Texts in Computer Science

More information about this series at http://www.springer.com/series/3191

Wilhelm Burger • Mark J. Burge

Digital Image Processing

An Algorithmic Introduction Using Java

Second Edition

Wilhelm Burger
School of Informatics/
Communications/Media
Upper Austria University
of Applied Sciences
Hagenberg, Austria

Mark J. Burge
Noblis, Inc.
Washington, DC, USA

Series Editors
David Gries
Department of Computer Science
Cornell University
Ithaca, NY, USA

Fred B. Schneider
Department of Computer Science
Cornell University
Ithaca, NY, USA

ISSN 1868-0941 ISSN 1868-095X (electronic)
Texts in Computer Science
ISBN 978-1-4471-6683-2 ISBN 978-1-4471-6684-9 (eBook)
DOI 10.1007/978-1-4471-6684-9

Library of Congress Control Number: 2016933770

Printed on acid-free paper

This Springer imprint is published by Springer Nature
The registered company is Springer-Verlag London Ltd.

Preface

This book provides a modern, self-contained introduction to digital image processing. We designed the book to be used both by learners desiring a firm foundation on which to build as well as practitioners in search of detailed analysis and transparent implementations of the most important techniques. This is the second English edition of the original German-language book, which has been widely used by:

- Scientists and engineers who use image processing as a tool and wish to develop a deeper understanding and create custom solutions to imaging problems in their field.
- IT professionals wanting a self-study course featuring easily adaptable code and completely worked out examples, enabling them to be productive right away.
- Faculty and students desiring an example-rich introductory textbook suitable for an advanced undergraduate or graduate level course that features exercises, projects, and examples that have been honed during our years of experience teaching this material.

While we concentrate on practical applications and concrete implementations, we do so without glossing over the important formal details and mathematics necessary for a deeper understanding of the algorithms. In preparing this text, we started from the premise that simply creating a recipe book of imaging solutions would not provide the deeper understanding needed to apply these techniques to novel problems, so instead our solutions are developed stepwise from three different perspectives: in mathematical form, as abstract pseudocode algorithms, and as complete Java programs. We use a common notation to intertwine all three perspectives—providing multiple, but linked, views of the problem and its solution.

Prerequisites

Instead of presenting digital image processing as a mathematical discipline, or strictly as signal processing, we present it from a practitioner's and programmer's perspective and with a view toward replacing many of the formalisms commonly used in other texts with constructs more readily understandable by our audience. To take full advantage of the *programming* components of this book, a knowledge of basic data structures and object-oriented programming, ideally in Java, is required. We selected Java for a number of reasons: it is the first programming language learned by students in a wide variety of engineering curricula, and professionals with knowledge of a related language, especially C# or C++, will find the programming examples easy to follow and extend.

The software in this book is designed to work with ImageJ, a widely used, programmer-extensible, imaging system developed, maintained, and distributed by the National Institutes of Health (NIH).[1] ImageJ is implemented completely in Java, and therefore runs on all major platforms, and is widely used because its "plugin"-based architecture enables it to be easily extended. While all examples run in ImageJ, they have been specifically designed to be easily ported to other environments and programming languages.

Use in research and development

This book has been especially designed for use as a textbook and as such features exercises and carefully constructed examples that supplement our detailed presentation of the fundamental concepts and techniques. As both practitioners and developers, we know that the details required to successfully understand, apply, and extend classical techniques are often difficult to find, and for this reason we have been very careful to provide the missing details, many gleaned over years of practical application. While this should make the text particularly valuable to those in research and development, it is not designed as a comprehensive, fully-cited scientific research text. On the contrary, we have carefully vetted our citations so that they can be obtained from easily accessible sources. While we have only briefly discussed the fundamentals of, or entirely omitted, topics such as hierarchical methods, wavelets, or eigenimages because of space limitations, other topics have been left out deliberately, including advanced issues such as object recognition, image understanding, and three-dimensional (3D) computer vision. So, while most techniques described in this book could be called "blind and dumb", it is our experience that straightforward, technically clean implementations of these simpler methods are essential to the success of any further domain-specific, or even "intelligent", approaches.

If you are only in search of a programming handbook for ImageJ or Java, there are certainly better sources. While the book includes many code examples, programming in and of itself is not our main focus. Instead Java serves as just one important element for describing each technique in a precise and immediately testable way.

Classroom use

Whether it is called signal processing, image processing, or media computation, the manipulation of digital images has been an integral part of most computer science and engineering curricula for many years. Today, with the omnipresence of all-digital work flows, it has become an integral part of the required skill set for professionals in many diverse disciplines.

Today the topic has migrated into the early stages of many curricula, where it is often a key foundation course. This migration uncovered a problem in that many of the texts relied on as standards

[1] http://rsb.info.nih.gov/ij/.

in the older graduate-level courses were not appropriate for beginners. The texts were usually too formal for novices, and at the same time did not provide detailed coverage of many of the most popular methods used in actual practice. The result was that educators had a difficult time selecting a single textbook or even finding a compact collection of literature to recommend to their students. Faced with this dilemma ourselves, we wrote this book in the sincere hope of filling this gap.

The contents of the following chapters can be presented in either a one- or two-semester sequence. Where feasible, we have added supporting material in order to make each chapter as independent as possible, providing the instructor with maximum flexibility when designing the course. Chapters 18–20 offer a complete introduction to the fundamental spectral techniques used in image processing and are essentially independent of the other material in the text. Depending on the goals of the instructor and the curriculum, they can be covered in as much detail as required or completely omitted. The following road map shows a possible partitioning of topics for a two-semester syllabus.

Road Map for a 1/2-Semester Syllabus	Sem. 1	2
1. Digital Images	■	□
2. ImageJ	■	□
3. Histograms and Image Statistics	■	□
4. Point Operations	■	□
5. Filters	■	□
6. Edges and Contours	■	□
7. Corner Detection	□	■
8. The Hough Transform: Finding Simple Curves	□	■
9. Morphological Filters	■	□
10. Regions in Binary Images	■	□
11. Automatic Thresholding	□	■
12. Color Images	■	□
13. Color Quantization	□	■
14. Colorimetric Color Spaces	□	■
15. Filters for Color Images	□	■
16. Edge Detection in Color Images	□	■
17. Edge-Preserving Smoothing Filters	□	■
18. Introduction to Spectral Techniques	□	■
19. The Discrete Fourier Transform in 2D	□	■
20. The Discrete Cosine Transform (DCT)	□	■
21. Geometric Operations	■	□
22. Pixel Interpolation	■	□
23. Image Matching and Registration	■	□
24. Non-Rigid Image Matching	□	■
25. Scale-Invariant Local Features (SIFT)	□	■
26. Fourier Shape Descriptors	□	■

Addendum to the second edition

This second edition is based on our completely revised German third edition and contains both additional material and several new chapters including: *automatic thresholding* (Ch. 11), *filters and edge detection for color images* (Chs. 15 and 16), *edge-preserving smoothing filters* (Ch. 17), *non-rigid image matching* (Ch. 24), and *Fourier shape descriptors* (Ch. 26). Much of this new material is presented for the first time at the level of detail necessary to completely understand and create a working implementation.

The two final chapters on SIFT and Fourier shape descriptors are particularly detailed to demonstrate the actual level of granularity and the special cases which must be considered when actually implementing complex techniques. Some other chapters have been rearranged or split into multiple parts for more clarity and easier use in teaching. The mathematical notation and programming examples were completely revised and almost all illustrations were adapted or created anew for this full-color edition.

For this edition, the *ImageJ Short Reference* and ancillary source code have been relocated from the Appendix and the most recently versions are freely available in electronic form from the book's website. The complete source code, consisting of the common `imagingbook` library, sample ImageJ plugins for each book chapter, and extended documentation are available from the book's SourceForge site.[2]

Online resources

Visit the website for this book

www.imagingbook.com

to download supplementary materials, including the complete Java source code for all examples and the underlying software library, full-size test images, useful references, and other supplements. Comments, questions, and corrections are welcome and may be addressed to

imagingbook@gmail.com

Exercises and solutions

Each chapter of this book contains a set of sample exercises, mainly for supporting instructors to prepare their own assignments. Most of these tasks are easy to solve after studying the corresponding chapter, while some others may require more elaborated reasoning or experimental work. We assume that scholars know best how to select and adapt individual assignments in order to fit the level and interest of their students. This is the main reason why we have abstained from publishing explicit solutions in the past. However, we are happy to answer any personal request if an exercise is unclear or seems to elude a simple solution.

Thank you!

This book would not have been possible without the understanding and support of our families. Our thanks go to Wayne Rasband at NIH for developing ImageJ and for his truly outstanding support of

[2] http://sourceforge.net/projects/imagingbook/.

the community and to all our readers of the previous editions who provided valuable input, suggestions for improvement, and encouragement. The use of open source software for such a project always carries an element of risk, since the long-term acceptance and continuity is difficult to assess. Retrospectively, choosing ImageJ as the software basis for this work was a good decision, and we would consider ourselves happy if our book has indirectly contributed to the success of the ImageJ project itself. Finally, we owe a debt of gratitude to the professionals at Springer, particularly to Wayne Wheeler and Simon Reeves who were responsible for the English edition.

Hagenberg / Washington D.C.
Fall 2015

Contents

1

Digital Images

For a long time, using a computer to manipulate a digital image (i.e., digital image processing) was something performed by only a relatively small group of specialists who had access to expensive equipment. Usually this combination of specialists and equipment was only to be found in research labs, and so the field of digital image processing has its roots in the academic realm. Now, however, the combination of a powerful computer on every desktop and the fact that nearly everyone has some type of device for digital image acquisition, be it their cell phone camera, digital camera, or scanner, has resulted in a plethora of digital images and, with that, for many digital image processing has become as common as word processing. It was not that many years ago that digitizing a photo and saving it to a file on a computer was a time-consuming task. This is perhaps difficult to imagine given today's powerful hardware and operating system level support for all types of digital media, but it is always sobering to remember that "personal" computers in the early 1990s were not powerful enough to even load into main memory a single image from a typical digital camera of today. Now powerful hardware and software packages have made it possible for amateurs to manipulate digital images and videos just as easily as professionals.

All of these developments have resulted in a large community that works productively with digital images while having only a basic understanding of the underlying mechanics. For the typical consumer merely wanting to create a digital archive of vacation photos, a deeper understanding is not required, just as a deep understanding of the combustion engine is unnecessary to successfully drive a car.

Today, IT professionals must be more then simply familiar with digital image processing. They are expected to be able to knowledgeably manipulate images and related digital media, which are an increasingly important part of the workflow not only of those involved in medicine and media but all industries. In the same way, software engineers and computer scientists are increasingly confronted with developing programs, databases, and related systems that must correctly deal with digital images. The simple lack of practical experience with this type of material, combined with an often unclear understanding of its basic foundations and a tendency to underestimate its difficulties, frequently leads to inefficient solutions, costly errors, and personal frustration.

1.1 Programming with Images

Even though the term "image processing" is often used interchangeably with that of "image editing", we introduce the following more precise definitions. Digital image editing, or as it is sometimes referred to, digital imaging, is the manipulation of digital images using an existing software application such as Adobe Photoshop® or Corel Paint®. Digital image processing, on the other hand, is the conception, design, development, and enhancement of digital imaging programs.

Modern programming environments, with their extensive APIs (application programming interfaces), make practically every aspect of computing, be it networking, databases, graphics, sound, or imaging, easily available to nonspecialists. The possibility of developing a program that can reach into an image and manipulate the individual elements at its very core is fascinating and seductive. You will discover that with the right knowledge, an image becomes ultimately no more than a simple array of values, that with the right tools you can manipulate in any way imaginable.

"Computer graphics", in contrast to digital image processing, concentrates on the *synthesis* of digital images from geometrical descriptions such as three-dimensional (3D) object models [75, 87, 247]. While graphics professionals today tend to be interested in topics such as realism and, especially in terms of computer games, rendering speed, the field does draw on a number of methods that originate in image processing, such as image transformation (morphing), reconstruction of 3D models from image data, and specialized techniques such as image-based and nonphotorealistic rendering [180, 248]. Similarly, image processing makes use of a number of ideas that have their origin in computational geometry and computer graphics, such as volumetric (voxel) models in medical image processing. The two fields perhaps work closest when it comes to digital postproduction of film and video and the creation of special effects [256]. This book provides a thorough grounding in the effective processing of not only images but also sequences of images; that is, videos.

1.2 Image Analysis and Computer Vision

Often it appears at first glance that a given image-processing task will have a simple solution, especially when it is something that is easily accomplished by our own visual system. Yet in practice it turns out that developing reliable, robust, and timely solutions is difficult or simply impossible. This is especially true when the problem involves image *analysis*; that is, where the ultimate goal is not to enhance or otherwise alter the appearance of an image but instead to extract meaningful information about its contents—be it distinguishing an object from its background, following a street on a map, or finding the bar code on a milk carton, tasks such as these often turn out to be much more difficult to accomplish than we would expect.

We expect technology to improve on what we can do by ourselves. Be it as simple as a lever to lift more weight or binoculars to see farther or as complex as an airplane to move us across continents—science has created so much that improves on, sometimes by unbelievable factors, what our biological systems are able to perform. So, it is perhaps humbling to discover that today's technology is nowhere near as capable, when it comes to image analysis, as our own visual system. While it is possible that this will always remain true, do not let this discourage you. Instead consider it a challenge to develop creative solutions. Using the tools, techniques, and fundamental knowledge available today, it is possible not only to solve many problems but to create robust, reliable, and fast applications.

While image analysis is not the main subject of this book, it often naturally intersects with image processing and we will explore this intersection in detail in these situations: finding simple curves (Ch. 8), segmenting image regions (Ch. 10), and comparing images (Ch. 23). In these cases, we present solutions that work directly on the pixel data in a *bottom-up* way without recourse to domain-specific knowledge (i.e., blind solutions). In this way, our solutions essentially embody the distinction between image processing, *pattern recognition*, and *computer vision*, respectively. While these two disciplines are firmly grounded in, and rely heavily on, image processing, their ultimate goals are much more lofty.

Pattern recognition is primarily a mathematical discipline and has been responsible for techniques such as clustering, hidden Markov models (HMMs), decision trees, and principal component analysis (PCA), which are used to discover patterns in data and signals. Methods from pattern recognition have been applied extensively to problems arising in computer vision and image analysis. A good example of their successful application is optical character recognition (OCR), where robust, highly accurate turnkey solutions are available for recognizing scanned text. Pattern recognition methods are truly universal and have been successfully applied not only to images but also speech and audio signals, text documents, stock trades, and finding trends in large databases, where it is often called data mining. Dimensionality reduction, statistical, and syntactical methods play important roles in pattern recognition (see, e.g., [64, 169, 228]).

Computer vision tackles the problem of engineering artificial visual systems capable of somehow comprehending and interpreting our real, 3D world. Popular topics in this field include scene understanding, object recognition, motion interpretation (tracking), autonomous navigation, and the robotic manipulation of objects in a scene. Since computer vision has its roots in artificial intelligence (AI), many AI methods were originally developed to either tackle or represent a problem in computer vision (see, e.g., [51, Ch. 13]). The fields still have much in common today, especially in terms of adaptive methods and machine learning. Further literature on computer vision includes [15, 78, 110, 214, 222, 232].

Ultimately you will find image processing to be both intellectually challenging and professionally rewarding, as the field is ripe with problems that were originally thought to be relatively simple to solve but have to this day refused to give up their secrets. With the background and techniques presented in this text, you will not only be able to develop complete image-processing solutions but will also have the prerequisite knowledge to tackle unsolved problems and the real possibility of expanding the horizons of science: for while image processing by itself may not change the world, it is likely to be the foundation that supports marvels of the future.

1.3 Types of Digital Images

Digital images are the central theme of this book, and unlike just a few years ago, this term is now so commonly used that there is really no reason to explain it further. Yet this book is not about all types of digital images, instead it focuses on images that are made up of *picture elements*, more commonly known as *pixels*, arranged in a regular rectangular grid.

Every day, people work with a large variety of digital raster images such as color photographs of people and landscapes, grayscale scans of printed documents, building plans, faxed documents, screenshots, medical images such as x-rays and ultrasounds, and a multitude of others (see Fig. 1.1 for examples). Despite all the different sources for these images, they are all, as a rule, ultimately represented as rectangular ordered arrays of image elements.

1.4 Image Acquisition

The process by which a scene becomes a digital image is varied and complicated, and, in most cases, the images you work with will already be in digital form, so we only outline here the essential stages in the process. As most image acquisition methods are essentially variations on the classical optical camera, we will begin by examining it in more detail.

1.4.1 The Pinhole Camera Model

The pinhole camera is one of the simplest camera models and has been in use since the 13th century, when it was known as the "Camera Obscura". While pinhole cameras have no practical use today except to hobbyists, they are a useful model for understanding the essential optical components of a simple camera. The pinhole camera consists of a closed box with a small opening on the front side through which light enters, forming an image on the opposing wall. The light forms a smaller, inverted image of the scene (Fig. 1.2).

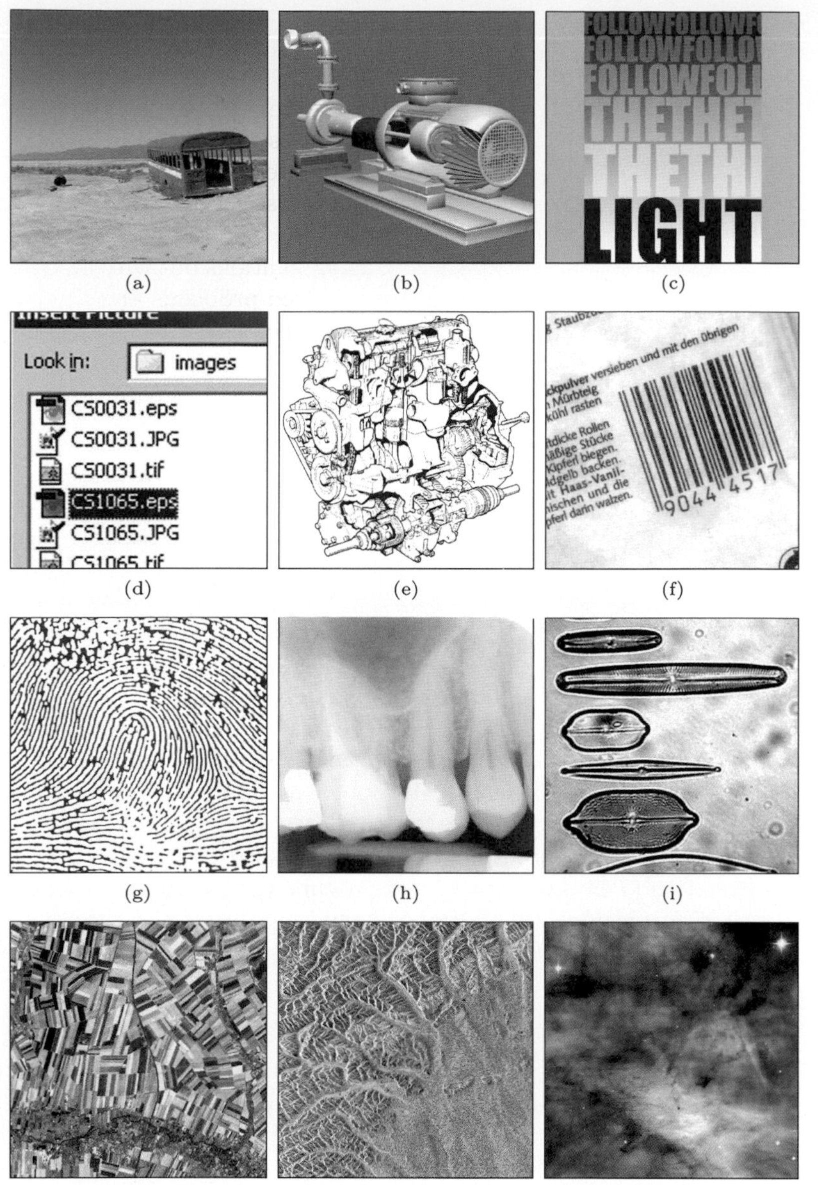

Fig. 1.1
Examples of digital images. Natural landscape (a), synthetically generated scene (b), poster graphic (c), computer screenshot (d), black and white illustration (e), barcode (f), fingerprint (g), x-ray (h), microscope slide (i), satellite image (j), radar image (k), astronomical object (l).

Perspective projection

The geometric properties of the pinhole camera are very simple. The optical axis runs through the pinhole perpendicular to the image plane. We assume a visible object, in our illustration the cactus, located at a horizontal distance Z from the pinhole and vertical distance Y from the optical axis. The height of the projection y is determined by two parameters: the fixed depth of the camera box f and the distance Z to the object from the origin of the coordinate system. By comparison we see that

$$x = -f \cdot \frac{X}{Z} \qquad \text{and} \qquad y = -f \cdot \frac{Y}{Z} \tag{1.1}$$

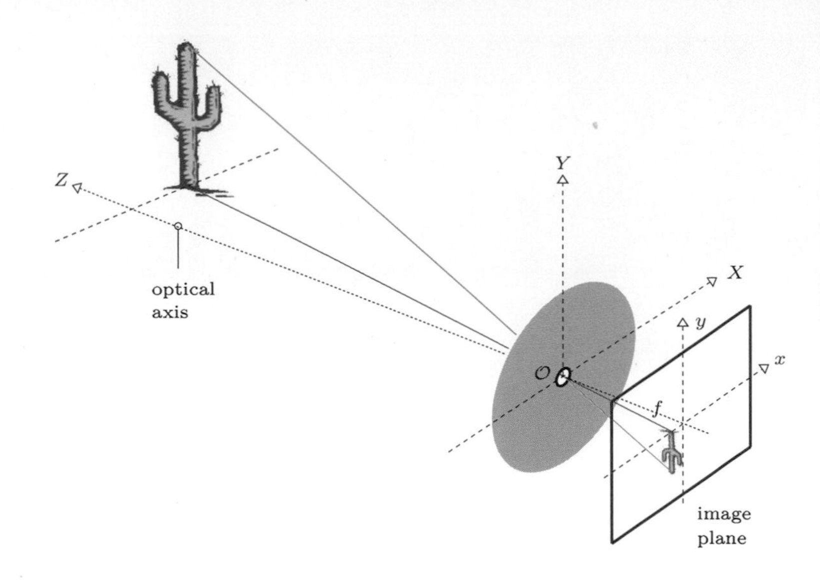

Fig. 1.2
Geometry of the pinhole camera. The pinhole opening serves as the origin ($\mathcal{O}$) of the 3D coordinate system (X, Y, Z) for the objects in the scene. The optical axis, which runs through the opening, is the Z axis of this coordinate system. A separate 2D coordinate system (x, y) describes the projection points on the image plane. The distance f ("focal length") between the opening and the image plane determines the scale of the projection.

change with the scale of the resulting image in proportion to the depth of the box (i.e., the distance f) in a way similar to how the focal length does in an everyday camera. For a fixed image, a small f (i.e., short focal length) results in a small image and a large viewing angle, just as occurs when a wide-angle lens is used, while increasing the "focal length" f results in a larger image and a smaller viewing angle, just as occurs when a telephoto lens is used. The negative sign in Eqn. (1.1) means that the projected image is flipped in the horizontal and vertical directions and rotated by 180°.

Equation (1.1) describes what is commonly known today as the *perspective transformation.*[1] Important properties of this theoretical model are that straight lines in 3D space always appear straight in 2D projections and that circles appear as ellipses.

1.4.2 The "Thin" Lens

While the simple geometry of the pinhole camera makes it useful for understanding its basic principles, it is never really used in practice. One of the problems with the pinhole camera is that it requires a very small opening to produce a sharp image. This in turn reduces the amount of light passed through and thus leads to extremely long exposure times. In reality, glass lenses or systems of optical lenses are used whose optical properties are greatly superior in many aspects but of course are also much more complex. Instead we can make our model more realistic, without unduly increasing its complexity, by replacing the pinhole with a "thin lens" as in Fig. 1.3.

In this model, the lens is assumed to be symmetric and infinitely thin, such that all light rays passing through it cross through a virtual plane in the middle of the lens. The resulting image geometry is the same as that of the pinhole camera. This model is not sufficiently complex to encompass the physical details of actual lens systems, such

[1] It is hard to imagine today that the rules of perspective geometry, while known to the ancient mathematicians, were only rediscovered in 1430 by the Renaissance painter Brunelleschi.

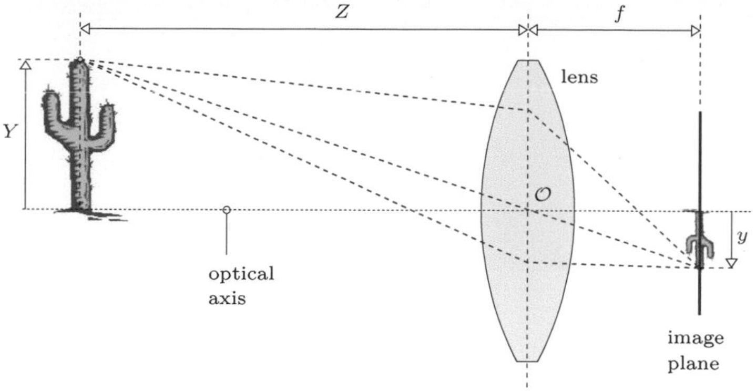

Fig. 1.3
Thin lens projection model.

as geometrical distortions and the distinct refraction properties of different colors. So, while this simple model suffices for our purposes (i.e., understanding the mechanics of image acquisition), much more detailed models that incorporate these additional complexities can be found in the literature (see, e.g., [126]).

1.4.3 Going Digital

What is projected on the image plane of our camera is essentially a two-dimensional (2D), time-dependent, continuous distribution of light energy. In order to convert this image into a digital image on our computer, the following three main steps are necessary:

1. The continuous light distribution must be spatially sampled.
2. This resulting function must then be sampled in time to create a single (still) image.
3. Finally, the resulting values must be quantized to a finite range of integers (or floating-point values) such that they can be represented by digital numbers.

Step 1: Spatial sampling

The spatial sampling of an image (i.e., the conversion of the continuous signal to its discrete representation) depends on the geometry of the sensor elements of the acquisition device (e.g., a digital or video camera). The individual sensor elements are arranged in ordered rows, almost always at right angles to each other, along the sensor plane (Fig. 1.4). Other types of image sensors, which include hexagonal elements and circular sensor structures, can be found in specialized products.

Step 2: Temporal sampling

Temporal sampling is carried out by measuring at regular intervals the amount of light incident on each individual sensor element. The CCD[2] in a digital camera does this by triggering the charging process and then measuring the amount of electrical charge that has built up during the specified amount of time that the CCD was illuminated.

[2] Charge-coupled device.

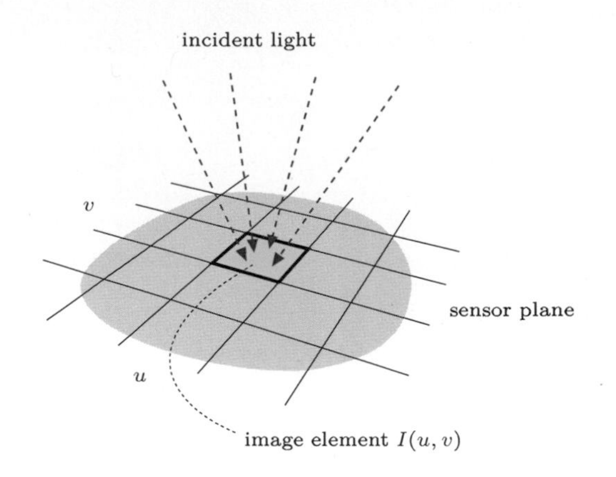

Fig. 1.4
The geometry of the sensor elements is directly responsible for the spatial sampling of the continuous image. In the simplest case, a plane of sensor elements are arranged in an evenly spaced grid, and each element measures the amount of light that falls on it.

Step 3: Quantization of pixel values

In order to store and process the image values on the computer they are commonly converted to an integer scale (e.g., $256 = 2^8$ or $4096 = 2^{12}$). Occasionally floating-point values are used in professional applications, such as medical imaging. Conversion is carried out using an analog to digital converter, which is typically embedded directly in the sensor electronics so that conversion occurs at image capture or is performed by special interface hardware.

Images as discrete functions

The result of these three stages is a description of the image in the form of a 2D, ordered matrix of integers (Fig. 1.5). Stated a bit more formally, a digital image I is a 2D function that maps from the domain of integer coordinates $\mathbb{N} \times \mathbb{N}$ to a range of possible pixel values $\mathbb{P}$ such that

$$I(u, v) \in \mathbb{P} \quad \text{and} \quad u, v \in \mathbb{N}.$$

Now we are ready to transfer the image to our computer so that we can save, compress, and otherwise manipulate it into the file format of our choice. At this point, it is no longer important to us how the image originated since it is now a simple 2D array of numerical data. Before moving on, we need a few more important definitions.

1.4.4 Image Size and Resolution

In the following, we assume rectangular images, and while that is a relatively safe assumption, exceptions do exist. The *size* of an image is determined directly from the *width* M (number of columns) and the *height* N (number of rows) of the image matrix I.

The *resolution* of an image specifies the spatial dimensions of the image in the real world and is given as the number of image elements per measurement; for example, *dots per inch* (dpi) or *lines per inch* (lpi) for print production, or in *pixels per kilometer* for satellite images. In most cases, the resolution of an image is the same in the horizontal and vertical directions, which means that the

$F(x, y)$

$\rightarrow$

148	123	52	107	123	162	172	123	64	89	···
147	130	92	95	98	130	171	155	169	163	···
141	118	121	148	117	107	144	137	136	134	···
82	106	93	172	149	131	138	114	113	129	···
57	101	72	54	109	111	104	135	106	125	···
138	135	114	82	121	110	34	76	101	111	···
138	102	128	159	168	147	116	129	124	117	···
113	89	89	109	106	126	114	150	164	145	···
120	121	123	87	85	70	119	64	79	127	···
145	141	143	134	111	124	117	113	64	112	···
⋮	⋮	⋮	⋮	⋮	⋮	⋮	⋮	⋮	⋮	

$I(u, v)$

Fig. 1.5
The transformation of a continuous grayscale image $F(x, y)$ to a discrete digital image $I(u, v)$ (left), image detail (below).

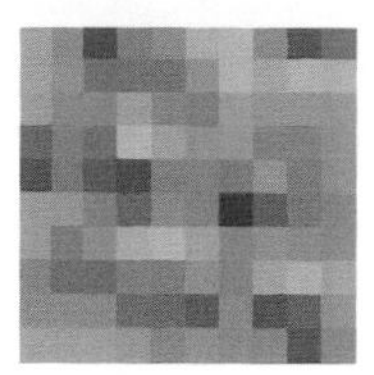

image elements are square. Note that this is not always the case as, for example, the image sensors of most current video cameras have non-square pixels!

The spatial resolution of an image may not be relevant in many basic image processing steps, such as point operations or filters. Precise resolution information is, however, important in cases where geometrical elements such as circles need to be drawn on an image or when distances within an image need to be measured. For these reasons, most image formats and software systems designed for professional applications rely on precise information about image resolution.

1.4.5 Image Coordinate System

In order to know which position on the image corresponds to which image element, we need to impose a coordinate system. Contrary to normal mathematical conventions, in image processing the coordinate system is usually flipped in the vertical direction; that is, the y-coordinate runs from top to bottom and the origin lies in the upper left corner (Fig. 1.6). While this system has no practical or theoretical advantage, and in fact may be a bit confusing in the context of geometrical transformations, it is used almost without exception in imaging software systems. The system supposedly has its roots in the original design of television broadcast systems, where the picture rows are numbered along the vertical deflection of the electron beam, which moves from the top to the bottom of the screen. We start the numbering of rows and columns at zero for practical reasons, since in Java array indexing also begins at zero.

1.4.6 Pixel Values

The information within an image element depends on the data type used to represent it. Pixel values are practically always binary words of length k so that a pixel can represent any of 2^k different values. The value k is called the bit depth (or just "depth") of the image. The exact bit-level layout of an individual pixel depends on the kind of

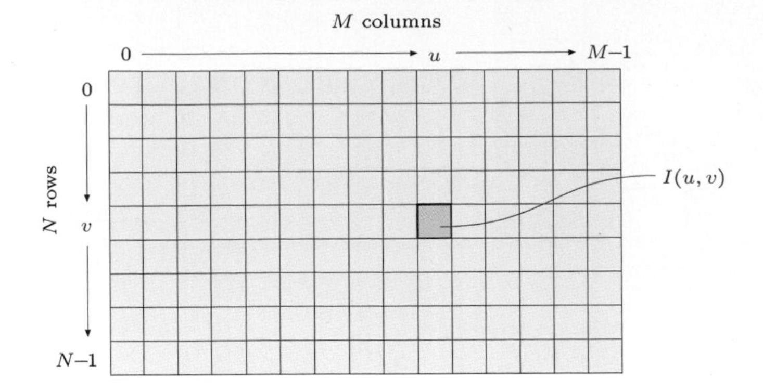

Fig. 1.6
Image coordinates. In digital image processing, it is common to use a coordinate system where the origin ($u = 0$, $v = 0$) lies in the upper left corner. The coordinates u, v represent the columns and the rows of the image, respectively. For an image with dimensions $M \times N$, the maximum column number is $u_{max} = M-1$ and the maximum row number is $v_{max} = N-1$.

Table 1.1
Bit depths of common image types and typical application domains.

Grayscale (Intensity Images):

Chan.	*Bits/Pix.*	*Range*	*Use*
1	1	$[0, 1]$	Binary image: document, illustration, fax
1	8	$[0, 255]$	Universal: photo, scan, print
1	12	$[0, 4095]$	High quality: photo, scan, print
1	14	$[0, 16383]$	Professional: photo, scan, print
1	16	$[0, 65535]$	Highest quality: medicine, astronomy

Color Images:

Chan.	*Bits/Pix.*	*Range*	*Use*
3	24	$[0, 255]^3$	RGB, universal: photo, scan, print
3	36	$[0, 4095]^3$	RGB, high quality: photo, scan, print
3	42	$[0, 16383]^3$	RGB, professional: photo, scan, print
4	32	$[0, 255]^4$	CMYK, digital prepress

Special Images:

Chan.	*Bits/Pix.*	*Range*	*Use*
1	16	$[-32768, 32767]$	Integer values pos./neg., increased range
1	32	$\pm 3.4 \cdot 10^{38}$	Floating-point values: medicine, astronomy
1	64	$\pm 1.8 \cdot 10^{308}$	Floating-point values: internal processing

image; for example, binary, grayscale, or RGB[3] color. The properties of some common image types are summarized below (also see Table 1.1).

Grayscale images (intensity images)

The image data in a grayscale image consist of a single channel that represents the intensity, brightness, or density of the image. In most cases, only positive values make sense, as the numbers represent the intensity of light energy or density of film and thus cannot be negative, so typically whole integers in the range $0, \ldots, 2^k-1$ are used. For example, a typical grayscale image uses $k = 8$ bits (1 byte) per pixel and intensity values in the range $0, \ldots, 255$, where the value 0 represents the minimum brightness (black) and 255 the maximum brightness (white).

For many professional photography and print applications, as well as in medicine and astronomy, 8 bits per pixel is not sufficient. Image depths of 12, 14, and even 16 bits are often encountered in these

[3] Red, green, and blue.

domains. Note that bit depth usually refers to the number of bits used to represent one color component, not the number of bits needed to represent an entire color pixel. For example, an RGB-encoded color image with an 8-bit depth would require 8 bits for each channel for a total of 24 bits, while the same image with a 12-bit depth would require a total of 36 bits.

Binary images

Binary images are a special type of intensity image where pixels can only take on one of two values, black or white. These values are typically encoded using a single bit (0/1) per pixel. Binary images are often used for representing line graphics, archiving documents, encoding fax transmissions, and of course in electronic printing.

Color images

Most color images are based on the primary colors red, green, and blue (RGB), typically making use of 8 bits for each color component. In these color images, each pixel requires $3 \times 8 = 24$ bits to encode all three components, and the range of each individual color component is $[0, 255]$. As with intensity images, color images with 30, 36, and 42 bits per pixel are commonly used in professional applications. Finally, while most color images contain three components, images with four or more color components are common in most prepress applications, typically based on the subtractive CMYK (**C**yan-**M**agenta-**Y**ellow-Blac**k**) color model (see Ch. 12).

Indexed or *palette* images constitute a very special class of color image. The difference between an indexed image and a *true color* image is the number of different colors (fewer for an indexed image) that can be used in a particular image. In an indexed image, the pixel values are only indices (with a maximum of 8 bits) onto a specific table of selected full-color values (see Sec. 12.1.1).

Special images

Special images are required if none of the above standard formats is sufficient for representing the image values. Two common examples of special images are those with negative values and those with floating-point values. Images with negative values arise during image-processing steps, such as filtering for edge detection (see Sec. 6.2.2), and images with floating-point values are often found in medical, biological, or astronomical applications, where extended numerical range and precision are required. These special formats are mostly application specific and thus may be difficult to use with standard image-processing tools.

1.5 Image File Formats

While in this book we almost always consider image data as being already in the form of a 2D array—ready to be accessed by a program—, in practice image data must first be loaded into memory from a file. Files provide the essential mechanism for storing, archiving, and exchanging image data, and the choice of the correct file format is an important decision. In the early days of digital image processing (i.e., before around 1985), most software developers created a new custom file format for almost every new application they developed.[4] Today there exist a wide range of standardized file formats, and developers can almost always find at least one existing format that is suitable for their application. Using standardized file formats vastly increases the ease with which images can be exchanged and the likelihood that the images will be readable by other software in the long term. Yet for many projects the selection of the right file format is not always simple, and compromises must be made. The following sub-sections outline a few of the typical criteria that need to be considered when selecting an appropriate file format.

1.5.1 Raster versus Vector Data

In the following, we will deal exclusively with file formats for storing *raster images*; that is, images that contain pixel values arranged in a regular matrix using discrete coordinates. In contrast, *vector graphics* represent geometric objects using continuous coordinates, which are only rasterized once they need to be displayed on a physical device such as a monitor or printer.

A number of standardized file formats exist for vector images, such as the ANSI/ISO standard format CGM (Computer Graphics Metafile) and SVG (Scalable Vector Graphics),[5] as well as proprietary formats such as DXF (Drawing Exchange Format from AutoDesk), AI (Adobe Illustrator), PICT (QuickDraw Graphics Metafile from Apple), and WMF/EMF (Windows Metafile and Enhanced Metafile from Microsoft). Most of these formats can contain both vector data and raster images in the same file. The PS (PostScript) and EPS (Encapsulated PostScript) formats from Adobe as well as the PDF (Portable Document Format) also offer this possibility, although they are typically used for printer output and archival purposes.[6]

1.5.2 Tagged Image File Format (TIFF)

This is a widely used and flexible file format designed to meet the professional needs of diverse fields. It was originally developed by Aldus and later extended by Microsoft and currently Adobe. The format supports a range of grayscale, indexed, and true color images, but also special image types with large-depth integer and floating-point elements. A TIFF file can contain a number of images with different properties. The TIFF specification provides a range of different compression methods (LZW, ZIP, CCITT, and JPEG) and color spaces,

[4] The result was a chaotic jumble of incompatible file formats that for a long time limited the practical sharing of images between research groups.

[5] www.w3.org/TR/SVG/.

[6] Special variations of PS, EPS, and PDF files are also used as (editable) exchange formats for raster and vector data; for example, both Adobe's Photoshop (Photoshop-EPS) and Illustrator (AI).

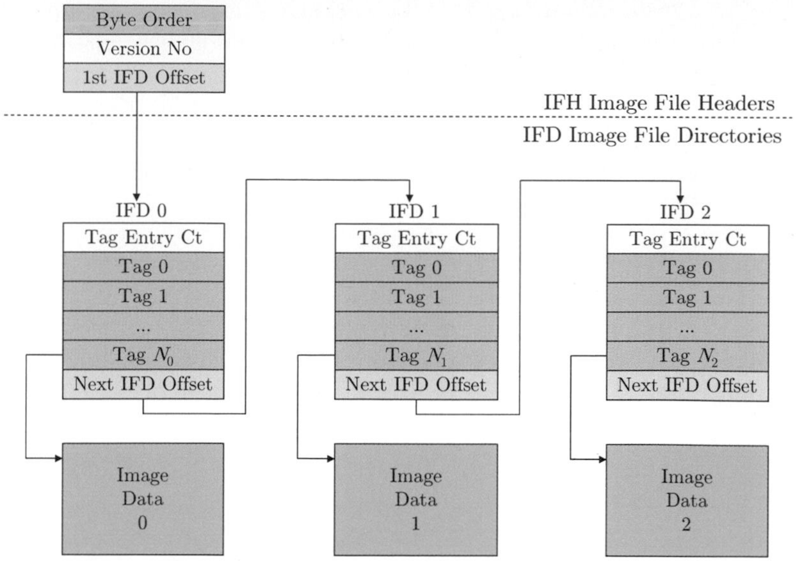

Fig. 1.7
Structure of a typical TIFF file. A TIFF file consists of a header and a linked list of image objects, three in this example. Each image object consists of a list of "tags" with their corresponding entries followed by a pointer to the actual image data.

so that it is possible, for example, to store a number of variations of an image in different sizes and representations together in a single TIFF file. The flexibility of TIFF has made it an almost universal exchange format that is widely used in archiving documents, scientific applications, digital photography, and digital video production.

The strength of this image format lies within its architecture (Fig. 1.7), which enables new image types and information blocks to be created by defining new "tags". In this flexibility also lies the weakness of the format, namely that proprietary tags are not always supported and so the "unsupported tag" error is sometimes still encountered when loading TIFF files. ImageJ also reads only a few uncompressed variations of TIFF formats,[7] and bear in mind that most popular Web browsers currently do not support TIFF either.

1.5.3 Graphics Interchange Format (GIF)

The Graphics Interchange Format (GIF) was originally designed by CompuServe in 1986 to efficiently encode the rich line graphics used in their dial-up Bulletin Board System (BBS). It has since grown into one of the most widely used formats for representing images on the Web. This popularity is largely due to its early support for indexed color at multiple bit depths, LZW[8] compression, interlaced image loading, and ability to encode simple animations by storing a number of images in a single file for later sequential display. GIF is essentially an indexed image file format designed for color and grayscale images with a maximum depth of 8 bits and consequently it does not support true color images. It offers efficient support for encoding palettes containing from 2 to 256 colors, one of which can be marked for transparency. GIF supports color tables in the range

[7] The `ImageIO` plugin offers support for a wider range of TIFF formats.
[8] Lempel-Ziv-Welch

of $2, \ldots, 256$, enabling pixels to be encoded using fewer bits. As an example, the pixels of an image using 16 unique colors require only 4 bits to store the 16 possible color values $0, \ldots, 15$. This means that instead of storing each pixel using 1 byte, as done in other bitmap formats, GIF can encode two 4-bit pixels into each 8-bit byte. This results in a 50% storage reduction over the standard 8-bit indexed color bitmap format.

The GIF file format is designed to efficiently encode "flat" or "iconic" images consisting of large areas of the same color. It uses lossy color quantization (see Ch. 13) as well as lossless LZW compression to efficiently encode large areas of the same color. Despite the popularity of the format, when developing new software, the PNG[9] format, presented in the next sub-section, should be preferred, as it outperforms GIF by almost every metric.

1.5.4 Portable Network Graphics (PNG)

PNG (pronounced "ping") was originally developed as a replacement for the GIF file format when licensing issues[10] arose because of its use of LZW compression. It was designed as a universal image format especially for use on the Internet, and, as such, PNG supports three different types of images:

- true color images (with up to 3×16 bits/pixel),
- grayscale images (with up to 16 bits/pixel),
- indexed color images (with up to 256 colors).

Additionally, PNG includes an *alpha* channel for transparency with a maximum depth of 16 bits. In comparison, the transparency channel of a GIF image is only a single bit deep. While the format only supports a single image per file, it is exceptional in that it allows images of up to $2^{30} \times 2^{30}$ pixels. The format supports lossless compression by means of a variation of PKZIP (Phil Katz's ZIP). No lossy compression is available, as PNG was not designed as a replacement for JPEG. Ultimately, the PNG format meets or exceeds the capabilities of the GIF format in every way except GIF's ability to include multiple images in a single file to create simple animations. Currently, PNG should be considered the format of choice for representing uncompressed, lossless, true color images for use on the Web.

1.5.5 JPEG

The JPEG standard defines a compression method for continuous grayscale and color images, such as those that would arise from nature photography. The format was developed by the Joint Photographic Experts Group (JPEG)[11] with the goal of achieving an average data reduction of a factor of 1:16 and was established in 1990 as ISO Standard IS-10918. Today it is the most widely used image file format. In practice, JPEG achieves, depending on the application, compression in the order of 1 bit per pixel (i.e., a compression factor of around

[9] Portable network graphics

[10] Unisys's U.S. LZW Patent No. 4,558,302 expired on June 20, 2003.

[11] www.jpeg.org.

1:25) when compressing 24-bit color images to an acceptable quality for viewing. The JPEG standard supports images with up to 256 color components, and what has become increasingly important is its support for CMYK images (see Sec. 12.2.5).

The modular design of the JPEG compression algorithm [163] allows for variations of the "baseline" algorithm; for example, there exists an uncompressed version, though it is not often used. In the case of RGB images, the core of the algorithm consists of three main steps:

1. **Color conversion and down sampling:** A color transformation from RGB into the YC_bC_r space (see Ch. 12, Sec. 12.2.4) is used to separate the actual color components from the brightness Y component. Since the human visual system is less sensitive to rapid changes in color, it is possible to compress the color components more, resulting in a significant data reduction, without a subjective loss in image quality.
2. **Cosine transform and quantization in frequency space:** The image is divided up into a regular grid of 8 blocks, and for each independent block, the frequency spectrum is computed using the discrete cosine transformation (see Ch. 20). Next, the 64 spectral coefficients of each block are quantized into a quantization table. The size of this table largely determines the eventual compression ratio, and therefore the visual quality, of the image. In general, the high frequency coefficients, which are essential for the "sharpness" of the image, are reduced most during this step. During decompression these high frequency values will be approximated by computed values.
3. **Lossless compression:** Finally, the quantized spectral components data stream is again compressed using a lossless method, such as arithmetic or Huffman encoding, in order to remove the last remaining redundancy in the data stream.

The JPEG compression method combines a number of different compression methods and its should not be underestimated. Implementing even the baseline version is nontrivial, so application support for JPEG increased sharply once the Independent JPEG Group (IJG)[12] made available a reference implementation of the JPEG algorithm in 1991. Drawbacks of the JPEG compression algorithm include its limitation to 8-bit images, its poor performance on non-photographic images such as line art (for which it was not designed), its handling of abrupt transitions within an image, and the striking artifacts caused by the 8×8 pixel blocks at high compression rates. Figure 1.9 shows the results of compressing a section of a grayscale image using different quality factors (Photoshop $Q_{\mathrm{JPG}} = 10, 5, 1$).

JPEG File Interchange Format (JFIF)

Despite common usage, JPEG is *not* a file format; it is "only" a method of compressing image data. The actual JPEG standard only specifies the JPEG codec (compressor and decompressor) and by de-

[12] www.ijg.org.

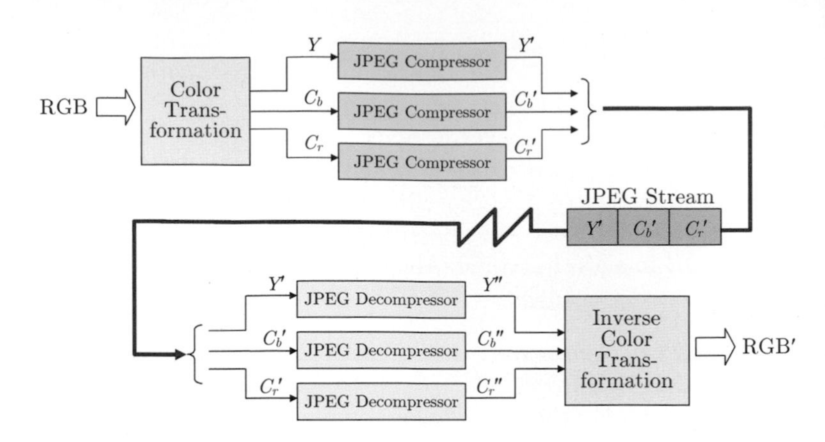

Fig. 1.8 JPEG compression of an RGB image. Using a color space transformation, the color components C_b, C_r are separated from the Y luminance component and subjected to a higher rate of compression. Each of the three components are then run independently through the JPEG compression pipeline and are merged into a single JPEG data stream. Decompression follows the same stages in reverse order.

sign leaves the wrapping, or file format, undefined.[13] What is normally referred to as a JPEG *file* is almost always an instance of a "JPEG File Interchange Format" (JFIF) file, originally developed by Eric Hamilton and the IJG. JFIF specifies a file format based on the JPEG standard by defining the remaining necessary elements of a file format. The JPEG standard leaves some parts of the codec undefined for generality, and in these cases JFIF makes a specific choice. As an example, in step 1 of the JPEG codec, the specific color space used in the color transformation is not part of the JPEG standard, so it is specified by the JFIF standard. As such, the use of different compression ratios for color and luminance is a practical implementation decision specified by JFIF and is not a part of the actual JPEG encoder.

Exchangeable Image File Format (EXIF)

The Exchangeable Image File Format (EXIF) is a variant of the JPEG (JFIF) format designed for storing image data originating on digital cameras, and to that end it supports storing metadata such as the type of camera, date and time, photographic parameters such as aperture and exposure time, as well as geographical (GPS) data. EXIF was developed by the Japan Electronics and Information Technology Industries Association (JEITA) as a part of the DCF[14] guidelines and is used today by practically all manufacturers as the standard format for storing digital images on memory cards. Internally, EXIF uses TIFF to store the metadata information and JPEG to encode a thumbnail preview image. The file structure is designed so that it can be processed by existing JPEG/JFIF readers without a problem.

JPEG-2000

JPEG-2000, which is specified by an ISO-ITU standard ("Coding of Still Pictures"),[15] was designed to overcome some of the better-known weaknesses of the traditional JPEG codec. Among the im-

[13] To be exact, the JPEG standard only defines how to compress the individual components and the structure of the JPEG stream.

[14] Design Rule for Camera File System.

[15] www.jpeg.org/JPEG2000.htm.

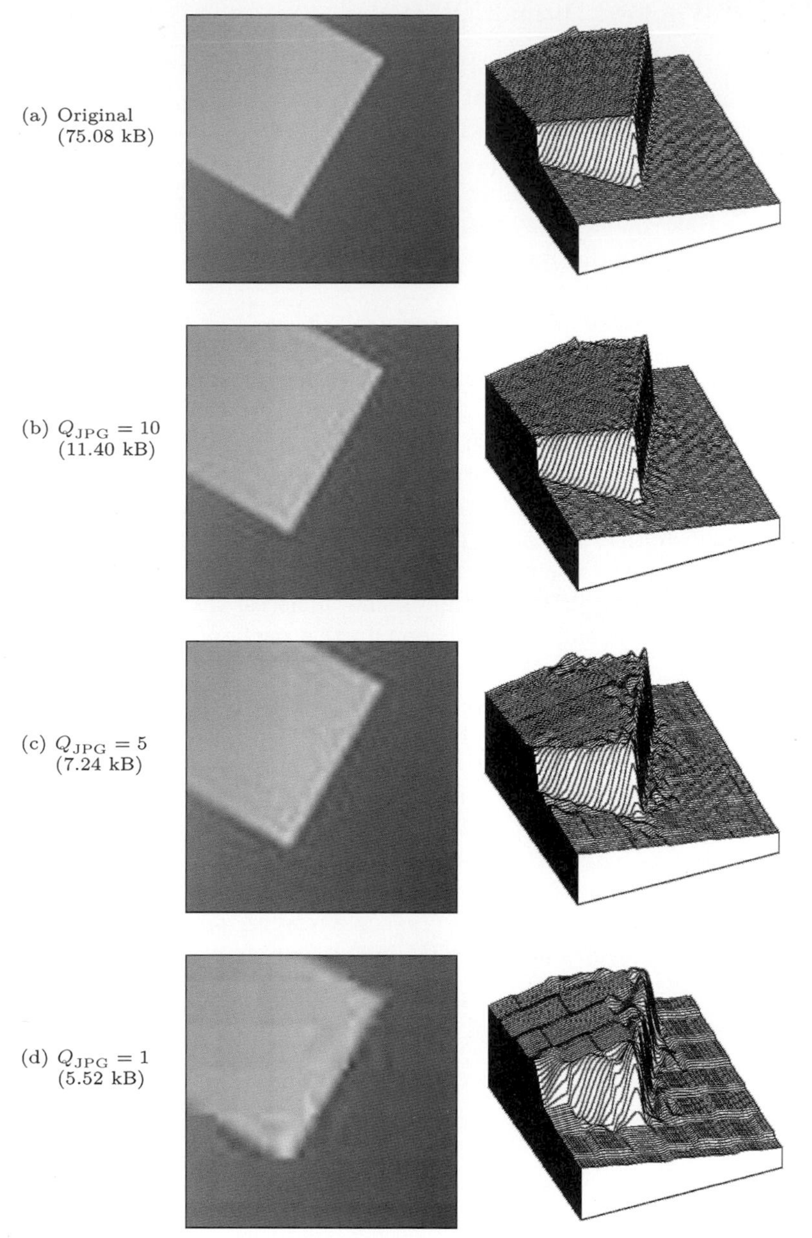

(a) Original (75.08 kB)

(b) $Q_{JPG} = 10$ (11.40 kB)

(c) $Q_{JPG} = 5$ (7.24 kB)

(d) $Q_{JPG} = 1$ (5.52 kB)

Fig. 1.9
Artifacts arising from JPEG compression. A section of the original image (a) and the results of JPEG compression at different quality factors: $Q_{JPG} = 10$ (b), $Q_{JPG} = 5$ (c), and $Q_{JPG} = 1$ (d). In parentheses are the resulting file sizes for the complete (dimensions 274×274) image.

provements made in JPEG-2000 are the use of larger, 64×64 pixel blocks and replacement of the discrete cosine transform by the *wavelet* transform. These and other improvements enable it to achieve significantly higher compression ratios than JPEG—up to 0.25 bits per pixel on RGB color images. Despite these advantages, JPEG-2000 is supported by only a few image-processing applications and Web browsers.[16]

[16] At this time, ImageJ does not offer JPEG-2000 support.

1.5.6 Windows Bitmap (BMP)

The Windows Bitmap (BMP) format is a simple, and under Windows widely used, file format supporting grayscale, indexed, and true color images. It also supports binary images, but not in an efficient manner, since each pixel is stored using an entire byte. Optionally, the format supports simple lossless, run-length-based compression. While BMP offers storage for a similar range of image types as TIFF, it is a much less flexible format.

1.5.7 Portable Bitmap Format (PBM)

The Portable Bitmap Format (PBM) family[17] consists of a series of very simple file formats that are exceptional in that they can be optionally saved in a human-readable text format that can be easily read in a program or simply edited using a text editor. A simple PGM image is shown in Fig. 1.10. The characters `P2` in the first line indicate that the image is a PGM ("plain") file stored in human-readable format. The next line shows how comments can be inserted directly into the file by beginning the line with the `#` symbol. Line three gives the image's dimensions, in this case width 17 and height 7, and line four defines the maximum pixel value, in this case 255. The remaining lines give the actual pixel values. This format makes it easy to create and store image data without any explicit imaging API, since it requires only basic text I/O that is available in any programming environment. In addition, the format supports a much more machine-optimized "raw" output mode in which pixel values are stored as bytes. PBM is widely used under Unix and supports the following formats: PBM (*portable bitmap*) for binary *bitmaps*, PGM (*portable graymap*) for grayscale images, and PNM (*portable any map*) for color images. PGM images can be opened by ImageJ.

```
P2
# oie.pgm
17 7
255
0 13 13 13 13 13 13 13  0  0  0  0  0  0  0  0  0
0 13  0  0  0  0  0 13  0  7  7  0  0 81 81 81 81
0 13  0  7  7  7  0 13  0  7  7  0  0 81  0  0  0
0 13  0  7  0  7  0 13  0  7  7  0  0 81 81 81  0
0 13  0  7  7  7  0 13  0  7  7  0  0 81  0  0  0
0 13  0  0  0  0  0 13  0  7  7  0  0 81 81 81 81
0 13 13 13 13 13 13 13  0  0  0  0  0  0  0  0  0
```

Fig. 1.10 Example of a PGM file in human-readable text format (top) and the corresponding grayscale image (below).

1.5.8 Additional File Formats

For most practical applications, one of the following file formats is sufficient: TIFF as a universal format supporting a wide variety of uncompressed images and JPEG/JFIF for digital color photos when storage size is a concern, and there is either PNG or GIF for when an image is destined for use on the Web. In addition, there exist

[17] http://netpbm.sourceforge.net.

countless other file formats, such as those encountered in legacy applications or in special application areas where they are traditionally used. A few of the more commonly encountered types are:

- **RGB**, a simple format from Silicon Graphics.
- **RAS** (Sun Raster Format), a simple format from Sun Microsystems.
- **TGA** (Truevision Targa File Format), the first 24-bit file format for PCs. It supports numerous image types with 8- to 32-bit depths and is still used in medicine and biology.
- **XBM/XPM** (X-Windows Bitmap/Pixmap), a group of ASCII-encoded formats used in the X-Windows system and similar to PBM/PGM.

1.5.9 Bits and Bytes

Today, opening, reading, and writing image files is mostly carried out by means of existing software libraries. Yet sometimes you still need to deal with the structure and contents of an image file at the byte level, for instance when you need to read an unsupported file format or when you receive a file where the format of the data is unknown.

Big endian and little endian

In the standard model of a computer, a file consists of a simple sequence of 8-bit bytes, and a byte is the smallest entry that can be read or written to a file. In contrast, the image elements as they are stored in memory are usually larger then a byte; for example, a 32-bit `int` value (= 4 bytes) is used for an RGB color pixel. The problem is that storing the four individual bytes that make up the image data can be done in different ways. In order to correctly recreate the original color pixel, we must naturally know the *order* in which bytes in the file are arranged.

Consider, for example, a 32-bit `int` number z with the binary and hexadecimal values[18]

$$z = \underbrace{00010010}_{\substack{12_H \\ \text{(MSB)}}}\,00110100\,01010110\,\underbrace{01111000}_{\substack{78_H \\ \text{(LSB)}}}{}_B \equiv 12345678_H, \quad (1.2)$$

then $00010010_B \equiv 12_H$ is the value of the *most significant byte* (MSB) and $01111000_B \equiv 78_H$ the *least significant byte* (LSB). When the individual bytes in the file are arranged in order from MSB to LSB when they are saved, we call the ordering "big endian", and when in the opposite direction, "little endian". Thus the 32-bit value z from Eqn. (1.2) could be stored in one of the following two modes:

Ordering	Byte Sequence	1	2	3	4
big endian	MSB $\rightarrow$ LSB	12_H	34_H	56_H	78_H
little endian	LSB $\rightarrow$ MSB	78_H	56_H	34_H	12_H

Even though correctly ordering the bytes should essentially be the responsibility of the operating and file systems, in practice it actually

[18] The decimal value of z is 305419896.

Table 1.2
Signatures of various image file formats. Most image file formats can be identified by inspecting the first bytes of the file. These byte sequences, or signatures, are listed in hexadecimal (0x..) form and as ASCII text (□ indicates a nonprintable character).

Format	*Signature*		*Format*	*Signature*	
PNG	0x89504e47	□PNG	BMP	0x424d	BM
JPEG/JFIF	0xffd8ffe0	□□□□	GIF	0x4749463839	GIF89
$\text{TIFF}_{\text{little}}$	0x49492a00	II*□	Photoshop	0x38425053	8BPS
TIFF_{big}	0x4d4d002a	MM□*	PS/EPS	0x25215053	%!PS

depends on the architecture of the processor.[19] Processors from the Intel family (e.g., x86, Pentium) are traditionally little endian, and processors from other manufacturers (e.g., IBM, MIPS, Motorola, Sun) are big endian.[20] Big endian is also called *network byte ordering* since in the IP protocol the data bytes are arranged in MSB to LSB order during transmission.

To correctly interpret image data with multi-byte pixel values, it is necessary to know the byte ordering used when creating it. In most cases, this is fixed and defined by the file format, but in some file formats, for example TIFF, it is variable and depends on a parameter given in the file header (see Table 1.2).

File headers and signatures

Practically all image file formats contain a data header consisting of important information about the layout of the image data that follows. Values such as the size of the image and the encoding of the pixels are usually present in the file header to make it easier for programmers to allocate the correct amount of memory for the image. The size and structure of this header are usually fixed, but in some formats, such as TIFF, the header can contain pointers to additional subheaders.

In order to interpret the information in the header, it is necessary to know the file type. In many cases, this can be determined by the *file name extension* (e.g., `.jpg` or `.tif`), but since these extensions are not standardized and can be changed at any time by the user, they are not a reliable way of determining the file type. Instead, many file types can be identified by their embedded "signature", which is often the first 2 bytes of the file. Signatures from a number of popular image formats are given in Table 1.2. Most image formats can be determined by inspecting the first few bytes of the file. These bytes, or signatures, are listed in hexadecimal (`0x..`) form and as ASCII text. A PNG file always begins with the 4-byte sequence `0x89`, `0x50`, `0x4e`, `0x47`, which is the "magic number" `0x89` followed by the ASCII sequence "PNG". Sometimes the signature not only identifies the type of image file but also contains information about its encoding; for instance, in TIFF the first two characters are either `II` for "Intel" or `MM` for "Motorola" and indicate the byte ordering (little endian or big endian, respectively) of the image data in the file.

[19] At least the ordering of the *bits* within a byte is almost universally uniform.

[20] In Java, this problem does not arise since internally all implementations of the *Java Virtual Machine* use big endian ordering.

1.6 Exercises

Exercise 1.1. Determine the actual physical measurement in millimeters of an image with 1400 rectangular pixels and a resolution of 72 dpi.

Exercise 1.2. A camera with a focal length of $f = 50\,\text{mm}$ is used to take a photo of a vertical column that is 12 m high and is 95 m away from the camera. Determine its height in the image in mm (a) and the number of pixels (b) assuming the camera has a resolution of 4000 dpi.

Exercise 1.3. The image sensor of a particular digital camera contains 2016×3024 pixels. The geometry of this sensor is identical to that of a traditional 35 mm camera (with an image size of 24×36 mm) except that it is 1.6 times smaller. Compute the resolution of this digital sensor in dpi.

Exercise 1.4. Assume the camera geometry described in Exercise 1.3 combined with a lens with focal length $f = 50$ mm. What amount of blurring (in pixels) would be caused by a uniform, $0.1°$ horizontal turn of the camera during exposure? Recompute this for $f = 300$ mm. Consider if the extent of the blurring also depends on the distance of the object.

Exercise 1.5. Determine the number of bytes necessary to store an uncompressed binary image of size 4000×3000 pixels.

Exercise 1.6. Determine the number of bytes necessary to store an uncompressed RGB color image of size 640×480 pixels using 8, 10, 12, and 14 bits per color channel.

Exercise 1.7. Given a black and white television with a resolution of 625×512 8-bit pixels and a frame rate of 25 images per second: (a) How may different images can this device ultimately display, and how long would you have to watch it (assuming no sleeping) in order to see every possible image at least once? (b) Perform the same calculation for a color television with 3×8 bits per pixel.

Exercise 1.8. Show that the projection of a 3D straight line in a pinhole camera (assuming perspective projection as defined in Eqn. (1.1)) is again a straight line in the resulting 2D image.

Exercise 1.9. Using Fig. 1.10 as a model, use a text editor to create a PGM file, `disk.pgm`, containing an image of a bright circle. Open your image with ImageJ and then try to find other programs that can open and display the image.

2

ImageJ

Until a few years ago, the image-processing community was a relatively small group of people who either had access to expensive commercial image-processing tools or, out of necessity, developed their own software packages. Usually such home-brew environments started out with small software components for loading and storing images from and to disk files. This was not always easy because often one had to deal with poorly documented or even proprietary file formats. An obvious (and frequent) solution was to simply design a *new* image file format from scratch, usually optimized for a particular field, application, or even a single project, which naturally led to a myriad of different file formats, many of which did not survive and are forgotten today [163, 168]. Nevertheless, writing software for *converting* between all these file formats in the 1980s and early 1990s was an important business that occupied many people. Displaying images on computer screens was similarly difficult, because there was only marginal support from operating systems, APIs, and display hardware, and capturing images or videos into a computer was close to impossible on common hardware. It thus may have taken many weeks or even months before one could do just elementary things with images on a computer and finally do some serious image processing.

Fortunately, the situation is much different today. Only a few common image file formats have survived (see also Sec. 1.5), which are readily handled by many existing tools and software libraries. Most standard APIs for C/C++, Java, and other popular programming languages already come with at least some basic support for working with images and other types of media data. While there is still much development work going on at this level, it makes our job a lot easier and, in particular, allows us to focus on the more interesting aspects of digital imaging.

2.1 Software for Digital Imaging

Traditionally, software for digital imaging has been targeted at either *manipulating* or *processing* images, either for practitioners and designers or software programmers, with quite different requirements.

Software packages for *manipulating* images, such as Adobe Photoshop, Corel Paint, and others, usually offer a convenient user interface and a large number of readily available functions and tools for working with images interactively. Sometimes it is possible to extend the standard functionality by writing scripts or adding self-programmed components. For example, Adobe provides a special API[1] for programming Photoshop "plugins" in C++, though this is a nontrivial task and certainly too complex for nonprogrammers.

In contrast to the aforementioned category of tools, digital image *processing* software primarily aims at the requirements of algorithm and software developers, scientists, and engineers working with images, where interactivity and ease of use are not the main concerns. Instead, these environments mostly offer comprehensive and well-documented software libraries that facilitate the implementation of new image-processing algorithms, prototypes, and working applications. Popular examples are Khoros/Accusoft,[2] MatLab,[3] ImageMagick,[4] among many others. In addition to the support for conventional programming (typically with C/C++), many of these systems provide dedicated scripting languages or visual programming aides that can be used to construct even highly complex processes in a convenient and safe fashion.

In practice, image manipulation and image processing are of course closely related. Although Photoshop, for example, is aimed at image manipulation by nonprogrammers, the software itself implements many traditional image-processing algorithms. The same is true for many Web applications using server-side image processing, such as those based on ImageMagick. Thus image processing is really at the base of any image manipulation software and certainly not an entirely different category.

2.2 ImageJ Overview

ImageJ, the software that is used for this book, is a combination of both worlds discussed in the previous section. It offers a set of ready-made tools for viewing and interactive manipulation of images but can also be extended easily by writing new software components in a "real" programming language. ImageJ is implemented entirely in Java and is thus largely platform-independent, running without modification under Windows, MacOS, or Linux. Java's dynamic execution model allows new modules ("plugins") to be written as independent pieces of Java code that can be compiled, loaded, and executed "on the fly" in the running system without the need to

[1] www.adobe.com/products/photoshop/.

[2] www.accusoft.com.

[3] www.mathworks.com.

[4] www.imagemagick.org.

even restart ImageJ. This quick turnaround makes ImageJ an ideal platform for developing and testing new image-processing techniques and algorithms. Since Java has become extremely popular as a first programming language in many engineering curricula, it is usually quite easy for students to get started in ImageJ without having to spend much time learning another programming language. Also, ImageJ is freely available, so students, instructors, and practitioners can install and use the software legally and without license charges on any computer. ImageJ is thus an ideal platform for education and self-training in digital image processing but is also in regular use for serious research and application development at many laboratories around the world, particularly in biological and medical imaging.

Wayne Rasband (right) at the 1st ImageJ Conference 2006 (picture courtesy of Marc Seil, CRP Henri Tudor, Luxembourg).

ImageJ was (and still *is*) developed by Wayne Rasband [193] at the U.S. National Institutes of Health (NIH), originally as a substitute for its predecessor, NIH-Image, which was only available for the Apple Macintosh platform. The current version of ImageJ, updates, documentation, the complete source code, test images, and a continuously growing collection of third-party plugins can be downloaded from the ImageJ website.[5] Installation is simple, with detailed instructions available online, in Werner Bailer's programming tutorial [12], and in the authors' *ImageJ Short Reference* [40].

In addition to ImageJ itself there are several popular software projects that build on or extend ImageJ. This includes in particular *Fiji*[6] ("Fiji Is Just ImageJ") which offers a consistent collection of numerous plugins, simple installation on various platforms and excellent documentation. All programming examples (plugins) shown in this book should also execute in Fiji without any modifications. Another important development is *ImgLib2*, which is a generic Java API for representing and processing n-dimensional images in a consistent fashion. ImgLib2 also provides the underlying data model for *ImageJ2*,[7] which is a complete reimplementation of ImageJ.

2.2.1 Key Features

As a pure Java application, ImageJ should run on any computer for which a current Java runtime environment (JRE) exists. ImageJ comes with its own Java runtime, so Java need not be installed separately on the computer. Under the usual restrictions, ImageJ can be run as a Java "applet" within a Web browser, though it is mostly used as a stand-alone application. It is sometimes also used on the server side in the context of Java-based Web applications (see [12] for details). In summary, the key features of ImageJ are:

- A set of ready-to-use, interactive tools for creating, visualizing, editing, processing, analyzing, loading, and storing images, with support for several common file formats. ImageJ also provides "deep" 16-bit integer images, 32-bit floating-point images, and image sequences ("stacks").

[5] http://rsb.info.nih.gov/ij/.

[6] http://fiji.sc.

[7] http://imagej.net/ImageJ2. To avoid confusion, the "classic" ImageJ platform is sometimes referred to as "ImageJ1" or simply "IJ1".

- A simple plugin mechanism for extending the core functionality of ImageJ by writing (usually small) pieces of Java code. All coding examples shown in this book are based on such plugins.
- A macro language and the corresponding interpreter, which make it easy to implement larger processing blocks by combining existing functions without any knowledge of Java. Macros are not discussed in this book, but details can be found in ImageJ's online documentation.[8]

2.2.2 Interactive Tools

When ImageJ starts up, it first opens its main window (Fig. 2.1), which includes the following menu entries:

- File: for opening, saving, and creating new images.
- Edit: for editing and drawing in images.
- Image: for modifying and converting images, geometric operations.
- Process: for image processing, including point operations, filters, and arithmetic operations between multiple images.
- Analyze: for statistical measurements on image data, histograms, and special display formats.
- Plugin: for editing, compiling, executing, and managing user-defined plugins.

The current version of ImageJ can open images in several common formats, including TIFF (uncompressed only), JPEG, GIF, PNG, and BMP, as well as the formats DICOM[9] and FITS,[10] which are popular in medical and astronomical image processing, respectively. As is common in most image-editing programs, all interactive operations are applied to the currently *active* image, i.e., the image most recently selected by the user. ImageJ provides a simple (single-step) "undo" mechanism for most operations, which can also revert modifications effected by user-defined plugins.

2.2.3 ImageJ Plugins

Plugins are small Java modules for extending the functionality of ImageJ by using a simple standardized interface (Fig. 2.2). Plugins can be created, edited, compiled, invoked, and organized through the Plugin menu in ImageJ's main window (Fig. 2.1). Plugins can be grouped to improve modularity, and plugin commands can be arbitrarily placed inside the main menu structure. Also, many of ImageJ's built-in functions are actually implemented as plugins themselves.

Program structure

Technically speaking, plugins are Java classes that implement a particular interface specification defined by ImageJ. There are two main types of plugins:

[8] http://rsb.info.nih.gov/ij/developer/macro/macros.html.

[9] Digital Imaging and Communications in Medicine.

[10] Flexible Image Transport System.

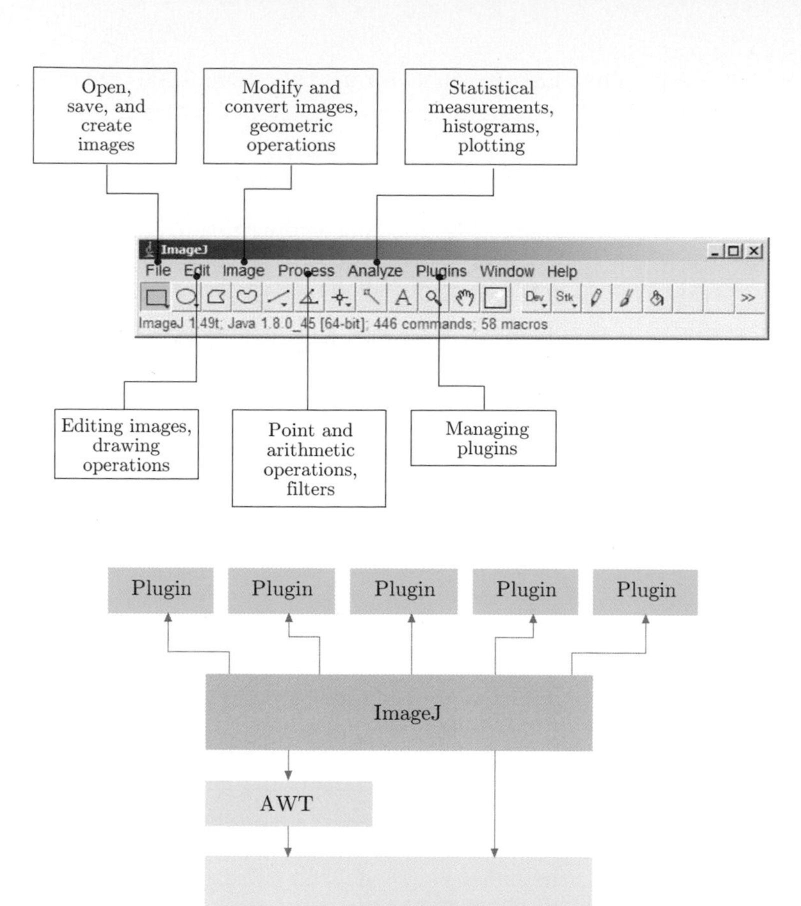

Fig. 2.1
ImageJ main window (under Windows).

Fig. 2.2
ImageJ software structure (simplified). ImageJ is based on the Java core system and depends in particular upon Java's Advanced Windowing Toolkit (AWT) for the implementation of the user interface and the presentation of image data. Plugins are small Java classes that extend the functionality of the basic ImageJ system.

- `PlugIn`: requires no image to be open to start a plugin.
- `PlugInFilter`: the currently active image is passed to the plugin when started.

Throughout the examples in this book, we almost exclusively use plugins of the second type (i.e., `PlugInFilter`) for implementing image-processing operations. The interface specification requires that any plugin of type `PlugInFilter` must at least implement two methods, `setup()` and `run()`, with the following signatures:

`int setup (String` *`args`*`, ImagePlus` *`im`*`)`
When the plugin is started, ImageJ calls this method first to verify that the capabilities of this plugin match the target image. `setup()` returns a vector of binary flags (packaged as a 32-bit `int` value) that describes the plugin's properties.

`void run (ImageProcessor` *`ip`*`)`
This method does the actual work for this plugin. It is passed a single argument *ip*, an object of type `ImageProcessor`, which contains the image to be processed and all relevant information

about it. The `run()` method returns no result value (`void`) but may modify the passed image and create new images.

2.2.4 A First Example: Inverting an Image

Let us look at a real example to quickly illustrate this mechanism. The task of our first plugin is to invert any 8-bit grayscale image to turn a positive image into a negative. As we shall see later, inverting the intensity of an image is a typical *point operation*, which is discussed in detail in Chapter 4. In ImageJ, 8-bit grayscale images have pixel values ranging from 0 (black) to 255 (white), and we assume that the width and height of the image are M and N, respectively. The operation is very simple: the value of each image pixel $I(u,v)$ is replaced by its inverted value,

$$I(u,v) \leftarrow 255 - I(u,v),$$

for all image coordinates (u,v), with $u = 0,\ldots,M-1$ and $v = 0,\ldots, N-1$.

2.2.5 Plugin `My_Inverter_A` (using `PlugInFilter`)

We decide to name our first plugin "`My_Inverter_A`", which is both the name of the Java class and the name of the source file[11] that contains it (see Prog. 2.1). The underscore characters ("`_`") in the name cause ImageJ to recognize this class as a plugin and to insert it automatically into the menu list at startup. The Java source code in file `My_Inverter.java` contains a few `import` statements, followed by the definition of the class `My_Inverter`, which implements the `PlugInFilter` interface (because it will be applied to an existing image).

The `setup()` method

When a plugin of type `PlugInFilter` is executed, ImageJ first invokes its `setup()` method to obtain information about the plugin itself. In this example, `setup()` only returns the value `DOES_8G` (a static `int` constant specified by the `PlugInFilter` interface), indicating that this plugin can handle 8-bit grayscale images. The parameters `arg` and `im` of the `setup()` method are not used in this example (see also Exercise 2.7).

The `run()` method

As mentioned already, the `run()` method of a `PlugInFilter` plugin receives an object (`ip`) of type `ImageProcessor`, which contains the image to be processed and all relevant information about it. First, we use the `ImageProcessor` methods `getWidth()` and `getHeight()` to query the size of the image referenced by `ip`. Then we use two nested `for` loops (with loop variables `u`, `v` for the horizontal and vertical coordinates, respectively) to iterate over all image pixels. For reading and writing the pixel values, we use two additional methods of the class `ImageProcessor`:

[11] File `My_Inverter_A.java`.

```
1 import ij.ImagePlus;
2 import ij.plugin.filter.PlugInFilter;
3 import ij.process.ImageProcessor;
4
5 public class My_Inverter_A implements PlugInFilter {
6
7   public int setup(String args, ImagePlus im) {
8     return DOES_8G; // this plugin accepts 8-bit grayscale images
9   }
10
11  public void run(ImageProcessor ip) {
12    int M = ip.getWidth();
13    int N = ip.getHeight();
14
15    // iterate over all image coordinates (u,v)
16    for (int u = 0; u < M; u++) {
17      for (int v = 0; v < N; v++) {
18        int p = ip.getPixel(u, v);
19        ip.putPixel(u, v, 255 - p);
20      }
21    }
22  }
23
24 }
```

Prog. 2.1
ImageJ plugin for inverting 8-bit grayscale images. This plugin implements the interface `PlugInFilter` and defines the required methods `setup()` and `run()`. The target image is received by the `run()` method as an instance of type `ImageProcessor`. ImageJ assumes that the plugin modifies the supplied image and automatically redisplays it after the plugin is executed. Program 2.2 shows an alternative implementation that is based on the `PlugIn` interface.

`int getPixel (int` u`, int` v`)`
: Returns the pixel value at the given position or zero if (u, v) is outside the image bounds.

`void putPixel (int` u`, int` v`, int` a`)`
: Sets the pixel value at position (u, v) to the new value a. Does nothing if (u, v) is outside the image bounds.

Both methods check the supplied image coordinates and pixel values to avoid unwanted errors. While this makes them more or less fail-safe it also makes them slow. If we are sure that no coordinates outside the image bounds are ever accessed (as in `My_Inverter` in Prog. 2.1) and the inserted pixel values are guaranteed not to exceed the image processor's range, we can use the significantly faster methods `get()` and `set()` in place of `getPixel()` and `putPixel()`, respectively. The most efficient way to process the image is to avoid read/write methods altogether and directly access the elements of the associated (1D) pixel array. Details on these and other methods can be found in the ImageJ API documentation.[12]

2.2.6 Plugin `My_Inverter_B` (using `PlugIn`)

Program 2.2 shows an alternative implementation of the inverter plugin based on ImageJ's `PlugIn` interface, which requires a `run()` method only. In this case the reference to the current image is not supplied directly but is obtained by invoking the (static) method

[12] http://rsbweb.nih.gov/ij/developer/api/index.html.

Prog. 2.2
Alternative implementation of the inverter plugin, based on ImageJ's PlugIn interface. In contrast to Prog. 2.1 this plugin has no setUp() method but defines a run() method only. The current image (im) is obtained as an instance of class ImagePlus by invoking the IJ.getImage() method. After checking for the proper image type the associated ImageProcessor (ip) is retrieved from im. The parameter string (args) is not used in this example. The remaining parts of the plugin are identical to Prog. 2.1, except that the (slightly faster) pixel access methods get() and set() are used. Also note that the modified image is not re-displayed automatically but by an explicit call to updateAndDraw().

```
1 import ij.IJ;
2 import ij.ImagePlus;
3 import ij.plugin.PlugIn;
4 import ij.process.ImageProcessor;
5
6 public class My_Inverter_B implements PlugIn {
7
8   public void run(String args) {
9     ImagePlus im = IJ.getImage();
10
11    if (im.getType() != ImagePlus.GRAY8) {
12      IJ.error("8-bit grayscale image required");
13      return;
14    }
15
16    ImageProcessor ip = im.getProcessor();
17    int M = ip.getWidth();
18    int N = ip.getHeight();
19
20    // iterate over all image coordinates (u,v)
21    for (int u = 0; u < M; u++) {
22      for (int v = 0; v < N; v++) {
23        int p = ip.get(u, v);
24        ip.set(u, v, 255 - p);
25      }
26    }
27
28    im.updateAndDraw();    // redraw the modified image
29  }
30 }
```

IJ.getImage(). If no image is currently open, getImage() automatically displays an error message and aborts the plugin. However, the subsequent test for the correct image type (GRAY8) and the corresponding error handling must be performed explicitly. The run() method accepts a single string argument that can be used to pass arbitrary information for controlling the plugin.

2.2.7 When to use PlugIn or PlugInFilter?

The choice of PlugIn or PlugInFilter is mostly a matter of taste, since both versions have their advantages and disadvantages. As a rule of thumb, we use the PlugIn type for tasks that do not require any image to be open but for tasks that create, load, or record images or perform operations without any images. Otherwise, if one or more open images should be processed, PlugInFilter is the preferred choice and thus almost all plugins in this book are of type PlugInFilter.

Editing, compiling, and executing the plugin

The Java source file for our plugin should be stored in directory <ij>/plugins/[13] or an immediate subdirectory. New plugin files

[13] <ij> denotes ImageJ's installation directory.

can be created with ImageJ's Plugins ▷ New... menu. ImageJ even provides a built-in Java editor for writing plugins, which is available through the Plugins ▷ Edit... menu but unfortunately is of little use for serious programming. A better alternative is to use a modern editor or a professional Java programming environment, such as Eclipse,[14] NetBeans,[15] or JBuilder,[16] all of which are freely available.

For compiling plugins (to Java bytecode), ImageJ comes with its own Java compiler as part of its runtime environment. To compile and execute the new plugin, simply use the menu

Plugins ▷ Compile and Run...

Compilation errors are displayed in a separate log window. Once the plugin is compiled, the corresponding `.class` file is automatically loaded and the plugin is applied to the currently active image. An error message is displayed if no images are open or if the current image cannot be handled by that plugin.

At startup, ImageJ automatically loads all correctly named plugins found in the `<ij>/plugins/` directory (or any immediate subdirectory) and installs them in its Plugins menu. These plugins can be executed immediately without any recompilation. References to plugins can also be placed manually with the

Plugins ▷ Shortcuts ▷ Install Plugin...

command at any other position in the ImageJ menu tree. Sequences of plugin calls and other ImageJ commands may be recorded as macro programs with Plugins ▷ Macros ▷ Record.

Displaying and "undoing" results

Our first plugins in Prog. 2.1–2.2 did not create a new image but "destructively" modified the target image. This is not always the case, but plugins can also create additional images or compute only statistics, without modifying the original image at all. It may be surprising, though, that our plugin contains no commands for displaying the modified image. This is done automatically by ImageJ whenever it can be assumed that the image passed to a plugin was modified.[17] In addition, ImageJ automatically makes a copy ("snapshot") of the image before passing it to the `run()` method of a `PlugInFilter`-type plugin. This feature makes it possible to restore the original image (with the Edit ▷ Undo menu) after the plugin has finished without any explicit precautions in the plugin code.

Logging and debugging

The usual console output from Java via `System.out` is not available in ImageJ by default. Instead, a separate logging window can be used which facilitates simple text output by the method

```
IJ.log(String s).
```

[14] www.eclipse.org.

[15] www.netbeans.org.

[16] www.borland.com.

[17] No automatic redisplay occurs if the `NO_CHANGES` flag is set in the return value of the plugin's `setup()` method.

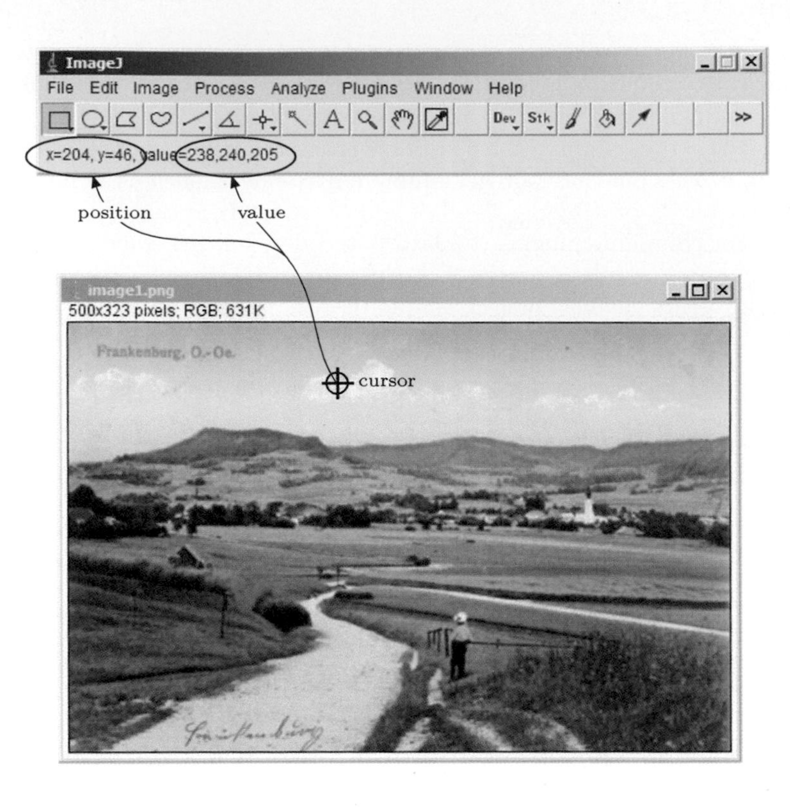

Fig. 2.3 Information displayed in ImageJ's main window is extremely helpful for debugging image-processing operations. The current cursor position is displayed in pixel coordinates unless the associated image is spatially calibrated. The way pixel *values* are displayed depends on the image type; in the case of a color image (as shown here) integer RGB component values are shown.

Such calls may be placed at any position in the plugin code for quick and simple debugging at runtime. However, because of the typically large amounts of data involved, they should be used with caution in real image-processing operations. Particularly, when placed in the body of inner processing loops that could execute millions of times, text output may produce an enormous overhead compared to the time used for the actual calculations.

ImageJ itself does not offer much support for "real" debugging, i.e., for setting breakpoints, inspecting local variables etc. However, it is possible to launch ImageJ from within a programming environment (IDE) such as Eclipse or *Netbeans* and then use all debugging options that the given environment provides.[18] According to experience, this is only needed in rare and exceptionally difficult situations. In most cases, inspection of pixel values displayed in ImageJ's main window (see Fig. 2.3) is much simpler and more effective. In general, many errors (in particular those related to image coordinates) can be easily avoided by careful planning in advance.

2.2.8 Executing ImageJ "Commands"

If possible, it is wise in most cases to re-use existing (and extensively tested) functionality instead of re-implementing it oneself. In particuar, the Java library that comes with ImageJ covers many standard image-processing operations, many of which are used throughout this

[18] For details see the "HowTo" section at http://imagejdocu.tudor.lu.

```
1 import ij.IJ;
2 import ij.ImagePlus;
3 import ij.plugin.PlugIn;
4
5 public class Run_Command_From_PlugIn implements PlugIn {
6
7     public void run(String args) {
8       ImagePlus im = IJ.getImage();
9       IJ.run(im, "Invert", ""); // run the "Invert" command on im
10      // ... continue with this plugin
11    }
12 }
```

Prog. 2.3
Executing the ImageJ command "Invert" within a Java plugin of type `PlugIn`.

```
1 public class Run_Command_From_PlugInFilter implements
      PlugInFilter {
2   ImagePlus im;
3
4   public int setup(String args, ImagePlus im) {
5     this.im = im;
6     return DOES_ALL;
7   }
8
9   public void run(ImageProcessor ip) {
10    im.unlock();                 // unlock im to run other commands
11    IJ.run(im, "Invert", "");    // run "Invert" command on im
12    im.lock();                   // lock im again (to be safe)
13    // ... continue with this plugin
14  }
15 }
```

Prog. 2.4
Executing the ImageJ command "Invert" within a Java plugin of type `PlugInFilter`. In this case the current image is automatically locked during plugin execution, such that no other operation may be applied to it. However, the image can be temporarily unlocked by calling `unlock()` and `lock()`, respectively, to run the external command.

book. Additional classes and methods for specific operations are contained in the associated (**imagingbook**) library.

In the context of ImageJ, the term "command" refers to any composite operation implemented as a (Java) plugin, a macro command or as a script.[19] ImageJ itself includes numerous commands which can be listed with the menu Plugins ▷ Utilities ▷ Find Commands.... They are usually referenced "by name", i.e., by a unique string. For example, the standard operation for inverting an image (Edit ▷ Invert) is implemented by the Java class **ij.plugin.filter.Filters** (with the argument **"invert"**).

An existing command can also be executed from within a Java plugin with the method **IJ.run()**, as demonstrated for the "Invert" command in Prog. 2.3. Some caution is required with plugins of type **PlugInFilter**, since these lock the current image during execution, such that no other operation can be applied to it. The example in Prog. 2.4 shows how this can be resolved by a pair of calls to **unlock()** and **lock()**, respectively, to temporarily release the current image.

A convenient tool for putting together complex commands is ImageJ's built-in *Macro Recorder*. Started with Plugins ▷ Macros ▷

[19] Scripting languages for ImageJ currently include *JavaScript*, *BeanShell*, and *Python*.

Record..., it logs all subsequent commands in a text file for later use. It can be set up to record commands in various modes, including *Java*, *JavaScript*, *BeanShell*, or ImageJ macro code. Of course it does record the application of self-defined plugins as well.

2.3 Additional Information on ImageJ and Java

In the following chapters, we mostly use concrete plugins and Java code to describe algorithms and data structures. This not only makes these examples immediately applicable, but they should also help in acquiring additional skills for using ImageJ in a step-by-step fashion. To keep the text compact, we often describe only the `run()` method of a particular plugin and additional class and method definitions if they are relevant in the given context. The complete source code for these examples can of course be downloaded from the book's supporting website.[20]

2.3.1 Resources for ImageJ

The complete and most current API reference, including source code, tutorials, and many example plugins, can be found on the official ImageJ website. Another great source for any serious plugin programming is the tutorial by Werner Bailer [12].

2.3.2 Programming with Java

While this book does not require extensive Java skills from its readers, some elementary knowledge is essential for understanding or extending the given examples. There is a huge and still-growing number of introductory textbooks on Java, such as [8, 29, 66, 70, 208] and many others. For readers with programming experience who have not worked with Java before, we particularly recommend some of the tutorials on Oracle's Java website.[21] Also, in Appendix F of this book, readers will find a small compilation of specific Java topics that cause frequent problems or programming errors.

2.4 Exercises

Exercise 2.1. Install the current version of ImageJ on your computer and make yourself familiar with the built-in commands (open, convert, edit, and save images).

Exercise 2.2. Write a new ImageJ plugin that reflects a grayscale image horizontally (or vertically) using `My_Inverter.java` (Prog. 2.1) as a template. Test your new plugin with appropriate images of different sizes (odd, even, extremely small) and inspect the results carefully.

[20] www.imagingbook.com.

[21] http://docs.oracle.com/javase/.

Exercise 2.3. The `run()` method of plugin `Inverter_Plugin_A` (see Prog. 2.1) iterates over all pixels of the given image. Find out in which order the pixels are visited: along the (horizontal) lines or along the (vertical) columns? Make a drawing to illustrate this process.

Exercise 2.4. Create an ImageJ plugin for 8-bit grayscale images of arbitrary size that paints a white frame (with pixel value 255) 10 pixels wide *into* the image (without increasing its size). Make sure this plugin also works for very small images.

Exercise 2.5. Create a plugin for 8-bit grayscale images that calculates and prints the result (with `IJ.log()`). Use a variable of type `int` or `long` for accumulating the pixel values. What is the maximum image size for which we can be certain that the result of summing with an `int` variable is correct?

Exercise 2.6. Create a plugin for 8-bit grayscale images that calculates and prints the minimum and maximum pixel values in the current image (with `IJ.log()`). Compare your output to the results obtained with Analyze ▷ Measure.

Exercise 2.7. Write a new ImageJ plugin that shifts an 8-bit grayscale image horizontally and circularly until the original state is reached again. To display the modified image after each shift, a reference to the corresponding `ImagePlus` object is required (`ImageProcessor` has no display methods). The `ImagePlus` object is only accessible to the plugin's `setup()` method, which is automatically called before the `run()` method. Modify the definition in Prog. 2.1 to keep a reference and to redraw the `ImagePlus` object as follows:

```
public class XY_Plugin implements PlugInFilter {
  ImagePlus im;       // new variable!

  public int setup(String args, ImagePlus im) {
    this.im = im;     // reference to the associated ImagePlus object
    return DOES_8G;
  }

  public void run(ImageProcessor ip) {
    // ... modify ip
    im.updateAndDraw(); // redraw the associated ImagePlus object
    // ...
  }
}
```

3

Histograms and Image Statistics

Histograms are used to depict image statistics in an easily interpreted visual format. With a histogram, it is easy to determine certain types of problems in an image, for example, it is simple to conclude if an image is properly exposed by visual inspection of its histogram. In fact, histograms are so useful that modern digital cameras often provide a real-time histogram overlay on the viewfinder (Fig. 3.1) to help prevent taking poorly exposed pictures. It is important to catch errors like this at the image capture stage because poor exposure results in a permanent loss of information, which it is not possible to recover later using image-processing techniques. In addition to their usefulness during image capture, histograms are also used later to improve the visual appearance of an image and as a "forensic" tool for determining what type of processing has previously been applied to an image. The final part of this chapter shows how to calculate simple image statistics from the original image, its histogram, or the so-called integral image.

Fig. 3.1
Digital camera back display showing the associated RGB histograms.

3.1 What is a Histogram?

Histograms in general are frequency distributions, and histograms of images describe the frequency of the intensity values that occur in an image. This concept can be easily explained by considering an old-fashioned grayscale image like the one shown in Fig. 3.2.

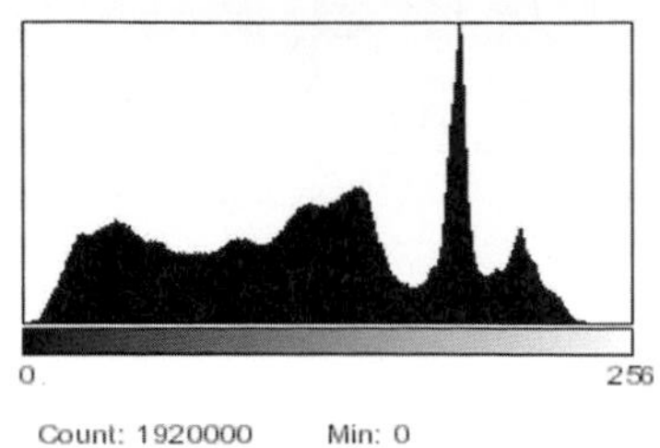

Fig. 3.2
An 8-bit grayscale image and a histogram depicting the frequency distribution of its 256 intensity values.

The histogram h for a grayscale image I with intensity values in the range $I(u,v) \in [0, K-1]$ holds exactly K entries, where $K = 2^8 = 256$ for a typical 8-bit grayscale image. Each single histogram entry is defined as

$$\mathsf{h}(i) = \text{the } \textit{number} \text{ of pixels in } I \text{ with the intensity value } i,$$

for all $0 \leq i < K$. More formally stated,[1]

$$\mathsf{h}(i) = \text{card}\{(u,v) \mid I(u,v) = i\}. \tag{3.1}$$

Therefore, $\mathsf{h}(0)$ is the number of pixels with the value 0, $\mathsf{h}(1)$ the number of pixels with the value 1, and so forth. Finally, $\mathsf{h}(255)$ is the number of all white pixels with the maximum intensity value $255 = K-1$. The result of the histogram computation is a 1D vector h of length K. Figure 3.3 gives an example for an image with $K = 16$ possible intensity values.

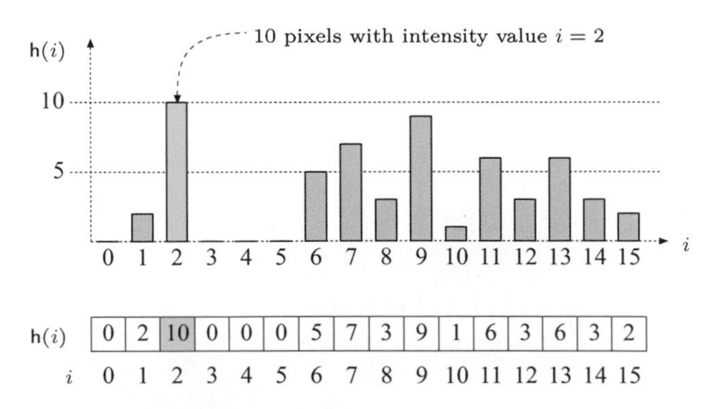

Fig. 3.3
Histogram vector for an image with $K = 16$ possible intensity values. The indices of the vector element $i = 0 \ldots 15$ represent intensity values. The value of 10 at index 2 means that the image contains 10 pixels of intensity value 2.

Since the histogram encodes no information about *where* each of its individual entries originated in the image, it contains no information about the spatial arrangement of pixels in the image. This

[1] card{...} denotes the number of elements ("cardinality") in a set (see also Sec. A.1 in the Appendix).

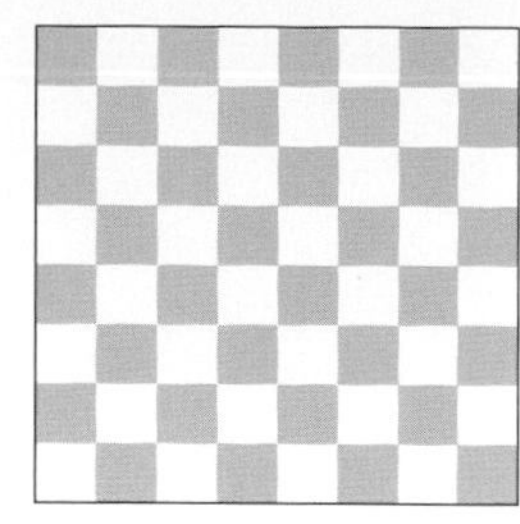

Fig. 3.4
Three very different images with identical histograms.

is intentional, since the main function of a histogram is to provide statistical information, (e.g., the distribution of intensity values) in a compact form. Is it possible to reconstruct an image using only its histogram? That is, can a histogram be somehow "inverted"? Given the loss of spatial information, in all but the most trivial cases, the answer is no. As an example, consider the wide variety of images you could construct using the same number of pixels of a specific value. These images would appear different but have exactly the same histogram (Fig. 3.4).

3.2 Interpreting Histograms

A histogram depicts problems that originate during image acquisition, such as those involving contrast and dynamic range, as well as artifacts resulting from image-processing steps that were applied to the image. Histograms are often used to determine if an image is making effective use of its intensity range (Fig. 3.5) by examining the size and uniformity of the histogram's distribution.

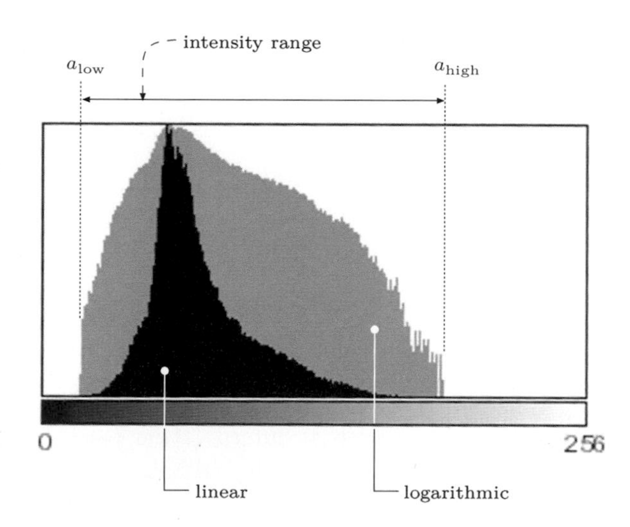

Fig. 3.5
Effective intensity range. The graph depicts the frequencies of pixel values *linearly* (black bars) and *logarithmically* (gray bars). The logarithmic form makes even relatively low occurrences, which can be very important in the image, readily apparent.

3.2.1 Image Acquisition

Histograms make typical exposure problems readily apparent. As an example, a histogram where a large section of the intensity range at one end is largely unused while the other end is crowded with

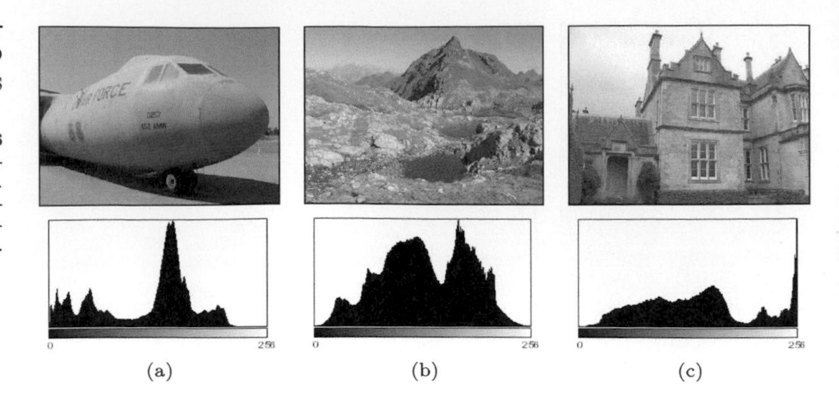

Fig. 3.6
Exposure errors are readily apparent in histograms. Underexposed (a), properly exposed (b), and overexposed (c) photographs.

high-value peaks (Fig. 3.6) is representative of an improperly exposed image.

Contrast

Contrast is understood as the range of intensity values *effectively* used within a given image, that is the difference between the image's maximum and minimum pixel values. A full-contrast image makes effective use of the entire range of available intensity values from $a = a_{\min}, \ldots, a_{\max}$ with $a_{\min} = 0$, $a_{\max} = K-1$ (black to white). Using this definition, image contrast can be easily read directly from the histogram. Figure 3.7 illustrates how varying the contrast of an image affects its histogram.

Dynamic range

The dynamic range of an image is, in principle, understood as the number of *distinct* pixel values in an image. In the ideal case, the dynamic range encompasses all K usable pixel values, in which case the value range is completely utilized. When an image has an available range of contrast $a = a_{\mathrm{low}}, \ldots, a_{\mathrm{high}}$, with

$$a_{\min} < a_{\mathrm{low}} \qquad \text{and} \qquad a_{\mathrm{high}} < a_{\max},$$

then the maximum possible dynamic range is achieved when all the intensity values lying in this range are utilized (i.e., appear in the image; Fig. 3.8).

While the contrast of an image can be increased by transforming its existing values so that they utilize more of the underlying value range available, the dynamic range of an image can only be increased by introducing artificial (that is, not originating with the image sensor) values using methods such as interpolation (see Ch. 22). An image with a high dynamic range is desirable because it will suffer less image-quality degradation during image processing and compression. Since it is not possible to increase dynamic range after image acquisition in a practical way, professional cameras and scanners work at depths of more than 8 bits, often 12–14 bits per channel, in order to provide high dynamic range at the acquisition stage. While most output devices, such as monitors and printers, are unable to actually reproduce more than 256 different shades, a high dynamic range is always beneficial for subsequent image processing or archiving.

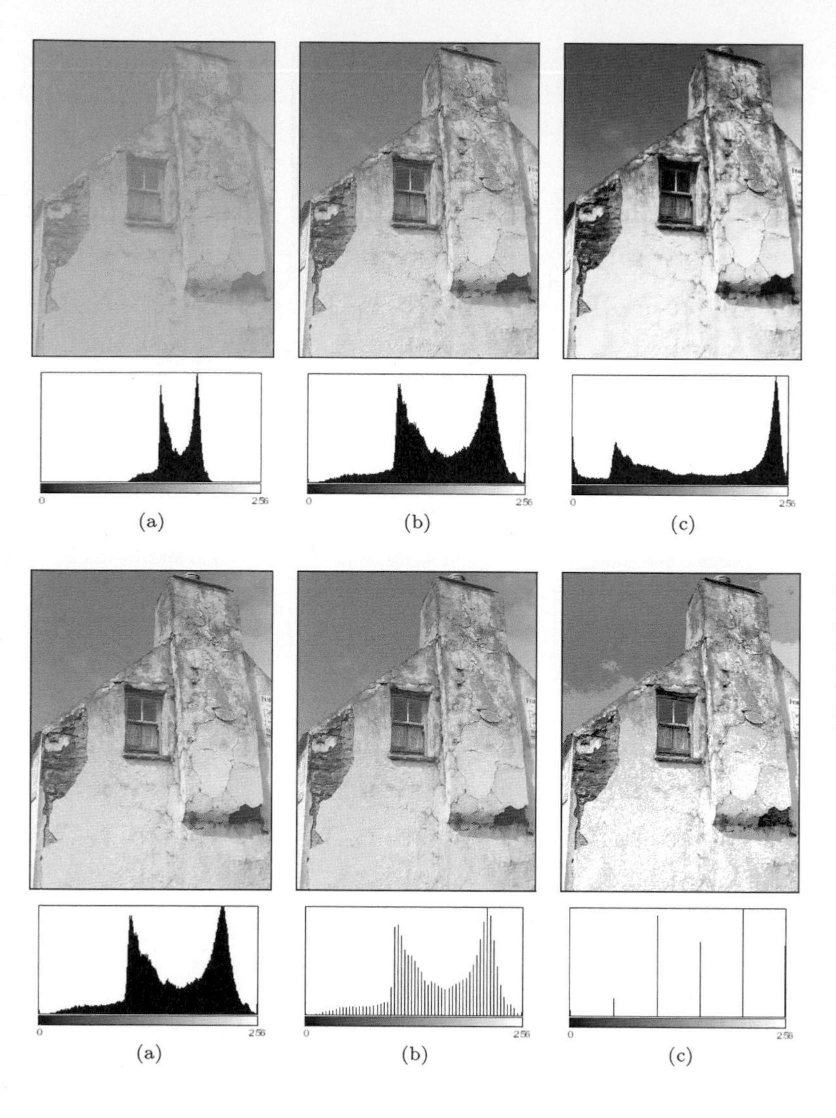

Fig. 3.7
How changes in contrast affect the histogram: low contrast (a), normal contrast (b), high contrast (c).

Fig. 3.8
How changes in dynamic range affect the histogram: high dynamic range (a), low dynamic range with 64 intensity values (b), extremely low dynamic range with only 6 intensity values (c).

3.2.2 Image Defects

Histograms can be used to detect a wide range of image defects that originate either during image acquisition or as the result of later image processing. Since histograms always depend on the visual characteristics of the scene captured in the image, no single "ideal" histogram exists. While a given histogram may be optimal for a specific scene, it may be entirely unacceptable for another. As an example, the ideal histogram for an astronomical image would likely be very different from that of a good landscape or portrait photo. Nevertheless, there are some general rules; for example, when taking a landscape image with a digital camera, you can expect the histogram to have evenly distributed intensity values and no isolated spikes.

Saturation

Ideally the contrast range of a sensor, such as that used in a camera, should be greater than the range of the intensity of the light that it receives from a scene. In such a case, the resulting histogram will be smooth at both ends because the light received from the very bright and the very dark parts of the scene will be less than the light received from the other parts of the scene. Unfortunately, this ideal is often not the case in reality, and illumination outside of the sensor's contrast range, arising for example from glossy highlights and especially dark parts of the scene, cannot be captured and is lost. The result is a histogram that is saturated at one or both ends of its range. The illumination values lying outside of the sensor's range are mapped to its minimum or maximum values and appear on the histogram as significant spikes at the tail ends. This typically occurs in an under- or overexposed image and is generally not avoidable when the inherent contrast range of the scene exceeds the range of the system's sensor (Fig. 3.9(a)).

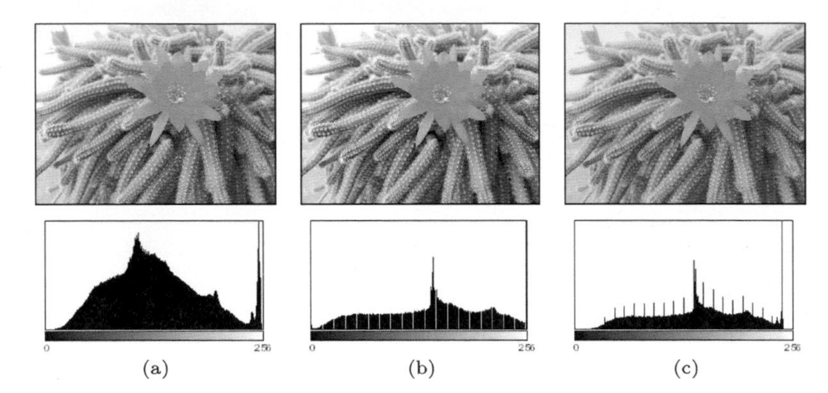

Fig. 3.9
Effect of image capture errors on histograms: saturation of high intensities (a), histogram gaps caused by a slight increase in contrast (b), and histogram spikes resulting from a reduction in contrast (c).

Spikes and gaps

As discussed already, the intensity value distribution for an unprocessed image is generally smooth; that is, it is unlikely that isolated spikes (except for possible saturation effects at the tails) or gaps will appear in its histogram. It is also unlikely that the count of any given intensity value will differ greatly from that of its neighbors (i.e., it is locally smooth). While artifacts like these are observed very rarely in original images, they will often be present after an image has been manipulated, for instance, by changing its contrast. Increasing the contrast (see Ch. 4) causes the histogram lines to separate from each other and, due to the discrete values, gaps are created in the histogram (Fig. 3.9(b)). Decreasing the contrast leads, again because of the discrete values, to the merging of values that were previously distinct. This results in increases in the corresponding histogram entries and ultimately leads to highly visible spikes in the histogram (Fig. 3.9(c)).[2]

Impacts of image compression

Image compression also changes an image in ways that are immediately evident in its histogram. As an example, during GIF compression, an image's dynamic range is reduced to only a few intensities

[2] Unfortunately, these types of errors are also caused by the internal contrast "optimization" routines of some image-capture devices, especially consumer-type scanners.

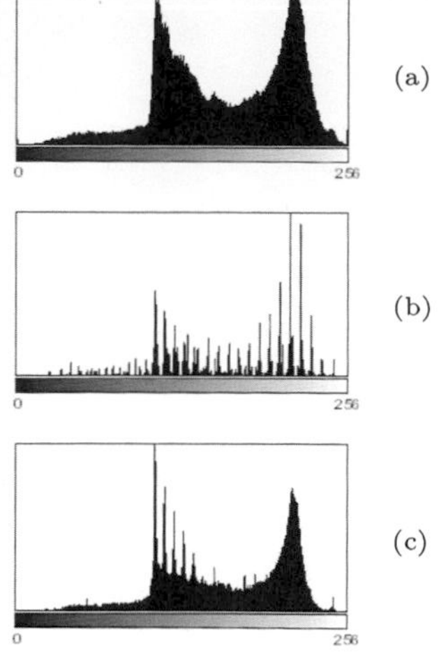

Fig. 3.10
Color quantization effects resulting from GIF conversion. The original image converted to a 256 color GIF image (left). Original histogram (a) and the histogram after GIF conversion (b). When the RGB image is scaled by 50%, some of the lost colors are recreated by interpolation, but the results of the GIF conversion remain clearly visible in the histogram (c).

or colors, resulting in an obvious line structure in the histogram that cannot be removed by subsequent processing (Fig. 3.10). Generally, a histogram can quickly reveal whether an image has ever been subjected to color quantization, such as occurs during conversion to a GIF image, even if the image has subsequently been converted to a full-color format such as TIFF or JPEG.

Figure 3.11 illustrates what occurs when a simple line graphic with only two gray values (128, 255) is subjected to a compression method such as JPEG, that is not designed for line graphics but instead for natural photographs. The histogram of the resulting image clearly shows that it now contains a large number of gray values that were not present in the original image, resulting in a poor-quality image[3] that appears dirty, fuzzy, and blurred.

3.3 Calculating Histograms

Computing the histogram of an 8-bit grayscale image containing intensity values between 0 and 255 is a simple task. All we need is a set of 256 counters, one for each possible intensity value. First, all counters are initialized to zero. Then we iterate through the image I, determining the pixel value p at each location (u, v), and incrementing the corresponding counter by one. At the end, each counter will contain the number of pixels in the image that have the corresponding intensity value.

An image with K possible intensity values requires exactly K counter variables; for example, since an 8-bit grayscale image can contain at most 256 different intensity values, we require 256 counters. While individual counters make sense conceptually, an actual

[3] Using JPEG compression on images like this, for which it was not designed, is one of the most egregious of imaging errors. JPEG is designed for photographs of natural scenes with smooth color transitions, and using it to compress iconic images with large areas of the same color results in strong visual artifacts (see, e.g., Fig. 1.9 on p. 17).

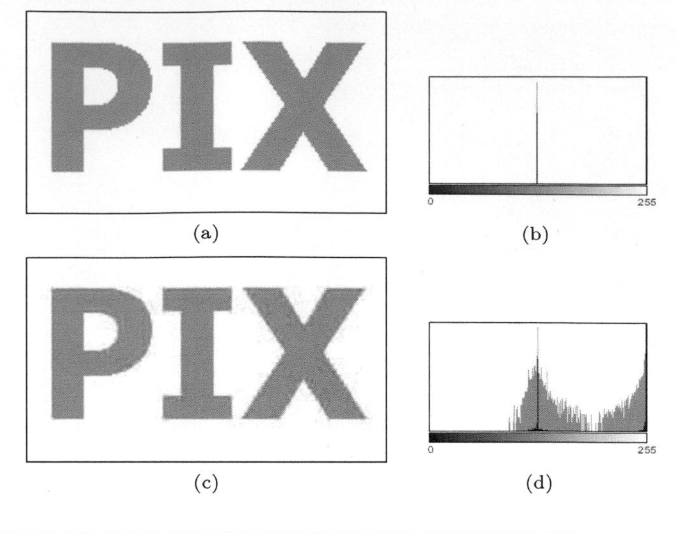

(a) (b) (c) (d)

Fig. 3.11 Effects of JPEG compression. The original image (a) contained only two different gray values, as its histogram (b) makes readily apparent. JPEG compression, a poor choice for this type of image, results in numerous additional gray values, which are visible in both the resulting image (c) and its histogram (d). In both histograms, the linear frequency (black bars) and the logarithmic frequency (gray bars) are shown.

Prog. 3.1 ImageJ plugin for computing the histogram of an 8-bit grayscale image. The `setup()` method returns `DOES_8G + NO_CHANGES`, which indicates that this plugin requires an 8-bit grayscale image and will not alter it (line 4). In Java, all elements of a newly instantiated numerical array are automatically initialized to zero (line 8).

```
1 public class Compute_Histogram implements PlugInFilter {
2
3     public int setup(String arg, ImagePlus img) {
4         return DOES_8G + NO_CHANGES;
5     }
6
7     public void run(ImageProcessor ip) {
8         int[] h = new int[256]; // histogram array
9         int w = ip.getWidth();
10        int h = ip.getHeight();
11
12        for (int v = 0; v < h; v++) {
13            for (int u = 0; u < w; u++) {
14                int i = ip.getPixel(u, v);
15                h[i] = h[i] + 1;
16            }
17        }
18        // ... histogram h can now be used
19    }
20 }
```

implementation would not use K individual *variables* to represent the counters but instead would use an *array* with K entries (`int[256]` in Java). In this example, the actual implementation as an array is straightforward. Since the intensity values begin at zero (like arrays in Java) and are all positive, they can be used directly as the indices $i \in [0, N-1]$ of the histogram array. Program 3.1 contains the complete Java source code for computing a histogram within the `run()` method of an ImageJ plugin.

At the start of Prog. 3.1, the array `h` of type `int[]` is created (line 8) and its elements are automatically initialized[4] to `0`. It makes no difference, at least in terms of the final result, whether the array is

[4] In Java, arrays of primitives such as `int`, `double` are initialized at creation to `0` in the case of integer types or `0.0` for floating-point types, while arrays of objects are initialized to `null`.

traversed in row or column order, as long as all pixels in the image are visited exactly once. In contrast to Prog. 2.1, in this example we traverse the array in the standard row-first order such that the outer `for` loop iterates over the *vertical* coordinates v and the inner loop over the *horizontal* coordinates u.[5] Once the histogram has been calculated, it is available for further processing steps or for being displayed.

Of course, histogram computation is already implemented in ImageJ and is available via the method `getHistogram()` for objects of the class `ImageProcessor`. If we use this built-in method, the `run()` method of Prog. 3.1 can be simplified to

```
public void run(ImageProcessor ip) {
    int[] h = ip.getHistogram();  // built-in ImageJ method
    ... // histogram h can now be used
}
```

3.4 Histograms of Images with More than 8 Bits

Normally histograms are computed in order to visualize the image's distribution on the screen. This presents no problem when dealing with images having $2^8 = 256$ entries, but when an image uses a larger range of values, for instance 16- and 32-bit or floating-point images (see Table 1.1), then the growing number of necessary histogram entries makes this no longer practical.

3.4.1 Binning

Since it is not possible to represent each intensity value with its own entry in the histogram, we will instead let a given entry in the histogram represent a *range* of intensity values. This technique is often referred to as "binning" since you can visualize it as collecting a range of pixel values in a container such as a bin or bucket. In a binned histogram of size B, each bin $\mathsf{h}(j)$ contains the number of image elements having values within the interval $[a_j, a_{j+1})$, and therefore (analogous to Eqn. (3.1))

$$\mathsf{h}(j) = \operatorname{card}\left\{(u,v) \mid a_j \leq I(u,v) < a_{j+1}\right\}, \tag{3.2}$$

for $0 \leq j < B$. Typically the range of possible values in B is divided into bins of equal size $k_B = K/B$ such that the starting value of the interval j is

$$a_j = j \cdot \frac{K}{B} = j \cdot k_B \,.$$

3.4.2 Example

In order to create a typical histogram containing $B = 256$ entries from a 14-bit image, one would divide the original value range $j =$

[5] In this way, image elements are traversed in exactly the same way that they are laid out in computer memory, resulting in more efficient memory access and with it the possibility of increased performance, especially when dealing with larger images (see also Appendix F).

$0, \ldots, 2^{14}-1$ into 256 equal intervals, each of length $k_B = 2^{14}/256 = 64$, such that $a_0 = 0$, $a_1 = 64$, $a_2 = 128$, ..., $a_{255} = 16320$ and $a_{256} = a_B = 2^{14} = 16320 = K$. This gives the following association between pixel values and histogram bins $\mathsf{h}(0), \ldots, \mathsf{h}(255)$:

$$\begin{array}{rcrl} 0, & \ldots, & 63 & \rightarrow \mathsf{h}(0), \\ 64, & \ldots, & 127 & \rightarrow \mathsf{h}(1), \\ 128, & \ldots, & 191 & \rightarrow \mathsf{h}(2), \\ \vdots & & \vdots & \quad \vdots \\ 16320, & \ldots, & 16383 & \rightarrow \mathsf{h}(255). \end{array}$$

3.4.3 Implementation

If, as in the previous example, the value range $0, \ldots, K-1$ is divided into equal length intervals $k_B = K/B$, there is naturally no need to use a mapping table to find a_j since for a given pixel value $a = I(u,v)$ the correct histogram element j is easily computed. In this case, it is enough to simply divide the pixel value $I(u,v)$ by the interval length k_B; that is,

$$\frac{I(u,v)}{k_B} = \frac{I(u,v)}{K/B} = \frac{I(u,v) \cdot B}{K}\,. \tag{3.3}$$

As an index to the appropriate histogram bin $\mathsf{h}(j)$, we require an integer value

$$j = \left\lfloor \frac{I(u,v) \cdot B}{K} \right\rfloor, \tag{3.4}$$

where $\lfloor \cdot \rfloor$ denotes the *floor* operator.[6] A Java method for computing histograms by "linear binning" is given in Prog. 3.2. Note that all the computations from Eqn. (3.4) are done with integer numbers without using any floating-point operations. Also there is no need to explicitly call the *floor* function because the expression

```
a * B / K
```

in line 11 uses integer division and in Java the fractional result of such an operation is truncated, which is equivalent to applying the floor function (assuming positive arguments).[7] The binning method can also be applied, in a similar way, to floating-point images.

3.5 Histograms of Color Images

When referring to histograms of color images, typically what is meant is a histogram of the image intensity (luminance) or of the individual color channels. Both of these variants are supported by practically every image-processing application and are used to objectively appraise the image quality, especially directly after image acquisition.

[6] $\lfloor x \rfloor$ rounds x down to the next whole number (see Appendix A).

[7] For a more detailed discussion, see the section on integer division in Java in Appendix F (p. 765).

```
1   int[] binnedHistogram(ImageProcessor ip) {
2     int K = 256; // number of intensity values
3     int B = 32; // size of histogram, must be defined
4     int[] H = new int[B]; // histogram array
5     int w = ip.getWidth();
6     int h = ip.getHeight();
7
8     for (int v = 0; v < h; v++) {
9       for (int u = 0; u < w; u++) {
10        int a = ip.getPixel(u, v);
11        int i = a * B / K; // integer operations only!
12        H[i] = H[i] + 1;
13      }
14    }
15    // return binned histogram
16    return H;
17  }
```

Prog. 3.2
Histogram computation using "binning" (Java method). Example of computing a histogram with $B = 32$ bins for an 8-bit grayscale image with $K = 256$ intensity levels. The method `binnedHistogram()` returns the histogram of the image object `ip` passed to it as an `int` array of size B.

3.5.1 Intensity Histograms

The intensity or *luminance* histogram $\mathsf{h}_{\mathrm{Lum}}$ of a color image is nothing more than the histogram of the corresponding grayscale image, so naturally all aspects of the preceding discussion also apply to this type of histogram. The grayscale image is obtained by computing the luminance of the individual channels of the color image. When computing the luminance, it is not sufficient to simply average the values of each color channel; instead, a weighted sum that takes into account color perception theory should be computed. This process is explained in detail in Chapter 12 (p. 304).

3.5.2 Individual Color Channel Histograms

Even though the luminance histogram takes into account all color channels, image errors appearing in single channels can remain undiscovered. For example, the luminance histogram may appear clean even when one of the color channels is oversaturated. In RGB images, the blue channel contributes only a small amount to the total brightness and so is especially sensitive to this problem.

Component histograms supply additional information about the intensity distribution within the individual color channels. When computing component histograms, each color channel is considered a separate intensity image and each histogram is computed independently of the other channels. Figure 3.12 shows the luminance histogram $\mathsf{h}_{\mathrm{Lum}}$ and the three component histograms h_{R}, h_{G}, and h_{B} of a typical RGB color image. Notice that saturation problems in all three channels (red in the upper intensity region, green and blue in the lower regions) are obvious in the component histograms but not in the luminance histogram. In this case it is striking, and not at all atypical, that the three component histograms appear completely different from the corresponding luminance histogram $\mathsf{h}_{\mathrm{Lum}}$ (Fig. 3.12(b)).

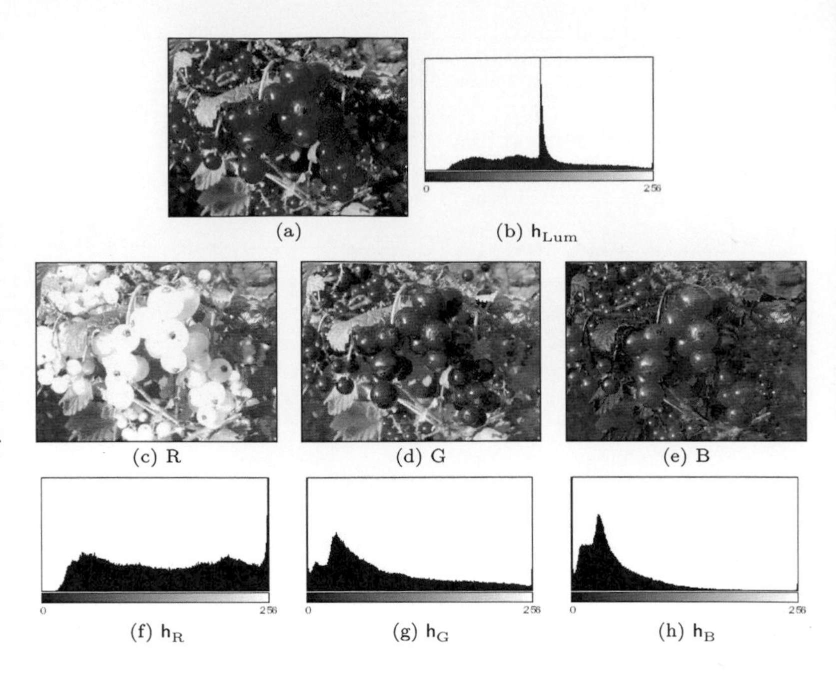

Fig. 3.12 Histograms of an RGB color image: original image (a), luminance histogram h_{Lum} (b), RGB color components as intensity images (c–e), and the associated component histograms h_R, h_G, h_B (f–h). The fact that all three color channels have saturation problems is only apparent in the individual component histograms. The spike in the distribution resulting from this is found in the middle of the luminance histogram (b).

3.5.3 Combined Color Histograms

Luminance histograms and component histograms both provide useful information about the lighting, contrast, dynamic range, and saturation effects relative to the individual color components. It is important to remember that they provide no information about the distribution of the actual *colors* in the image because they are based on the individual color channels and not the combination of the individual channels that forms the color of an individual pixel. Consider, for example, when h_R, the component histogram for the red channel, contains the entry

$$h_R(200) = 24.$$

Then it is only known that the image has 24 pixels that have a red intensity value of 200. The entry does not tell us anything about the green and blue values of those pixels, which could be any valid value ($*$), that is,

$$(r, g, b) = (200, *, *).$$

Suppose further that the three component histograms included the following entries:

$$h_R(50) = 100, \quad h_G(50) = 100, \quad h_B(50) = 100.$$

Could we conclude from this that the image contains 100 pixels with the color combination

$$(r, g, b) = (50, 50, 50)$$

or that this color occurs at all? In general, no, because there is no way of ascertaining from these data if there exists a pixel in the image in which all three components have the value 50. The only thing we could really say is that the color value $(50, 50, 50)$ can occur at most 100 times in this image.

So, although conventional (intensity or component) histograms of color images depict important properties, they do not really provide any useful information about the composition of the actual colors in an image. In fact, a collection of color images can have very similar component histograms and still contain entirely different colors. This leads to the interesting topic of the *combined* histogram, which uses statistical information about the combined color components in an attempt to determine if two images are roughly similar in their color composition. Features computed from this type of histogram often form the foundation of color-based image retrieval methods. We will return to this topic in Chapter 12, where we will explore color images in greater detail.

3.6 The Cumulative Histogram

The cumulative histogram, which is derived from the ordinary histogram, is useful when performing certain image operations involving histograms; for instance, histogram equalization (see Sec. 4.5). The cumulative histogram H is defined as

$$\mathsf{H}(i) = \sum_{j=0}^{i} \mathsf{h}(j) \qquad \text{for } 0 \leq i < K. \tag{3.5}$$

A particular value $\mathsf{H}(i)$ is thus the sum of all histogram values $\mathsf{h}(j)$, with $j \leq i$. Alternatively, we can define H recursively (as implemented in Prog. 4.2 on p. 66):

$$\mathsf{H}(i) = \begin{cases} \mathsf{h}(0) & \text{for } i = 0, \\ \mathsf{H}(i-1) + \mathsf{h}(i) & \text{for } 0 < i < K. \end{cases} \tag{3.6}$$

The cumulative histogram $\mathsf{H}(i)$ is a monotonically increasing function with the maximum value

$$\mathsf{H}(K-1) = \sum_{j=0}^{K-1} \mathsf{h}(j) = M \cdot N, \tag{3.7}$$

that is, the total number of pixels in an image of width M and height N. Figure 3.13 shows a concrete example of a cumulative histogram.

The cumulative histogram is useful not primarily for viewing but as a simple and powerful tool for capturing statistical information from an image. In particular, we will use it in the next chapter to compute the parameters for several common point operations (see Sec. 4.4–4.6).

3.7 Statistical Information from the Histogram

Some common statistical parameters of the image can be conveniently calculated directly from its histogram. For example, the minimum and maximum pixel value of an image I can be obtained by simply

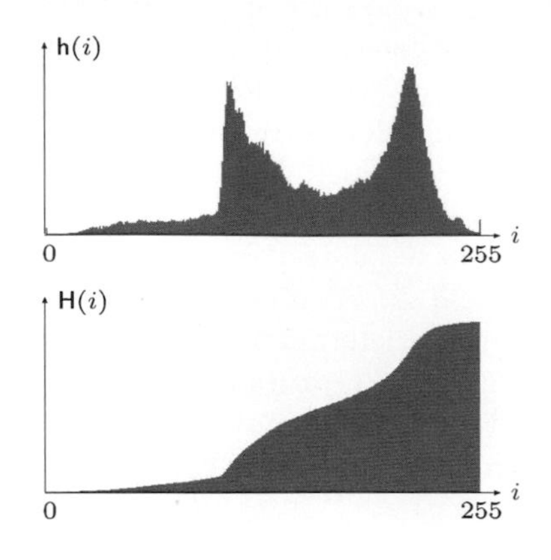

Fig. 3.13
The ordinary histogram $\mathsf{h}(i)$ and its associated cumulative histogram $\mathsf{H}(i)$.

finding the smallest and largest histogram index with nonzero value, i.e.,

$$\begin{aligned} \min(I) &= \min\{\, i \mid \mathsf{h}(i) > 0\}, \\ \max(I) &= \max\{\, i \mid \mathsf{h}(i) > 0\}. \end{aligned} \tag{3.8}$$

If we assume that the histogram is already available, the advantage is that the calculation does not include the entire image but only the relatively small set of histogram elements (typ. 256).

3.7.1 Mean and Variance

The *mean* value μ of an image I (of size $M \times N$) can be calculated as

$$\mu = \frac{1}{MN} \cdot \sum_{u=0}^{M-1} \sum_{v=0}^{N-1} I(u,v) = \frac{1}{MN} \cdot \sum_{i=0}^{K-1} \mathsf{h}(i) \cdot i, \tag{3.9}$$

i.e., either directly from the pixel values $I(u,v)$ or indirectly from the histogram h (of size K), where $MN = \sum_i \mathsf{h}(i)$ is the total number of pixels.

Analogously we can also calculate the *variance* of the pixel values straight from the histogram as

$$\sigma^2 = \frac{1}{MN} \cdot \sum_{u=0}^{M-1} \sum_{v=0}^{N-1} \bigl[I(u,v) - \mu\bigr]^2 = \frac{1}{MN} \cdot \sum_{i=0}^{K-1} (i-\mu)^2 \cdot \mathsf{h}(i). \tag{3.10}$$

As we see in the right parts of Eqns. (3.9) and (3.10), there is no need to access the original pixel values.

The formulation of the variance in Eqn. (3.10) assumes that the arithmetic mean μ has already been determined. This is not necessary though, since the mean and the variance can be calculated together in a single iteration over the image pixels or the associated histogram in the form

$$\mu = \frac{1}{MN} \cdot A \qquad \text{and} \tag{3.11}$$

$$\sigma^2 = \frac{1}{MN} \cdot \Bigl(B - \frac{1}{MN} \cdot A^2\Bigr), \tag{3.12}$$

with the quantities

$$A = \sum_{u=0}^{M-1}\sum_{v=0}^{N-1} I(u,v) \;= \sum_{i=0}^{K-1} i\cdot \mathsf{h}(i), \tag{3.13}$$

$$B = \sum_{u=0}^{M-1}\sum_{v=0}^{N-1} I^2(u,v) = \sum_{i=0}^{K-1} i^2\cdot \mathsf{h}(i). \tag{3.14}$$

The above formulation has the additional numerical advantage that all summations can be performed with integer values, in contrast to Eqn. (3.10) which requires the summation of floating-point values.

3.7.2 Median

The median m of an image is defined as the smallest pixel value that is greater or equal to one half of all pixel values, i.e., lies "in the middle" of the pixel values.[8] The median can also be easily calculated from the image's histogram.

To determine the median of an image I from the associated histogram it is sufficient to find the index i that separates the histogram into two halves, such that the sum of the histogram entries to the left and the right of i are approximately equal. In other words, i is the smallest index where the sum of the histogram entries below (and including) i corresponds to at least half of the image size, that is,

$$m = \min\Bigl\{ i \mid \sum_{j=0}^{i} \mathsf{h}(j) \geq \frac{MN}{2} \Bigr\}. \tag{3.15}$$

Since $\sum_{j=0}^{i} \mathsf{h}(j) = \mathsf{H}(i)$ (see Eqn. (3.5)), the median calculation can be formulated even simpler as

$$m = \min\Bigl\{ i \mid \mathsf{H}(i) \geq \frac{MN}{2} \Bigr\}, \tag{3.16}$$

given the cumulative histogram H.

3.8 Block Statistics

3.8.1 Integral Images

Integral images (also known as *summed area tables* [58]) provide a simple way for quickly calculating elementary statistics of arbitrary rectangular sub-images. They have found use in several interesting applications, such as fast filtering, adaptive thresholding, image matching, local feature extraction, face detection, and stereo reconstruction [20, 142, 244].

Given a scalar-valued (grayscale) image $I\colon M \times N \mapsto \mathbb{R}$ the associated *first-order* integral image is defined as

$$\Sigma_1(u,v) = \sum_{i=0}^{u}\sum_{j=0}^{v} I(i,j). \tag{3.17}$$

[8] See Sec. 5.4.2 for an alternative definition of the median.

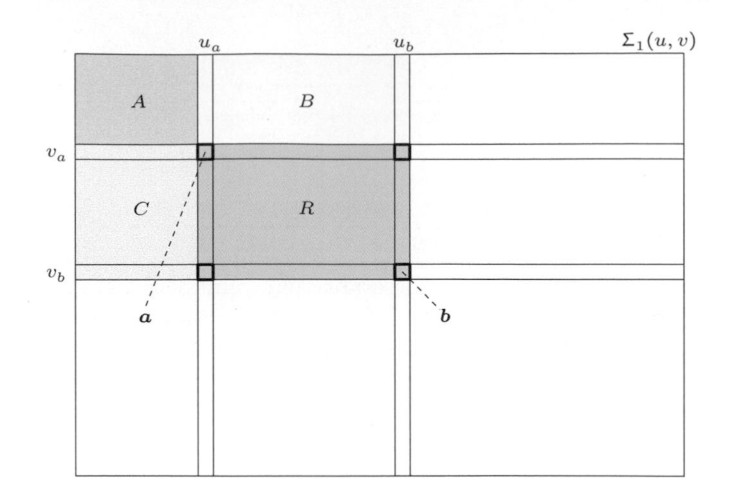

Fig. 3.14
Block-based calculations with integral images. Only four samples from the integral image Σ_1 are required to calculate the sum of the pixels inside the (green) rectangle $R = \langle \boldsymbol{a}, \boldsymbol{b} \rangle$, defined by the corner coordinates $\boldsymbol{a} = (u_a, v_a)$ and $\boldsymbol{b} = (u_b, v_b)$.

Thus a value in Σ_1 is the sum of all pixel values in the original image I located to the left and above the given position (u, v), inclusively. The integral image can be calculated efficiently with a single pass over the image I by using the recurrence relation

$$\Sigma_1(u,v) = \begin{cases} 0 & \text{for } u<0 \text{ or } v<0, \\ \Sigma_1(u-1,v) + \Sigma_1(u,v-1) - \\ \qquad \Sigma_1(u-1,v-1) + I(u,v) & \text{for } u,v \geq 0, \end{cases} \tag{3.18}$$

for positions $u = 0, \ldots, M-1$ and $v = 0, \ldots, N-1$ (see Alg. 3.1).

Suppose now that we wanted to calculate the sum of the pixel values in a given rectangular region R, defined by the corner positions $\boldsymbol{a} = (u_a, v_a)$, $\boldsymbol{b} = (u_b, v_b)$, that is, the *first-order block sum*

$$S_1(R) = \sum_{i=u_a}^{u_b} \sum_{j=v_a}^{v_b} I(i,j), \tag{3.19}$$

from the integral image Σ_1. As shown in Fig. 3.14, the quantity $\Sigma_1(u_a-1, v_a-1)$ corresponds to the pixel sum within rectangle A, and $\Sigma_1(u_b, v_b)$ is the pixel sum over all four rectangles A, B, C and R, that is,

$$\begin{aligned} \Sigma_1(u_a-1, v_a-1) &= S_1(A), \\ \Sigma_1(u_b, v_a-1) &= S_1(A) + S_1(B), \\ \Sigma_1(u_a-1, v_b) &= S_1(A) + S_1(C), \\ \Sigma_1(u_b, v_b) &= S_1(A) + S_1(B) + S_1(C) + S_1(R). \end{aligned} \tag{3.20}$$

Thus $S_1(R)$ can be calculated as

$$\begin{aligned} S_1(R) &= \underbrace{S_1(A)+S_1(B)+S_1(C)+S_1(R)}_{\Sigma_1(u_b,v_b)} + \underbrace{S_1(A)}_{\Sigma_1(u_a-1,v_a-1)} \\ &\quad - \underbrace{[S_1(A)+S_1(B)]}_{\Sigma_1(u_b,v_a-1)} - \underbrace{[S_1(A)+S_1(C)]}_{\Sigma_1(u_a-1,v_b)} \\ &= \Sigma_1(u_b,v_b) + \Sigma_1(u_a-1,v_a-1) - \Sigma_1(u_b,v_a-1) - \Sigma_1(u_a-1,v_b), \end{aligned} \tag{3.21}$$

that is, by taking only *four* samples from the integral image Σ_1.

3.8.2 Mean Intensity

Given the region size N_R and the sum of the pixel values $S_1(R)$, the average intensity value (*mean*) inside the rectangle R can now easily be found as

$$\mu_R = \frac{1}{N_R} \cdot S_1(R), \tag{3.22}$$

with $S_1(R)$ as defined in Eqn. (3.21) and the region size

$$N_R = |R| = (u_b - u_a + 1) \cdot (v_b - v_a + 1). \tag{3.23}$$

3.8.3 Variance

Calculating the *variance* inside a rectangular region R requires the summation of squared intensity values, that is, tabulating

$$\Sigma_2(u,v) = \sum_{i=0}^{u} \sum_{j=0}^{v} I^2(i,j), \tag{3.24}$$

which can be performed analogously to Eqn. (3.18) in the form

$$\Sigma_2(u,v) = \begin{cases} 0 & \text{for } u < 0 \text{ or } v < 0, \\ \Sigma_2(u{-}1,v) + \Sigma_2(u,v{-}1) - \\ \quad \Sigma_2(u{-}1,v{-}1) + I^2(u,v) & \text{for } u,v \geq 0. \end{cases} \tag{3.25}$$

As in Eqns. (3.19)–(3.21), the sum of the *squared* values inside a given rectangle R (i.e., the *second-order block sum*) can be obtained as

$$\begin{aligned} S_2(R) &= \sum_{i=u_0}^{u_1} \sum_{j=v_0}^{v_1} I^2(i,j) \qquad (3.26) \\ &= \Sigma_2(u_b,v_b) + \Sigma_2(u_a{-}1,v_a{-}1) - \Sigma_2(u_b,v_a{-}1) - \Sigma_2(u_a{-}1,v_b). \end{aligned}$$

From this, the variance inside the rectangular region R is finally calculated as

$$\sigma_R^2 = \frac{1}{N_R} \big[S_2(R) - \frac{1}{N_R} \cdot (S_1(R))^2 \big], \tag{3.27}$$

with N_R as defined in Eqn. (3.23). In addition, certain higher-order statistics can be efficiently calculated with summation tables in a similar fashion.

3.8.4 Practical Calculation of Integral Images

Algorithm 3.1 shows how Σ_1 and Σ_2 can be calculated in a single iteration over the original image I. Note that the accumulated values in the integral images Σ_1, Σ_2 tend to become quite large. Even with pictures of medium size and 8-bit intensity values, the range of 32-bit integers is quickly exhausted (particularly when calculating Σ_2). The use of 64-bit integers (type `long` in Java) or larger is recommended to avoid arithmetic overflow. A basic implementation of integral images is available as part of the `imagingbook` library.[9]

[9] Class `imagingbook.lib.image.IntegralImage`.

Alg. 3.1
Joint calculation of the integral images Σ_1 and Σ_2 for a scalar-valued image I.

```
1:  IntegralImage(I)
        Input: I, a scalar-valued input image with I(u,v) ∈ ℝ.
        Returns the first and second order integral images of I.
2:      (M, N) ← Size(I)
3:      Create maps Σ1, Σ2 : M × N ↦ ℝ
        Process the first image line (v = 0):
4:      Σ1(0,0) ← I(0,0)
5:      Σ2(0,0) ← I²(0,0)
6:      for u ← 1,...,M−1 do
7:          Σ1(u,0) ← Σ1(u−1,0) + I(u,0)
8:          Σ2(u,0) ← Σ2(u−1,0) + I²(u,0)
        Process the remaining image lines (v > 0):
9:      for v ← 1,...,N−1 do
10:         Σ1(0,v) ← Σ1(0,v−1) + I(0,v)
11:         Σ2(0,v) ← Σ2(0,v−1) + I²(0,v)
12:         for u ← 1,...,M−1 do
13:             Σ1(u,v) ← Σ1(u−1,v) + Σ1(u,v−1) −
                                  Σ1(u−1,v−1) + I(u,v)
14:             Σ2(u,v) ← Σ2(u−1,v) + Σ2(u,v−1) −
                                  Σ2(u−1,v−1) + I²(u,v)
15:     return (Σ1, Σ2)
```

3.9 Exercises

Exercise 3.1. In Prog. 3.2, B and K are constants. Consider if there would be an advantage to computing the value of B/K outside of the loop, and explain your reasoning.

Exercise 3.2. Develop an ImageJ plugin that computes the cumulative histogram of an 8-bit grayscale image and displays it as a new image, similar to $\mathsf{H}(i)$ in Fig. 3.13. *Hint:* Use the `ImageProcessor` method `int[] getHistogram()` to retrieve the original image's histogram values and then compute the cumulative histogram "in place" according to Eqn. (3.6). Create a new (blank) image of appropriate size (e.g., 256×150) and draw the scaled histogram data as black vertical bars such that the maximum entry spans the full height of the image. Program 3.3 shows how this plugin could be set up and how a new image is created and displayed.

Exercise 3.3. Develop a technique for nonlinear binning that uses a table of interval limits a_j (Eqn. (3.2)).

Exercise 3.4. Develop an ImageJ plugin that uses the Java methods `Math.random()` or `Random.nextInt(int n)` to create an image with random pixel values that are uniformly distributed in the range $[0, 255]$. Analyze the image's histogram to determine how equally distributed the pixel values truly are.

Exercise 3.5. Develop an ImageJ plugin that creates a random image with a Gaussian (normal) distribution with mean value $\mu = 128$ and standard deviation $\sigma = 50$. Use the standard Java method `double Random.nextGaussian()` to produce normally-distributed random numbers (with $\mu = 0$ and $\sigma = 1$) and scale them appropriately to pixel values. Analyze the resulting image histogram to see if it shows a Gaussian distribution too.

Exercise 3.6. Implement the calculation of the arithmetic *mean* μ and the *variance* σ^2 of a given grayscale image from its histogram h (see Sec. 3.7.1). Compare your results to those returned by ImageJ's Analyze ▷ Measure tool (they should match *exactly*).

Exercise 3.7. Implement the first-order integral image (Σ_1) calculation described in Eqn. (3.18) and calculate the sum of pixel values $S_1(R)$ inside a given rectangle R using Eqn. (3.21). Verify numerically that the results are the same as with the naive formulation in Eqn. (3.19).

Exercise 3.8. Values of integral images tend to become quite large. Assume that 32-bit signed integers (`int`) are used to calculate the integral of the squared pixel values, that is, Σ_2 (see Eqn. (3.24)), for an 8-bit grayscale image. What is the maximum image size that is guaranteed not to cause an arithmetic overflow? Perform the same analysis for 64-bit signed integers (`long`).

Exercise 3.9. Calculate the integral image Σ_1 for a given image I, convert it to a floating-point iamge (`FloatProcessor`) and display the result. You will realize that integral images are without any apparent structure and they all look more or less the same. Come up with an efficient method for reconstructing the original image I from Σ_1.

Prog. 3.3
Creating and displaying a new image (ImageJ plugin). First, we create a ByteProcessor object (histIp, line 20) that is subsequently filled. At this point, histIp has no screen representation and is thus not visible. Then, an associated ImagePlus object is created (line 33) and displayed by applying the show() method (line 34). Notice how the title (String) is retrieved from the original image inside the setup() method (line 10) and used to compose the new image's title (lines 30 and 33). If histIp is changed *after* calling show(), then the method updateAndDraw() could be used to redisplay the associated image again (line 34).

```
1 import ij.ImagePlus;
2 import ij.plugin.filter.PlugInFilter;
3 import ij.process.ByteProcessor;
4 import ij.process.ImageProcessor;
5
6 public class Create_New_Image implements PlugInFilter {
7   ImagePlus im;
8
9   public int setup(String arg, ImagePlus im) {
10    this.im = im;
11    return DOES_8G + NO_CHANGES;
12  }
13
14  public void run(ImageProcessor ip) {
15    // obtain the histogram of ip:
16    int[] hist = ip.getHistogram();
17    int K = hist.length;
18
19    // create the histogram image:
20    ImageProcessor hip = new ByteProcessor(K, 100);
21    hip.setValue(255);  // white = 255
22    hip.fill();
23
24    // draw the histogram values as black bars in hip here,
25    // for example, using hip.putpixel(u, v, 0)
26    // ...
27
28    // compose a nice title:
29    String imTitle = im.getShortTitle();
30    String histTitle = "Histogram of " + imTitle;
31
32    // display the histogram image:
33    ImagePlus him = new ImagePlus(title, hip);
34    him.show();
35  }
36 }
```

4

Point Operations

Point operations perform a modification of the pixel values without changing the size, geometry, or local structure of the image. Each new pixel value $b = I'(u, v)$ depends exclusively on the previous value $a = I(u, v)$ at the *same* position and is thus independent from any other pixel value, in particular from any of its neighboring pixels.[1] The original pixel values a are mapped to the new values b by some given function f, i.e.,

$$b = f\big(I(u, v)\big) \qquad \text{or} \qquad b = f(a). \tag{4.1}$$

If, as in this case, the function $f()$ is independent of the image coordinates (i.e., the same throughout the image), the operation is called "global" or "homogeneous". Typical examples of homogeneous point operations include, among others:

- modifying image brightness or contrast,
- applying arbitrary intensity transformations ("curves"),
- inverting images,
- quantizing (or "posterizing") images,
- global thresholding,
- gamma correction,
- color transformations
- etc.

We will look at some of these techniques in more detail in the following.

In contrast to Eqn. (4.1), the mapping function $g()$ for a *nonhomogeneous* point operation would also take into account the current image coordinate (u, v), that is,

$$b = g\big(I(u, v), u, v\big) \qquad \text{or} \qquad b = f(a, u, v). \tag{4.2}$$

A typical nonhomogeneous operation is the local adjustment of contrast or brightness used, for example, to compensate for uneven lighting during image acquisition.

[1] If the result depends on more than one pixel value, the operation is called a "filter", as described in Chapter 5.

4.1 Modifying Image Intensity

4.1.1 Contrast and Brightness

Let us start with a simple example. Increasing the image's contrast by 50% (i.e., by the factor 1.5) or raising the brightness by 10 units can be expressed by the mapping functions

$$f_{\mathrm{contr}}(a) = a \cdot 1.5 \qquad \text{or} \qquad f_{\mathrm{bright}}(a) = a + 10\,, \tag{4.3}$$

respectively. The first operation is implemented as an ImageJ plugin by the code shown in Prog. 4.1, which can easily be adapted to perform any other type of point operation. Rounding to the nearest integer values is accomplished by simply adding 0.5 before the truncation effected by the `(int)` typecast in line 8 (this only works for positive values). Also note the use of the more efficient image processor methods `get()` and `set()` (instead of `getPixel()` and `putPixel()`) in this example.

Prog. 4.1
Point operation to increase the contrast by 50% (ImageJ plugin). Note that in line 8 the result of the multiplication of the integer pixel value by the constant 1.5 (implicitly of type double) is of type double. Thus an explicit type cast (int) is required to assign the value to the int variable a. 0.5 is added in line 8 to round to the nearest integer values.

```
1  public void run(ImageProcessor ip) {
2    int w = ip.getWidth();
3    int h = ip.getHeight();
4
5    for (int v = 0; v < h; v++) {
6      for (int u = 0; u < w; u++) {
7        int a = ip.get(u, v);
8        int b = (int) (a * 1.5 + 0.5);
9        if (b > 255)
10         b = 255;   // clamp to the maximum value (a_max)
11       ip.set(u, v, b);
12     }
13   }
14 }
```

4.1.2 Limiting Values by Clamping

When implementing arithmetic operations on pixel values, we must keep in mind that the calculated results must not exceed the admissible range of pixel values for the given image type (e.g., $[0, 255]$ in the case of 8-bit grayscale images). This is commonly called "clamping" and can be expressed in the form

$$b = \min(\max(f(a), a_{\min}), a_{\max}) = \begin{cases} a_{\min} & \text{for } f(a) < a_{\min}, \\ a_{\max} & \text{for } f(a) > a_{\max}, \\ f(a) & \text{otherwise.} \end{cases} \tag{4.4}$$

For this purpose, line 10 of Prog. 4.1 contains the statement

```
if (b > 255) b = 255;
```

which limits the result to the maximum value 255. Similarly, one may also want to limit the results to the minimum value (0) to avoid negative pixel values (which cannot be represented by this type of 8-bit image), for example, by the statement

```
if (b < 0) b = 0;
```

The above statement is not needed in Prog. 4.1 because the intermediate results can never be negative in this particular operation.

4.1.3 Inverting Images

Inverting an intensity image is a simple point operation that reverses the ordering of pixel values (by multiplying by −1) and adds a constant value to map the result to the admissible range again. Thus for a pixel value $a = I(u,v)$ in the range $[0, a_{\max}]$, the corresponding point operation is

$$f_{\mathrm{inv}}(a) = -a + a_{\max} = a_{\max} - a. \tag{4.5}$$

The inversion of an 8-bit grayscale image with $a_{\max} = 255$ was the task of our first plugin example in Sec. 2.2.4 (Prog. 2.1). Note that in this case no clamping is required at all because the function always maps to the original range of values. In ImageJ, this operation is performed by the method `invert()` (for objects of type `ImageProcessor`) and is also available through the Edit ▷ Invert menu. Obviously, inverting an image mirrors its histogram, as shown in Fig. 4.5(c).

4.1.4 Threshold Operation

Thresholding an image is a special type of quantization that separates the pixel values in two classes, depending upon a given threshold value q that is usually constant. The threshold operation maps all pixels to one of two fixed intensity values a_0 or a_1, that is,

$$f_{\mathrm{threshold}}(a) = \begin{cases} a_0 & \text{for } a < q, \\ a_1 & \text{for } a \geq q, \end{cases} \tag{4.6}$$

with $0 < q \leq a_{\max}$. A common application is *binarizing* an intensity image with the values $a_0 = 0$ and $a_1 = 1$.

ImageJ does provide a special image type (`BinaryProcessor`) for binary images, but these are actually implemented as 8-bit intensity images (just like ordinary intensity images) using the values 0 and 255. ImageJ also provides the `ImageProcessor` method `threshold(int level)`, with *level* $\equiv q$, to perform this operation, which can also be invoked through the Image ▷ Adjust ▷ Threshold menu (see Fig. 4.1 for an example). Thresholding affects the histogram by separating the distribution into two entries at positions a_0 and a_1, as illustrated in Fig. 4.2.

4.2 Point Operations and Histograms

We have already seen that the effects of a point operation on the image's histogram are quite easy to predict in some cases. For example, increasing the brightness of an image by a constant value shifts the entire histogram to the right, raising the contrast widens

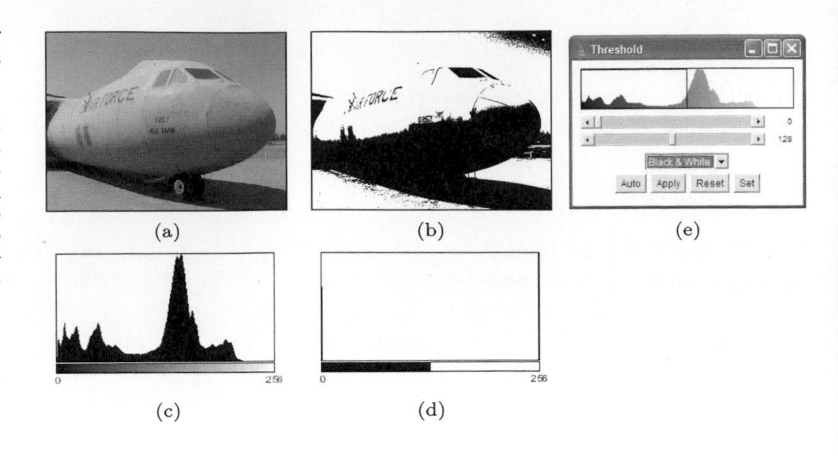

(a) (b) (e)

(c) (d)

Fig. 4.1
Threshold operation: original image (a) and corresponding histogram (c); result after thresholding with $a_{th} = 128$, $a_0 = 0$, $a_1 = 255$ (b) and corresponding histogram (d); ImageJ's interactive Threshold menu (e).

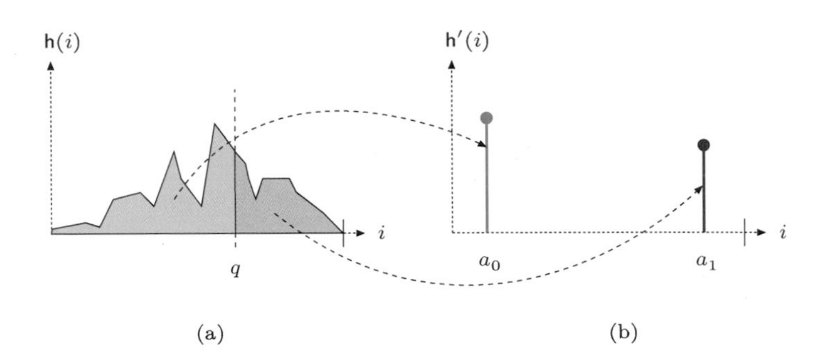

(a) (b)

Fig. 4.2
Effects of thresholding upon the histogram. The threshold value is a_{th}. The original distribution (a) is split and merged into two isolated entries at a_0 and a_1 in the resulting histogram (b).

it, and inverting the image flips the histogram. Although this appears rather simple, it may be useful to look a bit more closely at the relationship between point operations and the resulting changes in the histogram.

As the illustration in Fig. 4.3 shows, every entry (bar) at some position i in the histogram maps to a *set* (of size $\mathsf{h}(i)$) containing all image pixels whose values are exactly i.[2]

If a particular histogram line is *shifted* as a result of some point operation, then of course all pixels in the corresponding set are equally modified and vice versa. So what happens when a point operation (e.g., reducing image contrast) causes two previously separated histogram lines to fall together at the same position i? The answer is that the corresponding pixel sets are *merged* and the new common histogram entry is the sum of the two (or more) contributing entries (i.e., the size of the combined set). At this point, the elements in the merged set are no longer distinguishable (or separable), so this operation may have (perhaps unintentionally) caused an irreversible reduction of dynamic range and thus a permanent loss of information in that image.

[2] Of course this is only true for ordinary histograms with an entry for every single intensity value. If *binning* is used (see Sec. 3.4.1), each histogram entry maps to pixels within a certain *range* of values.

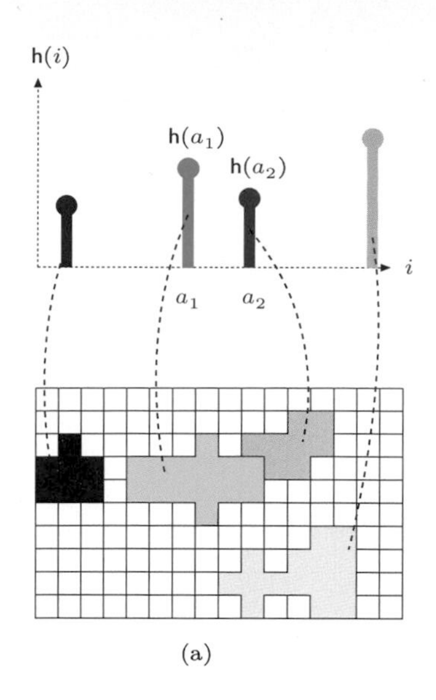

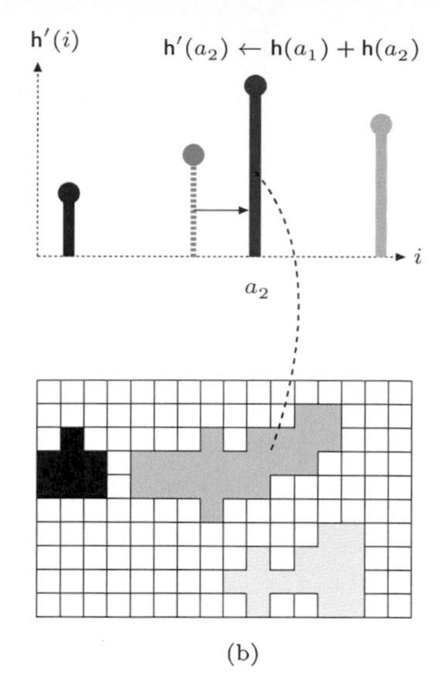

Fig. 4.3
Histogram entries represent *sets* of pixels of the same value. If a histogram line is moved as a result of some point operation, then all pixels in the corresponding set are equally modified (a). If, due to this operation, two histogram lines $h(a_1)$, $h(a_2)$ coincide on the same index, the two corresponding pixel sets merge and the contained pixels become undiscernable (b).

4.3 Automatic Contrast Adjustment

Automatic contrast adjustment (auto-contrast) is a point operation whose task is to modify the pixels such that the available range of values is fully covered. This is done by mapping the current darkest and brightest pixels to the minimum and maximum intensity values, respectively, and linearly distributing the intermediate values.

Let us assume that a_{lo} and a_{hi} are the lowest and highest pixel values found in the current image, whose full intensity range is $[a_{min}, a_{max}]$. To stretch the image to the full intensity range (see Fig. 4.4), we first map the smallest pixel value a_{lo} to zero, subsequently increase the contrast by the factor $(a_{max} - a_{min})/(a_{hi} - a_{lo})$, and finally shift to the target range by adding a_{min}. The mapping function for the auto-contrast operation is thus defined as

$$f_{ac}(a) = a_{min} + \big(a - a_{lo}\big) \cdot \frac{a_{max} - a_{min}}{a_{hi} - a_{lo}}, \tag{4.7}$$

provided that $a_{hi} \neq a_{lo}$; that is, the image contains at least *two* different pixel values. For an 8-bit image with $a_{min} = 0$ and $a_{max} = 255$, the function in Eqn. (4.7) simplifies to

$$f_{ac}(a) = (a - a_{lo}) \cdot \frac{255}{a_{hi} - a_{lo}}. \tag{4.8}$$

The target range $[a_{min}, a_{max}]$ need not be the maximum available range of values but can be any interval to which the image should be mapped. Of course the method can also be used to reduce the image contrast to a smaller range. Figure 4.5(b) shows the effects of an auto-contrast operation on the corresponding histogram, where the linear stretching of the intensity range results in regularly spaced gaps in the new distribution.

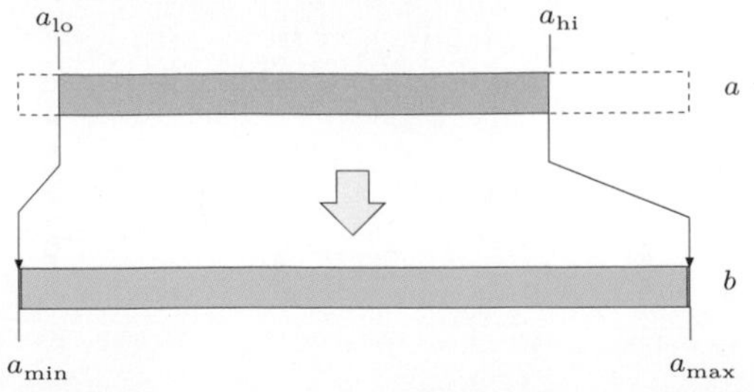

Fig. 4.4
Auto-contrast operation according to Eqn. (4.7). Original pixel values a in the range $[a_{\mathrm{lo}}, a_{\mathrm{hi}}]$ are mapped linearly to the target range $[a_{\min}, a_{\max}]$.

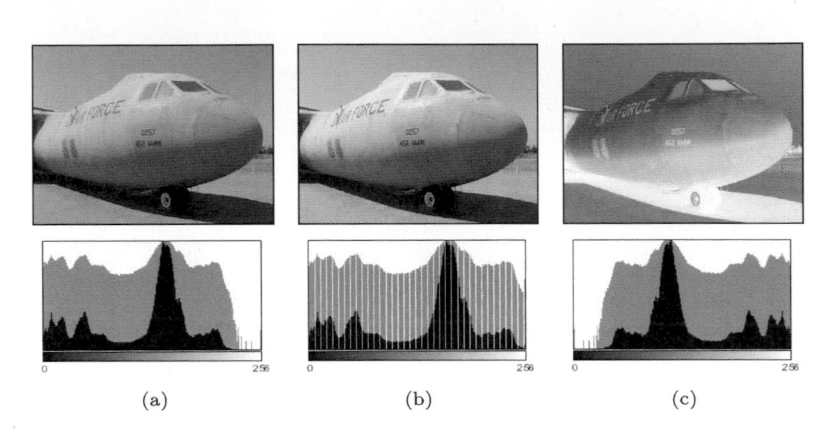

Fig. 4.5
Effects of auto-contrast and inversion operations on the resulting histograms. Original image (a), result of auto-contrast operation (b), and inversion (c). The histogram entries are shown both linearly (black bars) and logarithmically (gray bars).

4.4 Modified Auto-Contrast Operation

In practice, the mapping function in Eqn. (4.7) could be strongly influenced by only a few extreme (low or high) pixel values, which may not be representative of the main image content. This can be avoided to a large extent by "saturating" a fixed percentage $(p_{\mathrm{lo}}, p_{\mathrm{hi}})$ of pixels at the upper and lower ends of the target intensity range. To accomplish this, we determine two limiting values a'_{lo}, a'_{hi} such that a predefined quantile q_{lo} of all pixel values in the image I are smaller than a'_{lo} and another quantile q_{hi} of the values are greater than a'_{hi} (Fig. 4.6).

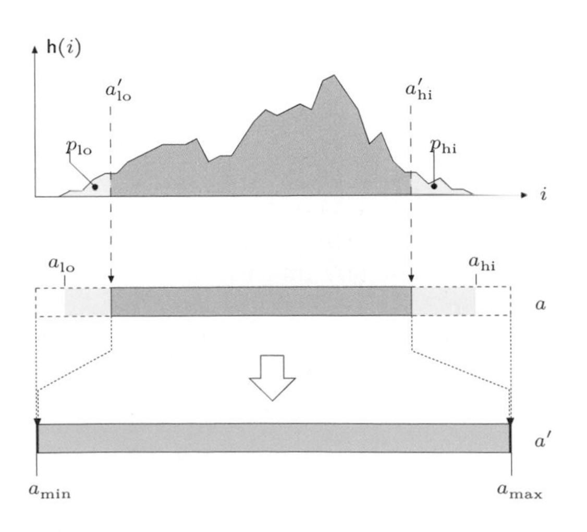

Fig. 4.6
Modified auto-contrast operation (Eqn. (4.11)). Predefined quantiles $(q_{\mathrm{lo}}, q_{\mathrm{hi}})$ of image pixels—shown as dark areas at the left and right ends of the histogram $\mathsf{h}(i)$—are "saturated" (i.e., mapped to the extreme values of the target range). The intermediate values $(a = a'_{\mathrm{lo}}, \ldots, a'_{\mathrm{hi}})$ are mapped linearly to the interval $a_{\min}, \ldots, a_{\max}$.

The values a'_{lo}, a'_{hi} depend on the image content and can be easily obtained from the image's cumulative histogram[3] H:

$$a'_{lo} = \min\{\, i \mid \mathsf{H}(i) \geq M \cdot N \cdot p_{lo} \},\tag{4.9}$$
$$a'_{hi} = \max\{\, i \mid \mathsf{H}(i) \leq M \cdot N \cdot (1 - p_{hi}) \},\tag{4.10}$$

where $0 \leq p_{lo}, p_{hi} \leq 1$, $p_{lo} + p_{hi} \leq 1$, and $M \cdot N$ is the number of pixels in the image. All pixel values *outside* (and including) a'_{lo} and a'_{hi} are mapped to the extreme values a_{min} and a_{max}, respectively, and intermediate values are mapped linearly to the interval $[a_{min}, a_{max}]$. Using this formulation, the mapping to minimum and maximum intensities does not depend on singular extreme pixels only but can be based on a representative set of pixels. The mapping function for the modified auto-contrast operation can thus be defined as

$$f_{mac}(a) = \begin{cases} a_{min} & \text{for } a \leq a'_{lo}, \\ a_{min} + \left(a - a'_{lo}\right) \cdot \dfrac{a_{max} - a_{min}}{a'_{hi} - a'_{lo}} & \text{for } a'_{lo} < a < a'_{hi}, \\ a_{max} & \text{for } a \geq a'_{hi}. \end{cases}\tag{4.11}$$

Usually the same value is taken for both upper and lower quantiles (i.e., $p_{lo} = p_{hi} = p$), with $p = 0.005, \ldots, 0.015$ ($0.5, \ldots, 1.5\,\%$) being common values. For example, the auto-contrast operation in Adobe Photoshop saturates $0.5\,\%$ ($p = 0.005$) of all pixels at both ends of the intensity range. Auto-contrast is a frequently used point operation and thus available in practically any image-processing software. ImageJ implements the modified auto-contrast operation as part of the Brightness/Contrast and Image ▷ Adjust menus (Auto button), shown in Fig. 4.7.

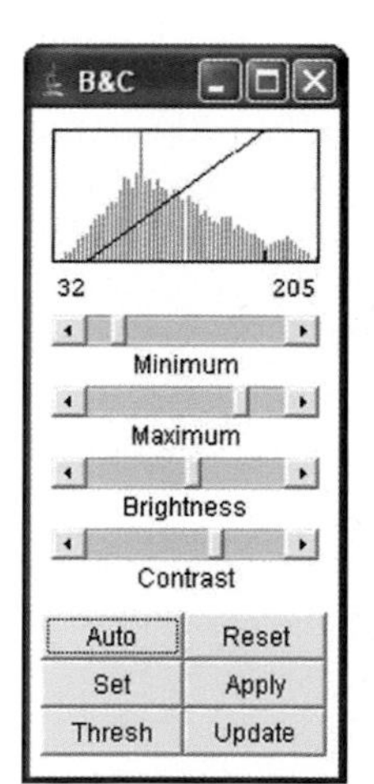

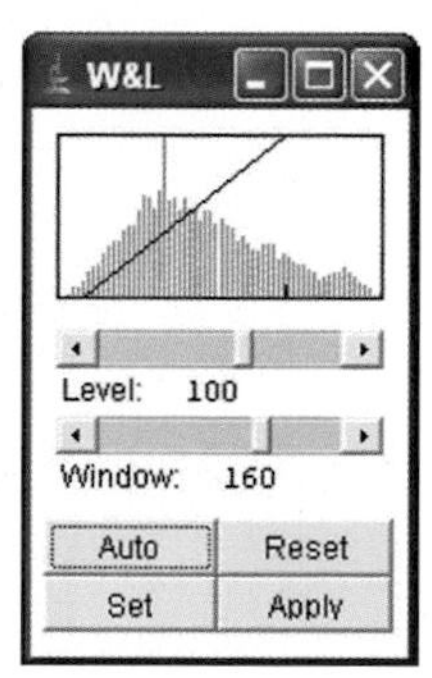

Fig. 4.7
ImageJ's Brightness/Contrast tool (left) and Window/Level tool (right) can be invoked through the Image ▷ Adjust menu. The Auto button displays the result of a modified auto-contrast operation. Apply must be hit to actually modify the image.

4.5 Histogram Equalization

A frequent task is to adjust two different images in such a way that their resulting intensity distributions are similar, for example, to use

[3] See Sec. 3.6.

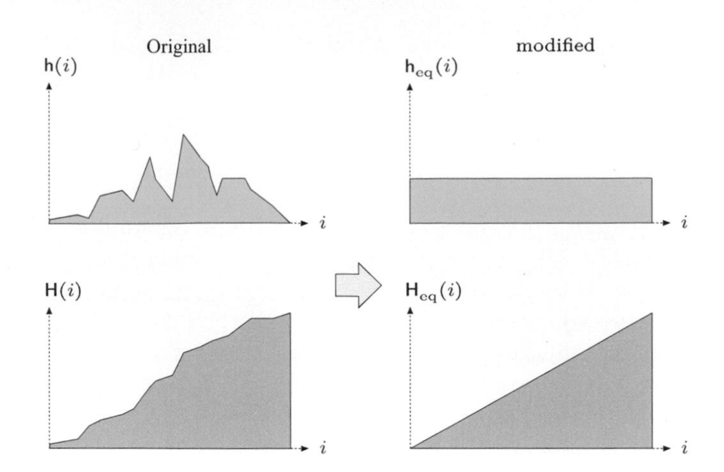

Fig. 4.8
Histogram equalization. The idea is to find and apply a point operation to the image (with original histogram h) such that the histogram h_{eq} of the modified image approximates a *uniform* distribution (top). The cumulative target histogram H_{eq} must thus be approximately wedge-shaped (bottom).

them in a print publication or to make them easier to compare. The goal of histogram equalization is to find and apply a point operation such that the histogram of the modified image approximates a *uniform* distribution (see Fig. 4.8). Since the histogram is a discrete distribution and homogeneous point operations can only shift and merge (but never split) histogram entries, we can only obtain an approximate solution in general. In particular, there is no way to eliminate or decrease individual peaks in a histogram, and a truly uniform distribution is thus impossible to reach. Based on point operations, we can thus modify the image only to the extent that the resulting histogram is *approximately* uniform. The question is how good this approximation can be and exactly which point operation (which clearly depends on the image content) we must apply to achieve this goal.

We may get a first idea by observing that the *cumulative* histogram (Sec. 3.6) of a uniformly distributed image is a linear ramp (wedge), as shown in Fig. 4.8. So we can reformulate the goal as finding a point operation that shifts the histogram lines such that the resulting cumulative histogram is approximately linear, as illustrated in Fig. 4.9.

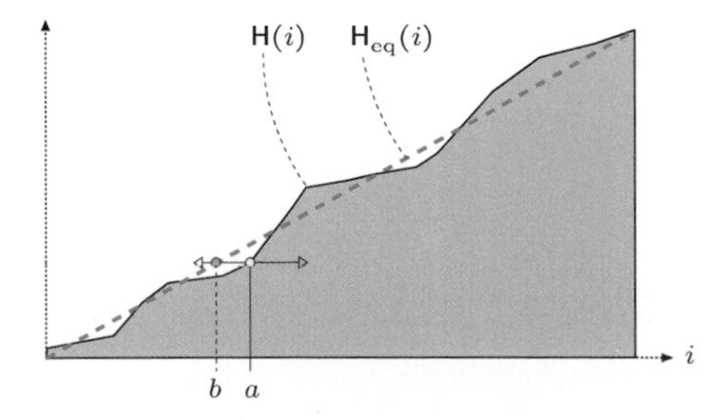

Fig. 4.9
Histogram equalization on the cumulative histogram. A suitable point operation $b \leftarrow f_{eq}(a)$ shifts each histogram line from its original position a to b (left or right) such that the resulting cumulative histogram H_{eq} is approximately linear.

The desired point operation $f_{eq}()$ is simply obtained from the cumulative histogram H of the original image as[4]

$$f_{eq}(a) = \left\lfloor \mathsf{H}(a) \cdot \frac{K-1}{M \cdot N} \right\rfloor, \tag{4.12}$$

[4] For a derivation, see, for example, [88, p. 173].

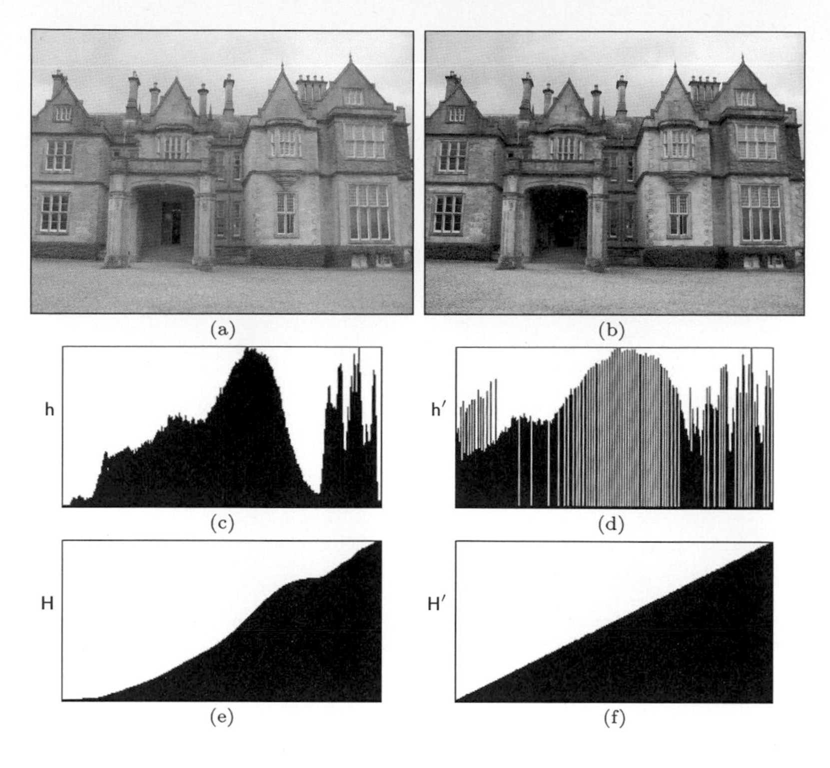

Fig. 4.10
Linear histogram equalization example. Original image I (a) and modified image I' (b), corresponding histograms h, h' (c, d), and cumulative histograms H, H' (e, f). The resulting cumulative histogram H' (f) approximates a uniformly distributed image. Notice that new peaks are created in the resulting histogram h' (d) by merging original histogram cells, particularly in the lower and upper intensity ranges.

for an image of size $M \times N$ with pixel values a in the range $[0, K-1]$. The resulting function $f_{\mathrm{eq}}(a)$ in Eqn. (4.12) is monotonically increasing, because $\mathsf{H}(a)$ is monotonic and K, M, N are all positive constants. In the (unusual) case where an image is already uniformly distributed, linear histogram equalization should not modify that image any further. Also, repeated applications of linear histogram equalization should not make any changes to the image after the first time. Both requirements are fulfilled by the formulation in Eqn. (4.12). Program 4.2 lists the Java code for a sample implementation of linear histogram equalization. An example demonstrating the effects on the image and the histograms is shown in Fig. 4.10.

Notice that for "inactive" pixel values i (i.e., pixel values that do not appear in the image, with $\mathsf{h}(i) = 0$), the corresponding entries in the cumulative histogram $\mathsf{H}(i)$ are either zero or identical to the neighboring entry $\mathsf{H}(i-1)$. Consequently a contiguous range of zero values in the histogram $\mathsf{h}(i)$ corresponds to a constant (i.e., flat) range in the cumulative histogram $\mathsf{H}(i)$, and the function $f_{\mathrm{eq}}(a)$ maps all "inactive" intensity values within such a range to the next lower "active" value. This effect is not relevant, however, since the image contains no such pixels anyway. Nevertheless, a linear histogram equalization may (and typically will) cause histogram lines to merge and consequently lead to a loss of dynamic range (see also Sec. 4.2).

This or a similar form of linear histogram equalization is implemented in almost any image-processing software. In ImageJ it can be invoked interactively through the Process ▷ Enhance Contrast menu (option Equalize). To avoid extreme contrast effects, the histogram

Prog. 4.2
Linear histogram equalization (ImageJ plugin). First the histogram of the image ip is obtained using the standard ImageJ method ip.getHistogram() in line 7. In line 9, the cumulative histogram is computed "in place" based on the recursive definition in Eqn. (3.6). The int division in line 16 implicitly performs the required floor ($\lfloor\ \rfloor$) operation by truncation.

```
1  public void run(ImageProcessor ip) {
2    int M = ip.getWidth();
3    int N = ip.getHeight();
4    int K = 256; // number of intensity values
5
6    // compute the cumulative histogram:
7    int[] H = ip.getHistogram();
8    for (int j = 1; j < H.length; j++) {
9      H[j] = H[j - 1] + H[j];
10   }
11
12   // equalize the image:
13   for (int v = 0; v < N; v++) {
14     for (int u = 0; u < M; u++) {
15       int a = ip.get(u, v);
16       int b = H[a] * (K - 1) / (M * N); // s. Equation (4.12)
17       ip.set(u, v, b);
18     }
19   }
20 }
```

equalization in ImageJ by default[5] cumulates the *square root* of the histogram entries using a modified cumulative histogram of the form

$$\tilde{\mathsf{H}}(i) = \sum_{j=0}^{i} \sqrt{\mathsf{h}(j)}\,. \tag{4.13}$$

4.6 Histogram Specification

Although widely implemented, the goal of linear histogram equalization—a uniform distribution of intensity values (as described in the previous section)—appears rather ad hoc, since good images virtually never show such a distribution. In most real images, the distribution of the pixel values is not even remotely uniform but is usually more similar, if at all, to perhaps a Gaussian distribution. The images produced by linear equalization thus usually appear quite unnatural, which renders the technique practically useless.

Histogram *specification* is a more general technique that modifies the image to match an arbitrary intensity distribution, including the histogram of a given image. This is particularly useful, for example, for adjusting a set of images taken by different cameras or under varying exposure or lighting conditions to give a similar impression in print production or when displayed. Similar to histogram equalization, this process relies on the alignment of the cumulative histograms by applying a homogeneous point operation. To be independent of the image size (i.e., the number of pixels), we first define *normalized* distributions, which we use in place of the original histograms.

[5] The "classic" linear approach (see Eqn. (3.5)) is used when simultaneously keeping the Alt key pressed.

4.6.1 Frequencies and Probabilities

The value in each histogram cell describes the observed frequency of the corresponding intensity value, i.e., the histogram is a discrete *frequency distribution*. For a given image I of size $M \times N$, the sum of all histogram entries $\mathsf{h}(i)$ equals the number of image pixels,

$$\sum_i \mathsf{h}(i) = M \cdot N\,. \tag{4.14}$$

The associated *normalized* histogram,

$$\mathsf{p}(i) = \frac{\mathsf{h}(i)}{M \cdot N}, \quad \text{for } 0 \leq i < K, \tag{4.15}$$

is usually interpreted as the *probability distribution* or *probability density function* (pdf) of a random process, where $\mathsf{p}(i)$ is the probability for the occurrence of the pixel value i. The cumulative probability of i being any possible value is 1, and the distribution p must thus satisfy

$$\sum_{i=0}^{K-1} \mathsf{p}(i) = 1\,. \tag{4.16}$$

The statistical counterpart to the cumulative histogram H (Eqn. (3.5)) is the discrete *distribution function* $\mathsf{P}()$ (also called the *cumulative distribution function* or cdf),

$$\mathsf{P}(i) = \frac{\mathsf{H}(i)}{\mathsf{H}(K-1)} = \frac{\mathsf{H}(i)}{M \cdot N} = \sum_{j=0}^{i} \frac{\mathsf{h}(j)}{M \cdot N} = \sum_{j=0}^{i} \mathsf{p}(j), \tag{4.17}$$

for $i = 0, \ldots, K-1$. The computation of the cdf from a given histogram h is outlined in Alg. 4.1. The resulting function $\mathsf{P}(i)$ is (as the cumulative histogram) monotonically increasing and, in particular,

$$\mathsf{P}(0) = \mathsf{p}(0) \quad \text{and} \quad \mathsf{P}(K-1) = \sum_{i=0}^{K-1} \mathsf{p}(i) = 1\,. \tag{4.18}$$

This statistical formulation implicitly treats the generation of images as a random process whose exact properties are mostly unknown.[6] However, the process is usually assumed to be homogeneous (independent of the image position); that is, each pixel value is the result of a "random experiment" on a single random variable i. The observed frequency distribution given by the histogram $\mathsf{h}(i)$ serves as a (coarse) estimate of the probability distribution $\mathsf{p}(i)$ of this random variable.

4.6.2 Principle of Histogram Specification

The goal of histogram specification is to modify a given image I_A by some point operation such that its distribution function P_A matches

[6] Statistical modeling of the image generation process has a long tradition (see, e.g., [128, Ch. 2]).

Alg. 4.1
Calculation of the cumulative distribution function (cdf) $\mathsf{P}(i)$ from a given histogram h of length K. See Prog. 4.3 (p. 73) for the corresponding Java implementation.

```
1: Cdf(h)
     Returns the cumulative distribution function P(i) ∈ [0,1] for a
     given histogram h(i), with i = 0,...,K−1.
2:   Let K ← Size(h)
3:   Let n ← Σ_{i=0}^{K−1} h(i)
4:   Create map P: [0, K−1] ↦ ℝ
5:   Let c ← 0
6:   for i ← 0,...,K−1 do
7:       c ← c + h(i)                         ▷ cumulate histogram values
8:       P(i) ← c/n
9:   return P.
```

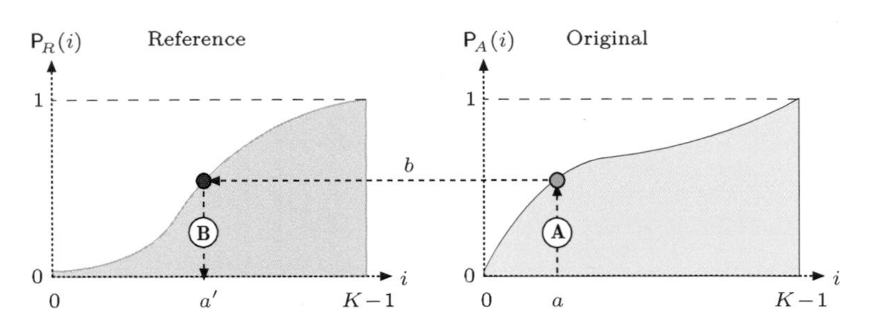

Fig. 4.11
Principle of histogram specification. Given is the reference distribution P_R (left) and the distribution function for the original image P_A (right). The result is the mapping function $f_{\mathrm{hs}}: a \rightarrow a'$ for a point operation, which replaces each pixel a in the original image I_A by a modified value a'. The process has two main steps: Ⓐ For each pixel value a, determine $b = \mathsf{P}_A(a)$ from the right distribution function. Ⓑ a' is then found by inverting the left distribution function as $a' = \mathsf{P}_R^{-1}(b)$. In summary, the result is $f_{\mathrm{hs}}(a) = a' = \mathsf{P}_R^{-1}\big(\mathsf{P}_A(a)\big)$.

a *reference distribution* P_R as closely as possible. We thus look for a mapping function

$$a' = f_{\mathrm{hs}}(a) \tag{4.19}$$

to convert the original image I_A by a point operation to a new image $I_{A'}$ with pixel values a', such that its distribution function P'_A matches P_R, that is,

$$\mathsf{P}'_A(i) \approx \mathsf{P}_R(i)\,, \quad \text{for } 0 \leq i < K. \tag{4.20}$$

As illustrated in Fig. 4.11, the desired mapping f_{hs} is found by combining the two distribution functions P_R and P_A (see [88, p. 180] for details). For a given pixel value a in the original image, we obtain the new pixel value a' as

$$a' = \mathsf{P}_R^{-1}\big(\mathsf{P}_A(a)\big) = \mathsf{P}_R^{-1}(b) \tag{4.21}$$

and thus the mapping f_{hs} (Eqn. (4.19)) is defined as

$$f_{\mathrm{hs}}(a) = \mathsf{P}_R^{-1}\big(\mathsf{P}_A(a)\big)\,, \quad \text{for } 0 \leq a < K. \tag{4.22}$$

This of course assumes that $\mathsf{P}_R(i)$ is invertible, that is, that the function $\mathsf{P}_R^{-1}(b)$ exists for $b \in [0,1]$.

4.6.3 Adjusting to a Piecewise Linear Distribution

If the reference distribution P_R is given as a continuous, invertible function, then the mapping function f_{hs} can be obtained from Eqn. (4.22) without any difficulty. In practice, it is convenient to specify the (synthetic) reference distribution as a *piecewise linear* function $\mathsf{P}_L(i)$; that is, as a sequence of $N+1$ coordinate pairs

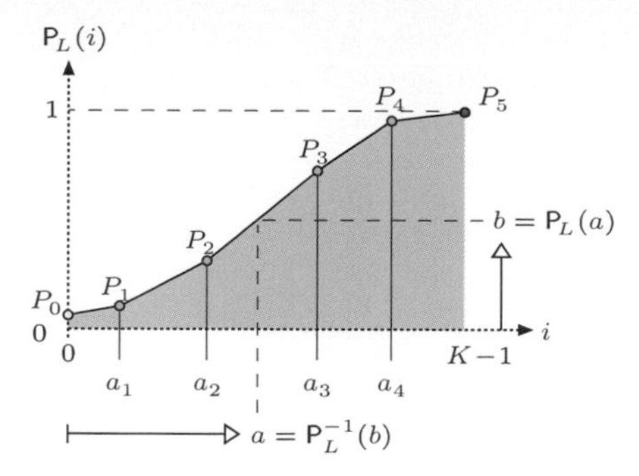

Fig. 4.12
Piecewise linear reference distribution. The function $P_L(i)$ is specified by $N = 5$ control points $(0, P_0)$, $(a_1, P_1), \ldots, (a_4, P_4)$, with $a_k < a_{k+1}$ and $P_k < P_{k+1}$. The final point P_5 is fixed at $(K-1, 1)$.

$$L = \big((a_0, P_0), (a_1, P_1), \ldots, (a_k, P_k), \ldots, (a_N, P_N)\big),$$

each consisting of an intensity value a_k and the corresponding cumulative probability P_k. We assert that $0 \leq a_k < K$, $a_k < a_{k+1}$, and $0 \leq P_k < 1$. Also, the two endpoints (a_0, P_0) and (a_N, P_N) are fixed at

$$(0, P_0) \quad \text{and} \quad (K-1, 1),$$

respectively. To be invertible, the function must also be strictly monotonic, that is, $P_k < P_{k+1}$ for $0 \leq k < N$. Figure 4.12 shows an example of such a function, which is specified by $N = 5$ variable points $(P_0, \ldots, P_4)$ and a fixed end point P_5 and thus consists of $N = 5$ linear segments. The reference distribution can of course be specified at an arbitrary accuracy by inserting additional control points.

The intermediate values of $\mathsf{P}_L(i)$ are obtained by linear interpolation between the control points as

$$\mathsf{P}_L(i) = \begin{cases} P_m + (i - a_m) \cdot \frac{(P_{m+1} - P_m)}{(a_{m+1} - a_m)} & \text{for } 0 \leq i < K-1, \\ 1 & \text{for } i = K-1. \end{cases} \tag{4.23}$$

where $m = \max\{j \in [0, N-1] \mid a_j \leq i\}$ is the index of the line segment $(a_m, P_m) \rightarrow (a_{m+1}, P_{m+1})$, which overlaps the position i. For instance, in the example in Fig. 4.12, the point a lies within the segment that starts at point (a_2, P_2); i.e., $m = 2$.

For the histogram specification according to Eqn. (4.22), we also need the *inverse* distribution function $\mathsf{P}_L^{-1}(b)$ for $b \in [0, 1]$. As we see from the example in Fig. 4.12, the function $\mathsf{P}_L(i)$ is in general not invertible for values $b < \mathsf{P}_L(0)$. We can fix this problem by mapping all values $b < \mathsf{P}_L(0)$ to zero and thus obtain a "semi-inverse" of the reference distribution in Eqn. (4.23) as

$$\mathsf{P}_L^{-1}(b) = \begin{cases} 0 & \text{for } 0 \leq b < \mathsf{P}_L(0), \\ a_n + (b - P_n) \cdot \frac{(a_{n+1} - a_n)}{(P_{n+1} - P_n)} & \text{for } \mathsf{P}_L(0) \leq b < 1, \\ K-1 & \text{for } b \geq 1. \end{cases} \tag{4.24}$$

Here $n = \max\{j \in \{0, \ldots N-1\} \mid P_j \leq b\}$ is the index of the line segment $(a_n, P_n) \rightarrow (a_{n+1}, P_{n+1})$, which overlaps the argument value b. The required mapping function f_{hs} for adapting a given image with intensity distribution P_A is finally specified, analogous to Eqn. (4.22), as

Alg. 4.2
Histogram specification using a piecewise linear reference distribution. Given is the histogram h of the original image and a piecewise linear reference distribution function, specified as a sequence of N control points L. The discrete mapping f_{hs} for the corresponding point operation is returned.

```
1:  MatchPiecewiseLinearHistogram(h, L)
    Input: h, histogram of the original image I; L, reference distri-
    bution function, given as a sequence of N+1 control points L =
    [(a_0, P_0), (a_1, P_1), ..., (a_N, P_N)], with 0 ≤ a_k < K, 0 ≤ P_k ≤ 1,
    and P_k < P_{k+1}. Returns a discrete mapping f_hs(a) to be applied
    to the original image I.
2:      N ← Size(L) + 1
3:      Let K ← Size(h)
4:      Let P ← Cdf(h)                            ▷ cdf for h (see Alg. 4.1)
5:      Create map f_hs: [0, K−1] ↦ ℝ                        ▷ function f_hs
6:      for a ← 0, ..., K−1 do
7:          b ← P(a)
8:          if (b ≤ P_0) then
9:              a' ← 0
10:         else if (b ≥ 1) then
11:             a' ← K−1
12:         else
13:             n ← N−1
14:             while (n ≥ 0) ∧ (P_n > b) do       ▷ find line segment in L
15:                 n ← n − 1
16:             a' ← a_n + (b−P_n)·(a_{n+1} − a_n)/(P_{n+1} − P_n)   ▷ see Eqn. 4.24
17:         f_hs[a] ← a'
18:     return f_hs.
```

$$f_{\mathrm{hs}}(a) = \mathsf{P}_L^{-1}(\mathsf{P}_A(a)), \quad \text{for } 0 \leq a < K. \tag{4.25}$$

The whole process of computing the pixel mapping function for a given image (histogram) and a piecewise linear target distribution is summarized in Alg. 4.2. A real example is shown in Fig. 4.14 (Sec. 4.6.5).

4.6.4 Adjusting to a Given Histogram (Histogram Matching)

If we want to adjust one image to the histogram of another image, the reference distribution function $\mathsf{P}_R(i)$ is not continuous and thus, in general, cannot be inverted (as required by Eqn. (4.22)). For example, if the reference distribution contains zero entries (i.e., pixel values k with probability $\mathsf{p}(k) = 0$), the corresponding cumulative distribution function P (just like the cumulative histogram) has intervals of constant value on which no inverse function value can be determined.

In the following, we describe a simple method for histogram matching that works with discrete reference distributions. The principal idea is graphically illustrated in Fig. 4.13. The mapping function f_{hs} is not obtained by inverting but by "filling in" the reference distribution function $\mathsf{P}_R(i)$. For each possible pixel value a, starting with $a = 0$, the corresponding probability $\mathsf{p}_A(a)$ is stacked layer by layer "under" the reference distribution P_R. The thickness of each horizontal bar for a equals the corresponding probability $\mathsf{p}_A(a)$. The bar for a particular intensity value a with thickness $\mathsf{p}_A(a)$ runs from

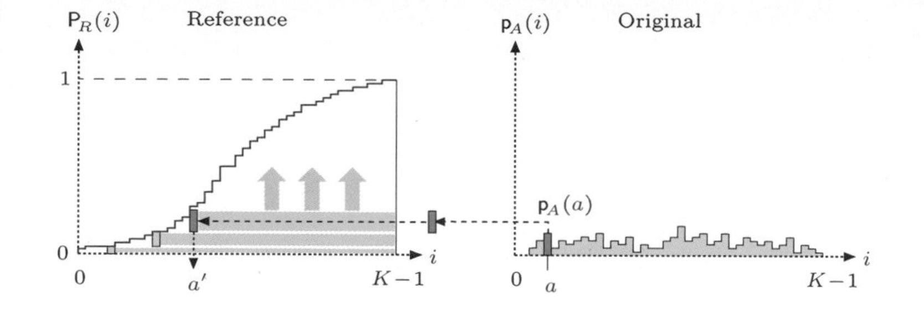

Fig. 4.13
Discrete histogram specification. The reference distribution P_R (left) is "filled" layer by layer from bottom to top and from right to left. For every possible intensity value a (starting from $a = 0$), the associated probability $p_A(a)$ is added as a horizontal bar to a stack accumulated 'under" the reference distribution P_R. The bar with thickness $p_A(a)$ is drawn from right to left down to the position a', where the reference distribution P_R is reached. The function $f_{hs}()$ must map a to a'.

right to left, down to position a', where it hits the reference distribution P_R. This position a' corresponds to the new pixel value to which a should be mapped.

Since the sum of all probabilities p_A and the maximum of the distribution function P_R are both 1 (i.e., $\sum_i p_A(i) = \max_i P_R(i) = 1$), all horizontal bars will exactly fit underneath the function P_R. One may also notice in Fig. 4.13 that the distribution value resulting at a' is identical to the cumulated probability $P_A(a)$. Given some intensity value a, it is therefore sufficient to find the minimum value a', where the reference distribution $P_R(a')$ is greater than or equal to the cumulative probability $P_A(a)$, that is,

$$f_{hs}(a) = \min\{ j \mid (0 \leq j < K) \wedge \big(P_A(a) \leq P_R(j)\big) \} . \tag{4.26}$$

This results in a very simple method, which is summarized in Alg. 4.3. The corresponding Java implementation in Prog. 4.3, consists of the method `matchHistograms()`, which accepts the original histogram (`hA`) and the reference histogram (`hR`) and returns the resulting mapping function (`fhs`) specifying the required point operation.

Due to the use of normalized distribution functions, the *size* of the associated images is not relevant. The following code fragment demonstrates the use of the `matchHistograms()` method from Prog. 4.3 in ImageJ:

```
ImageProcessor ipA = ...      // target image I_A (to be modified)
ImageProcessor ipR = ...      // reference image I_R

int[] hA = ipA.getHistogram(); // get histogram for I_A
int[] hR = ipR.getHistogram(); // get histogram for I_R

int[] fhs = matchHistograms(hA, hR); // mapping function f_hs(a)

ipA.applyTable(fhs);              // modify the target image I_A
```

The original image `ipA` is modified in the last line by applying the mapping function f_{hs} (`fhs`) with the method `applyTable()` (see also p. 83).

4.6.5 Examples

Adjusting to a piecewise linear reference distribution

The first example in Fig. 4.14 shows the results of histogram specification for a continuous, piecewise linear reference distribution, as

Alg. 4.3
Histogram matching. Given are two histograms: the histogram h_A of the target image I_A and a reference histogram h_R, both of size K. The result is a discrete mapping function $f_{\mathrm{hs}}()$ that, when applied to the target image, produces a new image with a distribution function similar to the reference histogram.

```
1:  MatchHistograms(h_A, h_R)
        Input: h_A, histogram of the target image I_A; h_R, reference his-
        togram (the same size as h_A). Returns a discrete mapping f_hs(a)
        to be applied to the target image I_A.
2:      K ← Size(h_A)
3:      P_A ← CDF(h_A)                          ▷ c.d.f. for h_A (Alg. 4.1)
4:      P_R ← CDF(h_R)                          ▷ c.d.f. for h_R (Alg. 4.1)
5:      Create map f_hs : [0, K−1] ↦ ℝ          ▷ pixel mapping function f_hs
6:      for a ← 0, ..., K−1 do
7:          j ← K−1
8:          repeat
9:              f_hs[a] ← j
10:             j ← j − 1
11:         while (j ≥ 0) ∧ (P_A(a) ≤ P_R(j))
12:     return f_hs.
```

described in Sec. 4.6.3. Analogous to Fig. 4.12, the actual distribution function P_R (Fig. 4.14(f)) is specified as a polygonal line consisting of five control points $\langle a_k, q_k \rangle$ with coordinates

$k =$	0	1	2	3	4	5
$a_k =$	0	28	75	150	210	255
$q_k =$	0.002	0.050	0.250	0.750	0.950	1.000

The resulting reference histogram (Fig. 4.14(c)) is a step function with ranges of constant values corresponding to the linear segments of the probability density function. As expected, the *cumulative* probability function for the modified image (Fig. 4.14(h)) is quite close to the reference function in Fig. 4.14(f), while the resulting *histogram* (Fig. 4.14(e)) shows little similarity with the reference histogram (Fig. 4.14(c)). However, as discussed earlier, this is all we can expect from a homogeneous point operation.

Adjusting to an arbitrary reference histogram

The example in Fig. 4.15 demonstrates this technique using synthetic reference histograms whose shape is approximately Gaussian. In this case, the reference distribution is not given as a continuous function but specified by a discrete histogram. We thus use the method described in Sec. 4.6.4 to compute the required mapping functions.

The target image used here was chosen intentionally for its poor quality, manifested by an extremely unbalanced histogram. The histograms of the modified images thus naturally show little resemblance to a Gaussian. However, the resulting *cumulative* histograms match nicely with the integral of the corresponding Gaussians, apart from the unavoidable irregularity at the center caused by the dominant peak in the original histogram.

Adjusting to another image

The third example in Fig. 4.16 demonstrates the adjustment of two images by matching their intensity histograms. One of the images is selected as the reference image I_R (Fig. 4.16(b)) and supplies the

```
1   int[] matchHistograms (int[] hA, int[] hR) {
2     // hA ... histogram h_A of the target image I_A (to be modified)
3     // hR ... reference histogram h_R
4     // returns the mapping f_hs() to be applied to image I_A
5
6     int K = hA.length;
7     double[] PA = Cdf(hA);        // get CDF of histogram h_A
8     double[] PR = Cdf(hR);        // get CDF of histogram h_R
9     int[] fhs = new int[K];       // mapping f_hs()
10
11    // compute mapping function f_hs():
12    for (int a = 0; a < K; a++) {
13      int j = K - 1;
14      do {
15        fhs[a] = j;
16        j--;
17      } while (j >= 0 && PA[a] <= PR[j]);
18    }
19    return fhs;
20  }
```

```
22  double[] Cdf (int[] h) {
23    // returns the cumul. distribution function for histogram h
24    int K = h.length;
25
26    int n = 0;                    // sum all histogram values
27    for (int i = 0; i < K; i++) {
28      n += h[i];
29    }
30
31    double[] P = new double[K];   // create CDF table P
32    int c = h[0];                 // cumulate histogram values
33    P[0] = (double) c / n;
34    for (int i = 1; i < K; i++) {
35      c += h[i];
36      P[i] = (double) c / n;
37    }
38    return P;
39  }
```

Prog. 4.3
Histogram matching (Java implementation of Alg. 4.3). The method `matchHistograms()` computes the mapping function `fhs` from the target histogram `hA` and the reference histogram `hR` (see Eqn. (4.26)). The method `Cdf()` computes the cumulative distribution function (cdf) for a given histogram (Eqn. (4.17)).

reference histogram h_R (Fig. 4.16(e)). The second (target) image I_A (Fig. 4.16(a)) is modified such that the resulting cumulative histogram matches the cumulative histogram of the reference image I_R. It can be expected that the final image $I_{A'}$ (Fig. 4.16(c)) and the reference image give a similar visual impression with regard to tonal range and distribution (assuming that both images show similar content).

Of course this method may be used to adjust multiple images to the same reference image (e.g., to prepare a series of similar photographs for a print project). For this purpose, one could either select a single representative image as a common reference or, alternatively, compute an "average" reference histogram from a set of typical images (see also Exercise 4.7).

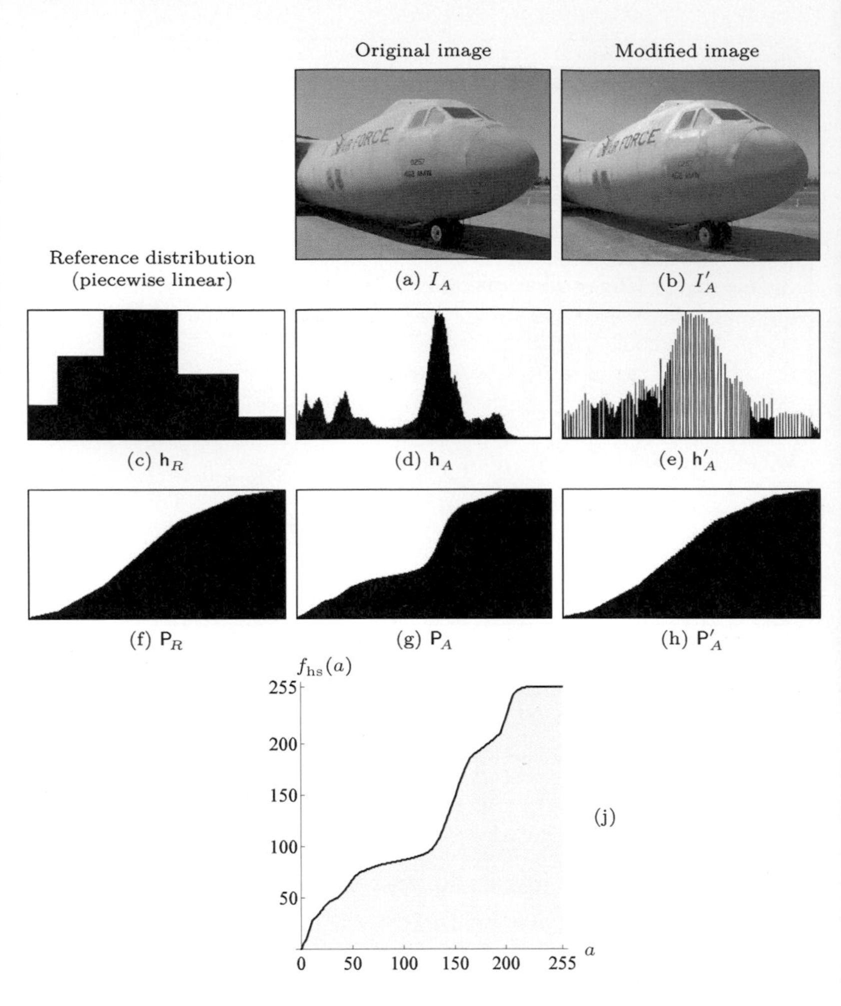

Fig. 4.14
Histogram specification with a piecewise linear reference distribution. The target image I_A (a), its histogram (d), and distribution function P_A (g); the reference histogram h_R (c) and the corresponding distribution P_R (f); the modified image $I_{A'}$ (b), its histogram $\mathsf{h}_{A'}$ (e), and the resulting distribution $\mathsf{P}_{A'}$ (h). Associated mapping function f_{hs} (j).

4.7 Gamma Correction

We have been using the terms "intensity" and "brightness" many times without really bothering with how the numeric pixel values in our images relate to these physical concepts, if at all. A pixel value may represent the amount of light falling onto a sensor element in a camera, the photographic density of film, the amount of light to be emitted by a monitor, the number of toner particles to be deposited by a printer, or any other relevant physical magnitude. In practice, the relationship between a pixel value and the corresponding physical quantity is usually complex and almost always nonlinear. In many imaging applications, it is important to know this relationship, at least approximately, to achieve consistent and reproducible results.

When applied to digital intensity images, the ideal is to have some kind of "calibrated intensity space" that optimally matches the human perception of intensity and requires a minimum number of bits to represent the required intensity range. Gamma correction denotes a simple point operation to compensate for the transfer characteristics of different input and output devices and to map them to a unified intensity space.

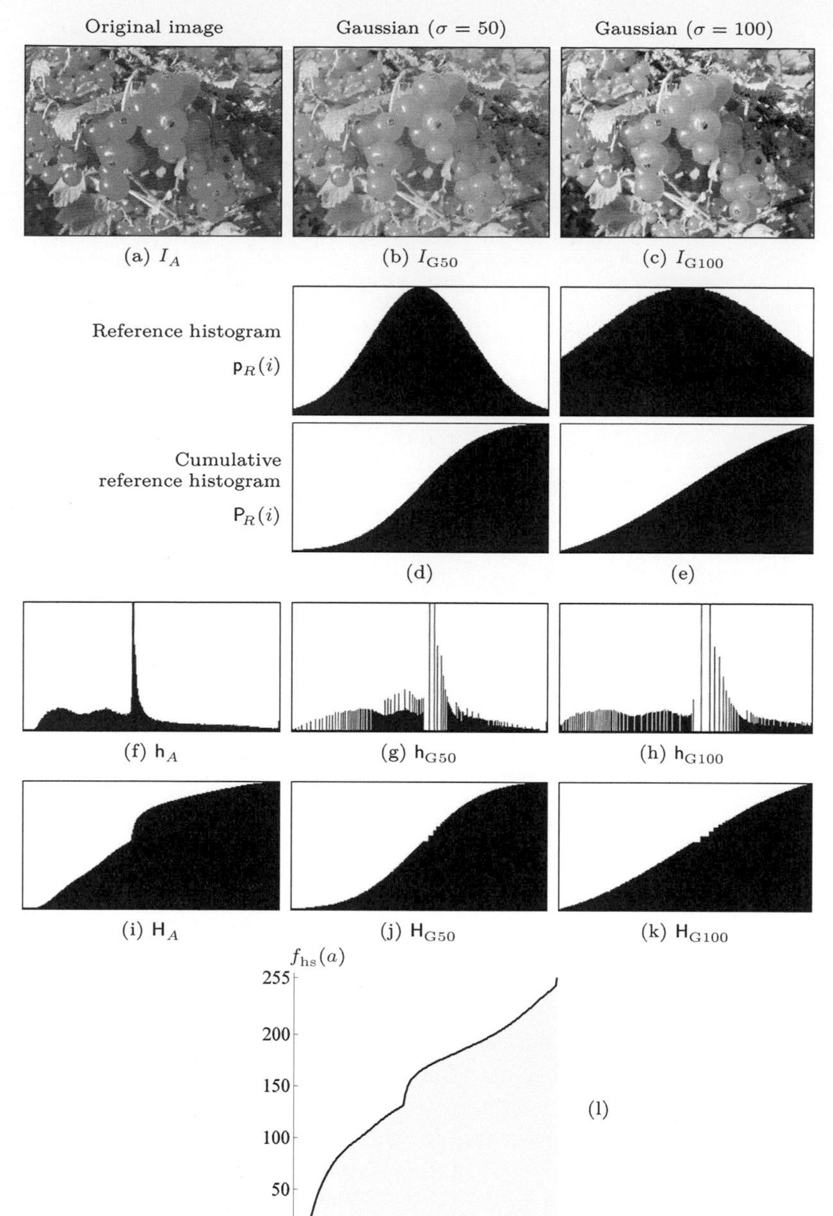

Fig. 4.15
Histogram matching: adjusting to a synthetic histogram. Original image I_A (a), corresponding histogram (f), and cumulative histogram (i). Gaussian-shaped reference histograms with center $\mu = 128$ and $\sigma = 50$ (d) and $\sigma = 100$ (e), respectively. Resulting images after histogram matching, I_{G50} (b) and I_{G100} (c) with the corresponding histograms (g, h) and cumulative histograms (j, k). Associated mapping function f_{hs} (l).

4.7.1 Why Gamma?

The term "gamma" originates from analog photography, where the relationship between the light energy and the resulting film density is approximately logarithmic. The "exposure function" (Fig. 4.17), specifying the relationship between the *logarithmic* light intensity and the resulting film density, is therefore approximately *linear* over a wide range of light intensities. The slope of this function within this linear range is traditionally referred to as the "gamma" of the photographic material. The same term was adopted later in televi-

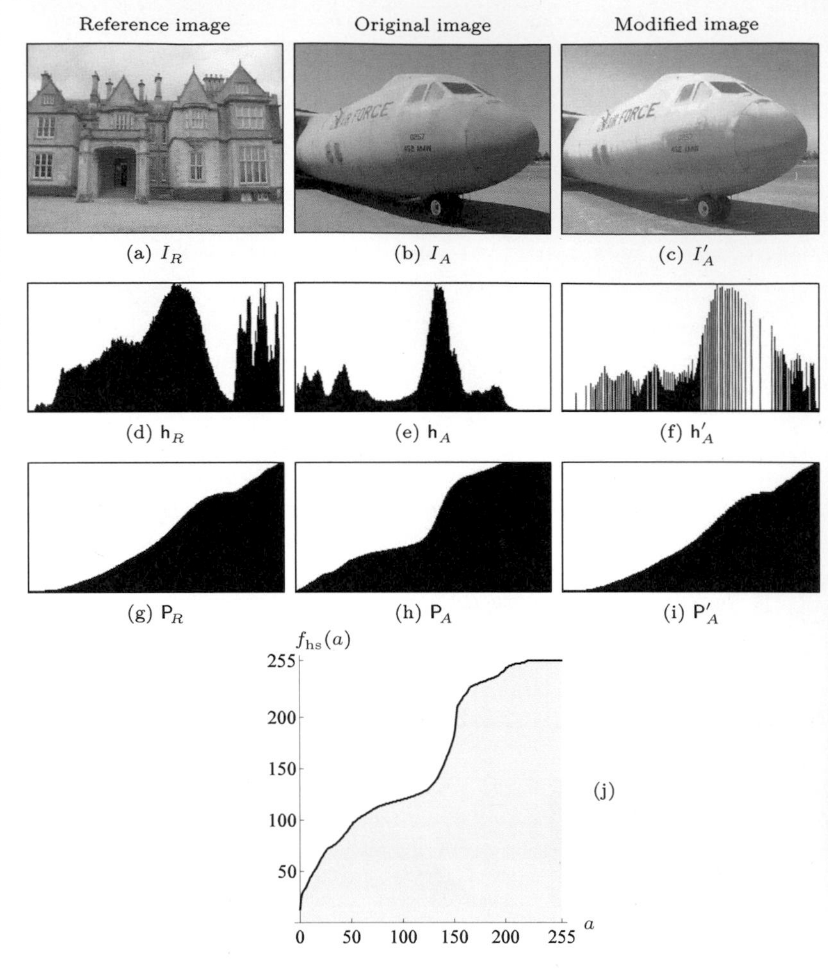

Fig. 4.16 Histogram matching: adjusting to a reference image. The target image I_A (a) is modified by matching its histogram to the reference image I_R (b), resulting in the new image $I_{A'}$ (c). The corresponding histograms h_A, h_R, $\mathsf{h}_{A'}$ (d–f) and cumulative histograms H_A, H_R, $\mathsf{P}_{A'}$ (g–i) are shown. Notice the good agreement between the cumulative histograms of the reference and adjusted images (h, i). Associated mapping function f_{hs} (j).

sion broadcasting to describe the nonlinearities of the cathode ray tubes used in TV receivers, that is, to model the relationship between the amplitude (voltage) of the video signal and the emitted light intensity. To compensate for the nonlinearities of the receivers, a "gamma correction" was (and is) applied to the TV signal once before broadcasting in order to avoid the need for costly correction measures on the receiver side.

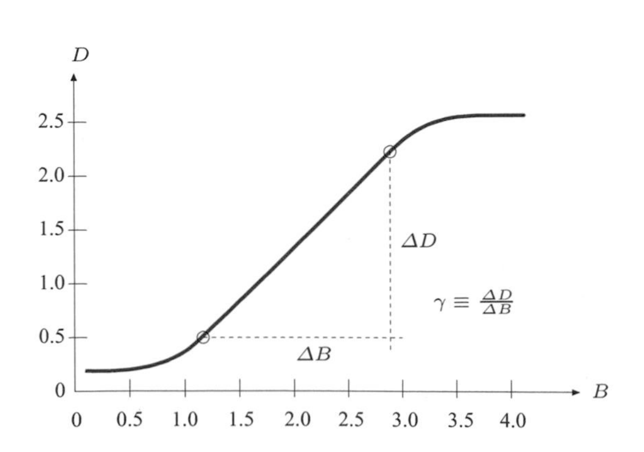

Fig. 4.17 Exposure function of photographic film. With respect to the *logarithmic* light intensity B, the resulting film density D is approximately *linear* over a wide intensity range. The slope ($\Delta D/\Delta B$) of this linear section of the function specifies the "gamma" (γ) value for a particular type of photographic material.

4.7.2 Mathematical Definition

Gamma correction is based on the exponential function

$$f_\gamma(a) = a^\gamma, \tag{4.27}$$

where the parameter $\gamma \in \mathbb{R}$ is called the *gamma* value. If a is constrained to the interval $[0,1]$, then—independent of γ—the value of $f_\gamma(a)$ also stays within $[0,1]$, and the function always runs through the points $(0,0)$ and $(1,1)$. In particular, $f_\gamma(a)$ is the identity function for $\gamma = 1$, as shown in Fig. 4.18. The function runs *above* the diagonal for gamma values $\gamma < 1$, and *below* it for $\gamma > 1$. Controlled by a single continuous parameter (γ), the power function can thus "imitate" both logarithmic and exponential types of functions. Within the interval $[0,1]$, the function is continuous and strictly monotonic, and also very simple to invert as

$$a = f_\gamma^{-1}(b) = b^{1/\gamma}, \tag{4.28}$$

since $b^{1/\gamma} = (a^\gamma)^{1/\gamma} = a^1 = a$. The inverse of the exponential function $f_\gamma^{-1}(b)$ is thus again an exponential function,

$$f_\gamma^{-1}(b) = f_{\bar{\gamma}}(b) = f_{1/\gamma}(b), \tag{4.29}$$

with the parameter $\bar{\gamma} = 1/\gamma$.

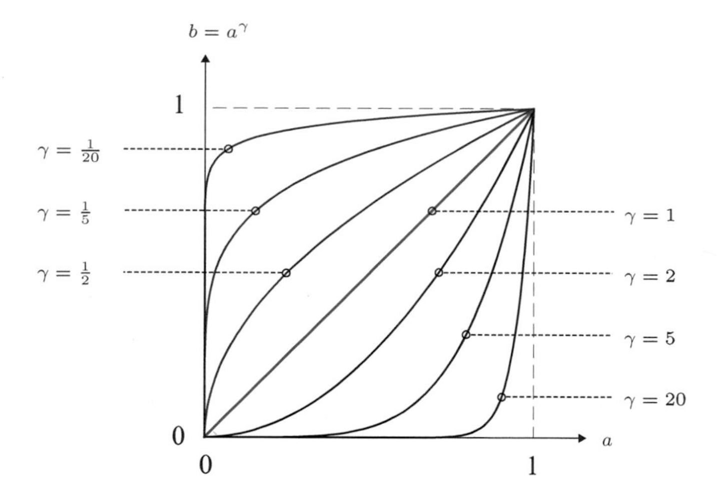

Fig. 4.18
Gamma correction function $f_\gamma(a) = a^\gamma$ for $a \in [0,1]$ and different gamma values.

4.7.3 Real Gamma Values

The actual gamma values of individual devices are usually specified by the manufacturers based on real measurements. For example, common gamma values for CRT monitors are in the range 1.8 to 2.8, with 2.4 as a typical value. Most LCD monitors are internally adjusted to similar values. Digital video and still cameras also emulate the transfer characteristics of analog film and photographic cameras by making internal corrections to give the resulting images an accustomed "look".

In TV receivers, gamma values are standardized with 2.2 for analog NTSC and 2.8 for the PAL system (these values are theoretical; results of actual measurements are around 2.35). A gamma value of $1/2.2 \approx 0.45$ is the norm for cameras in NTSC as well as the EBU[7] standards. The current international standard ITU-R BT.709[8] calls for uniform gamma values of 2.5 in receivers and $1/1.956 \approx 0.51$ for cameras [76, 122]. The ITU 709 standard is based on a slightly modified version of the gamma correction (see Sec. 4.7.6).

Computers usually allow adjustment of the gamma value applied to the video output signals to adapt to a wide range of different monitors. Note, however, that the power function $f_\gamma()$ is only a coarse approximation to the actual transfer characteristics of any device, which may also not be the same for different color channels. Thus significant deviations may occur in practice, despite the careful choice of gamma settings. Critical applications, such as prepress or high-end photography, usually require additional calibration efforts based on exactly measured device profiles (see Sec. 14.7.4).

4.7.4 Applications of Gamma Correction

Let us first look at the simple example illustrated in Fig. 4.19. Assume that we use a digital camera with a nominal gamma value γ_c, meaning that its output signal s relates to the incident light intensity L as

$$S = L^{\gamma_c}. \tag{4.30}$$

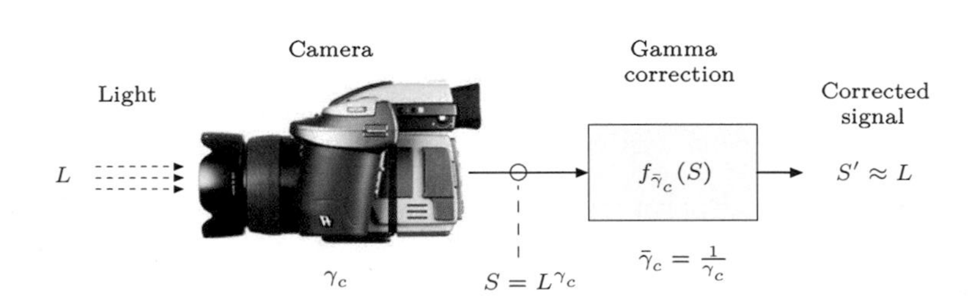

Fig. 4.19 Principle of gamma correction. To compensate the output signal S produced by a camera with nominal gamma value γ_c, a gamma correction is applied with $\bar{\gamma}_c = 1/\gamma_c$. The corrected signal S' is proportional to the received light intensity L.

To compensate the transfer characteristic of this camera (i.e., to obtain a measurement S' that is proportional to the original light intensity L), the camera signal S is subject to a gamma correction with the inverse of the camera's gamma value $\bar{\gamma}_c = 1/\gamma_c$ and thus

$$S' = f_{\bar{\gamma}_c}(S) = S^{1/\gamma_c}. \tag{4.31}$$

The resulting signal

$$S' = S^{1/\gamma_c} = \left(L^{\gamma_c}\right)^{1/\gamma_c} = L^{(\gamma_c \frac{1}{\gamma_c})} = L^1$$

is obviously proportional (in theory even identical) to the original light intensity L. Although this example is quite simplistic, it still demonstrates the general rule, which holds for output devices as well:

[7] European Broadcast Union (EBU).

[8] International Telecommunications Union (ITU).

The transfer characteristic of an input or output device with specified gamma value γ is compensated for by a gamma correction with $\bar{\gamma} = 1/\gamma$.

In the aforementioned, we have implicitly assumed that all values are strictly in the range $[0, 1]$, which usually is not the case in practice. When working with digital images, we have to deal with discrete pixel values, for example, in the range $[0, 255]$ for 8-bit images. In general, performing a gamma correction

$$b \leftarrow f_{\mathrm{gc}}(a, \gamma),$$

on a pixel value $a \in [0, a_{\max}]$ and a gamma value $\gamma > 0$ requires the following three steps:

1. Scale a linearly to $\hat{a} \in [0, 1]$.
2. Apply the gamma correction function to $\hat{a}$: $\hat{b} \leftarrow \hat{a}^{\gamma}$.
3. Scale $\hat{b} \in [0, 1]$ linearly back to $b \in [0, a_{\max}]$.

Formulated in a more compact way, the corrected pixel value b is obtained from the original value a as

$$b \leftarrow \left(\frac{a}{a_{\max}}\right)^{\gamma} \cdot a_{\max}. \tag{4.32}$$

Figure 4.20 illustrates the typical role of gamma correction in the digital work flow with two input (camera, scanner) and two output devices (monitor, printer), each with its individual gamma value. The central idea is to correct all images to be processed and stored in a device-independent, standardized intensity space.

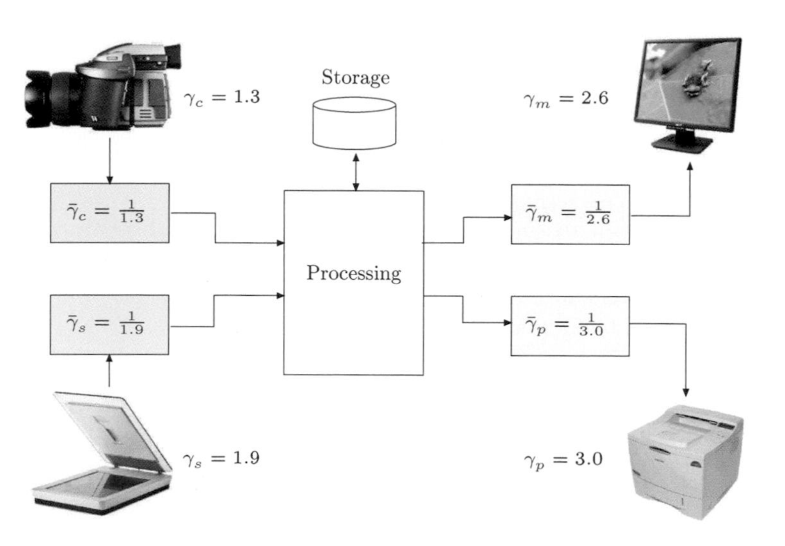

Fig. 4.20
Gamma correction in the digital imaging work flow. Images are processed and stored in a "linear" intensity space, where gamma correction is used to compensate for the transfer characteristic of each input and output device. (The gamma values shown are examples only.)

4.7.5 Implementation

Program 4.4 shows the implementation of gamma correction as an ImageJ plugin for 8-bit grayscale images. The mapping function $f_{\mathrm{gc}}(a, \gamma)$ is computed as a lookup table (`Fgc`), which is then applied to the image using the method `applyTable()` to perform the actual point operation (see also Sec. 4.8.1).

Prog. 4.4
Implementation of gamma correction in the run() method of an ImageJ plugin. The corrected intensity values b are only computed once and stored in the lookup table Fgc (line 15). The gamma value GAMMA is constant. The actual point operation is performed by calling the ImageJ method applyTable(Fgc) on the image object ip (line 18).

```
1  public void run(ImageProcessor ip) {
2      // works for 8-bit images only
3      int K = 256;
4      int aMax = K - 1;
5      double GAMMA = 2.8;
6
7      // create and fill the lookup table:
8      int[] Fgc = new int[K];
9
10     for (int a = 0; a < K; a++) {
11         double aa = (double) a / aMax;        // scale to [0, 1]
12         double bb = Math.pow(aa, GAMMA);    // power function
13         // scale back to [0, 255]:
14         int b = (int) Math.round(bb * aMax);
15         Fgc[a] = b;
16     }
17
18     ip.applyTable(Fgc);  // modify the image
19 }
```

4.7.6 Modified Gamma Correction

A subtle problem with the simple power function $f_\gamma(a) = a^\gamma$ (Eqn. (4.27)) appears if we take a closer look at the *slope* of this function, expressed by its first derivative,

$$f'_\gamma(a) = \gamma \cdot a^{(\gamma-1)},$$

which for $a = 0$ has the values

$$f'_\gamma(0) = \begin{cases} 0 & \text{for } \gamma > 1, \\ 1 & \text{for } \gamma = 1, \\ \infty & \text{for } \gamma < 1. \end{cases} \tag{4.33}$$

The tangent to the function at the origin is thus horizontal ($\gamma > 1$), diagonal ($\gamma = 1$), or vertical ($\gamma < 1$), with no intermediate values. For $\gamma < 1$, this causes extremely high amplification of small intensity values and thus increased noise in dark image regions. Theoretically, this also means that the power function is generally not invertible at the origin.

A common solution to this problem is to replace the lower part ($0 \leq a \leq a_0$) of the power function by a *linear* segment with constant slope and to continue with the ordinary power function for $a > a_0$. The resulting modified gamma correction function,

$$\bar{f}_{\gamma,a_0}(a) = \begin{cases} s \cdot a & \text{for } 0 \leq a \leq a_0, \\ (1+d) \cdot a^\gamma - d & \text{for } a_0 < a \leq 1, \end{cases} \tag{4.34}$$

$$\text{with} \quad s = \frac{\gamma}{a_0(\gamma-1) + a_0^{(1-\gamma)}} \quad \text{and} \quad d = \frac{1}{a_0^\gamma(\gamma-1)+1} - 1 \tag{4.35}$$

thus consists of a *linear* section (for $0 \leq a \leq a_0$) and a *nonlinear* section (for $a_0 < a \leq 1$) that connect smoothly at the transition point

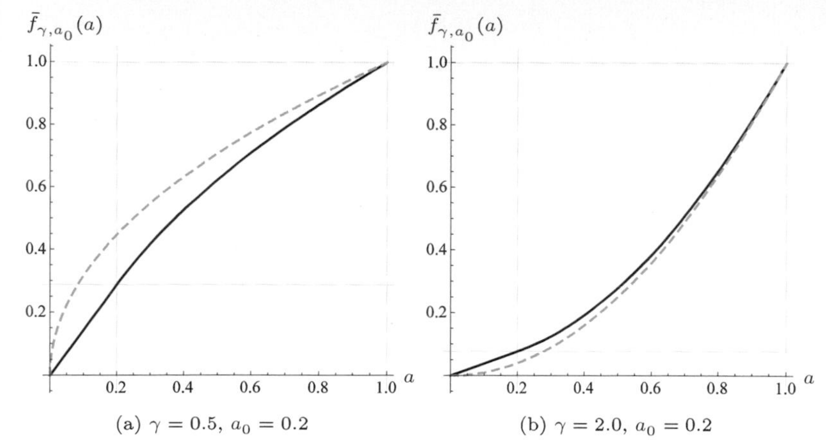

(a) $\gamma = 0.5$, $a_0 = 0.2$

(b) $\gamma = 2.0$, $a_0 = 0.2$

Fig. 4.21
Modified gamma correction. The mapping $\bar{f}_{\gamma,a_0}(a)$ consists of a linear segment with fixed slope s between $a = 0$ and $a = a_0$, followed by a power function with parameter γ (Eqn. (4.34)). The dashed lines show the ordinary power functions for the same gamma values.

$a = a_0$. The linear slope s and the parameter d are determined by the requirement that the two function segments must have identical values as well as identical slopes (first derivatives) at $a = a_0$ to produce a continuous function. The function in Eqn. (4.34) is thus fully specified by the two parameters a_0 and γ.

Figure 4.21 shows two examples of the modified gamma correction $\bar{f}_{\gamma,a_0}()$ with values $\gamma = 0.5$ and $\gamma = 2.0$, respectively. In both cases, the transition point is at $a_0 = 0.2$. For comparison, the figure also shows the ordinary gamma correction $f_\gamma(a)$ for the same gamma values (dashed lines), whose slope at the origin is ∞ (Fig. 4.21(a)) and zero (Fig. 4.21(b)), respectively.

Gamma correction in common standards

The modified gamma correction is part of several modern imaging standards. In practice, however, the values of a_0 are considerably smaller than the ones used for the illustrative examples in Fig. 4.21, and γ is chosen to obtain a good overall match to the desired correction function. For example, the ITU-BT.709 specification [122] mentioned in Sec. 4.7.3 specifies the parameters

$$\gamma = \frac{1}{2.222} \approx 0.45 \quad \text{and} \quad a_0 = 0.018\,, \tag{4.36}$$

with the corresponding slope and offset values $s = 4.50681$ and $d = 0.0991499$, respectively (Eqn. (4.35)). The resulting correction function $\bar{f}_{\mathrm{ITU}}(a)$ has a *nominal* gamma value of 0.45, which corresponds to the *effective* gamma value $\gamma_{\mathrm{eff}} = 1/1.956 \approx 0.511$. The gamma correction in the sRGB standard [224] is specified on the same basis (with different parameters; see Sec. 14.4).

Figure 4.22 shows the actual correction functions for the ITU and sRGB standards, respectively, each in comparison with the equivalent ordinary gamma correction. The ITU function (Fig. 4.22(a)) with $\gamma = 0.45$ and $a_0 = 0.018$ corresponds to an ordinary gamma correction with effective gamma value $\gamma_{\mathrm{eff}} = 0.511$ (dashed line). The curves for sRGB (Fig. 4.22(b)) differ only by the parameters γ and a_0, as summarized in Table 4.1.

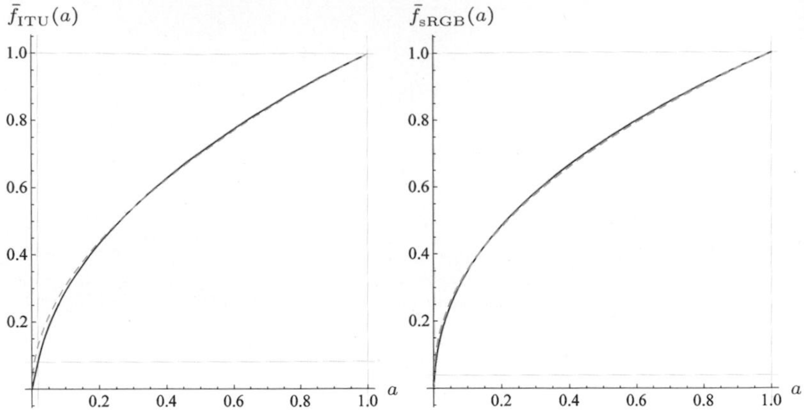

Fig. 4.22
Gamma correction functions specified by the ITU-R BT.709 (a) and sRGB (b) standards. The continuous plot shows the modified gamma correction with the nominal γ values and transition points a_0.

(a) $\gamma \approx 0.450$, $a_0 = 0.018$

(b) $\gamma \approx 0.417$, $a_0 = 0.0031308$

Table 4.1
Gamma correction parameters for the ITU and sRGB standards based on the modified mapping in Eqns. (4.34) and (4.35).

Standard	Nominal gamma value γ	a_0	s	d	Effective gamma value γ_{eff}
ITU-R BT.709	$1/2.222 \approx 0.450$	0.018	4.50	0.099	$1/1.956 \approx 0.511$
sRGB	$1/2.400 \approx 0.417$	0.0031308	12.92	0.055	$1/2.200 \approx 0.455$

Inverting the modified gamma correction

To invert the modified gamma correction of the form $b = \bar{f}_{\gamma,a_0}(a)$ (Eqn. (4.34)), we need the inverse of the function $\bar{f}_{\gamma,a_0}()$, which is again defined in two parts,

$$\bar{f}^{-1}_{\gamma,a_0}(b) = \begin{cases} b/s & \text{for } 0 \leq b \leq s \cdot a_0, \\ \left(\frac{b+d}{1+d}\right)^{1/\gamma} & \text{for } s \cdot a_0 < b \leq 1. \end{cases} \tag{4.37}$$

s and d are the quantities defined in Eqn. (4.35) and thus

$$a = \bar{f}^{-1}_{\gamma,a_0}\left(\bar{f}_{\gamma,a_0}(a)\right) \qquad \text{for } a \in [0,1], \tag{4.38}$$

with the *same* value γ being used in both functions. The inverse gamma correction function is required in particular for transforming between different color spaces if nonlinear (i.e., gamma-corrected) component values are involved (see also Sec. 14.2).

4.8 Point Operations in ImageJ

Several important types of point operations are already implemented in ImageJ, so there is no need to program every operation manually (as shown in Prog. 4.4). In particular, it is possible in ImageJ to apply point operations efficiently by using tabulated functions, to use built-in standard functions for point operations on single images, and to apply arithmetic operations on pairs of images. These issues are described briefly in the remaining parts of this section.

4.8.1 Point Operations with Lookup Tables

Some point operations require complex computations for each pixel, and the processing of large images may be quite time-consuming. If the point operation is *homogeneous* (i.e., independent of the pixel coordinates), the value of the mapping function can be precomputed for every possible pixel value and stored in a lookup table, which may then be applied very efficiently to the image. A lookup table **L** represents a discrete mapping (function f) from the original to the new pixel values,

$$\mathsf{F} : [0, K-1] \overset{f}{\longmapsto} [0, K-1] \,. \tag{4.39}$$

For a point operation specified by a particular pixel mapping function $a' = f(a)$, the table **L** is initialized with the values

$$\mathsf{F}[a] \leftarrow f(a), \quad \text{for } 0 \leq a < K. \tag{4.40}$$

Thus the K table elements of **F** need only be computed once, where typically $K = 256$. Performing the actual point operation only requires a simple (and quick) table lookup in **F** at each pixel, that is,

$$I'(u,v) \leftarrow \mathsf{F}[I(u,v)] \,, \tag{4.41}$$

which is much more efficient than any individual function call. ImageJ provides the method

```
void applyTable(int[] F)
```

for objects of type `ImageProcessor`, which requires a lookup table *F* as a 1D `int` array of size K (see Prog. 4.4 on page 80 for an example). The advantage of this approach is obvious: for an 8-bit image, for example, the mapping function is evaluated only 256 times (independent of the image size) and not a million times or more as in the case of a large image. The use of lookup tables for implementing point operations thus always makes sense if the number of image pixels ($M \times N$) is greater than the number of possible pixel values K (which is usually the case).

4.8.2 Arithmetic Operations

ImageJ implements a set of common arithmetic operations as methods for the class `ImageProcessor`, which are summarized in Table 4.2. In the following example, the image is multiplied by a scalar constant (1.5) to increase its contrast:

```
ImageProcessor ip = ... //some image
ip.multiply(1.5);
```

The image `ip` is destructively modified by all of these methods, with the results being limited (clamped) to the minimum and maximum pixel values, respectively.

4.8.3 Point Operations Involving Multiple Images

Point operations may involve more than one image at once, with arithmetic operations on the pixels of *pairs* of images being a special but important case. For example, we can express the pointwise *addition* of two images I_1 and I_2 (of identical size) to create a new image I' as

Table 4.2
ImageJ methods for arithmetic operations applicable to objects of type ImageProcessor.

`void abs()`	$I(u,v) \leftarrow \lvert I(u,v)\rvert$
`void add(int p)`	$I(u,v) \leftarrow I(u,v)+p$
`void gamma(double g)`	$I(u,v) \leftarrow \bigl(I(u,v)/255\bigr)^{g} \cdot 255$
`void invert(int p)`	$I(u,v) \leftarrow 255 - I(u,v)$
`void log()`	$I(u,v) \leftarrow \log_{10}\bigl(I(u,v)\bigr)$
`void max(double s)`	$I(u,v) \leftarrow \max\bigl(I(u,v),s\bigr)$
`void min(double s)`	$I(u,v) \leftarrow \min\bigl(I(u,v),s\bigr)$
`void multiply(double s)`	$I(u,v) \leftarrow \mathrm{round}\bigl(I(u,v)\cdot s\bigr)$
`void sqr()`	$I(u,v) \leftarrow I(u,v)^2$
`void sqrt()`	$I(u,v) \leftarrow \sqrt{I(u,v)}$

$$I'(u,v) \;\leftarrow\; I_1(u,v) + I_2(u,v) \tag{4.42}$$

for all positions (u,v). In general, any function $f(a_1, a_2, \ldots, a_n)$ over n pixel values a_i may be defined to perform pointwise combinations of n images, that is,

$$I'(u,v) \;\leftarrow\; f\bigl(I_1(u,v), I_2(u,v), \ldots, I_n(u,v)\bigr). \tag{4.43}$$

Of course, most arithmetic operations on multiple images can also be implemented as successive binary operations on pairs of images.

4.8.4 Methods for Point Operations on Two Images

ImageJ supplies a single method for implementing arithmetic operations on pairs of images,

`copyBits(ImageProcessor ip2, int u, int v, int mode)`,

which applies the binary operation specified by the transfer mode parameter ***mode*** to all pixel pairs taken from the *source image* ***ip2*** and the *target image* (the image on which this method is invoked) and stores the result in the target image. ***u***, ***v*** are the coordinates where the source image is inserted into the target image (usually ***u*** $=$ ***v*** $= 0$). The following code segment demonstrates the addition of two images:

```
ImageProcessor ip1 = ... // target image (I1)
ImageProcessor ip2 = ... // source image (I2)
...
ip1.copyBits(ip2, 0, 0, Blitter.ADD); // I1 <- I1 + I2
// ip1 holds the result, ip2 is unchanged
...
```

In this operation, the target image `ip1` is destructively modified, while the source image `ip2` remains unchanged. The constant `ADD` is one of several arithmetic transfer modes defined by the `Blitter` interface (see Table 4.3). In addition, `Blitter` defines (bitwise) logical operations, such as `OR` and `AND`. For arithmetic operations, the `copyBits()` method limits the results to the admissible range of pixel values (of the target image). Also note that (except for target images of type `FloatProcessor`) the results are *not* rounded but truncated to integer values.

ADD	$I_1(u,v) \leftarrow I_1(u,v) + I_2(u,v)$
AVERAGE	$I_1(u,v) \leftarrow \big(I_1(u,v) + I_2(u,v)\big) \,/\, 2$
COPY	$I_1(u,v) \leftarrow I_2(u,v)$
DIFFERENCE	$I_1(u,v) \leftarrow \lvert I_1(u,v) - I_2(u,v) \rvert$
DIVIDE	$I_1(u,v) \leftarrow I_1(u,v) \,/\, I_2(u,v)$
MAX	$I_1(u,v) \leftarrow \max\big(I_1(u,v), I_2(u,v)\big)$
MIN	$I_1(u,v) \leftarrow \min\big(I_1(u,v), I_2(u,v)\big)$
MULTIPLY	$I_1(u,v) \leftarrow I_1(u,v) \cdot I_2(u,v)$
SUBTRACT	$I_1(u,v) \leftarrow I_1(u,v) - I_2(u,v)$

Table 4.3
Arithmetic operations and corresponding transfer mode constants for ImageProcessor's `copyBits()` method. Example: `ip1.copyBits(ip2, 0, 0, Blitter.ADD)`.

4.8.5 ImageJ Plugins Involving Multiple Images

ImageJ provides two types of plugin: a generic plugin (`PlugIn`), which can be run without any open image, and plugins of type `PlugInFilter`, which apply to a single image. In the latter case, the currently active image is passed as an object of type `ImageProcessor` (or any of its subclasses) to the plugin's `run()` method (see also Sec. 2.2.3).

If two or more images $I_1, I_2, \ldots, I_k$ are to be combined by a plugin program, only a single image I_1 can be passed directly to the plugin's `run()` method, but not the additional images $I_2, \ldots, I_k$. The usual solution is to make the plugin open a dialog window to let the user select the remaining images interactively. This is demonstrated in the following example plugin for transparently blending two images.

Example: Linear blending

Linear blending is a simple method for continuously mixing two images, I_{BG} and I_{FG}. The background image I_{BG} is covered by the foreground image I_{FG}, whose transparency is controlled by the value α in the form

$$I'(u,v) = \alpha \cdot I_{BG}(u,v) \;+\; (1-\alpha) \cdot I_{FG}(u,v)\,, \tag{4.44}$$

with $0 \leq \alpha \leq 1$. For $\alpha = 0$, the foreground image I_{FG} is nontransparent (opaque) and thus entirely hides the background image I_{BG}. Conversely, the image I_{FG} is fully transparent for $\alpha = 1$ and only I_{BG} is visible. All α values between 0 and 1 result in a weighted sum of the corresponding pixel values taken from I_{BG} and I_{FG} (Eqn. (4.44)).

Figure 4.23 shows the results of linear blending for different α values. The Java code for the corresponding implementation (as an ImageJ plugin) is listed in Prog. 4.5. The background image (`bgIp`) is passed directly to the plugin's `run()` method. The second (foreground) image and the α value are specified interactively by creating an instance of the ImageJ class `GenericDialog`, which allows the simple implementation of dialog windows with various types of input fields.

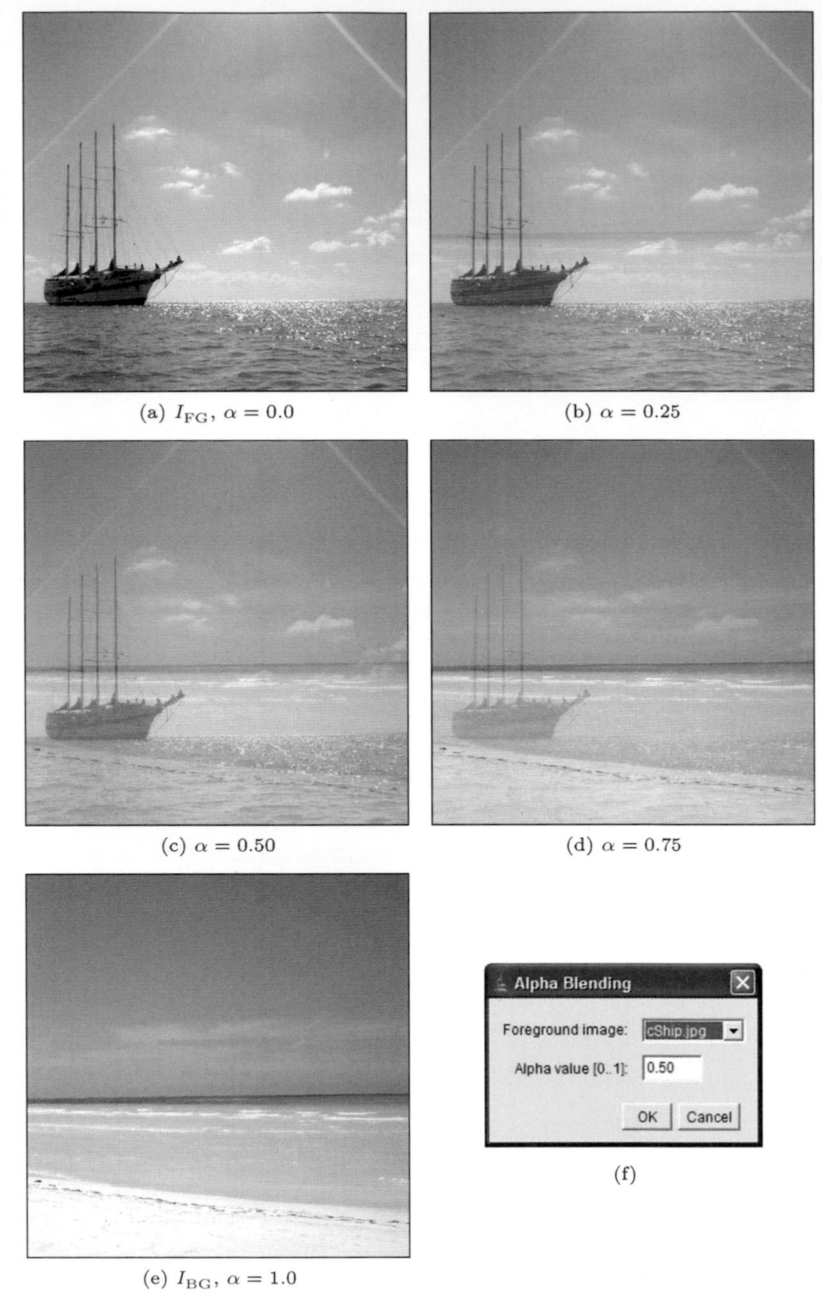

(a) I_{FG}, $\alpha = 0.0$

(b) $\alpha = 0.25$

(c) $\alpha = 0.50$

(d) $\alpha = 0.75$

(e) I_{BG}, $\alpha = 1.0$

(f)

Fig. 4.23
Linear blending example. Foreground image I_{FG} (a) and background image (I_{BG}) (e); blended images for transparency values $\alpha = 0.25, 0.50$, and 0.75 (b–d) and dialog window (f) produced by `GenericDialog` (see Prog. 4.5).

4.9 Exercises

Exercise 4.1. Implement the auto-contrast operation as defined in Eqns. (4.9)–(4.11) as an ImageJ plugin for an 8-bit grayscale image. Set the quantile p of pixels to be saturated at both ends of the intensity range (0 and 255) to $p = p_{lo} = p_{hi} = 1\%$.

Exercise 4.2. Modify the histogram equalization plugin in Prog. 4.2 to use a lookup table (Sec. 4.8.1) for computing the point operation.

Exercise 4.3. Implement the histogram equalization as defined in Eqn. (4.12), but use the *modified* cumulative histogram defined in Eqn. (4.13), cumulating the square root of the histogram entries. Compare the results to the standard (linear) approach by plotting the resulting histograms and cumulative histograms as shown in Fig. 4.10.

Exercise 4.4. Show formally that (a) a linear histogram equalization (Eqn. (4.12)) does not change an image that already has a uniform intensity distribution and (b) that any repeated application of histogram equalization to the same image causes no more changes.

Exercise 4.5. Show that the linear histogram equalization (Sec. 4.5) is only a special case of histogram specification (Sec. 4.6).

Exercise 4.6. Implement the histogram specification using a piecewise linear reference distribution function, as described in Sec. 4.6.3. Define a new object class with all necessary instance variables to represent the distribution function and implement the required functions $\mathsf{P}_L(i)$ (Eqn. (4.23)) and $\mathsf{P}_L^{-1}(b)$ (Eqn. (4.24)) as methods of this class.

Exercise 4.7. Using a histogram specification for adjusting *multiple* images (Sec. 4.6.4), one could either use one typical image as the reference or compute an "average" reference histogram from a set of images. Implement the second approach and discuss its possible advantages (or disadvantages).

Exercise 4.8. Implement the modified gamma correction (see Eqn. (4.34)) as an ImageJ plugin with variable values for γ and a_0 using a lookup table as shown in Prog. 4.4.

Exercise 4.9. Show that the modified gamma correction function $\bar{f}_{\gamma,a_0}(a)$, with the parameters defined in Eqns. (4.34)–(4.35), is C1-continuous (i.e., both the function itself and its first derivative are continuous).

Prog. 4.5
ImageJ-Plugin (Linear_Blending). A background image is transparently blended with a selected foreground image. The plugin is applied to the (currently active) background image, and the foreground image must also be open when the plugin is started. The background image (bgIp), which is passed to the plugin's run() method, is multiplied with α (line 22). The foreground image (fgIP, selected in part 2) is first duplicated (line 20) and then multiplied with $(1-\alpha)$ (line 21). Thus the original foreground image is not modified. The final result is obtained by adding the two weighted images (line 23). To select the foreground image, a list of currently open images and image titles is obtained (lines 30–32). Then a dialog object (of type GenericDialog) is created and opened for specifying the foreground image (fgIm) and the α value (lines 36–46).

```
1 import ij.ImagePlus;
2 import ij.gui.GenericDialog;
3 import ij.plugin.filter.PlugInFilter;
4 import ij.process.Blitter;
5 import ij.process.ImageProcessor;
6 import imagingbook.lib.ij.IjUtils;
7
8 public class Linear_Blending implements PlugInFilter {
9   static double alpha = 0.5;  // transparency of foreground image
10  ImagePlus fgIm;             // foreground image (to be selected)
11
12  public int setup(String arg, ImagePlus im) {
13    return DOES_8G;
14  }
15
16  public void run(ImageProcessor ipBG) {  // ipBG = I_BG
17    if(runDialog()) {
18      ImageProcessor ipFG =                      // ipFG = I_FG
19            fgIm.getProcessor().convertToByte(false);
20      ipFG = ipFG.duplicate();
21      ipFG.multiply(1 - alpha); // I_FG ← I_FG · (1 − α)
22      ipBG.multiply(alpha);     // I_BG ← I_BG · α
23      ipBG.copyBits(ipFG,0,0,Blitter.ADD); // I_BG ← I_BG + I_FG
24    }
25  }
26
27  boolean runDialog() {
28    // get list of open images and their titles:
29    ImagePlus[] openImages = IjUtils.getOpenImages(true);
30    String[] imageTitles = new String[openImages.length];
31    for (int i = 0; i < openImages.length; i++) {
32      imageTitles[i] = openImages[i].getShortTitle();
33    }
34    // create the dialog and show:
35    GenericDialog gd =
36            new GenericDialog("Linear Blending");
37    gd.addChoice("Foreground image:",
38        imageTitles, imageTitles[0]);
39    gd.addNumericField("Alpha value [0..1]:", alpha, 2);
40    gd.showDialog();
41
42    if (gd.wasCanceled())
43      return false;
44    else {
45      fgIm = openImages[gd.getNextChoiceIndex()];
46      alpha = gd.getNextNumber();
47      return true;
48    }
49  }
50 }
```

5

Filters

The essential property of point operations (discussed in the previous chapter) is that each new pixel value only depends on the original pixel at the *same* position. The capabilities of point operations are limited, however. For example, they cannot accomplish the task of *sharpening* or *smoothing* an image (Fig. 5.1). This is what filters can do. They are similar to point operations in the sense that they also produce a 1:1 mapping of the image coordinates, that is, the geometry of the image does not change.

Fig. 5.1
No point operation can blur or sharpen an image. This is an example of what filters can do. Like point operations, filters do not modify the geometry of an image.

5.1 What is a Filter?

The main difference between filters and point operations is that filters generally use more than one pixel from the source image for computing each new pixel value. Let us first take a closer look at the task of smoothing an image. Images look sharp primarily at places where the local intensity rises or drops sharply (i.e., where the difference between neighboring pixels is large). On the other hand, we perceive an image as blurred or fuzzy where the local intensity function is smooth.

A first idea for smoothing an image could thus be to simply replace every pixel by the *average* of its neighboring pixels. To determine the new pixel value in the smoothed image $I'(u, v)$, we use the original pixel $I(u,v) = p_0$ at the same position plus its eight neighboring pixels $p_1, p_2, \ldots, p_8$ to compute the arithmetic mean of these nine values,

$$I'(u,v) \leftarrow \frac{p_0 + p_1 + p_2 + p_3 + p_4 + p_5 + p_6 + p_7 + p_8}{9} . \tag{5.1}$$

Expressed in relative image coordinates this is

$$\begin{aligned} I'(u,v) \leftarrow \tfrac{1}{9} \cdot [\ & I(u-1,v-1) + I(u,v-1) + I(u+1,v-1) + \\ & I(u-1,v) \quad\;\; + I(u,v) \quad\;\; + I(u+1,v) \quad\;\; + \\ & I(u-1,v+1) + I(u,v+1) + I(u+1,v+1)\ ] , \end{aligned} \tag{5.2}$$

which we can write more compactly in the form

$$I'(u,v) \leftarrow \frac{1}{9} \cdot \sum_{j=-1}^{1} \sum_{i=-1}^{1} I(u+i, v+j) . \tag{5.3}$$

This simple local averaging already exhibits all the important elements of a typical filter. In particular, it is a so-called *linear* filter, which is a very important class of filters. But how are filters defined in general? First they differ from point operations mainly by using not a single source pixel but a *set* of them for computing each resulting pixel. The coordinates of the source pixels are fixed relative to the current image position (u,v) and usually form a contiguous region, as illustrated in Fig. 5.2.

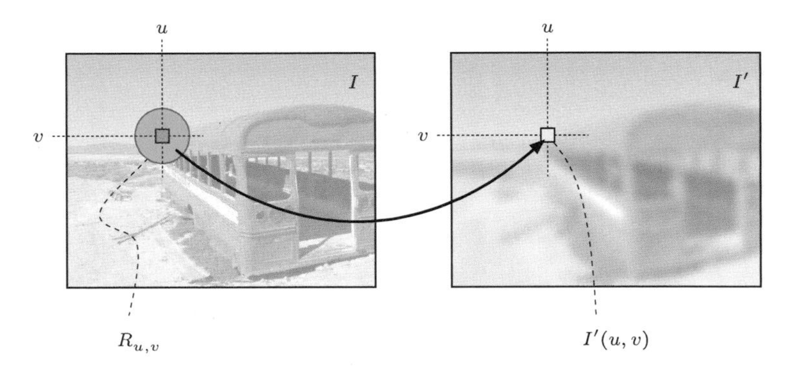

Fig. 5.2
Principal filter operation. Each new pixel value $I'(u,v)$ is calculated as a function of the pixel values within a specified region of source pixels $R_{u,v}$ in the original image I.

The *size* of the filter region is an important parameter of the filter because it specifies how many original pixels contribute to each resulting pixel value and thus determines the spatial extent (support) of the filter. For example, the smoothing filter in Eqn. (5.2) uses a 3×3 region of support that is centered at the current coordinate (u,v). Similar filters with larger support, such as 5×5, 7×7, or even 21×21 pixels, would obviously have stronger smoothing effects.

The *shape* of the filter region is not necessarily quadratic or even rectangular. In fact, a circular (disk-shaped) region would be preferred to obtain an *isotropic* blur effect (i.e., one that is the same in all image directions). Another option is to assign different *weights* to the pixels in the support region, such as to give stronger emphasis to pixels that are closer to the center of the region. Furthermore, the support region of a filter does not need to be contiguous and may not even contain the original pixel itself (imagine a ring-shaped filter region, for example). Theoretically the filter region could even be of infinite size.

It is probably confusing to have so many options—a more systematic method is needed for specifying and applying filters in a targeted manner. The traditional and proven classification into *linear* and *nonlinear* filters is based on the mathematical properties of the filter function; that is, whether the result is computed from the source pixels by a *linear* or a *nonlinear* expression. In the following, we discuss both classes of filters and show several practical examples.

5.2 Linear Filters

Linear filters are denoted that way because they combine the pixel values in the support region in a linear fashion, that is, as a weighted summation. The local averaging process discussed in the beginning (Eqn. (5.3)) is a special example, where all nine pixels in the 3×3 support region are added with identical weights ($1/9$). With the same mechanism, a multitude of filters with different properties can be defined by simply modifying the distribution of the individual weights.

5.2.1 The Filter Kernel

For any linear filter, the size and shape of the support region, as well as the individual pixel weights, are specified by the "filter kernel" or "filter matrix" $H(i,j)$. The size of the kernel H equals the size of the filter region, and every element $H(i,j)$ specifies the weight of the corresponding pixel in the summation. For the 3×3 smoothing filter in Eqn. (5.3), the filter kernel is

$$H = \begin{bmatrix} 1/9 & 1/9 & 1/9 \\ 1/9 & 1/9 & 1/9 \\ 1/9 & 1/9 & 1/9 \end{bmatrix} = \frac{1}{9} \cdot \begin{bmatrix} 1 & 1 & 1 \\ 1 & 1 & 1 \\ 1 & 1 & 1 \end{bmatrix}, \tag{5.4}$$

because each of the nine pixels contributes one-ninth of its value to the result.

In principle, the filter kernel $H(i,j)$ is, just like the image itself, a discrete, 2D, real-valued function, $H\colon \mathbb{Z} \times \mathbb{Z} \mapsto \mathbb{R}$. The filter has its own coordinate system with the origin—often referred to as the "hot spot"— mostly (but not necessarily) located at the center. Thus, filter coordinates are generally positive and negative (Fig. 5.3). The filter function is of infinite extent and considered zero outside the region defined by the matrix H.

5.2.2 Applying the Filter

For a linear filter, the result is unambiguously and completely specified by the coefficients of the filter matrix. Applying the filter to an image is a simple process that is illustrated in Fig. 5.4. The following steps are performed at each image position (u,v):

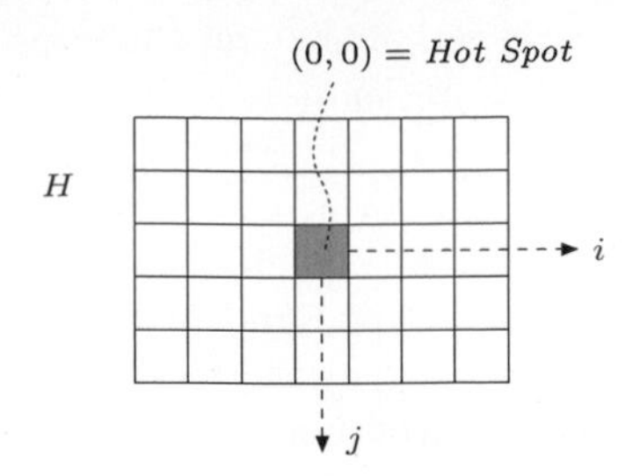

Fig. 5.3
Filter matrix and its coordinate system. i is the horizontal (column) index, j is the vertical (row) index.

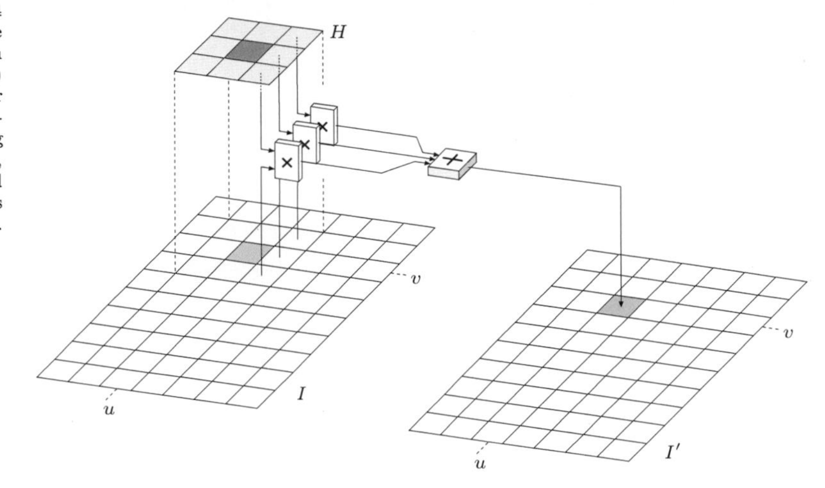

Fig. 5.4
Linear filter operation. The filter kernel H is placed with its origin at position (u, v) on the image I. Each filter coefficient $H(i, j)$ is multiplied with the corresponding image pixel $I(u+i, v+j)$, the results are added, and the final sum is inserted as the new pixel value $I'(u, v)$.

1. The filter kernel H is moved over the original image I such that its origin $H(0, 0)$ coincides with the current image position (u, v).
2. All filter coefficients $H(i, j)$ are multiplied with the corresponding image element $I(u+i, v+j)$, and the results are added up.
3. Finally, the resulting sum is stored at the current position in the new image $I'(u, v)$.

Described formally, the pixel values of the new image $I'(u, v)$ are computed by the operation

$$I'(u, v) \leftarrow \sum_{(i,j) \in R_H} I(u+i, v+j) \cdot H(i, j), \tag{5.5}$$

where R_H denotes the set of coordinates covered by the filter H. For a typical 3×3 filter with centered origin, this is

$$I'(u, v) \leftarrow \sum_{i=-1}^{i=1} \sum_{j=-1}^{j=1} I(u+i, v+j) \cdot H(i, j), \tag{5.6}$$

for all image coordinates (u, v). Not quite for *all* coordinates, to be exact. There is an obvious problem at the image borders where the filter reaches outside the image and finds no corresponding pixel values to use in computing a result. For the moment, we ignore this border problem, but we will attend to it again in Sec. 5.5.2.

5.2.3 Implementing the Filter Operation

Now that we understand the principal operation of a filter (Fig. 5.4) and know that the borders need special attention, we go ahead and program a simple linear filter in ImageJ. But before we do this, we may want to consider one more detail. In a point operation (e.g., in Progs. 4.1 and 4.2), each new pixel value depends only on the corresponding pixel value in the original image, and it was thus no problem simply to store the results back to the same image—the computation is done "in place" without the need for any intermediate storage. In-place computation is generally not possible for a filter since any original pixel contributes to more than one resulting pixel and thus may not be modified before all operations are complete.

We therefore require additional storage space for the resulting image, which subsequently could be copied back to the source image again (if desired). Thus the complete filter operation can be implemented in two different ways (Fig. 5.5):

A. The result of the filter computation is initially stored in a new image whose content is eventually copied back to the original image.
B. The original image is first copied to an intermediate image that serves as the source for the actual filter operation. The result replaces the pixels in the original image.

The same amount of storage is required for both versions, and thus none of them offers a particular advantage. In the following examples, we generally use version B.

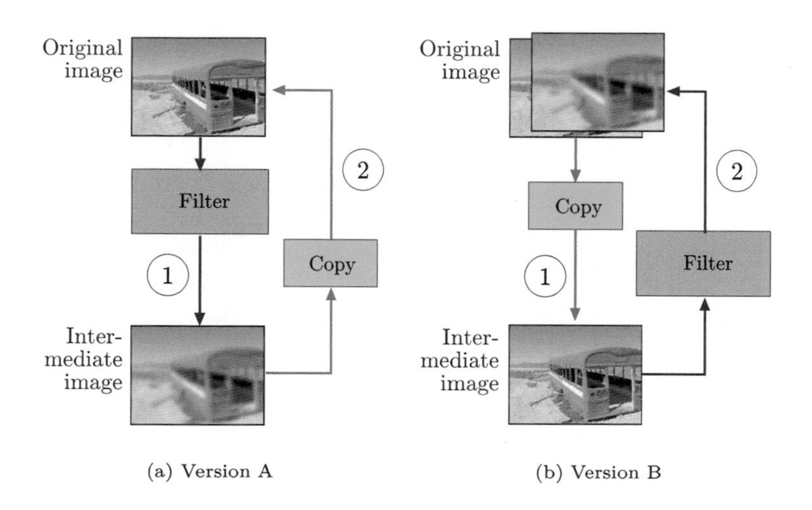

Fig. 5.5 Practical implementation of in-place filter operations. **Version A**: The result of the filter operation is first stored in an intermediate image and subsequently copied back to the original image (a). **Version B**: The original image is first copied to an intermediate image that serves as the source for the filter operation. The results are placed in the original image (b).

5.2.4 Filter Plugin Examples

The following examples demonstrate the implementation of two very basic filters that are nevertheless often used in practice.

Simple 3 × 3 averaging filter ("box" filter)

Program 5.1 shows the ImageJ code for a simple 3×3 smoothing filter based on local averaging (Eqn. (5.4)), which is often called a

Prog. 5.1
3 × 3 averaging "box" filter (Filter_Box_3x3). First (in line 10) a duplicate (copy) of the original image (orig) is created, which is used as the source image in the subsequent filter computation (line 18). In line 23, the resulting value is placed in the original image (line 23). Notice that the border pixels remain unchanged because they are not reached by the iteration over (u, v).

```
1 import ij.ImagePlus;
2 import ij.plugin.filter.PlugInFilter;
3 import ij.process.ImageProcessor;
4
5 public class Filter_Box_3x3 implements PlugInFilter {
6     ...
7     public void run(ImageProcessor ip) {
8         int M = ip.getWidth();
9         int N = ip.getHeight();
10        ImageProcessor copy = ip.duplicate();
11
12        for (int u = 1; u <= M - 2; u++) {
13          for (int v = 1; v <= N - 2; v++) {
14              //compute filter result for position (u, v):
15              int sum = 0;
16              for (int i = -1; i <= 1; i++) {
17                for (int j = -1; j <= 1; j++) {
18                    int p = copy.getPixel(u + i, v + j);
19                    sum = sum + p;
20                }
21              }
22              int q = (int) (sum / 9.0);
23              ip.putPixel(u, v, q);
24          }
25        }
26    }
27 }
```

"box" filter because of its box-like shape. No explicit filter matrix is required in this case, since all filter coefficients are identical ($^1/_9$). Also, no *clamping* (see Sec. 4.1.2) of the results is needed because the sum of the filter coefficients is 1 and thus no pixel values outside the admissible range can be created.

Although this example implements an extremely simple filter, it nevertheless demonstrates the general structure of a 2D filter program. In particular, *four* nested loops are needed: *two* (outer) loops for moving the filter over the image coordinates (u, v) and *two* (inner) loops to iterate over the (i, j) coordinates within the rectangular filter region. The required amount of computation thus depends not only upon the size of the image but equally on the size of the filter.

Another 3 × 3 smoothing filter

Instead of the constant weights applied in the previous example, we now use a real filter matrix with variable coefficients. For this purpose, we apply a bell-shaped 3×3 filter function $H(i, j)$, which puts more emphasis on the center pixel than the surrounding pixels:

$$H = \begin{bmatrix} 0.075 & 0.125 & 0.075 \\ 0.125 & 0.200 & 0.125 \\ 0.075 & 0.125 & 0.075 \end{bmatrix}. \tag{5.7}$$

Notice that all coefficients in H are positive and sum to 1 (i.e., the matrix is normalized) such that all results remain within the origi-

```
1   ...
2   public void run(ImageProcessor ip) {
3     int M = ip.getWidth();
4     int N = ip.getHeight();
5
6     //3x3 filter matrix:
7     double[][] H = {
8         {0.075, 0.125, 0.075},
9         {0.125, 0.200, 0.125},
10        {0.075, 0.125, 0.075}};
11
12    ImageProcessor copy = ip.duplicate();
13
14    for (int u = 1; u <= M - 2; u++) {
15      for (int v = 1; v <= N - 2; v++) {
16        // compute filter result for position (u,v):
17        double sum = 0;
18        for (int i = -1; i <= 1; i++) {
19          for (int j = -1; j <= 1; j++) {
20            int p = copy.getPixel(u + i, v + j);
21            // get the corresponding filter coefficient:
22            double c = H[j + 1][i + 1];
23            sum = sum + c * p;
24          }
25        }
26        int q = (int) Math.round(sum);
27        ip.putPixel(u, v, q);
28      }
29    }
30  }
```


Prog. 5.2
3×3 smoothing filter (`Filter_Smooth_3x3`). The filter matrix is defined as a 2D array of type `double` (line 7). The coordinate origin of the filter is assumed to be at the center of the matrix (i.e., at the array position [1, 1]), which is accounted for by an offset of 1 for the i, j coordinates in line 22. The results are rounded (line 26) and stored in the original image (line 27).

nal range of pixel values. Again no clamping is necessary and the program structure in Prog. 5.2 is virtually identical to the previous example. The filter matrix (**filter**) is represented by a 2D array[1] of type `double`. Each pixel is multiplied by the corresponding coefficient of the filter matrix, the resulting sum being also of type `double`. Accessing the filter coefficients, it must be considered that the coordinate origin of the filter matrix is assumed to be at its center (i.e., at position $(1, 1)$) in the case of a 3×3 matrix. This explains the offset of 1 for the i and j coordinates (see Prog. 5.2, line 22).

5.2.5 Integer Coefficients

Instead of using floating-point coefficients (as in the previous examples), it is often simpler and usually more efficient to work with integer coefficients in combination with some common scale factor s, that is,

$$H(i,j) = s \cdot H'(i,j), \tag{5.8}$$

with $H'(i,j) \in \mathbb{Z}$ and $s \in \mathbb{R}$. If all filter coefficients are positive (which is the case for any smoothing filter), then s is usually taken

[1] See the additional comments regarding 2D arrays in Java in Sec. F.2.4 in the Appendix.

as the reciprocal of the sum of the coefficients,

$$s = \frac{1}{\sum_{i,j} H'(i,j)}, \tag{5.9}$$

to obtain a normalized filter matrix. In this case, the results are bounded to the original range of pixel values. For example, the filter matrix in Eqn. (5.7) could be defined equivalently as

$$H = \begin{bmatrix} 0.075 & 0.125 & 0.075 \\ 0.125 & 0.200 & 0.125 \\ 0.075 & 0.125 & 0.075 \end{bmatrix} = \frac{1}{40} \cdot \begin{bmatrix} 3 & 5 & 3 \\ 5 & 8 & 5 \\ 3 & 5 & 3 \end{bmatrix} \tag{5.10}$$

with the common scale factor $s = \frac{1}{40} = 0.025$. A similar scaling is used for the filter operation in Prog. 5.3.

In Adobe Photoshop, linear filters can be specified with the "Custom Filter" tool (Fig. 5.6) using integer coefficients and a common scale factor *Scale* (which corresponds to the reciprocal of s). In addition, a constant *Offset* value can be specified; for example, to shift negative results (caused by negative coefficients) into the visible range of values. In summary, the operation performed by the 5×5 Photoshop custom filter can be expressed as

$$I'(u,v) \leftarrow \mathit{Offset} + \frac{1}{\mathit{Scale}} \cdot \sum_{j=-2}^{j=2} \sum_{i=-2}^{i=2} I(u+i, v+j) \cdot H(i,j). \tag{5.11}$$

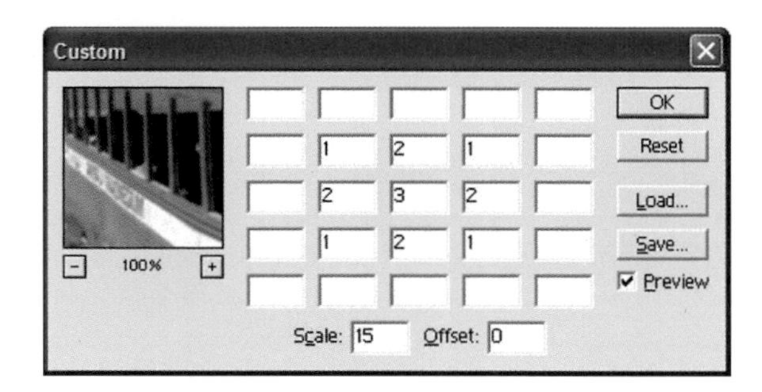

Fig. 5.6
Adobe Photoshop's "Custom Filter" implements linear filters up to a size of 5×5. The filter's coordinate origin ("hot spot") is assumed to be at the center (value set to 3 in this example), and empty cells correspond to zero coefficients. In addition to the (integer) coefficients, common *Scale* and *Offset* values can be specified (see Eqn. (5.11)).

5.2.6 Filters of Arbitrary Size

Small filters of size 3×3 are frequently used in practice, but sometimes much larger filters are required. Let us assume that the filter matrix H is centered and has an odd number of $(2K+1)$ columns and $(2L+1)$ rows, with $K, L \geq 0$. If the image is of size $M \times N$, that is

$$I(u,v) \quad \text{with} \quad 0 \leq u < M \quad \text{and} \quad 0 \leq v < N, \tag{5.12}$$

then the result of the filter can be calculated for all image coordinates (u', v') with

$$K \leq u' \leq (M-K-1) \qquad \text{and} \qquad L \leq v' \leq (N-L-1), \tag{5.13}$$

as illustrated in Fig. 5.7. Program 5.3 (which is adapted from Prog. 5.2) shows a 7×5 smoothing filter as an example for implementing

```
1   public void run(ImageProcessor ip) {
2       int M = ip.getWidth();
3       int N = ip.getHeight();
4
5       // filter matrix H of size (2K + 1) × (2L + 1)
6       int[][] H = {
7           {0,0,1,1,1,0,0},
8           {0,1,1,1,1,1,0},
9           {1,1,1,1,1,1,1},
10          {0,1,1,1,1,1,0},
11          {0,0,1,1,1,0,0}};
12
13      double s = 1.0 / 23;   // sum of filter coefficients is 23
14
15      // H[L][K] is the center element of H:
16      int K = H[0].length / 2;   // K = 3
17      int L = H.length / 2; // L = 2
18
19      ImageProcessor copy = ip.duplicate();
20
21      for (int u = K; u <= M - K - 1; u++) {
22        for (int v = L; v <= N - L - 1; v++) {
23          // compute filter result for position (u, v):
24          int sum = 0;
25          for (int i = -K; i <= K; i++) {
26            for (int j = -L; j <= L; j++) {
27              int p = copy.getPixel(u + i, v + j);
28              int c = H[j + L][i + K];
29              sum = sum + c * p;
30              }
31          }
32          int q = (int) Math.round(s * sum);
33          // clamp result:
34          if (q < 0)   q = 0;
35          if (q > 255) q = 255;
36          ip.putPixel(u, v, q);
37        }
38      }
39  }
```

Prog. 5.3
Linear filter of arbitrary size using integer coefficients (`Filter_Arbitrary`). The filter matrix is an integer array of size $(2K+1) \times (2L+1)$ with the origin at the center element. The summation variable `sum` is also defined as an integer (`int`), which is scaled by a constant factor `s` and rounded in line 32. The border pixels are not modified.

linear filters of arbitrary size. This example uses integer-valued filter coefficients (line 6) in combination with a common scale factor s, as described already. As usual, the "hot spot" of the filter is assumed to be at the matrix center, and the range of all iterations depends on the dimensions of the filter matrix. In this case, clamping of the results is included (in lines 34–35) as a preventive measure.

5.2.7 Types of Linear Filters

Since the effects of a linear filter are solely specified by the filter matrix (which can take on arbitrary values), an infinite number of different linear filters exists, at least in principle. So how can these filters be used and which filters are suited for a given task? In the following, we briefly discuss two broad classes of linear filters that are

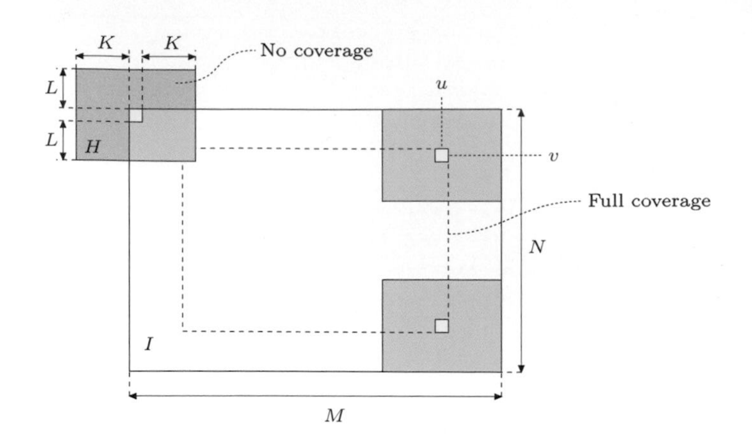

Fig. 5.7
Border geometry. The filter can be applied only at locations where the kernel H of size $(2K+1) \times (2L+1)$ is fully contained in the image (inner rectangle).

of key importance in practice: smoothing filters and difference filters (Fig. 5.8).

Smoothing filters

Every filter we have discussed so far causes some kind of smoothing. In fact, any linear filter with positive-only coefficients is a smoothing filter in a sense, because such a filter computes merely a weighted average of the image pixels within a certain image region.

Box filter

This simplest of all smoothing filters, whose 3D shape resembles a box (Fig. 5.8(a)), is a well-known friend already. Unfortunately, the box filter is far from an optimal smoothing filter due to its wild behavior in frequency space, which is caused by the sharp cutoff around its sides. Described in frequency terms, smoothing corresponds to low-pass filtering, that is, effectively attenuating all signal components above a given cutoff frequency (see also Chs. 18–19). The box filter, however, produces strong "ringing" in frequency space and is therefore not considered a high-quality smoothing filter. It may also appear rather ad hoc to assign the same weight to all image pixels in the filter region. Instead, one would probably expect to have stronger emphasis given to pixels near the center of the filter than to the more distant ones. Furthermore, smoothing filters should possibly operate "isotropically" (i.e., uniformly in each direction), which is certainly not the case for the rectangular box filter.

Gaussian filter

The filter matrix (Fig. 5.8(b)) of this smoothing filter corresponds to a 2D Gaussian function,

$$H^{\mathrm{G},\sigma}(x,y) = e^{-\frac{x^2+y^2}{2\sigma^2}}, \tag{5.14}$$

where σ denotes the width (standard deviation) of the bell-shaped function and r is the distance (radius) from the center. The pixel at the center receives the maximum weight (1.0, which is scaled to the integer value 9 in the matrix shown in Fig. 5.8(b)), and the remaining coefficients drop off smoothly with increasing distance from the

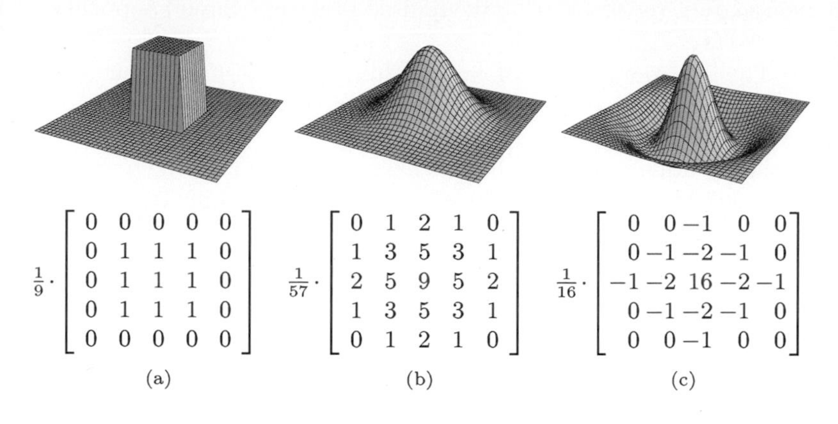

$$\frac{1}{9}\cdot\begin{bmatrix} 0 & 0 & 0 & 0 & 0 \\ 0 & 1 & 1 & 1 & 0 \\ 0 & 1 & 1 & 1 & 0 \\ 0 & 1 & 1 & 1 & 0 \\ 0 & 0 & 0 & 0 & 0 \end{bmatrix}$$
(a)
$$\frac{1}{57}\cdot\begin{bmatrix} 0 & 1 & 2 & 1 & 0 \\ 1 & 3 & 5 & 3 & 1 \\ 2 & 5 & 9 & 5 & 2 \\ 1 & 3 & 5 & 3 & 1 \\ 0 & 1 & 2 & 1 & 0 \end{bmatrix}$$
(b)
$$\frac{1}{16}\cdot\begin{bmatrix} 0 & 0 & -1 & 0 & 0 \\ 0 & -1 & -2 & -1 & 0 \\ -1 & -2 & 16 & -2 & -1 \\ 0 & -1 & -2 & -1 & 0 \\ 0 & 0 & -1 & 0 & 0 \end{bmatrix}$$
(c)

Fig. 5.8
Typical examples of linear filters, illustrated as 3D plots (top), profiles (center), and approximations by discrete filter matrices (bottom). The "box" filter (a) and the Gauss filter (b) are both *smoothing filters* with all-positive coefficients. The "Laplacian" or "Mexican hat" filter (c) is a *difference filter*. It computes the weighted difference between the center pixel and the surrounding pixels and thus reacts most strongly to local intensity peaks.

center. The Gaussian filter is isotropic if the discrete filter matrix is large enough for a sufficient approximation (at least 5×5). As a low-pass filter, the Gaussian is "well-behaved" in frequency space and thus clearly superior to the box filter. The 2D Gaussian filter is separable into a pair of 1D filters (see Sec. 5.3.3), which facilitates its efficient implementation.[2]

Difference filters

If some of the filter coefficients are negative, the filter calculation can be interpreted as the difference of two sums: the weighted sum of all pixels with associated positive coefficients minus the weighted sum of pixels with negative coefficients in the filter region R_H, that is,

$$I'(u,v) = \sum_{(i,j)\in R^+} I(u+i,v+j)\cdot|H(i,j)| - \sum_{(i,j)\in R^-} I(u+i,v+j)\cdot|H(i,j)|\,, \tag{5.15}$$

where R_H^+ and R_H^- denote the partitions of the filter with positive coefficients $H(i,j) > 0$ and negative coefficients $H(i,j) < 0$, respectively. For example, the 5×5 Laplace filter in Fig. 5.8(c) computes the difference between the center pixel (with weight 16) and the weighted sum of 12 surrounding pixels (with weights -1 or -2). The remaining 12 pixels have associated zero coefficients and are thus ignored in the computation.

While local intensity variations are *smoothed* by averaging, we can expect the exact contrary to happen when differences are taken: local intensity changes are *enhanced*. Important applications of difference filters thus include edge detection (Sec. 6.2) and image sharpening (Sec. 6.6).

5.3 Formal Properties of Linear Filters

In the previous sections, we have approached the concept of filters in a rather casual manner to quickly get a grasp of how filters are defined and used. While such a level of treatment may be sufficient for most practical purposes, the power of linear filters may not really

[2] See also Sec. E in the Appendix.

be apparent yet considering the limited range of (simple) applications seen so far.

The real importance of linear filters (and perhaps their formal elegance) only becomes visible when taking a closer look at some of the underlying theoretical details. At this point, it may be surprising to the experienced reader that we have not mentioned the term "convolution" in this context yet. We make up for this in the remaining parts of this section.

5.3.1 Linear Convolution

The operation associated with a linear filter, as described in the previous section, is not an invention of digital image processing but has been known in mathematics for a long time. It is called *linear convolution*[3] and in general combines two functions of the same dimensionality, either continuous or discrete. For discrete, 2D functions I and H, the convolution operation is defined as

$$I'(u,v) = \sum_{i=-\infty}^{\infty} \sum_{j=-\infty}^{\infty} I(u-i, v-j) \cdot H(i,j)\,, \tag{5.16}$$

or, expressed with the designated *convolution operator* ($*$) in the form

$$I' = I * H. \tag{5.17}$$

This almost looks the same as Eqn. (5.5), with two differences: the range of the variables i, j in the summation and the negative signs in the coordinates of $I(u-i, v-j)$. The first point is easy to explain: because the coefficients outside the filter matrix $H(i,j)$, also referred to as a filter *kernel*, are assumed to be zero, the positions outside the matrix are irrelevant in the summation. To resolve the coordinate issue, we modify Eqn. (5.16) by replacing the summation variables i, j to

$$I'(u,v) = \sum_{(i,j)\in R_H} I(u-i, v-j) \cdot H(i,j) \tag{5.18}$$

$$= \sum_{(i,j)\in R_H} I(u+i, v+j) \cdot H(-i,-j) \tag{5.19}$$

$$= \sum_{(i,j)\in R_H} I(u+i, v+j) \cdot H^*(i,j). \tag{5.20}$$

The result is identical to the linear filter in Eqn. (5.5), with the $H^*(i,j) = H(-i,-j)$ being the horizontally and vertically *reflected* (i.e., rotated by 180°) kernel H. To be precise, the operation in Eqn. (5.5) actually defines the linear *correlation*, which is merely a convolution with a reflected filter matrix.[4]

[3] Oddly enough the simple concept of convolution is often (though unjustly) feared as an intractable mystery.

[4] Of course this is the same in the 1D case. Linear correlation is typically used for comparing images or subpatterns (see Sec. 23.1 for details).

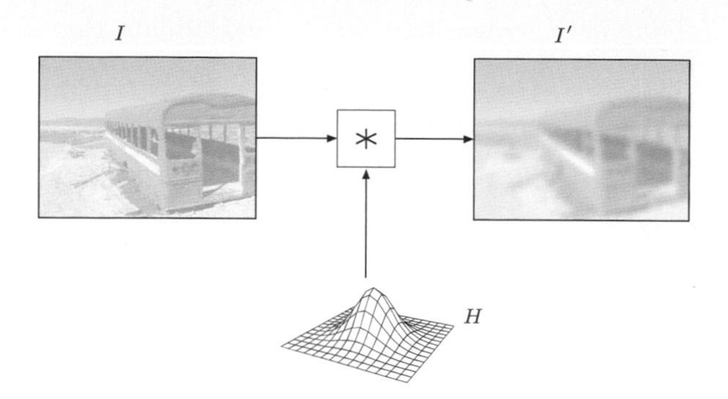

Fig. 5.9
Convolution as a "black box" operation. The original image I is subjected to a linear convolution ($*$) with the convolution kernel H, producing the output image I'.

Thus the mathematical concept underlying all linear filters is the convolution operation ($*$) and its results are completely and sufficiently specified by the convolution matrix (or kernel) H. To illustrate this relationship, the convolution is often pictured as a "black box" operation, as shown in Fig. 5.9.

5.3.2 Formal Properties of Linear Convolution

The importance of linear convolution is based on its simple mathematical properties as well as its multitude of manifestations and applications. Linear convolution is a suitable model for many types of natural phenomena, including mechanical, acoustic, and optical systems. In particular (as shown in Ch. 18), there are strong formal links to the Fourier representation of signals in the frequency domain that are extremely valuable for understanding complex phenomena, such as sampling and aliasing. In the following, however, we first look at some important properties of linear convolution in the accustomed "signal" or image space.

Commutativity

Linear convolution is *commutative*; that is, for any image I and filter kernel H,

$$I * H = H * I. \tag{5.21}$$

Thus the result is the same if the image and filter kernel are interchanged, and it makes no difference if we convolve the image I with the kernel H or the other way around. The two functions I and H are interchangeable and may assume either role.

Linearity

Linear filters are so called because of the linearity properties of the convolution operation, which manifests itself in various aspects. For example, if an image is multiplied by a scalar constant $s \in \mathbb{R}$, then the result of the convolution multiplies by the same factor, that is,

$$(s \cdot I) * H \;=\; I * (s \cdot H) \;=\; s \cdot (I * H)\;. \tag{5.22}$$

Similarly, if we add two images I_1, I_2 pixel by pixel and convolve the resulting image with some kernel H, the same outcome is obtained by convolving each image individually and adding the two results afterward, that is,

$$(I_1 + I_2) * H \;=\; (I_1 * H) + (I_2 * H). \tag{5.23}$$

It may be surprising, however, that simply *adding* a constant (scalar) value b to the image does *not* add to the convolved result by the same amount,

$$(b + I) * H \;\neq\; b + (I * H), \tag{5.24}$$

and is thus not part of the linearity property. While linearity is an important theoretical property, one should note that in practice "linear" filters are often only partially linear because of rounding errors or a limited range of output values.

Associativity

Linear convolution is associative, meaning that the order of successive filter operations is irrelevant, that is,

$$(I * H_1) * H_2 = I * (H_1 * H_2). \tag{5.25}$$

Thus multiple successive filters can be applied in any order, and multiple filters can be arbitrarily combined into new filters.

5.3.3 Separability of Linear Filters

A direct consequence of associativity is the separability of linear filters. If a convolution kernel H can be expressed as the convolution of multiple kernels H_i in the form

$$H = H_1 * H_2 * \ldots * H_n, \tag{5.26}$$

then (as a consequence of Eqn. (5.25)) the filter operation $I * H$ may be performed as a sequence of convolutions with the constituting kernels H_i,

$$\begin{aligned} I * H &= I * (H_1 * H_2 * \ldots * H_n) \\ &= (\ldots((I * H_1) * H_2) * \ldots * H_n). \end{aligned} \tag{5.27}$$

Depending upon the type of decomposition, this may result in significant computational savings.

x/y separability

The possibility of separating a 2D kernel H into a pair of 1D kernels h_x, h_y is of particular relevance and is used in many practical applications. Let us assume, as a simple example, that the filter is composed of the 1D kernels h_x and h_y, with

$$h_x = \begin{bmatrix} 1 & 1 & 1 & 1 & 1 \end{bmatrix} \quad \text{and} \quad h_y = \begin{bmatrix} 1 \\ 1 \\ 1 \end{bmatrix}, \tag{5.28}$$

respectively. If these filters are applied sequentially to the image I,

$$I' = (I * h_x) * h_y, \tag{5.29}$$

then (according to Eqn. (5.27)) this is equivalent to applying the composite filter

$$H = h_x * h_y = \begin{bmatrix} 1 & 1 & 1 & 1 & 1 \\ 1 & 1 & 1 & 1 & 1 \\ 1 & 1 & 1 & 1 & 1 \end{bmatrix}. \qquad (5.30)$$

Thus the 2D 5×3 "box" filter H can be constructed from two 1D filters of lengths 5 and 3, respectively (which is obviously true for box filters of any size). But what is the advantage of this? In the aforementioned case, the required amount of processing is $5 \cdot 3 = 15$ steps per image pixel for the 2D filter H as compared with $5 + 3 = 8$ steps for the two separate 1D filters, a reduction of almost 50%. In general, the number of operations for a 2D filter grows *quadratically* with the filter size (side length) but only *linearly* if the filter is x/y-separable. Clearly, separability is an eminent bonus for the implementation of large linear filters (see also Sec. 5.5.1).

Separable Gaussian filters

In general, a 2D filter is x/y-separable if (as in the earlier example) the filter function $H(i, j)$ can be expressed as the outer product ($\otimes$) of two 1D functions,

$$H(i, j) = h_x(i) \cdot h_y(j), \qquad (5.31)$$

because in this case the resulting function also corresponds to the convolution product $H = H_x * H_y$. A prominent example is the widely employed 2D Gaussian function $G_\sigma(x, y)$ (Eqn. (5.14)), which can be expressed as the product

$$G_\sigma(x, y) = e^{-\frac{x^2+y^2}{2\sigma^2}} \qquad (5.32)$$

$$= \exp(-\tfrac{x^2}{2\sigma^2}) \cdot \exp(-\tfrac{y^2}{2\sigma^2}) = g_\sigma(x) \cdot g_\sigma(y). \qquad (5.33)$$

Thus a 2D Gaussian filter H^G_σ can be implemented by a pair of 1D Gaussian filters $h^G_{x,\sigma}$ and $h^G_{y,\sigma}$ as

$$I * H^G_\sigma = I * h^G_{x,\sigma} * h^G_{y,\sigma}. \qquad (5.34)$$

The ordering of the two 1D filters is not relevant in this case. With different σ-values along the x and y axes, elliptical 2D Gaussians can be realized as separable filters in the same fashion.

The Gaussian function decays relatively slowly with increasing distance from the center. To avoid visible truncation errors, discrete approximations of the Gaussian should have a sufficiently large extent of about $\pm 2.5\,\sigma$ to $\pm 3.5\,\sigma$ samples. For example, a discrete 2D Gaussian with "radius" $\sigma = 10$ requires a minimum filter size of 51×51 pixels, in which case the x/y-separable version can be expected to run about 50 times faster than the full 2D filter. The Java method `makeGaussKernel1d()` in Prog. 5.4 shows how to dynamically create a 1D Gaussian filter kernel with an extent of $\pm 3\,\sigma$ (i.e., a vector of odd length $6\,\sigma + 1$). As an example, this method is used for implementing "unsharp masking" filters where relatively large Gaussian kernels may be required (see Prog. 6.1 in Sec. 6.6.2).

Prog. 5.4
Dynamic creation of 1D Gaussian filter kernels. For a given σ, the Java method makeGaussKernel1d() returns a discrete 1D Gaussian filter kernel (float array) large enough to avoid truncation effects.

```
1  float[] makeGaussKernel1d(double sigma) {
2    // create the 1D kernel h:
3    int center = (int) (3.0 * sigma);
4    float[] h = new float[2 * center + 1]; // odd size
5    // fill the 1D kernel h:
6    double sigma2 = sigma * sigma;     // σ²
7    for (int i = 0; i < h.length; i++) {
8      double r  = center - i;
9      h[i] = (float) Math.exp(-0.5 * (r * r) / sigma2);
10   }
11   return h;
12 }
```

5.3.4 Impulse Response of a Filter

Linear convolution is a binary operation involving two functions as its operands; it also has a "neutral element", which of course is a function, too. The *impulse* or *Dirac* function $\delta()$ is neutral under convolution, that is,

$$I * \delta = I. \tag{5.35}$$

In the 2D, discrete case, the impulse function is defined as

$$\delta(u,v) = \begin{cases} 1 & \text{for } u = v = 0, \\ 0 & \text{otherwise.} \end{cases} \tag{5.36}$$

Interpreted as an image, this function is merely a single bright pixel (with value 1) at the coordinate origin contained in a dark (zero value) plane of infinite extent (Fig. 5.10).

When the Dirac function is used as the filter kernel in a linear convolution as in Eqn. (5.35), the result is identical to the original image (Fig. 5.11). The reverse situation is more interesting, however, where some filter H is applied to the impulse δ as the input function. What happens? Since convolution is commutative (Eqn. (5.21)) it is evident that

$$H * \delta = \delta * H = H \tag{5.37}$$

and thus the result of this filter operation is identical to the filter H itself (Fig. 5.12)! While sending an impulse into a linear filter to obtain its filter function may seem paradoxical at first, it makes sense if the properties (coefficients) of the filter H are unknown. Assuming that the filter is actually linear, complete information about this filter is obtained by injecting only a single impulse and measuring the result, which is called the "impulse response" of the filter. Among

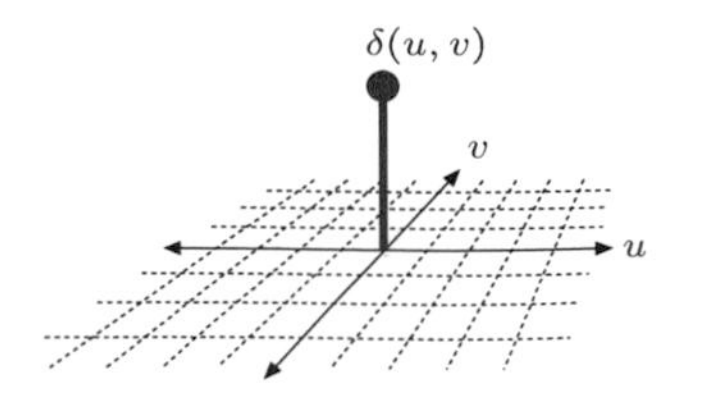

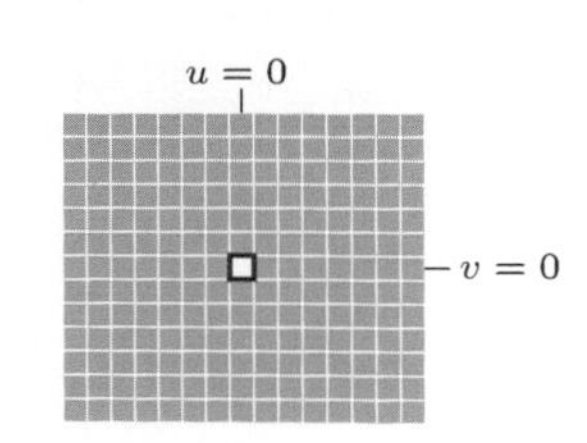

Fig. 5.10
Discrete 2D *impulse* or *Dirac* function $\delta(u, v)$.

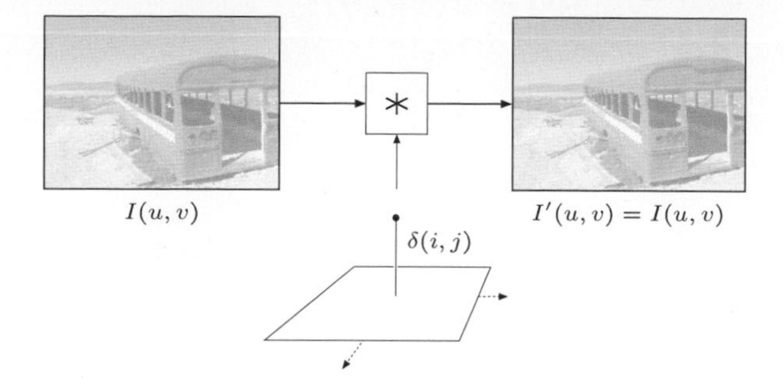

Fig. 5.11
Convolving the image I with the impulse δ returns the original (unmodified) image.

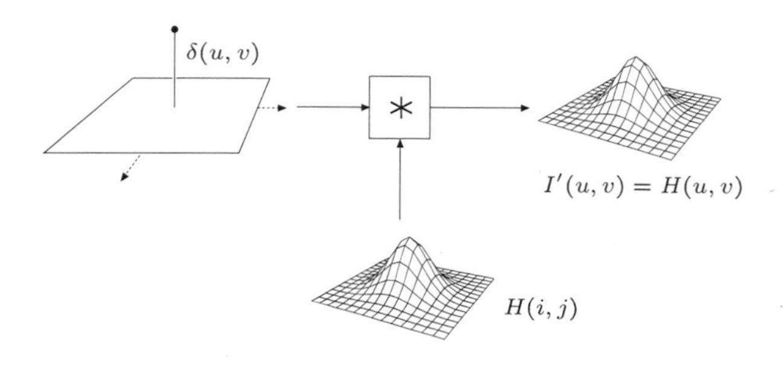

Fig. 5.12
The linear filter H with the impulse δ as the input yields the filter kernel H as the result.

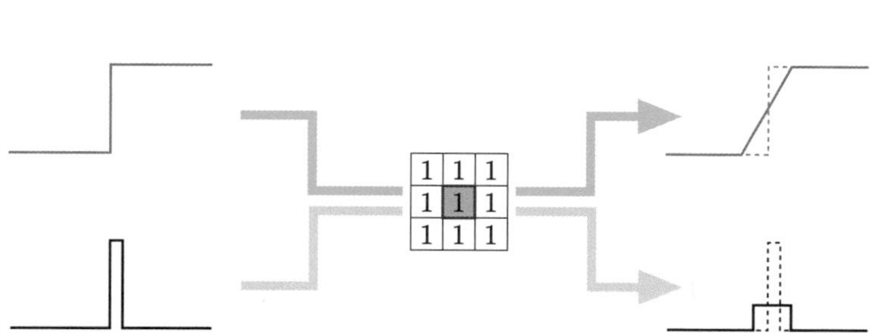

Fig. 5.13
Any image structure is blurred by a linear smoothing filter. Important image structures such as step edges (top) or thin lines (bottom) are widened, and local contrast is reduced.

other applications, this technique is used for measuring the behavior of optical systems (e.g., lenses), where a point light source serves as the impulse and the result—a distribution of light energy—is called the "point spread function" (PSF) of the system.

5.4 Nonlinear Filters

5.4.1 Minimum and Maximum Filters

Like all other filters, nonlinear filters calculate the result at a given image position (u, v) from the pixels inside the moving region $R_{u,v}$ of the original image. The filters are called "nonlinear" because the source pixel values are combined by some nonlinear function. The simplest of all nonlinear filters are the *minimum* and *maximum* filters, defined as

$$I'(u, v) = \min_{(i,j)\in R} \{I(u+i, v+j)\}, \tag{5.38}$$

$$I'(u, v) = \max_{(i,j)\in R} \{I(u+i, v+j)\}, \tag{5.39}$$

Fig. 5.14
3 × 3 linear box filter applied to a grayscale image corrupted with salt-and-pepper noise. Original (a), filtered image (b), enlarged details (c, d). Note that the individual noise pixels are only flattened but not removed.

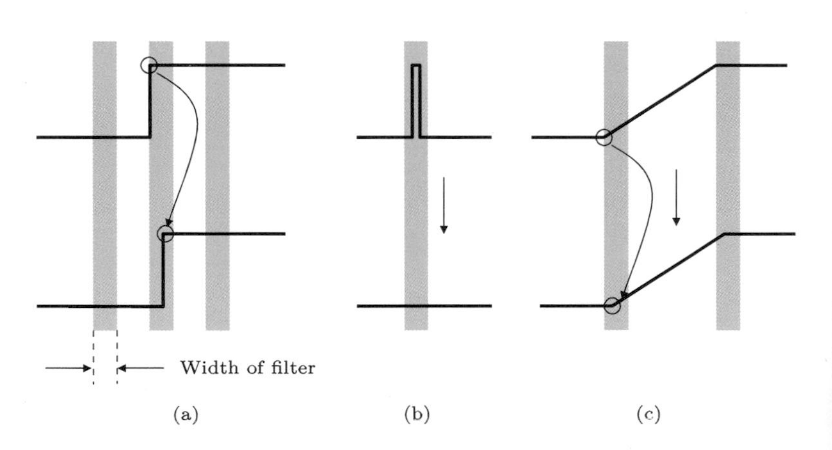

Fig. 5.15
Effects of a 1D minimum filter on different local signal structures. Original signal (top) and result after filtering (bottom), where the colored bars indicate the extent of the filter. The step edge (a) and the linear ramp (c) are shifted to the right by half the filter width, and the narrow pulse (b) is completely removed.

where R denotes the filter region (set of filter coordinates, usually a square of size 3×3 pixels). Figure 5.15 illustrates the effects of a 1D minimum filter on various local signal structures.

Figure 5.16 shows the results of applying 3×3 pixel minimum and maximum filters to a grayscale image corrupted with "salt-and-pepper" noise (i.e., randomly placed white and black dots), respectively. Obviously the minimum filter removes the white (salt) dots, because any single white pixel within the 3×3 filter region is replaced

Fig. 5.16
Minimum and maximum filters applied to a grayscale image corrupted with "salt-and-pepper" noise (see original in Fig. 5.14(a)). The 3×3 *minimum* filter eliminates the bright dots and widens all dark image structures (a, c). The *maximum* filter shows the exact opposite effects (b, d).

by one of its surrounding pixels with a smaller value. Notice, however, that the minimum filter at the same time widens all the dark structures in the image.

The reverse effects can be expected from the *maximum* filter. Any single bright pixel is a local maximum as soon as it is contained in the filter region R. White dots (and all other bright image structures) are thus widened to the size of the filter, while now the dark ("pepper") dots disappear.[5]

5.4.2 Median Filter

It is impossible of course to design a filter that removes any noise but keeps all the important image structures intact, because no filter can discriminate which image content is important to the viewer and which is not. The popular median filter is at least a good step in this direction.

[5] The image shown in Figs. 5.14 and 5.16, called "Lena" (or "Lenna"), is one of the most popular test images in digital image processing ever and thus of historic interest. The picture of the Swedish "playmate" Lena Sjööblom (Söderberg?), published in *Playboy* in 1972, was included in a collection of test images at the University of Southern California and was subsequently used by researches throughout the world (presumably without knowledge of its delicate origin) [115].

The median filter replaces every image pixel by the *median* of the pixels in the current filter region R, that is,

$$I'(u,v) = \underset{(i,j)\in R}{\text{median}}\{I(u+i, v+j)\}. \tag{5.40}$$

The median of a set of $2n+1$ values $A = \{a_0, \ldots, a_{2n}\}$ can be defined as the *center* value a_n after arranging (sorting) A to an ordered sequence, that is,

$$\text{median}(\underbrace{a_0, a_1, \ldots, a_{n-1}}_{n \text{ values}}, a_n, \underbrace{a_{n+1}, \ldots, a_{2n}}_{n \text{ values}}) = a_n, \tag{5.41}$$

where $a_i \leq a_{i+1}$. Figure 5.17 demonstrates the calculation of the median filter of size 3×3 (i.e., $n = 4$).

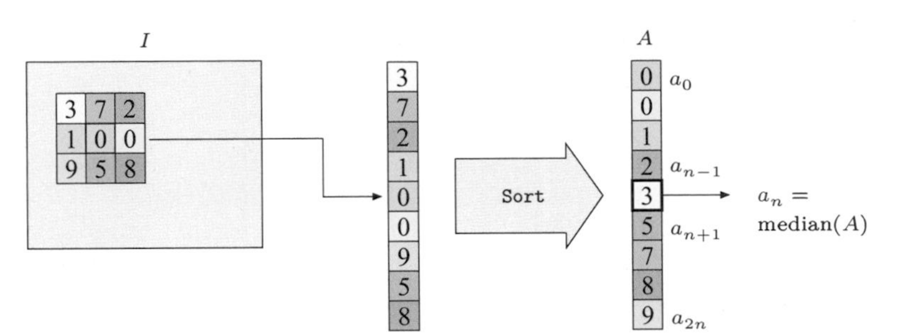

Fig. 5.17 Calculation of the median. The nine pixel values collected from the 3×3 image region are arranged as a vector that is subsequently sorted (A). The center value of A is taken as the median.

Equation (5.41) defines the median of an *odd*-sized set of values, and if the side length of the rectangular filters is odd (which is usually the case), then the number of elements in the filter region is odd as well. In this case, the median filter does not create any new pixel values that did not exist in the original image. If, however, the number of elements is *even*, then the median of the sorted sequence $A = (a_0, \ldots, a_{2n-1})$ is defined as the arithmetic mean of the two adjacent center values a_{n-1} and a_n,

$$\text{median}\big(\underbrace{a_0, \ldots, a_{n-1}}_{\substack{n \text{ values} \\ a_i \leq a_n}}, \underbrace{a_n, \ldots, a_{2n-1}}_{\substack{n \text{ values} \\ a_i \geq a_n}}\big) = \frac{a_{n-1} + a_n}{2}. \tag{5.42}$$

By averaging a_{n-1} and a_n, new pixel values are generally introduced by the median filter if the region is of even size.

Figure 5.18 compares the results of median filtering with a linear-smoothing filter. Finally, Fig. 5.19 illustrates the effects of a 3×3 pixel median filter on selected 2D image structures. In particular, very small structures (smaller than half the filter size) are eliminated, but all other structures remain largely unchanged. A sample Java implementation of the median filter of arbitrary size is shown in Prog. 5.5. The constant K specifies the side length of the filter region R of size $(2r+1) \times (2r+1)$. The number of elements in R (equal to the length of the vector A) is

$$(2r+1)^2 = 4(r^2 + r) + 1, \tag{5.43}$$

and thus the index of the middle vector element is $n = 2(r^2 + r)$. Setting $r = 1$ gives a 3×3 median filter ($n = 4$), $r = 2$ gives a 5×5

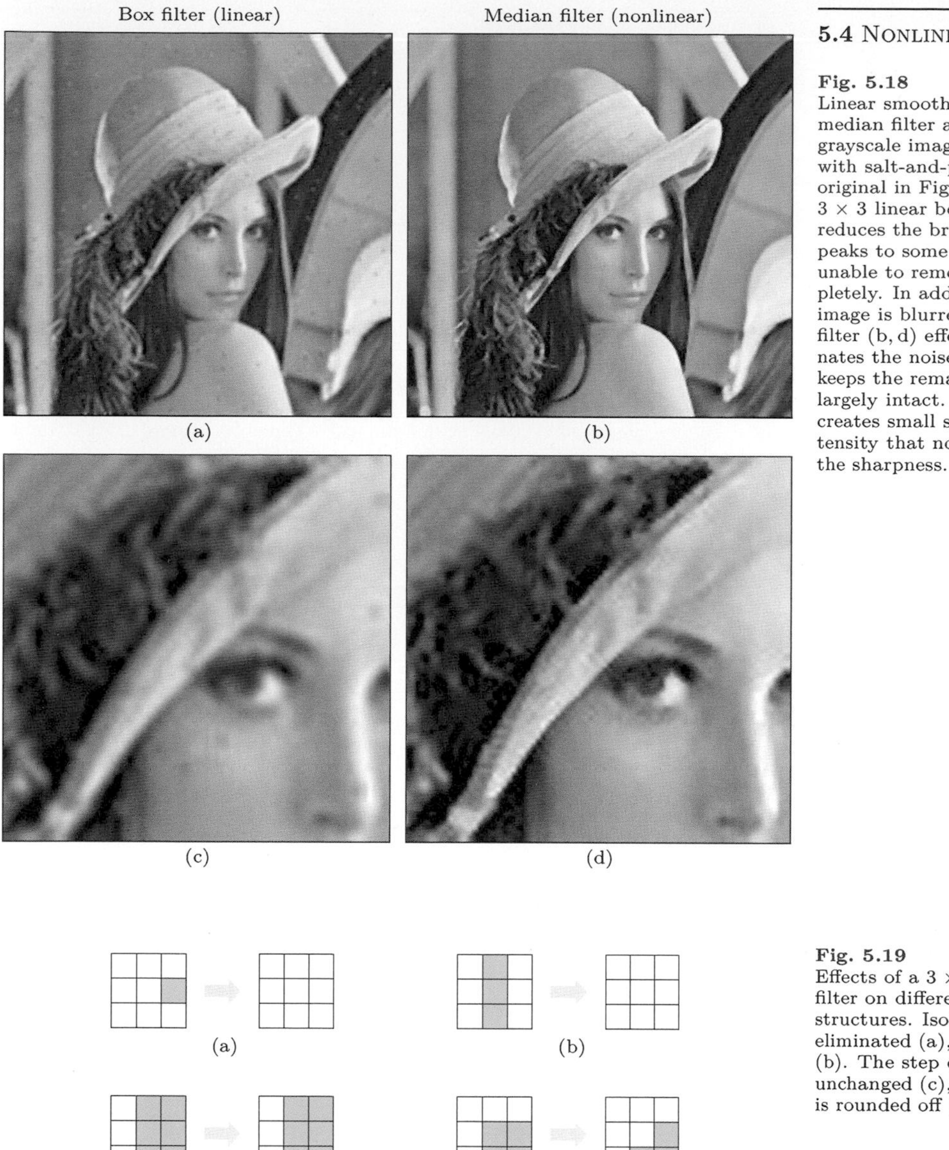

Fig. 5.18
Linear smoothing filter vs. median filter applied to a grayscale image corrupted with salt-and-pepper noise (see original in Fig. 5.14(a)). The 3×3 linear box filter (a, c) reduces the bright and dark peaks to some extent but is unable to remove them completely. In addition, the entire image is blurred. The median filter (b, d) effectively eliminates the noise dots and also keeps the remaining structures largely intact. However, it also creates small spots of flat intensity that noticeably affect the sharpness.

Fig. 5.19
Effects of a 3×3 pixel median filter on different 2D image structures. Isolated dots are eliminated (a), as are thin lines (b). The step edge remains unchanged (c), while a corner is rounded off (d).

filter ($n = 12$), etc. The structure of this plugin is similar to the arbitrary size linear filter in Prog. 5.3.

5.4.3 Weighted Median Filter

The median is a rank order statistic, and in a sense the “majority” of the pixel values involved determine the result. A single exceptionally high or low value (an “outlier”) cannot influence the result much but only shift the result up or down to the next value. Thus the median (in contrast to the linear average) is considered a “robust” measure. In an ordinary median filter, each pixel in the filter region has the same influence, regardless of its distance from the center.

Prog. 5.5
Median filter of arbitrary size (Plugin Filter_Median). An array A of type int is defined (line 16) to hold the region's pixel values for each filter position (u, v). This array is sorted by using the Java utility method Arrays.sort() in line 32. The center element of the sorted vector (A[n]) is taken as the median value and stored in the original image (line 33).

```
1 import ij.ImagePlus;
2 import ij.plugin.filter.PlugInFilter;
3 import ij.process.ImageProcessor;
4 import java.util.Arrays;
5
6 public class Filter_Median implements PlugInFilter {
7
8   final int r = 4;   // specifies the size of the filter
9
10  public void run(ImageProcessor ip) {
11    int M = ip.getWidth();
12    int N = ip.getHeight();
13    ImageProcessor copy = ip.duplicate();
14
15    // vector to hold pixels from (2r+1)x(2r+1) neighborhood:
16    int[] A = new int[(2 * r + 1) * (2 * r + 1)];
17
18    // index of center vector element n = 2(r^2 + r):
19    int n = 2 * (r * r + r);
20
21    for (int u = r; u <= M - r - 2; u++) {
22      for (int v = r; v <= N - r - 2; v++) {
23        // fill the pixel vector A for filter position (u,v):
24        int k = 0;
25        for (int i = -r; i <= r; i++) {
26          for (int j = -r; j <= r; j++) {
27            A[k] = copy.getPixel(u + i, v + j);
28            k++;
29          }
30        }
31        // sort vector A and take the center element A[n]:
32        Arrays.sort(A);
33        ip.putPixel(u, v, A[n]);
34      }
35    }
36  }
37 }
```

The weighted median filter assigns individual weights to the positions in the filter region, which can be interpreted as the "number of votes" for the corresponding pixel values. Similar to the coefficient matrix H of a linear filter, the distribution of weights is specified by a *weight matrix* W, with $W(i, j) \in \mathbb{N}$. To compute the result of the modified filter, each pixel value $I(u + i, v + j)$ involved is inserted $W(i, j)$ times into the extended pixel vector

$$A = (a_0, \ldots, a_{L-1}) \qquad \text{of length} \qquad L = \sum_{(i,j)\in R} W(i, j). \tag{5.44}$$

This vector is again sorted, and the resulting center value is taken as the median, as in the standard median filter. Figure 5.21 illustrates the computation of the weighted median filter using the 3×3 weight matrix

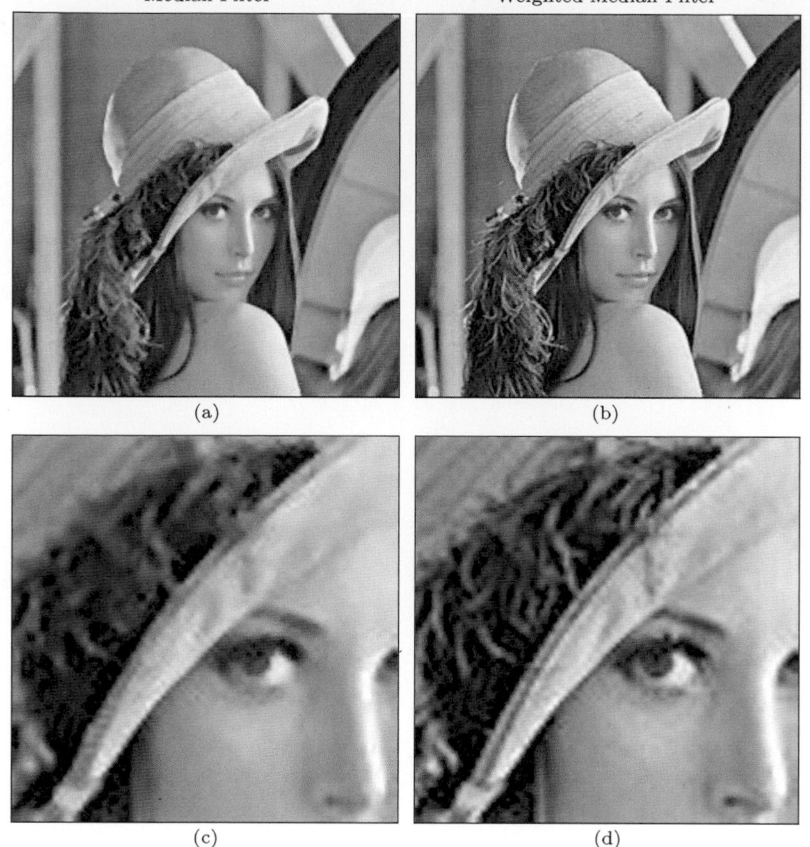

Fig. 5.20
Ordinary vs. weighted median filter. Compared to the ordinary median filter (a, c), the weighted median (b, d) shows superior preservation of structural details. Both filters are of size 3 × 3; the weight matrix in Eqn. (5.45) was used for the weighted median filter.

$$W = \begin{bmatrix} 1 & 2 & 1 \\ 2 & 3 & 2 \\ 1 & 2 & 1 \end{bmatrix}, \tag{5.45}$$

which requires an extended pixel vector of length $L = 15$, equal to the sum of the weights in W. If properly used, the weighted median filter yields effective noise removal with good preservation of structural details (see Fig. 5.20 for an example).

Of course this method may also be used to implement ordinary median filters of nonrectangular shape; for example, a *cross-shaped* median filter can be defined with the weight matrix

$$W^{+} = \begin{bmatrix} 0 & 1 & 0 \\ 1 & 1 & 1 \\ 0 & 1 & 0 \end{bmatrix}. \tag{5.46}$$

Not every arrangement of weights is useful, however. In particular, if the weight assigned to the center pixel is greater than the sum of all other weights, then that pixel would always have the "majority vote" and dictate the resulting value, thus inhibiting any filter effect.

5.4.4 Other Nonlinear Filters

Median and weighted median filters are two examples of nonlinear filters that are easy to describe and frequently used. Since "nonlin-

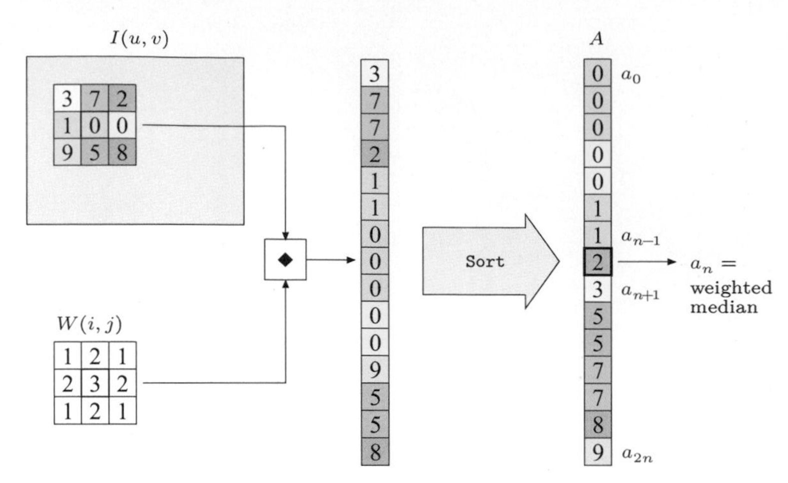

Fig. 5.21
Weighted median example. Each pixel value is inserted into the extended pixel vector multiple times, as specified by the weight matrix W. For example, the value 0 from the center pixel is inserted three times (since $W(0,0) = 3$) and the pixel value 7 twice. The pixel vector is sorted and the center value (2) is taken as the median.

ear" refers to anything that is not linear, there are a multitude of filters that fall into this category, including the morphological filters for binary and grayscale images, which are discussed in Ch. 9. Other types of nonlinear filters, such as the corner detector described in Ch. 7, are often described algorithmically and thus defy a simple, compact description.

In contrast to the linear case, there is usually no "strong theory" for nonlinear filters that could, for example, describe the relationship between the sum of two images and the results of a median filter, as does Eqn. (5.23) for linear convolution. Similarly, not much (if anything) can be stated in general about the effects of nonlinear filters in frequency space.

5.5 Implementing Filters

5.5.1 Efficiency of Filter Programs

Computing the results of filters is computationally expensive in most cases, especially with large images, large filter kernels, or both. Given an image of size $M \times N$ and a filter kernel of size $(2K+1) \times (2L+1)$, a direct implementation requires

$$2K \cdot 2L \cdot M \cdot N = 4\,KLMN \tag{5.47}$$

operations, namely multiplications and additions (in the case of a linear filter). Thus if both the image and the filter are simply assumed to be of size $N \times N$, the time complexity of direct filtering is $\mathcal{O}(N^4)$. As described in Sec. 5.3.3, substantial savings are possible when large, 2D filters can be decomposed (separated) into smaller, possibly 1D filters.

The programming examples in this chapter are deliberately designed to be simple and easy to understand, and none of the solutions shown is particularly efficient. Possibilities for tuning and code optimization exist in many places. It is particularly important to move all unnecessary instructions out of inner loops if possible because these are executed most often. This applies especially to "expensive" instructions, such as method invocations, which may be relatively time-consuming.

In the examples, we have intentionally used the ImageJ standard methods `getPixel()` for reading and `putPixel()` for writing image pixels, which is the simplest and safest approach to accessing image data but also the slowest, of course. Substantial speed can be gained by using the quicker read and write methods `get()` and `set()` defined for class `ImageProcessor` and its subclasses. Note, however, that these methods do not check if the passed image coordinates are valid. Maximum performance can be obtained by accessing the pixel arrays directly.

5.5.2 Handling Image Borders

As mentioned briefly in Sec. 5.2.2, the image borders require special attention in most filter implementations. We have argued that theoretically no filter results can be computed at positions where the filter matrix is not fully contained in the image array. Thus any filter operation would reduce the size of the resulting image, which is not acceptable in most applications. While no formally correct remedy exists, there are several more or less practical methods for handling the remaining border regions:

Method 1: Set the unprocessed pixels at the borders to some constant value (e.g., "black"). This is certainly the simplest method, but not acceptable in many situations because the image size is incrementally reduced by every filter operation.

Method 2: Set the unprocessed pixels to the original (unfiltered) image values. Usually the results are unacceptable, too, due to the noticeable difference between filtered and unprocessed image parts.

Method 3: Expand the image by "padding" additional pixels around it and apply the filter to the border regions as well. Fig. 5.22 shows different options for padding images.

A. The pixels outside the image have a *constant value* (e.g., "black" or"gray", see Fig. 5.22(a)). This may produce strong artifacts at the image borders, particularly when large filters are used.
B. The *border pixels extend* beyond the image boundaries (Fig. 5.22(b)). Only minor artifacts can be expected at the borders. The method is also simple to compute and is thus often considered the method of choice.
C. The *image is mirrored* at each of its four boundaries (Fig. 5.22(c)). The results will be similar to those of the previous method unless very large filters are used.
D. The *image repeats periodically* in the horizontal and vertical directions (Fig. 5.22(d)). This may seem strange at first, and the results are generally not satisfactory. However, in discrete spectral analysis, the image is implicitly treated as a periodic function, too (see Ch. 18). Thus, if the image is filtered in the frequency domain, the results will be equal to filtering in the space domain under this repetitive model.

Fig. 5.22
Methods for padding the image to facilitate filtering along the borders. The assumption is that the (nonexisting) pixels outside the original image are either set to some constant value (a), take on the value of the closest border pixel (b), are mirrored at the image boundaries (c), or repeat periodically along the coordinate axes (d).

None of these methods is perfect and, as usual, the right choice depends upon the type of image and the filter applied. Notice also that the special treatment of the image borders may sometimes require more programming effort (and computing time) than the processing of the interior image.

5.5.3 Debugging Filter Programs

Experience shows that programming errors can hardly ever be avoided, even by experienced practitioners. Unless errors occur during execution (usually caused by trying to access nonexistent array elements), filter programs always "do something" to the image that may be similar but not identical to the expected result. To assure that the code operates correctly, it is not advisable to start with full, large images but first to experiment with small test cases for which the outcome can easily be predicted. Particularly when implementing linear filters, a first "litmus test" should always be to inspect the impulse response of the filter (as described in Sec. 5.3.4) before processing any real images.

5.6 Filter Operations in ImageJ

ImageJ offers a collection of readily available filter operations, many of them contributed by other authors using different styles of implementation. Most of the available operations can also be invoked via ImageJ's Process menu.

5.6.1 Linear Filters

Filters based on linear convolution are implemented by the ImageJ plugin class `ij.plugin.filter.Convolver`, which offers useful "public" methods in addition to the standard `run()` method. Usage of this class is illustrated by the following example that convolves an 8-bit grayscale image with the filter kernel from Eqn. (5.7):

$$H = \begin{bmatrix} 0.075 & 0.125 & 0.075 \\ 0.125 & 0.200 & 0.125 \\ 0.075 & 0.125 & 0.075 \end{bmatrix} .$$

In the following `run()` method, we first define the filter matrix `H` as a 1D `float` array (notice the syntax for the `float` constants "`0.075f`", etc.) and then create a new instance (`cv`) of class `Convolver` in line 8:

```
import ij.plugin.filter.Convolver;
...
public void run(ImageProcessor I) {
  float[] H = { // coefficient array H is one-dimensional!
      0.075f, 0.125f, 0.075f,
      0.125f, 0.200f, 0.125f,
      0.075f, 0.125f, 0.075f };
  Convolver cv = new Convolver();
  cv.setNormalize(true);      // turn on filter normalization
  cv.convolve(I, H, 3, 3);    // apply the filter H to I
}
```

The invocation of the method `convolve()` applies the filter `H` to the image `I`. It requires two additional arguments for the dimensions of the filter matrix since `H` is passed as a 1D array. The image `I` is destructively modified by the `convolve` operation.

In this case, one could have also used the nonnormalized, integer-valued filter matrix given in Eqn. (5.10) because `convolve()` normalizes the given filter automatically (after `cv.setNormalize(true)`).

5.6.2 Gaussian Filters

The ImageJ class `ij.plugin.filter.GaussianBlur` implements a simple Gaussian blur filter with arbitrary radius (σ). The filter uses separable 1D Gaussians as described in Sec. 5.3.3. Here is an example showing its application with $\sigma = 2.5$:

```
import ij.plugin.filter.GaussianBlur;
...
public void run(ImageProcessor I) {
  GaussianBlur gb = new GaussianBlur();
```

```
    double sigmaX = 2.5;
    double sigmaY = sigmaX;
    double accuracy = 0.01;
    gb.blurGaussian(I, sigmaX, sigmaY, accuracy);
    ...
  }
```

The `accuracy` value specifies the size of the discrete filter kernel. Higher accuracy reduces truncation errors but requires larger kernels and more processing time.

An alternative implementation of separable Gaussian filters can be found in Prog. 6.1 (see p. 145), which uses the method `makeGaussKernel1d()` defined in Prog. 5.4 (page 104) for dynamically calculating the required 1D filter kernels.

5.6.3 Nonlinear Filters

A small set of nonlinear filters is implemented in the ImageJ class `ij.plugin.filter.RankFilters`, including the minimum, maximum, and standard median filters. The filter region is (approximately) circular with variable radius. Here is an example that applies three different filters with the same radius in sequence:

```
import ij.plugin.filter.RankFilters;
...
public void run(ImageProcessor I) {
  RankFilters rf = new RankFilters();
  double radius = 3.5;
  rf.rank(I, radius, RankFilters.MIN);       // minimum filter
  rf.rank(I, radius, RankFilters.MAX);       // maximum filter
  rf.rank(I, radius, RankFilters.MEDIAN);    // median filter
}
```

5.7 Exercises

Exercise 5.1. Explain why the "custom filter" in Adobe Photoshop (Fig. 5.6) is not strictly a linear filter.

Exercise 5.2. Determine the possible maximum and minimum results (pixel values) for the following linear filter, when applied to an 8-bit grayscale image (with pixel values in the range $[0, 255]$):

$$H = \begin{bmatrix} -1 & -2 & 0 \\ -2 & 0 & 2 \\ 0 & 2 & 1 \end{bmatrix}.$$

Assume that no clamping of the results occurs.

Exercise 5.3. Modify the ImageJ plugin shown in Prog. 5.3 such that the image borders are processed as well. Use one of the methods for extending the image outside its boundaries as described in Sec. 5.5.2.

Exercise 5.4. Show that a standard box filter is not isotropic (i.e., does not smooth the image identically in all directions).

Exercise 5.5. Explain why the clamping of results to a limited range of pixel values may violate the linearity property (Sec. 5.3.2) of linear filters.

Exercise 5.6. Verify the properties of the *impulse* function with respect to linear filters (see Eqn. (5.37)). Create a black image with a white pixel at its center and use this image as the 2D impulse. See if linear filters really deliver the filter matrix H as their impulse response.

Exercise 5.7. Describe the effects of the linear filters with the following kernels:

$$H_1 = \begin{bmatrix} 0 & 0 & 0 \\ 0 & 0 & 1 \\ 0 & 0 & 0 \end{bmatrix}, \qquad H_2 = \begin{bmatrix} 0 & 0 & 0 \\ 0 & 2 & 0 \\ 0 & 0 & 0 \end{bmatrix}, \qquad H_3 = \frac{1}{3} \cdot \begin{bmatrix} 0 & 0 & 1 \\ 0 & 1 & 0 \\ 1 & 0 & 0 \end{bmatrix}.$$

Exercise 5.8. Design a linear filter (kernel) that creates a horizontal blur over a length of 7 pixels, thus simulating the effect of camera movement during exposure.

Exercise 5.9. Compare the number of processing steps required for non-separable linear filters and x/y-separable filters sized 5×5, 11×11, 25×25, and 51×51 pixels. Compute the speed gain resulting from separability in each case.

Exercise 5.10. Program your own ImageJ plugin that implements a Gaussian smoothing filter with variable filter width (radius σ). The plugin should dynamically create the required filter kernels with a size of at least 5σ in both directions. Make use of the fact that the Gaussian function is x/y-separable (see Sec. 5.3.3).

Exercise 5.11. The *Laplacian of Gaussian* (LoG) filter (see Fig. 5.8) is based on the sum of the second derivatives of the 2D Gaussian. It is defined as

$$L_\sigma(x, y) = -\Big(\frac{x^2 + y^2 - 2 \cdot \sigma^2}{\sigma^4}\Big) \cdot e^{-\frac{x^2+y^2}{2 \cdot \sigma^2}}. \tag{5.48}$$

Implement the LoG filter as an ImageJ plugin of variable width (σ), analogous to Exercise 5.10. Find out if the LoG function is x/y-separable.

Exercise 5.12. Implement a circular (i.e., disk-shaped) median filter for grayscale images. Make the filter's radius r adjustable in the range from 1 to 10 pixels (e.g., using ImageJ's `GenericDialog` class). Use a binary (0/1) disk-shaped *mask* to represent the filter's support region R, with a minimum size of $(2r+1) \times (2r+1)$, as shown in Fig. 5.23(a). Create this mask dynamically for the chosen filter radius r (see Fig. 5.23(c–h) for typical results).

Exercise 5.13. Implement a weighted median filter (see Sec. 5.4.3) as an ImageJ plugin, specifying the weights as a constant, 2D `int` array. Test the filter on suitable images and compare the results with those from a standard median filter. Explain why, for example, the following weight matrix does *not* make sense:

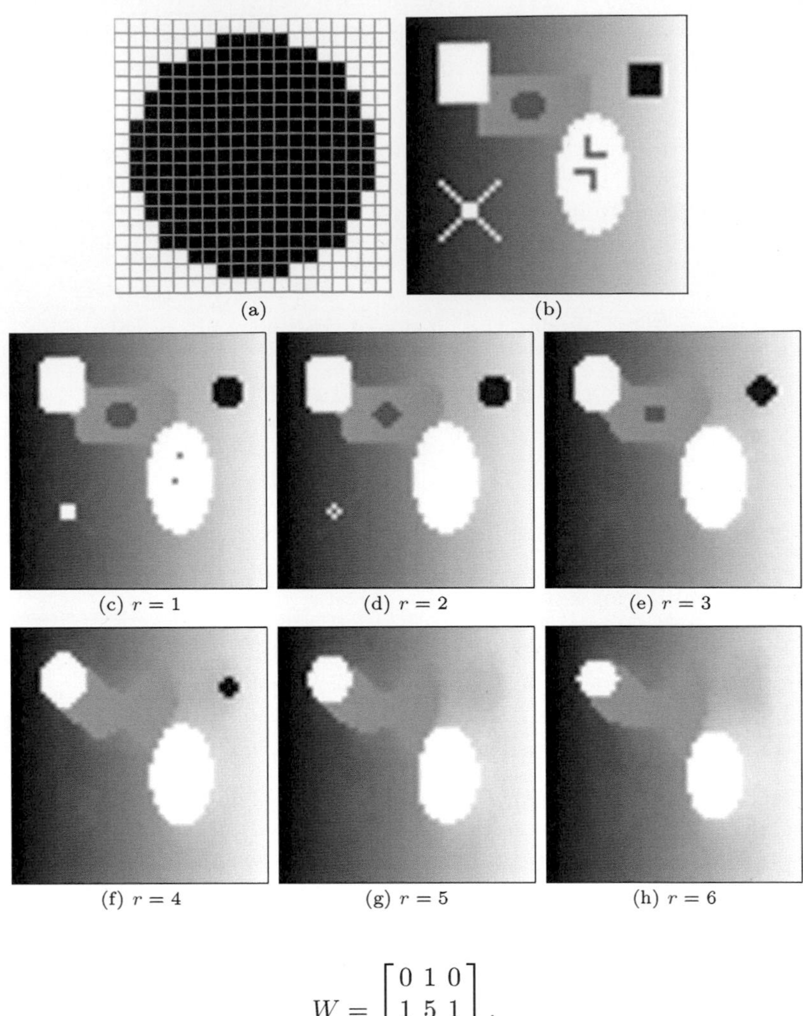

Fig. 5.23
Disk-shaped median filter. Example of a binary mask to represent the support region R with radius $r = 8$ (a). The origin of the filter region is located at its center. Synthetic test image (b). Results of the median filter for $r = 1, \ldots, 6$ (c–h).

(a) (b) (c) $r = 1$ (d) $r = 2$ (e) $r = 3$ (f) $r = 4$ (g) $r = 5$ (h) $r = 6$

$$W = \begin{bmatrix} 0 & 1 & 0 \\ 1 & 5 & 1 \\ 0 & 1 & 0 \end{bmatrix}.$$

Exercise 5.14. The "jitter" filter is a (quite exotic) example for a *nonhomogeneous filter*. For each image position, it selects a space-variant filter kernel (of size $2r + 1$) containing a single, randomly placed impulse (1); for example,

$$H_{u,v} = \begin{bmatrix} 0 & 0 & 0 & \mathbf{1} & 0 \\ 0 & 0 & 0 & 0 & 0 \\ 0 & 0 & 0 & 0 & 0 \\ 0 & 0 & 0 & 0 & 0 \\ 0 & 0 & 0 & 0 & 0 \end{bmatrix} \tag{5.49}$$

for $r = 2$. The position of the 1-value in the kernel $H_{u,v}$ is uniformly distributed in the range $i, j \in [-r, r]$; thus the filter effectively picks a random pixel value from the surrounding $(2r+1) \times (2r+1)$ neighborhood. Implement this filter for $r = 3, 5, 10$, as shown in Fig. 5.24. Is this filter linear or nonlinear? Develop another version using a Gaussian random distribution.

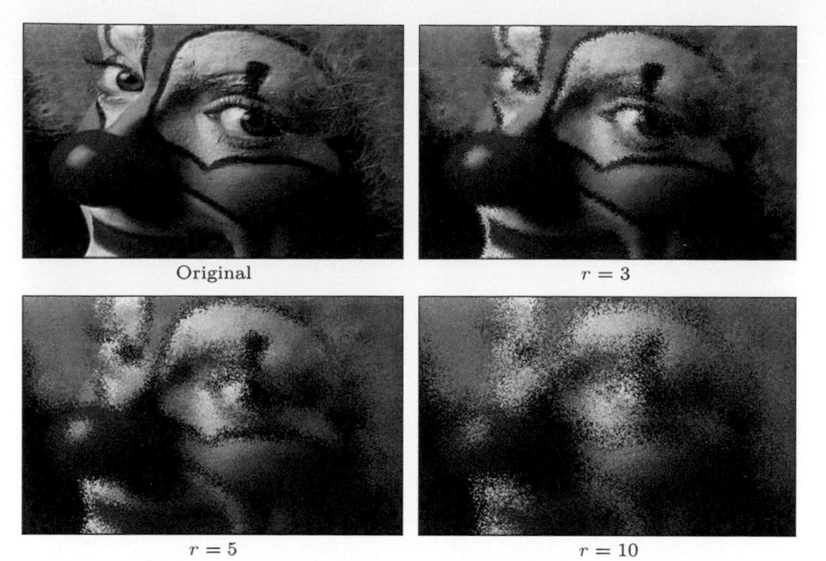

Fig. 5.24
Jitter filter example.

6

Edges and Contours

Prominent image "events" originating from local changes in intensity or color, such as edges and contours, are of high importance for the visual perception and interpretation of images. The perceived amount of information in an image appears to be directly related to the distinctiveness of the contained structures and discontinuities. In fact, edge-like structures and contours seem to be so important for our human visual system that a few lines in a caricature or illustration are often sufficient to unambiguously describe an object or a scene. It is thus no surprise that the enhancement and detection of edges has been a traditional and important topic in image processing as well. In this chapter, we first look at simple methods for localizing edges and then attend to the related issue of image sharpening.

6.1 What Makes an Edge?

Edges and contours play a dominant role in human vision and probably in many other biological vision systems as well. Not only are edges visually striking, but it is often possible to describe or reconstruct a complete figure from a few key lines, as the example in Fig. 6.1 shows. But how do edges arise, and how can they be technically localized in an image?

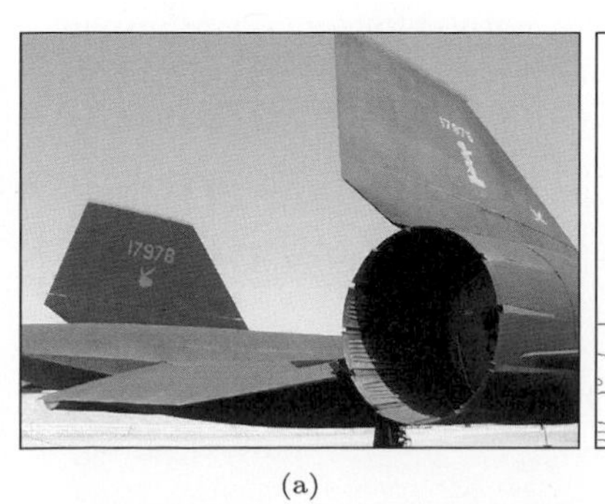

(a)

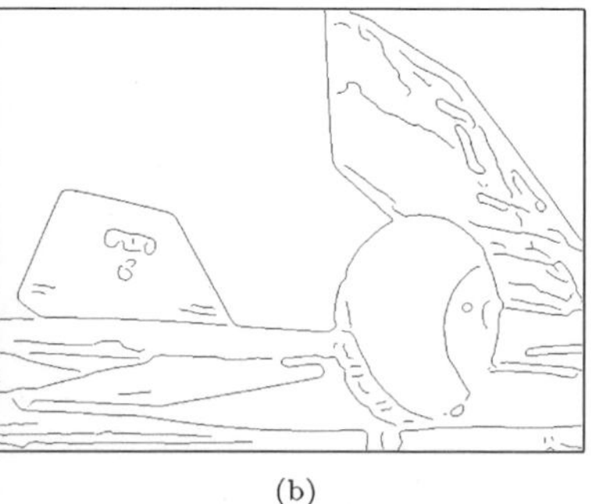

(b)

Fig. 6.1
Edges play an important role in human vision. Original image (a) and edge image (b).

Edges can roughly be described as image positions where the local intensity changes distinctly along a particular orientation. The stronger the local intensity change, the higher is the evidence for an edge at that position. In mathematics, the amount of change with respect to spatial distance is known as the first derivative of a function, and we thus start with this concept to develop our first simple edge detector.

6.2 Gradient-Based Edge Detection

For simplicity, we first investigate the situation in only one dimension, assuming that the image contains a single bright region at the center surrounded by a dark background (Fig. 6.2(a)). In this case, the intensity profile along one image line would look like the 1D function $f(x)$, as shown in Fig. 6.2(b). Taking the first derivative of the function f,

$$f'(x) = \frac{df}{dx}(x), \tag{6.1}$$

results in a positive swing at those positions where the intensity rises and a negative swing where the value of the function drops (Fig. 6.2(c)).

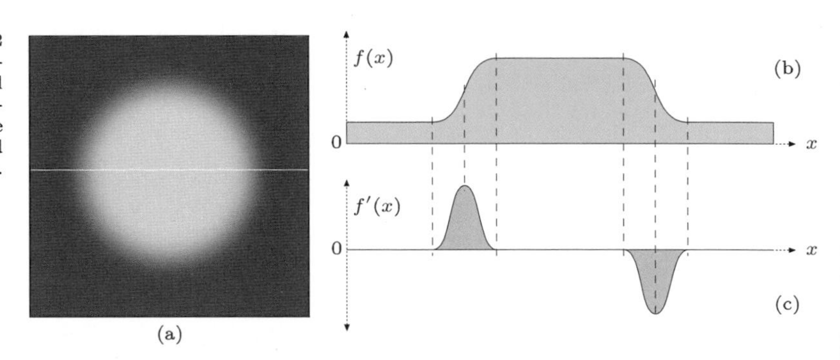

Fig. 6.2 Sample image and first derivative in one dimension: original image (a), horizontal intensity profile $f(x)$ along the center image line (b), and first derivative $f'(x)$ (c).

Unlike in the continuous case, however, the first derivative is undefined for a *discrete* function $f(u)$ (such as the line profile of a real image), and some method is needed to estimate it. Figure 6.3 shows the basic idea, again for the 1D case: the first derivative of a continuous function at position x can be interpreted as the slope of its *tangent* at this position. One simple method for roughly approximating the slope of the tangent for a *discrete* function $f(u)$ at position u is to fit a straight line through the neighboring function values $f(u–1)$ and $f(u+1)$,

$$\frac{df}{dx}(u) \approx \frac{f(u+1) - f(u-1)}{(u+1) - (u-1)} = \frac{f(u+1) - f(u-1)}{2}. \tag{6.2}$$

Of course, the same method can be applied in the vertical direction to estimate the first derivative along the y-axis, thats is, along the image columns.

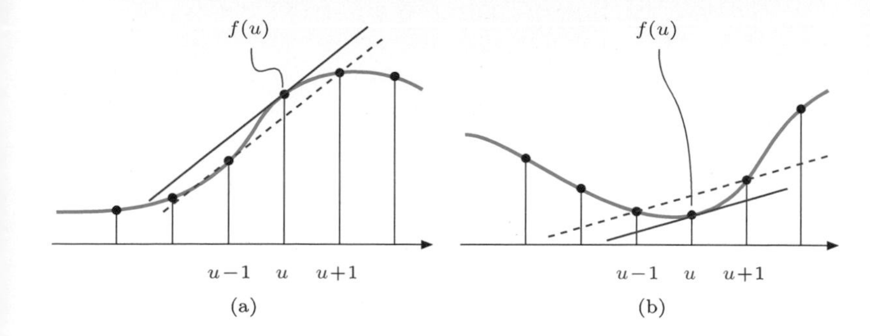

Fig. 6.3
Estimating the first derivative of a discrete function. The slope of the straight (dashed) line between the neighboring function values $f(u-1)$ and $f(u+1)$ is taken as the estimate for the slope of the tangent (i.e., the first derivative) at $f(u)$.

6.2.1 Partial Derivatives and the Gradient

A derivative of a multi-dimensional function taken along one of its coordinate axes is called a *partial derivative*; for example,

$$I_x = \frac{\partial I}{\partial x}(u,v) \qquad \text{and} \qquad I_y = \frac{\partial I}{\partial y}(u,v) \tag{6.3}$$

are the partial derivatives of the 2D image function $I(u,v)$ along the u and v axes, respectively.[1] The vector

$$\nabla I(u,v) = \begin{pmatrix} I_x(u,v) \\ I_y(u,v) \end{pmatrix} = \begin{pmatrix} \frac{\partial I}{\partial x}(u,v) \\ \frac{\partial I}{\partial y}(u,v) \end{pmatrix} \tag{6.4}$$

is called the *gradient* of the function I at position (u,v). The *magnitude* of the gradient,

$$|\nabla I| = \sqrt{I_x^2 + I_y^2}, \tag{6.5}$$

is invariant under image rotation and thus independent of the orientation of the underlying image structures. This property is important for isotropic localization of edges, and thus $|\nabla I|$ is the basis of many practical edge detection methods.

6.2.2 Derivative Filters

The components of the gradient function (Eqn. (6.4)) are simply the first derivatives of the image lines (Eqn. (6.1)) and columns along the horizontal and vertical axes, respectively. The approximation of the first horizontal derivatives (Eqn. (6.2)) can be easily implemented by a linear filter (see Sec. 5.2) with the 1D kernel

$$H_x^{\mathrm{D}} = \begin{bmatrix} -0.5 & 0 & 0.5 \end{bmatrix} = 0.5 \cdot \begin{bmatrix} -1 & 0 & 1 \end{bmatrix}, \tag{6.6}$$

where the coefficients -0.5 and $+0.5$ apply to the image elements $I(u-1,v)$ and $I(u+1,v)$, respectively. Notice that the center pixel $I(u,v)$ itself is weighted with the zero coefficient and is thus ignored. Analogously, the vertical component of the gradient is obtained with the linear filter

[1] ∂ denotes the *partial derivative* or "del" operator.

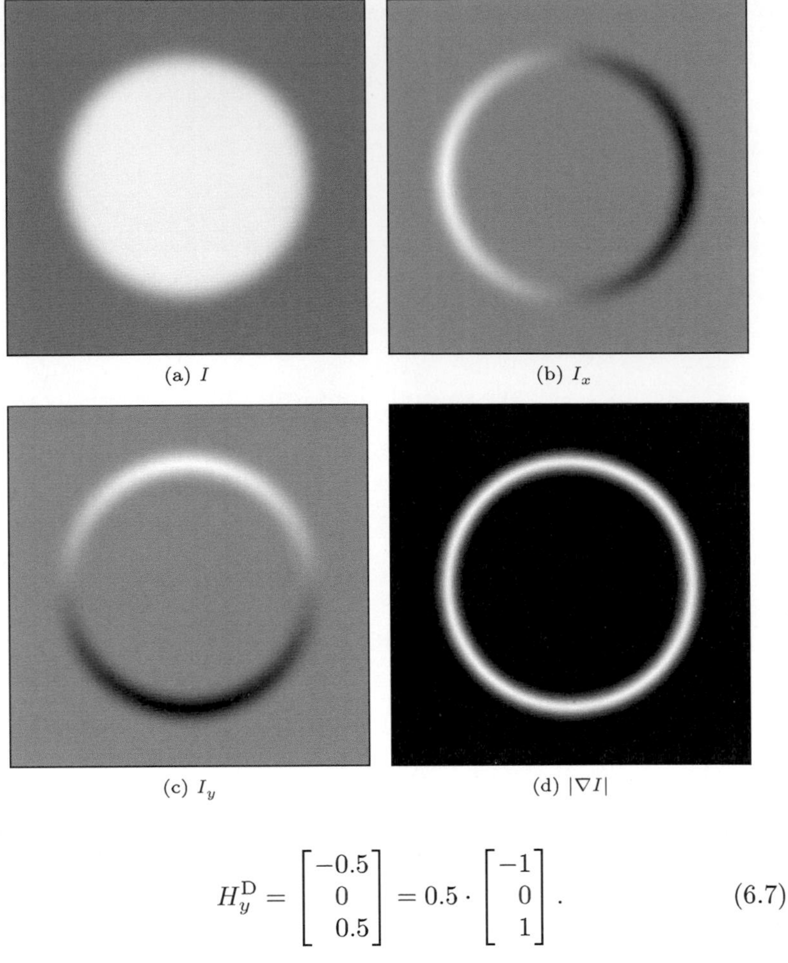

Fig. 6.4
Partial derivatives of a 2D function: synthetic image function I (a); approximate first derivatives in the horizontal direction $\partial I/\partial u$ (b) and the vertical direction $\partial I/\partial v$ (c); magnitude of the resulting gradient $|\nabla I|$ (d). In (b) and (c), the lowest (negative) values are shown black, the maximum (positive) values are white, and zero values are gray.

$$H_y^{\mathrm{D}} = \begin{bmatrix} -0.5 \\ 0 \\ 0.5 \end{bmatrix} = 0.5 \cdot \begin{bmatrix} -1 \\ 0 \\ 1 \end{bmatrix}. \tag{6.7}$$

Figure 6.4 shows the results of applying the gradient filters defined in Eqn. (6.6) and Eqn. (6.7) to a synthetic test image.

The orientation dependence of the filter responses can be seen clearly. The horizontal gradient filter H_x^D reacts most strongly to rapid changes along the horizontal direction, (i.e., to *vertical* edges); analogously the vertical gradient filter H_y^D reacts most strongly to *horizontal* edges. The filter response is zero in flat image regions (shown as gray in Fig. 6.4(b, c)).

6.3 Simple Edge Operators

The local gradient of the image function is the basis of many classical edge-detection operators. Practically, they only differ in the type of filter used for estimating the gradient components and the way these components are combined. In many situations, one is not only interested in the *strength* of edge points but also in the local *direction* of the edge. Both types of information are contained in the gradient function and can be easily computed from the directional components. The following small collection describes some frequently used, simple edge operators that have been around for many years and are thus interesting from a historic perspective as well.

6.3.1 Prewitt and Sobel Operators

The edge operators by Prewitt [191] and Sobel [61] are two classic methods that differ only marginally in the derivative filters they use.

Gradient filters

Both operators use linear filters that extend over three adjacent lines and columns, respectively, to counteract the noise sensitivity of the simple (single line/column) gradient operators (Eqns. (6.6) and (6.7)). The Prewitt operator uses the filter kernels

$$H_x^{\mathrm{P}} = \begin{bmatrix} -1 & 0 & 1 \\ -1 & 0 & 1 \\ -1 & 0 & 1 \end{bmatrix} \quad \text{and} \quad H_y^{\mathrm{P}} = \begin{bmatrix} -1 & -1 & -1 \\ 0 & 0 & 0 \\ 1 & 1 & 1 \end{bmatrix}, \tag{6.8}$$

which compute the average gradient components across three neighboring lines or columns, respectively. When the filters are written in separated form,

$$H_x^{\mathrm{P}} = \begin{bmatrix} 1 \\ 1 \\ 1 \end{bmatrix} * \begin{bmatrix} -1 & 0 & 1 \end{bmatrix} \quad \text{and} \quad H_y^{\mathrm{P}} = \begin{bmatrix} 1 & 1 & 1 \end{bmatrix} * \begin{bmatrix} -1 \\ 0 \\ 1 \end{bmatrix}, \tag{6.9}$$

respectively, it becomes obvious that H_x^P performs a simple (box) smoothing over three lines before computing the x gradient (Eqn. (6.6)), and analogously H_y^P smooths over three columns before computing the y gradient (Eqn. (6.7)).[2] Because of the commutativity property of linear convolution, this could equally be described the other way around, with smoothing being applied *after* the computation of the gradients.

The filters for the Sobel operator are almost identical; however, the smoothing part assigns higher weight to the current center line and column, respectively:

$$H_x^{\mathrm{S}} = \begin{bmatrix} -1 & 0 & 1 \\ -2 & 0 & 2 \\ -1 & 0 & 1 \end{bmatrix} \quad \text{and} \quad H_y^{\mathrm{S}} = \begin{bmatrix} -1 & -2 & -1 \\ 0 & 0 & 0 \\ 1 & 2 & 1 \end{bmatrix}. \tag{6.10}$$

The estimates for the local gradient components are obtained from the filter results by appropriate scaling, that is,

$$\nabla I(u,v) \approx \frac{1}{6} \cdot \begin{pmatrix} (I * H_x^{\mathrm{P}})(u,v) \\ (I * H_y^{\mathrm{P}})(u,v) \end{pmatrix} \tag{6.11}$$

for the *Prewitt* operator and

$$\nabla I(u,v) \approx \frac{1}{8} \cdot \begin{pmatrix} (I * H_x^{\mathrm{S}})(u,v) \\ (I * H_y^{\mathrm{S}})(u,v) \end{pmatrix} \tag{6.12}$$

for the *Sobel* operator.

[2] In Eqn. (6.9), $*$ is the linear convolution operator (see Sec. 5.3.1).

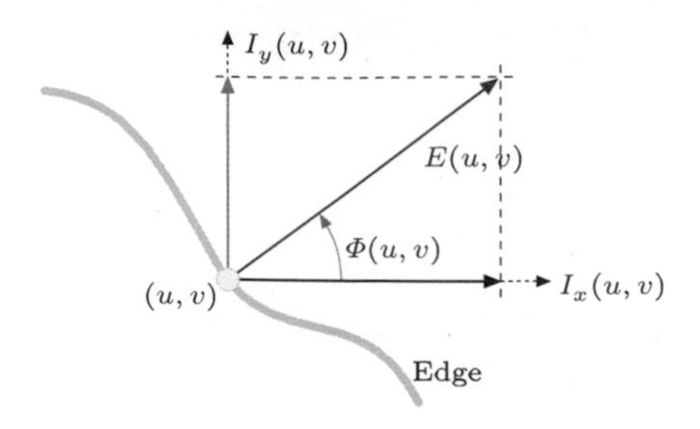

Fig. 6.5
Calculation of edge magnitude and orientation (geometry).

Edge strength and orientation

In the following, we denote the scaled filter results (obtained with either the Prewitt or Sobel operator) as

$$I_x = I * H_x \qquad \text{and} \qquad I_y = I * H_y.$$

In both cases, the local edge strength $E(u,v)$ is defined as the gradient magnitude

$$E(u,v) = \sqrt{I_x^2(u,v) + I_y^2(u,v)} \tag{6.13}$$

and the local edge orientation angle $\Phi(u,v)$ is[3]

$$\Phi(u,v) = \tan^{-1}\Bigl(\frac{I_y(u,v)}{I_x(u,v)}\Bigr) = \mathrm{ArcTan}\bigl(I_x(u,v), I_y(u,v)\bigr), \tag{6.14}$$

as illustrated in Fig. 6.5.

The whole process of extracting the edge magnitude and orientation is summarized in Fig. 6.6. First, the original image I is independently convolved with the two gradient filters H_x and H_y, and subsequently the edge strength E and orientation Φ are computed from the filter results. Figure 6.7 shows the edge strength and orientation for two test images, obtained with the Sobel filters in Eqn. (6.10).

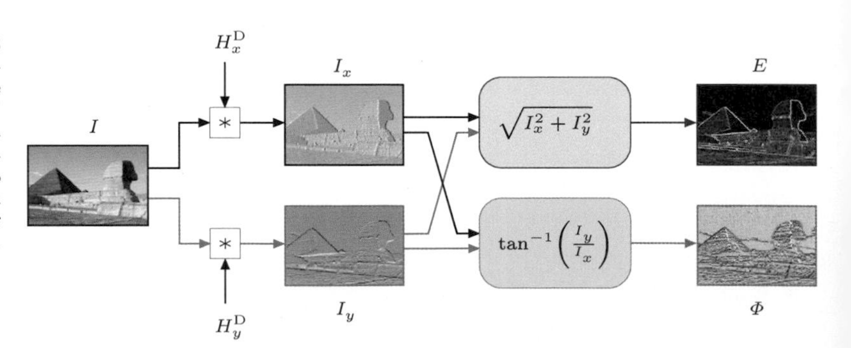

Fig. 6.6
Typical process of gradient-based edge extraction. The linear derivative filters H_x^{D} and H_y^{D} produce two gradient images, I_x and I_y, respectively. They are used to compute the edge strength E and orientation Φ for each image position (u,v).

The estimate of the edge orientation based on the original Prewitt and Sobel filters is relatively inaccurate, and improved versions of the Sobel filters were proposed in [126, p. 353] to minimize the orientation errors:

[3] See the hints in Sec. F.1.6 in the Appendix for computing the inverse tangent $\tan^{-1}(y/x)$ with the $\mathrm{ArcTan}(x,y)$ function.

(a)

(b)

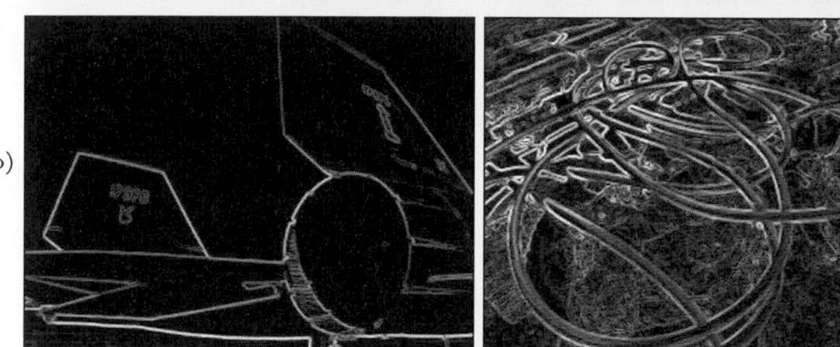

(c)

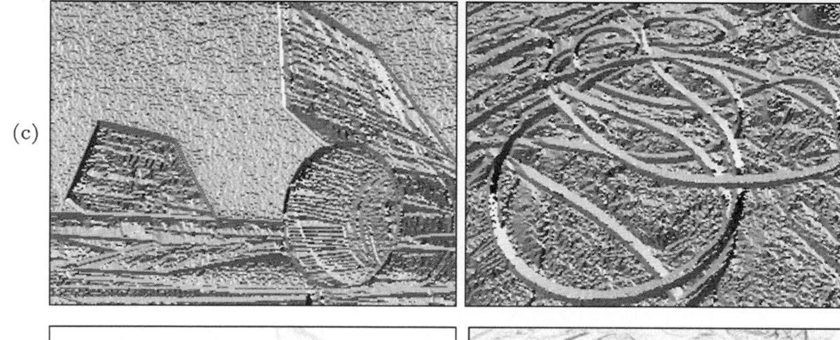

(d)

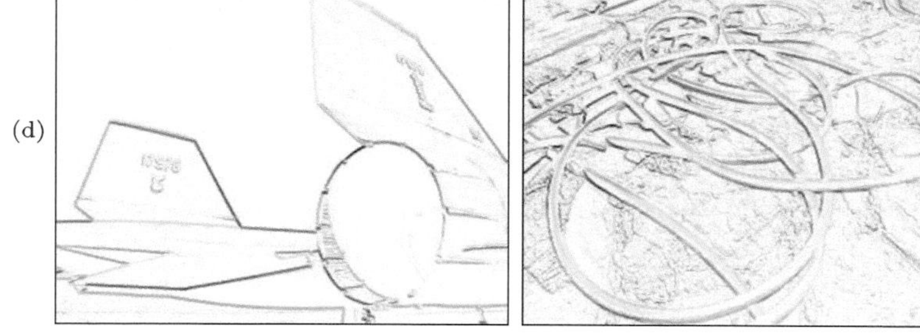

Fig. 6.7
Edge strength and orientation obtained with a Sobel operator. Original images (a), the edge strength $E(u,v)$ (b), and the local edge orientation $\Phi(u,v)$ (c). The images in (d) show the orientation angles coded as color hues, with the edge strength controlling the color saturation (see Sec. 12.2.3 for the corresponding definitions).

$$H_x^{S'} = \frac{1}{32} \cdot \begin{bmatrix} -3 & 0 & 3 \\ -10 & 0 & 10 \\ -3 & 0 & 3 \end{bmatrix} \quad \text{and} \quad H_y^{S'} = \frac{1}{32} \cdot \begin{bmatrix} -3 & -10 & -3 \\ 0 & 0 & 0 \\ 3 & 10 & 3 \end{bmatrix}. \tag{6.15}$$

These edge operators are frequently used because of their good results (see also Fig. 6.11) and simple implementation. The Sobel operator, in particular, is available in many image-processing tools and software packages (including ImageJ).

6.3.2 Roberts Operator

As one of the simplest and oldest edge finders, the Roberts operator [199] today is mainly of historic interest. It employs two extremely small filters of size 2×2 for estimating the directional gradient along

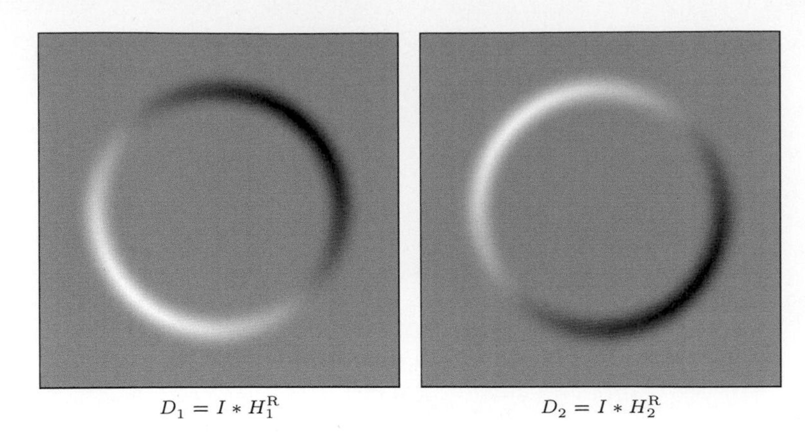

Fig. 6.8
Diagonal gradient components produced by the two Roberts filters.

the image diagonals:

$$H_1^R = \begin{bmatrix} 0 & 1 \\ -1 & 0 \end{bmatrix} \quad \text{and} \quad H_2^R = \begin{bmatrix} -1 & 0 \\ 0 & 1 \end{bmatrix}. \tag{6.16}$$

These filters naturally respond to diagonal edges but are not highly selective to orientation; that is, both filters show strong results over a relatively wide range of angles (Fig. 6.8). The local edge strength is calculated by measuring the length of the resulting 2D vector, similar to the gradient computation but with its components rotated 45° (Fig. 6.9).

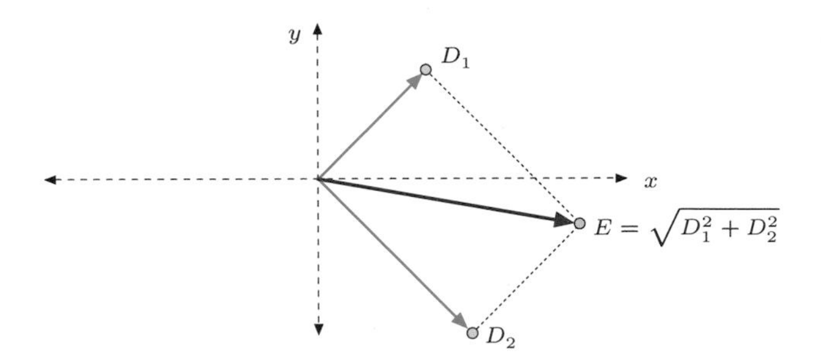

Fig. 6.9
Definition of edge strength for the Roberts operator. The edge strength $E(u, v)$ corresponds to the length of the vector obtained by adding the two orthogonal gradient components (filter results) $D_1(u, v)$ and $D_2(u, v)$.

6.3.3 Compass Operators

The design of linear edge filters involves a trade-off: the stronger a filter responds to edge-like structures, the more sensitive it is to orientation. In other words, filters that are orientation insensitive tend to respond to nonedge structures, while the most discriminating edge filters only respond to edges in a narrow range of orientations. One solution is to use not only a single pair of relatively "wide" filters for two directions (such as the Prewitt and the simple Sobel operator discussed in Sec. 6.3.1) but a larger set of filters with narrowly spaced orientations.

Extended Sobel operator

Classic examples are the edge operator proposed by *Kirsch* [136] and the "extended Sobel" or *Robinson* operator [200], which employs the following eight filters with orientations spaced at 45°:

$$H_0^{\mathrm{ES}} = \begin{bmatrix} -1 & 0 & 1 \\ -2 & 0 & 2 \\ -1 & 0 & 1 \end{bmatrix}, \qquad H_1^{\mathrm{ES}} = \begin{bmatrix} -2 & -1 & 0 \\ -1 & 0 & 1 \\ 0 & 1 & 2 \end{bmatrix}, \tag{6.17}$$

$$H_2^{\mathrm{ES}} = \begin{bmatrix} -1 & -2 & -1 \\ 0 & 0 & 0 \\ 1 & 2 & 1 \end{bmatrix}, \qquad H_3^{\mathrm{ES}} = \begin{bmatrix} 0 & -1 & -2 \\ 1 & 0 & -1 \\ 2 & 1 & 0 \end{bmatrix}, \tag{6.18}$$

$$H_4^{\mathrm{ES}} = \begin{bmatrix} 1 & 0 & -1 \\ 2 & 0 & -2 \\ 1 & 0 & -1 \end{bmatrix}, \qquad H_5^{\mathrm{ES}} = \begin{bmatrix} 2 & 1 & 0 \\ 1 & 0 & -1 \\ 0 & -1 & -2 \end{bmatrix}, \tag{6.19}$$

$$H_6^{\mathrm{ES}} = \begin{bmatrix} 1 & 2 & 1 \\ 0 & 0 & 0 \\ -1 & -2 & -1 \end{bmatrix}, \qquad H_7^{\mathrm{ES}} = \begin{bmatrix} 0 & 1 & 2 \\ -1 & 0 & 1 \\ -2 & -1 & 0 \end{bmatrix}. \tag{6.20}$$

Only the results of four of these eight filters ($H_0^{\mathrm{ES}}, H_1^{\mathrm{ES}}, \ldots, H_7^{\mathrm{ES}}$) must actually be computed since the remaining four are identical except for the reversed sign. For example, from the fact that $H_4^{\mathrm{ES}} = -H_0^{\mathrm{ES}}$ and the convolution being linear (Eqn. (5.22)), it follows that

$$I * H_4^{\mathrm{ES}} = I * -H_0^{\mathrm{ES}} = -(I * H_0^{\mathrm{ES}}), \tag{6.21}$$

that is, the result for filter H_4^S is simply the negative result for filter H_0^S. The directional outputs $D_0, D_1, \ldots D_7$ for the eight Sobel filters can thus be computed as follows:

$$\begin{array}{llll} D_0 \leftarrow I * H_0^{\mathrm{ES}}, & D_1 \leftarrow I * H_1^{\mathrm{ES}}, & D_2 \leftarrow I * H_2^{\mathrm{ES}}, & D_3 \leftarrow I * H_3^{\mathrm{ES}}, \\ D_4 \leftarrow -D_0, & D_5 \leftarrow -D_1, & D_6 \leftarrow -D_2, & D_7 \leftarrow -D_3. \end{array} \tag{6.22}$$

The edge strength E^S at position (u, v) is defined as the maximum of the eight filter outputs; that is,

$$\begin{aligned} E^{\mathrm{ES}}(u,v) &= \max\bigl(D_0(u,v), D_1(u,v), \ldots, D_7(u,v)\bigr) \\ &= \max\bigl(|D_0(u,v)|, |D_1(u,v)|, |D_2(u,v)|, |D_3(u,v)|\bigr), \end{aligned} \tag{6.23}$$

and the strongest-responding filter also determines the local edge orientation as

$$\Phi^{\mathrm{ES}}(u,v) = \frac{\pi}{4} j, \quad \text{with } j = \operatorname*{argmax}_{0 \le i \le 7} D_i(u,v). \tag{6.24}$$

Kirsch operator

Another classic compass operator is the one proposed by Kirsch [136], which also employs eight oriented filters with the following kernels:

$$H_0^{\mathrm{K}} = \begin{bmatrix} -5 & 3 & 3 \\ -5 & 0 & 3 \\ -5 & 3 & 3 \end{bmatrix}, \qquad H_4^{\mathrm{K}} = \begin{bmatrix} 3 & 3 & -5 \\ 3 & 0 & -5 \\ 3 & 3 & -5 \end{bmatrix}, \tag{6.25}$$

$$H_1^{\mathrm{K}} = \begin{bmatrix} -5 & -5 & 3 \\ -5 & 0 & 3 \\ 3 & 3 & 3 \end{bmatrix}, \qquad H_5^{\mathrm{K}} = \begin{bmatrix} 3 & 3 & 3 \\ 3 & 0 & -5 \\ 3 & -5 & -5 \end{bmatrix}, \tag{6.26}$$

$$H_2^{\mathrm{K}} = \begin{bmatrix} -5 & -5 & -5 \\ 3 & 0 & 3 \\ 3 & 3 & 3 \end{bmatrix}, \qquad H_6^{\mathrm{K}} = \begin{bmatrix} 3 & 3 & 3 \\ 3 & 0 & 3 \\ -5 & -5 & -5 \end{bmatrix}, \tag{6.27}$$

$$H_3^{\mathrm{K}} = \begin{bmatrix} 3 & -5 & -5 \\ 3 & 0 & -5 \\ 3 & 3 & 3 \end{bmatrix}, \qquad H_7^{\mathrm{K}} = \begin{bmatrix} 3 & 3 & 3 \\ -5 & 0 & 3 \\ -5 & -5 & 3 \end{bmatrix}. \tag{6.28}$$

Again, because of the symmetries, only four of the eight filters need to be applied and the results may be combined in the same way as already described for the extended Sobel operator.

In practice, this and other "compass operators" show only minor benefits over the simpler operators described earlier, including the small advantage of not requiring the computation of square roots (which is considered a relatively "expensive" operation).

6.3.4 Edge Operators in ImageJ

The current version of ImageJ implements the Sobel operator (as described in Eqn. (6.10)) for practically any type of image. It can be invoked via the

Process ▷ Find Edges

menu and is also available through the method `void findEdges()` for objects of type `ImageProcessor`.

6.4 Other Edge Operators

One problem with edge operators based on first derivatives (as described in the previous section) is that each resulting edge is as wide as the underlying intensity transition and thus edges may be difficult to localize precisely. An alternative class of edge operators makes use of the second derivatives of the image function, including some popular modern edge operators that also address the problem of edges appearing at various levels of scale. These issues are briefly discussed in the following.

6.4.1 Edge Detection Based on Second Derivatives

The second derivative of a function measures its local curvature. The idea is that edges can be found at zero positions or—even better—at the zero crossings of the second derivatives of the image function, as illustrated in Fig. 6.10 for the 1D case. Since second derivatives generally tend to amplify image noise, some sort of presmoothing is usually applied with suitable low-pass filters.

A popular example is the "Laplacian-of-Gaussian" (LoG) operator [161], which combines gGussian smoothing and computing the second derivatives (see the *Laplace Filter* in Sec. 6.6.1) into a single linear filter. The example in Fig. 6.11 shows that the edges produced by the LoG operator are more precisely localized than the ones delivered by the Prewitt and Sobel operators, and the amount of "clutter" is comparably small. Details about the LoG operator and a comprehensive survey of common edge operators can be found in [203, Ch. 4] and [165].

6.4.2 Edges at Different Scales

Unfortunately, the results of the simple edge operators we have discussed so far often deviate from what we as humans perceive as important edges. The two main reasons for this are:

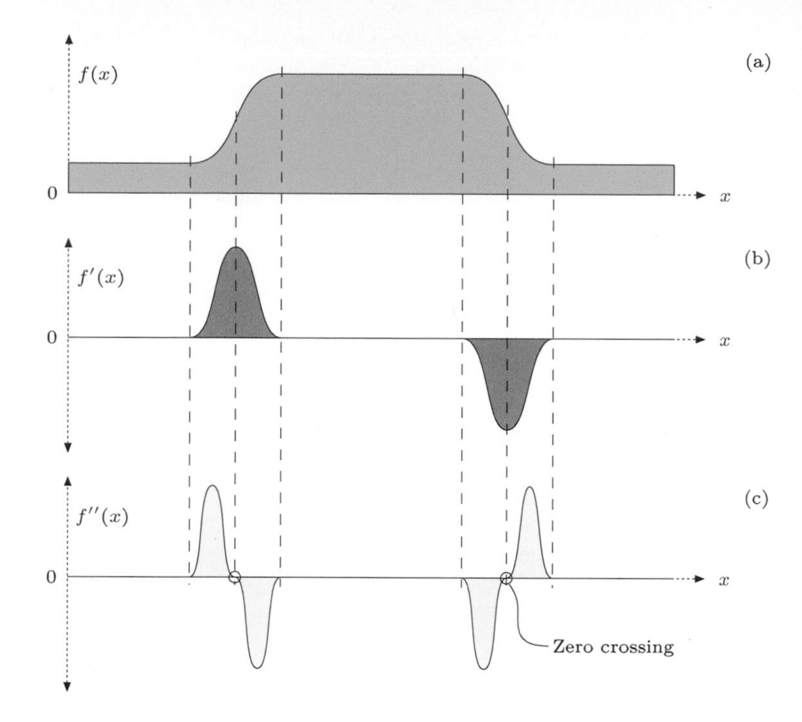

Fig. 6.10
Principle of edge detection with the second derivative: original function (a), first derivative (b), and second derivative (c). Edge points are located where the second derivative crosses through zero and the first derivative has a high magnitude.

- First, edge operators only respond to local intensity differences, while our visual system is able to extend edges across areas of minimal or vanishing contrast.
- Second, edges exist not at a single fixed resolution or at a certain scale but over a whole range of different scales.

Typical small edge operators, such as the Sobel operator, can only respond to intensity differences that occur within their 3×3 pixel filter regions. To recognize edge-like events over a greater horizon, we would either need larger edge operators (with correspondingly large filters) or to use the original (small) operators on reduced (i.e., scaled) images. This is the principal idea of "multiresolution" techniques (also referred to as "hierarchical" or "pyramid" techniques), which have traditionally been used in many image-processing applications [41, 151]. In the context of edge detection, this typically amounts to detecting edges at various scale levels first and then deciding which edge (if any) at which scale level is dominant at each image position.

6.4.3 From Edges to Contours

Whatever method is used for edge detection, the result is usually a continuous value for the edge strength for each image position and possibly also the angle of local edge orientation. How can this information be used, for example, to find larger image structures and contours of objects in particular?

Binary edge maps

In many situations, the next step after edge enhancement (by some edge operator) is the selection of edge points, a binary decision about whether an image pixel is an edge point or not. The simplest method is to apply a *threshold* operation to the edge strength delivered by the edge operator using either a fixed or adaptive threshold value, which results in a binary edge image or "edge map".

In practice, edge maps hardly ever contain perfect contours but instead many small, unconnected contour fragments, interrupted at positions of insufficient edge strength. After thresholding, the empty positions of course contain no edge information at all that could possibly be used in a subsequent step, such as for linking adjacent edge segments. Despite this weakness, global thresholding is often used at this point because of its simplicity, and some common postprocessing methods, such as the Hough transform (see Ch. 8), can cope well with incomplete edge maps.

Contour following

The idea of tracing contours sequentially along the discovered edge points is not uncommon and appears quite simple in principle. Starting from an image point with high edge strength, the edge is followed iteratively in both directions until the two traces meet and a closed contour is formed. Unfortunately, there are several obstacles that make this task more difficult than it seems at first, including the following:

- edges may end in regions of vanishing intensity gradient,
- crossing edges lead to ambiguities, and
- contours may branch into several directions.

Because of these problems, contour following usually is not applied to original images or continuous-valued edge images except in very simple situations, such as when there is a clear separation between objects (foreground) and the background. Tracing contours in segmented binary images is much simpler, of course (see Ch. 10).

6.5 Canny Edge Operator

The operator proposed by Canny [42] is widely used and still considered "state of the art" in edge detection. The method tries to reach three main goals: (a) to minimize the number of false edge points, (b) achieve good localization of edges, and (c) deliver only a single mark on each edge. These properties are usually not achieved with simple edge operators (mostly based on first derivatives and subsequent thresholding).

At its core, the Canny "filter" is a gradient method (based on first derivatives; see Sec. 6.2), but it uses the zero crossings of second derivatives for precise edge localization.[4] In this regard, the method is similar to edge detectors that are based on the second derivatives of the image function [161].

Fully implemented, the Canny detector uses a set of relatively large, oriented filters at multiple image resolutions and merges the

[4] The zero crossings of a function's second derivative are found where the first derivates exhibit a local maximum or minimum.

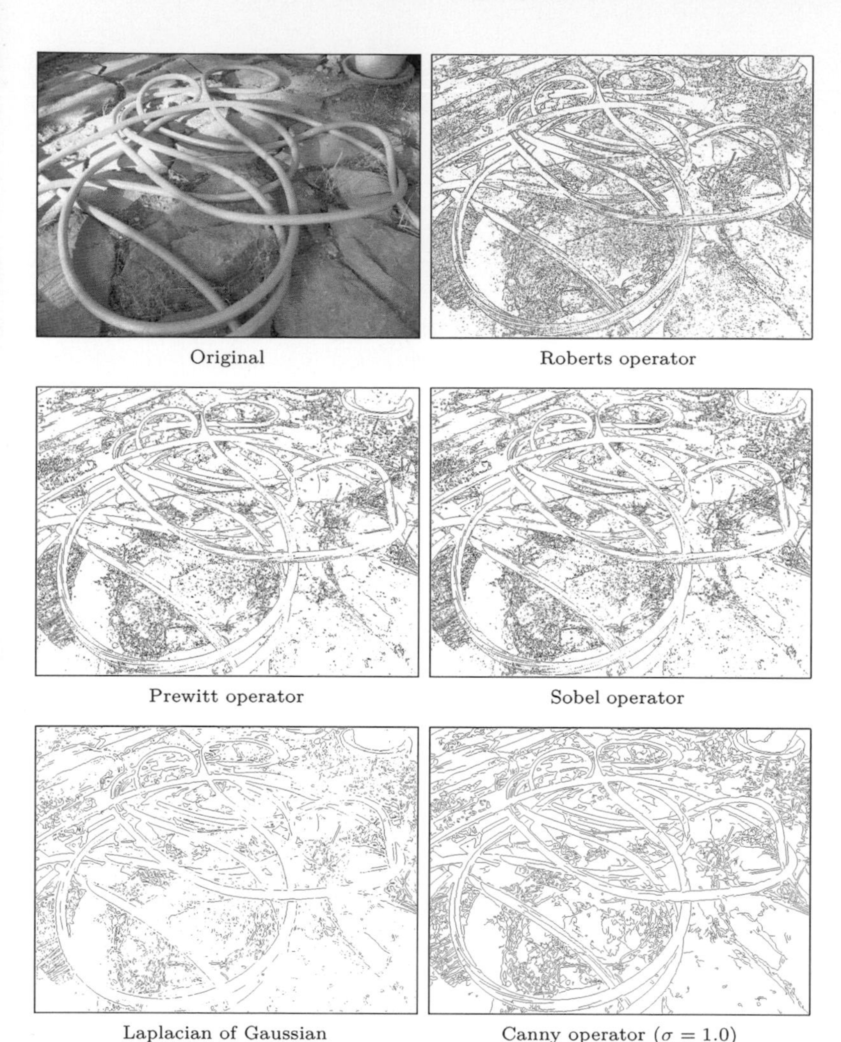

Fig. 6.11
Comparison of various edge operators. Important criteria for the quality of edge results are the amount of "clutter" (irrelevant edge elements) and the connectedness of dominant edges. The Roberts operator responds to very small edge structures because of the small size of its filters. The similarity of the Prewitt and Sobel operators is manifested in the corresponding results. The edge map produced by the Canny operator is substantially cleaner than those of the simpler operators, even for a fixed and relatively small scale value σ.

individual results into a common *edge map*. It is quite common, however, to use only a single-scale implementation of the algorithm with an adjustable filter radius (smoothing parameter σ), which is nevertheless superior to most of the simple edge operators (see Fig. 6.11). In addition, the algorithm not only yields a binary edge map but connected chains of edge pixels, which greatly simplifies the subsequent processing steps. Thus, even in its basic (single-scale) form, the Canny operator is often preferred over other edge detection methods.

In its basic (single-scale) form, the Canny operator performs the following steps (stated more precisely in Algs. 6.1–6.2):

1. **Pre-processing:** Smooth the image with a Gaussian filter of width σ, which specifies the scale level of the edge detector. Calculate the x/y gradient vector at each position of the filtered image and determine the local gradient magnitude and orientation.
2. **Edge localization:** Isolate local maxima of gradient magnitude by "non-maximum suppression" along the local gradient direction.

3. **Edge tracing and hysteresis thresholding:** Collect sets of connected edge pixels from the local maxima by applying "hysteresis thresholding".

6.5.1 Pre-processing

The original intensity image I is first smoothed with a Gaussian filter kernel $H^{G,\sigma}$; its width σ specifies the spatial scale at which edges are to be detected (see Alg. 6.1, lines 2–10). Subsequently, first-order difference filters are applied to the smoothed image $\bar{I}$ to calculate the components $\bar{I}_x, \bar{I}_y$ of the local gradient vectors (Alg. 6.1, line 3–3).[5] Then the local magnitude E_{mag} is calculated as the norm of the corresponding gradient vector (Alg. 6.1, line 11). In view of the subsequent thresholding it may be helpful to normalize the edge magnitude values to a standard range (e.g., to $[0, 100]$).

6.5.2 Edge localization

Candidate edge pixels are isolated by local "non-maximum suppression" of the edge magnitude E_{mag}. In this step, only those pixels are preserved that represent a local maximum along the 1D profile in the direction of the gradient, that is, perpendicular to the edge tangent (see Fig. 6.12). While the gradient may point in any continuous direction, only *four* discrete directions are typically used to facilitate efficient processing. The pixel at position (u, v) is only retained as an edge candidate if its gradient magnitude is greater than both its immediate neighbors in the direction specified by the gradient vector (d_x, d_y) at position (u, v). If a pixel is not a local maximum, its edge magnitude value is set to zero (i.e., "suppressed"). In Alg. 6.1, the non-maximum suppressed edge values are stored in the map E_{nms}.

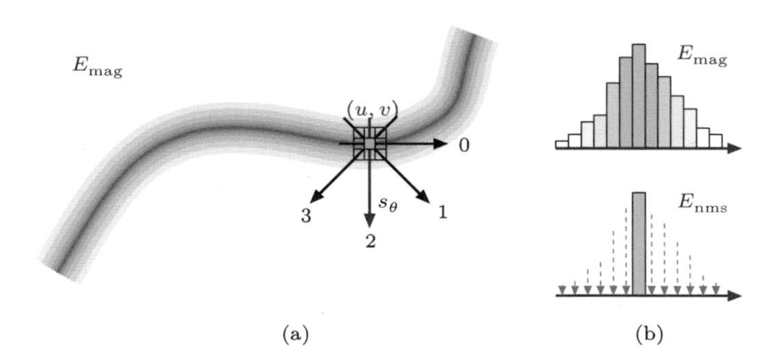

Fig. 6.12
Non-maximum suppression of gradient magnitude. The gradient direction at position (u, v) is coarsely quantized to four discrete orientations $s_\theta \in \{0, 1, 2, 3\}$ (a). Only pixels where the gradient magnitude $E_{\mathrm{mag}}(u, v)$ is a local maximum in the gradient direction (i.e., perpendicular to the edge tangent) are taken as candidate edge points (b). The gradient magnitude at all other points is set (suppressed) to zero.

The problem of finding the discrete orientation $s_\theta = 0, ..., 3$ for a given gradient vector $\boldsymbol{q} = (d_x, d_y)$ is illustrated in Fig. 6.13. This task is simple if the corresponding angle $\theta = \tan^{-1}(d_y/d_x)$ is known, but at this point the use of the trigonometric functions is typically avoided for efficiency reasons. The octant that corresponds to $\boldsymbol{q}$ can be inferred directly from the signs and magnitude of the components d_x, d_y, however, the necessary decision rules are quite complex. Much simpler rules apply if the coordinate system and gradient vector $\boldsymbol{q}$ are

[5] See also Sec. C.3.1 in the Appendix.

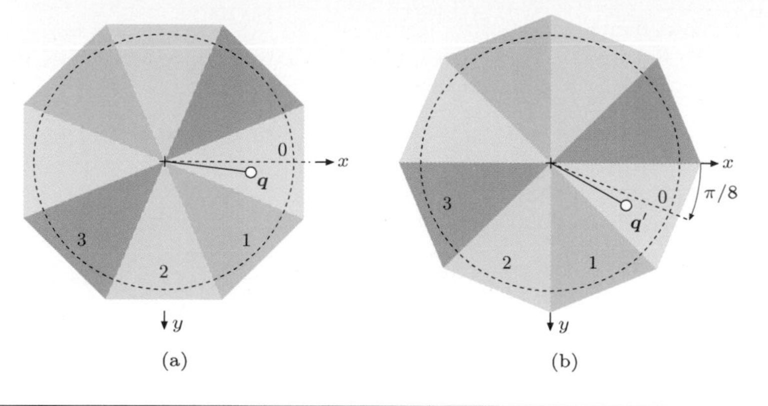

Fig. 6.13
Discrete gradient directions. In (a), calculating the octant for a given orientation vector $\boldsymbol{q} = (d_x, d_y)$ requires a relatively complex decision. Alternatively (b), if $\boldsymbol{q}$ is rotated by $\frac{\pi}{8}$ to $\boldsymbol{q}'$, the corresponding octant can be found directly from the components of $\boldsymbol{q}' = (d'_x, d'_y)$ without the need to calculate the actual angle. Orientation vectors in the other octants are mirrored to octants $s_\theta = 0, 1, 2, 3$.

Alg. 6.1
Canny edge detector for grayscale images.

```
1:  CannyEdgeDetector(I, σ, t_hi, t_lo)
      Input: I, a grayscale image of size M×N; σ, scale (radius of
      Gaussian filter H^{G,σ}); t_hi, t_lo, hysteresis thresholds (t_hi > t_lo).
      Returns a binary edge map of size M×N.
2:    Ī ← I ∗ H^{G,σ}                        ▷ blur with Gaussian of width σ
3:    Ī_x ← Ī ∗ [−0.5  0  0.5]                               ▷ x-gradient
4:    Ī_y ← Ī ∗ [−0.5  0  0.5]^T                             ▷ y-gradient
5:    (M, N) ← Size(I)
6:    Create maps:
7:       E_mag : M×N ↦ ℝ                              ▷ gradient magnitude
8:       E_nms : M×N ↦ ℝ                               ▷ maximum magnitude
9:       E_bin : M×N ↦ {0, 1}                          ▷ binary edge pixels
10:   for all image coordinates (u, v) ∈ M×N do
11:      E_mag(u, v) ← [Ī_x²(u, v) + Ī_y²(u, v)]^{1/2}
12:      E_nms(u, v) ← 0
13:      E_bin(u, v) ← 0
14:   for u ← 1, ..., M−2 do
15:      for v ← 1, ..., N−2 do
16:         d_x ← Ī_x(u, v),   d_y ← Ī_y(u, v)
17:         s_θ ← GetOrientationSector(d_x, d_y)               ▷ Alg. 6.2
18:         if IsLocalMax(E_mag, u, v, s_θ, t_lo) then         ▷ Alg. 6.2
19:            E_nms(u, v) ← E_mag(u, v)         ▷ only keep local maxima
20:   for u ← 1, ..., M−2 do
21:      for v ← 1, ..., N−2 do
22:         if (E_nms(u, v) ≥ t_hi) ∧ (E_bin(u, v) = 0) then
23:            TraceAndThreshold(E_nms, E_bin, u, v, t_lo)
                                                                ▷ Alg. 6.2
24:   return E_bin.
```

rotated by $\frac{\pi}{8}$, as illustrated in Fig. 6.13(b). This step is implemented by the function GetOrientationSector() in Alg. 6.2.[6]

6.5.3 Edge tracing and hysteresis thresholding

In the final step, sets of connected edge points are collected from the magnitude values that remained unsuppressed in the previous oper-

[6] Note that the elements of the rotation matrix in Alg. 6.2 (line 2) are constants and thus no repeated use of trigonometric functions is required.

Alg. 6.2
Procedures used in Alg. 6.1 (Canny edge detector).

1: **GetOrientationSector**(d_x, d_y)
Returns the discrete octant s_θ for the orientation vector $(d_x, d_y)^\mathsf{T}$. See Fig. 6.13 for an illustration.

2: $\begin{pmatrix} d'_x \\ d'_y \end{pmatrix} \leftarrow \begin{pmatrix} \cos(\pi/8) & -\sin(\pi/8) \\ \sin(\pi/8) & \cos(\pi/8) \end{pmatrix} \cdot \begin{pmatrix} d_x \\ d_y \end{pmatrix}$ ▷ rotate $\begin{pmatrix} d_x \\ d_y \end{pmatrix}$ by $\pi/8$

3: **if** $d'_y < 0$ **then**

4: $d'_x \leftarrow -d'_x, \quad d'_y \leftarrow -d'_y$ ▷ mirror to octants $0, \ldots, 3$

5: $s_\theta \leftarrow \begin{cases} 0 & \text{if } (d'_x \geq 0) \wedge (d'_x \geq d'_y) \\ 1 & \text{if } (d'_x \geq 0) \wedge (d'_x < d'_y) \\ 2 & \text{if } (d'_x < 0) \wedge (-d'_x < d'_y) \\ 3 & \text{if } (d'_x < 0) \wedge (-d'_x \geq d'_y) \end{cases}$

6: **return** s_θ. ▷ sector index $s_\theta \in \{0, 1, 2, 3\}$

7: **IsLocalMax**$(E_{\mathrm{mag}}, u, v, s_\theta, \mathsf{t}_{\mathrm{lo}})$
Determines if the gradient magnitude E_{mag} is a local maximum at position (u, v) in direction $s_\theta \in \{0, 1, 2, 3\}$.

8: $m_\mathrm{C} \leftarrow E_{\mathrm{mag}}(u, v)$

9: **if** $m_\mathrm{C} < \mathsf{t}_{\mathrm{lo}}$ **then**

10: **return** false

11: **else**

12: $m_\mathrm{L} \leftarrow \begin{cases} E_{\mathrm{mag}}(u{-}1, v) & \text{if } s_\theta = 0 \\ E_{\mathrm{mag}}(u{-}1, v{-}1) & \text{if } s_\theta = 1 \\ E_{\mathrm{mag}}(u, v{-}1) & \text{if } s_\theta = 2 \\ E_{\mathrm{mag}}(u{-}1, v{+}1) & \text{if } s_\theta = 3 \end{cases}$

13: $m_\mathrm{R} \leftarrow \begin{cases} E_{\mathrm{mag}}(u{+}1, v) & \text{if } s_\theta = 0 \\ E_{\mathrm{mag}}(u{+}1, v{+}1) & \text{if } s_\theta = 1 \\ E_{\mathrm{mag}}(u, v{+}1) & \text{if } s_\theta = 2 \\ E_{\mathrm{mag}}(u{+}1, v{-}1) & \text{if } s_\theta = 3 \end{cases}$

14: **return** $(m_\mathrm{L} \leq m_\mathrm{C}) \wedge (m_\mathrm{C} \geq m_\mathrm{R})$.

15: **TraceAndThreshold**$(E_{\mathrm{nms}}, E_{\mathrm{bin}}, u_0, v_0, \mathsf{t}_{\mathrm{lo}})$
Recursively collects and marks all pixels of an edge that are 8-connected to (u_0, v_0) and have a gradient magnitude above t_{lo}.

16: $E_{\mathrm{bin}}(u_0, v_0) \leftarrow 1$ ▷ mark (u_0, v_0) as an edge pixel

17: $u_\mathrm{L} \leftarrow \max(u_0{-}1, 0)$ ▷ limit to image bounds

18: $u_\mathrm{R} \leftarrow \min(u_0{+}1, M{-}1)$

19: $v_\mathrm{T} \leftarrow \max(v_0{-}1, 0)$

20: $v_\mathrm{B} \leftarrow \min(v_0{+}1, N{-}1)$

21: **for** $u \leftarrow u_\mathrm{L}, \ldots, u_\mathrm{R}$ **do**

22: **for** $v \leftarrow v_\mathrm{T}, \ldots, v_\mathrm{B}$ **do**

23: **if** $(E_{\mathrm{nms}}(u, v) \geq \mathsf{t}_{\mathrm{lo}}) \wedge (E_{\mathrm{bin}}(u, v) = 0)$ **then**

24: TraceAndThreshold$(E_{\mathrm{nms}}, E_{\mathrm{bin}}, u, v, \mathsf{t}_{\mathrm{lo}})$

25: **return**

ation. This is done with a technique called "hysteresis thresholding" using two different threshold values , t_{lo} (with $\mathsf{t}_{\mathrm{hi}} > \mathsf{t}_{\mathrm{lo}}$). The image is scanned for pixels with edge magnitude $E_{\mathrm{nms}}(u, v) \geq \mathsf{t}_{\mathrm{hi}}$. Whenever such a (previously unvisited) location is found, a new *edge trace* is started and all connected edge pixels (u', v') are added to it as long as $E_{\mathrm{nms}}(u', v') \geq \mathsf{t}_{\mathrm{lo}}$. Only those edge traces remain that contain at least one pixel with edge magnitude greater than t_{hi} and no pixels with edge magnitude less than t_{lo}. This process (which is similar to

(a)

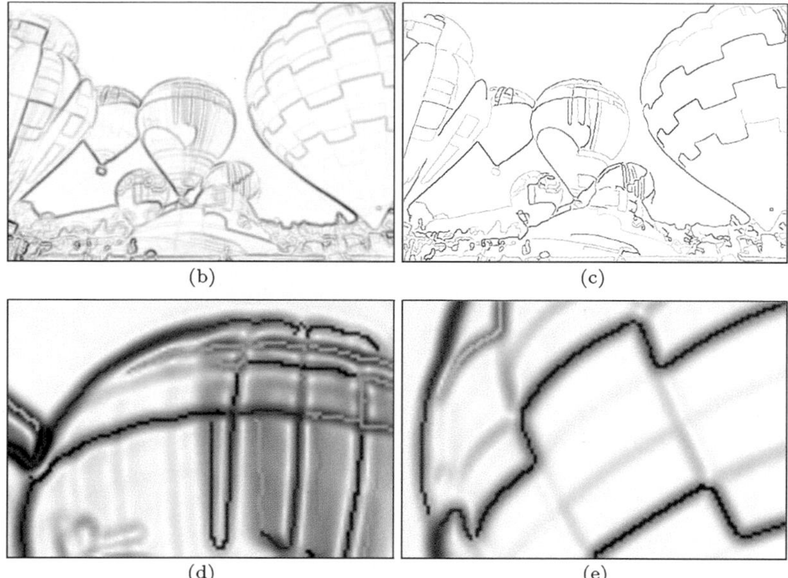

(b) (c)

(d) (e)

Fig. 6.14
Grayscale Canny edge operator details. Inverted gradient magnitude (a), detected edge points with connected edge tracks shown in distinctive colors (b). Details with gradient magnitude and detected edge points overlaid (c, d). Settings: $\sigma = 2.0$, $\mathsf{t}_{\mathrm{hi}} = 20\%$, $\mathsf{t}_{\mathrm{lo}} = 5\%$ (of the max. edge magnitude).

flood-fill region growing) is detailed in procedure GetOrientationSector in Alg. 6.2. Typical threshold values for 8-bit grayscale images are $\mathsf{t}_{\mathrm{hi}} = 5.0$ and $\mathsf{t}_{\mathrm{lo}} = 2.5$.

Figure 6.14 illustrates the effectiveness of non-maximum suppression for localizing the edge centers and edge-linking with hysteresis thresholding. Results from the single-scale Canny detector are shown in Fig. 6.15 for different settings of σ and fixed upper/lower threshold values $\mathsf{t}_{\mathrm{hi}} = 20\%$, $\mathsf{t}_{\mathrm{lo}} = 5\%$ (relative to the maximum gradient magnitude).

6.5.4 Additional Information

Due to the long-lasting popularity of the Canny operator, additional descriptions and some excellent illustrations can be found at various places in the literature, including [89, p. 719], [232, pp. 71–80], and [166, pp. 548–549]. An edge operator similar to the Canny detector, but based on a set of recursive filters, is described in [62]. While the Canny detector was originally designed for grayscale images, modified versions for color images exist, including the one we describe in the next section.

Fig. 6.15
Results from the single-scale grayscale Canny edge operator (Algs. 6.1–6.2) for different values of $\sigma = 0.5, \ldots, 5.0$. Inverted gradient magnitude (left column) and detected edge points (right column). The detected edge points (right column) are linked to connected edge chains.

6.5.5 Implementation

A complete implementation of the Canny edge detector for both grayscale and RGB color images can be found in the Java library for this book.[7] A basic usage example Prog. 16.1 is shown in Prog. 16.1 on p. 411.

[7] Class `CannyEdgeDetector` in package `imagingbook.pub.coloredge`.

6.6 Edge Sharpening

Making images look sharper is a frequent task, such as to make up for a lack of sharpness after scanning or scaling an image or to pre-compensate for a subsequent loss of sharpness in the course of printing or displaying an image. A common approach to image sharpening is to amplify the high-frequency image components, which are mainly responsible for the perceived sharpness of an image and for which the strongest occur at rapid intensity transitions. In the following, we describe two methods for artificial image sharpening that are based on techniques similar to edge detection and thus fit well in this chapter. In the following, we describe two methods for artificial image sharpening that are based on techniques similar to edge detection and thus fit well in this chapter.

6.6.1 Edge Sharpening with the Laplacian Filter

A common method for localizing rapid intensity changes are filters based on the second derivatives of the image function. Figure 6.16 illustrates this idea on a 1D, continuous function $f(x)$. The second derivative $f''(x)$ of the step function shows a positive pulse at the lower end of the transition and a negative pulse at the upper end. The edge is sharpened by subtracting a certain fraction w of the second derivative $f''(x)$ from the original function $f(x)$,

$$\hat{f}(x) = f(x) - w \cdot f''(x)\,. \tag{6.29}$$

Depending upon the weight factor $w \geq 0$, the expression in Eqn. (6.29) causes the intensity function to overshoot at both sides of an edge, thus exaggerating edges and increasing the perceived sharpness.

Laplacian operator

Sharpening of a 2D function can be accomplished with the second derivatives in the horizontal and vertical directions combined by the so-called Laplacian operator. The Laplacian operator ∇^2 of a 2D function $f(x,y)$ is defined as the sum of the second partial derivatives along the x and y directions:

$$(\nabla^2 f)(x,y) = \frac{\partial^2 f}{\partial^2 x}(x,y) + \frac{\partial^2 f}{\partial^2 y}(x,y). \tag{6.30}$$

Similar to the first derivatives (see Sec. 6.2.2), the second derivatives of a discrete image function can also be estimated with a set of simple linear filters. Again, several versions, have been proposed. For example, the two 1D filters

$$\frac{\partial^2 f}{\partial^2 x} \approx H_x^{\mathrm{L}} = \begin{bmatrix} 1 & -2 & 1 \end{bmatrix} \quad \text{and} \quad \frac{\partial^2 f}{\partial^2 y} \approx H_y^{\mathrm{L}} = \begin{bmatrix} 1 \\ -2 \\ 1 \end{bmatrix}, \tag{6.31}$$

for estimating the second derivatives along the x and y directions, respectively, combine to make the 2D Laplacian filter

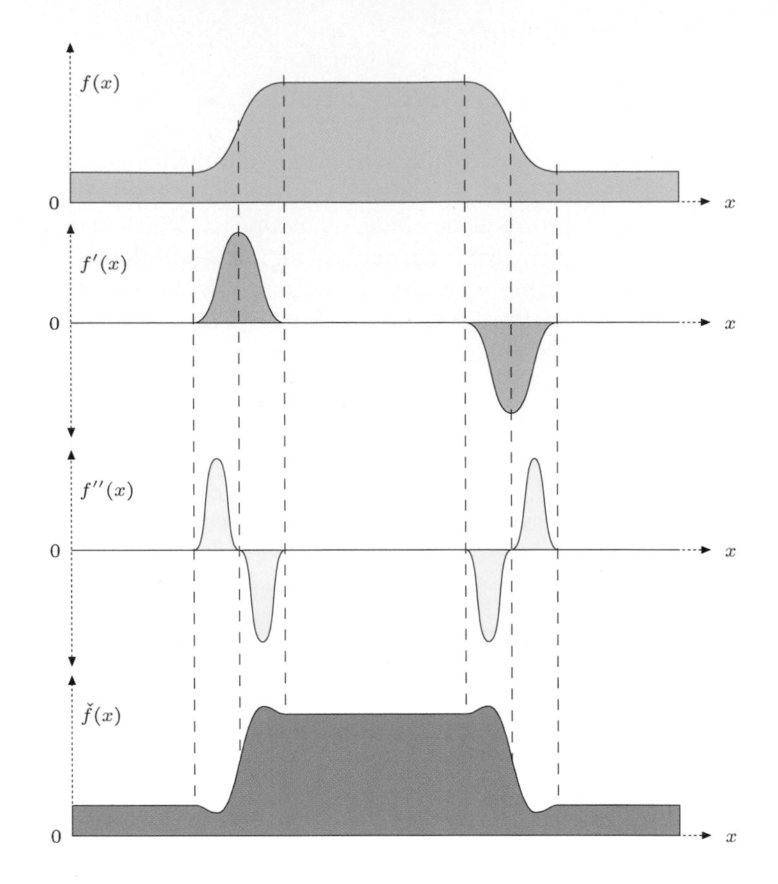

Fig. 6.16
Edge sharpening with the second derivative. The original intensity function $f(x)$, first derivative $f'(x)$, second derivative $f''(x)$, and sharpened intensity function $\hat{f}(x) = f(x) - w \cdot f''(x)$ are shown (w is a weighting factor).

$$H^{\mathrm{L}} = H_x^{\mathrm{L}} + H_y^{\mathrm{L}} = \begin{bmatrix} 0 & 1 & 0 \\ 1 & -4 & 1 \\ 0 & 1 & 0 \end{bmatrix}. \quad (6.32)$$

Figure 6.17 shows an example of applying the Laplacian filter H^{L} to a grayscale image, where the pairs of positive-negative peaks at both sides of each edge are clearly visible. The filter appears almost isotropic despite the coarse approximation with the small filter kernels.

Notice that H^{L} in Eqn. (6.32) is not *separable* in the usual sense (as described in Sec. 5.3.3) but, because of the linearity property of convolution (Eqns. (5.21) and (5.23)), it can be expressed (and computed) as the *sum* of two 1D filters,

$$I * H^{\mathrm{L}} = I * (H_x^{\mathrm{L}} + H_y^{\mathrm{L}}) = (I * H_x^{\mathrm{L}}) + (I * H_y^{\mathrm{L}}) = I_{xx} + I_{yy}. \quad (6.33)$$

Analogous to the gradient filters (for estimating the first derivatives), the sum of the coefficients is zero in any Laplace filter, such that its response is zero in areas of constant (flat) intensity (Fig. 6.17). Other common variants of 3 × 3 pixel Laplace filters are

$$H_8^{\mathrm{L}} = \begin{bmatrix} 1 & 1 & 1 \\ 1 & -8 & 1 \\ 1 & 1 & 1 \end{bmatrix} \quad \text{oder} \quad H_{12}^{\mathrm{L}} = \begin{bmatrix} 1 & 2 & 1 \\ 2 & -12 & 2 \\ 1 & 2 & 1 \end{bmatrix}. \quad (6.34)$$

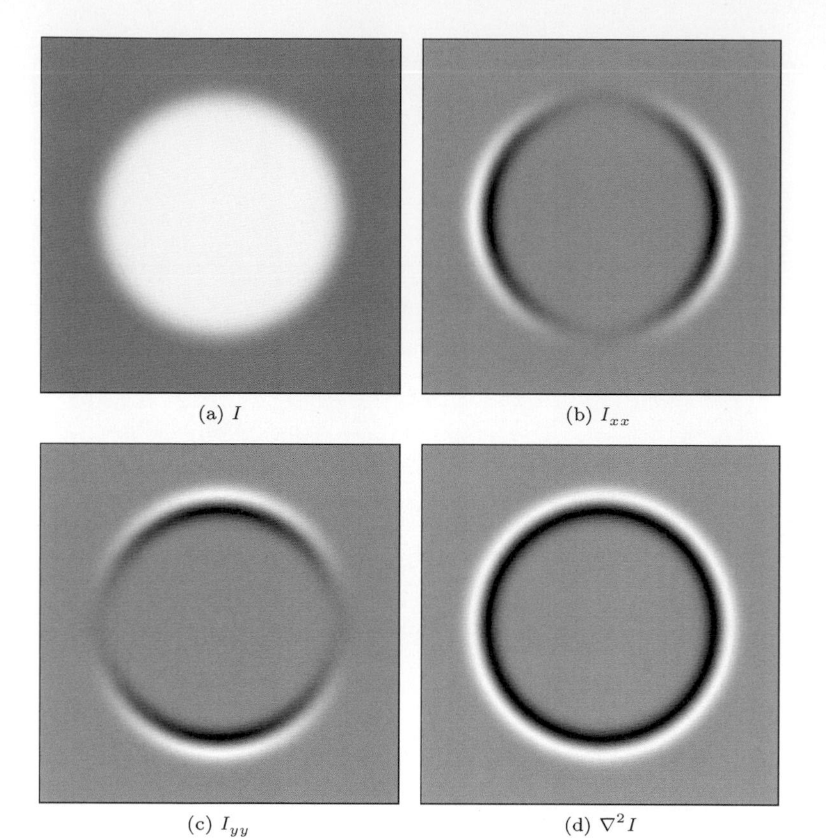

Fig. 6.17
Results of Laplace filter H^{L}: synthetic test image I (a), second partial derivative $I_{xx} = \partial^2 I/\partial^2 x$ in the horizontal direction (b), second partial derivative $I_{yy} = \partial^2 I/\partial^2 y$ in the vertical direction (c), and Laplace filter $\nabla^2 I = I_{xx} + I_{yy}$ (d). Intensities in (b–d) are scaled such that maximally negative and positive values are shown as black and white, respectively, and zero values are gray.

Sharpening

To perform the actual sharpening, as described by Eqn. (6.29) for the 1D case, we first apply a Laplacian filter H^{L} to the image I and then subtract a fraction of the result from the original image,

$$I' \leftarrow I - w \cdot (H^{\mathrm{L}} * I). \tag{6.35}$$

The factor w specifies the proportion of the Laplacian component and thus the sharpening strength. The proper choice of w also depends on the specific Laplacian filter used in Eqn. (6.35) since none of the aforementioned filters is normalized.

Figure 6.17 shows the result of applying a Laplacian filter (with the kernel given in Eqn. (6.32)) to a synthetic test image where the pairs of positive/negative peaks at both sides of each edge are clearly visible. The filter appears almost isotropic despite the coarse approximation with the small filter kernels. The application to a real grayscale image using the filter H^{L} (Eqn. (6.32)) and $w = 1.0$ is shown in Fig. 6.18.

As we can expect from second-order derivatives, the Laplacian filter is fairly sensitive to image noise, which can be reduced (as is commonly done in edge detection with first derivatives) by previous smoothing, such as with a Gaussian filter (see also Sec. 6.4.1).

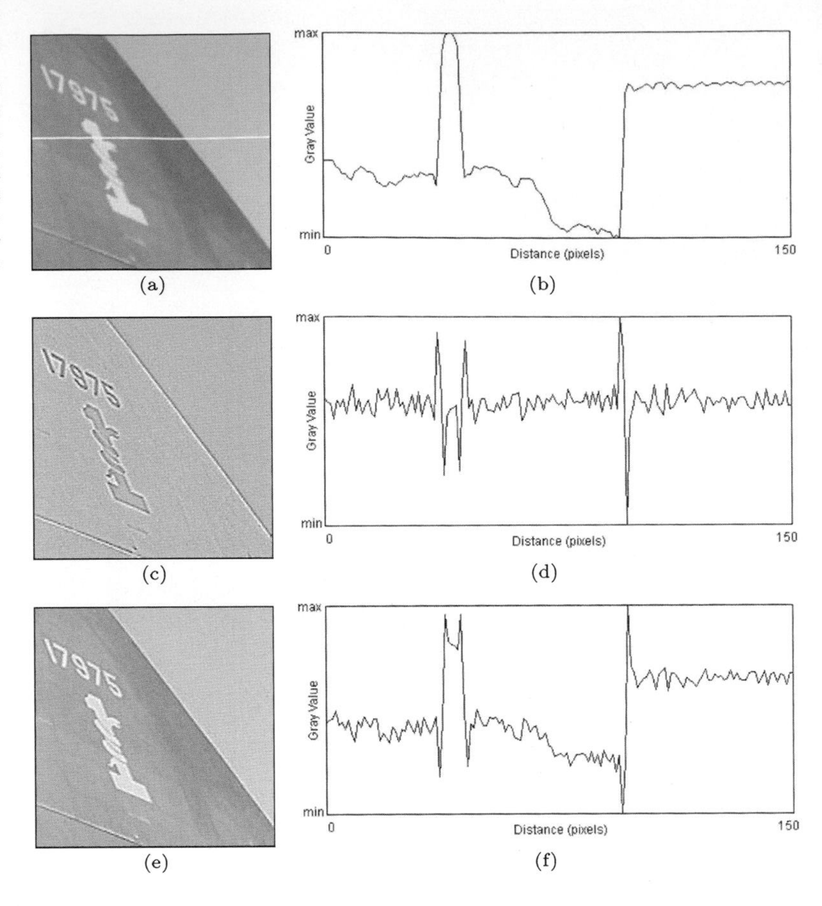

Fig. 6.18 Edge sharpening with the Laplacian filter. Original image with a horizontal profile taken from the marked line (a, b), result of Laplacian filter H^{L} (c, d), and sharpened image with sharpening factor $w = 1.0$ (e, f).

6.6.2 Unsharp Masking

"Unsharp masking" (USM) is a technique for edge sharpening that is particularly popular in astronomy, digital printing, and many other areas of image processing. The term originates from classical photography, where the sharpness of an image was optically enhanced by combining it with a smoothed ("unsharp") copy. This process is in principle the same for digital images.

Process

The first step in the USM filter is to subtract a smoothed version of the image from the original, which enhances the edges. The result is called the "mask". In analog photography, the required smoothing was achieved by simply defocusing the lens. Subsequently, the mask is again added to the original, such that the edges in the image are sharpened. In summary, the steps involved in USM filtering are:

1. The mask image M is generated by subtracting (from the original image I) a smoothed version of I, obtained by filtering with $\tilde{H}$, that is,
$$M \leftarrow I - (I * \tilde{H}) = I - \tilde{I}. \tag{6.36}$$
The kernel $\tilde{H}$ of the smoothing filter is assumed to be normalized (see Sec. 5.2.5).

2. To obtain the sharpened image $\check{I}$, the mask M is added to the original image I, weighted by the factor a, which controls the amount of sharpening,

$$\check{I} \leftarrow I + a \cdot M, \tag{6.37}$$

and thus (by inserting from Eqn. (6.36))

$$\check{I} \leftarrow I + a \cdot (I - \tilde{I}) = (1 + a) \cdot I - a \cdot \tilde{I}. \tag{6.38}$$

Smoothing filter

In principle, any smoothing filter could be used for the kernel $\tilde{H}$ in Eqn. (6.36), but Gaussian filters $H^{\mathrm{G},\sigma}$ with variable radius σ are most common (see also Sec. 5.2.7). Typical parameter values are 1 to 20 for σ and 0.2 to 4.0 (equivalent to 20% to 400%) for the sharpening factor a.

Figure 6.19 shows two examples of USM filters using Gaussian smoothing filters with different radii σ.

Extensions

The advantages of the USM filter over the Laplace filter are reduced noise sensitivity due to the involved smoothing and improved controllability through the parameters σ (spatial extent) and a (sharpening strength).

Of course the USM filter responds not only to real edges but to some extent to any intensity transition, and thus potentially increases any visible noise in continuous image regions. Some implementations (e.g., Adobe Photoshop) therefore provide an additional *threshold* parameter t_c to specify the *minimum local contrast* required to perform edge sharpening. Sharpening is only applied if the local contrast at position (u, v), expressed, for example, by the gradient magnitude $|\nabla I|$ (Eqn. (6.5)), is greater than that threshold. Otherwise, that pixel remains unmodified, that is,

$$\check{I}(u,v) \leftarrow \begin{cases} I(u,v) + a \cdot M(u,v) & \text{for } |\nabla I|(u,v) \geq t_c, \\ I(u,v) & \text{otherwise.} \end{cases} \tag{6.39}$$

Different to the original USM filter (Eqn. (6.37)), this extended version is no longer a *linear* filter. On color images, the USM filter is usually applied to all color channels with identical parameter settings.

Implementation

The USM filter is available in virtually any image-processing software and, due to its simplicity and flexibility, has become an indispensable tool for many professional users. In ImageJ, the USM filter is implemented by the plugin class **`UnsharpMask`**[8] and can be applied through the menu

Process ▷ Filter ▷ Unsharp Mask...

[8] In package `ij.plugin.filter`.

Fig. 6.19
Unsharp masking filters with varying smoothing radii $\sigma = 2.5$ and 10.0. The sharpening strength a is set to 1.0 (100%). The profiles show the intensity function for the image line marked in the original image (top-left).

max
Gray Value
min
0
Distance (pixels)
150

(a) Original

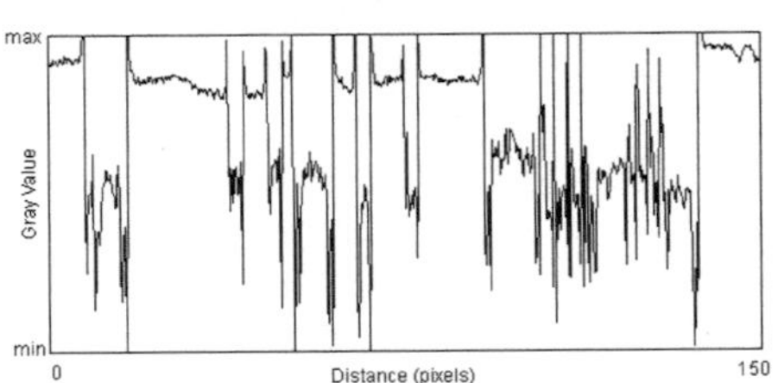

b) $\sigma = 2.5$

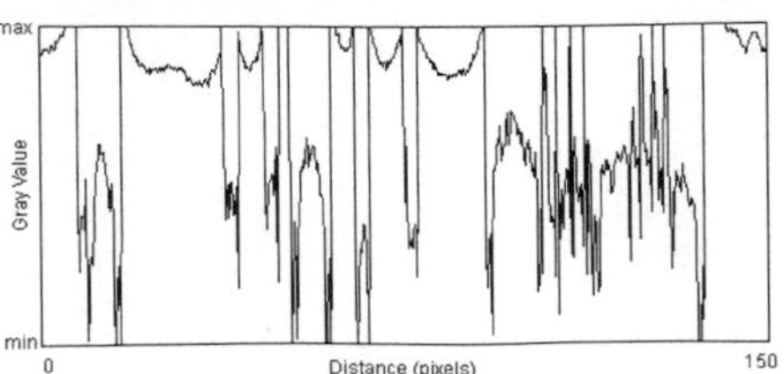

(c) $\sigma = 10.0$

This filter can also be used from other plugin classes, for example, in the following way:

```
import ij.plugin.filter.UnsharpMask;
...
public void run(ImageProcessor ip) {
  UnsharpMask usm = new UnsharpMask();
  double r = 2.0; // standard settings for radius
  double a = 0.6; // standard settings for weight
  usm.sharpen(ip, r, a);
  ...
}
```

ImageJ's `UnsharpMask` implementation uses the class `GaussianBlur` for the required smoothing operation. The alternative implementation shown in Prog. 6.1 follows the definition in Eqn. (6.38) and uses Gaussian filter kernels that are created with the method `makeGaussKernel1d()`, as defined in Prog. 5.4.

```
1  double radius = 1.0;  // radius (sigma of Gaussian)
2  double amount = 1.0;  // amount of sharpening (1 = 100%)
3  ...
4  public void run(ImageProcessor ip) {
5    ImageProcessor I = ip.convertToFloat(); // I
6
7    // create a blurred version of the image:
8    ImageProcessor J = I.duplicate();           // Ĩ
9    float[] H = GaussianFilter.makeGaussKernel1d(sigma);
10   Convolver cv = new Convolver();
11   cv.setNormalize(true);
12   cv.convolve(J, H, 1, H.length);
13   cv.convolve(J, H, H.length, 1);
14
15   I.multiply(1 + a);      // I ← (1 + a) · I
16   J.multiply(a);          // Ĩ ← a · Ĩ
17   I.copyBits(J,0,0,Blitter.SUBTRACT); //Ĩ ← (1+a) · I−a · Ĩ
18
19   //copy result back into original byte image
20   ip.insert(I.convertToByte(false), 0, 0);
21 }
```

Prog. 6.1
Unsharp masking (Java implementation). First the original image is converted to a `FloatProcessor` object `I` (I) in line 5, which is duplicated to hold the blurred image `J` ($\tilde{I}$) in line 8. The method `makeGaussKernel1d()`, defined in Prog. 5.4, is used to create the 1D Gaussian filter kernel applied in the horizontal and vertical directions (lines 12–13). The remaining calculations follow Eqn. (6.38).

Laplace vs. USM filter

A closer look at these two methods reveals that sharpening with the Laplace filter (Sec. 6.6.1) can be viewed as a special case of the USM filter. If the Laplace filter in Eqn. (6.32) is decomposed as

$$H^{\mathrm{L}} = \begin{bmatrix} 0 & 1 & 0 \\ 1 & -4 & 1 \\ 0 & 1 & 0 \end{bmatrix} = \begin{bmatrix} 0 & 1 & 0 \\ 1 & 1 & 1 \\ 0 & 1 & 0 \end{bmatrix} - 5 \begin{bmatrix} 0 & 0 & 0 \\ 0 & 1 & 0 \\ 0 & 0 & 0 \end{bmatrix} = 5 \cdot \big(\tilde{H}^{\mathrm{L}} - \delta\big), \quad (6.40)$$

one can see that H^L consists of a simple 3×3 pixel smoothing filter $\tilde{H}$ minus the impulse function δ. Laplace sharpening with the weight factor w as defined in Eqn. (6.35) can therefore (by a little manipulation) be expressed as

$$\begin{aligned}\check{I}_L &\leftarrow I - w \cdot (H^{\mathrm{L}} * I) &&= I - w \cdot \bigl(5(\tilde{H}^{\mathrm{L}} - \delta) * I\bigr) \\ &= I - 5w \cdot (\tilde{H}^{\mathrm{L}} * I - I) &&= I + 5w \cdot (I - \tilde{H}^{\mathrm{L}} * I) \\ &= I + 5w \cdot M^{\mathrm{L}}, \end{aligned} \tag{6.41}$$

that is, in the form of a USM filter $\check{I} \leftarrow I + a \cdot M$ (Eqn. (6.37)). Laplacian sharpening is thus a special case of a USM filter with the mask $M = M^L = (I - \tilde{H}^L * I)$, the specific smoothing filter

$$\tilde{H}^{\mathrm{L}} = \frac{1}{5} \begin{bmatrix} 0 & 1 & 0 \\ 1 & 1 & 1 \\ 0 & 1 & 0 \end{bmatrix}$$

and the sharpening factor $a = 5w$.

6.7 Exercises

Exercise 6.1. Calculate (manually) the gradient and the Laplacian (using the discrete approximations in Eqn. (6.2) and Eqn. (6.32), respectively) for the following "image":

$$I = \begin{bmatrix} 14 & 10 & 19 & 16 & 14 & 12 \\ 18 & 9 & 11 & 12 & 10 & 19 \\ 9 & 14 & 15 & 26 & 13 & 6 \\ 21 & 27 & 17 & 17 & 19 & 16 \\ 11 & 18 & 18 & 19 & 16 & 14 \\ 16 & 10 & 13 & 7 & 22 & 21 \end{bmatrix}.$$

Exercise 6.2. Implement the Sobel edge operator as defined in Eqn. (6.10) (and illustrated in Fig. 6.6) as an ImageJ plugin. The plugin should generate two new images for the edge magnitude $E(u, v)$ and the edge orientation $\Phi(u, v)$. Come up with a suitable way to display local edge orientation.

Exercise 6.3. Express the Sobel operator (Eqn. (6.10)) in x/y-separable form analogous to the decomposition of the Prewitt operator in Eqn. (6.9).

Exercise 6.4. Implement the Kirsch operator (Eqns. (6.25)–(6.28)) analogous to the two-directional Sobel operator in Exercise 6.2 and compare the results from both methods, particularly the edge orientation estimates.

Exercise 6.5. Devise and implement a compass edge operator with more than eight (16?) differently oriented filters.

Exercise 6.6. Compare the results of the unsharp masking filters in ImageJ and Adobe Photoshop using a suitable test image. How should the parameters for σ (*radius*) and a (*weight*) be defined in both implementations to obtain similar results?

7

Corner Detection

Corners are prominent structural elements in an image and are therefore useful in a wide variety of applications, including following objects across related images (*tracking*), determining the correspondence between stereo images, serving as reference points for precise geometrical measurements, and calibrating camera systems for machine vision applications. Thus corner points are not only important in human vision but they are also "robust" in the sense that they do not arise accidentally in 3D scenes and, furthermore, can be located quite reliably under a wide range of viewing angles and lighting conditions.

7.1 Points of Interest

Despite being easily recognized by our visual system, accurately and precisely detecting corners automatically is not a trivial task. A good corner detector must satisfy a number of criteria, including distinguishing between true and accidental corners, reliably detecting corners in the presence of realistic image noise, and precisely and accurately determining the locations of corners. Finally, it should also be possible to implement the detector efficiently enough so that it can be utilized in real-time applications such as video tracking.

Numerous methods for finding corners or similar interest points have been proposed and most of them take advantage of the following basic principle. While an *edge* is usually defined as a location in the image at which the gradient is especially high in *one* direction and low in the direction normal to it, a *corner point* is defined as a location that exhibits a strong gradient value in *multiple* directions at the same time.

Most methods take advantage of this observation by examining the first or second derivative of the image in the x and y directions to find corners (e.g., [77, 102, 137, 154]). In the next section, we describe in detail the Harris detector, also known as the "Plessey feature point detector" [102], since it turns out that even though more efficient detectors are known (see, e.g., [210, 220]), the Harris detector, and other detectors based on it, are the most widely used in practice.

7.2 Harris Corner Detector

This operator, developed by Harris and Stephens [102], is one of a group of related methods based on the same premise: a corner point exists where the gradient of the image is especially strong in more than one direction at the same time. In addition, locations along edges, where the gradient is strong in only one direction, should not be considered as corners, and the detector should be isotropic, that is, independent of the orientation of the local gradients.

7.2.1 Local Structure Matrix

The Harris corner detector is based on the first partial derivatives (gradient) of the image function $I(u,v)$, that is,

$$I_x(u,v) = \frac{\partial I}{\partial x}(u,v) \quad \text{and} \quad I_y(u,v) = \frac{\partial I}{\partial y}(u,v). \tag{7.1}$$

For each image position (u,v), we first calculate the three quantities

$$A(u,v) = I_x^2(u,v), \tag{7.2}$$
$$B(u,v) = I_y^2(u,v), \tag{7.3}$$
$$C(u,v) = I_x(u,v) \cdot I_y(u,v) \tag{7.4}$$

that constitute the elements of the *local structure matrix* $\mathbf{M}(u,v)$:[1]

$$\mathbf{M} = \begin{pmatrix} I_x^2 & I_x I_y \\ I_x I_y & I_y^2 \end{pmatrix} = \begin{pmatrix} A & C \\ C & B \end{pmatrix}. \tag{7.5}$$

Next, each of the three scalar fields $A(u,v)$, $B(u,v)$, $C(u,v)$ is individually smoothed by convolution with a linear Gaussian filter $H^{G,\sigma}$ (see Sec. 5.2.7),

$$\bar{\mathbf{M}} = \begin{pmatrix} A * H_\sigma^G & C * H_\sigma^G \\ C * H_\sigma^G & B * H_\sigma^G \end{pmatrix} = \begin{pmatrix} \bar{A} & \bar{C} \\ \bar{C} & \bar{B} \end{pmatrix}. \tag{7.6}$$

The *eigenvalues*[2] of the matrix $\bar{\mathbf{M}}$, defined as[3]

$$\begin{aligned} \lambda_{1,2} &= \frac{\mathrm{trace}(\bar{\mathbf{M}})}{2} \pm \sqrt{\left(\frac{\mathrm{trace}(\bar{\mathbf{M}})}{2}\right)^2 - \det(\bar{\mathbf{M}})} \\ &= \frac{1}{2} \cdot \left(\bar{A} + \bar{B} \pm \sqrt{\bar{A}^2 - 2 \cdot \bar{A} \cdot \bar{B} + \bar{B}^2 + 4 \cdot \bar{C}^2}\right), \end{aligned} \tag{7.7}$$

[1] For improved legibility, we simplify the notation used in the following by omitting the function coordinates (u,v); for example, the function $I_x(u,v)$ is abbreviated as I_x or $A(u,v)$ is simply denoted A etc.

[2] See also Sec. B.4 in the Appendix.

[3] $\det(\bar{\mathbf{M}})$ denotes the *determinant* and $\mathrm{trace}(\bar{\mathbf{M}})$ denotes the *trace* of the matrix $\bar{\mathbf{M}}$ (see, e.g., [35, pp. 252 and 259]).

are (because the matrix is symmetric) positive and real. They contain essential information about the local image structure. Within an image region that is uniform (that is, appears flat), $\bar{\mathbf{M}} = 0$ and therefore $\lambda_1 = \lambda_2 = 0$. On an ideal ramp, however, the eigenvalues are $\lambda_1 > 0$ and $\lambda_2 = 0$, independent of the orientation of the edge. The eigenvalues thus encode an edge's *strength*, and their associated *eigenvectors* correspond to the local edge *orientation*.

A corner should have a strong edge in the main direction (corresponding to the larger of the two eigenvalues), another edge normal to the first (corresponding to the smaller eigenvalues), and both eigenvalues must be significant. Since $\bar{A}, \bar{B} \geq 0$, we can assume that $\mathrm{trace}(\bar{\mathbf{M}}) > 0$ and thus $|\lambda_1| \geq |\lambda_2|$. Therefore only the smaller of the two eigenvalues, $\lambda_2 = \mathrm{trace}(\bar{\mathbf{M}})/2 - \sqrt{\ldots}$, is relevant when determining a corner.

7.2.2 Corner Response Function (CRF)

From Eqn. (7.7) we see that the difference between the two eigenvalues of the local structure matrix is

$$\lambda_1 - \lambda_2 = 2 \cdot \sqrt{0.25 \cdot \big(\mathrm{trace}(\bar{\mathbf{M}})\big)^2 - \det(\bar{\mathbf{M}})}, \tag{7.8}$$

where the expression under the square root is always non-negative. At a good corner position, the difference between the two eigenvalues λ_1, λ_2 should be as small as possible and thus the expression under the root in Eqn. (7.8) should be a minimum. To avoid the explicit calculation of the eigenvalues (and the square root) the Harris detector defines the function

$$\begin{aligned} Q(u, v) &= \det(\bar{\mathbf{M}}(u, v)) - \alpha \cdot \big(\mathrm{trace}(\bar{\mathbf{M}}(u, v))\big)^2 \\ &= \bar{A}(u, v) \cdot \bar{B}(u, v) - \bar{C}^2(u, v) - \alpha \cdot [\bar{A}(u, v) + \bar{B}(u, v)]^2 \end{aligned} \tag{7.9}$$

as a measure of "corner strength", where the parameter α determines the sensitivity of the detector. $Q(u, v)$ is called the "corner response function" and returns maximum values at isolated corners. In practice, α is assigned a fixed value in the range of 0.04 to 0.06 (max. $0.25 = \frac{1}{4}$). The larger the value of α, the less sensitive the detector is and the fewer corners detected.

7.2.3 Determining Corner Points

An image location (u, v) is selected as a potential candidate for a corner point if

$$Q(u, v) > \mathsf{t}_H,$$

where the threshold t_H is selected based on image content and typically lies within the range of 10,000 to 1,000,000. Once selected, the corners $\boldsymbol{c}_i = \langle u_i, v_i, q_i \rangle$ are inserted into the sequence

$$\mathcal{C} = (\boldsymbol{c}_1, \boldsymbol{c}_2, \ldots, \boldsymbol{c}_N),$$

which is then sorted in descending order (i.e., $q_i \geq q_{i+1}$) according to *corner strength* $q_i = Q(u_i, v_i)$, as defined in Eqn. (7.9). To suppress

Table 7.1
Harris corner detector—typical parameter settings for Alg. 7.1.

Prefilter (Alg. 7.1, line 2–3): Smoothing with a small xy-separable filter $H_p = H_{px} * H_{py}$, where

$$H_{px} = \frac{1}{9} \cdot \begin{bmatrix} 2 & 5 & 2 \end{bmatrix} \qquad \text{and} \qquad H_{py} = H_{px}^{\mathsf{T}} = \frac{1}{9} \cdot \begin{bmatrix} 2 \\ 5 \\ 2 \end{bmatrix}.$$

Gradient filter (Alg. 7.1, line 3): Computing the first partial derivative in the x and y directions with

$$h_{dx} = \begin{bmatrix} -0.5 & 0 & 0.5 \end{bmatrix} \qquad \text{and} \qquad h_{dy} = h_{dx}^{\mathsf{T}} = \begin{bmatrix} -0.5 \\ 0 \\ 0.5 \end{bmatrix}.$$

Blur filter (Alg. 7.1, line 10): Smoothing the individual components of the structure matrix M with separable Gaussian filters $H_b = H_{bx} * H_{by}$ with

$$h_{bx} = \frac{1}{64} \cdot \begin{bmatrix} 1 & 6 & 15 & 20 & 15 & 6 & 1 \end{bmatrix} \quad \text{and} \quad h_{by} = h_{bx}^{\mathsf{T}} = \frac{1}{64} \cdot \begin{bmatrix} 1 \\ 6 \\ 15 \\ 20 \\ 15 \\ 6 \\ 1 \end{bmatrix}.$$

Control parameter (Alg. 7.1, line 14): $\alpha = 0.04, \ldots, 0.06$ (default 0.05).

Response threshold (Alg. 7.1, line 19): $\mathtt{t}_H = 10\,000, \ldots, 1\,000\,000$ (default 20 000).

Neighborhood radius (Alg. 7.1, line 37): $d_{\mathrm{min}} = 10$ Pixel.

the false corners that tend to arise in densely packed groups around true corners, all except the strongest corner in a specified vicinity are eliminated. To accomplish this, the list $\mathcal{C}$ is traversed from the front to the back, and the weaker corners toward the end of the list, which lie in the surrounding neighborhood of a stronger corner, are deleted.

The complete algorithm for the Harris detector is summarized again in Alg. 7.1; the associated parameters are listed in Table 7.1.

7.2.4 Examples

Figure 7.1 uses a simple synthetic image to illustrate the most important steps in corner detection using the Harris detector. The figure shows the result of the gradient computation, the three components of the structure matrix $\mathbf{M}(u,v) = \left(\begin{smallmatrix} A & C \\ C & B \end{smallmatrix}\right)$, and the values of the *corner response function* $Q(u,v)$ for each image position (u,v). This example was calculated with the standard settings as given in Table 7.1.

The second example (Fig. 7.2) illustrates the detection of corner points in a grayscale representation of a natural scene. It demonstrates how weak corners are eliminated in favor of the strongest corner in a region.

Alg. 7.1
Harris corner detector. This algorithm takes an intensity image I and creates a sorted list of detected corner points. $*$ is the convolution operator used for linear filter operations. Details for the parameters H_p, H_{dx}, H_{dy}, H_b, α, and t_H can be found in Table 7.1.

1: **HarrisCorners**($I, \alpha, \mathrm{t}_H, d_{\min}$)
Input: I, the source image; α, sensitivity parameter (typ. 0.05); t_H, response threshold (typ. 20 000); $d_{\min}$, minimum distance between final corners. Returns a sequence of the strongest corners detected in I.

Step 1 – calculate the corner response function:

2: $I_x \leftarrow (I * h_{px}) * h_{dx}$ ▷ horizontal prefilter and derivative
3: $I_y \leftarrow (I * h_{py}) * h_{dy}$ ▷ vertical prefilter and derivative
4: $(M, N) \leftarrow \mathsf{Size}(I)$
5: Create maps $A, B, C, Q\colon M \times N \mapsto \mathbb{R}$
6: **for all** image coordinates (u, v) **do**
Compute the local structure matrix $\mathbf{M} = \left(\begin{smallmatrix} A & C \\ C & B \end{smallmatrix}\right)$:
7: $A(u,v) \leftarrow (I_x(u,v))^2$
8: $B(u,v) \leftarrow (I_y(u,v))^2$
9: $C(u,v) \leftarrow I_x(u,v) \cdot I_y(u,v)$
Blur the components of the local structure matrix ($\bar{\mathbf{M}}$):
10: $\bar{A} \leftarrow A * H_b$
11: $\bar{B} \leftarrow B * H_b$
12: $\bar{C} \leftarrow C * H_b$
13: **for all** image coordinates (u, v) **do** ▷ calc. corner response:
14: $Q(u,v) \leftarrow \bar{A}(u,v) \cdot \bar{B}(u,v) - \bar{C}^2(u,v) - \alpha \cdot [\bar{A}(u,v) + \bar{B}(u,v)]^2$

Step 2 – collect the corner points:

15: $\mathcal{C} \leftarrow (\,)$ ▷ start with an empty corner sequence
16: **for all** image coordinates (u, v) **do**
17: **if** $Q(u,v) > \mathrm{t}_H \wedge \mathsf{IsLocalMax}(Q, u, v)$ **then**
18: $\boldsymbol{c} \leftarrow \langle u, v, Q(u,v) \rangle$ ▷ create a new corner $\boldsymbol{c}$
19: $\mathcal{C} \leftarrow \mathcal{C} \smile (\boldsymbol{c})$ ▷ add $\boldsymbol{c}$ to corner sequence $\mathcal{C}$
20: $\mathcal{C}_{\text{clean}} \leftarrow \mathsf{CleanUpCorners}(\mathcal{C}, d_{\min})$
21: **return** $\mathcal{C}_{\text{clean}}$

22: **IsLocalMax**(Q, u, v) ▷ determine if $Q(u,v)$ is a local maximum
23: $\mathcal{N} \leftarrow \mathsf{GetNeighbors}(Q, u, v)$ ▷ se below
24: **return** $Q(u,v) > \max(\mathcal{N})$ ▷ true or false

25: **GetNeighbors**(Q, u, v)
Returns the 8 neighboring values around $Q(u,v)$.
26: $\mathcal{N} \leftarrow (Q(u{+}1,v), Q(u{+}1,v{-}1), Q(u,v{-}1), Q(u-1,v{-}1),$
$Q(u{-}1,v), Q(u{-}1,v{+}1), Q(u,v{+}1), Q(u{+}1,v{+}1))$
27: **return** $\mathcal{N}$

28: **CleanUpCorners**($\mathcal{C}, d_{\min}$)
29: $\mathsf{Sort}(\mathcal{C})$ ▷ sort $\mathcal{C}$ by desc. q_i (strongest corners first)
30: $\mathcal{C}_{\text{clean}} \leftarrow (\,)$ ▷ empty "clean" corner sequence
31: **while** $\mathcal{C}$ is not empty **do**
32: $\boldsymbol{c}_0 \leftarrow \mathsf{GetFirst}(\mathcal{C})$ ▷ get the strongest corner from $\mathcal{C}$
33: $\mathcal{C} \leftarrow \mathsf{Delete}(\boldsymbol{c}_0, \mathcal{C})$ ▷ the 1st element is removed from $\mathcal{C}$
34: $\mathcal{C}_{\text{clean}} \leftarrow \mathcal{C}_{\text{clean}} \smile (\boldsymbol{c}_0)$ ▷ add $\boldsymbol{c}_0$ to $\mathcal{C}_{\text{clean}}$
35: **for all** $\boldsymbol{c}_j$ in $\mathcal{C}$ **do**
36: **if** $\mathsf{Dist}(\boldsymbol{c}_0, \boldsymbol{c}_j) < d_{\min}$ **then**
37: $\mathcal{C} \leftarrow \mathsf{Delete}(\boldsymbol{c}_j, \mathcal{C})$ ▷ remove element $\boldsymbol{c}_j$ from $\mathcal{C}$
38: **return** $\mathcal{C}_{\text{clean}}$

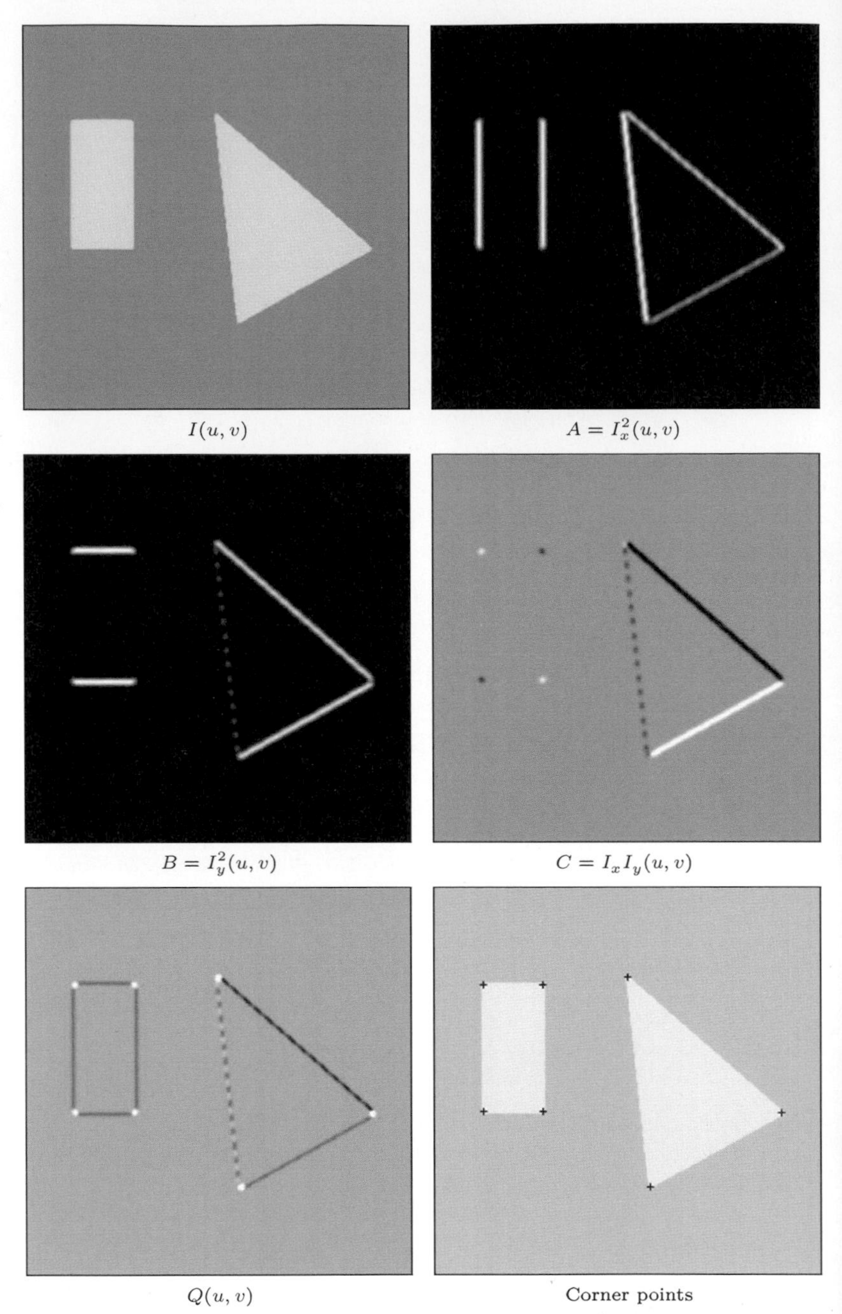

Fig. 7.1
Harris corner detector—Example 1. Starting with the original image $I(u, v)$, the first derivative is computed, and then from it the components of the structure matrix $\mathbf{M}(u, v)$, with $A(u, v) = I_x^2(u, v)$, $B = I_y^2(u, v)$, $C = I_x(u, v) \cdot I_y(u, v)$. $A(u, v)$ and $B(u, v)$ represent, respectively, the strength of the horizontal and vertical edges. In $C(u, v)$, the values are strongly positive (white) or strongly negative (black) only where the edges are strong in both directions (null values are shown in gray). The corner response function, $Q(u, v)$, exhibits noticeable positive peaks at the corner positions.

7.3 Implementation

Since the Harris detector algorithm is more complex than the algorithms we presented earlier, in the following sections we explain its implementation in greater detail. While reading the following you may wish to refer to the complete source code for the class `HarrisCornerDetector`, which is available online as part of the `imagingbook` library.[4]

[4] Package `imagingbook.pub.corners`.

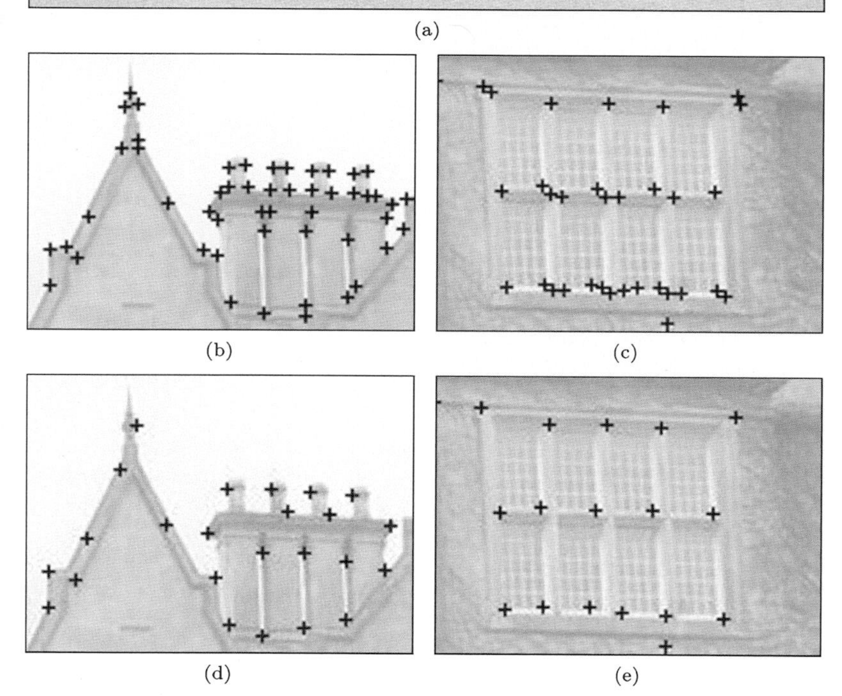

Fig. 7.2
Harris corner detector—Example 2. A complete result with the final corner points marked (a). After selecting the strongest corner points within a 10-pixel radius, only 335 of the original 615 candidate corners remain. Details *before* (b, c) and *after* selection (d, e).

7.3.1 Step 1: Calculating the Corner Response Function

To handle the range of the positive and negative values generated by the filters used in this step, we will need to use floating-point images to store the intermediate results, which also assures sufficient range and precision for small values. The kernels of the required filters, that is, the presmoothing filter H_p, the gradient filters H_{dx}, H_{dy}, and the smoothing filter for the structure matrix H_b, are defined as 1D `float` arrays:

```
1 float[] hp = {2f/9, 5f/9, 2f/9};
```

```
2 float[] hd = {0.5f, 0, -0.5f};
3 float[] hb =
4     {1f/64, 6f/64, 15f/64, 20f/64, 15f/64, 6f/64, 1f/64};
```

From the original 8-bit image (of type `ByteProcessor`), we first create two copies, `Ix` and `Iy`, of type `FloatProcessor`:

```
5   FloatProcessor Ix = I.convertToFloatProcessor();
6   FloatProcessor Iy = I.convertToFloatProcessor();
```

The first processing step is to presmooth the image with the 1D filter kernel `hp` ($= h_{px} = h_{py}^{\mathsf{T}}$, see Alg. 7.1, line 2). Subsequently the 1D gradient filter `hd` ($= h_{dx} = h_{dy}^{\mathsf{T}}$) is used to calculate the horizontal and vertical derivatives (see Alg. 7.1, line 3). To perform the convolution with the corresponding 1D kernels we use the (static) methods `convolveX()` and `convolveY()` defined in class `Filter`:[5]

```
7   Filter.convolveX(Ix, hp);          // I_x ← I_x * h_px
8   Filter.convolveX(Ix, hd);          // I_x ← I_x * h_dx
9   Filter.convolveY(Iy, hp);          // I_y ← I_y * h_py
10  Filter.convolveY(Iy, hd);          // I_y ← I_y * h_dy
```

Now the components $A(u,v)$, $B(u,v)$, $C(u,v)$ of the structure matrix $\mathbf{M}$ are calculated for all image positions (u,v):

```
11  A = ImageMath.sqr(Ix);             // A(u,v) ← I_x^2(u,v)
12  B = ImageMath.sqr(Iy);             // B(u,v) ← I_y^2(u,v)
13  C = ImageMath.mult(Ix, Iy);        // C(u,v) ← I_x(u,v) · I_y(u,v)
14
```

The components of the structure matrix are then smoothed with a separable filter kernel $H_b = h_{bx} * h_{by}$:

```
15  Filter.convolveXY(A, hb);          // A ← (A * h_bx) * h_by
16  Filter.convolveXY(B, hb);          // B ← (B * h_bx) * h_by
17  Filter.convolveXY(C, hb);          // C ← (C * h_bx) * h_by
```

The variables `A`, `B`, `C` of type `FloatProcessor` are declared in the class `HarrisCornerDetector`. `sqr()` and `mult()` are static methods of class `ImageMath` for squaring an image and multiplying two images, respectively. The method `convolveXY(I, h)` is used to apply a x/y-separable 2D convolution with the 1D kernel `h` to the image `I`.

Finally, the corner response function (Alg. 7.1, line 14) is calculated by the method `makeCrf()` and a new image (of type `FloatProcessor`) is created:

```
18 private FloatProcessor  makeCrf(float alpha) {
19   FloatProcessor Q = new FloatProcessor(M, N);
20   final float[] pA = (float[]) A.getPixels();
21   final float[] pB = (float[]) B.getPixels();
22   final float[] pC = (float[]) C.getPixels();
23   final float[] pQ = (float[]) Q.getPixels();
24   for (int i = 0; i < M * N; i++) {
25     float a = pA[i], b = pB[i], c = pC[i];
26     float det = a * b - c * c;   // det(M̄)
27     float trace = a + b;             // trace(M̄)
28     pQ[i] = det - alpha * (trace * trace);
```

[5] Package `imagingbook.lib.image`.

```
29    }
30    return Q;
31 }
```

7.3.2 Step 2: Selecting "Good" Corner Points

The result of the first stage of Alg. 7.1 is the corner response function $Q(u, v)$, which in our implementation is stored as a floating-point image (`FloatProcessor`). In the second stage, the dominant corner points are selected from Q. For this we need (a) an object type to describe the corners and (b) a flexible container, in which to store these objects. In this case, the container should be a dynamic data structure since the number of objects to be stored is not known beforehand.

The `Corner` class

Next we define a new class `Corner`[6] to represent individual corner points $\boldsymbol{c} = \langle x, y, q \rangle$ and a single constructor (in line 35) with `float` parameters x, y for the position and corner strength q:

```
32 public class Corner implements Comparable<Corner> {
33    final float x, y, q;
34
35    public Corner (float x, float y, float q) {
36      this.x = x;
37      this.y = y;
38      this.q = q;
39    }
40
41    public int compareTo (Corner c2) {
42      if (this.q > c2.q) return -1;
43      if (this.q < c2.q) return 1;
44      else return 0;
45    }
46    ...
47 }
```

The class `Corner` implements Java's `Comparable` interface, such that objects of type `Corner` can be compared with each other and thereby sorted into an ordered sequence. The `compareTo()` method required by the `Comparable` interface is defined (in line 41) to sort corners by descending `q` values.

Choosing a suitable container

In Alg. 7.1, we used the notion of a *sequence* or *lists* to organize and manipulate the collections of potential corner points generated at various stages. One solution would be to utilize *arrays*, but since the size of arrays must be declared before they are used, we would have to allocate memory for extremely large arrays in order to store all the possible corner points that might be identified. Instead, we make use of the `ArrayList` class, which is one of many dynamic data structures conveniently provided by Java's *Collections Framework*.[7]

[6] Package `imagingbook.pub.corners`.
[7] Package `java.util`.

The collectCorners() method

The method `collectCorners()` outlined here selects the dominant corner points from the corner response function $Q(u, v)$. The parameter *border* specifies the width of the image's border, within which corner points should be ignored.

```
48 List<Corner> collectCorners(FloatProcessor Q, float tH, int
       border) {
49   List<Corner> C = new ArrayList<Corner>();
50   for (int v = border; v < N - border; v++) {
51     for (int u = border; u < M - border; u++) {
52       float q = Q.getf(u, v);
53       if (q > tH && isLocalMax(Q, u, v)) {
54         Corner c = new Corner(u, v, q);
55         C.add(c);
56       }
57     }
58   }
59   return C;
60 }
```

First (in line 49), a new instance of `ArrayList`[8] is created and assigned to the variable `C`. Then the CRF image `Q` is traversed, and when a potential corner point is located, a new `Corner` is instantiated (line 54) and added to `C` (line 55). The Boolean method `isLocalMax()` (defined in class `HarrisCornerDetector`) determines if the 2D function `Q` is a local maximum at the given position `u, v`:

```
61 boolean isLocalMax (FloatProcessor Q, int u, int v) {
62   if (u <= 0 || u >= M - 1 || v <= 0 || v >= N - 1) {
63     return false;
64   }
65   else {
66     float[] q = (float[]) Q.getPixels();
67     int i0 = (v - 1) * M + u;
68     int i1 = v * M + u;
69     int i2 = (v + 1) * M + u;
70     float q0 = q[i1];
71     return  // check 8 neighbors of q0:
72       q0 >= q[i0 - 1] && q0 >= q[i0] && q0 >= q[i0 + 1] &&
73       q0 >= q[i1 - 1] &&                    q0 >= q[i1 + 1]
      &&
74       q0 >= q[i2 - 1] && q0 >= q[i2] && q0 >= q[i2 + 1] ;
75   }
76 }
```

7.3.3 Step 3: Cleaning up

The final step is to remove the weakest corners in a limited area where the size of this area is specified by the radius $d_{\min}$ (Alg. 7.1, lines 29–38). This process is outlined in Fig. 7.3 and implemented by the following method `cleanupCorners()`.

[8] The specification `ArrayList<Corner>` indicates that the list `C` may only contain objects of type `Corner`.

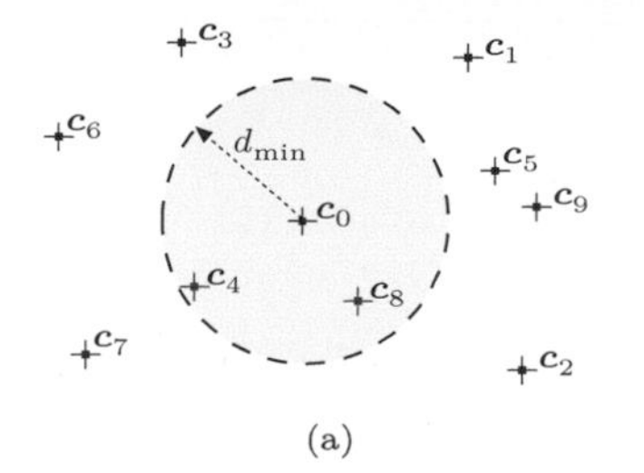

(a)

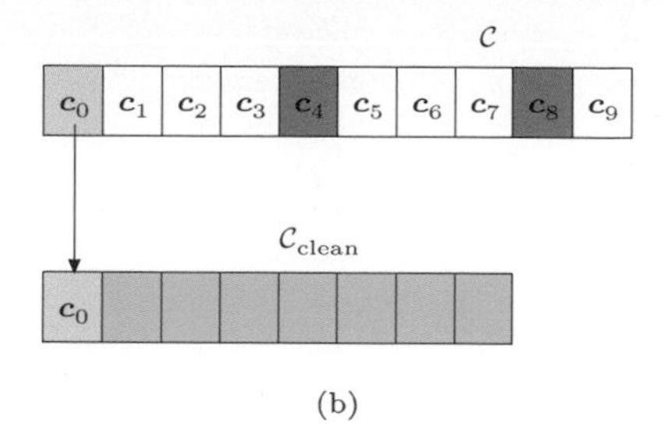

(b)

Fig. 7.3
Selecting the strongest corners within a given spatial distance. (a) Sample corner positions in the 2D plane. (b) The original list of corners ($\mathcal{C}$) is sorted by "corner strength" (q) in descending order; that is, $\boldsymbol{c}_0$ is the strongest corner. First, corner $\boldsymbol{c}_0$ is added to a new list $\mathcal{C}_{clean}$, while the weaker corners $\boldsymbol{c}_4$ and $\boldsymbol{c}_8$ (which are both within distance d_{min} from $\boldsymbol{c}_0$) are removed from the original list $\mathcal{C}$. The following corners $\boldsymbol{c}_1, \boldsymbol{c}_2, \ldots$ are treated similarly until no more elements remain in $\mathcal{C}$. None of the corners in the resulting list $\mathcal{C}_{clean}$ is closer to another corner than d_{min}.

```
77 List<Corner> cleanupCorners(List<Corner> C, double dmin) {
78   double dmin2 = dmin * dmin;
79   // sort corners by descending q-value:
80   Collections.sort(C);
81   // we use an array of corners for efficiency reasons:
82   Corner[] Ca = C.toArray(new Corner[C.size()]);
83   List<Corner> Cclean = new ArrayList<Corner>(C.size());
84   for (int i = 0; i < Ca.length; i++) {
85     Corner c0 = Ca[i];        // get next strongest corner
86     if (c0 != null) {
87       Cclean.add(c0);
88       // delete all remaining corners cj too close to c0:
89       for (int j = i + 1; j < Ca.length; j++) {
90         Corner cj = Ca[j];
91         if (cj != null && c0.dist2(cj) < dmin2)
92           Ca[j] = null;       //delete corner cj from Ca
93       }
94     }
95   }
96   return Cclean;
97 }
```

Initially (in line 80) the corner list `C` is sorted by decreasing corner strenth q by calling the static method `sort()`.[9] The sorted sequence is then converted to an array (line 82) which is traversed from start to end (line 84–95). For each selected corner (`c0`), all subsequent corners (`cj`) with a distance d_{min} are deleted from the sequence (line 92). The "surviving" corners are then transferred to the final corner sequence `Cclean`.

Note that the call `c0.dist2(cj)` in line 91 returns the *squared* Euclidean distance between the corner points $\boldsymbol{c}_0$ and $\boldsymbol{c}_j$, that is, the quantity $d^2 = (x_0 - x_j)^2 + (y_0 - y_j)^2$. Since the square of the distance suffices for the comparison, we do not need to compute the actual distance, and consequently we avoid calling the expensive square root function. This is a common trick when comparing distances.

7.3.4 Summary

Most of the implementation steps we have just described are initiated through calls from the method `findCorners()` in class `HarrisCornerDetector`:

```
98 public List<Corner> findCorners() {
```

[9] Defined in class `java.util.Collections`.

```
 99   FloatProcessor Q = makeCrf((float)params.alpha);
100   List<Corner> corners =
101       collectCorners(Q, (float)params.tH, params.border);
102   if (params.doCleanUp) {
103     corners = cleanupCorners(corners, params.dmin);
104   }
105   return corners;
106 }
```

An example of how to use the class `HarrisCornerDetector` is shown by the associated ImageJ plugin `Find_Corners` whose `run()` consists of only a few lines of code. This method simply creates a new object of the class `HarrisCornerDetector`, calls the `findCorners()` method, and finally displays the results in a new image (`R`):

```
107 public class Find_Corners implements PlugInFilter {
108
109   public void run(ImageProcessor ip) {
110     HarrisCornerDetector cd = new HarrisCornerDetector(ip);
111     List<Corner> corners = cd.findCorners();
112     ColorProcessor R = ip.convertToColorProcessor();
113     drawCorners(R, corners);
114     (new ImagePlus("Result", R)).show();
115   }
116
117   void drawCorners(ImageProcessor ip,
118                    List<Corner> corners) {
119     ip.setColor(cornerColor);
120     for (Corner c : corners) {
121       drawCorner(ip, c);
122     }
123   }
124
125   void drawCorner(ImageProcessor ip, Corner c) {
126     int size = cornerSize;
127     int x = Math.round(c.getX());
128     int y = Math.round(c.getY());
129     ip.drawLine(x - size, y, x + size, y);
130     ip.drawLine(x, y - size, x, y + size);
131   }
132 }
```

For completeness, the definition of the `drawCorners()` method has been included here; the complete source code can be found online. Again, when writing this code, the focus is on understandability and not necessarily speed and memory usage. Many elements of the code can be optimized with relatively little effort (perhaps as an exercise?) if efficiency becomes important.

7.4 Exercises

Exercise 7.1. Adapt the `draw()` method in the class `Corner` (see p. 155) so that the strength (q-value) of the corner points can also be visualized. This could be done, for example, by manipulating the size, color, or intensity of the markers drawn in relation to the strength of the corner.

Exercise 7.2. Conduct a series of experiments to determine how image contrast affects the performance of the Harris detector, and then develop an idea for how you might automatically determine the parameter t_H depending on image content.

Exercise 7.3. Explore how rotation and distortion of the image affect the performance of the Harris corner detector. Based on your experiments, is the operator truly isotropic?

Exercise 7.4. Determine how image noise affects the performance of the Harris detector in terms of the positional accuracy of the detected corners and the omission of actual corners. Remark: ImageJ's menu command Process ▷ Noise ▷ Add Specified Noise... can be used to easily add certain types of random noise to a given image.

8

Finding Simple Curves: The Hough Transform

In Chapter 6 we demonstrated how to use appropriately designed filters to detect edges in images. These filters compute both the edge strength and orientation at every position in the image. In the following sections, we explain how to decide (e.g., by using a threshold operation on the edge strength) if a curve is actually present at a given image location. The result of this process is generally represented as a binary *edge map*. Edge maps are considered preliminary results, since with an edge filter's limited ("myopic") view it is not possible to accurately ascertain if a point belongs to a true edge. Edge maps created using simple threshold operations contain many edge points that do not belong to true edges (false positives), and, on the other hand, many edge points are not detected and hence are missing from the map (false negatives).

8.1 Salient Image Structures

An intuitive approach to locating large image structures is to first select an arbitrary edge point, systematically examine its neighboring pixels and add them if they belong to the object's contour, and repeat. In principle, such an approach could be applied to either a continuous edge map consisting of edge strengths and orientations or a simple binary *edge map*. Unfortunately, with either input, such an approach is likely to fail due to image noise and ambiguities that arise when trying to follow the contours. Additional constraints and information about the type of object sought are needed in order to handle pixel-level problems such as branching, as well as interruptions. This type of local sequential *contour tracing* makes for an interesting optimization problem [128] (see also Sec. 10.2).

A completely different approach is to search for globally apparent structures that consist of certain simple shape features. As an example, Fig. 8.1 shows that certain structures are readily apparent to the human visual system, even when they overlap in noisy images. The biological basis for why the human visual system spontaneously

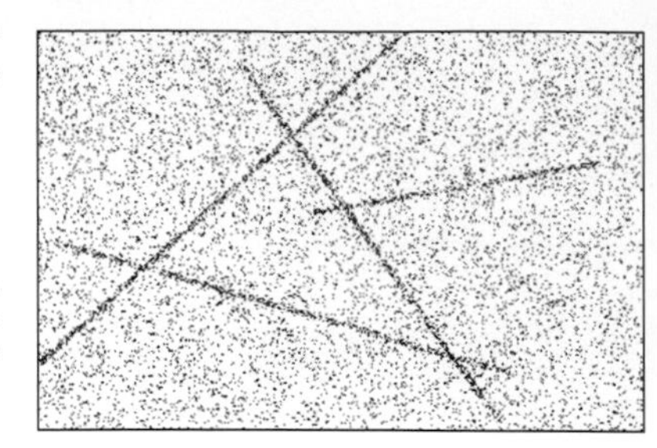

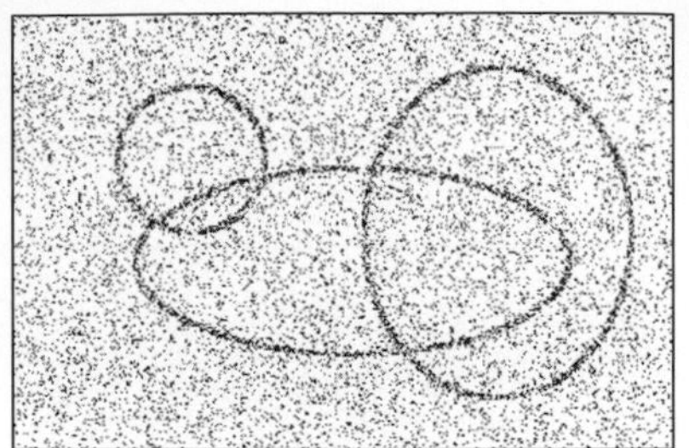

Fig. 8.1 The human visual system is capable of instantly recognizing prominent image structures even under difficult conditions.

recognizes four lines or three ellipses in Fig. 8.1 instead of a larger number of disjoint segments and arcs is not completely known. At the cognitive level, theories such as "Gestalt" grouping have been proposed to address this behavior. The next sections explore one technique, the Hough transform, that provides an algorithmic solution to this problem.

8.2 The Hough Transform

The method from Paul Hough—originally published as a US Patent [111] and often referred to as the "Hough transform" (HT)—is a general approach to localizing any shape that can be defined parametrically within a distribution of points [64, 117]. For example, many geometrical shapes, such as lines, circles, and ellipses, can be readily described using simple equations with only a few parameters. Since simple geometric forms often occur as part of man-made objects, they are especially useful features for analysis of these types of images (Fig. 8.2).

The Hough transform is perhaps most often used for detecting straight line segments in edge maps. A line segment in 2D can be described with two real-valued parameters using the classic slope-intercept form

$$y = k \cdot x + d, \tag{8.1}$$

Fig. 8.2 Simple geometrical forms such as sections of lines, circles, and ellipses are often found in man-made objects.

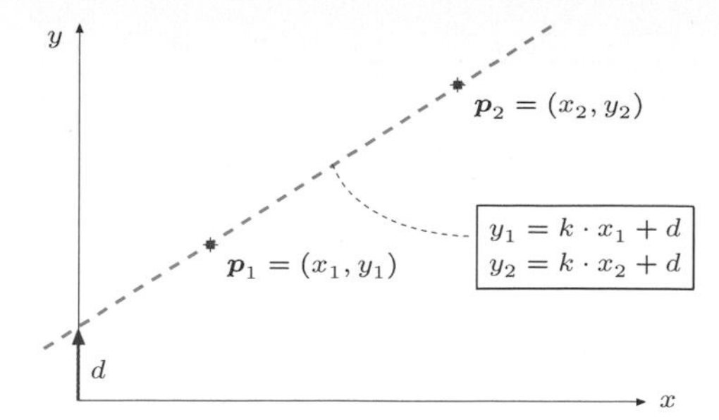

Fig. 8.3
Two points, p_1 and p_2, lie on the same line when $y_1 = kx_1 + d$ and $y_2 = kx_2 + d$ for a particular pair of parameters k and d.

where k is the slope and d the intercept—that is, the height at which the line would intercept the y axis (Fig. 8.3). A line segment that passes through two given edge points $p_1 = (x_1, y_1)$ and $p_2 = (x_2, y_2)$ must satisfy the conditions

$$y_1 = k \cdot x_1 + d \qquad \text{and} \qquad y_2 = k \cdot x_2 + d, \tag{8.2}$$

for $k, d \in \mathbb{R}$. The goal is to find values of k and d such that as many edge points as possible lie on the line they describe; in other words, the line that fits the most edge points. But how can you determine the number of edge points that lie on a given line segment? One possibility is to exhaustively "draw" every possible line segment into the image while counting the number of points that lie exactly on each of these. Even though the discrete nature of pixel images (with only a finite number of different lines) makes this approach possible in theory, generating such a large number of lines is infeasible in practice.

8.2.1 Parameter Space

The Hough transform approaches the problem from another direction. It examines all the possible line segments that run through a single given point in the image. Every line $L_j = \langle k_j, d_j \rangle$ that runs through a point $p_0 = (x_0, y_0)$ must satisfy the condition

$$L_j \colon y_0 = k_j x_0 + d_j \tag{8.3}$$

for suitable values k_j, d_j. Equation 8.3 is underdetermined and the possible solutions for k_j, d_j correspond to an infinite set of lines passing through the given point p_0 (Fig. 8.4). Note that for a given k_j, the solution for d_j in Eqn. (8.3) is

$$d_j = -x_0 \cdot k_j + y_0, \tag{8.4}$$

which is another equation for a line, where now k_j, d_j are the *variables* and x_0, y_0 are the constant *parameters* of the equation. The solution set $\{(k_j, d_j)\}$ of Eqn. (8.4) describes the parameters of all possible lines L_j passing through the image point $p_0 = (x_0, y_0)$.

For an *arbitrary* image point $p_i = (x_i, y_i)$, Eqn. (8.4) describes the line

$$M_i : d = -x_i \cdot k + y_i \tag{8.5}$$

with the parameters $-x_i, y_i$ in the so-called *parameter* or *Hough space*, spanned by the coordinates k, d. The relationship between

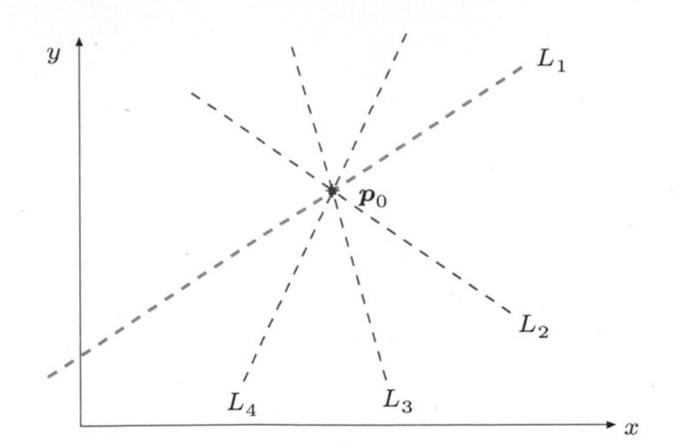

Fig. 8.4
A set of lines passing through an image point. For all possible lines L_j passing through the point $p_0 = (x_0, y_0)$, the equation $y_0 = k_j x_0 + d_j$ holds for appropriate values of the parameters k_j, d_j.

(x, y) *image* space and (k, d) *parameter* space can be summarized as follows:

	Image Space (x, y)		*Parameter Space* (k, d)	
Point	$\boldsymbol{p}_i = (x_i, y_i)$	$\longleftrightarrow$	$M_i\colon d = -x_i \cdot k + y_i$	Line
Line	$L_j\colon y = k_j \cdot x + d_j$	$\longleftrightarrow$	$\boldsymbol{q}_j = (k_j, d_j)$	Point

Each image point $\boldsymbol{p}_i$ and its associated line bundle correspond to exactly one line M_i in parameter space. Therefore we are interested in those places in the parameter space where lines *intersect*. The example in Fig. 8.5 illustrates how the lines M_1 and M_2 intersect at the position $\boldsymbol{q}_{12} = (k_{12}, d_{12})$ in the parameter space, which means (k_{12}, d_{12}) are the parameters of the line in the image space that runs through both image points $\boldsymbol{p}_1$ and $\boldsymbol{p}_2$. The more lines M_i that intersect at a single point in the parameter space, the more image space points lie on the corresponding line in the image! In general, we can state:

> If N lines intersect at position (k', d') in *parameter space*, then N image points lie on the corresponding line $y = k'x + d'$ in *image space.*

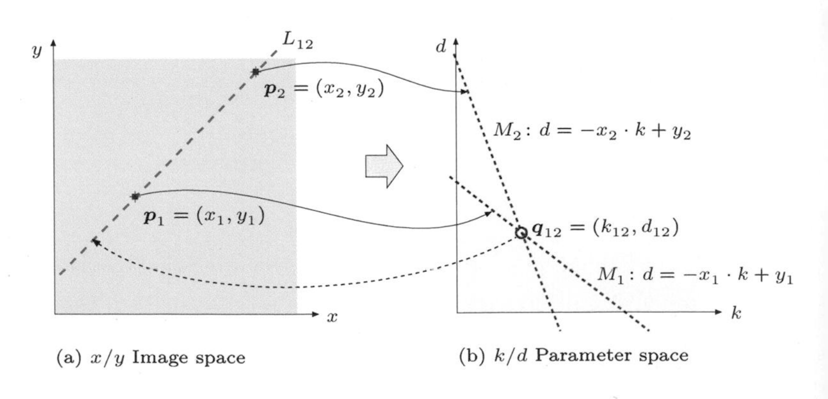

(a) x/y Image space

(b) k/d Parameter space

Fig. 8.5
Relationship between image space and parameter space. The parameter values for all possible lines passing through the image point $\boldsymbol{p}_i = (x_i, y_i)$ in image space (a) lie on a single line M_i in parameter space (b). This means that each point $\boldsymbol{q}_j = (k_j, d_j)$ in parameter space corresponds to a single line L_j in image space. The intersection of the two lines M_1, M_2 at the point $\boldsymbol{q}_{12} = (k_{12}, d_{12})$ in parameter space indicates that a line L_{12} through the two points k_{12} and d_{12} exists in the image space.

8.2.2 Accumulator Map

Finding the dominant lines in the image can now be reformulated as finding all the locations in parameter space where a significant number of lines intersect. This is basically the goal of the HT. In order

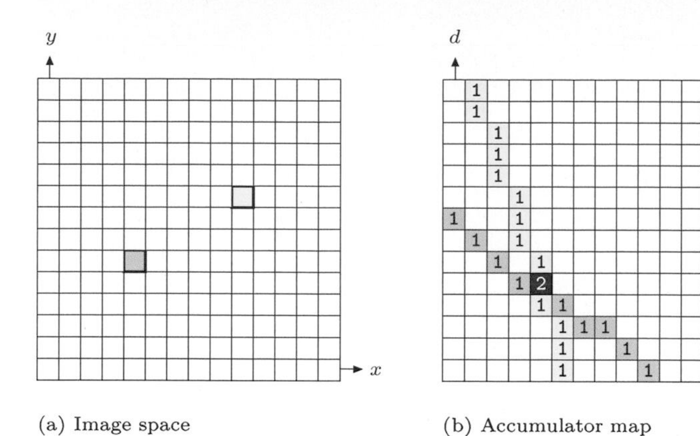

(a) Image space (b) Accumulator map

Fig. 8.6
The accumulator map is a discrete representation of the parameter space (k, d). For each image point found (a), a discrete line in the parameter space (b) is drawn. This operation is performed *additively* so that the values of the array through which the line passes are incremented by 1. The value at each cell of the accumulator array is the number of parameter space lines that intersect it (in this case 2).

to compute the HT, we must first decide on a discrete representation of the continuous parameter space by selecting an appropriate step size for the k and d axes. Once we have selected step sizes for the coordinates, we can represent the space naturally using a 2D array. Since the array will be used to keep track of the number of times parameter space lines intersect, it is called an "accumulator" array. Each parameter space line is painted into the accumulator array and the cells through which it passes are incremented, so that ultimately each cell accumulates the total number of lines that intersect at that cell (Fig. 8.6).

8.2.3 A Better Line Representation

The line representation in Eqn. (8.1) is not used in practice because for vertical lines the slope is infinite, that is, $k = \infty$. A more practical representation is the so-called *Hessian normal form* (HNF)[1] for representing lines,

$$x \cdot \cos(\theta) + y \cdot \sin(\theta) = r, \tag{8.6}$$

which does not exhibit such singularities and also provides a natural linear quantization for its parameters, the angle θ and the radius r (Fig. 8.7).

With the HNF representation, the parameter space is defined by the coordinates θ, r, and a point $\boldsymbol{p} = (x, y)$ in image space corresponds to the relation

$$r(\theta) = x \cdot \cos(\theta) + y \cdot \sin(\theta), \tag{8.7}$$

for angles in the range $0 \leq \theta < \pi$ (see Fig. 8.8). Thus, for a given image point $\boldsymbol{p}$, the associated radius r is simply a function of the angle θ. If we use the center of the image (of size $M \times N$),

$$\boldsymbol{x}_r = \begin{pmatrix} x_r \\ y_r \end{pmatrix} = \frac{1}{2} \cdot \begin{pmatrix} M \\ N \end{pmatrix}, \tag{8.8}$$

[1] The Hessian normal form is a normalized version of the general ("algebraic") line equation $Ax + By + C = 0$, with $A = \cos(\theta)$, $B = \sin(\theta)$, and $C = -r$ (see, e.g., [35, p. 194]).

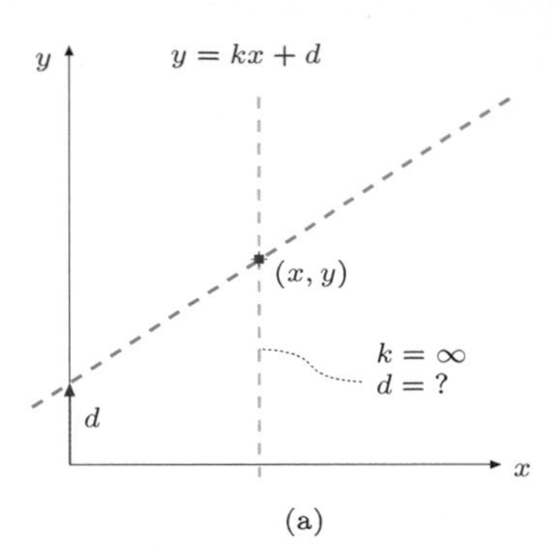

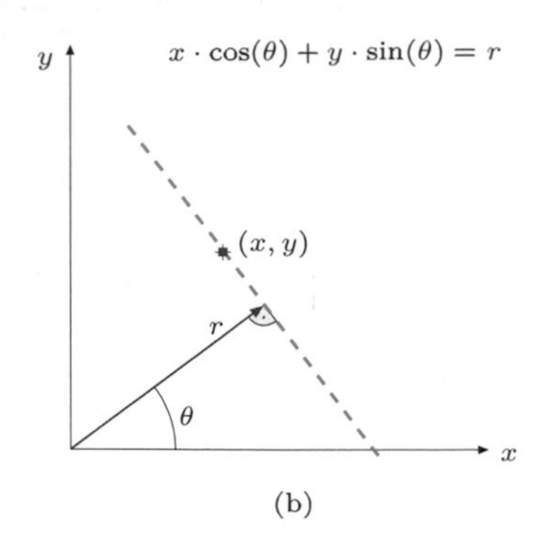

Fig. 8.7 Representation of lines in 2D. In the common k, d representation (a), vertical lines pose a problem because $k = \infty$. The Hessian normal form (b) avoids this problem by representing a line by its angle θ and distance r from the origin.

as the reference point for the x/y image coordinates, then it is possible to limit the range of the radius to half the diagonal of the image, that is,

$$-r_{\max} \leq r(\theta) \leq r_{\max}, \qquad \text{with} \qquad r_{\max} = \tfrac{1}{2}\sqrt{M^2 + N^2}. \tag{8.9}$$

We can see that the function $r(\theta)$ in Eqn. (8.7) is the sum of a cosine and a sine function on θ, each being weighted by the x and y coordinates of the image point (assumed to be constant for the moment). The result is again a sinusoidal function whose magnitude and phase depend only on the weights (coefficients) x, y. Thus, with the Hessian parameterization θ/r, an image point (x, y) does not create a straight line in the accumulator map $\mathsf{A}(i, j)$ but a unique sinusoidal curve, as shown in Fig. 8.8. Again, each image point adds a curve to the accumulator and each resulting cluster point corresponds to to a dominant line in the image with a proportional number of points on it.[2]

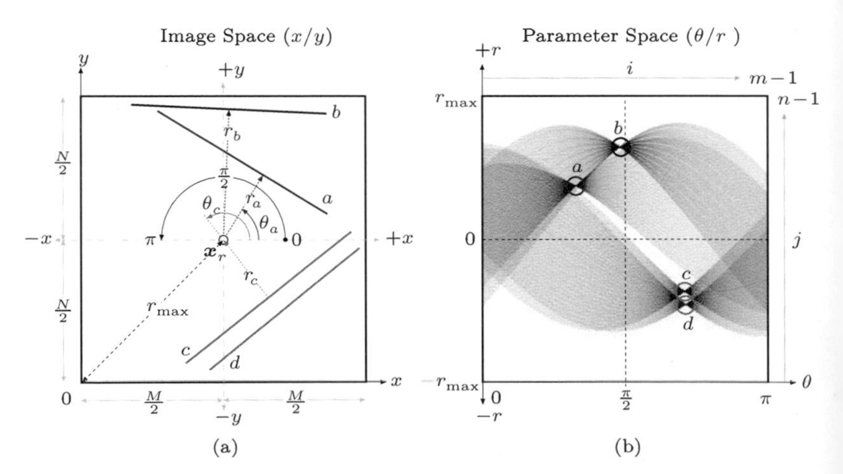

Fig. 8.8 Image space and parameter space using the HNF representation. The image (a) of size $M \times N$ contains four straight lines $L_a, \ldots, L_d$. Each point on an image line creates a sinusoidal curve in the θ/r parameter space (b) and the corresponding line parameters are indicated by the clearly visible cluster points in the accumulator map. The reference point $\boldsymbol{x}_r$ for the x/y coordinates lies at the center of the image. The line angles θ_i are in the range $[0, \pi)$ and the associated radii r_i are in $[-r_{\max}, r_{\max}]$ (the length $r_{\max}$ is half of the image diagonal). For example, the the angle θ_a of line L_a is approximately $\pi/3$, with the (positive) radius $r_a \approx 0.4\, r_{\max}$. Note that, with this parameterization, line L_c has the angle $\theta_c \approx 2\pi/3$ and the *negative* radius $r_c \approx -0.4\, r_{\max}$.

[2] Note that, in Fig. 8.8(a), the positive direction of the y-coordinate runs *upwards* (unlike our usual convention for image coordinates) to stay in line with the previous illustrations (and high school geometry). In practice, the consequences are minor: only the rotation angle runs in the opposite direction and thus the accumulator image in Fig. 8.8(b) was mirrored horizontally for proper display.

8.3 Hough Algorithm

The fundamental Hough algorithm using the HNF line representation (Eqn. (8.6)) is given in Alg. 8.1. Starting with a binary image $I(u, v)$ where the edge pixels have been assigned a value of 1, the first stage creates a 2D accumulator array and then iterates over the image to fill it. The resulting increments are

$$d_\theta = \pi/m \qquad \text{and} \qquad d_r = \sqrt{M^2 + N^2}/n \tag{8.10}$$

for the angle θ and the radius r, respectively. The discrete indices of the accumulators cells are denoted i and j, with $j_0 = n \div 2$ as the center index (for $r = 0$).

For each relevant image point (u, v), a sinusoidal curve is added to the accumulator map by stepping over the discrete angles $\theta_i = \theta_0, \ldots, \theta_{m-1}$, calculating the corresponding radius[3]

$$r(\theta_i) = (u - x_r) \cdot \cos(\theta_i) + (v - y_r) \cdot \sin(\theta_i) \tag{8.11}$$

(see Eqn. (8.7)) and its discrete index

$$j = j_0 + \operatorname{round}\left(\frac{r(\theta_i)}{d_r}\right), \tag{8.12}$$

and subsequently incrementing the accumulator cell $\mathsf{A}(i, j)$ by one (see Alg. 8.1, lines 10–17). The line parameters θ_i and r_j for a given accumulator position (i, j) can be calculated as

$$\theta_i = i \cdot d_\theta \qquad \text{and} \qquad r_j = (j - j_0) \cdot d_r. \tag{8.13}$$

In the second stage of Alg. 8.1, the accumulator array is searched for local peaks above a given minimum Values $a_{\min}$. For each detected peak, a line object is created of the form

$$L_k = \langle \theta_k, r_k, a_k \rangle, \tag{8.14}$$

consisting of the angle θ_k, the radius r_k (relative to the reference point $\boldsymbol{x}_r$), and the corresponding accumulator value a_k. The resulting sequence of lines $\mathcal{L} = (L_1, L_2, \ldots)$ is then sorted by descending a_k and returned.

Figure 8.9 shows the result of applying the Hough transform to a very noisy binary image, which obviously contains four straight lines. They appear clearly as cluster points in the corresponding accumulator map in Fig. 8.9 (b). Figure 8.9 (c) shows the reconstruction of these lines from the extracted parameters. In this example, the resolution of the discrete parameter space is set to 256×256.[4]

[3] The frequent (and expensive) calculation of $\cos(\theta_i)$ and $\sin(\theta_i)$ in Eqn. (8.11) and Alg. 8.1 (line 15) can be easily avoided by initially tabulating the function values for all m possible angles $\theta_i = \theta_0, \ldots, \theta_{m-1}$, which should yield a significant performance gain.

[4] Note that *drawing* a straight line given in Hessian normal form is not really a trivial task (see Excercises 8.1–8.2 for details).

Alg. 8.1
Hough algorithm for detecting straight lines. The algorithm returns a sorted list of straight lines of the form $L_k = \langle \theta_k, r_k, a_k \rangle$ for the binary input image I of size $M \times N$. The resolution of the discrete Hough accumulator map (and thus the step size for the angle and radius) is specified by parameters m and n, respectively. $a_{\min}$ defines the minimum accumulator value, that is, the minimum number of image point on any detected line. The function IsLocalMax() used in line 20 is the same as in Alg. 7.1 (see p. 151).

```
1: HoughTransformLines(I, m, n, a_min)
   Input: I, a binary image of size M × N; m, angular accumulator
   steps; n, radial accumulator steps; a_min, minimum accumulator
   count per line. Returns a sorted sequence 𝓛 = (L_1, L_2, ...) of the
   most dominant lines found.
2:    (M, N) ← Size(I)
3:    (x_r, y_r) ← 1/2 · (M, N)           ▷ reference point x_r (image center)
4:    d_θ ← π/m                                      ▷ angular step size
5:    d_r ← √(M² + N²)/n                              ▷ radial step size
6:    j_0 ← n ÷ 2                                 ▷ map index for r = 0
      Step 1 – set up and fill the Hough accumulator:
7:    Create map A: [0, m−1] × [0, n−1] ↦ ℤ               ▷ accumulator
8:    for all accumulator cells (i, j) do
9:        A(i, j) ← 0                           ▷ initialize accumulator
10:   for all (u, v) ∈ M×N do                           ▷ scan the image
11:       if I(u, v) > 0 then               ▷ I(u, v) is a foreground pixel
12:           (x, y) ← (u−x_r, v−y_r)                ▷ shift to reference
13:           for i ← 0, ..., m−1 do              ▷ angular coordinate i
14:               θ ← d_θ · i                          ▷ angle, 0 ≤ θ < π
15:               r ← x · cos(θ) + y · sin(θ)              ▷ see Eqn. 8.7
16:               j ← j_0 + round(r/d_r)            ▷ radial coordinate j
17:               A(i, j) ← A(i, j) + 1               ▷ increment A(i, j)
      Step 2 – extract the most dominant lines:
18:   𝓛 ← ()                      ▷ start with empty sequence of lines
19:   for all accumulator cells (i, j) do          ▷ collect local maxima
20:       if (A(i, j) ≥ a_min) ∧ IsLocalMax(A, i, j) then
21:           θ ← i · d_θ                                      ▷ angle θ
22:           r ← (j − j_0) · d_r                              ▷ radius r
23:           a ← A(i, j)                         ▷ accumulated value a
24:           L ← ⟨θ, r, a⟩                         ▷ create a new line L
25:           𝓛 ← 𝓛 ⌣ (L)                     ▷ add line L to sequence 𝓛
26:   Sort(𝓛)             ▷ sort 𝓛 by descending accumulator count a
27:   return 𝓛
```

8.3.1 Processing the Accumulator Array

The reliable detection and precise localization of peaks in the accumulator map $\mathsf{A}(i, j)$ is not a trivial problem. As can readily be seen in Fig. 8.9(b), even in the case where the lines in the image are geometrically "straight", the parameter space curves associated with them do not intersect at *exactly* one point in the accumulator array but rather their intersection points are distributed within a small area. This is primarily caused by the rounding errors introduced by the discrete coordinate grid used for the accumulator array. Since the maximum points are really maximum *areas* in the accumulator array, simply traversing the array and returning the positions of its largest values is not sufficient. Since this is a critical step in the algorithm, we examine two different approaches below (see Fig. 8.10).

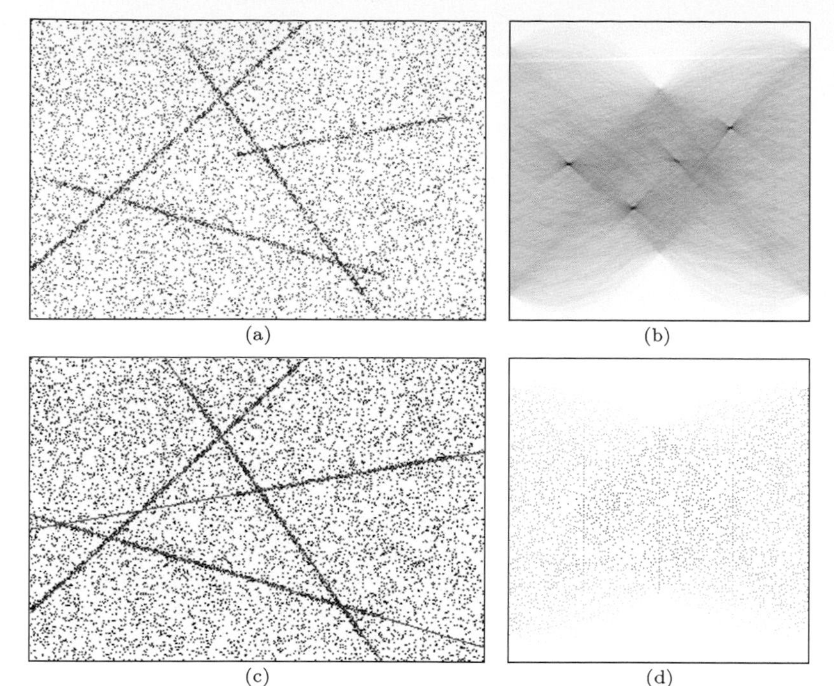

Fig. 8.9
Hough transform for straight lines. The dimensions of the original image (a) are 360×240 pixels, so the maximal radius (measured from the image center) is $r_{\max} \approx 216$. For the parameter space (b), a step size of 256 is used for both the angle $\theta = 0, \ldots, \pi$ (horizontal axis) and the radius $r = -r_{\max}, \ldots, r_{\max}$ (vertical axis). The four (dark) clusters in (b) surround the maximum values in the accumulator array, and their parameters correspond to the four lines in the original image. Intensities are shown inverted in all images to improve legibility.

Approach A: Thresholding

First the accumulator is thresholded to the value of t_a by setting all accumulator values $\mathsf{A}(i,j) < t_a$ to 0. The resulting scattering of points, or point clouds, are first coalesced into regions (Fig. 8.10(b)) using a technique such as a morphological *closing* operation (see Sec. 9.3.2). Next the remaining regions must be localized, for instance using the region-finding technique from Sec. 10.1, and then each region's centroid (see Sec. 10.5) can be utilized as the (noninteger) coordinates for the potential image space line. Often the sum of the accumulator's values within a region is used as a measure of the strength (number of image points) of the line it represents.

Approach B: Nonmaximum suppression

In this method, local maxima in the accumulator array are found by suppressing nonmaximal values.[5] This is carried out by determining for every accumulator cell $\mathsf{A}(i,j)$ whether the value is higher than the value of all of its neighboring cells. If this is the case, then the value remains the same; otherwise it is set to 0 (Fig. 8.10(c)). The (integer) coordinates of the remaining peaks are potential line parameters, and their respective heights correlate with the strength of the image space line they represent. This method can be used in conjunction with a threshold operation to reduce the number of candidate points that must be considered. The result for Fig. 8.9(a) is shown in Fig. 8.10(d).

[5] Nonmaximum suppression is also used in Sec. 7.2.3 for isolating corner points.

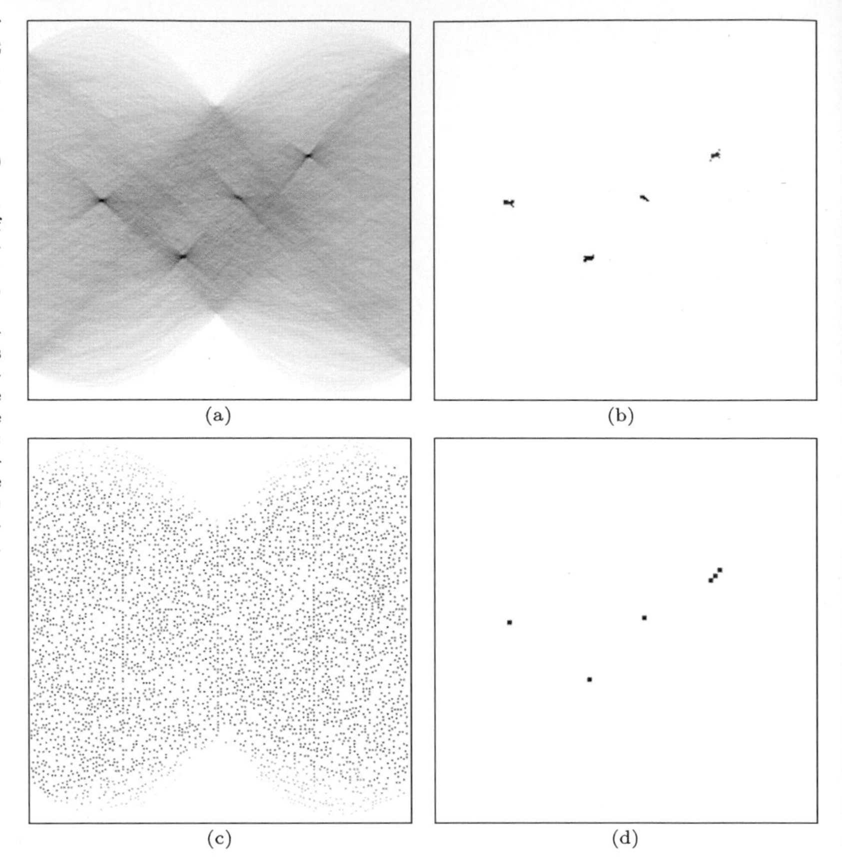

Fig. 8.10 Finding local maximum values in the accumulator array. Original distribution of the values in the Hough accumulator (a). **Variant A:** *Threshold operation* using 50% of the maximum value (b). The remaining regions represent the four dominant lines in the image, and the coordinates of their centroids are a good approximation to the line parameters. **Variant B:** Using *non-maximum suppression* results in a large number of local maxima (c) that must then be reduced using a threshold operation (d).

Mind the vertical lines!

Special consideration should be given to *vertical* lines (once more!) when processing the contents of the accumulator map. The parameter pairs for these lines lie near $\theta = 0$ and $\theta = \pi$ at the left and right borders, respectively, of the accumulator map (see Fig. 8.8(b)). Thus, to locate peak clusters in this part of the parameter space, the horizontal coordinate along the θ axis must be treated circularly, that is, modulo m. However, as can be seen clearly in Fig. 8.8(b), the sinusoidal traces in the parameter space do not continue smoothly at the transition $\theta = \pi \to 0$, but are vertically mirrored! Evaluating such neighborhoods near the borders of the parameter space thus requires special treatment of the vertical (r) accumulator coordinate.

8.3.2 Hough Transform Extensions

So far, we have presented the Hough transform only in its most basic formulation. The following is a list of some of the more common methods of improving and refining the method.

Modified accumulation

The purpose of the accumulator map is to locate the intersections of multiple 2D curves. Due to the discrete nature of the image and accumulator coordinates, rounding errors usually cause the parameter curves not to intersect in a single accumulator cell, even when the associated image lines are exactly straight. A common remedy is, for a given angle $\theta = i_\theta \cdot \Delta_\theta$ (Alg. 8.1), to increment not only the *main* accumulator cell $\mathsf{A}(i,j)$ but also the *neighboring* cells $\mathsf{A}(i,j-1)$ and $\mathsf{A}(i,j+1)$, possibly with different weights. This makes the Hough transform more tolerant against inaccurate point coordinates and rounding errors.

Considering edge strength and orientation

Until now, the raw data for the Hough transform was typically an edge map that was interpreted as a binary image with ones at potential edge points. Yet edge maps contain additional information, such as the edge strength $E(u,v)$ and local edge orientation $\Phi(u,v)$ (see Sec. 6.3), which can be used to improve the results of the HT.

The *edge strength* $E(u,v)$ is especially easy to take into consideration. Instead of incrementing visited accumulator cells by 1, add the strength of the respective edge, that is,

$$\mathsf{A}(i,j) \leftarrow \mathsf{A}(i,j) + E(u,v). \tag{8.15}$$

In this way, strong edge points will contribute more to the accumulated values than weak ones (see also Exercise 8.6).

The local *edge orientation* $\Phi(u,v)$ is also useful for limiting the range of possible orientation angles for the line at (u,v). The angle $\Phi(u,v)$ can be used to increase the efficiency of the algorithm by reducing the number of accumulator cells to be considered along the θ axis. Since this also reduces the number of irrelevant "votes" in the accumulator, it increases the overall sensitivity of the Hough transform (see, e.g., [125, p. 483]).

Bias compensation

Since the value of a cell in the Hough accumulator represents the number of image points falling on a line, longer lines naturally have higher values than shorter lines. This may seem like an obvious point to make, but consider when the image only contains a small section of a "long" line. For instance, if a line only passes through the corner of an image then the cells representing it in the accumulator array will naturally have lower values than a "shorter" line that lies entirely within the image (Fig. 8.11). It follows then that if we only search the accumulator array for maximal values, it is likely that we will completely miss short line segments. One way to compensate for

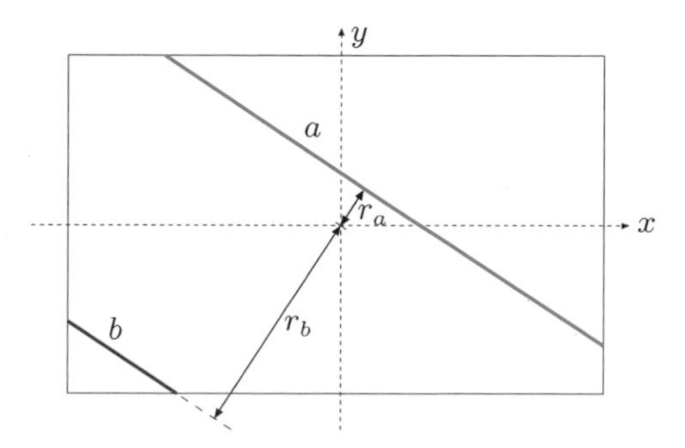

Fig. 8.11
Hough transform bias problem. When an image represents only a finite section of an object, then those lines nearer the center (smaller r values) will have higher values than those farther away (larger r values). As an example, the maximum value of the accumulator for line a will be higher than that of line b.

this inherent bias is to compute for each accumulator entry $\mathsf{A}(i,j)$ the maximum number of image points $\mathsf{A}_{\mathrm{max}}(i,j)$ possible for a line with the corresponding parameters and then normalize the result, for example, in the form

$$\mathsf{A}(i,j) \leftarrow \frac{\mathsf{A}(i,j)}{\max(1, \mathsf{A}_{\mathrm{max}}(i,j))}. \tag{8.16}$$

The normalization map $\mathsf{A}_{\mathrm{max}}(i,j)$ can be determined analytically (by calculating the intersecting length of each line) or by simulation; for example, by computing the Hough transform of an image with the same dimensions in which all pixels are edge pixels or by using a random image in which the pixels are uniformly distributed.

Line endpoints

Our simple version of the Hough transform determines the parameters of the line in the image but not their endpoints. These could be found in a subsequent step by determining which image points belong to any detected line (e.g., by applying a threshold to the perpendicular distance between the ideal line—defined by its parameters—and the actual image points). An alternative solution is to calculate the extreme point of the line during the computation of the accumulator array. For this, every cell of the accumulator array is supplemented with four addition coordinates to

$$\mathsf{A}(i,j) = (a, u_{\mathrm{min}}, v_{\mathrm{min}}, u_{\mathrm{max}}, v_{\mathrm{max}}), \tag{8.17}$$

where component a denotes the original accumulator value and u_{min}, v_{min}, u_{max}, v_{max} are the coordinates of the line's bounding box. After the additional coordinates are initialized, they are updated simultaneously with the positions along the parameter trace for every image point (u,v). After completion of the process, the accumulator cell (i,j) contains the bounding box for all image points that contributed it. When finding the maximum values in the second stage, care should be taken so that the merged cells contain the correct endpoints (see also Exercise 8.4).

Hierarchical Hough transform

The accuracy of the results increases with the size of the parameter space used; for example, a step size of 256 along the θ axis is equivalent to searching for lines at every $\frac{\pi}{256} \approx 0.7°$. While increasing the number of accumulator cells provides a finer result, bear in mind that it also increases the computation time and especially the amount of memory required.

Instead of increasing the resolution of the entire parameter space, the idea of the hierarchical HT is to gradually "zoom" in and refine the parameter space. First, the regions containing the most important lines are found using a relatively low-resolution parameter space, and then the parameter spaces of those regions are recursively passed to the HT and examined at a higher resolution. In this way, a relatively exact determination of the parameters can be found using a limited (in comparison) parameter space.

Line intersections

It may be useful in certain applications not to find the lines themselves but their intersections, for example, for precisely locating the corner points of a polygon-shaped object. The Hough transform delivers the parameters of the recovered lines in Hessian normal form (that is, as pairs $L_k = \langle \theta_k, r_k \rangle$). To compute the point of intersection $\boldsymbol{x}_{12} = (x_{12}, y_{12})^\mathsf{T}$ for two lines $L_1 = \langle \theta_1, r_1 \rangle$ and $L_2 = \langle \theta_2, r_2 \rangle$ we need to solve the system of linear equations

$$\begin{aligned} x_{12} \cdot \cos(\theta_1) + y_{12} \cdot \sin(\theta_1) &= r_1, \\ x_{12} \cdot \cos(\theta_2) + y_{12} \cdot \sin(\theta_2) &= r_2, \end{aligned} \tag{8.18}$$

for the unknowns x_{12}, y_{12}. The solution is

$$\begin{aligned} \begin{pmatrix} x_{12} \\ y_{12} \end{pmatrix} &= \frac{1}{\cos(\theta_1)\sin(\theta_2) - \cos(\theta_2)\sin(\theta_1)} \cdot \begin{pmatrix} r_1 \sin(\theta_2) - r_2 \sin(\theta_1) \\ r_2 \cos(\theta_1) - r_1 \cos(\theta_2) \end{pmatrix} \\ &= \frac{1}{\sin(\theta_2 - \theta_1)} \cdot \begin{pmatrix} r_1 \sin(\theta_2) - r_2 \sin(\theta_1) \\ r_2 \cos(\theta_1) - r_1 \cos(\theta_2) \end{pmatrix}, \end{aligned} \tag{8.19}$$

for $\sin(\theta_2 - \theta_1) \neq 0$. Obviously $\boldsymbol{x}_0$ is undefined (no intersection point exists) if the lines L_1, L_2 are parallel to each other (i.e., if $\theta_1 \equiv \theta_2$).

Figure 8.12 shows an illustrative example using *ARToolkit*[6] markers. After automatic thresholding (see Ch. 11) the straight line segments along the outer boundary of the largest binary region are analyzed with the Hough transform. Subsequently, the corners of the marker are calculated precisely as the intersection points of the involved line segments.

8.4 Java Implementation

The complete Java source code for the straight line Hough transform is available online in class `HoughTransformLines`.[7] Detailed usage of this class is shown in the ImageJ plugin `Find_Straight_Lines` (see also Prog. 8.1 for a minimal example).[8]

`HoughTransformLines` (class)

This class is a direct implementation of the Hough transform for straight lines, as outlined in Alg. 8.1. The sin/cos function calls (see Alg. 8.1, line 15) are substituted by precalculated tables for improved efficiency. The class defines the following constructors:

`HoughTransformLines (ImageProcessor I, Parameters params)`
: `I` denotes the input image, where all pixel values > 0 are assumed to be relevant (edge) points; `params` is an instance of the (inner) class `HoughTransformLines.Parameters`, which allows to specify the accumulator size (`nAng`, `nRad`) etc.

[6] Used for augmented reality applications, see www.hitl.washington.edu/artoolkit/.

[7] Package `imagingbook.pub.hough`.

[8] Note that the current implementation has no bias compensation (see Sec. 8.3.2, Fig. 8.11).

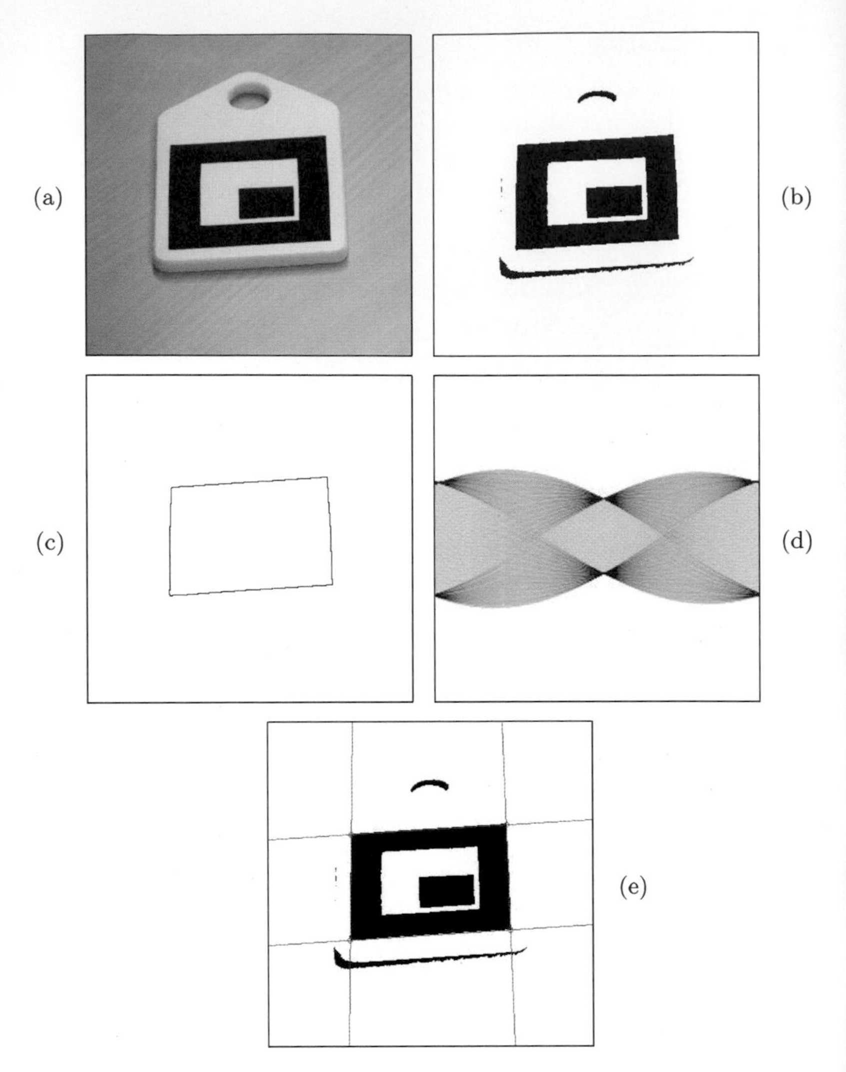

Fig. 8.12
Hough transform used for precise calculation of corner points. Original image showing a typical ARToolkit marker (a), result after automatic thresholding (b). The outer contour pixels of the largest binary region (c) are used as input points to the Hough transform. Hough accumulator map (d), detected lines and marked intersection points (e).

`HoughTransformLines (Point2D[] points, int M, int N, Parameters params)`
: In this case the Hough transform is calculated for a sequence of 2D points (`points`); `M`, `N` specify the associated coordinate frame (for calculating the reference point $\boldsymbol{x}_r$), which is typically the original image size; `params` is a parameter object (as described before).

The most important public methods of the class ClassHoughTransformLines are:

`HoughLine[] getLines (int amin, int maxLines)`
: Returns a sorted sequence of line objects[9] whose accumulator value is `amin` or greater. The sequence is sorted by accumulator values and contains up to `maxLines` elements

`int[][] getAccumulator ()`
: Returns a reference to the accumulator map A (of size $m \times n$ for angles and radii, respectively).

[9] Of type `HoughTransformLines.HoughLine`.

```
1 import imagingbook... .HoughTransformLines;
2 import imagingbook... .HoughTransformLines.HoughLine;
3 import imagingbook... .HoughTransformLines.Parameters;
4 ...
5
6   public void run(ImageProcessor ip) {
7     Parameters params = new Parameters();
8     params.nAng = 256;     // = m
9     params.nRad = 256;     // = n
10
11     // compute the Hough Transform:
12     HoughTransformLines ht =
13           new HoughTransformLines(ip, params);
14
15     // retrieve the 5 strongest lines with min. 50 accumulator votes
16     HoughLine[] lines = ht.getLines(50, 5);
17
18     if (lines.length > 0) {
19       IJ.log("Lines found:");
20       for (HoughLine L : lines) {
21         IJ.log(L.toString()); // list the resulting lines
22       }
23     }
24     else
25       IJ.log("No lines found!");
26   }
```

Prog. 8.1
Minimal example for the usage of class `HoughTransformLines` (`run()` method for an ImageJ plugin of type `PlugInFilter`). First (in lines 7–9) a parameter ob ect is created and configured; `nAng` ($= m$) and `nRad` ($= n$) specify the number of discrete angular and radial steps in the Hough accumulator map. In lines 12-13 an instance of `HoughTransformLines` is created for the image `ip`. The accumulator map is calculated in this step. In line 16, `getLines()` is called to retrieve the sequence of the 5 strongest detected lines, with at least 50 image points each. Unless empty, this sequence is subsequently listed.

`int[][] getAccumulatorMax()`
: Returns a copy of accumulator array in which all non-maxima are replaced by zero values.

`FloatProcessor getAccumulatorImage()`
: Returns a floating-point image of the accumulator array, analogous to `getAccumulator()`. Angles θ_i run horizontally, radii r_j vertically.

`FloatProcessor getAccumulatorMaxImage()`
: Returns a floating-point image of the accumulator array with suppressed non-maximum values, analogous to `getAccumulatorMax()`.

`double angleFromIndex(int i)`
: Returns the angle $\theta_i \in [0, \pi)$ for the given index `i` in the range $0, \ldots, m-1$.

`double radiusFromIndex(int j)`
: Returns the radius $r_j \in [-r_{\mathrm{max}}, r_{\mathrm{max}}]$ for the given index j in the range $0, \ldots, n-1$.

`Point2D getReferencePoint()`
: Returns the (fixed) reference point $\boldsymbol{x}_r$ for this Hough transform instance.

HoughLine (class)

HoughLine represents a straight line in Hessian normal form. It is implemented as an inner class of **HoughTransformLines**. It offers no public constructor but the following methods:

`double getAngle ()`
: Returns the angle $\theta \in [0, \pi)$ of this line.

`double getRadius ()`
: Returns the radius $r \in [-r_{\mathrm{max}}, r_{\mathrm{max}}]$ of this line, relative to the associated Hough transform's reference point $\boldsymbol{x}_r$.

`int getCount ()`
: Returns the Hough transform's accumulator value (number of registered image points) for this line.

`Point2D getReferencePoint ()`
: Returns the (fixed) reference point $\boldsymbol{x}_r$ for this line. Note that all lines associated with a given Hough transform share the same reference point.

`double getDistance (Point2D p)`
: Returns the Euclidean distance of point `p` to this line. The result may be positive or negative, depending on which side of the line `p` is located.

8.5 Hough Transform for Circles and Ellipses

8.5.1 Circles and Arcs

Since lines in 2D have two degrees of freedom, they could be completely specified using two real-valued parameters. In a similar fashion, representing a circle in 2D requires *three* parameters, for example

$$C = \langle \bar{x}, \bar{y}, r \rangle,$$

where $\bar{x}$, $\bar{y}$ are the coordinates of the center and ρ is the radius of the circle (Fig. 8.13).

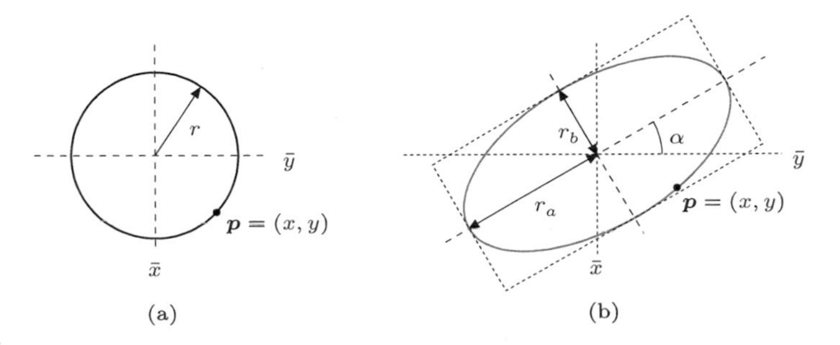

Fig. 8.13 Representation of circles and ellipses in 2D. A circle (a) requires three parameters (e.g., $\bar{x}, \bar{y}, r$). An arbitrary ellipse (b) takes five parameters (e.g., $\bar{x}, \bar{y}, r_a, r_b, \alpha$).

A point $\boldsymbol{p} = (x, y)$ lies exactly on the circle C if the condition

$$(x - \bar{x})^2 + (x - \bar{y})^2 = r^2 \tag{8.20}$$

holds. Therefore the Hough transform for circles requires a 3D parameter space $\mathsf{A}(i, j, k)$ to find the position and radius of circles (and circular arcs) in an image. Unlike the HT for lines, there does not exist a simple functional dependency between the coordinates in parameter space; so how can we find every parameter combination $(\bar{x}, \bar{y}, r)$ that satisfies Eqn. (8.20) for a given image point (x, y)? A "brute force" is to a exhaustively test all cells of the parameter space to see if the relation in Eqn. (8.20) holds, which is computationally quite expensive, of course.

If we examine Fig. 8.14, we can see that a better idea might be to make use of the fact that the coordinates of the center points also form a circle in Hough space. It is not necessary therefore to search the entire 3D parameter space for each image point. Instead we need only increase the cell values along the edge of the appropriate circle on each r plane of the accumulator array. To do this, we can adapt any of the standard algorithms for generating circles. In this case, the integer math version of the well-known *Bresenham* algorithm [33] is particularly well-suited.

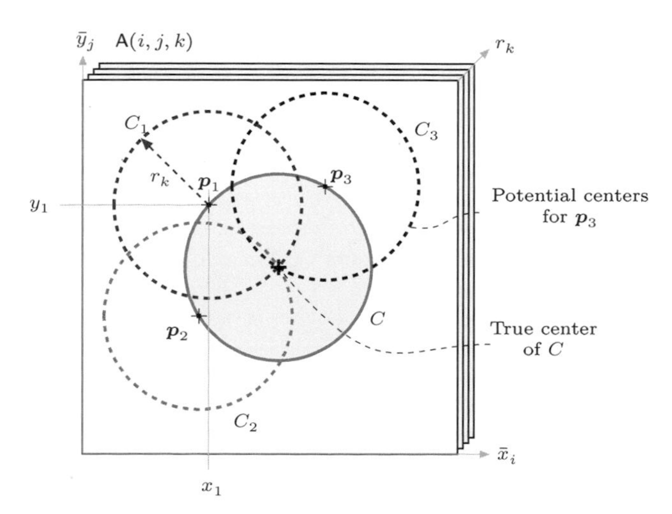

Fig. 8.14
Hough transform for circles. The illustration depicts a single slice of the 3D accumulator array $\mathsf{A}(i, j, k)$ at a given circle radius r_k. The center points of all the circles running through a given image point $\boldsymbol{p}_1 = (x_1, y_1)$ form a circle C_1 with a radius of r_k centered around $\boldsymbol{p}_1$, just as the center points of the circles that pass through $\boldsymbol{p}_2$ and $\boldsymbol{p}_3$ lie on the circles C_2, C_3. The cells along the edges of the three circles C_1, C_2, C_3 of radius r_k are traversed and their values in the accumulator array incremented. The cell in the accumulator array contains a value of 3 where the circles intersect at the true center of the image circle C.

Figure 8.15 shows the spatial structure of the 3D parameter space for circles. For a given image point $\boldsymbol{p}_m = (u_m, v_m)$, at each plane along the r axis (for $r_k = r_{\min}, \ldots, r_{\max}$), a circle centered at (u_m, v_m) with the radius r_k is traversed, ultimately creating a 3D cone-shaped surface in the parameter space. The coordinates of the dominant circles can be found by searching the accumulator space for the cells with the highest values; that is, the cells where the most cones intersect. Just as in the linear HT, the *bias* problem (see Sec. 8.3.2) also occurs in the circle HT. Sections of circles (i.e., arcs) can be found in a similar way, in which case the maximum value possible for a given cell is proportional to the arc length.

8.5.2 Ellipses

In a perspective image, most circular objects originating in our real, 3D world will actually appear in 2D images as ellipses, except in the case where the object lies on the optical axis and is observed from the front. For this reason, perfectly circular structures seldom occur

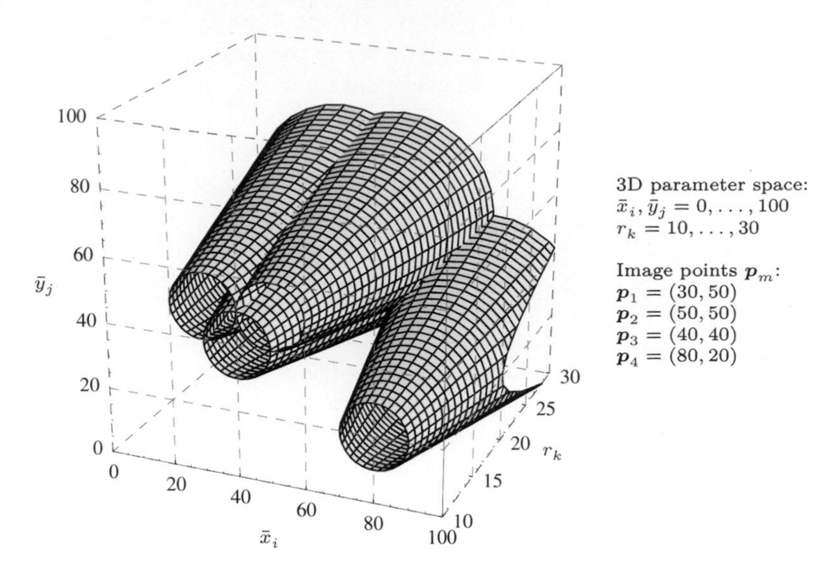

Fig. 8.15
3D parameter space for circles. For each image point $p = (u, v)$, the cells lying on a cone (with its axis at (u, v) and varying radius r_k) in the 3D accumulator $\mathsf{A}(i, j, k)$ are traversed and incremented. The size of the discrete accumulator is set to $100 \times 100 \times 30$. Candidate center points are found where many of the 3D surfaces intersect.

in photographs. While the Hough transform can still be used to find ellipses, the larger parameter space required makes it substantially more expensive.

A general ellipse in 2D has five degrees of freedom and therefore requires five parameters to represent it, for example,

$$E = \langle \bar{x}, \bar{y}, r_a, r_b, \alpha \rangle, \tag{8.21}$$

where $(\bar{x}, \bar{y})$ are the coordinates of the center points, (r_a, r_b) are the two radii, and α is the orientation of the principal axis (Fig. 8.13).[10] In order to find ellipses of any size, position, and orientation using the Hough transform, a 5D parameter space with a suitable resolution in each dimension is required. A simple calculation illustrates the enormous expense of representing this space: using a resolution of only $128 = 2^7$ steps in every dimension results in 2^{35} accumulator cells, and implementing these using 4-byte `int` values thus requires 2^{37} bytes (128 gigabytes) of memory. Moreover, the amount of processing required for filling and evaluating such a huge parameter space makes this method unattractive for real applications.

An interesting alternative in this case is the *generalized Hough transform*, which in principle can be used for detecting any arbitrary 2D shape [15,117]. Using the generalized Hough transform, the shape of the sought-after contour is first encoded point by point in a table and then the associated parameter space is related to the position (x_c, y_c), scale S, and orientation θ of the shape. This requires a 4D space, which is smaller than that of the Hough method for ellipses described earlier.

[10] See Chapter 10, Eqn. (10.39) for a parametric equation of this ellipse.

8.6 Exercises

Exercise 8.1. Drawing a straight line given in Hessian normal (HNF) form is not directly possible because typical graphics environments can only draw lines between two specified end points.[11] An HNF line $L = \langle\theta, r\rangle$, specified relative to a reference point $\boldsymbol{x}_r = (x_r, y_r)$, can be drawn into an image I in several ways (implement both versions):

Version 1: Iterate over all image points (u, v); if Eqn. (8.11), that is,

$$r = (u - x_r) \cdot \cos(\theta) + (v - y_r) \cdot \sin(\theta), \tag{8.22}$$

is satisfied for position (u, v), then mark the pixel $I(u, v)$. Of course, this "brute force" method will only show those (few) line pixels whose positions satisfy the line equation *exactly*. To obtain a more "tolerant" drawing method, we first reformulate Eqn. (8.22) to

$$(u - x_r) \cdot \cos(\theta) + (v - y_r) \cdot \sin(\theta) - r = d. \tag{8.23}$$

Obviously, Eqn. (8.22) is only then exactly satisfied if $d = 0$ in Eqn. (8.23). If, however, Eqn. (8.22) is *not* satisfied, then the magnitude of $d \neq 0$ equals the distance of the point (u, v) from the line. Note that d itself may be positive or negative, depending on which side of the line (u, v) is located. This suggests the following version.

Version 2: Define a constant $w > 0$. Iterate over all image positions (u, v); whenever the inequality

$$|(u - x_r) \cdot \cos(\theta) + (v - y_r) \cdot \sin(\theta) - r| \leq w \tag{8.24}$$

is satisfied for position (u, v), mark the pixel $I(u, v)$. For example, all line points should show with $w = 1$. What is the geometric meaning of w?

Exercise 8.2. Develop a less "brutal" method (compared to Exercise 8.1) for drawing a straight line $L = \langle\theta, r\rangle$ in Hessian normal form (HNF). First, set up the HNF equations for the four border lines of the image, A, B, C, D. Now determine the intersection points of the given line L with each border line $A, \ldots, D$ and use the built-in `drawLine()` method or a similar routine to draw L by connecting the intersection points. Consider which special situations may appear and how they could be handled.

Exercise 8.3. Implement (or extend) the Hough transform for straight lines by including measures against the bias problem, as discussed in Sec. 8.3.2 (Eqn. (8.16)).

Exercise 8.4. Implement (or extend) the Hough transform for finding lines that takes into account line endpoints, as described in Sec. 8.3.2 (Eqn. (8.17)).

Exercise 8.5. Calculate the pairwise intersection points of all detected lines (see Eqns. (8.18)–(8.19)) and show the results graphically.

[11] For example, with `drawLine(x1, y1, x2, y2)` in ImageJ.

Exercise 8.6. Extend the Hough transform for straight lines so that updating the accumulator map takes into account the intensity (edge magnitude) of the current pixel, as described in Eqn. (8.15).

Exercise 8.7. Implement a *hierarchical* Hough transform for straight lines (see p. 172) capable of accurately determining line parameters.

Exercise 8.8. Implement the Hough transform for finding circles and circular arcs with varying radii. Make use of a fast algorithm for drawing circles in the accumulator array, such as described in Sec. 8.5.

9

Morphological Filters

In the discussion of the median filter in Chapter 5 (Sec. 5.4.2), we noticed that this type of filter can somehow alter 2D image structures. Figure 9.1 illustrates once more how corners are rounded off, holes of a certain size are filled, and small structures, such as single dots or thin lines, are removed. The median filter thus responds selectively to the local shape of image structures, a property that might be useful for other purposes if it can be applied not just randomly but in a controlled fashion. Altering the local structure in a predictable way is exactly what "morphological" filters can do, which we focus on in this chapter.

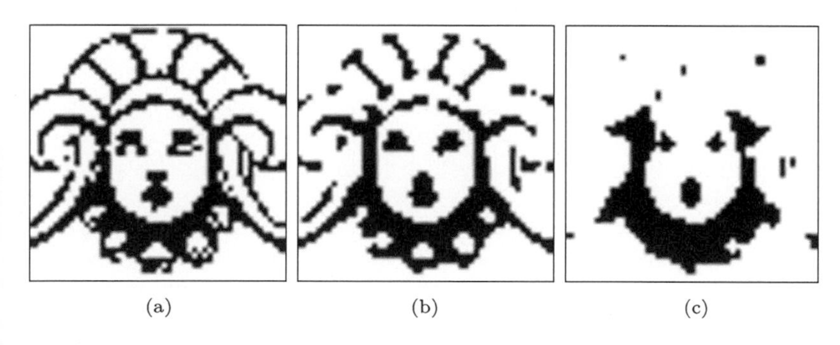

Fig. 9.1
Median filter applied to a binary image: original image (a) and results from a 3×3 pixel median filter (b) and a 5×5 pixel median filter (c).

In their original form, morphological filters are aimed at binary images, images with only two possible pixel values, 0 and 1 or *black* and *white*, respectively. Binary images are found in many places, in particular in digital printing, document transmission (FAX) and storage, or as selection masks in image and video editing. Binary images can be obtained from grayscale images by simple thresholding (see Sec. 4.1.4) using either a global or a locally varying threshold value. We denote binary pixels with values 1 and 0 as *foreground* and *background* pixels, respectively. In most of the following examples, the foreground pixels are shown in black and background pixels are shown in white, as is common in printing.

At the end of this chapter, we will see that morphological filters are applicable not only to binary images but also to grayscale and

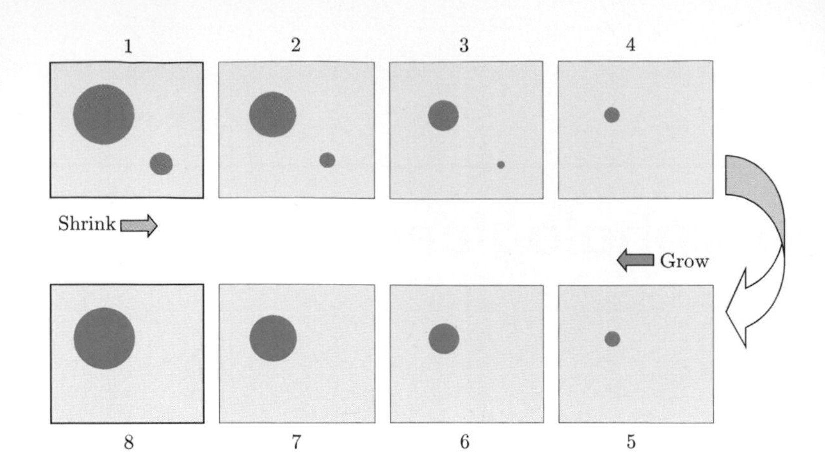

Fig. 9.2
Basic idea of size-dependent removal of image structures. Small structures may be eliminated by iterative shrinking and subsequent growing. Ideally, the "surviving" structures should be restored to their original shape.

even color images, though these operations differ significantly from their binary counterparts.

9.1 Shrink and Let Grow

Our starting point was the observation that a simple 3×3 pixel median filter can round off larger image structures and remove smaller structures, such as points and thin lines, in a binary image. This could be useful to eliminate structures that are below a certain size (e.g., to clean an image from noise or dirt). But how can we control the size and possibly the shape of the structures affected by such an operation?

Although its structural effects may be interesting, we disregard the median filter at this point and start with this task again from the beginning. Let's assume that we want to remove small structures from a binary image without significantly altering the remaining larger structures. The key idea for accomplishing this could be the following (Fig. 9.2):

1. First, all structures in the image are iteratively "shrunk" by peeling off a layer of a certain thickness around the boundaries.
2. Shrinking removes the smaller structures step by step, and only the larger structures remain.
3. The remaining structures are then grown back by the same amount.
4. Eventually the larger regions should have returned to approximately their original shapes, while the smaller regions have disappeared from the image.

All we need for this are two types of operations. "Shrinking" means to remove a layer of pixels from a foreground region around all its borders against the background (Fig. 9.3). The other way around, "growing", adds a layer of pixels around the border of a foreground region (Fig. 9.4).

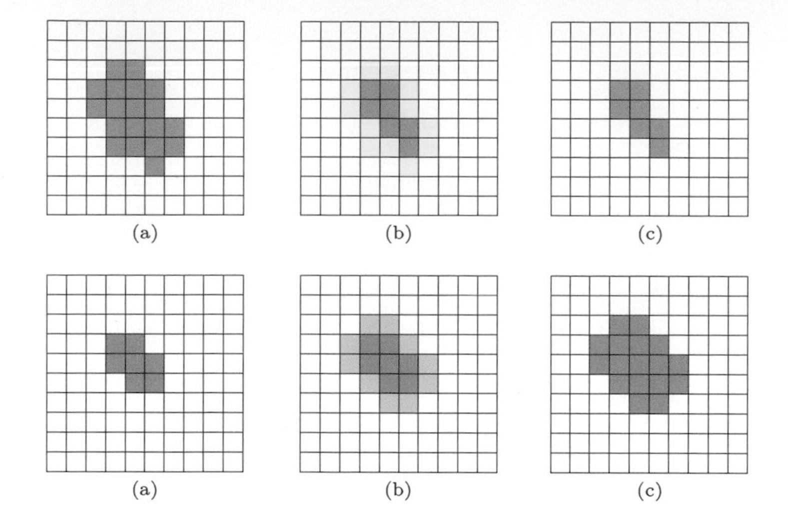

Fig. 9.3
"Shrinking" a foreground region by removing a layer of border pixels: original image (a), identified foreground pixels that are in direct contact with the background (b), and result after shrinking (c).

Fig. 9.4
"Growing" a foreground region by attaching a layer of pixels: original image (a), identified background pixels that are in direct contact with the region (b), and result after growing (c).

9.1.1 Neighborhood of Pixels

For both operations, we must define the meaning of two pixels being adjacent (i.e., being "neighbors"). Two definitions of "neighborhood" are commonly used for rectangular pixel grids (Fig. 9.5):

- **4-neighborhood** ($\mathcal{N}_4$): the four pixels adjacent to a given pixel in the horizontal and vertical directions;
- **8-neighborhood** ($\mathcal{N}_8$): the pixels contained in $\mathcal{N}_4$ plus the four adjacent pixels along the diagonals.

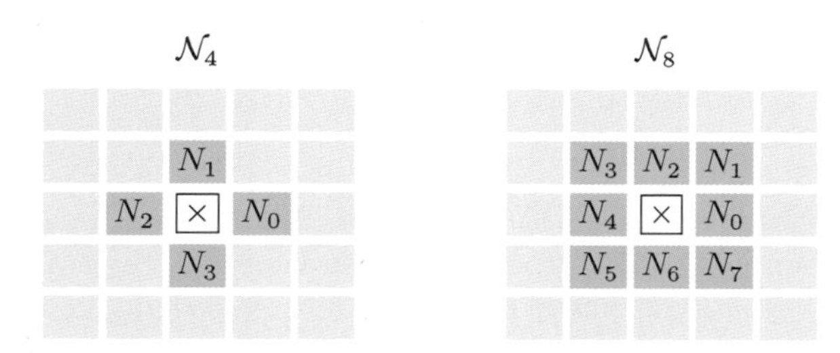

Fig. 9.5
Definitions of "neighborhood" on a rectangular pixel grid: *4-neighborhood* $\mathcal{N}_4 = \{N_1, \ldots, N_4\}$ and *8-neighborhood* $\mathcal{N}_8 = \mathcal{N}_4 \cup \{N_5, \ldots, N_8\}$.

9.2 Basic Morphological Operations

Shrinking and growing are indeed the two most basic morphological operations, which are referred to as "erosion" and "dilation", respectively. These morphological operations, however, are much more general than illustrated in the example in Sec. 9.1. They go well beyond removing or attaching single pixel layers and—in combination—can perform much more complex operations.

9.2.1 The Structuring Element

Similar to the coefficient matrix of a linear filter (see Sec. 5.2), the properties of a morphological filter are specified by elements in a matrix called a "structuring element". In binary morphology, the structuring element (just like the image itself) contains only the values 0 and 1,

$$H(i, j) \in \{0, 1\},$$

and the *hot spot* marks the origin of the coordinate system of H (Fig. 9.6). Notice that the hot spot is not necessarily located at the center of the structuring element, nor must its value be 1.

Fig. 9.6
Binary structuring element (example). 1–elements are marked with •; 0–cells are empty. The hot spot (boxed) is not necessarily located at the center.

9.2.2 Point Sets

For the formal specification of morphological operations, it is sometimes helpful to describe binary images as *sets* of 2D coordinate points.[1]

For a binary image $I(u, v) \in \{0, 1\}$, the corresponding point set $\mathcal{Q}_I$ consists of the coordinate pairs $\boldsymbol{p} = (u, v)$ of all foreground pixels,

$$\mathcal{Q}_I = \{\boldsymbol{p} \mid I(\boldsymbol{p}) = 1\}. \tag{9.1}$$

Of course, as shown in Fig. 9.7, not only a binary image I but also a structuring element H can be described as a point set.

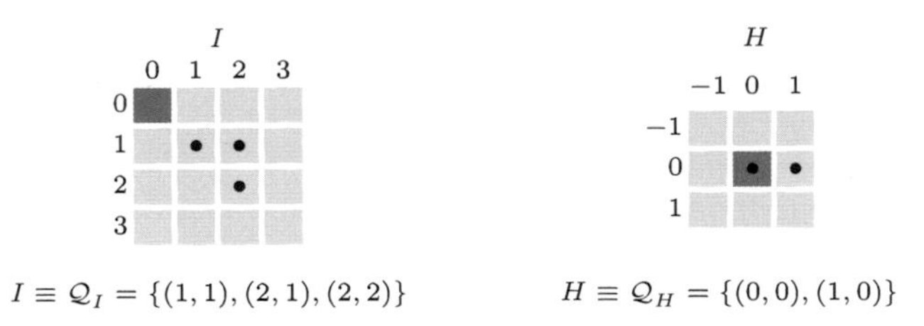

Fig. 9.7
A binary image I or a structuring element H can each be described as a set of coordinate pairs, $\mathcal{Q}_I$ and $\mathcal{Q}_H$, respectively. The dark shaded element in H marks the coordinate origin (hot spot).

With the description as point sets, fundamental operations on binary images can also be expressed as simple set operations. For example, *inverting* a binary image $I \rightarrow \bar{I}$ (i.e., exchanging foreground and background) is equivalent to building the *complementary* set

$$\mathcal{Q}_{\bar{I}} = \bar{\mathcal{Q}}_I = \{\boldsymbol{p} \in \mathbb{Z}^2 \mid \boldsymbol{p} \notin \mathcal{Q}_I\}. \tag{9.2}$$

Combining two binary images I_1 and I_2 by an OR operation between corresponding pixels, the resulting point set is the *union* of the individual point sets $\mathcal{Q}_{I_1}$ and $\mathcal{Q}_{I_2}$; that is,

$$\mathcal{Q}_{I_1 \vee I_2} = \mathcal{Q}_{I_1} \cup \mathcal{Q}_{I_2}. \tag{9.3}$$

Since a point set $\mathcal{Q}_I$ is only an alternative representation of the binary image I (i.e., $I \equiv \mathcal{Q}_I$), we will use both image and set notations synonymously in the following. For example, we simply write $\bar{I}$ instead of $\bar{\mathcal{Q}}_I$ for an inverted image as in Eqn. (9.2) or $I_1 \cup I_2$ instead of $\mathcal{Q}_{I_1} \cup \mathcal{Q}_{I_2}$ in Eqn. (9.3). The meaning should always be clear in the given context.

[1] *Morphology* is a mathematical discipline dealing with the algebraic analysis of geometrical structures and shapes, with strong roots in set theory.

Translating (shifting) a binary image I by some coordinate vector $\boldsymbol{d}$ creates a new image with the content

$$I_{\boldsymbol{d}}(\boldsymbol{p}+\boldsymbol{d}) = I(\boldsymbol{p}) \qquad \text{oder} \qquad I_{\boldsymbol{d}}(\boldsymbol{p}) = I(\boldsymbol{p}-\boldsymbol{d}), \tag{9.4}$$

which is equivalent to changing the coordinates of the original point set in the form

$$I_{\boldsymbol{d}} \equiv \big\{(\boldsymbol{p}+\boldsymbol{d}) \mid \boldsymbol{p} \in I\big\}. \tag{9.5}$$

In some cases, it is also necessary to *reflect* (mirror) a binary image or point set about its origin, which we denote as

$$I^{*} \equiv \{-\boldsymbol{p} \mid \boldsymbol{p} \in I\}. \tag{9.6}$$

9.2.3 Dilation

A *dilation* is the morphological operation that corresponds to our intuitive concept of "growing" as discussed already. As a set operation, it is defined as

$$I \oplus H \equiv \big\{(\boldsymbol{p}+\boldsymbol{q}) \mid \text{for all } \boldsymbol{p} \in I, \boldsymbol{q} \in H\big\}. \tag{9.7}$$

Thus the point set produced by a dilation is the (vector) sum of all possible pairs of coordinate points from the original sets I and H, as illustrated by a simple example in Fig. 9.8. Alternatively, one could view the dilation as the structuring element H being *replicated* at each foreground pixel of the image I or, conversely, the image I being replicated at each foreground element of H. Expressed in set notation,[2] this is

$$I \oplus H \equiv \bigcup_{\boldsymbol{p}\in I} H_{\boldsymbol{p}} = \bigcup_{\boldsymbol{q}\in H} I_{\boldsymbol{q}}, \tag{9.8}$$

with $H_{\boldsymbol{p}}, I_{\boldsymbol{q}}$ denoting the sets H, I shifted by $\boldsymbol{p}$ and $\boldsymbol{q}$, respectively (see Eqn. (9.5)).

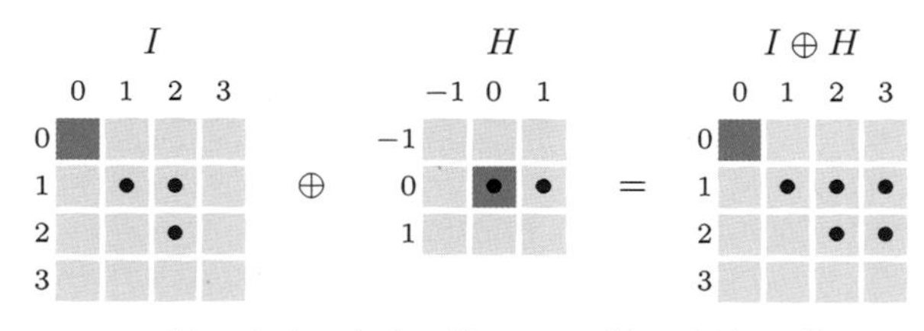

$$I \equiv \{(1,1),(2,1),(2,2)\}, \;\; H \equiv \{(\mathbf{0},\mathbf{0}),(\mathbf{1},\mathbf{0})\}$$

$$\begin{aligned} I \oplus H \;\equiv\; \{\; & (1,1)+(\mathbf{0},\mathbf{0}),\; (1,1)+(\mathbf{1},\mathbf{0}),\\ & (2,1)+(\mathbf{0},\mathbf{0}),\; (2,1)+(\mathbf{1},\mathbf{0}),\\ & (2,2)+(\mathbf{0},\mathbf{0}),\; (2,2)+(\mathbf{1},\mathbf{0})\,\} \end{aligned}$$

Fig. 9.8
Binary dilation example. The binary image I is subject to dilation with the structuring element H. In the result $I \oplus H$ the structuring element H is replicated at every foreground pixel of the original image I.

[2] See also Sec. A.2 in the Appendix.

9.2.4 Erosion

The quasi-inverse of dilation is the *erosion* operation, again defined in set notation as

$$I \ominus H \equiv \big\{ \boldsymbol{p} \in \mathbb{Z}^2 \mid (\boldsymbol{p}+\boldsymbol{q}) \in I, \text{ for all } \boldsymbol{q} \in H \big\}. \tag{9.9}$$

This operation can be interpreted as follows. A position $\boldsymbol{p}$ is contained in the result $I \ominus H$ if (and only if) the structuring element H—when placed at this position $\boldsymbol{p}$—is *fully contained* in the foreground pixels of the original image; that is, if $H_{\boldsymbol{p}}$ is a subset of I. Equivalent to Eqn. (9.9), we could thus define binary erosion as

$$I \ominus H \equiv \{ \boldsymbol{p} \in \mathbb{Z}^2 \mid H_{\boldsymbol{p}} \subseteq I \}. \tag{9.10}$$

Figure 9.9 shows a simple example for binary erosion.

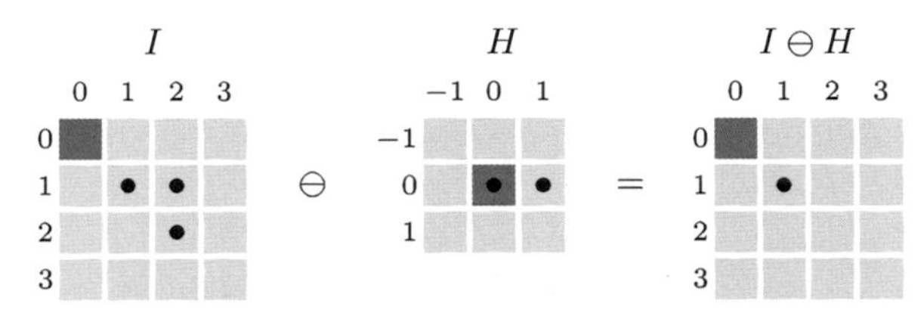

$$I \equiv \{(1,1),(2,1),(2,2)\},\ H \equiv \{(\mathbf{0},\mathbf{0}),(\mathbf{1},\mathbf{0})\}$$

$$I \ominus H \equiv \{\,(1,1)\,\} \text{ because}$$
$$(1,1)+(\mathbf{0},\mathbf{0}) = (1,1) \in I \quad \textbf{and} \quad (1,1)+(\mathbf{1},\mathbf{0}) = (2,1) \in I$$

Fig. 9.9 Binary erosion example. The binary image I is subject to erosion with H as the structuring element. H is only covered by I when placed at position $\boldsymbol{p} = (1,1)$, thus the resulting points set contains only the single coordinate $(1,1)$.

9.2.5 Formal Properties of Dilation and Erosion

The dilation operation is *commutative*,

$$I \oplus H = H \oplus I, \tag{9.11}$$

and therefore—just as in linear convolution—the image and the structuring element (filter) can be exchanged to get the same result. Dilation is also *associative*, that is,

$$(I_1 \oplus I_2) \oplus I_3 = I_1 \oplus (I_2 \oplus I_3), \tag{9.12}$$

and therefore the ordering of multiple dilations is not relevant. This also means—analogous to linear filters (cf. Eqn. (5.25))—that a dilation with a large structuring element of the form $H_{\text{big}} = H_1 \oplus H_2 \oplus \ldots \oplus H_K$ can be efficiently implemented as a sequence of multiple dilations with smaller structuring elements by

$$I \oplus H_{\text{big}} = (\ldots((I \oplus H_1) \oplus H_2) \oplus \ \ldots\ \oplus H_K) \tag{9.13}$$

There is also a *neutral element* (δ) for the dilation operation, similar to the Dirac function for the linear convolution (see Sec. 5.3.4),

$$I \oplus \delta = \delta \oplus I = I, \quad \text{with } \delta = \{(0,0)\}. \tag{9.14}$$

The *erosion* operation is, in contrast to dilation (but similar to arithmetic subtraction), *not* commutative, that is,

$$I \ominus H \neq H \ominus I, \tag{9.15}$$

in general. However, if erosion and dilation are combined, then—again in analogy with arithmetic subtraction and addition—the following chain rule holds:

$$(I_1 \ominus I_2) \ominus I_3 = I_1 \ominus (I_2 \oplus I_3). \tag{9.16}$$

Although dilation and erosion are not mutually inverse (in general, the effects of dilation cannot be undone by a subsequent erosion), there are still some strong formal relations between these two operations. For one, dilation and erosion are *dual* in the sense that a dilation of the *foreground* (I) can be accomplished by an erosion of the *background* ($\bar{I}$) and subsequent inversion of the result,

$$I \oplus H \;=\; \overline{(\bar{I} \ominus H^*)}, \tag{9.17}$$

where H^* denotes the *reflection* of H (Eqn. (9.6)). This works similarly the other way, too, namely

$$\bar{I} \ominus H = \overline{(I \oplus H^*)}, \tag{9.18}$$

effectively eroding the foreground by dilating the background with the mirrored structuring element, as illustrated by the example in Fig. 9.10 (see [88, pp. 521–524] for a formal proof).

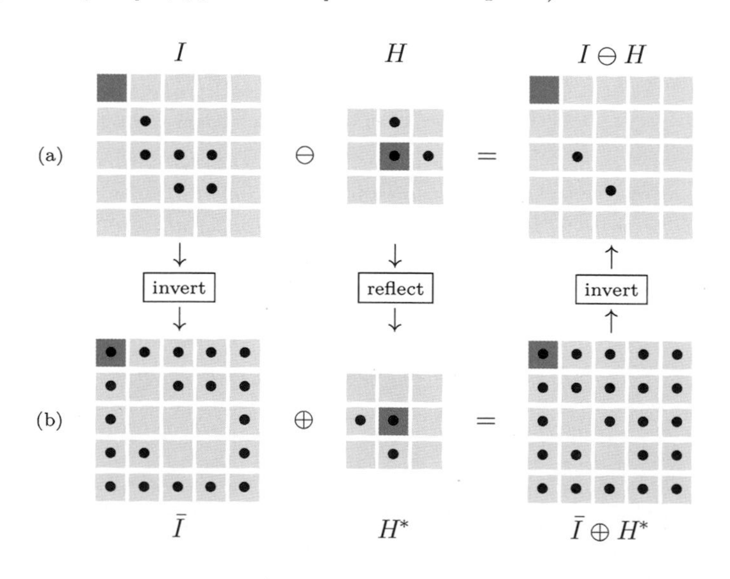

Fig. 9.10
Implementing erosion via dilation. The binary erosion of the foreground $I \ominus H$ (a) can be implemented by dilating the inverted (background) image $\bar{I}$ with the reflected structuring element H^* and subsequently inverting the result again (b).

Equation (9.18) is interesting because it shows that we only need to implement either dilation or erosion for computing both, considering that the foreground–background inversion is a very simple task. Algorithm 9.1 gives a simple algorithmic description of dilation and erosion based on the aforementioned relationships.

Alg. 9.1
Binary dilation and erosion. Procedure Dilate() implements the binary dilation as suggested by Eqn. (9.8). The original image I is displaced to each foreground coordinate of H and then copied into the resulting image I'. The hot spot of the structuring element H is assumed to be at coordinate $(0,0)$. Procedure Erode() implements the binary erosion by dilating the inverted image $\bar{I}$ with the reflected structuring element H^*, as described by Eqn. (9.18).

```
1: Dilate(I, H)
       Input: I, a binary image of size M × N;
       H, a binary structuring element.
       Returns the dilated image I' = I ⊕ H.
2:     Create map I': M × N ↦ {0,1}            ▷ new binary image I'
3:     for all (p) ∈ M × N do
4:         I'(p) ← 0                            ▷ I' ← { }
5:     for all q ∈ H do
6:         for all p ∈ I do
7:             I'(p + q) ← 1                    ▷ I' ← I' ∪ {(p+q)}
8:     return I'                                ▷ I' = I ⊕ H

9: Erode(I, H)
       Input: I, a binary image of size M × N;
       H, a binary structuring element.
       Returns the eroded image I' = I ⊖ H.
10:    Ī ← Invert(I)                            ▷ Ī ← ¬I
11:    H* ← Reflect(H)
12:    I' ← Invert(Dilate(Ī, H*))               ▷ I' = I ⊖ H = ‾(Ī ⊕ H*)‾
13:    return I'
```

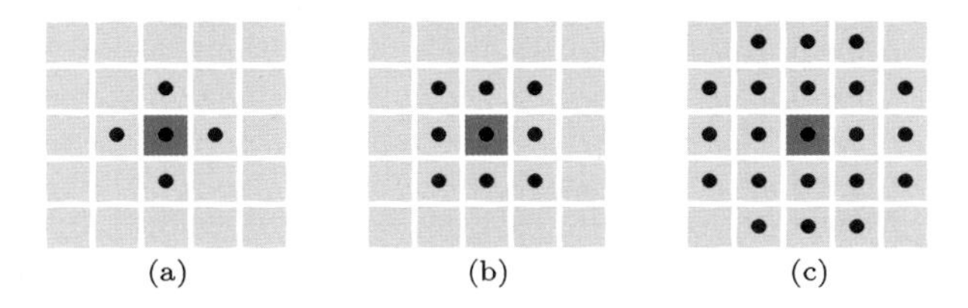

Fig. 9.11
Typical binary structuring elements of various sizes. 4-neighborhood (a), 8-neighborhood (b), "small disk" (c).

9.2.6 Designing Morphological Filters

A morphological filter is unambiguously specified by (a) the type of operation and (b) the contents of the structuring element. The appropriate size and shape of the structuring element depends upon the application, image resolution, etc. In practice, structuring elements of quasi-circular shape are frequently used, such as the examples shown in Fig. 9.11.

A dilation with a circular (disk-shaped) structuring element with radius r adds a layer of thickness r to any foreground structure in the image. Conversely, an erosion with that structuring element peels off layers of the same thickness. Figure 9.13 shows the results of dilation and erosion with disk-shaped structuring elements of different diameters applied to the original image in Fig. 9.12. Dilation and erosion results for various other structuring elements are shown in Fig. 9.14.

Disk-shaped structuring elements are commonly used to implement *isotropic* filters, morphological operations that have the same effect in every direction. Unlike linear filters (e.g., the 2D Gaussian filter in Sec. 5.3.3), it is generally not possible to compose an isotropic 2D structuring element H° from 1D structuring elements H_x and H_y since the dilation $H_x \oplus H_y$ always results in a rectangular (i.e., non-isotropic) structure. A remedy for approximating large disk-shaped filters is to alternately apply smaller disk-shaped operators of differ-

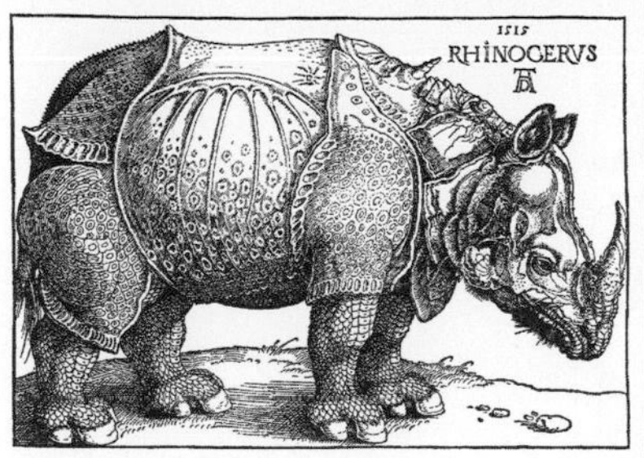

Fig. 9.12
Original binary image and the section used in the following examples (illustration by Albrecht Dürer, 1515).

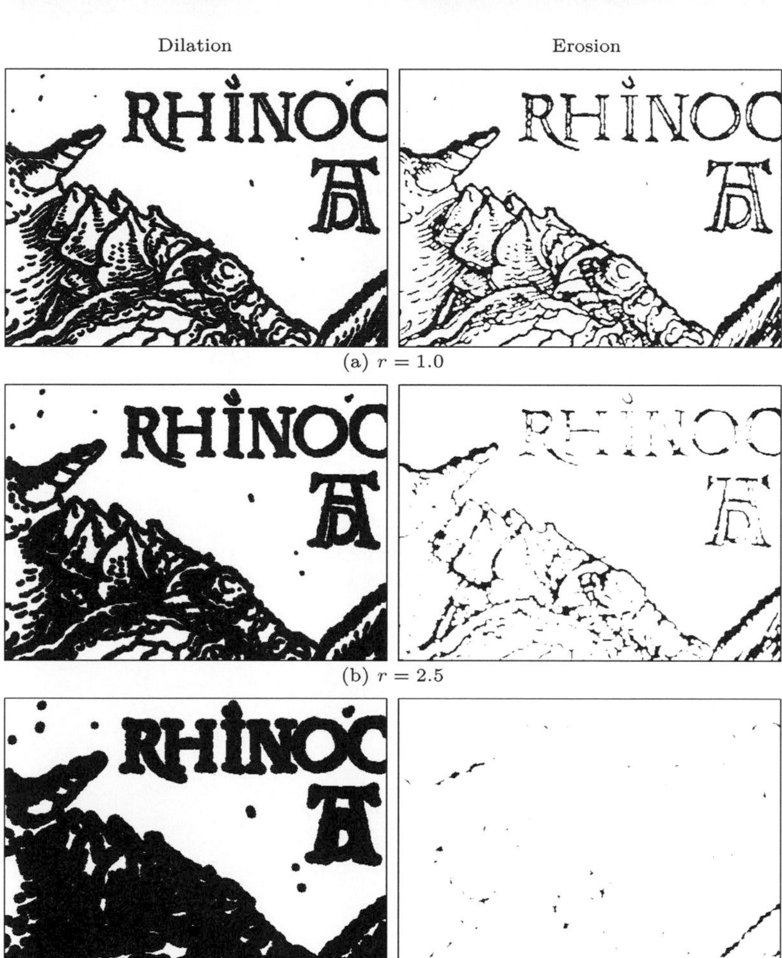

Fig. 9.13
Results of binary dilation and erosion with disk-shaped structuring elements. The radius of the disk (r) is 1.0 (a), 2.5 (b), and 5.0 (c).

ent shapes, as illustrated in Fig. 9.15. The resulting filter is generally not fully isotropic but can be implemented efficiently as a sequence of small filters.

9.2.7 Application Example: Outline

A typical application of morphological operations is to extract the boundary pixels of the foreground structures. The process is very simple. First, we apply an erosion on the original image I to remove the boundary pixels of the foreground,

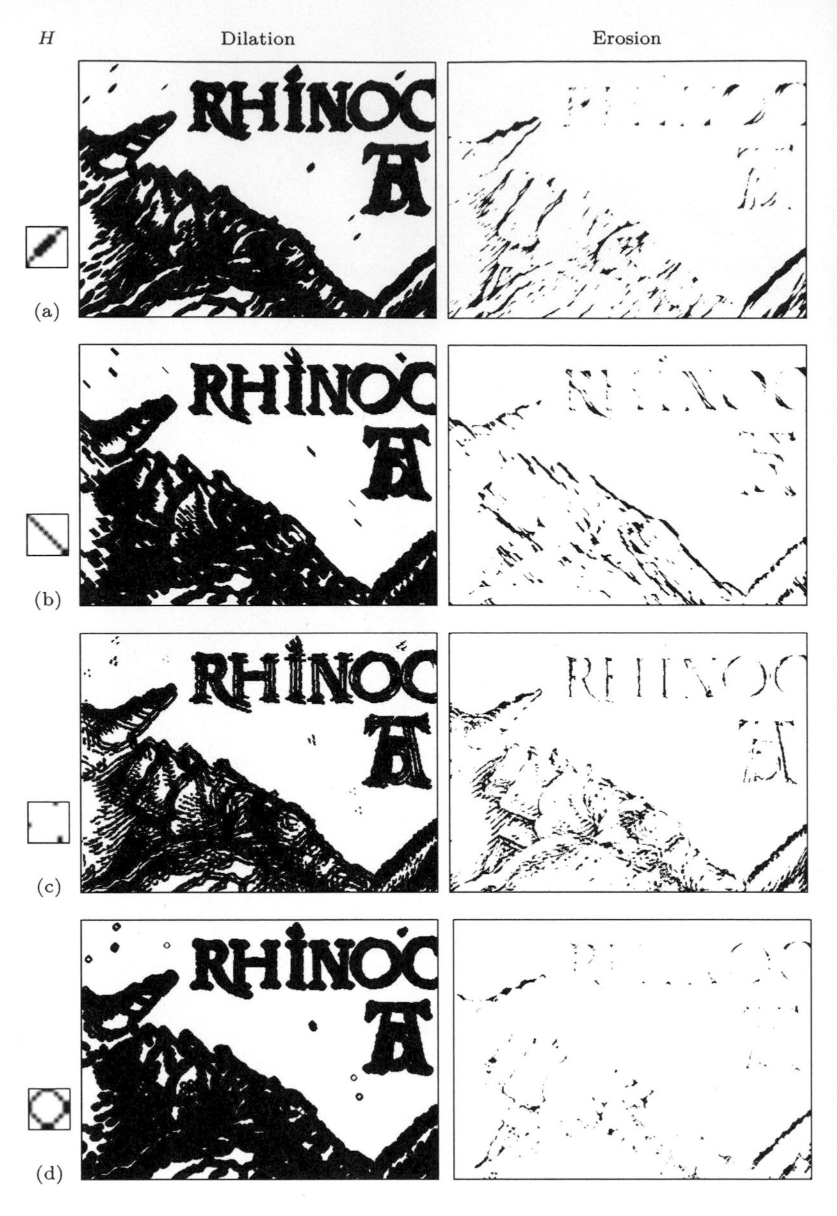

Fig. 9.14
Examples of binary dilation and erosion with various free-form structuring elements. The structuring elements H are shown in the left column (enlarged). Notice that the dilation expands every isolated foreground point to the shape of the structuring element, analogous to the *impulse response* of a linear filter. Under erosion, only those elements where the structuring element is fully contained in the original image survive.

$$I' = I \ominus H_n,$$

where H_n is a structuring element, for example, for a 4- or 8-neighborhood (Fig. 9.11) as the structuring element H_n. The actual boundary pixels B are those contained in the original image but *not* in the eroded image, that is, the *intersection* of the original image I and the inverted result $\bar{I}'$, or

$$B \leftarrow I \cap \overline{I'} = I \cap \overline{(I \ominus H_n)}\,. \tag{9.19}$$

Figure 9.17 shows an example for the extraction of region boundaries. Notice that using the 4-neighborhood as the structuring element H_n produces "8-connected" contours and vice versa [125, p. 504].

The process of boundary extraction is illustrated on a simple example in Fig. 9.16. As can be observed in this figure, the result B

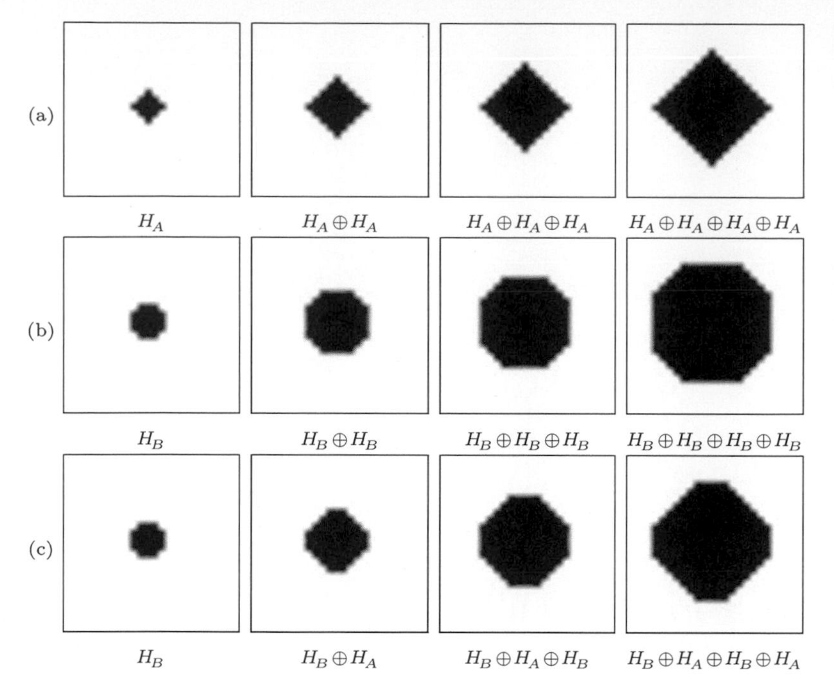

Fig. 9.15
Composition of large morphological filters by repeated application of smaller filters: repeated application of the structuring element H_A (a) and structuring element H_B (b); alternating application of H_B and H_A (c).

contains exactly those pixels that are *different* in the original image I and the eroded image $I' = I \ominus H_n$, which can also be obtained by an exclusive-OR (XOR) operation between pairs of pixels; that is, boundary extraction from a binary image can be implemented as

$$B(\boldsymbol{p}) \leftarrow I(\boldsymbol{p}) \operatorname{XOR} (I \ominus H_n)(\boldsymbol{p}), \quad \text{for all } \boldsymbol{p}. \tag{9.20}$$

Figure 9.17 shows a more complex example for isolating the boundary pixels in a real image.

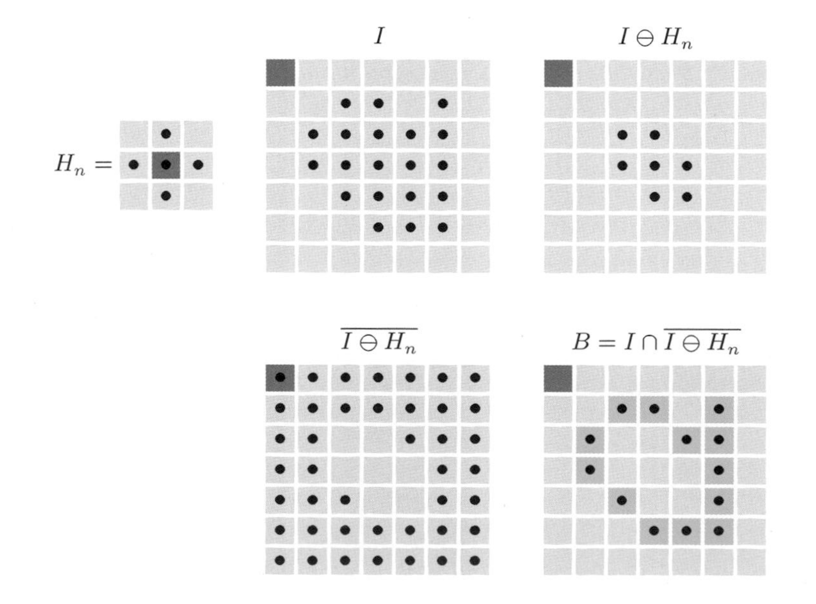

Fig. 9.16
Outline example using a 4-neighborhood structuring element H_n. The image I is first eroded ($I \ominus H_n$) and subsequently inverted ($\overline{I \ominus H_n}$). The boundary pixels are finally obtained as the intersection $I \cap \overline{I \ominus H_n}$.

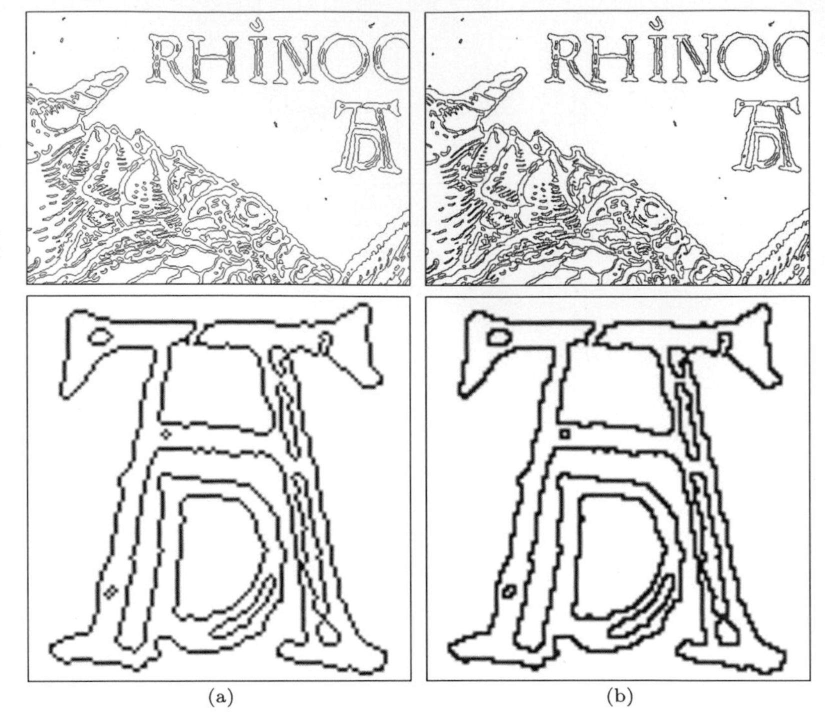

Fig. 9.17
Extraction of boundary pixels using morphological operations. The 4-neighborhood structuring element used in (a) produces 8-connected contours. Conversely, using the 8-neighborhood as the structuring element gives 4-connected contours (b).

9.3 Composite Morphological Operations

Due to their semiduality, dilation and erosion are often used together in composite operations, two of which are so important that they even carry their own names and symbols: "opening" and "closing". They are probably the most frequently used morphological operations in practice.

9.3.1 Opening

A binary opening $I \circ H$ denotes an erosion followed by a dilation with the *same* structuring element H,

$$I \circ H = (I \ominus H) \oplus H. \tag{9.21}$$

The main effect of an opening is that all foreground structures that are smaller than the structuring element are eliminated in the first step (erosion). The remaining structures are smoothed by the subsequent dilation and grown back to approximately their original size, as demonstrated by the examples in Fig. 9.18. This process of shrinking and subsequent growing corresponds to the idea for eliminating small structures that we had initially sketched in Sec. 9.1.

9.3.2 Closing

When the sequence of erosion and dilation is reversed, the resulting operation is called a closing and denoted $I \bullet H$,

$$I \bullet H = (I \oplus H) \ominus H. \tag{9.22}$$

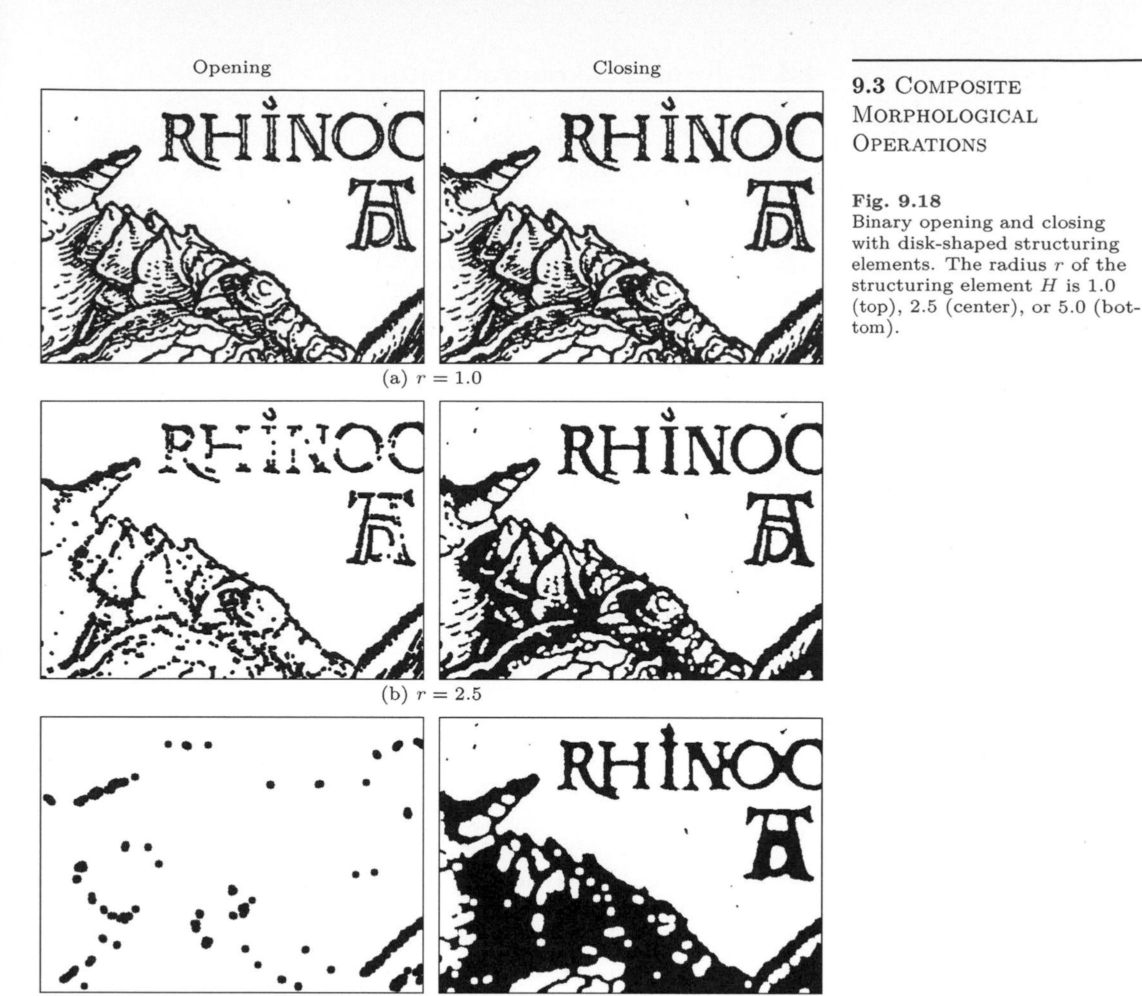

Fig. 9.18
Binary opening and closing with disk-shaped structuring elements. The radius r of the structuring element H is 1.0 (top), 2.5 (center), or 5.0 (bottom).

A *closing* removes (closes) holes and fissures in the foreground structures that are smaller than the structuring element H. Some examples with typical disk-shaped structuring elements are shown in Fig. 9.18.

9.3.3 Properties of Opening and Closing

Both operations, opening as well as closing, are *idempotent*, meaning that their results are "final" in the sense that any subsequent application of the same operation no longer changes the result, that is,

$$\begin{aligned} I \circ H &= (I \circ H) \circ H = ((I \circ H) \circ H) \circ H = \ldots, \\ I \bullet H &= (I \bullet H) \bullet H = ((I \bullet H) \bullet H) \bullet H = \ldots. \end{aligned} \tag{9.23}$$

Also, opening and closing are "duals" in the sense that opening the foreground is equivalent to closing the background and vice versa, that is,

$$I \circ H = \overline{\overline{I} \bullet H} \quad \text{and} \quad I \bullet H = \overline{\overline{I} \circ H}. \tag{9.24}$$

9.4 Thinning (Skeletonization)

Thinning is a common morphological technique which aims at shrinking binary structures down to a maximum thickness of one pixel without splitting them into multiple parts. This is accomplished by iterative "conditional" erosion. It is applied to a local neighborhood only if a sufficiently thick structure remains and the operation does not cause a separation to occur. This requires that, depending on the local image structure, a decision must be made at every image position whether another erosion step may be applied or not. The operation continues until no more changes appear in the resulting image. It follows that, compared to the ordinary ("homogeneous") morphological discussed earlier, thinning is computationally expensive in general. A frequent application of thinning is to calculate the "skeleton" of a binary region, for example, for structural matching of 2D shapes.

Thinning is also known by the terms *center line detection* and *medial axis transform*. Many different implementations of varied complexity and efficiency exist (see, e.g., [2, 7, 68, 108, 201]). In the following, we describe the classic algorithm by Zhang and Suen [265] and its implementation as a representative example.[3]

9.4.1 Thinning Algorithm by Zhang and Suen

The input to this algorithm is a binary image I, with foreground pixels carrying the value 1 and background pixels with value 0. The algorithm scans the image and at each position (u, v) examines a 3×3 neighborhood with the central element P and the surrounding values $\mathsf{N} = (N_0, N_1, \ldots, N_7)$, as illustrated in Fig. 9.5(b). The complete process is summarized in Alg. 9.2.

For classifying the contents of the local neighborhood N we first define the function

$$B(\mathsf{N}) = N_0 + N_1 + \cdots + N_7 = \sum_{i=0}^{7} N_i, \tag{9.25}$$

which simply counts surrounding foreground pixels. We also define the so-called "connectivity number" to express how many binary components are connected via the current center pixel at position (u, v). This quantity is equivalent to the number of $1 \rightarrow 0$ transitions in the sequence $(N_0, \ldots, N_7, N_0)$, or expressed in arithmetic terms,

$$C(\mathsf{N}) = \sum_{i=0}^{7} N_i \cdot [N_i - N_{(i+1) \bmod 8}]. \tag{9.26}$$

Figure 9.19 shows some selected examples for the neighborhood N and the associated values for the functions $B(\mathsf{N})$ and $C(\mathsf{N})$. Based on the above functions, we finally define two Boolean predicates R_1, R_2 on the neighborhood N,

[3] The built-in thinning operation in ImageJ is also based on this algorithm.

$$R_1(\mathsf{N}) := [\,2 \leq B(\mathsf{N}) \leq 6\,] \wedge [\,C(\mathsf{N}) = 1\,] \wedge [\,N_6 \cdot N_0 \cdot N_2 = 0\,] \wedge [\,N_4 \cdot N_6 \cdot N_0 = 0\,], \qquad (9.27)$$

$$R_2(\mathsf{N}) := [\,2 \leq B(\mathsf{N}) \leq 6\,] \wedge [\,C(\mathsf{N}) = 1\,] \wedge [\,N_0 \cdot N_2 \cdot N_4 = 0\,] \wedge [\,N_2 \cdot N_4 \cdot N_6 = 0\,]. \qquad (9.28)$$

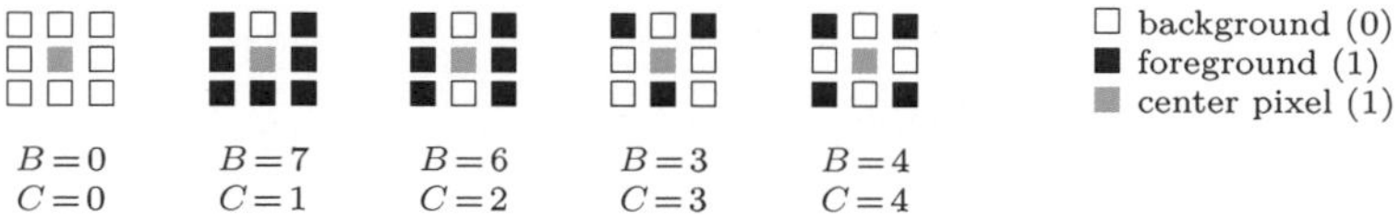

$B=0$, $C=0$ | $B=7$, $C=1$ | $B=6$, $C=2$ | $B=3$, $C=3$ | $B=4$, $C=4$

□ background (0)
■ foreground (1)
■ center pixel (1)

Fig. 9.19 Selected binary neighborhood patterns N and associated function values $B(\mathsf{N})$ and $C(\mathsf{N})$ (see Eqns. (9.25)–(9.26)).

Depending on the outcome of $R_1(\mathsf{N})$ and $R_2(\mathsf{N})$, the foreground pixel at the center position of N is either deleted (i.e., eroded) or marked as non-removable (see Alg. 9.2, lines 16 and 27).

Figure 9.20 illustrates the effect of layer-by-layer thinning performed by procedure ThinOnce(). In every iteration, only one "layer" of foreground pixels is selectively deleted. An example of thinning applied to a larger binary image is shown in Fig. 9.21.

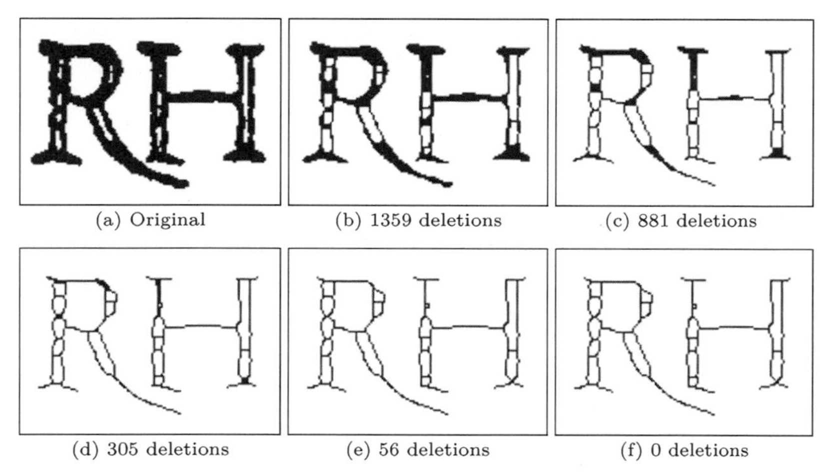

(a) Original (b) 1359 deletions (c) 881 deletions (d) 305 deletions (e) 56 deletions (f) 0 deletions

Fig. 9.20 Iterative application of the ThinOnce() procedure. The "deletions" indicated in (b–f) denote the number of pixels that were removed from the previous image. No deletions occurred in the final iteration (from (e) to (f)). Thus five iterations were required to thin this image.

9.4.2 Fast Thinning Algorithm

In a binary image, only $2^8 = 256$ different combinations of zeros and ones are possible inside any 8-neighborhood. Since the expressions in Eqns. (9.27)–(9.27) are relatively costly to evaluate it makes sense to pre-calculate and tabulate all 256 instances (see Fig. 9.22). This is the basis of the fast version of Zhang and Suen's algorithm, summarized in Alg. 9.3. It uses a decision table Q, which is constant and calculated only once by procedure MakeDeletionCodeTable() in Alg. 9.3 (lines 34–45). The table contains the binary codes

$$\mathsf{Q}(i) \in \{0, 1, 2, 3\} = \{00_{\mathrm{b}}, 01_{\mathrm{b}}, 10_{\mathrm{b}}, 11_{\mathrm{b}}\}, \qquad (9.29)$$

for $i = 0, \ldots, 255$, where the two bits correspond to the predicates R_1 and R_2, respectively. The associated test is found in procedure ThinOnceFast() in line 19. The two passes are in this case controlled by a separate loop variable ($p = 1, 2$). In the concrete implementation, the map Q is not calculated at the start but defined as a constant array (see Prog. 9.1 for the actual Java code).

Alg. 9.2
Iterative thinning algorithm by Zhang und Suen [265]. Procedure ThinOnce() performs a single thinning step on the supplied binary image I_b and returns the number of deleted foreground pixels. It is iteratively invoked by Thin() until no more pixels are deleted. The required pixel deletions are only registered in the binary map D and executed en-bloc at the end of every iteration. Lines 40–42 define the functions $R_1()$, $R_2()$, $B()$ and $C()$ used to characterize the local pixel neighborhoods. Note that the order of processing the image positions (u, v) in the **for all** loops in **Pass 1** and **Pass 2** is completely arbitrary. In particular, positions could be processed simultaneously, so the algorithm may be easily parallelized (and thereby accelerated).

```
1:  Thin(I_b, i_max)
      Input: I_b, binary image with background = 0, foreground > 0;
      i_max, max. number of iterations. Returns the number of iterations
      performed and modifies I_b.
2:    (M, N) ← Size(I_b)
3:    Create a binary map D: M × N ↦ {0, 1}
4:    i ← 0
5:    do
6:        n_d ← ThinOnce(I_b, D)
7:        i ← i + 1
8:    while (n_d > 0 ∧ i < i_max)      ▷ do ... while more deletions required
9:    return i

10: ThinOnce(I_b, D)
      Pass 1:
11:   n_1 ← 0                                            ▷ deletion counter
12:   for all image positions (u, v) ∈ M×N do
13:       D(u, v) ← 0
14:       if I_b(u, v) > 0 then
15:           N ← GetNeighborhood(I_b, u, v)
16:           if R_1(N) then                              ▷ see Eq. 9.27
17:               D(u, v) ← 1                ▷ mark pixel (u, v) for deletion
18:               n_1 ← n_1 + 1
19:   if n_1 > 0 then                          ▷ at least 1 deletion required
20:       for all image positions (u, v) ∈ M×N do
21:           I_b(u, v) ← I_b(u, v) − D(u, v)     ▷ delete all marked pixels
      Pass 2:
22:   n_2 ← 0
23:   for all image positions (u, v) ∈ M×N do
24:       D(u, v) ← 0
25:       if I_b(u, v) > 0 then
26:           N ← GetNeighborhood(I_b, u, v)
27:           if R_2(N) then                              ▷ see Eq. 9.28
28:               D(u, v) ← 1                ▷ mark pixel (u, v) for deletion
29:               n_2 ← n_2 + 1
30:   if n_2 > 0 then                          ▷ at least 1 deletion required
31:       for all image positions (u, v) ∈ M×N do
32:           I_b(u, v) ← I_b(u, v) − D(u, v)     ▷ delete all marked pixels
33:   return n_1 + n_2

34: GetNeighborhood(I_b, u, v)
35:   N_0 ← I_b(u+1, v),       N_1 ← I_b(u+1, v−1)
36:   N_2 ← I_b(u, v−1),       N_3 ← I_b(u−1, v−1)
37:   N_4 ← I_b(u−1, v),       N_5 ← I_b(u−1, v+1)
38:   N_6 ← I_b(u, v+1),       N_7 ← I_b(u+1, v+1)
39:   return (N_0, N_1, ..., N_7)
```

40: $R_1(\mathsf{N}) := [2 \leq B(\mathsf{N}) \leq 6] \wedge [C(\mathsf{N}) = 1] \wedge [N_6 \cdot N_0 \cdot N_2 = 0] \wedge [N_4 \cdot N_6 \cdot N_0 = 0]$

41: $R_2(\mathsf{N}) := [2 \leq B(\mathsf{N}) \leq 6] \wedge [C(\mathsf{N}) = 1] \wedge [N_0 \cdot N_2 \cdot N_4 = 0] \wedge [N_2 \cdot N_4 \cdot N_6 = 0]$

42: $B(\mathsf{N}) := \sum_{i=0}^{7} N_i, \qquad C(\mathsf{N}) := \sum_{i=0}^{7} N_i \cdot [N_i - N_{(i+1) \bmod 8}]$

Alg. 9.3
Thinning algorithm by Zhang und Suen (accelerated version of Alg. 9.2). This algorithm employs a pre-calculated table of "deletion codes" (Q). Procedure GetNeighborhood() has been replaced by GetNeighborhoodIndex(), which does not return the neighboring pixel values themselves but the associated 8-bit index c with possible values in $0, \ldots, 255$ (see Fig. 9.22). For completeness, the calculation of table Q is included in procedure MakeDeletionCodeTable(), although this table is fixed and may be simply defined as a constant array (see Prog. 9.1).

```
1:  ThinFast(I_b, i_max)
        Input: I_b, binary image with background = 0, foreground > 0;
        i_max, max. number of iterations. Returns the number of iterations
        performed and modifies I_b.
2:      (M, N) ← Size(I_b)
3:      Q ← MakeDeletionCodeTable()
4:      Create a binary map D: M × N ↦ {0, 1}
5:      i ← 0
6:      do
7:          n_d ← ThinOnce(I_b, D)
8:      while (n_d > 0 ∧ i < i_max)  ▷ do ... while more deletions required
9:      return i

10: ThinOnceFast(I_b, D)                  ▷ performs a single thinning iteration
11:     n_d ← 0                            ▷ number of deletions in both passes
12:     for p ← 1, 2 do                            ▷ pass counter (2 passes)
13:         n ← 0                          ▷ number of deletions in current pass
14:         for all image positions (u, v) do
15:             D(u, v) ← 0
16:             if I_b(u, v) = 1 then                        ▷ I_b(u, v) = P
17:                 c ← GetNeighborhoodIndex(I_b, u, v)
18:                 q ← Q(c)        ▷ q ∈ {0, 1, 2, 3} = {00_b, 01_b, 10_b, 11_b}
19:                 if (p and q) ≠ 0 then          ▷ bitwise 'and' operation
20:                     D(u, v) ← 1            ▷ mark pixel (u, v) for deletion
21:                     n ← n + 1
22:         if n > 0 then                      ▷ at least 1 deletion is required
23:             n_d ← n_d + n
24:             for all image positions (u, v) do
25:                 I_b(u, v) ← I_b(u, v) − D(u, v)         ▷ delete all marked
                    pixels
26:     return n_d

27: GetNeighborhoodIndex(I_b, u, v)
28:     N_0 ← I_b(u+1, v),       N_1 ← I_b(u+1, v−1)
29:     N_2 ← I_b(u, v−1),       N_3 ← I_b(u−1, v−1)
30:     N_4 ← I_b(u−1, v),       N_5 ← I_b(u−1, v+1)
31:     N_6 ← I_b(u, v+1),       N_7 ← I_b(u+1, v+1)
32:     c ← N_0 + N_1·2 + N_2·4 + N_3·8 + N_4·16 + N_5·32 + N_6·64 + N_7·128
33:     return c                                             ▷ c ∈ [0, 255]

34: MakeDeletionCodeTable()
35:     Create maps Q: [0, 255] ↦ {0, 1, 2, 3},   N: [0, 7] ↦ {0, 1}
36:     for i ← 0, ..., 255 do                 ▷ list all possible neighborhoods
37:         for k ← 0, ..., 7 do                       ▷ check neighbors 0, ..., 7
38:             N(k) ← { 1  if (i and 2^k) ≠ 0         ▷ test the k^th bit of i
                       { 0  otherwise
39:         q ← 0
40:         if R_1(N) then                           ▷ see Alg. 9.2, line 40
41:             q ← q + 1                                  ▷ set bit 0 of q
42:         if R_2(N) then                           ▷ see Alg. 9.2, line 41
43:             q ← q + 2                                  ▷ set bit 1 of q
44:         Q(i) ← q              ▷ q ∈ {0, 1, 2, 3} = {00_b, 01_b, 10_b, 11_b}
45:     return Q
```

Fig. 9.21
Thinning a binary image (Alg. 9.2 or 9.3). Original image with enlarged detail (a, c) and results after thinning (b, d). The original foreground pixels are marked green, the resulting pixels are black.

Prog. 9.1
Java definition for the "deletion code" table Q (see Fig. 9.22).

```
static final byte[] Q = {
  0, 0, 0, 3, 0, 0, 3, 3, 0, 0, 0, 0, 3, 0, 3, 3,
  0, 0, 0, 0, 0, 0, 0, 0, 3, 0, 0, 0, 3, 0, 3, 1,
  0, 0, 0, 0, 0, 0, 0, 0, 0, 0, 0, 0, 0, 0, 0, 0,
  3, 0, 0, 0, 0, 0, 0, 0, 3, 0, 0, 0, 3, 0, 3, 1,
  0, 0, 0, 0, 0, 0, 0, 0, 0, 0, 0, 0, 0, 0, 0, 0,
  0, 0, 0, 0, 0, 0, 0, 0, 0, 0, 0, 0, 0, 0, 0, 0,
  3, 0, 0, 0, 0, 0, 0, 0, 0, 0, 0, 0, 0, 0, 0, 0,
  3, 0, 0, 0, 0, 0, 0, 0, 3, 0, 0, 0, 1, 0, 1, 0,
  0, 3, 0, 3, 0, 0, 0, 3, 0, 0, 0, 0, 0, 0, 0, 3,
  0, 0, 0, 0, 0, 0, 0, 0, 0, 0, 0, 0, 0, 0, 0, 1,
  0, 0, 0, 0, 0, 0, 0, 0, 0, 0, 0, 0, 0, 0, 0, 0,
  0, 0, 0, 0, 0, 0, 0, 0, 0, 0, 0, 0, 0, 0, 0, 0,
  3, 3, 0, 3, 0, 0, 0, 2, 0, 0, 0, 0, 0, 0, 0, 2,
  0, 0, 0, 0, 0, 0, 0, 0, 0, 0, 0, 0, 0, 0, 0, 0,
  3, 3, 0, 3, 0, 0, 0, 2, 0, 0, 0, 0, 0, 0, 0, 0,
  3, 2, 0, 2, 0, 0, 0, 0, 3, 2, 0, 0, 1, 0, 0, 0
};
```

9.4.3 Java Implementation

The complete Java source code for the morphological operations on binary images is available online as part of the `imagingbook`[4] library.

[4] Package `imagingbook.pub.morphology`.

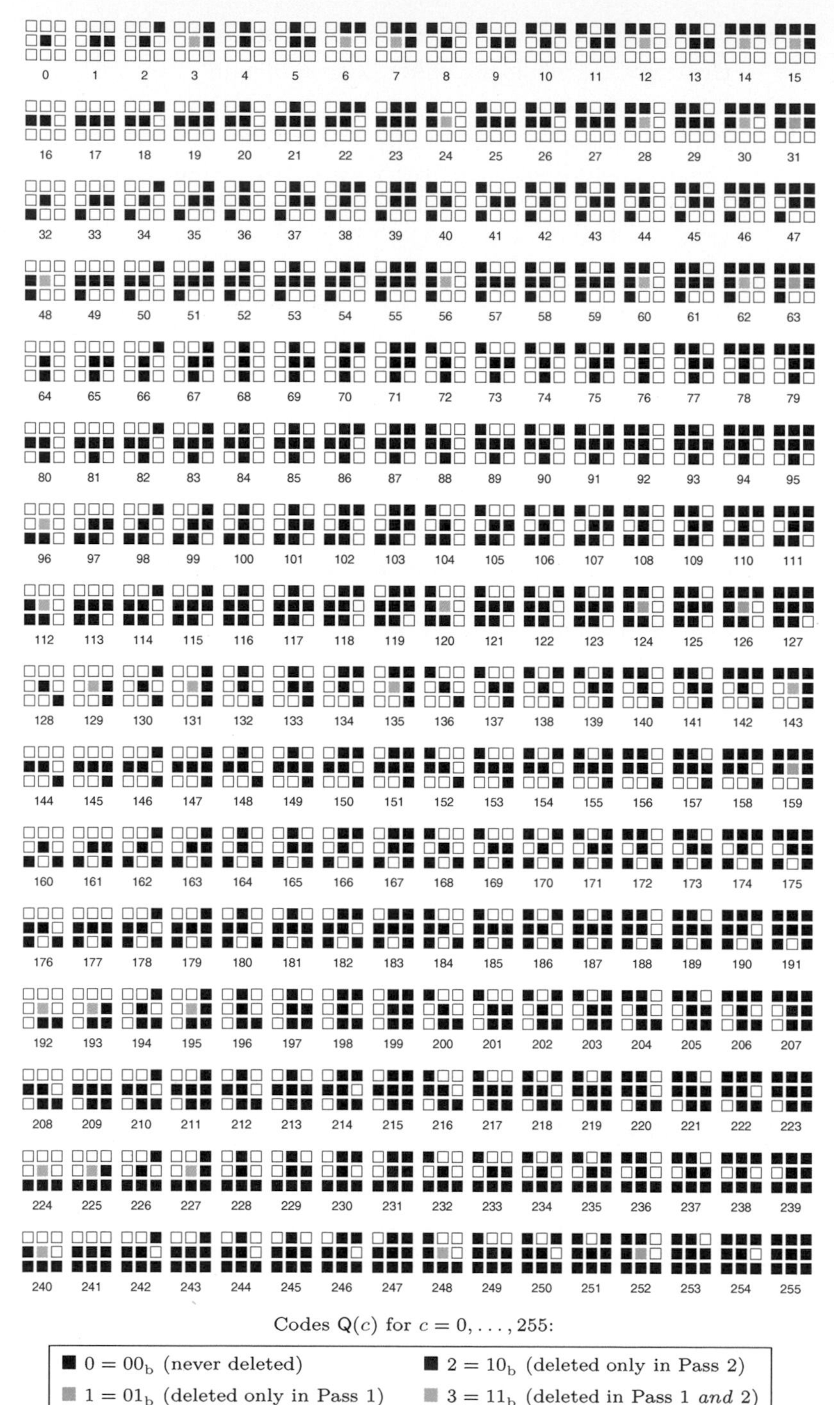

Fig. 9.22
"Deletion codes" for the 256 possible binary 8-neighborhoods tabulated in map $Q(c)$ of Alg. 9.3. □ = 0 and ■ = 1 denote background and foreground pixels, respectively. The 2-bit codes are color coded as indicated at the bottom.

BinaryMorphologyFilter class

This class implements several morphological operators for binary images of type `ByteProcessor`. It defines the sub-classes `Box` and `Disk` with different structuring elements. The class provides the following constructors:

`BinaryMorphologyFilter ()`
Creates a morphological filter with a (default) structuring element of size 3×3 as depicted in Fig. 9.11(b).

`BinaryMorphologyFilter (int[][] H)`
Creates a morphological filter with a structuring element specified by the 2D array `H`, which may contain 0/1 values only (all values > 0 are treated as 1).

`BinaryMorphologyFilter.Box (int rad)`
Creates a morphological filter with a square structuring element of radius $\texttt{rad} \geq 1$ and side length $2 \cdot \texttt{rad} + 1$ pixels.

`BinaryMorphologyFilter.Disk (double rad)`
Creates a morphological filter with a disk-shaped structuring element with radius $\texttt{rad} \geq 1$ and diameter $2 \cdot \text{round}(\texttt{rad}) + 1$ pixels.

The key methods[5] of `BinaryMorphologyFilter` are:

`void applyTo (ByteProcessor I, OpType op)`
Destructively applies the morphological operator `op` to the image `I`. Possible arguments for `op` are `Dilate`, `Erode`, `Open`, `Close`, `Outline`, `Thin`.

`void dilate (ByteProcessor I)`
Performs (destructive) *dilation* on the binary image `I` with the initial structuring element of this filter.

`void erode (ByteProcessor I)`
Performs (destructive) *erosion* on the binary image `I`.

`void open (ByteProcessor I)`
Performs (destructive) *opening* on the binary image `I`.

`void close (ByteProcessor I)`
Performs (destructive) *closing* on the binary image `I`.

`void outline (ByteProcessor I)`
Performs a (destructive) *outline* operation on the binary image `I` using a 3×3 structuring element (see Sec. 9.2.7).

`void thin (ByteProcessor I)`
Performs a (destructive) *thinning* operation on the binary image `I` using a 3×3 structuring element (with at most $i_{\max} = 1500$ iterations, see Alg. 9.3).

`void thin (ByteProcessor I, int iMax)`
Performs a thinning operation with at most `iMax` iterations (see Alg. 9.3).

`int thinOnce (ByteProcessor I)`
Performs a single iteration of the thinning operation and returns the number of pixel deletions (see Alg. 9.3).

The methods listed here *always* treat image pixels with value 0 as background and values > 0 as foreground. Unlike ImageJ's built-in implementation of morphological operations (described in Sec. 9.4.4), the display lookup table (LUT, typically only used for display purposes) of the image is *not* taken into account at all.

[5] See the online documentation for additional methods.

```
1 import ij.ImagePlus;
2 import ij.plugin.filter.PlugInFilter;
3 import ij.process.ByteProcessor;
4 import ij.process.ImageProcessor;
5 import imagingbook.pub.morphology.BinaryMorphologyFilter;
6 import imagingbook.pub.morphology.BinaryMorphologyFilter.
      OpType;
7
8 public class Bin_Dilate_Disk_Demo implements PlugInFilter {
9   static double radius = 5.0;
10  static OpType op = OpType.Dilate; // Erode, Open, Close, ...
11
12  public int setup(String arg, ImagePlus imp) {
13    return DOES_8G;
14  }
15
16  public void run(ImageProcessor ip) {
17    BinaryMorphologyFilter bmf =
18            new BinaryMorphologyFilter.Disk(radius);
19    bmf.applyTo((ByteProcessor) ip, op);
20  }
21 }
```

Prog. 9.2
Example for using class `BinaryMorphologyFilter` (see Sec. 9.4.3) inside a ImageJ plugin. The actual filter operator is instantiated in line 18 and subsequently (in line 19) applied to the image `ip` of type `ByteProcessor`. Available operations (`OpType`) are `Dilate`, `Erode`, `Open`, `Close`, `Outline` and `Thin`. Note that the results depend strictly on the pixel values of the input image, with values 0 taken as background and values > 0 taken as foreground. The display lookup-table (LUT) is irrelevant.

The example in Prog. 9.2 shows the use of class `BinaryMorphologyFilter` in a complete ImageJ plugin that performs dilation with a disk-shaped structuring element of radius 5 (pixel units). Other examples can be found in the online code repository.

9.4.4 Built-in Morphological Operations in ImageJ

Apart from the implementation described in the previous section, the ImageJ API provides built-in methods for basic morphological operations, such as `dilate()` and `erode()`. These methods use a 3×3 structuring element (analogous to Fig. 9.11(b)) and are only defined for images of type `ByteProcessor` and `ColorProcessor`. In the case of RGB color images (`ColorProcessor`) the morphological operation is applied individually to the three color channels. All these and other morphological operations can be applied interactively through ImageJ's Process ▷ Binary menu (see Fig. 9.23(a)).

Note that ImageJ's `dilate()` and `erode()` methods use the current settings of display lookup table (LUT) to discriminate between background and foreground pixels. Thus the results of morphological operations depend not only on the stored pixel values but how they are being displayed (in addition to the settings in Process ▷ Binary ▷ Options..., see Fig. 9.23(b)).[6] It is therefore recommended to use the methods (defined for `ByteProcessor` only)

```
dilate(int count, int background),
erode(int count, int background)
```

[6] These dependencies may be quite confusing because the same program will produce different results under different user setups.

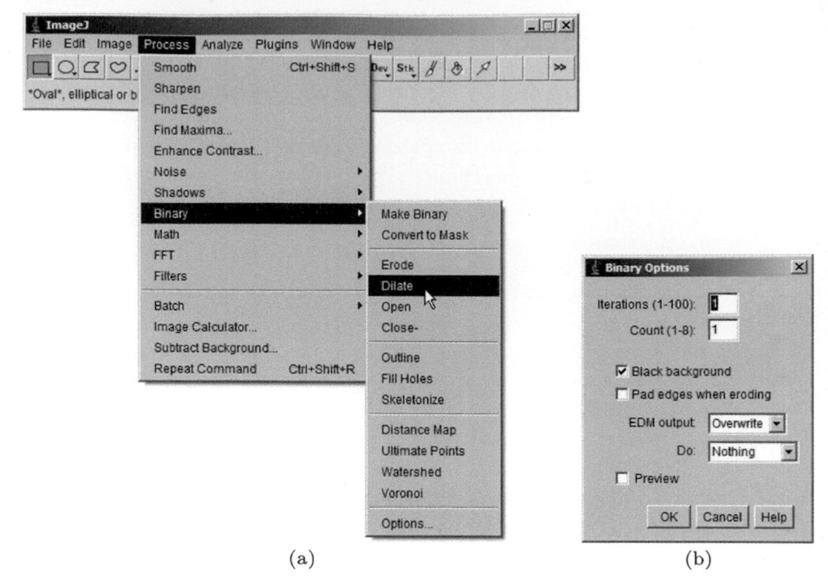

Fig. 9.23 Morphological operations in ImageJ's built-in standard menu Process ▷ Binary (a) and optional settings with Process ▷ Binary ▷ Options... (b). The choice "Black background" specifies if background pixels are bright or dark, which is taken into account by ImageJ's morphological operations.

instead, since they provide explicit control of the background pixel value and are thus independent from other settings. ImageJ's `ByteProcessor` class defines additional methods for morphological operations on binary images, such as `outline()` and `skeletonize()`. The method `outline()` implements the extraction of region boundaries using an 8-neighborhood structuring element, as described in Sec. 9.2.7. The method `skeletonize()`, on the other hand, implements a thinning process similar to Alg. 9.3.

9.5 Grayscale Morphology

Morphological operations are not confined to binary images but are also for intensity (grayscale) images. In fact, the definition of grayscale morphology is a *generalization* of binary morphology, with the binary OR and AND operators replaced by the arithmetic MAX and MIN operators, respectively. As a consequence, procedures designed for grayscale morphology can also perform binary morphology (but not the other way around).[7] In the case of color images, the grayscale operations are usually applied individually to each color channel.

9.5.1 Structuring Elements

Unlike in the binary scheme, the structuring elements for grayscale morphology are not defined as point sets but as real-valued 2D functions, that is,

$$H(i,j) \in \mathbb{R}, \qquad \text{for } (i,j) \in \mathbb{Z}^2. \tag{9.30}$$

The values in H may be negative or zero. Notice, however, that, in contrast to linear convolution (Sec. 5.3.1), zero elements in grayscale

[7] ImageJ provides a single implementation of morphological operations that handles both binary and grayscale images (see Sec. 9.4.4).

morphology generally *do* contribute to the result.[8] The design of structuring elements for grayscale morphology must therefore distinguish explicitly between cells containing the value 0 and empty ("don't care") cells, for example,

$$\begin{array}{|c|c|c|}\hline 0 & 1 & 0 \\ \hline 1 & \mathbf{2} & 1 \\ \hline 0 & 1 & 0 \\ \hline \end{array} \;\neq\; \begin{array}{|c|c|c|}\hline & 1 & \\ \hline 1 & \mathbf{2} & 1 \\ \hline & 1 & \\ \hline \end{array}\,. \qquad (9.31)$$

9.5.2 Dilation and Erosion

The result of grayscale *dilation* $I \oplus H$ is defined as the *maximum* of the values in H added to the values of the current subimage of I, that is,

$$(I \oplus H)\,(u,v) = \max_{(i,j)\in H} \bigl(I(u+i,v+j) + H(i,j)\bigr)\,. \qquad (9.32)$$

Similarly, the result of grayscale *erosion* is the *minimum* of the differences,

$$(I \ominus H)(u,v) = \min_{(i,j)\in H} \bigl(I(u+i,v+j) - H(i,j)\bigr)\,. \qquad (9.33)$$

Figures 9.24 and 9.25 demonstrate the basic process of grayscale dilation and erosion, respectively, on a simple example.

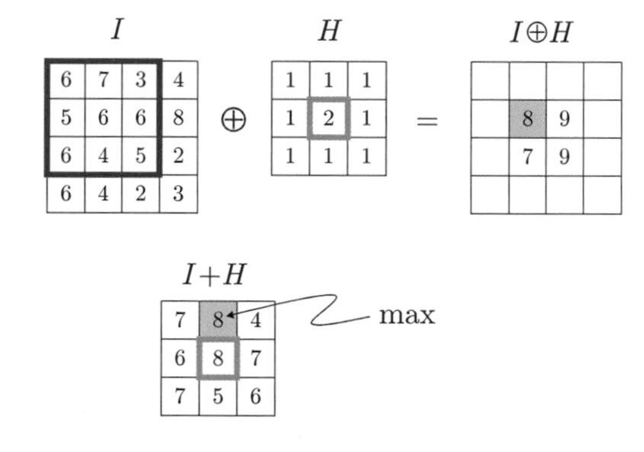

Fig. 9.24
Grayscale dilation $I \oplus H$. The 3×3 pixel structuring element H is placed on the image I in the upper left position. Each value of H is added to the corresponding element of I; the intermediate result $(I + H)$ for this particular position is shown below. Its maximum value $8 = 7 + 1$ is inserted into the result $(I \oplus H)$ at the current position of the filter origin. The results for three other filter positions are also shown.

In general, either operation may produce *negative* results that must be considered if the range of pixel values is restricted, for example, by clamping the results (see Ch. 4, Sec. 4.1.2). Some examples of grayscale dilation and erosion on natural images using disk-shaped structuring elements of various sizes are shown in Fig. 9.26. Figure 9.28 demonstrates the same operations with some freely designed structuring elements.

9.5.3 Grayscale Opening and Closing

Opening and closing on grayscale images are defined, identical to the binary case (Eqns. (9.21) and (9.22)), as operations composed

[8] While a zero coefficient in a linear convolution matrix simply means that the corresponding image pixel is ignored.

I

6	7	3	4
5	6	6	8
6	4	5	2
6	4	2	3

$\ominus$ H

1	1	1
1	2	1
1	1	1

$=$ $I \ominus H$

	2	1	
	1	1	

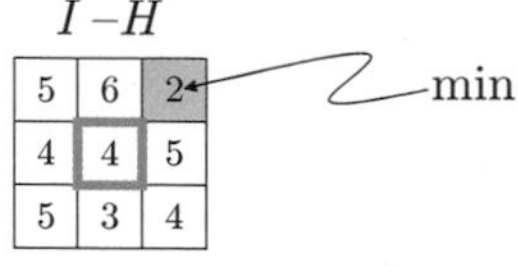

Fig. 9.25
Grayscale erosion $I \ominus H$. The 3×3 pixel structuring element H is placed on the image I in the upper left position. Each value of H is subtracted from the corresponding element of I; the intermediate result $(I - H)$ for this particular position is shown below. Its minimum value $3-1 = 2$ is inserted into the result $(I \ominus H)$ at the current position of the filter origin. The results for three other filter positions are also shown.

Original

Dilation

Erosion

(a) $r = 2.5$

(b) $r = 5.0$

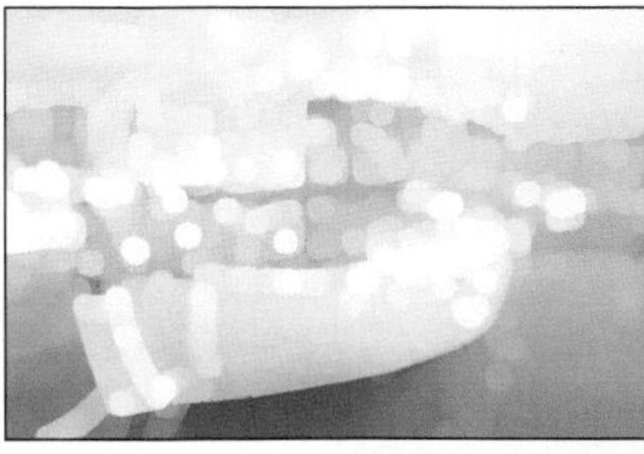

(c) $r = 10.0$

Fig. 9.26
Grayscale dilation and erosion with disk-shaped structuring elements. The radius r of the structuring element is 2.5 (a), 5.0 (b), and 10.0 (c).

Opening Closing

(a) $r = 2.5$

(b) $r = 5.0$

(c) $r = 10.0$

Fig. 9.27
Grayscale opening and closing with disk-shaped structuring elements. The radius r of the structuring element is 2.5 (a), 5.0 (b), and 10.0 (c).

of dilation and erosion with the same structuring element. Some examples are shown in Fig. 9.27 for disk-shaped structuring elements and in Fig. 9.29 for various nonstandard structuring elements. Notice that interesting effects can be obtained, particularly from structuring elements resembling the shape of brush or other stroke patterns.

As mentioned in Sec. 9.4.4, the morphological operations avaiable in ImageJ can be applied to binary images as well as grayscale images. In addition, several additional plugins and complete morphological packages are available online,[9] including the morphology operators by Gabriel Landini and the Grayscale Morphology package by Dimiter Prodanov, which allows structuring elements to be interactively specified (a modified version was used for some examples in this chapter).

9.6 Exercises

Exercise 9.1. Manually calculate the results of dilation and erosion for the following image I and the structuring elements H_1 and H_2:

[9] See http://rsb.info.nih.gov/ij/plugins/.

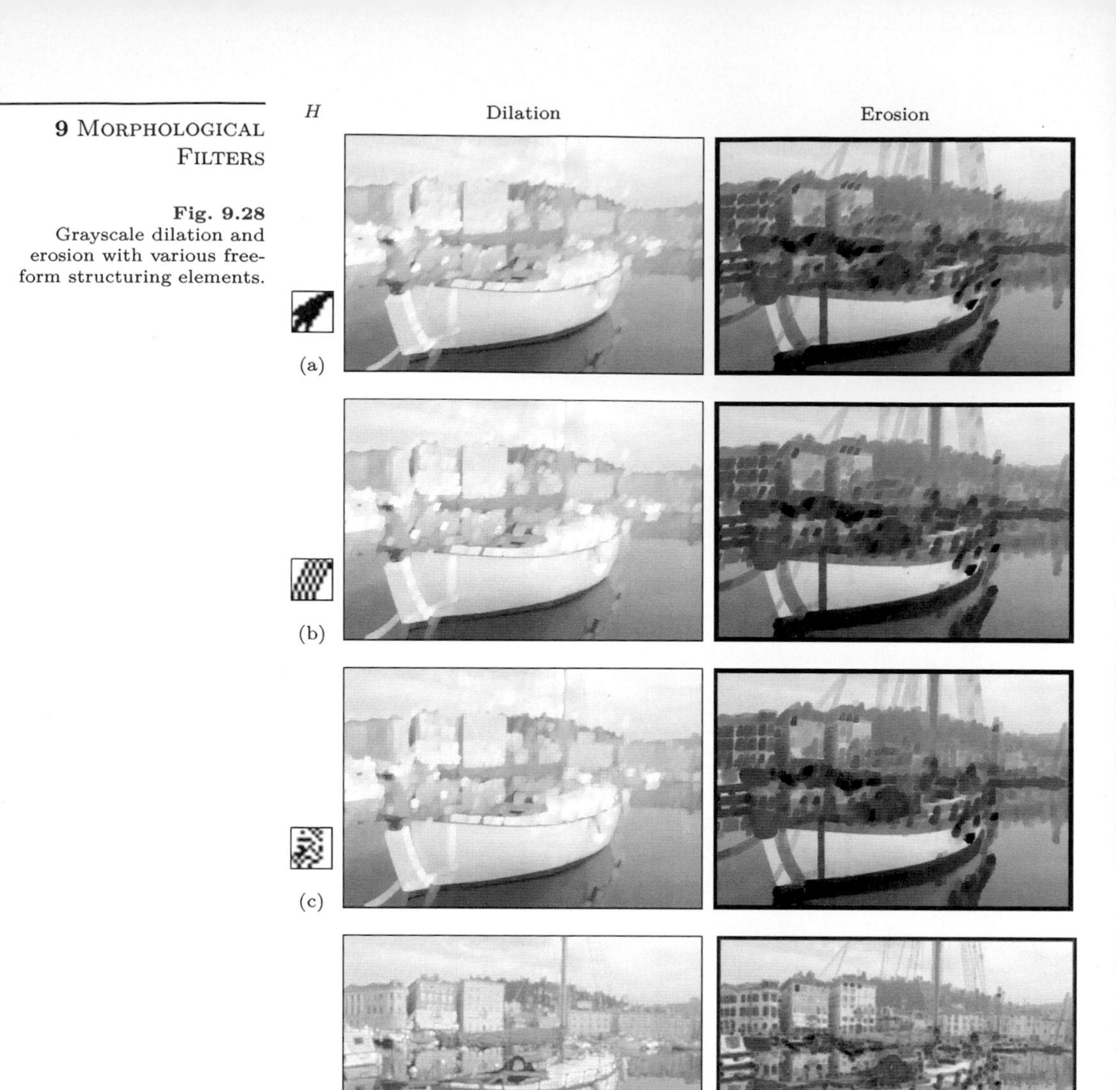

Fig. 9.28 Grayscale dilation and erosion with various free-form structuring elements.

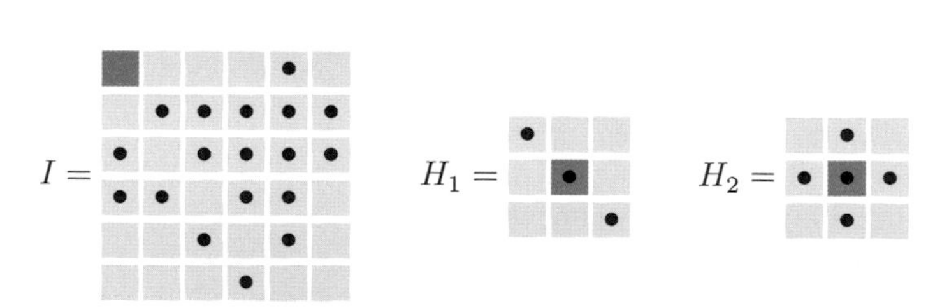

Exercise 9.2. Assume that a binary image I contains unwanted foreground spots with a maximum diameter of 5 pixels that should be removed without damaging the remaining structures. Design a suitable morphological procedure, and evaluate its performance on appropriate test images.

Exercise 9.3. Investigate if the results of the thinning operation described in Alg. 9.2 (and implemented by the `thin()` method of class `BinaryMorphologyFilter`) are invariant against rotating the image

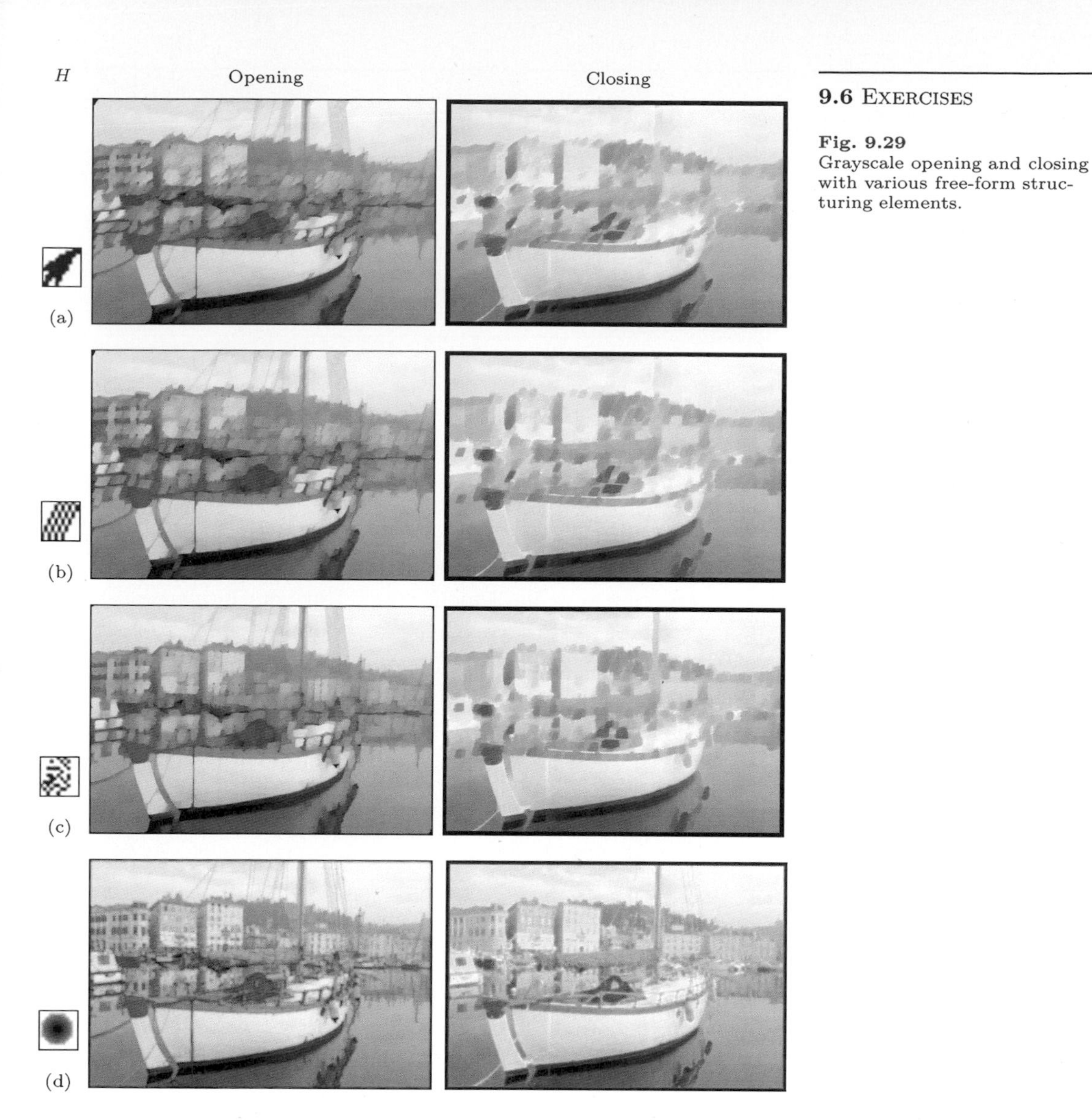

Fig. 9.29
Grayscale opening and closing with various free-form structuring elements.

by 90° and horizontal or vertical mirroring. Use appropriate test images to see if the results are identical.

Exercise 9.4. Show that, in the special case of the structuring elements with the contents

•	•	•
•	•	•
•	•	•

for *binary* and

0	0	0
0	0	0
0	0	0

for *grayscale* images,

dilation is equivalent to a 3×3 pixel maximum filter and erosion is equivalent to a 3×3 pixel minimum filter (see Ch. 5, Sec. 5.4.1).

Exercise 9.5. Thinning can be applied to extract the "skeleton" of a binary region, which in turn can be used to characterize the shape of the region. A common approach is to partition the skeleton into a graph, consisting of nodes and connecting segments, as a

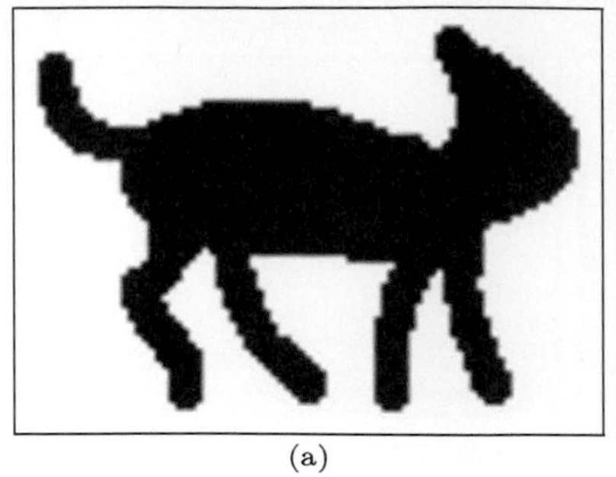

(a)

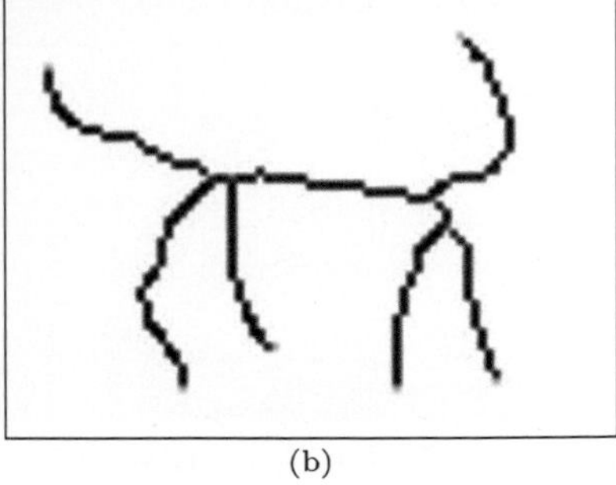

(b)

Fig. 9.30
Segmentation of a region skeleton. Original binary image (a) and the skeleton obtained by thinning (b). Terminal nodes are marked green, connecting (inner) nodes are marked red.

shape representation (see Fig. 9.30 for an example). Use ImageJ's `skeletonize()` method or the `thin()` methode of class `BinaryMorphologyFilter` (see Sec. 9.4.3) to generate the skeleton, then locate and mark the connecting and terminal nodes of this structure. Define precisely the properties of each type of node and use this definition in your implementation. Test your implementation on different examples. How would you generally judge the robustness of this approach as a 2D shape representation?

10

Regions in Binary Images

In a binary image, pixels can take on exactly one of two values. These values are often thought of as representing the "foreground" and "background" in the image, even though these concepts often are not applicable to natural scenes. In this chapter we focus on connected regions in images and how to isolate and describe such structures.

Let us assume that our task is to devise a procedure for finding the number and type of objects contained in an image as shown in Fig. 10.1. As long as we continue to consider each pixel in isolation, we will not be able to determine how many objects there are overall in the image, where they are located, and which pixels belong to which objects. Therefore our first step is to find each object by grouping together all the pixels that belong to it. In the simplest case, an object is a group of touching foreground pixels, that is, a connected *binary region* or "component".

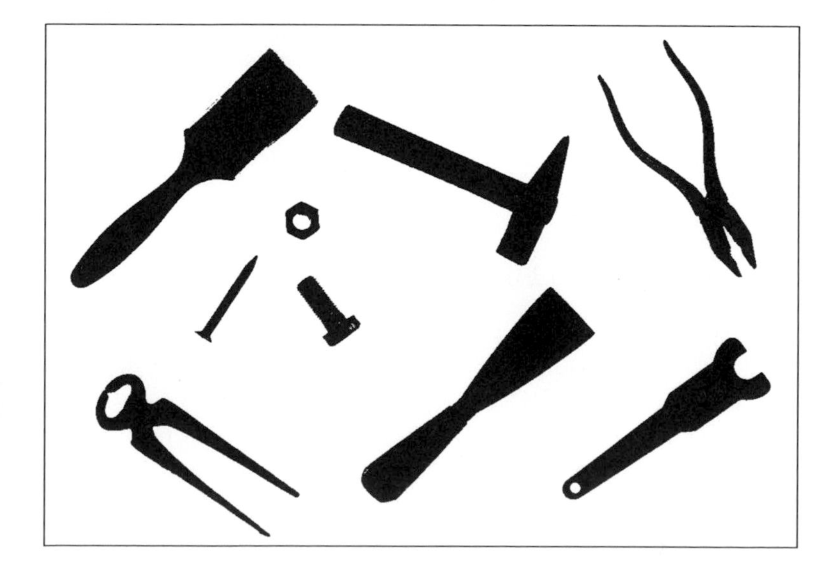

Fig. 10.1
Binary image with nine components. Each component corresponds to a connected region of (black) foreground pixels.

10.1 Finding Connected Image Regions

In the search for binary regions, the most important tasks are to find out which pixels belong to which regions, how many regions are in the image, and where these regions are located. These steps usually take place as part of a process called *region labeling* or *region coloring*. During this process, neighboring pixels are pieced together in a stepwise manner to build regions in which all pixels within that region are assigned a unique number ("label") for identification. In the following sections, we describe two variations on this idea. In the first method, region marking through *flood filling*, a region is filled in all directions starting from a single point or "seed" within the region. In the second method, *sequential region marking*, the image is traversed from top to bottom, marking regions as they are encountered. In Sec. 10.2.2, we describe a third method that combines two useful processes, region labeling and contour finding, in a single algorithm.

Independent of which of these methods we use, we must first settle on either the 4- or 8-connected definition of neighboring (see Ch. 9, Fig. 9.5) for determining when two pixels are "connected" to each other, since under each definition we can end up with different results. In the following region-marking algorithms, we use the following convention: the original binary image $I(u, v)$ contains the values 0 and 1 to mark the *background* and *foreground*, respectively; any other value is used for numbering (labeling) the regions, that is, the pixel values are

$$I(u,v) = \begin{cases} 0 & \text{background,} \\ 1 & \text{foreground,} \\ 2, 3, \ldots & \text{region label.} \end{cases} \tag{10.1}$$

10.1.1 Region Labeling by Flood Filling

The underlying algorithm for region marking by *flood filling* is simple: search for an unmarked foreground pixel and then fill (visit and mark) all the rest of the neighboring pixels in its region. This operation is called a "flood fill" because it is as if a flood of water erupts at the start pixel and flows out across a flat region. There are various methods for carrying out the fill operation that ultimately differ in how to select the coordinates of the next pixel to be visited during the fill. We present three different ways of performing the FloodFill() procedure: a recursive version, an iterative *depth-first* version, and an iterative *breadth-first* version (see Alg. 10.1):

A. Recursive Flood Filling: The recursive version (Alg. 10.1, line 8) does not make use of explicit data structures to keep track of the image coordinates but uses the local variables that are implicitly allocated by recursive procedure calls.[1] Within each region, a tree structure, rooted at the starting point, is defined by the neighborhood relation between pixels. The recursive step corresponds to a *depth-first traversal* [54] of this tree and results

[1] In Java, and similar imperative programming languages such as C and C++, local variables are automatically stored on the *call stack* at each procedure call and restored from the stack when the procedure returns.

Alg. 10.1
Region marking by *flood filling*. The binary input image I uses the value 0 for background pixels and 1 for foreground pixels. Unmarked foreground pixels are searched for, and then the region to which they belong is filled. Procedure FloodFill() is defined in three different versions: *recursive*, emphdepth-first and *breadth-first*.

```
1: RegionLabeling(I)
   Input: I, an integer-valued image with initial values 0 = background,
   1 = foreground. Returns nothing but modifies the image I.
2:   label ← 2                          ▷ value of the next label to be assigned
3:   for all image coordinates u, v do
4:     if I(u, v) = 1 then                             ▷ a foreground pixel
5:       FloodFill(I, u, v, label)             ▷ any of the 3 versions below
6:       label ← label + 1.
7:   return
─────────────────────────────────────────────────────────────────────────
8: FloodFill(I, u, v, label)                          ▷ Recursive Version
9:   if u, v is within the image boundaries and I(u, v) = 1 then
10:    I(u, v) ← label
11:    FloodFill(I, u+1, v, label)          ▷ recursive call to FloodFill()
12:    FloodFill(I, u, v+1, label)
13:    FloodFill(I, u, v−1, label)
14:    FloodFill(I, u−1, v, label)
15:  return
─────────────────────────────────────────────────────────────────────────
16: FloodFill(I, u, v, label)                       ▷ Depth-First Version
17:  S ← ()                                       ▷ create an empty stack S
18:  S ← (u, v) ⌣ S                   ▷ push seed coordinate (u, v) onto S
19:  while S ≠ () do                                ▷ while S is not empty
20:    (x, y) ← GetFirst(S)
21:    S ← Delete((x, y), S)           ▷ pop first coordinate off the stack
22:    if x, y is within the image boundaries and I(x, y) = 1 then
23:      I(x, y) ← label
24:      S ← (x+1, y) ⌣ S                           ▷ push (x+1, y) onto S
25:      S ← (x, y+1) ⌣ S                           ▷ push (x, y+1) onto S
26:      S ← (x, y−1) ⌣ S                           ▷ push (x, y−1) onto S
27:      S ← (x−1, y) ⌣ S                           ▷ push (x−1, y) onto S
28:  return
─────────────────────────────────────────────────────────────────────────
29: FloodFill(I, u, v, label)                     ▷ Breadth-First Version
30:  Q ← ()                                       ▷ create an empty queue Q
31:  Q ← Q ⌣ (u, v)                  ▷ append seed coordinate (u, v) to Q
32:  while Q ≠ () do                                ▷ while Q is not empty
33:    (x, y) ← GetFirst(Q)
34:    Q ← Delete((x, y), Q)                  ▷ dequeue first coordinate
35:    if x, y is within the image boundaries and I(x, y) = 1 then
36:      I(x, y) ← label
37:      Q ← Q ⌣ (x+1, y)                          ▷ append (x+1, y) to Q
38:      Q ← Q ⌣ (x, y+1)                          ▷ append (x, y+1) to Q
39:      Q ← Q ⌣ (x, y−1)                          ▷ append (x, y−1) to Q
40:      Q ← Q ⌣ (x−1, y)                          ▷ append (x−1, y) to Q
41:  return
```

in very short and elegant program code. Unfortunately, since the maximum depth of the recursion—and thus the size of the required stack memory—is proportional to the size of the region, stack memory is quickly exhausted. Therefore this method is risky and really only practical for very small images.

B. **Iterative Flood Filling (*depth-first*):** Every recursive algorithm can also be reformulated as an iterative algorithm (Alg. 10.1, line 16) by implementing and managing its own *stacks*. In this case, the stack records the "open" (that is, the adjacent but not yet visited) elements. As in the recursive version (A), the corresponding tree of pixels is traversed in *depth-first* order. By making use of its own dedicated stack (which is created in the much larger *heap* memory), the depth of the tree is no longer limited to the size of the call stack.

C. **Iterative Flood Filling (*breadth-first*):** In this version, pixels are traversed in a way that resembles an expanding wave front propagating out from the starting point (Alg. 10.1, line 29). The data structure used to hold the as yet unvisited pixel coordinates is in this case a *queue* instead of a stack, but otherwise it is identical to version B.

Java implementation

The recursive version (A) of the algorithm is an exact blueprint of the Java implementation. However, a normal Java runtime environment does not support more than about 10,000 recursive calls of the FloodFill() procedure (Alg. 10.1, line 8) before the memory allocated for the call stack is exhausted. This is only sufficient for relatively small images with fewer than approximately 200×200 pixels.

Program 10.1 (line 1–17) gives the complete Java implementation for both variants of the iterative FloodFill() procedure. The stack (S) in the *depth-first* Version (B) and the queue (Q) in the *breadth-first* variant (C) are both implemented as instances of type `LinkedList`.[2] `<Point>` is specified as a type parameter for both generic container classes so they can only contain objects of type `Point`.[3]

Figure 10.2 illustrates the progress of the region marking in both variants within an example region, where the start point (i.e., seed point), which would normally lie on a contour edge, has been placed arbitrarily within the region in order to better illustrate the process. It is clearly visible that the *depth-first* method first explores *one* direction (in this case horizontally to the left) completely (that is, until it reaches the edge of the region) and only then examines the remaining directions. In contrast the *breadth-first* method markings proceed outward, layer by layer, equally in all directions.

Due to the way exploration takes place, the memory requirement of the *breadth-first* variant of the *flood-fill* version is generally much lower than that of the *depth-first* variant. For example, when flood filling the region in Fig. 10.2 (using the implementation given Prog. 10.1), the stack in the *depth-first* variant grows to a maximum of 28,822 elements, while the queue used by the *breadth-first* variant never exceeds a maximum of 438 nodes.

[2] The class `LinkedList` is part of Java's *collections framework*.

[3] Note that the depth-first and breadth-first implementations in Prog. 10.1 typically run slower than the recursive version described in Alg. 10.1, since they allocate (and immediately discard) large numbers of `Point` objects. A better solution is to use a queue or stack with elements of a primitive type (e.g., `int`) instead. See also Exercise 10.3.

Prog. 10.1
Java implementation of iterative flood filling (*depth-first* and *breadth-first* variants).

Depth-first version (using a *stack*):

```
1 void floodFill(int u, int v, int label) {
2   Deque<Point> S = new LinkedList<Point>(); // stack S
3   S.push(new Point(u, v));
4   while (!S.isEmpty()) {
5     Point p = S.pop();
6     int x = p.x;
7     int y = p.y;
8     if ((x >= 0) && (x < width) && (y >= 0) && (y < height)
9         && ip.getPixel(x, y) == 1) {
10      ip.putPixel(x, y, label);
11      S.push(new Point(x + 1, y));
12      S.push(new Point(x, y + 1));
13      S.push(new Point(x, y - 1));
14      S.push(new Point(x - 1, y));
15    }
16  }
17 }
```

Breadth-first version (using a *queue*):

```
18 void floodFill(int u, int v, int label) {
19   Queue<Point> Q = new LinkedList<Point>(); // queue Q
20   Q.add(new Point(u, v));
21   while (!Q.isEmpty()) {
22     Point p = Q.remove(); // get the next point to process
23     int x = p.x;
24     int y = p.y;
25     if ((x >= 0) && (x < width) && (y >= 0) && (y < height)
26         && ip.getPixel(x, y) == 1) {
27       ip.putPixel(x, y, label);
28       Q.add(new Point(x + 1, y));
29       Q.add(new Point(x, y + 1));
30       Q.add(new Point(x, y - 1));
31       Q.add(new Point(x - 1, y));
32     }
33   }
34 }
```

10.1.2 Sequential Region Labeling

Sequential region marking is a classical, nonrecursive technique that is known in the literature as "region labeling". The algorithm consists of two steps: (1) preliminary labeling of the image regions and (2) resolving cases where more than one label occurs (i.e., has been assigned in the previous step) in the same connected region. Even though this algorithm is relatively complex, especially its second stage, its moderate memory requirements make it a good choice under limited memory conditions. However, this is not a major issue on modern computers and thus, in terms of overall efficiency, sequential labeling offers no clear advantage over the simpler methods described earlier. The sequential technique is nevertheless interesting (not only from a historic perspective) and inspiring. The complete process is summarized in Alg. 10.2, with the following main steps:

Fig. 10.2
Iterative *flood filling*—comparison between the *depth-first* and *breadth-first* approach. The starting point, marked + in the top two image (a), was arbitrarily chosen. Intermediate results of the *flood fill* process after 1000 (a), 5000 (b), and 10,000 (c) marked pixels are shown. The image size is 250 × 242 pixels.

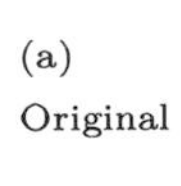

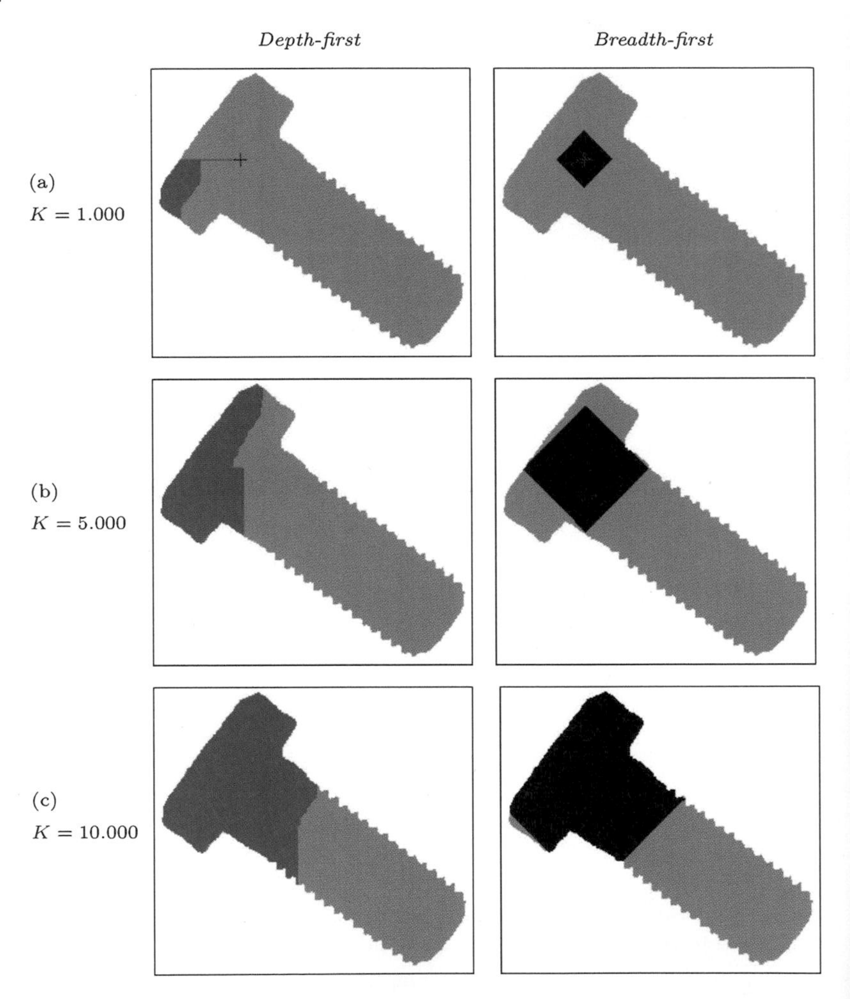

Step 1: Initial labeling

In the first stage of region labeling, the image is traversed from top left to bottom right sequentially to assign a preliminary label to every foreground pixel. Depending on the definition of neighborhood (either 4- or 8-connected) used, the following neighbors in the direct vicinity of each pixel must be examined (× marks the current pixel at the position (u, v)):

Alg. 10.2
Sequential region labeling. The binary input image I uses the value $I(u,v) = 0$ for background pixels and $I(u,v) = 1$ for foreground (region) pixels. The resulting labels have the values $2, \ldots, label - 1$.

```
 1: SequentialLabeling(I)
    Input: I, an integer-valued image with initial values 0 = background,
    1 = foreground. Returns nothing but modifies the image I.

    Step 1 – Assign initial labels:
 2: (M, N) ← Size(I)
 3: label ← 2                      ▷ value of the next label to be assigned
 4: C ← ()                                  ▷ empty list of label collisions
 5: for v ← 0, ..., N−1 do
 6:     for u ← 0, ..., M−1 do
 7:         if I(u,v) = 1 then              ▷ I(u,v) is a foreground pixel
 8:             𝒩 ← GetNeighbors(I, u, v)                 ▷ see Eqn. 10.2
 9:             if N_i = 0 for all N_i ∈ 𝒩 then
10:                 I(u,v) ← label.
11:                 label ← label + 1.
12:             else if exactly one N_j ∈ 𝒩 has a value > 1 then
13:                 set I(u,v) ← N_j
14:             else if more than one N_k ∈ 𝒩 have values > 1 then
15:                 I(u,v) ← N_k       ▷ select one N_k > 1 as the new
                      label
16:                 for all N_l ∈ 𝒩, with l ≠ k and N_l > 1 do
17:                     C ← C ⌣ (N_k, N_l) ▷ register collision (N_k, N_l)
    Remark: The image I now contains labels 0, 2, ..., label−1.

    Step 2 – Resolve label collisions:
    Create a partitioning of the label set (sequence of 1-element sets):
18: R ← ({2}, {3}, {4}, ..., {label−1})
19: for all collisions (A, B) in C do
        Find the sets R(a), R(b) holding labels A, B:
20:     a ← index of the set R(a) that contains label A
21:     b ← index of the set R(b) that contains label B
22:     if a ≠ b then          ▷ A and B are contained in different sets
23:         R(a) ← R(a) ∪ R(b)      ▷ merge elements of R(b) into R(a)
24:         R(b) ← {}
    Remark: All equivalent labels (i.e., all labels of pixels in the same
    connected component) are now contained in the same subset of R.

25: Step 3: Relabel the image:
26: for all (u,v) ∈ M × N do
27:     if I(u,v) > 1 then        ▷ this is a labeled foreground pixel
28:         j ← index of the set R(j) that contains label I(u,v)
            Choose a representative element k from the set R(j):
29:         k ← min(R(j))                ▷ e.g., pick the minimum value
30:         I(u,v) ← k                        ▷ replace the image label
31: return
```

$$\mathcal{N}_4 = \begin{array}{|c|c|c|} \hline & N_1 & \\ \hline N_2 & \times & N_0 \\ \hline & N_3 & \\ \hline \end{array} \quad \text{or} \quad \mathcal{N}_8 = \begin{array}{|c|c|c|} \hline N_3 & N_2 & N_1 \\ \hline N_4 & \times & N_0 \\ \hline N_5 & N_6 & N_7 \\ \hline \end{array}. \qquad (10.2)$$

When using the 4-connected neighborhood $\mathcal{N}_4$, only the two neighbors $N_1 = I(u-1, v)$ and $N_2 = I(u, v-1)$ need to be considered, but when using the 8-connected neighborhood $\mathcal{N}_8$, all four neighbors $N_1 \ldots N_4$ must be examined. In the following examples (Figs. 10.3–10.5), we use an 8-connected neighborhood and a very simple test image (Fig. 10.3(a)) to demonstrate the sequential region labeling process.

Propagating region labels

Again we assume that, in the image, the value $I(u, v) = 0$ represents background pixels and the value $I(u, v) = 1$ represents foreground pixels. We will also consider neighboring pixels that lie outside of the image matrix (e.g., on the array borders) to be part of the background. The neighborhood region $\mathcal{N}(u, v)$ is slid over the image horizontally and then vertically, starting from the top left corner. When the current image element $I(u, v)$ is a foreground pixel, it is either assigned a new region number or, in the case where one of its previously examined neighbors in $\mathcal{N}(u, v)$ was a foreground pixel, it takes on the region number of the neighbor. In this way, existing region numbers propagate in the image from the left to the right and from the top to the bottom, as shown in (Fig. 10.3(b–c)).

Label collisions

In the case where two or more neighbors have labels belonging to *different* regions, then a label collision has occurred; that is, pixels within a single connected region have different labels. For example, in a U-shaped region, the pixels in the left and right arms are at first assigned different labels since it is not immediately apparent that they are actually part of a single region. The two labels will propagate down independently from each other until they eventually collide in the lower part of the "U" (Fig. 10.3(d)).

When two labels a, b collide, then we know that they are actually "equivalent"; that is, they are contained in the same image region. These collisions are registered but otherwise not dealt with during the first step. Once all collisions have been registered, they are then resolved in the second step of the algorithm. The number of collisions depends on the content of the image. There can be only a few or very many collisions, and the exact number is only known at the end of the first step, once the whole image has been traversed. For this reason, collision management must make use of dynamic data structures such as lists or hash tables.

Upon the completion of the first steps, all the original foreground pixels have been provisionally marked, and all the collisions between labels within the same regions have been registered for subsequent processing. The example in Fig. 10.4 illustrates the state upon completion of step 1: all foreground pixels have been assigned preliminary labels (Fig. 10.4(a)), and the following collisions (depicted by circles) between the labels $(2, 4)$, $(2, 5)$, and $(2, 6)$ have been registered. The labels $\mathcal{L} = \{2, 3, 4, 5, 6, 7\}$ and collisions $\mathcal{C} = \{(2,4), (2,5), (2,6)\}$ correspond to the nodes and edges of an undirected graph (Fig. 10.4(b)).

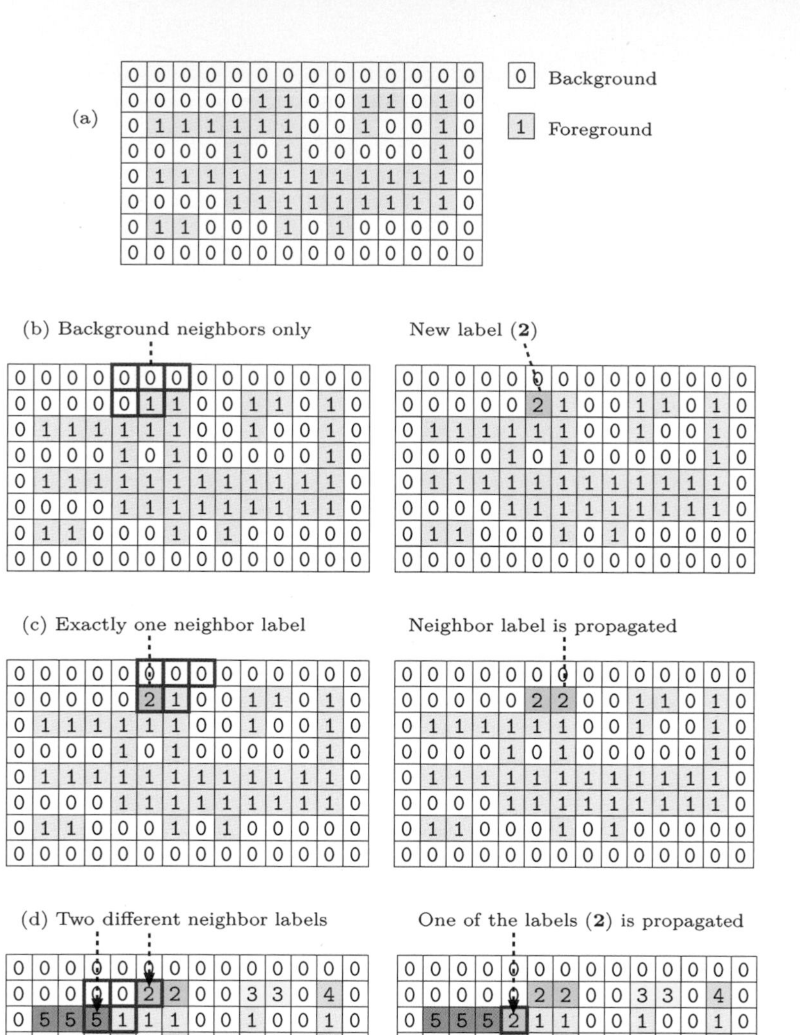

Fig. 10.3
Sequential region labeling—label propagation. Original image (a). The first foreground pixel (marked **1**) is found in (b): all neighbors are background pixels (marked **0**), and the pixel is assigned the first label (**2**). In the next step (c), there is exactly *one* neighbor pixel marked with the label **2**, so this value is propagated. In (d) there are *two* neighboring pixels, and they have differing labels (**2** and **5**); one of these values is propagated, and the collision (**2**, **5**) is registered.

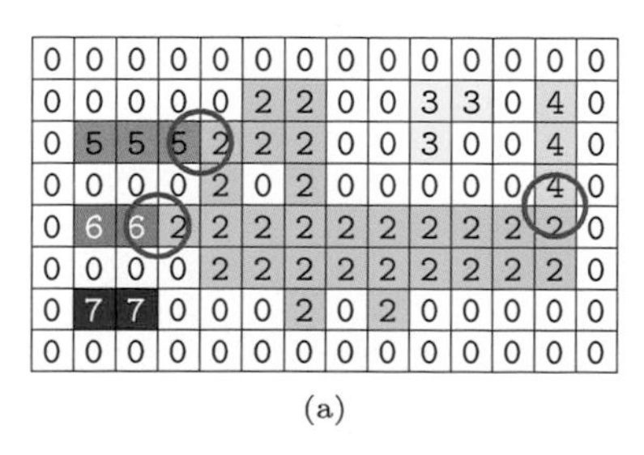

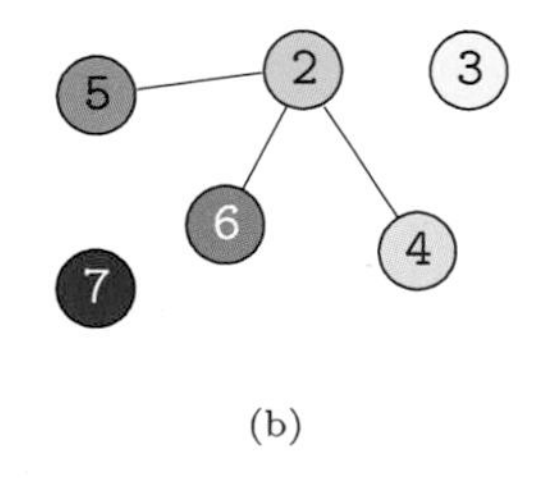

Fig. 10.4
Sequential region labeling—intermediate result after step 1. Label collisions indicated by circles (a); the nodes of the undirected graph (b) correspond to the labels, and its edges correspond to the collisions.

Step 2: Resolving label collisions

The task in the second step is to resolve the label collisions that arose in the first step in order to merge the corresponding "partial" regions. This process is nontrivial since it is possible for two regions with different labels to be connected transitively (e.g., $(a,b) \cap (b,c) \Rightarrow (a,c)$) through a third region or, more generally, through a series of regions. In fact, this problem is identical to the problem of finding the *connected components* of a graph [54], where the labels $\mathcal{L}$ determined in step 1 constitute the "nodes" of the graph and the registered collisions $\mathcal{C}$ make up its "edges" (Fig. 10.4(b)).

Once all the distinct labels within a single region have been collected, the labels of all the pixels in the region are updated so they carry the same label (e.g., choosing the smallest label number in the region), as depicted in Fig. 10.5. Figure 10.6 shows the complete segmentation with some region statistics that can be easily calculated from the labeling data.

Fig. 10.5
Sequential region labeling—final result after step 2. All equivalent labels have been replaced by the smallest label within that region.

0	0	0	0	0	0	0	0	0	0	0	0	0	0
0	0	0	0	0	2	2	0	0	3	3	0	2	0
0	2	2	2	2	2	2	0	0	3	0	0	2	0
0	0	0	0	2	0	2	0	0	0	0	0	2	0
0	2	2	2	2	2	2	2	2	2	2	2	2	0
0	0	0	0	2	2	2	2	2	2	2	2	2	0
0	7	7	0	0	0	2	0	2	0	0	0	0	0
0	0	0	0	0	0	0	0	0	0	0	0	0	0

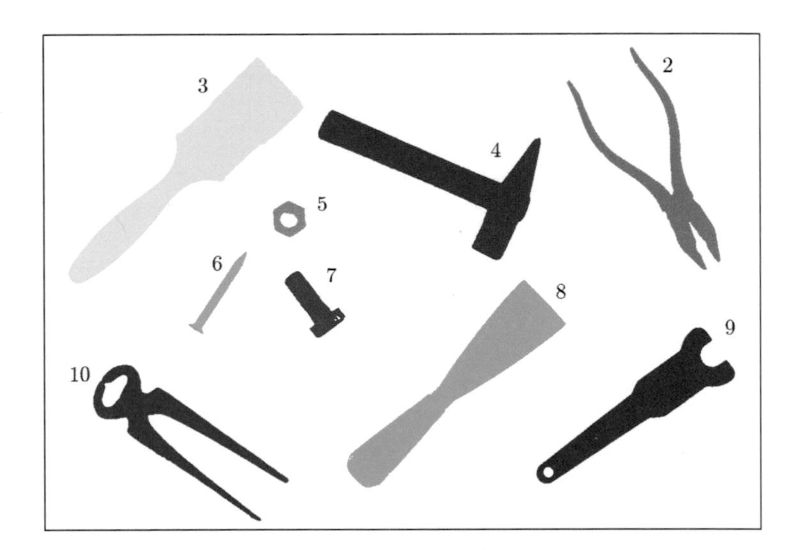

Fig. 10.6
Example of a complete region labeling. The pixels within each region have been colored according to the consecutive label values $2, 3, \ldots, 10$ they were assigned. The corresponding region statistics are shown in the table (total image size is 1212×836).

Label	Area (*pixels*)	Bounding Box (*left, top, right, bottom*)	Centroid (x_c, y_c)
2	14978	(887, 21, 1144, 399)	(1049.7, 242.8)
3	36156	(40, 37, 438, 419)	(261.9, 209.5)
4	25904	(464, 126, 841, 382)	(680.6, 240.6)
5	2024	(387, 281, 442, 341)	(414.2, 310.6)
6	2293	(244, 367, 342, 506)	(294.4, 439.0)
7	4394	(406, 400, 507, 512)	(454.1, 457.3)
8	29777	(510, 416, 883, 765)	(704.9, 583.9)
9	20724	(833, 497, 1168, 759)	(1016.0, 624.1)
10	16566	(82, 558, 411, 821)	(208.7, 661.6)

10.1.3 Region Labeling—Summary

In this section, we have described a selection of algorithms for finding and labeling connected regions in images. We discovered that the elegant idea of labeling individual regions using a simple recursive flood-filling method (Sec. 10.1.1) was not useful because of practical limitations on the depth of recursion and the high memory costs associated with it. We also saw that classical sequential region labeling (Sec. 10.1.2) is relatively complex and offers no real advantage over iterative implementations of the *depth-first* and *breadth-first* methods. In practice, the iterative breadth-first method is generally the best choice for large and complex images. In the following section we present a modern and efficient algorithm that performs region labeling and also delineates the regions' contours. Since contours are required in many applications, this combined approach is highly practical.

10.2 Region Contours

Once the regions in a binary image have been found, the next step is often to find the contours (that is, the outlines) of the regions. Like so many other tasks in image processing, at first glance this appears to be an easy one: simply follow along the edge of the region. We will see that, in actuality, describing this apparently simple process algorithmically requires careful thought, which has made contour finding one of the classic problems in image analysis.

10.2.1 External and Internal Contours

As we discussed in Chapter 9, Sec. 9.2.7, the pixels along the edge of a binary region (i.e., its border) can be identified using simple morphological operations and difference images. It must be stressed, however, that this process only *marks* the pixels along the contour, which is useful, for instance, for display purposes. In this section, we will go one step further and develop an algorithm for obtaining an *ordered sequence* of border pixel coordinates for describing a region's contour. Note that connected image regions contain exactly one *outer* contour, yet, due to holes, they can contain arbitrarily many *inner* contours. Within such holes, smaller regions may be found, which will again have their own outer contours, and in turn these regions may themselves contain further holes with even smaller regions, and so on in a recursive manner (Fig. 10.7). An additional complication arises when regions are connected by parts that taper down to the width of a single pixel. In such cases, the contour can run through the same pixel more than once and from different directions (Fig. 10.8). Therefore, when tracing a contour from a start point $\boldsymbol{x}_{\mathrm{s}}$, returning to the start point is *not* a sufficient condition for terminating the contour-tracing process. Other factors, such as the current direction along which contour points are being traversed, must be taken into account.

One apparently simple way of determining a contour is to proceed in analogy to the two-stage process presented in Sec. 10.1; that is,

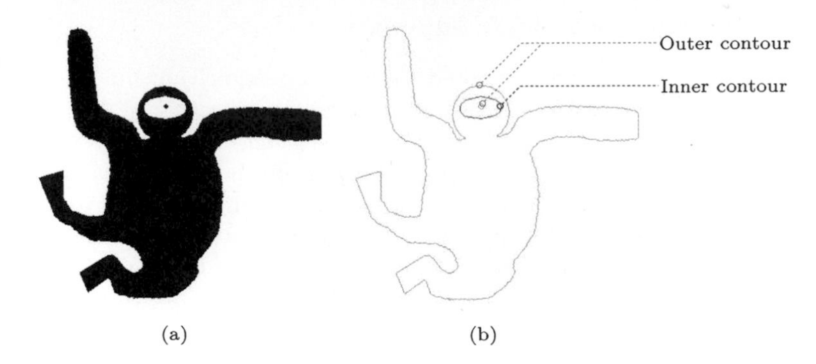

Fig. 10.7
Binary image with outer and inner contours. The outer contour lies along the outside of the foreground region (dark). The inner contour surrounds the space within the region, which may contain further regions (holes), and so on.

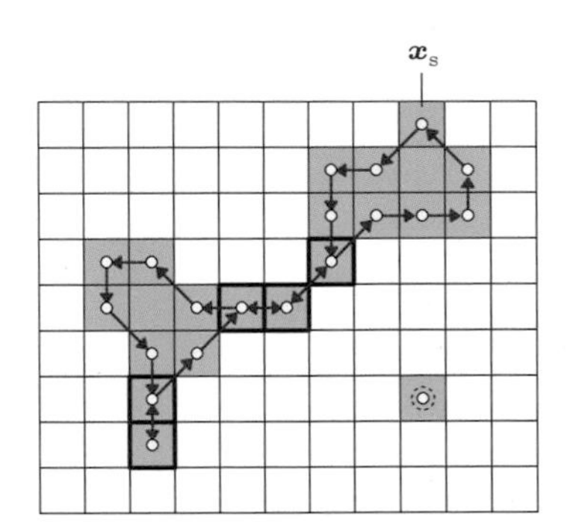

Fig. 10.8
The path along a contour as an ordered sequence of pixel coordinates with a given start point $\boldsymbol{x}_s$. Individual pixels may occur (be visited) more than once within the path, and a region consisting of a single isolated pixel will also have a contour (bottom right).

to *first* identify the connected regions in the image and *second*, for each region, proceed around it, starting from a pixel selected from its border. In the same way, an internal contour can be found by starting at a border pixel of a region's hole. A wide range of algorithms based on first finding the regions and then following along their contours have been published, including [202], [180, pp. 142–148], and [214, p. 296].

As a modern alternative, we present the following *combined* algorithm that, in contrast to the aforementioned classical methods, combines contour finding and region labeling in a single process.

10.2.2 Combining Region Labeling and Contour Finding

This method, based on [47], combines the concepts of sequential region labeling (Sec. 10.1) and traditional contour tracing into a single algorithm able to perform both tasks simultaneously during a single pass through the image. It identifies and labels regions and at the same time traces both their inner and outer contours. The algorithm does not require any complicated data structures and is relatively efficient when compared to other methods with similar capabilities. The key steps of this method are described here and illustrated in Fig. 10.9:

1. As in the sequential region labeling (Alg. 10.2), the binary image I is traversed from the top left to the bottom right. Such a traversal ensures that all pixels in the image are eventually examined and assigned an appropriate label.

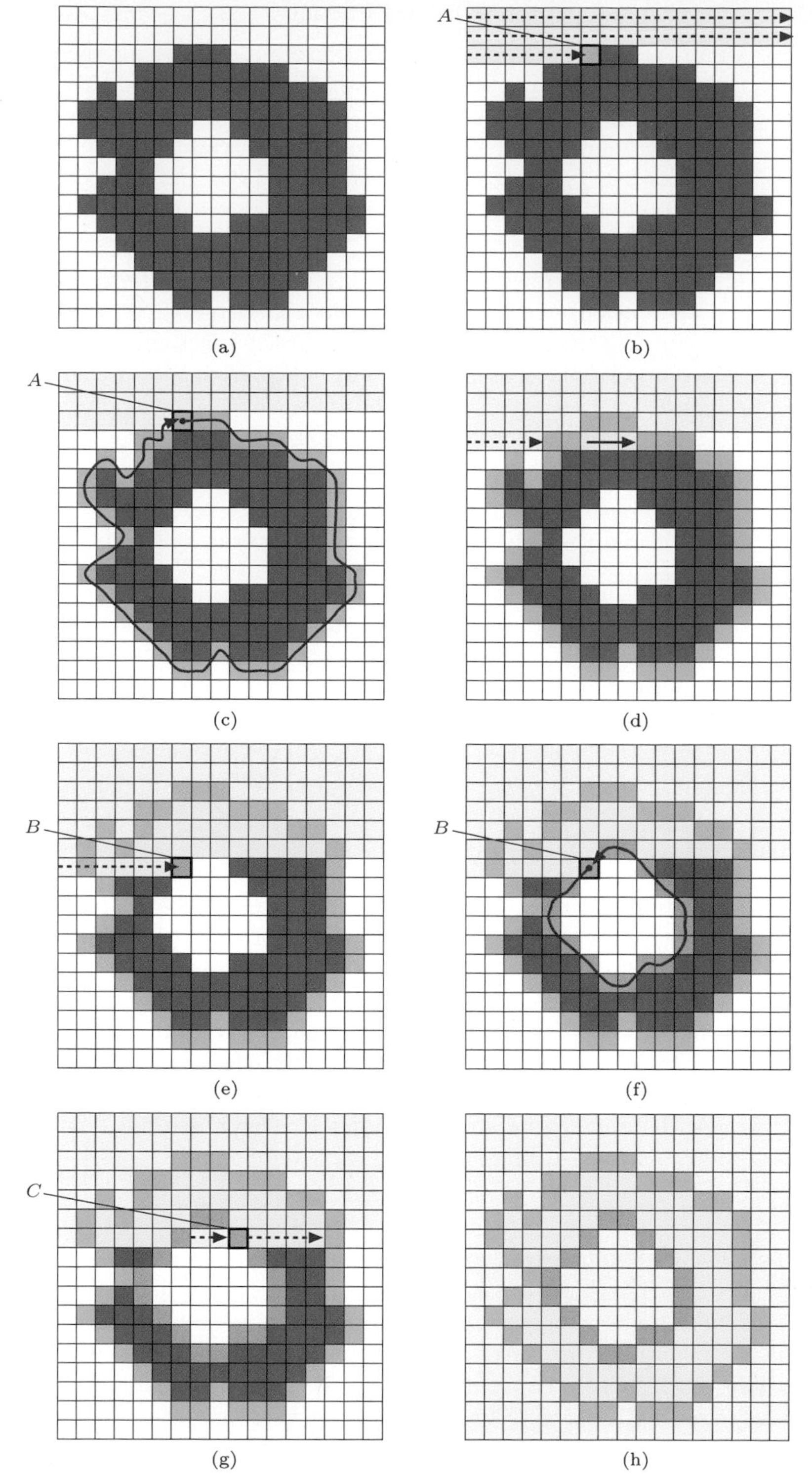

Fig. 10.9
Combined region labeling and contour following (after [47]). The image in (a) is traversed from the top left to the lower right, one row at a time. In (b), the first foreground pixel A on the outer edge of the region is found. Starting from point A, the pixels on the edge along the outer contour are visited and labeled until A is reached again (c). Labels picked up at the outer contour are propagated along the image line inside the region (d). In (e), B was found as the first point on the *inner contour*. Now the inner contour is traversed in clock-wise direction, marking the contour pixels until point B is reached again (f). The same tracing process is used as in step (c), with the inside of the region always lying to the right of the contour path. In (g) a previously marked point C on an inner contour is detected. Its label is again propagated along the image line inside the region. The final result is shown in (h).

2. At a given position in the image, the following cases may occur:

   Case A: The transition from a background pixel to a previously unmarked foreground pixel means that this pixel lies on the outer edge of a new region. A new *label* is assigned and the associated *outer* contour is traversed and marked by calling the method TraceContour (see Alg. 10.3 and Fig. 10.9(a)). Furthermore, all background pixels directly bordering the region are marked with the special label -1.

   Case B: The transition from a foreground pixel B to an unmarked background pixel means that this pixel lies on an *inner* contour (Fig. 10.9(b)). Starting from B, the inner contour is traversed and its pixels are marked with labels from the surrounding region (Fig. 10.9(c)). Also, all bordering background pixels are again assigned the special label value -1.

   Case C: When a foreground pixel does not lie on a contour, then the neighboring pixel to the left has already been labeled (Fig. 10.9(d)) and this label is propagated to the current pixel.

In Algs. 10.3–10.4, the entire procedure is presented again and explained precisely. Procedure RegionContourLabeling traverses the image line-by-line and calls procedure TraceContour whenever a new inner or outer contour must be traced. The labels of the image elements along the contour, as well as the neighboring foreground pixels, are stored in the "label map" L (a rectangular array of the same size as the image) by procedure FindNextContourPoint in Alg. 10.4.

10.2.3 Java Implementation

The Java implementation of the combined region labeling and contour tracing algorithm can be found online in class `RegionContourLabeling`[4] (for details see Sec. 10.9). It almost exactly follows Algs. 10.3–10.4, only the image I and the associated *label map* L are initially *padded* (i.e., enlarged) by a surrounding layer of background pixels. This simplifies the process of tracing the outer region contours, since no special treatment is needed at the image borders. Program 10.2 shows a minimal example of its usage within the `run()` method of an ImageJ plugin (class `Trace_Contours`).

Examples

This combined algorithm for region marking and contour following is particularly well suited for processing large binary images since it is efficient and has only modest memory requirements. Figure 10.10 shows a synthetic test image that illustrates a number of special situations, such as isolated pixels and thin sections, which the algorithm must deal with correctly when following the contours. In the resulting plot, outer contours are shown as black polygon lines running trough the centers of the contour pixels, and inner contours are drawn white. Contours of single-pixel regions are marked by small circles filled with the corresponding color. Figure 10.11 shows the results for a larger section taken from a real image (Fig. 9.12).

[4] Package `imagingbook.pub.regions`.

```
1: RegionContourLabeling(I)
     Input: I, a binary image with 0 = background, 1 = foreground.
     Returns sequences of outer and inner contours and a map of
     region labels.
2:   (M, N) ← Size(I)
3:   C_out ← ()                              ▷ empty list of outer contours
4:   C_in ← ()                               ▷ empty list of inner contours
5:   Create map L: M × N ↦ ℤ                 ▷ create the label map L
6:   for all (u, v) do
7:       L(u, v) ← 0                         ▷ initialize L to zero
8:   r ← 0                                   ▷ region counter
9:   for v ← 0, ..., N−1 do                  ▷ scan the image top to bottom
10:      label ← 0
11:      for u ← 0, ..., M−1 do              ▷ scan the image left to right
12:          if I(u, v) > 0 then             ▷ I(u, v) is a foreground pixel
13:              if (label ≠ 0) then         ▷ continue existing region
14:                  L(u, v) ← label
15:              else
16:                  label ← L(u, v)
17:                  if (label = 0) then     ▷ hit a new outer contour
18:                      r ← r + 1
19:                      label ← r
20:                      x_s ← (u, v)
21:                      C ← TraceContour(x_s, 0, label, I, L)  ▷ outer c.
22:                      C_out ← C_out ⌣ (C)        ▷ collect outer contour
23:                      L(u, v) ← label
24:          else                            ▷ I(u, v) is a background pixel
25:              if (label ≠ 0) then
26:                  if (L(u, v) = 0) then   ▷ hit new inner contour
27:                      x_s ← (u−1, v)
28:                      C ← TraceContour(x_s, 1, label, I, L)    ▷ inner
                             cntr.
29:                      C_in ← C_in ⌣ (C)          ▷ collect inner contour
30:                  label ← 0
31:  return (C_out, C_in, L)
```

continued in Alg. 10.4 ▷▷

Alg. 10.3
Combined contour tracing and region labeling (part 1). Given a binary image I, the application of RegionContourLabeling(I) returns a set of contours and an array containing region labels for all pixels in the image. When a new point on either an outer or inner contour is found, then an ordered list of the contour's points is constructed by calling procedure TraceContour (line 21 and line 28). TraceContour itself is described in Alg. 10.4.

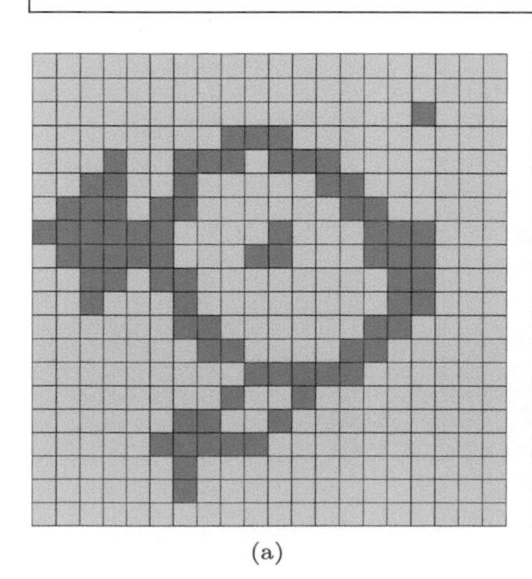

(a)

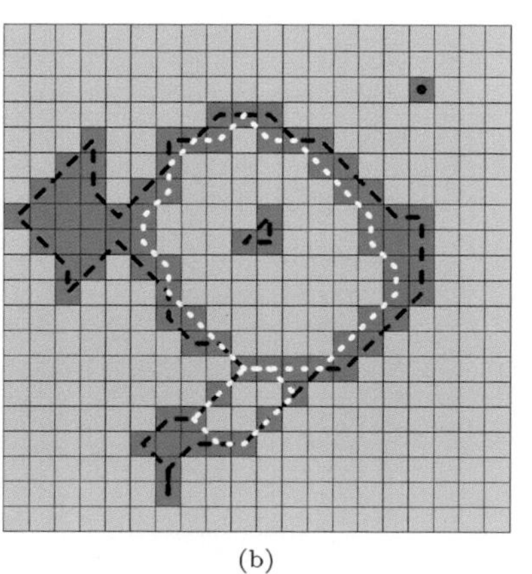

(b)

Fig. 10.10
Combined contour and region marking. Original image, with foreground pixels marked green (a); located contours with black lines for *outer* and white lines for *inner* contours (b). Contour polygons pass through the pixel centers. Outer contours of single-pixel regions (e.g., in the upper-right of (b)) are marked by a single dot.

Alg. 10.4
Combined contour finding and region labeling (part 2, continued from Alg. 10.3). Starting from $\boldsymbol{x}_s$, the procedure TraceContour traces along the contour in the direction $d_S = 0$ for outer contours or $d_S = 1$ for inner contours. During this process, all contour points as well as neighboring background points are marked in the label array L. Given a point $\boldsymbol{x}_c$, TraceContour uses FindNextContourPoint() to determine the next point along the contour (line 9). The function Delta() returns the next coordinate in the sequence, taking into account the search direction d.

```
1:  TraceContour(x_s, d_s, label, I, L)
        Input: x_s, start position; d_s, initial search direction; label, the
        label assigned to this contour; I, the binary input image; L, label
        map. Returns a new outer or inner contour (sequence of points)
        starting at x_s.
2:      (x, d) ← FindNextContourPoint(x_s, d_s, I, L)
3:      c ← (x)                      ▷ new contour with the single point x
4:      x_p ← x_s                    ▷ previous position x_p = (u_p, v_p)
5:      x_c ← x                      ▷ current position x_c = (u_c, v_c)
6:      done ← (x_s ≡ x)                               ▷ isolated pixel?
7:      while (¬done) do
8:          L(u_c, v_c) ← label
9:          (x_n, d) ← FindNextContourPoint(x_c, (d + 6) mod 8, I, L)
10:         x_p ← x_c
11:         x_c ← x_n
12:         done ← (x_p ≡ x_s ∧ x_c ≡ x)     ▷ back at starting position?
13:         if (¬done) then
14:             c ← c ⌣ (x_n)                ▷ add point x_n to contour c
15:     return c                                    ▷ return this contour

16: FindNextContourPoint(x, d, I, L)
        Input: x, initial position; d ∈ [0, 7], search direction, I, binary
        input image; L, the label map.
        Returns the next point on the contour and the modified search
        direction.
17:     for i ← 0, ..., 6 do                      ▷ search in 7 directions
18:         x_n ← x + Delta(d)
19:         if I(x_n) = 0 then            ▷ I(u_n, v_n) is a background pixel
20:             L(x_n) ← −1              ▷ mark background as visited (−1)
21:             d ← (d + 1) mod 8
22:         else                    ▷ found a non-background pixel at x_n
23:             return (x_n, d)
24:     return (x, d)       ▷ found no next node, return start position
```

25: $\mathsf{Delta}(d) := \begin{pmatrix} \Delta x \\ \Delta y \end{pmatrix}$, with

d	0	1	2	3	4	5	6	7
Δx	1	1	0	−1	−1	−1	0	1
Δy	0	1	1	1	0	−1	−1	−1

Prog. 10.2
Example of using the class ContourTracer. (plugin Trace_Contours). First (in line 9) a new instance of RegionContourLabeling is created for the input image I. The segmentation into regions and contours is done by the constructor. In lines 11–12 the outer and inner contours are retrieved as (possibly empty) lists of type Contour. Finally, the list of connected regions is obtained in line 14.

```
1  import imagingbook.pub.regions.BinaryRegion;
2  import imagingbook.pub.regions.Contour;
3  import imagingbook.pub.regions.RegionContourLabeling;
4  import java.util.List;
5  ...
6  public void run(ImageProcessor ip) {
7    // Make sure we have a proper byte image:
8    ByteProcessor I = ip.convertToByteProcessor();
9    // Create the region labeler / contour tracer:
10   RegionContourLabeling seg = new RegionContourLabeling(I);
11   // Get all outer/inner contours and connected regions:
12   List<Contour> outerContours = seg.getOuterContours();
13   List<Contour> innerContours = seg.getInnerContours();
14   List<BinaryRegion> regions = seg.getRegions();
15   ...
16 }
```

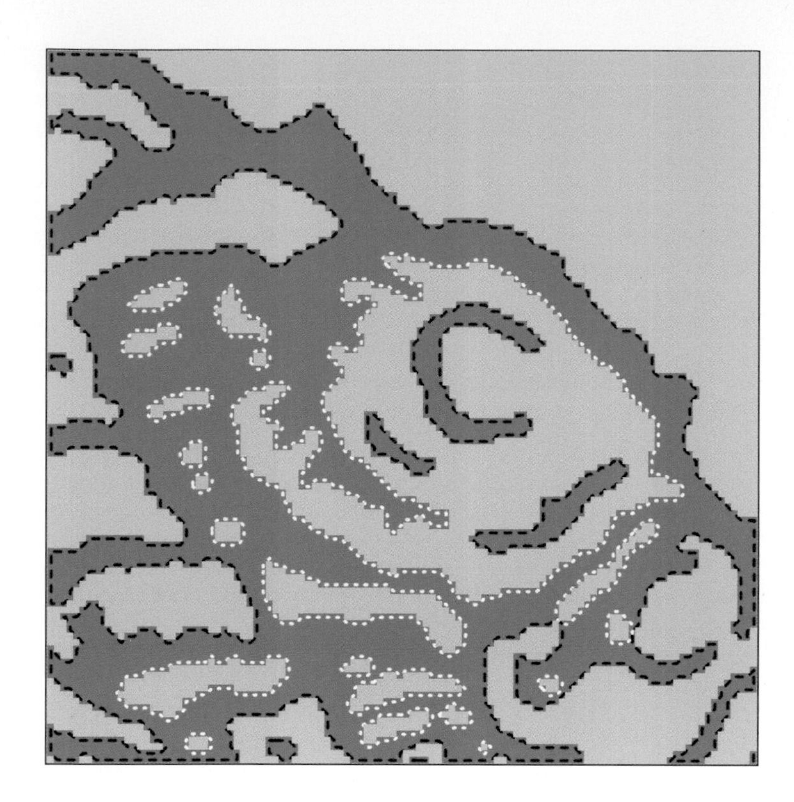

Fig. 10.11
Example of a complex contour (original image in Ch. 9, Fig. 9.12). Outer contours are marked in black and inner contours in white.

10.3 Representing Image Regions

10.3.1 Matrix Representation

A natural representation for images is a matrix (i.e., a two-dimensional array) in which elements represent the intensity or the color at a corresponding position in the image. This representation lends itself, in most programming languages, to a simple and elegant mapping onto two-dimensional arrays, which makes possible a very natural way to work with raster images. One possible disadvantage with this representation is that it does not depend on the content of the image. In other words, it makes no difference whether the image contains only a pair of lines or is of a complex scene because the amount of memory required is constant and depends only on the dimensions of the image.

Regions in an image can be represented using a logical mask in which the area within the region is assigned the value *true* and the area without the value *false* (Fig. 10.12). Since these values can be represented by a single bit, such a matrix is often referred to as a "bitmap".[5]

10.3.2 Run Length Encoding

In *run length encoding* (RLE), sequences of adjacent foreground pixels can be represented compactly as "runs". A run, or contiguous

[5] Java does not provide a genuine 1-bit data type. Even variables of type `boolean` are represented internally (i.e., within the Java virtual machine) as 32-bit `int`s.

Fig. 10.12
Use of a binary mask to specify a region of an image: original image (a), logical (bit) mask (b), and masked image (c).

block, is a maximal length sequence of adjacent pixels of the same type within either a row or a column. Runs of arbitrary length can be encoded compactly using three integers,

$$Run_i = \langle row_i, column_i, length_i \rangle,$$

as illustrated in Fig. 10.13. When representing a sequence of runs within the same row, the number of the row is redundant and can be left out. Also, in some applications, it is more useful to record the coordinate of the end column instead of the length of the run.

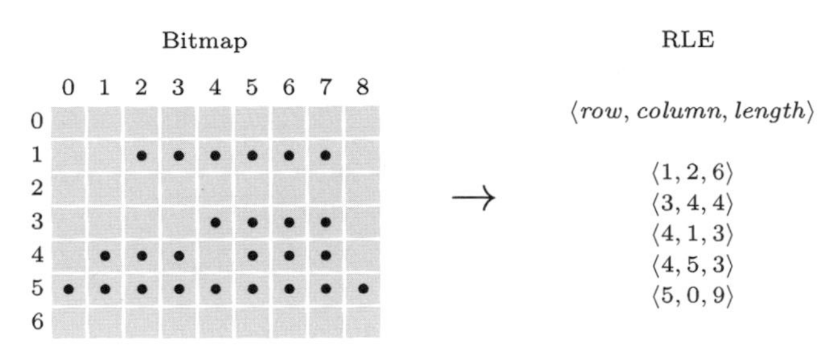

Fig. 10.13
Run length encoding in row direction. A run of pixels can be represented by its starting point $(1, 2)$ and its length (6).

Since the RLE representation can be easily implemented and efficiently computed, it has long been used as a simple lossless compression method. It forms the foundation for fax transmission and can be found in a number of other important codecs, including TIFF, GIF, and JPEG. In addition, RLE provides precomputed information about the image that can be used directly when computing certain properties of the image (for example, statistical moments; see Sec. 10.5.2).

10.3.3 Chain Codes

Regions can be represented not only using their interiors but also by their contours. Chain codes, which are often referred to as Freeman codes [79], are a classical method of contour encoding. In this encoding, the contour beginning at a given start point $\boldsymbol{x}_\mathrm{s}$ is represented by the sequence of directional changes it describes on the discrete image grid (Fig. 10.14).

Absolute chain code

For a closed contour of a region $\mathcal{R}$, described by the sequence of points $\boldsymbol{c}_\mathcal{R} = (\boldsymbol{x}_0, \boldsymbol{x}_1, \ldots \boldsymbol{x}_{M-1})$ with $\boldsymbol{x}_i = \langle u_i, v_i \rangle$, we create the elements of its chain code sequence $\boldsymbol{c}'_\mathcal{R} = (c'_0, c'_1, \ldots c'_{M-1})$ with

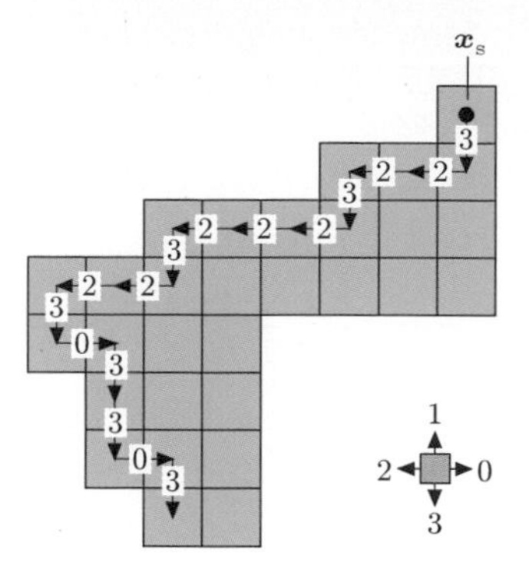

(a) 4-Chain Code
3223222322303303...111
length = 28

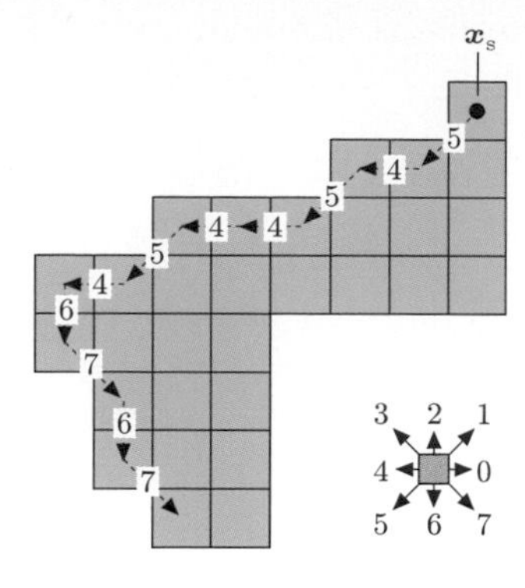

(b) 8-Chain Code
54544546767...222
length $= 16 + 6\sqrt{2} \approx 24.5$

Fig. 10.14
Chain codes with 4- and 8-connected neighborhoods. To compute a chain code, begin traversing the contour from a given starting point x_s. Encode the relative position between adjacent contour points using the directional code for either 4-connected (left) or 8-connected (right) neighborhoods. The length of the resulting path, calculated as the sum of the individual segments, can be used to approximate the true length of the contour.

$$c'_i = \mathsf{Code}(u', v'), \tag{10.3}$$

where

$$(u', v') = \begin{cases} (u_{i+1} - u_i, v_{i+1} - v_i) & \text{for } 0 \leq i < M-1, \\ (u_0 - u_i, v_0 - v_i) & \text{for } i = M-1, \end{cases} \tag{10.4}$$

and $\mathsf{Code}(u', v')$ being defined (assuming an 8-connected neighborhood) by the following table:

u'	1	1	0	−1	−1	−1	0	1
v'	0	1	1	1	0	−1	−1	−1
$\mathsf{Code}(u', v')$	0	1	2	3	4	5	6	7

Chain codes are compact since instead of storing the absolute coordinates for every point on the contour, only that of the starting point is recorded. The remaining points are encoded relative to the starting point by indicating in which of the eight possible directions the next point lies. Since only 3 bits are required to encode these eight directions the values can be stored using a smaller numeric type.

Differential chain code

Directly comparing two regions represented using chain codes is difficult since the description depends on the starting point selected x_s, and for instance simply rotating the region by 90° results in a completely different chain code. When using a *differential* chain code, the situation improves slightly. Instead of encoding the difference in the *position* of the next contour point, the change in the *direction* along the discrete contour is encoded. A given *absolute* chain code $c'_{\mathcal{R}} = (c'_0, c'_1, \ldots c'_{M-1})$ can be converted element by element to a *differential* chain code $c''_{\mathcal{R}} = (c''_0, c''_1, \ldots c''_{M-1})$, with[6]

$$c''_i = \begin{cases} (c'_{i+1} - c'_i) \bmod 8 & \text{for } 0 \leq i < M-1, \\ (c'_0 - c'_i) \bmod 8 & \text{for } i = M-1, \end{cases} \tag{10.5}$$

[6] For the implementation of the mod operator see Sec. F.1.2 in the Appendix.

again under the assumption of an 8-connected neighborhood. The element c''_i thus describes the change in direction (curvature) of the contour between two successive segments c'_i and c'_{i+1} of the original chain code $\boldsymbol{c}'_{\mathcal{R}}$. For the contour in Fig. 10.14(b), for example, the result is

$$\begin{aligned} \boldsymbol{c}'_{\mathcal{R}} &= (5,4,5,4,4,5,4,6,7,6,7,\ldots,2,2,2), \\ \boldsymbol{c}''_{\mathcal{R}} &= (7,1,7,0,1,7,2,1,7,1,1,\ldots,0,0,3). \end{aligned}$$

Given the start position $\boldsymbol{x}_{\mathrm{s}}$ and the (absolute) initial direction c_0, the original contour can be unambiguously reconstructed from the differential chain code.

Shape numbers

While the differential chain code remains the same when a region is rotated by 90°, the encoding is still dependent on the selected starting point. If we want to determine the similarity of two contours of the same length M using their differential chain codes $\boldsymbol{c}''_1$, $\boldsymbol{c}''_2$, we must first ensure that the same start point was used when computing the codes. A method that is often used [15,88] is to interpret the elements c''_i in the differential chain code as the digits of a number to the base b ($b = 8$ for an 8-connected contour or $b = 4$ for a 4-connected contour) and the numeric value

$$\mathsf{Val}(\boldsymbol{c}''_{\mathcal{R}}) = c''_0 \cdot b^0 + c''_1 \cdot b^1 + \;\ldots\; + c''_{M-1} \cdot b^{M-1} = \sum_{i=0}^{M-1} c''_i \cdot b^i. \tag{10.6}$$

Then the sequence $\boldsymbol{c}''_{\mathcal{R}}$ is shifted circularly until the numeric value of the corresponding number reaches a maximum. We use the expression $\boldsymbol{c}''_{\mathcal{R}} \rhd k$ to denote the sequence $\boldsymbol{c}''_{\mathcal{R}}$ being circularly shifted by k positions to the right.[7] For example, for $k = 2$ this is

$$\begin{aligned} \boldsymbol{c}''_{\mathcal{R}} &= (0,1,3,2,\ldots,5,3,7,4), \\ \boldsymbol{c}''_{\mathcal{R}} \rhd 2 &= (7,4,0,1,3,2,\ldots,5,3), \end{aligned}$$

and

$$k_{\mathrm{max}} = \underset{0 \le k < M}{\operatorname{argmax}} \, \mathsf{Val}(\boldsymbol{c}''_{\mathcal{R}} \rhd k), \tag{10.7}$$

denotes the shift required to maximize the corresponding arithmetic value. The resulting code sequence or *shape number*,

$$\boldsymbol{s}_{\mathcal{R}} = \boldsymbol{c}''_{\mathcal{R}} \rhd k_{\mathrm{max}}, \tag{10.8}$$

is *normalized* with respect to the starting point and can thus be directly compared element by element with other normalized code sequences. Since the function $\mathsf{Val}()$ in Eqn. (10.6) produces values that are in general too large to be actually computed, in practice the relation

$$\mathsf{Val}(\boldsymbol{c}''_1) > \mathsf{Val}(\boldsymbol{c}''_2)$$

[7] That is, $(\boldsymbol{c}''_{\mathcal{R}} \rhd k)(i) = \boldsymbol{c}''_{\mathcal{R}}((i-k) \bmod M)$.

is determined by comparing the *lexicographic ordering* between the sequences $\boldsymbol{c}_1''$ and $\boldsymbol{c}_2''$ so that the arithmetic values need not be computed at all.

Unfortunately, comparisons based on chain codes are generally not very useful for determining the similarity between regions simply because rotations at arbitrary angles ($\neq 90°$) have too great of an impact (change) on a region's code. In addition, chain codes are not capable of handling changes in size (scaling) or other distortions. Section 10.4 presents a number of tools that are more appropriate in these types of cases.

Fourier shape descriptors

An elegant approach to describing contours are so-called Fourier shape descriptors, which interpret the two-dimensional contour $C = (\boldsymbol{x}_0, \boldsymbol{x}_1, \ldots, \boldsymbol{x}_{M-1})$ with $\boldsymbol{x}_i = (u_i, v_i)$ as a sequence of values in the complex plane, where

$$z_i = (u_i + \mathrm{i} \cdot v_i) \in \mathbb{C}. \tag{10.9}$$

From this sequence, one obtains (using a suitable method of interpolation in case of an 8-connected contour), a discrete, one-dimensional periodic function $f(s) \in \mathbb{C}$ with a constant sampling interval over s, the path length around the contour. The coefficients of the 1D *Fourier spectrum* (see Sec. 18.3) of this function $f(s)$ provide a shape description of the contour in frequency space, where the lower spectral coefficients deliver a gross description of the shape. The details of this classical method can be found, for example, in [88, 97, 126, 128, 222]. This technique is described in considerable detail in Chapter 26.

10.4 Properties of Binary Regions

Imagine that you have to describe the contents of a digital image to another person over the telephone. One possibility would be to call out the value of each pixel in some agreed upon order. A much simpler way of course would be to describe the image on the basis of its properties—for example, "a red rectangle on a blue background", or at an even higher level such as "a sunset at the beach with two dogs playing in the sand". While using such a description is simple and natural for us, it is not (yet) possible for a computer to generate these types of descriptions without human intervention. For computers, it is of course simpler to calculate the mathematical properties of an image or region and to use these as the basis for further classification. Using features to classify, be they images or other items, is a fundamental part of the field of pattern recognition, a research area with many applications in image processing and computer vision [64, 169, 228].

10.4.1 Shape Features

The comparison and classification of binary regions is widely used, for example, in optical character recognition (OCR) and for automating processes ranging from blood cell counting to quality control inspection of manufactured products on assembly lines. The analysis of binary regions turns out to be one of the simpler tasks for which many efficient algorithms have been developed and used to implement reliable applications that are in use every day.

By a *feature* of a region, we mean a specific numerical or qualitative measure that is computable from the values and coordinates of the pixels that make up the region. As an example, one of the simplest features is its *size* or *area*; that is the number of pixels that make up a region. In order to describe a region in a compact form, different features are often combined into a *feature vector*. This vector is then used as a sort of "signature" for the region that can be used for classification or comparison with other regions. The best features are those that are simple to calculate and are not easily influenced (robust) by irrelevant changes, particularly translation, rotation, and scaling.

10.4.2 Geometric Features

A region $\mathcal{R}$ of a binary image can be interpreted as a two-dimensional distribution of foreground points $\boldsymbol{p}_i = (u_i, v_i)$ on the discrete plane $\mathbb{Z}^2$, that is, as a set

$$\mathcal{R} = \{\boldsymbol{x}_0, \ldots, \boldsymbol{x}_{N-1}\} = \{(u_0, v_0), (u_1, v_1), \ldots, (u_{N-1}, v_{N-1})\}.$$

Most geometric properties are defined in such a way that a region is considered to be a set of pixels that, in contrast to the definition in Sec. 10.1, does not necessarily have to be connected.

Perimeter

The perimeter (or circumference) of a region $\mathcal{R}$ is defined as the length of its outer contour, where $\mathcal{R}$ must be connected. As illustrated in Fig. 10.14, the type of neighborhood relation must be taken into account for this calculation. When using a 4-neighborhood, the measured length of the contour (except when that length is 1) will be larger than its actual length.

In the case of 8-neighborhoods, a good approximation is reached by weighing the horizontal and vertical segments with 1 and diagonal segments with $\sqrt{2}$. Given an 8-connected chain code $\boldsymbol{c}'_{\mathcal{R}} = (c'_0, c'_1, \ldots c'_{M-1})$, the perimeter of the region is arrived at by

$$\mathsf{Perimeter}(\mathcal{R}) = \sum_{i=0}^{M-1} \text{length}(c'_i), \tag{10.10}$$

with

$$\text{length}(c) = \begin{cases} 1 & \text{for } c = 0, 2, 4, 6, \\ \sqrt{2} & \text{for } c = 1, 3, 5, 7. \end{cases} \tag{10.11}$$

However, with this conventional method of calculation, the real perimeter $P(\mathcal{R})$ is systematically overestimated. As a simple remedy, an empirical correction factor of 0.95 works satisfactorily even for relatively small regions, that is,

$$P(\mathcal{R}) \approx 0.95 \cdot \mathsf{Perimeter}(\mathcal{R}). \tag{10.12}$$

Area

The area of a binary region $\mathcal{R}$ can be found by simply counting the image pixels that make up the region, that is,

$$A(\mathcal{R}) = N = |\mathcal{R}|. \qquad (10.13)$$

The area of a connected region without holes can also be approximated from its closed contour, defined by M coordinate points $(\boldsymbol{x}_0, \boldsymbol{x}_1, \ldots \boldsymbol{x}_{M-1})$, where $\boldsymbol{x}_i = (u_i, v_i)$, using the Gaussian area formula for polygons:

$$A(\mathcal{R}) \approx \frac{1}{2} \cdot \left| \sum_{i=0}^{M-1} \left(u_i \cdot v_{(i+1) \bmod M} \; - \; u_{(i+1) \bmod M} \cdot v_i \right) \right|. \qquad (10.14)$$

When the contour is already encoded as a chain code $\boldsymbol{c}'_{\mathcal{R}} = (c'_0, c'_1, \ldots c'_{M-1})$, then the region's area can be computed (trivially) with Eqn. (10.14) by expanding $\mathsf{C}_{\mathrm{abs}}$ into a sequence of contour points from an arbitrary starting point (e.g., $(0,0)$). However, the area can also be calculated directly from the chain code representation without expanding the contour [263] (see also Exercise 10.12).

While simple region properties such as area and perimeter are not influenced (except for quantization errors) by translation and rotation of the region, they are definitely affected by changes in size; for example, when the object to which the region corresponds is imaged from different distances. However, as will be described, it is possible to specify combined features that are *invariant* to translation, rotation, *and* scaling as well.

Compactness and roundness

Compactness is understood as the relation between a region's area and its perimeter. We can use the fact that a region's perimeter P increases linearly with the enlargement factor while the area A increases quadratically to see that, for a particular shape, the ratio A/P^2 should be the same at any scale. This ratio can thus be used as a feature that is invariant under translation, rotation, and scaling. When applied to a circular region of any diameter, this ratio has a value of $\frac{1}{4\pi}$, so by normalizing it against a filled circle, we create a feature that is sensitive to the *roundness* or *circularity* of a region,

$$\mathsf{Circularity}(\mathcal{R}) = 4\pi \cdot \frac{A(\mathcal{R})}{P^2(\mathcal{R})}, \qquad (10.15)$$

which results in a maximum value of 1 for a perfectly round region $\mathcal{R}$ and a value in the range $[0,1]$ for all other shapes (Fig. 10.15). If an absolute value for a region's roundness is required, the corrected perimeter estimate (Eqn. (10.12)) should be employed. Figure 10.15 shows the circularity values of different regions as computed with the formulation in Eqn. (10.15).

Bounding box

The bounding box of a region $\mathcal{R}$ is the minimal axis-parallel rectangle that encloses all points of $\mathcal{R}$,

Fig. 10.15
Circularity values for different shapes. Shown are the corresponding estimates for Circularity($\mathcal{R}$) as defined in Eqn. (10.15). Corrected values calculated with Eqn. (10.12) are shown in parentheses.

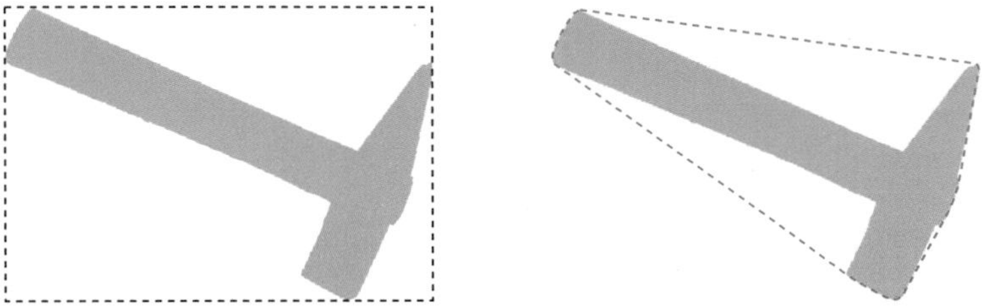

(a) 0.904 (1.001)

(b) 0.607 (0.672)

(c) 0.078 (0.086)

Fig. 10.16
Example bounding box (a) and convex hull (b) of a binary image region.

(a)

(b)

$$\mathsf{BoundingBox}(\mathcal{R}) = \langle u_{\min}, u_{\max}, v_{\min}, v_{\max} \rangle, \tag{10.16}$$

where $u_{\min}, u_{\max}$ and $v_{\min}, v_{\max}$ are the minimal and maximal coordinate values of all points $(u_i, v_i) \in \mathcal{R}$ in the x and y directions, respectively (Fig. 10.16(a)).

Convex hull

The convex hull is the smallest convex polygon that contains all points of the region $\mathcal{R}$. A physical analogy is a board in which nails stick out in correspondence to each of the points in the region. If you were to place an elastic band around *all* the nails, then, when you release it, it will contract into a convex hull around the nails (see Figs. 10.16(b) and 10.21(c)). Given N contour points, the convex hull can be computed in time $\mathcal{O}(N \log V)$, where V is the number vertices in the polygon of the resulting convex hull [17].

The convex hull is useful, for example, for determining the convexity or the *density* of a region. The *convexity* is defined as the relationship between the length of the convex hull and the original perimeter of the region. *Density* is then defined as the ratio between the area of the region and the area of its convex hull. The *diameter*, on the other hand, is the maximal distance between any two nodes on the convex hull.

10.5 Statistical Shape Properties

When computing statistical shape properties, we consider a region $\mathcal{R}$ to be a collection of coordinate points distributed within a two-dimensional space. Since statistical properties can be computed for point distributions that do not form a connected region, they can be applied before segmentation. An important concept in this context are the *central moments* of the region's point distribution, which measure characteristic properties with respect to its midpoint or *centroid*.

10.5.1 Centroid

The centroid or center of gravity of a connected region can be easily visualized. Imagine drawing the region on a piece of cardboard or tin and then cutting it out and attempting to balance it on the tip of your finger. The location on the region where you must place your finger in order for the region to balance is the *centroid* of the region.[8]

The centroid $\bar{\boldsymbol{x}} = (\bar{x}, \bar{y})^\mathsf{T}$ of a binary (not necessarily connected) region is the arithmetic mean of the pont coordinates $\boldsymbol{x}_i = (u_i, v_i)$, that is,

$$\bar{\boldsymbol{x}} = \frac{1}{|\mathcal{R}|} \cdot \sum_{\boldsymbol{x}_i \in \mathcal{R}} \boldsymbol{x}_i \qquad (10.17)$$

or

$$\bar{x} = \frac{1}{|\mathcal{R}|} \cdot \sum_{(u_i,v_i)} u_i \quad \text{and} \quad \bar{y} = \frac{1}{|\mathcal{R}|} \cdot \sum_{(u_i,v_i)} v_i \,. \qquad (10.18)$$

10.5.2 Moments

The formulation of the region's centroid in Eqn. (10.18) is only a special case of the more general statistical concept of a *moment*. Specifically, the expression

$$m_{pq}(\mathcal{R}) = \sum_{(u,v)\in\mathcal{R}} I(u,v) \cdot u^p \cdot v^q \qquad (10.19)$$

describes the (ordinary) moment of order p, q for a discrete (image) function $I(u,v) \in \mathbb{R}$; for example, a grayscale image. All the following definitions are also generally applicable to regions in grayscale images. The moments of connected binary regions can also be calculated directly from the coordinates of the contour points [212, p. 148].

In the special case of a binary image $I(u,v) \in \{0,1\}$, only the foreground pixels with $I(u,v) = 1$ in the region $\mathcal{R}$ need to be considered, and therefore Eqn. (10.19) can be simplified to

$$m_{pq}(\mathcal{R}) = \sum_{(u,v)\in\mathcal{R}} u^p \cdot v^q \,. \qquad (10.20)$$

In this way, the *area* of a binary region can be expressed as its *zero-order moment*,

$$A(\mathcal{R}) = |\mathcal{R}| = \sum_{(u,v)} 1 = \sum_{(u,v)} u^0 \cdot v^0 = m_{00}(\mathcal{R}) \qquad (10.21)$$

and similarly the *centroid* $\bar{\boldsymbol{x}}$ Eqn. (10.18) can be written as

[8] Assuming you did not imagine a region where the centroid lies outside of the region or within a hole in the region, which is of course possible.

$$\bar{x} = \frac{1}{|\mathcal{R}|} \cdot \sum_{(u,v)} u^1 \cdot v^0 = \frac{m_{10}(\mathcal{R})}{m_{00}(\mathcal{R})},$$
$$\bar{y} = \frac{1}{|\mathcal{R}|} \cdot \sum_{(u,v)} u^0 \cdot v^1 = \frac{m_{01}(\mathcal{R})}{m_{00}(\mathcal{R})}. \tag{10.22}$$

These moments thus represent concrete physical properties of a region. Specifically, the area m_{00} is in practice an important basis for characterizing regions, and the centroid $(\bar{x}, \bar{y})$ permits the reliable and (within a fraction of a pixel) exact specification of a region's position.

10.5.3 Central Moments

To compute position-independent (translation-invariant) region features, the region's centroid, which can be determined precisely in any situation, can be used as a reference point. In other words, we can shift the origin of the coordinate system to the region's centroid $\bar{\boldsymbol{x}} = (\bar{x}, \bar{y})$ to obtain the *central* moments of order p, q:

$$\mu_{pq}(\mathcal{R}) = \sum_{(u,v)\in\mathcal{R}} I(u,v) \cdot (u - \bar{x})^p \cdot (v - \bar{y})^q. \tag{10.23}$$

For a binary image (with $I(u, v) = 1$ within the region $\mathcal{R}$), Eqn. (10.23) can be simplified to

$$\mu_{pq}(\mathcal{R}) = \sum_{(u,v)\in\mathcal{R}} (u - \bar{x})^p \cdot (v - \bar{y})^q. \tag{10.24}$$

10.5.4 Normalized Central Moments

Central moment values of course depend on the absolute size of the region since the value depends directly on the distance of all region points to its centroid. So, if a 2D shape is scaled uniformly by some factor $s \in \mathbb{R}$, its central moments multiply by the factor

$$s^{(p+q+2)}. \tag{10.25}$$

Thus size-invariant "normalized" moments are obtained by scaling with the reciprocal of the area $A = \mu_{00} = m_{00}$ raised to the required power in the form

$$\bar{\mu}_{pq}(\mathcal{R}) = \mu_{pq} \cdot \left(\frac{1}{\mu_{00}(\mathcal{R})}\right)^{(p+q+2)/2}, \tag{10.26}$$

for $(p + q) \geq 2$ [126, p. 529].

10.5.5 Java Implementation

Program 10.3 gives a direct (brute force) Java implementation for computing the ordinary, central, and normalized central moments for binary images (`BACKGROUND = 0`). This implementation is only meant to clarify the computation, and naturally much more efficient implementations are possible (see, e.g., [131]).

```
1 // Ordinary moment:
2
3 double moment(ImageProcessor I, int p, int q) {
4   double Mpq = 0.0;
5   for (int v = 0; v < I.getHeight(); v++) {
6     for (int u = 0; u < I.getWidth(); u++) {
7       if (I.getPixel(u, v) > 0) {
8         Mpq+= Math.pow(u, p) * Math.pow(v, q);
9       }
10    }
11  }
12  return Mpq;
13 }
14
15 // Central moments:
16
17 double centralMoment(ImageProcessor I, int p, int q) {
18   double m00  = moment(I, 0, 0);  // region area
19   double xCtr = moment(I, 1, 0) / m00;
20   double yCtr = moment(I, 0, 1) / m00;
21   double cMpq = 0.0;
22   for (int v = 0; v < I.getHeight(); v++) {
23     for (int u = 0; u < I.getWidth(); u++) {
24       if (I.getPixel(u, v) > 0) {
25         cMpq+= Math.pow(u-xCtr, p) * Math.pow(v-yCtr, q);
26       }
27     }
28   }
29   return cMpq;
30 }
31
32 // Normalized central moments:
33
34 double nCentralMoment(ImageProcessor I, int p, int q) {
35   double m00 = moment(I, 0, 0);
36   double norm = Math.pow(m00, 0.5 * (p + q + 2));
37   return centralMoment(I, p, q) / norm;
38 }
```

Prog. 10.3
Example of directly computing moments in Java. The methods `moment()`, `centralMoment()`, and `nCentralMoment()` compute for a binary image the moments m_{pq}, μ_{pq}, and $\bar{\mu}_{pq}$ (Eqns. (10.20), (10.24), and (10.26)).

10.6 Moment-Based Geometric Properties

While normalized moments can be directly applied for classifying regions, further interesting and geometrically relevant features can be elegantly derived from statistical region moments.

10.6.1 Orientation

Orientation describes the direction of the major axis, that is, the axis that runs through the centroid and along the widest part of the region (Fig. 10.18(a)). Since rotating the region around the major axis requires less effort (smaller moment of inertia) than spinning it around any other axis, it is sometimes referred to as the major axis of rotation. As an example, when you hold a pencil between your hands and twist it around its major axis (that is, around the lead),

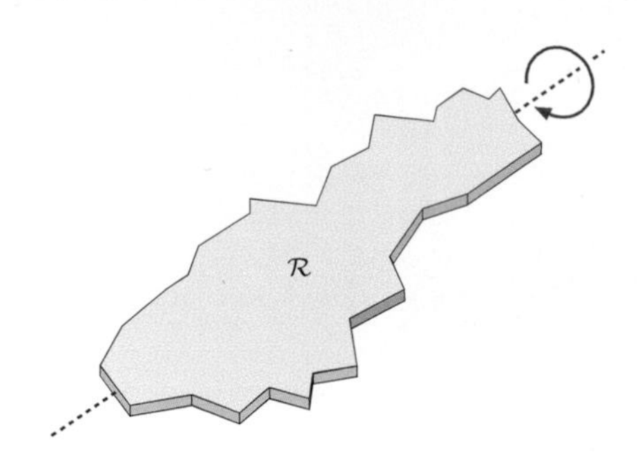

Fig. 10.17
Major axis of a region. Rotating an elongated region $\mathcal{R}$, interpreted as a physical body, around its major axis requires less effort (least moment of inertia) than rotating it around any other axis.

the pencil exhibits the least mass inertia (Fig. 10.17). As long as a region exhibits an orientation at all ($\mu_{20}(\mathcal{R}) \neq \mu_{02}(\mathcal{R})$), the direction $\theta_{\mathcal{R}}$ of the major axis can be found directly from the central moments μ_{pq} as

$$\tan(2\,\theta_{\mathcal{R}}) = \frac{2 \cdot \mu_{11}(\mathcal{R})}{\mu_{20}(\mathcal{R}) - \mu_{02}(\mathcal{R})} \tag{10.27}$$

and thus the corresponding angle is

$$\theta_{\mathcal{R}} = \frac{1}{2} \cdot \tan^{-1}\Bigl(\frac{2 \cdot \mu_{11}(\mathcal{R})}{\mu_{20}(\mathcal{R}) - \mu_{02}(\mathcal{R})}\Bigr) \tag{10.28}$$

$$= \frac{1}{2} \cdot \operatorname{ArcTan}\bigl(\mu_{20}(\mathcal{R}) - \mu_{02}(\mathcal{R}), 2 \cdot \mu_{11}(\mathcal{R})\bigr). \tag{10.29}$$

The resulting angle $\theta_{\mathcal{R}}$ is in the range $[-\frac{\pi}{2}, \frac{\pi}{2}]$.[9] Orientation measurements based on region moments are very accurate in general.

Calculating orientation vectors

When visualizing region properties, a frequent task is to plot the region's orientation as a line or arrow, usually anchored at the center of gravity $\bar{\boldsymbol{x}} = (\bar{x}, \bar{y})^{\mathsf{T}}$; for example, by a parametric line of the form

$$\boldsymbol{x} = \bar{\boldsymbol{x}} + \lambda \cdot \boldsymbol{x}_{\mathrm{d}} = \begin{pmatrix} \bar{x} \\ \bar{y} \end{pmatrix} + \lambda \cdot \begin{pmatrix} \cos(\theta_{\mathcal{R}}) \\ \sin(\theta_{\mathcal{R}}) \end{pmatrix}, \tag{10.30}$$

with the normalized orientation vector $\boldsymbol{x}_{\mathrm{d}}$ and the length variable $\lambda > 0$. To find the unit orientation vector $\boldsymbol{x}_d = (\cos\theta, \sin\theta)^{\mathsf{T}}$, we could first compute the inverse tangent to get 2θ (Eqn. (10.28)) and then compute the cosine and sine of θ. However, the vector $\boldsymbol{x}_d$ can also be obtained without using trigonometric functions as follows. Rewriting Eqn. (10.27) as

$$\tan(2\theta_{\mathcal{R}}) = \frac{2 \cdot \mu_{11}(\mathcal{R})}{\mu_{20}(\mathcal{R}) - \mu_{02}(\mathcal{R})} = \frac{a}{b} = \frac{\sin(2\theta_{\mathcal{R}})}{\cos(2\theta_{\mathcal{R}})}, \tag{10.31}$$

we get (by Pythagora's theorem)

[9] See Sec. A.1 in the Appendix for the computation of angles with the ArcTan() (inverse tangent) function and Sec. F.1.6 for the corresponding Java method `Math.atan2()`.

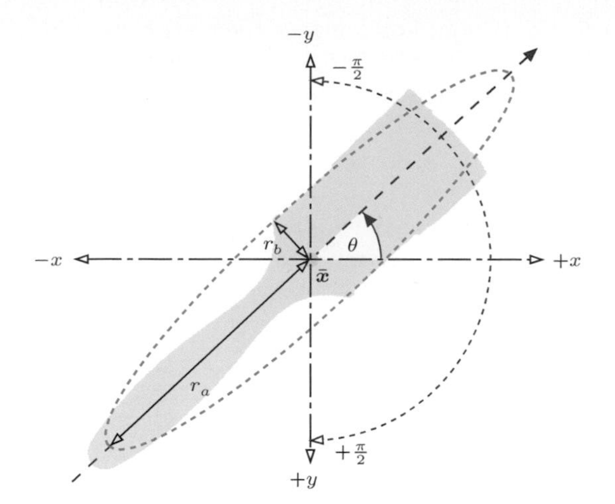

Fig. 10.18
Region orientation and eccentricity. The major axis of the region extends through its center of gravity $\bar{\boldsymbol{x}}$ at the orientation θ. Note that angles are in the range $[-\frac{\pi}{2}, +\frac{\pi}{2}]$ and increment in the *clockwise* direction because the y axis of the image coordinate system points downward (in this example, $\theta \approx -0.759 \approx -43.5°$). The eccentricity of the region is defined as the ratio between the lengths of the major axis (r_a) and the minor axis (r_b) of the "equivalent" ellipse.

$$\sin(2\theta_{\mathcal{R}}) = \frac{a}{\sqrt{a^2+b^2}} \qquad \text{and} \qquad \cos(2\theta_{\mathcal{R}}) = \frac{b}{\sqrt{a^2+b^2}},$$

where $A = 2\mu_{11}(\mathcal{R})$ and $B = \mu_{20}(\mathcal{R}) - \mu_{02}(\mathcal{R})$. Using the relations $\cos^2\alpha = \frac{1}{2}[1+\cos(2\alpha)]$ and $\sin^2\alpha = \frac{1}{2}[1-\cos(2\alpha)]$, we can compute the normalized orientation vector $\boldsymbol{x}_\mathrm{d} = (x_\mathrm{d}, y_\mathrm{d})^\mathsf{T}$ as

$$x_\mathrm{d} = \cos(\theta_{\mathcal{R}}) = \begin{cases} 0 & \text{for } a = b = 0, \\ \left[\frac{1}{2} \cdot \left(1 + \frac{b}{\sqrt{a^2+b^2}}\right)\right]^{\frac{1}{2}} & \text{otherwise,} \end{cases} \tag{10.32}$$

$$y_\mathrm{d} = \sin(\theta_{\mathcal{R}}) = \begin{cases} 0 & \text{for } a = b = 0, \\ \left[\frac{1}{2} \cdot \left(1 - \frac{b}{\sqrt{a^2+b^2}}\right)\right]^{\frac{1}{2}} & \text{for } a \geq 0, \\ -\left[\frac{1}{2} \cdot \left(1 - \frac{b}{\sqrt{a^2+b^2}}\right)\right]^{\frac{1}{2}} & \text{for } a < 0, \end{cases} \tag{10.33}$$

straight from the central region moments $\mu_{11}(\mathcal{R})$, $\mu_{20}(\mathcal{R})$, and $\mu_{02}(\mathcal{R})$, as defined in Eqn. (10.31). The horizontal component (x_d) in Eqn. (10.32) is always positive, while the case switch in Eqn. (10.33) corrects the sign of the vertical component (y_d) to map to the same angular range $[-\frac{\pi}{2}, +\frac{\pi}{2}]$ as Eqn. (10.28). The resulting vector $\boldsymbol{x}_d$ is normalized (i.e., $\|(x_d, y_d)\| = 1$) and could be scaled arbitrarily for display purposes by a suitable length λ, for example, using the region's eccentricity value described in Sec. 10.6.2 (see also Fig. 10.19).

10.6.2 Eccentricity

Similar to the region orientation, moments can also be used to determine the "elongatedness" or *eccentricity* of a region. A naive approach for computing the eccentricity could be to rotate the region until we can fit a bounding box (or enclosing ellipse) with a maximum aspect ratio. Of course this process would be computationally intensive simply because of the many rotations required. If we know the orientation of the region (Eqn. (10.28)), then we may fit a bounding box that is parallel to the region's major axis. In general, the proportions of the region's bounding box is not a good eccentricity measure anyway because it does not consider the distribution of pixels inside the box.

Based on region moments, highly accurate and stable measures can be obtained without any iterative search or optimization. Also, moment-based methods do not require knowledge of the boundary length (as required for computing the circularity feature in Sec. 10.4.2), and they can also handle nonconnected regions or point clouds. Several different formulations of region eccentricity can be found in the literature [15, 126, 128] (see also Exercise 10.17). We adopt the following definition because of its simple geometrical interpretation:

$$\mathsf{Ecc}(\mathcal{R}) = \frac{a_1}{a_2} = \frac{\mu_{20} + \mu_{02} + \sqrt{(\mu_{20} - \mu_{02})^2 + 4 \cdot \mu_{11}^2}}{\mu_{20} + \mu_{02} - \sqrt{(\mu_{20} - \mu_{02})^2 + 4 \cdot \mu_{11}^2}}, \tag{10.34}$$

where $a_1 = 2\lambda_1$, $a_2 = 2\lambda_2$ are proportional to the eigenvalues λ_1, λ_2 (with $\lambda_1 \geq \lambda_2$) of the symmetric 2×2 matrix

$$\mathbf{A} = \begin{pmatrix} \mu_{20} & \mu_{11} \\ \mu_{11} & \mu_{02} \end{pmatrix}, \tag{10.35}$$

with the region's central moments $\mu_{11}, \mu_{20}, \mu_{02}$ (see Eqn. (10.23)).[10] The values of Ecc are in the range $[1, \infty)$, where $\mathsf{Ecc} = 1$ corresponds to a circular disk and elongated regions have values > 1.

The value returned by $\mathsf{Ecc}(\mathcal{R})$ is invariant to the region's orientation and size, that is, this quantity has the important property of being rotation and scale invariant. However, the values a_1, a_2 contain relevant information about the spatial structure of the region. Geometrically, the eigenvalues λ_1, λ_2 (and thus a_1, a_2) directly relate to the proportions of the "equivalent" ellipse, positioned at the region's center of gravity $(\bar{x}, \bar{y})$ and oriented at $\theta = \theta_{\mathcal{R}}$ Eqn. (10.28). The lengths of the major and minor axes, r_a and r_b, are

$$r_a = 2 \cdot \left(\frac{\lambda_1}{|\mathcal{R}|}\right)^{\frac{1}{2}} = \left(\frac{2\, a_1}{|\mathcal{R}|}\right)^{\frac{1}{2}}, \tag{10.36}$$

$$r_b = 2 \cdot \left(\frac{\lambda_2}{|\mathcal{R}|}\right)^{\frac{1}{2}} = \left(\frac{2\, a_2}{|\mathcal{R}|}\right)^{\frac{1}{2}}, \tag{10.37}$$

respectively, with a_1, a_2 as defined in Eqn. (10.34) and $|\mathcal{R}|$ being the number of pixels in the region. Given the axes' lengths r_a, r_b and the centroid $(\bar{x}, \bar{y})$, the parametric equation of this ellipse is

$$\boldsymbol{x}(t) = \begin{pmatrix} \bar{x} \\ \bar{y} \end{pmatrix} + \begin{pmatrix} \cos(\theta) & -\sin(\theta) \\ \sin(\theta) & \cos(\theta) \end{pmatrix} \cdot \begin{pmatrix} r_a \cdot \cos(t) \\ r_b \cdot \sin(t) \end{pmatrix} \tag{10.38}$$

$$= \begin{pmatrix} \bar{x} + \cos(\theta) \cdot r_a \cdot \cos(t) - \sin(\theta) \cdot r_b \cdot \sin(t) \\ \bar{y} + \sin(\theta) \cdot r_a \cdot \cos(t) + \cos(\theta) \cdot r_b \cdot \sin(t) \end{pmatrix}, \tag{10.39}$$

for $0 \leq t < 2\pi$. If entirely *filled*, the region described by this ellipse would have the same central moments as the original region $\mathcal{R}$. Figure 10.19 shows a set of regions with overlaid orientation and eccentricity results.

[10] **A** is actually the *covariance matrix* for the distribution of pixel positions inside the region (see Sec. D.2 in the Appendix).

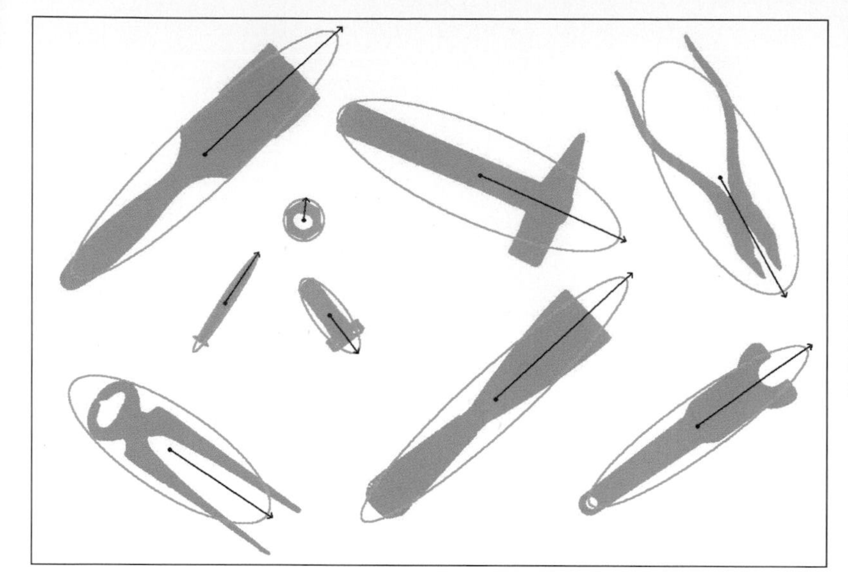

Fig. 10.19
Orientation and eccentricity examples. The orientation θ (Eqn. (10.28)) is displayed for each connected region as a vector with the length proportional to the region's eccentricity value $\mathsf{Ecc}(\mathcal{R})$ (Eqn. (10.34)). Also shown are the ellipses (Eqns. (10.36) and (10.37)) corresponding to the orientation and eccentricity parameters.

10.6.3 Bounding Box Aligned to the Major Axis

While the ordinary, x/y axis-aligned bounding box (see Sec. 10.4.2) is of little practical use (because it is sensitive to rotation), it may be interesting to see how to find a region's bounding box that is aligned with its major axis, as defined in Sec. 10.6.1. Given a region's orientation angle $\theta_{\mathcal{R}}$,

$$\boldsymbol{e}_a = \begin{pmatrix} x_a \\ y_a \end{pmatrix} = \begin{pmatrix} \cos(\theta_{\mathcal{R}}) \\ \sin(\theta_{\mathcal{R}}) \end{pmatrix} \tag{10.40}$$

is the unit vector parallel to its major axis; thus

$$\boldsymbol{e}_b = \boldsymbol{e}_a^{\perp} = \begin{pmatrix} y_a \\ -x_a \end{pmatrix} \tag{10.41}$$

is the unit vector orthogonal to $\boldsymbol{e}_a$.[11] The bounding box can now be determined as follows (see Fig. 10.20):

1. Project each region point[12] $\boldsymbol{u}_i = (u_i, v_i)$ onto the vector $\boldsymbol{e}_a$ (parallel to the region's major axis) by calculating the dot product[13]

$$a_i = \boldsymbol{u}_i \cdot \boldsymbol{e}_a \tag{10.42}$$

and keeping the minimum and maximum values

$$a_{\min} = \min_{\boldsymbol{u}_i \in \mathcal{R}} a_i, \qquad a_{\max} = \max_{\boldsymbol{u}_i \in \mathcal{R}} a_i. \tag{10.43}$$

2. Analogously, project each region point $\boldsymbol{u}_i$ onto the *orthogonal axis* (specified by the vector $\boldsymbol{e}_b$) by

[11] $\boldsymbol{x}^{\perp} = \mathrm{perp}(\boldsymbol{x}) = \left(\begin{smallmatrix} 0 & 1 \\ -1 & 0 \end{smallmatrix}\right) \cdot \boldsymbol{x}$.

[12] Of course, if the region's contour is available, it is sufficient to iterate over the contour points only.

[13] See Sec. B.3.1, Eqn. (B.19) in the Appendix.

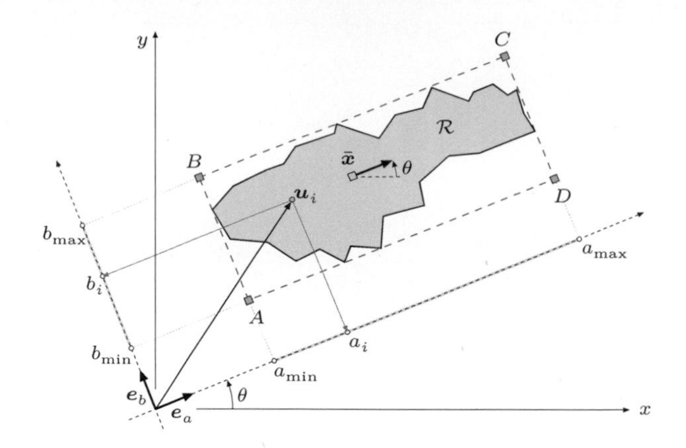

Fig. 10.20 Calculation of a region's major axis-aligned bounding box. The unit vector $\boldsymbol{e}_a$ is parallel to the region's major axis (oriented at angle θ); $\boldsymbol{e}_b$ is perpendicular to $\boldsymbol{e}_a$. The projection of a region point $\boldsymbol{u}_i$ onto the lines defined by $\boldsymbol{e}_a$ and $\boldsymbol{e}_b$ yields the lengths a_i and b_i, respectively (measured from the coordinate origin). The resulting quantities a_{min}, a_{max}, b_{min}, b_{max} define the corner points (A, B, C, D) of the axis-aligned bounding box. Note that the position of the region's centroid ($\bar{\boldsymbol{x}}$) is not required in this calculation.

$$b_i = \boldsymbol{u}_i \cdot \boldsymbol{e}_b \tag{10.44}$$

and keeping the minimum and maximum values, that is,

$$b_{\mathrm{min}} = \min_{\boldsymbol{u}_i \in \mathcal{R}} b_i, \qquad b_{\mathrm{max}} = \max_{\boldsymbol{u}_i \in \mathcal{R}} b_i. \tag{10.45}$$

Note that steps 1 and 2 can be performed in a single iteration over all region points.

3. Finally, from the resulting quantities a_{min}, a_{max}, b_{min}, b_{max}, calculate the four corner points A, B, C, D of the bounding box as

$$\begin{aligned} A &= a_{\mathrm{min}} \cdot \boldsymbol{e}_a + b_{\mathrm{min}} \cdot \boldsymbol{e}_b, & B &= a_{\mathrm{min}} \cdot \boldsymbol{e}_a + b_{\mathrm{max}} \cdot \boldsymbol{e}_b, \\ C &= a_{\mathrm{max}} \cdot \boldsymbol{e}_a + b_{\mathrm{max}} \cdot \boldsymbol{e}_b, & D &= a_{\mathrm{max}} \cdot \boldsymbol{e}_a + b_{\mathrm{min}} \cdot \boldsymbol{e}_b. \end{aligned} \tag{10.46}$$

The complete calculation is summarized in Alg. 10.20; a typical example is shown in Fig. 10.21(d).

Alg. 10.5 Calculation of the major axis-aligned bounding box for a binary region $\mathcal{R}$. If the region's contour is available, it is sufficient to use the contour points only.

```
1: MajorAxisAlignedBoundingBox(R)
       Input: R = {u_i}, a binary region containing points u_i ∈ ℝ².
       Returns the four corner points of the region's bounding box.
2:     θ ← 0.5 · ArcTan(μ20(R) − μ02(R), 2 · μ11(R))      ▷ see Eq. 10.28
3:     e_a ← (cos(θ), sin(θ))ᵀ  ▷ unit vector parall. to region's major axis
4:     e_b ← (sin(θ), −cos(θ))ᵀ   ▷ unit vector perpendic. to major axis
5:     a_min ← ∞,   a_max ← −∞
6:     b_min ← ∞,   b_max ← −∞
7:     for all u ∈ R do
8:         a ← u · e_a                          ▷ project u onto e_a (Eq. 10.42)
9:         a_min ← min(a_min, a)
10:        a_max ← max(a_max, a)
11:        b ← u · e_b                          ▷ project u onto e_b (Eq. 10.44)
12:        b_min ← min(b_min, b)
13:        b_max ← max(b_max, b)
14:    A ← a_min · e_a + b_min · e_b
15:    B ← a_min · e_a + b_max · e_b
16:    C ← a_max · e_a + b_max · e_b
17:    D ← a_max · e_a + b_min · e_b
18:    return (A, B, C, D)                      ▷ corners of the bounding box
```

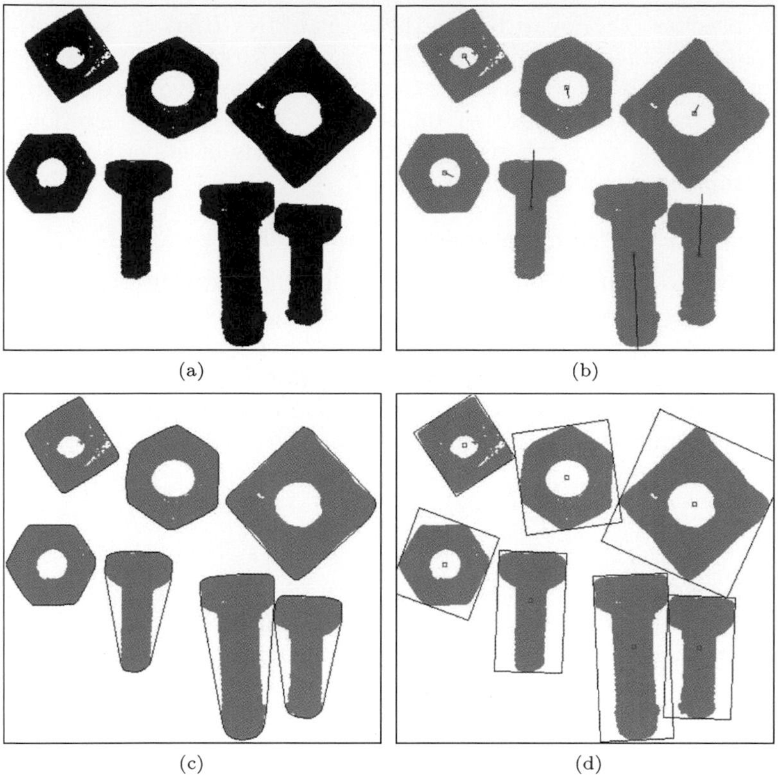

(a) (b) (c) (d)

Fig. 10.21
Geometric region properties. Original binary image (a), centroid and orientation vector (length determined by the region's eccentricity) of the major axis (b), convex hull (c), and major axis-aligned bounding box (d).

10.6.4 Invariant Region Moments

Normalized central moments are not affected by the translation or uniform scaling of a region (i.e., the values are invariant), but in general rotating the image will change these values.

Hu's invariant moments

A classical solution to this problem is a clever combination of simpler features known as "Hu's Moments" [112]:[14]

$$\begin{aligned}
\phi_1 &= \bar{\mu}_{20} + \bar{\mu}_{02}, \qquad (10.47)\\
\phi_2 &= (\bar{\mu}_{20} - \bar{\mu}_{02})^2 + 4\,\bar{\mu}_{11}^2,\\
\phi_3 &= (\bar{\mu}_{30} - 3\,\bar{\mu}_{12})^2 + (3\,\bar{\mu}_{21} - \bar{\mu}_{03})^2,\\
\phi_4 &= (\bar{\mu}_{30} + \bar{\mu}_{12})^2 + (\bar{\mu}_{21} + \bar{\mu}_{03})^2,\\
\phi_5 &= (\bar{\mu}_{30} - 3\,\bar{\mu}_{12}) \cdot (\bar{\mu}_{30} + \bar{\mu}_{12}) \cdot [(\bar{\mu}_{30} + \bar{\mu}_{12})^2 - 3(\bar{\mu}_{21} + \bar{\mu}_{03})^2] +\\
&\quad (3\,\bar{\mu}_{21} - \bar{\mu}_{03}) \cdot (\bar{\mu}_{21} + \bar{\mu}_{03}) \cdot [3\,(\bar{\mu}_{30} + \bar{\mu}_{12})^2 - (\bar{\mu}_{21} + \bar{\mu}_{03})^2],\\
\phi_6 &= (\bar{\mu}_{20} - \bar{\mu}_{02}) \cdot [(\bar{\mu}_{30} + \bar{\mu}_{12})^2 - (\bar{\mu}_{21} + \bar{\mu}_{03})^2] +\\
&\quad 4\,\bar{\mu}_{11} \cdot (\bar{\mu}_{30} + \bar{\mu}_{12}) \cdot (\bar{\mu}_{21} + \bar{\mu}_{03}),\\
\phi_7 &= (3\,\bar{\mu}_{21} - \bar{\mu}_{03}) \cdot (\bar{\mu}_{30} + \bar{\mu}_{12}) \cdot [(\bar{\mu}_{30} + \bar{\mu}_{12})^2 - 3\,(\bar{\mu}_{21} + \bar{\mu}_{03})^2] +\\
&\quad (3\,\bar{\mu}_{12} - \bar{\mu}_{30}) \cdot (\bar{\mu}_{21} + \bar{\mu}_{03}) \cdot [3\,(\bar{\mu}_{30} + \bar{\mu}_{12})^2 - (\bar{\mu}_{21} + \bar{\mu}_{03})^2].
\end{aligned}$$

[14] In order to improve the legibility of Eqn. (10.47) the argument for the region ($\mathcal{R}$) has been dropped; as an example, with the region argument, the first line would read $H_1(\mathcal{R}) = \bar{\mu}_{20}(\mathcal{R}) + \bar{\mu}_{02}(\mathcal{R})$, and so on.

In practice, the logarithm of these quantities (that is, $\log(\phi_k)$) is used since the raw values may have a very large range. These features are also known as *moment invariants* since they are invariant under translation, rotation, and scaling. While defined here for binary images, they are also applicable to parts of grayscale images; examples can be found in [88, p. 517].

Flusser's invariant moments

It was shown in [72,73] that Hu's moments, as listed in Eqn. (10.47), are partially redundant and incomplete. Based on so-called *complex moments* $c_{pq} \in \mathbb{C}$, Flusser designed an improved set of 11 rotation and scale-invariant features $\psi_1, \ldots, \psi_{11}$ (see Eqn. (10.51)) for characterizing 2D shapes. For grayscale images (with $I(u,v) \in \mathbb{R}$), the complex moments of order p, q are defined as

$$c_{pq}(\mathcal{R}) = \sum_{(u,v)\in\mathcal{R}} I(u,v) \cdot (x + \mathrm{i}\cdot y)^p \cdot (x - \mathrm{i}\cdot y)^q, \tag{10.48}$$

with centered positions $x = u - \bar{x}$ and $y = v - \bar{y}$, and $(\bar{x}, \bar{y})$ being the *centroid* of $\mathcal{R}$ (i denotes the imaginary unit). In the case of binary images (with $I(u,v) \in [0,1]$) Eqn. (10.48) simplifies to

$$c_{pq}(\mathcal{R}) = \sum_{(u,v)\in\mathcal{R}} (x + \mathrm{i}\cdot y)^p \cdot (x - \mathrm{i}\cdot y)^q. \tag{10.49}$$

Analogous to Eqn. (10.26), the complex moments can be *scale-normalized* to

$$\hat{c}_{p,q}(\mathcal{R}) = \frac{1}{A^{(p+q+2)/2}} \cdot c_{p,q}, \tag{10.50}$$

with A being the area of $\mathcal{R}$ [74, p. 29]. Finally, the derived rotation and scale invariant region moments of 2nd to 4th order are[15]

$$\begin{aligned}
\psi_1 &= \mathrm{Re}(\hat{c}_{1,1}), & \psi_2 &= \mathrm{Re}(\hat{c}_{2,1} \cdot \hat{c}_{1,2}), & \psi_3 &= \mathrm{Re}(\hat{c}_{2,0} \cdot \hat{c}_{1,2}^2), \\
\psi_4 &= \mathrm{Im}(\hat{c}_{2,0} \cdot \hat{c}_{1,2}^2), & \psi_5 &= \mathrm{Re}(\hat{c}_{3,0} \cdot \hat{c}_{1,2}^3), & \psi_6 &= \mathrm{Im}(\hat{c}_{3,0} \cdot \hat{c}_{1,2}^3), \\
\psi_7 &= \mathrm{Re}(\hat{c}_{2,2}), & \psi_8 &= \mathrm{Re}(\hat{c}_{3,1} \cdot \hat{c}_{1,2}^2), & \psi_9 &= \mathrm{Im}(\hat{c}_{3,1} \cdot \hat{c}_{1,2}^2), \\
\psi_{10} &= \mathrm{Re}(\hat{c}_{4,0} \cdot \hat{c}_{1,2}^4), & \psi_{11} &= \mathrm{Im}(\hat{c}_{4,0} \cdot \hat{c}_{1,2}^4).
\end{aligned} \tag{10.51}$$

Table 10.1 lists the normalized Flusser moments for five binary shapes taken from the Kimia dataset [134].

Shape matching with region moments

One obvious use of invariant region moments is shape matching and classification. Given two binary shapes A and B, with associated moment ("feature") vectors

$$\boldsymbol{f}_A = (\psi_1(A), \ldots, \psi_{11}(A)) \quad \text{and} \quad \boldsymbol{f}_B = (\psi_1(B), \ldots, \psi_{11}(B)),$$

respectively, one approach could be to simply measure the difference between shapes by the Euclidean distance of these vectors in the form

[15] In Eqn. (10.51), the use of Re( ) for the quantities ψ_1, ψ_2, ψ_7 (which are real-valued *per se*) is redundant.

ψ_1	0.3730017575	0.2545476083	0.2154034257	0.2124041195	0.3600613700
ψ_2	0.0012699373	0.0004247053	0.0002068089	0.0001089652	0.0017187073
ψ_3	0.0004041515	0.0000644829	0.0000274491	0.0000014248	-0.0003853999
ψ_4	0.0000097827	-0.0000076547	0.0000071688	-0.0000022103	-0.0001944121
ψ_5	0.0000012672	0.0000002327	0.0000000637	0.0000000083	-0.0000078073
ψ_6	0.0000001090	-0.0000000483	0.0000000041	0.0000000153	-0.0000061997
ψ_7	0.2687922057	0.1289708408	0.0814034374	0.0712567626	0.2340886626
ψ_8	0.0003192443	0.0000414818	0.0000134036	0.0000003020	-0.0002878997
ψ_9	0.0000053208	-0.0000032541	0.0000030880	-0.0000008365	-0.0001628669
ψ_{10}	0.0000103461	0.0000000091	0.0000000019	-0.0000000003	0.0000001922
ψ_{11}	0.0000000120	-0.0000000020	0.0000000008	-0.0000000000	0.0000003015

Table 10.1
Binary shapes and associated normalized Flusser moments $\psi_1, \ldots, \psi_{11}$. Notice the magnitude of the moments varies by a large factor.

	0.000	0.183	0.245	0.255	0.037
	0.183	0.000	0.062	0.071	0.149
	0.245	0.062	0.000	0.011	0.210
	0.255	0.071	0.011	0.000	0.220
	0.037	0.149	0.210	0.220	0.000

Table 10.2
Inter-class (Euclidean) distances $\mathrm{d_E}(A, B)$ between normalized shape feature vectors for the five reference shapes (see Eqn. (10.52)). Off-diagonal values should be consistently large to allow good shape discrimination.

$$\mathrm{d_E}(A, B) = \|\boldsymbol{f}_A - \boldsymbol{f}_B\| = \Big[\sum_{i=1}^{11} |\psi_i(A) - \psi_i(B)|^2\Big]^{1/2}. \tag{10.52}$$

Concrete distances between the five sample shapes are listed in Table 10.2. Since the moment vectors are rotation and scale invariant,[16] shape comparisons should remain unaffected by such transformations. Note, however, that the magnitude of the individual moments varies over a very large range. Thus, if the Euclidean distance is used as we have just suggested, the comparison (matching) of shapes is typically dominated by a few moments (or even a single moment) of relatively large magnitude, while the small-valued moments play virtually no role in the distance calculation. This is because the Euclidean distance treats the multi-dimensional feature space uniformly along all dimensions.

As a consequence, moment-based shape discrimination with the ordinary Euclidean distance is typically not very selective. A simple solution is to replace Eqn. (10.52) by a *weighted distance* measure of the form

$$\mathrm{d'_E}(A, B) = \Big[\sum_{i=1}^{11} w_i \cdot |\psi_i(A) - \psi_i(B)|^2\Big]^{1/2}, \tag{10.53}$$

with fixed weights $w_1, \ldots, w_{11} \geq 0$ assigned to each each moment feature to compensate for the differences in magnitude.

A more elegant approach is to use of the *Mahalanobis* distance [24, 157] for comparing the moment vectors, which accounts for the statistical distribution of each vector component and avoids large-magnitude components dominating the smaller ones. In this case,

[16] Although the invariance property holds perfectly for continuous shapes, rotating and scaling *discrete* binary images may significantly affect the associated region moments.

the distance calculation becomes

$$d_M(A,B) = \big[(\boldsymbol{f}_A - \boldsymbol{f}_B)^\mathsf{T} \cdot \boldsymbol{\Sigma}^{-1} \cdot (\boldsymbol{f}_A - \boldsymbol{f}_B)\big]^{1/2}, \tag{10.54}$$

where $\boldsymbol{\Sigma}$ is the 11×11 *covariance matrix* for the moment vectors $\boldsymbol{f}$. Note that the expression under the root in Eqn. (10.54) is the dot product of a row vector and a column vector, that is, the result is a non-negative scalar value. The Mahalanobis distance can be viewed as a special form of the weighted Euclidean distance (Eqn. (10.53)), where the weights are determined by the variability of the individual vector components. See Sec. D.3 in the Appendix and Exercise 10.16 for additional details.

10.7 Projections

Image projections are 1D representations of the image contents, usually calculated parallel to the coordinate axis. In this case, the horizontal and vertical projections of a scalar-valued image $I(u,v)$ of size $M \times N$ are defined as

$$\mathsf{P}_{\mathrm{hor}}(v) = \sum_{u=0}^{M-1} I(u,v) \quad \text{for } 0 < v < N, \tag{10.55}$$

$$\mathsf{P}_{\mathrm{ver}}(u) = \sum_{v=0}^{N-1} I(u,v) \quad \text{for } 0 < u < M. \tag{10.56}$$

The *horizontal* projection $P_{\mathrm{hor}}(v_0)$ (Eqn. (10.55)) is the sum of the pixel values in the image *row* v_0 and has length N corresponding to the height of the image. On the other hand, a *vertical* projection P_{ver} of length M is the sum of all the values in the image *column* u_0 (Eqn. (10.56)). In the case of a binary image with $I(u,v) \in 0,1$, the projection contains the count of the foreground pixels in the corresponding image row or column.

Program 10.4 gives a direct implementation of the projection calculations as the `run()` method for an ImageJ plugin, where projections in both directions are computed during a single traversal of the image.

Projections in the direction of the coordinate axis are often utilized to quickly analyze the structure of an image and isolate its component parts; for example, in document images it is used to separate graphic elements from text blocks as well as to isolate individual lines (see the example in Fig. 10.22). In practice, especially to account for document skew, projections are often computed along the major axis of an image region Eqn. (10.28). When the projection vectors of a region are computed in reference to the centroid of the region along the major axis, the result is a rotation-invariant vector description (often referred to as a "signature") of the region.

10.8 Topological Region Properties

Topological features do not describe the shape of a region in continuous terms; instead, they capture its structural properties. Topological

```
1  public void run(ImageProcessor I) {
2    int M = I.getWidth();
3    int N = I.getHeight();
4    int[] pHor = new int[N]; // = Phor(v)
5    int[] pVer = new int[M]; // = Pver(u)
6    for (int v = 0; v < N; v++) {
7      for (int u = 0; u < M; u++) {
8        int p = I.getPixel(u, v);
9        pHor[v] += p;
10       pVer[u] += p;
11     }
12   }  // use projections pHor, pVer now
13   // ...
14 }
```

Prog. 10.4
Calculation of horizontal and vertical projections. The `run()` method for an ImageJ plugin (`ip` is of type `ByteProcessor` or `ShortProcessor`) computes the projections in x and y directions simultaneously in a a single traversal of the image. The projections are represented by the one-dimensional arrays `horProj` and `verProj` with elements of type `int`.

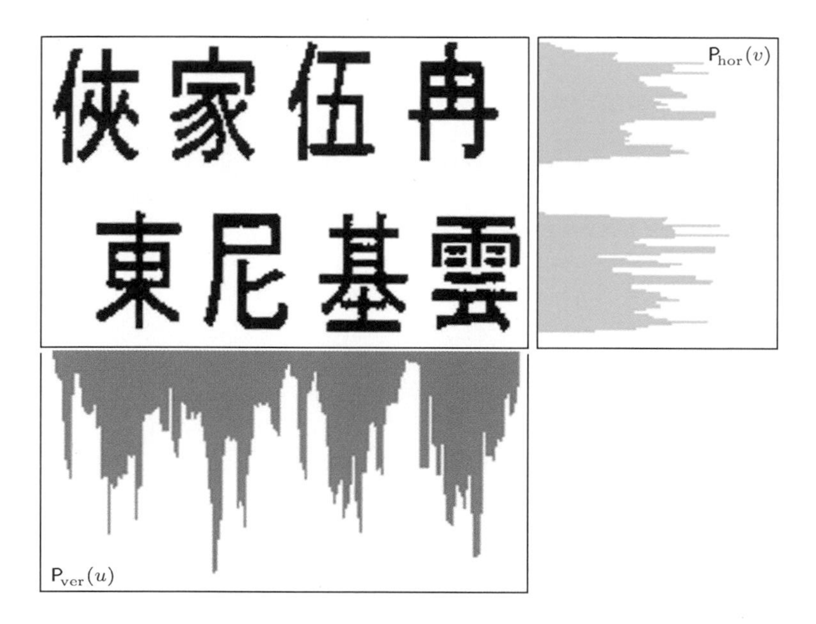

Fig. 10.22
Horizontal and vertical projections of a binary image.

properties are typically invariant even under strong image transformations. The convexity of a region, which can be calculated from the convex hull (Sec. 10.4.2), is also a topological property.

A simple and robust topological feature is the *number of holes* $N_L(\mathcal{R})$ in a region. This feature is easily determined while finding the inner contours of a region, as described in Sec. 10.2.2.

A useful topological feature that can be derived directly from the number of holes is the so-called *Euler number* N_E, which is the difference between the number of connected regions N_R and the number of their holes N_L, that is,

$$N_E(\mathcal{R}) = N_R(\mathcal{R}) - N_L(\mathcal{R}). \tag{10.57}$$

In the case of a single connected region this is simply $1 - N_L$. For a picture of the number "**8**", for example, $N_E = 1 - 2 = -1$ and for the letter "**D**" we get $N_E = 1 - 1 = 0$.

Topological features are often used in combination with numerical features for classification. A classic example of this combination is OCR (optical character recognition) [38]. Figure 10.23 shows an

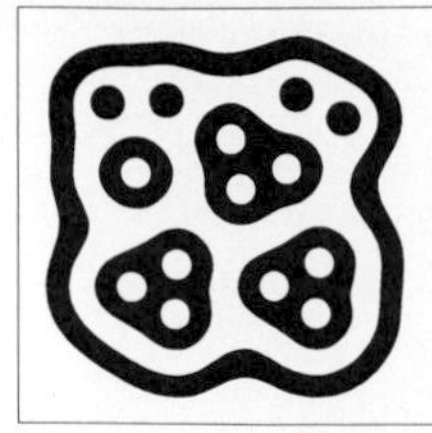
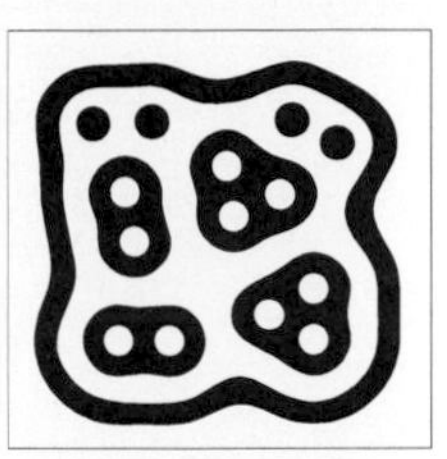

Fig. 10.23 Visual identification markers composed of recursively nested regions [22].

interesting use of topological structures for coding optical markers used in augmented reality applications [22].[17] The recursive nesting of outer and inner regions is equivalent to a tree structure that allows fast and unique identification of a larger number of known patterns (see also Exercise 10.21).

10.9 Java Implementation

Most algorithms described in this chapter are implemented as part of the `imagingbook` library.[18] The key classes are `BinaryRegion` and `Contour`, the abstract class `RegionLabeling` and its concrete subclasses `RecursiveLabeling`, `BreadthFirstLabeling`, `DepthFirstLabeling` (Alg. 10.1) and `SequentialLabeling` (Alg. 10.2). The combined region labeling and contour tracing method (Algs. 10.3 and 10.4) is implemented by class `RegionContourLabeling`. Additional details can be found in the online documentation.

Example

A complete example for the use of this API is shown in Prog. 10.5. Particularly useful is the facility for visiting all positions of a specific region using the iterator returned by method `getRegionPoints()`, as demonstrated by this code segment:

```
RegionLabeling segmenter =  ....
// Get the largest region:
BinaryRegion r = segmenter.getRegions(true).get(0);
// Loop over all points of region r:
for (Point p : r.getRegionPoints()) {
  int u = p.x;
  int v = p.y;
  // do something with position u, v
}
```

10.10 Exercises

Exercise 10.1. Manually simulate the execution of both variations (*depth-first* and *breadth-first*) of the flood-fill algorithm using the image in Fig. 10.24 and starting at position $(5, 1)$.

[17] http://reactivision.sourceforge.net/.

[18] Package `imagingbook.pub.regions`.

```
1 ...
2 import imagingbook.pub.regions.BinaryRegion;
3 import imagingbook.pub.regions.Contour;
4 import imagingbook.pub.regions.ContourOverlay;
5 import imagingbook.pub.regions.RegionContourLabeling;
6 import java.awt.geom.Point2D;
7 import java.util.List;
8
9 public class Region_Contours_Demo implements PlugInFilter {
10
11   public int setup(String arg, ImagePlus im) {
12     return DOES_8G + NO_CHANGES;
13   }
14
15   public void run(ImageProcessor ip) {
16     // Make sure we have a proper byte image:
17     ByteProcessor bp = ip.convertToByteProcessor();
18
19     // Create the region labeler / contour tracer:
20     RegionContourLabeling segmenter =
21             new RegionContourLabeling(bp);
22
23     // Get the list of detected regions (sort by size):
24     List<BinaryRegion> regions =
25             segmenter.getRegions(true);
26     if (regions.isEmpty()) {
27       IJ.error("No regions detected!");
28       return;
29     }
30
31     // List all regions:
32     IJ.log("Detected regions: " + regions.size());
33     for (BinaryRegion r: regions) {
34       IJ.log(r.toString());
35     }
36
37     // Get the outer contour of the largest region:
38     BinaryRegion largestRegion = regions.get(0);
39     Contour oc =  largestRegion.getOuterContour();
40     IJ.log("Points on outer contour of largest region:");
41     Point2D[] points = oc.getPointArray();
42     for (int i = 0; i < points.length; i++) {
43       Point2D p = points[i];
44       IJ.log("Point " + i + ": " + p.toString());
45     }
46
47     // Get all inner contours of the largest region:
48     List<Contour> ics = largestRegion.getInnerContours();
49     IJ.log("Inner regions (holes): " + ics.size());
50   }
51 }
```

Prog. 10.5
Complete example for the use of the regions API. The ImageJ plugin `Region_Contours_Demo` segments the binary (8-bit grayscale) image `ip` into connected components. This is done with an instance of class `RegionContourLabeling` (see line 21), which also extracts the regions' contours. In line 25, a list of regions (sorted by size) is produced which is subsequently traversed (line 33). The treatment of outer and inner contours as well as the iteration over individual contour points is shown in lines 38–49.

	0	1	2	3	4	5	6	7	8	9	10	11	12	13
0	0	0	0	0	0	0	0	0	0	0	0	0	0	0
1	0	0	0	0	0	1	1	0	0	1	1	0	1	0
2	0	1	1	1	1	1	1	0	0	1	0	0	1	0
3	0	0	0	0	1	0	1	0	0	0	0	0	1	0
4	0	1	1	1	1	1	1	1	1	1	1	1	1	0
5	0	0	0	0	1	1	1	1	1	1	1	1	1	0
6	0	1	1	0	0	0	1	0	1	0	0	0	0	0
7	0	0	0	0	0	0	0	0	0	0	0	0	0	0

0 Background
1 Foreground

Fig. 10.24
Binary image for Exercise 10.1.

Exercise 10.2. The implementation of the flood-fill algorithm in Prog. 10.1 places all the neighboring pixels of each visited pixel into either the *stack* or the *queue* without ensuring they are foreground pixels and that they lie within the image boundaries. The number of items in the stack or the queue can be reduced by ignoring (not inserting) those neighboring pixels that do not meet the two conditions given. Modify the *depth-first* and *breadth-first* variants given in Prog. 10.1 accordingly and compare the new running times.

Exercise 10.3. The implementations of depth-first and breadth-first labeling shown in Prog. 10.1 will run significantly slower than the recursive version because the frequent creation of new `Point` objects is quite time consuming. Modify the *depth-first* version of Prog. 10.1 to use a stack with elements of a *primitive type* (e.g., `int`) instead. Note that (at least in Java)[19] it is not possible to specify a built-in list structure (such as `Deque` or `LinkedList`) for a primitive element type. Implement you own stack class that internally uses an `int`-array to store the (u, v) coordinates. What is the maximum number of stack entries needed for a given image of size $M \times N$? Compare the performance of your solution to the original version in Prog. 10.1.

Exercise 10.4. Implement an ImageJ plugin that encodes a given binary image by run length encoding (Sec. 10.3.2) and stores it in a file. Develop a second plugin that reads the file and reconstructs the image.

Exercise 10.5. Calculate the amount of memory required to represent a contour with 1000 points in the following ways: (a) as a sequence of coordinate points stored as pairs of `int` values; (b) as an 8-chain code using Java `byte` elements, and (c) as an 8-chain code using only 3 bits per element.

Exercise 10.6. Implement a Java class for describing a binary image region using chain codes. It is up to you, whether you want to use an absolute or differential chain code. The implementation should be able to encode closed contours as chain codes and also reconstruct the contours given a chain code.

Exercise 10.7. The *Graham Scan* method [91] is an efficient algorithm for calculating the convex hull of a 2D point set (of size n), with time complexity $\mathcal{O}(n \cdot \log(n))$.[20] Implement this algorithm and show that it is sufficient to consider only the outer contour points of a region to calculate its convex hull.

[19] Other languages like C# allow this.

[20] See also http://en.wikipedia.org/wiki/Graham_scan.

Exercise 10.8. While computing the convex hull of a region, the maximal diameter (maximum distance between two arbitrary points) can also be simply found. Devise an alternative method for computing this feature without using the convex hull. Determine the running time of your algorithm in terms of the number of points in the region.

Exercise 10.9. Implement an algorithm for comparing contours using their shape numbers Eqn. (10.6). For this purpose, develop a metric for measuring the distance between two normalized chain codes. Describe if, and under which conditions, the results will be reliable.

Exercise 10.10. Sketch the contour equivalent to the *absolute* chain code sequence $\boldsymbol{c}'_{\mathcal{R}} = (6, 7, 7, 1, 2, 0, 2, 3, 5, 4, 4)$. (a) Choose an arbitrary starting point and determine if the resulting contour is closed. (b) Find the associated *differential* chain code $\boldsymbol{c}''_{\mathcal{R}}$ (Eqn. (10.5)).

Exercise 10.11. Calculate (under assumed 8-neighborhood) the *shape number* of base $b = 8$ (see Eqn. (10.6)) for the differential chain code $\boldsymbol{c}''_{\mathcal{R}} = (1, 0, 2, 1, 6, 2, 1, 2, 7, 0, 2)$ and all possible circular shifts of this code. Which shift yields the maximum arithmetic value?

Exercise 10.12. Using Eqn. (10.14) as the basis, develop and implement an algorithm that computes the area of a region from its 8-chain-encoded contour (see also [263], [127, Sec. 19.5]).

Exercise 10.13. Modify Alg. 10.3 such that the outer and inner contours are not returned as individual lists $(\mathsf{C}_{\text{out}}, \mathsf{C}_{\text{in}})$ but as a composite tree structure. An outer contour thus represents a region that may contain zero, one, or more inner contours (i.e., holes). Each inner contour may again contain other regions (i.e., outer contours), and so on.

Exercise 10.14. Sketch an example binary region where the centroid does not lie inside the region itself.

Exercise 10.15. Implement the binary region moment features proposed by *Hu* (Eqn. (10.47)) and/or *Flusser* (Eqn. (10.51)) and verify that they are invariant under image scaling and rotation. Use the test image in Fig. 10.25[21] (or create your own), which contains rotated and mirrored instances of the reference shapes, in addition to other (unknown) shapes.

Exercise 10.16. Implement the Mahalanobis distance calculation, as defined in Eqn. (10.54), for measuring the similarity between shape moment vectors.

A. Compute the covariance matrix $\boldsymbol{\Sigma}$ (see Sec. D.3 in the Appendix) for the $m = 11$ Flusser shape features $\psi_1, \ldots, \psi_{11}$ of the reference images in Table 10.1. Calculate and tabulate the inter-class Mahalanobis distances for the reference shapes, analogous to the example in Table 10.2.

[21] Images are available on the book's website.

Fig. 10.25
Test image for moment-based shape matching. Reference shapes (top) and test image (bottom) composed of rotated and/or scaled shapes from the Kimia database and additional (unclassified) shapes.

B. Extend your analysis to a larger set of 500–1000 shapes (e.g., from the Kimia dataset [134], which contains more than 20 000 binary shape images). Calculate the normalized moment features and the covariance matrix $\boldsymbol{\Sigma}$ for the entire image set. Calculate the inter-class distance matrices for (a) the Euclidean and (b) the Mahalanobis distance. Display the distance matrices as grayscale images (`FloatProcessor`) and interpret them.

Exercise 10.17. There are alternative definitions for the *eccentricity* of a region Eqn. (10.34); for example [128, p. 394],

$$\mathsf{Ecc}_2(\mathcal{R}) = \frac{[\mu_{20}(\mathcal{R}) - \mu_{02}(\mathcal{R})]^2 + 4\cdot\mu_{11}^2(\mathcal{R})}{[\mu_{20}(\mathcal{R}) + \mu_{02}(\mathcal{R})]^2}. \tag{10.58}$$

Implement this version as well as the one in Eqn. (10.34) and contrast the results using suitably designed regions. Determine the numeric range of these quantities and test if they are really rotation and scale-invariant.

Exercise 10.18. Write an ImageJ plugin that (a) finds (labels) all regions in a binary image, (b) computes the orientation and eccentricity for each region, and (c) shows the results as a direction vector and the equivalent ellipse on top of each region (as exemplified in Fig. 10.19). Hint: Use Eqn. (10.39) to develop a method for drawing ellipses at arbitrary orientations (not available in ImageJ).

Exercise 10.19. The Java method in Prog. 10.4 computes an image's horizontal and vertical projections. The scheme described in Sec. 10.6.3 and illustrated in Fig. 10.20 can be used to calculate projections along arbitrary directions θ. Develop and implement such a process and display the resulting projections.

Exercise 10.20. Text recognition (OCR) methods are likely to fail if the document image is not perfectly axis-aligned. One method for estimating the skew angle of a text document is to perform binary segmentation and connected components analysis (see Fig. 10.26):

- *Smear* the original binary image by applying a disk-shaped morphological dilation with a specified radius (see Chapter 9, Sec. 9.2.3). The aim is to close the gaps between neighboring glyphs without closing the space between adjacent text lines (Fig. 10.26(b))
- Apply region segmentation to the resulting image and calculate the orientation $\theta(\mathcal{R})$ and the eccentricity $E(\mathcal{R})$ of each region $\mathcal{R}$ (see Secs. 10.6.1 and 10.6.2). Ignore all regions that are either too small or not sufficiently elongated.
- Estimate the global skew angle by averaging the regions' orientations θ_i. Note that, since angles are *circular*, they cannot be averaged in the usual way (see Chapter 15, Eqn. (15.14) for how to calculate the mean of a circular quantity). Consider using the eccentricity as a weight for the contribution of the associated region to the global average.
- Obviously, this scheme is sensitive to *outliers*, that is, against angles that deviate strongly from the average orientation. Try to improve this estimate (i.e., make it more robust and accurate) by iteratively removing angles that are "too far" from the average orientation and then recalculating the result.

Exercise 10.21. Draw the tree structure, defined by the recursive nesting of outer and inner regions, for each of the markers shown in Fig. 10.23. Based on this graph structure, suggest an algorithm for matching pairs of markers or, alternatively, for retrieving the best-matching marker from a database of markers.

Fig. 10.26
Document skew estimation example (see Exercise 10.20). Original binary image (a); result of applying a disk-shaped morphological dilation with radius 3.0 (b); region orientation vectors (c); histogram of the orientation angle θ (d). The real skew angle in this scan is approximately 1.1°.

As President Eisenhower once said, nuclear weapons are the only thing that can destroy the United States. Americans want to hear how the next president plans to control the thousands of these weapons of mass destruction that exist in the world.

It's worth remembering that in October 1986 President Ronald Reagan was meeting with Soviet President Mikhail Gorbachev in Reykjavik, Iceland, to discuss eliminating nuclear weapons.

The two leaders focused on nuclear weapons testing. If you are serious about total nuclear disarmament, you have to end testing first. As Reagan wrote then, "I am committed to the ultimate attainment of a total ban on nuclear testing, a goal that has been endorsed by every U.S. president since President Eisenhower."

But Reagan had some prerequisites. In 1986 the United States Senate had yet to ratify two treaties that had been negotiated with the Soviets: the Threshold Test Ban, which limited the size of underground lowed underground tests for peaceful purposes. Reagan wanted to get these treaties ratified first, and that meant making sure the agreements could not be cheated on by secret tests. As Reagan like to say "Trust, but verify."

In 1990, after Reagan had left office, both the Threshold Test Ban and the Peaceful Nuclear Explosions Treaty were ratified by the Senate after satisfactory review of the verification provisions. Reagan's first requirement on the road to a nuclear test ban was complete.

Reagan's second requirement for ending nuclear testing was that the Soviets and the Americans should reduce their nuclear stockpiles. That effort started with the 1987 Intermediate-Range Nuclear Forces Treaty, which eliminated medium- and short-range nuclear missiles. The Strategic Arms Reduction Talks (START) treaties subsequently continued U.S. and Russian reductions, although thousands still remain.

In 1996 the Comprehensive Nuclear Test Ban Treaty was crafted to ban all nuclear test

(a)

(b)

(c)

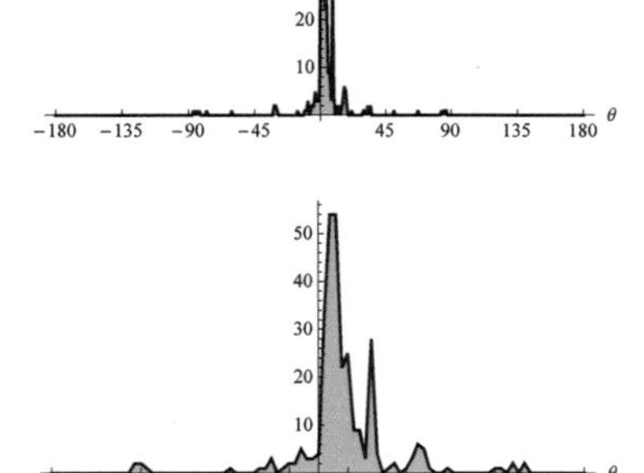

(d)

11

Automatic Thresholding

Although techniques based on binary image regions have been used for a very long time, they still play a major role in many practical image processing applications today because of their simplicity and efficiency. To obtain a binary image, the first and perhaps most critical step is to convert the initial grayscale (or color) image to a binary image, in most cases by performing some form of thresholding operation, as described in Chapter 4, Sec. 4.1.4.

Anyone who has ever tried to convert a scanned document image to a readable binary image has experienced how sensitively the result depends on the proper choice of the threshold value. This chapter deals with finding the best threshold automatically only from the information contained in the image, i.e., in an "unsupervised" fashion. This may be a single, "global" threshold that is applied to the whole image or different thresholds for different parts of the image. In the latter case we talk about "adaptive" thresholding, which is particularly useful when the image exhibits a varying background due to uneven lighting, exposure, or viewing conditions.

Automatic thresholding is a traditional and still very active area of research that had its peak in the 1980s and 1990s. Numerous techniques have been developed for this task, ranging from simple ad-hoc solutions to complex algorithms with firm theoretical foundations, as documented in several reviews and evaluation studies [86, 178, 204, 213, 231]. Binarization of images is also considered a "segmentation" technique and thus often categorized under this term. In the following, we describe some representative and popular techniques in greater detail, starting in Sec. 11.1 with global thresholding methods and continuing with adaptive methods in Sec. 11.2.

11.1 Global Histogram-Based Thresholding

Given a grayscale image I, the task is to find a single "optimal" threshold value for binarizing this image. Applying a particular threshold q is equivalent to classifying each pixel as being either part

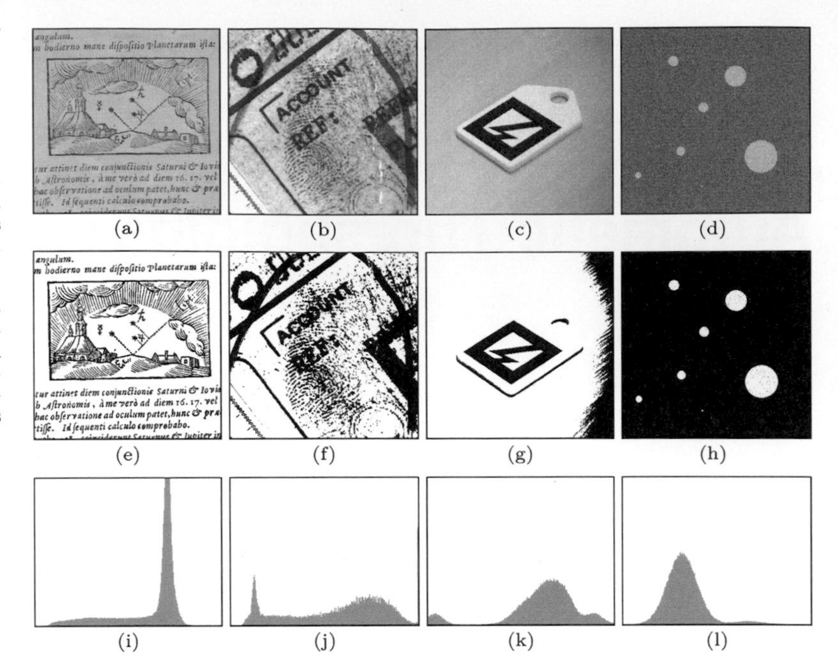

Fig. 11.1 Test images used for subsequent thresholding experiments. Detail from a manuscript by Johannes Kepler (a), document with fingerprint (b), ARToolkit marker (c), synthetic two-level Gaussian mixture image (d). Results of thresholding with the fixed threshold value $q = 128$ (e–h). Histograms of the original images (i–l) with intensity values from 0 (left) to 255 (right).

of the *background* or the *foreground*. Thus the set of all image pixels is partitioned into two disjoint sets $\mathcal{C}_0$ and $\mathcal{C}_1$, where $\mathcal{C}_0$ contains all elements with values in $[0, 1, \ldots, q]$ and $\mathcal{C}_1$ collects the remaining elements with values in $[q+1, \ldots, K-1]$, that is,

$$(u, v) \in \begin{cases} \mathcal{C}_0 & \text{if } I(u, v) \leq q \text{ (background)}, \\ \mathcal{C}_1 & \text{if } I(u, v) > q \text{ (foreground)}. \end{cases} \tag{11.1}$$

Of course, the meaning of *background* and *foreground* may differ from one application to another. For example, the aforementioned scheme is quite natural for astronomical or thermal images, where the relevant "foreground" pixels are bright and the background is dark. Conversely, in document analysis, for example, the objects of interest are usually the *dark* letters or artwork printed on a bright background. This should not be confusing and of course one can always *invert* the image to adapt to this scheme, so there is no loss of generality here.

Figure 11.1 shows several test images used in this chapter and the result of thresholding with a fixed threshold value. The synthetic image in Fig. 11.1(d) is the mixture of two Gaussian random distributions $\mathcal{N}_0, \mathcal{N}_1$ for the background and foreground, respectively, with $\mu_0 = 80$, $\mu_1 = 170$, $\sigma_0 = \sigma_1 = 20$. The corresponding histograms of the test images are shown in Fig. 11.1(i–l). Note that all histograms are normalized to constant area (not to maximum values, as usual), with intensity values ranging from 0 (left) to 255 (right).

The key question is how to find a suitable (or even "optimal") threshold value for binarizing the image. As the name implies, histogram-based methods calculate the threshold primarily from the information contained in the image's histogram, without inspecting the actual image pixels. Other methods process individual pixels for finding the threshold and there are also hybrid methods that rely both on the histogram and the local image content. Histogram-based techniques are usually simple and efficient, because they operate on a small set of data (256 values in case of an 8-bit histogram); they can be grouped into two main categories: *shape-based* and *statistical* methods.

Shape-based methods analyze the structure of the histogram's distribution, for example by trying to locate peaks, valleys and other "shape" features. Usually the histogram is first smoothed to eliminate narrow peaks and gaps. While shape-based methods were quite popular early on, they are usually not as robust as their statistical counterparts or at least do not seem to offer any distinct advantages. A classic representative of this category is the "triangle" (or "chord") algorithm described in [261]. References to numerous other shape-based methods can be found in [213].

Statistical methods, as their name suggests, rely on statistical information derived from the image's histogram (which of course is a statistic itself), such as the mean, variance, or entropy. In the next section, we discuss a few elementary parameters that can be obtained from the histogram, followed by a description of concrete algorithms that use this information. Again there are a vast number of similar methods and we have selected four representative algorithms to be described in more detail: (a) iterative threshold selection by Ridler and Calvard [198], (b) Otsu's clustering method [177], (c) the minimum error method by Kittler and Illingworth [116], and (d) the maximum entropy thresholding method by Kapur, Sahoo, and Wong [133].

11.1.1 Image Statistics from the Histogram

As described in Chapter 3, Sec. 3.7, several statistical quantities, such as the arithmetic mean, variance and median, can be calculated directly from the histogram, without reverting to the original image data. If we *threshold* the image at level q $(0 \leq q < K)$, the set of pixels is partitioned into the disjoint subsets $\mathcal{C}_0$, $\mathcal{C}_1$, corresponding to the background and the foreground. The number of pixels assigned to each subset is

$$n_0(q) = |\mathcal{C}_0| = \sum_{g=0}^{q} \mathsf{h}(g) \quad \text{and} \quad n_1(q) = |\mathcal{C}_1| = \sum_{g=q+1}^{K-1} \mathsf{h}(g), \tag{11.2}$$

respectively. Also, because all pixels are assigned to either the *background* set $\mathcal{C}_0$ or the *foreground* set $\mathcal{C}_1$,

$$n_0(q) + n_1(q) = |\mathcal{C}_0| + |\mathcal{C}_1| = |\mathcal{C}_0 \cup \mathcal{C}_1| = MN. \tag{11.3}$$

For any threshold q, the *mean* values of the associated partitions $\mathcal{C}_0, \mathcal{C}_1$ can be calculated from the image histogram as

$$\mu_0(q) = \frac{1}{n_0(q)} \cdot \sum_{g=0}^{q} g \cdot \mathsf{h}(g), \tag{11.4}$$

$$\mu_1(q) = \frac{1}{n_1(q)} \cdot \sum_{g=q+1}^{K-1} g \cdot \mathsf{h}(g) \tag{11.5}$$

and these quantities relate to the image's overall mean μ_I (Eqn. (3.9)) by[1]

$$\mu_I = \frac{1}{MN} \cdot \big[n_0(q) \cdot \mu_0(q) + n_1(q) \cdot \mu_1(q) \big] = \mu_0(K-1). \tag{11.6}$$

Analogously, the *variances* of the background and foreground partitions can be extracted from the histogram as[2]

$$\begin{aligned} \sigma_0^2(q) &= \frac{1}{n_0(q)} \cdot \sum_{g=0}^{q} (g - \mu_0(q))^2 \cdot \mathsf{h}(g) \\ \sigma_1^2(q) &= \frac{1}{n_1(q)} \cdot \sum_{g=q+1}^{K-1} (g - \mu_1(q))^2 \cdot \mathsf{h}(g). \end{aligned} \tag{11.7}$$

(Of course, as in Eqn. (3.12), this calculation can also be performed in a single iteration and without knowing $\mu_0(q), \mu_1(q)$ in advance.) The overall variance σ_I^2 for the whole image is identical to the variance of the background for $q = K-1$,

$$\sigma_I^2 = \frac{1}{MN} \cdot \sum_{g=0}^{K-1} (g - \mu_I)^2 \cdot \mathsf{h}(g) = \sigma_0^2(K-1), \tag{11.8}$$

that is, for all pixels being assigned to the background partition. Note that, unlike the simple relation of the means given in Eqn. (11.6),

$$\sigma_I^2 \neq \frac{1}{MN} \big[n_0(q) \cdot \sigma_0^2(q) + n_1(q) \cdot \sigma_1^2(q) \big] \tag{11.9}$$

in general (see also Eqn. (11.20)).

We will use these basic relations in the discussion of histogram-based threshold selection algorithms in the following and add more specific ones as we go along.

11.1.2 Simple Threshold Selection

Clearly, the choice of the threshold value should not be fixed but somehow based on the content of the image. In the simplest case, we could use the *mean* of all image pixels,

$$q \leftarrow \text{mean}(I) = \mu_I, \tag{11.10}$$

as the threshold value q, or the *median*, (see Sec. 3.7.2),

$$q \leftarrow \text{median}(I) = m_I, \tag{11.11}$$

or, alternatively, the average of the *minimum* and the *maximum* (mid-range value), that is,

$$q \leftarrow \frac{\max(I) + \min(I)}{2}. \tag{11.12}$$

[1] Note that $\mu_0(q), \mu_1(q)$ are meant to be functions over q and thus $\mu_0(K-1)$ in Eqn. (11.6) denotes the mean of partition $\mathcal{C}_0$ for the threshold $K-1$.

[2] $\sigma_0^2(q)$ and $\sigma_1^2(q)$ in Eqn. (11.7) are also functions over q.

1:	**QuantileThreshold**(h, p) Input: h : $[0, K{-}1] \mapsto \mathbb{N}$, a grayscale histogram. p, the proportion of expected background pixels ($0 < p < 1$). Returns the optimal threshold value or -1 if no threshold is found.
2:	$K \leftarrow$ Size(h) ▷ number of intensity levels
3:	$MN \leftarrow \sum_{i=0}^{K-1} \mathsf{h}(i)$ ▷ number of image pixels
4:	$i \leftarrow 0$
5:	$c \leftarrow \mathsf{h}(0)$
6:	**while** $(i < K) \wedge (c < MN \cdot p)$ **do** ▷ quantile calc. (Eq. 11.13)
7:	$i \leftarrow i + 1$
8:	$c \leftarrow c + \mathsf{h}(j)$
9:	**if** $c < MN$ **then** ▷ foreground is non-empty
10:	$q \leftarrow i$
11:	**else** ▷ foreground is empty, all pixels are background
12:	$q \leftarrow -1$
13:	**return** q

Alg. 11.1
Quantile thresholding. The optimal threshold value $q \in [0, K-2]$ is returned, or -1 if no valid threshold was found. Note the test in line 9 to check if the foreground is empty or not (the background is always non-empty by definition).

Like the image mean μ_I (see Eqn. (3.9)), all these quantities can be obtained directly from the histogram h.

Thresholding at the median segments the image into approximately equal-sized background and foreground sets, that is, $|\mathcal{C}_0| \approx |\mathcal{C}_1|$, which assumes that the "interesting" (foreground) pixels cover about half of the image. This may be appropriate for certain images, but completely wrong for others. For example, a scanned text image will typically contain a lot more white than black pixels, so using the median threshold would probably be unsatisfactory in this case. If the approximate fraction p ($0 < p < 1$) of expected background pixels is known in advance, the threshold could be set to that *quantile* instead. In this case, q is simply chosen as

$$q \leftarrow \min\Big\{i \mid \sum_{j=0}^{i} \mathsf{h}(i) \geq M{\cdot}N{\cdot}p\Big\}, \tag{11.13}$$

where N is the total number of pixels. We see that the *median* is only a special case of a quantile measure, with $p = 0.5$. This simple thresholding method is summarized in Alg. 11.1.

For the *mid-range* technique (Eqn. (11.12)), the limiting intensity values $\min(I)$ and $\max(I)$ can be found by searching for the smallest and largest non-zero entries, respectively, in the histogram h. The mid-range threshold segments the image at 50 % (or any other percentile) of the contrast range. In this case, nothing can be said in general about the relative sizes of the resulting background and foreground partitions. Because a single extreme pixel value (outlier) may change the contrast range dramatically, this approach is not very robust. Here too it is advantageous to define the contrast range by specifying pixel *quantiles*, analogous to the calculation of the quantities a'_{low} and a'_{high} in the modified auto-contrast function (see Ch. 4, Sec. 4.4).

In the pathological (but nevertheless possible) case that all pixels in the image have the *same* intensity g, all the aforementioned meth-

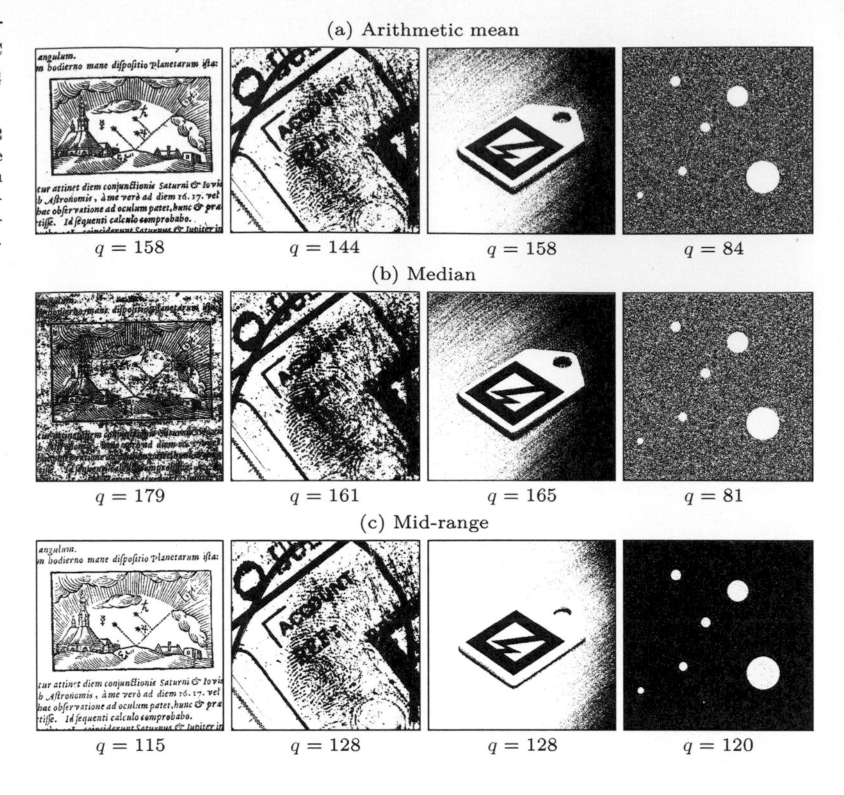

Fig. 11.2
Results from various simple thresholding schemes. Mean (a–d), median (e–h), and mid-range (i–l) threshold, as specified in Eqns. (11.10)–(11.12).

ods will return the threshold $q = g$, which assigns all pixels to the background partition and leaves the foreground empty. Algorithms should try to detect this situation, because thresholding a uniform image obviously makes no sense. Results obtained with these simple thresholding techniques are shown in Fig. 11.2. Despite the obvious limitations, even a simple automatic threshold selection (such as the quantile technique in Alg. 11.1) will typically yield more reliable results than the use of a fixed threshold.

11.1.3 Iterative Threshold Selection (Isodata Algorithm)

This classic iterative algorithm for finding an optimal threshold is attributed to Ridler and Calvard [198] and was related to Isodata clustering by Velasco [242]. It is thus sometimes referred to as the "isodata" or "intermeans" algorithm. Like in many other global thresholding schemes it is assumed that the image's histogram is a mixture of two separate distributions, one for the intensities of the background pixels and the other for the foreground pixels. In this case, the two distributions are assumed to be Gaussian with approximately identical spreads (variances).

The algorithm starts by making an initial guess for the threshold, for example, by taking the mean or the median of the whole image. This splits the set of pixels into a background and a foreground set, both of which should be non-empty. Next, the means of both sets are calculated and the threshold is repositioned to their average, that is, centered between the two means. The means are then re-calculated for the resulting background and foreground sets, and so on, until

```
1:  IsodataThreshold(h)
        Input: h : [0, K−1] ↦ ℕ, a grayscale histogram.
        Returns the optimal threshold value or −1 if no threshold is
        found.
2:      K ← Size(h)                          ▷ number of intensity levels
3:      q ← Mean(h, 0, K−1)       ▷ set initial threshold to overall mean
4:      repeat
5:          n_0 ← Count(h, 0, q)                  ▷ background population
6:          n_1 ← Count(h, q+1, K−1)              ▷ foreground population
7:          if (n_0 = 0) ∨ (n_1 = 0) then  ▷ backgrd. or foregrd. is empty
8:              return −1
9:          μ_0 ← Mean(h, 0, q)                         ▷ background mean
10:         μ_1 ← Mean(h, q+1, K−1)                     ▷ foreground mean
11:         q' ← q                              ▷ keep previous threshold
12:         q ← ⌊(μ_0 + μ_1)/2⌋             ▷ calculate the new threshold
13:     until q = q'                          ▷ terminate if no change
14:     return q

15: Count(h, a, b) := Σ_{g=a}^{b} h(g)

16: Mean(h, a, b) := [Σ_{g=a}^{b} g · h(g)] / [Σ_{g=a}^{b} h(g)]
```

Alg. 11.2
"Isodata" threshold selection based on the iterative method by Ridler and Calvard [198].

the threshold does not change any longer. In practice, it takes only a few iterations for the threshold to converge.

This procedure is summarized in Alg. 11.2. The initial threshold is set to the overall mean (line 3). For each threshold q, separate mean values μ_0, μ_1 are computed for the corresponding foreground and background partitions. The threshold is repeatedly set to the average of the two means until no more change occurs. The clause in line 7 tests if either the background or the foreground partition is empty, which will happen, for example, if the image contains only a single intensity value. In this case, no valid threshold exists and the procedure returns -1. The functions $\mathsf{Count}(\mathsf{h}, a, b)$ and $\mathsf{Mean}(\mathsf{h}, a, b)$ in lines 15–16 return the number of pixels and the mean, respectively, of the image pixels with intensity values in the range $[a, b]$. Both can be computed directly from the histogram h without inspecting the image itself.

The performance of this algorithm can be easily improved by using tables $\boldsymbol{\mu}_0(q), \boldsymbol{\mu}_1(q)$ for the background and foreground means, respectively. The modified, table-based version of the iterative threshold selection procedure is shown in Alg. 11.3. It requires two passes over the histogram to initialize the tables $\boldsymbol{\mu}_0$, $\boldsymbol{\mu}_1$ and only a small, constant number of computations for each iteration in its main loop. Note that the image's overall mean μ_I, used as the initial guess for the threshold q (Alg. 11.3, line 4), need not be calculated separately but can be obtained as $\mu_I = \boldsymbol{\mu}_0(K{-}1)$, given that threshold $q = K{-}1$ assigns all image pixels to the background. The time complexity of this algorithm is thus $\mathcal{O}(K)$, that is, linear w.r.t. the size of the

Alg. 11.3 Fast version of "isodata" threshold selection. Precalculated tables are used for the foreground and background means μ_0 and μ_1, respectively.

1: **FastIsodataThreshold**(h)
  Input: $\mathsf{h} : [0, K-1] \mapsto \mathbb{N}$, a grayscale histogram.
  Returns the optimal threshold value or -1 if no threshold is found.
2: $K \leftarrow \mathsf{Size}(\mathsf{h})$ ▷ number of intensity levels
3: $\langle \boldsymbol{\mu}_0, \boldsymbol{\mu}_1, N \rangle \leftarrow \mathsf{MakeMeanTables}(\mathsf{h})$
4: $q \leftarrow \lfloor \boldsymbol{\mu}_0(K-1) \rfloor$ ▷ take the overall mean μ_I as initial threshold
5: **repeat**
6:   **if** $(\boldsymbol{\mu}_0(q) < 0) \vee (\boldsymbol{\mu}_1(q) < 0)$ **then**
7:     **return** -1 ▷ background or foreground is empty
8:   $q' \leftarrow q$ ▷ keep previous threshold
9:   $q \leftarrow \left\lfloor \frac{\boldsymbol{\mu}_0(q) + \boldsymbol{\mu}_1(q)}{2} \right\rfloor$ ▷ calculate the new threshold
10: **until** $q = q'$ ▷ terminate if no change
11: **return** q

12: **MakeMeanTables**(h)
13: $K \leftarrow \mathsf{Size}(\mathsf{h})$
14: Create maps $\boldsymbol{\mu}_0, \boldsymbol{\mu}_1 : [0, K-1] \mapsto \mathbb{R}$
15: $n_0 \leftarrow 0, \quad s_0 \leftarrow 0$
16: **for** $q \leftarrow 0, \ldots, K-1$ **do** ▷ tabulate background means $\mu_0(q)$
17:   $n_0 \leftarrow n_0 + \mathsf{h}(q)$
18:   $s_0 \leftarrow s_0 + q \cdot \mathsf{h}(q)$
19:   $\boldsymbol{\mu}_0(q) \leftarrow \begin{cases} s_0/n_0 & \text{if } n_0 > 0 \\ -1 & \text{otherwise} \end{cases}$
20: $N \leftarrow n_0$
21: $n_1 \leftarrow 0, \quad s_1 \leftarrow 0$
22: $\boldsymbol{\mu}_1(K-1) \leftarrow 0$
23: **for** $q \leftarrow K-2, \ldots, 0$ **do** ▷ tabulate foreground means $\mu_1(q)$
24:   $n_1 \leftarrow n_1 + \mathsf{h}(q+1)$
25:   $s_1 \leftarrow s_1 + (q+1) \cdot \mathsf{h}(q+1)$
26:   $\boldsymbol{\mu}_1(q) \leftarrow \begin{cases} s_1/n_1 & \text{if } n_1 > 0 \\ -1 & \text{otherwise} \end{cases}$
27: **return** $\langle \boldsymbol{\mu}_0, \boldsymbol{\mu}_1, N \rangle$

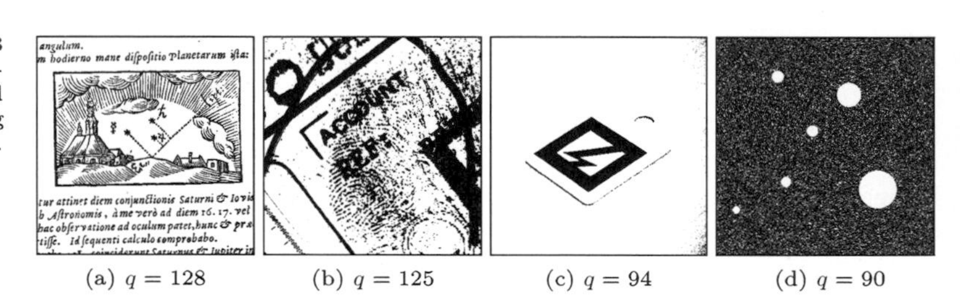

(a) $q = 128$ (b) $q = 125$ (c) $q = 94$ (d) $q = 90$

Fig. 11.3 Thresholding with the isodata algorithm. Binarized images and the corresponding optimal threshold values (q).

histogram. Figure 11.3 shows the results of thresholding with the isodata algorithm applied to the test images in Fig. 11.1.

11.1.4 Otsu's Method

The method proposed by Otsu [147, 177] also assumes that the original image contains pixels from two classes, whose intensity distributions are unknown. The goal is to find a threshold q such that the resulting background and foreground distributions are maximally separated, which means that they are (a) each as narrow as possible (have minimal variances) and (b) their centers (means) are most distant from each other.

For a given threshold q, the variances of the corresponding background and foreground partitions can be calculated straight from the image's histogram (see Eqn. (11.7)). The combined width of the two distributions is measured by the *within-class* variance

$$\sigma_{\mathrm{w}}^2(q) = \mathsf{P}_0(q) \cdot \sigma_0^2(q) + \mathsf{P}_1(q) \cdot \sigma_1^2(q) \tag{11.14}$$

$$= \frac{1}{MN} \cdot \big[n_0(q) \cdot \sigma_0^2(q) + n_1(q) \cdot \sigma_1^2(q)\big], \tag{11.15}$$

where

$$\mathsf{P}_0(q) = \sum_{i=0}^{q} \mathsf{p}(i) = \frac{1}{MN} \cdot \sum_{i=0}^{q} \mathsf{h}(i) = \frac{n_0(q)}{MN}, \tag{11.16}$$

$$\mathsf{P}_1(q) = \sum_{i=q+1}^{K-1} \mathsf{p}(i) = \frac{1}{MN} \cdot \sum_{i=q+1}^{K-1} \mathsf{h}(i) = \frac{n_1(q)}{MN} \tag{11.17}$$

are the class probabilities for $\mathcal{C}_0$, $\mathcal{C}_1$, respectively. Thus the within-class variance in Eqn. (11.15) is simply the sum of the individual variances weighted by the corresponding class probabilities or "populations". Analogously, the *between-class* variance,

$$\sigma_{\mathrm{b}}^2(q) = \mathsf{P}_0(q) \cdot \big(\mu_0(q) - \mu_I\big)^2 + \mathsf{P}_1(q) \cdot \big(\mu_1(q) - \mu_I\big)^2 \tag{11.18}$$

$$= \frac{1}{MN} \big[n_0(q) \cdot \big(\mu_0(q) - \mu_I\big)^2 + n_1(q) \cdot \big(\mu_1(q) - \mu_I\big)^2\big] \tag{11.19}$$

measures the distances between the cluster means μ_0, μ_1 and the overall mean μ_I. The total image variance σ_I^2 is the sum of the within-class variance and the between-class variance, that is,

$$\sigma_I^2 = \sigma_{\mathrm{w}}^2(q) + \sigma_{\mathrm{b}}^2(q), \tag{11.20}$$

for $q = 0, \ldots, K-1$. Since σ_I^2 is constant for a given image, the threshold q can be found by either *minimizing* the within-variance σ_{w}^2 or *maximizing* the between-variance σ_{b}^2. The natural choice is to maximize σ_{b}^2, because it only relies on first-order statistics (i.e., the within-class means μ_0, μ_1). Since the overall mean μ_I can be expressed as the weighted sum of the partition means μ_0 and μ_1 (Eqn. (11.6)), we can simplify Eqn. (11.19) to

$$\sigma_{\mathrm{b}}^2(q) = \mathsf{P}_0(q) \cdot \mathsf{P}_1(q) \cdot \big[\mu_0(q) - \mu_1(q)\big]^2 \tag{11.21}$$

$$= \frac{1}{(MN)^2} \cdot n_0(q) \cdot n_1(q) \cdot \big[\mu_0(q) - \mu_1(q)\big]^2. \tag{11.22}$$

The optimal threshold is finally found by *maximizing* the expression for the between-class variance in Eqn. (11.22) with respect to q, thereby *minimizing* the within-class variance in Eqn. (11.15).

Noting that $\sigma_{\mathrm{b}}^2(q)$ only depends on the means (and *not* on the variances) of the two partitions for a given threshold q allows for a very efficient implementation, as outlined in Alg. 11.4. The algorithm assumes a grayscale image with a total of N pixels and K intensity

Alg. 11.4
Finding the optimal threshold using Otsu's method [177]. Initially (outside the **for**-loop), the threshold q is assumed to be -1, which corresponds to the background class being empty ($n_0 = 0$) and all pixels are assigned to the foreground class ($n_1 = N$). The **for**-loop (lines 7–14) examines each possible threshold $q = 0, \ldots, K-2$. The factor $1/(MN)^2$ in line 11 is constant and thus not relevant for the optimization. The optimal threshold value is returned, or -1 if no valid threshold was found. The function MakeMeanTables() is defined in Alg. 11.3.

```
1:  OtsuThreshold(h)
      Input: h : [0, K−1] ↦ ℕ, a grayscale histogram. Returns the
      optimal threshold value or −1 if no threshold is found.
2:    K ← Size(h)                                  ▷ number of intensity levels
3:    (μ0, μ1, MN) ← MakeMeanTables(h)                        ▷ see Alg. 11.3
4:    σ²bmax ← 0
5:    qmax ← −1
6:    n0 ← 0
7:    for q ← 0, ..., K−2 do  ▷ examine all possible threshold values q
8:        n0 ← n0 + h(q)
9:        n1 ← MN − n0
10:       if (n0 > 0) ∧ (n1 > 0) then
11:           σ²b ← 1/(MN)² · n0 · n1 · [μ0(q) − μ1(q)]²     ▷ see Eq. 11.22
12:           if σ²b > σ²bmax then                            ▷ maximize σ²b
13:               σ²bmax ← σ²b
14:               qmax ← q
15:   return qmax
```

levels. As in Alg. 11.3, precalculated tables $\boldsymbol{\mu}_0(q), \boldsymbol{\mu}_1(q)$ are used for the background and foreground means for all possible threshold values $q = 0, \ldots, K-1$.

Possible threshold values are $q = 0, \ldots, K-2$ (with $q = K-1$, all pixels are assigned to the background). Initially (before entering the main **for**-loop in line 7) $q = -1$; at this point, the set of background pixels ($\leq q$) is empty and all pixels are classified as foreground ($n_0 = 0$ and $n_1 = N$). Each possible threshold value is examined inside the body of the **for**-loop.

As long as any of the two classes is empty ($n_0(q) = 0$ or $n_1(q) = 0$),[3] the resulting between-class variance $\sigma_{\mathrm{b}}^2(q)$ is zero. The threshold that yields the maximum between-class variance (σ_{bmax}^2) is returned, or -1 if no valid threshold could be found. This occurs when all image pixels have the same intensity, that all pixels are in either the background or the foreground class.

Note that in line 11 of Alg. 11.4, the factor $\frac{1}{N^2}$ is constant (independent of q) and can thus be ignored in the optimization. However, care must be taken at this point because the computation of σ_{b}^2 may produce intermediate values that exceed the range of typical (32-bit) integer variables, even for medium-size images. Variables of type `long` should be used or the computation be performed with floating-point values.

The absolute "goodness" of the final thresholding by q_{max} could be measured as the ratio

$$\eta = \frac{\sigma_{\mathrm{b}}^2(q_{\mathrm{max}})}{\sigma_I^2} \in [0, 1] \tag{11.23}$$

[3] This is the case if the image contains no pixels with values $I(u,v) \leq q$ or $I(u,v) > q$, that is, the histogram h is empty either below or above the index q.

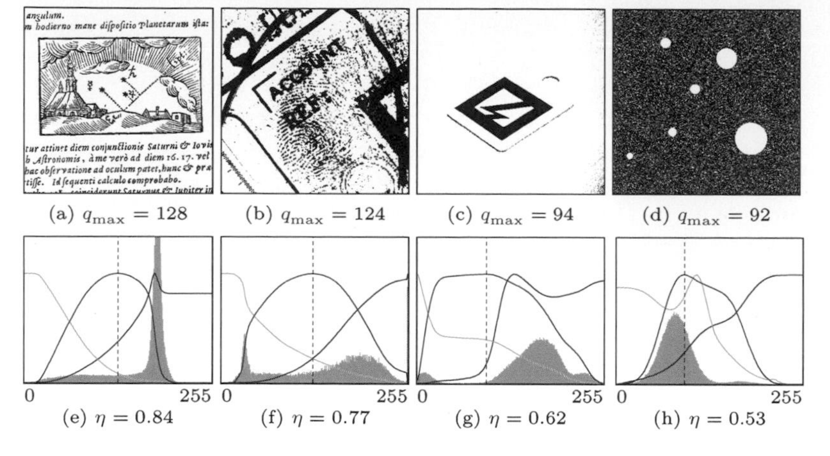

Fig. 11.4
Results of thresholding with Otsu's method. Calculated threshold values q and resulting binary images (a–d). Graphs in (e–h) show the corresponding within-background variance σ_0^2 (green), the within-foreground variance σ_1^2 (blue), and the between-class variance σ_b^2 (red), for varying threshold values $q = 0, \ldots, 255$. The optimal threshold q_{max} (dashed vertical line) is positioned at the maximum of σ_b^2. The value η denotes the "goodness" estimate for the thresholding, as defined in Eqn. (11.23).

(see Eqn. (11.8)), which is invariant under linear changes of contrast and brightness [177]. Greater values of η indicate better thresholding.

Results of automatic threshold selection with Otsu's method are shown in Fig. 11.4, where q_{max} denotes the optimal threshold and η is the corresponding "goodness" estimate, as defined in Eqn. (11.23). The graph underneath each image shows the original histogram (gray) overlaid with the variance within the background σ_0^2 (green), the variance within the foreground σ_1^2 (blue), and the between-class variance σ_b^2 (red) for varying threshold values q. The dashed vertical line marks the position of the optimal threshold q_{max}.

Due to the pre-calculation of the mean values, Otsu's method requires only three passes over the histogram and is thus very fast ($\mathcal{O}(K)$), in contrast to opposite accounts in the literature. The method is frequently quoted and performs well in comparison to other approaches [213], despite its long history and its simplicity. In general, the results are very similar to the ones produced by the iterative threshold selection ("isodata") algorithm described in Sec. 11.1.3.

11.1.5 Maximum Entropy Thresholding

Entropy is an important concept in information theory and particularly in data compression. It is a statistical measure that quantifies the average amount of information contained in the "messages" generated by a stochastic data source [99, 101]. For example, the MN pixels in an image I can be interpreted as a message of MN symbols, each taken independently from a finite alphabet of K (e.g., 256) different intensity values. Every pixel is assumed to be statically independent. Knowing the probability of each intensity value g to occur, entropy measures how likely it is to observe a particular image, or, in other words, how much we should be surprised to see such an image. Before going into further details, we briefly review the notion of probabilities in the context of images and histograms (see also Ch. 4, Sec. 4.6.1).

For modeling the image generation as a random process, we first need to define an "alphabet", that is, a set of symbols

$$Z = \{0, 1, \ldots, K-1\}, \tag{11.24}$$

which in this case is simply the set of possible intensity values $g = 0, \ldots, K-1$, together with the probability $p(g)$ that a particular intensity value g occurs. These probabilities are supposed to be known in advance, which is why they are called *a priori* (or *prior*) probabilities. The vector of probabilities,

$$\big(p(0), p(1), \ldots, p(K-1)\big),$$

is a *probability distribution* or *probability density function* (pdf). In practice, the *a priori* probabilities are usually *un*known, but they can be estimated by observing how often the intensity values actually occur in one or more images, assuming that these are representative instances of the images typically produced by that source. An estimate $\mathsf{p}(g)$ of the image's probability density function $p(g)$ is obtained by normalizing its histogram h in the form

$$p(g) \approx \mathsf{p}(g) = \frac{\mathsf{h}(g)}{MN}, \tag{11.25}$$

for $0 \leq g < K$, such that $0 \leq \mathsf{p}(g) \leq 1$ and $\sum_{g=0}^{K-1} \mathsf{p}(g) = 1$. The associated *cumulative distribution function* (cdf) is

$$\mathsf{P}(g) = \sum_{i=0}^{g} \frac{\mathsf{h}(g)}{MN} = \sum_{i=0}^{g} \mathsf{p}(i), \tag{11.26}$$

where $\mathsf{P}(0) = \mathsf{p}(0)$ and $\mathsf{P}(K-1) = 1$. This is simply the normalized *cumulative histogram*.[4]

Entropy of images

Given an estimate of its intensity probability distribution $\mathsf{p}(g)$, the *entropy* of an image is defined as[5]

$$H(Z) = \sum_{g \in Z} \mathsf{p}(g) \cdot \log_b\Big(\frac{1}{\mathsf{p}(g)}\Big) = -\sum_{g \in Z} \mathsf{p}(g) \cdot \log_b\left(\mathsf{p}(g)\right), \tag{11.27}$$

where $g = I(u,v)$ and $\log_b(x)$ denotes the logarithm of x to the base b. If $b = 2$, the entropy (or "information content") is measured in *bits*, but proportional results are obtained with any other logarithm (such as ln or $\log_{10}$). Note that the value of $H()$ is always positive, because the probabilities $\mathsf{p}()$ are in $[0, 1]$ and thus the terms $\log_b[\mathsf{p}()]$ are negative or zero for any b.

Some other properties of the entropy are also quite intuitive. For example, if all probabilities $\mathsf{p}(g)$ are zero except for one intensity g', then the entropy $H(I)$ is *zero*, indicating that there is no uncertainty (or "surprise") in the messages produced by the corresponding data source. The (rather boring) images generated by this source will contain nothing but pixels of intensity g', since all other intensities are

[4] See also Chapter 3, Sec. 3.6.

[5] Note the subtle difference in notation for the cumulative histogram H and the entropy H.

impossible. Conversely, the entropy is a maximum if all K intensities have the same probability (uniform distribution),

$$\mathsf{p}(g) = \frac{1}{K}, \qquad \text{for } 0 \le g < K, \tag{11.28}$$

and therefore (from Eqn. (11.27)) the entropy in this case is

$$H(Z) = -\sum_{i=0}^{K-1} \frac{1}{K} \cdot \log_b\!\left(\frac{1}{K}\right) = \frac{1}{K} \cdot \underbrace{\sum_{i=0}^{K-1} \log_b(K)}_{K\cdot\log_b(K)} \tag{11.29}$$

$$= \frac{1}{K} \cdot \big(K \cdot \log_b(K)\big) = \log_b(K). \tag{11.30}$$

This is the maximum possible entropy of a stochastic source with an alphabet Z of size K. Thus the entropy $H(Z)$ is always in the range $[0, \log(K)]$.

Using image entropy for threshold selection

The use of image entropy as a criterion for threshold selection has a long tradition and numerous methods have been proposed. In the following, we describe the early (but still popular) technique by Kapur et al. [100, 133] as a representative example.

Given a particular threshold q (with $0 \le q < K{-}1$), the estimated probability distributions for the resulting partitions $\mathcal{C}_0$ and $\mathcal{C}_1$ are

$$\begin{array}{l} \mathcal{C}_0 : \left(\frac{\mathsf{p}(0)}{\mathsf{P}_0(q)} \;\; \frac{\mathsf{p}(1)}{\mathsf{P}_0(q)} \;\; \cdots \;\; \frac{\mathsf{p}(q)}{\mathsf{P}_0(q)} \;\; 0 \;\; 0 \;\; \cdots \;\; 0 \right), \\ \mathcal{C}_1 : \left(0 \;\; 0 \;\; \cdots \;\; 0 \;\; \frac{\mathsf{p}(q+1)}{\mathsf{P}_1(q)} \;\; \frac{\mathsf{p}(q+2)}{\mathsf{P}_1(q)} \;\; \cdots \;\; \frac{\mathsf{p}(K-1)}{\mathsf{P}_1(q)} \right), \end{array} \tag{11.31}$$

with the associated cumulated probabilities (see Eqn. (11.26))

$$\mathsf{P}_0(q) = \sum_{i=0}^{q} \mathsf{p}(i) = \mathsf{P}(q) \qquad \text{and} \qquad \mathsf{P}_1(q) = \sum_{i=q+1}^{K-1} \mathsf{p}(i) = 1 - \mathsf{P}(q). \tag{11.32}$$

Note that $\mathsf{P}_0(q) + \mathsf{P}_1(q) = 1$, since the background and foreground partitions are disjoint. The entropies *within* each partition are defined as

$$H_0(q) = -\sum_{i=0}^{q} \frac{\mathsf{p}(i)}{\mathsf{P}_0(q)} \cdot \log\!\left(\frac{\mathsf{p}(i)}{\mathsf{P}_0(q)}\right), \tag{11.33}$$

$$H_1(q) = -\sum_{i=q+1}^{K-1} \frac{\mathsf{p}(i)}{\mathsf{P}_1(q)} \cdot \log\!\left(\frac{\mathsf{p}(i)}{\mathsf{P}_1(q)}\right), \tag{11.34}$$

and the *overall* entropy for the threshold q is

$$H_{01}(q) = H_0(q) + H_1(q). \tag{11.35}$$

This expression is to be maximized over q, also called the "information between the classes" $\mathcal{C}_0$ and $\mathcal{C}_1$. To allow for an efficient computation, the expression for $H_0(q)$ in Eqn. (11.33) can be rearranged to

$$H_0(q) = -\sum_{i=0}^{q} \frac{\mathsf{p}(i)}{\mathsf{P}_0(q)} \cdot \big[\log(\mathsf{p}(i)) - \log(\mathsf{P}_0(q))\big] \tag{11.36}$$

$$= -\frac{1}{\mathsf{P}_0(q)} \cdot \sum_{i=0}^{q} \mathsf{p}(i) \cdot \big[\log(\mathsf{p}(i)) - \log(\mathsf{P}_0(q))\big] \tag{11.37}$$

$$= -\frac{1}{\mathsf{P}_0(q)} \cdot \underbrace{\sum_{i=0}^{q} \mathsf{p}(i) \cdot \log(\mathsf{p}(i))}_{S_0(q)} + \frac{1}{\mathsf{P}_0(q)} \cdot \underbrace{\sum_{i=0}^{q} \mathsf{p}(i)}_{=\mathsf{P}_0(q)} \cdot \log(\mathsf{P}_0(q))$$

$$= -\frac{1}{\mathsf{P}_0(q)} \cdot S_0(q) + \log(\mathsf{P}_0(q)). \tag{11.38}$$

Similarly $H_1(q)$ in Eqn. (11.34) becomes

$$H_1(q) = -\sum_{i=q+1}^{K-1} \frac{\mathsf{p}(i)}{\mathsf{P}_1(q)} \cdot \big[\log(\mathsf{p}(i)) - \log(\mathsf{P}_1(q))\big] \tag{11.39}$$

$$= -\frac{1}{1-\mathsf{P}_0(q)} \cdot S_1(q) + \log(1-\mathsf{P}_0(q)). \tag{11.40}$$

Given the estimated probability distribution $\mathsf{p}(i)$, the cumulative probability P_0 and the summation terms S_0, S_1 (see Eqns. (11.38)–(11.40)) can be calculated from the recurrence relations

$$\mathsf{P}_0(q) = \begin{cases} \mathsf{p}(0) & \text{for } q = 0, \\ \mathsf{P}_0(q-1) + \mathsf{p}(q) & \text{for } 0 < q < K, \end{cases}$$

$$S_0(q) = \begin{cases} \mathsf{p}(0) \cdot \log(\mathsf{p}(0)) & \text{for } q = 0, \\ S_0(q-1) + \mathsf{p}(q) \cdot \log(\mathsf{p}(q)) & \text{for } 0 < q < K, \end{cases}$$

$$S_1(q) = \begin{cases} 0 & \text{for } q = K-1, \\ S_1(q+1) + \mathsf{p}(q+1) \cdot \log(\mathsf{p}(q+1)) & \text{for } 0 \leq q < K-1. \end{cases} \tag{11.41}$$

The complete procedure is summarized in Alg. 11.5, where the values $S_0(q), S_1(q)$ are obtained from precalculated tables $\mathsf{S}_0, \mathsf{S}_1$. The algorithm performs three passes over the histogram of length K (two for filling the tables $\mathsf{S}_0, \mathsf{S}_1$ and one in the main loop), so its time complexity is $\mathcal{O}(K)$, like the algorithms described before.

Results obtained with this technique are shown in Fig. 11.5. The technique described in this section is simple and efficient, because it again relies entirely on the image's histogram. More advanced entropy-based thresholding techniques exist that, among other improvements, take into account the spatial structure of the original image. An extensive review of entropy-based methods can be found in [46].

11.1.6 Minimum Error Thresholding

The goal of minimum error thresholding is to optimally fit a combination (mixture) of Gaussian distributions to the image's histogram. Before we proceed, we briefly look at some additional concepts from statistics. Note, however, that the following material is only intended

1: **MaximumEntropyThreshold**(h)
Input: $\mathsf{h} : [0, K-1] \mapsto \mathbb{N}$, a grayscale histogram. Returns the optimal threshold value or -1 if no threshold is found.

2: $K \leftarrow \mathsf{Size}(\mathsf{h})$ ▷ number of intensity levels
3: $\mathsf{p} \leftarrow \mathsf{Normalize}(\mathsf{h})$ ▷ normalize histogram
4: $(\mathsf{S}_0, \mathsf{S}_1) \leftarrow \mathsf{MakeTables}(\mathsf{p}, K)$ ▷ tables for $S_0(q), S_1(q)$
5: $\mathsf{P}_0 \leftarrow 0$ ▷ $\mathsf{P}_0 \in [0,1]$
6: $q_{\max} \leftarrow -1$
7: $H_{\max} \leftarrow -\infty$ ▷ maximum joint entropy
8: **for** $q \leftarrow 0, \ldots, K-2$ **do** ▷ check all possible threshold values q
9: $\quad \mathsf{P}_0 \leftarrow \mathsf{P}_0 + \mathsf{p}(q)$
10: $\quad \mathsf{P}_1 \leftarrow 1 - \mathsf{P}_0$ ▷ $\mathsf{P}_1 \in [0,1]$
11: $\quad H_0 \leftarrow \begin{cases} -\frac{1}{\mathsf{P}_0} \cdot \mathsf{S}_0(q) + \log(\mathsf{P}_0) & \text{if } \mathsf{P}_0 > 0 \\ 0 & \text{otherwise} \end{cases}$ ▷ *BG* entropy
12: $\quad H_1 \leftarrow \begin{cases} -\frac{1}{\mathsf{P}_1} \cdot \mathsf{S}_1(q) + \log(\mathsf{P}_1) & \text{if } \mathsf{P}_1 > 0 \\ 0 & \text{otherwise} \end{cases}$ ▷ *FG* entropy
13: $\quad H_{01} = H_0 + H_1$ ▷ *overall* entropy for q
14: $\quad$ **if** $H_{01} > H_{\max}$ **then** ▷ maximize $H_{01}(q)$
15: $\quad\quad H_{\max} \leftarrow H_{01}$
16: $\quad\quad q_{\max} \leftarrow q$
17: **return** $q_{\max}$

18: **MakeTables**(p, K)
19: Create maps $\mathsf{S}_0, \mathsf{S}_1 : [0, K-1] \mapsto \mathbb{R}$
20: $s_0 \leftarrow 0$
21: **for** $i \leftarrow 0, \ldots, K-1$ **do** ▷ initialize table S_0
22: $\quad$ **if** $\mathsf{p}(i) > 0$ **then**
23: $\quad\quad s_0 \leftarrow s_0 + \mathsf{p}(i) \cdot \log\big(\mathsf{p}(i)\big)$
24: $\quad \mathsf{S}_0(i) \leftarrow s_0$
25: $s_1 \leftarrow 0$
26: **for** $i \leftarrow K-1, \ldots, 0$ **do** ▷ initialize table S_1
27: $\quad \mathsf{S}_1(i) \leftarrow s_1$
28: $\quad$ **if** $\mathsf{p}(i) > 0$ **then**
29: $\quad\quad s_1 \leftarrow s_1 + \mathsf{p}(i) \cdot \log\big(\mathsf{p}(i)\big)$
30: **return** $(\mathsf{S}_0, \mathsf{S}_1)$

Alg. 11.5
Maximum entropy threshold selection after Kapur et al. [133]. Initially (outside the **for**-loop), the threshold q is assumed to be -1, which corresponds to the background class being empty ($n_0 = 0$) and all pixels assigned to the foreground class ($n_1 = N$). The **for**-loop (lines 8–16) examines each possible threshold $q = 0, \ldots, K-2$. The optimal threshold value $(0, \ldots, K-2)$ is returned, or -1 if no valid threshold was found.

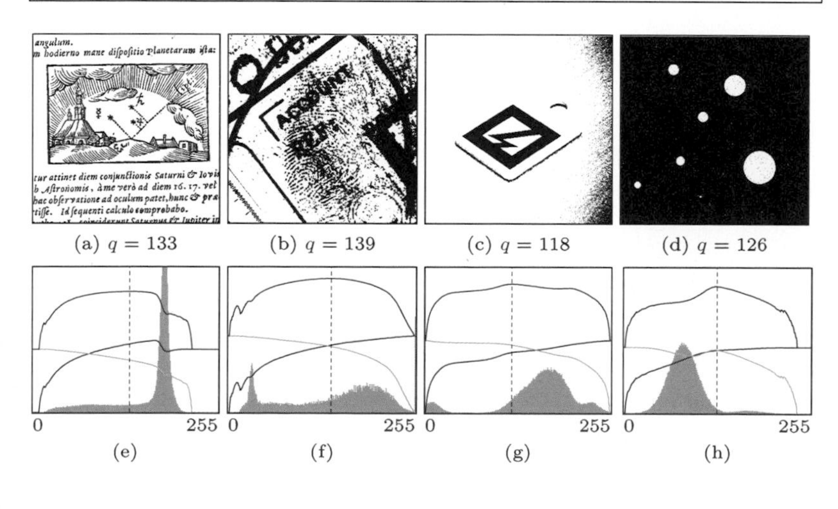

Fig. 11.5
Thresholding with the Maximum-entropy method. Calculated threshold values q and resulting binary images (a–d). Graphs in (e–h) show the *background* entropy $H_0(q)$ (green), *foreground* entropy $H_1(q)$ (blue) and *overall* entropy $H_{01}(q) = H_0(q) + H_1(q)$ (red), for varying threshold values q. The optimal threshold $q_{\max}$ is found at the maximum of H_{01} (dashed vertical line).

as a superficial outline to explain the elementary concepts. For a solid grounding of these and related topics readers are referred to the excellent texts available on statistical pattern recognition, such as [24, 64].

Bayesian decision making

The assumption is again that the image pixels originate from one of two classes, $\mathcal{C}_0$ and $\mathcal{C}_1$, or background and foreground, respectively. Both classes generate random intensity values following unknown statistical distributions. Typically, these are modeled as Gaussian distributions with unknown parameters μ and σ^2, as will be described. The task is to decide for each pixel value x to which of the two classes it most likely belongs. Bayesian reasoning is a classic technique for making such decisions in a probabilistic context.

The *probability*, that a certain intensity value x originates from a background pixel is denoted

$$p(x \mid \mathcal{C}_0).$$

This is called a "conditional probability".[6] It tells us how likely it is to observe the gray value x when a pixel is a member of the background class $\mathcal{C}_0$. Analogously, $p(x \mid \mathcal{C}_1)$ is the conditional probability of observing the value x when a pixel is known to be of the foreground class $\mathcal{C}_1$.

For the moment, let us assume that the conditional probability functions $p(x \mid \mathcal{C}_0)$ and $p(x \mid \mathcal{C}_1)$ are known. Our problem is reversed though, namely to decide which class a pixel most likely belongs to, given that its intensity is x. This means that we are actually interested in the conditional probabilities

$$p(\mathcal{C}_0 \mid x) \qquad \text{and} \qquad p(\mathcal{C}_1 \mid x), \tag{11.42}$$

also called *a posteriori* (or *posterior*) probabilities. If we knew these, we could simply select the class with the higher probability in the form

$$\mathcal{C} = \begin{cases} \mathcal{C}_0 & \text{if } p(\mathcal{C}_0 \mid x) > p(\mathcal{C}_1 \mid x), \\ \mathcal{C}_1 & \text{otherwise.} \end{cases} \tag{11.43}$$

Bayes' theorem provides a method for estimating these *posterior* probabilities, that is,

$$p(\mathcal{C}_j \mid x) = \frac{p(x \mid \mathcal{C}_j) \cdot p(\mathcal{C}_j)}{p(x)}, \tag{11.44}$$

where $p(\mathcal{C}_j)$ is the *prior* probability of class $\mathcal{C}_j$. While, in theory, the prior probabilities are also unknown, they can be easily estimated from the image histogram (see also Sec. 11.1.5). Finally, $p(x)$ in Eqn. (11.44) is the overall probability of observing the intensity value x,

[6] In general, $p(A \mid B)$ denotes the (conditional) probability of observing the event A in a given situation B. It is usually read as "the probability of A, given B".

which is typically estimated from its relative frequency in one or more images.[7]

Note that for a particular intensity x, the corresponding evidence $p(x)$ only *scales* the posterior probabilities and is thus not relevant for the classification itself. Consequently, we can reformulate the binary decision rule in Eqn. (11.43) to

$$\mathcal{C} = \begin{cases} \mathcal{C}_0 & \text{if } p(x \mid \mathcal{C}_0) \cdot p(\mathcal{C}_0) > p(x \mid \mathcal{C}_1) \cdot p(\mathcal{C}_1), \\ \mathcal{C}_1 & \text{otherwise.} \end{cases} \tag{11.45}$$

This is called Bayes' decision rule. It minimizes the probability of making a classification error if the involved probabilities are known and is also called the "minimum error" criterion.

Gaussian probability distributions

If the probability distributions $p(x \mid \mathcal{C}_j)$ are modeled as *Gaussian*[8] distributions $\mathcal{N}(x \mid \mu_j, \sigma_j^2)$, where μ_j, σ_j^2 denote the *mean* and *variance* of class C_j, we can rewrite the scaled posterior probabilities in Eqn. (11.45) as

$$p(x \mid \mathcal{C}_j) \cdot p(\mathcal{C}_j) = \frac{1}{\sqrt{2\pi\sigma_j^2}} \cdot \exp\Bigl(-\frac{(x-\mu_j)^2}{2\sigma_j^2}\Bigr) \cdot p(\mathcal{C}_j). \tag{11.46}$$

As long as the ordering between the resulting class scores remains unchanged, these quantities can be scaled or transformed arbitrarily. In particular, it is common to use the *logarithm* of the above expression to avoid repeated multiplications of small numbers. For example, applying the natural logarithm[9] to both sides of Eqn. (11.46) yields

$$\ln\bigl(p(x \mid \mathcal{C}_j) \cdot p(\mathcal{C}_j)\bigr) = \ln\bigl(p(x \mid \mathcal{C}_j)\bigr) + \ln\bigl(p(\mathcal{C}_j)\bigr) \tag{11.47}$$

$$= \ln\Bigl(\frac{1}{\sqrt{2\pi\sigma_j^2}}\Bigr) + \ln\Bigl(\exp\Bigl(-\frac{(x-\mu_j)^2}{2\sigma_j^2}\Bigr)\Bigr) + \ln\bigl(p(\mathcal{C}_j)\bigr) \tag{11.48}$$

$$= -\frac{1}{2} \cdot \ln(2\pi) - \frac{1}{2} \cdot \ln(\sigma_j^2) - \frac{(x-\mu_j)^2}{2\sigma_j^2} + \ln\bigl(p(\mathcal{C}_j)\bigr) \tag{11.49}$$

$$= -\frac{1}{2} \cdot \Bigl[\ln(2\pi) + \frac{(x-\mu_j)^2}{\sigma_j^2} + \ln(\sigma_j^2) - 2 \cdot \ln\bigl(p(\mathcal{C}_j)\bigr)\Bigr]. \tag{11.50}$$

Since $\ln(2\pi)$ in Eqn. (11.50) is constant, it can be ignored for the classification decision, as well as the factor $\frac{1}{2}$ at the front. Thus, to find the class $\mathcal{C}_j$ that maximizes $p(x \mid \mathcal{C}_j) \cdot p(\mathcal{C}_j)$ for a given intensity value x, it is sufficient to *maximize* the quantity

$$-\Bigl[\frac{(x-\mu_j)^2}{\sigma_j^2} + 2 \cdot \bigl[\ln(\sigma_j) - \ln\bigl(p(\mathcal{C}_j)\bigr)\bigr]\Bigr] \tag{11.51}$$

or, alternatively, to *minimize*

[7] $p(x)$ is also called the "evidence" for the event x.

[8] See also Sec. D.4 in the Appendix.

[9] Any logarithm could be used but the natural logarithm complements the exponential function of the Gaussian.

$$\varepsilon_j(x) = \frac{(x-\mu_j)^2}{\sigma_j^2} + 2\cdot\big[\ln(\sigma_j) - \ln\big(p(\mathcal{C}_j)\big)\big]. \tag{11.52}$$

The quantity $\varepsilon_j(x)$ can be viewed as a *measure of the potential error* involved in classifying the observed value x as being of class $\mathcal{C}_j$. To obtain the decision associated with the minimum risk, we can modify the binary decision rule in Eqn. (11.45) to

$$\mathcal{C} = \begin{cases} \mathcal{C}_0 & \text{if } \varepsilon_0(x) \le \varepsilon_1(x), \\ \mathcal{C}_1 & \text{otherwise.} \end{cases} \tag{11.53}$$

Remember that this rule tells us how to correctly classify the observed intensity value x as being either of the background class $\mathcal{C}_0$ or the foreground class $\mathcal{C}_1$, assuming that the underlying distributions are really Gaussian and their parameters are well estimated.

Goodness of classification

If we apply a threshold q, all pixel values $g \le q$ are implicitly classified as $\mathcal{C}_0$ (background) and all $g > q$ as $\mathcal{C}_1$ (foreground). The goodness of this classification by q over all N image pixels $I(u,v)$ can be measured with the criterion function

$$\begin{aligned} \mathsf{e}(q) &= \frac{1}{MN}\cdot\sum_{u,v}\begin{cases} \varepsilon_0(I(u,v)) & \text{for } I(u,v) \le q \\ \varepsilon_1(I(u,v)) & \text{for } I(u,v) > q \end{cases} && (11.54) \\ &= \frac{1}{MN}\cdot\sum_{g=0}^{q}\mathsf{h}(g)\cdot\varepsilon_0(g) \;+\; \frac{1}{MN}\cdot\sum_{g=q+1}^{K-1}\mathsf{h}(g)\cdot\varepsilon_1(g) && (11.55) \\ &= \sum_{g=0}^{q}\mathsf{p}(g)\cdot\varepsilon_0(g) \;+\; \sum_{g=q+1}^{K-1}\mathsf{p}(g)\cdot\varepsilon_1(g), && (11.56) \end{aligned}$$

with the normalized frequencies $\mathsf{p}(g) = \mathsf{h}(g)/N$ and the function $\varepsilon_j(g)$ as defined in Eqn. (11.52). By substituting $\varepsilon_j(g)$ from Eqn. (11.52) and some mathematical gymnastics, $\mathsf{e}(q)$ can be written as

$$\begin{aligned} \mathsf{e}(q) = 1 &+ P_0(q)\cdot\ln\big(\sigma_0^2(q)\big) + P_1(q)\cdot\ln\big(\sigma_1^2(q)\big) \\ &- 2\cdot P_0(q)\cdot\ln\big(P_0(q)\big) - 2\cdot P_1(q)\cdot\ln\big(P_1(q)\big). \end{aligned} \tag{11.57}$$

The remaining task is to find the threshold q that *minimizes* $\mathsf{e}(q)$ (where the constant 1 in Eqn. (11.57) can be omitted, of course). For each possible threshold q, we only need to estimate (from the image's histogram, as in Eqn. (11.31)) the "prior" probabilities $P_0(q)$, $P_1(q)$ and the corresponding within-class variances $\sigma_0(q), \sigma_1(q)$. The *prior* probabilities for the background and foreground classes are estimated as

$$P_0(q) \approx \sum_{g=0}^{q}\mathsf{p}(g) = \frac{1}{MN}\cdot\sum_{g=0}^{q}\mathsf{h}(g) = \frac{n_0(q)}{MN}, \tag{11.58}$$

$$P_1(q) \approx \sum_{g=q+1}^{K-1}\mathsf{p}(g) = \frac{1}{MN}\cdot\sum_{g=q+1}^{K-1}\mathsf{h}(g) = \frac{n_1(q)}{MN}, \tag{11.59}$$

where $n_0(q) = \sum_{i=0}^{q} \mathsf{h}(i)$, $n_1(q) = \sum_{i=q+1}^{K-1} \mathsf{h}(i)$, and $MN = n_0(q) + n_1(q)$ is the total number of image pixels. Estimates for background and foreground variances ()$\sigma_0^2(q)$ and $\sigma_1^2(q)$, respectively) defined in Eqn. (11.7), can be calculated efficiently by expressing them in the form

$$\begin{aligned}
\sigma_0^2(q) &\approx \frac{1}{n_0(q)} \cdot \Bigl[\underbrace{\sum_{g=0}^{q} \mathsf{h}(g) \cdot g^2}_{B_0(q)} - \frac{1}{n_0(q)} \cdot \bigl(\underbrace{\sum_{g=0}^{q} \mathsf{h}(g) \cdot g}_{A_0(q)}\bigr)^2\Bigr] \\
&= \frac{1}{n_0(q)} \cdot \Bigl[B_0(q) - \frac{1}{n_0(q)} \cdot A_0^2(q)\Bigr], \qquad (11.60) \\
\sigma_1^2(q) &\approx \frac{1}{n_1(q)} \cdot \Bigl[\underbrace{\sum_{g=q+1}^{K-1} \mathsf{h}(g) \cdot g^2}_{B_1(q)} - \frac{1}{n_1(q)} \cdot \bigl(\underbrace{\sum_{g=q+1}^{K-1} \mathsf{h}(g) \cdot g}_{A_1(q)}\bigr)^2\Bigr] \\
&= \frac{1}{n_1(q)} \cdot \Bigl[B_1(q) - \frac{1}{n_1(q)} \cdot A_1^2(q)\Bigr], \qquad (11.61)
\end{aligned}$$

with the quantities

$$\begin{aligned}
A_0(q) &= \sum_{g=0}^{q} \mathsf{h}(g) \cdot g, & B_0(q) &= \sum_{g=0}^{q} \mathsf{h}(g) \cdot g^2, \\
A_1(q) &= \sum_{g=q+1}^{K-1} \mathsf{h}(g) \cdot g, & B_1(q) &= \sum_{g=q+1}^{K-1} \mathsf{h}(g) \cdot g^2.
\end{aligned} \qquad (11.62)$$

Furthermore, the values $\sigma_0^2(q)$, $\sigma_1^2(q)$ can be tabulated for every possible q in only two passes over the histogram, using the recurrence relations

$$A_0(q) = \begin{cases} 0 & \text{for } q = 0, \\ A_0(q-1) + \mathsf{h}(q) \cdot q & \text{for } 1 \le q \le K-1, \end{cases} \qquad (11.63)$$

$$B_0(q) = \begin{cases} 0 & \text{for } q = 0, \\ B_0(q-1) + \mathsf{h}(q) \cdot q^2 & \text{for } 1 \le q \le K-1, \end{cases} \qquad (11.64)$$

$$A_1(q) = \begin{cases} 0 & \text{for } q = K-1, \\ A_1(q+1) + \mathsf{h}(q+1) \cdot (q+1) & \text{for } 0 \le q \le K-2, \end{cases} \qquad (11.65)$$

$$B_1(q) = \begin{cases} 0 & \text{for } q = K-1, \\ B_1(q+1) + \mathsf{h}(q+1) \cdot (q+1)^2 & \text{for } 0 \le q \le K-2. \end{cases} \qquad (11.66)$$

The complete minimum-error threshold calculation is summarized in Alg. 11.6. First, the tables $\mathsf{S}_0, \mathsf{S}_1$ are set up and initialized with the values of $\sigma_0^2(q), \sigma_1^2(q)$, respectively, for $0 \le q < K$, following the recursive scheme in Eqns. (11.63–11.66). Subsequently, the error value $\mathsf{e}(q)$ is calculated for every possible threshold value q to find the global minimum. Again $\mathsf{e}(q)$ can only be calculated for those values of q, for which both resulting partitions are non-empty (i.e., with $n_0(q), n_1(q) > 0$). Note that, in lines 27 and 37 of Alg. 11.6, a small constant ($\frac{1}{12}$) is added to the variance to avoid zero values when the corresponding class population is homogeneous (i.e., only contains a single intensity value).[10] This ensures that the algorithm works properly on images with only two distinct gray values. The algorithm computes the optimal threshold by performing three passes over the histogram (two for initializing the tables and one for finding the minimum); it thus has the same time complexity of $\mathcal{O}(K)$ as the algorithms described before.

Figure 11.6 shows the results of minimum-error thresholding on our set of test images. It also shows the fitted pair of Gaussian distributions for the background and the foreground pixels, respectively, for the optimal threshold as well as the graphs of the error function $\mathsf{e}(q)$, which is minimized over all threshold values q. Obviously the error function is quite flat in certain cases, indicating that similar scores are obtained for a wide range of threshold values and the optimal threshold is not very distinct. We can also see that the estimate is quite accurate in case of the synthetic test image in Fig. 11.6(d), which is actually generated as a mixture of two Gaussians (with parameters $\mu_0 = 80$, $\mu_1 = 170$ and $\sigma_0 = \sigma_1 = 20$). Note that the histograms in Fig. 11.6 have been properly normalized (to constant area) to illustrate the curves of the Gaussians, that is, properly scaled by their prior probabilities (P_0, P_1), while the original histograms are scaled with respect to their maximum values.

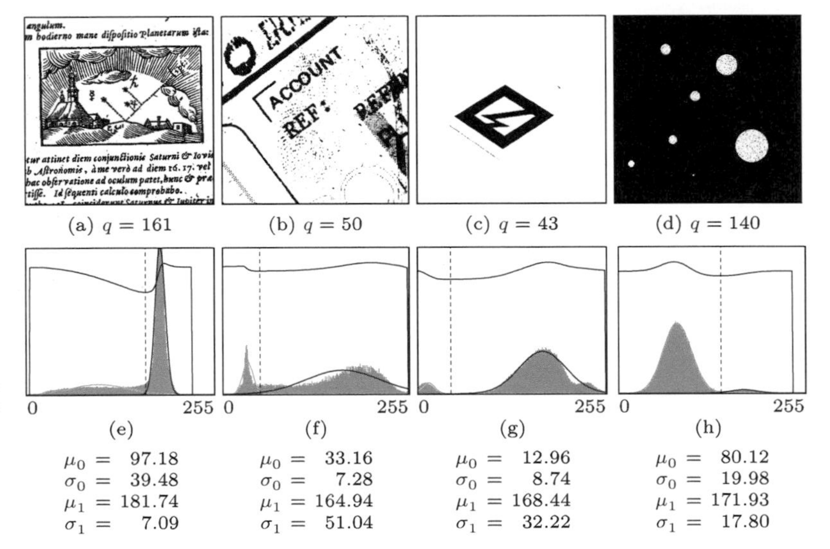

Fig. 11.6
Results from minimum-error thresholding. Calculated threshold values q and resulting binary images (a–d). The green and blue graphs in (e–h) show the fitted Gaussian background and foreground distributions $\mathcal{N}_0 = (\mu_0, \sigma_0)$ and $\mathcal{N}_1 = (\mu_1, \sigma_1)$, respectively. The red graph corresponds to the error quantity $\mathsf{e}(q)$ for varying threshold values $q = 0, \ldots, 255$ (see Eqn. (11.57)). The optimal threshold $q_{\min}$ is located at the *minimum* of $\mathsf{e}(q)$ (dashed vertical line). The estimated parameters of the background/foreground Gaussians are listed at the bottom.

A minor theoretical problem with the minimum error technique is that the parameters of the Gaussian distributions are always estimated from *truncated* samples. This means that, for any threshold q, only the intensity values smaller than q are used to estimate the parameters of the background distribution, and only the intensities greater than q contribute to the foreground parameters. In practice, this problem is of minor importance, since the distributions are typically not strictly Gaussian either.

[10] This is explained by the fact that each histogram bin $\mathsf{h}(i)$ represents intensities in the continuous range $[i \pm 0.5]$ and the variance of uniformly distributed values in the unit interval is $\frac{1}{12}$.

```
1:  MinimumErrorThreshold(h)
    Input: h : [0, K−1] ↦ ℕ, a grayscale histogram. Returns the
    optimal threshold value or −1 if no threshold is found.
2:  K ← Size(h)
3:  (S_0, S_1, N) ← MakeSigmaTables(h, K)
4:  n_0 ← 0
5:  q_min ← −1
6:  e_min ← ∞
7:  for q ← 0, ..., K−2 do           ▷ evaluate all possible thresholds q
8:      n_0 ← n_0 + h(q)             ▷ background population
9:      n_1 ← N − n_0                ▷ foreground population
10:     if (n_0 > 0) ∧ (n_1 > 0) then
11:         P_0 ← n_0/N              ▷ prior probability of C_0
12:         P_1 ← n_1/N              ▷ prior probability of C_1
13:         e ← P_0 · ln(S_0(q)) + P_1 · ln(S_1(q))
                − 2 · (P_0 · ln(P_0) + P_1 · ln(P_1))   ▷ Eq. 11.57
14:         if e < e_min then        ▷ minimize error for q
15:             e_min ← e
16:             q_min ← q
17: return q_min

18: MakeSigmaTables(h, K)
19:     Create maps S_0, S_1 : [0, K−1] ↦ ℝ
20:     n_0 ← 0
21:     A_0 ← 0
22:     B_0 ← 0
23:     for q ← 0, ..., K−1 do       ▷ tabulate σ_0^2(q)
24:         n_0 ← n_0 + h(q)
25:         A_0 ← A_0 + h(q) · q     ▷ Eq. 11.63
26:         B_0 ← B_0 + h(q) · q^2   ▷ Eq. 11.64
27:         S_0(q) ← { 1/12 + (B_0 − A_0^2/n_0)/n_0   for n_0 > 0
                     { 0                             otherwise   ▷ Eq. 11.60
28:     N ← n_0
29:     n_1 ← 0
30:     A_1 ← 0
31:     B_1 ← 0
32:     S_1(K−1) ← 0
33:     for q ← K−2, ..., 0 do       ▷ tabulate σ_1^2(q)
34:         n_1 ← n_1 + h(q+1)
35:         A_1 ← A_1 + h(q+1) · (q+1)     ▷ Eq. 11.65
36:         B_1 ← B_1 + h(q+1) · (q+1)^2   ▷ Eq. 11.66
37:         S_1(q) ← { 1/12 + (B_1 − A_1^2/n_1)/n_1   for n_1 > 0
                     { 0                             otherwise   ▷ Eq. 11.61
38:     return (S_0, S_1, N)
```

Alg. 11.6
Minimum error threshold selection based on a Gaussian mixture model (after [116]). Tables S_0, S_1 are intialized with values $\sigma_0^2(q)$ and $\sigma_1^2(q)$, respectively (see Eqns. (11.60)–(11.61)), for all possible threshold values $q = 0, \ldots, K-1$. N is the number of image pixels. Initially (outside the **for**-loop), the threshold q is assumed to be -1, which corresponds to the background class being empty ($n_0 = 0$) and all pixels assigned to the foreground class ($n_1 = N$). The **for**-loop (lines 8–16) examines each possible threshold $q = 0, \ldots, K-2$. The optimal threshold is returned, or -1 if no valid threshold was found.

11.2 Local Adaptive Thresholding

In many situations, a fixed threshold is not appropriate to classify the pixels in the entire image, for example, when confronted with stained backgrounds or uneven lighting or exposure. Figure 11.7 shows a typical, unevenly exposed document image and the results obtained with some global thresholding methods described in the previous sections.

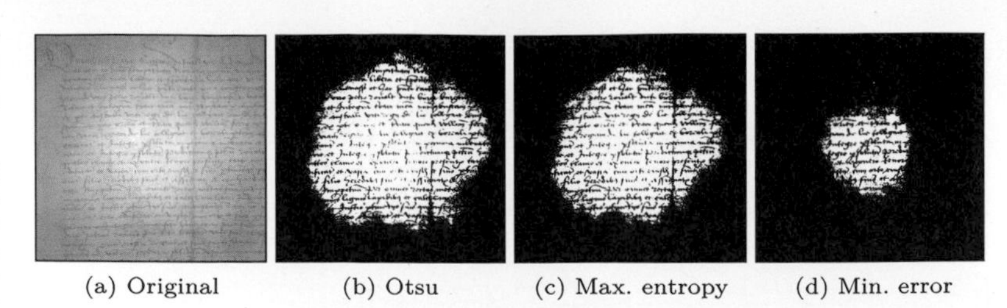
(a) Original (b) Otsu (c) Max. entropy (d) Min. error

Fig. 11.7 Global thresholding methods fail under uneven lighting or exposure. Original image (a), results from global thresholding with various methods described above (b–d).

Instead of using a single threshold value for the whole image, adaptive thresholding specifies a *varying* threshold value $Q(u, v)$ for each image position that is used to classify the corresponding pixel $I(u, v)$ in the same way as described in Eqn. (11.1) for a global threshold. The following approaches differ only with regard to how the threshold "surface" Q is derived from the input image.

11.2.1 Bernsen's Method

The method proposed by Bernsen [23] specifies a dynamic threshold for each image position (u, v), based on the minimum and maximum intensity found in a local neighborhood $R(u, v)$. If

$$I_{\min}(u,v) = \min_{\substack{(i,j)\in\\ R(u,v)}} I(i,j), \tag{11.67}$$

$$I_{\max}(u,v) = \max_{\substack{(i,j)\in\\ R(u,v)}} I(i,j) \tag{11.68}$$

are the minimum and maximum intensity values within a fixed-size neighborhood region R centered at position (u, v), the space-varying threshold is simply calculated as the *mid-range* value

$$Q(u,v) = \frac{I_{\min}(u,v) + I_{\max}(u,v)}{2}. \tag{11.69}$$

This is done as long as the local contrast $c(u, v) = I_{\max}(u, v) - I_{\min}(u, v)$ is above some predefined limit $c_{\min}$. If $c(u, v) < c_{\min}$, the pixels in the corresponding image region are assumed to belong to a single class and are (by default) assigned to the background.

The whole process is summarized in Alg. 11.7. Note that the meaning of "background" in terms of intensity levels depends on the application. For example, in astronomy, the image background is usually darker than the objects of interest. In typical OCR applications, however, the background (paper) is brighter than the foreground objects (print). The main function provides a control parameter bg to select the proper default threshold $\bar{q}$, which is set to K in case of a dark background (bg = dark) and to 0 for a bright background (bg = bright). The support region R may be square or circular, typically with a radius $r = 15$. The choice of the minimum contrast limit $c_{\min}$ depends on the type of imagery and the noise level ($c_{\min} = 15$ is a suitable value to start with).

Figure 11.8 shows the results of Bernsen's method on the uneven test image used in Fig. 11.7 for different settings of the region's radius r. Due to the nonlinear min- and max-operation, the resulting

Alg. 11.7
Adaptive thresholding using local contrast (after Bernsen [23]). The argument to bg should be set to dark if the image background is darker than the structures of interest, and to bright if the background is brighter than the objects.

```
1: BernsenThreshold(I, r, c_min, bg)
     Input: I, intensity image of size M × N; r, radius of support
     region; c_min, minimum contrast; bg, background type (dark or
     bright). Returns a map with an individual threshold value for
     each image position.
2:   (M, N) ← Size(I)
3:   Create map Q : M × N ↦ ℝ
4:   q̄ ← { K  if bg = dark
          { 0  if bg = bright
5:   for all image coordinates (u, v) ∈ M × N do
6:       R ← MakeCircularRegion(u, v, r)
7:       I_min ← min_{(i,j)∈R} I(i, j)
8:       I_max ← max_{(i,j)∈R} I(i, j)
9:       c ← I_max − I_min
10:      Q(u, v) ← { (I_min + I_max)/2   if c ≥ c_min
                   { q̄                   otherwise
11:  return Q

12: MakeCircularRegion(u, v, r)
     Returns the set of pixel coordinates within the circle of radius r,
     centered at (u, v)
13:  return {(i, j) ∈ ℤ² | (u − i)² + (v − j)² ≤ r²}
```

threshold surface is not smooth. The minimum contrast is set to $c_{\min} = 15$, which is too low to avoid thresholding low-contrast noise visible along the left image margin. By increasing the minimum contrast $c_{\min}$, more neighborhoods are considered "flat" and thus ignored, that is, classified as background. This is demonstrated in Fig. 11.9. While larger values of $c_{\min}$ effectively eliminate low-contrast noise, relevant structures are also lost, which illustrates the difficulty of finding a suitable global value for $c_{\min}$. Additional examples, using the test images previously used for global thresholding, are shown in Fig. 11.10.

What Alg. 11.7 describes formally can be implemented quite efficiently, noting that the calculation of local minima and maxima over a sliding window (lines 6–8) corresponds to a simple nonlinear filter operation (see Ch. 5, Sec. 5.4). To perform these calculations, we can use a *minimum* and *maximum* filter with radius r, as provided by virtually every image processing environment. For example, the Java implementation of the Bernsen thresholder in Prog. 11.1 uses ImageJ's built-in `RankFilters` class for this purpose. The complete implementation can be found on the book's website (see Sec. 11.3 for additional details on the corresponding API).

11.2.2 Niblack's Method

In this approach, originally presented in [172, Sec. 5.1], the threshold $Q(u, v)$ is varied across the image as a function of the local intensity average $\mu_R(u, v)$ and standard deviation[11] $\sigma_R(u, v)$ in the form

[11] The standard deviation σ is the square root of the variance σ^2.

Prog. 11.1
Bernsen's thresholder (ImageJ plugin implementation of Alg. 11.7). Note the use of ImageJ's RankFilters class (lines 30–32) for calculating the local minimum (Imin) and maximum (Imax) maps inside the getThreshold() method. The resulting threshold surface $Q(u, v)$ is returned as an 8-bit image of type ByteProcessor.

```
1 package imagingbook.pub.threshold.adaptive;
2 import ij.plugin.filter.RankFilters;
3 import ij.process.ByteProcessor;
4 import imagingbook.pub.threshold.BackgroundMode;
5
6 public class BernsenThresholder extends AdaptiveThresholder
      {
7
8   public static class Parameters {
9     public int radius = 15;
10    public int cmin = 15;
11    public BackgroundMode bgMode = BackgroundMode.DARK;
12  }
13
14  private final Parameters params;
15
16  public BernsenThresholder() {
17    this.params = new Parameters();
18  }
19
20  public BernsenThresholder(Parameters params) {
21    this.params = params;
22  }
23
24  public ByteProcessor getThreshold(ByteProcessor I) {
25    int M = I.getWidth();
26    int N = I.getHeight();
27    ByteProcessor Imin = (ByteProcessor) I.duplicate();
28    ByteProcessor Imax = (ByteProcessor) I.duplicate();
29
30    RankFilters rf = new RankFilters();
31    rf.rank(Imin,params.radius,RankFilters.MIN); // Imin(u,v)
32    rf.rank(Imax,params.radius,RankFilters.MAX); // Imax(u,v)
33
34    int q = (params.bgMode == BackgroundMode.DARK) ?
35             256 : 0;
36    ByteProcessor Q = new ByteProcessor(M, N);  // Q(u,v)
37
38    for (int v = 0; v < N; v++) {
39      for (int u = 0; u < M; u++) {
40        int gMin = Imin.get(u, v);
41        int gMax = Imax.get(u, v);
42        int c = gMax - gMin;
43        if (c >= params.cmin)
44          Q.set(u, v, (gMin + gMax) / 2);
45        else
46          Q.set(u, v, q);
47      }
48    }
49    return Q;
50  }
51 }
```

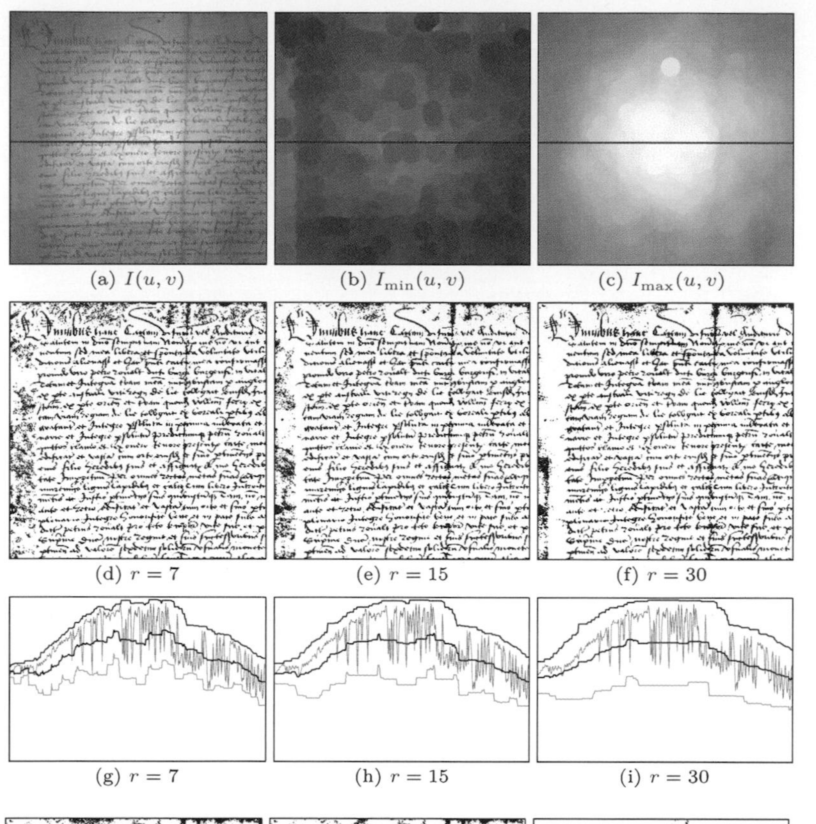

Fig. 11.8 Adaptive thresholding using Bernsen's method. Original image (a), local minimum (b), and maximum (c). The center row shows the binarized images for different settings of r (d–f). The corresponding curves in (g–i) show the original intensity (gray), local minimum (green), maximum (red), and the actual threshold (blue) along the horizontal line marked in (a–c). The region radius r is 15 pixels, the minimum contrast $c_{\min}$ is 15 intensity units.

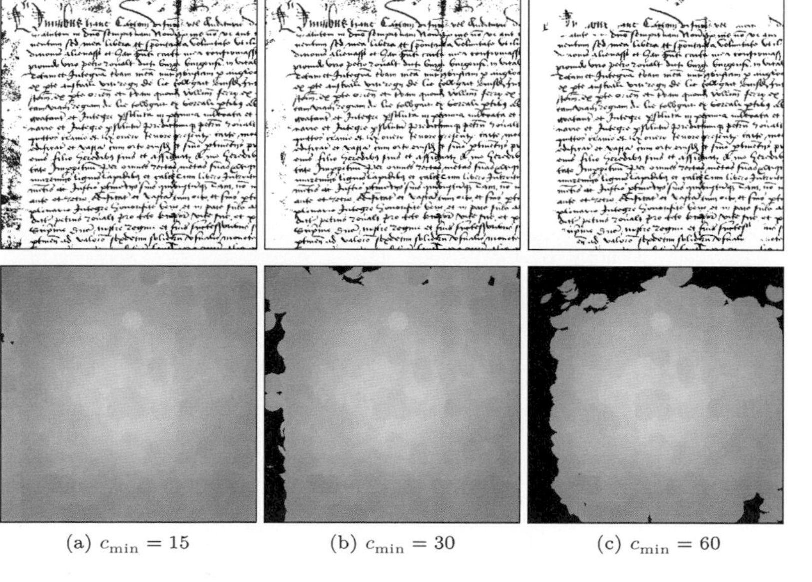

Fig. 11.9 Adaptive thresholding using Bernsen's method with different settings of $c_{\min}$. Binarized images (top row) and threshold surface $Q(u, v)$ (bottom row). Black areas in the threshold functions indicate that the local contrast is below $c_{\min}$; the corresponding pixels are classified as background (white in this case).

$$Q(u, v) := \mu_R(u, v) + \kappa \cdot \sigma_R(u, v). \tag{11.70}$$

Thus the local threshold $Q(u, v)$ is determined by adding a constant portion ($\kappa \geq 0$) of the local standard deviation σ_R to the local mean μ_R. μ_R and σ_R are calculated over a square support region R centered at (u, v). The size (radius) of the averaging region R should be as large as possible, at least larger than the size of the structures to be detected, but small enough to capture the variations (unevenness)

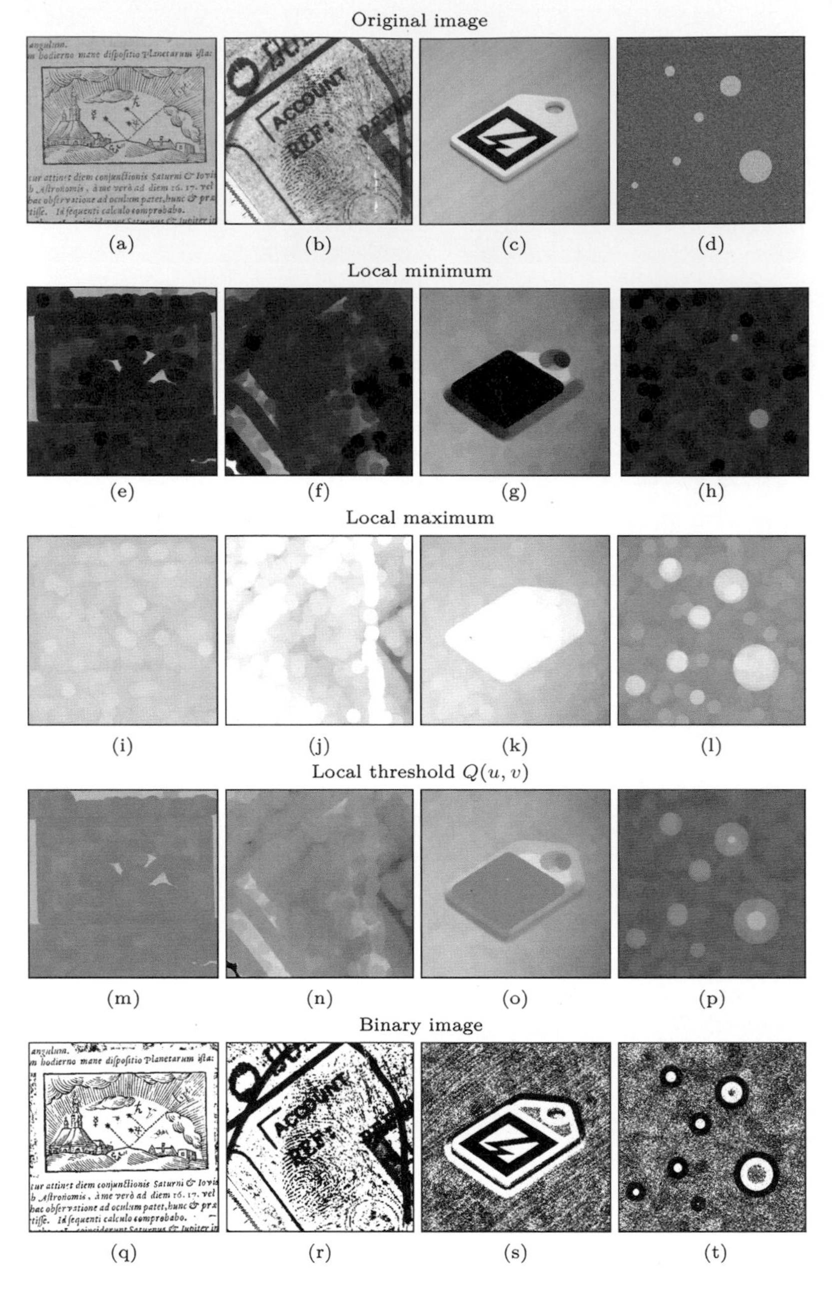

Fig. 11.10 Additional examples for Bernsen's method. Original images (a–d), local minimum I_{min} (e–h), maximum I_{max} (i–l), and threshold map Q (m–p); results after thresholding the images (q–t). Settings are $r = 15$, $c_{\text{min}} = 15$. A bright background is assumed for all images (bg = bright), except for image (d).

of the background. A size of 31×31 pixels (or radius $r = 15$) is suggested in [172] and $\kappa = 0.18$, though the latter does not seem to be critical.

One problem is that, for small values of σ_R (as obtained in "flat" image regions of approximately constant intensity), the threshold will be close to the local average, which makes the segmentation quite sensitive to low-amplitude noise ("ghosting"). A simple improvement is to secure a minimum distance from the mean by adding a constant offset d, that is, replacing Eqn. (11.70) by

$$Q(u, v) := \mu_R(u, v) + \kappa \cdot \sigma_R(u, v) + d, \tag{11.71}$$

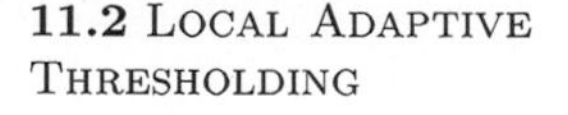

with $d \geq 0$, in the range $2, \ldots, 20$ for typical 8-bit images.

The original formulation (Eqn. (11.70)) is aimed at situations where the foreground structures are *brighter* than the background (Fig. 11.11(a)) but does not work if the images are set up the other way round (Fig. 11.11(b)). In the case that the structures of interest are *darker* than the background (as, e.g., in typical OCR applications), one could either work with inverted images or modify the calculation of the threshold to

$$Q(u,v) := \begin{cases} \mu_R(u,v) + (\kappa \cdot \sigma_R(u,v) + d) & \text{for } \textit{dark} \text{ BG}, \\ \mu_R(u,v) - (\kappa \cdot \sigma_R(u,v) + d) & \text{for } \textit{bright} \text{ BG}. \end{cases} \tag{11.72}$$

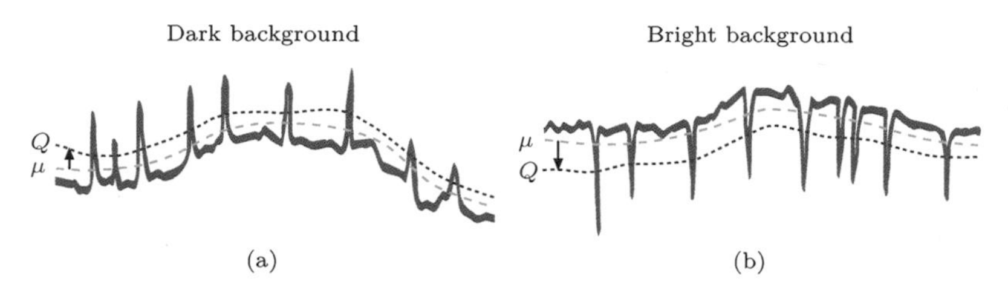

Fig. 11.11
Adaptive thresholding based on average local intensity. The illustration shows a line profile as typically found in document imaging. The space-variant threshold Q (dotted blue line) is chosen as the local average μ_R (dashed green line) offset by a multiple of the local intensity variation σ_R. The offset is chosen to be *positive* for images with a dark background and bright structures (a) and *negative* if the background is brighter than the contained structures of interest (b).

The modified procedure is detailed in Alg. 11.8. The example in Fig. 11.12 shows results obtained with this method on an image with a bright background containing dark structures, for $\kappa = 0.3$ and varying settings of d. Note that setting $d = 0$ (Fig. 11.12(d, g)) corresponds to Niblack's original method. For these examples, a circular window of radius $r = 15$ was used to compute the local mean $\mu_R(u,v)$ and variance $\sigma_R(u,v)$. Additional examples are shown in Fig. 11.13. Note that the selected radius r is obviously too small for the structures in the images in Fig. 11.13(c, d), which are thus not segmented cleanly. Better results can be expected with a larger radius.

With the intent to improve upon Niblack's method, particularly for thresholding deteriorated text images, Sauvola and Pietikäinen [207] proposed setting the threshold to

$$Q(u,v) := \begin{cases} \mu_R(u,v) \cdot \left[1 - \kappa \cdot \left(\frac{\sigma_R(u,v)}{\sigma_{\max}} - 1\right)\right] & \text{for } \textit{dark} \text{ BG}, \\ \mu_R(u,v) \cdot \left[1 + \kappa \cdot \left(\frac{\sigma_R(u,v)}{\sigma_{\max}} - 1\right)\right] & \text{for } \textit{bright} \text{ BG}, \end{cases} \tag{11.73}$$

with $\kappa = 0.5$ and $\sigma_{\max} = 128$ (the "dynamic range of the standard deviation" for 8-bit images) as suggested parameter values. In this approach, the offset between the threshold and the local average not only depends on the local variation σ_R (as in Eqn. (11.70)), but also on the magnitude of the local *mean* μ_R! Thus, changes in absolute brightness lead to modified relative threshold values, even when the image contrast remains constant. Though this technique is frequently referenced in the literature, it appears questionable if this behavior is generally desirable.

Calculating local mean and variance

Algorithm 11.8 shows the principle operation of Niblack's method and also illustrates how to efficiently calculate the local average and

Fig. 11.12
Adaptive thresholding using Niblack's method (with $r = 15$, $\kappa = 0.3$). Original image (a), local mean μ_R (b), and standard deviation σ_R (c). The result for $d = 0$ in (d) corresponds to Niblack's original formulation. Increasing the value of d reduces the amount of clutter in regions with low variance (e, f). The curves in (g–i) show the local intensity (gray), mean (green), variance (red), and the actual threshold (blue) along the horizontal line marked in (a–c).

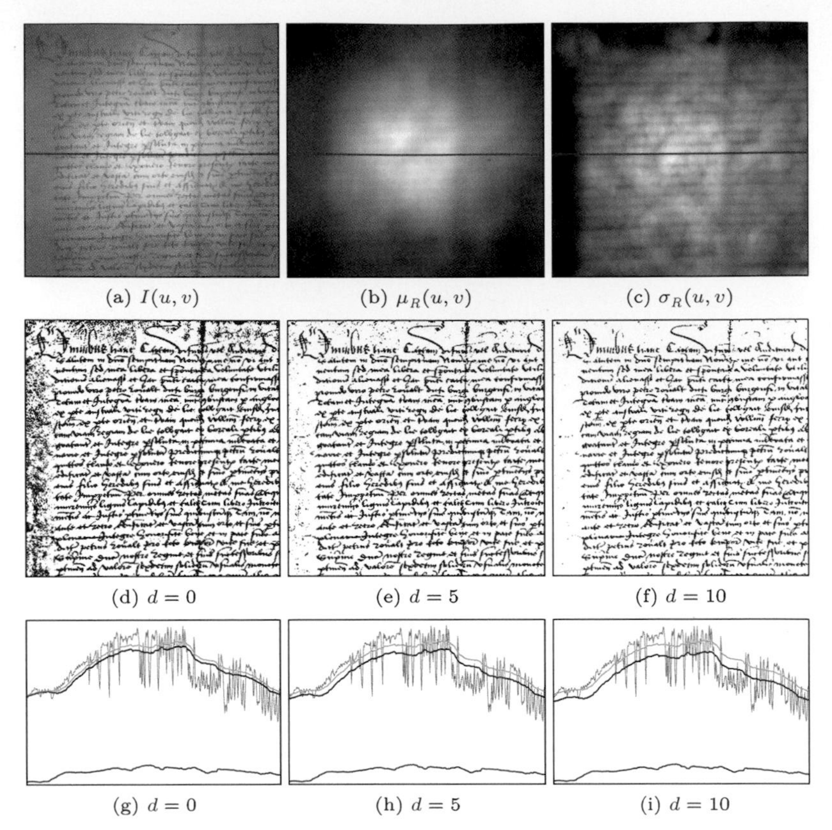

(a) $I(u, v)$ (b) $\mu_R(u, v)$ (c) $\sigma_R(u, v)$

(d) $d = 0$ (e) $d = 5$ (f) $d = 10$

(g) $d = 0$ (h) $d = 5$ (i) $d = 10$

variance. Given the image I and the averaging region R, we can use the shortcut suggested in Eqn. (3.12) to obtain these quantities as

$$\mu_R = \frac{1}{n} \cdot A \qquad \text{and} \qquad \sigma_R^2 = \frac{1}{n} \cdot \bigl(B - \frac{1}{n} \cdot A^2\bigr), \tag{11.74}$$

with

$$A = \sum_{(i,j)\in R} I(i,j), \qquad B = \sum_{(i,j)\in R} I^2(i,j), \qquad n = |R|. \tag{11.75}$$

Procedure GetLocalMeanAndVariance() in Alg. 11.8 shows this calculation in full detail.

When computing the local average and variance, attention must be paid to the situation at the image borders, as illustrated in Fig. 11.14. Two approaches are frequently used. In the first approach (following the common practice for implementing filter operations), all outside pixel values are replaced by the closest inside pixel, which is always a border pixel. Thus the border pixel values are effectively replicated outside the image boundaries and thus these pixels have a strong influence on the local results. The second approach is to perform the calculation of the average and variance on only those image pixels that are actually covered by the support region. In this case, the number of pixels (N) is reduced at the image borders to about 1/4 of the full region size.

Although the calculation of the local mean and variance outlined by function GetLocalMeanAndVariance() in Alg. 11.8 is definitely more

1: $\textsf{NiblackThreshold}(I, r, \kappa, d, bg)$
Input: I, intensity image of size $M \times N$; r, radius of support region; κ, variance control parameter; d, minimum offset; $bg \in \{\textsf{dark}, \textsf{bright}\}$, background type. Returns a map with an individual threshold value for each image position.
2: $(M, N) \leftarrow \text{Size}(I)$
3: Create map $Q : M \times N \mapsto \mathbb{R}$
4: **for all** image coordinates $(u, v) \in M \times N$ **do**
Define a support region of radius r, centered at (u, v):
5: $(\mu, \sigma^2) \leftarrow \textsf{GetLocalMeanAndVariance}(I, u, v, r)$
6: $\sigma \leftarrow \sqrt{\sigma^2}$ ▷ local std. deviation σ_R
7: $Q(u,v) \leftarrow \begin{cases} \mu + (\kappa \cdot \sigma + d) & \text{if } bg = \textsf{dark} \\ \mu - (\kappa \cdot \sigma + d) & \text{if } bg = \textsf{bright} \end{cases}$ ▷ Eq. 11.72
8: **return** Q

9: $\textsf{GetLocalMeanAndVariance}(I, u, v, r)$
Returns the local mean and variance of the image pixels $I(i, j)$ within the disk-shaped region with radius r around position (u, v).
10: $R \leftarrow \textsf{MakeCircularRegion}(u, v, r)$ ▷ see Alg. 11.7
11: $n \leftarrow 0$
12: $A \leftarrow 0$
13: $B \leftarrow 0$
14: **for all** $(i, j) \in R$ **do**
15: $n \leftarrow n + 1$
16: $A \leftarrow A + I(i, j)$
17: $B \leftarrow B + I^2(i, j)$
18: $\mu \leftarrow \frac{1}{n} \cdot A$
19: $\sigma^2 \leftarrow \frac{1}{n} \cdot (B - \frac{1}{n} \cdot A^2)$
20: **return** (μ, σ^2)

Alg. 11.8
Adaptive thresholding using local mean and variance (modified version of Niblack's method [172]). The argument to bg should be dark if the image background is darker than the structures of interest, bright if the background is brighter than the objects. The function MakeCircularRegion() is defined in Alg. 11.7.

efficient than a brute-force approach, additional optimizations are possible. Most image processing environments have suitable routines already built in. With ImageJ, for example, we can again use the `RankFilters` class (as with the *min*- and *max*-filters in the *Bernsen* approach, see Sec. 11.2.1). Instead of performing the computation for each pixel individually, the following ImageJ code segment uses predefined filters to compute two separate images `Imean` (μ_R) and `Ivar` (σ_R^2) containing the local mean and variance values, respectively, with a disk-shaped support region of radius 15:

```
ByteProcessor I; // original image I(u,v)
int radius = 15;

FloatProcessor Imean = I.convertToFloatProcessor();
FloatProcessor Ivar =  Imean.duplicate();

RankFilters rf = new RankFilters();
rf.rank(Imean, radius, RankFilters.MEAN);          // μ_R(u,v)
rf.rank(Ivar, radius, RankFilters.VARIANCE);       // σ_R^2(u,v)
...
```

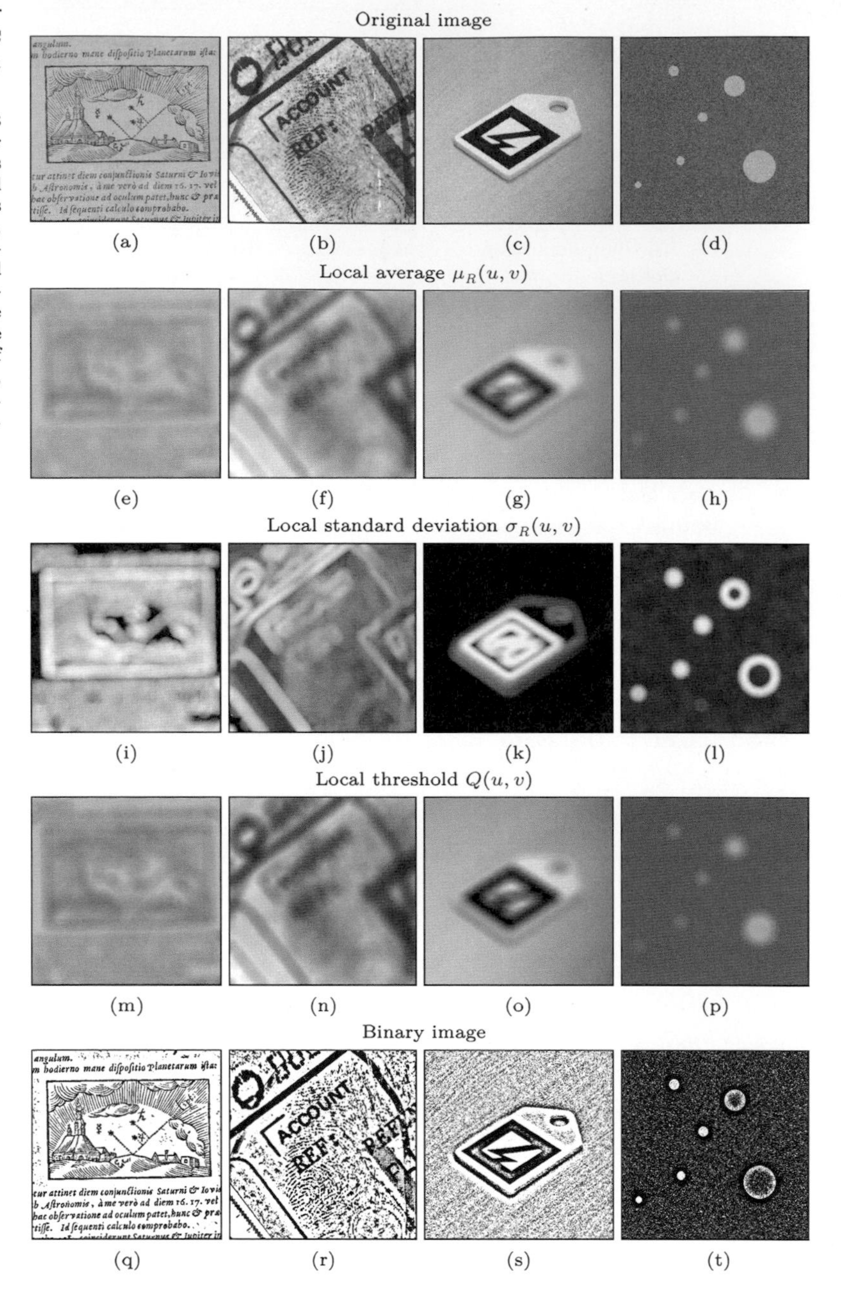

Fig. 11.13
Additional examples for thresholding with Niblack's method using a disk-shaped support region of radius $r = 15$. Original images (a–d), local mean μ_R (e–h), std. deviation σ_R (i–l), and threshold Q (m–p); results after thresholding the images (q–t). The background is assumed to be brighter than the structures of interest, except for image (d), which has a dark background. Settings are $\kappa = 0.3$, $d = 5$.

See Sec. 11.3 and the online code for additional implementation details. Note that the filter methods implemented in `RankFilters` perform replication of border pixels as the border handling strategy, as discussed earlier.

Local average and variance with Gaussian kernels

The purpose of taking the local average is to smooth the image to obtain an estimate of the varying background intensity. In case of a square or circular region, this is equivalent to convolving the image with a box- or disk-shaped kernel, respectively. Kernels of this

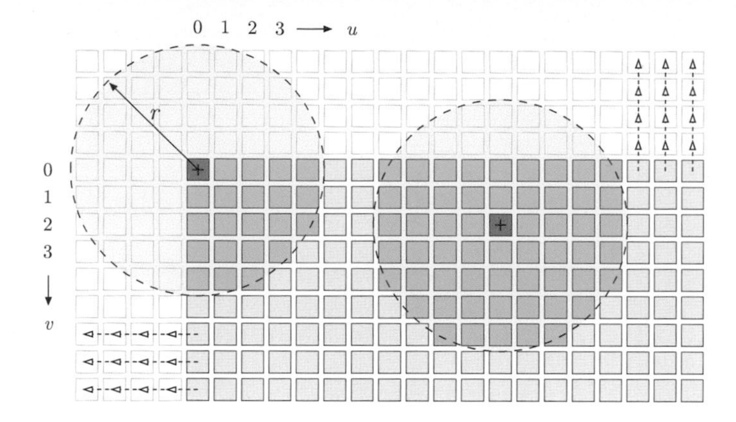

Fig. 11.14
Calculating local statistics at image boundaries. The illustration shows a disk-shaped support region with radius r, placed at the image border. Pixel values outside the image can be replaced ("filled-in") by the closest border pixel, as is common in many filter operations. Alternatively, the calculation of the local statistics can be confined to include only those pixels inside the image that are actually covered by the support region. At any border pixel, the number of covered elements (N) is still more than $\approx 1/4$ of the full region size. In this particular case, the circular region covers a maximum of $N = 69$ pixels when fully embedded and $N = 22$ when positioned at an image corner.

type, however, are not well suited for image smoothing, because they create strong ringing and truncating effects, as demonstrated in Fig. 11.15. Moreover, convolution with a box-shaped (rectangular) kernel is a non-isotropic operation, that is, the results are orientation-dependent. From this perspective alone it seems appropriate to consider other smoothing kernels, Gaussian kernels in particular.

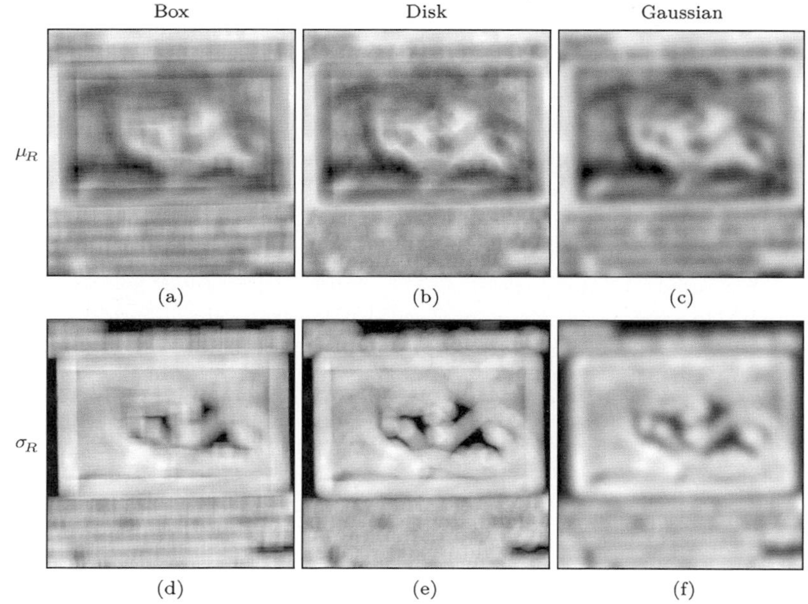

Box | Disk | Gaussian

μ_R: (a) (b) (c)

σ_R: (d) (e) (f)

Fig. 11.15
Local average (a–c) and variance (d–f) obtained with different smoothing kernels. 31×31 box filter (a, d), disk filter with radius $r = 15$ (b, e), Gaussian kernel with $\sigma = 0.6 \cdot 15 = 9.0$ (c, f). Both the box and disk filter show strong truncation effects (ringing), the box filter is also highly non-isotropic. All images are contrast-enhanced for better visibility.

Using a Gaussian kernel H^{G} for smoothing is equivalent to calculating a weighted average of the corresponding image pixels, with the weights being the coefficients of the kernel. Thus calculating this weighted local average can be expressed by

$$\mu_{\mathrm{G}}(u,v) = \frac{1}{\Sigma H^{\mathrm{G}}} \cdot \big(I * H^{\mathrm{G}}\big)(u,v), \tag{11.76}$$

where ΣH^{G} is the sum of the coefficients in the kernel H^{G} and $*$ denotes the linear convolution operator.[12] Analogously, there is also

[12] See Chapter 5, Sec. 5.3.1.

a weighted *variance* σ_G^2 which can be calculated jointly with the local average μ_G (as in Eqn. (11.74)) in the form

$$\mu_G(u,v) = \frac{1}{\Sigma H^G} \cdot A_G(u,v), \tag{11.77}$$

$$\sigma_G^2(u,v) = \frac{1}{\Sigma H^G} \cdot \big(B_G(u,v) - \frac{1}{\Sigma H^G} \cdot A_G^2(u,v)\big), \tag{11.78}$$

with $A_G = I * H^G$ and $B_G = I^2 * H^G$.

Thus all we need is two filter operations, one applied to the original image ($I * H^G$) and another applied to the squared image ($I^2 * H^G$), using the same 2D Gaussian kernel H^G (or any other suitable smoothing kernel). If the kernel H^G is *normalized* (i.e., $\Sigma H^G = 1$), Eqns. (11.77)–(11.78) reduce to

$$\mu_G(u,v) = A_G(u,v), \tag{11.79}$$

$$\sigma_G^2(u,v) = B_G(u,v) - A_G^2(u,v), \tag{11.80}$$

with A_G, B_G as defined already.

This suggests a very simple process for computing the local average and variance by Gaussian filtering, as summarized in Alg. 11.9. The width (standard deviation σ) of the Gaussian kernel is set to 0.6 times the radius r of the corresponding disk filter to produce a similar effect as Alg. 11.8. The Gaussian approach has two advantages: First, the Gaussian makes a much superior low-pass filter, compared to the box or disk kernels. Second, the 2D Gaussian is (unlike the circular disk kernel) separable in the x- and y-direction, which permits a very efficient implementation of the 2D filter using only a pair of 1D convolutions (see Ch. 5, Sec. 5.2).

For practical calculation, A_G, B_G can be represented as (floating-point) images, and most modern image-processing environments provide efficient (multi-scale) implementations of Gaussian filters with large-size kernels. In ImageJ, fast Gaussian filtering is implemented by the class `GaussianBlur` with the public methods `blur()`, `blurGaussian()`, and `blurFloat()`, which all use normalized filter kernels by default. Programs 11.2–11.3 show the complete ImageJ implementation of Niblack's thresholder using Gaussian smoothing kernels.

11.3 Java Implementation

All thresholding methods described in this chapter have been implemented as part of the `imagingbook` library that is available with full source code at the book's website. The top class in this library[13] is `Thresholder` with the sub-classes `GlobalThresholder` and `AdaptiveThresholder` for the methods described in Secs. 11.1 and 11.2, respectively. Class `Thresholder` itself is abstract and only defines a set of (non-public) utility methods for histogram analysis.

[13] Package `imagingbook.pub.threshold`.

Alg. 11.9
Adaptive thresholding using Gaussian averaging (extended from Alg. 11.8). Parameters are the original image I, the radius r of the Gaussian kernel, variance control k, and minimum offset d. The argument to bg should be dark if the image background is darker than the structures of interest, bright if the background is brighter than the objects. The procedure MakeGaussianKernel2D(σ) creates a discrete, normalized 2D Gaussian kernel with standard deviation σ.

```
1:  AdaptiveThresholdGauss(I, r, κ, d, bg)
      Input: I, intensity image of size M × N; r, support region radius; κ, variance control parameter; d, minimum offset; bg ∈ {dark, bright}, background type.
      Returns a map Q of local thresholds for the grayscale image I.
2:    (M, N) ← Size(I)
3:    Create maps A, B, Q : M × N ↦ ℝ
4:    for all image coordinates (u, v) ∈ M × N do
5:        A(u, v) ← I(u, v)
6:        B(u, v) ← (I(u, v))^2
7:    H^G ← MakeGaussianKernel2D(0.6 · r)
8:    A ← A ∗ H^G                    ▷ filter the original image with H^G
9:    B ← B ∗ H^G                    ▷ filter the squared image with H^G
10:   for all image coordinates (u, v) ∈ M × N do
11:       μ_G ← A(u, v)                                  ▷ Eq. 11.79
12:       σ_G ← sqrt(B(u, v) − A^2(u, v))                ▷ Eq. 11.80
13:       Q(u, v) ← { μ_G + (κ · σ_G + d)   if bg = dark
                    { μ_G − (κ · σ_G + d)   if bg = bright  ▷ Eq. 11.72
14:   return Q

15: MakeGaussianKernel2D(σ)
      Returns a discrete 2D Gaussian kernel H with std. deviation σ, sized sufficiently large to avoid truncation effects.
16:   r ← max(1, ⌈3.5 · σ⌉)          ▷ size the kernel sufficiently large
17:   Create map H : [−r, r]^2 ↦ ℝ
18:   s ← 0
19:   for x ← −r, ..., r do
20:       for y ← −r, ..., r do
21:           H(x, y) ← e^(−(x^2+y^2)/(2·σ^2))   ▷ unnormalized 2D Gaussian
22:           s ← s + H(x, y)
23:   for x ← −r, ..., r do
24:       for y ← −r, ..., r do
25:           H(x, y) ← (1/s) · H(x, y)           ▷ normalize H
26:   return H
```

11.3.1 Global Thresholding Methods

The thresholding methods covered in Sec. 11.1 are implemented by the following classes:

- `MeanThresholder`, `MedianThresholder` (Sec. 11.1.2),
- `QuantileThresholder` (Alg. 11.1),
- `IsodataThresholder` (Alg. 11.2–11.3),
- `OtsuThresholder` (Alg. 11.4),
- `MaxEntropyThresholder` (Alg. 11.5), and
- `MinErrorThresholder` (Alg. 11.6).

These are sub-classes of the (abstract) class `GlobalThresholder`. The following example demonstrates the typical use of this method for a given `ByteProcessor` object `I`:

```
...
GlobalThresholder thr = new IsodataThresholder();
int q = thr.getThreshold(I);
```

Prog. 11.2
Niblack's thresholder using Gaussian smoothing kernels (ImageJ implementation of Alg. 11.9, part 1).

```
1 package threshold;
2
3 import ij.plugin.filter.GaussianBlur;
4 import ij.plugin.filter.RankFilters;
5 import ij.process.ByteProcessor;
6 import ij.process.FloatProcessor;
7 import imagingbook.pub.threshold.BackgroundMode;
8
9 public abstract class NiblackThresholder extends
      AdaptiveThresholder {
10
11   // parameters for this thresholder
12   public static class Parameters {
13     public int radius = 15;
14     public double kappa =  0.30;
15     public int dMin = 5;
16     public BackgroundMode bgMode = BackgroundMode.DARK;
17   }
18
19   private final Parameters params;  // parameter object
20
21   protected FloatProcessor Imean;   // = μ_G(u,v)
22   protected FloatProcessor Isigma;  // = σ_G(u,v)
23
24   public ByteProcessor getThreshold(ByteProcessor I) {
25     int w = I.getWidth();
26     int h = I.getHeight();
27
28     makeMeanAndVariance(I, params);
29     ByteProcessor Q = new ByteProcessor(w, h);
30
31     final double kappa = params.kappa;
32     final int dMin = params.dMin;
33     final boolean darkBg =
34         (params.bgMode == BackgroundMode.DARK);
35
36     for (int v = 0; v < h; v++) {
37       for (int u = 0; u < w; u++) {
38         double sigma = Isigma.getf(u, v);
39         double mu = Imean.getf(u, v);
40         double diff = kappa * sigma + dMin;
41         int q = (int)
42             Math.rint((darkBg) ? mu + diff : mu - diff);
43         if (q < 0)   q = 0;
44         if (q > 255) q = 255;
45         Q.set(u, v, q);
46       }
47     }
48     return Q;
49   }
50
51   // continues in Prog. 11.3
```

```
52 // continued from Prog. 11.2
53
54   public static class Gauss extends NiblackThresholder {
55
56     protected void makeMeanAndVariance(ByteProcessor I,
       Parameters params) {
57       int width = I.getWidth();
58       int height = I.getHeight();
59
60       Imean = new FloatProcessor(width,height);
61       Isigma = new FloatProcessor(width,height);
62
63       FloatProcessor A = I.convertToFloatProcessor(); // = I
64       FloatProcessor B = I.convertToFloatProcessor(); // = I
65       B.sqr();     // = I^2
66
67       GaussianBlur gb = new GaussianBlur();
68       double sigma = params.radius * 0.6;
69       gb.blurFloat(A, sigma, sigma, 0.002); // = A
70       gb.blurFloat(B, sigma, sigma, 0.002); // = B
71
72       for (int v = 0; v < height; v++) {
73         for (int u = 0; u < width; u++) {
74           float a = A.getf(u, v);
75           float b = B.getf(u, v);
76           float sigmaG =
77               (float) Math.sqrt(b - a*a); // Eq. 11.80
78           Imean.setf(u, v, a);        // = μ_G(u,v)
79           Isigma.setf(u, v, sigmaG); // = σ_G(u,v)
80         }
81       }
82     }
83   } // end of inner class NiblackThresholder.Gauss
84 } // end of class NiblackThresholder
```


Prog. 11.3
Niblack's thresholder using Gaussian smoothing kernels (part 2). The floating-point images AG and BG correspond to the maps A_G (filtered original image) and B_G (filtered squared image) in Alg. 11.9. An instance of the ImageJ class GaussianBlur is created in line 67 and subsequently used to filter both images in lines 69–70. The last argument to the ImageJ method blurFloat (0.002) specifies the accuracy of the Gaussian kernel.

```
if (q > 0) I.threshold(q);
else ...
```

Here `threshold()` is the built-in ImageJ's method defined by class `ImageProcessor`.

11.3.2 Adaptive Thresholding

The techniques described in Sec. 11.2 are implemented by the following classes:

- `BernsenThresholder` (Alg. 11.7),
- `NiblackThresholder` (Alg. 11.8, multiple versions), and
- `SauvolaThresholder` (Eqn. (11.73)).

These are sub-classes of the (abstract) class `AdaptiveThresholder`. The following example demonstrates the typical use of these methods for a given `ByteProcessor` object `I`:

```
...
AdaptiveThresholder thr = new BernsenThresholder();
```

```
ByteProcessor Q = thr.getThreshold(I);
thr.threshold(I, Q);
...
```

The 2D threshold surface is represented by the image `Q`; the method `threshold(I, Q)` is defined by class `AdaptiveThresholder`. Alternatively, the same operation can be performed without making `Q` explicit, as demonstrated by the following code segment:

```
...
// Create and set up a parameter object:
Parameters params = new BernsenThresholder.Parameters();
params.radius = 15;
params.cmin = 15;
params.bgMode = BackgroundMode.DARK;

// Create the thresholder:
AdaptiveThresholder thr = new BernsenThresholder(params);

// Perform the threshold operation:
thr.threshold(I);
...
```

This example also shows how to specify a parameter object (`params`) for the instantiation of the thresholder.

11.4 Summary and Further Reading

The intention of this chapter was to give an overview of established methods for automatic image thresholding. A vast body of relevant literature exists, and thus only a fraction of the proposed techniques could be discussed here. For additional approaches and references, several excellent surveys are available, including [86, 178, 204, 231] and [213].

Given the obvious limitations of global techniques, adaptive thresholding methods have received continued interest and are still a focus of ongoing research. Another popular approach is to calculate an adaptive threshold through image decomposition. In this case, the image is partitioned into (possibly overlapping) tiles, an "optimal" threshold is calculated for each tile and the adaptive threshold is obtained by interpolation between adjacent tiles. Another interesting idea, proposed in [260], is to specify a "threshold surface" by sampling the image at specific points that exhibit a high gradient, with the assumption that these points are at transitions between the background and the foreground. From these irregularly spaced point samples, a smooth surface is interpolated that passes through the sample points. Interpolation between these irregularly spaced point samples is done by solving a Laplacian difference equation to obtain a continuous "potential surface". This is accomplished with the so-called "successive over-relaxation" method, which requires about N scans over an image of size $N \times N$ to converge, so its time complexity is an expensive $\mathcal{O}(N^3)$. A more efficient approach was proposed in [26], which uses a hierarchical, multi-scale algorithm for interpolating the threshold surface. Similarly, a quad-tree representation was used for this purpose in [49]. Another interesting concept is "kriging" [175], which was originally developed for interpolating 2D geological data [190, Ch. 3, Sec. 3.7.4].

In the case of color images, simple thresholding is often applied individually to each color channel and the results are subsequently merged using a suitable logical operation. Transformation to a non-RGB color space (such as HSV or CIELAB) might be helpful for this purpose. For a binarization method aimed specifically at vector-valued images, see [159], for example. Since thresholding can be viewed as a specific form of segmentation, color segmentation methods [50, 53, 85, 216] are also relevant for binarizing color images.

11.5 Exercises

Exercise 11.1. Define a procedure for estimating the minimum and maximum pixel value of an image from its histogram. Threshold the image at the resulting mid-range value (see Eqn. (11.12)). Can anything be said about the size of the resulting partitions?

Exercise 11.2. Define a procedure for estimating the median of an image from its histogram. Threshold the image at the resulting median value (see Eqn. (11.11)) and verify that the foreground and background partitions are of approximately equal size.

Exercise 11.3. The algorithms described in this chapter assume 8-bit grayscale input images (of type `ByteProcessor` in ImageJ). Adopt the current implementations to work with 16-bit integer image (of type `ShortProcessor`). Images of this type may contain pixel values in the range $[0, 2^{16}-1]$ and the `getHistogram()` method returns the histogram as an integer array of length 65536.

Exercise 11.4. Implement simple thresholding for RGB color images by thresholding each (scalar-valued) color channel individually and then merging the results by performing a pixel-wise AND operation. Compare the results to those obtained by thresholding the corresponding grayscale (luminance) images.

Exercise 11.5. Re-implement the Bernsen and/or Niblack thresholder (classes `BernsenThresholder` and `NiblackThresholder`) using integral images (see Ch. 3, Sec. 3.8) for efficiently calculating the required local mean and variance of the input image over a rectangular support region R.

12

Color Images

Color images are involved in every aspect of our lives, where they play an important role in everyday activities such as television, photography, and printing. Color perception is a fascinating and complicated phenomenon that has occupied the interests of scientists, psychologists, philosophers, and artists for hundreds of years [211, 217]. In this chapter, we focus on those technical aspects of color that are most important for working with digital color images. Our emphasis will be on understanding the various representations of color and correctly utilizing them when programming. Additional color-related issues, such as colorimetric color spaces, color quantization, and color filters, are covered in subsequent chapters.

12.1 RGB Color Images

The RGB color schema encodes colors as combinations of the three primary colors: red, green, and blue (R, G, B). This scheme is widely used for transmission, representation, and storage of color images on both analog devices such as television sets and digital devices such as computers, digital cameras, and scanners. For this reason, many image-processing and graphics programs use the RGB schema as their internal representation for color images, and most language libraries, including Java's imaging APIs, use it as their standard image representation.

RGB is an *additive* color system, which means that all colors start with black and are created by adding the primary colors. You can think of color formation in this system as occurring in a dark room where you can overlay three beams of light—one red, one green, and one blue—on a sheet of white paper. To create different colors, you would modify the intensity of each of these beams independently. The distinct intensity of each primary color beam controls the shade and brightness of the resulting color. The colors gray and white are created by mixing the three primary color beams at the same intensity. A similar operation occurs on the screen of a color television or

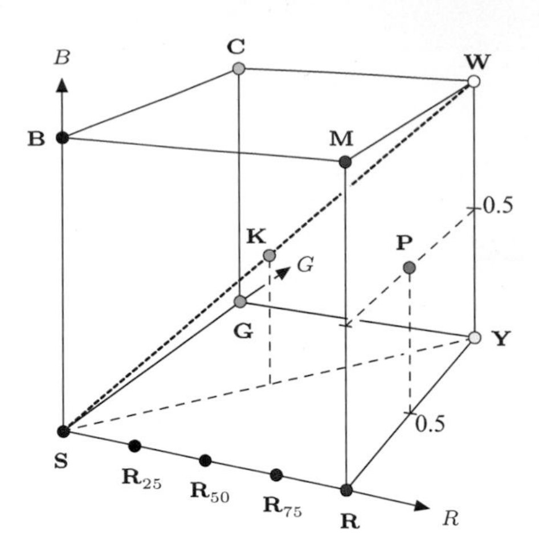

Pt.	Color	RGB values R	G	B
S	Black	0.00	0.00	0.00
R	Red	1.00	0.00	0.00
Y	Yellow	1.00	1.00	0.00
G	Green	0.00	1.00	0.00
C	Cyan	0.00	1.00	1.00
B	Blue	0.00	0.00	1.00
M	Magenta	1.00	0.00	1.00
W	White	1.00	1.00	1.00
K	50% Gray	0.50	0.50	0.50
$\mathbf{R}_{75}$	75% Red	0.75	0.00	0.00
$\mathbf{R}_{50}$	50% Red	0.50	0.00	0.00
$\mathbf{R}_{25}$	25% Red	0.25	0.00	0.00
P	Pink	1.00	0.50	0.50

Fig. 12.1 Representation of the RGB color space as a 3D unit cube. The primary colors red (R), green (G), and blue (B) form the coordinate system. The "pure" red color (**R**), green (**G**), blue (**B**), cyan (**C**), magenta (**M**), and yellow (**Y**) lie on the vertices of the color cube. All the shades of gray, of which **K** is an example, lie on the diagonal between black **S** and white **W**.

CRT[1]-based computer monitor, where tiny, close-lying dots of red, green, and blue phosphorous are simultaneously excited by a stream of electrons to distinct energy levels (intensities), creating a seemingly continuous color image.

The RGB color space can be visualized as a 3D unit cube in which the three primary colors form the coordinate axis. The RGB values are positive and lie in the range $[0, C_{\max}]$; for most digital images, $C_{\max} = 255$. Every possible color $\mathbf{C}_i$ corresponds to a point within the RGB color cube of the form

$$\mathbf{C}_i = (R_i, G_i, B_i),$$

where $0 \leq R_i, G_i, B_i \leq C_{\max}$. RGB values are often normalized to the interval $[0, 1]$ so that the resulting color space forms a unit cube (Fig. 12.1). The point $\mathbf{S} = (0, 0, 0)$ corresponds to the color black, $\mathbf{W} = (1, 1, 1)$ corresponds to the color white, and all the points lying on the diagonal between **S** and **W** are shades of gray created from equal color components $R = G = B$.

Figure 12.2 shows a color test image and its corresponding RGB color components, displayed here as intensity images. We will refer to this image in a number of examples that follow in this chapter.

RGB is a very simple color system, and as demonstrated in Sec. 12.2, a basic knowledge of it is often sufficient for processing color images or transforming them into other color spaces. At this point, we will not be able to determine what color a particular RGB pixel corresponds to in the real world, or even what the primary colors red, green, and blue truly mean in a physical (i.e., colorimetric) sense. For now we rely on our intuitive understanding of color and will address colorimetry and color spaces later in the context of the CIE color system (see Ch. 14).

12.1.1 Structure of Color Images

Color images are represented in the same way as grayscale images, by using an array of pixels in which different models are used to order the

[1] Cathode ray tube.

Fig. 12.2
A color image and its corresponding RGB channels. The fruits depicted are mainly yellow and red and therefore have high values in the R and G channels. In these regions, the B content is correspondingly lower (represented here by darker gray values) except for the bright highlights on the apple, where the color changes gradually to white. The tabletop in the foreground is purple and therefore displays correspondingly higher values in its B channel.

individual color components. In the next sections we will examine the difference between *true color* images, which utilize colors uniformly selected from the entire color space, and so-called *palleted* or *indexed* images, in which only a select set of distinct colors are used. Deciding which type of image to use depends on the requirements of the application. Farbbilder werden üblicherweise, genau wie Grauwertbilder, als Arrays von Pixeln dargestellt, wobei unterschiedliche Modelle für die Anordnung der einzelnen Farbkomponenten verwendet werden. Zunächst ist zu unterscheiden zwischen *Vollfarbenbildern*, die den gesamten Farbraum gleichförmig abdecken können, und so genannten *Paletten-* oder *Indexbildern*, die nur eine beschränkte Zahl unterschiedlicher Farben verwenden. Beide Bildtypen werden in der Praxis häufig eingesetzt.

True color images

A pixel in a true color image can represent any color in its color space, as long as it falls within the (discrete) range of its individual color components. True color images are appropriate when the image contains many colors with subtle differences, as occurs in digital photography and photo-realistic computer graphics. Next we look at two methods of ordering the color components in true color images: *component ordering* and *packed ordering*.

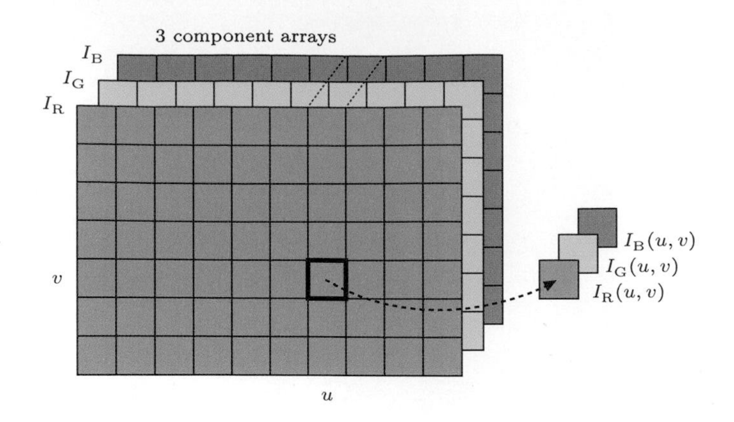

Fig. 12.3
RGB color image in component ordering. The three color components are laid out in separate arrays I_R, I_G, I_B of the same size.

Component ordering

In *component ordering* (also referred to as *planar ordering*) the color components are laid out in separate arrays of identical dimensions. In this case, the color image

$$\boldsymbol{I}_{\mathrm{comp}} = (I_{\mathrm{R}}, I_{\mathrm{G}}, I_{\mathrm{B}}) \tag{12.1}$$

can be thought of as a vector of related intensity images I_R, I_G, and I_B (Fig. 12.3), and the RGB values of the color image I at position (u, v) are obtained by accessing the three component images in the form

$$\begin{pmatrix} R(u,v) \\ G(u,v) \\ B(u,v) \end{pmatrix} = \begin{pmatrix} I_{\mathrm{R}}(u,v) \\ I_{\mathrm{G}}(u,v) \\ I_{\mathrm{B}}(u,v) \end{pmatrix}. \tag{12.2}$$

Packed ordering

In *packed ordering*, the component values that represent the color of a particular pixel are packed together into a single element of the image array (Fig. 12.4) such that

$$\boldsymbol{I}_{\mathrm{pack}}(u,v) = (R, G, B). \tag{12.3}$$

The RGB value of a packed image I at the location (u, v) is obtained by accessing the individual components of the color pixel as

$$\begin{pmatrix} R(u,v) \\ G(u,v) \\ B(u,v) \end{pmatrix} = \begin{pmatrix} \mathsf{Red}(\boldsymbol{I}_{\mathrm{pack}}(u,v)) \\ \mathsf{Green}(\boldsymbol{I}_{\mathrm{pack}}(u,v)) \\ \mathsf{Blue}(\boldsymbol{I}_{\mathrm{pack}}(u,v)) \end{pmatrix}. \tag{12.4}$$

The access functions Red(), Green(), Blue(), will depend on the specific implementation used for encoding the color pixels.

Indexed images

Indexed images permit only a limited number of distinct colors and therefore are used mostly for illustrations and graphics that contain large regions of the same color. Often these types of images are stored in indexed GIF or PNG files for use on the Web. In these indexed

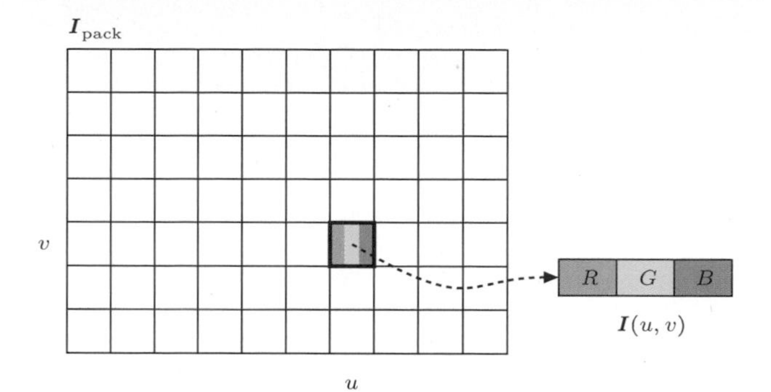

Fig. 12.4
RGB-color image using packed ordering. The three color components R, G, and B are placed together in a single array element.

images, the pixel array does not contain color or brightness data but instead consists of integer numbers k that are used to index into a *color table* or "palette"

$$\mathsf{P} = (\mathsf{P}_r, \mathsf{P}_g, \mathsf{P}_g) : [0, Q-1]^3 \mapsto [0, K-1]. \tag{12.5}$$

Here Q denotes the size of the color table, equal to the maximum number of distinct image colors (typically $Q = 2, \ldots, 256$). K is the number of distinct component values (typ. $K = 256$). This table contains a specific color vector $\mathsf{P}(q) = (R_q, G_q, B_q)$ for every color index $q = 0, \ldots, Q-1$ (see Fig. 12.5). The RGB component values of an indexed image I_{idx} at position (u, v) are obtained as

$$\begin{pmatrix} R(u,v) \\ G(u,v) \\ B(u,v) \end{pmatrix} = \begin{pmatrix} R_q \\ G_q \\ B_q \end{pmatrix} = \begin{pmatrix} \mathsf{P}_r(q) \\ \mathsf{P}_g(q) \\ \mathsf{P}_b(q) \end{pmatrix}, \tag{12.6}$$

with the index $q = I_{idx}(u, v)$. To allow proper reconstruction, the color table P must of course be stored and/or transmitted along with the indexed image.

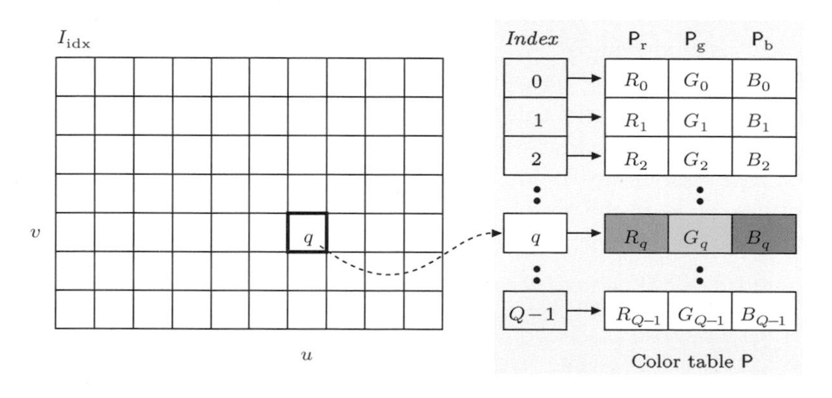

Fig. 12.5
RGB indexed image. The image array I_{idx} itself does not contain any color component values. Instead, each cell contains an index $q \in [0, Q-1]$. into the associated color table ("palette") P. The actual color value is specified by the table entry $\mathsf{P}_q = (R_q, G_q, B_q)$.

During the transformation from a true color image to an indexed image (e.g., from a JPEG image to a GIF image), the problem of optimal color reduction, or *color quantization*, arises. Color quantization is the process of determining an optimal color table and then mapping it to the original colors. This process is described in detail in Chapter 13.

12.1.2 Color Images in ImageJ

ImageJ provides two simple types of color images:

- RGB full-color images (24-bit "RGB color").
- Indexed images ("8-bit color").

RGB true color images

RGB color images in ImageJ use a packed order (see Sec. 12.1.1), where each color pixel is represented by a 32-bit `int` value. As Fig. 12.6 illustrates, 8 bits are used to represent each of the RGB components, which limits the range of the individual components to 0–255. The remaining 8 bits are reserved for the transparency,[2] or *alpha* (α), component. This is also the usual ordering in Java[3] for RGB color images.

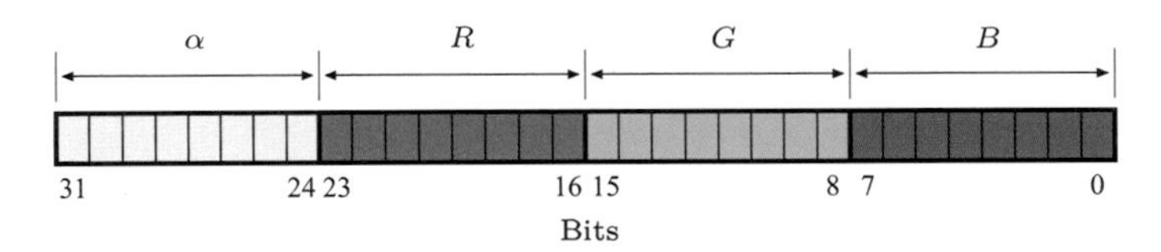

Fig. 12.6
Structure of a packed RGB color pixel in Java. Within a 32-bit `int`, 8 bits are allocated, in the following order, for each of the color components R, G, B, and the transparency value α (unused in ImageJ).

Accessing RGB pixel values

RGB color images are represented by an array of pixels, the elements of which are standard Java `int`s. To disassemble the packed `int` value into the three color components, you apply the appropriate bitwise shifting and masking operations. In the following example, we assume that the image processor `ip` (of type `ColorProcessor`) contains an RGB color image:

```
int c = ip.getPixel(u,v);          // a packed RGB color pixel
int r = (c & 0xff0000) >> 16;  // red component
int g = (c & 0x00ff00) >> 8;   // green component
int b = (c & 0x0000ff);            // blue component
```

In this example, each of the RGB components of the packed pixel `c` are isolated using a bitwise AND operation (`&`) with an appropriate bit mask (following convention, bit masks are given in hexadecimal[4] notation), and afterwards the extracted bits are shifted right by 16 (for R) or 8 (for G) bit positions (see Fig. 12.7).

The "assembly" of an RGB pixel from separate R, G, and B values works in the opposite direction using the bitwise OR operator (`|`) and shifting the bits left (`<<`):

```
int r = 169;   // red component
int g = 212;   // green component
int b = 17;    // blue component
int c = ((r & 0xff) << 16) | ((g & 0xff) << 8) | b & 0xff;
ip.putPixel(u, v, c);
```

[2] The transparency value α (alpha) represents the ability to see through a color pixel onto the background. At this time, the α channel is unused in ImageJ.

[3] Java Advanced Window Toolkit (AWT).

[4] The mask `0xff0000` is of type `int` and represents the 32-bit binary pattern `00000000111111110000000000000000`.

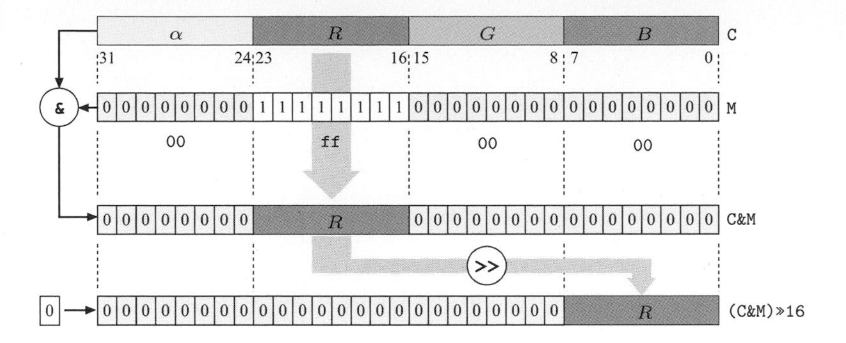

Fig. 12.7
Decomposition of a 32-bit RGB color pixel using bit operations. The R component (bits 16–23) of the RGB pixels `C` (above) is isolated using a bitwise AND operation (`&`) together with a bit mask `M = 0xff0000`. All bits except the R component are set to the value 0, while the bit pattern within the R component remains unchanged. This bit pattern is subsequently shifted 16 positions to the right (`>>`), so that the R component is moved into the lowest 8 bits and its value lies in the range of $0, \ldots, 255$. During the shift operation, zeros are filled in from the left.

```
1 // File Brighten_RGB_1.java
2 import ij.ImagePlus;
3 import ij.plugin.filter.PlugInFilter;
4 import ij.process.ImageProcessor;
5
6 public class Brighten_RGB_1 implements PlugInFilter {
7
8   public int setup(String arg, ImagePlus imp) {
9     return DOES_RGB;   // this plugin works on RGB images
10  }
11
12  public void run(ImageProcessor ip) {
13    int[] pixels = (int[]) ip.getPixels();
14
15    for (int i = 0; i < pixels.length; i++) {
16      int c = pixels[i];
17      // split color pixel into rgb-components:
18      int r = (c & 0xff0000) >> 16;
19      int g = (c & 0x00ff00) >> 8;
20      int b = (c & 0x0000ff);
21      // modify colors:
22      r = r + 10; if (r > 255) r = 255;
23      g = g + 10; if (g > 255) g = 255;
24      b = b + 10; if (b > 255) b = 255;
25      // reassemble color pixel and insert into pixel array:
26      pixels[i]
27        = ((r & 0xff)<<16) | ((g & 0xff)<<8) | b & 0xff;
28    }
29  }
30 }
```

Prog. 12.1
Processing RGB color data with the use of bit operations (ImageJ plugin, version 1). This plugin increases the values of all three color components by 10 units. It demonstrates the use of direct access to the pixel array (line 16), the separation of color components using bit operations (lines 18–20), and the reassembly of color pixels after modification (line 27). The value `DOES_RGB` (defined in the interface `PlugInFilter`) returned by the `setup()` method indicates that this plugin is designed to work on RGB formatted true color images (line 9).

Masking the component values with `0xff` works in this case because, except for the bits in positions $0, \ldots, 7$ (values in the range 0–255), all the other bits are already set to zero. A complete example of manipulating an RGB color image using bit operations is presented in Prog. 12.1. Instead of accessing color pixels using ImageJ's access functions, these programs directly access the pixel array for increased efficiency

The ImageJ class `ColorProcessor` provides an easy to use alternative which returns the separated RGB components (as an `int` array

Prog. 12.2
Working with RGB color images without bit operations (ImageJ plugin, version 2). This plugin increases the values of all three color components by 10 units using the access methods `getPixel(int, int, int[])` and `putPixel(int, int, int[])` from the class `ColorProcessor` (lines 21 and 25, respectively). Exection time is approximately four times higher than that of version 1 (Prog. 12.1) because of the additional method calls.

```
1  // File Brighten_RGB_2.java
2  import ij.ImagePlus;
3  import ij.plugin.filter.PlugInFilter;
4  import ij.process.ColorProcessor;
5  import ij.process.ImageProcessor;
6
7  public class Brighten_RGB_2 implements PlugInFilter {
8    static final int R = 0, G = 1, B = 2; // component indices
9
10   public int setup(String arg, ImagePlus imp) {
11     return DOES_RGB;  // this plugin works on RGB images
12   }
13
14   public void run(ImageProcessor ip) {
15     // typecast the image to ColorProcessor (no duplication):
16     ColorProcessor cp = (ColorProcessor) ip;
17     int[] RGB = new int[3];
18
19     for (int v = 0; v < cp.getHeight(); v++) {
20       for (int u = 0; u < cp.getWidth(); u++) {
21         cp.getPixel(u, v, RGB);
22         RGB[R] = Math.min(RGB[R] + 10, 255);  // add 10 and
23         RGB[G] = Math.min(RGB[G] + 10, 255);  // limit to 255
24         RGB[B] = Math.min(RGB[B] + 10, 255);
25         cp.putPixel(u, v, RGB);
26       }
27     }
28   }
29 }
```

with three elements). In the following example, which demonstrates its use, `ip` is of type `ColorProcessor`:

```
int[] RGB = new int[3];
...
ip.getPixel(u, v, RGB); // modifies RGB
int r = RGB[0];
int g = RGB[1];
int b = RGB[2];
...
ip.putPixel(u, v, RGB);
```

A more detailed and complete example is shown by the simple plugin in Prog. 12.2, which increases the value of all three color components of an RGB image by 10 units. Notice that the plugin limits the resulting component values to 255, because the `putPixel()` method only uses the lowest 8 bits of each component and does not test if the value passed in is out of the permitted 0–255 range. Without this test, arithmetic overflow errors can occur. The price for using this access method, instead of direct array access, is a noticeably longer running time (approximately a factor of 4 when compared to the version in Prog. 12.1).

Opening and saving RGB images

ImageJ supports the following types of image formats for RGB true color images:

- **TIFF** (uncompressed only): 3 × 8-bit RGB. TIFF color images with 16-bit depth are opened as an image stack consisting of three 16-bit intensity images.
- **BMP**, **JPEG**: 3 × 8-bit RGB.
- **PNG**: 3 × 8-bit RGB.
- **RAW**: using the ImageJ menu File ▷ Import ▷ Raw, RGB images can be opened whose format is not directly supported by ImageJ. It is then possible to select different arrangements of the color components.

Creating RGB color images

The simplest way to create a new RGB image using ImageJ is to use an instance of the class `ColorProcessor`, as the following example demonstrates:

```
int w = 640, h = 480;
ColorProcessor cp = new ColorProcessor(w, h);
(new ImagePlus("My New Color Image", cp)).show();
```

When needed, the color image can be displayed by creating an instance of the class `ImagePlus` and calling its `show()` method. Since `cip` is of type `ColorProcessor`, the resulting `ImagePlus` object `cimg` is also a color image.

Indexed color images

The structure of an indexed image in ImageJ is given in Fig. 12.5, where each element of the index array is 8 bits and therefore can represent a maximum of 256 different colors. When programming, indexed images are similar to grayscale images, as both make use of a color table to determine the actual color of the pixel. Indexed images differ from grayscale images only in that the contents of the color table are not intensity values but RGB values.

Opening and saving indexed images

ImageJ supports the indexed images in GIF, PNG, BMP, and TIFF format with index values of 1–8 bits (i.e., 2–256 distinct colors) and 3 × 8-bit color values.

Processing indexed images

The indexed format is mostly used as a space-saving means of image storage and is not directly useful as a processing format since an index value in the pixel array is arbitrarily related to the actual color, found in the color table, that it represents. When working with indexed images it usually makes no sense to base any numerical interpretations on the pixel values or to apply any filter operations designed for 8-bit intensity images. Figure 12.8 illustrates an example of applying a Gaussian filter and a median filter to the pixels of an indexed image. Since there is no meaningful quantitative relation between the actual colors and the index values, the results are erratic.

Fig. 12.8
Improper application of smoothing filters to an indexed color image. Indexed image with 16 colors (a) and results of applying a linear smoothing filter (b) and a 3×3 median filter (c) to the pixel array (that is, the *index* values). The application of a linear filter makes no sense, of course, since no meaningful relation exists between the index values in the pixel array and the actual image intensities. While the median filter (c) delivers seemingly plausible results in this case, its use is also inadmissible because no meaningful ordering relation exists between the index values.

Note that even the use of the median filter is inadmissible because no ordering relation exists between the index values. Thus, with few exceptions, ImageJ functions do not permit the application of such operations to indexed images. Generally, when processing an indexed image, you first convert it into a true color RGB image and then after processing convert it back into an indexed image.

When an ImageJ plugin is supposed to process indexed images, its `setup()` method should return the `DOES_8C` ("8-bit color") flag. The plugin in Prog. 12.3 shows how to increase the intensity of the three color components of an indexed image by 10 units (analogously to Progs. 12.1 and 12.2 for RGB images). Notice how in indexed images only the palette is modified and the original pixel data, the index values, remain the same. The color table of `ImageProcessor` is accessible through a `ColorModel`[5] object, which can be read using the method `getColorModel()` and modified using `setColorModel()`.

The `ColorModel` object for indexed images (as well as 8-bit grayscale images) is a subtype of `IndexColorModel`, which contains three color tables (*maps*) representing the red, green, and blue components as separate `byte` arrays. The size of these tables $(2, \ldots, 256)$ can be determined by calling the method `getMapSize()`. Note that the elements of the palette should be interpreted as *unsigned* bytes with values ranging from $0, \ldots, 255$. Just as with grayscale pixel values, during the conversion to `int` values, these color component values must also be bitwise masked with `0xff` as shown in Prog. 12.3 (lines 30–32).

As a further example, Prog. 12.4 shows how to convert an indexed image to a true color RGB image of type `ColorProcessor`. Conversion in this direction poses no problems because the RGB component values for a particular pixel are simply taken from the corresponding color table entry, as described by Eqn. (12.6). On the other hand,

[5] Defined in the standard Java class `java.awt.image.ColorModel`.

Prog. 12.3
Working with indexed images (ImageJ plugin). This plugin increases the brightness of an image by 10 units by modifying the image's color table (palette). The actual values in the pixel array, which are indices into the palette, are not changed.

```
1 // File Brighten_Index_Image.java
2
3 import ij.ImagePlus;
4 import ij.plugin.filter.PlugInFilter;
5 import ij.process.ImageProcessor;
6 import java.awt.image.IndexColorModel;
7
8 public class Brighten_Index_Image implements PlugInFilter {
9
10   public int setup(String arg, ImagePlus imp) {
11     return DOES_8C; // this plugin works on indexed color images
12   }
13
14   public void run(ImageProcessor ip) {
15     IndexColorModel icm =
16       (IndexColorModel) ip.getColorModel();
17     int pixBits = icm.getPixelSize();
18     int nColors = icm.getMapSize();
19
20     //retrieve the current lookup tables (maps) for R, G, B:
21     byte[] pRed = new byte[nColors];
22     byte[] pGrn = new byte[nColors];
23     byte[] pBlu = new byte[nColors];
24     icm.getReds(pRed);
25     icm.getGreens(pGrn);
26     icm.getBlues(pBlu);
27
28     //modify the lookup tables:
29     for (int idx = 0; idx < nColors; idx++){
30       int r = 0xff & pRed[idx]; // mask to treat as unsigned byte
31       int g = 0xff & pGrn[idx];
32       int b = 0xff & pBlu[idx];
33       pRed[idx] = (byte) Math.min(r + 10, 255);
34       pGrn[idx] = (byte) Math.min(g + 10, 255);
35       pBlu[idx] = (byte) Math.min(b + 10, 255);
36     }
37     //create a new color model and apply to the image:
38     IndexColorModel icm2 =
39       new IndexColorModel(pixBits,nColors,pRed,pGrn,pBlu);
40     ip.setColorModel(icm2);
41   }
42 }
```

conversion in the other direction requires *quantization* of the RGB color space and is as a rule more difficult and involved (see Ch. 13 for details). In practice, most applications make use of existing conversion methods such as those provided by the ImageJ API.

Creating indexed images

In ImageJ, no special method is provided for the creation of indexed images, so in almost all cases they are generated by converting an existing image. The following method demonstrates how to directly create an indexed image if required:

```
ByteProcessor makeIndexColorImage(int w, int h, int nColors) {
```

Prog. 12.4
Converting an indexed image to a true color RGB image (ImageJ plugin).

```
1 // File Index_To_Rgb.java
2
3 import ij.ImagePlus;
4 import ij.plugin.filter.PlugInFilter;
5 import ij.process.ColorProcessor;
6 import ij.process.ImageProcessor;
7 import java.awt.image.IndexColorModel;
8
9 public class Index_To_Rgb implements PlugInFilter {
10   static final int R = 0, G = 1, B = 2;
11   ImagePlus imp;
12
13   public int setup(String arg, ImagePlus imp) {
14     this.imp = imp;
15     return DOES_8C + NO_CHANGES;   // does not alter original image
16   }
17
18   public void run(ImageProcessor ip) {
19     int w = ip.getWidth();
20     int h = ip.getHeight();
21
22     // retrieve the lookup tables (maps) for R, G, B:
23     IndexColorModel icm =
24         (IndexColorModel) ip.getColorModel();
25     int nColors = icm.getMapSize();
26     byte[] pRed = new byte[nColors];
27     byte[] pGrn = new byte[nColors];
28     byte[] pBlu = new byte[nColors];
29     icm.getReds(pRed);
30     icm.getGreens(pGrn);
31     icm.getBlues(pBlu);
32
33     // create a new 24-bit RGB image:
34     ColorProcessor cp = new ColorProcessor(w, h);
35     int[] RGB = new int[3];
36     for (int v = 0; v < h; v++) {
37       for (int u = 0; u < w; u++) {
38         int idx = ip.getPixel(u, v);
39         RGB[R] = 0xFF & pRed[idx];
40         RGB[G] = 0xFF & pGrn[idx];
41         RGB[B] = 0xFF & pBlu[idx];
42         cp.putPixel(u, v, RGB);
43       }
44     }
45     ImagePlus cwin =
46         new ImagePlus(imp.getShortTitle() + " (RGB)", cp);
47     cwin.show();
48   }
49 }
```

```
byte[] rMap = new byte[nColors]; // red, green, blue color maps
byte[] gMap = new byte[nColors];
byte[] bMap = new byte[nColors];
// color maps need to be filled here
byte[] pixels = new byte[w * h];
```

```
  IndexColorModel cm
    = new IndexColorModel(8, nColors, rMap, gMap, bMap);
  return new ByteProcessor(w, h, pixels, cm);
}
```

The parameter **nColors** defines the number of colors (and thus the size of the palette) and must be a value in the range of $2, \ldots, 256$. To use the above template, you would complete it with code that filled the three **byte** arrays for the RGB components (**rMap**, **gMap**, **bMap**) and the index array (**pixels**) with the appropriate values.

Transparency

Transparency is one of the reasons indexed images are often used for Web graphics. In an indexed image, it is possible to define one of the index values so that it is displayed in a transparent manner and at selected image locations the background beneath the image shows through. In Java this can be controlled when creating the image's color model (IndexColorModel). As an example, to make color index 2 in Prog. 12.3 transparent, line 39 would need to be modified as follows:

```
int tidx = 2; // index of transparent color
IndexColorModel icm2 =
  new IndexColorModel(pixBits,nColors,pRed,pGrn,pBlu,tidx);
ip.setColorModel(icm2);
```

At this time, however, ImageJ does not support the transparency property; it is not considered during display, and it is lost when the image is saved.

12.2 Color Spaces and Color Conversion

The RGB color system is well-suited for use in programming, as it is simple to manipulate and maps directly to the typical display hardware. When modifying colors within the RGB space, it is important to remember that the *metric*, or *measured distance* within this color space, does not proportionally correspond to our perception of color (e.g., doubling the value of the red component does not necessarily result in a color which appears to be twice as red). In general, in this space, modifying different color points by the same amount can cause very different changes in color. In addition, brightness changes in the RGB color space are also perceived as nonlinear.

Since changing any component modifies color tone, saturation, and brightness all at once, color selection in RGB space is difficult and quite non-intuitive. Color selection is more intuitive in other color spaces, such as the HSV space (see Sec. 12.2.3), since perceptual color features, such as saturation, are represented individually and can be modified independently. Alternatives to the RGB color space are also used in applications such as the automatic separation of objects from a colored background (the *blue box* technique in television), encoding television signals for transmission, or in printing, and are thus also relevant in digital image processing.

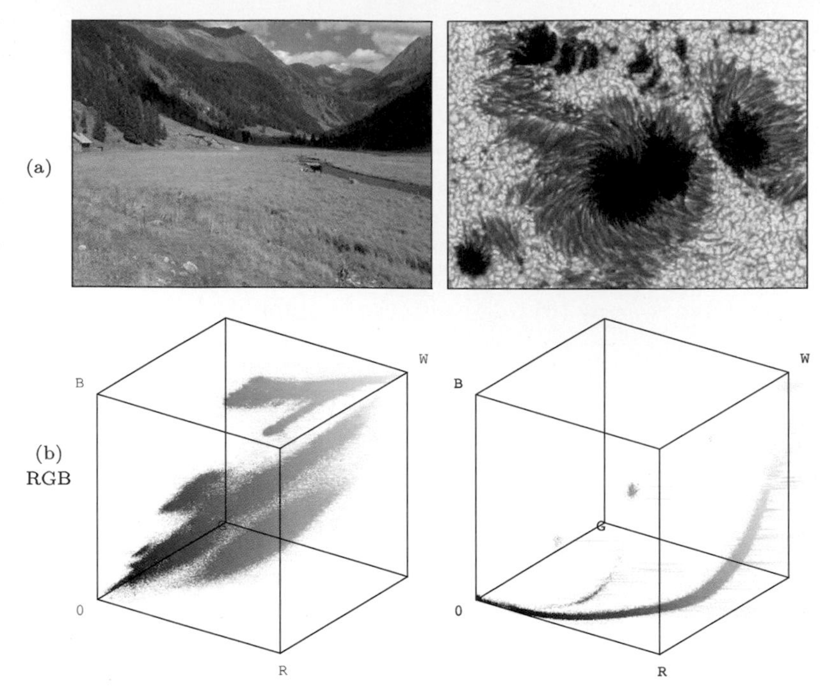

Fig. 12.9
Examples of the color distribution of natural images. Original images: landscape photograph with dominant green and blue components and sun-spot image with rich red and yellow components (a). Distribution of image colors in RGB-space (b).

Figure 12.9 shows the distribution of the colors from natural images in the RGB color space. The first half of this section introduces alternative color spaces and the methods of converting between them, and later discusses the choices that need to be made to correctly convert a color image to grayscale. In addition to the classical color systems most widely used in programming, precise reference systems, such as the CIEXYZ color space, gain increasing importance in practical color processing.

12.2.1 Conversion to Grayscale

The conversion of an RGB color image to a grayscale image proceeds by computing the equivalent gray or *luminance* value Y for each RGB pixel. In its simplest form, Y could be computed as the average

$$Y = \mathrm{Avg}(R, G, B) = \frac{R + G + B}{3} \tag{12.7}$$

of the three color components R, G, and B. Since we perceive both red and green as being substantially brighter than blue, the resulting image will appear to be too dark in the red and green areas and too bright in the blue ones. Therefore, a weighted sum of the color components is typically used for calculating the equivalent brightness or *luminance* in the form

$$Y = \mathrm{Lum}(R, G, B) = w_{\mathrm{R}} \cdot R + w_{\mathrm{G}} \cdot G + w_{\mathrm{B}} \cdot B \tag{12.8}$$

The weights most often used were originally developed for encoding analog color television signals (see Sec. 12.2.4) are

$$w_{\mathrm{R}} = 0.299, \qquad w_{\mathrm{G}} = 0.587, \qquad w_{\mathrm{B}} = 0.114, \tag{12.9}$$

and the weights recommended in ITU-BT.709 [122] for digital color encoding are

$$w_{\mathrm{R}} = 0.2126, \qquad w_{\mathrm{G}} = 0.7152, \qquad w_{\mathrm{B}} = 0.0722. \tag{12.10}$$

If each color component is assigned the same weight, as in Eqn. (12.7), this is of course just a special case of Eqn. (12.8).

Note that, although these weights were developed for use with TV signals, they are optimized for *linear* RGB component values, that is, signals with no gamma correction. In many practical situations, however, the RGB components are actually *nonlinear*, particularly when we work with sRGB images (see Ch. 14, Sec. 14.4). In this case, the RGB components must first be linearized to obtain the correct luminance values with the aforementioned weights.

In some color systems, instead of a weighted sum of the RGB color components, a nonlinear brightness function, for example the *value* V in HSV (Eqn. (12.14) in Sec. 12.2.3) or the *luminance* L in HLS (Eqn. (12.25)), is used as the intensity value Y.

Hueless (gray) color images

An RGB image is hueless or gray when the RGB components of each pixel $\boldsymbol{I}(u,v) = (R,G,B)$ are the same; that is, if

$$R = G = B.$$

Therefore, to completely remove the color from an RGB image, simply replace the R, G, and B component of each pixel with the equivalent gray value Y,

$$\begin{pmatrix} R_{\mathrm{gray}} \\ G_{\mathrm{gray}} \\ B_{\mathrm{gray}} \end{pmatrix} = \begin{pmatrix} Y \\ Y \\ Y \end{pmatrix}, \tag{12.11}$$

by using $Y = \mathrm{Lum}(R,G,B)$ from Eqns. (12.8) and (12.9), for example. The resulting grayscale image should have the same subjective brightness as the original color image.

Grayscale conversion in ImageJ

In ImageJ, the simplest way to convert an RGB color image (of type `ColorProcessor`) into an 8-bit grayscale image is to use the `ImageProcessor`-method

```
convertToByteProcessor(),
```

which returns a new image of type `ByteProcessor`. ImageJ uses the default weights $w_R = w_G = w_B = \frac{1}{3}$ (as in Eqn. (12.7)) for the RGB components, or alternatively $w_R = 0.299$, $w_G = 0.587$, $w_B = 0.114$ (as in Eqn. (12.9)) if the "Weighted RGB Conversions" option is selected in the Edit ▷ Options ▷ Conversions dialog. Arbitrary component weights can be specified for subsequent conversion operations through the static `ColorProcessor` method

```
setRGBWeights(double wR, double wG, double wB).
```

Similarly, the static method `getWeightingFactors()` of class `ColorProcessor` can be used to retrieve the current component weights as a 3-element `double`-array. Note that no *linearization* is performed on the color components, which should be considered when working with (nonlinear) sRGB colors (see Ch. 14, Sec. 14.4 for details).

12.2.2 Desaturating RGB Color Images

Desaturation is the uniform reduction of the amount of color in an RGB image in a *continuous* manner. It is done by replacing each RGB pixel by a desaturated color obtained by linear interpolation between the pixel's original color and the corresponding (Y, Y, Y) gray point in the RGB space, that is,

$$\begin{pmatrix} R_{\mathrm{desat}} \\ G_{\mathrm{desat}} \\ B_{\mathrm{desat}} \end{pmatrix} = \begin{pmatrix} Y \\ Y \\ Y \end{pmatrix} + s \cdot \begin{pmatrix} R - Y \\ G - Y \\ B - Y \end{pmatrix}, \tag{12.12}$$

again with $Y = \mathrm{Lum}(R, G, B)$ from Eqns. (12.8) and (12.9), where the factor $s \in [0, 1]$ controls the remaining amount of color saturation (Fig. 12.10). A value of $s = 0$ completely eliminates all color, resulting in a true grayscale image, and with $s = 1$ the color values will be unchanged. In Prog. 12.5, continuous desaturation as defined in Eqn. (12.12) is implemented as an ImageJ plugin.

In color spaces where color saturation is represented by an explicit component (such as HSV and HLS, for example), desaturation is of course much easier to accomplish (by simply reducing the saturation value to zero).

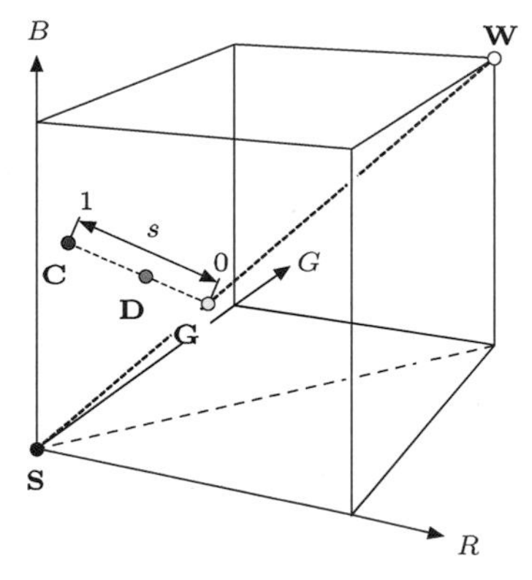

Fig. 12.10 Desaturation in RGB space: original color point $\mathbf{C} = (R, G, B)$, its corresponding gray point $\mathbf{G} = (Y, Y, Y)$, and the desaturated color point $\mathbf{D}$. Saturation is controlled by the factor s.

12.2.3 HSV/HSB and HLS Color Spaces

In the **HSV** color space, colors are specified by the components *hue*, *saturation*, and *value*. Often, such as in Adobe products and the Java API, the **HSV** space is called **HSB**. While the acronym is different (in this case B = *brightness*),[6] it denotes the same color space. The HSV color space is traditionally shown as an upside-down, six-sided pyramid (Fig. 12.11(a)), where the vertical axis represents the V (brightness) value, the horizontal distance from the axis the S (saturation) value, and the angle the H (hue) value. The black point is at the tip of the pyramid and the white point lies in the center of the base. The three primary colors *red*, *green*, and *blue* and the pairwise mixed colors *yellow*, *cyan*, and *magenta* are the corner points of the

[6] Sometimes the HSV space is also referred to as the "HSI" space, where "I" stands for *intensity*.

```
1 // File Desaturate_Rgb.java
2
3 import ij.ImagePlus;
4 import ij.plugin.filter.PlugInFilter;
5 import ij.process.ImageProcessor;
6
7 public class Desaturate_Rgb implements PlugInFilter {
8   double s = 0.3; // color saturation value
9
10   public int setup(String arg, ImagePlus imp) {
11     return DOES_RGB;
12   }
13
14   public void run(ImageProcessor ip) {
15     //iterate over all pixels:
16     for (int v = 0; v < ip.getHeight(); v++) {
17       for (int u = 0; u < ip.getWidth(); u++) {
18
19         // get int-packed color pixel:
20         int c = ip.get(u, v);
21
22         //extract RGB components from color pixel
23         int r = (c & 0xff0000) >> 16;
24         int g = (c & 0x00ff00) >> 8;
25         int b = (c & 0x0000ff);
26
27         // compute equiv. gray value:
28         double y = 0.299 * r + 0.587 * g + 0.114 * b;
29
30         // linear interpolate (yyy) ↔ (rgb):
31         r = (int) (y + s * (r - y));
32         g = (int) (y + s * (g - y));
33         b = (int) (y + s * (b - y));
34
35         // reassemble the color pixel:
36         c = ((r & 0xff)<<16) | ((g & 0xff)<<8) | b & 0xff;
37         ip.set(u, v, c);
38       }
39     }
40   }
41
42 }
```

Prog. 12.5
Continuous desaturation of an RGB color image (ImageJ plugin). The amount of color saturation is controlled by the variable s defined in line 8 (see Eqn. (12.12)).

base. While this space is often represented as a pyramid, according to its mathematical definition, the space is actually a *cylinder*, as shown in Fig. 12.12.

The **HLS** color space[7] (*hue*, *luminance*, *saturation*) is very similar to the HSV space, and the *hue* component is in fact completely identical in both spaces. The *luminance* and *saturation* values also correspond to the vertical axis and the radius, respectively, but are defined differently than in HSV space. The common representation of the HLS space is as a double pyramid (Fig. 12.11(b)), with black

[7] The acronyms HLS and HSL are used interchangeably.

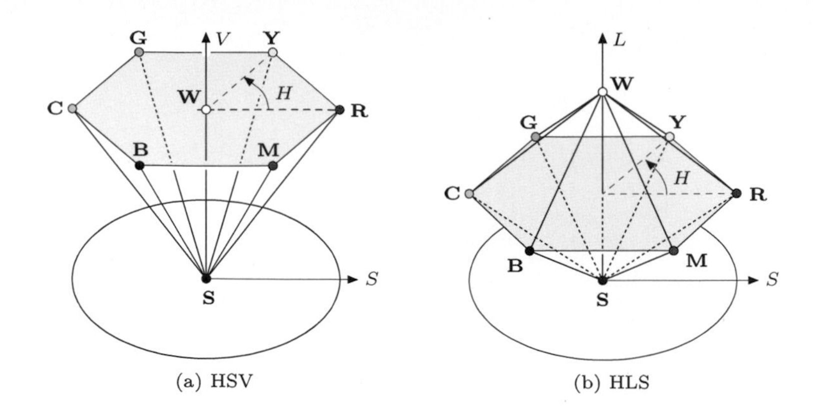

Fig. 12.11
HSV and HLS color space are traditionally visualized as a single or double hexagonal pyramid. The brightness V (or L) is represented by the vertical dimension, the color saturation S by the radius from the pyramid's axis, and the hue h by the angle. In both cases, the primary colors red (**R**), green (**G**), and blue (**B**) and the mixed colors yellow (**Y**), cyan (**C**), and magenta (**M**) lie on a common plane with black (**S**) at the tip. The essential difference between the HSV and HLS color spaces is the location of the white point (**W**).

on the bottom tip and white on the top. The primary colors lie on the corner points of the hexagonal base between the two pyramids. Even though it is often portrayed in this intuitive way, mathematically the HLS space is again a cylinder (see Fig. 12.15).

RGB→HSV conversion

To convert from RGB to the HSV color space, we first find the *saturation* of the RGB color components $R, G, B \in [0, C_{\mathrm{max}}]$, with C_{max} being the maximum component value (typically 255), as

$$S_{\mathrm{HSV}} = \begin{cases} \frac{C_{\mathrm{rng}}}{C_{\mathrm{high}}} & \text{for } C_{\mathrm{high}} > 0, \\ 0 & \text{otherwise} \end{cases} \tag{12.13}$$

and the brightness (*value*)

$$V_{\mathrm{HSV}} = \frac{C_{\mathrm{high}}}{C_{\mathrm{max}}}, \tag{12.14}$$

with

$$\begin{aligned} C_{\mathrm{low}} &= \min(R, G, B), \qquad C_{\mathrm{high}} = \max(R, G, B), \\ C_{\mathrm{rng}} &= C_{\mathrm{high}} - C_{\mathrm{low}}. \end{aligned} \tag{12.15}$$

Finally, we need to specify the *hue* value H_{HSV}. When all three RGB color components have the same value ($R = G = B$), then we are dealing with an *achromatic* (gray) pixel. In this particular case $C_{\mathrm{rng}} = 0$ and thus the saturation value $S_{\mathrm{HSV}} = 0$, consequently the hue is undefined. To calculate H_{HSV} when $C_{\mathrm{rng}} > 0$, we first normalize each component using

$$R' = \frac{C_{\mathrm{high}} - R}{C_{\mathrm{rng}}}, \qquad G' = \frac{C_{\mathrm{high}} - G}{C_{\mathrm{rng}}}, \qquad B' = \frac{C_{\mathrm{high}} - B}{C_{\mathrm{rng}}}. \tag{12.16}$$

Then, depending on which of the three original color components had the maximal value, we compute a preliminary hue H' as

$$H' = \begin{cases} B' - G' & \text{for } R = C_{\mathrm{high}}, \\ R' - B' + 2 & \text{for } G = C_{\mathrm{high}}, \\ G' - R' + 4 & \text{for } B = C_{\mathrm{high}}. \end{cases} \tag{12.17}$$

Since the resulting value for H' lies on the interval $[-1, 5]$, we obtain the final hue value by normalizing to the interval $[0, 1]$ as

$$H_{\mathrm{HSV}} = \frac{1}{6} \cdot \begin{cases} (H' + 6) & \text{for } H' < 0, \\ H' & \text{otherwise.} \end{cases} \tag{12.18}$$

Hence all three components H_{HSV}, S_{HSV}, and V_{HSV} will lie within the interval $[0, 1]$. The hue value H_{HSV} can naturally also be computed in another angle interval, for example, in the 0 to 360° interval using

$$H^{\circ}_{\mathrm{HSV}} = H_{\mathrm{HSV}} \cdot 360. \tag{12.19}$$

Under this definition, the RGB space unit cube is mapped to a *cylinder* with height and radius of length 1 (Fig. 12.12). In contrast to the traditional representation (Fig. 12.11), all HSB points within the entire cylinder correspond to valid color coordinates in RGB space. The mapping from RGB to the HSV space is nonlinear, as can be noted by examining how the black point stretches completely across the cylinder's base. Figure 12.12 plots the location of some notable color points and compares them with their locations in RGB space (see also Fig. 12.1). Figure 12.13 shows the individual HSV components (in grayscale) of the test image in Fig. 12.2.

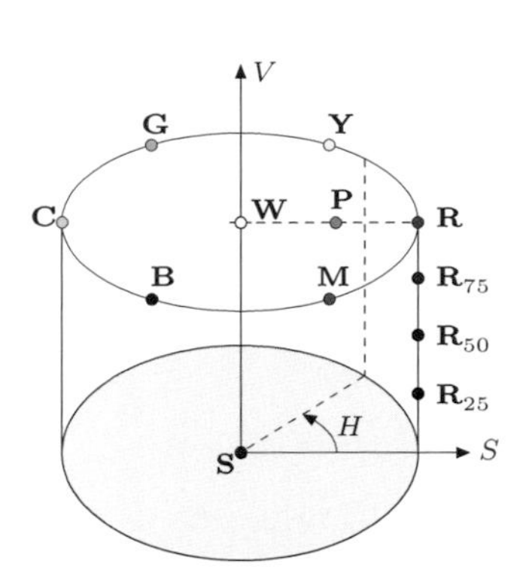

RGB/HSV values

Pt.	Color	R	G	B	H	S	V
S	Black	0.00	0.00	0.00	—	0.00	0.00
R	Red	1.00	0.00	0.00	0	1.00	1.00
Y	Yellow	1.00	1.00	0.00	1/6	1.00	1.00
G	Green	0.00	1.00	0.00	2/6	1.00	1.00
C	Cyan	0.00	1.00	1.00	3/6	1.00	1.00
B	Blue	0.00	0.00	1.00	4/6	1.00	1.00
M	Magenta	1.00	0.00	1.00	5/6	1.00	1.00
W	White	1.00	1.00	1.00	—	0.00	1.00
$\mathbf{R}_{75}$	75% Red	0.75	0.00	0.00	0	1.00	0.75
$\mathbf{R}_{50}$	50% Red	0.50	0.00	0.00	0	1.00	0.50
$\mathbf{R}_{25}$	25% Red	0.25	0.00	0.00	0	1.00	0.25
P	Pink	1.00	0.50	0.50	0	0.5	1.00

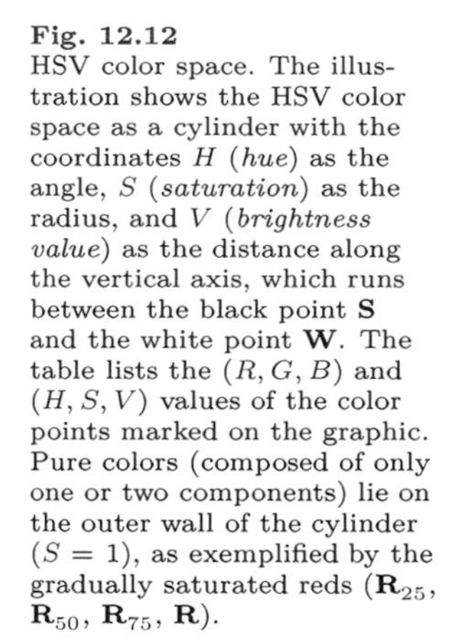

Fig. 12.12
HSV color space. The illustration shows the HSV color space as a cylinder with the coordinates H (*hue*) as the angle, S (*saturation*) as the radius, and V (*brightness value*) as the distance along the vertical axis, which runs between the black point **S** and the white point **W**. The table lists the (R, G, B) and (H, S, V) values of the color points marked on the graphic. Pure colors (composed of only one or two components) lie on the outer wall of the cylinder ($S = 1$), as exemplified by the gradually saturated reds ($\mathbf{R}_{25}$, $\mathbf{R}_{50}$, $\mathbf{R}_{75}$, $\mathbf{R}$).

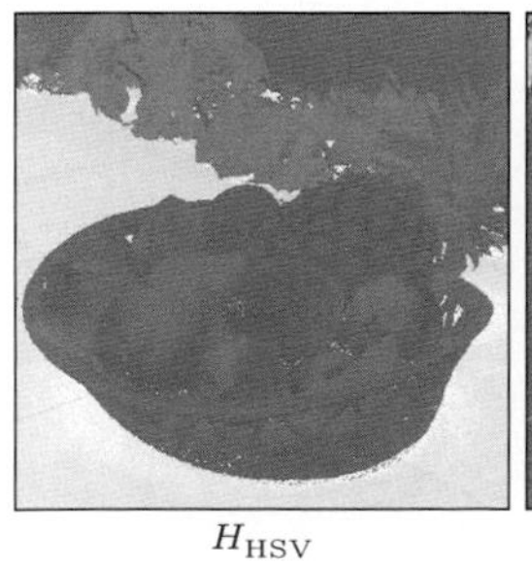

H_{HSV}

S_{HSV}

V_{HSV}

Fig. 12.13
HSV components for the test image in Fig. 12.2. The darker areas in the h_{HSV} component correspond to the red and yellow colors, where the *hue* angle is near zero.

Java implementation

In Java, the RGB→HSV conversion is implemented in the standard AWT `Color` class by the static method `float[] RGBtoHSB (int r, int g, int b, float[] hsv)`

(HSV and HSB denote the same color space). The method takes three `int` arguments `r`, `g`, `b` (within the range $[0, 255]$) and returns a `float` array with the resulting H, S, V values in the interval $[0, 1]$. When an existing `float` array is passed as the argument ***hsv***, then the result is placed in it; otherwise (when ***hsv*** $=$ `null`) a new array is created. Here is a simple example:

```
import java.awt.Color;
...
float[] hsv = new float[3];
int red = 128, green = 255, blue = 0;
hsv = Color.RGBtoHSB (red, green, blue, hsv);
float h = hsv[0];
float s = hsv[1];
float v = hsv[2];
...
```

A possible implementation of the Java method `RGBtoHSB()` using the definition in Eqns. (12.14)–(12.18) is given in Prog. 12.6.

HSV→RGB conversion

To convert an HSV tuple $(H_{\mathrm{HSV}}, S_{\mathrm{HSV}}, V_{\mathrm{HSV}})$, where H_{HSV}, S_{HSV}, and $V_{\mathrm{HSV}} \in [0, 1]$, into the corresponding (R, G, B) color values, the appropriate color sector

$$H' = (6 \cdot H_{\mathrm{HSV}}) \bmod 6 \tag{12.20}$$

(with $0 \leq H' < 6$) is determined first, followed by computing the intermediate values

$$\begin{aligned} c_1 &= \lfloor H' \rfloor, & x &= (1 - S_{\mathrm{HSV}}) \cdot V_{\mathrm{HSV}}, \\ c_2 &= H' - c_1, & y &= (1 - (S_{\mathrm{HSV}} \cdot c_2)) \cdot V_{\mathrm{HSV}}, \\ & & z &= (1 - (S_{\mathrm{HSV}} \cdot (1 - c_2))) \cdot V_{\mathrm{HSV}}. \end{aligned} \tag{12.21}$$

Depending on the value of c_1, the normalized RGB values $R', G', B' \in [0, 1]$ are then calculated from $v = V_{\mathrm{HSV}}$, x, y, and z as follows:[8]

$$(R', G', B') \leftarrow \begin{cases} (v, z, x) & \text{for } c_1 = 0, \\ (y, v, x) & \text{for } c_1 = 1, \\ (x, v, z) & \text{for } c_1 = 2, \\ (x, y, v) & \text{for } c_1 = 3, \\ (z, x, v) & \text{for } c_1 = 4, \\ (v, x, y) & \text{for } c_1 = 5. \end{cases} \tag{12.22}$$

Scaling the RGB components back to integer values in the range $[0, 255]$ is carried out as follows:

$$\begin{aligned} R &\leftarrow \min\bigl(\mathrm{round}(K \cdot R'), 255\bigr), \\ G &\leftarrow \min\bigl(\mathrm{round}(K \cdot G'), 255\bigr), \\ B &\leftarrow \min\bigl(\mathrm{round}(K \cdot B'), 255\bigr). \end{aligned} \tag{12.23}$$

[8] The variables x, y, z used here are not related to the CIEXYZ color space (see Ch. 14, Sec. 14.1).

Prog. 12.6
RGB→HSV conversion (Java implementation). This Java method for RGB→HSV conversion follows the process given in the text to compute a single color tuple. It takes the same arguments and returns results identical to the standard `Color.RGBtoHSB()` method.

```
1   float[] RGBtoHSV (int[] RGB) {
2     int R = RGB[0], G = RGB[1], B = RGB[2]; // R, G, B ∈ [0, 255]
3     int cHi = Math.max(R,Math.max(G,B));  // max. comp. value
4     int cLo = Math.min(R,Math.min(G,B));  // min. comp. value
5     int cRng = cHi - cLo;    // component range
6     float H = 0, S = 0, V = 0;
7     float cMax = 255.0f;
8
9     // compute value V
10    V = cHi / cMax;
11
12    // compute saturation S
13    if (cHi > 0)
14      S = (float) cRng / cHi;
15
16    // compute hue H
17    if (cRng > 0) { // hue is defined only for color pixels
18      float rr = (float)(cHi - R) / cRng;
19      float gg = (float)(cHi - G) / cRng;
20      float bb = (float)(cHi - B) / cRng;
21      float hh;
22      if (R == cHi)                 // R is largest component value
23        hh = bb - gg;
24      else if (G == cHi)            // G is largest component value
25        hh = rr - bb + 2.0f;
26      else                          // B is largest component value
27        hh = gg - rr + 4.0f;
28      if (hh < 0)
29        hh = hh + 6;
30      H = hh / 6;
31    }
32    return new float[] {H, S, V};
33  }
```

Java implementation

HSV→RGB conversion is implemented in Java's standard AWT `Color` class by the static method

`int HSBtoRGB (float h, float s, float v)`,

which takes three `float` arguments $h, s, v \in [0, 1]$ and returns the corresponding RGB color as an `int` value with 3×8 bits arranged in the standard Java RGB format (see Fig. 12.6). One possible implementation of this method is shown in Prog. 12.7.

RGB→HLS conversion

In the HLS model, the *hue* value H_{HLS} is computed in the same way as in the HSV model (Eqns. (12.16)–(12.18)), that is,

$$H_{\mathrm{HLS}} = H_{\mathrm{HSV}}. \tag{12.24}$$

The other values, L_{HLS} and S_{HLS}, are calculated as follows (for C_{high}, C_{low}, and C_{rng}, see Eqn. (12.15)):

Prog. 12.7
HSV→RGB conversion (Java implementation).

```
1  int HSVtoRGB (float[] HSV) {
2    float H = HSV[0], S = HSV[1], V = HSV[2]; // H,S,V ∈ [0,1]
3    float r = 0, g = 0, b = 0;
4    float hh = (6 * H) % 6;     // h' ← (6·h) mod 6
5    int   c1 = (int) hh;        // c1 ← ⌊h'⌋
6    float c2 = hh - c1;
7    float x = (1 - S) * V;
8    float y = (1 - (S * c2)) * V;
9    float z = (1 - (S * (1 - c2))) * V;
10   switch (c1) {
11     case 0: r = V; g = z; b = x; break;
12     case 1: r = y; g = V; b = x; break;
13     case 2: r = x; g = V; b = z; break;
14     case 3: r = x; g = y; b = V; break;
15     case 4: r = z; g = x; b = V; break;
16     case 5: r = V; g = x; b = y; break;
17   }
18   int R = Math.min((int)(r * 255), 255);
19   int G = Math.min((int)(g * 255), 255);
20   int B = Math.min((int)(b * 255), 255);
21   return new int[] {R, G, B};
22 }
```

$$L_{\mathrm{HLS}} = \frac{(C_{\mathrm{high}} + C_{\mathrm{low}})/255}{2}, \tag{12.25}$$

$$S_{\mathrm{HLS}} = \begin{cases} 0 & \text{for } L_{\mathrm{HLS}} = 0, \\ 0.5 \cdot \frac{C_{\mathrm{rng}}/255}{L_{\mathrm{HLS}}} & \text{for } 0 < L_{\mathrm{HLS}} \leq 0.5, \\ 0.5 \cdot \frac{C_{\mathrm{rng}}/255}{1-L_{\mathrm{HLS}}} & \text{for } 0.5 < L_{\mathrm{HLS}} < 1, \\ 0 & \text{for } L_{\mathrm{HLS}} = 1. \end{cases} \tag{12.26}$$

Using the aforementioned definitions, the RGB color cube is again mapped to a cylinder with height and radius 1 (see Fig. 12.15). In contrast to the HSV space (Fig. 12.12), the primary colors lie together in the horizontal plane at $L_{\mathrm{HLS}} = 0.5$ and the white point lies outside of this plane at $L_{\mathrm{HLS}} = 1.0$. Using these nonlinear transformations, the black and the white points are mapped to the top and the bottom planes of the cylinder, respectively. All points inside HLS cylinder correspond to valid colors in RGB space. Figure 12.14 shows the individual HLS components of the test image as grayscale images.

Fig. 12.14
HLS color components H_{HLS} (*hue*), S_{HLS} (*saturation*), and L_{HLS} (*luminance*).

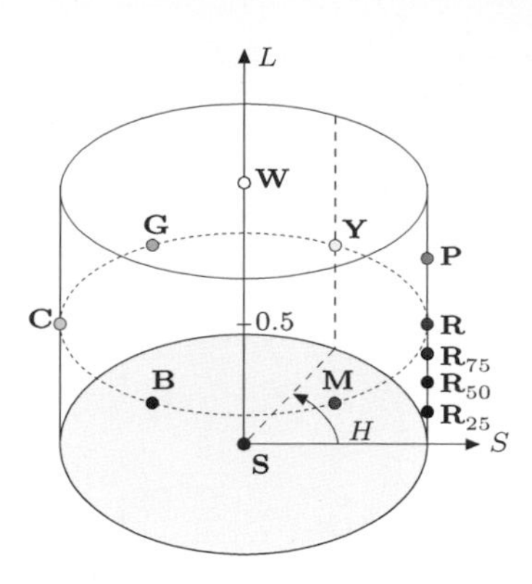

RGB/HLS values

Pt.	Color	R	G	B	H	S	L
S	Black	0.00	0.00	0.00	—	0.00	0.00
R	Red	1.00	0.00	0.00	0	1.00	0.50
Y	Yellow	1.00	1.00	0.00	1/6	1.00	0.50
G	Green	0.00	1.00	0.00	2/6	1.00	0.50
C	Cyan	0.00	1.00	1.00	3/6	1.00	0.50
B	Blue	0.00	0.00	1.00	4/6	1.00	0.50
M	Magenta	1.00	0.00	1.00	5/6	1.00	0.50
W	White	1.00	1.00	1.00	—	0.00	1.00
$\mathbf{R}_{75}$	75% Red	0.75	0.00	0.00	0	1.00	0.375
$\mathbf{R}_{50}$	50% Red	0.50	0.00	0.00	0	1.00	0.250
$\mathbf{R}_{25}$	25% Red	0.25	0.00	0.00	0	1.00	0.125
P	Pink	1.00	0.50	0.50	0/6	1.00	0.75

Fig. 12.15
HLS color space. The illustration shows the HLS color space visualized as a cylinder with the coordinates H (*hue*) as the angle, S (*saturation*) as the radius, and L (*lightness*) as the distance along the vertical axis, which runs between the black point **S** and the white point **W**. The table lists the (R, G, B) and (H, S, L) values where "pure" colors (created using only one or two color components) lie on the lower half of the outer cylinder wall ($S = 1$), as illustrated by the gradually saturated reds ($\mathbf{R}_{25}$, $\mathbf{R}_{50}$, $\mathbf{R}_{75}$, $\mathbf{R}$). Mixtures of all three primary colors, where at least one of the components is completely saturated, lie along the upper half of the outer cylinder wall; for example, the point **P** (pink).

HLS→RGB conversion

When converting from HLS to the RGB space, we assume that H_{HLS}, S_{HLS}, $L_{\mathrm{HLS}} \in [0, 1]$. In the case where $L_{\mathrm{HLS}} = 0$ or $L_{\mathrm{HLS}} = 1$, the result is

$$(R', G', B') = \begin{cases} (0, 0, 0) & \text{for } L_{\mathrm{HLS}} = 0, \\ (1, 1, 1) & \text{for } L_{\mathrm{HLS}} = 1. \end{cases} \tag{12.27}$$

Otherwise, we again determine the appropriate color sector

$$H' = (6 \cdot H_{\mathrm{HLS}}) \bmod 6, \tag{12.28}$$

such that $0 \leq H' < 6$, and from this

$$c_1 = \lfloor H' \rfloor, \qquad c_2 = H' - c_1, \tag{12.29}$$

$$d = \begin{cases} S_{\mathrm{HLS}} \cdot L_{\mathrm{HLS}} & \text{for } L_{\mathrm{HLS}} \leq 0.5, \\ S_{\mathrm{HLS}} \cdot (1 - L_{\mathrm{HLS}}) & \text{for } L_{\mathrm{HLS}} > 0.5, \end{cases} \tag{12.30}$$

and the quantities

$$w = L_{\mathrm{HLS}} + d, \qquad x = L_{\mathrm{HLS}} - d, \tag{12.31}$$

$$y = w - (w - x) \cdot c_2, \qquad z = x + (w - x) \cdot c_2. \tag{12.32}$$

The final mapping to the RGB values is (similar to Eqn. (12.22))

$$(R', G', B') = \begin{cases} (w, z, x) & \text{for } c_1 = 0, \\ (y, w, x) & \text{for } c_1 = 1, \\ (x, w, z) & \text{for } c_1 = 2, \\ (x, y, w) & \text{for } c_1 = 3, \\ (z, x, w) & \text{for } c_1 = 4, \\ (w, x, y) & \text{for } c_1 = 5. \end{cases} \tag{12.33}$$

Finally, scaling the normalized R', G', B' ($\in [0, 1]$) color components back to the $[0, 255]$ range is done as in Eqn. (12.23).

Prog. 12.8
RGB→HLS conversion (Java implementation).

```
1  float[] RGBtoHLS (int[] RGB) {
2    int R = RGB[0], G = RGB[1], B = RGB[2]; // R,G,B in [0,255]
3    float cHi = Math.max(R, Math.max(G, B));
4    float cLo = Math.min(R, Math.min(G, B));
5    float cRng = cHi - cLo;    // component range
6
7    // compute lightness L
8    float L = ((cHi + cLo) / 255f) / 2;
9
10   // compute saturation S
11   float S = 0;
12   if (0 < L && L < 1) {
13     float d = (L <= 0.5f) ? L : (1 - L);
14     S = 0.5f * (cRng / 255f) / d;
15   }
16
17   // compute hue H (same as in HSV)
18   float H = 0;
19   if (cHi > 0 && cRng > 0) {          // this is a color pixel!
20     float r = (float)(cHi - R) / cRng;
21     float g = (float)(cHi - G) / cRng;
22     float b = (float)(cHi - B) / cRng;
23     float h;
24     if (R == cHi)                      // R is largest component
25       h = b - g;
26     else if (G == cHi)                 // G is largest component
27       h = r - b + 2.0f;
28     else                               // B is largest component
29       h = g - r + 4.0f;
30     if (h < 0)
31       h = h + 6;
32     H = h / 6;
33   }
34   return new float[] {H, L, S};
35 }
```

Java implementation

Currently there is no method in either the standard Java API or ImageJ for converting color values between RGB and HLS. Program 12.8 gives one possible implementation of the RGB→HLS conversion that follows the definitions in Eqns. (12.24)–(12.26). The HLS→RGB conversion is shown in Prog. 12.9.

HSV and HLS compared

Despite the obvious similarity between the two color spaces, as Fig. 12.16 illustrates, substantial differences in the V/L and S components do exist. The essential difference between the HSV and HLS spaces is the ordering of the colors that lie between the white point **W** and the "pure" colors (**R**, **G**, **B**, **Y**, **C**, **M**), which consist of at most two primary colors, at least one of which is completely saturated.

The difference in how colors are distributed in RGB, HSV, and HLS space is readily apparent in Fig. 12.17. The starting point was a distribution of 1331 ($11 \times 11 \times 11$) color tuples obtained by uniformly

Prog. 12.9
HLS→RGB conversion (Java implementation).

```
1   float[] HLStoRGB (float[] HLS) {
2     float H = HLS[0], L = HLS[1], S = HLS[2]; // H,L,S in [0, 1]
3     float r = 0, g = 0, b = 0;
4     if (L <= 0)        // black
5       r = g = b = 0;
6     else if (L >= 1)      // white
7       r = g = b = 1;
8     else {
9       float hh = (6 * H) % 6;        // = H'
10      int   c1 = (int) hh;
11      float c2 = hh - c1;
12      float d = (L <= 0.5f) ? (S * L) : (S * (1 - L));
13      float w = L + d;
14      float x = L - d;
15      float y = w - (w - x) * c2;
16      float z = x + (w - x) * c2;
17      switch (c1) {
18        case 0: r = w; g = z; b = x; break;
19        case 1: r = y; g = w; b = x; break;
20        case 2: r = x; g = w; b = z; break;
21        case 3: r = x; g = y; b = w; break;
22        case 4: r = z; g = x; b = w; break;
23        case 5: r = w; g = x; b = y; break;
24      }
25    } // r, g, b in [0, 1]
26    int R = Math.min(Math.round(r * 255), 255);
27    int G = Math.min(Math.round(g * 255), 255);
28    int B = Math.min(Math.round(b * 255), 255);
29    return new int[] {R, G, B};
30  }
```

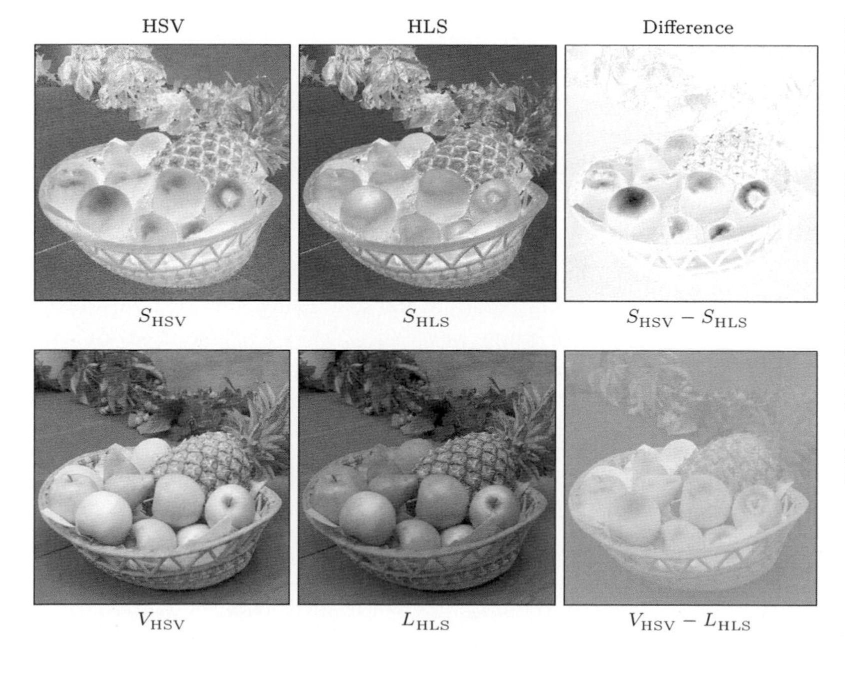

Fig. 12.16
HSV and HLS components compared. *Saturation* (top row) and *intensity* (bottom row). In the color *saturation* difference image $S_{\mathrm{HSV}} - S_{\mathrm{HLS}}$ (top), light areas correspond to positive values and dark areas to negative values. Saturation in the HLS representation, especially in the brightest sections of the image, is notably higher, resulting in negative values in the difference image. For the *intensity* (*value* and *luminance*, respectively) in general, $V_{\mathrm{HSV}} \geq L_{\mathrm{HLS}}$ and therefore the difference $V_{\mathrm{HSV}} - L_{\mathrm{HLS}}$ (bottom) is always positive. The *hue* component H (not shown) is identical in both representations.

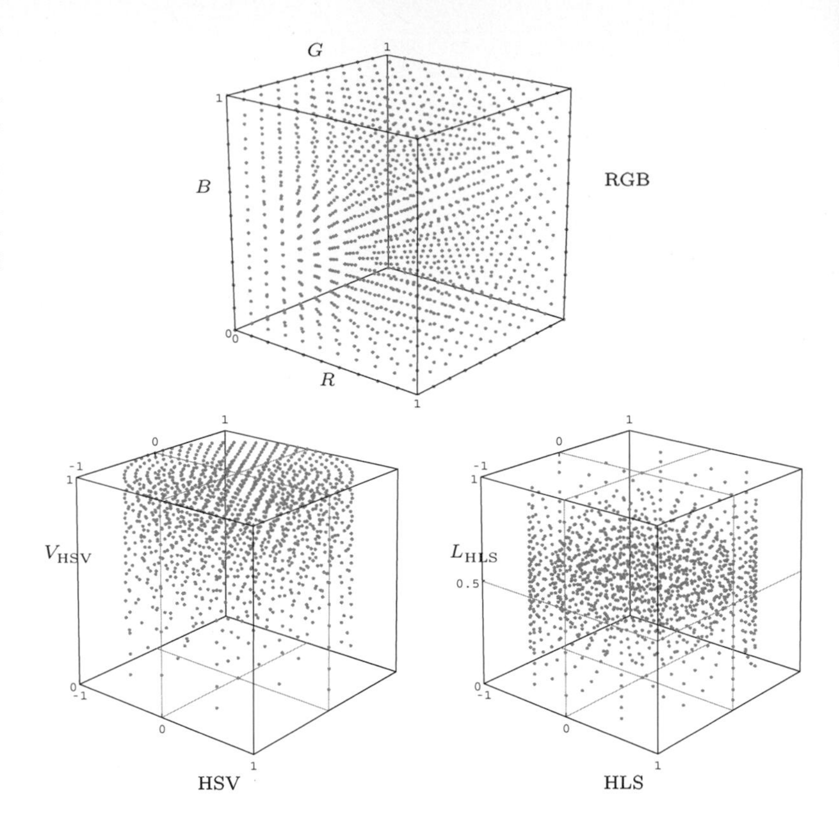

Fig. 12.17
Distribution of colors in the RGB, HSV, and HLS spaces. The starting point is the uniform distribution of colors in RGB space (top). The corresponding colors in the cylindrical spaces are distributed nonsymmetrically in HSV and symmetrically in HLS.

sampling the RGB space at an interval of 0.1 in each dimension. We can see clearly that in HSV space the maximally saturated colors ($s = 1$) form circular rings with increasing density toward the upper plane of the cylinder. In HLS space, however, the color samples are spread out symmetrically around the center plane and the density is significantly lower, particularly in the region near white. A given coordinate shift in this part of the color space leads to relatively small color changes, which allows the specification of very fine color grades in HLS space, especially for colors located in the upper half of the HLS cylinder.

Both the HSV and HLS color spaces are widely used in practice; for instance, for selecting colors in image editing and graphics design applications. In digital image processing, they are also used for *color keying* (i.e., isolating objects according to their *hue*) on a homogeneously colored background where the brightness is not necessarily constant.

Desaturation in HSV/HLS color space

Desaturation of color images (cf. Sec. 12.2.2) represented in HSV or HLS color space is trivial since color saturation is available as a separate component. In particular, pixels with zero saturation are uncolored or gray. For example, HSV colors can be gradually or fully desaturated by simply multiplying the component S by a fixed saturation factor $s \in [0, 1]$ and keeping H, V unchanged, that is,

$$\begin{pmatrix} H_{\text{desat}} \\ S_{\text{desat}} \\ V_{\text{desat}} \end{pmatrix} = \begin{pmatrix} H \\ s \cdot S \\ V \end{pmatrix}, \tag{12.34}$$

which works analogously with HLS colors. While Eqn. (12.34) applies equally to all colors, it might be interesting to *selectively* modify only colors with certain hues. This is easily accomplished by replacing the fixed saturation factor s by a hue-dependent function $f(H)$ (see also Exercise 12.6).

12.2.4 TV Component Color Spaces—YUV, YIQ, and $\text{YC}_\text{b}\text{C}_\text{r}$

These color spaces are an integral part of the standards surrounding the recording, storage, transmission, and display of television signals. YUV and YIQ are the fundamental color-encoding methods for the analog NTSC and PAL systems, and $\text{YC}_\text{b}\text{C}_\text{r}$ is a part of the international standards governing digital television [114]. All of these color spaces have in common the idea of separating the luminance component Y from two chroma components and, instead of directly encoding colors, encoding color differences. In this way, compatibility with legacy black and white systems is maintained while at the same time the bandwidth of the signal can be optimized by using different transmission bandwidths for the brightness and the color components. Since the human visual system is not able to perceive detail in the color components as well as it does in the intensity part of a video signal, the amount of information, and consequently bandwidth, used in the color channel can be reduced to approximately 1/4 of that used for the intensity component. This fact is also used when compressing digital still images and is why, for example, the JPEG codec converts RGB images to $\text{YC}_\text{b}\text{C}_\text{r}$. That is why these color spaces are important in digital image processing, even though raw YIQ or YUV images are rarely encountered in practice.

YUV

YUV is the basis for the color encoding used in analog television in both the North American NTSC and the European PAL systems. The luminance component Y is computed, just as in Eqn. (12.9), from the RGB components as

$$Y = 0.299 \cdot R + 0.587 \cdot G + 0.114 \cdot B \tag{12.35}$$

under the assumption that the RGB values have already been gamma corrected according to the TV encoding standard ($\gamma_{\text{NTSC}} = 2.2$ and $\gamma_{\text{PAL}} = 2.8$, see Ch. 4, Sec. 4.7) for playback. The UV components are computed from a weighted difference between the luminance and the blue or red components as

$$U = 0.492 \cdot (B - Y) \qquad \text{und} \qquad V = 0.877 \cdot (R - Y), \tag{12.36}$$

and the entire transformation from RGB to YUV is

$$\begin{pmatrix} Y \\ U \\ V \end{pmatrix} = \begin{pmatrix} 0.299 & 0.587 & 0.114 \\ -0.147 & -0.289 & 0.436 \\ 0.615 & -0.515 & -0.100 \end{pmatrix} \cdot \begin{pmatrix} R \\ G \\ B \end{pmatrix}. \tag{12.37}$$

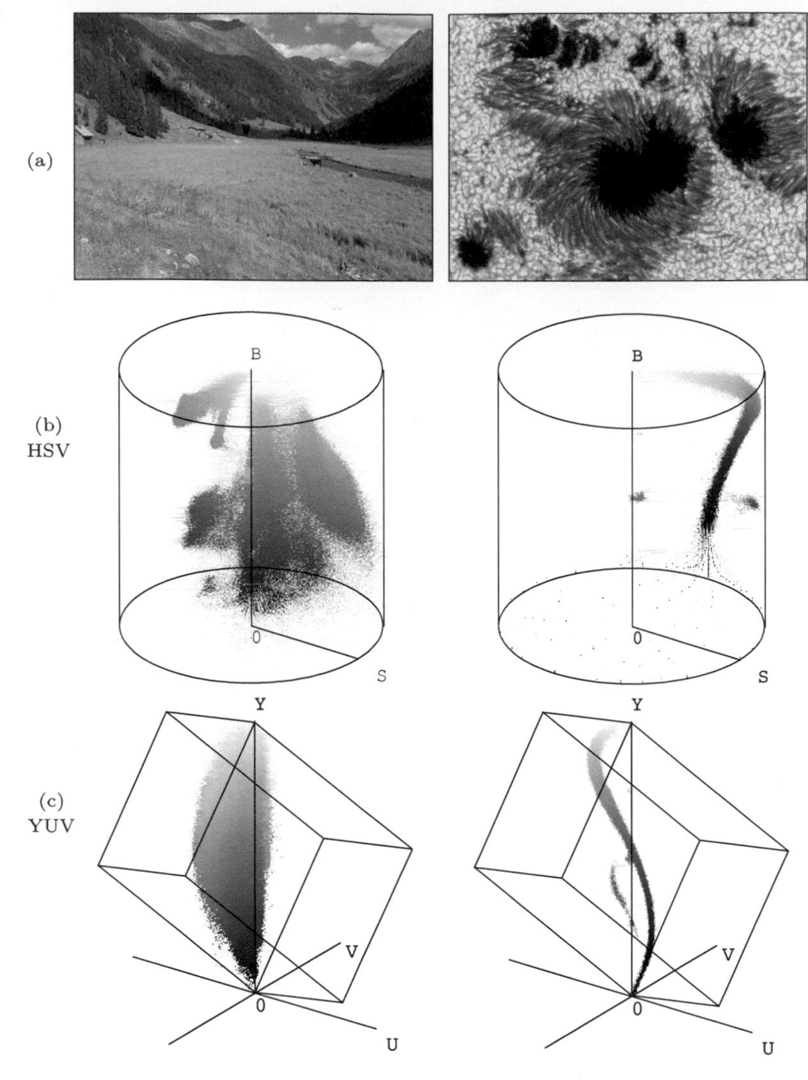

Fig. 12.18
Examples of the color distribution of natural images in different color spaces. Original images (a); color distribution in HSV- (b), and YUV-space (c). See Fig. 12.9 for the corresponding distributions in RGB color space.

The transformation from YUV back to RGB is found by inverting the matrix in Eqn. (12.37):

$$\begin{pmatrix} R \\ G \\ B \end{pmatrix} = \begin{pmatrix} 1.000 & 0.000 & 1.140 \\ 1.000 & -0.395 & -0.581 \\ 1.000 & 2.032 & 0.000 \end{pmatrix} \cdot \begin{pmatrix} Y \\ U \\ V \end{pmatrix}. \tag{12.38}$$

The color distributions in YUV-space for a set of natural images are shown in Fig. 12.18.

YIQ

The original NTSC system used a variant of YUV called YIQ (I for "in-phase", Q for "quadrature"), where both the U and V color vectors were rotated and mirrored such that

$$\begin{pmatrix} I \\ Q \end{pmatrix} = \begin{pmatrix} 0 & 1 \\ 1 & 0 \end{pmatrix} \cdot \begin{pmatrix} \cos\beta & \sin\beta \\ -\sin\beta & \cos\beta \end{pmatrix} \cdot \begin{pmatrix} U \\ V \end{pmatrix}, \tag{12.39}$$

where $\beta = 0.576$ (33°). The Y component is the same as in YUV. Although the YIQ has certain advantages with respect to bandwidth requirements it has been completely replaced by YUV [124, p. 240].

$\mathrm{YC_bC_r}$

The $\mathrm{YC_bC_r}$ color space is an internationally standardized variant of YUV that is used for both digital television and image compression (e.g., in JPEG). The chroma components C_b, C_r are (similar to U, V) difference values between the luminance and the blue and red components, respectively. In contrast to YUV, the weights of the RGB components for the luminance Y depend explicitly on the coefficients used for the chroma values C_b and C_r [197, p. 16]. For arbitrary weights w_B, w_R, the transformation is defined as

$$Y = w_\mathrm{R} \cdot R + (1 - w_\mathrm{B} - w_\mathrm{R}) \cdot G + w_\mathrm{B} \cdot B, \tag{12.40}$$

$$C_\mathrm{b} = \frac{0.5}{1 - w_\mathrm{B}} \cdot (B - Y), \tag{12.41}$$

$$C_\mathrm{r} = \frac{0.5}{1 - w_\mathrm{R}} \cdot (R - Y), \tag{12.42}$$

with $w_\mathrm{R} = 0.299$ and $w_\mathrm{B} = 0.114$ ($w_\mathrm{G} = 0.587$)[9] according to ITU[10] recommendation BT.601 [123]. Analogously, the reverse mapping from $\mathrm{YC_bC_r}$ to RGB is

$$R = Y + \frac{(1 - w_\mathrm{R}) \cdot C_\mathrm{r}}{0.5}, \tag{12.43}$$

$$G = Y - \frac{w_\mathrm{B} \cdot (1 - w_\mathrm{B}) \cdot C_\mathrm{b} + w_\mathrm{R} \cdot (1 - w_\mathrm{R}) \cdot C_\mathrm{r}}{0.5 \cdot (1 - w_\mathrm{B} - w_\mathrm{R})}, \tag{12.44}$$

$$B = Y + \frac{(1 - w_\mathrm{B}) \cdot C_\mathrm{b}}{0.5}. \tag{12.45}$$

In matrix-vector notation this gives the linear transformation

$$\begin{pmatrix} Y \\ C_\mathrm{b} \\ C_\mathrm{r} \end{pmatrix} = \begin{pmatrix} 0.299 & 0.587 & 0.114 \\ -0.169 & -0.331 & 0.500 \\ 0.500 & -0.419 & -0.081 \end{pmatrix} \cdot \begin{pmatrix} R \\ G \\ B \end{pmatrix}, \tag{12.46}$$

$$\begin{pmatrix} R \\ G \\ B \end{pmatrix} = \begin{pmatrix} 1.000 & 0.000 & 1.403 \\ 1.000 & -0.344 & -0.714 \\ 1.000 & 1.773 & 0.000 \end{pmatrix} \cdot \begin{pmatrix} Y \\ C_\mathrm{b} \\ C_\mathrm{r} \end{pmatrix}. \tag{12.47}$$

Different weights are recommended based on how the color space is used; for example, ITU-BT.709 [122] recommends $w_R = 0.2125$ and $w_B = 0.0721$ to be used in digital HDTV production. The values of U, V, I, Q, and C_b, C_r may be both positive or negative. To encode C_b, C_r values to digital numbers, a suitable offset is typically added to obtain positive-only values, for example, $128 = 2^7$ in case of 8-bit components.

Figure 12.19 shows the three color spaces YUV, YIQ, and $\mathrm{YC_bC_r}$ together for comparison. The U, V, I, Q, and C_b, C_r values in the

[9] $w_\mathrm{R} + w_\mathrm{G} + w_\mathrm{B} = 1$.

[10] International Telecommunication Union (www.itu.int).

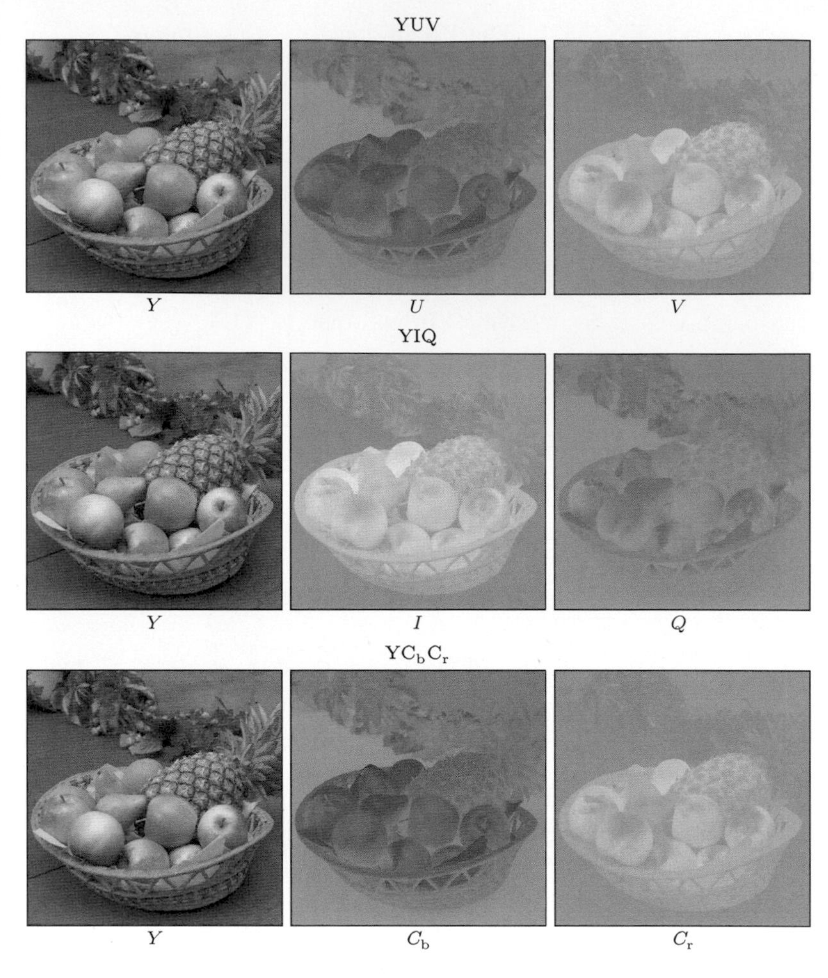

Fig. 12.19
Comparing YUV-, YIQ-, and YC_bC_r values. The Y values are identical in all three color spaces.

right two frames have been offset by 128 so that the negative values are visible. Thus a value of zero is represented as medium gray in these images. The YC_bC_r encoding is practically indistinguishable from YUV in these images since they both use very similar weights for the color components.

12.2.5 Color Spaces for Printing—CMY and CMYK

In contrast to the *additive* RGB color scheme (and its various color models), color printing makes use of a *subtractive* color scheme, where each printed color reduces the intensity of the reflected light at that location. Color printing requires a minimum of three primary colors; traditionally *cyan* (C), *magenta* (M), and *yellow* (Y)[11] have been used.

Using subtractive color mixing on a white background, $C = M = Y = 0$ (no ink) results in the color *white* and $C = M = Y = 1$ (complete saturation of all three inks) in the color *black*. A cyan-colored ink will absorb *red* (R) most strongly, magenta absorbs *green*

[11] Note that in this case Y stands for *yellow* and is unrelated to the Y luma or luminance component in YUV or YC_bC_r.

(G), and yellow absorbs *blue* (B). The simplest form of the CMY model is defined as

$$C = 1 - R, \qquad M = 1 - G, \qquad Y = 1 - B. \tag{12.48}$$

In practice, the color produced by fully saturating all three inks is not physically a true black. Therefore, the three primary colors C, M, Y are usually supplemented with a black ink (K) to increase the color range and coverage (gamut). In the simplest case, the amount of black is

$$K = \min(C, M, Y). \tag{12.49}$$

With rising levels of black, however, the intensity of the C, M, Y components can be gradually reduced. Many methods for reducing the primary dyes have been proposed and we look at three of them in the following.

CMY→CMYK conversion (version 1)

In this simple variant the C, M, Y values are reduced linearly with increasing K (Eqn. (12.49)), which yields the modified components as

$$\begin{pmatrix} C_1 \\ M_1 \\ Y_1 \\ K_1 \end{pmatrix} = \begin{pmatrix} C - K \\ M - K \\ Y - K \\ K \end{pmatrix}. \tag{12.50}$$

CMY→CMYK conversion (version 2)

The second variant corrects the color by reducing the C, M, Y components by $s = \frac{1}{1-K}$, resulting in stronger colors in the dark areas of the image:

$$\begin{pmatrix} C_2 \\ M_2 \\ Y_2 \\ K_2 \end{pmatrix} = \begin{pmatrix} (C-K) \cdot s \\ (M-K) \cdot s \\ (Y-K) \cdot s \\ K \end{pmatrix}, \quad \text{with } s = \begin{cases} \frac{1}{1-K} & \text{for } K < 1, \\ 1 & \text{otherwise.} \end{cases} \tag{12.51}$$

In both versions, the K component (as defined in Eqn. (12.49)) is used directly without modification, and all gray tones (that is, when $R = G = B$) are printed using black ink K, without any contribution from C, M, or Y.

While both of these simple definitions are widely used, neither one produces high quality results. Figure 12.20(a) compares the result from version 2 with that produced with Adobe Photoshop (Fig. 12.20(c)). The difference in the cyan component C is particularly noticeable and also the exceeding amount of black (K) in the brighter areas of the image.

In practice, the required amounts of black K and C, M, Y depend so strongly on the printing process and the type of paper used that print jobs are routinely calibrated individually.

Fig. 12.20
RGB→CMYK conversion comparison. Simple conversion using Eqn. (12.51) (a), applying the *undercolor-removal* and *black-generation* functions of Eqn. (12.52) (b), and results obtained with Adobe Photoshop (c). The color intensities are shown inverted, that is, darker areas represent higher CMYK color values. The simple conversion (a), in comparison with Photoshop's result (c), shows strong deviations in all color components, C and K in particular. The results in (b) are close to Photoshop's and could be further improved by tuning the corresponding function parameters.

CMY→CMYK conversion (version 3)

In print production, special transfer functions are applied to tune the results. For example, the Adobe PostScript interpreter [135, p. 345] specifies an *undercolor-removal function* $f_{\mathrm{UCR}}(K)$ for gradually reducing the CMY components and a separate *black-generation function* $f_{\mathrm{BG}}(K)$ for controlling the amount of black. These functions are used in the form

$$\begin{pmatrix} C_3 \\ M_3 \\ Y_3 \\ K_3 \end{pmatrix} = \begin{pmatrix} C - f_{\mathrm{UCR}}(K) \\ M - f_{\mathrm{UCR}}(K) \\ Y - f_{\mathrm{UCR}}(K) \\ f_{\mathrm{BG}}(K) \end{pmatrix}, \qquad (12.52)$$

where $K = \min(C, M, Y)$, as defined in Eqn. (12.49). The functions f_{UCR} and f_{BG} are usually nonlinear, and the resulting values

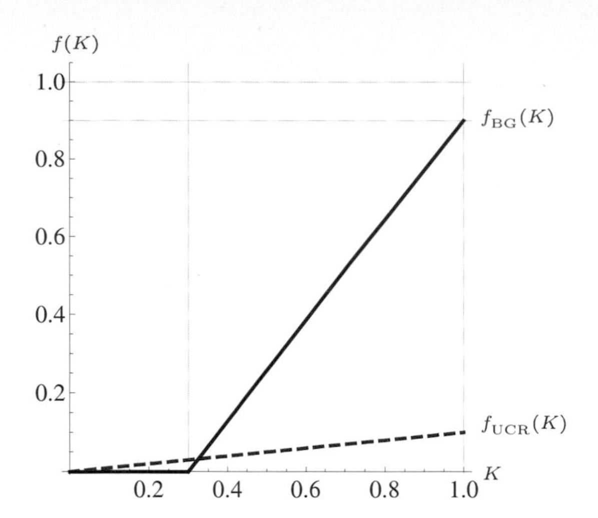

Fig. 12.21
Examples of *undercolor-removal function* f_{UCR} (Eqn. (12.53)) and *black generation function* f_{BG} (Eqn. (12.54)). The parameter settings are $s_K = 0.1$, $K_0 = 0.3$, and $K_{max} = 0.9$.

C_3, M_3, Y_3, K_3 are scaled (typically by means of *clamping*) to the interval $[0, 1]$. The example shown in Fig. 12.20(b) was produced to approximate the results of Adobe Photoshop using the definitions

$$f_{UCR}(K) = s_K \cdot K, \qquad (12.53)$$

$$f_{BG}(K) = \begin{cases} 0 & \text{for } K < K_0, \\ K_{max} \cdot \frac{K-K_0}{1-K_0} & \text{for } K \geq K_0, \end{cases} \qquad (12.54)$$

where $s_K = 0.1$, $K_0 = 0.3$, and $K_{max} = 0.9$ (see Fig. 12.21). With this definition, f_{UCR} reduces the CMY components by 10% of the K value (by Eqn. (12.52)), which mostly affects the dark areas of the image with high K values. The effect of the function f_{BG} (Eqn. (12.54)) is that for values of $K < K_0$ (i.e., in the light areas of the image) no black ink is added at all. In the interval $K = K_0, \ldots, 1.0$, the black component is increased linearly up to the maximum value K_{max}. The result in Fig. 12.20(b) is relatively close to the CMYK component values produced by Photoshop[12] in Fig. 12.20(c). It could be further improved by adjusting the function parameters s_K, K_0, and K_{max} (Eqn. (12.52)).

Even though the results of this last variant (3) for converting RGB to CMYK are better, it is only a gross approximation and still too imprecise for professional work. As we discuss in Chapter 14, technically correct color conversions need to be based on precise, "colorimetric" grounds.

12.3 Statistics of Color Images

12.3.1 How Many Different Colors are in an Image?

A minor but frequent task in the context of color images is to determine how many different colors are contained in a given image.

[12] Actually Adobe Photoshop does not convert directly from RGB to CMYK. Instead, it first converts to, and then from, the CIELAB color space (see Ch. 14, Sec. 14.1).

One way of doing this would be to create and fill a histogram array with one integer element for each color and subsequently count all histogram cells with values greater than zero. But since a 24-bit RGB color image potentially contains $2^{24} = 16,777,216$ colors, the resulting histogram array (with a size of 64 megabytes) would be larger than the image itself in most cases!

A simple solution to this problem is to *sort* the pixel values in the (1D) pixel array such that all identical colors are placed next to each other. The sorting order is of course completely irrelevant, and the number of contiguous color blocks in the sorted pixel vector corresponds to the number of different colors in the image. This number can be obtained by simply counting the transitions between neighboring color blocks, as shown in Prog. 12.10. Of course, we do not want to sort the original pixel array (which would destroy the image) but a copy of it, which can be obtained with Java's `clone()` method.[13] Sorting of the 1D array in Prog. 12.10 is accomplished (in line 4) with the generic Java method `Arrays.sort()`, which is implemented very efficiently.

Prog. 12.10
Counting the colors contained in an RGB image. The method `countColors()` first creates a copy of the 1D RGB (`int`) pixel array (line 3), then sorts that array, and finally counts the transitions between contiguous blocks of identical colors.

```
1  int countColors (ColorProcessor cp) {
2    // duplicate the pixel array and sort it
3    int[] pixels = ((int[]) cp.getPixels()).clone();
4    Arrays.sort(pixels);  // requires java.util.Arrays
5
6    int k = 1;  // color count (image contains at least 1 color)
7    for (int i = 0; i < pixels.length-1; i++) {
8      if (pixels[i] != pixels[i + 1])
9        k = k + 1;
10   }
11   return k;
12 }
```

12.3.2 Color Histograms

We briefly touched on histograms of color images in Chapter 3, Sec. 3.5, where we only considered the 1D distributions of the image intensity and the individual color channels. For instance, the built-in ImageJ method `getHistogram()`, when applied to an object of type `ColorProcessor`, simply computes the intensity histogram of the corresponding gray values:

```
ColorProcessor cp;
int[] H = cp.getHistogram();
```

As an alternative, one could compute the individual intensity histograms of the three color channels, although (as discussed in Chapter 3, Sec. 3.5.2) these do not provide any information about the actual colors in this image. Similarly, of course, one could compute the distributions of the individual components of any other color space, such as HSV or CIELAB.

[13] Java arrays implement the `Cloneable` interface.

A *full* histogram of an RGB image is 3D and, as noted earlier, consists of $256 \times 256 \times 256 = 2^{24}$ cells of type `int` (for 8-bit color components). Such a histogram is not only very large[14] but also difficult to visualize.

2D color histograms

A useful alternative to the full 3D RGB histogram are 2D histogram projections (Fig. 12.22). Depending on the axis of projection, we obtain 2D histograms with coordinates red-green (h_{RG}), red-blue (h_{RB}), or green-blue (h_{GB}), respectively, with the values

$$\begin{aligned} \mathsf{h}_{\mathrm{RG}}(r,g) &:= \text{number of pixels with } \boldsymbol{I}(u,v) = (r,g,*), \\ \mathsf{h}_{\mathrm{RB}}(r,b) &:= \text{number of pixels with } \boldsymbol{I}(u,v) = (r,*,b), \\ \mathsf{h}_{\mathrm{GB}}(g,b) &:= \text{number of pixels with } \boldsymbol{I}(u,v) = (*,g,b), \end{aligned} \tag{12.55}$$

where $*$ denotes an arbitrary component value. The result is, independent of the original image size, a set of 2D histograms of size 256×256 (for 8-bit RGB components), which can easily be visualized as images. Note that it is not necessary to obtain the full RGB histogram in order to compute the combined 2D histograms (see Prog. 12.11).

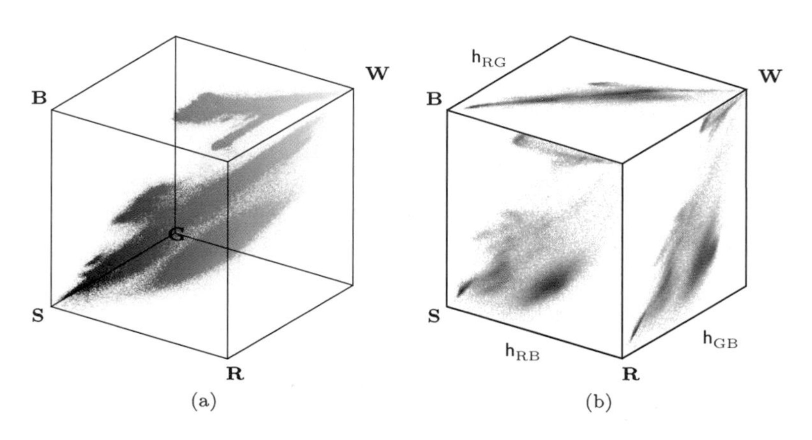

Fig. 12.22
2D RGB histogram projections. 3D RGB cube illustrating an image's color distribution (a). The color points indicate the corresponding pixel colors and not the color frequency. The combined histograms for red-green (h_{RG}), red-blue (h_{RB}), and green-blue (h_{GB}) are 2D projections of the 3D histogram. The corresponding image is shown in Fig. 12.9(a).

As the examples in Fig. 12.23 show, the combined color histograms do, to a certain extent, express the color characteristics of an image. They are therefore useful, for example, to identify the coarse type of the depicted scene or to estimate the similarity between images (see also Exercise 12.8).

12.4 Exercises

Exercise 12.1. Create an ImageJ plugin that rotates the individual components of an RGB color image; that is, $R \to G \to B \to R$.

Exercise 12.2. Pseudocolors are sometimes used for displaying grayscale images (i.e., for viewing medical images with high dynamic

[14] It may seem a paradox that, although the RGB histogram is usually much larger than the image itself, the histogram is not sufficient in general to reconstruct the original image.

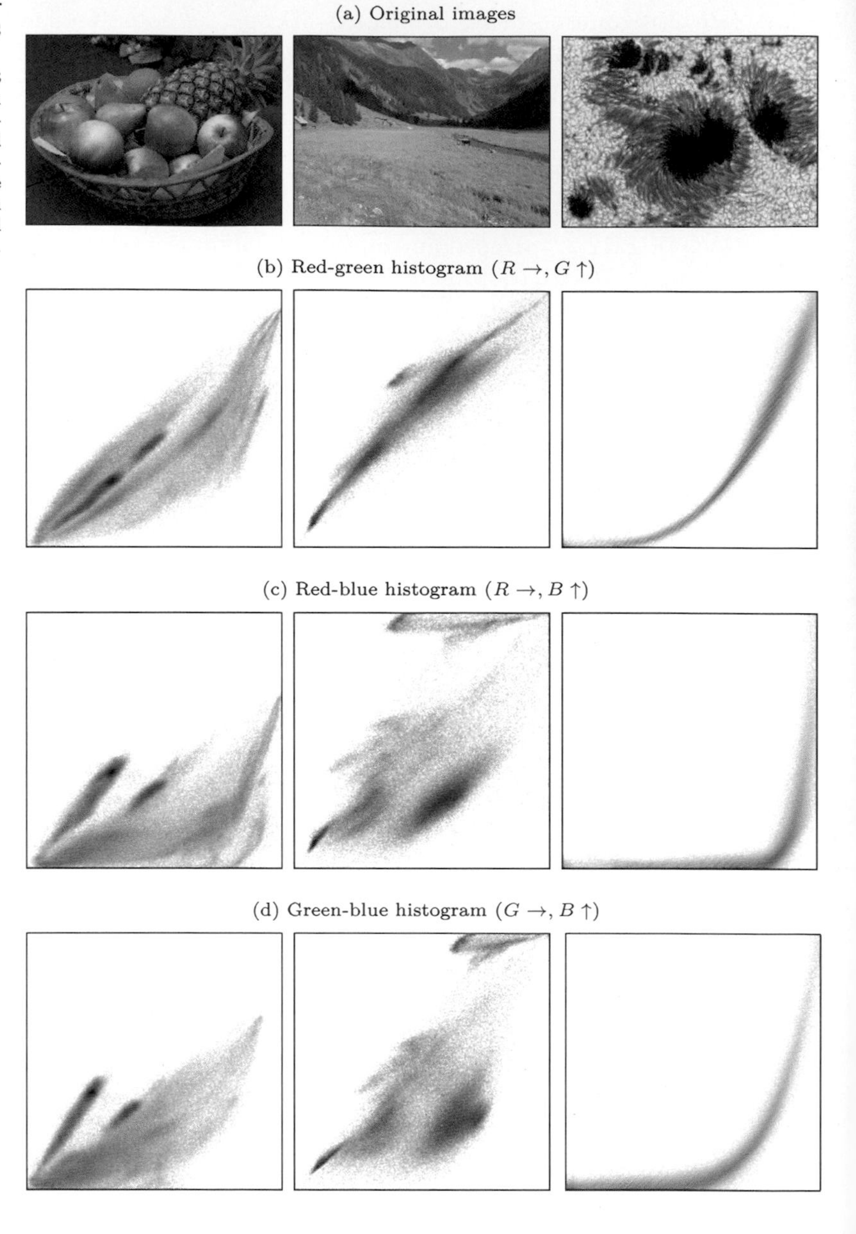

(a) Original images
(b) Red-green histogram ($R \rightarrow, G \uparrow$)
(c) Red-blue histogram ($R \rightarrow, B \uparrow$)
(d) Green-blue histogram ($G \rightarrow, B \uparrow$)

Fig. 12.23
Combined color histogram examples. For better viewing, the images are inverted (dark regions indicate high frequencies) and the gray value corresponds to the logarithm of the histogram entries (scaled to the maximum entries).

range). Create an ImageJ plugin for converting 8-bit grayscale images to an indexed image with 256 colors, simulating the hues of glowing iron (from dark red to yellow and white).

Exercise 12.3. Create an ImageJ plugin that shows the color table of an 8-bit indexed image as a new image with 16×16 rectangular color fields. Mark all unused color table entries in a suitable way. Look at Prog. 12.3 as a starting point.

Exercise 12.4. Show that a "desaturated" RGB pixel produced in the form $(r, g, b) \rightarrow (y, y, y)$, where y is the equivalent luminance value (see Eqn. (12.11)), has the luminance y as well.

```
1 int[][] get2dHistogram
2     (ColorProcessor cp, int c1, int c2) {
3   // c1, c2: component index R = 0, G = 1, B = 2
4
5   int[] RGB = new int[3];
6   int[][] h = new int[256][256];  // 2D histogram h[c1][c2]
7
8   for (int v = 0; v < cp.getHeight(); v++) {
9     for (int u = 0; u < cp.getWidth(); u++) {
10      cp.getPixel(u, v, RGB);
11      int i1 = RGB[c1];
12      int i2 = RGB[c2];
13      // increment the associated histogram cell
14      h[i1][i2]++;
15    }
16  }
17  return h;
18 }
```


Prog. 12.11
Java method `get2dHistogram()` for computing a combined 2D color histogram. The color components (histogram axes) are specified by the parameters `c1` and `c2`. The color distribution `H` is returned as a 2D `int` array. The method is defined in class `ColorStatistics` (Prog. 12.10).

Exercise 12.5. Extend the ImageJ plugin for desaturating color images in Prog. 12.5 such that the image is only modified inside the user-selected region of interest (ROI).

Exercise 12.6. Write an ImageJ plugin that *selectively desaturates* an RGB image, preserving colors with a hue close to a given *reference color* $\boldsymbol{c}_{\mathrm{ref}} = (R_{\mathrm{ref}}, G_{\mathrm{ref}}, B_{\mathrm{ref}})$, with (HSV) hue H_{ref} (see the example in Fig. 12.24). Transform the image to HSV and modify the colors (cf. Eqn. (12.34)) in the form

$$\begin{pmatrix} H_{\mathrm{desat}} \\ S_{\mathrm{desat}} \\ V_{\mathrm{desat}} \end{pmatrix} = \begin{pmatrix} H \\ f(H)\cdot S \\ V \end{pmatrix}, \qquad (12.56)$$

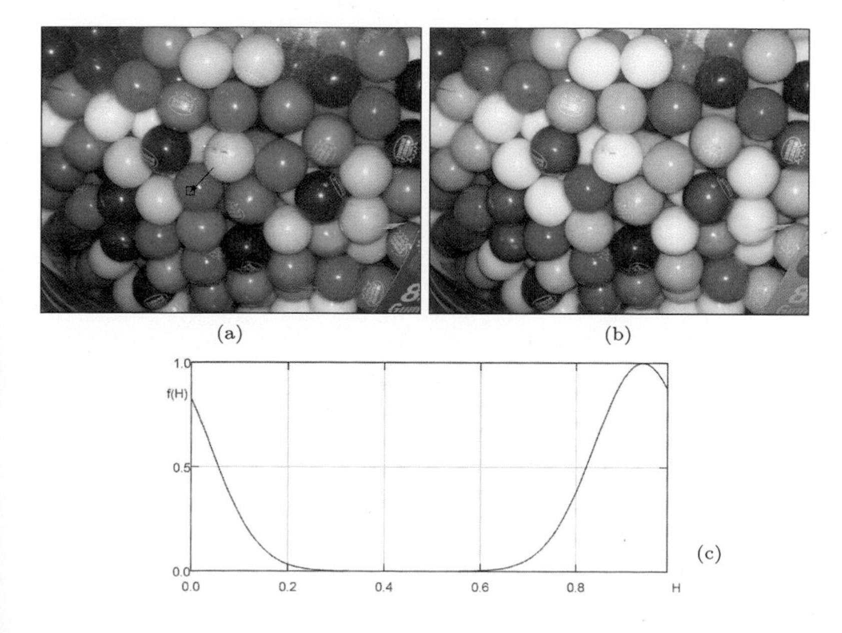

Fig. 12.24
Selective desaturation example. Original image with selected reference color $\boldsymbol{c}_{\mathrm{ref}} = (250, 92, 150)$ (a), desaturated image (b). Gaussian saturation function $f(H)$ (see Eqn. (12.58)) with reference hue $H_{\mathrm{ref}} = 0.9388$ and $\sigma = 0.1$ (c).

where $f(H)$ is a smooth saturation function, for example, a Gaussian function of the form

$$f(H) = e^{-\frac{(H-H_{\mathrm{ref}})^2}{2\cdot\sigma^2}} = g_\sigma(H-H_{\mathrm{ref}}), \qquad (12.57)$$

with center H_{ref} and variance σ^2 (see Fig. 12.24(c)). Recall that the H component is circular in $[0, 1)$. To obtain a continuous and periodic saturation function we note that $H' = H - H_{\mathrm{ref}}$ is in the range $[-1, 1]$ and reformulate $f(H)$ as

$$f(H) = \begin{cases} g_\sigma(H') & \text{for } -0.5 \leq H' \leq 0.5, \\ g_\sigma(H'+1) & \text{for } H' < -0.5, \\ g_\sigma(H'-1) & \text{for } H' > 0.5. \end{cases} \qquad (12.58)$$

Verify the values of the function $f(H)$, check in particular that it is 1 for the reference color! What would be a good (synthetic) color image for validating the saturation function? Use ImageJ's color picker (pipette) tool to specify the reference color $\boldsymbol{c}_{\mathrm{ref}}$ interactively.[15]

Exercise 12.7. Calculate (analogous to Eqns. (12.46)–(12.47)) the complete transformation matrices for converting from (linear) RGB colors to $\mathrm{YC_bC_r}$ for the ITU-BT.709 (HDTV) standard with the coefficients $w_\mathrm{R} = 0.2126$, $w_\mathrm{B} = 0.0722$ and $w_\mathrm{G} = 0.7152$.

Exercise 12.8. Determining the similarity between images of different sizes is a frequent problem (e.g., in the context of image data bases). Color statistics are commonly used for this purpose because they facilitate a coarse classification of images, such as landscape images, portraits, etc. However, 2D color histograms (as described in Sec. 12.3.2) are usually too large and thus cumbersome to use for this purpose. A simple idea could be to split the 2D histograms or even the full RGB histogram into K regions (*bins*) and to combine the corresponding entries into a K-dimensional feature vector, which could be used for a coarse comparison. Develop a concept for such a procedure, and also discuss the possible problems.

Exercise 12.9. Write a program (plugin) that generates a sequence of colors with constant hue and saturation but different brightness (value) in HSV space. Transform these colors to RGB and draw them into a new image. Verify (visually) if the hue really remains constant.

Exercise 12.10. When applying any type of filter in HSV or HLS color space one must keep in mind that the *hue* component H is circular in $[0, 1)$ and thus shows a discontinuity at the $1 \rightarrow 0$ ($360 \rightarrow 0°$) transition. For example, a linear filter would not take into account that $H = 0.0$ and $H = 1.0$ refer to the same hue (red) and thus cannot be applied directly to the H component. One solution is to filter the *cosine* and *sine* values of the H component (which really is an angle) instead, and composing the filtered hue array from the filtered cos / sin values (see Ch. 15, Sec. 15.1.3 for details). Based on this idea, implement a variable-sized linear Gaussian filter (see Ch. 5, Sec. 5.2.7) for the HSV color space.

---

[15] The current color pick is returned by the ImageJ method `Toolbar.getForegroundColor()`.

13

Color Quantization

The task of color quantization is to select and assign a limited set of colors for representing a given color image with maximum fidelity. Assume, for example, that a graphic artist has created an illustration with beautiful shades of color, for which he applied 150 different crayons. His editor likes the result but, for some technical reason, instructs the artist to draw the picture again, this time using only 10 different crayons. The artist now faces the problem of color quantization—his task is to select a "palette" of the 10 best suited from his 150 crayons and then choose the most similar color to redraw each stroke of his original picture.

In the general case, the original image I contains a set of m different colors $\mathcal{C} = \{\mathbf{C}_1, \mathbf{C}_2, \ldots, \mathbf{C}_m\}$, where m could be only a few or several thousand, but at most 2^{24} for a 3×8-bit color image. The goal is to replace the original colors by a (usually much smaller) set of colors $\mathcal{C}' = \{\mathbf{C}'_1, \mathbf{C}'_2, \ldots, \mathbf{C}'_n\}$, with $n < m$. The difficulty lies in the proper choice of the reduced color palette $\mathcal{C}'$ such that damage to the resulting image is minimized.

In practice, this problem is encountered, for example, when converting from full-color images to images with lower pixel depth or to index ("palette") images, such as the conversion from 24-bit TIFF to 8-bit GIF images with only 256 (or fewer) colors. Until a few years ago, a similar problem had to be solved for displaying full-color images on computer screens because the available display memory was often limited to only 8 bits. Today, even the cheapest display hardware has at least 24-bit depth and therefore this particular need for (fast) color quantization no longer exists.

13.1 Scalar Color Quantization

Scalar (or *uniform*) quantization is a simple and fast process that is independent of the image content. Each of the original color components c_i (e.g., R_i, G_i, B_i) in the range $[0, \ldots, m-1]$ is independently converted to the new range $[0, \ldots, n-1]$, in the simplest case by a

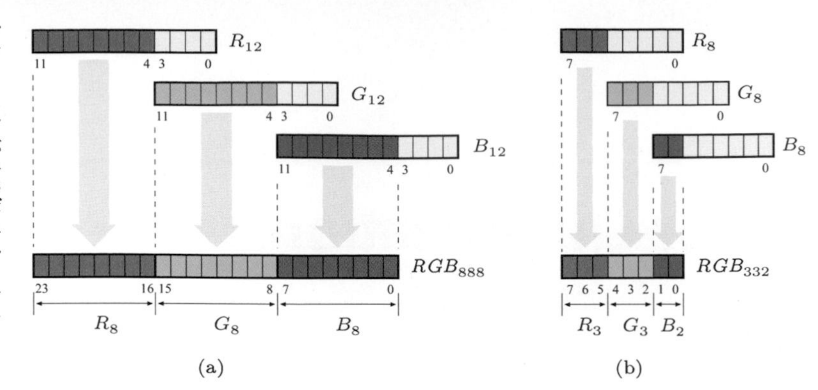

Fig. 13.1 Scalar quantization of color components by truncating lower bits. Quantization of 3 × 12-bit to 3 × 8-bit colors (a). Quantization of 3 × 8-bit to 3:3:2-packed 8-bit colors (b). The Java code segment in Prog. 13.1 shows the corresponding sequence of bit operations.

linear quantization in the form

$$c_i' \leftarrow \left\lfloor c_i \cdot \frac{n}{m} \right\rfloor \tag{13.1}$$

for all color components c_i. A typical example would be the conversion of a color image with 3 × 12-bit components ($m = 4096$) to an RGB image with 3 × 8-bit components ($n = 256$). In this case, each original component value is multiplied by $n/m = 256/4096 = 1/16 = 2^{-4}$ and subsequently truncated, which is equivalent to an integer division by 16 or simply ignoring the lower 4 bits of the corresponding binary values (see Fig. 13.1(a)). m and n are usually the same for all color components but not always.

An extreme (today rarely used) approach is to quantize 3 × 8 color vectors to single-byte (8-bit) colors, where 3 bits are used for red and green and only 2 bits for blue, as shown in Prog. 13.1(b). In this case, $m = 256$ for all color components, $n_{\text{red}} = n_{\text{green}} = 8$, and $m_{\text{blue}} = 4$.

Prog. 13.1 Quantization of a 3 × 8-bit RGB color pixel to 8 bits by 3:3:2 packing.

```
1 ColorProcessor cp = (ColorProcessor) ip;
2 int C = cp.getPixel(u, v);
3 int R = (C & 0x00ff0000) >> 16;
4 int G = (C & 0x0000ff00) >> 8;
5 int B = (C & 0x000000ff);
6 // 3:3:2 uniform color quantization
7 byte RGB =
8     (byte) ((R & 0xE0) | (G & 0xE0)>>3 | ((B & 0xC0)>>6));
```

Unlike the techniques described in the following, scalar quantization does not take into account the distribution of colors in the original image. Scalar quantization is an optimal solution only if the image colors are *uniformly* distributed within the RGB cube. However, the typical color distribution in natural images is anything but uniform, with some regions of the color space being densely populated and many colors entirely missing. In this case, scalar quantization is not optimal because the interesting colors may not be sampled with sufficient density while at the same time colors are represented that do not appear in the image at all.

(a)

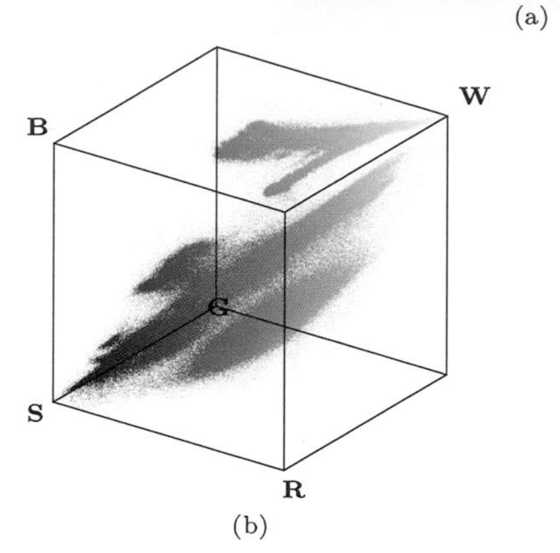

(b)

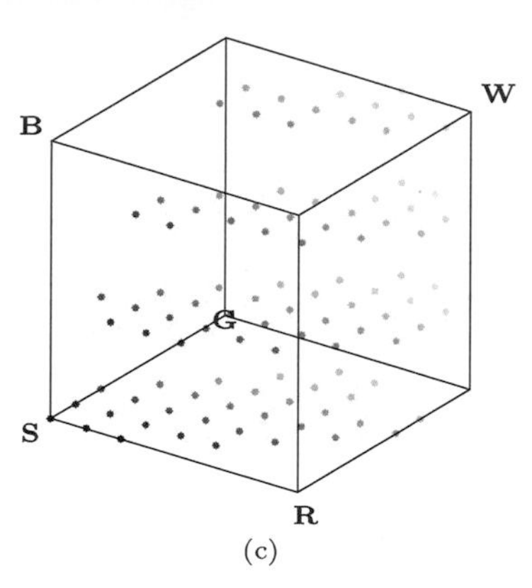

(c)

Fig. 13.2
Color distribution after a scalar 3:3:2 quantization. Original color image (a). Distribution of the original 226,321 colors (b) and the remaining $8 \times 8 \times 4 = 256$ colors after 3:3:2 quantization (c) in the RGB color cube.

13.2 Vector Quantization

Vector quantization does not treat the individual color components separately as does scalar quantization, but each color vector $\mathbf{C}_i = (r_i, g_i, b_i)$ or pixel in the image is treated as a single entity. Starting from a set of original color tuples $\mathcal{C} = \{\mathbf{c}_1, \mathbf{c}_2, \ldots, \mathbf{c}_m\}$, the task of vector quantization is

a) to find a set of n representative color vectors $\mathcal{C}' = \{\mathbf{c}'_1, \mathbf{c}'_2, \ldots, \mathbf{c}'_n\}$ and
b) to replace each original color $\mathbf{C}_i$ by one of the new color vectors $\mathbf{C}'_j \in \mathcal{C}'$,

where n is usually predetermined $(n < m)$ and the resulting deviation from the original image shall be minimal. This is a combinatorial optimization problem in a rather large search space, which usually makes it impossible to determine a global optimum in adequate time. Thus all of the following methods only compute a "local" optimum at best.

13.2.1 Populosity Algorithm

The populosity algorithm[1] [104] selects the n most frequent colors in the image as the representative set of color vectors $\mathcal{C}'$. Being very easy to implement, this procedure is quite popular. The method described in Sec. 12.3.1, based on sorting the image pixels, can be used to determine the n most frequent image colors. Each original

[1] Sometimes also called the "popularity" algorithm.

pixel $\mathbf{C}_i$ is then replaced by the closest representative color vector in $\mathcal{C}'$; that is, the quantized color vector with the smallest distance in the 3D color space.

The algorithm performs sufficiently only as long as the original image colors are not widely scattered through the color space. Some improvement is possible by grouping similar colors into larger cells first (by scalar quantization). However, a less frequent (but possibly important) color may get lost whenever it is not sufficiently similar to any of the n most frequent colors.

13.2.2 Median-Cut Algorithm

The median-cut algorithm [104] is considered a classical method for color quantization that is implemented in many applications (including ImageJ). As in the populosity method, a color histogram is first computed for the original image, traditionally with a reduced number of histogram cells (such as $32 \times 32 \times 32$) for efficiency reasons.[2] The initial histogram volume is then recursively split into smaller boxes until the desired number of representative colors is reached. In each recursive step, the color box representing the largest number of pixels is selected for splitting. A box is always split across the longest of its three axes at the median point, such that half of the contained pixels remain in each of the resulting subboxes (Fig. 13.3).

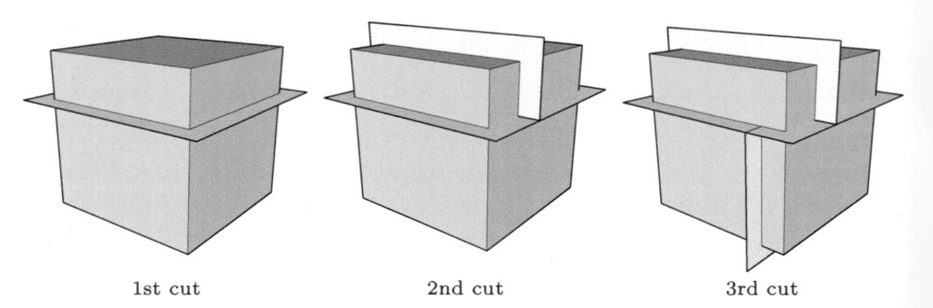

Fig. 13.3 Median-cut algorithm. The RGB color space is recursively split into smaller cubes along one of the color axes.

The result of this recursive splitting process is a partitioning of the color space into a set of disjoint boxes, with each box ideally containing the same number of image pixels. In the last step, a representative color vector (e.g., the mean vector of the contained colors) is computed for each color cube, and all the image pixels it contains are replaced by that color.

The advantage of this method is that color regions of high pixel density are split into many smaller cells, thus reducing the overall quantization error. In color regions of low density, however, relatively large cubes and thus large color deviations may occur for individual pixels.

The median-cut method is described in detail in Algorithms 13.1–13.3 and a corresponding Java implementation can be found in the source code section of this book's website (see Sec. 13.2.5).

[2] This corresponds to a scalar prequantization on the color components, which leads to additional quantization errors and thus produces suboptimal results. This step seems unnecessary on modern computers and should be avoided.

```
1:  MedianCut(I, K_max)
        I: color image, K_max: max. number of quantized colors
        Returns a new quantized image with at most K_max colors.
2:      C_q ← FindRepresentativeColors(I, K_max)
3:      return QuantizeImage(I, C_q)                    ▷ see Alg. 13.3

4:  FindRepresentativeColors(I, K_max)
        Returns a set of up to K_max representative colors for the image
        I.
5:      Let C = {c_1, c_2, ..., c_K} be the set of distinct colors in I. Each of
          the K color elements in C is a tuple c_i = ⟨red_i, grn_i, blu_i, cnt_i⟩
          consisting of the RGB color components (red, grn, blu) and
          the number of pixels (cnt) in I with that particular color.
6:      if |C| ≤ K_max then
7:          return C
8:      else
            Create a color box b_0 at level 0 that contains all image colors
            C and make it the initial element in the set of color boxes B:
9:          b_0 ← CreateColorBox(C, 0)                  ▷ see Alg. 13.2
10:         B ← {b_0}                                   ▷ initial set of color boxes
11:         k ← 1
12:         done ← false
13:         while k < N_max and ¬done do
14:             b ← FindBoxToSplit(B)                   ▷ see Alg. 13.2
15:             if b ≠ nil then
16:                 (b_1, b_2) ← SplitBox(b)            ▷ see Alg. 13.2
17:                 B ← B − {b}                         ▷ remove b from B
18:                 B ← B ∪ {b_1, b_2}                  ▷ insert b_1, b_2 into B
19:                 k ← k + 1
20:             else                                    ▷ no more boxes to split
21:                 done ← true
            Collect the average colors of all color boxes in B:
22:         C_q ← {AverageColor(b_j) | b_j ∈ B}         ▷ see Alg. 13.3
23:     return C_q
```

Alg. 13.1
Median-cut color quantization (part 1). The input image $\boldsymbol{I}$ is quantized to up to K_{max} representative colors and a new, quantized image is returned. The main work is done in procedure FindRepresentativeColors(), which iteratively partitions the color space into increasingly smaller boxes. It returns a set of representative colors ($\mathcal{C}_{\mathrm{q}}$) that are subsequently used by procedure QuantizeImage() to quantize the original image $\boldsymbol{I}$. Note that (unlike in most common implementations) no prequantization is applied to the original image colors.

13.2.3 Octree Algorithm

Similar to the median-cut algorithm, this method is also based on partitioning the 3D color space into cells of varying size. The octree algorithm [82] utilizes a hierarchical structure, where each cube in color space may contain eight subcubes. This partitioning is represented by a tree structure (octree) with a cube at each node that may again link to up to eight further nodes. Thus each node corresponds to a subrange of the color space that reduces to a single color point at a certain tree depth d (e.g., $d = 8$ for a 3×8-bit RGB color image).

When an image is processed, the corresponding quantization tree, which is initially empty, is created dynamically by evaluating all pixels in a sequence. Each pixel's color tuple is inserted into the quantization tree, while at the same time the number of nodes is limited to a predefined value K (typically 256). When a new color tuple $\mathbf{C}_i$ is inserted and the tree does not contain this color, one of the following situations can occur:

Alg. 13.2
Median-cut color quantization (part 2).

```
1: CreateColorBox(C, m)
      Creates and returns a new color box containing the colors C and
      level m. A color box b is a tuple ⟨colors, level, rmin, rmax, gmin,
      gmax, bmin, bmax⟩, where colors is the set of image colors repre-
      sented by the box, level denotes the split-level, and rmin, ..., bmax
      describe the color boundaries of the box in RGB space.
      Find the RGB extrema of all colors in C:
2:    r_min, g_min, b_min ← +∞
3:    r_max, g_max, b_max ← −∞
4:    for all c ∈ C do
          r_min ← min (r_min, red(c))
          r_max ← max (r_max, red(c))
5:        g_min ← min (g_min, grn(c))
          g_max ← max (g_max, grn(c))
          b_min ← min (b_min, blu(c))
          b_max ← max (b_max, blu(c))
6:    b ← ⟨C, m, r_min, r_max, g_min, g_max, b_min, b_max⟩
7:    return b

8: FindBoxToSplit(B)
      Searches the set of boxes B for a box to split and returns this
      box, or nil if no splittable box can be found.
      Find the set of color boxes that can be split (i.e., contain at least
      2 different colors):
9:    B_s ← { b | b ∈ B ∧ |colors(b)| ≥ 2}
10:   if B_s = {} then                      ▷ no splittable box was found
11:       return nil
12:   else
          Select a box b_x from B_s, such that level(b_x) is a minimum:
13:       b_x ← argmin_{b∈B_s}(level(b))
14:       return b_x

15: SplitBox(b)
      Splits the color box b at the median plane perpendicular to its
      longest dimension and returns a pair of new color boxes.
16:   m ← level(b)
17:   d ← FindMaxBoxDimension(b)                          ▷ see Alg. 13.3
18:   C ← colors(b)                          ▷ the set of colors in box b
          From all colors in C determine the median of the color dis-
          tribution along dimension d and split C into C_1, C_2:
                 ⎧ {c ∈ C | red(c) ≤ median_{c∈C}(red(c))}   for d = Red
19:   C_1 ←     ⎨ {c ∈ C | grn(c) ≤ median_{c∈C}(grn(c))}   for d = Green
                 ⎩ {c ∈ C | blu(c) ≤ median_{c∈C}(blu(c))}   for d = Blue
20:   C_2 ← C \ C_1
21:   b_1 ← CreateColorBox(C_1, m + 1)
22:   b_2 ← CreateColorBox(C_2, m + 1)
23:   return (b_1, b_2)
```

1. If the number of nodes is less than K and no suitable node for the color $\mathbf{c}_i$ exists already, then a new node is created for $\mathbf{C}_i$.
2. Otherwise (i.e., if the number of nodes is K), the existing nodes at the maximum tree depth (which represent similar colors) are merged into a common node.

Alg. 13.3
Median-cut color quantization (part 3).

```
1:  AverageColor(b)
      Returns the average color c_avg for the pixels represented by the
      color box b.
2:    C ← colors(b)                          ▷ the set of colors in box b
3:    n ← 0
4:    Σ_r ← 0,  Σ_g ← 0,  Σ_b ← 0
5:    for all c ∈ C do
6:        k ← cnt(c)
7:        n ← n + k
8:        Σ_r ← Σ_r + k · red(c)
9:        Σ_g ← Σ_g + k · grn(c)
10:       Σ_b ← Σ_b + k · blu(c)
11:   c̄ ← (Σ_r/n, Σ_g/n, Σ_b/n)
12:   return c̄

13: FindMaxBoxDimension(b)
      Returns the largest dimension of the color box b (Red, Green, or
      Blue).
14:   d_r = rmax(b) − rmin(b)
15:   d_g = gmax(b) − gmin(b)
16:   d_b = bmax(b) − bmin(b)
17:   d_max = max(d_r, d_g, d_b)
18:   if d_max = d_r then
19:       return Red.
20:   else if d_max = d_g then
21:       return Green
22:   else
23:       return Blue

24: QuantizeImage(I, C_q)
      Returns a new image with color pixels from I replaced by their
      closest representative colors in C_q.
25:   I' ← duplicate(I)                      ▷ create a new image
26:   for all image coordinates (u, v) do
          Find the quantization color in C_q that is "closest" to the cur-
          rent pixel color (e.g., using the Euclidean distance in RGB
          space):
27:       I'(u, v) ← argmin_{c∈C_q} ||I(u, v) − c||
28:   return I'
```

A key advantage of the iterative octree method is that the number of color nodes remains limited to K in any step and thus the amount of required storage is small. The final replacement of the image pixels by the quantized color vectors can also be performed easily and efficiently with the octree structure because only up to eight comparisons (one at each tree layer) are necessary to locate the best-matching color for each pixel.

Figure 13.4 shows the resulting color distributions in RGB space after applying the median-cut and octree algorithms. In both cases, the original image (Fig. 13.2(a)) is quantized to 256 colors. Notice in particular the dense placement of quantized colors in certain regions of the green hues. For both algorithms and the (scalar) 3:3:2 quan-

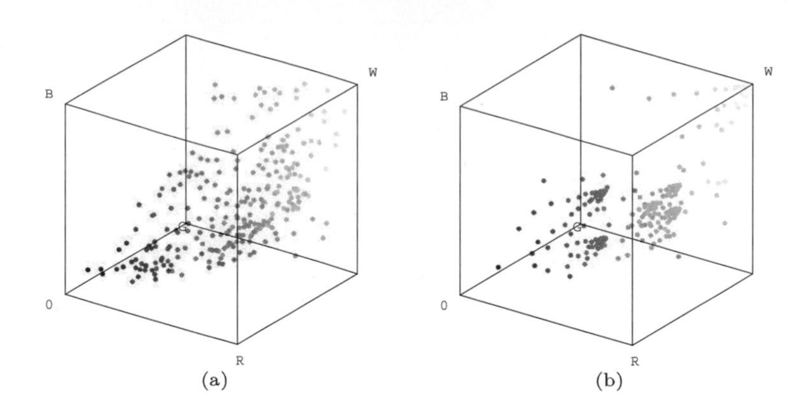

(a) (b)

Fig. 13.4 Color distribution after application of the median-cut (a) and octree (b) algorithms. In both cases, the set of 226,321 colors in the original image (Fig. 13.2(a)) was reduced to 256 representative colors.

tization, the resulting distances between the original pixels and the quantized colors are shown in Fig. 13.5. The greatest error naturally results from 3:3:2 quantization, because this method does not consider the contents of the image at all. Compared with the median-cut method, the overall error for the octree algorithm is smaller, although the latter creates several large deviations, particularly inside the colored foreground regions and the forest region in the background. In general, however, the octree algorithm does not offer significant advantages in terms of the resulting image quality over the simpler median-cut algorithm.

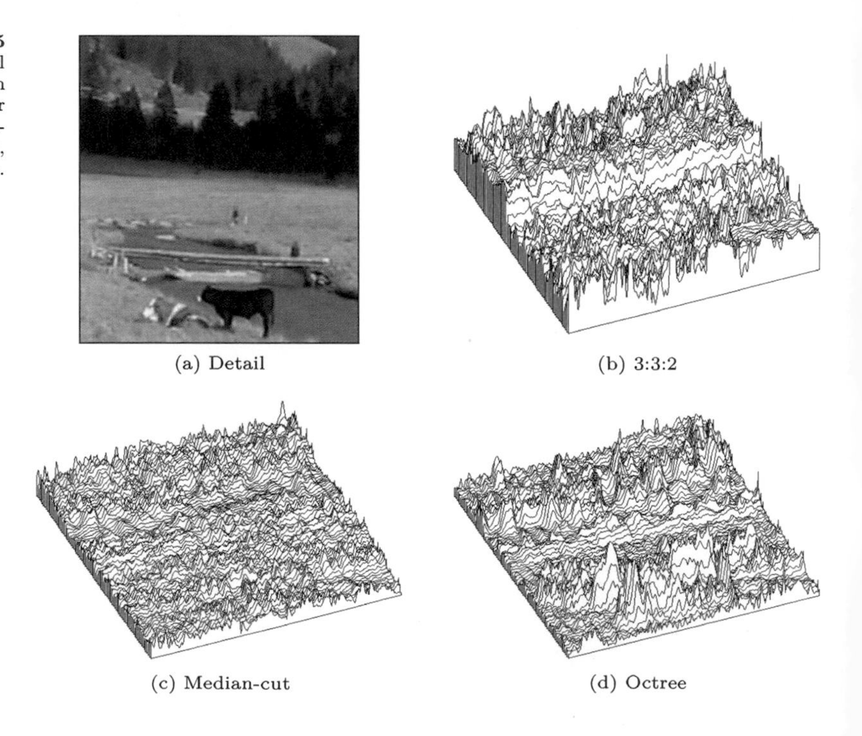

(a) Detail (b) 3:3:2 (c) Median-cut (d) Octree

Fig. 13.5 Quantization errors. Original image (a), distance between original and quantized color pixels for scalar 3:3:2 quantization (b), median-cut (c), and octree (d) algorithms.

13.2.4 Other Methods for Vector Quantization

A suitable set of representative color vectors can usually be determined without inspecting all pixels in the original image. It is often sufficient to use only 10% of randomly selected pixels to obtain a high probability that none of the important colors is lost.

In addition to the color quantization methods described already, several other procedures and refined algorithms have been proposed. This includes statistical and clustering methods, such as the classical *k-means* algorithm, but also the use of neural networks and genetic algorithms. A good overview can be found in [219].

13.2.5 Java Implementation

The Java implementation[3] of the algorithms described in this chapter consists of a common interface `ColorQuantizer` and the concrete classes

- `MedianCutQuantizer`,
- `OctreeQuantizer`.

Program 13.2 shows a complete ImageJ plugin that employs the class `MedianCutQuantizer` for quantizing an RGB full-color image to an indexed image. The choice of data structures for the representation of color sets and the implementation of the associated set operations are essential to achieve good performance. The data structures used in this implementation are illustrated in Fig. 13.6.

Initially, the set of all colors contained in the original image (`ip` of type `ColorProcessor`) is computed by `new ColorHistogram()`. The result is an array `imageColors` of size K Each cell of `imageColors` refers to a `colorNode` object ($\boldsymbol{c}_i$) that holds the associated color (`red`, `green`, `blue`) and its frequency (`cnt`) in the image. Each `colorBox` object (corresponding to a color box $\boldsymbol{b}$ in Alg. 13.1) selects a contiguous range of image colors, bounded by the indices `lower` and `upper`. The ranges of elements in `imageColors`, indexed by different `colorBox` objects, never overlap. Each element in `imageColors` is contained in exactly one `colorBox`; that is, the color boxes held in `colorSet` ($\mathcal{B}$ in Alg. 13.1) form a partitioning of `imageColors` (`colorSet` is implemented as a list of `ColorBox` objects). To split a particular `colorBox` along a color dimension $d =$ Red, Green, or Blue, the corresponding subrange of elements in `imageColors` is *sorted* with the property `red`, `green`, or `blue`, respectively, as the sorting key. In Java, this is quite easy to implement using the standard `Arrays.sort()` method and a dedicated `Comparator` object for each color dimension. Finally, the method `quantize()` replaces each pixel in `ip` by the closest color in `colorSet`.

13.3 Exercises

Exercise 13.1. Simplify the 3:3:2 quantization given in Prog. 13.1 such that only a single bit mask/shift step is performed for each color component.

[3] Package `imagingbook.pub.color.quantize`.

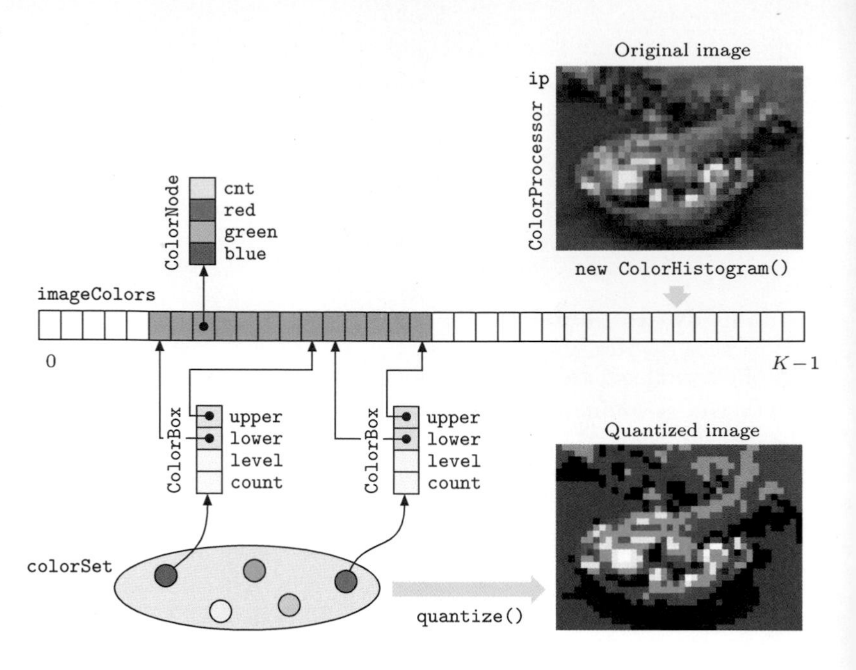

Fig. 13.6
Data structures used in the implementation of the median-cut quantization algortihm (class `MedianCutQuantizer`).

Exercise 13.2. The median-cut algorithm for color quantization (Sec. 13.2.2) is implemented in the *Independent JPEG Group's*[4] `libjpeg` open source software with the following modification: the choice of the cube to be split next depends alternately on (a) the number of contained image pixels and (b) the cube's geometric volume. Consider the possible motives and discuss examples where this approach may offer an improvement over the original algorithm.

Exercise 13.3. The *signal-to-noise ratio* (SNR) is a common measure for quantifying the loss of image quality introduced by color quantization. It is defined as the ratio between the average *signal energy* P_{signal} and the average *noise energy* P_{noise}. For example, given an original color image $\boldsymbol{I}$ and the associated quantized image $\boldsymbol{I}'$, this ratio could be calculated as

$$\mathrm{SNR}(\boldsymbol{I},\boldsymbol{I}') = \frac{P_{\mathrm{signal}}}{P_{\mathrm{noise}}} = \frac{\sum_{u=0}^{M-1}\sum_{v=0}^{N-1} \|\boldsymbol{I}(u,v)\|^2}{\sum_{u=0}^{M-1}\sum_{v=0}^{N-1} \left\|\boldsymbol{I}(u,v)-\boldsymbol{I}'(u,v)\right\|^2} . \tag{13.2}$$

Thus all deviations between the original and the quantized image are considered "noise". The signal-to-noise ratio is usually specified on a logarithmic scale with the unit *decibel* (dB), that is,

$$\mathrm{SNR}_{\log}(\boldsymbol{I},\boldsymbol{I}') = 10 \cdot \log_{10}(\mathrm{SNR}(\boldsymbol{I},\boldsymbol{I}')) \;\; [\mathrm{dB}]. \tag{13.3}$$

Implement the calculation of the SNR, as defined in Eqns. (13.2)–(13.3), for color images and compare the results for the median-cut and the octree algorithms for the same number of target colors.

[4] www.ijg.org.

```
1 import ij.ImagePlus;
2 import ij.plugin.filter.PlugInFilter;
3 import ij.process.ByteProcessor;
4 import ij.process.ColorProcessor;
5 import ij.process.ImageProcessor;
6 import imagingbook.pub.color.quantize.ColorQuantizer;
7 import imagingbook.pub.color.quantize.MedianCutQuantizer;
8
9 public class Median_Cut_Quantization implements
      PlugInFilter {
10   static int NCOLORS = 32;
11
12   public int setup(String arg, ImagePlus imp) {
13     return DOES_RGB + NO_CHANGES;
14   }
15
16   public void run(ImageProcessor ip) {
17     ColorProcessor cp = ip.convertToColorProcessor();
18     int w = ip.getWidth();
19     int h = ip.getHeight();
20
21     // create a quantizer:
22     ColorQuantizer q =
23         new MedianCutQuantizer(cp, NCOLORS);
24
25     // quantize cp to an indexed image:
26     ByteProcessor idxIp = q.quantize(cp);
27     (new ImagePlus("Quantized Index Image", idxIp)).show();
28
29     // quantize cp to an RGB image:
30     int[] rgbPix = q.quantize((int[]) cp.getPixels());
31     ImageProcessor rgbIp =
32           new ColorProcessor(w, h, rgbPix);
33     (new ImagePlus("Quantized RGB Image", rgbIp)).show();
34   }
35 }
```

Prog. 13.2
Color quantization by the median-cut method (ImageJ plugin). This example uses the class `MedianCutQuantizer` to quantize the original full-color RGB image into (a) an indexed color image (of type `ByteProcessor`) and (b) another RGB image (of type `ColorProcessor`). Both images are finally displayed.

14

Colorimetric Color Spaces

In any application that requires precise, reproducible, and device-independent presentation of colors, the use of calibrated color systems is an absolute necessity. For example, color calibration is routinely used throughout the digital print work flow but also in digital film production, professional photography, image databases, etc. One may have experienced how difficult it is, for example, to render a good photograph on a color laser printer, and even the color reproduction on monitors largely depends on the particular manufacturer and computer system.

All the color spaces described in Chapter 12, Sec. 12.2, somehow relate to the physical properties of some media device, such as the specific colors of the phosphor coatings inside a CRT tube or the colors of the inks used for printing. To make colors appear similar or even identical on different media modalities, we need a representation that is independent of how a particular device reproduces these colors. Color systems that describe colors in a measurable, device-independent fashion are called *colorimetric* or *calibrated*, and the field of *color science* is traditionally concerned with the properties and application of these color systems (see, e.g., [258] or [215] for an overview). While several colorimetric standards exist, we focus on the most widely used CIE systems in the remaining part of this section.

14.1 CIE Color Spaces

The XYZ color system, developed by the CIE (Commission Internationale d'Èclairage)[1] in the 1920s and standardized in 1931, is the foundation of most colorimetric color systems that are in use today [195, p. 22].

[1] International Commission on Illumination (www.cie.co.at).

14.1.1 CIE XYZ Color Space

The CIE XYZ color scheme was developed after extensive measurements of human visual perception under controlled conditions. It is based on three imaginary primary colors X, Y, Z, which are chosen such that all visible colors can be described as a summation of positive-only components, where the Y component corresponds to the perceived lightness or *luminosity* of a color. All visible colors lie inside a 3D cone-shaped region (Fig. 14.1(a)), which interestingly enough does not include the primary colors themselves.

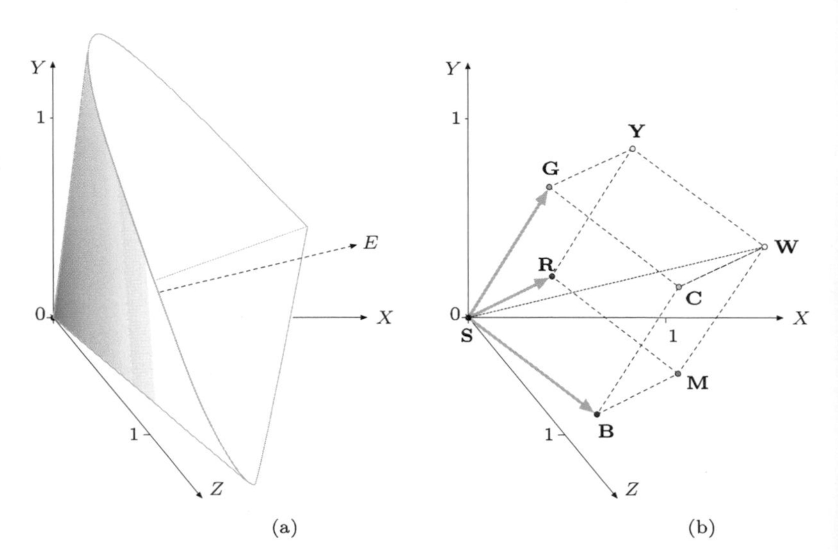

Fig. 14.1
The XYZ color space is defined by the three imaginary primary colors X, Y, Z, where the Y dimension corresponds to the perceived luminance. All visible colors are contained inside an open, cone-shaped volume that originates at the black point **S** (a), where E denotes the axis of neutral (gray) colors. The RGB color space maps to the XYZ space as a linearly distorted cube (b). See also Fig. 14.5(a).

Some common color spaces, and the RGB color space in particular, conveniently relate to XYZ space by a *linear* coordinate transformation, as described in Sec. 14.4. Thus, as shown in Fig. 14.1(b), the RGB color space is embedded in the XYZ space as a distorted cube, and therefore straight lines in RGB space map to straight lines in XYZ again. The CIE XYZ scheme is (similar to the RGB color space) *nonlinear* with respect to human visual perception, that is, a particular fixed distance in XYZ is not perceived as a uniform color change throughout the entire color space. The XYZ coordinates of the RGB color cube (based on the primary colors defined by ITU-R BT.709) are listed in Table 14.1.

14.1.2 CIE x, y Chromaticity

As mentioned, the luminance in XYZ color space increases along the Y axis, starting at the black point **S** located at the coordinate origin $(X = Y = Z = 0)$. The color hue is independent of the luminance and thus independent of the Y value. To describe the corresponding "pure" color hues and saturation in a convenient manner, the CIE system also defines the three *chromaticity* values

$$x = \frac{X}{X+Y+Z}, \quad y = \frac{Y}{X+Y+Z}, \quad z = \frac{Z}{X+Y+Z}, \tag{14.1}$$

where (obviously) $x+y+z = 1$ and thus one of the three values (e.g., z) is redundant. Equation (14.1) describes a central projection from

Pt.	Color	R	G	B	X	Y	Z	x	y
S	Black	0.00	0.00	0.00	0.0000	0.0000	0.0000	0.3127	0.3290
R	Red	1.00	0.00	0.00	0.4125	0.2127	0.0193	0.6400	0.3300
Y	Yellow	1.00	1.00	0.00	0.7700	0.9278	0.1385	0.4193	0.5052
G	Green	0.00	1.00	0.00	0.3576	0.7152	0.1192	0.3000	0.6000
C	Cyan	0.00	1.00	1.00	0.5380	0.7873	1.0694	0.2247	0.3288
B	Blue	0.00	0.00	1.00	0.1804	0.0722	0.9502	0.1500	0.0600
M	Magenta	1.00	0.00	1.00	0.5929	0.2848	0.9696	0.3209	0.1542
W	White	1.00	1.00	1.00	0.9505	1.0000	1.0888	0.3127	0.3290

Table 14.1
Coordinates of the RGB color cube in CIE XYZ space. The X, Y, Z values refer to standard (ITU-R BT. 709) primaries and white point D65 (see Table 14.2), x, y denote the corresponding CIE chromaticity coordinates.

X, Y, Z coordinates onto the 3D plane

$$X + Y + Z = 1, \qquad (14.2)$$

with the origin **S** as the projection center (Fig. 14.2). Thus, for an arbitrary XYZ color point $\mathbf{A} = (X_a, Y_a, Z_a)$, the corresponding chromaticity coordinates $\mathbf{a} = (x_a, y_a, z_a)$ are found by intersecting the line $\overline{\mathbf{SA}}$ with the $X + Y + Z = 1$ plane (Fig. 14.2(a)). The final x, y coordinates are the result of projecting these intersection points onto the X/Y-plane (Fig. 14.2(b)) by simply dropping the Z component z_a.

The result is the well-known horseshoe-shaped *CIE* x, y *chromaticity diagram*, which is shown in Fig. 14.2(c). Any x, y point in this diagram defines the hue and saturation of a particular color, but only the colors inside the horseshoe curve are potentially visible. Obviously an infinite number of X, Y, Z colors (with different luminance values) project to the same x, y, z chromaticity values, and the XYZ color coordinates thus cannot be uniquely reconstructed from given chromaticity values. Additional information is required. For example, it is common to specify the visible colors of the CIE system in the form Yxy, where Y is the original luminance component of the XYZ color. Given a pair of chromaticity values x, y (with $y > 0$) and an arbitrary Y value, the missing X, Z coordinates are obtained (using the definitions in Eqn. (14.1)) as

$$X = x \cdot \frac{Y}{y}, \qquad Z = z \cdot \frac{Y}{y} = (1 - x - y) \cdot \frac{Y}{y}. \qquad (14.3)$$

The CIE diagram not only yields an intuitive layout of color hues but exhibits some remarkable formal properties. The xy values along the outer horseshoe boundary correspond to monochromatic ("spectrally pure"), maximally saturated colors with wavelengths ranging from below 400 nm (purple) up to 780 nm (red). Thus the position of any color inside the xy diagram can be specified with respect to any of the primary colors at the boundary, except for the points on the connecting line ("purple line") between 380 and 780 nm, whose purple hues do not correspond to primary colors but can only be generated by mixing other colors.

The *saturation* of colors falls off continuously toward the "neutral point" (**E**) at the center of the horseshoe, with $x = y = \frac{1}{3}$ (or $X = Y = Z = 1$, respectively) and zero saturation. All other colorless (i.e., gray) values also map to the neutral point, just as any set of colors

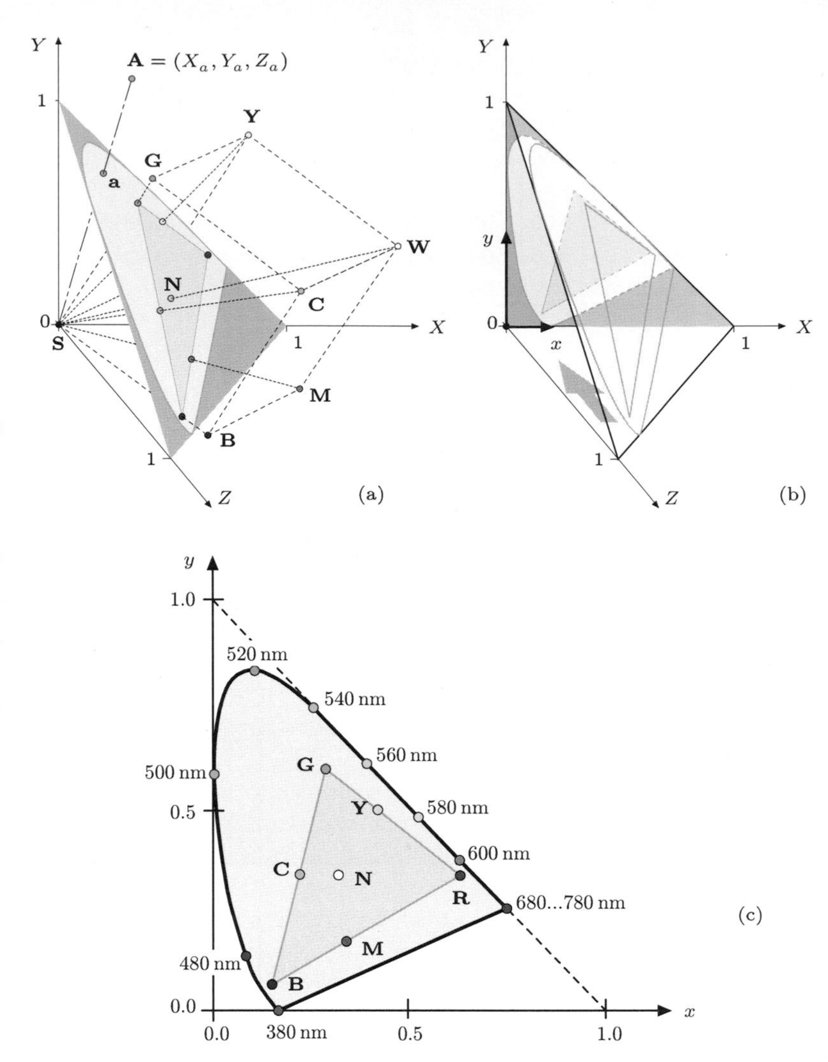

Fig. 14.2 CIE x, y chromaticity diagram. For an arbitrary XYZ color point $\mathbf{A} = (X_a, Y_a, Z_a)$, the chromaticity values $\mathbf{a} = (x_a, y_a, z_a)$ are obtained by a central projection onto the 3D plane $X + Y + Z = 1$ (a). The corner points of the RGB cube map to a triangle, and its white point **W** maps to the (colorless) neutral point **E**. The intersection points are then projected onto the X/Y plane (b) by simply dropping the Z component, which produces the familiar CIE chromaticity diagram shown in (c). The CIE diagram contains all visible color tones (hues and saturations) but no luminance information, with wavelengths in the range 380–780 nanometers. A particular color space is specified by at least three primary colors (tristimulus values; e.g., **R**, **G**, **B**), which define a triangle (linear hull) containing all representable colors.

with the same hue but different brightness corresponds to a single x, y point. All possible composite colors lie inside the convex hull specified by the coordinates of the primary colors of the CIE diagram and, in particular, complementary colors are located on straight lines that run diagonally through the white point.

14.1.3 Standard Illuminants

A central goal of colorimetry is the quantitative measurement of colors in physical reality, which strongly depends on the color properties of the illumination. The CIE system specifies a number of standard illuminants for a variety of real and hypothetical light sources, each specified by a spectral radiant power distribution and the "correlated color temperature" (expressed in degrees Kelvin) [258, Sec. 3.3.3]. The following daylight (D) illuminants are particularly important for the design of digital color spaces (Table 14.2):

D50 emulates the spectrum of natural (direct) sunlight with an equivalent color temperature of approximately 5000° K. D50 is the recommended illuminant for viewing reflective images, such as paper prints. In practice, D50 lighting is commonly implemented with fluorescent lamps using multiple phosphors to approximate the specified color spectrum.

D65 has a correlated color temperature of approximately 6500° K and is designed to emulate the average (indirect) daylight observed under an overcast sky on the northern hemisphere. D65 is also used as the reference white for emittive devices, such as display screens.

The standard illuminants serve to specify the ambient viewing light but also to define the reference white points in various color spaces in the CIE color system. For example, the sRGB standard (see Sec. 14.4) refers to D65 as the media white point and D50 as the ambient viewing illuminant. In addition, the CIE system also specifies the range of admissible viewing angles (commonly at ±2°).

	°K	X	Y	Z	x	y
D50	5000	0.96429	1.00000	0.82510	0.3457	0.3585
D65	6500	0.95045	1.00000	1.08905	0.3127	0.3290
N	—	1.00000	1.00000	1.00000	$0.333\dot{3}$	$0.333\dot{3}$

Table 14.2
CIE color parameters for the standard illuminants **D50** and **D65**. **E** denotes the absolute neutral point in CIE XYZ space.

14.1.4 Gamut

The set of all colors that can be handled by a certain media device or can be represented by a particular color space is called "gamut". This is usually a contiguous region in the 3D CIE XYZ color space or, reduced to the representable color hues and ignoring the luminance component, a convex region in the 2D CIE chromaticity diagram.

Figure 14.3 illustrates some typical gamut regions inside the CIE diagram. The gamut of an output device mainly depends on the technology employed. For example, ordinary color monitors are typically not capable of displaying all colors of the gamut covered by the corresponding color space (usually sRGB). Conversely, it is also possible that devices would reproduce certain colors that cannot be represented in the utilized color space. Significant deviations exist, for example, between the RGB color space and the gamuts associated with CMYK-based printers. Also, media devices with very large gamuts exist, as demonstrated by the laser display system in Fig. 14.3. Representing such large gamuts and, in particular, transforming between different color representations requires adequately sized color spaces, such as the Adobe-RGB color space or CIELAB (described in Sec. 14.2), which covers the entire visible portion of the CIE diagram.

14.1.5 Variants of the CIE Color Space

The original CIEXYZ color space and the derived xy chromaticity diagram have the disadvantage that color differences are not perceived equally in different regions of the color space. For example,

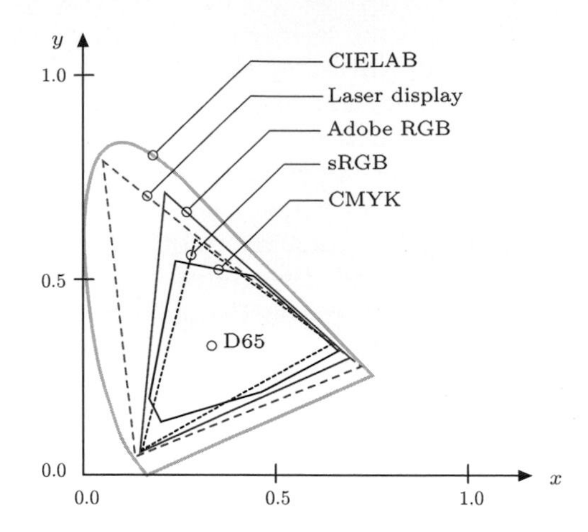

Fig. 14.3
Gamut regions for different color spaces and output devices inside the CIE diagram.

large color changes are perceived in the *magenta* region for a given shift in XYZ while the change is relatively small in the *green* region for the same coordinate distance. Several variants of the CIE color space have been developed for different purposes, primarily with the goal of creating perceptually uniform color representations without sacrificing the formal qualities of the CIE reference system. Popular CIE-derived color spaces include CIE YUV, YU′V′, YC_bC_r, and particularly CIELAB and CIELUV, which are described in the following sections. In addition, CIE-compliant specifications exist for most common color spaces (see Ch. 12, Sec. 12.2), which allow more or less dependable conversions between almost any pair of color spaces.

14.2 CIELAB

The CIELAB color model (specified by CIE in 1976) was developed with the goal of linearizing the representation with respect to human color perception and at the same time creating a more intuitive color system. Since then, CIELAB[2] has become a popular and widely used color model, particularly for high-quality photographic applications. It is used, for example, inside Adobe Photoshop as the standard model for converting between different color spaces. The dimensions in this color space are the luminosity L^* and the two color components a^*, b^*, which specify the color hue and saturation along the *green-red* and *blue-yellow* axes, respectively. All three components are *relative* values and refer to the specified reference white point $\mathbf{C}_{\mathrm{ref}} = (X_{\mathrm{ref}}, Y_{\mathrm{ref}}, Z_{\mathrm{ref}})$. In addition, a nonlinear correction function (similar to the modified gamma correction described in Ch. 4, Sec. 4.7.6) is applied to all three components, as will be detailed further.

14.2.1 CIEXYZ→CIELAB Conversion

Several specifications for converting to and from CIELAB space exist that, however, differ marginally and for very small L values only. The

[2] Often CIELAB is simply referred to as the “Lab” color space.

L^*

a^*

b^*

Fig. 14.4
CIELAB components shown as grayscale images. The contrast of the a^* and b^* images has been increased by 40% for better viewing.

current specification for converting between CIEXYZ and CIELAB colors is defined by ISO Standard 13655 [120] as follows:

$$L^* = 116 \cdot Y' - 16, \tag{14.4}$$
$$a^* = 500 \cdot (X' - Y'), \tag{14.5}$$
$$b^* = 200 \cdot (Y' - Z'), \tag{14.6}$$

with

$$X' = f_1\big(\tfrac{X}{X_{\mathrm{ref}}}\big), \quad Y' = f_1\big(\tfrac{Y}{Y_{\mathrm{ref}}}\big), \quad Z' = f_1\big(\tfrac{Z}{Z_{\mathrm{ref}}}\big), \tag{14.7}$$
$$f_1(c) = \begin{cases} c^{1/3} & \text{for } c > \epsilon, \\ \kappa \cdot c + \frac{16}{116} & \text{for } c \leq \epsilon, \end{cases} \tag{14.8}$$

and

$$\epsilon = \big(\tfrac{6}{29}\big)^3 = \tfrac{216}{24389} \approx 0.008856, \tag{14.9}$$
$$\kappa = \tfrac{1}{116}\,\big(\tfrac{29}{3}\big)^3 = \tfrac{841}{108} \approx 7.787. \tag{14.10}$$

For the conversion in Eqn. (14.7), D65 is usually specified as the reference white point $\mathbf{C}_{\mathrm{ref}} = (X_{\mathrm{ref}}, Y_{\mathrm{ref}}, Z_{\mathrm{ref}})$, that is, $X_{\mathrm{ref}} = 0.95047$, $Y_{\mathrm{ref}} = 1.0$ and $Z_{\mathrm{ref}} = 1.08883$ (see Table 14.2). The L^* values are positive and typically in the range $[0, 100]$ (often scaled to $[0, 255]$), but may theoretically be greater. Values for a^* and b^* are in the range $[-127, +127]$. Figure 14.4 shows the separation of a color image into the corresponding CIELAB components. Table 14.3 lists the relation between CIELAB and XYZ coordinates for selected RGB colors. The given $R'G'B'$ values are (nonlinear) sRGB coordinates with D65 as the reference white point.[3] Figure 14.5(c) shows the transformation of the RGB color cube into the CIELAB color space.

14.2.2 CIELAB→CIEXYZ Conversion

The reverse transformation from CIELAB space to CIEXYZ coordinates is defined as follows:

$$X = X_{\mathrm{ref}} \cdot f_2\big(L' + \tfrac{a^*}{500}\big), \tag{14.11}$$
$$Y = Y_{\mathrm{ref}} \cdot f_2\big(L'\big), \tag{14.12}$$
$$Z = Z_{\mathrm{ref}} \cdot f_2\big(L' - \tfrac{b^*}{200}\big), \tag{14.13}$$

[3] Note that sRGB colors in Java are specified with respect to white point D50, which explains certain numerical deviations (see Sec. 14.7).

Table 14.3
CIELAB coordinates for selected color points in sRGB. The sRGB components R', G', B' are nonlinear (i.e., gamma-corrected), white point is D65 (see Table 14.2).

		sRGB			CIEXYZ (D65)			CIELAB		
Pt.	Color	R'	G'	B'	X_{65}	Y_{65}	Z_{65}	L^*	a^*	b^*
S	Black	0.00	0.00	0.00	0.0000	0.0000	0.0000	0.00	0.00	0.00
R	Red	1.00	0.00	0.00	0.4125	0.2127	0.0193	53.24	80.09	67.20
Y	Yellow	1.00	1.00	0.00	0.7700	0.9278	0.1385	97.14	−21.55	94.48
G	Green	0.00	1.00	0.00	0.3576	0.7152	0.1192	87.74	−86.18	83.18
C	Cyan	0.00	1.00	1.00	0.5380	0.7873	1.0694	91.11	−48.09	−14.13
B	Blue	0.00	0.00	1.00	0.1804	0.0722	0.9502	32.30	79.19	−107.86
M	Magenta	1.00	0.00	1.00	0.5929	0.2848	0.9696	60.32	98.24	−60.83
W	White	1.00	1.00	1.00	0.9505	1.0000	1.0888	100.00	0.00	0.00
K	50% Gray	0.50	0.50	0.50	0.2034	0.2140	0.2330	53.39	0.00	0.00
$\mathbf{R}_{75}$	75% Red	0.75	0.00	0.00	0.2155	0.1111	0.0101	39.77	64.51	54.13
$\mathbf{R}_{50}$	50% Red	0.50	0.00	0.00	0.0883	0.0455	0.0041	25.42	47.91	37.91
$\mathbf{R}_{25}$	25% Red	0.25	0.00	0.00	0.0210	0.0108	0.0010	9.66	29.68	15.24
P	Pink	1.00	0.50	0.50	0.5276	0.3812	0.2482	68.11	48.39	22.83

with

$$L' = \tfrac{L^*+16}{116} \quad \text{and} \tag{14.14}$$

$$f_2(c) = \begin{cases} c^3 & \text{for } c^3 > \epsilon, \\ \frac{c-16/116}{\kappa} & \text{for } c^3 \leq \epsilon, \end{cases} \tag{14.15}$$

and ϵ, κ as defined in Eqns. (14.9–14.10). The complete Java code for the CIELAB→XYZ conversion and the implementation of the associated `ColorSpace` class can be found in Progs. 14.1 and 14.2 (pp. 363–364).

14.3 CIELUV

14.3.1 CIEXYZ→CIELUV Conversion

The CIELUV component values L^*, u^*, v^* are calculated from given X, Y, Z color coordinates as follows:

$$L^* = 116 \cdot Y' - 16, \tag{14.16}$$

$$u^* = 13 \cdot L^* \cdot (u' - u'_{\mathrm{ref}}), \tag{14.17}$$

$$v^* = 13 \cdot L^* \cdot (v' - v'_{\mathrm{ref}}), \tag{14.18}$$

with Y' as defined in Eqn. (14.7) (identical to CIELAB) and

$$\begin{aligned} u' &= f_u(X,Y,Z), & u'_{\mathrm{ref}} &= f_u(X_{\mathrm{ref}}, Y_{\mathrm{ref}}, Z_{\mathrm{ref}}), \\ v' &= f_v(X,Y,Z), & v'_{\mathrm{ref}} &= f_v(X_{\mathrm{ref}}, Y_{\mathrm{ref}}, Z_{\mathrm{ref}}), \end{aligned} \tag{14.19}$$

with the correction functions

$$f_u(X,Y,Z) = \begin{cases} 0 & \text{for } X = 0, \\ \frac{4X}{X+15Y+3Z} & \text{for } X > 0, \end{cases} \tag{14.20}$$

$$f_v(X,Y,Z) = \begin{cases} 0 & \text{for } Y = 0, \\ \frac{9Y}{X+15Y+3Z} & \text{for } Y > 0. \end{cases} \tag{14.21}$$

Linear RGB sRGB

(a) **XYZ** (b)

(c) **CIELAB** (d)

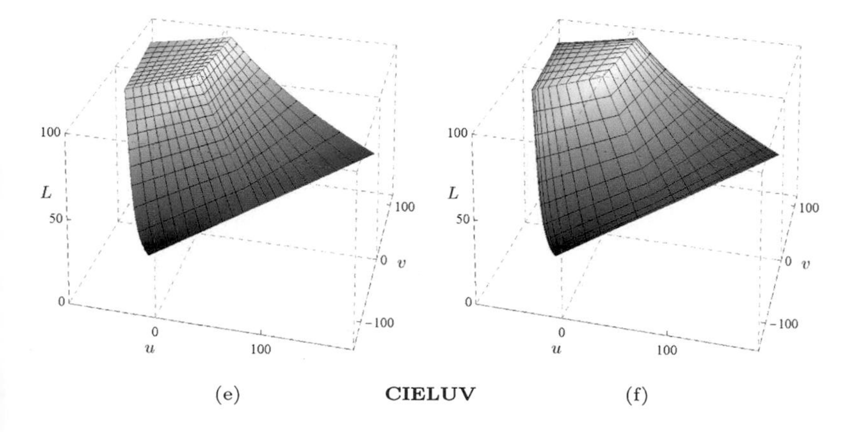

(e) **CIELUV** (f)

Fig. 14.5
Transformation of the RGB color cube to the XYZ, CIELAB, and CIELUV color space. The left column shows the color cube in *linear* RGB space, the right column in *nonlinear* sRGB space. Both RGB volumes were uniformly subdivided into 10 × 10 × 10 cubes of equal size. In both cases, the transformation to XYZ space (a, b) yields a distorted cube with straight edges and planar faces. Due to the linear transformation from RGB to XYZ, the subdivision of the RGB cube remains uniform (a). However, the nonlinear transformation (due to gamma correction) from sRGB to XYZ makes the tesselation strongly nonuniform in XYZ space (b). Since CIELAB uses gamma correction as well, the transformation of the linear RGB cube in (c) appears much less uniform than the nonlinear sRGB cube in (d), although this appears to be the other way round in CIELUV (e, f). Note that the RGB/sRGB color cube maps to a *non-convex* volume in both the CIELAB and the CLIELUV space.

Note that the checks for zero X, Y in Eqns. (14.20)–(14.21) are not part of the original definitions but are essential in any real implementation to avoid divisions by zero.[4]

[4] Remember though that floating-point values (`double`, `float`) should never be strictly tested against zero but compared to a sufficiently small (epsilon) quantity (see Sec. F.1.8 in the Appendix).

Table 14.4
CIELUV coordinates for selected color points in sRGB. Reference white point is D65. The L^* values are identical to CIELAB (see Table 14.3).

		sRGB			CIEXYZ (D65)			CIELUV		
Pt.	Color	R'	G'	B'	X_{65}	Y_{65}	Z_{65}	L^*	u^*	v^*
S	Black	0.00	0.00	0.00	0.0000	0.0000	0.0000	0.00	0.00	0.00
R	Red	1.00	0.00	0.00	0.4125	0.2127	0.0193	53.24	175.01	37.75
Y	Yellow	1.00	1.00	0.00	0.7700	0.9278	0.1385	97.14	7.70	106.78
G	Green	0.00	1.00	0.00	0.3576	0.7152	0.1192	87.74	−83.08	107.39
C	Cyan	0.00	1.00	1.00	0.5380	0.7873	1.0694	91.11	−70.48	−15.20
B	Blue	0.00	0.00	1.00	0.1804	0.0722	0.9502	32.30	−9.40	−130.34
M	Magenta	1.00	0.00	1.00	0.5929	0.2848	0.9696	60.32	84.07	−108.68
W	White	1.00	1.00	1.00	0.9505	1.0000	1.0888	100.00	0.00	0.00
K	50% Gray	0.50	0.50	0.50	0.2034	0.2140	0.2330	53.39	0.00	0.00
$\mathbf{R}_{75}$	75% Red	0.75	0.00	0.00	0.2155	0.1111	0.0101	39.77	130.73	28.20
$\mathbf{R}_{50}$	50% Red	0.50	0.00	0.00	0.0883	0.0455	0.0041	25.42	83.56	18.02
$\mathbf{R}_{25}$	25% Red	0.25	0.00	0.00	0.0210	0.0108	0.0010	9.66	31.74	6.85
P	Pink	1.00	0.50	0.50	0.5276	0.3812	0.2482	68.11	92.15	19.88

14.3.2 CIELUV→CIEXYZ Conversion

The reverse mapping from L^*, u^*, v^* components to X, Y, Z coordinates is defined as follows:

$$Y = Y_{\mathrm{ref}} \cdot f_2\big(\tfrac{L^*+16}{116}\big), \tag{14.22}$$

with $f_2()$ as defined in Eqn. (14.15), and

$$X = Y \cdot \frac{9u'}{4v'}, \qquad Z = Y \cdot \frac{12 - 3u' - 20v'}{4v'}, \tag{14.23}$$

with

$$(u', v') = \begin{cases} (u'_{\mathrm{ref}}, v'_{\mathrm{ref}}) & \text{for } L^* = 0, \\ (u'_{\mathrm{ref}}, v'_{\mathrm{ref}}) + \frac{1}{13 \cdot L^*} \cdot (u^*, v^*) & \text{for } L^* > 0, \end{cases} \tag{14.24}$$

and $u'_{\mathrm{ref}}, v'_{\mathrm{ref}}$ as in Eqn. (14.19).[5]

14.3.3 Measuring Color Differences

Due to its high uniformity with respect to human color perception, the CIELAB color space is a particularly good choice for determining the difference between colors (the same holds for the CIELUV space) [94, p. 57]. The difference between two color points $\mathbf{c}_1 = (L^*_1, a^*_1, b^*_1)$ and $\mathbf{c}_2 = (L^*_2, a^*_2, b^*_2)$ can be found by simply measuring the *Euclidean distance* in CIELAB or CIELUV space, for example,

$$\mathrm{ColorDist}(\mathbf{c}_1, \mathbf{c}_2) = \|\mathbf{c}_1 - \mathbf{c}_2\| \tag{14.25}$$

$$= \sqrt{(L^*_1 - L^*_2)^2 + (a^*_1 - a^*_2)^2 + (b^*_1 - b^*_2)^2}. \tag{14.26}$$

14.4 Standard RGB (sRGB)

CIE-based color spaces such as CIELAB (and CIELUV) are device-independent and have a gamut sufficiently large to represent virtually

[5] No explicit check for zero denominators is required in Eqn. (14.23) since v' can be assumed to be greater than zero.

all visible colors in the CIEXYZ system. However, in many computer-based, display-oriented applications, such as computer graphics or multimedia, the direct use of CIE-based color spaces may be too cumbersome or inefficient.

sRGB ("standard RGB" [119]) was developed (jointly by Hewlett-Packard and Microsoft) with the goal of creating a precisely specified color space for these applications, based on standardized mappings with respect to the colorimetric CIEXYZ color space. This includes precise specifications of the three primary colors, the white reference point, ambient lighting conditions, and gamma values. Interestingly, the sRGB color specification is the same as the one specified many years before for the European PAL/SECAM television standards. Compared to CIELAB, sRGB exhibits a relatively small gamut (see Fig. 14.3), which, however, includes most colors that can be reproduced by current computer and video monitors. Although sRGB was not designed as a universal color space, its CIE-based specification at least permits more or less exact conversions to and from other color spaces.

Several standard image formats, including EXIF (JPEG) and PNG are based on sRGB color data, which makes sRGB the de facto standard for digital still cameras, color printers, and other imaging devices at the consumer level [107]. sRGB is used as a relatively dependable archive format for digital images, particularly in less demanding applications that do not require (or allow) explicit color management [225]. Thus, in practice, working with any RGB color data almost always means dealing with sRGB. It is thus no coincidence that sRGB is also the common color scheme in Java and is extensively supported by the Java standard API (see Sec. 14.7 for details).

Table 14.5 lists the key parameters of the sRGB color space (i.e., the XYZ coordinates for the primary colors **R**, **G**, **B** and the white point **W** (D65)), which are defined according to ITU-R BT.709 [122] (see Tables 14.1 and 14.2). Together, these values permit the unambiguous mapping of all other colors in the CIE diagram.

Pt.	R	G	B	X_{65}	Y_{65}	Z_{65}	x_{65}	y_{65}
R	1.0	0.0	0.0	0.412453	0.212671	0.019334	0.6400	0.3300
G	0.0	1.0	0.0	0.357580	0.715160	0.119193	0.3000	0.6000
B	0.0	0.0	1.0	0.180423	0.072169	0.950227	0.1500	0.0600
W	1.0	1.0	1.0	0.950456	1.000000	1.088754	0.3127	0.3290

Table 14.5
sRGB tristimulus values **R**, **G**, **B** with reference to the white point D65 (**W**).

14.4.1 Linear vs. Nonlinear Color Components

sRGB is a *nonlinear* color space with respect to the XYZ coordinate system, and it is important to carefully distinguish between the *linear* and *nonlinear* RGB component values. The nonlinear values (denoted R', G', B') represent the actual color tuples, the data values read from an image file or received from a digital camera. These values are pre-corrected with a fixed Gamma (≈ 2.2) such that they can be easily viewed on a common color monitor without any additional conversion. The corresponding *linear* components (denoted

R, G, B) relate to the CIEXYZ color space by a linear mapping and can thus be computed from X, Y, Z coordinates and vice versa by simple matrix multiplication, that is,

$$\begin{pmatrix} R \\ G \\ B \end{pmatrix} = \boldsymbol{M}_{\mathrm{RGB}} \cdot \begin{pmatrix} X \\ Y \\ Z \end{pmatrix} \quad \text{and} \quad \begin{pmatrix} X \\ Y \\ Z \end{pmatrix} = \boldsymbol{M}_{\mathrm{RGB}}^{-1} \cdot \begin{pmatrix} R \\ G \\ B \end{pmatrix}, \tag{14.27}$$

with

$$\boldsymbol{M}_{\mathrm{RGB}} = \begin{pmatrix} 3.240479 & -1.537150 & -0.498535 \\ -0.969256 & 1.875992 & 0.041556 \\ 0.055648 & -0.204043 & 1.057311 \end{pmatrix}, \tag{14.28}$$

$$\boldsymbol{M}_{\mathrm{RGB}}^{-1} = \begin{pmatrix} 0.412453 & 0.357580 & 0.180423 \\ 0.212671 & 0.715160 & 0.072169 \\ 0.019334 & 0.119193 & 0.950227 \end{pmatrix}. \tag{14.29}$$

Notice that the three column vectors of $\boldsymbol{M}_{\mathrm{RGB}}^{-1}$ (Eqn. (14.29)) are the coordinates of the primary colors $\mathbf{R}$, $\mathbf{G}$, $\mathbf{B}$ (tristimulus values) in XYZ space (cf. Table 14.5) and thus

$$\mathbf{R} = \boldsymbol{M}_{\mathrm{RGB}}^{-1} \cdot \begin{pmatrix} 1 \\ 0 \\ 0 \end{pmatrix}, \quad \mathbf{G} = \boldsymbol{M}_{\mathrm{RGB}}^{-1} \cdot \begin{pmatrix} 0 \\ 1 \\ 0 \end{pmatrix}, \quad \mathbf{B} = \boldsymbol{M}_{\mathrm{RGB}}^{-1} \cdot \begin{pmatrix} 0 \\ 0 \\ 1 \end{pmatrix}. \tag{14.30}$$

14.4.2 CIEXYZ→sRGB Conversion

To transform a given XYZ color to sRGB (Fig. 14.6), we first compute the *linear* R, G, B values by multiplying the (X, Y, Z) coordinate vector with the matrix $\boldsymbol{M}_{\mathrm{RGB}}$ (Eqn. (14.28)),

$$\begin{pmatrix} R \\ G \\ B \end{pmatrix} = \boldsymbol{M}_{\mathrm{RGB}} \cdot \begin{pmatrix} X \\ Y \\ Z \end{pmatrix}. \tag{14.31}$$

Subsequently, a modified gamma correction (see Ch. 4, Sec. 4.7.6) with $\gamma = 2.4$ (which corresponds to an effective gamma value of ca. 2.2) is applied to the linear R, G, B values,

$$R' = f_1(R), \qquad G' = f_1(G), \qquad B' = f_1(B), \tag{14.32}$$

with

$$f_1(c) = \begin{cases} 12.92 \cdot c & \text{for } c \leq 0.0031308, \\ 1.055 \cdot c^{1/2.4} - 0.055 & \text{for } c > 0.0031308. \end{cases} \tag{14.33}$$

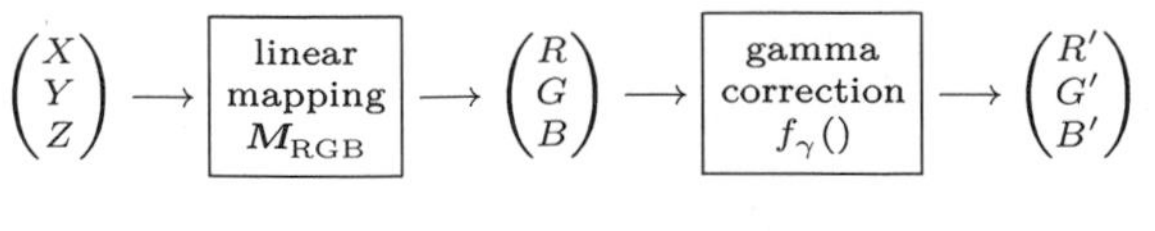

$$\begin{pmatrix} X \\ Y \\ Z \end{pmatrix} \longrightarrow \boxed{\begin{matrix}\text{linear} \\ \text{mapping} \\ \boldsymbol{M}_{\mathrm{RGB}}\end{matrix}} \longrightarrow \begin{pmatrix} R \\ G \\ B \end{pmatrix} \longrightarrow \boxed{\begin{matrix}\text{gamma} \\ \text{correction} \\ f_\gamma()\end{matrix}} \longrightarrow \begin{pmatrix} R' \\ G' \\ B' \end{pmatrix}$$

Fig. 14.6
Color transformation from CIEXYZ to sRGB.

The resulting sRGB components R', G', B' are limited to the interval $[0, 1]$ (see Table 14.6). To obtain discrete numbers, the R', G', B' values are finally scaled linearly to the 8-bit integer range $[0, 255]$.

Pt.	Color	sRGB (*nonlinear*) R'	G'	B'	RGB (*linear*) R	G	B	CIEXYZ X_{65}	Y_{65}	Z_{65}
S	Black	0.00	0.00	0.00	0.0000	0.0000	0.0000	0.0000	0.0000	0.0000
R	Red	1.00	0.00	0.00	1.0000	0.0000	0.0000	0.4125	0.2127	0.0193
Y	Yellow	1.00	1.00	0.00	1.0000	1.0000	0.0000	0.7700	0.9278	0.1385
G	Green	0.00	1.00	0.00	0.0000	1.0000	0.0000	0.3576	0.7152	0.1192
C	Cyan	0.00	1.00	1.00	0.0000	1.0000	1.0000	0.5380	0.7873	1.0694
B	Blue	0.00	0.00	1.00	0.0000	0.0000	1.0000	0.1804	0.0722	0.9502
M	Magenta	1.00	0.00	1.00	1.0000	0.0000	1.0000	0.5929	0.2848	0.9696
W	White	1.00	1.00	1.00	1.0000	1.0000	1.0000	0.9505	1.0000	1.0888
K	50% Gray	0.50	0.50	0.50	0.2140	0.2140	0.2140	0.2034	0.2140	0.2330
$\mathbf{R}_{75}$	75% Red	0.75	0.00	0.00	0.5225	0.0000	0.0000	0.2155	0.1111	0.0101
$\mathbf{R}_{50}$	50% Red	0.50	0.00	0.00	0.2140	0.0000	0.0000	0.0883	0.0455	0.0041
$\mathbf{R}_{25}$	25% Red	0.25	0.00	0.00	0.0509	0.0000	0.0000	0.0210	0.0108	0.0010
P	Pink	1.00	0.50	0.50	1.0000	0.2140	0.2140	0.5276	0.3812	0.2482

Table 14.6
CIEXYZ coordinates for selected sRGB colors. The table lists the *nonlinear* R', G', and B' components, the *linearized* R, G, and B values, and the corresponding X, Y, and Z coordinates (for white point D65). The linear and nonlinear RGB values are identical for the extremal points of the RGB color cube $\mathbf{S}, \ldots, \mathbf{W}$ (top rows) because the gamma correction does not affect 0 and 1 component values. However, *intermediate* colors ($\mathbf{K}, \ldots, \mathbf{P}$, shaded rows) may exhibit large differences between the nonlinear and linear components (e.g., compare the R' and R values for $\mathbf{R}_{25}$).

14.4.3 sRGB→CIEXYZ Conversion

To calculate the reverse transformation from sRGB to XYZ, the given (nonlinear) $R'G'B'$ values (in the range $[0, 1]$) are first linearized by inverting the gamma correction (Eqn. (14.33)), that is,

$$R = f_2(R'), \quad G = f_2(G'), \quad B = f_2(B'), \tag{14.34}$$

with

$$f_2(c') = \begin{cases} \frac{c'}{12.92} & \text{for } c' \leq 0.04045, \\ \left(\frac{c'+0.055}{1.055}\right)^{2.4} & \text{for } c' > 0.04045. \end{cases} \tag{14.35}$$

Subsequently, the linearized (R, G, B) vector is transformed to XYZ coordinates by multiplication with the inverse of the matrix $\boldsymbol{M}_{\mathrm{RGB}}$ (Eqn. (14.29)),

$$\begin{pmatrix} X \\ Y \\ Z \end{pmatrix} = \boldsymbol{M}_{\mathrm{RGB}}^{-1} \cdot \begin{pmatrix} R \\ G \\ B \end{pmatrix}. \tag{14.36}$$

14.4.4 Calculations with Nonlinear sRGB Values

Due to the wide use of sRGB in digital photography, graphics, multimedia, Internet imaging, etc., there is a probability that a given image is encoded in sRGB colors. If, for example, a JPEG image is opened with ImageJ or Java, the pixel values in the resulting data array are media-oriented (i.e., nonlinear R', G', B' components of the sRGB color space). Unfortunately, this fact is often overlooked by programmers, with the consequence that colors are incorrectly manipulated and reproduced.

As a general rule, any arithmetic operation on color values should always be performed on the *linearized* R, G, B components, which are obtained from the nonlinear R', G', B' values through the inverse gamma function f_γ^{-1} (Eqn. (14.35)) and converted back again with f_γ (Eqn. (14.33)).

Example: color to grayscale conversion

The principle of converting RGB colors to grayscale values by computing a weighted sum of the color components was described already in Chapter 12, Sec. 12.2.1, where we had simply ignored the issue of possible nonlinearities. As one may have guessed, however, the variables R, G, B, and Y in Eqn. (12.10) on p. 305,

$$Y = 0.2125 \cdot R + 0.7154 \cdot G + 0.072 \cdot B \tag{14.37}$$

implicitly refer to *linear* color and gray values, respectively, and not the raw sRGB values! Based on Eqn. (14.37), the *correct* grayscale conversion from raw (nonlinear) sRGB components R', G', B' is

$$Y' = f_1\big(0.2125 \cdot f_2(R') + 0.7154 \cdot f_2(G') + 0.0721 \cdot f_2(B')\big), \tag{14.38}$$

with $f_\gamma()$ and $f_\gamma^{-1}()$ as defined in Eqns. (14.33) and (14.35). The result (Y') is again a nonlinear, sRGB-compatible gray value; that is, the sRGB color tuple (Y', Y', Y') should have the same perceived luminance as the original color (R', G', B').

Note that setting the components of an sRGB color pixel to three arbitrary but identical values Y',

$$(R', G', B') \leftarrow (Y', Y', Y')$$

always creates a gray (colorless) pixel, despite the nonlinearities of the sRGB space. This is due to the fact that the gamma correction (Eqns. (14.33) and (14.35)) applies evenly to all three color components and thus any three identical values map to a (linearized) color on the straight gray line between the black point $\mathbf{S}$ and the white point $\mathbf{W}$ in XYZ space (cf. Fig. 14.1(b)).

For many applications, however, the following *approximation* to the exact grayscale conversion in Eqn. (14.38) is sufficient. It works without converting the sRGB values (i.e., directly on the nonlinear R', G', B' components) by computing a linear combination

$$Y' \approx w'_R \cdot R' + w'_G \cdot G' + w'_B \cdot B', \tag{14.39}$$

with a slightly different set of weights; for example, $w'_R = 0.309$, $w'_G = 0.609$, $w'_B = 0.082$, as proposed in [188]. The resulting quantity from Eqn. (14.39) is sometimes called *luma* (compared to *luminance* in Eqn. (14.37)).

14.5 Adobe RGB

A distinct weakness of sRGB is its relatively small gamut, which is limited to the range of colors reproducible by ordinary color monitors. This causes problems, for example, in printing, where larger gamuts are needed, particularly in the green regions. The "Adobe RGB (1998)" [1] color space, developed by Adobe as their own standard, is based on the same general concept as sRGB but exhibits a significantly larger gamut (Fig. 14.3), which extends its use particularly to print applications. Figure 14.7 shows the noted difference between the sRGB and Adobe RGB gamuts in 3D CIEXYZ color space.

The neutral point of Adobe RGB corresponds to the D65 standard (with $x = 0.3127$, $y = 0.3290$), and the gamma value is 2.199

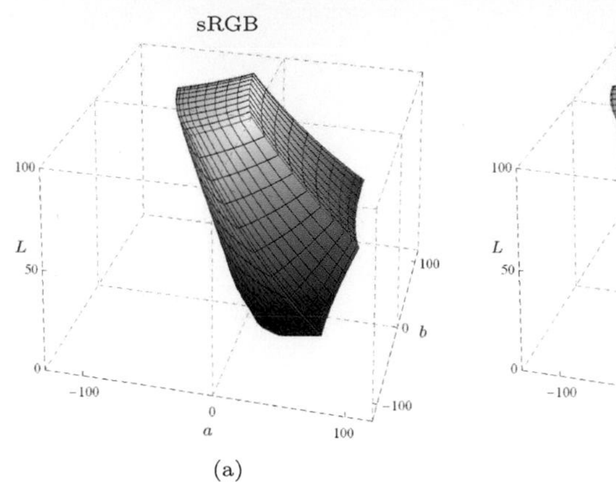

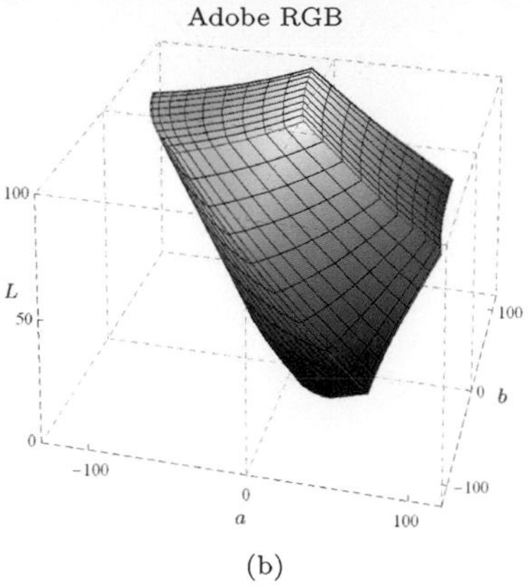

(a) (b)

Fig. 14.7
Gamuts of sRGB and Adobe RGB shown in CIELAB color space. The volume of the sRGB gamut (a) is significantly smaller than the Adobe RGB gamut (b), particularly in the green color region. The tesselation corresponds to a uniform subdivision of the original RGB cubes (in the respective color spaces).

(compared with 2.4 for sRGB) for the forward correction and $\frac{1}{2.199}$ for the inverse correction, respectively. The associated file specification provides for a number of different codings (8- to 16-bit integer and 32-bit floating point) for the color components. Adobe RGB is frequently used in professional photography as an alternative to the CIELAB color space and for picture archive applications.

14.6 Chromatic Adaptation

The human eye has the capability to interpret colors as being constant under varying viewing conditions and illumination in particular. A white sheet of paper appears white to us in bright daylight as well as under fluorescent lighting, although the spectral composition of the light that enters the eye is completely different in both situations. The CIE color system takes into account the color temperature of the ambient lighting because the exact interpretation of XYZ color values also requires knowledge of the corresponding reference white point. For example, a color value (X, Y, Z) specified with respect to the D50 reference white point is generally perceived differently when reproduced by a D65-based media device, although the absolute (i.e., measured) color is the same. Thus the actual meaning of XYZ values cannot be known without knowing the corresponding white point. This is known as *relative colorimetry*.

If colors are specified with respect to *different* white points, for example $\mathbf{W}_1 = (X_{\mathrm{W1}}, Y_{\mathrm{W1}}, Z_{\mathrm{W1}})$ and $\mathbf{W}_2 = (X_{\mathrm{W2}}, Y_{\mathrm{W2}}, Z_{\mathrm{W2}})$, they can be related by first applying a so-called *chromatic adaptation transformation* (CAT) [114, Ch. 34] in XYZ color space. This transformation determines, for given color coordinates (X_1, Y_1, Z_1) and the associated white point $\mathbf{W}_1$, the new color coordinates (X_2, Y_2, Z_2) relative to another white point $\mathbf{W}_2$.

14.6.1 XYZ Scaling

The simplest chromatic adaptation method is XYZ scaling, where the individual color coordinates are individually multiplied by the ratios of the corresponding white point coordinates, that is,

$$X_2 = X_1 \cdot \frac{\hat{X}_2}{\hat{X}_1}, \qquad Y_2 = Y_1 \cdot \frac{\hat{Y}_2}{\hat{Y}_1}, \qquad Z_2 = Z_1 \cdot \frac{\hat{Z}_2}{\hat{Z}_1}. \tag{14.40}$$

For example, for converting colors (X_{65}, Y_{65}, Z_{65}) related to the white point **D65** $= (\hat{X}_{65}, \hat{Y}_{65}, \hat{Z}_{65})$ to the corresponding colors for white point **D50** $= (\hat{X}_{50}, \hat{Y}_{50}, \hat{Z}_{50})$,[6] the concrete scaling is

$$\begin{aligned} X_{50} &= X_{65} \cdot \tfrac{\hat{X}_{50}}{\hat{X}_{65}} = X_{65} \cdot \tfrac{0.964296}{0.950456} = X_{65} \cdot 1.01456, \\ Y_{50} &= Y_{65} \cdot \tfrac{\hat{Y}_{50}}{\hat{Y}_{65}} = Y_{65} \cdot \tfrac{1.000000}{1.000000} = Y_{65}, \\ Z_{50} &= Z_{65} \cdot \tfrac{\hat{Z}_{50}}{\hat{Z}_{65}} = Z_{65} \cdot \tfrac{0.825105}{1.088754} = Z_{65} \cdot 0.757843\,. \end{aligned} \tag{14.41}$$

This form of scaling the color coordinates in XYZ space is usually not considered a good color adaptation model and is not recommended for high-quality applications.

14.6.2 Bradford Adaptation

The most common chromatic adaptation models are based on scaling the color coordinates not directly in XYZ but in a "virtual" $R^*G^*B^*$ color space obtained from the XYZ values by a linear transformation

$$\begin{pmatrix} R^* \\ G^* \\ B^* \end{pmatrix} = \boldsymbol{M}_{\mathrm{CAT}} \cdot \begin{pmatrix} X \\ Y \\ Z \end{pmatrix}, \tag{14.42}$$

where $\boldsymbol{M}_{\mathrm{CAT}}$ is a 3×3 transformation matrix (defined in Eqn. (14.45)). After appropriate scaling, the $R^*G^*B^*$ coordinates are transformed back to XYZ, so the complete adaptation transform from color coordinates X_1, Y_1, Z_1 (w.r.t. white point $\mathbf{W}_1 = (X_{\mathrm{W1}}, Y_{\mathrm{W1}}, Z_{\mathrm{W1}})$) to the new color coordinates X_2, Y_2, Z_2 (w.r.t. white point $\mathbf{W}_2 = (X_{\mathrm{W2}}, Y_{\mathrm{W2}}, Z_{\mathrm{W2}})$) takes the form

$$\begin{pmatrix} X_2 \\ Y_2 \\ Z_2 \end{pmatrix} = \boldsymbol{M}_{\mathrm{CAT}}^{-1} \cdot \begin{pmatrix} \frac{R^*_{\mathrm{W2}}}{R^*_{\mathrm{W1}}} & 0 & 0 \\ 0 & \frac{G^*_{\mathrm{W2}}}{G^*_{\mathrm{W1}}} & 0 \\ 0 & 0 & \frac{B^*_{\mathrm{W2}}}{B^*_{\mathrm{W1}}} \end{pmatrix} \cdot \boldsymbol{M}_{\mathrm{CAT}} \cdot \begin{pmatrix} X_1 \\ Y_1 \\ Z_1 \end{pmatrix}, \tag{14.43}$$

where the diagonal elements $\frac{R^*_{\mathrm{W2}}}{R^*_{\mathrm{W1}}}, \frac{G^*_{\mathrm{W2}}}{G^*_{\mathrm{W1}}}, \frac{B^*_{\mathrm{W2}}}{B^*_{\mathrm{W1}}}$ are the (constant) ratios of the $R^*G^*B^*$ values of the white points $\mathbf{W}_2$, $\mathbf{W}_1$, respectively; that is,

$$\begin{pmatrix} R^*_{\mathrm{W1}} \\ G^*_{\mathrm{W1}} \\ B^*_{\mathrm{W1}} \end{pmatrix} = \boldsymbol{M}_{\mathrm{CAT}} \cdot \begin{pmatrix} X_{\mathrm{W1}} \\ Y_{\mathrm{W1}} \\ Z_{\mathrm{W1}} \end{pmatrix}, \quad \begin{pmatrix} R^*_{\mathrm{W2}} \\ G^*_{\mathrm{W2}} \\ B^*_{\mathrm{W2}} \end{pmatrix} = \boldsymbol{M}_{\mathrm{CAT}} \cdot \begin{pmatrix} X_{\mathrm{W2}} \\ Y_{\mathrm{W2}} \\ Z_{\mathrm{W2}} \end{pmatrix}. \tag{14.44}$$

The "Bradford" model [114, p. 590] specifies for Eqn. (14.43) the particular transformation matrix

$$\boldsymbol{M}_{\mathrm{CAT}} = \begin{pmatrix} 0.8951 & 0.2664 & -0.1614 \\ -0.7502 & 1.7135 & 0.0367 \\ 0.0389 & -0.0685 & 1.0296 \end{pmatrix}. \tag{14.45}$$

[6] See Table 14.2.

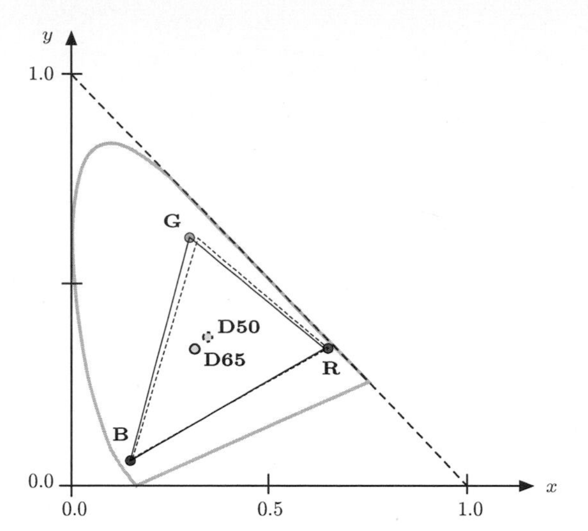

Fig. 14.8
Bradford chromatic adaptation from white point D65 to D50. The solid triangle represents the original RGB gamut for white point D65, with the primaries $(\mathbf{R}, \mathbf{G}, \mathbf{B})$ located at the corner points. The dashed triangle is the corresponding gamut after chromatic adaptation to white point D50.

Inserting $\boldsymbol{M}_{\mathrm{CAT}}$ matrix in Eqn. (14.43) gives the complete chromatic adaptation. For example, the resulting transformation for converting from D65-based to D50-based colors (i.e., $\mathbf{W}_1 = \mathbf{D65}$, $\mathbf{W}_2 = \mathbf{D50}$, as listed in Table 14.2) is

$$\begin{pmatrix} X_{50} \\ Y_{50} \\ Z_{50} \end{pmatrix} = \boldsymbol{M}_{50|65} \cdot \begin{pmatrix} X_{65} \\ Y_{65} \\ Z_{65} \end{pmatrix}$$
$$= \begin{pmatrix} 1.047884 & 0.022928 & -0.050149 \\ 0.029603 & 0.990437 & -0.017059 \\ -0.009235 & 0.015042 & 0.752085 \end{pmatrix} \cdot \begin{pmatrix} X_{65} \\ Y_{65} \\ Z_{65} \end{pmatrix}, \quad (14.46)$$

and conversely from D50-based to D65-based colors (i.e., $\mathbf{W}_1 = \mathbf{D50}$, $\mathbf{W}_2 = \mathbf{D65}$),

$$\begin{pmatrix} X_{65} \\ Y_{65} \\ Z_{65} \end{pmatrix} = \boldsymbol{M}_{65|50} \cdot \begin{pmatrix} X_{50} \\ Y_{50} \\ Z_{50} \end{pmatrix} = \boldsymbol{M}_{50|65}^{-1} \cdot \begin{pmatrix} X_{50} \\ Y_{50} \\ Z_{50} \end{pmatrix}$$
$$= \begin{pmatrix} 0.955513 & -0.023079 & 0.063190 \\ -0.028348 & 1.009992 & 0.021019 \\ 0.012300 & -0.020484 & 1.329993 \end{pmatrix} \cdot \begin{pmatrix} X_{50} \\ Y_{50} \\ Z_{50} \end{pmatrix}. \quad (14.47)$$

Figure 14.8 illustrates the effects of adaptation from the D65 white point to D50 in the CIE x, y chromaticity diagram. A short list of corresponding color coordinates is given in Table 14.7.

The Bradford model is a widely used chromatic adaptation scheme but several similar procedures have been proposed (see also Exercise 14.1). Generally speaking, chromatic adaptation and related problems have a long history in color engineering and are still active fields of scientific research [258, Ch. 5, Sec. 5.12].

Table 14.7
Bradford chromatic adaptation from white point D65 to D50 for selected sRGB colors. The XYZ coordinates X_{65}, Y_{65}, Z_{65} relate to the original white point D65 ($\mathbf{W}_1$). X_{50}, Y_{50}, Z_{50} are the corresponding coordinates for the new white point D50 ($\mathbf{W}_2$), obtained with the Bradford adaptation according to Eqn. (14.46).

		sRGB			XYZ (**D65**)			XYZ (**D50**)		
Pt.	Color	R'	G'	B'	X_{65}	Y_{65}	Z_{65}	X_{50}	Y_{50}	Z_{50}
S	Black	0.00	0.0	0.0	0.0000	0.0000	0.0000	0.0000	0.0000	0.0000
R	Red	1.00	0.0	0.0	0.4125	0.2127	0.0193	0.4361	0.2225	0.0139
Y	Yellow	1.00	1.0	0.0	0.7700	0.9278	0.1385	0.8212	0.9394	0.1110
G	Green	0.00	1.0	0.0	0.3576	0.7152	0.1192	0.3851	0.7169	0.0971
C	Cyan	0.00	1.0	1.0	0.5380	0.7873	1.0694	0.5282	0.7775	0.8112
B	Blue	0.00	0.0	1.0	0.1804	0.0722	0.9502	0.1431	0.0606	0.7141
M	Magenta	1.00	0.0	1.0	0.5929	0.2848	0.9696	0.5792	0.2831	0.7280
W	White	1.00	1.0	1.0	0.9505	1.0000	1.0888	0.9643	1.0000	0.8251
K	50% Gray	0.50	0.5	0.5	0.2034	0.2140	0.2330	0.2064	0.2140	0.1766
$\mathbf{R}_{75}$	75% Red	0.75	0.0	0.0	0.2155	0.1111	0.0101	0.2279	0.1163	0.0073
$\mathbf{R}_{50}$	50% Red	0.50	0.0	0.0	0.0883	0.0455	0.0041	0.0933	0.0476	0.0030
$\mathbf{R}_{25}$	25% Red	0.25	0.0	0.0	0.0210	0.0108	0.0010	0.0222	0.0113	0.0007
P	Pink	1.00	0.5	0.5	0.5276	0.3812	0.2482	0.5492	0.3889	0.1876

14.7 Colorimetric Support in Java

sRGB is the standard color space in Java; that is, the components of color objects and RGB color images are gamma-corrected, *nonlinear* R', G', B' values (see Fig. 14.6). The nonlinear R', G', B' values are related to the linear R, G, B values by a modified gamma correction, as specified by the sRGB standard (Eqns. (14.33) and (14.35)).

14.7.1 Profile Connection Space (PCS)

The Java API (AWT) provides classes for representing color objects and color spaces, together with a rich set of corresponding methods. Java's color system is designed after the ICC[7] "color management architecture", which uses a CIEXYZ-based device-independent color space called the "profile connection space" (PCS) [118, 121]. The PCS color space is used as the intermediate reference for converting colors between different color spaces. The ICC standard defines device profiles (see Sec. 14.7.4) that specify the transforms to convert between a device's color space and the PCS. The advantage of this approach is that for any given device only a single color transformation (profile) must be specified to convert between device-specific colors and the unified, colorimetric profile connection space. Every `ColorSpace` class (or subclass) provides the methods `fromCIEXYZ()` and `toCIEXYZ()` to convert device color values to XYZ coordinates in the standardized PCS. Figure 14.9 illustrates the principal application of `ColorSpace` objects for converting colors between different color spaces in Java using the XYZ space as a common "hub".

Different to the sRGB specification, the ICC specifies **D50** (and *not* D65) as the illuminant white point for its default PCS color space (see Table 14.2). The reason is that the ICC standard was developed primarily for color management in photography, graphics, and printing, where D50 is normally used as the reflective media white point. The Java methods `fromCIEXYZ()` and `toCIEXYZ()` thus take and return X, Y, Z color coordinates that are relative to the D50 white point. The resulting coordinates for the primary colors (listed in Table 14.8) are different from the ones given for white point D65 (see Table 14.5)! This is a frequent cause of confusion since the sRGB

[7] International Color Consortium (ICC, www.color.org).

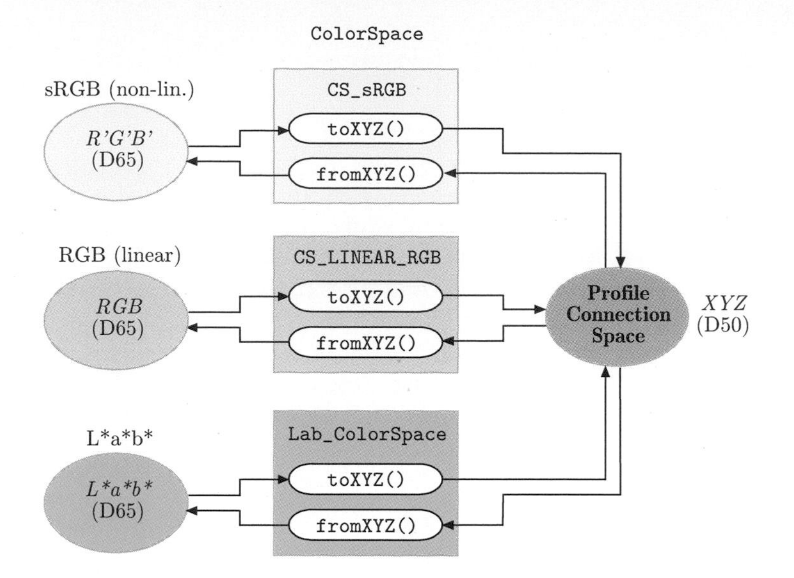

Fig. 14.9
XYZ-based color conversion in Java. ColorSpace objects implement the methods fromCIEXYZ() and toCIEXYZ() to convert color vectors from and to the CIEXYZ color space, respectively. Colorimetric transformations between color spaces can be accomplished as a two-step process via the XYZ space. For example, to convert from sRGB to CIELAB, the sRGB color is first converted to XYZ and subsequently from XYZ to CIELAB. Notice that Java's standard XYZ color space is based on the D50 white point, while most common color spaces refer to D65.

Pt.	R	G	B	X_{50}	Y_{50}	Z_{50}	x_{50}	y_{50}
R	1.0	0.0	0.0	0.436108	0.222517	0.013931	0.6484	0.3309
G	0.0	1.0	0.0	0.385120	0.716873	0.097099	0.3212	0.5978
B	0.0	0.0	1.0	0.143064	0.060610	0.714075	0.1559	0.0660
W	1.0	1.0	1.0	0.964296	1.000000	0.825106	0.3457	0.3585

Table 14.8
Color coordinates for sRGB primaries and the white point in Java's default XYZ color space. Color coordinates for sRGB primaries and the white point in Java's default XYZ color space. The white point **W** is equal to D50.

component values are D65-based (as specified by the sRGB standard) but Java's XYZ values are relative to the D50.

Chromatic adaptation (see Sec. 14.6) is used to convert between XYZ color coordinates that are measured with respect to different white points. The ICC specification [118] recommends a linear chromatic adaptation based on the Bradford model to convert between the D65-related XYZ coordinates (X_{65}, Y_{65}, Z_{65}) and D50-related values (X_{50}, Y_{50}, Z_{50}). This is also implemented by the Java API.

The complete mapping between the linearized sRGB color values (R, G, B) and the D50-based (X_{50}, Y_{50}, Z_{50}) coordinates can be expressed as a linear transformation composed of the RGB→XYZ$_{65}$ transformation by matrix $\boldsymbol{M}_{\mathrm{RGB}}$ (Eqns. (14.28) and (14.29)) and the chromatic adaptation transformation XYZ$_{65}$→XYZ$_{50}$ defined by the matrix $\boldsymbol{M}_{50|65}$ (Eqn. (14.46)),

$$\begin{aligned}\begin{pmatrix} X_{50} \\ Y_{50} \\ Z_{50} \end{pmatrix} &= \boldsymbol{M}_{50|65} \cdot \boldsymbol{M}_{\mathrm{RGB}}^{-1} \cdot \begin{pmatrix} R \\ G \\ B \end{pmatrix} = \left(\boldsymbol{M}_{\mathrm{RGB}} \cdot \boldsymbol{M}_{65|50}\right)^{-1} \cdot \begin{pmatrix} R \\ G \\ B \end{pmatrix} \\ &= \begin{pmatrix} 0.436131 & 0.385147 & 0.143033 \\ 0.222527 & 0.716878 & 0.060600 \\ 0.013926 & 0.097080 & 0.713871 \end{pmatrix} \cdot \begin{pmatrix} R \\ G \\ B \end{pmatrix},\end{aligned} \tag{14.48}$$

and, in the reverse direction,

$$\begin{pmatrix} R \\ G \\ B \end{pmatrix} = \boldsymbol{M}_{\mathrm{RGB}} \cdot \boldsymbol{M}_{65|50} \cdot \begin{pmatrix} X_{50} \\ Y_{50} \\ Z_{50} \end{pmatrix}$$

$$= \begin{pmatrix} 3.133660 & -1.617140 & -0.490588 \\ -0.978808 & 1.916280 & 0.033444 \\ 0.071979 & -0.229051 & 1.405840 \end{pmatrix} \cdot \begin{pmatrix} X_{50} \\ Y_{50} \\ Z_{50} \end{pmatrix}. \quad (14.49)$$

Equations (14.48) and (14.49) are the transformations implemented by the methods `toCIEXYZ()` and `fromCIEXYZ()`, respectively, for Java's default sRGB `ColorSpace` class. Of course, these methods must also perform the necessary gamma correction between the linear R, G, B components and the actual (nonlinear) sRGB values R', G', B'. Figure 14.10 illustrates the complete transformation from D50-based PCS coordinates to nonlinear sRGB values.

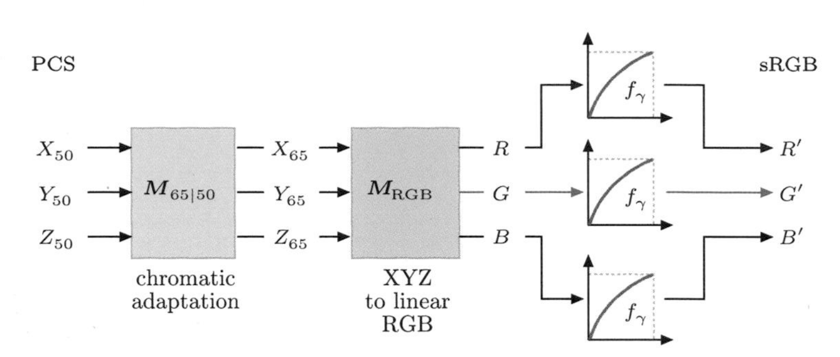

Fig. 14.10 Transformation from **D50**-based CIEXYZ coordinates (X_{50}, Y_{50}, Z_{50}) in Java's *Profile Connection Space* (PCS) to nonlinear sRGB values (R', G', B'). The first step ist chromatic adaptation from **D50** to **D65** (by $\boldsymbol{M}_{65|50}$), followed by mapping the CIEXYZ coordinates to linear RGB values (by $\boldsymbol{M}_{\mathrm{RGB}}$). Finally, gamma correction is applied individually to all three color components.

14.7.2 Color-Related Java Classes

The Java standard API offers extensive support for working with colors and color images. The most important classes contained in the Java AWT package are:

- `Color`: defines individual color objects.
- `ColorSpace`: specifies entire color spaces.
- `ColorModel`: describes the structure of color images; e.g., full-color images or indexed-color images (see Prog. 12.3 on p. 301).

Class `Color` *(*`java.awt.Color`*)*

An object of class `Color` describes a particular color in the associated color space, which defines the number and type of the color components. `Color` objects are primarily used for graphic operations, such as to specify the color for drawing or filling graphic objects. Unless the color space is not explicitly specified, new `Color` objects are created as sRGB colors. The arguments passed to the `Color` constructor methods may be either `float` components in the range [0, 1] or integers in the range [0, 255], as demonstrated by the following example:

```
Color pink = new Color(1.0f, 0.5f, 0.5f);
Color blue = new Color(0, 0, 255);
```

Note that in both cases the arguments are interpreted as *nonlinear* sRGB values (R', G', B'). Other constructor methods exist for class

Color that also accept alpha (transparency) values. In addition, the Color class offers two useful static methods, RGBtoHSB() and HSBtoRGB(), for converting between sRGB and HSV[8] colors (see Ch. 12, Sec. 12.2.3).

Class ColorSpace (java.awt.color.ColorSpace)

An object of type ColorSpace represents an entire color space, such as sRGB or CMYK. Every subclass of ColorSpace (which itself is an abstract class) provides methods for converting its native colors to the CIEXYZ and sRGB color space and vice versa, such that conversions between arbitrary color spaces can easily be performed (through Java's XYZ-based profile connection space). In the following example, we first create an instance of the default sRGB color space by invoking the static method ColorSpace.getInstance() and subsequently convert an sRGB color object (R', B', G') to the corresponding (X, Y, Z) coordinates in Java's (D50-based) profile connection space:

```
// create an sRGB color space object:
  ColorSpace sRGBcsp
        = ColorSpace.getInstance(ColorSpace.CS_sRGB);
  float[] pink_RGB = new float[] {1.0f, 0.5f, 0.5f};
  // convert from sRGB to XYZ:
  float[] pink_XYZ = sRGBcsp.toCIEXYZ(pink_RGB);
```

Notice that color vectors are represented as float[] arrays for color conversions with ColorSpace objects. If required, the method getComponents() can be used to convert Color objects to float[] arrays. In summary, the types of color spaces that can be created with the ColorSpace.getInstance() method include:

- CS_sRGB: the standard (D65-based) RGB color space with *non-linear* R', G', B' components, as specified in [119].
- CS_LINEAR_RGB: color space with *linear* R, G, B components (i.e., no gamma correction applied).
- CS_GRAY: single-component color space with linear grayscale values.
- CS_PYCC: Kodak's Photo YCC color space.
- CS_CIEXYZ: the default XYZ profile connection space (based on the D50 white point).

Other color spaces can be implemented by creating additional implementations (subclasses) of ColorSpace, as demonstrated for CIELAB in the example in Sec. 14.7.3.

14.7.3 Implementation of the CIELAB Color Space (Example)

In the following, we show a complete implementation of the CIELAB color space, which is not available in the current Java API, based on the specification given in Sec. 14.2. For this purpose, we define a

[8] The HSV color space is referred to as "HSB" (hue, saturation, *brightness*) in the Java API.

subclass of ColorSpace (defined in the package java.awt.color) named Lab_ColorSpace, which implements the required methods toCIEXYZ(), fromCIEXYZ() for converting to and from Java's default profile connection space, respectively, and toRGB(), fromRGB() for converting between CIELAB and sRGB (Progs. 14.1 and 14.2). These conversions are performed in two steps via XYZ coordinates, where care must be taken regarding the right choice of the associated white point (CIELAB is based on D65 and Java XYZ on D50). The following examples demonstrate the principal use of the new Lab_ColorSpace class:[9]

```
ColorSpace labCs = new LabColorSpace();
float[] cyan_sRGB = {0.0f, 1.0f, 1.0f};
float[] cyan_LAB = labCs.fromRGB(cyan_sRGB) // sRGB→LAB
float[] cyan_XYZ = labCs.toXYZ(cyan_LAB);   // LAB→XYZ (D50)
```

14.7.4 ICC Profiles

Even with the most precise specification, a standard color space may not be sufficient to accurately describe the transfer characteristics of some input or output device. ICC[10] profiles are standardized descriptions of individual device transfer properties that warrant that an image or graphics can be reproduced accurately on different media. The contents and the format of ICC profile files is specified in [118], which is identical to ISO standard 15076 [121]. Profiles are thus a key element in the process of digital color management [246].

The Java graphics API supports the use of ICC profiles mainly through the classes ICC_ColorSpace and ICC_Profile, which allow application designers to create various standard profiles and read ICC profiles from data files.

Assume, for example, that an image was recorded with a calibrated scanner and shall be displayed accurately on a monitor. For this purpose, we need the ICC profiles for the scanner and the monitor, which are often supplied by the manufacturers as .icc data files.[11] For standard color spaces, the associated ICC profiles are often available as part of the computer installation, such as CIERGB.icc or NTSC1953.icc. With these profiles, a color space object can be specified that converts the image data produced by the scanner into corresponding CIEXYZ or sRGB values, as illustrated by the following example:

```
// load the scanner's ICC profile and create a corresponding color space:
ICC_ColorSpace scannerCs = new
  ICC_ColorSpace(ICC_ProfileRGB.getInstance("scanner.icc"));

// specify a device-specific color:
float[] deviceColor = {0.77f, 0.13f, 0.89f};
```

[9] Classes LabColorSpace, LuvColorSpace (analogous implementation of the CIELUV color space) and associated auxiliary classes are found in package imagingbook.pub.colorimage.

[10] International Color Consortium ICC (www.color.org).

[11] ICC profile files may also come with extensions .icm or .pf (as in the Java distribution).

```
1 package imagingbook.pub.color.image;
2
3 import static imagingbook.pub.color.image.Illuminant.D50;
4 import static imagingbook.pub.color.image.Illuminant.D65;
5
6 import java.awt.color.ColorSpace;
7
8 public class LabColorSpace extends ColorSpace {
9
10   // D65 reference white point and chromatic adaptation objects:
11   static final double Xref = D65.X; // 0.950456
12   static final double Yref = D65.Y; // 1.000000
13   static final double Zref = D65.Z; // 1.088754
14
15   static final ChromaticAdaptation catD65toD50 =
16         new BradfordAdaptation(D65, D50);
17   static final ChromaticAdaptation catD50toD65 =
18         new BradfordAdaptation(D50, D65);
19
20   // the only constructor:
21   public LabColorSpace() {
22     super(TYPE_Lab,3);
23   }
24
25   // XYZ (Profile Connection Space, D50) → CIELab conversion:
26   public float[] fromCIEXYZ(float[] XYZ50) {
27     float[] XYZ65 = catD50toD65.apply(XYZ50);
28     return fromCIEXYZ65(XYZ65);
29   }
30
31   // XYZ (D65) → CIELab conversion (Eqn. (14.6)–14.10):
32   public float[] fromCIEXYZ65(float[] XYZ65) {
33     double xx = f1(XYZ65[0] / Xref);
34     double yy = f1(XYZ65[1] / Yref);
35     double zz = f1(XYZ65[2] / Zref);
36     float L = (float)(116.0 * yy - 16.0);
37     float a = (float)(500.0 * (xx - yy));
38     float b = (float)(200.0 * (yy - zz));
39     return new float[] {L, a, b};
40   }
41   // CIELab→XYZ (Profile Connection Space, D50) conversion:
42   public float[] toCIEXYZ(float[] Lab) {
43     float[] XYZ65 = toCIEXYZ65(Lab);
44     return catD65toD50.apply(XYZ65);
45   }
46
47   // CIELab→XYZ (D65) conversion (Eqn. (14.13)–14.15):
48   public float[] toCIEXYZ65(float[] Lab) {
49     double ll = ( Lab[0] + 16.0 ) / 116.0;
50     float Y65 = (float) (Yref * f2(ll));
51     float X65 = (float) (Xref * f2(ll + Lab[1] / 500.0));
52     float Z65 = (float) (Zref * f2(ll - Lab[2] / 200.0));
53     return new float[] {X65, Y65, Z65};
54   }
```

Prog. 14.1
Java implementation of the CIELAB color space as a sub-class of `ColorSpace` (part 1). The conversion from D50-based profile connection space XYZ coordinates to CIELAB (Eqn. (14.6)) and back is implemented by the required methods `fromCIEXYZ()` and `toCIEXYZ()`, respectively. The auxiliary methods `fromCIEXYZ65()` and `toCIEXYZ65()` are used for converting D65-based XYZ coordinates (see Eqn. (14.6)). Chromatic adaptation from D50 to D65 is performed by the objects `catD65toD50` and `catD50toD65` of type `ChromaticAdaptation`. The gamma correction functions f_1 (Eqn. (14.8)) and f_2 (Eqn. (14.15)) are implemented by the methods `f1()` and `f2()`, respectively (see Prog. 14.2).

Prog. 14.2
Java implementation of the CIELAB color space as a subclass of `ColorSpace` (part 2). The methods `fromRGB()` and `toRGB()` perform direct conversion between CIELAB and sRGB via D65-based XYZ coordinates, i.e., without conversion to Java's *Profile Connection Space*. Gamma correction (for mapping between linear RGB and sRGB component values) is implemented by the methods `gammaFwd()` and `gammaInv()` in class `sRgbUtil` (not shown). The methods `f1()` and `f2()` implement the forward and inverse gamma correction of CIELAB components (see Eqns. (14.6) and (14.13)).

```
55    // sRGB→CIELab conversion:
56    public float[] fromRGB(float[] srgb) {
57      // get linear rgb components:
58      double r = sRgbUtil.gammaInv(srgb[0]);
59      double g = sRgbUtil.gammaInv(srgb[1]);
60      double b = sRgbUtil.gammaInv(srgb[2]);
61      // convert to XYZ (D65-based, Eqn. (14.29)):
62      float X =
63        (float) (0.412453 * r + 0.357580 * g + 0.180423 * b);
64      float Y =
65        (float) (0.212671 * r + 0.715160 * g + 0.072169 * b);
66      float Z =
67        (float) (0.019334 * r + 0.119193 * g + 0.950227 * b);
68      float[] XYZ65 = new float[] {X, Y, Z};
69      return fromCIEXYZ65(XYZ65);
70    }
71
72    // CIELab→sRGB conversion:
73    public float[] toRGB(float[] Lab) {
74      float[] XYZ65 = toCIEXYZ65(Lab);
75      double X = XYZ65[0];
76      double Y = XYZ65[1];
77      double Z = XYZ65[2];
78      // XYZ→RGB (linear components, Eqn. (14.28)):
79      double r = ( 3.240479*X + -1.537150*Y + -0.498535*Z);
80      double g = (-0.969256*X +  1.875992*Y +  0.041556*Z);
81      double b = ( 0.055648*X + -0.204043*Y +  1.057311*Z);
82      // RGB→sRGB (nonlinear components):
83      float rr = (float) sRgbUtil.gammaFwd(r);
84      float gg = (float) sRgbUtil.gammaFwd(g);
85      float bb = (float) sRgbUtil.gammaFwd(b);
86      return new float[] {rr, gg, bb};
87    }
88
89    static final double epsilon = 216.0 / 24389;  // Eqn. (14.9)
90    static final double kappa = 841.0 / 108;  // Eqn. (14.10)
91
92    // Gamma correction for L* (forward, Eqn. (14.8)):
93    double f1 (double c) {
94      if (c > epsilon) // 0.008856
95        return Math.cbrt(c);
96      elses
97        return (kappa * c) + (16.0 / 116);
98    }
99
100   // Gamma correction for L* (inverse, Eqn. (14.15)):
101   double f2 (double c) {
102     double c3 = c * c * c;
103     if (c3 > epsilon)
104       return c3;
105     else
106       return (c - 16.0 / 116) / kappa;
107   }
108
109 } // end of class LabColorSpace
```

```
// convert to sRGB:
float[] RGBColor = scannerCs.toRGB(deviceColor);

// convert to (D50-based) XYZ:
float[] XYZColor = scannerCs.toCIEXYZ(deviceColor);
```

Similarly, we can calculate the accurate color values to be sent to the monitor by creating a suitable color space object from this device's ICC profile.

14.8 Exercises

Exercise 14.1. For chromatic adaptation (defined in Eqn. (14.43)), transformation matrices other than the Bradford model (Eqn. (14.45)) have been proposed; for example, [225],

$$\boldsymbol{M}_{\mathrm{CAT}}^{(2)} = \begin{pmatrix} 1.2694 & -0.0988 & -0.1706 \\ -0.8364 & 1.8006 & 0.0357 \\ 0.0297 & -0.0315 & 1.0018 \end{pmatrix} \quad \text{or} \tag{14.50}$$

$$\boldsymbol{M}_{\mathrm{CAT}}^{(3)} = \begin{pmatrix} 0.7982 & 0.3389 & -0.1371 \\ -0.5918 & 1.5512 & 0.0406 \\ 0.0008 & -0.0239 & 0.9753 \end{pmatrix}. \tag{14.51}$$

Derive the complete chromatic adaptation transformations $\boldsymbol{M}_{50|65}$ and $\boldsymbol{M}_{65|50}$ for converting between D65 and D50 colors, analogous to Eqns. (14.46) and (14.47), for each of the above transformation matrices.

Exercise 14.2. Implement the conversion of an sRGB color image to a colorless (grayscale) sRGB image using the three methods in Eqn. (14.37) (incorrectly applying standard weights to nonlinear $R'G'B'$ components), Eqn. (14.38) (exact computation), and Eqn. (14.39) (approximation using nonlinear components and modified weights). Compare the results by computing difference images, and also determine the total errors.

Exercise 14.3. Write a program to evaluate the errors that are introduced by using *nonlinear* instead of linear color components for grayscale conversion. To do this, compute the diffence between the Y values obtained with the linear variant (Eqn. (14.38)) and the nonlinear variant (Eqn. (14.39) with $w'_R = 0.309$, $w'_G = 0.609$, $w'_B = 0.082$) for all possible 2^{24} RGB colors. Let your program return the maximum gray value difference and the sum of the absolute differences for all colors.

Exercise 14.4. Determine the virtual primaries $\mathbf{R}^*, \mathbf{G}^*, \mathbf{B}^*$ obtained by Bradford adaptation (Eqn. (14.42)), with $\boldsymbol{M}_{\mathrm{CAT}}$ as defined in Eqn. (14.45). What are the resulting coordinates in the xy chromaticity diagram? Are the primaries inside the visible color range?

15

Filters for Color Images

Color images are everywhere and filtering them is such a common task that it does not seem to require much attention at all. In this chapter, we describe how classical linear and nonlinear filters, which we covered before in the context of grayscale images (see Ch. 5), can be either used directly or adapted for the processing of color images. Often color images are treated as stacks of intensity images and existing monochromatic filters are simply applied independently to the individual color channels. While this is straightforward and performs satisfactorily in many situations, it does not take into account the vector-valued nature of color pixels as samples taken in a specific, multi-dimensional color space. As we show in this chapter, the outcome of filter operations depends strongly on the working color space and the variations between different color spaces may be substantial. Although this may not be apparent in many situations, it should be of concern if high-quality color imaging is an issue.

15.1 Linear Filters

Linear filters are important in many applications, such as smoothing, noise removal, interpolation for geometric transformations, decimation in scale-space transformations, image compression, reconstruction and edge enhancement. The general properties of linear filters and their use on scalar-valued grayscale images are detailed in Chapter 5, Sec. 5.2. For color images, it is common practice to apply these monochromatic filters separately to each color channel, thereby treating the image as a stack of scalar-valued images. As we describe in the following section, this approach is simple as well as efficient, since existing implementations for grayscale images can be reused without any modification. However, the outcome depends strongly on the choice of the color space in which the filter operation is performed. For example, it makes a great difference if the channels of an RGB image contain linear or nonlinear component values. This issue is discussed in more detail in Sec. 15.1.2.

15.1.1 Monochromatic Application of Linear Filters

Given a discrete *scalar* (grayscale) image with elements $I(u,v) \in \mathbb{R}$, the application of a linear filter can be expressed as a linear 2D convolution[1]

$$\bar{I}(u,v) = (I * H)(u,v) = \sum_{(i,j)\in\mathcal{R}_H} I(u-i, v-j) \cdot H(i,j), \tag{15.1}$$

where H denotes a discrete filter kernel defined over the (usually rectangular) region $\mathcal{R}_H$, with $H(i,j) \in \mathbb{R}$. For a *vector*-valued image $\boldsymbol{I}$ with K components, the individual picture elements are vectors, that is,

$$\boldsymbol{I}(u,v) = \begin{pmatrix} I_1(u,v) \\ I_2(u,v) \\ \vdots \\ I_K(u,v) \end{pmatrix}, \tag{15.2}$$

with $\boldsymbol{I}(u,v) \in \mathbb{R}^K$ or $I_k(u,v) \in \mathbb{R}$, respectively. In this case, the linear filter operation can be generalized to

$$\bar{\boldsymbol{I}}(u,v) = (\boldsymbol{I} * H)(u,v) = \sum_{(i,j)\in\mathcal{R}_H} \boldsymbol{I}(u-i, v-j) \cdot H(i,j), \tag{15.3}$$

with the same scalar-valued filter kernel H as in Eqn. (15.1). Thus the kth element of the resulting pixels,

$$\bar{I}_k(u,v) = \sum_{(i,j)\in\mathcal{R}_H} I_k(u-i, v-j) \cdot H(i,j) = (I_k * H)\,(u,v), \tag{15.4}$$

is simply the result of scalar convolution (Eqn. (15.1)) applied to the corresponding component plane I_k. In the case of an RGB color image (with $K = 3$ components), the filter kernel H is applied separately to the scalar-valued R, G, and B planes $(I_\mathrm{R}, I_\mathrm{G}, I_\mathrm{B})$, that is,

$$\bar{\boldsymbol{I}}(u,v) = \begin{pmatrix} \bar{I}_\mathrm{R}(u,v) \\ \bar{I}_\mathrm{G}(u,v) \\ \bar{I}_\mathrm{B}(u,v) \end{pmatrix} = \begin{pmatrix} (I_\mathrm{R} * H)\,(u,v) \\ (I_\mathrm{G} * H)\,(u,v) \\ (I_\mathrm{B} * H)\,(u,v) \end{pmatrix}. \tag{15.5}$$

Figure 15.1 illustrates how linear filters for color images are typically implemented by individually filtering the three scalar-valued color components.

Linear smoothing filters

Smoothing filters are a particular class of linear filters that are found in many applications and characterized by positive-only filter coefficients. Let $C_{u,v} = (\boldsymbol{c}_1, \ldots, \boldsymbol{c}_n)$ denote the vector of color pixels $\boldsymbol{c}_m \in \mathbb{R}^K$ contained in the spatial support region of the kernel H, placed at position (u,v) in the original image $\boldsymbol{I}$, where n is the size of H. With arbitrary kernel coefficients $H(i,j) \in \mathbb{R}$, the resulting

[1] See also Chapter 5, Sec. 5.3.1.

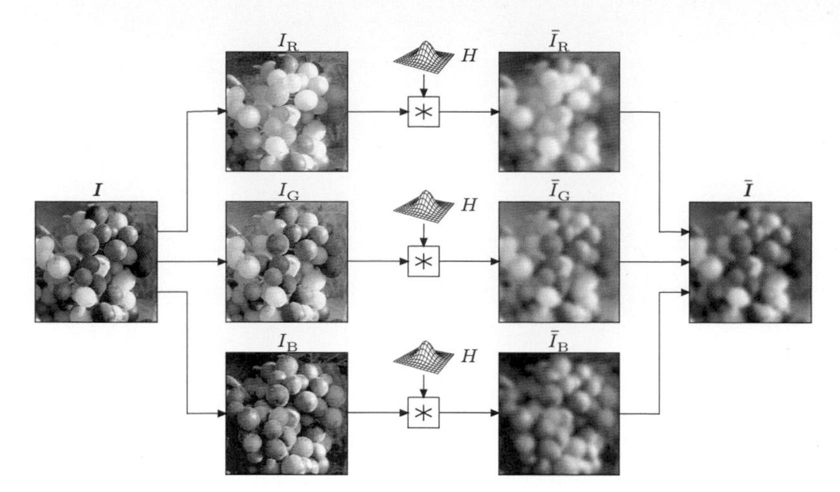

Fig. 15.1
Monochromatic application of a linear filter. The filter, specified by the kernel H, is applied separately to each of the scalar-valued color channels I_R, I_G, I_B. Combining the filtered component channels $\bar{I}_R$, $\bar{I}_G$, $\bar{I}_B$ produces the filtered color image $\bar{\boldsymbol{I}}$.

color pixel $\bar{\boldsymbol{I}}(u, v) = \bar{\boldsymbol{c}}$ in the filtered image is a *linear combination* of the original colors in $C_{u,v}$, that is,

$$\bar{\boldsymbol{c}} = w_1 \cdot \boldsymbol{c}_1 + w_2 \cdot \boldsymbol{c}_2 + \cdots + w_n \cdot \boldsymbol{c}_n = \sum_{i=1}^{n} w_i \cdot \boldsymbol{c}_i, \qquad (15.6)$$

where w_m is the coefficient in H that corresponds to pixel $\boldsymbol{c}_m$. If the kernel is *normalized* (i.e., $\sum H(i, j) = \sum \alpha_m = 1$), the result is an *affine combination* of the original colors. In case of a typical smoothing filter, with H normalized and all coefficients $H(i, j)$ being *positive*, any resulting color $\bar{\boldsymbol{c}}$ is a *convex combination* of the original color vectors $\boldsymbol{c}_1, \ldots, \boldsymbol{c}_n$.

Geometrically this means that the mixed color $\bar{\boldsymbol{c}}$ is contained within the *convex hull* of the contributing colors $\boldsymbol{c}_1, \ldots, \boldsymbol{c}_n$, as illustrated in Fig. 15.2. In the special case that only *two* original colors $\boldsymbol{c}_1, \boldsymbol{c}_2$ are involved, the result $\bar{\boldsymbol{c}}$ is located on the straight line segment connecting $\boldsymbol{c}_1$ and $\boldsymbol{c}_2$ (Fig. 15.2(b)).[2]

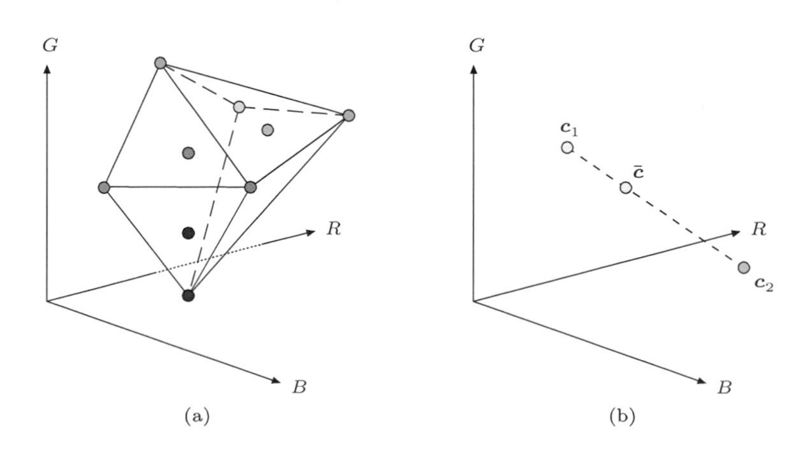

Fig. 15.2
Convex linear color mixtures. The result of the convex combination (mixture) of n color vectors $\mathcal{C} = \{\boldsymbol{c}_1, \ldots, \boldsymbol{c}_n\}$ is confined to the convex hull of $\mathcal{C}$ (a). In the special case of only two initial colors $\boldsymbol{c}_1$ and $\boldsymbol{c}_2$, any mixed color $\bar{\boldsymbol{c}}$ is located on the straight line segment connecting $\boldsymbol{c}_1$ and $\boldsymbol{c}_2$ (b).

[2] The convex hull of *two* points $\boldsymbol{c}_1, \boldsymbol{c}_2$ consists of the straight line segment between them.

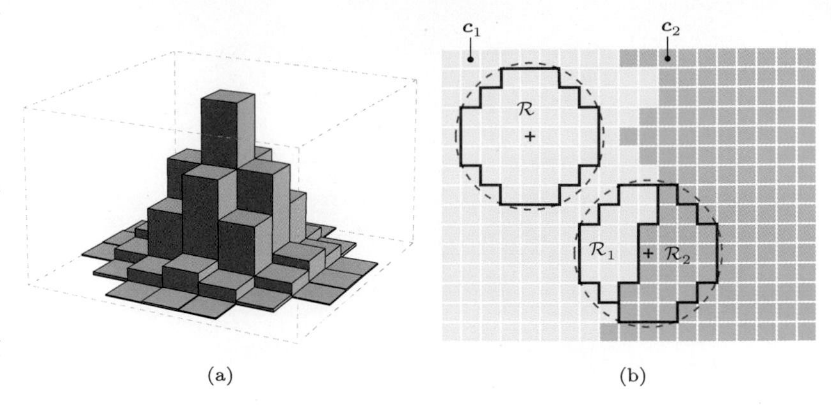

Fig. 15.3
Linear smoothing filter at a color edge. Discrete filter kernel with positive-only elements and support region $\mathcal{R}$ (a). Filter kernel positioned over a region of constant color $\boldsymbol{c}_1$ and over a color step edge $\boldsymbol{c}_1/\boldsymbol{c}_2$, respectively (b). If the (normalized) filter kernel of extent $\mathcal{R}$ is completely embedded in a region of constant color ($\boldsymbol{c}_1$), the result of filtering is exactly that same color. At a step edge between two colors $\boldsymbol{c}_1, \boldsymbol{c}_2$, one part of the kernel ($\mathcal{R}_1$) covers pixels of color $\boldsymbol{c}_1$ and the remaining part ($\mathcal{R}_2$) covers pixels of color $\boldsymbol{c}_2$. In this case, the result is a *linear mixture* of the colors $\boldsymbol{c}_1, \boldsymbol{c}_2$, as illustrated in Fig. 15.2(b).

Response to a color step edge

Assume, as a special case, that the original RGB image $\boldsymbol{I}$ contains a *step edge* separating two regions of constant colors $\boldsymbol{c}_1$ and $\boldsymbol{c}_2$, respectively, as illustrated in Fig. 15.3(b). If the normalized smoothing kernel H is placed at some position (u, v), where it is fully supported by pixels of identical color $\boldsymbol{c}_1$, the (trivial) response of the filter is

$$\bar{\boldsymbol{I}}(u,v) = \sum_{(i,j)\in\mathcal{R}_H} \boldsymbol{c}_1 \cdot H(i,j) = \boldsymbol{c}_1 \cdot \sum_{(i,j)\in\mathcal{R}_H} H(i,j) = \boldsymbol{c}_1 \cdot 1 = \boldsymbol{c}_1. \tag{15.7}$$

Thus the result at this position is the original color $\boldsymbol{c}_1$. Alternatively, if the filter kernel is placed at some position *on* a color edge (between two colors $\boldsymbol{c}_1, \boldsymbol{c}_2$, see again Fig. 15.3(b)), a subset of its coefficients ($\mathcal{R}_1$) is supported by pixels of color $\boldsymbol{c}_1$, while the other coefficients ($\mathcal{R}_2$) overlap with pixels of color $\boldsymbol{c}_2$. Since $\mathcal{R}_1 \cup \mathcal{R}_2 = \mathcal{R}$ and the kernel is normalized, the resulting color is

$$\bar{\boldsymbol{c}} = \sum_{(i,j)\in\mathcal{R}_1} \boldsymbol{c}_1 \cdot H(i,j) + \sum_{(i,j)\in\mathcal{R}_2} \boldsymbol{c}_2 \cdot H(i,j) \tag{15.8}$$

$$= \boldsymbol{c}_1 \cdot \underbrace{\sum_{(i,j)\in\mathcal{R}_1} H(i,j)}_{1-s} + \boldsymbol{c}_2 \cdot \underbrace{\sum_{(i,j)\in\mathcal{R}_2} H(i,j)}_{s} \tag{15.9}$$

$$= \boldsymbol{c}_1 \cdot (1-s) + \boldsymbol{c}_2 \cdot s = \boldsymbol{c}_1 + s \cdot (\boldsymbol{c}_2 - \boldsymbol{c}_1), \tag{15.10}$$

for some $s \in [0, 1]$. As we see, the resulting color coordinate $\bar{\boldsymbol{c}}$ lies on the straight line segment connecting the original colors $\boldsymbol{c}_1$ and $\boldsymbol{c}_2$ in the respective color space. Thus, at a step edge between two colors $\boldsymbol{c}_1, \boldsymbol{c}_2$, the intermediate colors produced by a (normalized) smoothing filter are located on the straight line between the two original color coordinates. Note that this relationship between linear filtering and linear color mixtures is independent of the particular color space in which the operation is performed.

15.1.2 Color Space Considerations

Since a linear filter always yields a convex linear mixture of the involved colors it should make a difference in which color space the

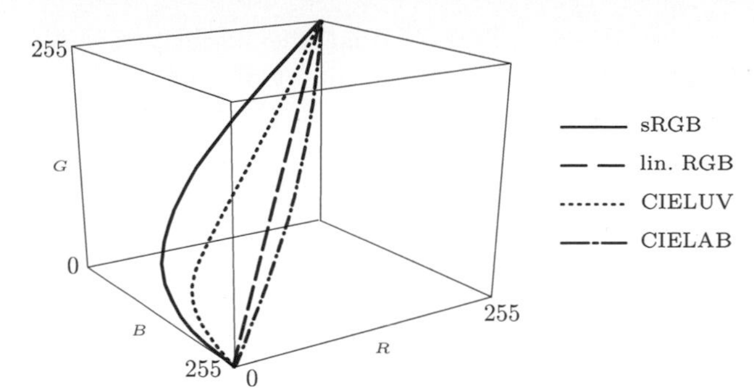

Fig. 15.4
Intermediate colors produced by linear interpolation between *yellow* and *blue*, performed in different color spaces. The 3D plot shows the resulting colors in linear RGB space.

filter operation is performed. For example, Fig. 15.4 shows the intermediate colors produced by a smoothing filter being applied to the same blue/yellow step edge but in different color spaces: sRGB, linear RGB, CIELUV, and CIELAB. As we see, the differences between the various color spaces are substantial. To obtain dependable and standardized results it might be reasonable to first transform the input image to a particular operating color space, perform the required filter operation, and finally transform the result back to the original color space, as illustrated in Fig. 15.5.

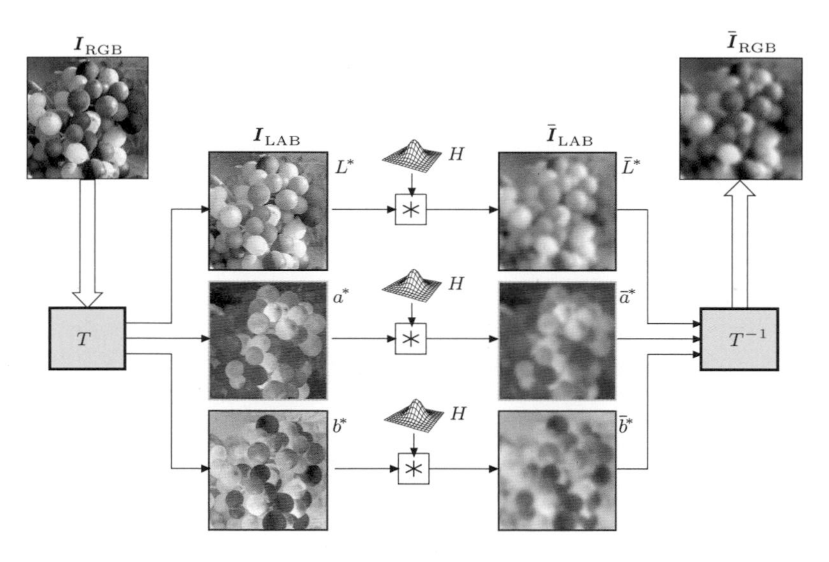

Fig. 15.5
Linear filter operation performed in a "foreign" color space. The original RGB image I_{RGB} is first transformed to CIELAB (by T), where the linear filter is applied separately to the three channels L^*, a^*, b^*. The filtered RGB image $\bar{I}_{RGB}$ is obtained by transforming back from CIELAB to RGB (by T^{-1}).

Obviously, a linear filter implies certain "metric" properties of the underlying color space. If we assume that a certain color space S_A has this property, then this is also true for any color space S_B that is related to S_A by a linear transformation, such as CIEXYZ and linear RGB space (see Ch. 14, Sec. 14.4.1). However, many color spaces used in practice (sRGB in particular) are related to these reference color spaces by highly nonlinear mappings, and thus significant deviations can be expected.

Preservation of brightness (luminance)

Apart from the intermediate colors produced by interpolation, another important (and easily measurable) aspect is the resulting change of *brightness* or *luminance* across the filter region. In particular it should generally hold that the luminance of the filtered color image is identical to the result of filtering only the (scalar) luminance channel of the original image with the same kernel H. Thus, if $\mathrm{Lum}(\boldsymbol{I})$ denotes the luminance of the original color image and $\mathrm{Lum}(\boldsymbol{I} * H)$ is the luminance of the filtered image, it should hold that

$$\mathrm{Lum}(\boldsymbol{I} * H) = \mathrm{Lum}(\boldsymbol{I}) * H. \tag{15.11}$$

This is only possible if $\mathrm{Lum}(\cdot)$ is linearly related to the components of the associated color space, which is mostly not the case. From Eqn. (15.11) we also see that, when filtering a step edge with colors $\boldsymbol{c}_1$ and $\boldsymbol{c}_2$, the resulting brightness should also change *monotonically* from $\mathrm{Lum}(\boldsymbol{c}_1)$ to $\mathrm{Lum}(\boldsymbol{c}_2)$ and, in particular, none of the intermediate brightness values should fall outside this range.

Figure 15.6 shows the results of filtering a synthetic test image with a normalized Gaussian kernel (of radius $\sigma = 3$) in different color spaces. Differences are most notable at the *red–blue* and *green–magenta* transitions, with particularly large deviations in the sRGB space. The corresponding luminance values Y (calculated from linear RGB components as in Eqn. (12.35)) are shown in Fig. 15.6(g–j). Again conspicuous is the result for sRGB (Fig. 15.6(c, g)), which exhibits transitions at the *red–blue*, *magenta–blue*, and *magenta–green* edges, where the resulting brightness drops below the original brightness of both contributing colors. Thus Eqn. (15.11) is not satisfied in this case. On the other hand, filtering in linear RGB space has the tendency to produce too high brightness values, as can be seen at the *black–white* markers in Fig. 15.6(d, h).

Out-of-gamut colors

If we apply a linear filter in RGB or sRGB space, the resulting intermediate colors are always valid RGB colors again and contained in the original RGB gamut volume. However, transformed to CIELUV or CIELAB, the set of possible RGB or sRGB colors forms a nonconvex shape (see Ch. 14, Fig. 14.5), such that linearly interpolated colors may fall outside the RGB gamut volume. Particularly critical (in both CIELUV and CIELAB) are the *red–white*, *red–yellow*, and *red–magenta* transitions, as well as *yellow–green* in CIELAB, where the resulting distances from the gamut surface can be quite large (see Fig. 15.7). During back-transformation to the original color space, such "out-of-gamut" colors must receive special treatment, since simple clipping of the affected components may cause unacceptable color distortions [167].

Implications and further reading

Applying a linear filter to the individual component channels of a color image presumes a certain "linearity" of the underlying color space. Smoothing filters implicitly perform additive linear mixing and interpolation. Despite common practice (and demonstrated by the results), there is no justification for performing a linear filter operation directly on gamma-mapped sRGB components. However, contrary to expectation, filtering in linear RGB does not yield better

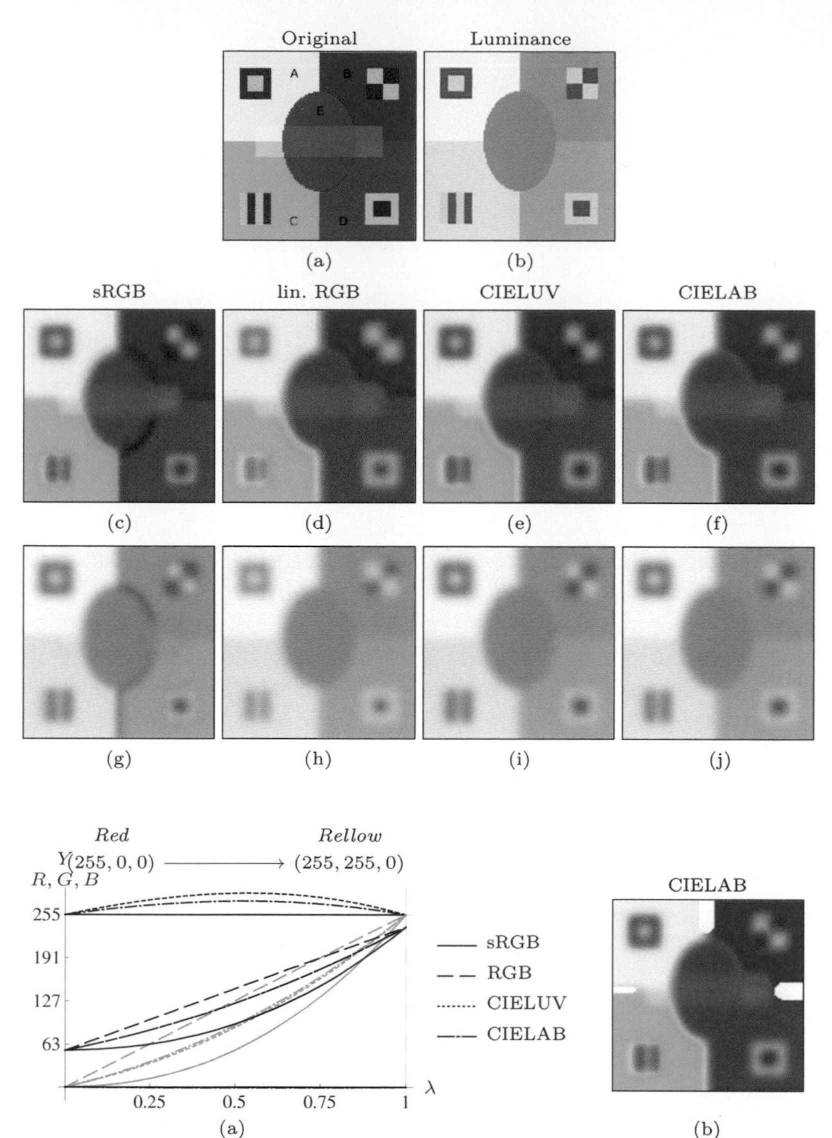

Fig. 15.6
Gaussian smoothing performed in different color spaces. Synthetic color image (a) and corresponding luminance image (b). The test image contains a horizontal bar with reduced color saturation but the same luminance as its surround, i.e., it is invisible in the luminance image. Gaussian filter applied in different color spaces: sRGB (c), linear RGB (d), CIELUV (e), and CIELAB (f). The bottom row (g–j) shows the corresponding luminance (Y) images. Note the dark bands in the sRGB result (b), particularly along the color boundaries between regions B–E, C–D, and D–E, which stand out clearly in the corresponding luminance image (g). Filtering in linear RGB space (d, h) gives good results between highly saturated colors, but subjectively too high luminance in unsaturated regions, which is apparent around the gray markers. Results with CIELUV (e, i) and CIELAB color spaces (f, j) appear most consistent as far as the preservation of luminance is concerned.

Fig. 15.7
Out-of-gamut colors produced by linear interpolation between *red* and *yellow* in "foreign" color spaces. The graphs in (a) show the (linear) R, G, B component values and the luminance Y (gray curves) resulting from a linear filter between *red* and *yellow* performed in different color spaces. The graphs show that the red component runs significantly outside the RGB gamut for both CIELUV and CIELAB. In (b) all pixels with any component outside the RGB gamut by more than 1% are marked white (for filtering in CIELAB).

overall results either. In summary, both nonlinear sRGB and linear RGB color spaces are unsuitable for linear filtering if perceptually accurate results are desired. Perceptually uniform color spaces, such as CIELUV and CIELAB, are good choices for linear filtering because of their metric properties, with CIELUV being perhaps slightly superior when it comes to interpolation over large color distances. When using CIELUV or CIELAB as intermediate color spaces for filtering RGB images, one must consider that out-of-gamut colors may be produced that must be handled properly. Thus none of the existing standard color spaces is universally suited or even "ideal" with respect to linear filtering.

The proper choice of the working color space is relevant not only to smoothing filters, but also to other types of filters, such as linear interpolation filters for geometric image transformations, decimation filters used in multi-scale techniques, and also nonlinear filters that involve averaging colors or calculation of color distances, such as the vector median filter (see Sec. 15.2.2). While complex color space transformations in the context of filtering (e.g., sRGB $\leftrightarrow$ CIELUV) are usually avoided for performance reasons, they should certainly be considered when high-quality results are important.

Although the issues related to color mixtures and interpolation have been investigated for some time (see, e.g., [149, 258]), their relevance to image filtering has not received much attention in the literature. Most image processing tools (including commercial software) apply linear filters directly to color images, without proper linearization or color space conversion. Lindbloom [149] was among the first to describe the problem of accurate color reproduction, particularly in the context of computer graphics and photo-realistic imagery. He also emphasized the relevance of perceptual uniformity for color processing and recommended the use of CIELUV as a suitable (albeit not perfect) processing space. Tomasi and Manduchi [229] suggested the use of the Euclidean distance in CIELAB space as "most natural" for bilateral filtering applied to color images (see also Ch. 17, Sec. 17.2) and similar arguments are put forth in [109]. De Weijer [239] notes that the additional chromaticities introduced by linear smoothing are "visually unacceptable" and argues for the use of nonlinear operators as an alternative. Lukac et al. [156] mention "certain inaccuracies" and color artifacts related to the application of scalar filters and discuss the issue of choosing a proper distance metric for vector-based filters. The practical use of alternative color spaces for image filtering is described in [141, Ch. 5].

15.1.3 Linear Filtering with Circular Values

If any of the color components is a *circular* quantity, such as the hue component in the HSV and HLS color spaces (see Ch. 12, Sec. 12.2.3), linear filters cannot be applied directly without additional provisions. As described in the previous section, a linear filter effectively calculates a weighted average over the values inside the filter region. Since the hue component represents a revolving angle and exhibits a discontinuity at the $1 \rightarrow 0$ (i.e., $360° \rightarrow 0°$) transition, simply averaging this quantity is not admissible (see Fig. 15.8).

However, correct interpolation of angular data is possible by utilizing the corresponding cosine and sine values, without any special treatment of discontinuities [69]. Given two angles α_1, α_2, the average angle α_{12} can be calculated as[3]

$$\alpha_{12} = \tan^{-1}\left(\frac{\sin(\alpha_1) + \sin(\alpha_2)}{\cos(\alpha_1) + \cos(\alpha_2)}\right) \tag{15.12}$$

$$= \mathrm{ArcTan}\big(\cos(\alpha_1) + \cos(\alpha_2), \sin(\alpha_1) + \sin(\alpha_2)\big) \tag{15.13}$$

and, in general, multiple angular values $\alpha_1, \ldots, \alpha_n$ can be correctly averaged in the form

$$\bar{\alpha} = \mathrm{ArcTan}\Big(\sum_{i=1}^{n} \cos(\alpha_i), \sum_{i=1}^{n} \sin(\alpha_i)\Big). \tag{15.14}$$

[3] See Sec. A.1 in the Appendix for the definition of the ArcTan() function.

Fig. 15.8
Naive linear filtering in HSV color space. Original RGB color image (a) and the associated HSV hue component I_h (b), with values in the range $[0, 1)$. Hue component after direct application of a Gaussian blur filter H with $\sigma = 3.0$ (c). Reconstructed RGB image $\tilde{\boldsymbol{I}}$ after filtering all components in HSV space (d). Note the false colors introduced around the $0 \rightarrow 1$ discontinuity (near red) of the hue component.

Also, the calculation of a *weighted* average is possible in the same way, that is,

$$\bar{\alpha} = \mathrm{ArcTan}\Bigl(\sum_{i=1}^{n} w_i \cdot \cos(\alpha_i), \sum_{i=1}^{n} w_i \cdot \sin(\alpha_i)\Bigr), \tag{15.15}$$

without any additional provisions, even the weights w_i need not be normalized. This approach can be used for linearly filtering circular data in general.

Filtering the hue component in HSV color space

To apply a linear filter H to the circular hue component I_h (with original values in $[0, 1)$) of a HSV or HLS image (see Ch. 12, Sec. 12.2.3), we first calculate the corresponding cosine and sine parts $I_h^{\sin}$ and $I_h^{\cos}$ by

$$\begin{aligned} I_h^{\sin}(u, v) &= \sin(2\pi \cdot I_h(u, v)), \\ I_h^{\cos}(u, v) &= \cos(2\pi \cdot I_h(u, v)), \end{aligned} \tag{15.16}$$

with resulting values in the range $[-1, 1]$. These are then filtered individually, that is,

$$\begin{aligned} \bar{I}_h^{\sin} &= I_h^{\sin} * H, \\ \bar{I}_h^{\cos} &= I_h^{\cos} * H. \end{aligned} \tag{15.17}$$

Finally, the filtered hue component $\bar{I}_h$ is obtained in the form

$$\bar{I}_h(u, v) = \frac{1}{2\pi} \cdot \bigl[\mathrm{ArcTan}\bigl(\bar{I}_h^{\cos}(u, v), \bar{I}_h^{\sin}(u, v)\bigr) \bmod 2\pi\bigr], \tag{15.18}$$

with values again in the range $[0, 1]$.

Fig. 15.9 demonstrates the correct application of a Gaussian smoothing filter to the hue component of an HSV color image by

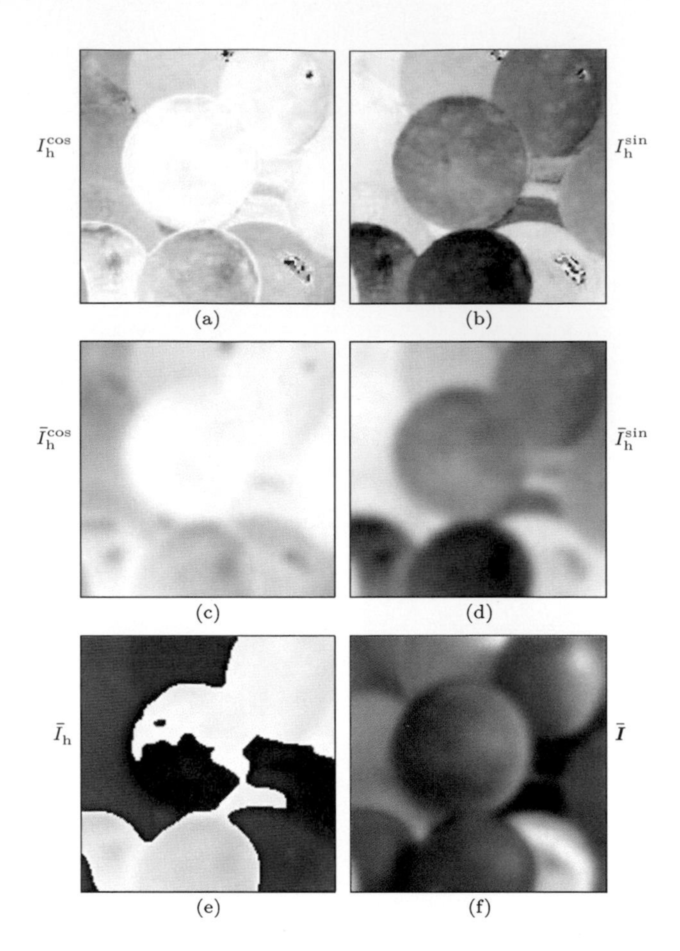

Fig. 15.9
Correct filtering of the HSV hue component by separation into cosine and sine parts (see Fig. 15.8(a) for the original image). Cosine and sine parts I_h^{sin}, I_h^{cos} of the hue component before (a, b) and after the application of a Gaussian blur filter with $\sigma = 3.0$ (c, d). Smoothed hue component $\bar{I}_h$ after merging the filtered cosine and sine parts $\bar{I}_h^{sin}$, $\bar{I}_h^{cos}$ (e). Reconstructed RGB image $\bar{\boldsymbol{I}}$ after filtering all HSV components (f). It is apparent that the hard 0/1 hue transitions in (e) are in fact only gradual color changes around the red hues. The other HSV components (S, V, which are non-circular) were filtered in the usual way. The reconstructed RGB image (f) shows no false colors and all hues correctly filtered.

separation into cosine and sine parts. The other two HSV components (S, V) are non-circular and were filtered as usual. In contrast to the result in Fig. 15.8(d), no false colors are produced at the $0 \rightarrow 1$ boundary. In this context it is helpful to look at the *distribution* of the hue values, which are clustered around 0/1 in the sample image (see Fig. 15.10(a)). In Fig. 15.10(b) we can clearly see how naive filtering of the hue component produces new (false) colors in the middle of the histogram. This does not occur when the hue component is filtered correctly (see Fig. 15.10(c)).

Saturation-weighted filtering

The method just described does not take into account that in HSV (and HLS) the hue and saturation components are closely related. In particular, the hue angle may be very inaccurate (or even indeterminate) if the associated saturation value goes to zero. For example, the test image in Fig. 15.8(a) contains a bright patch in the lower right-hand corner, where the *saturation* is low and the *hue* value is quite unstable, as seen in Fig. 15.9(a, b). However, the circular filter defined in Eqns. (15.16)–(15.18) takes all color samples as equally significant.

A simple solution is to use the saturation value $I_s(u, v)$ as a weight factor for the associated pixel [98], by modifying Eqn. (15.16) to

1: **HsvLinearFilter**$(\boldsymbol{I}_{\mathrm{hsv}}, H)$
Input: $\boldsymbol{I}_{\mathrm{hsv}} = (I_{\mathrm{h}}, I_{\mathrm{s}}, I_{\mathrm{v}})$, a HSV color image of size $M \times N$, with all components in $[0, 1]$; H, a 2D filter kernel. Returns a new (filtered) HSV color image of size $M \times N$.

2: $(M, N) \leftarrow \mathsf{Size}(\boldsymbol{I}_{\mathrm{hsv}})$
3: Create 2D maps $I_{\mathrm{h}}^{\sin}, I_{\mathrm{h}}^{\cos}, \bar{I}_{\mathrm{h}} \colon M \times N \mapsto \mathbb{R}$
Split the *hue* channel into sine/cosine parts:
4: **for all** $(u, v) \in M \times N$ **do**
5: $\quad \theta \leftarrow 2\pi \cdot I_{\mathrm{h}}(u, v)$ $\quad \triangleright$ hue angle $\theta \in [0, 2\pi]$
6: $\quad s \leftarrow I_{\mathrm{s}}(u, v)$ $\quad \triangleright$ saturation $s \in [0, 1]$
7: $\quad I_{\mathrm{h}}^{\sin}(u, v) \leftarrow s \cdot \sin(\theta)$ $\quad \triangleright$ $I_{\mathrm{h}}^{\sin}(u, v) \in [-1, 1]$
8: $\quad I_{\mathrm{h}}^{\cos}(u, v) \leftarrow s \cdot \cos(\theta)$ $\quad \triangleright$ $I_{\mathrm{h}}^{\cos}(u, v) \in [-1, 1]$
Filter all components with the same kernel:
9: $\bar{I}_{\mathrm{h}}^{\sin} \leftarrow I_{\mathrm{h}}^{\sin} * H$
10: $\bar{I}_{\mathrm{h}}^{\cos} \leftarrow I_{\mathrm{h}}^{\cos} * H$
11: $\bar{I}_{\mathrm{s}} \leftarrow I_{\mathrm{s}} * H$
12: $\bar{I}_{\mathrm{v}} \leftarrow I_{\mathrm{v}} * H$
Reassemble the filtered hue channel:
13: **for all** $(u, v) \in M \times N$ **do**
14: $\quad \theta \leftarrow \mathrm{ArcTan}\big(\bar{I}_{\mathrm{h}}^{\cos}(u, v), \bar{I}_{\mathrm{h}}^{\sin}(u, v)\big)$ $\quad \triangleright$ $\theta \in [-\pi, \pi]$
15: $\quad \bar{I}_{\mathrm{h}}(u, v) \leftarrow \frac{1}{2\pi} \cdot (\theta \bmod 2\pi)$ $\quad \triangleright$ $\bar{I}_{\mathrm{h}}(u, v) \in [0, 1]$
16: $\bar{\boldsymbol{I}}_{\mathrm{hsv}} \leftarrow (\bar{I}_{\mathrm{h}}, \bar{I}_{\mathrm{s}}, \bar{I}_{\mathrm{v}})$
17: **return** $\bar{\boldsymbol{I}}_{\mathrm{hsv}}$

Alg. 15.1
Linear filtering in HSV color space. All component values of the original HSV image are in the range $[0, 1]$. The algorithm considers the circular nature of the hue component and uses the saturation component (in line 6) as a weight factor, as defined in Eqn. (15.19). The same filter kernel H is applied to all three color components (lines 9–12).

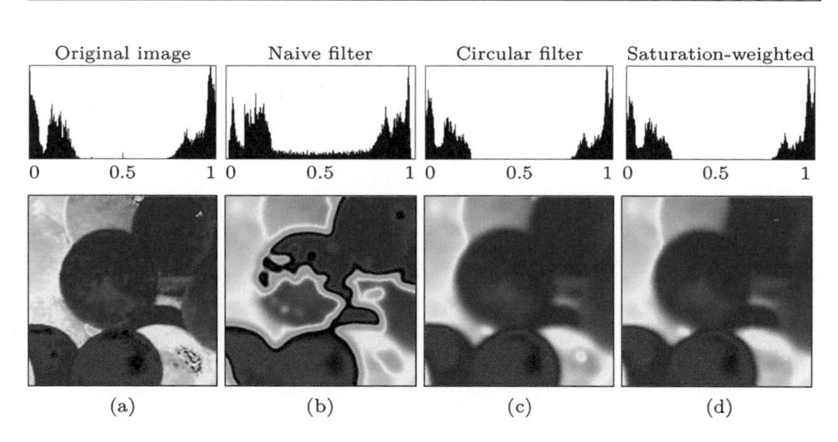

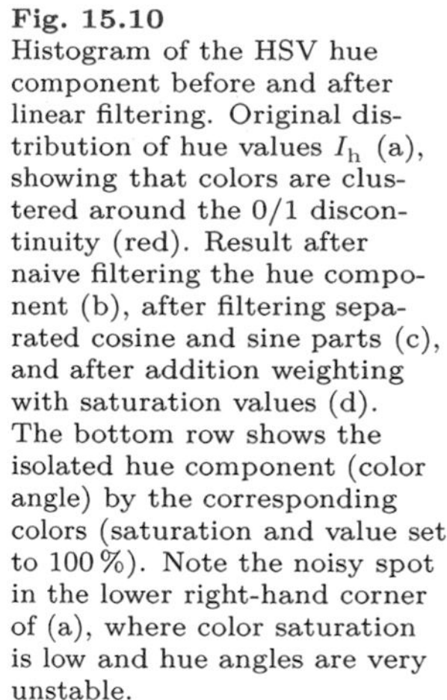

Fig. 15.10
Histogram of the HSV hue component before and after linear filtering. Original distribution of hue values I_{h} (a), showing that colors are clustered around the 0/1 discontinuity (red). Result after naive filtering the hue component (b), after filtering separated cosine and sine parts (c), and after addition weighting with saturation values (d). The bottom row shows the isolated hue component (color angle) by the corresponding colors (saturation and value set to 100 %). Note the noisy spot in the lower right-hand corner of (a), where color saturation is low and hue angles are very unstable.

$$
\begin{aligned}
I_{\mathrm{h}}^{\sin}(u, v) &= I_{\mathrm{s}}(u, v) \cdot \sin(2\pi \cdot I_{\mathrm{h}}(u, v)), \\
I_{\mathrm{h}}^{\cos}(u, v) &= I_{\mathrm{s}}(u, v) \cdot \cos(2\pi \cdot I_{\mathrm{h}}(u, v)).
\end{aligned}
\tag{15.19}
$$

All other steps in Eqns. (15.17)–(15.18) remain unchanged. The complete process is summarized in Alg. 15.1. The result in Fig. 15.10(d) shows that, particularly in regions of low color saturation, more stable hue values can be expected. Note that no normalization of the weights is required because the calculation of the hue angles (with the ArcTan() function in Eqn. (15.18)) only considers the ratio of the resulting sine and cosine parts.

15.2 Nonlinear Color Filters

In many practical image processing applications, linear filters are of limited use and nonlinear filters, such as the median filter, are applied instead.[4] In particular, for effective noise removal, nonlinear filters are usually the better choice. However, as with linear filters, the techniques originally developed for scalar (grayscale) images do not transfer seamlessly to vector-based color data. One reason is that, unlike in scalar data, no natural ordering relation exists for multi-dimensional data. As a consequence, nonlinear filters of the scalar type are often applied separately to the individual color channels, and again one must be cautious about the intermediate colors being introduced by these types of filters.

In the remainder of this section we describe the application of the classic (scalar) median filter to color images, a vector-based version of the median filter, and edge-preserving smoothing filters designed for color images. Additional filters for color images are presented in Chapter 17.

15.2.1 Scalar Median Filter

Applying a median filter with support region $\mathcal{R}$ (e.g., a disk-shaped region) at some image position (u, v) means to select one pixel value that is the most representative of the pixels in $\mathcal{R}$ to replace the current center pixel (*hot spot*). In case of a median filter, the statistical *median* of the pixels in $\mathcal{R}$ is taken as that representative. Since we always select the value of one of the existing image pixels, the median filter does not introduce any new pixel values that were not contained in the original image.

If a median filter is applied independently to the components of a color image, each channel is treated as a scalar image, like a single grayscale image. In this case, with the support region $\mathcal{R}$ centered at some point (u, v), the median for each color channel will typically originate from a *different* spatial position in $\mathcal{R}$, as illustrated in Fig. 15.11. Thus the components of the resulting color vector are generally collected from more than one pixel in $\mathcal{R}$, therefore the color placed in the filtered image may not match any of the original colors and new colors may be generated that were not contained in the original image. Despite its obvious deficiencies, the scalar (monochromatic) median filter is used in many popular image processing environments (including Photoshop and ImageJ) as the standard median filter for color images.

15.2.2 Vector Median Filter

The scalar median filter is based on the concept of *rank ordering*, that is, it assumes that the underlying data can be ordered and sorted. However, no such natural ordering exists for data elements that are vectors. Although vectors can be sorted in many different ways, for example by length or lexicographically along their dimensions, it is

[4] See also Chapter 5, Sec. 5.4.

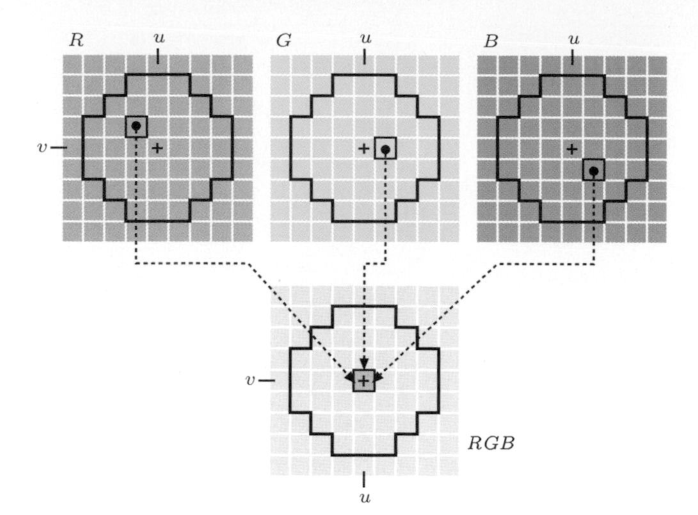

Fig. 15.11
Scalar median filter applied separately to color channels. With the filter region $\mathcal{R}$ centered at some point (u, v), the median pixel value is generally found at different locations in the R, G, B channels of the original image. The components of the resulting RGB color vector are collected from spatially separated pixels. It thus may not match any of the colors in the original image.

usually impossible to define a useful greater-than relation between any pair of vectors.

One can show, however, that the median of a sequence of n scalar values $P = (p_1, \ldots, p_n)$ can also be defined as the value p_{m} selected from P, such that

$$\sum_{i=1}^{n} |p_{\mathrm{m}} - p_i| \leq \sum_{i=1}^{n} |p_j - p_i| , \tag{15.20}$$

holds for any $p_j \in P$. In other words, the median value $p_{\mathrm{m}} = \mathrm{median}(P)$ is the one for which the sum of the differences to *all other* elements in the sequence P is the smallest.

With this definition, the concept of the median can be easily extended from the scalar situation to the case of multi-dimensional data. Given a sequence of vector-valued samples $\boldsymbol{P} = (\boldsymbol{p}_1, \ldots, \boldsymbol{p}_n)$, with $\boldsymbol{p}_i \in \mathbb{R}^K$, we define the median element $\boldsymbol{p}_{\mathrm{m}}$ to satisfy

$$\sum_{i=1}^{n} \|\boldsymbol{p}_{\mathrm{m}} - \boldsymbol{p}_i\| \leq \sum_{i=1}^{n} \|\boldsymbol{p}_j - \boldsymbol{p}_i\| , \tag{15.21}$$

for every possible $\boldsymbol{p}_j \in \boldsymbol{P}$. This is analogous to Eqn. (15.20), with the exception that the scalar difference $|\cdot|$ has been replaced by the vector norm $\|\cdot\|$ for measuring the distance between two points in the K-dimensional space.[5] We call

$$D_{\mathrm{L}}(\boldsymbol{p}, \boldsymbol{P}) = \sum_{\boldsymbol{p}_i \in \boldsymbol{P}} \|\boldsymbol{p} - \boldsymbol{p}_i\|_{\mathrm{L}} \tag{15.22}$$

the "aggregate distance" of the sample vector $\boldsymbol{p}$ with respect to all samples $\boldsymbol{p}_i$ in $\boldsymbol{P}$ under the distance norm L. Common choices for the distance norm are the L_1, L_2 and L_∞ norms, that is,

[5] K denotes the dimensionality of the samples in $\boldsymbol{p}_i$, for example, $K = 3$ for RGB color samples.

Fig. 15.12
Noisy test image (a) with enlarged details (b, c), used in the following examples.

$$\text{L}_1: \quad \|\boldsymbol{p}-\boldsymbol{q}\|_1 = \sum_{k=1}^{K} |p_k - q_k|, \tag{15.23}$$

$$\text{L}_2: \quad \|\boldsymbol{p}-\boldsymbol{q}\|_2 = \Big[\sum_{k=1}^{K} |p_k - q_k|^2\Big]^{1/2}, \tag{15.24}$$

$$\text{L}_\infty: \quad \|\boldsymbol{p}-\boldsymbol{q}\|_\infty = \max_{1\leq k\leq K} |p_k - q_k|. \tag{15.25}$$

The vector median of the sequence $\boldsymbol{P}$ can thus be defined as

$$\text{median}(\boldsymbol{P}) = \underset{\boldsymbol{p}\in\boldsymbol{P}}{\text{argmin}}\, D_{\text{L}}(\boldsymbol{p},\boldsymbol{P}), \tag{15.26}$$

that is, the sample $\boldsymbol{p}$ with the smallest aggregate distance to all other elements in $\boldsymbol{P}$.

A straight forward implementation of the vector median filter for RGB images is given in Alg. 15.2. The calculation of the aggregate distance $D_{\text{L}}(\boldsymbol{p},\boldsymbol{P})$ is performed by the function AggregateDistance $(\boldsymbol{p},\boldsymbol{P})$. At any position (u, v), the center pixel is replaced by the neighborhood pixel with the smallest aggregate distance D_{min}, but only if it is smaller than the center pixel's aggregate distance D_{ctr} (line 15). Otherwise, the center pixel is left unchanged (line 17). This is to prevent that the center pixel is unnecessarily changed to another color, which incidentally has the same aggregate distance.

The optimal choice of the norm L for calculating the distances between color vectors in Eqn. (15.22) depends on the assumed noise distribution of the underlying signal [10]. The effects of using different norms (L_1, L_2, L_∞) are shown in Fig. 15.13 (see Fig. 15.12 for the original images). Although the results for these norms show numerical differences, they are hardly noticeable in real images (particularly in print). Unless otherwise noted, the L_1 norm is used in all subsequent examples.

Results of the scalar median filter and the vector median filter are compared in Fig. 15.14. Note how new colors are introduced by the scalar filter at certain locations (Fig. 15.14(a, c)), as illustrated in Fig. 15.11. In contrast, the vector median filter (Fig. 15.14(b, d)) can only produce colors that already exist in the original image. Figure

Alg. 15.2
Vector median filter for color images.

```
1: VectorMedianFilter(I, r)
     Input: I = (I_R, I_G, I_B), a color image of size M × N;
     r, filter radius (r ≥ 1).
     Returns a new (filtered) color image of size M × N.
2:   (M, N) ← Size(I)
3:   I' ← Duplicate(I)
4:   for all image coordinates (u, v) ∈ M × N do
5:       p_ctr ← I(u, v)                      ▷ center pixel of support region
6:       P ← GetSupportRegion(I, u, v, r)
7:       d_ctr ← AggregateDistance(p_ctr, P)
8:       d_min ← ∞
9:       for all p ∈ P do
10:          d ← AggregateDistance(p, P)
11:          if d < d_min then
12:              p_min ← p
13:              d_min ← d
14:      if d_min < d_ctr then
15:          I'(u, v) ← p_min                  ▷ modify this pixel
16:      else
17:          I'(u, v) ← I(u, v)               ▷ keep the original pixel value
18:  return I'

19: GetSupportRegion(I, u, v, r)
     Returns a vector of n pixel values P = (p_1, p_2, ..., p_n) from
     image I that are inside a disk of radius r, centered at position
     (u, v).
20:  P ← ()
21:  for i ← ⌊u−r⌋, ..., ⌈u+r⌉ do
22:      for j ← ⌊v−r⌋, ..., ⌈v+r⌉ do
23:          if (u − i)^2 + (v − j)^2 ≤ r^2 then
24:              p ← I(i, j)
25:              P ← P ⌣ (p)
26:  return P                                  ▷ P = (p_1, p_2, ..., p_n)

27: AggregateDistance(p, P)
     Returns the aggregate distance D_L(p, P) of the sample vector p
     over all elements p_i ∈ P (see Eq. 15.22).
28:  d ← 0
29:  for all q ∈ P do
30:      d ← d + ||p − q||_L                   ▷ choose any distance norm L
31:  return d
```

15.15 shows the results of applying the vector median filter to real color images while varying the filter radius.

Since the vector median filter relies on measuring the distance between pairs of colors, the considerations in Sec. 15.1.2 regarding the metric properties of the color space do apply here as well. It is thus not uncommon to perform this filter operation in a perceptual uniform color space, such as CIELUV or CIELAB, rather than in RGB [132, 240, 254].

The vector median filter is computationally expensive. Calculating the aggregate distance for all sample vectors $\boldsymbol{p}_i$ in $\boldsymbol{P}$ requires $\mathcal{O}(n^2)$ steps, for a support region of size n. Finding the candidate neighborhood pixel with the minimum aggregate distance in $\boldsymbol{P}$ can

Fig. 15.13
Results of vector median filtering using different color distance norms: L_1 norm (a), L_2 norm (b), L_∞ norm (c). Filter radius $r = 2.0$.

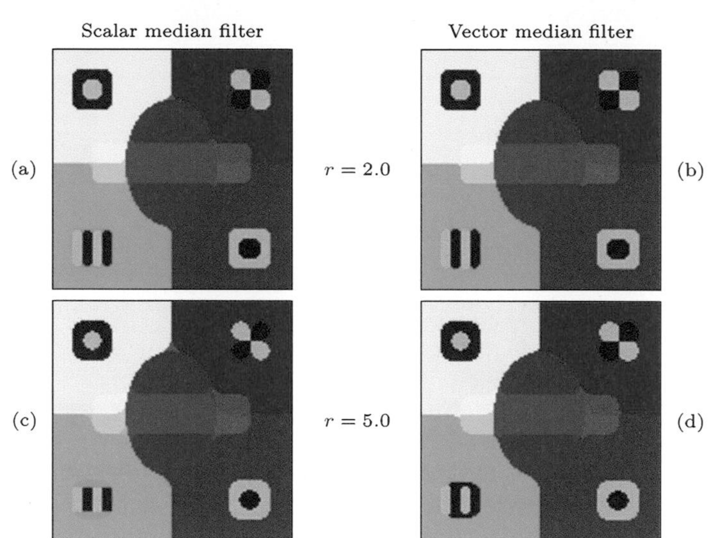

Fig. 15.14
Scalar median vs. vector median filter applied to a color test image, with filter radius $r = 2.0$ (a, b) and $r = 5.0$ (c, d). Note how the scalar median filter (a, c) introduces new colors that are not contained in the original image.

be done in $\mathcal{O}(n)$. Since n is proportional to the square of the filter radius r, the number of steps required for calculating a single image pixel is roughly $\mathcal{O}(r^4)$. While faster implementations have been proposed [10, 18, 221], calculating the vector median filter remains computationally demanding.

15.2.3 Sharpening Vector Median Filter

Although the vector median filter is a good solution for suppressing impulse noise and additive Gaussian noise in color images, it does tend to blur or even eliminate relevant structures, such as lines and edges. The *sharpening* vector median filter, proposed in [155], aims at improving the edge preservation properties of the standard vector median filter described earlier. The key idea is not to calculate the aggregate distances against *all* other samples in the neighborhood but only against the *most similar* ones. The rationale is that

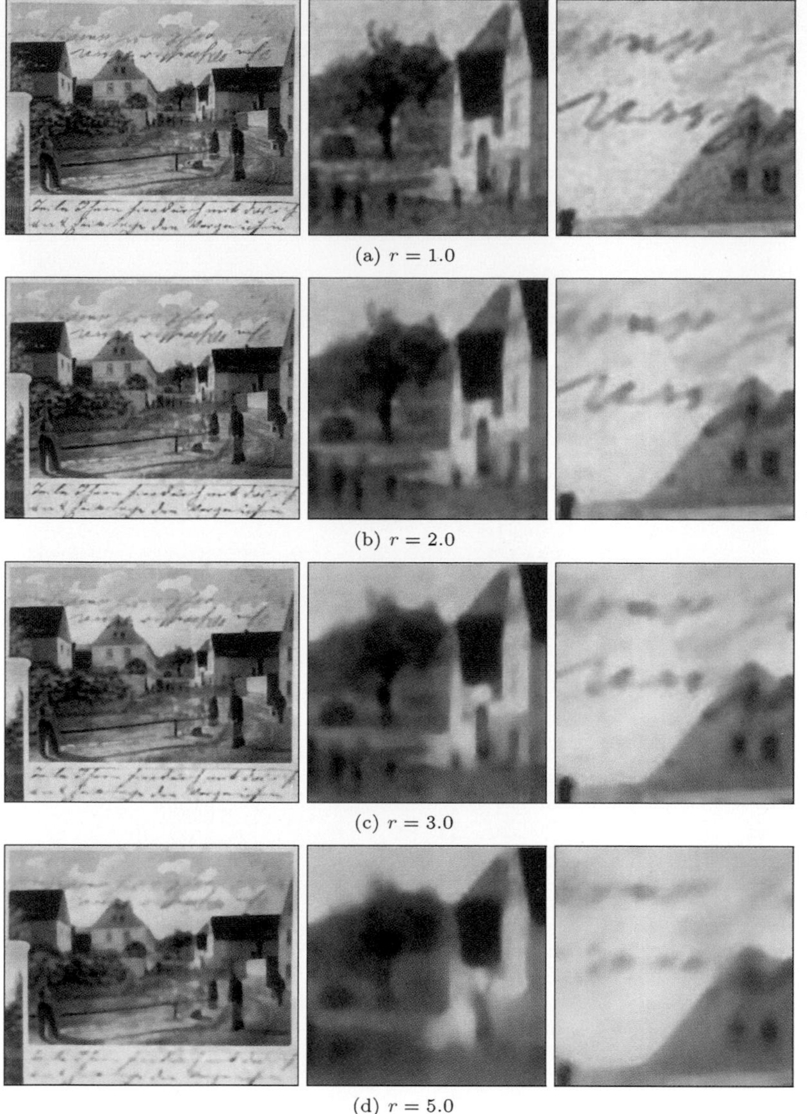

(a) $r = 1.0$

(b) $r = 2.0$

(c) $r = 3.0$

(d) $r = 5.0$

Fig. 15.15
Vector median filter with varying radii applied to a real color image (L_1 norm).

the samples deviating strongly from their neighbors tend to be *outliers* (e.g., caused by nearby edges) and should be excluded from the median calculation to avoid blurring of structural details.

The operation of the sharpening vector median filter is summarized in Alg. 15.3. For calculating the aggregate distance $D_L(\boldsymbol{p}, \boldsymbol{P})$ of a given sample vector $\boldsymbol{p}$ (see Eqn. (15.22)), not all samples in $\boldsymbol{P}$ are considered, but only those a samples that are *closest* to $\boldsymbol{p}$ in the 3D color space (a being a fixed fraction of the support region size). The subsequent minimization is performed over what is called the "trimmed aggregate distance". Thus, only a fixed number (a) of neighborhood pixels is included in the calculation of the aggregate distances. As a consequence, the sharpening vector median filter provides good noise removal while at the same time leaving edge structures intact.

Alg. 15.3
Sharpening vector median filter for RGB color images (extension of Alg. 15.2). The *sharpening parameter* $s \in [0,1]$ controls the number of most-similar neighborhood pixels included in the median calculation. For $s = 0$, all pixels in the given support region are included and no sharpening occurs; setting $s = 1$ leads to maximum sharpening. The *threshold parameter* t controls how much smaller the aggregate distance of any neighborhood pixel must be to replace the current center pixel.

```
1: SharpeningVectorMedianFilter(I, r, s, t)
   Input: I, a color image of size M×N, I(u,v) ∈ ℝ³; r, filter
   radius (r ≥ 1); s, sharpening parameter (0 ≤ s ≤ 1); t, threshold
   (t ≥ 0). Returns a new (filtered) color image of size M×N.
2:   (M, N) ← Size(I)
3:   I' ← Duplicate(I)
4:   for all image coordinates (u, v) ∈ M×N do
5:       P ← GetSupportRegion(I, u, v, r)            ▷ see Alg. 15.2
6:       n ← |P|                                     ▷ size of P
7:       a ← round (n − s·(n − 2))                   ▷ a = 2, ..., n
8:       d_ctr ← TrimmedAggregateDistance(I(u, v), P, a)
9:       d_min ← ∞
10:      for all p ∈ P do
11:          d ← TrimmedAggregateDistance(p, P, a)
12:          if d < d_min then
13:              p_min ← p
14:              d_min ← d
15:      if (d_ctr − d_min) > t·a then
16:          I'(u, v) ← p_min                        ▷ replace the center pixel
17:      else
18:          I'(u, v) ← I(u, v)                      ▷ keep the original center pixel
19:  return I'

20: TrimmedAggregateDistance(p, P, a)
    Returns the aggregate distance from p to the a most similar ele-
    ments in P = (p_1, p_2, ..., p_n).
21:  n ← |P|                                         ▷ size of P
22:  Create map D : [1, n] ↦ ℝ
23:  for i ← 1, ..., n do
24:      D(i) ← ‖p − P(i)‖_L                         ▷ choose any distance norm L
25:  D' ← Sort(D)                                    ▷ D'(1) ≤ D'(2) ≤ ... ≤ D'(n)
26:  d ← 0
27:  for i ← 2, ..., a do                            ▷ D'(1) = 0, thus skipped
28:      d ← d + D'(i)
29:  return d
```

Typically, the aggregate distance of $\boldsymbol{p}$ to the a closest neighborhood samples is found by first calculating the distances between $\boldsymbol{p}$ and all other samples in $\boldsymbol{P}$, then sorting the result, and finally adding up only the a initial elements of the sorted distances (see procedure TrimmedAggregateDistance($\boldsymbol{p}, \boldsymbol{P}, a$) in Alg. 15.3). Thus the sharpening median filter requires an additional sorting step over $n \propto r^2$ elements at each pixel, which again adds to its time complexity.

The parameter s in Alg. 15.3 specifies the fraction of region pixels included in the calculation of the median and thus controls the amount of sharpening. The number of incorporated pixels a is determined as $a = \text{round}(n - s \cdot (n-2))$ (see Alg. 15.3, line 7), so that $a = n, \ldots, 2$ for $s \in [0,1]$. With $s = 0$, all $a = |\boldsymbol{P}| = n$ pixels in the filter region are included in the median calculation and the filter behaves like the ordinary vector-median filter described in Alg. 15.2. At maximum sharpening (i.e., with $s = 1$) the calculation of the aggregate distance includes only the single most similar color pixel in the neighborhood $\boldsymbol{P}$.

The calculation of the "trimmed aggregate distance" is shown in Alg. 15.3 (lines 20–29). The function TrimmedAggregateDistance $(\boldsymbol{p}, \boldsymbol{P}, a)$ calculates the aggregate distance for a given vector (color sample) $\boldsymbol{p}$ over the a closest samples in the support region $\boldsymbol{P}$. Initially (in line 24), the n distances $D(i)$ between $\boldsymbol{p}$ and all elements in $\boldsymbol{P}$ are calculated, with $D(i) = \|\boldsymbol{p} - \boldsymbol{P}(i)\|_{\mathrm{L}}$ (see Eqns. (15.23)–(15.25)). These are subsequently sorted by increasing value (line 25) and the sum of the a smallest values $D'(1), \ldots, D'(a)$ (line 28) is returned.[6]

The effects of varying the sharpen parameter s are shown in Fig. 15.16, with a fixed filter radius $r = 2.0$ and threshold $\mathtt{t} = 0$. For $s = 0.0$ (Fig. 15.16(a)), the result is the same as that of the ordinary vector median filter (see Fig. 15.15(b)).

The value of the current center pixel is only replaced by a neighboring pixel value if the corresponding minimal (trimmed) aggregate distance d_{min} is significantly smaller than the center pixel's aggregate distance d_{ctr}. In Alg. 15.3, this is controlled by the threshold t. The center pixel is replaced only if the condition

$$(d_{\mathrm{ctr}} - d_{\mathrm{min}}) > \mathtt{t} \cdot a \qquad (15.27)$$

holds; otherwise it remains unmodified. Note that the distance limit is proportional to a and thus t really specifies the minimum "average" pixel distance; it is independent of the filter radius r and the sharpening parameter s.

Results for typical values of t (in the range $0, \ldots, 10$) are shown in Figs. 15.17–15.18. To illustrate the effect, the images in Fig. 15.18 only display those pixels that were *not* replaced by the filter, while all modified pixels are set to black. As one would expect, increasing the threshold t leads to fewer pixels being modified. Of course, the same thresholding scheme may also be used with the ordinary vector median filter (see Exercise 15.2).

15.3 Java Implementation

Implementations of the scalar and vector median filter as well as the sharpening vector median filter are available with full Java source code at the book's website.[7] The corresponding classes

- `ScalarMedianFilter`,
- `VectorMedianFilter`, and
- `VectorMedianFilterSharpen`

are based on the common super-class `GenericFilter`, which provides the abstract methods

```
void applyTo (ImageProcessor ip),
```

which greatly simplifies the use of these filters. The code segment in Prog. 15.1 demonstrates the use of the class `VectorMedianFilter` (with radius 3.0 and L_1-norm) for RGB color images in an ImageJ plugin. For the specific filters described in this chapter, the following constructors are provided:

[6] $D'(1)$ is zero because it is the distance between $\boldsymbol{p}$ and itself.

[7] Package `imagingbook.pub.color.filters`.

(a) $s = 0.0$

(b) $s = 0.2$

(c) $s = 0.5$

(d) $s = 0.8$

Fig. 15.16
Sharpening vector median filter with different sharpness values s. The filter radius is $r = 2.0$ and the corresponding filter mask contains $n = 21$ pixels. At each pixel, only the $a = 21, 17, 12, 6$ closest color samples (for sharpness $s = 0.0, 0.2, 0.5, 0.8$, respectively) are considered when calculating the local vector median.

`ScalarMedianFilter (Parameters params)`
Creates a scalar median filter, as described in Sec. 15.2.1, with parameter `radius` = 3.0 (default).

`VectorMedianFilter (Parameters params)`
Creates a vector median filter, as described in Sec. 15.2.2, with parameters `radius` = 3.0 (default), `distanceNorm` = `NormType.L1` (default), `L2`, `Lmax`.

`VectorMedianFilterSharpen (Parameters params)`
Creates a sharpening vector median filter (see Sec. 15.2.3) with parameters `radius` = 3.0 (default), `distanceNorm` = `NormType.L1` (default), `L2`, `Lmax`, sharpening factor `sharpen` = 0.5 (default), `threshold` = 0.0 (default).

The listed default values pertain to the parameterless constructors that are also available. See the online API documentation or the

(a) t = 0

(b) t = 2

(c) t = 5

(d) t = 10

Fig. 15.17
Sharpening vector median filter with different threshold values t = 0, 2, 5, 10. The filter radius and sharpening factor are fixed at $r = 2.0$ and $s = 0.0$, respectively.

source code for additional details. Note that the created filter objects are generic and can be applied to both grayscale and color images without any modification.

15.4 Further Reading

A good overview of different linear and nonlinear filtering techniques for color images can be found in [141]. In [186, Ch. 2], the authors give a concise treatment of color image filtering, including statistical noise models, vector ordering schemes, and different color similarity measures. Several variants of weighted median filters for color images and multi-channel data in general are described in [6, Ch. 2, Sec. 2.4]. A very readable and up-to-date survey of important color issues in computer vision, such as color constancy, photometric invariance, and

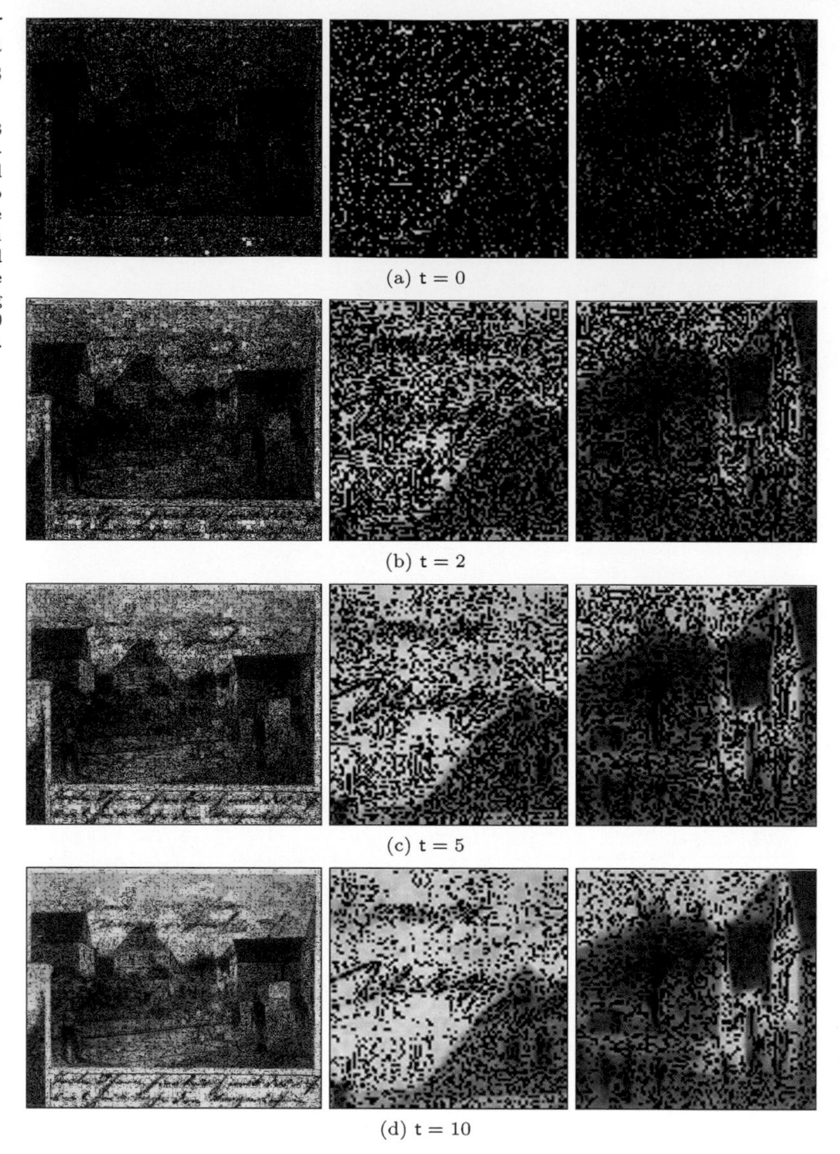

Fig. 15.18
Sharpening vector median filter with different threshold values t = 0, 2, 5, 10 (also see Fig. 15.17). Only the *unmodified* pixels are shown in color, while all modified pixels are set to *black*. The filter radius and sharpening factor are fixed at $r = 2.0$ and $s = 0.0$, respectively.

(a) t = 0

(b) t = 2

(c) t = 5

(d) t = 10

color feature extraction, can be found in [83]. A vector median filter operating in HSV color space is proposed in [240]. In addition to the techniques discussed in this chapter, most of the filters described in Chapter 17 can either be applied directly to color images or easily modified for this purpose.

15.5 Exercises

Exercise 15.1. Verify Eqn. (15.20) by showing (formally or experimentally) that the usual calculation of the scalar median (by sorting a sequence and selecting the center value) indeed gives the value with the smallest sum of differences from all other values in the same sequence. Is the result independent of the type of distance norm used?

```
1 import ij.ImagePlus;
2 import ij.plugin.filter.PlugInFilter;
3 import ij.process.ImageProcessor;
4 import imagingbook.lib.math.VectorNorm.NormType;
5 import imagingbook.lib.util.Enums;
6 import imagingbook.pub.colorfilters.VectorMedianFilter;
7 import imagingbook.pub.colorfilters.VectorMedianFilter.*;
8
9 public class MedianFilter_Color_Vector implements
      PlugInFilter
10 {
11     public int setup(String arg, ImagePlus imp) {
12       return DOES_RGB;
13     }
14
15     public void run(ImageProcessor ip) {
16       Parameters params =
17           new VectorMedianFilter.Parameters();
18       params.distanceNorm = NormType.L1;
19       params.radius = 3.0;
20
21       VectorMedianFilter filter =
22           new VectorMedianFilter(params);
23
24       filter.applyTo(ip);
25     }
26 }
```

Prog. 15.1
Color median filter using class `VectorMedianFilter`. In line 17, a suitable parameter object (with default values) is created, then modified and passed to the constructor of the filter (in line 22). The filter itself is applied to the input image, which is destructively modified (in line 24).

Exercise 15.2. Modify the ordinary vector median filter described in Alg. 15.2 to incorporate a threshold t for deciding whether to modify the current center pixel or not, analogous to the approach taken in the sharpening vector median filter in Alg. 15.3.

Exercise 15.3. Implement a dedicated median filter (analogous to Alg. 15.1) for the HSV color space. The filter should process the color components independently but consider the circular nature of the hue component, as discussed in Sec. 15.1.3. Compare the results to the vector-median filter in Sec. 15.2.2.

16

Edge Detection in Color Images

Edge information is essential in many image analysis and computer vision applications and thus the ability to locate and characterize edges robustly and accurately is an important task. Basic techniques for edge detection in *grayscale* images are discussed in Chapter 6. *Color* images contain richer information than grayscale images and it appears natural to assume that edge detection methods based on color should outperform their monochromatic counterparts. For example, locating an edge between two image regions of different hue but similar brightness is difficult with an edge detector that only looks for changes in image intensity. In this chapter, we first look at the use of "ordinary" (i.e., monochromatic) edge detectors for color images and then discuss dedicated detectors that are specifically designed for color images.

Although the problem of color edge detection has been pursued for a long time (see [140,266] for a good overview), most image processing texts do not treat this subject in much detail. One reason could be that, in practice, edge detection in color images is often accomplished by using "monochromatic" techniques on the intensity channel or the individual color components. We discuss these simple methods—which nevertheless give satisfactory results in many situations—in Sec. 16.1.

Unfortunately, monochromatic techniques do not extend naturally to color images and other "multi-channel" data, since edge information in the different color channels may be ambiguous or even contradictory. For example, multiple edges running in different directions may coincide at a given image location, edge gradients may cancel out, or edges in different channels may be slightly displaced. In Sec. 16.2, we describe how local gradients can be calculated for edge detection by treating the color image as a 2D *vector field*. In Sec. 16.3, we show how the popular Canny edge detector, originally designed for monochromatic images, can be adapted for color images, and Sec. 16.4 goes on to look at other color edge operators. Implementations of the discussed algorithms are described in Sec. 16.5, with complete source code available on the book's website.

16.1 Monochromatic Techniques

Linear filters are the basis of most edge enhancement and edge detection operators for scalar-valued grayscale images, particularly the gradient filters described in Chapter 15, Sec. 6.3. Again, it is quite common to apply these scalar filters separately to the individual color channels of RGB images. A popular example is the Sobel operator with the filter kernels

$$H_x^{\mathrm{S}} = \frac{1}{8} \cdot \begin{bmatrix} -1 & 0 & 1 \\ -2 & 0 & 2 \\ -1 & 0 & 1 \end{bmatrix} \quad \text{and} \quad H_y^{\mathrm{S}} = \frac{1}{8} \cdot \begin{bmatrix} -1 & -2 & -1 \\ 0 & 0 & 0 \\ 1 & 2 & 1 \end{bmatrix} \tag{16.1}$$

for the x- and y-direction, respectively. Applied to a grayscale image I, with $I_x = I * H_x^{\mathrm{S}}$ and $I_y = I * H_y^{\mathrm{S}}$, these filters give a reasonably good estimate of the local gradient vector,

$$\nabla I(\boldsymbol{u}) = \begin{pmatrix} I_x(\boldsymbol{u}) \\ I_y(\boldsymbol{u}) \end{pmatrix}, \tag{16.2}$$

at position $\boldsymbol{u} = (u, v)$. The local edge *strength* of the grayscale image is then taken as

$$E_{\mathrm{gray}}(\boldsymbol{u}) = \|\nabla I(\boldsymbol{u})\| = \sqrt{I_x^2(\boldsymbol{u}) + I_y^2(\boldsymbol{u})}, \tag{16.3}$$

and the corresponding edge *orientation* is calculated as

$$\Phi(\boldsymbol{u}) = \angle \nabla I(\boldsymbol{u}) = \tan^{-1}\Bigl(\frac{I_y(\boldsymbol{u})}{I_x(\boldsymbol{u})}\Bigr). \tag{16.4}$$

The angle $\Phi(\boldsymbol{u})$ gives the direction of maximum intensity change on the 2D image surface at position $(\boldsymbol{u})$, which is the normal to the edge tangent.

Analogously, to apply this technique to a color image $\boldsymbol{I} = (I_{\mathrm{R}}, I_{\mathrm{G}}, I_{\mathrm{B}})$, each color plane is first filtered individually with the two gradient kernels given in Eqn. (16.1), resulting in

$$\begin{aligned} \nabla I_{\mathrm{R}} &= \begin{pmatrix} I_{\mathrm{R},x} \\ I_{\mathrm{R},y} \end{pmatrix} = \begin{pmatrix} I_{\mathrm{R}} * H_x^{\mathrm{S}} \\ I_{\mathrm{R}} * H_y^{\mathrm{S}} \end{pmatrix}, \\ \nabla I_{\mathrm{G}} &= \begin{pmatrix} I_{\mathrm{G},x} \\ I_{\mathrm{G},y} \end{pmatrix} = \begin{pmatrix} I_{\mathrm{G}} * H_x^{\mathrm{S}} \\ I_{\mathrm{G}} * H_y^{\mathrm{S}} \end{pmatrix}, \\ \nabla I_{\mathrm{B}} &= \begin{pmatrix} I_{\mathrm{B},x} \\ I_{\mathrm{B},y} \end{pmatrix} = \begin{pmatrix} I_{\mathrm{B}} * H_x^{\mathrm{S}} \\ I_{\mathrm{B}} * H_y^{\mathrm{S}} \end{pmatrix}. \end{aligned} \tag{16.5}$$

The local edge strength is calculated separately for each color channel which yields a vector

$$\boldsymbol{E}(\boldsymbol{u}) = \begin{pmatrix} E_{\mathrm{R}}(\boldsymbol{u}) \\ E_{\mathrm{G}}(\boldsymbol{u}) \\ E_{\mathrm{B}}(\boldsymbol{u}) \end{pmatrix} = \begin{pmatrix} \|\nabla I_{\mathrm{R}}(\boldsymbol{u})\| \\ \|\nabla I_{\mathrm{G}}(\boldsymbol{u})\| \\ \|\nabla I_{\mathrm{B}}(\boldsymbol{u})\| \end{pmatrix} \tag{16.6}$$

$$= \begin{pmatrix} [I_{\mathrm{R},x}^2(\boldsymbol{u}) + I_{\mathrm{R},y}^2(\boldsymbol{u})]^{1/2} \\ [I_{\mathrm{G},x}^2(\boldsymbol{u}) + I_{\mathrm{G},y}^2(\boldsymbol{u})]^{1/2} \\ [I_{\mathrm{B},x}^2(\boldsymbol{u}) + I_{\mathrm{B},y}^2(\boldsymbol{u})]^{1/2} \end{pmatrix} \tag{16.7}$$

for each image position $\boldsymbol{u}$. These vectors could be combined into a new color image $\boldsymbol{E} = (E_{\mathrm{R}}, E_{\mathrm{G}}, E_{\mathrm{B}})$, although such a "color edge image" has no particularly useful interpretation.[1] Finally, a scalar quantity of *combined edge strength* (C) over all color planes can be obtained, for example, by calculating the Euclidean (L_2) norm of $\boldsymbol{E}$ as

$$C_2(\boldsymbol{u}) = \|\boldsymbol{E}(\boldsymbol{u})\|_2 = \left[E_{\mathrm{R}}^2(\boldsymbol{u}) + E_{\mathrm{G}}^2(\boldsymbol{u}) + E_{\mathrm{B}}^2(\boldsymbol{u})\right]^{1/2}$$
$$= \left[I_{\mathrm{R},x}^2 + I_{\mathrm{R},y}^2 + I_{\mathrm{G},x}^2 + I_{\mathrm{G},y}^2 + I_{\mathrm{B},x}^2 + I_{\mathrm{B},y}^2\right]^{1/2} \tag{16.8}$$

(coordinates ($\boldsymbol{u}$) are omitted in the second line) or, using the L_1 norm,

$$C_1(\boldsymbol{u}) = \|\boldsymbol{E}(\boldsymbol{u})\|_1 = |E_{\mathrm{R}}(\boldsymbol{u})| + |E_{\mathrm{G}}(\boldsymbol{u})| + |E_{\mathrm{B}}(\boldsymbol{u})|\,. \tag{16.9}$$

Another alternative for calculating a combined edge strength is to take the *maximum* magnitude of the RGB gradients (i.e., the L_∞ norm),

$$C_\infty(\boldsymbol{u}) = \|\boldsymbol{E}(\boldsymbol{u})\|_\infty = \max\left(|E_{\mathrm{R}}(\boldsymbol{u})|, |E_{\mathrm{G}}(\boldsymbol{u})|, |E_{\mathrm{B}}(\boldsymbol{u})|\right). \tag{16.10}$$

An example using the test image from Chapter 15 is given in Fig. 16.1. It shows the edge magnitude of the corresponding grayscale image and the combined color edge magnitude calculated with the different norms defined in Eqns. (16.8)–(16.10).[2]

As far as edge *orientation* is concerned, there is no simple extension of the grayscale case. While edge orientation can easily be calculated for each individual color component (using Eqn. (16.4)), the gradients, three color channels are generally different (or even contradictory) and there is no obvious way of combining them.

A simple ad hoc approach is to choose, at each image position $\boldsymbol{u}$, the gradient direction from the color channel of maximum edge strength, that is,

$$\Phi_{\mathrm{col}}(\boldsymbol{u}) = \tan^{-1}\left(\frac{I_{m,y}(\boldsymbol{u})}{I_{m,x}(\boldsymbol{u})}\right), \tag{16.11}$$

with $m = \operatorname*{argmax}_{k=\mathrm{R,G,B}} E_k(\boldsymbol{u})$.

This simple (monochromatic) method for calculating edge strength and orientation in color images is summarized in Alg. 16.1 (see Sec. 16.5 for the corresponding Java implementation). Two sample results are shown in Fig. 16.2. For comparison, these figures also show the edge maps obtained by first converting the color image to a grayscale

[1] Such images are nevertheless produced by the "Find Edges" command in ImageJ and the filter of the same name in Photoshop (showing inverted components).

[2] In this case, the grayscale image in (c) was calculated with the *direct* conversion method (see Chapter 14, Eqn. (14.39)) from nonlinear sRGB components. With *linear* grayscale conversion (Ch. 14, Eqn. (14.37)), the desaturated bar at the center would exhibit no grayscale edges along its borders, since the luminance is the same inside and outside.

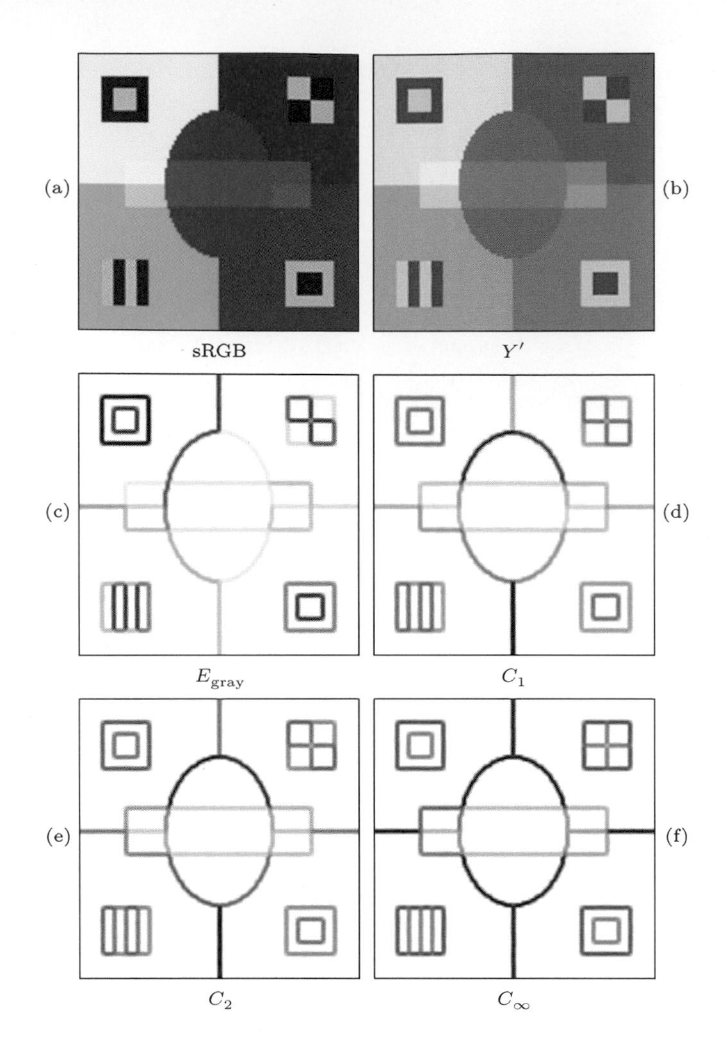

Fig. 16.1
Color edge enhancement with monochromatic methods. Original color image (a) and corresponding grayscale image (b); edge magnitude from the grayscale image (c). Color edge magnitude calculated with different norms: L_1 (d), L_2 (e), and L_∞ (f). The images in (c–f) are inverted for better viewing.

image and then applying the Sobel operator[3] (Fig. 16.2(b)). The edge magnitude in all examples is normalized; it is shown inverted and contrast-enhanced to increase the visibility of low-contrast edges. As expected and apparent from the examples, even simple monochromatic techniques applied to color images perform better than edge detection on the corresponding grayscale images. In particular, edges between color regions of similar brightness are not detectable in this way, so using color information for edge detection is generally more powerful than relying on intensity alone. Among the simple color techniques, the maximum channel edge strength C_∞ (Eqn. (16.10)) seems to give the most consistent results with the fewest edges getting lost.

However, none of the monochromatic detection techniques can be expected to work reliably under these circumstances. While the threshold for binarizing the edge magnitude could be tuned manually to give more pleasing results on specific images, it is difficult in practice to achieve consistently good results over a wide range of images. Methods for determining the optimal edge threshold dynam-

[3] See Chapter 6, Sec. 6.3.1.

```
1: MonochromaticColorEdge(I)
     Input: I = (I_R, I_G, I_B), an RGB color image of size M×N. Returns a pair of maps (E_2, Φ) for edge magnitude and orientation.
2:   H_x^S ← 1/8 · [−1 0 1; −2 0 2; −1 0 1]
3:   H_y^S ← 1/8 · [−1 −2 −1; 0 0 0; 1 2 1]          ▷ x/y gradient kernels
4:   (M, N) ← Size(I)
5:   Create maps E, Φ : M×N → ℝ          ▷ edge magnitude/orientation
6:   I_R,x ← I_R*H_x^S,   I_R,y ← I_R*H_y^S          ▷ apply gradient filters
7:   I_G,x ← I_G*H_x^S,   I_G,y ← I_G*H_y^S
8:   I_B,x ← I_B*H_x^S,   I_B,y ← I_B*H_y^S
9:   for all image coordinates u ∈ M×N do
10:      (r_x, g_x, b_x) ← (I_R,x(u), I_G,x(u), I_B,x(u))
11:      (r_y, g_y, b_y) ← (I_R,y(u), I_G,y(u), I_B,y(u))
12:      e_R^2 ← r_x^2 + r_y^2
13:      e_G^2 ← g_x^2 + g_y^2
14:      e_B^2 ← b_x^2 + b_y^2
15:      e_max^2 ← e_R^2          ▷ find maximum gradient channel
16:      c_x ← r_x,   c_y ← r_y
17:      if e_G^2 > e_max^2 then
18:          e_max^2 ← e_G^2,   c_x ← g_x,   c_y ← g_y
19:      if e_B^2 > e_max^2 then
20:          e_max^2 ← e_B^2,   c_x ← b_x,   c_y ← b_y
21:      E(u) ← sqrt(e_R^2 + e_G^2 + e_B^2)          ▷ edge magnitude (L_2 norm)
22:      Φ(u) ← ArcTan(c_x, c_y)          ▷ edge orientation
23:  return (E, Φ).
```

Alg. 16.1
Monochromatic color edge operator. A pair of Sobel-type filter kernels (H_x^S, H_y^S) is used to estimate the local x/y gradients of each component of the RGB input image $\boldsymbol{I}$. Color edge magnitude is calculated as the L_2 norm of the color gradient vector (see Eqn. (16.8)). The procedure returns a pair of maps, holding the edge magnitude E_2 and the edge orientation Φ, respectively.

ically, that is, depending on the image content, have been proposed, typically based on the statistical variability of the color gradients. Additional details can be found in [84, 171, 192].

16.2 Edges in Vector-Valued Images

In the "monochromatic" scheme described in Sec. 16.1, the edge magnitude in each color channel is calculated separately and thus no use is made of the potential coupling between color channels. Only in a subsequent step are the individual edge responses in the color channels combined, albeit in an ad hoc fashion. In other words, the color data are not treated as vectors, but merely as separate and unrelated scalar values.

To obtain better insight into this problem it is helpful to treat the color image as a *vector field*, a standard construct in vector calculus [32, 223].[4] A three-channel RGB color image $\boldsymbol{I}(\boldsymbol{u}) = (I_R(\boldsymbol{u}), I_G(\boldsymbol{u}), I_B(\boldsymbol{u}))$ can be modeled as a discrete 2D vector field, that is, a function whose coordinates $\boldsymbol{u} = (u, v)$ are 2D and whose values are 3D vectors.

[4] See Sec. C.2 in the Appendix for some general properties of vector fields.

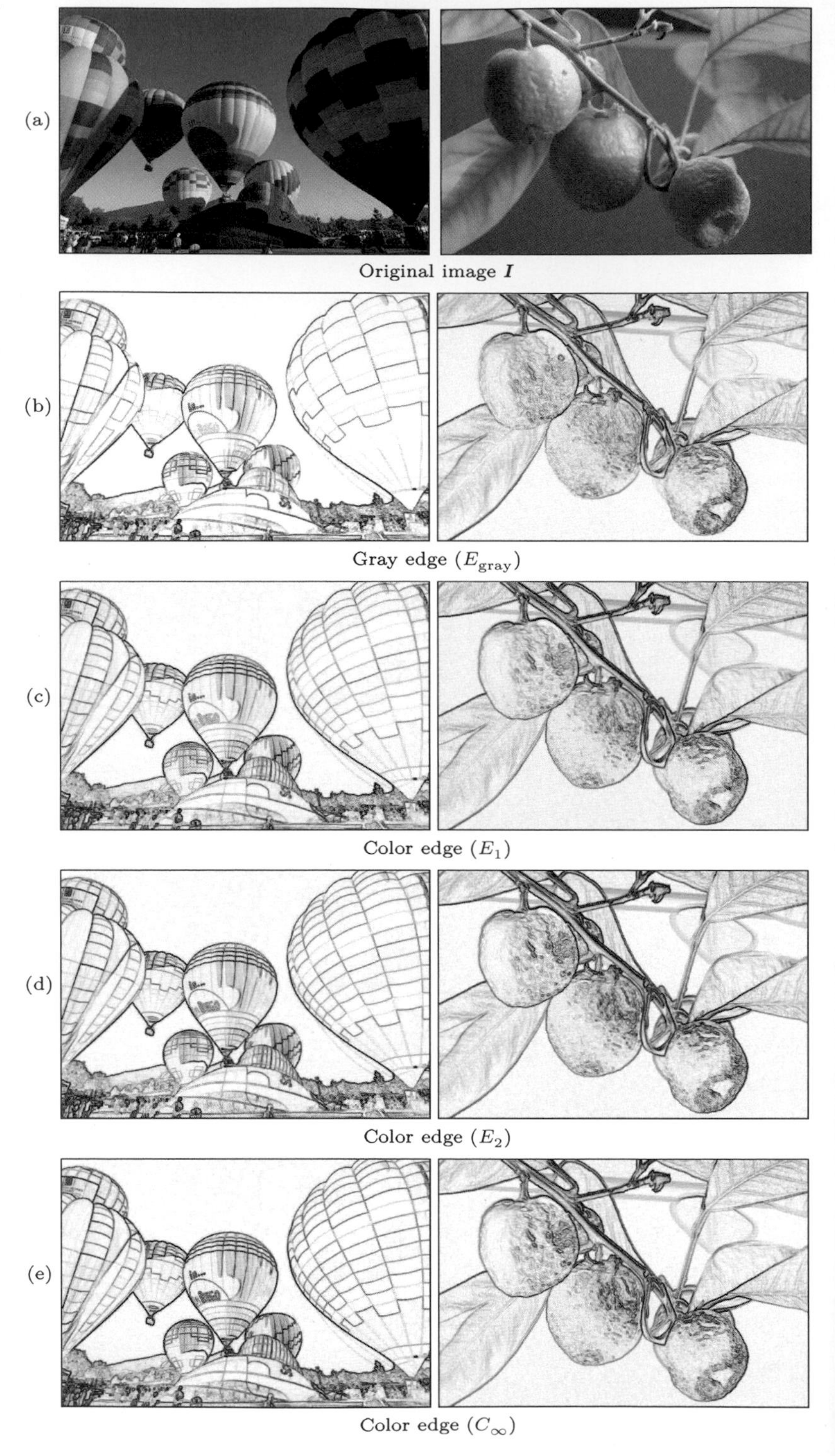

Fig. 16.2
Example of color edge enhancement with monochromatic techniques (balloons image). Original color image and corresponding grayscale image (a), edge magnitude obtained from the grayscale image (b), color edge magnitude calculated with the L_2 norm (c), and the L_∞ norm (d). Differences between the grayscale edge detector (b) and the color-based detector (c–e) are particularly visible inside the right balloon and at the lower borders of the tangerines.

Similarly, a grayscale image can be described as a discrete *scalar field*, since its pixel values are only 1D.

16.2.1 Multi-Dimensional Gradients

As noted in the previous section, the gradient of a scalar image I at a specific position $\boldsymbol{u}$ is defined as

$$\nabla I(\boldsymbol{u}) = \begin{pmatrix} \frac{\partial I}{\partial x}(\boldsymbol{u}) \\ \frac{\partial I}{\partial y}(\boldsymbol{u}) \end{pmatrix}, \tag{16.12}$$

that is, the vector of the partial derivatives of the function I in the x- and y-direction, respectively.[5] Obviously, the gradient of a scalar image is a 2D vector field.

In the case of a color image $\boldsymbol{I} = (I_R, I_G, I_B)$, we can treat the three color channels as separate scalar images and obtain their gradients analogously as

$$\nabla I_R(\boldsymbol{u}) = \begin{pmatrix} \frac{\partial I_R}{\partial x}(\boldsymbol{u}) \\ \frac{\partial I_R}{\partial y}(\boldsymbol{u}) \end{pmatrix}, \quad \nabla I_G(\boldsymbol{u}) = \begin{pmatrix} \frac{\partial I_G}{\partial x}(\boldsymbol{u}) \\ \frac{\partial I_G}{\partial y}(\boldsymbol{u}) \end{pmatrix}, \quad \nabla I_B(\boldsymbol{u}) = \begin{pmatrix} \frac{\partial I_B}{\partial x}(\boldsymbol{u}) \\ \frac{\partial I_B}{\partial y}(\boldsymbol{u}) \end{pmatrix}, \tag{16.13}$$

which is equivalent to what we did in Eqn. (16.5). Before we can take the next steps, we need to introduce a standard tool for the analysis of vector fields.

16.2.2 The Jacobian Matrix

The *Jacobian* matrix[6] $\mathbf{J}_{\boldsymbol{I}}(\boldsymbol{u})$ combines all first partial derivatives of a vector field $\boldsymbol{I}$ at a given position $\boldsymbol{u}$, its row vectors being the gradients of the scalar component functions. In particular, for an RGB color image $\boldsymbol{I}$, the Jacobian matrix is defined as

$$\mathbf{J}_{\boldsymbol{I}}(\boldsymbol{u}) = \begin{pmatrix} (\nabla I_R)^\intercal(\boldsymbol{u}) \\ (\nabla I_G)^\intercal(\boldsymbol{u}) \\ (\nabla I_B)^\intercal(\boldsymbol{u}) \end{pmatrix} = \begin{pmatrix} \frac{\partial I_R}{\partial x}(\boldsymbol{u}) & \frac{\partial I_R}{\partial y}(\boldsymbol{u}) \\ \frac{\partial I_G}{\partial x}(\boldsymbol{u}) & \frac{\partial I_G}{\partial y}(\boldsymbol{u}) \\ \frac{\partial I_B}{\partial x}(\boldsymbol{u}) & \frac{\partial I_B}{\partial y}(\boldsymbol{u}) \end{pmatrix} = (\boldsymbol{I}_x(\boldsymbol{u}) \mid \boldsymbol{I}_y(\boldsymbol{u})), \tag{16.14}$$

with $\nabla I_R, \nabla I_G, \nabla I_B$ as defined in Eqn. (16.13). We see that the 2D gradient vectors $(\nabla I_R)^\intercal$, $(\nabla I_G)^\intercal$, $(\nabla I_B)^\intercal$ constitute the rows of the resulting 3×2 matrix $\mathbf{J}_{\boldsymbol{I}}$. The two 3D column vectors of this matrix,

$$\boldsymbol{I}_x(\boldsymbol{u}) = \frac{\partial \boldsymbol{I}}{\partial x}(\boldsymbol{u}) = \begin{pmatrix} \frac{\partial I_R}{\partial x}(\boldsymbol{u}) \\ \frac{\partial I_G}{\partial x}(\boldsymbol{u}) \\ \frac{\partial I_B}{\partial x}(\boldsymbol{u}) \end{pmatrix}, \qquad \boldsymbol{I}_y(\boldsymbol{u}) = \frac{\partial \boldsymbol{I}}{\partial y}(\boldsymbol{u}) = \begin{pmatrix} \frac{\partial I_R}{\partial y}(\boldsymbol{u}) \\ \frac{\partial I_G}{\partial y}(\boldsymbol{u}) \\ \frac{\partial I_B}{\partial y}(\boldsymbol{u}) \end{pmatrix}, \tag{16.15}$$

are the partial derivatives of the color components along the x- and y-axes, respectively. At a given position $\boldsymbol{u}$, the total amount of change over all three color channels in the horizontal direction can be quantified by the norm of the corresponding column vector $\|\boldsymbol{I}_x(\boldsymbol{u})\|$. Analogously, $\|\boldsymbol{I}_y(\boldsymbol{u})\|$ gives the total amount of change over all three color channels along the vertical axis.

[5] Of course, images are discrete functions and the partial derivatives are estimated from finite differences (see Sec. C.3.1 in the Appendix).

[6] See also Sec. C.2.1 in the Appendix.

16.2.3 Squared Local Contrast

Now that we can quantify the change along the horizontal and vertical axes at any position $\boldsymbol{u}$, the next task is to find out the direction of the *maximum* change to find the angle of the edge normal, which we then use to derive the local edge strength. How can we calculate the gradient in some direction θ other than horizontal and vertical? For this purpose, we use the product of the unit vector oriented at angle θ,

$$\mathbf{e}_\theta = \begin{pmatrix} \cos(\theta) \\ \sin(\theta) \end{pmatrix}, \tag{16.16}$$

and the Jacobian matrix $\mathbf{J}_{\boldsymbol{I}}$ (Eqn. (16.14)) in the form

$$\begin{aligned} (\mathrm{grad}_\theta\, \boldsymbol{I})(\boldsymbol{u}) = \mathbf{J}_{\boldsymbol{I}}(\boldsymbol{u}) \cdot \mathbf{e}_\theta &= \big(\boldsymbol{I}_x(\boldsymbol{u}) \,\vdots\, \boldsymbol{I}_y(\boldsymbol{u})\big) \cdot \begin{pmatrix} \cos(\theta) \\ \sin(\theta) \end{pmatrix} \\ &= \boldsymbol{I}_x(\boldsymbol{u}) \cdot \cos(\theta) + \boldsymbol{I}_y(\boldsymbol{u}) \cdot \sin(\theta). \end{aligned} \tag{16.17}$$

The resulting 3D vector $(\mathrm{grad}_\theta\, \boldsymbol{I})(\boldsymbol{u})$ is called the *directional gradient*[7] of the color image $\boldsymbol{I}$ in the direction θ at position $\boldsymbol{u}$. By taking the squared norm of this vector,

$$\begin{aligned} S_\theta(\boldsymbol{I}, \boldsymbol{u}) &= \|(\mathrm{grad}_\theta\, \boldsymbol{I})(\boldsymbol{u})\|_2^2 \\ &= \big\|\boldsymbol{I}_x(\boldsymbol{u})\cdot\cos(\theta) + \boldsymbol{I}_y(\boldsymbol{u})\cdot\sin(\theta)\big\|_2^2 \\ &= \boldsymbol{I}_x^2(\boldsymbol{u}) \cdot \cos^2(\theta) + 2\cdot\boldsymbol{I}_x(\boldsymbol{u})\cdot\boldsymbol{I}_y(\boldsymbol{u})\cdot\cos(\theta)\cdot\sin(\theta) + \boldsymbol{I}_y^2(\boldsymbol{u})\cdot\sin^2(\theta), \end{aligned} \tag{16.18}$$

we obtain what is called the *squared local contrast* of the vector-valued image $\boldsymbol{I}$ at position $\boldsymbol{u}$ in direction θ.[8] For an RGB image $\boldsymbol{I} = (I_\mathrm{R}, I_\mathrm{G}, I_\mathrm{B})$, the squared local contrast in Eqn. (16.18) is, explicitly written,

$$S_\theta(\boldsymbol{I}, \boldsymbol{u}) = \left\| \begin{pmatrix} I_{\mathrm{R},x}(\boldsymbol{u}) \\ I_{\mathrm{G},x}(\boldsymbol{u}) \\ I_{\mathrm{B},x}(\boldsymbol{u}) \end{pmatrix} \cdot \cos(\theta) + \begin{pmatrix} I_{\mathrm{R},y}(\boldsymbol{u}) \\ I_{\mathrm{G},y}(\boldsymbol{u}) \\ I_{\mathrm{B},y}(\boldsymbol{u}) \end{pmatrix} \cdot \sin(\theta) \right\|_2^2 \tag{16.19}$$

$$\begin{aligned} &= \big[I_{\mathrm{R},x}^2(\boldsymbol{u}) + I_{\mathrm{G},x}^2(\boldsymbol{u}) + I_{\mathrm{B},x}^2(\boldsymbol{u})\big] \cdot \cos^2(\theta) \\ &+ \big[I_{\mathrm{R},y}^2(\boldsymbol{u}) + I_{\mathrm{G},y}^2(\boldsymbol{u}) + I_{\mathrm{B},y}^2(\boldsymbol{u})\big] \cdot \sin^2(\theta) \\ &+ 2 \cdot \cos(\theta) \cdot \sin(\theta) \cdot \\ &\quad \big[I_{\mathrm{R},x}(\boldsymbol{u})\cdot I_{\mathrm{R},y}(\boldsymbol{u}) + I_{\mathrm{G},x}(\boldsymbol{u})\cdot I_{\mathrm{G},y}(\boldsymbol{u}) + I_{\mathrm{B},x}(\boldsymbol{u})\cdot I_{\mathrm{B},y}(\boldsymbol{u})\big]. \end{aligned} \tag{16.20}$$

Note that, in the case that I is a *scalar* image, the squared local contrast reduces to

$$S_\theta(I, \boldsymbol{u}) = \|(\mathrm{grad}_\theta\, I)(\boldsymbol{u})\|^2 = \left\| \begin{pmatrix} I_x(\boldsymbol{u}) \\ I_y(\boldsymbol{u}) \end{pmatrix}^{\!\mathsf{T}} \cdot \begin{pmatrix} \cos(\theta) \\ \sin(\theta) \end{pmatrix} \right\|_2^2 \tag{16.21}$$

$$= \big[I_x(\boldsymbol{u}) \cdot \cos(\theta) + I_y(\boldsymbol{u}) \cdot \sin(\theta)\big]^2. \tag{16.22}$$

We will return to this result again later in Sec. 16.2.6. In the following, we use the root of the squared local contrast, that is, $\sqrt{S_\theta(I, \boldsymbol{u})}$, under the term *local contrast*.

[7] See also Sec. C.2.2 in the Appendix (Eqn. (C.18)).

[8] Note that $\boldsymbol{I}_x^2 = \boldsymbol{I}_x \cdot \boldsymbol{I}_x$, $\boldsymbol{I}_y^2 = \boldsymbol{I}_y \cdot \boldsymbol{I}_y$ and $\boldsymbol{I}_x \cdot \boldsymbol{I}_y$ in Eqn. (16.18) are dot products and thus the results are scalar values.

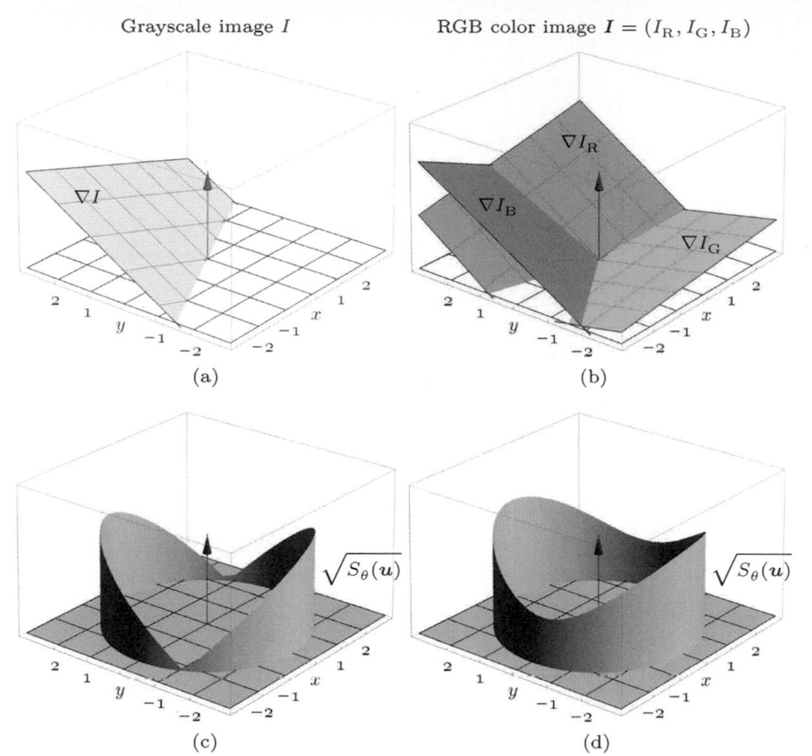

Fig. 16.3
Local image gradients and local contrast. In case of a scalar (grayscale) image I (a), the local gradient ∇I defines a single plane that is tangential to the image function I at position $\boldsymbol{u} = (u, v)$. In case of an RGB color image $\boldsymbol{I} = (I_R, I_G, I_B)$ (b), the local gradients ∇I_R, ∇I_G, ∇I_B for each color channel define three tangent planes. The vertical axes in graphs (c, d) show the corresponding local contrast values $\sqrt{S_\theta(\boldsymbol{I}, \boldsymbol{u})}$ (see Eqns. (16.18) and (16.19)) for all possible directions $\theta = 0, \ldots, 2\pi$.

Figure 16.3 illustrates the meaning of the squared local contrast in relation to the local image gradients. At a given image position $\boldsymbol{u}$, the local gradient $\nabla I(\boldsymbol{u})$ in a grayscale image (Fig. 16.3(a)) defines a single plane that is tangential to the image function I at position $\boldsymbol{u}$. In case of a *color* image (Fig. 16.3(b)), each color channel defines an individual tangent plane. In Fig. 16.3(c, d) the *local contrast* values are shown as the height of cylindrical surfaces for all directions θ. For a *grayscale* image (Fig. 16.3(c)), the local contrast changes *linearly* with the orientation θ, while the relation is *quadratic* for a color image (Fig. 16.3(d)). To calculate the strength and orientation of edges we need to determine the direction of the *maximum* local contrast, which is described in the following.

16.2.4 Color Edge Magnitude

The directions that *maximize* $S_\theta(\boldsymbol{I}, \boldsymbol{u})$ in Eqn. (16.18) can be found analytically as the roots of the first partial derivative of S with respect to the angle θ, as originally suggested by Di Zenzo [63], and the resulting quantity is called *maximum local contrast*. As shown in [59], the maximum local contrast can also be found from the Jacobian matrix $\mathbf{J}_{\boldsymbol{I}}$ (Eqn. (16.14)) as the largest eigenvalue of the (symmetric) 2×2 matrix

$$\mathbf{M}(\boldsymbol{u}) = \mathbf{J}_{\boldsymbol{I}}^\mathsf{T}(\boldsymbol{u}) \cdot \mathbf{J}_{\boldsymbol{I}}(\boldsymbol{u}) = \begin{pmatrix} \boldsymbol{I}_x^\mathsf{T}(\boldsymbol{u}) \\ \boldsymbol{I}_y^\mathsf{T}(\boldsymbol{u}) \end{pmatrix} \cdot \big(\boldsymbol{I}_x(\boldsymbol{u}) \; \boldsymbol{I}_y(\boldsymbol{u}) \big) \tag{16.23}$$

$$= \begin{pmatrix} \boldsymbol{I}_x^2(\boldsymbol{u}) & \boldsymbol{I}_x(\boldsymbol{u}) \cdot \boldsymbol{I}_y(\boldsymbol{u}) \\ \boldsymbol{I}_y(\boldsymbol{u}) \cdot \boldsymbol{I}_x(\boldsymbol{u}) & \boldsymbol{I}_y^2(\boldsymbol{u}) \end{pmatrix} = \begin{pmatrix} A(\boldsymbol{u}) & C(\boldsymbol{u}) \\ C(\boldsymbol{u}) & B(\boldsymbol{u}) \end{pmatrix}, \tag{16.24}$$

with the elements

$$\begin{aligned} A(\boldsymbol{u}) &= \boldsymbol{I}_x^2(\boldsymbol{u}) = \boldsymbol{I}_x(\boldsymbol{u})\cdot\boldsymbol{I}_x(\boldsymbol{u}), \\ B(\boldsymbol{u}) &= \boldsymbol{I}_y^2(\boldsymbol{u}) = \boldsymbol{I}_y(\boldsymbol{u})\cdot\boldsymbol{I}_y(\boldsymbol{u}), \\ C(\boldsymbol{u}) &= \boldsymbol{I}_x(\boldsymbol{u})\cdot\boldsymbol{I}_y(\boldsymbol{u}) = \boldsymbol{I}_y(\boldsymbol{u})\cdot\boldsymbol{I}_x(\boldsymbol{u}). \end{aligned} \tag{16.25}$$

The matrix $\mathbf{M}(\boldsymbol{u})$ could be considered as the color equivalent to the *local structure matrix* used for corner detection on grayscale images in Chapter 7, Sec. 7.2.1. The two eigenvalues λ_1, λ_2 of $\mathbf{M}$ can be found in closed form as[9]

$$\begin{aligned} \lambda_1(\boldsymbol{u}) &= \big(A + B + \sqrt{(A-B)^2 + 4\cdot C^2}\big)/2, \\ \lambda_2(\boldsymbol{u}) &= \big(A + B - \sqrt{(A-B)^2 + 4\cdot C^2}\big)/2. \end{aligned} \tag{16.26}$$

Since $\mathbf{M}$ is symmetric, the expression under the square root in Eqn. (16.26) is positive and thus all eigenvalues are real. In addition, A, B are both positive and therefore λ_1 is always the *larger* of the two eigenvalues. It is equivalent to the maximum squared local contrast (Eqn. (16.18)), that is,

$$\lambda_1(\boldsymbol{u}) \equiv \max_{0 \le \theta < 2\pi} S_\theta(\boldsymbol{I}, \boldsymbol{u}), \tag{16.27}$$

and thus $\sqrt{\lambda_1}$ can be used directly to quantify the local edge strength. The eigen*vector* associated with $\lambda_1(\boldsymbol{u})$ is

$$\boldsymbol{q}_1(\boldsymbol{u}) = \begin{pmatrix} A - B + \sqrt{(A-B)^2 + 4\cdot C^2} \\ 2\cdot C \end{pmatrix}, \tag{16.28}$$

or, equivalently, any multiple of $\boldsymbol{q}_1$.[10] Thus the rate of change along the vector $\boldsymbol{q}_1$ is the same as in the opposite direction $-\boldsymbol{q}_1$, and it follows that the local contrast $S_\theta(\boldsymbol{I}, \boldsymbol{u})$ at orientation θ is the same at orientation $\theta + k\pi$ (for any $k \in \mathbb{Z}$).[11] As usual, the *unit vector* corresponding to $\boldsymbol{q}_1$ is obtained by scaling $\boldsymbol{q}_1$ by its magnitude, that is,

$$\hat{\boldsymbol{q}}_1 = \frac{1}{\|\boldsymbol{q}_1\|}\cdot\boldsymbol{q}_1. \tag{16.29}$$

An alternative method, proposed in [60], is to calculate the unit eigenvector $\hat{\boldsymbol{q}}_1 = (\hat{x}_1, \hat{y}_1)^\intercal$ in the form

$$\hat{\boldsymbol{q}}_1 = \Big(\sqrt{\tfrac{1+\alpha}{2}},\, \operatorname{sgn}(C)\cdot\sqrt{\tfrac{1-\alpha}{2}}\Big)^\intercal, \tag{16.30}$$

with $\alpha = (A-B)/\sqrt{(A-B)^2 + 4\,C^2}$, directly from the matrix elements A, B, C defined in Eqn. (16.25).

While $\boldsymbol{q}_1$ (the eigenvector associated with the greater eigenvalue of $\mathbf{M}$) points in the direction of maximum change, the second eigenvector $\boldsymbol{q}_2$ (associated with λ_2) is *orthogonal* to $\boldsymbol{q}_1$, that is, has the same direction as the local edge tangent.

[9] See Sec. B.4 in the Appendix for details.

[10] The eigenvalues of a matrix are unique, but the corresponding eigenvectors are not.

[11] Thus the orientation of maximum change is inherently ambiguous [60].

16.2.5 Color Edge Orientation

The local orientation of the edge (i.e., the *normal* to the edge tangent) at a given position $\boldsymbol{u}$ can be obtained directly from the associated eigenvector $\boldsymbol{q}_1(\boldsymbol{u}) = (q_x(\boldsymbol{u}), q_y(\boldsymbol{u}))^\mathsf{T}$ using the relation

$$\tan(\theta_1(\boldsymbol{u})) = \frac{q_x(\boldsymbol{u})}{q_y(\boldsymbol{u})} = \frac{2 \cdot C}{A - B + \sqrt{(A-B)^2 + 4 \cdot C^2}}, \qquad (16.31)$$

which can be simplified[12] to

$$\tan(2 \cdot \theta_1(\boldsymbol{u})) = \frac{2 \cdot C}{A - B}. \qquad (16.32)$$

Unless both $A = B$ *and* $C = 0$ (in which case the edge orientation is undetermined) the angle of maximum local contrast or color edge orientation can be calculated as

$$\theta_1(\boldsymbol{u}) = \frac{1}{2} \cdot \tan^{-1}\Bigl(\frac{2 \cdot C}{A - B}\Bigr) = \frac{1}{2} \cdot \mathrm{ArcTan}(A - B, 2 \cdot C). \qquad (16.33)$$

The above steps are summarized in Alg. 16.2, which is a color edge operator based on the first derivatives of the image function (see Sec. 16.5 for the corresponding Java implementation). It is similar to the algorithm proposed by Di Zenzo [63] but uses the eigenvalues of the local structure matrix for calculating edge magnitude and orientation, as suggested in [59] (see Eqn. (16.24)).

Results of the monochromatic edge operator in Alg. 16.1 and the Di Zenzo-Cumani multi-gradient operator in Alg. 16.2 are compared in Fig. 16.4. The synthetic test image in Fig. 16.4(a) has constant luminance (brightness) and thus no gray-value operator should be able to detect edges in this image. The local edge strength $E(\boldsymbol{u})$ produced by the two operators is very similar (Fig. 16.4(b)). The vectors in Fig. 16.4(c–f) show the orientation of the edge tangents that are normals to the direction of maximum color contrast, $\Phi(\boldsymbol{u})$. The length of each tangent vector is proportional to the local edge strength $E(\boldsymbol{u})$.

Figure 16.5 shows two examples of applying the Di Zenzo-Cumani-style color edge operator (Alg. 16.2) to real images. Note that the multi-gradient edge magnitude (calculated from the eigenvalue λ_1 in Eqn. (16.27)) in Fig. 16.5(b) is virtually identical to the monochromatic edge magnitude E_{mag} under the L_2 norm in Fig. 16.2(d). The larger difference to the result for the L_∞ norm in Fig. 16.2(e) is shown in Fig. 16.5(c).

Thus, considering only edge *magnitude*, the Di Zenzo-Cumani operator has no significant advantage over the simpler, monochromatic operator in Sec. 16.1. However, if edge *orientation* is important (as in the color version of the Canny operator described in Sec. 16.3), the Di Zenzo-Cumani technique is certainly more reliable and consistent.

16.2.6 Grayscale Gradients Revisited

As one might have guessed, the usual gradient-based calculation of the edge orientation (see Ch. 6, Sec. 6.2) is only a special case of the

[12] Using the relation $\tan(2\theta) = [2 \cdot \tan(\theta)] \,/\, [1 - \tan^2(\theta)]$.

Alg. 16.2
Di Zenzo/Cumani-style multi-gradient color edge operator. A pair of Sobel-type filters (H_x^S, H_y^S) is used for estimating the local x/y gradients in each component of the RGB input image $\boldsymbol{I}$. The procedure returns a pair of maps, holding the edge magnitude $E(\boldsymbol{u})$ and the edge orientation $\Phi(\boldsymbol{u})$, respectively.

```
1:  MultiGradientColorEdge(I)
      Input: I = (I_R, I_G, I_B), an RGB color image of size M × N.
      Returns a pair of maps (E, Φ) for edge magnitude and orientation.
2:    H_x^S := 1/8 · [−1 0 1; −2 0 2; −1 0 1]
3:    H_y^S := 1/8 · [−1 −2 −1; 0 0 0; 1 2 1]          ▷ x/y gradient kernels
4:    (M, N) ← Size(I)
5:    Create maps E, Φ : M×N ↦ ℝ              ▷ edge magnitude/orientation
6:    I_R,x ← I_R * H_x^S,   I_R,y ← I_R * H_y^S       ▷ apply gradient filters
7:    I_G,x ← I_G * H_x^S,   I_G,y ← I_G * H_y^S
8:    I_B,x ← I_B * H_x^S,   I_B,y ← I_B * H_y^S
9:    for all u ∈ M×N do
10:       (r_x, g_x, b_x) ← (I_R,x(u), I_G,x(u), I_B,x(u))
11:       (r_y, g_y, b_y) ← (I_R,y(u), I_G,y(u), I_B,y(u))
12:       A ← r_x^2 + g_x^2 + b_x^2                              ▷ A = I_x·I_x
13:       B ← r_y^2 + g_y^2 + b_y^2                              ▷ B = I_y·I_y
14:       C ← r_x·r_y + g_x·g_y + b_x·b_y                        ▷ C = I_x·I_y
15:       λ1 ← (A+B + sqrt((A−B)^2 + 4·C^2)) / 2                 ▷ Eq. 16.26
16:       E(u) ← sqrt(λ1)                                        ▷ Eq. 16.27
17:       Φ(u) ← 1/2 · ArcTan(A−B, 2·C)                          ▷ Eq. 16.33
18:   return (E, Φ).
```

multi-dimensional gradient calculation described already. Given a scalar image I, the intensity gradient vector $(\nabla I)(\boldsymbol{u}) = (I_x(\boldsymbol{u}), I_y(\boldsymbol{u}))^\intercal$ defines a single plane that is tangential to the image function at position $\boldsymbol{u}$, as illustrated in Fig. 16.3(a). With

$$A = I_x^2(\boldsymbol{u}), \qquad B = I_y^2(\boldsymbol{u}), \qquad C = I_x(\boldsymbol{u})\cdot I_y(\boldsymbol{u}) \tag{16.34}$$

(analogous to Eqn. (16.25)) the squared local contrast at position $\boldsymbol{u}$ in direction θ (as defined in Eqn. (16.18)) is

$$S_\theta(I, \boldsymbol{u}) = \big(I_x(\boldsymbol{u})\cdot\cos(\theta) + I_y(\boldsymbol{u})\cdot\sin(\theta)\big)^2. \tag{16.35}$$

From Eqn. (16.26), the eigenvalues of the local structure matrix $\mathbf{M} = \left(\begin{smallmatrix} A & C \\ C & B \end{smallmatrix}\right)$ at position $\boldsymbol{u}$ are (see Eqn. (16.26))

$$\lambda_{1,2}(\boldsymbol{u}) = \big(A + B \pm \sqrt{(A-B)^2 + 4C^2}\big)\,/\,2, \tag{16.36}$$

but here, with I_x, I_y not being vectors but scalar values, we get $C^2 = (I_x \cdot I_y)^2 = I_x^2 \cdot I_y^2$, such that $(A-B)^2 + 4C^2 = (A+B)^2$, and therefore

$$\lambda_{1,2}(\boldsymbol{u}) = \big(A + B \pm (A + B)\big)\,/\,2\,. \tag{16.37}$$

We see that, for a scalar-valued image, the dominant eigenvalue,

$$\lambda_1(\boldsymbol{u}) = A + B = I_x^2(\boldsymbol{u}) + I_y^2(\boldsymbol{u}) = \|\nabla I(\boldsymbol{u})\|_2^2\,, \tag{16.38}$$

is simply the squared L_2 norm of the local gradient vector, while the smaller eigenvalue λ_2 is always zero. Thus, for a grayscale image, the

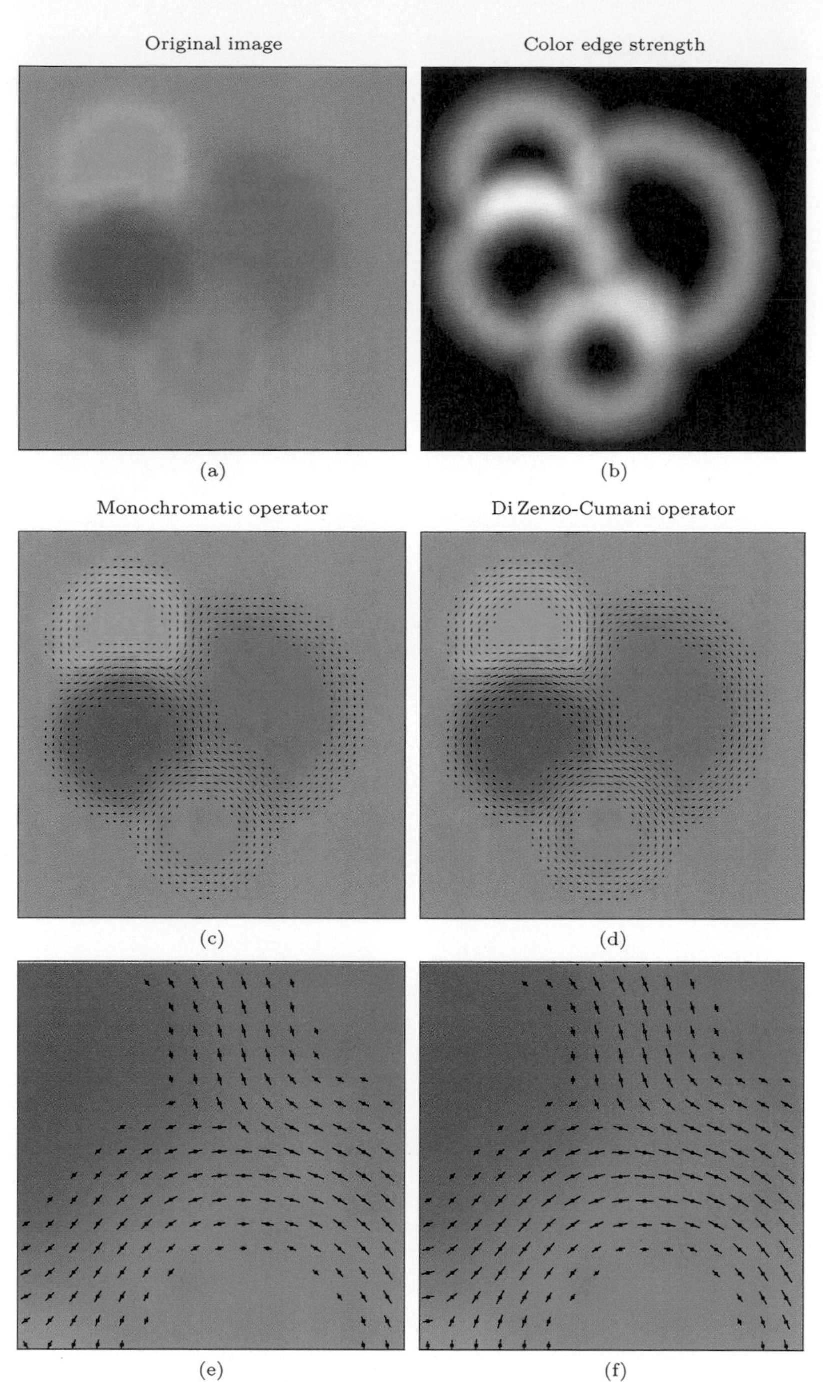

Fig. 16.4
Results from the monochromatic (Alg. 16.1) and the Di Zenzo-Cumani color edge operators (Alg. 16.2). The original color image (a) has *constant luminance*, that is, the intensity gradient is zero and thus a simple grayscale operator would not detect any edges at all. The local edge strength $E(\boldsymbol{u})$ is almost identical for both color edge operators (b). Edge tangent orientation vectors (normal to $\Phi(\boldsymbol{u})$) for the monochromatic and multi-gradient operators (c, d); enlarged details in (e, f).

maximum edge strength $\sqrt{\lambda_1(\boldsymbol{u})} = \|\nabla I(\boldsymbol{u})\|_2$ is equivalent to the *magnitude* of the local intensity gradient.[13] The fact that $\lambda_2 = 0$ indicates that the local contrast in the orthogonal direction (i.e., along the edge tangent) is zero (see Fig. 16.3(c)).

To calculate the local edge *orientation*, at position $\boldsymbol{u}$ we use Eqn. (16.31) to get

[13] See Eqns. (6.5) and (6.13) in Chapter 6, Sec. 6.2.

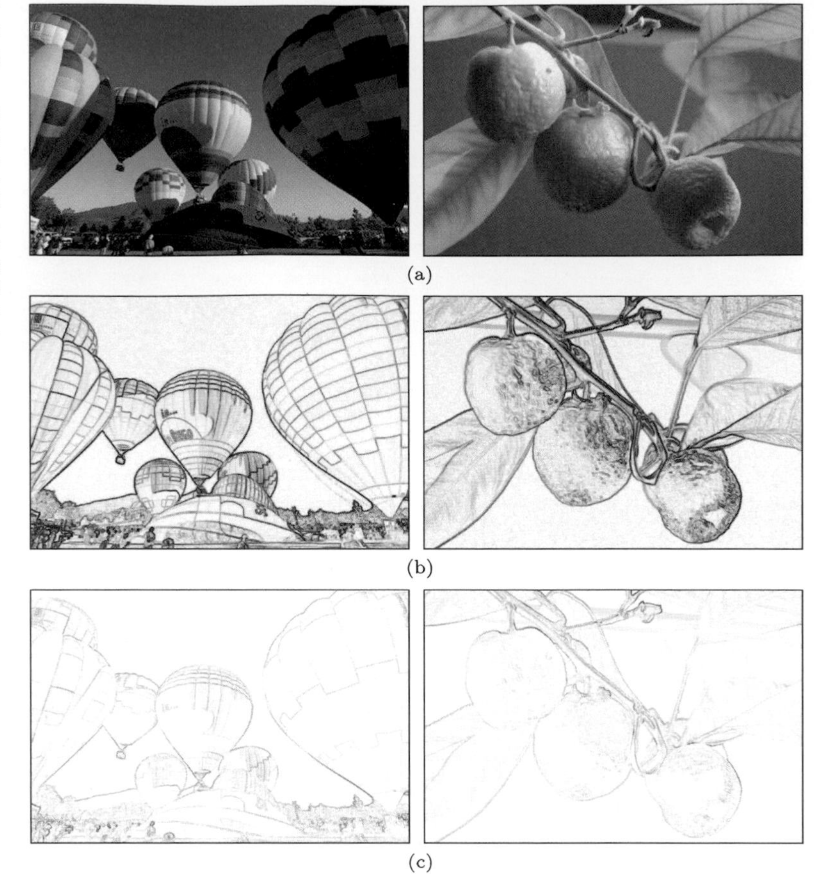
(a)
(b)
(c)

Fig. 16.5
Results of Di Zenzo-Cumani color edge operator (Alg. 16.2) on real images. Original image (a) and inverted color edge magnitude (b). The images in (c) show the differences to the edge magnitude returned by the monochromatic operator (Alg. 16.1, using the L_∞ norm).

$$\tan(\theta_1(\boldsymbol{u})) = \frac{2C}{A-B+(A+B)} = \frac{2C}{2A} = \frac{I_x(\boldsymbol{u})\cdot I_y(\boldsymbol{u})}{I_x^2(\boldsymbol{u})} = \frac{I_y(\boldsymbol{u})}{I_x(\boldsymbol{u})} \qquad (16.39)$$

and the direction of maximum contrast[14] is then found as

$$\theta_1(\boldsymbol{u}) = \tan^{-1}\Bigl(\frac{I_y(\boldsymbol{u})}{I_x(\boldsymbol{u})}\Bigr) = \mathrm{ArcTan}(I_x(\boldsymbol{u}), I_y(\boldsymbol{u})). \qquad (16.40)$$

Thus, for scalar-valued images, the general (multi-dimensional) technique based on the eigenvalues of the structure matrix leads to exactly the same result as the conventional grayscale edge detection approach described in Chapter 6, Sec. 6.3.

16.3 Canny Edge Detector for Color Images

Like most other edge operators, the Canny detector was originally designed for grayscale (i.e., scalar-valued) images. To use it on color images, a trivial approach is to apply the monochromatic operator separately to each of the color channels and subsequently merge the results into a single edge map. However, since edges within the different color channels rarely occur in the same places, the result will

[14] See Eqn. (6.14) in Chapter 6.

usually contain multiple edge marks and undesirable clutter (see Fig. 16.8 for an example).

Fortunately, the original grayscale version of the Canny edge detector can be easily adapted to color imagery using the multi-gradient concept described in Sec. 16.2.1. The only changes required in Alg. 6.1 are the calculation of the local gradients and the edge magnitude E_{mag}. The modified procedure is shown in Alg. 16.3 (see Sec. 16.5 for the corresponding Java implementation).

1: **ColorCannyEdgeDetector**$(\boldsymbol{I}, \sigma, \mathsf{t_{hi}}, \mathsf{t_{lo}})$
   Input: $\boldsymbol{I} = (I_R, I_G, I_B)$, an RGB color image of size $M \times N$; σ, radius of Gaussian filter $H^{G,\sigma}$; $\mathsf{t_{hi}}$, $\mathsf{t_{lo}}$, hysteresis thresholds ($\mathsf{t_{hi}} > \mathsf{t_{lo}}$). Returns a binary edge image of size $M \times N$.
2: $\bar{I}_R \leftarrow I_R * H^{G,\sigma}$ ▷ blur components with Gaussian of width σ
3: $\bar{I}_G \leftarrow I_G * H^{G,\sigma}$
4: $\bar{I}_B \leftarrow I_B * H^{G,\sigma}$
5: $H_x^{\nabla} \leftarrow [-0.5\ \ \mathbf{0}\ \ 0.5]$ ▷ x gradient filter
6: $H_y^{\nabla} \leftarrow [-0.5\ \ \mathbf{0}\ \ 0.5]^\intercal$ ▷ y gradient filter
7: $\bar{I}_{R,x} \leftarrow \bar{I}_R * H_x^{\nabla}$, $\bar{I}_{R,y} \leftarrow \bar{I}_R * H_y^{\nabla}$
8: $\bar{I}_{G,x} \leftarrow \bar{I}_G * H_x^{\nabla}$, $\bar{I}_{G,y} \leftarrow \bar{I}_G * H_y^{\nabla}$
9: $\bar{I}_{B,x} \leftarrow \bar{I}_B * H_x^{\nabla}$, $\bar{I}_{B,y} \leftarrow \bar{I}_B * H_y^{\nabla}$
10: $(M, N) \leftarrow \mathsf{Size}(\boldsymbol{I})$
11: Create maps:
12: $E_{mag}, E_{nms}, E_x, E_y : M \times N \rightarrow \mathbb{R}$
13: $E_{bin} : M \times N \rightarrow \{0, 1\}$
14: **for all** image coordinates $\boldsymbol{u} \in M \times N$ **do**
15: $(\mathsf{r}_x, \mathsf{g}_x, \mathsf{b}_x) \leftarrow (I_{R,x}(\boldsymbol{u}), I_{G,x}(\boldsymbol{u}), I_{B,x}(\boldsymbol{u}))$
16: $(\mathsf{r}_y, \mathsf{g}_y, \mathsf{b}_y) \leftarrow (I_{R,y}(\boldsymbol{u}), I_{G,y}(\boldsymbol{u}), I_{B,y}(\boldsymbol{u}))$
17: $A \leftarrow \mathsf{r}_x^2 + \mathsf{g}_x^2 + \mathsf{b}_x^2$,
18: $B \leftarrow \mathsf{r}_y^2 + \mathsf{g}_y^2 + \mathsf{b}_y^2$
19: $C \leftarrow \mathsf{r}_x \cdot \mathsf{r}_y + \mathsf{g}_x \cdot \mathsf{g}_y + \mathsf{b}_x \cdot \mathsf{b}_y$
20: $D \leftarrow \left[(A-B)^2 + 4C^2\right]^{1/2}$
21: $E_{mag}(\boldsymbol{u}) \leftarrow \left[0.5 \cdot \left(A + B + D\right)\right]^{1/2}$ ▷ $\sqrt{\lambda_1}$, Eq. 16.27
22: $E_x(\boldsymbol{u}) \leftarrow A - B + D$ ▷ $\boldsymbol{q}_1$, Eq. 16.28
23: $E_y(\boldsymbol{u}) \leftarrow 2C$
24: $E_{nms}(\boldsymbol{u}) \leftarrow 0$
25: $E_{bin}(\boldsymbol{u}) \leftarrow 0$
26: **for** $u \leftarrow 1, \ldots, M-2$ **do**
27: **for** $v \leftarrow 1, \ldots, N-2$ **do**
28: $\boldsymbol{u} \leftarrow (u, v)$
29: $d_x \leftarrow E_x(\boldsymbol{u})$
30: $d_y \leftarrow E_y(\boldsymbol{u})$
31: $s \leftarrow \mathsf{GetOrientationSector}(d_x, d_y)$ ▷ Alg. 6.2
32: **if** $\mathsf{IsLocalMax}(E_{mag}, \boldsymbol{u}, s, \mathsf{t_{lo}})$ **then** ▷ Alg. 6.2
33: $E_{nms}(\boldsymbol{u}) \leftarrow E_{mag}(\boldsymbol{u})$
34: **for** $u \leftarrow 1, \ldots, M-2$ **do**
35: **for** $v \leftarrow 1, \ldots, N-2$ **do**
36: $\boldsymbol{u} \leftarrow (u, v)$
37: **if** $(E_{nms}(\boldsymbol{u}) \geq \mathsf{t_{hi}} \wedge E_{bin}(\boldsymbol{u}) = 0)$ **then**
38: $\mathsf{TraceAndThreshold}(E_{nms}, E_{bin}, u, v, \mathsf{t_{lo}})$ ▷ Alg. 6.2
39: **return** E_{bin}.

Alg. 16.3
Canny edge detector for color images. Structure and parameters are identical to the grayscale version in Alg. 6.1 (p. 135). In the algorithm below, edge magnitude (E_{mag}) and orientation (E_x, E_y) are obtained from the gradients of the individual color channels (as described in Sec. 16.2.1).

In the pre-processing step, each of the three color channels is individually smoothed by a Gaussian filter of width σ, before calculating the gradient vectors (Alg. 16.3, lines 2–9). As in Alg. 16.2, the color edge magnitude is calculated as the squared local contrast, obtained from the dominant eigenvalue of the structure matrix $\mathbf{M}$ (Eqns. (16.24)–(16.27)). The local gradient vector (E_x, E_y) is calculated from the elements A, B, C, of the matrix $\mathbf{M}$, as given in Eqn. (16.28). The corresponding steps are found in Alg. 16.3, lines 14–22. The remaining steps, including non-maximum suppression, edge tracing and thresholding, are exactly the same as in Alg. 6.1.

Results from the grayscale and color version of the Canny edge detector are compared in Figs. 16.6 and 16.7 for varying values of σ and t_{hi}, respectively. In all cases, the gradient magnitude was normalized and the threshold values $\mathsf{t}_{\mathrm{hi}}, \mathsf{t}_{\mathrm{lo}}$ are given as a percentage of the maximum edge magnitude. Evidently, the color detector gives the more consistent results, particularly at color edges with low intensity difference.

For comparison, Fig. 16.8 shows the results of applying the monochromatic Canny operator separately to each color channel and subsequently merging the edge pixels into a combined edge map, as mentioned at the beginning of this section. We see that this leads to multiple responses and cluttered edges, since maximum gradient positions in the different color channels are generally not collocated.

In summary, the Canny edge detector is superior to simpler schemes based on first-order gradients and global thresholding, in terms of extracting clean and well-located edges that are immediately useful for subsequent processing. The results in Figs. 16.6 and 16.7 demonstrate that the use of color gives additional improvements over the grayscale approach, since edges with insufficient brightness gradients can still be detected from local color differences. Essential for the good performance of the color Canny edge detector, however, is the reliable calculation of the gradient direction, based on the multi-dimensional local contrast formulation given in Sec. 16.2.3. Quite a few variations of Canny detectors for color images have been proposed in the literature, including the one attributed to Kanade (in [140]), which is similar to the algorithm described here.

16.4 Other Color Edge Operators

The idea of using a vector field model in the context of color edge detection was first presented by Di Zenzo [63], who suggested finding the orientation of maximum change by maximizing $S(\boldsymbol{u}, \theta)$ in Eqn. (16.18) over the angle θ. Later Cumani [59, 60] proposed directly using the eigenvalues and eigenvectors of the local structure matrix $\mathbf{M}$ (Eqn. (16.24)) for calculating edge strength and orientation. He also proposed using the zero-crossings of the second-order gradients along the direction of maximum contrast to precisely locate edges, which is a general problem with first-order techniques. Both Di Zenzo and Cumani used only the dominant eigenvalue, indicating the edge strength perpendicular to the edge (if an edge existed at all), and then discarded the smaller eigenvalue proportional to the edge strength in

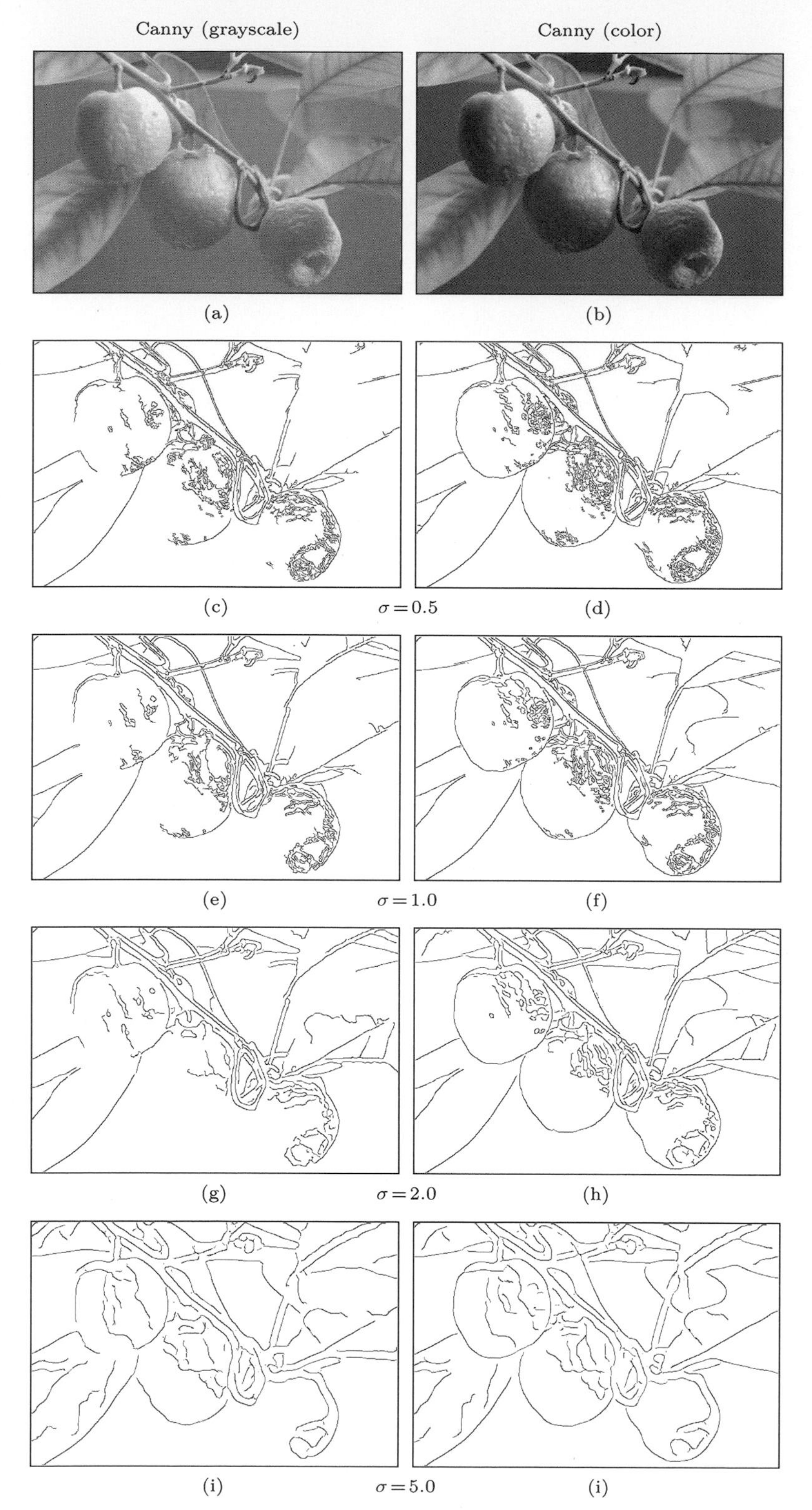

Fig. 16.6
Canny grayscale vs. color version. Results from the grayscale (left) and the color version (right) of the Canny operator for different values of σ ($\mathsf{t}_{\mathrm{hi}} = 20\%$, $\mathsf{t}_{\mathrm{lo}} = 5\%$ of max. edge magnitude).

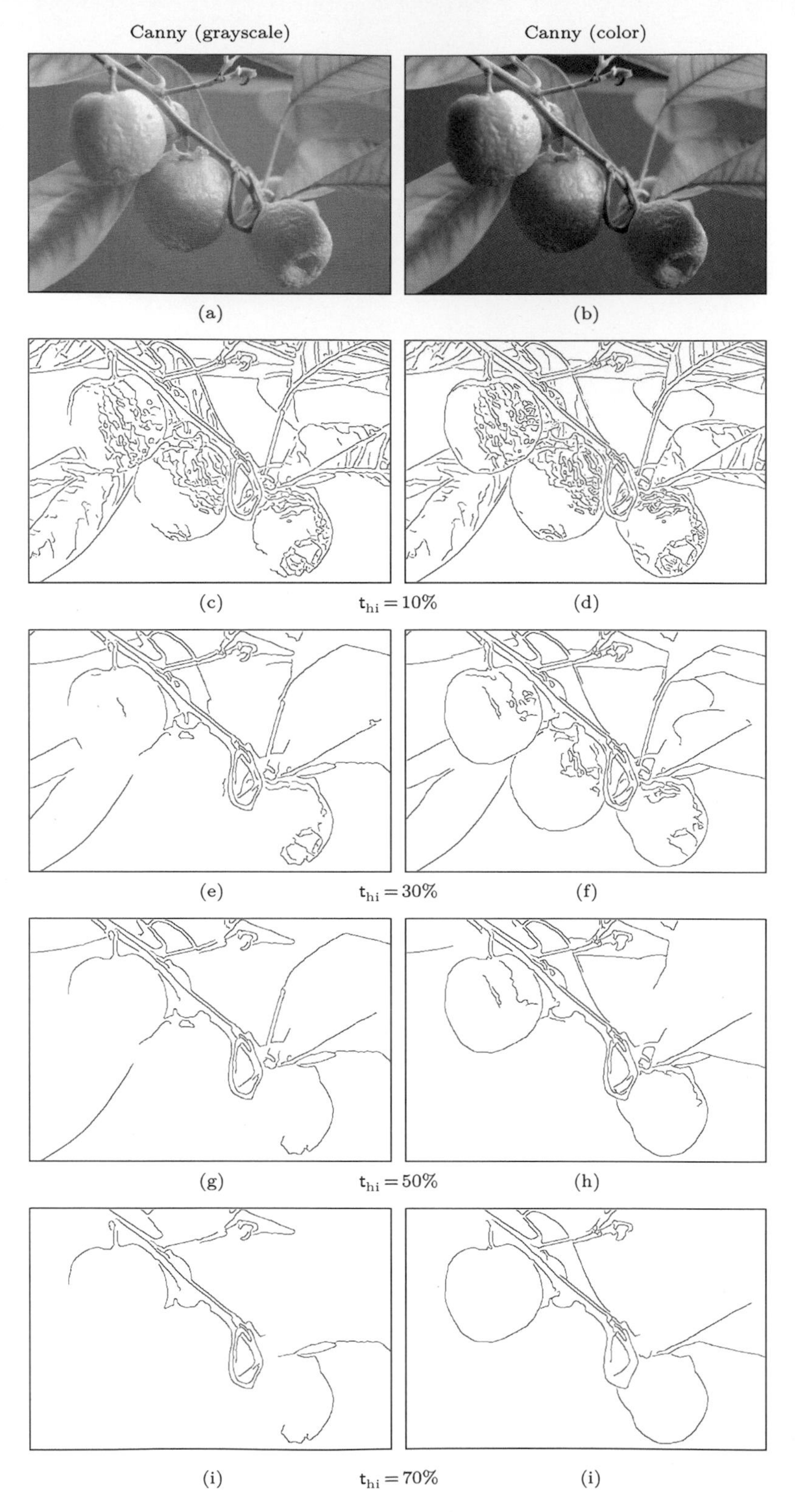

Fig. 16.7
Canny grayscale vs. color version. Results from the grayscale (left) and the color version (right) of the Canny operator for different threshold values t_{hi}, given in % of max. edge magnitude ($t_{lo} = 5\%$, $\sigma = 2.0$).

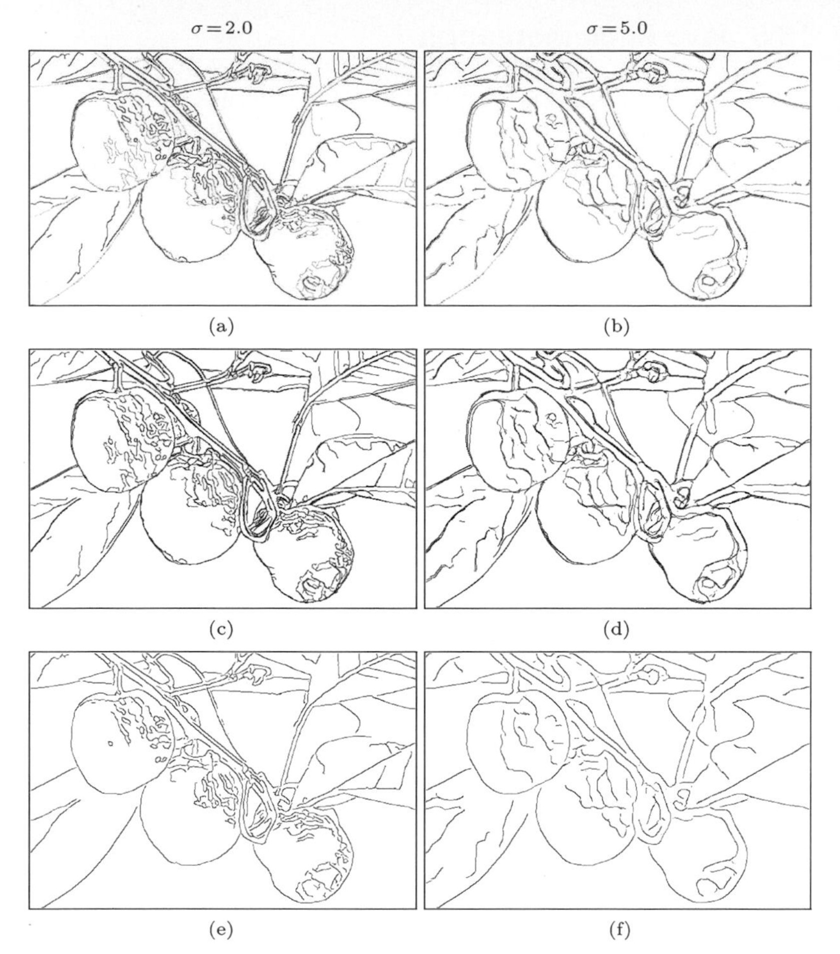

Fig. 16.8
Scalar vs. vector-based color Canny operator. Results from the scalar Canny operator applied separately to each color channel (a, b). Channel edges are shown in corresponding colors, with mixed colors indicating that edge points were detected in multiple channels (e.g., yellow marks overlapping points from the red and the green channel). A black pixel indicates that an edge point was detected in all three color channels. Channel edges combined into a joint edge map (c, d). For comparison, the result of the vector-based color Canny operator (e, f). Common parameter settings are $\sigma = 2.0$ and 5.0, $t_{hi} = 20\%$, $t_{lo} = 5\%$ of max. edge magnitude.

the perpendicular (i.e., tangential) direction. Real edges only exist where the larger eigenvalue is considerably greater than the smaller one. If both eigenvalues have similar values, this indicates that the local image surface exhibits change in all directions, which is not typically true at an edge but quite characteristic of flat, noisy regions and corners. One solution therefore is to use the difference between the eigenvalues, $\lambda_1 - \lambda_2$, to quantify edge strength [206].

Several color versions of the Canny edge detector can be found in the literature, such as the one proposed by Kanade (in [140]) which is very similar to the algorithm presented here. Other approaches of adapting the Canny detector for color images can be found in [85]. In addition to Canny's scheme, other types of color edge detectors have been used successfully, including techniques based on vector order statistics and color difference vectors. Excellent surveys of the various color edge detection approaches can be found in [266] and [141, Ch. 6].

16.5 Java Implementation

The following Java implementations of the algorithms described in this chapter can be found in the source code section[15] of the book's website. The common (abstract) super-class for all color edge detectors is `ColorEdgeDetector`, which mainly provides the following methods:

`FloatProcessor getEdgeMagnitude ()`
Returns the resulting edge magnitude map $E(\boldsymbol{u})$ as a `FloatProcessor` object.

`FloatProcessor getEdgeOrientation ()`
Returns the resulting edge orientation map $\Phi(\boldsymbol{u})$ as a `FloatProcessor` object, with values in the range $[-\pi, \pi]$.

The following edge detectors are defined as concrete sub-classes of `ColorEdgeDetector`:

`GrayscaleEdgeDetector`: Implements an edge detector that uses only the intensity (brightness) of the supplied color image.
`MonochromaticEdgeDetector`: Implements the monochromatic color edge detector described in Alg. 16.1.
`DiZenzoCumaniEdgeDetector`: Implements the Di Zenzo-Cumani type color edge detector described in Alg. 16.2.
`CannyEdgeDetector`: Implements the canny edge detector for grayscale and color images described in Alg. 16.3. This class defines the additional methods

```
ByteProcessor getEdgeBinary(),
List<List<java.awt.Point>> getEdgeTraces().
```

Program 16.1 shows a complete example for the use of the class `CannyEdgeDetector` in the context of an ImageJ plugin.

[15] Package `imagingbook.pub.color.edge`.

```
 1 import ij.ImagePlus;
 2 import ij.plugin.filter.PlugInFilter;
 3 import ij.process.ByteProcessor;
 4 import ij.process.FloatProcessor;
 5 import ij.process.ImageProcessor;
 6 import imagingbook.pub.coloredge.CannyEdgeDetector;
 7
 8 import java.awt.Point;
 9 import java.util.List;
10
11 public class Canny_Edge_Demo implements PlugInFilter {
12
13   public int setup(String arg0, ImagePlus imp) {
14     return DOES_ALL + NO_CHANGES;
15   }
16
17   public void run(ImageProcessor ip) {
18
19     CannyEdgeDetector.Parameters params =
20               new CannyEdgeDetector.Parameters();
21
22     params.gSigma = 3.0f; // σ of Gaussian
23     params.hiThr  = 20.0f;  // 20% of max. edge magnitude
24     params.loThr  = 5.0f;   //  5% of max. edge magnitude
25
26     CannyEdgeDetector detector =
27         new CannyEdgeDetector(ip, params);
28
29     FloatProcessor eMag = detector.getEdgeMagnitude();
30     FloatProcessor eOrt = detector.getEdgeOrientation();
31     ByteProcessor  eBin = detector.getEdgeBinary();
32     List<List<Point>> edgeTraces =
33         detector.getEdgeTraces();
34
35     (new ImagePlus("Canny Edges", eBin)).show();
36
37     // process edge detection results ...
38   }
39 }
```

Prog. 16.1
Use of the `CannyEdgeDetector` class in an ImageJ plugin. A parameter object (`params`) is created in line 20, subsequently configured (in lines 22–24) and finally used to construct a `CannyEdgeDetector` object in line 27. Note that edge detection is performed within the constructor method. Lines 29–33 demonstrate how different types of edge detection results can be retrieved. The binary edge map `eBin` is displayed in line 35. As indicated in the `setup()` method (by returning `DOES_ALL`), this plugin works with any type of image.

17

Edge-Preserving Smoothing Filters

Noise reduction in images is a common objective in image processing, not only for producing pleasing results for human viewing but also to facilitate easier extraction of meaningful information in subsequent steps, for example, in segmentation or feature detection. Simple smoothing filters, such as the Gaussian filter[1] and the filters discussed in Chapter 15 effectively perform low-pass filtering and thus remove high-frequency noise. However, they also tend to suppress high-rate intensity variations that are part of the original signal, thereby destroying image structures that are visually important. The filters described in this chapter are "edge preserving" in the sense that they change their smoothing behavior adaptively depending upon the local image structure. In general, maximum smoothing is performed over "flat" (uniform) image regions, while smoothing is reduced near or across edge-like structures, typically characterized by high intensity gradients.

In the following, three classical types of edge preserving filters are presented, which are largely based on different strategies. The *Kuwahara-type* filters described in Sec. 17.1 partition the filter kernel into smaller sub-kernels and select the most "homogeneous" of the underlying image regions for calculating the filter's result. In contrast, the *bilateral* filter in Sec. 17.2 uses the differences between pixel *values* to control how much each individual pixel in the filter region contributes to the local average. Pixels which are similar to the current center pixel contribute strongly, while highly different pixels add little to the result. Thus, in a sense, the bilateral filter is a non-homogeneous linear filter with a convolution kernel that is adaptively controlled by the local image content. Finally, the *anisotropic diffusion* filters in Sec. 17.3 iteratively smooth the image similar to the process of thermal diffusion, using the image gradient to block the local diffusion at edges and similar structures. It should be noted that all filters described in this chapter are nonlinear and can be applied to either grayscale or color images.

[1] See Chapter 5, Sec. 5.2.

17.1 Kuwahara-Type Filters

The filters described in this section are all based on a similar concept that has its early roots in the work of Kuwahara et al. [144]. Although many variations have been proposed by other authors, we summarize them here under the term "Kuwahara-type" to indicate their origin and algorithmic similarities.

In principle, these filters work by calculating the mean and variance within neighboring image regions and selecting the mean value of the most "homogeneous" region, that is, the one with the smallest variance, to replace the original (center) pixel. For this purpose, the filter region R is divided into K partially overlapping subregions $R_1, R_2, \ldots, R_K$. At every image position (u, v), the *mean* μ_k and the *variance* σ_k^2 of each subregion R_k are calculated from the corresponding pixel values in I as

$$\mu_k(I,u,v) = \frac{1}{|R_k|} \cdot \sum_{(i,j)\in R_k} I(u+i,v+j) = \frac{1}{n_k} \cdot S_{1,k}(I,u,v), \tag{17.1}$$

$$\sigma_k^2(I,u,v) = \frac{1}{|R_k|} \cdot \sum_{(i,j)\in R_k} \bigl(I(u+i,v+j) - \mu_k(I,u,v)\bigr)^2 \tag{17.2}$$

$$= \frac{1}{|R_k|} \cdot \Bigl(S_{2,k}(I,u,v) - \frac{S_{1,k}^2(I,u,v)}{|R_k|}\Bigr), \tag{17.3}$$

for $k = 1, \ldots, K$, with[2]

$$S_{1,k}(I,u,v) = \sum_{(i,j)\in R_k} I(u+i,v+j), \tag{17.4}$$

$$S_{2,k}(I,u,v) = \sum_{(i,j)\in R_k} I^2(u+i,v+j). \tag{17.5}$$

The mean (μ) of the subregion with the smallest variance (σ^2) is selected as the update value, that is,

$$I'(u,v) \leftarrow \mu_{k'}(u,v), \quad \text{with } k' = \operatorname*{argmin}_{k=1,\ldots,K} \sigma_k^2(I,u,v). \tag{17.6}$$

The subregion structure originally proposed by Kuwahara et al. [144] is shown in Fig. 17.1(a) for a 3×3 filter ($r = 1$). It uses four square subregions of size $(r+1) \times (r+1)$ that overlap at the center. In general, the size of the whole filter is $(2r+1) \times (2r+1)$. This particular filter process is summarized in Alg. 17.1.

Note that this filter does not have a centered subregion, which means that the center pixel is always replaced by the mean of one of the neighboring regions, even if it had perfectly fit the surrounding values. Thus the filter always performs a spatial shift, which introduces jitter and banding artifacts in regions of smooth intensity change. This effect is reduced with the filter proposed by Tomita and Tsuji [230], which is similar but includes a fifth subregion at its center (Fig. 17.1(b)). Filters of arbitrary size can be built by simply scaling the corresponding structure. In case of the *Tomita-Tsuji* filter, the side length of the subregions should be odd.

[2] $|R_k|$ denotes the size (number of pixels) of the subregion R_k.

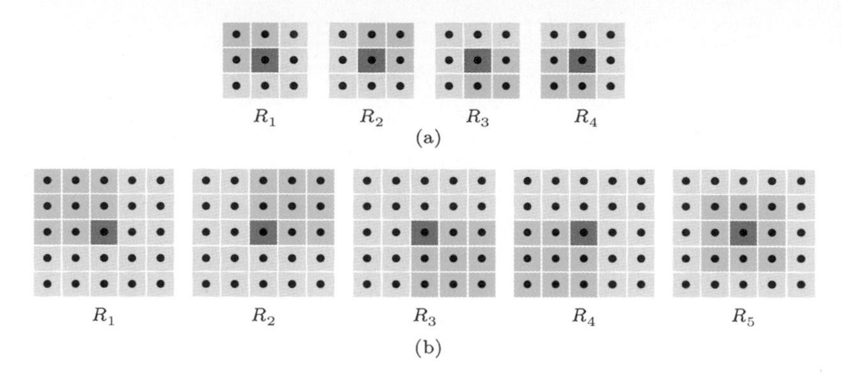

Fig. 17.1
Subregion structures for Kuwahara-type filters. The orginal *Kuwahara-Hachimura* filter (a) considers *four* square, overlapping subregions [144]. *Tomita-Tsuji* filter (b) with *five* subregions ($r = 2$). The current center pixel (red) is contained in all subregions. Das aktuelle Zentralpixel (rot) ist in allen Subregionen enthalten.

Note that replacing a pixel value by the mean of a square neighborhood is equivalent to linear filtering with a simple box kernel, which is not an optimal smoothing operator. To reduce the artifacts caused by the square subregions, alternative filter structures have been proposed, such as the 5×5 *Nagao-Matsuyama* filter [170] shown in Fig. 17.2.

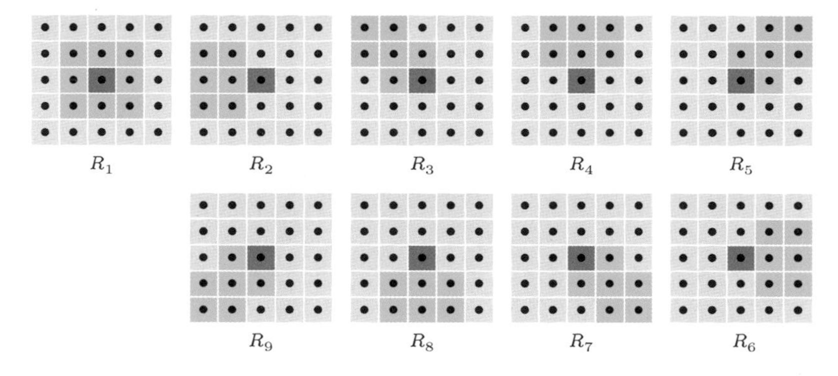

Fig. 17.2
Subregions for the 5×5 ($r = 2$) *Nagao-Matsuyama* filter [170]. Note that the centered subregion (R_1) has a different size than the remaining subregions ($R_2, \ldots, R_9$).

If all subregions are of identical *size* $|R_k| = n$, the quantities

$$\sigma_k^2(I,u,v) \cdot n = S_{2,k}(I,u,v) - S_{1,k}^2(I,u,v)/n \quad \text{or} \tag{17.7}$$

$$\sigma_k^2(I,u,v) \cdot n^2 = S_{2,k}(I,u,v) \cdot n - S_{1,k}^2(I,u,v) \tag{17.8}$$

can be used to measure the amount of variation within the corresponding subregion. Both expressions require calculating one multiplication less for each pixel than the "real" variance σ_k^2 in Eqn. (17.3). Moreover, if all subregions have the same *shape* (such as the filters in Fig. 17.1), additional optimizations are possible that substantially improve the performance. In this case, the local mean and variance need to be calculated only once over a fixed neighborhood for each image position. This type of filter can be efficiently implemented by using a set of pre-calculated maps for the local variance and mean values, as described in Alg. 17.2. As before, the parameter r specifies the radius of the composite filter, with subregions of size $(r+1) \times (r+1)$ and overall size $(2r+1) \times (2r+1)$. The individual subregions are of size $(r+1) \times (r+1)$; for example, $r = 2$ for the 5×5 filter shown in Fig. 17.1(b).

All these filters tend to generate banding artifacts in smooth image regions due to erratic spatial displacements, which become worse

Alg. 17.1
Simple *Kuwahara-Hachimura* filter.

```
1: KuwaharaFilter(I)
       Input: I, a grayscale image of size M×N.
       Returns a new (filtered) image of size M×N.
2:     R_1 ← {(−1,−1), (0,−1), (−1,0), (0,0)}
3:     R_2 ← {(0,−1), (1,−1), (0,0), (1,0)}
4:     R_3 ← {(0,0), (1,0), (1,0), (1,1)}
5:     R_4 ← {(−1,0), (0,0), (−1,1), (1,0)}
6:     I' ← Duplicate(I)
7:     (M,N) ← Size(I)
8:     for all image coordinates (u,v) ∈ M×N do
9:         σ²_min ← ∞
10:        for R ← R_1,...,R_4 do
11:            (σ², μ) ← EvalSubregion(I, R, u, v)
12:            if σ² < σ²_min then
13:                σ²_min ← σ²
14:                μ_min ← μ
15:        I'(u,v) ← μ_min
16:    return I'

17: EvalSubregion(I, R, u, v)
       Returns the variance and mean of the grayscale image I for the
       subregion R positioned at (u,v).
18:    n ← Size(R)
19:    S_1 ← 0,   S_2 ← 0
20:    for all (i,j) ∈ R do
21:        a ← I(u+i, v+j)
22:        S_1 ← S_1 + a                         ▷ Eq. 17.4
23:        S_2 ← S_2 + a²                        ▷ Eq. 17.5
24:    σ² ← (S_2 − S_1²/n)/n     ▷ variance of subregion R, see Eq. 17.1
25:    μ ← S_1/n                    ▷ mean of subregion R, see Eq. 17.3
26:    return (σ², μ)
```

with increasing filter size. If a centered subregion is used (such as R_5 in Fig. 17.1 or R_1 in Fig. 17.2), one could reduce this effect by applying a threshold (t_σ) to select any off-center subregion R_k *only* if its variance is significantly smaller than the variance of the center region R_1 (see Alg. 17.2, line 13).

17.1.1 Application to Color Images

While all of the aforementioned filters were originally designed for grayscale images, they are easily modified to work with color images. We only need to specify how to calculate the variance and mean for any subregion; the decision and replacement mechanisms then remain the same.

Given an RGB color image $\boldsymbol{I} = (I_\mathrm{R}, I_\mathrm{G}, I_\mathrm{B})$ with a subregion R_k, we can calculate the local mean and variance for each color channel as

$$\boldsymbol{\mu}_k(\boldsymbol{I},u,v) = \begin{pmatrix} \mu_k(I_\mathrm{R},u,v) \\ \mu_k(I_\mathrm{G},u,v) \\ \mu_k(I_\mathrm{B},u,v) \end{pmatrix}, \qquad \boldsymbol{\sigma}_k^2(\boldsymbol{I},u,v) = \begin{pmatrix} \sigma_k^2(I_\mathrm{R},u,v) \\ \sigma_k^2(I_\mathrm{G},u,v) \\ \sigma_k^2(I_\mathrm{B},u,v) \end{pmatrix}, \tag{17.9}$$

Alg. 17.2
Fast Kuwahara-type (Tomita-Tsuji) filter with variable size and fixed subregion structure. The filter uses five square subregions of size $(r+1) \times (r+1)$, with a composite filter of $(2r+1) \times (2r+1)$, as shown in Fig. 17.1(b). The purpose of the variance threshold t_σ is to reduce banding effects in smooth image regions (typically $t_\sigma = 5, \ldots, 50$ for 8-bit images).

```
1: FastKuwaharaFilter(I, r, t_σ)
     Input: I, a grayscale image of size M×N; r, filter radius (r ≥ 1);
     t_σ, variance threshold.
     Returns a new (filtered) image of size M × N.
2:   (M, N) ← Size(I)
3:   Create maps:
       S : M×N → ℝ             ▷ local variance S(u, v) ≡ n · σ²(I, u, v)
       A : M×N → ℝ                   ▷ local mean A(u, v) ≡ μ(I, u, v)
4:   d_min ← (r ÷ 2) − r                 ▷ subregions' left/top position
5:   d_max ← d_min + r               ▷ subregions' right/bottom position
6:   for all image coordinates (u, v) ∈ M×N do
7:       (s, μ) ← EvalSquareSubregion(I, u, v, d_min, d_max)
8:       S(u, v) ← s
9:       A(u, v) ← μ
10:  n ← (r + 1)²                                ▷ fixed subregion size
11:  I' ← Duplicate(I)
12:  for all image coordinates (u, v) ∈ M×N do
13:      s_min ← S(u, v) − t_σ · n            ▷ variance of center region
14:      μ_min ← A(u, v)                          ▷ mean of center region
15:      for p ← d_min, ..., d_max do
16:          for q ← d_min, ..., d_max do
17:              if S(u + p, v + q) < s_min then
18:                  s_min ← S(u + p, v + q)
19:                  μ_min ← A(u + p, v + q)
20:      I'(u, v) ← μ_min
21:  return I'

22: EvalSquareSubregion(I, u, v, d_min, d_max)
     Returns the variance and mean of the grayscale image I for a
     square subregion positioned at (u, v).
23:  S_1 ← 0,  S_2 ← 0
24:  for i ← d_min, ..., d_max do
25:      for j ← d_min, ..., d_max do
26:          a ← I(u + i, v + j)
27:          S_1 ← S_1 + a                                    ▷ Eq. 17.4
28:          S_2 ← S_2 + a²                                   ▷ Eq. 17.5
29:  s ← S_2 − S_1²/n                ▷ subregion variance (s ≡ n · σ²)
30:  μ ← S_1/n                                     ▷ subregion mean (μ)
31:  return (s, μ)
```

with $\mu_k()$, $\sigma_k^2()$ as defined in Eqns. (17.1) and (17.3), respectively. Analogous to the grayscale case, each pixel is then replaced by the average color in the subregion with the smallest variance, that is,

$$\boldsymbol{I}'(u,v) \leftarrow \boldsymbol{\mu}_{k'}(\boldsymbol{I},u,v), \quad \text{with } k' = \underset{k=1,\ldots,K}{\operatorname{argmin}}\ \sigma^2_{k,\mathrm{RGB}}(\boldsymbol{I},u,v). \tag{17.10}$$

The overall variance $\sigma^2_{k,\mathrm{RGB}}$, used to determine k' in Eqn. (17.10), can be defined in different ways, for example, as the sum of the variances in the individual color channels, that is,

$$\sigma^2_{k,\mathrm{RGB}}(\boldsymbol{I},u,v) = \sigma^2_k(I_\mathrm{R},u,v) + \sigma^2_k(I_\mathrm{G},u,v) + \sigma^2_k(I_\mathrm{B},u,v). \tag{17.11}$$

Alg. 17.3
Color version of the *Kuwahara-type* filter (adapted from Alg. 17.1). The algorithm uses the definition in Eqn. (17.11) for the total variance σ^2 in the subregion R (see line 25). The vector $\boldsymbol{\mu}$ (calculated in line 26) is the average color of the subregion.

```
1:  KuwaharaFilterColor(I)
        Input: I, an RGB image of size M×N.
        Returns a new (filtered) color image of size M×N.
2:      R1 ← {(−1,−1), (0,−1), (−1,0), (0,0)}
3:      R2 ← {(0,−1), (1,−1), (0,0), (1,0)}
4:      R3 ← {(0,0), (1,0), (1,0), (1,1)}
5:      R4 ← {(−1,0), (0,0), (−1,1), (1,0)}
6:      I' ← Duplicate(I)
7:      (M,N) ← Size(I)
8:      for all image coordinates (u,v) ∈ M×N do
9:          σ²_min ← ∞
10:         for R ← R1,...,R4 do
11:             (σ², μ) ← EvalSubregion(I, Rk, u, v)
12:             if σ² < σ²_min then
13:                 σ²_min ← σ²
14:                 μ_min ← μ
15:         I'(u,v) ← μ_min
16:     return I'

17: EvalSubregion(I, R, u, v)
        Returns the total variance and the mean vector of the color image
        I for the subregion R positioned at (u,v).
18:     n ← Size(R)
19:     S1 ← 0,   S2 ← 0                        ▷ S1, S2 ∈ ℝ³
20:     for all (i,j) ∈ R do
21:         a ← I(u+i, v+j)                     ▷ a ∈ ℝ³
22:         S1 ← S1 + a
23:         S2 ← S2 + a²                        ▷ a² = a · a (dot product)
24:     S ← (S2 − S1² · 1/n) · 1/n              ▷ S = (σ²_R, σ²_G, σ²_B)
25:     σ²_RGB ← ΣS                 ▷ σ²_RGB = σ²_R+σ²_G+σ²_B, total variance in R
26:     μ ← 1/n · S1                ▷ μ ∈ ℝ³, avg. color vector for subregion R
27:     return (σ²_RGB, μ)
```

This is sometimes called the "total variance". The resulting filter process is summarized in Alg. 17.3 and color examples produced with this algorithm are shown in Figs. 17.3 and 17.4.

Alternatively [109], one could define the combined color variance as the norm of the *color covariance matrix*[3] for the subregion R_k,

$$\Sigma_k(\boldsymbol{I},u,v) = \begin{pmatrix} \sigma_{k,\mathrm{RR}} & \sigma_{k,\mathrm{RG}} & \sigma_{k,\mathrm{RB}} \\ \sigma_{k,\mathrm{GR}} & \sigma_{k,\mathrm{GG}} & \sigma_{k,\mathrm{GB}} \\ \sigma_{k,\mathrm{BR}} & \sigma_{k,\mathrm{BG}} & \sigma_{k,\mathrm{BB}} \end{pmatrix}, \tag{17.12}$$

with

$$\sigma_{k,pq} = \frac{1}{|R_k|} \cdot \sum_{(i,j)\in R_k} \big[I_p(u+i,v+j) - \mu_k(I_p,u,v)\big] \cdot \big[I_q(u+i,v+j) - \mu_k(I_q,u,v)\big], \tag{17.13}$$

for all possible color pairs $(p,q) \in \{\mathrm{R},\mathrm{G},\mathrm{B}\}^2$. Note that $\sigma_{k,pp} = \sigma^2_{k,p}$ and $\sigma_{k,pq} = \sigma_{k,qp}$, and thus the matrix Σ_k is symmetric and only 6 of its 9 entries need to be calculated. The (Frobenius) *norm* of the 3×3 color covariance matrix is defined as

[3] See Sec. D.2 in the Appendix for details.

(a) RGB test image with selected details

(b) $r = 1$ (3×3 filter)

(c) $r = 2$ (5×5 filter)

(d) $r = 3$ (7×7 filter)

(e) $r = 4$ (9×9 filter)

Fig. 17.3
Kuwahara-type (*Tomita-Tsuji*) filter—color example using the variance definition in Eqn. (17.11). The filter radius is varied from $r = 1$ (b) to $r = 4$ (e).

$$\sigma^2_{k,\mathrm{RGB}} = \|\Sigma_k(\boldsymbol{I}, u, v)\|_2^2 = \sum_{\substack{p,q \in \\ \{\mathrm{R,G,B}\}}} (\sigma_{k,pq})^2. \qquad (17.14)$$

Note that the total variance in Eqn. (17.11)—which is simpler to calculate than this norm—is equivalent to the *trace* of the covariance matrix Σ_k.

(a) 5 × 5 *Tomita-Tsuji* filter ($r = 2$)

(b) 5 × 5 *Nagao-Matsuyama* filter

Fig. 17.4
Color versions of the *Tomita-Tsuji* (Fig. 17.1(b)) and *Nagao-Matsuyama* filter (Fig. 17.2). Both filters are of size 5 × 5 and use the variance definition in Eqn. (17.11). Results are visually similar, but in general the *Nagao-Matsuyama* filter is slightly less destructive on diagonal structures. Original image in Fig. 17.3(a).

Since each pixel of the filtered image is calculated as the *mean* (i.e., a linear combination) of a set of original color pixels, the results depend on the color space used, as discussed in Chapter 15, Sec. 15.1.2.

17.2 Bilateral Filter

Traditional linear smoothing filters operate by convolving the image with a kernel, whose coefficients act as weights for the corresponding image pixels and only depend on the spatial distance from the center coordinate. Pixels close to the filter center are typically given larger weights while pixels at a greater distance carry smaller weights. Thus the convolution kernel effectively encodes the closeness of the underlying pixels in space. In the following, a filter whose weights depend only on the distance in the spatial domain is called a *domain filter*.

To make smoothing filters less destructive on edges, a typical strategy is to exclude individual pixels from the filter operation or to reduce the weight of their contribution if they are very dissimilar *in value* to the pixel found at the center position. This operation too can be formulated as a filter, but this time the kernel coefficients depend only upon the differences in pixel *values* or *range*. Therefore this is called a *range filter*, as explained in more detail Sec. 17.2.2. The idea of the *bilateral* filter, proposed by Tomasi and Manduchi in [229], is to *combine* both domain and range filtering into a common, edge-preserving smoothing filter.

17.2.1 Domain Filter

In an ordinary 2D linear filter (or "convolution") operation,[4]

[4] See also Chapter 5, Eqn. (5.5) on page 92.

$$I'(u,v) \leftarrow \sum_{\substack{m=\\-\infty}}^{\infty} \sum_{\substack{n=\\-\infty}}^{\infty} I(u+m,v+n) \cdot H(m,n) \tag{17.15}$$

$$= \sum_{\substack{i=\\-\infty}}^{\infty} \sum_{\substack{j=\\-\infty}}^{\infty} I(i,j) \cdot H(i-u,j-v), \tag{17.16}$$

every new pixel value $I'(u,v)$ is the weighted average of the original image pixels I in a certain neighborhood, with the weights specified by the elements of the filter kernel H.[5] The weight assigned to each pixel only depends on its spatial position relative to the current center coordinate (u,v). In particular, $H(0,0)$ specifies the weight of the center pixel $I(u,v)$, and $H(m,n)$ is the weight assigned to a pixel displaced by (m,n) from the center. Since only the spatial image coordinates are relevant, such a filter is called a *domain filter*. Obviously, ordinary filters as we know them are *all* domain filters.

17.2.2 Range Filter

Although the idea may appear strange at first, one could also apply a linear filter to the pixel *values* or *range* of an image in the form

$$I'_r(u,v) \leftarrow \sum_{\substack{i=\\-\infty}}^{\infty} \sum_{\substack{j=\\-\infty}}^{\infty} I(i,j) \cdot H_\mathrm{r}\big(I(i,j) - I(u,v)\big). \tag{17.17}$$

The contribution of each pixel is specified by the function H_r and depends on the difference between its own *value* $I(i,j)$ and the value at the current center pixel $I(u,v)$. The operation in Eqn. (17.17) is called a *range filter*, where the spatial position of a contributing pixel is irrelevant and only the difference in values is considered. For a given position (u,v), all surrounding image pixels $I(i,j)$ with the same value contribute equally to the result $I'_r(u,v)$. Consequently, the application of a *range* filter has no *spatial* effect upon the image—in contrast to a *domain* filter, no blurring or sharpening will occur. Instead, a range filter effectively performs a global *point operation* by remapping the intensity or color values. However, a global *range filter* by itself is of little use, since it combines pixels from the entire image and only changes the intensity or color map of the image, equivalent to a nonlinear, image-dependent point operation.

17.2.3 Bilateral Filter—General Idea

The key idea behind the bilateral filter is to *combine* domain filtering (Eqn. (17.16)) *and* range filtering (Eqn. (17.17)) in the form

$$I'(u,v) = \frac{1}{W_{u,v}} \cdot \sum_{\substack{i=\\-\infty}}^{\infty} \sum_{\substack{j=\\-\infty}}^{\infty} \underbrace{I(i,j) \cdot H_\mathrm{d}(i-u,j-v) \cdot H_\mathrm{r}\big(I(i,j)-I(u,v)\big)}_{w_{i,j}}, \tag{17.18}$$

[5] In Eqn. (17.16), functions I and H are assumed to be zero outside their domains of definition.

where H_d, H_r are the *domain* and *range* kernels, respectively, $w_{i,j}$ are the composite weights, and

$$W_{u,v} = \sum_{i=-\infty}^{\infty} \sum_{j=-\infty}^{\infty} w_{i,j} = \sum_{i=-\infty}^{\infty} \sum_{j=-\infty}^{\infty} H_\mathrm{d}(i-u, j-v) \cdot H_\mathrm{r}\big(I(i,j)-I(u,v)\big) \tag{17.19}$$

is the (position-dependent) sum of the weights $w_{i,j}$ used to normalize the combined filter kernel.

In this form, the scope of range filtering is constrained to the spatial neighborhood defined by the domain kernel H_d. At a given filter position (u,v), the weight $w_{i,j}$ assigned to each contributing pixel depends upon (1) its spatial position relative to (u,v), and (2) the similarity of its pixel value to the value at the center position (u,v). In other words, the resulting pixel is the weighted average of pixels that are nearby *and* similar to the original pixel. In a flat image region, where most surrounding pixels have values similar to the center pixel, the bilateral filter acts as a conventional smoothing filter, controlled only by the domain kernel H_d. However, when placed near a step edge or on an intensity ridge, only those pixels are included in the smoothing process that are similar in value to the center pixel, thus avoiding blurring the edges.

If the domain kernel H_d has a limited radius D, or size $(2D+1) \times (2D+1)$, the bilateral filter defined in Eqn. (17.18) can be written as

$$I'(u,v) = \frac{\sum\limits_{i=u-D}^{u+D} \sum\limits_{j=v-D}^{v+D} I(i,j) \cdot H_\mathrm{d}(i-u, j-v) \cdot H_\mathrm{r}\left(I(i,j)-I(u,v)\right)}{\sum\limits_{i=u-D}^{u+D} \sum\limits_{j=v-D}^{v+D} H_\mathrm{d}(i-u, j-v) \cdot H_\mathrm{r}\left(I(i,j)-I(u,v)\right)} \tag{17.20}$$

$$= \frac{\sum\limits_{m=-D}^{D} \sum\limits_{n=-D}^{D} I(u+m, v+n) \cdot H_\mathrm{d}(m,n) \cdot H_\mathrm{r}\left(I(u+m,v+n)-I(u,v)\right)}{\sum\limits_{m=-D}^{D} \sum\limits_{n=-D}^{D} H_\mathrm{d}(m,n) \cdot H_\mathrm{r}\left(I(u+m,v+n)-I(u,v)\right)} \tag{17.21}$$

(by substituting $(i-u) \rightarrow m$ and $(j-v) \rightarrow n$). The effective, space variant filter kernel for the image I at position (u,v) then is

$$\bar{H}_{I,u,v}(i,j) = \frac{H_\mathrm{d}(i,j) \cdot H_\mathrm{r}\big(I(u+i,v+j)-I(u,v)\big)}{\sum\limits_{m=-D}^{D} \sum\limits_{n=-D}^{D} H_\mathrm{d}(m,n) \cdot H_\mathrm{r}\left(I(u+m,v+n)-I(u,v)\right)}, \tag{17.22}$$

for $-D \leq i,j \leq D$, whereas $\bar{H}_{I,u,v}(i,j) = 0$ otherwise. This quantity specifies the contribution of the original image pixels $I(u+i, v+j)$ to the resulting new pixel value $I'(u,v)$.

17.2.4 Bilateral Filter with Gaussian Kernels

A special (but common) case is the use of Gaussian kernels for both the domain and the range parts of the bilateral filter. The discrete 2D Gaussian *domain kernel* of width σ_d is defined as

$$H_\mathrm{d}^{\mathrm{G},\sigma_\mathrm{d}}(m,n) = \frac{1}{2\pi\sigma_\mathrm{d}^2} \cdot e^{-\frac{\rho^2}{2\sigma_\mathrm{d}^2}} = \frac{1}{2\pi\sigma_\mathrm{d}^2} \cdot e^{-\frac{m^2+n^2}{2\sigma_\mathrm{d}^2}} \qquad (17.23)$$

$$= \frac{1}{\sqrt{2\pi}\,\sigma_\mathrm{d}} \cdot \exp\!\left(-\tfrac{m^2}{2\sigma_\mathrm{d}^2}\right) \cdot \frac{1}{\sqrt{2\pi}\,\sigma_\mathrm{d}} \cdot \exp\!\left(-\tfrac{n^2}{2\sigma_\mathrm{d}^2}\right), \qquad (17.24)$$

for $m, n \in \mathbb{Z}$. It has its maximum at the center ($m = n = 0$) and declines smoothly and isotropically with increasing radius $\rho = \sqrt{m^2+n^2}$; for $\rho > 3.5\sigma_\mathrm{d}$, $H_\mathrm{d}^{\mathrm{G},\sigma_\mathrm{d}}(m,n)$ is practically zero. The factorization in Eqn. (17.24) indicates that the Gaussian 2D kernel can be separated into 1D Gaussians, allowing for a more efficient implementation.[6] The constant factors $1/(\sqrt{2\pi}\,\sigma_\mathrm{d})$ can be omitted in practice, since the bilateral filter requires individual normalization at each image position (Eqn. (17.19)).

Similarly, the corresponding *range filter kernel* is defined as a (continuous) *1*D Gaussian of width σ_r,

$$H_\mathrm{r}^{\mathrm{G},\sigma_\mathrm{r}}(x) = \frac{1}{\sqrt{2\pi}\,\sigma_\mathrm{r}} \cdot e^{-\frac{x^2}{2\sigma_\mathrm{r}^2}} = \frac{1}{\sqrt{2\pi}\,\sigma_\mathrm{r}} \cdot \exp\!\left(-\tfrac{x^2}{2\sigma_\mathrm{r}^2}\right), \qquad (17.25)$$

for $x \in \mathbb{R}$. The constant factor $1/(\sqrt{2\pi}\,\sigma_\mathrm{r})$ may again be omitted and the resulting composite filter (Eqn. (17.18)) can thus be written as

$$I'(u,v) = \frac{1}{W_{u,v}} \cdot \sum_{\substack{i=\\u-D}}^{u+D} \sum_{\substack{j=\\v-D}}^{v+D} \Big[I(i,j) \cdot H_\mathrm{d}^{\mathrm{G},\sigma_\mathrm{d}}(i-u, j-v) \cdot H_\mathrm{r}^{\mathrm{G},\sigma_\mathrm{r}}(I(i,j) - I(u,v)) \Big] \qquad (17.26)$$

$$= \frac{1}{W_{u,v}} \cdot \sum_{\substack{m=\\-D}}^{D} \sum_{\substack{n=\\-D}}^{D} \Big[I(u+m, v+n) \cdot H_\mathrm{d}^{\mathrm{G},\sigma_\mathrm{d}}(m,n) \cdot H_\mathrm{r}^{\mathrm{G},\sigma_\mathrm{r}}(I(u+m,v+n) - I(u,v)) \Big] \qquad (17.27)$$

$$= \frac{1}{W_{u,v}} \cdot \sum_{\substack{m=\\-D}}^{D} \sum_{\substack{n=\\-D}}^{D} \Big[I(u+m, v+n) \cdot \exp\!\left(-\tfrac{m^2+n^2}{2\sigma_\mathrm{d}^2}\right) \cdot \exp\!\left(-\tfrac{(I(u+m,v+n)-I(u,v))^2}{2\sigma_\mathrm{r}^2}\right) \Big], \qquad (17.28)$$

with $D = \lceil 3.5 \cdot \sigma_\mathrm{d} \rceil$ and

$$W_{u,v} = \sum_{\substack{m=\\-D}}^{D} \sum_{\substack{n=\\-D}}^{D} \exp\!\left(-\tfrac{m^2+n^2}{2\sigma_\mathrm{d}^2}\right) \cdot \exp\!\left(-\tfrac{(I(u+m,v+n)-I(u,v))^2}{2\sigma_\mathrm{r}^2}\right). \qquad (17.29)$$

For 8-bit grayscale images, with pixel values in the range [0, 255], the width of the range kernel is typically set to $\sigma_\mathrm{r} = 10, \ldots, 50$. The width of the domain kernel (σ_d) depends on the desired amount of spatial smoothing. Algorithm 17.4 gives a summary of the steps involved in bilateral filtering for grayscale images.

[6] See also Chapter 5, Sec. 5.3.3.

Alg. 17.4
Bilateral filter with Gaussian kernels (grayscale version).

```
1: BilateralFilterGray(I, σd, σr)
     Input: I, a grayscale image of size M×N; σd, width of the 2D
     Gaussian domain kernel; σr, width of the 1D Gaussian range
     kernel. Returns a new filtered image of size M×N.
2:   (M, N) ← Size(I)
3:   D ← ⌈3.5 · σd⌉                        ▷ width of domain filter kernel
4:   I' ← Duplicate(I)
5:   for all image coordinates (u, v) ∈ M×N do
6:       S ← 0                             ▷ sum of weighted pixel values
7:       W ← 0                             ▷ sum of weights
8:       a ← I(u, v)                       ▷ center pixel value
9:       for m ← −D, ..., D do
10:          for n ← −D, ..., D do
11:              b ← I(u + m, v + n)        ▷ off-center pixel value
12:              wd ← exp(−(m²+n²)/(2σd²))  ▷ domain coefficient
13:              wr ← exp(−(a−b)²/(2σr²))   ▷ range coefficient
14:              w ← wd · wr                ▷ composite coefficient
15:              S ← S + w · b
16:              W ← W + w
17:      I'(u, v) ← S/W
18:  return I'
```

Figures 17.5–17.9 show the effective, space-variant filter kernels (see Eqn. (17.22)) and the results of applying a bilateral filter with Gaussian domain and range kernels in different situations. Uniform noise was applied to the original images to demonstrate the filtering effect. One can see clearly how the range part makes the combined filter kernel adapt to the local image structure. Only those surrounding parts that have brightness values similar to the center pixel are included in the filter operation. The filter parameters were set to $\sigma_{\mathrm{d}} = 2.0$ and $\sigma_{\mathrm{r}} = 50$; the domain kernel is of size 15×15.

17.2.5 Application to Color Images

Linear smoothing filters are typically used on color images by separately applying the same filter to the individual color channels. As discussed in Chapter 15, Sec. 15.1, this is legitimate if a suitable working color space is used to avoid the introduction of unnatural intensity and chromaticity values. Thus, for the domain-part of the bilateral filter, the same considerations apply as for any linear smoothing filter. However, as will be described, the bilateral filter as a whole cannot be implemented by filtering the color channels separately.

In the *range* part of the filter, the weight assigned to each contributing pixel depends on its difference to the value of the center pixel. Given a suitable distance measure $\mathrm{dist}(\boldsymbol{a}, \boldsymbol{b})$ between two color vectors $\boldsymbol{a}, \boldsymbol{b}$, the bilateral filter in Eqn. (17.18) can be easily modified for a color image $\boldsymbol{I}$ to

$$\boldsymbol{I}'(u,v) = \frac{1}{W_{u,v}} \cdot \sum_{\substack{i= \\ -\infty}}^{\infty} \sum_{\substack{j= \\ -\infty}}^{\infty} \boldsymbol{I}(i,j) \cdot H_{\mathrm{d}}(i-u, j-v) \cdot H_{\mathrm{r}}\bigl(\mathrm{dist}(\boldsymbol{I}(i,j), \boldsymbol{I}(u,v))\bigr), \tag{17.30}$$

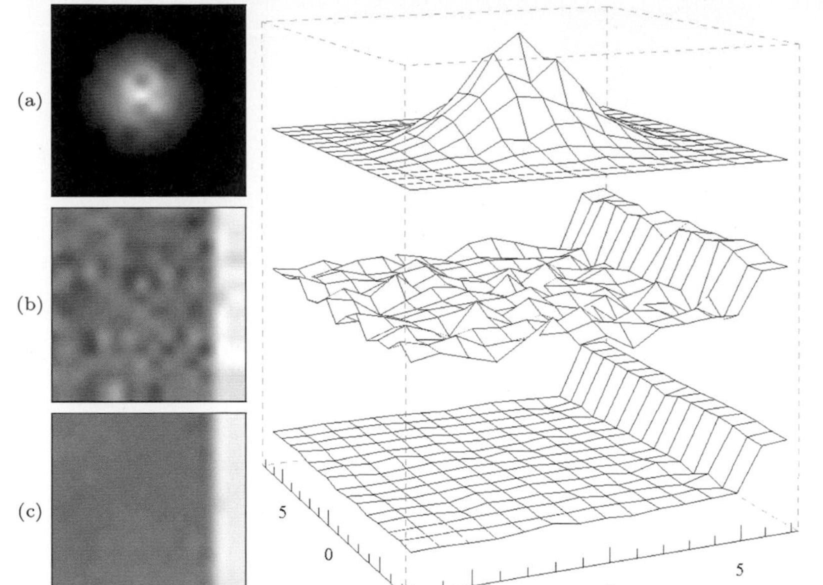

Fig. 17.5
Bilateral filter response when positioned in a flat, noisy image region. Original image function (b), filtered image (c), combined impulse response (a) of the filter at the given position.

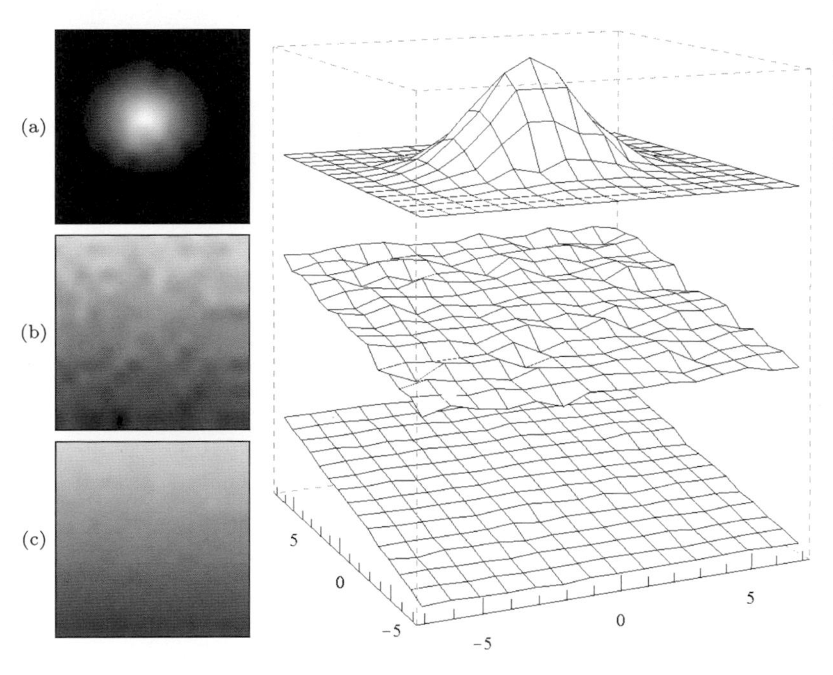

Fig. 17.6
Bilateral filter response when positioned on a linear ramp. Original image function (b), filtered image (c), combined impulse response (a) of the filter at the given position.

with

$$W_{u,v} = \sum_{i,j} H_{\mathrm{d}}(i-u, j-v) \cdot H_{\mathrm{r}}\bigl(\mathrm{dist}(\boldsymbol{I}(i,j), \boldsymbol{I}(u,v))\bigr). \qquad (17.31)$$

It is common to use one of the popular norms for measuring color distances, such as the L_1, L_2 (Euclidean), or the L_∞ (maximum) norms, for example,

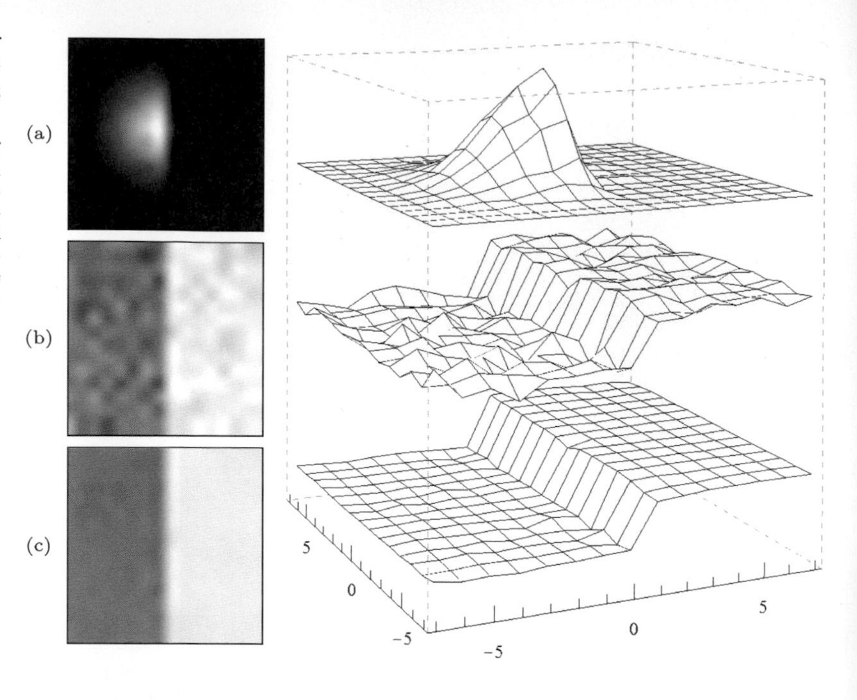

Fig. 17.7
Bilateral filter response when positioned left to a vertical step edge. Original image function (b), filtered image (c), combined impulse response (a) of the filter at the given position.

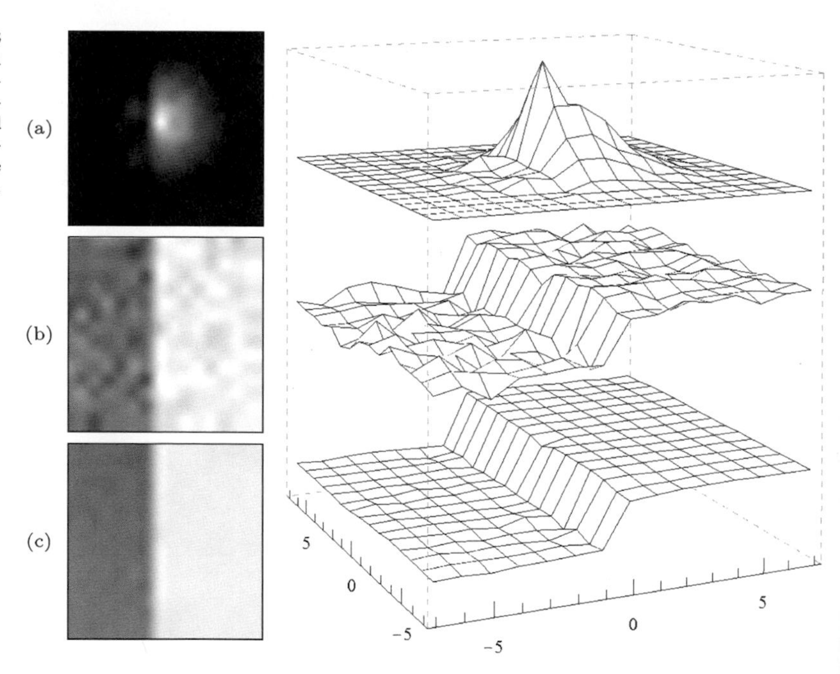

Fig. 17.8
Bilateral filter response when positioned right to a vertical step edge. Original image function (b), filtered image (c), combined impulse response (a) of the filter at the given position.

$$\mathrm{dist}_1(\boldsymbol{a},\boldsymbol{b}) := \tfrac{1}{3}\cdot\|\boldsymbol{a}-\boldsymbol{b}\|_1 = \tfrac{1}{3}\cdot\textstyle\sum_{k=1}^{K}|a_k-b_k|\,, \tag{17.32}$$

$$\mathrm{dist}_2(\boldsymbol{a},\boldsymbol{b}) := \tfrac{1}{\sqrt{3}}\cdot\|\boldsymbol{a}-\boldsymbol{b}\|_2 = \tfrac{1}{\sqrt{3}}\cdot\big(\textstyle\sum_{k=1}^{K}(a_k-b_k)^2\big)^{1/2}, \tag{17.33}$$

$$\mathrm{dist}_\infty(\boldsymbol{a},\boldsymbol{b}) := \|\boldsymbol{a}-\boldsymbol{b}\|_\infty = \max_k\,|a_k-b_k|\,. \tag{17.34}$$

The normalizing factors $1/3$ and $1/\sqrt{3}$ in Eqns. (17.32)–(17.33) are necessary to obtain results comparable in magnitude to those of

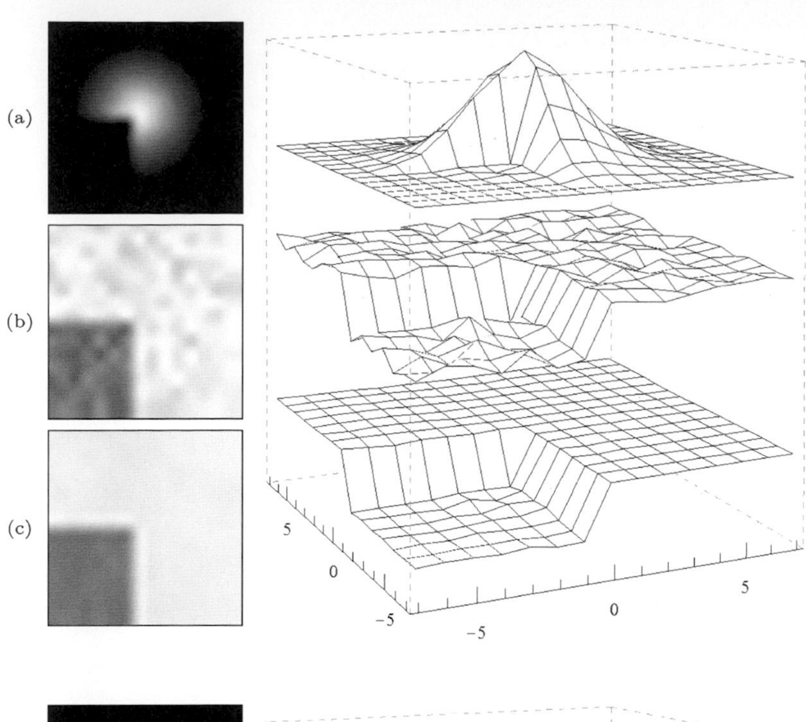

Fig. 17.9
Bilateral filter response when positioned at a corner. Original image function (b), filtered image (c), combined impulse response (a) of the filter at the given position.

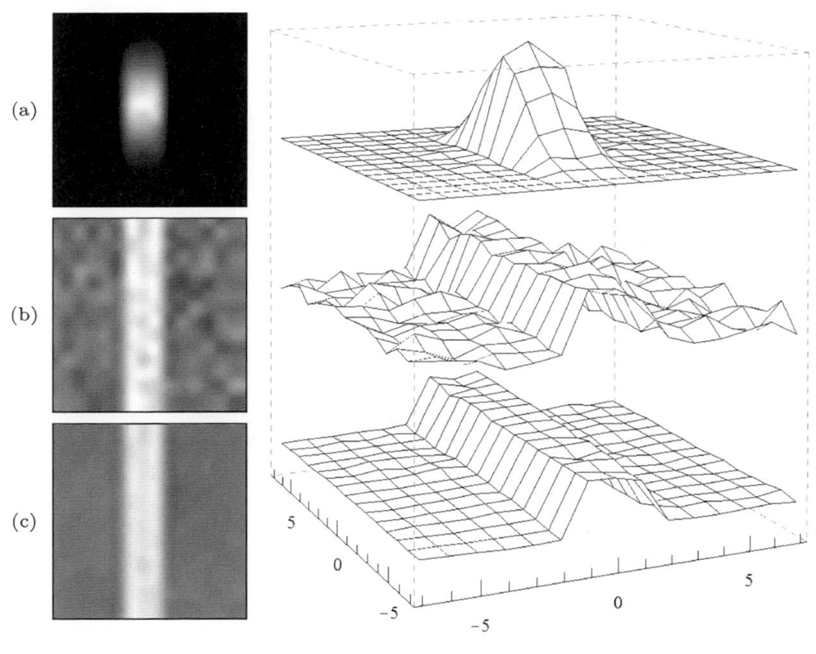

Fig. 17.10
Bilateral filter response when positioned on a vertical ridge. Original image function (b), filtered image (c), combined impulse response (a) of the filter at the given position.

grayscale images when using the same range kernel H_r.[7] Of course in most color spaces, none of these norms measures perceived color difference.[8] However, the distance function itself is not really critical since it only affects the relative *weights* assigned to the contributing

[7] For example, with 8-bit RGB color images, $\mathrm{dist}(\boldsymbol{a}, \boldsymbol{b})$ is always in the range $[0, 255]$.

[8] The CIELAB and CIELUV color spaces are designed to use the Euclidean distance (L_2 norm) as a valid metric for color difference.

Alg. 17.5
Bilateral filter with Gaussian kernels (color version). The function $\mathrm{dist}(\boldsymbol{a},\boldsymbol{b})$ measures the distance between two colors $\boldsymbol{a}$ and $\boldsymbol{b}$, for example, using the L_2 norm (line 5, see Eqns. (17.32)–(17.34) for other options).

```
1:  BilateralFilterColor(I, σ_d, σ_r)
      Input: I, a color image of size M×N; σ_d, width of the 2D
      Gaussian domain kernel; σ_r, width of the 1D Gaussian range
      kernel. Returns a new filtered color image of size M×N.
2:    (M, N) ← Size(I)
3:    D ← ⌈3.5 · σ_d⌉                      ▷ width of domain filter kernel
4:    I' ← Duplicate(I)
5:    dist(a, b) := (1/√3) · ‖a − b‖_2     ▷ color distance (e.g., Euclidean)
6:    for all image coordinates (u, v) ∈ (M×N) do
7:        S ← 0               ▷ S ∈ ℝ³, sum of weighted pixel vectors
8:        W ← 0                      ▷ sum of pixel weights (scalar)
9:        a ← I(u, v)                  ▷ a ∈ ℝ³, center pixel vector
10:       for m ← −D, ..., D do
11:           for n ← −D, ..., D do
12:               b ← I(u + m, v + n)  ▷ b ∈ ℝ³, off-center pixel vector
13:               w_d ← exp(−(m² + n²)/(2σ_d²))     ▷ domain coefficient
14:               w_r ← exp(−(dist(a, b))²/(2σ_r²))   ▷ range coefficient
15:               w ← w_d · w_r                  ▷ composite coefficient
16:               S ← S + w · b
17:               W ← W + w
18:       I'(u, v) ← (1/W) · S
19:   return I'
```

color pixels. Regardless of the distance function used, the resulting chromaticities are linear, convex combinations of the original colors in the filter region, and thus the choice of the working color space is more important (see Chapter 15, Sec. 15.1).

The process of bilateral filtering for color images (again using Gaussian kernels for the domain and the range filters) is summarized in Alg. 17.5. The Euclidean distance (L_2 norm) is used to measure the difference between color vectors. The examples in Fig. 17.11 were produced using sRGB as the color working space.

17.2.6 Efficient Implementation by x/y Separation

The bilateral filter, if implemented in the way described in Algs. 17.4–17.5, is computationally expensive, with a time complexity of $\mathcal{O}(K^2)$ for each pixel, where K denotes the radius of the filter. Some mild speedup is possible by tabulating the domain and range kernels, but the performance of the brute-force implementation is usually not acceptable for practical applications. In [185] a separable *approximation* of the bilateral filter is proposed that brings about a significant performance increase. In this implementation, a 1D bilateral filter is first applied in the horizontal direction only, which uses 1D domain and range kernels h_d and h_r, respectively, and produces the intermediate image $I^{\triangleright}$, that is,

(a) $\sigma_r = 10$

(b) $\sigma_r = 20$

(c) $\sigma_r = 50$

(d) $\sigma_r = 100$

Fig. 17.11
Bilateral filter—color example. A Gaussian kernel with $\sigma_d = 2.0$ (kernel size 15×15) is used for the domain part of the filter; working color space is sRGB. The width of the range filter is varied from $\sigma_r = 10$ to 100. The filter was applied in sRGB color space.

$$I^{\triangleright}(u,v) = \frac{\sum_{m=-D}^{D} I(u+m,v) \cdot h_{\mathrm{d}}(m) \cdot h_{\mathrm{r}}\big(I(u+m,v) - I(u,v)\big)}{\sum_{m=-D}^{D} h_{\mathrm{d}}(m) \cdot h_{\mathrm{r}}\big(I(u+m,v) - I(u,v)\big)} . \tag{17.35}$$

In the second pass, the *same* filter is applied to the intermediate result $I^{\triangleright}$ in the vertical direction to obtain the final result I' as

$$I'(u,v) = \frac{\sum_{n=-D}^{D} I^{\triangleright}(u,v+n) \cdot h_{\mathrm{d}}(n) \cdot h_{\mathrm{r}}\left(I^{\triangleright}(u,v+n) - I^{\triangleright}(u,v)\right)}{\sum_{n=-D}^{D} h_{\mathrm{d}}(n) \cdot h_{\mathrm{r}}\left(I^{\triangleright}(u,v+n) - I^{\triangleright}(u,v)\right)} , \tag{17.36}$$

for all (u,v), using the same 1D domain and range kernels h_{d} and h_{r}, respectively, as in Eqn. (17.35).

For the *horizontal* part of the filter, the effective space-variant kernel at image position (u, v) is

$$\bar{h}^{\triangleright}_{I,u,v}(i) = \frac{h_{\mathrm{d}}(i) \cdot h_{\mathrm{r}}\left(I(u{+}i, v) - I(u, v)\right)}{\sum\limits_{m=-D}^{D} h_{\mathrm{d}}(i) \cdot h_{\mathrm{r}}\left(I(u{+}m, v) - I(u, v)\right)}, \tag{17.37}$$

for $-D \leq i \leq D$ (zero otherwise). Analogously, the effective kernel for the *vertical* part of the filter is

$$\bar{h}^{\triangledown}_{I,u,v}(j) = \frac{h_{\mathrm{d}}(i) \cdot h_{\mathrm{r}}\left(I(u, v{+}j) - I(u, v)\right)}{\sum\limits_{n=-D}^{D} h_{\mathrm{d}}(j) \cdot h_{\mathrm{r}}\left(I(u, v{+}j) - I(u, v)\right)}, \tag{17.38}$$

again for $-D \leq j \leq D$. For the *combined* filter, the effective 2D kernel at position (u, v) then is

$$\bar{H}_{I,u,v}(i,j) = \begin{cases} \bar{h}^{\triangleright}_{I,u,v}(i) \cdot \bar{h}^{\triangledown}_{I^{\triangleright},u,v}(j) & \text{for } -D \leq i,j \leq D, \\ 0 & \text{otherwise,} \end{cases} \tag{17.39}$$

where I is the original image and $I^{\triangleright}$ denotes the intermediate image, as defined in Eqn. (17.35).

Alternatively, the vertical filter could be applied first, followed by the horizontal filter. Algorithm 17.6 shows a direct implementation of the separable bilateral filter for grayscale images, using Gaussian kernels for both the domain and the range parts of the filter. Again, the extension to color images is straightforward (see Eqn. (17.31) and Exercise 17.3).

As intended, the advantage of the separable filter is performance. For a given kernel radius D, the original (non-separable) requires $\mathcal{O}(D^2)$ calculations for each pixel, while the separable version takes only $\mathcal{O}(D)$ steps. This means a substantial saving and speed increase, particularly for large filters.

Figure 17.12 shows the response of the 1D separable bilateral filter in various situations. The results produced by the separable filter are very similar to those obtained with the original filter in Figs. 17.5–17.9, partly because the local structures in these images are parallel to the coordinate axes. In general, the results are different, as demonstrated for a diagonal step edge in Fig. 17.13. The effective filter kernels are shown in Fig. 17.13(g, h) for an anchor point positioned on the bright side of the edge. It can be seen that, while the kernel of the full filter Fig. 17.13(g) is orientation-insensitive, the upper part of the separable kernel is clearly truncated in Fig. 17.13(h). But although the separable bilateral filter is sensitive to local structure orientation, it performs well and is usually a sufficient substitute for the non-separable version [185]. The color examples shown in Fig. 17.14 demonstrate the effects of 1D bilateral filtering in the x- and y-directions. Note that the results are not exactly the same if the filter is first applied in the x- or in y-direction, but usually the differences are negligible.

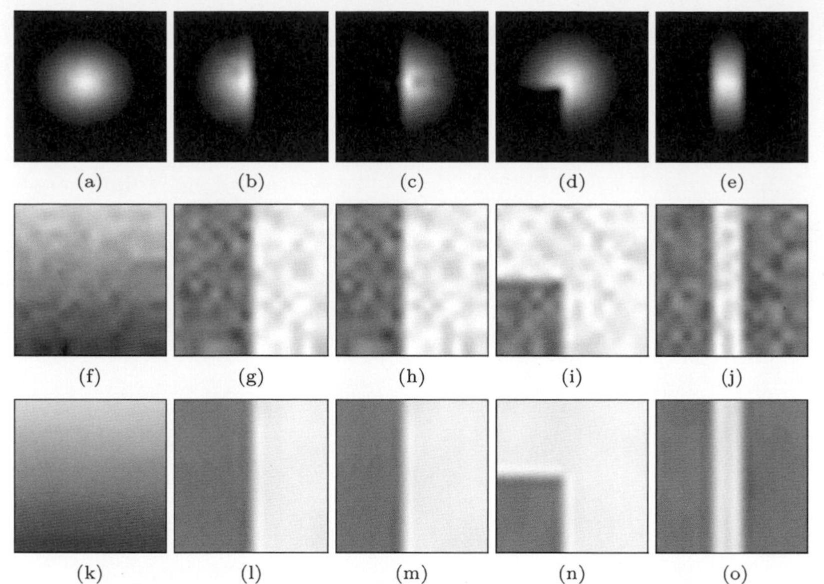

Fig. 17.12
Response of a *separable* bilateral filter in various situations. Effective kernel $\bar{H}_{I,u,v}$ (Eqn. (17.39)) at the center pixel (a–e), original image data (f–j), filtered image data (k–o). Settings are the same as in Figs. 17.5–17.9.

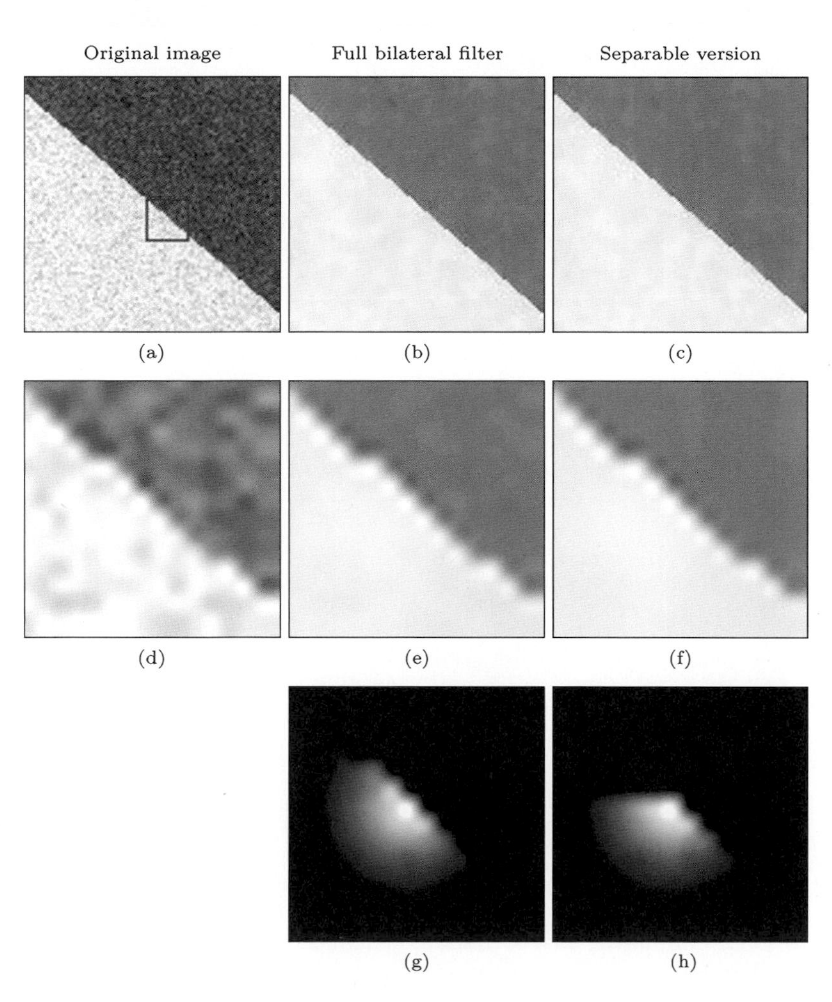

Fig. 17.13
Bilateral filter—full vs. separable version. Original image (a) and enlarged detail (d). Results of the full bilateral filter (b, e) and the separable version (c, f). The corresponding local filter kernels for the center pixel (positioned on the bright side of the step edge) for the full filter (g) and the separable version (h). Note how the upper part of the kernel in (h) is truncated along the horizontal axis, which shows that the separable filter is orientation-sensitive. In both cases, $\sigma_\mathrm{d} = 2.0$, $\sigma_\mathrm{r} = 25$.

Alg. 17.6
Separable bilateral filter with Gaussian kernels (adapted from Alg. 17.4). The input image is processed in two passes. In each pass, a 1D kernel is applied in horizontal or vertical direction, respectively (see Eqns. (17.35)–(17.36)). Note that results of the separable filter are similar (but not identical) to the full (2D) bilateral filter in Alg. 17.4.

```
1: BilateralFilterGraySeparable(I, σ_d, σ_r)
     Input: I, a grayscale image of size M×N; σ_d, width of the 2D
     Gaussian domain kernel; σ_r, width of the 1D Gaussian range
     kernel. Returns a new filtered image of size M×N.
2:   (M, N) ← Size(I)
3:   D ← ⌈3.5 · σ_d⌉                          ▷ width of domain filter kernel
4:   I^▷ ← Duplicate(I)
     Pass 1 (horizontal):
5:   for all coordinates (u, v) ∈ M×N do
6:       a ← I(u, v)
7:       S ← 0,   W ← 0
8:       for m ← −D, ..., D do
9:           b ← I(u + m, v)
10:          w_d ← exp(−m²/(2σ_d²))            ▷ domain kernel coeff. h_d(m)
11:          w_r ← exp(−(a−b)²/(2σ_r²))        ▷ range kernel coeff. h_r(b)
12:          w ← w_d · w_r                     ▷ composite filter coeff.
13:          S ← S + w · b
14:          W ← W + w
15:      I^▷(u, v) ← S/W                       ▷ see Eq. 17.35
16:  I' ← Duplicate(I)
     Pass 2 (vertical):
17:  for all coordinates (u, v) ∈ M×N do
18:      a ← I^▷(u, v)
19:      S ← 0,   W ← 0
20:      for n ← −D, ..., D do
21:          b ← I^▷(u, v + n)
22:          w_d ← exp(−n²/(2σ_d²))            ▷ domain kernel coeff. H_d(n)
23:          w_r ← exp(−(a−b)²/(2σ_r²))        ▷ range kernel coeff. H_r(b)
24:          w ← w_d · w_r                     ▷ composite filter coeff.
25:          S ← S + w · b
26:          W ← W + w
27:      I'(u, v) ← S/W                        ▷ see Eq. 17.36
28:  return I'
```

17.2.7 Further Reading

A thorough analysis of the bilateral filter as well as its relationship to adaptive smoothing and nonlinear diffusion can be found in [16] and [67]. In addition to the simple separable implementation described, several other fast versions of the bilateral filter have been proposed. For example, the method described in [65] approximates the bilateral filter by filtering sub-sampled copies of the image with discrete intensity kernels and recombining the results using linear interpolation. An improved and theoretically well-grounded version of this method was presented in [179]. The fast technique proposed in [253] eliminates the redundant calculations performed in partly overlapping image regions, albeit being restricted to the use of box-shaped domain kernels. As demonstrated in [187, 259], real-time performance using arbitrary-shaped kernels can be obtained by decomposing the filter into a set of smaller spatial filters.

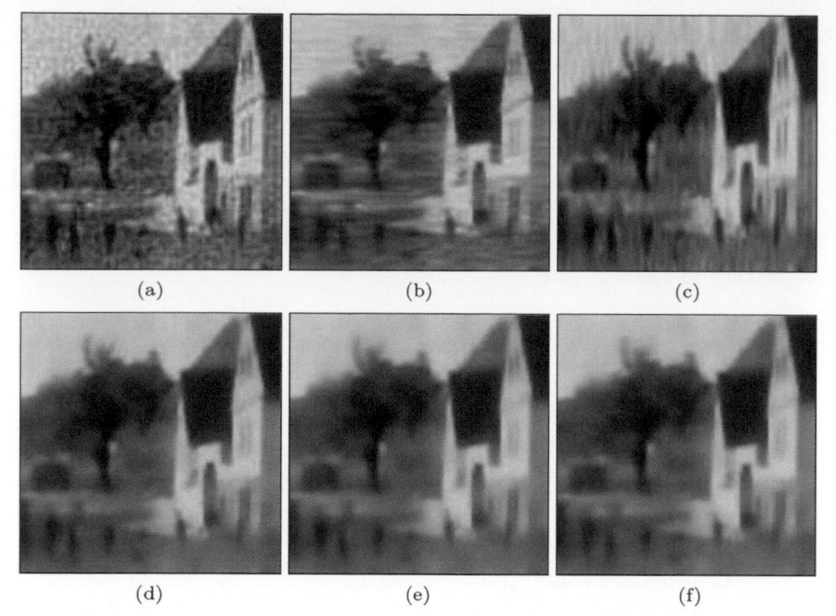

Fig. 17.14
Separable bilateral filter (color example). Original image (a), bilateral filter applied only in the x-direction (b) and only in the y-direction (c). Result of applying the *full* bilateral filter (d) and the *separable* bilateral filter applied in x/y order (e) and y/x order (f). Settings: $\sigma_d = 2.0$, $\sigma_r = 50$, L_2 color distance.

17.3 Anisotropic Diffusion Filters

Diffusion is a concept adopted from physics that models the spatial propagation of particles or state properties within substances. In the real world, certain physical properties (such as temperature) tend to diffuse homogeneously through a physical body, that is, equally in all directions. The idea viewing image smoothing as a diffusion process has a long history in image processing (see, e.g., [11,139]). To smooth an image and, at the same time, preserve edges or other "interesting" image structures, the diffusion process must somehow be made locally *non*-homogeneous; otherwise the entire image would come out equally blurred. Typically, the dominant smoothing direction is chosen to be *parallel* to nearby image contours, while smoothing is inhibited in the perpendicular direction, that is, *across* the contours.

Since the pioneering work by Perona and Malik [182], anisotropic diffusion has seen continued interest in the image processing community and research in this area is still strong today. The main elements of their approach are outlined in Sec. 17.3.2. While various other formulations have been proposed since, a key contribution by Weickert [250, 251] and Tschumperlé [233, 236] unified them into a common framework and demonstrated their extension to color images. They also proposed to separate the actual smoothing process from the smoothing geometry in order to obtain better control of the local smoothing behavior. In Sec. 17.3.4 we give a brief introduction to the approach proposed by Tschumperlé and Deriche, as initially described in [233]. Beyond these selected examples, a vast literature exists on this topic, including excellent reviews [95, 250], textbook material [125, 205], and journal articles (see [3, 45, 52, 173, 206, 226], for example).

17.3.1 Homogeneous Diffusion and the Heat Equation

Assume that in a homogeneous, 3D volume some physical property (e.g., temperature) is specified by a continuous function $f(\boldsymbol{x}, t)$ at position $\boldsymbol{x} = (x, y, z)$ and time t. With the system left to itself, the local differences in the property f will equalize over time until a global equilibrium is reached. This *diffusion process* in 3D space (x, y, z) and time (t) can be expressed using a partial differential equation (PDE),

$$\frac{\partial f}{\partial t} = c \cdot (\nabla^2 f) = c \cdot \Big(\frac{\partial^2 f}{\partial x^2} + \frac{\partial^2 f}{\partial y^2} + \frac{\partial^2 f}{\partial z^2} \Big). \tag{17.40}$$

This is the so-called *heat equation*, where $\nabla^2 f$ denotes the *Laplace operator*[9] applied to the scalar-valued function f, and c is a constant which describes the (thermal) *conductivity* or *conductivity coefficient* of the material. Since the conductivity is independent of position and orientation (c is constant), the resulting process is *isotropic*, that is, the heat spreads evenly in all directions.

For simplicity, we assume $c = 1$. Since f is a multi-dimensional function in space and time, we make this fact a bit more transparent by attaching explicit space and time coordinates $\boldsymbol{x}$ and τ to Eqn. (17.40), that is,

$$\frac{\partial f}{\partial t}(\boldsymbol{x}, \tau) = \frac{\partial^2 f}{\partial x^2}(\boldsymbol{x}, \tau) + \frac{\partial^2 f}{\partial y^2}(\boldsymbol{x}, \tau) + \frac{\partial^2 f}{\partial z^2}(\boldsymbol{x}, \tau), \tag{17.41}$$

or, written more compactly,

$$f_t(\boldsymbol{x}, \tau) = f_{\mathrm{xx}}(\boldsymbol{x}, \tau) \; + \; f_{\mathrm{yy}}(\boldsymbol{x}, \tau) \; + \; f_{zz}(\boldsymbol{x}, \tau). \tag{17.42}$$

Diffusion in images

A continuous, time-varying image I may be treated analogously to the function $f(\boldsymbol{x}, \tau)$, with the local intensities taking on the role of the temperature values in Eqn. (17.42). In this 2D case, the isotropic diffusion equation can be written as[10]

$$\frac{\partial I}{\partial t} = \nabla^2 I = \frac{\partial^2 I}{\partial x^2} + \frac{\partial^2 I}{\partial y^2} \qquad \text{or} \tag{17.43}$$

$$I_t(\boldsymbol{x}, \tau) = I_{\mathrm{xx}}(\boldsymbol{x}, \tau) + I_{\mathrm{yy}}(\boldsymbol{x}, \tau), \tag{17.44}$$

with the derivatives $I_t = \partial I / \partial t$, $I_{\mathrm{xx}} = \partial^2 I / \partial x^2$, and $I_{\mathrm{yy}} = \partial^2 I / \partial y^2$. An approximate, numerical solution of such a PDE can be obtained by replacing the derivatives with finite differences.

Starting with the initial (typically noisy) image $I^{(0)} = I$, the solution to the differential equation in Eqn. (17.44) can be calculated iteratively in the form

[9] Remember that ∇f denotes the *gradient* of the function f, which is a vector for any multi-dimensional function. The Laplace operator (or *Laplacian*) $\nabla^2 f$ corresponds to the *divergence* of the *gradient* of f, denoted $\operatorname{div} \nabla f$, which is a scalar value (see Secs. C.2.5 and C.2.4 in the Appendix). Other notations for the Laplacian are $\nabla \cdot (\nabla f)$, $(\nabla \cdot \nabla) f$, $\nabla \cdot \nabla f$, $\nabla^2 f$, or Δf.

[10] Function arguments $(\boldsymbol{\xi}, \tau)$ are omitted here for better readability.

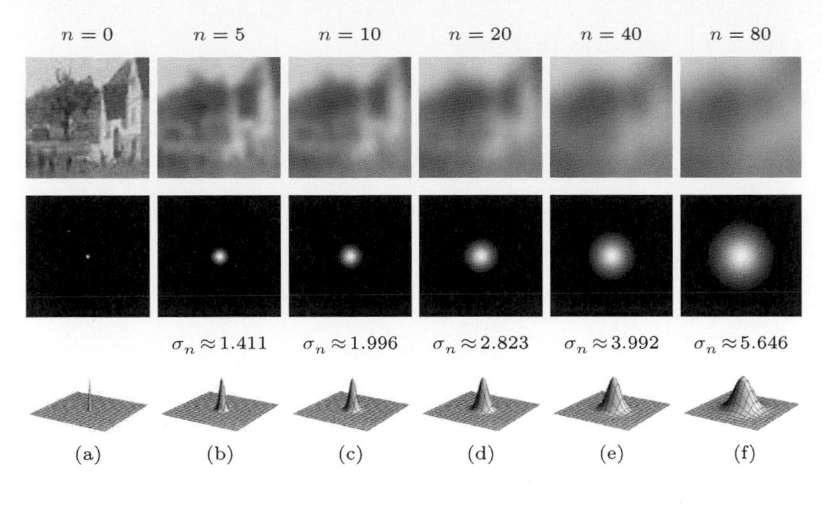

Fig. 17.15
Discrete isotropic diffusion. Blurred images and impulse response obtained after n iterations, with $\alpha = 0.20$ (see Eqn. (17.45)). The size of the images is 50×50. The width of the equivalent Gaussian kernel (σ_n) grows with the square root of n (the number of iterations). Impulse response plots are normalized to identical peak values.

$$I^{(n)}(\boldsymbol{u}) \leftarrow \begin{cases} I(\boldsymbol{u}) & \text{for } n = 0, \\ I^{(n-1)}(\boldsymbol{u}) + \alpha \cdot \left[\nabla^2 I^{(n-1)}(\boldsymbol{u})\right] & \text{for } n > 0, \end{cases} \tag{17.45}$$

for each image position $\boldsymbol{u} = (u, v)$, with n denoting the iteration number. This is called the "direct" solution method (there are other methods but this is the simplest). The constant α in Eqn. (17.45) is the time increment, which controls the speed of the diffusion process. Its value should be in the range $(0, 0.25]$ for the numerical scheme to be stable. At each iteration n, the variations in the image function are reduced and (depending on the boundary conditions) the image function should eventually flatten out to a constant plane as n approaches infinity.

For a discrete image I, the Laplacian $\nabla^2 I$ in Eqn. (17.45) can be approximated by a linear 2D filter,

$$\nabla^2 I \approx I * H^L, \qquad \text{with } H^L = \begin{bmatrix} 0 & 1 & 0 \\ 1 & -4 & 1 \\ 0 & 1 & 0 \end{bmatrix}, \tag{17.46}$$

as described earlier.[11] An essential property of isotropic diffusion is that it has the same effect as a Gaussian filter whose width grows with the elapsed time. For a discrete 2D image, in particular, the result obtained after n diffusion steps (Eqn. (17.45)), is the same as applying a linear filter to the original image I,

$$I^{(n)} \equiv I * H^{\mathrm{G},\sigma_n}, \tag{17.47}$$

with the normalized Gaussian kernel

$$H^{\mathrm{G},\sigma_n}(x, y) = \frac{1}{2\pi\sigma_n^2} \cdot e^{-\frac{x^2+y^2}{2\sigma_n^2}} \tag{17.48}$$

of width $\sigma_n = \sqrt{2t} = \sqrt{2n/\alpha}$. The example in Fig. 17.15 illustrates this Gaussian smoothing behavior obtained with discrete isotropic diffusion.

[11] See also Chapter 6, Sec. 6.6.1 and Sec. C.3.1 in the Appendix.

17.3.2 Perona-Malik Filter

Isotropic diffusion, as we have described, is a homogeneous operation that is independent of the underlying image content. Like any Gaussian filter, it effectively suppresses image noise but also tends to blur away sharp boundaries and detailed structures, a property that is often undesirable. The idea proposed in [182] is to make the conductivity coefficient *variable* and dependent on the local image structure. This is done by replacing the conductivity constant c in Eqn. (17.40), which can be written as

$$\frac{\partial I}{\partial t}(\boldsymbol{x},\tau) = c \cdot [\nabla^2 I](\boldsymbol{x},\tau), \tag{17.49}$$

by a *function* $c(\boldsymbol{x},t)$ that *varies* over space $\boldsymbol{x}$ and time t, that is,

$$\frac{\partial I}{\partial t}(\boldsymbol{x},\tau) = c(\boldsymbol{x},\tau) \cdot [\nabla^2 I](\boldsymbol{x},\tau). \tag{17.50}$$

If the conductivity function $c()$ is constant, then the equation reduces to the isotropic diffusion model in Eqn. (17.44).

Different behaviors can be implemented by selecting a particular function $c()$. To achieve edge-preserving smoothing, the conductivity $c()$ is chosen as a function of the magnitude of the local gradient vector ∇I, that is,

$$c(\boldsymbol{x},\tau) := g(d) = g\big(\|[\nabla I^{(\tau)}](\boldsymbol{x})\|\big). \tag{17.51}$$

To preserve edges, the function $g(d) : \mathbb{R} \rightarrow [0,1]$ should return high values in areas of low image gradient, enabling smoothing of homogeneous regions, but return low values (and thus inhibit smoothing) where the local brightness changes rapidly. Commonly used conductivity functions $g(d)$ are, for example [48, 182],

$$g_1(d) = e^{-(d/\kappa)^2}, \qquad g_2(d) = \frac{1}{1+(d/\kappa)^2}, \tag{17.52}$$

$$g_3(d) = \frac{1}{\sqrt{1+(d/\kappa)^2}}, \qquad g_4(d) = \begin{cases} (1-(d/2\kappa)^2)^2 & \text{for } d \leq 2\kappa, \\ 0 & \text{otherwise,} \end{cases}$$

where $\kappa > 0$ is a constant that is either set manually (typically in the range $[5, 50]$ for 8-bit images) or adjusted to the amount of image noise. Graphs of the four functions in Eqn. (17.52) are shown in Fig. 17.16 for selected values of κ. The Gaussian conductivity function g_1 tends to promote high-contrast edges, whereas g_2 and even more g_3 prefer wide, flat regions over smaller ones. Function g_4, which corresponds to Tuckey's *biweight* function known from robust statistics [205, p. 230], is strictly zero for any argument $d > 2\kappa$. The exact shape of the function $g()$ does not appear to be critical; other functions with similar properties (e.g., with a linear cutoff) are sometimes used instead.

As an approximate discretization of Eqn. (17.50), Perona and Malik [182] proposed the simple iterative scheme

$$I^{(n)}(\boldsymbol{u}) \leftarrow I^{(n-1)}(\boldsymbol{u}) + \alpha \cdot \sum_{i=0}^{3} g\big(|\delta_i(I^{(n-1)}, \boldsymbol{u})|\big) \cdot \delta_i(I^{(n-1)}, \boldsymbol{u}), \tag{17.53}$$

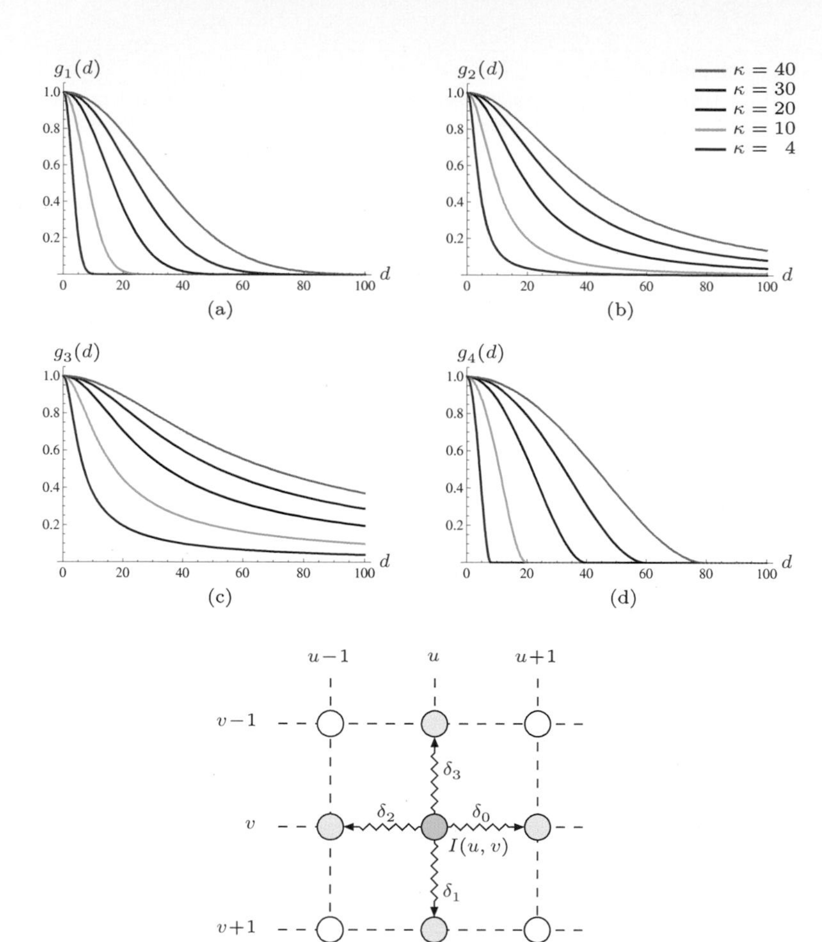

Fig. 17.16
Typical conductivity functions $g_1(), \ldots, g_4()$ for $\kappa = 4, 10, 20, 30, 40$ (see Eqn. (17.52)). If the magnitude of the local gradient d is small (near zero), smoothing amounts to a maximum (1.0), whereas diffusion is reduced where the gradient is high, for example, at or near edges. Smaller values of κ result in narrower curves, thereby restricting the smoothing operation to image areas with only small variations.

Fig. 17.17
Discrete lattice used for implementing diffusion filters in the Perona-Malik algorithm. The green element represents the current image pixel at position $\boldsymbol{u} = (u, v)$ and the yellow elements are its direct 4-neighbors.

where $I^{(0)} = I$ is the original image and

$$\delta_i(I, \boldsymbol{u}) = I(\boldsymbol{u} + \boldsymbol{d}_i) - I(\boldsymbol{u}) \tag{17.54}$$

denotes the difference between the pixel value $I(\boldsymbol{u})$ and its direct neighbor $i = 0, \ldots, 3$ (see Fig. 17.17), with

$$\boldsymbol{d}_0 = \left(\begin{smallmatrix}1\\0\end{smallmatrix}\right), \quad \boldsymbol{d}_1 = \left(\begin{smallmatrix}0\\1\end{smallmatrix}\right), \quad \boldsymbol{d}_2 = -\left(\begin{smallmatrix}1\\0\end{smallmatrix}\right), \quad \boldsymbol{d}_3 = -\left(\begin{smallmatrix}0\\1\end{smallmatrix}\right). \tag{17.55}$$

The procedure for computing the Perona-Malik filter for scalar-valued images is summarized in Alg. 17.7. The examples in Fig. 17.18 demonstrate how this filter performs along a step edge in a noisy grayscale image compared to isotropic (i.e., Gaussian) filtering.

In summary, the principle operation of this filter is to inhibit smoothing in the direction of strong local gradient vectors. Wherever the local contrast (and thus the gradient) is small, diffusion occurs uniformly in all directions, effectively implementing a Gaussian smoothing filter. However, in locations of high gradients, smoothing is inhibited along the gradient direction and allowed only in the direction perpendicular to it. If viewed as a heat diffusion process, a high-gradient brightness edge in an image acts like an insulating layer between areas of different temperatures. While temperatures

Alg. 17.7
Perona-Malik anisotropic diffusion filter for scalar (grayscale) images. The input image I is assumed to be real-valued (floating-point). Temporary real-valued maps $\mathsf{D}_x, \mathsf{D}_y$ are used to hold the directional gradient values, which are then re-calculated in every iteration. The conductivity function $g(d)$ can be one of the functions defined in Eqn. (17.52), or any similar function.

```
1: PeronaMalikGray(I, α, κ, T)
   Input: I, a grayscale image of size M × N; α, update rate; κ,
   smoothness parameter; T, number of iterations. Returns the
   modified image I.

   Specify the conductivity function:
2:    g(d) := e^(−(d/κ)^2)      ▷ for example, see alternatives in Eq. 17.52
3:    (M, N) ← Size(I)
4:    Create maps D_x, D_y: M×N → ℝ
5:    for n ← 1, ..., T do                          ▷ perform T iterations
6:        for all coordinates (u, v) ∈ M×N do            ▷ re-calculate
          gradients
7:            D_x(u, v) ← { I(u+1, v) − I(u, v)   if u < M−1
                          { 0                     otherwise
8:            D_y(u, v) ← { I(u, v+1) − I(u, v)   if v < N−1
                          { 0                     otherwise
9:        for all coordinates (u, v) ∈ M×N do     ▷ update the image
10:           δ0 ← D_x(u, v)
11:           δ1 ← D_y(u, v)
12:           δ2 ← { −D_x(u−1, v)   if u > 0
                   { 0              otherwise
13:           δ3 ← { −D_y(u, v−1)   if v > 0
                   { 0              otherwise
14:           I(u, v) ← I(u, v) + α · Σ_{k=0}^{3} g(|δk|) · δk
15:   return I
```

continuously level out in the homogeneous regions on either side of an edge, thermal energy does not diffuse across the edge itself.

Note that the Perona-Malik filter (as defined in Eqn. (17.50)) is formally considered a *nonlinear* filter but not an *anisotropic* diffusion filter because the conductivity function $g()$ is only a scalar and not a (directed) vector-valued function [250]. However, the (inexact) discretization used in Eqn. (17.53), where each lattice direction is attenuated individually, makes the filter appear to perform in an anisotropic fashion.

17.3.3 Perona-Malik Filter for Color Images

The original Perona-Malik filter is not explicitly designed for color images or vector-valued images in general. The simplest way to apply this filter to a color image is (as usual) to treat the color channels as a set of independent scalar images and filter them separately. Edges should be preserved, since they occur only where at least one of the color channels exhibits a strong variation. However, different filters are applied to the color channels and thus new chromaticities may be produced that were not contained in the original image. Nevertheless, the results obtained (see the examples in Fig. 17.19(b–d)) are often satisfactory and the approach is frequently used because of its simplicity.

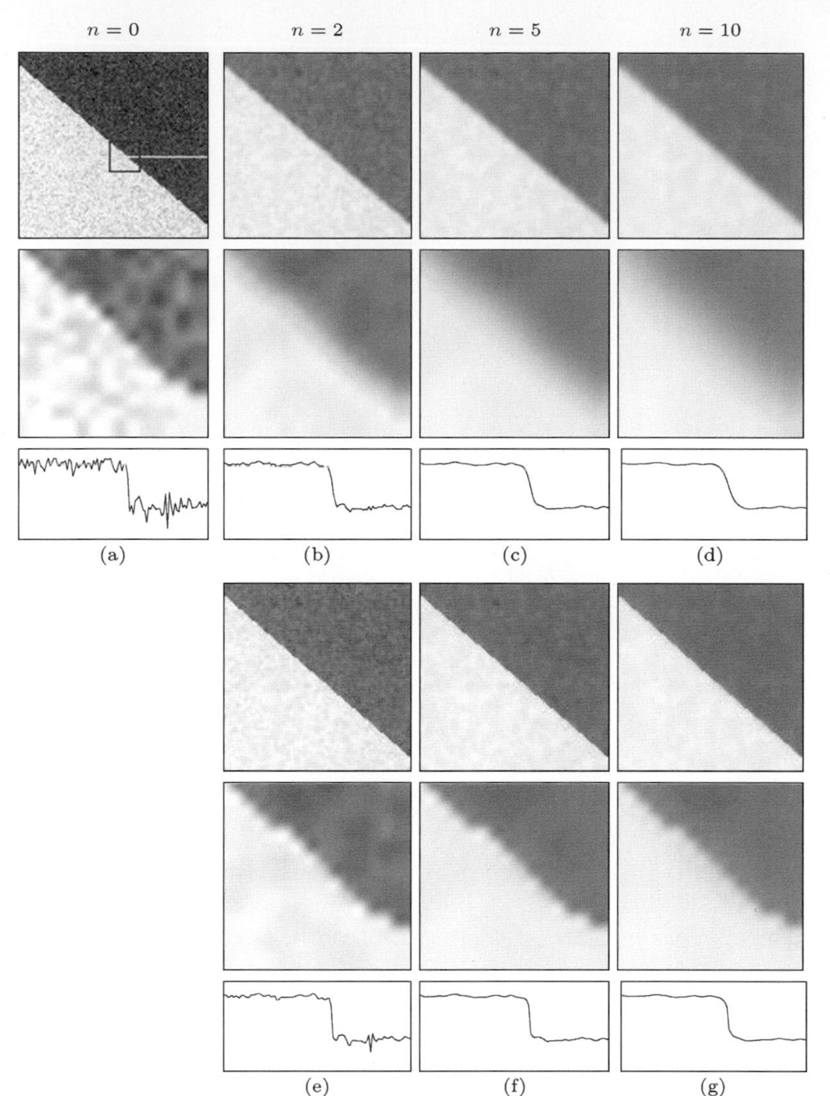

Fig. 17.18
Isotropic vs. anisotropic diffusion applied to a noisy step edge. Original image, enlarged detail, and horizontal profile (a), results of isotropic diffusion (b–d), results of anisotropic diffusion (e–g) after $n = 2, 5, 10$ iterations, respectively ($\alpha = 0.20$, $\kappa = 40$).

Color diffusion based on the brightness gradient

As an alternative to filtering each color channel separately, it has been proposed to use only the brightness (intensity) component to control the diffusion process of all color channels. Given an RGB color image $\boldsymbol{I} = (I_{\mathrm{R}}, I_{\mathrm{G}}, I_{\mathrm{B}})$ and a brightness function $\beta(\boldsymbol{I})$, the iterative scheme in Eqn. (17.53) could be modified to

$$\boldsymbol{I}^{(n)}(\boldsymbol{u}) \leftarrow \boldsymbol{I}^{(n-1)}(\boldsymbol{u}) + \alpha \cdot \sum_{i=0}^{3} g\bigl(|\beta_i(\boldsymbol{I}^{(n-1)}, \boldsymbol{u})|\bigr) \cdot \boldsymbol{\delta}_i(\boldsymbol{I}^{(n-1)}, \boldsymbol{u}), \tag{17.56}$$

where $\quad \beta_i(\boldsymbol{I}, \boldsymbol{u}) = \beta(\boldsymbol{I}(\boldsymbol{u} + \boldsymbol{d}_i)) - \beta(\boldsymbol{I}(\boldsymbol{u})), \qquad (17.57)$

$\boldsymbol{d}_i$ is the local brightness difference (as defined in Eqn. (17.55)) and

$$\boldsymbol{\delta}_i(\boldsymbol{I}, \boldsymbol{u}) = \begin{pmatrix} I_{\mathrm{R}}(\boldsymbol{u} + \boldsymbol{d}_i) - I_{\mathrm{R}}(\boldsymbol{u}) \\ I_{\mathrm{G}}(\boldsymbol{u} + \boldsymbol{d}_i) - I_{\mathrm{G}}(\boldsymbol{u}) \\ I_{\mathrm{B}}(\boldsymbol{u} + \boldsymbol{d}_i) - I_{\mathrm{B}}(\boldsymbol{u}) \end{pmatrix} = \begin{pmatrix} \delta_i(I_{\mathrm{R}}, \boldsymbol{u}) \\ \delta_i(I_{\mathrm{G}}, \boldsymbol{u}) \\ \delta_i(I_{\mathrm{B}}, \boldsymbol{u}) \end{pmatrix} \tag{17.58}$$

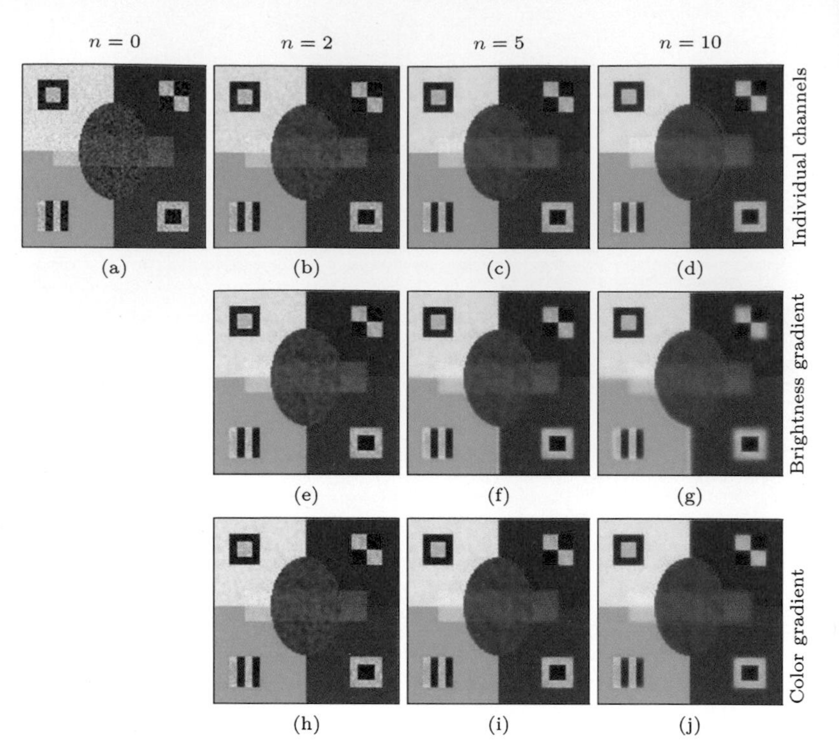

Fig. 17.19 Anisotropic diffusion filter (color). Noisy test image (a). Anisotropic diffusion filter applied separately to *individual color channels* (b–d), diffusion controlled by *brightness gradient* (e–g), diffusion controlled by *color gradient* (h–j), after 2, 5, and 10 iterations, respectively ($\alpha = 0.20$, $\kappa = 40$). With diffusion controlled by the brightness gradient, strong blurring occurs between regions of different color but similar brightness (e–g). The most consistent results are obtained by diffusion controlled by the *color gradient* (h–j). Filtering was performed in linear RGB color space.

is the local color difference vector for the neighboring pixels in directions $i = 0, \ldots, 3$ (see Fig. 17.17). Typical choices for the brightness function $\beta()$ are the *luminance* Y (calculated as a weighted sum of the linear R, G, B components), *luma* Y' (from nonlinear R', G', B' components), or the lightness component L of the CIELAB and CIELUV color spaces (see Chapter 15, Sec. 15.1 for a detailed discussion).

Algorithm 17.7 can be easily adapted to implement this type of color filter. An obvious disadvantage of this method is that it naturally blurs across color edges if the neighboring colors are of similar brightness, as the examples in Fig. 17.19(e–g)) demonstrate. This limits its usefulness for practical applications.

Using the color gradient

A better option for controlling the diffusion process in all three color channels is to use the color gradient (see Ch. 16, Sec. 16.2.1). As defined in Eqn. (16.17), the color gradient

$$(\mathrm{grad}_\theta\, \boldsymbol{I})(\boldsymbol{u}) = \boldsymbol{I}_x(\boldsymbol{u}) \cdot \cos(\theta) + \boldsymbol{I}_y(\boldsymbol{u}) \cdot \sin(\theta) \qquad (17.59)$$

is a 3D vector, representing the combined variations of the color image $\boldsymbol{I}$ at position $\boldsymbol{u}$ in a given direction θ. The squared norm of this vector, $S_\theta(\boldsymbol{I}, \boldsymbol{u}) = \|(\mathrm{grad}_\theta\, \boldsymbol{I})(\boldsymbol{u})\|^2$, called the *squared local contrast*, is a scalar quantity useful for color edge detection. Along the horizontal and vertical directions of the discrete diffusion lattice (see Fig. 17.17), the angle θ is a multiple of $\pi/2$, and thus one of the cosine/sine terms in Eqn. (17.59) vanishes, that is,

$$\begin{aligned}\|(\mathrm{grad}_\theta \boldsymbol{I})(\boldsymbol{u})\| &= \|(\mathrm{grad}_{i\pi/2} \boldsymbol{I})(\boldsymbol{u})\| \\ &= \begin{cases} \|\boldsymbol{I}_x(\boldsymbol{u})\| & \text{for } i = 0, 2, \\ \|\boldsymbol{I}_y(\boldsymbol{u})\| & \text{for } i = 1, 3. \end{cases}\end{aligned} \tag{17.60}$$

Taking $\boldsymbol{\delta}_i$ (Eqn. (17.58)) as an estimate for the horizontal and vertical derivatives $\boldsymbol{I}_x$, $\boldsymbol{I}_y$, the diffusion iteration (adapted from Eqn. (17.53)) thus becomes

$$\boldsymbol{I}^{(n)}(\boldsymbol{u}) \leftarrow \boldsymbol{I}^{(n-1)}(\boldsymbol{u}) + \alpha \cdot \sum_{i=0}^{3} g\big(\|\boldsymbol{\delta}_i(\boldsymbol{I}^{(n-1)}, \boldsymbol{u})\|\big) \cdot \boldsymbol{\delta}_i(\boldsymbol{I}^{(n-1)}, \boldsymbol{u}), \tag{17.61}$$

with $g()$ chosen from one of the conductivity functions in Eqn. (17.52). Note that this is almost identical to the formulation in Eqn. (17.53), except for the use of vector-valued images and the absolute values $|\cdot|$ being replaced by the vector norm $\|\cdot\|$. The diffusion process is coupled between color channels, because the local diffusion strength depends on the combined color difference vectors. Thus, unlike in the brightness-governed diffusion scheme in Eqn. (17.56), opposing variations in different color do not cancel out and edges between colors of similar brightness are preserved (see the examples in Fig. 17.19(h–j)).

The resulting process is summarized in Alg. 17.8. The algorithm assumes that the components of the color image $\boldsymbol{I}$ are real-valued. In practice, integer-valued images must be converted to floating point before this procedure can be applied and integer results should be recovered by appropriate rounding.

Examples

Figure 17.20 shows the results of applying the Perona-Malik filter to a color image, using different modalities to control the diffusion process. In Fig. 17.20(a) the *scalar* (grayscale) diffusion filter (described in Alg. 17.7) is applied *separately* to each color channel. In Fig. 17.20(b) the diffusion process is coupled over all three color channels and controlled by the *brightness gradient*, as specified in Eqn. (17.56). Finally, in Fig. 17.20(c) the *color gradient* is used to control the common diffusion process, as defined in Eqn. (17.61) and Alg. 17.8. In each case, $T = 10$ diffusion iterations were applied, with update rate $\alpha = 0.20$, smoothness $\kappa = 25$, and conductivity function $g_1(d)$. The example demonstrates that, under otherwise equal conditions, edges and line structures are best preserved by the filter if the diffusion process is controlled by the color gradient.

17.3.4 Geometry Preserving Anisotropic Diffusion

Historically, the seminal publication by Perona and Malik [182] was followed by increased interest in the use of diffusion filters based on partial differential equations. Numerous different schemes were proposed, mainly with the aim to better adapt the diffusion process to the underlying image geometry.

Alg. 17.8
Anistropic diffusion filter for color images based on the color gradient (see Ch. 16, Sec. 16.2.1). The conductivity function $g(d)$ may be chosen from the functions defined in Eqn. (17.52), or any similar function. Note that (unlike in Alg. 17.7) the maps $\mathsf{D}_x, \mathsf{D}_y$ are vector-valued.

1: PeronaMalikColor($\boldsymbol{I}, \alpha, \kappa, T$)
Input: $\boldsymbol{I}$, an RGB color image of size $M \times N$; α, update rate; κ, smoothness parameter; T, number of iterations. Returns the modified image $\boldsymbol{I}$.

Specify the conductivity function:
2: $g(d) := e^{-(d/\kappa)^2}$ ▷ for example, see alternatives in Eq. 17.52
3: $(M, N) \leftarrow \mathsf{Size}(\boldsymbol{I})$
4: Create maps $\mathsf{D}_x, \mathsf{D}_y$: $M \times N \to \mathbb{R}^3$; $\mathsf{S}_x, \mathsf{S}_y : M \times N \to \mathbb{R}$
5: **for** $n \leftarrow 1, \ldots, T$ **do** ▷ perform T iterations
6: **for all** $(u, v) \in M \times N$ **do** ▷ re-calculate gradients
7: $\mathsf{D}_x(u,v) \leftarrow \begin{cases} \boldsymbol{I}(u{+}1,v) - \boldsymbol{I}(u,v) & \text{if } u < M{-}1 \\ \boldsymbol{0} & \text{otherwise} \end{cases}$
8: $\mathsf{D}_y(u,v) \leftarrow \begin{cases} \boldsymbol{I}(u,v{+}1) - \boldsymbol{I}(u,v) & \text{if } v < N{-}1 \\ \boldsymbol{0} & \text{otherwise} \end{cases}$
9: $\mathsf{S}_x(u,v) \leftarrow (\mathsf{D}_x(u,v))^2$ ▷ $= I_{R,x}^2 + I_{G,x}^2 + I_{B,x}^2$
10: $\mathsf{S}_y(u,v) \leftarrow (\mathsf{D}_y(u,v))^2$ ▷ $= I_{R,y}^2 + I_{G,y}^2 + I_{B,y}^2$
11: **for all** $(u, v) \in M \times N$ **do** ▷ update the image
12: $s_0 \leftarrow \mathsf{S}_x(u,v)$, $\boldsymbol{\Delta}_0 \leftarrow \mathsf{D}_x(u,v)$
13: $s_1 \leftarrow \mathsf{S}_y(u,v)$, $\boldsymbol{\Delta}_1 \leftarrow \mathsf{D}_y(u,v)$
14: $s_2 \leftarrow 0$, $\boldsymbol{\Delta}_2 \leftarrow \boldsymbol{0}$
15: $s_3 \leftarrow 0$, $\boldsymbol{\Delta}_3 \leftarrow \boldsymbol{0}$
16: **if** $u > 0$ **then**
17: $s_2 \leftarrow \mathsf{S}_x(u{-}1,v)$
18: $\boldsymbol{\Delta}_2 \leftarrow -\mathsf{D}_x(u{-}1,v)$
19: **if** $v > 0$ **then**
20: $s_3 \leftarrow \mathsf{S}_y(u,v{-}1)$
21: $\boldsymbol{\Delta}_3 \leftarrow -\mathsf{D}_y(u,v{-}1)$
22: $\boldsymbol{I}(u,v) \leftarrow \boldsymbol{I}(u,v) + \alpha \cdot \sum_{k=0}^{3} g(|s_k|) \cdot \boldsymbol{\Delta}_k$
23: **return** $\boldsymbol{I}$

Generalized divergence-based formulation

Weickert [249, 250] generalized the divergence-based formulation of the Perona-Malik approach (see Eqn. (17.49)), that is,

$$\frac{\partial I}{\partial t} = \operatorname{div}\big(c \cdot \nabla I\big),$$

by replacing the time-varying, scalar diffusivity field $c(\boldsymbol{x}, \tau) \in \mathbb{R}$ by a *diffusion tensor* field $\boldsymbol{D}(\boldsymbol{x}, \tau) \in \mathbb{R}^{2\times 2}$ in the form

$$\frac{\partial I}{\partial t} = \operatorname{div}\big(\boldsymbol{D} \cdot \nabla I\big). \tag{17.62}$$

The time-varying tensor field $\boldsymbol{D}(\boldsymbol{x}, \tau)$ specifies a symmetric, positive-definite 2×2 matrix for each 2D image position $\boldsymbol{x}$ and time τ (i.e., $\boldsymbol{D} : \mathbb{R}^3 \to \mathbb{R}^{2\times 2}$ in the continuous case). Geometrically, $\boldsymbol{D}$ specifies an oriented, stretched ellipse which controls the local diffusion process. $\boldsymbol{D}$ may be independent of the image I but is typically derived from it. For example, the original Perona-Malik diffusion equation could be (trivially) written in the form[12]

[12] $\mathbf{I}_2$ denotes the 2×2 identity matrix.

(a) Color channels filtered separately

(b) Diffusion controlled by the local brightness gradient

(c) Diffusion controlled by the local color gradient

Fig. 17.20
Perona-Malik color example. Scalar diffusion filter applied separately to each color channel (a); diffusion controlled by the brightness gradient (b); diffusion controlled by color gradient (c). Common settings are $T = 10$, $\alpha = 0.20$, $g(d) = g_1(d)$, $\kappa = 25$; original image in Fig. 17.3(a).

$$\frac{\partial I}{\partial t} = \operatorname{div}\Big[\underbrace{(c \cdot \mathbf{I}_2)}_{\boldsymbol{D}} \cdot \nabla I\Big] = \operatorname{div}\Big[\begin{pmatrix} c & 0 \\ 0 & c \end{pmatrix} \cdot \nabla I\Big], \qquad (17.63)$$

where $c = g\left(\|\nabla I(\boldsymbol{x}, t)\|\right)$ (see Eqn. (17.51)), and thus $\boldsymbol{D}$ is coupled to the image content. In Weickert's approach, $\boldsymbol{D}$ is constructed from the eigenvalues of the local "image structure tensor" [251], which we have encountered under different names in several places. This approach was also adapted to work with color images [252].

Trace-based formulation

Similar to the work of Weickert, the approach proposed by Tschumperlé and Deriche [233, 235] also pursues a geometry-oriented generalization of anisotropic diffusion. The approach is directly aimed at vector-valued (color) images, but can also be applied to single-channel (scalar-valued) images. For a vector-valued image $\boldsymbol{I} = (I_1, \ldots, I_n)$, the smoothing process is specified as

$$\frac{\partial I_k}{\partial t} = \operatorname{trace}\left(\boldsymbol{A} \cdot \mathbf{H}_k\right), \qquad (17.64)$$

for each channel k, where $\mathbf{H}_k$ denotes the *Hessian* matrix of the scalar-valued image function of channel I_k, and $\boldsymbol{A}$ is a square (2×2 for 2D images) matrix that depends on the complete image $\boldsymbol{I}$ and adapts the smoothing process to the local image geometry. Note that $\boldsymbol{A}$ is the same for all image channels. Since the trace of the Hessian matrix[13] is the Laplacian of the corresponding function (i.e., $\mathrm{trace}(\mathbf{H}_I) = \nabla^2 I$) the diffusion equation for the Perona-Malik filter (Eqn. (17.49)) can be written as

$$\begin{aligned} \frac{\partial I}{\partial t} &= c\cdot(\nabla^2 I) = \mathrm{div}(c\cdot\nabla I) \\ &= \mathrm{trace}\left((c\cdot\mathbf{I}_2)\cdot\mathbf{H}_I\right) = \mathrm{trace}\left(c\cdot\mathbf{H}_I\right). \end{aligned} \tag{17.65}$$

In this case, $\boldsymbol{A} = c\cdot\mathbf{I}_2$, which merely applies the constant scalar factor c to the Hessian matrix $\mathbf{H}_I$ (and thus to the resulting Laplacian) that is derived from the local image (since $c = g\left(\|\nabla I(\boldsymbol{x},t)\|\right)$) and does not represent any geometric information.

17.3.5 Tschumperlé-Deriche Algorithm

This is different in the trace-based approach proposed by Tschumperlé and Deriche [233, 235], where the matrix $\boldsymbol{A}$ in Eqn. (17.64) is composed by the expression

$$\boldsymbol{A} = f_1(\lambda_1,\lambda_2)\cdot(\hat{\boldsymbol{q}}_2\cdot\hat{\boldsymbol{q}}_2^{\mathsf{T}}) + f_2(\lambda_1,\lambda_2)\cdot(\hat{\boldsymbol{q}}_1\cdot\hat{\boldsymbol{q}}_1^{\mathsf{T}}), \tag{17.66}$$

where λ_1, λ_2 and $\hat{\boldsymbol{q}}_1, \hat{\boldsymbol{q}}_2$ are the eigenvalues and normalized eigenvectors, respectively, of the (smoothed) 2×2 structure matrix

$$\boldsymbol{G} = \sum_{k=1}^{K} (\nabla I_k)\cdot(\nabla I_k)^{\mathsf{T}}, \tag{17.67}$$

with ∇I_k denoting the local gradient vector in image channel I_k. The functions $f_1(), f_2()$, which are defined in Eqn. (17.79), use the two eigenvalues to control the diffusion strength along the dominant direction of the contours (f_1) and perpendicular to it (f_2). Since the resulting algorithm is more involved than most previous ones, we describe it in more detail than usual.

Given a vector-valued image $\boldsymbol{I}\colon M\times N \to \mathbb{R}^n$, the following steps are performed in each iteration of the algorithm:

Step 1:

Calculate the gradient at each image position $\boldsymbol{u} = (u, v)$,

$$\nabla I_k(\boldsymbol{u}) = \begin{pmatrix} \frac{\partial I_k}{\partial x}(\boldsymbol{u}) \\ \frac{\partial I_k}{\partial y}(\boldsymbol{u}) \end{pmatrix} = \begin{pmatrix} I_{k,\mathrm{x}}(\boldsymbol{u}) \\ I_{k,\mathrm{y}}(\boldsymbol{u}) \end{pmatrix} = \begin{pmatrix} (I_k * H_x^{\nabla})(\boldsymbol{u}) \\ (I_k * H_y^{\nabla})(\boldsymbol{u}) \end{pmatrix}, \tag{17.68}$$

for each color channel $k = 1, \ldots, K$.[14] The first derivatives of the gradient vector ∇I_k are estimated by convolving the image with the kernels

[13] See Sec. C.2.6 in the Appendix for details.

[14] Note that $\nabla I_k(\boldsymbol{u})$ in Eqn. (17.68) is a 2D, vector-valued function, that is, a dedicated vector is calculated for every image position $\boldsymbol{u} = (u, v)$. For better readability, we omit the spatial coordinate ($\boldsymbol{u}$) in the following and simply write ∇I_k instead of $\nabla I_k(\boldsymbol{u})$. Analogously, all related vectors and matrices defined below (including the vectors $\mathbf{e}_1, \mathbf{e}_2$ and the matrices $\boldsymbol{G}$, $\bar{\boldsymbol{G}}$, $\boldsymbol{A}$, and $\mathbf{H}_k$) are also calculated for each image point $\boldsymbol{u}$, without the spatial coordinate being explicitly given.

$$H_x^{\nabla} = \begin{bmatrix} -a & 0 & a \\ -b & 0 & b \\ -a & 0 & a \end{bmatrix} \quad \text{and} \quad H_y^{\nabla} = \begin{bmatrix} -a & -b & -a \\ 0 & 0 & 0 \\ a & b & a \end{bmatrix}, \tag{17.69}$$

with $a = (2-\sqrt{2})/4$ and $b = (\sqrt{2}-1)/2$ (such that $2a+b = 1/2$).[15]

Step 2:

Smooth the channel gradients $I_{k,\mathrm{x}}, I_{k,\mathrm{y}}$ with an isotropic 2D Gaussian filter kernel $H^{\mathrm{G},\sigma_\mathrm{d}}$ of radius σ_d,

$$\overline{\nabla I}_k = \begin{pmatrix} \bar{I}_{k,\mathrm{x}} \\ \bar{I}_{k,\mathrm{y}} \end{pmatrix} = \begin{pmatrix} I_{k,\mathrm{x}} * H^{\mathrm{G},\sigma_\mathrm{d}} \\ I_{k,\mathrm{y}} * H^{\mathrm{G},\sigma_\mathrm{d}} \end{pmatrix}, \tag{17.70}$$

for each image channel $k = 1, \ldots, K$. In practice, this step is usually skipped by setting $\sigma_\mathrm{d} = 0$.

Step 3:

Calculate the *Hessian matrix* (see Sec. C.2.6 in the Appendix) for each image channel I_k, $k = 1, \ldots, K$, that is,

$$\mathbf{H}_k = \begin{pmatrix} \frac{\partial^2 I_k}{\partial x^2} & \frac{\partial^2 I_k}{\partial x \partial y} \\ \frac{\partial^2 I_k}{\partial y \partial x} & \frac{\partial^2 I_k}{\partial y^2} \end{pmatrix} = \begin{pmatrix} I_{k,\mathrm{xx}} & I_{k,\mathrm{xy}} \\ I_{k,\mathrm{xy}} & I_{k,\mathrm{yy}} \end{pmatrix} = \begin{pmatrix} I_k * H_{\mathrm{xx}}^{\nabla} & I_k * H_{\mathrm{xy}}^{\nabla} \\ I_k * H_{\mathrm{xy}}^{\nabla} & I_k * H_{\mathrm{yy}}^{\nabla} \end{pmatrix}, \tag{17.71}$$

using the filter kernels

$$H_{\mathrm{xx}}^{\nabla} = \begin{bmatrix} 1 & -2 & 1 \end{bmatrix}, \quad H_{\mathrm{yy}}^{\nabla} = \begin{bmatrix} 1 \\ -2 \\ 1 \end{bmatrix}, \quad H_{\mathrm{xy}}^{\nabla} = \frac{1}{4} \begin{bmatrix} 1 & 0 & -1 \\ 0 & 0 & 0 \\ -1 & 0 & 1 \end{bmatrix}. \tag{17.72}$$

Step 4:

Calculate the local variation (structure) matrix as

$$\begin{aligned} \boldsymbol{G} = \begin{pmatrix} G_0 & G_1 \\ G_1 & G_2 \end{pmatrix} &= \sum_{k=1}^{K} (\overline{\nabla I}_k) \cdot (\overline{\nabla I}_k)^\mathsf{T} \\ &= \sum_{k=1}^{K} \begin{pmatrix} \bar{I}_{k,\mathrm{x}}^2 & \bar{I}_{k,\mathrm{x}} \cdot \bar{I}_{k,\mathrm{y}} \\ \bar{I}_{k,\mathrm{x}} \cdot \bar{I}_{k,\mathrm{y}} & \bar{I}_{k,\mathrm{y}}^2 \end{pmatrix} = \begin{pmatrix} \sum_{k=1}^{K} \bar{I}_{k,\mathrm{x}}^2 & \sum_{k=1}^{K} \bar{I}_{k,\mathrm{x}} \cdot \bar{I}_{k,\mathrm{y}} \\ \sum_{k=1}^{K} \bar{I}_{k,\mathrm{x}} \cdot \bar{I}_{k,\mathrm{y}} & \sum_{k=1}^{K} \bar{I}_{k,\mathrm{y}}^2 \end{pmatrix}, \end{aligned} \tag{17.73}$$

for each image position $\boldsymbol{u}$. Note that the matrix $\boldsymbol{G}$ is symmetric (and positive semidefinite). In particular, for a RGB color image this is (coordinates $\boldsymbol{u}$ again omitted)

$$\begin{aligned} \boldsymbol{G} &= \begin{pmatrix} \bar{I}_{\mathrm{R},x}^2 & \bar{I}_{\mathrm{R},x}\bar{I}_{\mathrm{R},y} \\ \bar{I}_{\mathrm{R},x}\bar{I}_{\mathrm{R},y} & \bar{I}_{\mathrm{R},y}^2 \end{pmatrix} + \begin{pmatrix} \bar{I}_{\mathrm{G},x}^2 & \bar{I}_{\mathrm{G},x}\bar{I}_{\mathrm{G},y} \\ \bar{I}_{\mathrm{G},x}\bar{I}_{\mathrm{G},y} & \bar{I}_{\mathrm{G},y}^2 \end{pmatrix} + \begin{pmatrix} \bar{I}_{\mathrm{B},x}^2 & \bar{I}_{\mathrm{B},x}\bar{I}_{\mathrm{B},y} \\ \bar{I}_{\mathrm{B},x}\bar{I}_{\mathrm{B},y} & \bar{I}_{\mathrm{B},x}^2 \end{pmatrix} \\ &= \begin{pmatrix} \bar{I}_{\mathrm{R},x}^2 + \bar{I}_{\mathrm{G},x}^2 + \bar{I}_{\mathrm{B},x}^2 & \bar{I}_{\mathrm{R},x}\bar{I}_{\mathrm{R},y} + \bar{I}_{\mathrm{G},x}\bar{I}_{\mathrm{G},y} + \bar{I}_{\mathrm{B},x}\bar{I}_{\mathrm{B},y} \\ \bar{I}_{\mathrm{R},x}\bar{I}_{\mathrm{R},y} + \bar{I}_{\mathrm{G},x}\bar{I}_{\mathrm{G},y} + \bar{I}_{\mathrm{B},x}\bar{I}_{\mathrm{B},y} & \bar{I}_{\mathrm{R},y}^2 + \bar{I}_{\mathrm{G},y}^2 + \bar{I}_{\mathrm{B},y}^2 \end{pmatrix}. \end{aligned} \tag{17.74}$$

[15] Any other common set of x/y gradient kernels (e.g., Sobel masks) could be used instead, but these filters have better rotation invariance than their traditional counterparts. Similar kernels (with $a = 3/32$, $b = 10/32$) were proposed by Jähne in [126, p. 353].

Step 5:

Smooth the elements of the structure matrix $\boldsymbol{G}$ using an isotropic Gaussian filter kernel H^{G,σ_g} of radius σ_g, that is,

$$\bar{\boldsymbol{G}} = \begin{pmatrix} \bar{G}_0 & \bar{G}_1 \\ \bar{G}_1 & \bar{G}_2 \end{pmatrix} = \begin{pmatrix} G_0 * H^{G,\sigma_g} & G_1 * H^{G,\sigma_g} \\ G_1 * H^{G,\sigma_g} & G_2 * H^{G,\sigma_g} \end{pmatrix}. \tag{17.75}$$

Step 6:

For each image position $\boldsymbol{u}$, calculate the eigenvalues λ_1, λ_2 for the smoothed 2×2 matrix $\bar{\boldsymbol{G}}$, such that $\lambda_1 \geq \lambda_2$, and the corresponding normalized eigenvectors[16]

$$\hat{\boldsymbol{q}}_1 = \begin{pmatrix} \hat{x}_1 \\ \hat{y}_1 \end{pmatrix}, \quad \hat{\boldsymbol{q}}_2 = \begin{pmatrix} \hat{x}_2 \\ \hat{y}_2 \end{pmatrix},$$

such that $\|\hat{\boldsymbol{q}}_1\| = \|\hat{\boldsymbol{q}}_2\| = 1$. Note that $\hat{\boldsymbol{q}}_1$ points in the direction of maximum change and $\hat{\boldsymbol{q}}_2$ points in the perpendicular direction, that is, along the edge tangent. Thus, smoothing should occur predominantly along $\hat{\boldsymbol{q}}_2$. Since $\hat{\boldsymbol{q}}_1$ and $\hat{\boldsymbol{q}}_2$ are normal to each other, we can express $\hat{\boldsymbol{q}}_2$ in terms of $\hat{\boldsymbol{q}}_1$, for example,

$$\hat{\boldsymbol{q}}_2 \equiv \begin{pmatrix} 0 & -1 \\ 1 & 0 \end{pmatrix} \cdot \hat{\boldsymbol{q}}_1 = \begin{pmatrix} -\hat{y}_1 \\ \hat{x}_1 \end{pmatrix}. \tag{17.76}$$

Step 7:

From the eigenvalues (λ_1, λ_2) and the normalized eigenvectors $(\hat{\boldsymbol{q}}_1, \hat{\boldsymbol{q}}_2)$ of $\bar{\boldsymbol{G}}$, compose the symmetric matrix $\boldsymbol{A}$ in the form

$$\begin{aligned} \boldsymbol{A} = \begin{pmatrix} A_0 & A_1 \\ A_1 & A_2 \end{pmatrix} &= \underbrace{f_1(\lambda_1, \lambda_2)}_{c_1} \cdot (\hat{\boldsymbol{q}}_2 \cdot \hat{\boldsymbol{q}}_2^\intercal) + \underbrace{f_2(\lambda_1, \lambda_2)}_{c_2} \cdot (\hat{\boldsymbol{q}}_1 \cdot \hat{\boldsymbol{q}}_1^\intercal) \\ &= c_1 \cdot \begin{pmatrix} \hat{y}_1^2 & -\hat{x}_1 \cdot \hat{y}_1 \\ -\hat{x}_1 \cdot \hat{y}_1 & \hat{x}_1^2 \end{pmatrix} + c_2 \cdot \begin{pmatrix} \hat{x}_1^2 & \hat{x}_1 \cdot \hat{y}_1 \\ \hat{x}_1 \cdot \hat{y}_1 & \hat{y}_1^2 \end{pmatrix} & (17.77) \\ &= \begin{pmatrix} c_1 \cdot \hat{y}_1^2 + c_2 \cdot \hat{x}_1^2 & (c_2 - c_1) \cdot \hat{x}_1 \cdot \hat{y}_1 \\ (c_2 - c_1) \cdot \hat{x}_1 \cdot \hat{y}_1 & c_1 \cdot \hat{x}_1^2 + c_2 \cdot \hat{y}_1^2 \end{pmatrix}, & (17.78) \end{aligned}$$

using the conductivity coefficients

$$\begin{aligned} c_1 = f_1(\lambda_1, \lambda_2) &= \frac{1}{(1 + \lambda_1 + \lambda_2)^{a_1}}, \\ c_2 = f_2(\lambda_1, \lambda_2) &= \frac{1}{(1 + \lambda_1 + \lambda_2)^{a_2}}, \end{aligned} \tag{17.79}$$

with fixed parameters $a_1, a_2 > 0$ to control the non-isotropy of the filter: a_1 specifies the amount of smoothing along contours, a_2 in perpendicular direction (along the gradient). Small values of a_1, a_2 facilitate diffusion in the corresponding direction, while larger values inhibit smoothing. With a_1 close to zero, diffusion is practically unconstrained along the tangent direction. Typical default values are $a_1 = 0.5$ and $a_2 = 0.9$; results from other settings are shown in the examples.

[16] See Sec. B.4.1 in the Appendix for details on calculating the eigensystem of a 2×2 matrix.

Step 8:

Finally, each image channel I_k is updated using the recurrence relation

$$I_k \leftarrow I_k + \alpha \cdot \operatorname{trace}(\boldsymbol{A}\cdot\mathbf{H}_k) = I_k + \alpha \cdot \beta_k \qquad (17.80)$$
$$= I_k + \alpha \cdot \big(A_0 \cdot I_{k,\mathrm{xx}} + A_1 \cdot I_{k,\mathrm{xy}} + A_1 \cdot I_{k,yx} + A_2 \cdot I_{k,\mathrm{yy}}\big) \qquad (17.81)$$
$$= I_k + \alpha \cdot \underbrace{\big(A_0 \cdot I_{k,\mathrm{xx}} + 2 \cdot A_1 \cdot I_{k,\mathrm{xy}} + A_2 \cdot I_{k,\mathrm{yy}}\big)}_{\beta_k} \qquad (17.82)$$

(since $I_{k,\mathrm{xy}} = I_{k,\mathrm{yx}}$). The term $\beta_k = \operatorname{trace}(\boldsymbol{A}\cdot\mathbf{H}_k)$ represents the local image *velocity* in channel k. Note that, although a separate Hessian matrix $\mathbf{H}_k$ is calculated for each channel, the structure matrix $\boldsymbol{A}$ is the same for all image channels. The image is thus smoothed along a common image geometry which considers the correlation between color channels, since $\boldsymbol{A}$ is derived from the joint structure matrix $\boldsymbol{G}$ (Eqn. (17.74)) and therefore combines all K color channels.

In each iteration, the factor α in Eqn. (17.82) is adjusted dynamically to the maximum current velocity β_k in all channels in the form

$$\alpha = \frac{d_\mathrm{t}}{\max \beta_k} = \frac{d_\mathrm{t}}{\max\limits_{k,\boldsymbol{u}} |\operatorname{trace}(\boldsymbol{A}\cdot\mathbf{H}_k)|}\,, \qquad (17.83)$$

where d_t is the (constant) "time increment" parameter. Thus the time step α is kept small as long as the image gradients (vector field velocities) are large. As smoothing proceeds, image gradients are reduced and thus α typically increases over time. In the actual implementation, the values of I_k (in Eqn. (17.82)) are hard-limited to the initial minimum and maximum.

The steps (1–8) we have just outlined are repeated for the specified number of iterations. The complete procedure is summarized in Alg. 17.9 and a corresponding Java implementation can be found on the book's website (see Sec. 17.4).

Beyond this baseline algorithm, several variations and extensions of this filter exist, including the use of spatially-adaptive, oriented smoothing filters.[17] This type of filter has also been used with good results for *image inpainting* [234], where diffusion is applied to fill out only selected (masked) parts of the image where the content is unknown or should be removed.

Examples

The example in Fig. 17.21 demonstrates the influence of image geometry and how the non-isotropy of the Tschumperlé-Deriche filter can be controlled by varying the diffusion parameters a_1, a_2 (see Eqn. (17.79)). Parameter a_1, which specifies the diffusion in the direction of contours, is changed while a_2 (controlling the diffusion in the gradient direction) is held constant. In Fig. 17.21(a), smoothing along contours is modest and very small across edges with the default settings $a_1 = 0.5$ and $a_2 = 0.9$. With lower values of a_1, increased

[17] A recent version was released by the original authors as part of the "GREYC's Magic Image Converter" open-source framework, which is also available as a GIMP plugin (http://gmic.sourceforge.net).

Alg. 17.9
Tschumperlé-Deriche anisotropic diffusion filter for vector-valued (color) images. Typical settings are $T = 5, \ldots, 20$, $d_t = 20$, $\sigma_g = 0$, $\sigma_s = 0.5$, $a_1 = 0.5$, $a_2 = 0.9$. See Sec. B.4.1 for a description of the procedure RealEigenValues2x2 (used in line 12).

1: **TschumperleDericheFilter**$(\boldsymbol{I}, T, d_t, \sigma_g, \sigma_s, a_1, a_2)$
   Input: $\boldsymbol{I} = (I_1, \ldots, I_K)$, color image of size $M \times N$ with K channels; T, number of iterations; d_t, time increment; σ_g, width of the Gaussian kernel for smoothing the gradient; σ_s, width of the Gaussian kernel for smoothing the structure matrix; a_1, a_2, diffusion parameters for directions of min./max. variation, respectively. Returns the modified image $\boldsymbol{I}$.
2: Create maps:
   $\mathsf{D}: K \times M \times N \to \mathbb{R}^2$ ▷ $\mathsf{D}(k,u,v) \equiv \nabla I_k(u,v)$, grad. vector
   $\mathsf{H}: K \times M \times N \to \mathbb{R}^{2\times 2}$ ▷ $\mathsf{H}(k,u,v) \equiv \mathbf{H}_k(u,v)$, Hess. matrix
   $\mathsf{G}: M \times N \to \mathbb{R}^{2\times 2}$ ▷ $\mathsf{G}(u,v) \equiv \boldsymbol{G}(u,v)$, structure matrix
   $\mathsf{A}: M \times N \to \mathbb{R}^{2\times 2}$ ▷ $\mathsf{A}(u,v) \equiv \boldsymbol{A}(u,v)$, geometry matrix
   $\mathsf{B}: K \times M \times N \to \mathbb{R}$ ▷ $\mathsf{B}(k,u,v) \equiv \beta_k(u,v)$, velocity
3: **for** $t \leftarrow 1, \ldots, T$ **do** ▷ perform T iterations
4:   **for** $k \leftarrow 1, \ldots, K$ and **all** coordinates $(u,v) \in M \times N$ **do**
5:     $\mathsf{D}(k,u,v) \leftarrow \begin{pmatrix} (I_k * H_x^{\nabla})(u,v) \\ (I_k * H_y^{\nabla})(u,v) \end{pmatrix}$ ▷ Eq. 17.68–17.69
6:     $\mathsf{H}(k,u,v) \leftarrow \begin{pmatrix} (I_k * H_{xx}^{\nabla})(u,v) & (I_k * H_{xy}^{\nabla})(u,v) \\ (I_k * H_{xy}^{\nabla})(u,v) & (I_k * H_{yy}^{\nabla})(u,v) \end{pmatrix}$ ▷ Eq. 17.71–17.72
7:   $\mathsf{D} \leftarrow \mathsf{D} * H_G^{\sigma_d}$ ▷ smooth elements of D over (u,v)
8:   **for all** coordinates $(u,v) \in M \times N$ **do**
9:     $\mathsf{G}(u,v) \leftarrow \sum_{k=1}^{K} \begin{pmatrix} (\mathsf{D}_x(k,u,v))^2 & \mathsf{D}_x(k,u,v) \cdot \mathsf{D}_y(k,u,v) \\ \mathsf{D}_x(k,u,v) \cdot \mathsf{D}_y(k,u,v) & (\mathsf{D}_y(k,u,v))^2 \end{pmatrix}$
10:   $\mathsf{G} \leftarrow \mathsf{G} * H_G^{\sigma_g}$ ▷ smooth elements of G over (u,v)
11:   **for all** coordinates $(u,v) \in M \times N$ **do**
12:     $(\lambda_1, \lambda_2, \boldsymbol{q}_1, \boldsymbol{q}_2) \leftarrow \text{RealEigenValues2x2}(\mathsf{G}(u,v))$ ▷ p. 724
13:     $\hat{\boldsymbol{q}}_1 \leftarrow \begin{pmatrix} \hat{x}_1 \\ \hat{y}_1 \end{pmatrix} = \frac{\boldsymbol{q}_1}{\|\boldsymbol{q}_1\|}$ ▷ normalize 1st eigenvector $(\lambda_1 \geq \lambda_2)$
14:     $c_1 \leftarrow \frac{1}{(1+\lambda_1+\lambda_2)^{a_1}}, \quad c_2 \leftarrow \frac{1}{(1+\lambda_1+\lambda_2)^{a_2}}$ ▷ Eq. 17.79
15:     $\mathsf{A}(u,v) \leftarrow \begin{pmatrix} c_1 \cdot \hat{y}_1^2 + c_2 \cdot \hat{x}_1^2 & (c_2 - c_1) \cdot \hat{x}_1 \cdot \hat{y}_1 \\ (c_2 - c_1) \cdot \hat{x}_1 \cdot \hat{y}_1 & c_1 \cdot \hat{x}_1^2 + c_2 \cdot \hat{y}_1^2 \end{pmatrix}$ ▷ Eq. 17.78
16:   $\beta_{\max} \leftarrow -\infty$
17:   **for** $k \leftarrow 1, \ldots, K$ and **all** $(u,v) \in M \times N$ **do**
18:     $\mathsf{B}(k,u,v) \leftarrow \text{trace}(\mathsf{A}(u,v) \cdot \mathsf{H}(k,u,v))$ ▷ β_k, Eq. 17.82
19:     $\beta_{\max} \leftarrow \max(\beta_{\max}, |\mathsf{B}(k,u,v)|)$
20:   $\alpha \leftarrow d_t / \beta_{\max}$ ▷ Eq. 17.83
21:   **for** $k \leftarrow 1, \ldots, K$ and **all** $(u,v) \in M \times N$ **do**
22:     $I_k(u,v) \leftarrow I_k(u,v) + \alpha \cdot \mathsf{B}(k,u,v)$ ▷ update the image
23: **return** $\boldsymbol{I}$

blurring occurs in the direction of the contours, as shown in Figs. 17.21(b, c).

17.4 Java Implementation

Implementations of the filters described in this chapter are available as part of the `imagingbook`[18] library at the book's website. The associated classes `KuwaharaFilter`, `NagaoMatsuyamaFilter`, `PeronaMalikFilter` and `TschumperleDericheFilter` are based on

[18] Package `imagingbook.pub.edgepreservingfilters`.

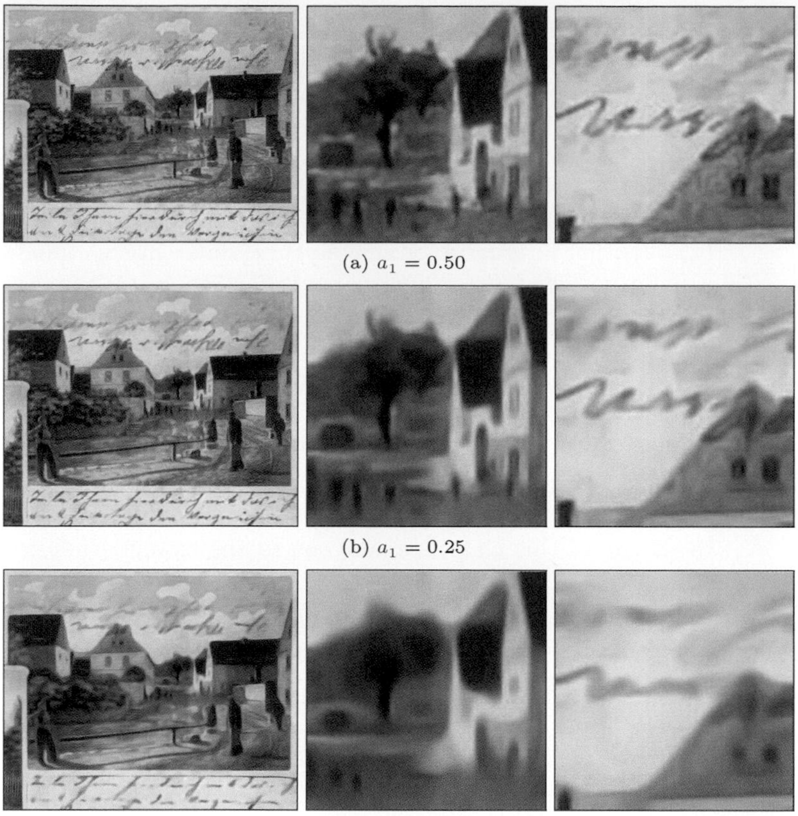

(a) $a_1 = 0.50$

(b) $a_1 = 0.25$

(c) $a_1 = 0.00$

Fig. 17.21
Tschumperlé-Deriche filter example. The non-isotropy of the filter can be adjusted by changing parameter a_1, which controls the diffusion along contours (see Eqn. (17.79)): $a_1 = 0.50, 0.25, 0.00$ (a–c). Parameter $a_2 = 0.90$ (constant) controls the diffusion in the direction of the gradient (perpendicular to contours). Remaining settings are $T = 20$, $d_t = 20$, $\sigma_g = 0.5$, $\sigma_s = 0.5$ (see the description of Alg. 17.9); original image in Fig. 17.3(a).

the common super-class `GenericFilter`[19] and define the following constructors:

`KuwaharaFilter (Parameters p)`
: Creates a Kuwahara-type filter for grayscale and color images, as described in Sec. 17.1 (Alg. 17.2), with radius `r` (default 2) and variance threshold `tsigma` (denoted t_σ in Alg. 17.2, default 0.0). The size of the resulting filter is $(2r+1) \times (2r+1)$.

`BilateralFilter (Parameters p)`
: Creates a bilateral filter for grayscale and color images using Gaussian kernels, as described in Sec. 17.2 (seeAlgs. 17.4 and 17.5). Parameters `sigmaD` (σ_d, default 2.0) and `sigmaR` (σ_r, default 50.0) specify the widths of the domain and the range kernels, respectively. The type of norm for measuring color distances is specified by `colorNormType` (default is `NormType.L2`).

`BilateralFilterSeparable (Parameters p)`
: Creates a x/y-separable bilateral filter for grayscale and color images, (see Alg. 17.6). Constructor parameters are the same as for the class `BilateralFilter` above.

[19] Package `imagingbook.lib.filters`. Filters of this type can be applied to images using the method `applyTo(ImageProcessor ip)`, as described in Chapter 15, Sec. 15.3.

`PeronaMalikFilter (Parameters p)`
Creates an anisotropic diffusion filter for grayscale and color images (see Algs. 17.7 and 17.8). The key parameters and their default values are **iterations** ($T = 10$), **alpha** ($\alpha = 0.2$), **kappa** ($\kappa = 25$), **smoothRegions** (**true**), **colorMode** (**SeparateChannels**). With **smoothRegions** = **true**, function $g_\kappa^{(2)}$ is used to control conductivity, otherwise $g_\kappa^{(1)}$ (see Eqn. (17.52)). For filtering color images, three different color modes can be specified for diffusion control: **SeparateChannels**, **BrightnessGradient**, or **ColorGradient**. See Prog. 17.1 for an example of using this class in a simple ImageJ plugin.

`TschumperleDericheFilter (Parameters p)`
Creates an anisotropic diffusion filter for color images, as described in Sec. 17.3.4 (Alg. 17.9). Parameters and default values are **iterations** ($T = 20$), **dt** ($d_\mathrm{t} = 20$), **sigmaG** ($\sigma_\mathrm{g} = 0.0$), **sigmaS** ($\sigma_\mathrm{s} = 0.5$), **a1** ($a_1 = 0.25$), **a2** ($a_2 = 0.90$). Otherwise the usage of this class is analogous to the example in Prog. 17.1.

All default values pertain to the parameterless constructors that are also available. Note that these filters are generic and can be applied to grayscale and color images without any modification.

17.5 Exercises

Exercise 17.1. Implement a pure *range filter* (Eqn. (17.17)) for grayscale images, using a 1D Gaussian kernel

$$H_\mathrm{r}(x) = \frac{1}{\sqrt{2\pi} \cdot \sigma} \cdot \exp(-\frac{x^2}{2\sigma^2}).$$

Investigate the effects of this filter upon the image and its histogram for $\sigma = 10$, 20, and 25.

Exercise 17.2. Modify the Kuwahara-type filter for color images in Alg. 17.3 to use the *norm of the color covariance matrix* (as defined in Eqn. (17.12)) for quantifying the amount of variation in each subregion. Estimate the number of additional calculations required for processing each image pixel. Implement the modified algorithm, compare the results and execution times.

Exercise 17.3. Modify the separable bilateral filter algorithm (given in Alg. 17.6) to handle color images, using Alg. 17.5 as a starting point. Implement and test your algorithm, compare the results (see also Fig. 17.14) and execution times.

Exercise 17.4. Verify (experimentally) that n iterations of the diffusion process defined in Eqn. (17.45) have the same effect as a Gaussian filter of width σ_n, as stated in Eqn. (17.48). To determine the impulse response of the resulting diffusion filter, use an "impulse" test image, that is, a black (zero-valued) image with a single bright pixel at the center.

```
1  import ij.ImagePlus;
2  import ij.plugin.filter.PlugInFilter;
3  import ij.process.ImageProcessor;
4  import imagingbook...PeronaMalikFilter;
5  import imagingbook...PeronaMalikFilter.ColorMode;
6  import imagingbook...PeronaMalikFilter.Parameters;
7
8  public class Perona_Malik_Demo implements PlugInFilter {
9
10   public int setup(String arg0, ImagePlus imp) {
11     return DOES_ALL + DOES_STACKS;
12   }
13
14   public void run(ImageProcessor ip) {
15     // create a parameter object:
16     Parameters params = new Parameters();
17
18     // modify filter settings if needed:
19     params.iterations = 20;
20     params.alpha = 0.15f;
21     params.kappa = 20.0f;
22     params.smoothRegions = true;
23     params.colorMode = ColorMode.ColorGradient;
24
25     // instantiate the filter object:
26     PeronaMalikFilter filter =
27         new PeronaMalikFilter(params);
28
29     // apply the filter:
30     filter.applyTo(ip);
31   }
32
33 }
```


Prog. 17.1
Perona-Malik filter (complete ImageJ plugin). Inside the run() method, a parameter object (instance of class PeronaMalikFilter.Parameters) is created in line 16. Individual parameters may then be modified, as shown in lines 19–23. This would typically be done be querying the user (e.g., with ImageJ's GenericDialog class). In line 27, a new instance of PeronaMalikFilter is created, the parameter object (params) being passed to the constructor as the only argument. Finally, in line 30, the filter is (destructively) applied to the input image, that is, ip is modified. ColorMode (in line 23) is implemented as an enumeration type within class PeronaMalikFilter, providing the options SeparateChannels (default), BrightnessGradient and ColorGradient. Note that, as specified in the setup() method, this plugin works for any type of image and image stacks.

Exercise 17.5. Use the signal-to-noise ratio (SNR) to measure the effectiveness of noise suppression by edge-preserving smoothing filters on grayscale images. Add synthetic Gaussian noise (see Sec. D.4.3 in the Appendix) to the original image I to create a corrupted image $\tilde{I}$. Then apply the filter to $\tilde{I}$ to obtain $\bar{I}$. Finally, calculate $\mathrm{SNR}(I, \bar{I})$ as defined in Eqn. (13.2). Compare the SNR values obtained with various types of filters and different parameter settings, for example, for the *Kuwahara filter* (Alg. 17.2), the *bilateral filter* (Alg. 17.4), and the *Perona-Malik* anisotropic diffusion filter (Alg. 17.7). Analyze if and how the SNR values relate to the perceived image quality.

18

Introduction to Spectral Techniques

The following three chapters deal with the representation and analysis of images in the frequency domain, based on the decomposition of image signals into sine and cosine functions using the well-known *Fourier transform*. Students often consider this a difficult topic, mainly because of its mathematical flavor and that its practical applications are not immediately obvious. Indeed, most common operations and methods in digital image processing can be sufficiently described in the original signal or image space without even mentioning spectral techniques. This is the reason why we pick up this topic relatively late in this text.

While spectral techniques were often used to improve the efficiency of image-processing operations, this has become increasingly less important due to the high power of modern computers. There exist, however, some important effects, concepts, and techniques in digital image processing that are considerably easier to describe in the frequency domain or cannot otherwise be understood at all. The topic should therefore not be avoided all together. Fourier analysis not only owns a very elegant (perhaps not always sufficiently appreciated) mathematical theory but interestingly enough also complements some important concepts we have seen earlier, in particular linear filters and linear *convolution* (see Chapter 5, Sec. 5.2). Equally important are applications of spectral techniques in many popular methods for image and video compression, and they provide valuable insight into the mechanisms of sampling (discretization) of continuous signals as well as the reconstruction and interpolation of discrete signals.

In the following, we first give a basic introduction to the concepts of frequency and spectral decomposition that tries to be minimally formal and thus should be easily "digestible" even for readers without previous exposure to this topic. We start with the representation of 1D signals and will then extend the discussion to 2D signals (images) in the next chapter. Subsequently, Chapter 20 briefly explains the *discrete cosine transform*, a popular variant of the discrete Fourier transform that is frequently used in image compression.

18.1 The Fourier Transform

The concept of frequency and the decomposition of waveforms into elementary "harmonic" functions first arose in the context of music and sound. The idea of describing acoustic events in terms of "pure" sinusoidal functions does not seem unreasonable, considering that sine waves appear naturally in every form of oscillation (e.g., on a free-swinging pendulum).

18.1.1 Sine and Cosine Functions

The well-known *cosine* function,

$$f(x) = \cos(x), \tag{18.1}$$

has the value 1 at the origin ($\cos(0) = 1$) and performs exactly *one* full cycle between the origin and the point $x = 2\pi$ (Fig. 18.1(a)). We say that the function is periodic with a cycle length (period) $T = 2\pi$; that is,

$$\cos(x) = \cos(x + 2\pi) = \cos(x + 4\pi) = \cdots = \cos(x + k2\pi), \tag{18.2}$$

for any $k \in \mathbb{Z}$. The same is true for the corresponding *sine* function, except that its value is zero at the origin (since $\sin(0) = 0$).

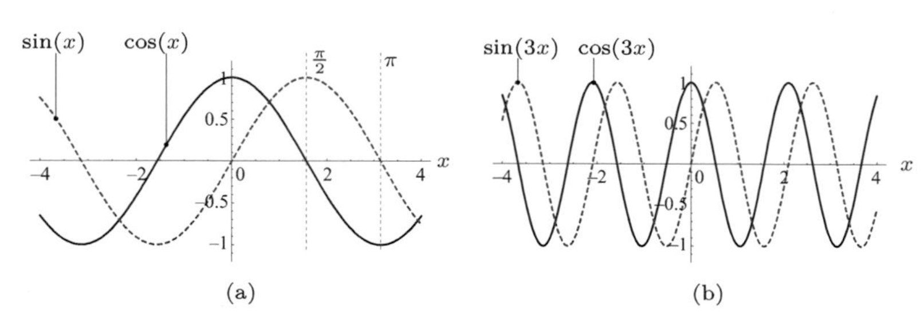

Fig. 18.1 Cosine and sine functions of different frequency. The expression $\cos(\omega x)$ describes a cosine function with angular frequency ω at position x. The angular frequency ω of this periodic function corresponds to a cycle length (period) $T = 2\pi/\omega$. For $\omega = 1$, the period is $T_1 = 2\pi$ (a), and for $\omega = 3$ it is $T_3 = 2\pi/3 \approx 2.0944$ (b). The same holds for the sine function $\sin(\omega x)$.

Frequency and amplitude

The number of oscillations of $\cos(x)$ over the distance $T = 2\pi$ is *one* and thus the value of the *angular frequency*

$$\omega = \frac{2\pi}{T} = 1\,. \tag{18.3}$$

If we modify the cosine function in Eqn. (18.1) to

$$f(x) = \cos(3x)\,, \tag{18.4}$$

we obtain a compressed cosine wave that oscillates three times faster than the original function $\cos(x)$ (see Fig. 18.1(b)). The function $\cos(3x)$ performs three full cycles over a distance of 2π and thus has the angular frequency $\omega = 3$ and a period $T = \frac{2\pi}{3}$. In general, the period T relates to the angular frequency ω as

$$T = \tfrac{2\pi}{\omega}, \tag{18.5}$$

for $\omega > 0$. A sine or cosine function oscillates between peak values $+1$ and -1, and its *amplitude* is 1. Multiplying by a constant $a \in \mathbb{R}$

changes the peak values of the function to $\pm a$ and its *amplitude* to a. In general, the expressions

$$a \cdot \cos(\omega x) \qquad \text{and} \qquad a \cdot \sin(\omega x)$$

denote a cosine or sine function, respectively, with amplitude a and angular frequency ω, evaluated at position (or point in time) x. The relation between the angular frequency ω and the "common" frequency f is given by

$$f = \frac{1}{T} = \frac{\omega}{2\pi} \quad \text{or} \quad \omega = 2\pi f, \tag{18.6}$$

respectively, where f is measured in cycles per length or time unit.[1] In the following, we use either ω or f as appropriate, and the meaning should always be clear from the symbol used.

Phase

Shifting a cosine function along the x axis by a distance φ,

$$\cos(x) \rightarrow \cos(x - \varphi),$$

changes the *phase* of the cosine wave, and φ denotes the *phase angle* of the resulting function. Thus a sine function is really just a cosine function shifted to the right[2] by a quarter period ($\varphi = \frac{2\pi}{4} = \frac{\pi}{2}$), so

$$\sin(\omega x) = \cos\left(\omega x - \tfrac{\pi}{2}\right). \tag{18.7}$$

If we take the cosine function as the reference with phase $\varphi_{\cos} = 0$, then the phase angle of the corresponding sine function is $\varphi_{\sin} = \frac{\pi}{2} = 90°$.

Cosine and sine functions are "orthogonal" in a sense and we can use this fact to create new "sinusoidal" functions with arbitrary frequency, phase, and amplitude. In particular, adding a cosine and a sine function with the identical frequencies ω and arbitrary amplitudes A and B, respectively, creates another sinusoid:

$$A \cdot \cos(\omega x) + B \cdot \sin(\omega x) = C \cdot \cos(\omega x - \varphi). \tag{18.8}$$

The resulting amplitude C and the phase angle φ are defined only by the two original amplitudes A and B as

$$C = \sqrt{A^2 + B^2} \quad \text{and} \quad \varphi = \tan^{-1}\left(\tfrac{B}{A}\right). \tag{18.9}$$

Figure 18.2(a) shows an example with amplitudes $A = B = 0.5$ and a resulting phase angle $\varphi = 45°$.

[1] For example, a temporal oscillation with frequency $f = 1000$ cycles/s (Hertz) has the period $T = 1/1000$ s and therefore the angular frequency $\omega = 2000\pi$. The latter is a unitless quantity.

[2] In general, the function $f(x-d)$ is the original function $f(x)$ shifted to the right by a distance d.

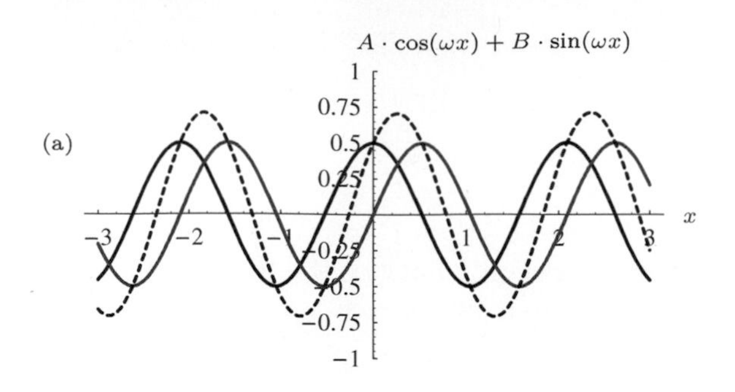

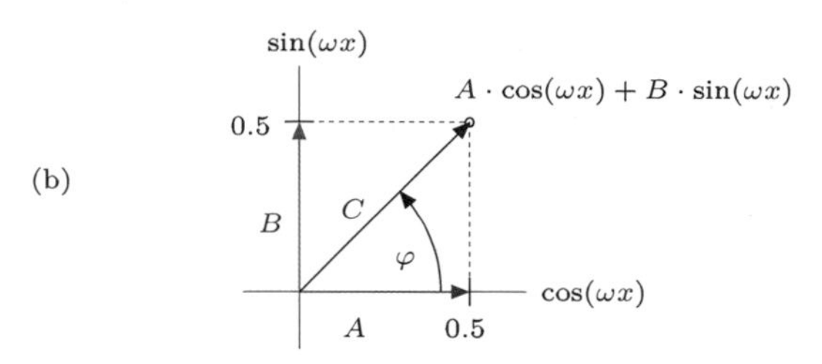

Fig. 18.2
Adding cosine and sine functions with identical frequencies, $A \cdot \cos(\omega x) + B \cdot \sin(\omega x)$, with $\omega = 3$ and $A = B = 0.5$. The result is a phase-shifted cosine function (dotted curve) with amplitude $C = \sqrt{0.5^2 + 0.5^2} \approx 0.707$ and phase angle $\varphi = 45°$ (a). If the cosine and sine components are treated as orthogonal vectors (A, B) in 2-space, the amplitude and phase of the resulting sinusoid (C) can be easily determined by vector summation (b).

Complex-valued sine functions—Euler's notation

Figure 18.2(b) depicts the contributing cosine and sine components of the new function as a pair of orthogonal vectors in 2-space whose *lengths* correspond to the amplitudes A and B. Not coincidentally, this reminds us of the representation of real and imaginary components of complex numbers,

$$z = a + \mathrm{i}\, b,$$

in the 2D plane $\mathbb{C}$, where i is the imaginary unit ($\mathrm{i}^2 = -1$). This association becomes even stronger if we look at Euler's famous notation of complex numbers along the unit circle,

$$z = e^{\mathrm{i}\theta} = \cos(\theta) + \mathrm{i} \cdot \sin(\theta), \tag{18.10}$$

where $e \approx 2.71828$ is the Euler number. If we take the expression $e^{\mathrm{i}\theta}$ as a function of the angle θ rotating around the unit circle, we obtain a "complex-valued sinusoid" whose real and imaginary parts correspond to a cosine and a sine function, respectively,

$$\begin{aligned} \mathrm{Re}(e^{\mathrm{i}\theta}) &= \cos(\theta), \\ \mathrm{Im}(e^{\mathrm{i}\theta}) &= \sin(\theta). \end{aligned} \tag{18.11}$$

Since $z = e^{\mathrm{i}\theta}$ is placed on the unit circle, the *amplitude* of the complex-valued sinusoid is $|z| = r = 1$. We can easily modify the amplitude of this function by multiplying it by some real value $a \geq 0$, that is,

$$|a \cdot e^{\mathrm{i}\theta}| = a \cdot |e^{\mathrm{i}\theta}| = a\,. \tag{18.12}$$

Similarly, we can alter the *phase* of a complex-valued sinusoid by adding a phase angle φ in the function's exponent or, equivalently, by multiplying it by a complex-valued constant $c = e^{\mathrm{i}\varphi}$,

$$e^{\mathrm{i}(\theta+\varphi)} = e^{\mathrm{i}\theta} \cdot e^{\mathrm{i}\varphi}. \tag{18.13}$$

In summary, multiplying by some real value affects only the *amplitude* of a sinusoid, while multiplying by some complex value c (with unit amplitude $|c| = 1$) modifies only the function's *phase* (without changing its amplitude). In general, of course, multiplying by some arbitrary complex value changes both the amplitude *and* the phase of the function (also see Sec. A.3 in the Appendix).

The complex notation makes it easy to combine orthogonal pairs of sine functions $\cos(\omega x)$ and $\sin(\omega x)$ with identical frequencies ω into a single expression,

$$e^{\mathrm{i}\theta} = e^{\mathrm{i}\omega x} = \cos(\omega x) + \mathrm{i} \cdot \sin(\omega x). \tag{18.14}$$

We will make more use of this notation later (in Sec. 18.1.4) to explain the Fourier transform.

18.1.2 Fourier Series Representation of Periodic Functions

As we demonstrated in Eqn. (18.8), sinusoidal functions of arbitrary frequency, amplitude, and phase can be described as the sum of suitably weighted cosine and sine functions. One may wonder if non-sinusoidal functions can also be decomposed into a sum of cosine and sine functions. The answer is yes, of course. It was Fourier[3] who first extended this idea to arbitrary functions and showed that (almost) any periodic function $g(x)$ with a fundamental frequency ω_0 can be described as a—possibly infinite—sum of "harmonic" sinusoids, that is,

$$g(x) = \sum_{k=0}^{\infty} A_k \cdot \cos(k\omega_0 x) + B_k \cdot \sin(k\omega_0 x). \tag{18.15}$$

This is called a *Fourier series*, and the constant factors A_k, B_k are the *Fourier coefficients* of the function $g(x)$. Notice that in Eqn. (18.15) the frequencies of the sine and cosine functions contributing to the Fourier series are integral multiples ("harmonics") of the fundamental frequency ω_0, including the zero frequency for $k = 0$. The corresponding coefficients A_k and B_k, which are initially unknown, can be uniquely derived from the original function $g(x)$. This process is commonly referred to as *Fourier analysis*.

18.1.3 Fourier Integral

Fourier did not want to limit this concept to periodic functions and postulated that nonperiodic functions, too, could be described as sums of sine and cosine functions. While this proved to be true in principle, it generally requires—beyond multiples of the fundamental frequency ($k\omega_0$)—infinitely many, densely spaced frequencies! The resulting decomposition,

$$g(x) = \int_0^{\infty} A_\omega \cdot \cos(\omega x) + B_\omega \cdot \sin(\omega x) \, d\omega, \tag{18.16}$$

is called a *Fourier integral* and the coefficients A_ω, B_ω are again the weights for the corresponding cosine and sine functions with the

[3] Jean-Baptiste Joseph de Fourier (1768–1830).

(continuous) frequency ω. The Fourier integral is the basis of the Fourier spectrum and the Fourier transform, as will be described (for details, see, e.g., [35, Ch. 15, Sec. 15.3]).

In Eqn. (18.16), every coefficient A_ω and B_ω specifies the *amplitude* of the corresponding cosine or sine function, respectively. The coefficients thus define "how much of each frequency" contributes to a given function or signal $g(x)$. But what are the proper values of these coefficients for a given function $g(x)$, and can they be determined uniquely? The answer is yes again, and the "recipe" for computing the coefficients is amazingly simple:

$$\begin{aligned} A_\omega = A(\omega) &= \frac{1}{\pi} \cdot \int_{-\infty}^{\infty} g(x) \cdot \cos(\omega x) \, \mathrm{d}x, \\ B_\omega = B(\omega) &= \frac{1}{\pi} \cdot \int_{-\infty}^{\infty} g(x) \cdot \sin(\omega x) \, \mathrm{d}x. \end{aligned} \tag{18.17}$$

Since this representation of the function $g(x)$ involves infinitely many densely spaced frequency values ω, the corresponding coefficients $A(\omega)$ and $B(\omega)$ are indeed continuous functions as well. They hold the continuous distribution of frequency components contained in the original signal, which is called a "spectrum".

Thus the Fourier integral in Eqn. (18.16) describes the original function $g(x)$ as a sum of infinitely many cosine and sine functions, with the corresponding Fourier coefficients contained in the functions $A(\omega)$ and $B(\omega)$. In addition, a signal $g(x)$ is uniquely and fully represented by the corresponding coefficient functions $A(\omega)$ and $B(\omega)$. We know from Eqn. (18.17) how to compute the spectrum for a given function $g(x)$, and Eqn. (18.16) explains how to reconstruct the original function from its spectrum if it is ever needed.

18.1.4 Fourier Spectrum and Transformation

There is now only a small remaining step from the decomposition of a function $g(x)$, as shown in Eqn. (18.17), to the "real" Fourier transform. In contrast to the Fourier *integral*, the Fourier *transform* treats both the original signal and the corresponding spectrum as *complex-valued* functions, which considerably simplifies the resulting notation.

Based on the functions $A(\omega)$ and $B(\omega)$ defined in the Fourier integral (Eqn. (18.17)), the *Fourier spectrum* $G(\omega)$ of a function $g(x)$ is given as

$$\begin{aligned} G(\omega) &= \sqrt{\tfrac{\pi}{2}} \cdot \big[A(\omega) - \mathrm{i} \cdot B(\omega)\big] \\ &= \sqrt{\tfrac{\pi}{2}} \cdot \Big[\frac{1}{\pi} \int_{-\infty}^{\infty} g(x) \cdot \cos(\omega x) \, \mathrm{d}x \;-\; \mathrm{i} \cdot \frac{1}{\pi} \int_{-\infty}^{\infty} g(x) \cdot \sin(\omega x) \, \mathrm{d}x\Big] \\ &= \frac{1}{\sqrt{2\pi}} \cdot \int_{-\infty}^{\infty} g(x) \cdot \big[\cos(\omega x) - \mathrm{i} \cdot \sin(\omega x)\big] \, \mathrm{d}x \,, \end{aligned} \tag{18.18}$$

with $g(x), G(\omega) \in \mathbb{C}$. Using Euler's notation of complex values (see Eqn. (18.14)) yields the continuous Fourier spectrum in Eqn. (18.18) in its common form:

$$G(\omega) = \frac{1}{\sqrt{2\pi}} \int_{-\infty}^{\infty} g(x) \cdot \big[\cos(\omega x) - \mathrm{i} \cdot \sin(\omega x)\big] \, \mathrm{d}x = \frac{1}{\sqrt{2\pi}} \int_{-\infty}^{\infty} g(x) \cdot e^{-\mathrm{i}\omega x} \, \mathrm{d}x \,. \tag{18.19}$$

The transition from the function $g(x)$ to its Fourier spectrum $G(\omega)$ is called the *Fourier transform*[4] ($\mathcal{F}$). Conversely, the original function $g(x)$ can be reconstructed completely from its Fourier spectrum $G(\omega)$ using the *inverse Fourier transform*[5] ($\mathcal{F}^{-1}$), defined as

$$g(x) = \frac{1}{\sqrt{2\pi}} \int_{-\infty}^{\infty} G(\omega) \cdot \big[\cos(\omega x) + \mathrm{i} \cdot \sin(\omega x)\big] \, \mathrm{d}\omega = \frac{1}{\sqrt{2\pi}} \int_{-\infty}^{\infty} G(\omega) \cdot e^{\mathrm{i}\omega x} \, \mathrm{d}\omega \,. \tag{18.20}$$

In general, even if one of the involved functions ($g(x)$ or $G(\omega)$) is real-valued (which is usually the case for physical signals $g(x)$), the other function is complex-valued. One may also note that the forward transformation $\mathcal{F}$ (Eqn. (18.19)) and the inverse transformation $\mathcal{F}^{-1}$ (Eqn. (18.20)) are almost completely symmetrical, the sign of the exponent being the only difference.[6] The spectrum produced by the Fourier transform is a new representation of the signal in a space of frequencies. Apparently, this "frequency space" and the original "signal space" are *dual* and interchangeable mathematical representations.

18.1.5 Fourier Transform Pairs

The relationship between a function $g(x)$ and its Fourier spectrum $G(\omega)$ is unique in both directions: the Fourier spectrum is uniquely defined for a given function, and for any Fourier spectrum there is only one matching signal—the two functions $g(x)$ and

$$g(x) \circ\!\!-\!\!\bullet\; G(\omega).$$

Table 18.1 lists the transform pairs for some selected analytical functions, which are also shown graphically in Figs. 18.3 and 18.4.

The Fourier spectrum of a *cosine function* $\cos(\omega_0 x)$, for example, consists of two separate thin pulses arranged symmetrically at a distance ω_0 from the origin (Fig. 18.3(a, c)). Intuitively, this corresponds to our physical understanding of a spectrum (e.g., if we think of a pure monophonic sound in acoustics or the thin line produced by some extremely pure color in the optical spectrum). Increasing the frequency ω_0 would move the corresponding pulses in the spectrum

[4] Also called the "direct" or "forward" transformation.

[5] Also called "backward" transformation.

[6] Various definitions of the Fourier transform are in common use. They are contrasted mainly by the constant factors outside the integral and the signs of the exponents in the forward and inverse transforms, but all versions are equivalent in principle. The symmetric variant shown here uses the same factor ($1/\sqrt{2\pi}$) in the forward and inverse transforms.

Table 18.1
Fourier transforms of selected analytical functions; $\delta()$ denotes the "impulse" or *Dirac* function (see Sec. 18.2.1).

Function	*Transform pair* $g(x) \circ\!\!-\!\!\bullet\, G(\omega)$	*Figure*
Cosine function with frequency ω_0	$g(x) = \cos(\omega_0 x)$ $G(\omega) = \sqrt{\frac{\pi}{2}} \cdot \big(\delta(\omega+\omega_0) + \delta(\omega-\omega_0)\big)$	18.3(a,c)
Sine function with frequency ω_0	$g(x) = \sin(\omega_0 x)$ $G(\omega) = \mathrm{i}\sqrt{\frac{\pi}{2}} \cdot \big(\delta(\omega+\omega_0) - \delta(\omega-\omega_0)\big)$	18.3(b,d)
Gaussian function of width σ	$g(x) = \frac{1}{\sigma} \cdot e^{-\frac{x^2}{2\sigma^2}}$ $G(\omega) = e^{-\frac{\sigma^2\omega^2}{2}}$	18.4(a,b)
Rectangular pulse of width $2b$	$g(x) = \Pi_b(x) = \begin{cases} 1 & \lvert x\rvert \leq b \\ 0 & \text{sonst} \end{cases}$ $G(\omega) = \frac{2b\sin(b\omega)}{\sqrt{2\pi}\omega}$	18.4(c,d)

away from the origin. Notice that the spectrum of the cosine function is real-valued, the imaginary part being zero. Of course, the same relation holds for the sine function (Fig. 18.3(b, d)), with the only difference being that the pulses have different polarities and appear in the imaginary part of the spectrum. In this case, the real part of the spectrum $G(\omega)$ is zero.

The *Gaussian function* is particularly interesting because its Fourier spectrum is also a Gaussian function (Fig. 18.4(a, b))! It is one of the few examples where the function type in frequency space is the same as in signal space. With the Gaussian function, it is also clear to see that *stretching* a function in signal space corresponds to *shortening* its spectrum and vice versa.

The Fourier transform of a *rectangular pulse* (Fig. 18.4(c, d)) is the "Sinc" function of type $\sin(x)/x$. With increasing frequencies, this function drops off quite slowly, which shows that the components contained in the original rectangular signal are spread out over a large frequency range. Thus a rectangular pulse function exhibits a very wide spectrum in general.

18.1.6 Important Properties of the Fourier Transform

Symmetry

The Fourier spectrum extends over positive and negative frequencies and could, in principle, be an arbitrary complex-valued function. However, in many situations, the spectrum is symmetric about its origin (see, e.g., [43, p. 178]). In particular, the Fourier transform of a real-valued signal $g(x) \in \mathbb{R}$ is a so-called *Hermite* function with the property

$$G(\omega) = G^*(-\omega), \tag{18.21}$$

where G^* denotes the complex conjugate of G (see also Sec. A.3 in the Appendix).

Fig. 18.3
Fourier transform pairs—cosine and sine functions.

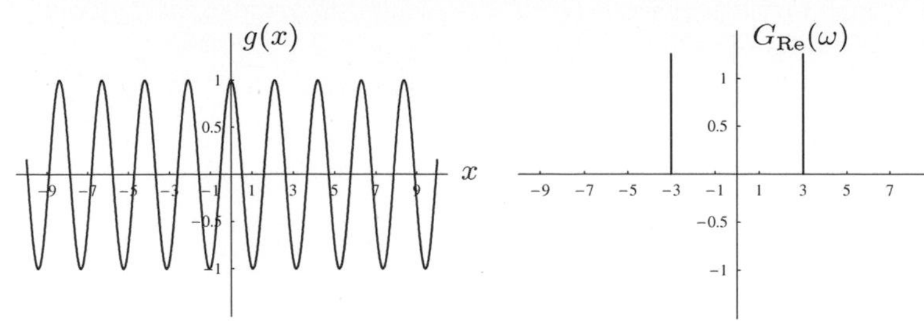

(a) Cosine ($\omega_0 = 3$): $g(x) = \cos(3x)$ ○—● $G(\omega) = \sqrt{\frac{\pi}{2}} \cdot \big(\delta(\omega+3) + \delta(\omega-3)\big)$

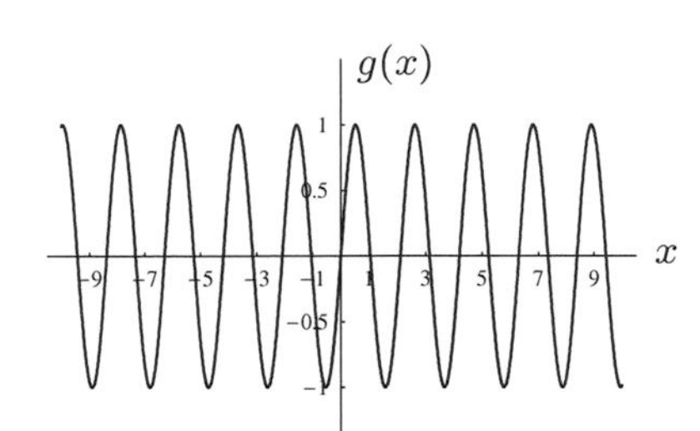

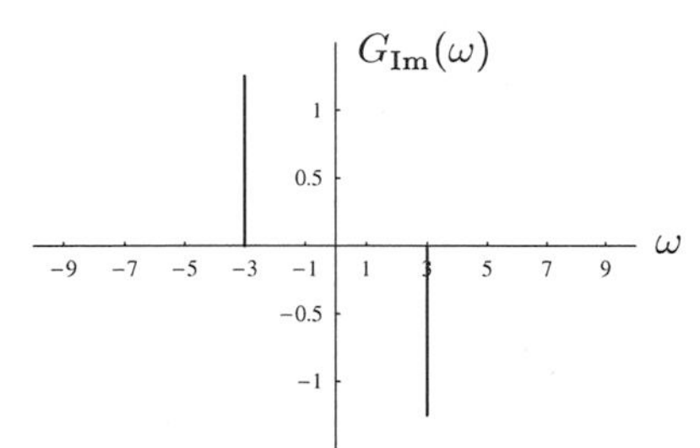

(b) Sine ($\omega_0 = 3$): $g(x) = \sin(3x)$ ○—● $G(\omega) = \mathrm{i}\sqrt{\frac{\pi}{2}} \cdot \big(\delta(\omega+3) - \delta(\omega-3)\big)$

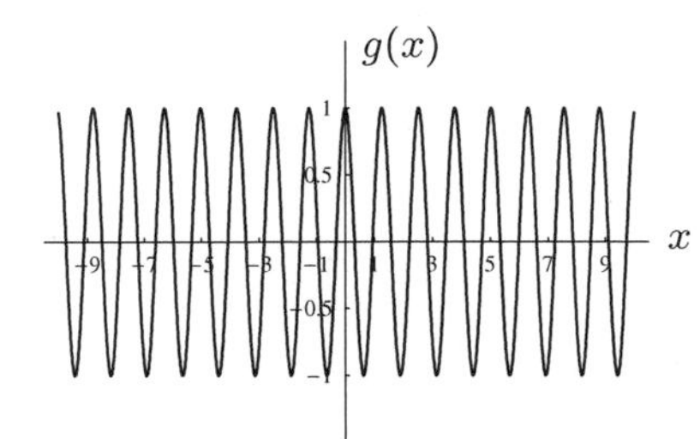

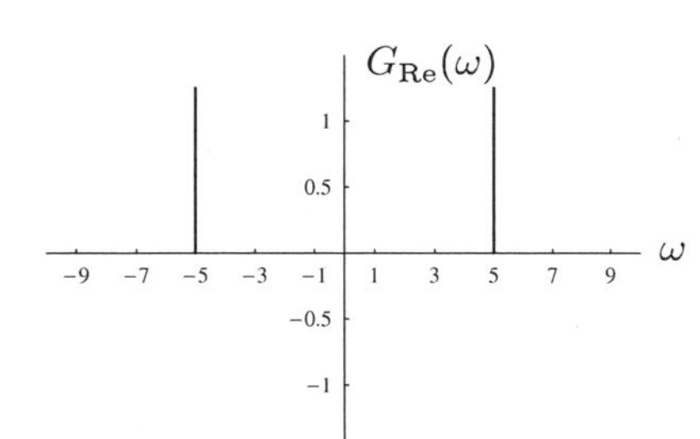

(c) Cosine ($\omega_0 = 5$): $g(x) = \cos(5x)$ ○—● $G(\omega) = \sqrt{\frac{\pi}{2}} \cdot \big(\delta(\omega+5) + \delta(\omega-5)\big)$

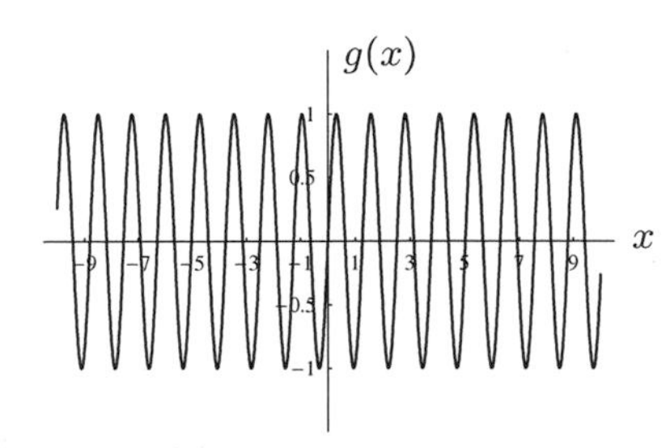

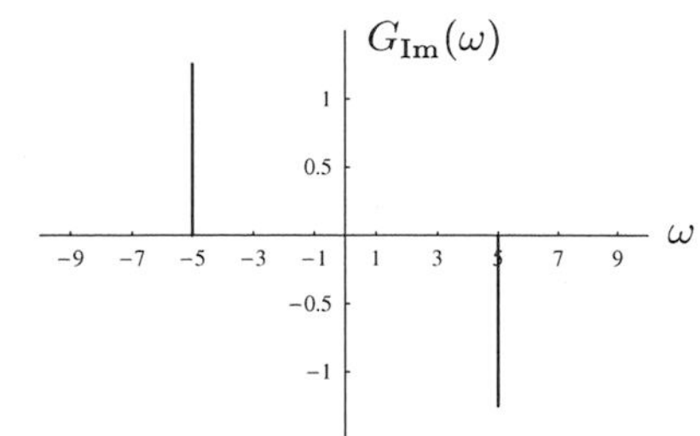

(d) Sine ($\omega_0 = 5$): $g(x) = \sin(5x)$ ○—● $G(\omega) = \mathrm{i}\sqrt{\frac{\pi}{2}} \cdot \big(\delta(\omega+5) - \delta(\omega-5)\big)$

Fig. 18.4
Fourier transform pairs—Gaussian functions and square pulses.

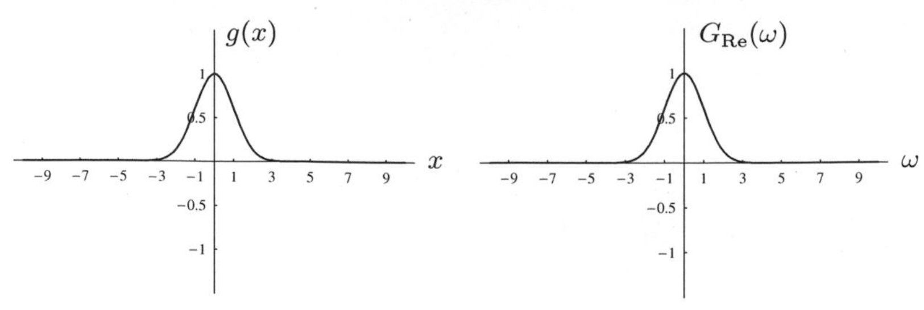

(a) Gauss. $(\sigma = 1)$: $g(x) = e^{-\frac{x^2}{2}}$ $\circ\!\!-\!\!\bullet$ $G(\omega) = e^{-\frac{\omega^2}{2}}$

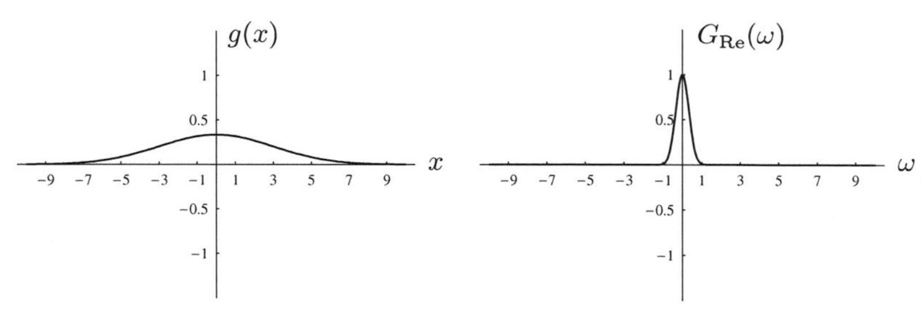

(b) Gauss. $(\sigma = 3)$: $g(x) = \frac{1}{3} \cdot e^{-\frac{x^2}{2 \cdot 9}}$ $\circ\!\!-\!\!\bullet$ $G(\omega) = e^{-\frac{9\omega^2}{2}}$

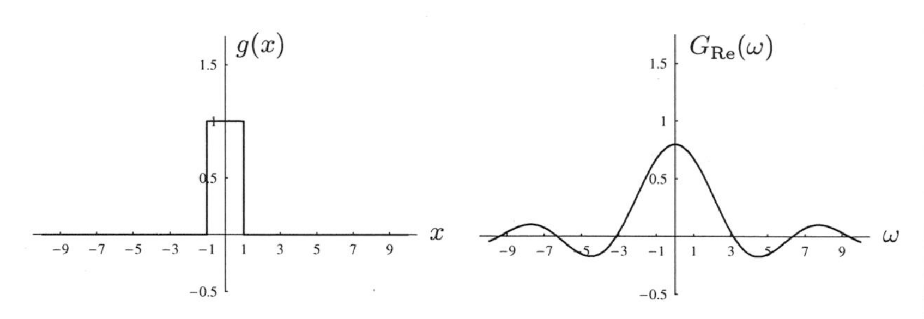

(c) Pulse $(b = 1)$: $g(x) = \Pi_1(x)$ $\circ\!\!-\!\!\bullet$ $G(\omega) = \frac{2\sin(\omega)}{\sqrt{2\pi}\omega}$

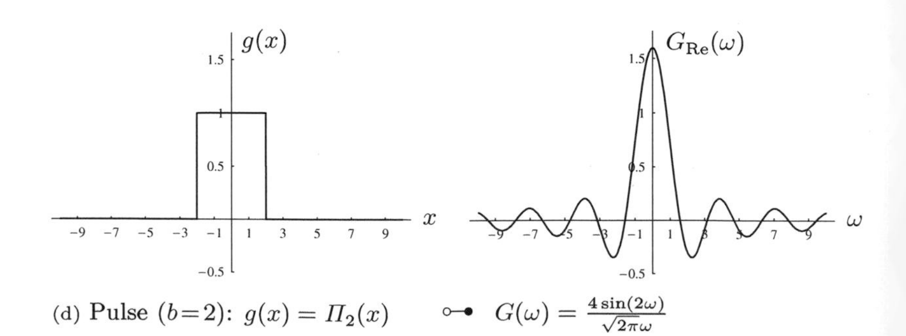

(d) Pulse $(b = 2)$: $g(x) = \Pi_2(x)$ $\circ\!\!-\!\!\bullet$ $G(\omega) = \frac{4\sin(2\omega)}{\sqrt{2\pi}\omega}$

Linearity

The Fourier transform is also a *linear* operation such that multiplying the signal by a constant value $c \in \mathbb{C}$ scales the corresponding spectrum by the same amount,

$$a \cdot g(x) \multimap a \cdot G(\omega). \tag{18.22}$$

Linearity also means that the transform of the sum of two signals $g(x) = g_1(x) + g_2(x)$ is identical to the sum of their individual transforms $G_1(\omega)$ and $G_2(\omega)$ and thus

$$g_1(x) + g_2(x) \multimap G_1(\omega) + G_2(\omega). \tag{18.23}$$

Similarity

If the original function $g(x)$ is scaled in space or time, the opposite effect appears in the corresponding Fourier spectrum. In particular, as observed on the Gaussian function in Fig. 18.4, *stretching* a signal by a factor s (i.e., $g(x) \rightarrow g(sx)$) leads to a *shortening* of the Fourier spectrum:

$$g(sx) \multimap \tfrac{1}{|s|} \cdot G\left(\tfrac{\omega}{s}\right). \tag{18.24}$$

Similarly, the signal is shortened if the corresponding spectrum is stretched.

Shift property

If the original function $g(x)$ is shifted by a distance d along its coordinate axis (i.e., $g(x) \rightarrow g(x-d)$), then the Fourier spectrum multiplies by the complex value $e^{-\mathrm{i}\omega d}$ dependent on ω:

$$g(x-d) \multimap e^{-\mathrm{i}\omega d} \cdot G(\omega). \tag{18.25}$$

Since $e^{-\mathrm{i}\omega d}$ lies on the unit circle, the multiplication causes a phase shift on the spectral values (i.e., a redistribution between the real and imaginary components) without altering the magnitude $|G(\omega)|$. Obviously, the amount (angle) of phase shift (ωd) is proportional to the angular frequency ω.

Convolution property

From the image-processing point of view, the most interesting property of the Fourier transform is its relation to linear convolution (see Ch. 5, Sec. 5.3.1). Let us assume that we have two functions $g(x)$ and $h(x)$ and their corresponding Fourier spectra $G(\omega)$ and $H(\omega)$, respectively. If the original functions are subject to linear convolution (i.e., $g(x) * h(x)$), then the Fourier transform of the result equals the (pointwise) product of the individual Fourier transforms $G(\omega)$ and $H(\omega)$:

$$g(x) * h(x) \multimap G(\omega) \cdot H(\omega). \tag{18.26}$$

Due to the duality of signal space and frequency space, the same also holds in the opposite direction; i.e., a pointwise multiplication of two signals is equivalent to convolving the corresponding spectra:

$$g(x) \cdot h(x) \multimap G(\omega) * H(\omega). \tag{18.27}$$

A multiplication of the functions in *one* space (signal or frequency space) thus corresponds to a linear convolution of the Fourier spectra in the *opposite* space.

18.2 Working with Discrete Signals

The definition of the continuous Fourier transform in Sec. 18.1 is of little use for numerical computation on a computer. Neither can arbitrary continuous (and possibly infinite) functions be represented in practice. Nor can the required integrals be computed. In reality, we must always deal with *discrete* signals, and we therefore need a new version of the Fourier transform that treats signals and spectra as finite data vectors—the "discrete" Fourier transform. Before continuing with this issue we want to use our existing wisdom to take a closer look at the process of discretizing signals in general.

18.2.1 Sampling

We first consider the question of how a continuous function can be converted to a discrete signal in the first place. This process is usually called "sampling" (i.e., taking samples of the continuous function at certain points in time (or in space), usually spaced at regular distances). To describe this step in a simple but formal way, we require an inconspicuous but nevertheless important piece from the mathematician's toolbox.

The impulse function $\delta(x)$

We casually encountered the impulse function (also called the *delta* or *Dirac* function) earlier when we looked at the impulse response of linear filters (see Ch. 5, Sec. 5.3.4) and in the Fourier transforms of the cosine and sine functions (Fig. 18.3). This function, which models a continuous "ideal" impulse, is unusual in several respects: its value is zero everywhere except at the origin, where it is nonzero (though undefined), but its integral is one, that is,

$$\delta(x) = 0 \ \text{ for } x \neq 0 \qquad \text{and} \qquad \int_{-\infty}^{\infty} \delta(x) \, \mathrm{d}x = 1 \, . \tag{18.28}$$

One could imagine $\delta(x)$ as a single pulse at position $x = 0$ that is infinitesimally narrow but still contains finite energy (1). Also remarkable is the impulse function's behavior under scaling along the time (or space) axis (i.e., $\delta(x) \rightarrow \delta(sx)$), with

$$\delta(sx) = \frac{1}{|s|} \cdot \delta(x), \tag{18.29}$$

for $s \neq 0$. Despite the fact that $\delta(x)$ does not exist in physical reality and cannot be plotted (the corresponding plots in Fig. 18.3 are for illustration only), this function is a useful mathematical tool for describing the sampling process, as will be shown.

Sampling with the impulse function

Using the concept of the ideal impulse, the sampling process can be described in a straightforward and intuitive way.[7] If a continuous

[7] The following description is intentionally a bit superficial (in a mathematical sense). See, for example, [43, 128] for more precise coverage of these topics.

function $g(x)$ is multiplied with the impulse function $\delta(x)$, we obtain a new function

$$\bar{g}(x) = g(x) \cdot \delta(x) = \begin{cases} g(0) & \text{for } x = 0, \\ 0 & \text{otherwise.} \end{cases} \tag{18.30}$$

The resulting function $\bar{g}(x)$ consists of a single pulse at position 0 whose height corresponds to the original function value $g(0)$ (at position 0). Thus, by multiplying the function $g(x)$ by the impulse function, we obtain a single discrete sample value of $g(x)$ at position $x = 0$. If the impulse function $\delta(x)$ is shifted by a distance x_0, we can sample $g(x)$ at an arbitrary position $x = x_0$,

$$\bar{g}(x) = g(x) \cdot \delta(x-x_0) = \begin{cases} g(x_0) & \text{for } x = x_0, \\ 0 & \text{otherwise.} \end{cases} \tag{18.31}$$

Here $\delta(x-x_0)$ is the impulse function shifted by x_0, and the resulting function $\bar{g}(x)$ is zero except at position x_0, where it contains the original function value $g(x_0)$. This relationship is illustrated in Fig. 18.5 for the sampling position $x_0 = 3$.

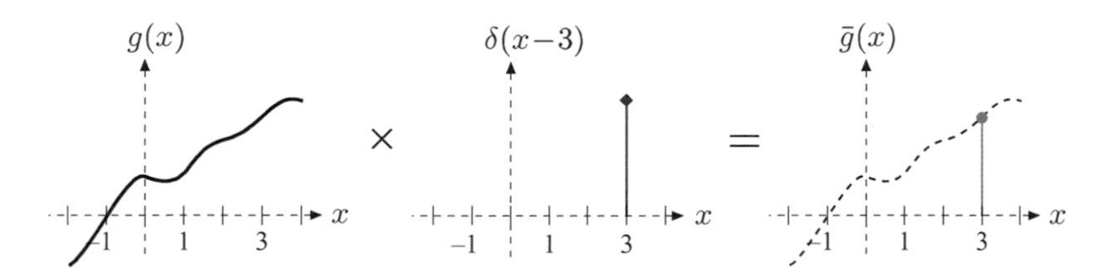

Fig. 18.5
Sampling with the impulse function. The continuous signal $g(x)$ is sampled at position $x_0 = 3$ by multiplying $g(x)$ by a shifted impulse function $\delta(x-3)$.

To sample the function $g(x)$ at more than one position simultaneously (e.g., at positions x_1 and x_2), we use two separately shifted versions of the impulse function, multiply $g(x)$ by both of them, and simply add the resulting function values. In this particular case, we get

$$\begin{aligned} \bar{g}(x) &= g(x) \cdot \delta(x-x_1) + g(x) \cdot \delta(x-x_2) && (18.32) \\ &= g(x) \cdot \big[\delta(x-x_1) + \delta(x-x_2)\big] && (18.33) \\ &= \begin{cases} g(x_1) & \text{for } x = x_1, \\ g(x_2) & \text{for } x = x_2, \\ 0 & \text{otherwise.} \end{cases} && (18.34) \end{aligned}$$

From Eqn. (18.33), sampling a continuous function $g(x)$ at N positions $x_i = 1, 2, \ldots, N$ can thus be described as the sum of the N individual samples, that is,

$$\begin{aligned} \bar{g}(x) &= g(x) \cdot \big[\delta(x-1) + \delta(x-2) + \ldots + \delta(x-N)\big] \\ &= g(x) \cdot \sum_{i=1}^{N} \delta(x-i) \, . \end{aligned} \tag{18.35}$$

The comb function

The sum of shifted impulses $\sum_{i=1}^{N} \delta(x-i)$ in Eqn. (18.35) is called a *pulse sequence* or *pulse train*. Extending this sequence to infinity in both directions, we obtain the "comb" or "Shah" function

$$\text{III}(x) = \sum_{i=-\infty}^{\infty} \delta(x-i)\,. \tag{18.36}$$

The process of discretizing a continuous function by taking samples at regular integral intervals can thus be written simply as

$$\bar{g}(x) = g(x) \cdot \text{III}(x), \tag{18.37}$$

that is, as a pointwise multiplication of the original signal $g(x)$ with the comb function $\text{III}(x)$. As Fig. 18.6 illustrates, the function values of $g(x)$ at integral positions $x_i \in \mathbb{Z}$ are transferred to the discrete function $\bar{g}(x_i)$ and ignored at all non-integer positions.

Of course, the sampling interval (i.e., the distance between adjacent samples) is not restricted to 1. To take samples at regular but *arbitrary* intervals τ, the sampling function $\text{III}(x)$ is simply scaled along the time or space axis; that is,

$$\bar{g}(x) = g(x) \cdot \text{III}\left(\tfrac{x}{\tau}\right), \quad \text{for } \tau > 0. \tag{18.38}$$

Effects of sampling in frequency space

Despite the elegant formulation made possible by the use of the comb function, one may still wonder why all this math is necessary to describe a process that appears intuitively to be so simple anyway. The Fourier spectrum gives one answer to this question. Sampling a continuous function has massive—though predictable—effects upon the frequency spectrum of the resulting (discrete) signal. Using the comb function as a formal model for the sampling process makes it relatively easy to estimate and interpret those spectral effects. Similar to the Gaussian (see Sec. 18.1.5), the comb function features the special property that its Fourier transform

$$\text{III}(x) \;\circ\!\!-\!\!\bullet\; \text{III}\left(\tfrac{1}{2\pi}\omega\right) \tag{18.39}$$

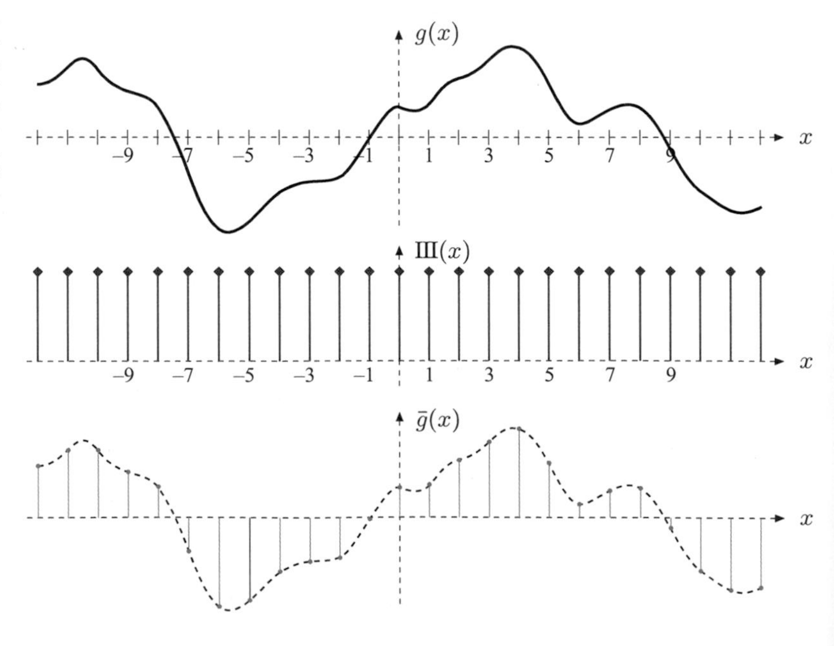

Fig. 18.6
Sampling with the comb function. The original continuous signal $g(x)$ is multiplied by the comb function $\text{III}(x)$. The function value $g(x)$ is transferred to the resulting function $\bar{g}(x)$ only at integral positions $x = x_i \in \mathbb{Z}$ and ignored at all non-integer positions.

is again a comb function (i.e., the same type of function). In general, the Fourier transform of a comb function scaled to an arbitrary sampling interval τ is

$$\text{Ш}(\tfrac{x}{\tau}) \;\circ\!\!-\!\!\bullet\; \tau\text{Ш}\left(\tfrac{\tau}{2\pi}\omega\right), \tag{18.40}$$

due to the similarity property of the Fourier transform (Eqn. (18.24)). Figure 18.7 shows two examples of the comb function $\text{Ш}_\tau(x)$ with sampling intervals $\tau = 1$ and $\tau = 3$ and the corresponding Fourier transforms.

Now, what happens to the Fourier spectrum during discretization, that is, when we multiply a function in signal space by the comb function $\text{Ш}(\frac{x}{\tau})$? We get the answer by recalling the convolution property of the Fourier transform (Eqn. (18.26)): the product of two functions in one space (signal or frequency space) corresponds to the linear convolution of the transformed functions in the opposite space, and thus

$$g(x) \cdot \text{Ш}(\tfrac{x}{\tau}) \;\circ\!\!-\!\!\bullet\; G(\omega) * \tau \cdot \text{Ш}\left(\tfrac{\tau}{2\pi}\omega\right). \tag{18.41}$$

We already know that the Fourier spectrum of the sampling function is a comb function again and therefore consists of a sequence of regularly spaced pulses (Fig. 18.7). In addition, we know that convolving an arbitrary function with the impulse $\delta(x)$ returns the original function; that is, $f(x) * \delta(x) = f(x)$ (see Ch. 5, Sec. 5.3.4). Convolving with a *shifted* pulse $\delta(x-d)$ also reproduces the original function $f(x)$, though shifted by the same distance d:

$$f(x) * \delta(x-d) = f(x-d). \tag{18.42}$$

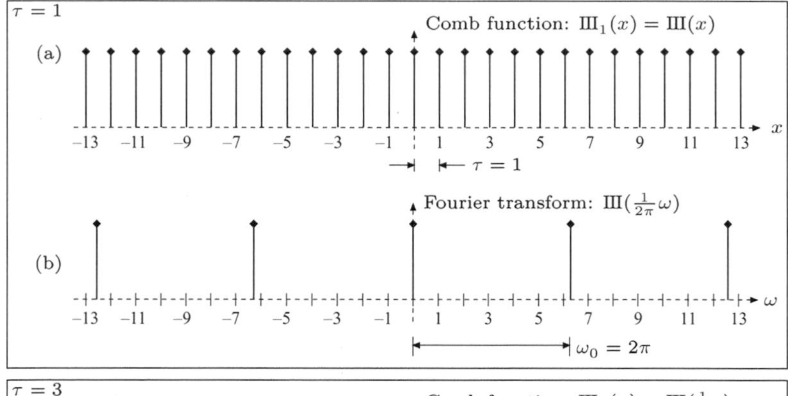

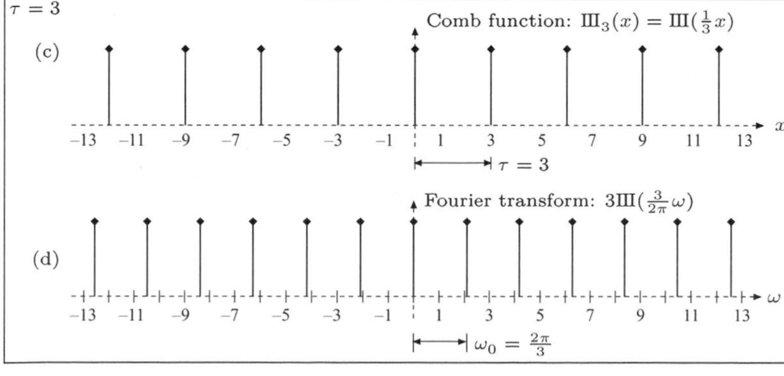

Fig. 18.7
Comb function and its Fourier transform. Comb function $\text{Ш}_\tau(x)$ for the sampling interval $\tau = 1$ (a) and its Fourier transform. Comb function for $\tau = 3$ (c) and its Fourier transform (d). Note that the actual height of the δ-pulses is undefined and shown only for illustration.

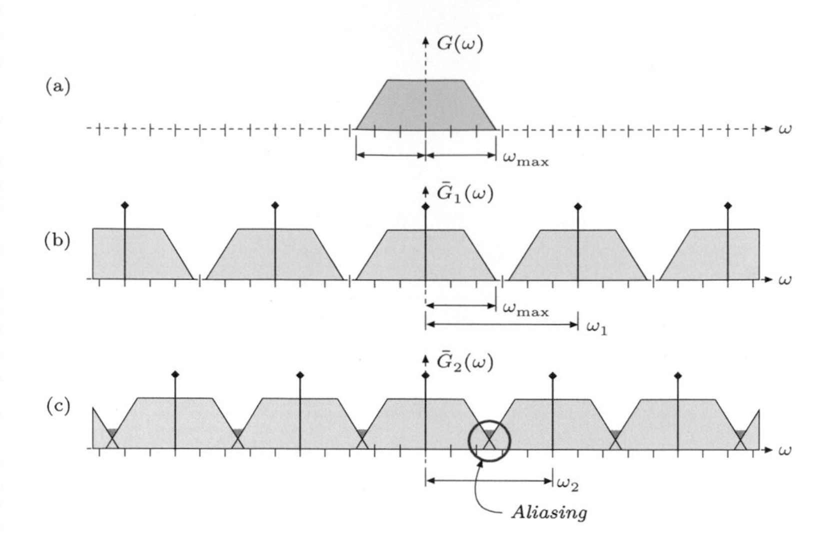

Fig. 18.8
Spectral effects of sampling. The spectrum $G(\omega)$ of the original continuous signal is assumed to be band-limited within the range $\pm\omega_{max}$ (a). Sampling the signal at a rate (sampling frequency) $\omega_s = \omega_1$ causes the signal's spectrum $G(\omega)$ to be replicated at multiples of ω_1 along the frequency (ω) axis (b). Obviously, the replicas in the spectrum do not overlap as long as $\omega_s > 2\omega_{max}$. In (c), the sampling frequency $\omega_s = \omega_2$ is less than $2\omega_{max}$, so there is overlap between the replicas in the spectrum, and frequency components are mirrored at $2\omega_{max}$ and superimpose the original spectrum. This effect is called "aliasing" because the original spectrum (and thus the original signal) cannot be reproduced from such a corrupted spectrum.

As a consequence, the spectrum $G(\omega)$ of the original continuous signal becomes *replicated* in the Fourier spectrum $\bar{G}(\omega)$ of a sampled signal at every pulse of the sampling function's spectrum; that is, infinitely many times (see Fig. 18.8(a, b))! Thus the resulting Fourier spectrum is repetitive with a period $\frac{2\pi}{\tau}$, which corresponds to the sampling frequency ω_s.

Aliasing and the sampling theorem

As long as the spectral replicas in $\bar{G}(\omega)$ created by the sampling process do not overlap, the original spectrum $G(\omega)$—and thus the original continuous function—can be reconstructed without loss from any isolated replica of $G(\omega)$ in the periodic spectrum $\bar{G}(\omega)$. As we can see in Fig. 18.8, this requires that the frequencies contained in the original signal $g(x)$ be within some upper limit ω_{max}; that is, the signal contains no components with frequencies greater than ω_{max}. The maximum allowed signal frequency ω_{max} depends upon the sampling frequency ω_s used to discretize the signal, with the requirement

$$\omega_{max} \leq \tfrac{1}{2} \cdot \omega_s \qquad \text{or} \qquad \omega_s \geq 2 \cdot \omega_{max}. \tag{18.43}$$

Discretizing a continuous signal $g(x)$ with frequency components in the range $0 \leq \omega \leq \omega_{max}$ thus requires a sampling frequency ω_s of at least twice the maximum signal frequency ω_{max}. If this condition is not met, the replicas in the spectrum of the sampled signal overlap (Fig. 18.8(c)) and the spectrum becomes corrupted. Consequently, the original signal cannot be recovered flawlessly from the sampled signal's spectrum. This effect is commonly called "aliasing".

What we just said in simple terms is nothing but the essence of the famous "sampling theorem" formulated by Shannon and Nyquist (see, e.g., [43, p. 256]). It actually states that the sampling frequency must be at least twice the *bandwidth*[8] of the continuous signal to avoid aliasing effects. However, if we assume that a signal's frequency range

[8] This may be surprising at first because it allows a signal with high frequency—but low bandwidth—to be sampled (and correctly recon-

starts at zero, then bandwidth and maximum frequency are the same anyway.

18.2.2 Discrete and Periodic Functions

Assume that we are given a continuous signal $g(x)$ that is periodic with a period of length T. In this case, the corresponding Fourier spectrum $G(\omega)$ is a sequence of thin spectral lines equally spaced at a distance $\omega_0 = 2\pi/T$. As discussed in Sec. 18.1.2, the Fourier spectrum of a periodic function can be represented as a Fourier series and is therefore *discrete*. Conversely, if a continuous signal $g(x)$ is *sampled* at regular intervals τ, then the corresponding Fourier spectrum becomes *periodic* with a period of length $\omega_s = 2\pi/\tau$.

Sampling in signal space thus leads to periodicity in frequency space and vice versa. Figure 18.9 illustrates this relationship and the transition from a continuous nonperiodic signal to a discrete periodic function, which can be represented as a finite vector of numbers and thus easily processed on a computer.

Thus, in general, the Fourier spectrum of a continuous, nonperiodic signal $g(x)$ is also continuous and nonperiodic (Fig. 18.9(a, b)). However, if the signal $g(x)$ is *periodic*, then the corresponding spectrum is *discrete* (Fig. 18.9(c,d)). Conversely, a discrete—but not necessarily periodic—signal leads to a periodic spectrum (Fig. 18.9(e, f)). Finally, if a signal is discrete *and* periodic with M samples per period, then its spectrum is also discrete and periodic with M values (Fig. 18.9(g, h)). Note that the particular signals and spectra in Fig. 18.9 were chosen for illustration only and do not really correspond with each other.

18.3 The Discrete Fourier Transform (DFT)

In the case of a discrete periodic signal, only a finite sequence of M sample values is required to completely represent either the signal $g(u)$ itself or its Fourier spectrum $G(m)$.[9] This representation as finite vectors makes it straightforward to store and process signals and spectra on a computer. What we still need is a version of the Fourier transform applicable to discrete signals.

18.3.1 Definition of the DFT

The discrete Fourier transform is, just like its continuous counterpart, identical in both directions. For a discrete signal $g(u)$ of length M ($u = 0 \ldots M-1$), the forward transform (DFT) is defined as

structed) at a relatively low sampling frequency, even well below the maximum signal frequency. This is possible because one can also use a filter with suitably low bandwidth for reconstructing the original signal. For example, it may be sufficient to strike (i.e., "sample") a church bell (a low-bandwidth oscillatory system with small internal damping) to uniquely generate a sound wave of relatively high frequency.

[9] Notation: We use $g(x)$, $G(\omega)$ for a *continuous* signal or spectrum, respectively, and $g(u)$, $G(m)$ for the *discrete* versions.

Fig. 18.9
Transition from continuous to discrete periodic functions (illustration only).

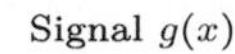

Spectrum $G(\omega)$

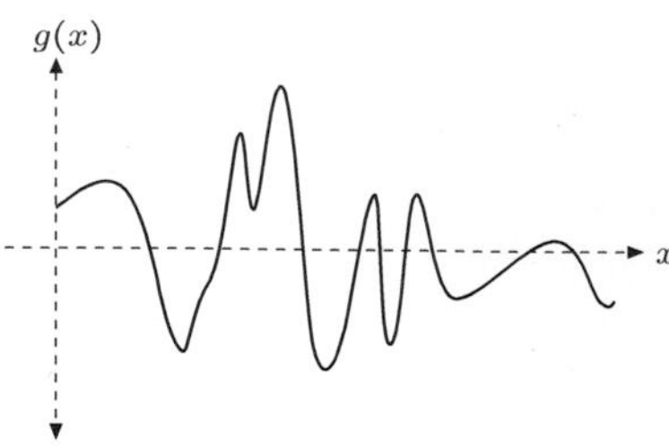

(a) Continuous nonperiodic signal

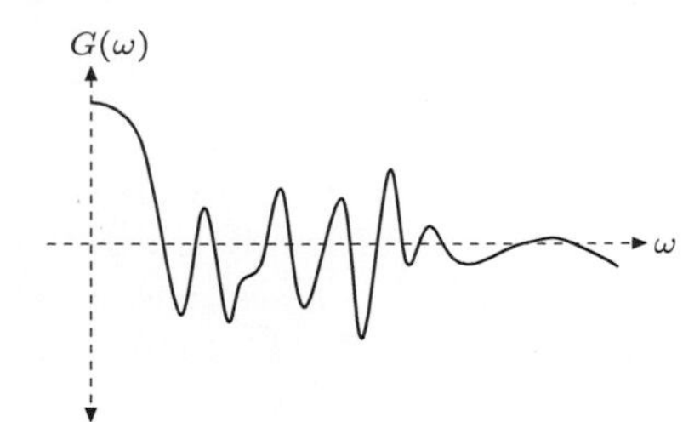

(b) Continuous nonperiodic spectrum

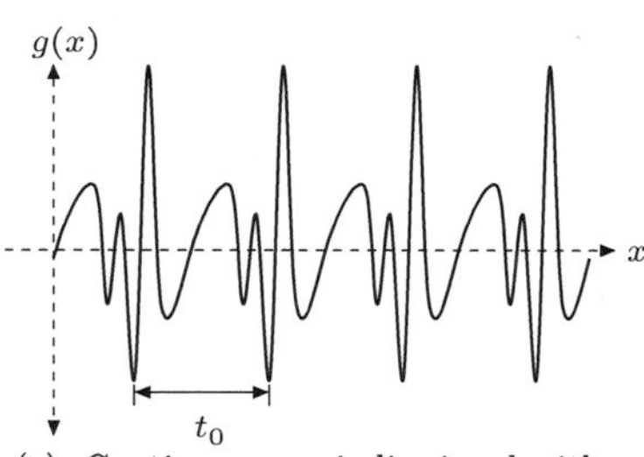

(c) Continuous periodic signal with period t_0

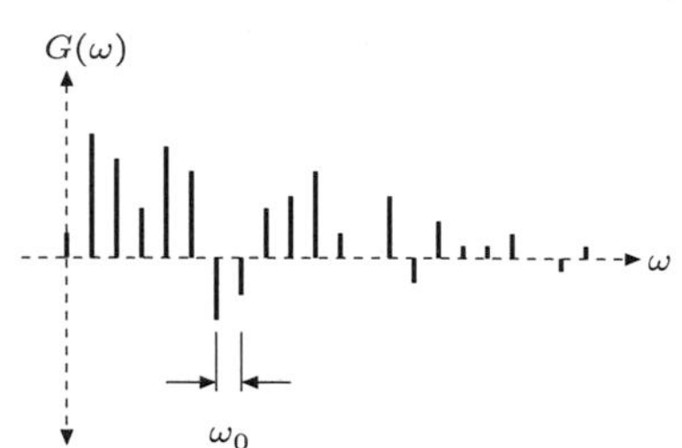

(d) Discrete nonperiodic spectrum with values spaced at $\omega_0 = 2\pi/t_0$

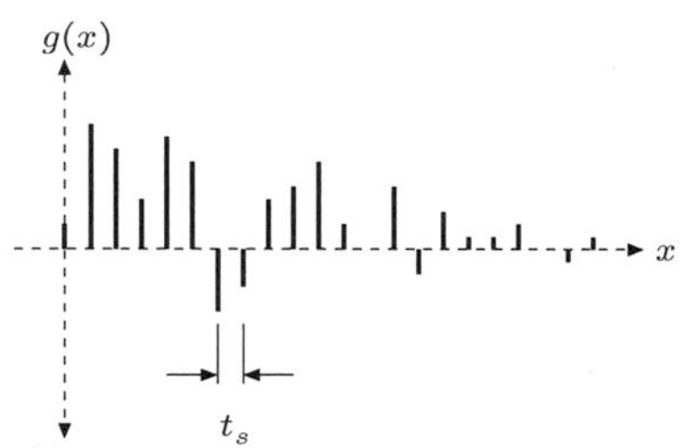

(e) Discrete nonperiodic signal with samples spaced at t_s

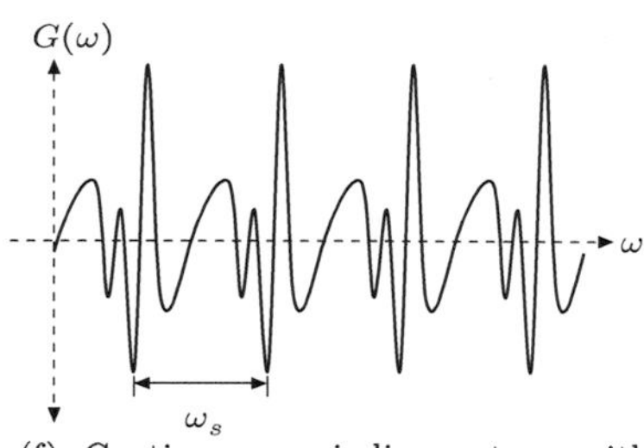

(f) Continuous periodic spectrum with period $\omega_s = 2\pi/t_s$

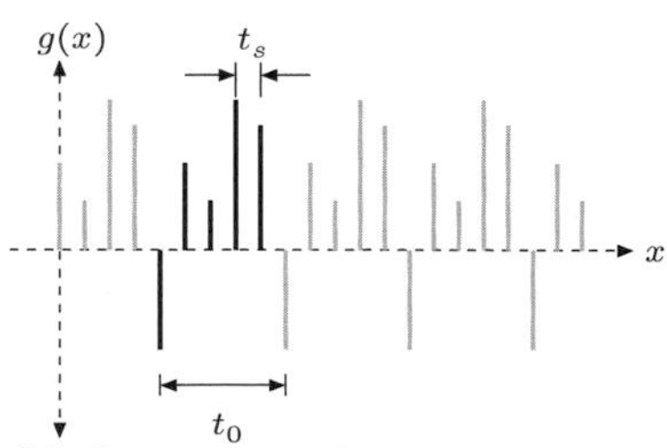

(g) Discrete periodic signal with samples spaced at $t_0 = t_s M$

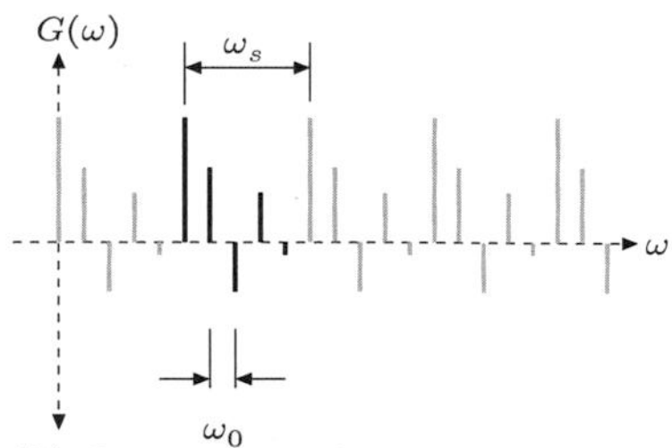

(h) Discrete periodic spectrum with values spaced at $\omega_0 = 2\pi/t_0$ and period $\omega_s = 2\pi/t_s = \omega_0 M$

$$G(m) = \frac{1}{\sqrt{M}} \sum_{u=0}^{M-1} g(u) \cdot \Bigl[\cos\Bigl(2\pi\frac{mu}{M}\Bigr) - \mathrm{i}\cdot\sin\Bigl(2\pi\frac{mu}{M}\Bigr)\Bigr] \quad (18.44)$$

$$= \frac{1}{\sqrt{M}} \sum_{u=0}^{M-1} g(u) \cdot e^{-\mathrm{i}2\pi\frac{mu}{M}}, \quad (18.45)$$

for $0 \leq m < M$, and the *inverse* transform (DFT^{-1}) is[10]

$$g(u) = \frac{1}{\sqrt{M}} \sum_{m=0}^{M-1} G(m) \cdot \Bigl[\cos\Bigl(2\pi \frac{mu}{M}\Bigr) + \mathrm{i} \cdot \sin\Bigl(2\pi \frac{mu}{M}\Bigr)\Bigr] \quad (18.46)$$

$$= \frac{1}{\sqrt{M}} \sum_{m=0}^{M-1} G(m) \cdot e^{\mathrm{i} 2\pi \frac{mu}{M}}, \quad (18.47)$$

for $0 \leq u < M$. Note that both the *signal* $g(u)$ and the discrete *spectrum* $G(m)$ are complex-valued vectors of length M, that is,

$$\begin{aligned} g(u) &= g_{\mathrm{Re}}(u) + \mathrm{i} \cdot g_{\mathrm{Im}}(u), \\ G(m) &= G_{\mathrm{Re}}(m) + \mathrm{i} \cdot G_{\mathrm{Im}}(m), \end{aligned} \quad (18.48)$$

for $u, m = 0, \ldots, M{-}1$. A numerical example for a DFT with $M = 10$ is shown in Fig. 18.10. Converting Eqn. (18.44) from Euler's exponential notation (Eqn. (18.10)) we obtain the discrete Fourier spectrum in component notation as

$$G(m) = \frac{1}{\sqrt{M}} \cdot \sum_{u=0}^{M-1} \Bigl[\underbrace{g_{\mathrm{Re}}(u) + \mathrm{i} \cdot g_{\mathrm{Im}}(u)}_{g(u)}\Bigr] \cdot \Bigl[\underbrace{\cos\bigl(2\pi \tfrac{mu}{M}\bigr)}_{\boldsymbol{C}_m^M(u)} - \mathrm{i} \cdot \underbrace{\sin\bigl(2\pi \tfrac{mu}{M}\bigr)}_{\boldsymbol{S}_m^M(u)}\Bigr], \quad (18.49)$$

where we denote as $\boldsymbol{C}_m^M$ and $\boldsymbol{S}_m^M$ the discrete (cosine and sine) basis functions, as described in the next section. Applying the usual complex multiplication,[11] we obtain the real and imaginary parts of the discrete Fourier spectrum as

$$G_{\mathrm{Re}}(m) = \frac{1}{\sqrt{M}} \cdot \sum_{u=0}^{M-1} g_{\mathrm{Re}}(u) \cdot \boldsymbol{C}_m^M(u) + g_{\mathrm{Im}}(u) \cdot \boldsymbol{S}_m^M(u), \quad (18.50)$$

$$G_{\mathrm{Im}}(m) = \frac{1}{\sqrt{M}} \cdot \sum_{u=0}^{M-1} g_{\mathrm{Im}}(u) \cdot \boldsymbol{C}_m^M(u) - g_{\mathrm{Re}}(u) \cdot \boldsymbol{S}_m^M(u), \quad (18.51)$$

for $m = 0, \ldots, M{-}1$. Analogously, the *inverse* DFT in Eqn. (18.46) expands to

$$g_{\mathrm{Re}}(u) = \frac{1}{\sqrt{M}} \cdot \sum_{m=0}^{M-1} G_{\mathrm{Re}}(m) \cdot \boldsymbol{C}_u^M(m) - G_{\mathrm{Im}}(m) \cdot \boldsymbol{S}_u^M(m), \quad (18.52)$$

$$g_{\mathrm{Im}}(u) = \frac{1}{\sqrt{M}} \cdot \sum_{m=0}^{M-1} G_{\mathrm{Im}}(m) \cdot \boldsymbol{C}_u^M(m) + G_{\mathrm{Re}}(m) \cdot \boldsymbol{S}_u^M(m), \quad (18.53)$$

for $u = 0, \ldots, M{-}1$.

[10] Compare these definitions with the corresponding expressions for the *continuous* forward and inverse Fourier transforms in Eqns. (18.19) and (18.20), respectively.

[11] See also Sec. A.3 in the Appendix.

u	$g(u)$ Re	$g(u)$ Im		$G(m)$ Re	$G(m)$ Im	m
0	1.0000	0.0000		14.2302	0.0000	0
1	3.0000	0.0000	DFT	−5.6745	−2.9198	1
2	5.0000	0.0000	$\longrightarrow$	*0.0000	*0.0000	2
3	7.0000	0.0000		−0.0176	−0.6893	3
4	9.0000	0.0000		*0.0000	*0.0000	4
5	8.0000	0.0000		0.3162	0.0000	5
6	6.0000	0.0000		*0.0000	*0.0000	6
7	4.0000	0.0000	DFT^{-1}	−0.0176	0.6893	7
8	2.0000	0.0000	$\longleftarrow$	*0.0000	*0.0000	8
9	0.0000	0.0000		−5.6745	2.9198	9

Fig. 18.10
Complex-valued result of the DFT for a signal of length $M = 10$ (example). In the discrete Fourier transform (DFT), both the original signal $g(u)$ and its spectrum $G(m)$ are complex-valued vectors of length M; $*$ indicates values with $|G(m)| < 10^{-15}$.

18.3.2 Discrete Basis Functions

The inverse DFT (Eqn. (18.46)) performs the decomposition of the discrete function $g(u)$ into a finite sum of M discrete cosine and sine functions ($\boldsymbol{C}_m^M, \boldsymbol{S}_m^M$) whose weights (or "amplitudes") are determined by the DFT coefficients in $G(m)$. Each of these 1D basis functions (first used in Eqn. (18.49)),

$$\boldsymbol{C}_m^M(u) = \boldsymbol{C}_u^M(m) = \cos\bigl(2\pi \tfrac{mu}{M}\bigr), \tag{18.54}$$

$$\boldsymbol{S}_m^M(u) = \boldsymbol{S}_u^M(m) = \sin\bigl(2\pi \tfrac{mu}{M}\bigr), \tag{18.55}$$

is periodic with M and has a discrete frequency (wave number) m, which corresponds to the angular frequency

$$\omega_m = 2\,\pi \cdot \frac{m}{M}. \tag{18.56}$$

For example, Figs. 18.11 and 18.12 show the discrete basis functions (with integer ordinate values $u \in \mathbb{Z}$) for the DFT of length $M = 8$ as well as their continuous counterparts (with ordinate values $x \in \mathbb{R}$).

For wave number $m = 0$, the cosine function $\boldsymbol{C}_0^M(u)$ (Eqn. (18.54)) has the constant value 1. The corresponding DFT coefficient $G_{\text{Re}}(0)$—the real part of $G(0)$—thus specifies the constant part of the signal or the average value of the signal $g(u)$ in Eqn. (18.52). In contrast, the zero-frequency sine function $\boldsymbol{S}_0^M(u)$ is zero for any value of u and thus cannot contribute anything to the signal. The corresponding DFT coefficients $G_{\text{Im}}(0)$ in Eqn. (18.52) and $G_{\text{Re}}(0)$ in Eqn. (18.53) are therefore of no relevance. For a real-valued signal (i.e., $g_{\text{Im}}(u) = 0$ for all u), the coefficient $G_{\text{Im}}(0)$ in the corresponding Fourier spectrum must also be zero.

As seen in Fig. 18.11, the wave number $m = 1$ relates to a cosine or sine function that performs exactly one full cycle over the signal length $M = 8$. Similarly, the wave numbers $m = 2, \ldots, 7$ correspond to $2, \ldots, 7$ complete cycles over the signal length M (see Figs. 18.11 and 18.12).

18.3.3 Aliasing Again!

A closer look at Figs. 18.11 and 18.12 reveals an interesting fact: the sampled (discrete) cosine and sine functions for $m = 3$ and $m = 5$ are *identical*, although their continuous counterparts are different! The same is true for the frequency pairs $m = 2, 6$ and $m = 1, 7$. What we

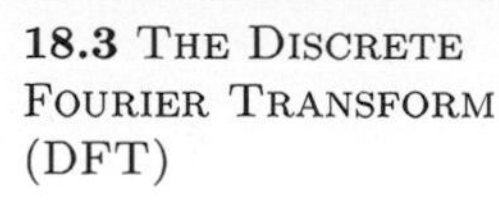

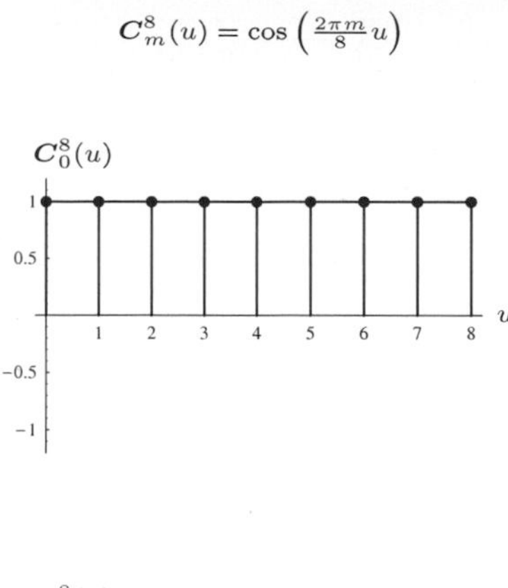

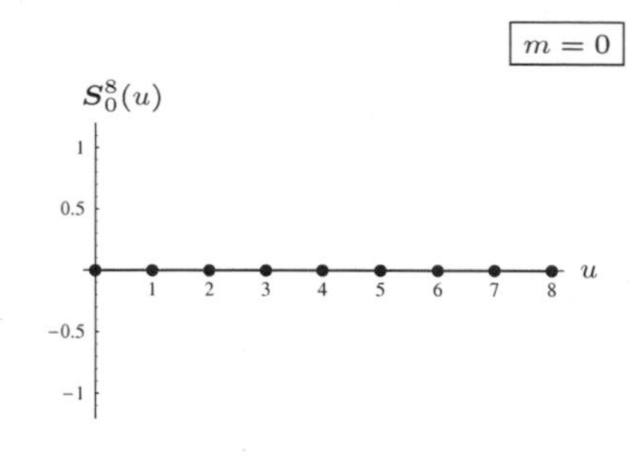

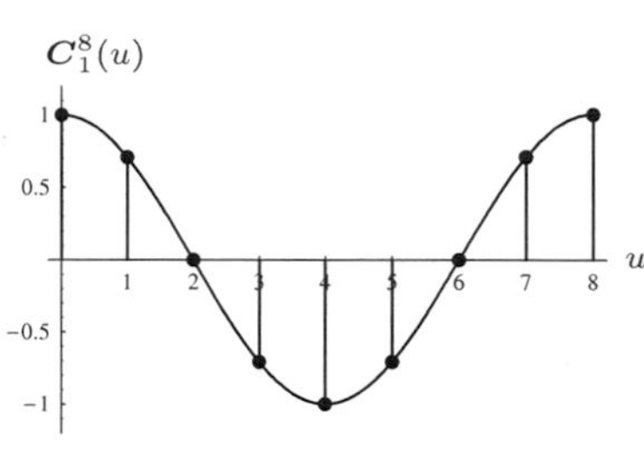

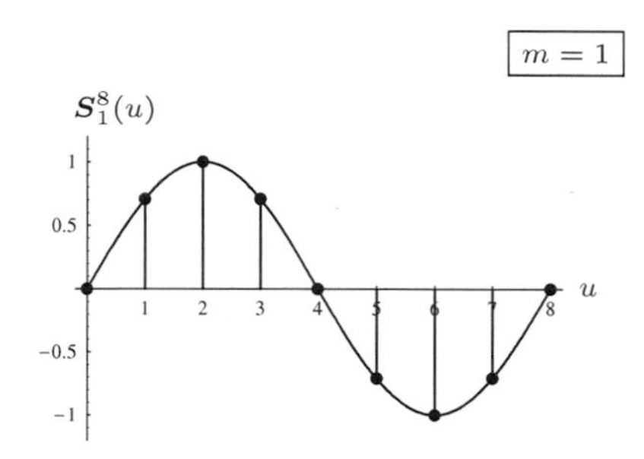

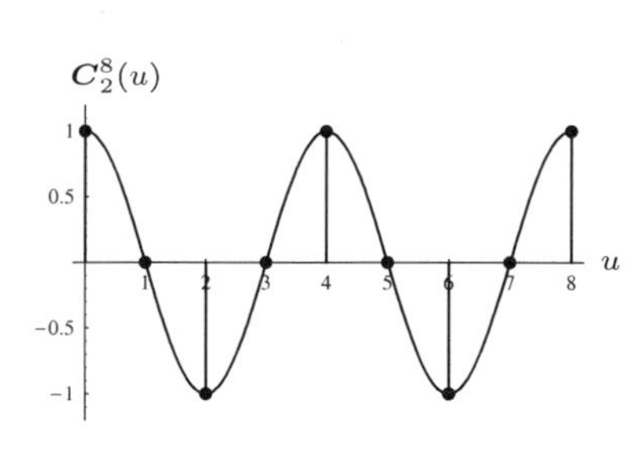

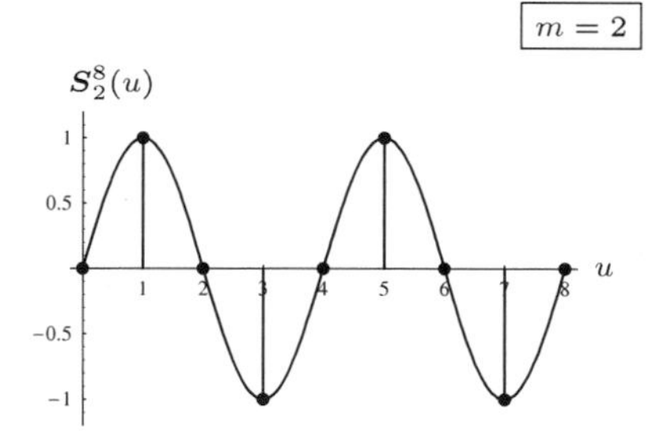

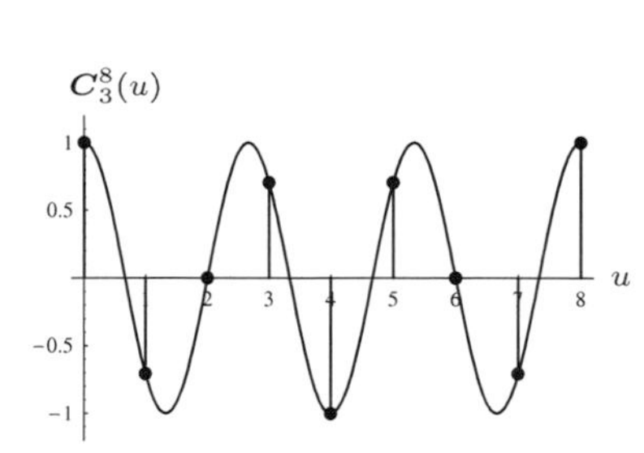

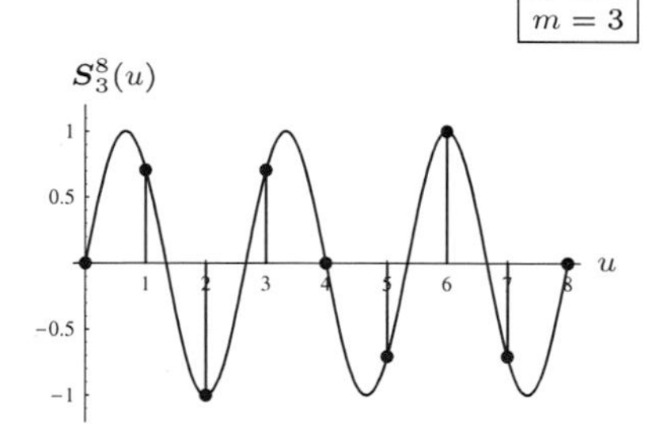

Fig. 18.11
Discrete basis functions $C_m^M(u)$ and $S_m^M(u)$ for the signal length $M = 8$ and wave numbers $m = 0, \ldots, 3$. Each plot shows both the discrete function (round dots) and the corresponding continuous function.

see here is another manifestation of the sampling theorem—which we had originally encountered (Sec. 18.2.1) in frequency space—in *signal space*. Obviously, $m = 4$ is the maximum frequency component that can be represented by a discrete signal of length $M = 8$. Any discrete function with a higher frequency ($m = 5, \ldots, 7$ in this case) has an identical counterpart with a lower wave number and thus cannot be reconstructed from the sampled signal (see also Fig. 18.13)!

If a continuous signal is sampled at a regular distance τ, the corresponding Fourier spectrum is repeated at multiples of $\omega_s = 2\pi/\tau$,

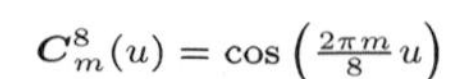

$S_m^8(u) = \sin\left(\frac{2\pi m}{8} u\right)$

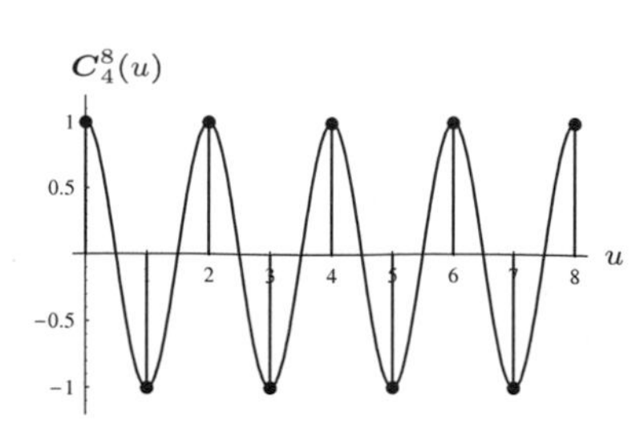

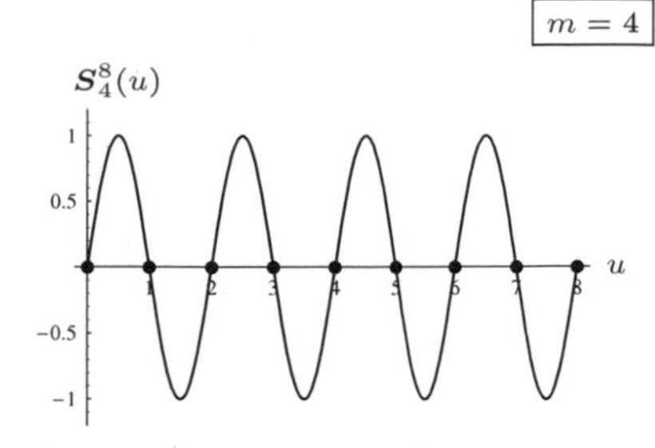

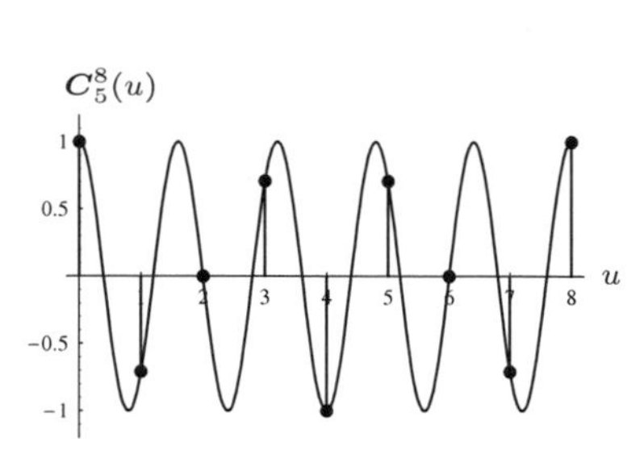

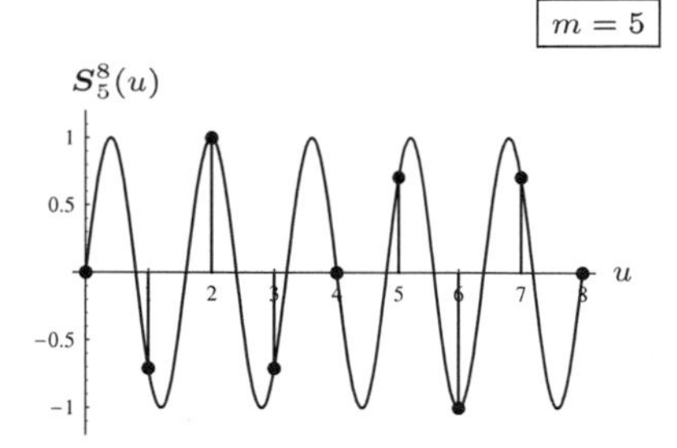

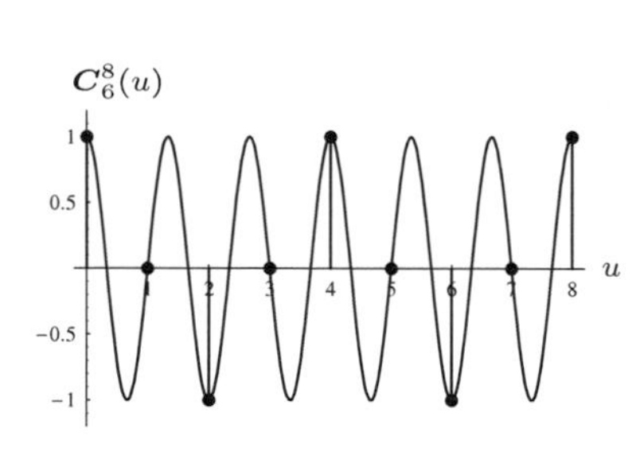

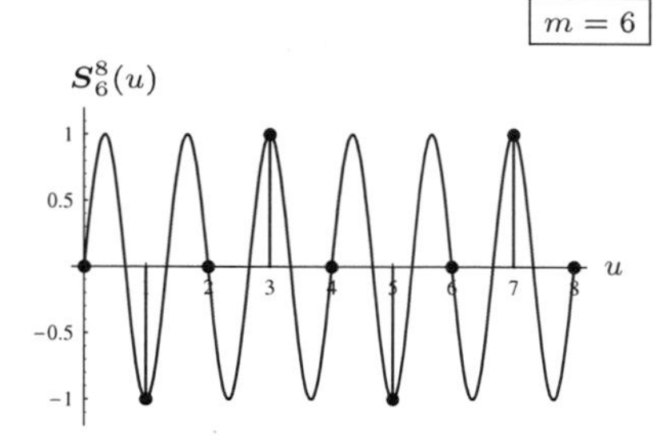

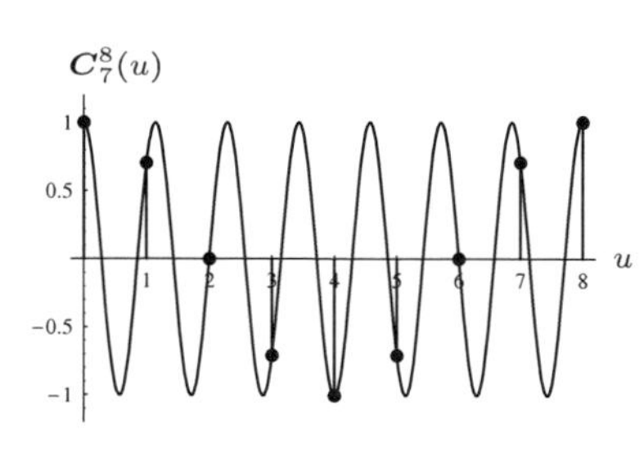

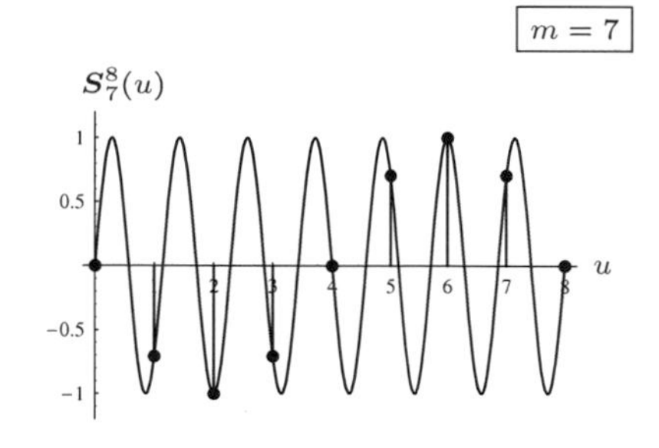

Fig. 18.12
Discrete basis functions (continued). Signal length $M = 8$ and wave numbers $m = 4, \ldots, 7$. Notice that, for example, the discrete functions for $m = 5$ and $m = 3$ (Fig. 18.11) are identical because $m = 4$ is the maximum wave number that can be represented in a discrete spectrum of length $M = 8$.

as we have shown earlier (Fig. 18.8). In the discrete case, the spectrum is periodic with length M. Since the Fourier spectrum of a real-valued signal is symmetric about the origin (Eqn. (18.21)), there is for every coefficient with wave number m an equal-sized duplicate with wave number $-m$. Thus the spectral components appear pairwise and mirrored at multiples of M; that is,

$C^8_m(u) = \cos\left(\frac{2\pi m}{8} u\right)$ $\quad$ $S^8_m(u) = \sin\left(\frac{2\pi m}{8} u\right)$

$m = 1$

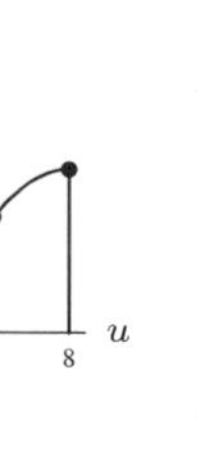

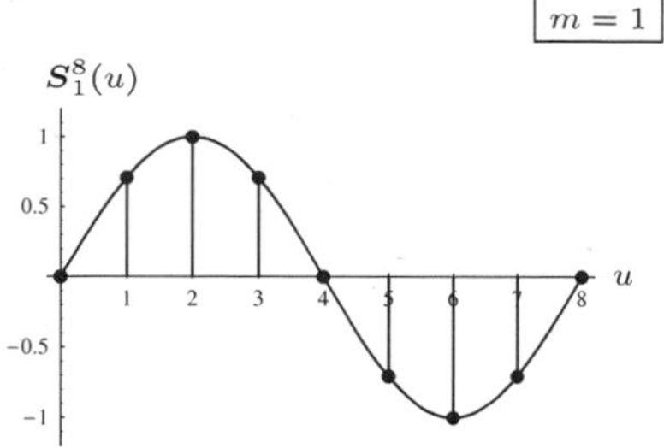

$m = 9$

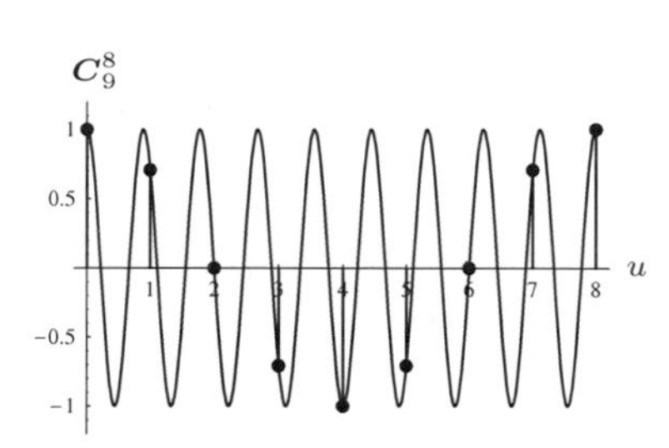

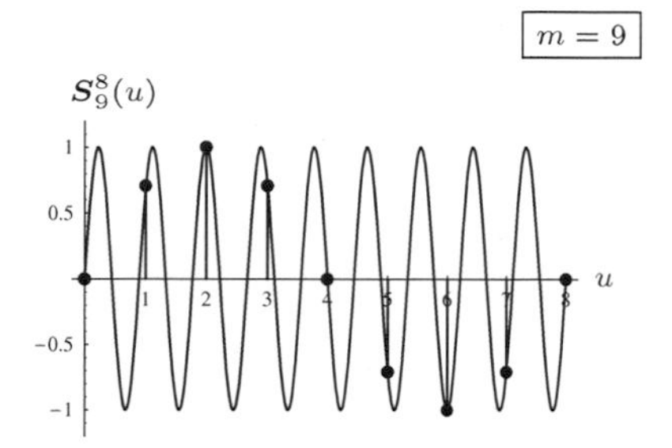

$m = 17$

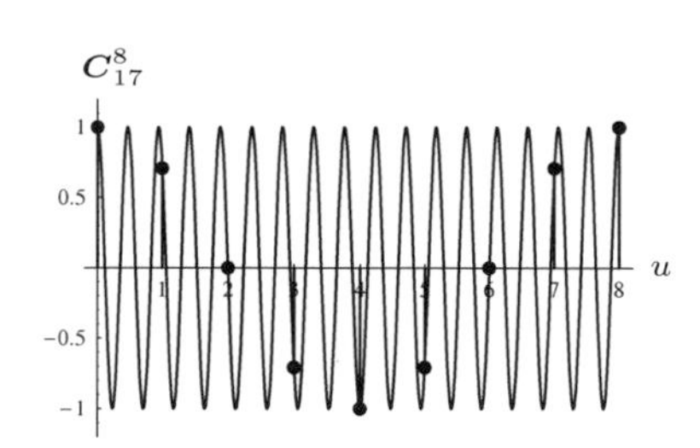

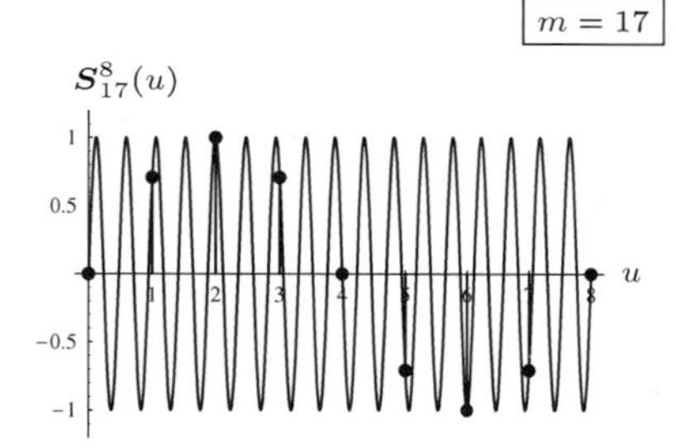

Fig. 18.13
Aliasing in signal space. For the signal length $M = 8$, the discrete cosine and sine basis functions for the wave numbers $m = 1, 9, 17, \ldots$ (round dots) are all identical. The sampling frequency itself corresponds to the wave number $m = 8$.

$$\begin{aligned} |G(m)| &= |G(M-m)| = |G(M+m)| \\ &= |G(2M-m)| = |G(2M+m)| \\ &\ldots \\ &= |G(kM-m)| = |G(kM+m)|, \end{aligned} \tag{18.57}$$

for all $k \in \mathbb{Z}$. If the original continuous signal contains "energy" at the frequencies

$$\omega_m > \omega_{M/2}$$

(i.e., signal components with wave numbers $m > M/2$), then, according to the sampling theorem, the overlapping parts of the spectra are superimposed in the resulting periodic spectrum of the discrete signal.

18.3.4 Units in Signal and Frequency Space

The relation between the units in signal and frequency space and the interpretation of wave numbers m is a common cause of confusion. While the discrete signal and its spectrum are simple numerical vectors and units of measurement are irrelevant for computing the DFT itself, it is nevertheless important to understand how the coordinates in the spectrum relate to physical dimensions in the real world.

Clearly, every complex-valued spectral coefficient $G(m)$ corresponds to one pair of cosine and sine functions with a particular frequency in signal space. Assume a continuous signal is sampled at M consecutive positions spaced at τ (an interval in time or distance in space). The *wave number* $m = 1$ then corresponds to the *fundamental period* of the discrete signal (which is now assumed to be periodic) with a period of length $M\tau$; that is, to the *frequency*

$$f_1 = \frac{1}{M\tau} . \tag{18.58}$$

In general, the wave number m of a discrete spectrum relates to the physical frequency as

$$f_m = m\frac{1}{M\tau} = m \cdot f_1 \tag{18.59}$$

for $0 \leq m < M$, which is equivalent to the angular frequency

$$\omega_m = 2\pi f_m = m\frac{2\pi}{M\tau} = m \cdot \omega_1 . \tag{18.60}$$

Obviously then, the sampling frequency $f_s = 1/\tau = M \cdot f_1$ corresponds to the wave number $m_s = M$. As expected, the maximum nonaliased wave number in the spectrum is

$$m_{\mathrm{max}} = \frac{M}{2} = \frac{m_s}{2} , \tag{18.61}$$

that is, half the sampling frequency index m_s.

Example 1: time-domain signal

We assume for this example that $g(u)$ is a signal in the time domain (e.g., a discrete sound signal) that contains $M = 500$ sample values taken at regular intervals $\tau = 1\text{ ms} = 10^{-3}\,\text{s}$. Thus the sampling frequency is $f_s = 1/\tau = 1000\,\text{Hertz}$ (cycles per second) and the total duration (fundamental period) of the signal is $M\tau = 0.5\,\text{s}$.

The signal is implicitly periodic, and from Eqn. (18.58) we obtain its fundamental frequency as $f_1 = \frac{1}{500 \cdot 10^{-3}} = \frac{1}{0.5} = 2\,\text{Hertz}$. The wave number $m = 2$ in this case corresponds to a real frequency $f_2 = 2f_1 = 4\,\text{Hertz}$, $f_3 = 6\,\text{Hertz}$, etc. The maximum frequency that can be represented by this discrete signal without aliasing is $f_{\mathrm{max}} = \frac{M}{2} f_1 = \frac{1}{2\tau} = 500\,\text{Hertz}$, exactly half the sampling frequency f_s.

Example 2: space-domain signal

Assume we have a 1D print pattern with a resolution (i.e., spatial sampling frequency) of 120 dots per cm, which equals approximately 300 dots per inch (dpi) and a total signal length of $M = 1800$ samples. This corresponds to a spatial sampling interval of $\tau = 1/120\,\text{cm} \approx 83\,\mu\text{m}$ and a physical signal length of $(1800/120)\,\text{cm} = 15\,\text{cm}$.

The fundamental frequency of this signal (again implicitly assumed to be periodic) is $f_1 = \frac{1}{15}$, expressed in cycles per cm. The sampling frequency is $f_s = 120$ cycles per cm and thus the maximum signal frequency is $f_{\mathrm{max}} = \frac{f_s}{2} = 60$ cycles per cm. The maximum signal frequency specifies the finest structure ($\frac{1}{60}$ cm) that can be reproduced by this print raster.

18.3.5 Power Spectrum

The *magnitude* of the complex-valued Fourier spectrum,

$$|G(m)| = \sqrt{G^2_{\mathrm{Re}}(m) + G^2_{\mathrm{Im}}(m)}, \tag{18.62}$$

is commonly called the "power spectrum" of a signal. It specifies the energy that individual frequency components in the spectrum contribute to the signal. The power spectrum is real-valued and positive and thus often used for graphically displaying the results of Fourier transforms (see also Ch. 19, Sec. 19.2).

Since all phase information is lost in the power spectrum, the original signal cannot be reconstructed from the power spectrum alone. However, because of the missing phase information, the power spectrum is insensitive to shifts of the original signal and can thus be efficiently used for comparing signals. To be more precise, the power spectrum of a circularly shifted signal is identical to the power spectrum of the original signal. Thus, given a discrete periodic signal $g_1(u)$ of length M and a second signal $g_2(u)$ shifted by some offset d, such that

$$g_2(u) = g_1(u-d) \tag{18.63}$$

the corresponding power spectra are the same, that is,

$$|G_2(m)| = |G_1(m)|, \tag{18.64}$$

although in general the complex-valued spectra $G_1(m)$ and $G_2(m)$ are different. Furthermore, from the symmetry property of the Fourier spectrum, it follows that

$$|G(m)| = |G(-m)|, \tag{18.65}$$

for real-valued signals $g(u) \in \mathbb{R}$.

18.4 Implementing the DFT

18.4.1 Direct Implementation

Based on the definitions in Eqns. (18.50) and (18.51) the DFT can be directly implemented, as shown in Prog. 18.1. The main method `DFT()` transforms a signal vector of arbitrary length M (not necessarily a power of 2). It requires roughly M^2 operations (multiplications and additions); that is, the time complexity of this DFT algorithm is $\mathcal{O}(M^2)$.

One way to improve the efficiency of the DFT algorithm is to use lookup tables for the sin and cos functions (which are relatively "expensive" to compute) since only function values for a set of M different angles φ_m are ever needed. The angles $\varphi_m = 2\pi\frac{m}{M}$ corresponding to $m = 0, \ldots, M-1$ are evenly distributed over the full 360° circle. Any integral multiple $\varphi_m \cdot u$ (for $u \in \mathbb{Z}$) can only fall onto one of these angles again because

Prog. 18.1
Direct implementation of the DFT based on the definition in Eqns. (18.50) and (18.51). The method `DFT()` returns a complex-valued vector with the same length as the complex-valued input (signal) vector `g`. This method implements both the forward and the inverse transforms, controlled by the Boolean parameter `forward`. The class `Complex` (bottom) defines the structure of the complex-valued vector elements.

```
1 class Complex {
2     double re, im;
3     Complex(double re, double im) { //constructor method
4         this.re = re;
5         this.im = im;
6     }
7 }
```

```
8 Complex[] DFT(Complex[] g, boolean forward) {
9     int M = g.length;
10    double s = 1 / Math.sqrt(M); //common scale factor
11    Complex[] G = new Complex[M];
12    for (int m = 0; m < M; m++) {
13        double sumRe = 0;
14        double sumIm = 0;
15        double phim = 2 * Math.PI * m / M;
16        for (int u = 0; u < M; u++) {
17            double gRe = g[u].re;
18            double gIm = g[u].im;
19            double cosw = Math.cos(phim * u);
20            double sinw = Math.sin(phim * u);
21            if (!forward) // inverse transform
22                sinw = -sinw;
23            // complex multiplication: [gRe+i·gIm]·[cos(ω)+i·sin(ω)]
24            sumRe += gRe * cosw + gIm * sinw;
25            sumIm += gIm * cosw - gRe * sinw;
26        }
27        G[m] = new Complex(s * sumRe, s * sumIm);
28    }
29    return G;
30 }
```

$$\varphi_m \cdot u = 2\pi \tfrac{mu}{M} \equiv \tfrac{2\pi}{M} \cdot (\underbrace{mu \mod M}_{0 \le k < M}) = 2\pi \tfrac{k}{M} = \varphi_k, \quad (18.66)$$

where mod denotes the "modulus" operator.[12] Thus we can set up two constant tables (floating-point arrays) W_C and W_S of size M with the values

$$\mathsf{W}_C(k) \leftarrow \cos(\omega_k) = \cos\left(2\pi \tfrac{k}{M}\right), \quad (18.67)$$

$$\mathsf{W}_S(k) \leftarrow \sin(\omega_k) = \sin\left(2\pi \tfrac{k}{M}\right), \quad (18.68)$$

for $0 \le k < M$. For computing the DFT, the necessary cosine and sine values (Eqn. (18.49)) can be read from these tables as

$$\boldsymbol{C}_k^M(u) = \cos\left(2\pi \tfrac{mu}{M}\right) \equiv \mathsf{W}_C(mu \bmod M), \quad (18.69)$$

$$\boldsymbol{S}_k^M(u) = \sin\left(2\pi \tfrac{mu}{M}\right) \equiv \mathsf{W}_S(mu \bmod M), \quad (18.70)$$

for arbitrary values of $m, u \in \mathbb{Z}$, without any additional computation. The necessary modification of the `DFT()` method in Prog. 18.1 is left as an exercise (Exercise 18.5).

Despite this significant improvement, the direct implementation of the DFT remains computationally intensive. As a matter of fact,

[12] See also Sec. F.1.2 in the Appendix.

it has been impossible for a long time to compute this form of DFT in sufficiently short time on off-the-shelf computers, and this is still true today for many real applications.

18.4.2 Fast Fourier Transform (FFT)

Fortunately, for computing the DFT in practice, fast algorithms exist that lay out the sequence of computations in such a way that intermediate results are only computed once and optimally reused many times. This "fast Fourier transform", which exists in many variations, generally reduces the time complexity of the computation from $\mathcal{O}(M^2)$ to $\mathcal{O}(M \log_2 M)$. The benefits are substantial, in particular for longer signals. For example, with a signal of length $M = 10^3$, the FFT leads to a speedup by a factor of 100 over the direct DFT implementation and an impressive gain of 10,000 times for a signal of length $M = 10^6$. Since its invention, the FFT has therefore become an indispensable tool in almost any application of spectral signal analysis [34].

Most FFT algorithms, including the one described in the famous publication by Cooley and Tukey in 1965 (see [88, p. 156] for a historic overview), are designed for signals of length $M = 2^k$ (i.e., powers of 2). However, FFT algorithms have also been developed for other lengths, including several small prime numbers [25]. Efficient Java implementations are available, for example, as part of the *JTransform* library[13] by Piotr Wendykier [255] or the *Apache Commons Math* libary.[14]

It is important to remember, though, that the DFT and FFT compute exactly the *same* result and the FFT is only a special—though ingenious—method for *implementing* the discrete Fourier transform (Eqn. (18.44)).

18.5 Exercises

Exercise 18.1. Calculate the values of the cosine function $f(x) = \cos(\omega x)$ with angular frequency $\omega = 5$ for the positions $x = -3, -2, \ldots, 2, 3$. What is the length of this function's period?

Exercise 18.2. Determine the phase angle φ of the function $f(x) = A \cdot \cos(\omega x) + B \cdot \sin(\omega x)$ for $A = -1$ and $B = 2$.

Exercise 18.3. Calculate the real part, the imaginary part, and the magnitude of the complex value $z = 1.5 \cdot e^{-\mathrm{i}\,2.5}$.

Exercise 18.4. A 1D optical scanner for sampling film transparencies is supposed to resolve image structures with a precision of 4,000 dpi. What spatial distance (in mm) between samples is required such that no aliasing occurs?

Exercise 18.5. Modify the direct implementation of the 1D DFT given in Prog. 18.1 by using lookup tables for the cos and sin functions as described in Eqns. (18.69)–(18.70).

[13] http://sites.google.com/site/piotrwendykier/software/jtransforms.

[14] http://commons.apache.org/math/ (class `FastFourierTransformer`).

19

The Discrete Fourier Transform in 2D

The Fourier transform is defined not only for 1D signals but for functions of arbitrary dimension. Thus, 2D images are nothing special from a mathematical point of view.

19.1 Definition of the 2D DFT

For a 2D, periodic function (e.g., an intensity image) $g(u, v)$ of size $M \times N$, the discrete Fourier transform (2D DFT) is defined as

$$G(m,n) = \frac{1}{\sqrt{MN}} \cdot \sum_{u=0}^{M-1} \sum_{v=0}^{N-1} g(u,v) \cdot e^{-\mathrm{i}2\pi \frac{mu}{M}} \cdot e^{-\mathrm{i}2\pi \frac{nv}{N}} \tag{19.1}$$

$$= \frac{1}{\sqrt{MN}} \cdot \sum_{u=0}^{M-1} \sum_{v=0}^{N-1} g(u,v) \cdot e^{-\mathrm{i}2\pi (\frac{mu}{M} + \frac{nv}{N})}, \tag{19.2}$$

for the spectral coordinates $m = 0, \ldots, M-1$ and $n = 0, \ldots, N-1$. As we see, the resulting Fourier transform is again a 2D function of the same size ($M \times N$) as the original signal. Similarly, the *inverse* 2D DFT is defined as

$$g(u,v) = \frac{1}{\sqrt{MN}} \cdot \sum_{m=0}^{M-1} \sum_{n=0}^{N-1} G(m,n) \cdot e^{\mathrm{i}2\pi \frac{mu}{M}} \cdot e^{\mathrm{i}2\pi \frac{nv}{N}} \tag{19.3}$$

$$= \frac{1}{\sqrt{MN}} \cdot \sum_{m=0}^{M-1} \sum_{n=0}^{N-1} G(m,n) \cdot e^{\mathrm{i}2\pi (\frac{mu}{M} + \frac{nv}{N})}, \tag{19.4}$$

for the image coordinates $u = 0, \ldots, M-1$ and $v = 0, \ldots, N-1$.

19.1.1 2D Basis Functions

Equation (19.4) shows that a discrete 2D, periodic function $g(u, v)$ can be represented as a linear combination (i.e., as a weighted sum) of 2D sinusoids of the form

$$e^{\mathrm{i}\cdot 2\pi(\frac{mu}{M}+\frac{nv}{N})} = e^{\mathrm{i}\cdot(\omega_m u+\omega_n v)} \tag{19.5}$$

$$= \underbrace{\cos\left[2\pi\left(\frac{mu}{M}+\frac{nv}{N}\right)\right]}_{\boldsymbol{C}^{M,N}_{m,n}(u,v)} + \mathrm{i}\cdot\underbrace{\sin\left[2\pi\left(\frac{mu}{M}+\frac{nv}{N}\right)\right]}_{\boldsymbol{S}^{M,N}_{m,n}(u,v)}. \tag{19.6}$$

$\boldsymbol{C}^{M,N}_{m,n}(u,v)$ and $\boldsymbol{S}^{M,N}_{m,n}(u,v)$ are discrete, 2D cosine and sine functions with horizontal and vertical wave numbers n and m, respectively, and the corresponding angular frequencies ω_m, ω_n, that is,

$$\boldsymbol{C}^{M,N}_{m,n}(u,v) = \cos\left[2\pi\left(\frac{mu}{M}+\frac{nv}{N}\right)\right] = \cos(\omega_m u+\omega_n v), \tag{19.7}$$

$$\boldsymbol{S}^{M,N}_{m,n}(u,v) = \sin\left[2\pi\left(\frac{mu}{M}+\frac{nv}{N}\right)\right] = \sin(\omega_m u+\omega_n v). \tag{19.8}$$

Each of these basis functions is periodic with M units in the horizontal direction and N units in the vertical direction.

Examples

Figures 19.1 and 19.2 show a set of 2D cosine functions $\boldsymbol{C}^{M,N}_{m,n}$ of size $M \times N = 16 \times 16$ for various combinations of wave numbers $m, n = 0, \ldots, 3$. As we can clearly see, these functions correspond to a directed, cosine-shaped waveform whose orientation is determined by the wave numbers m and n. For example, the wave numbers $m = n = 2$ specify a cosine function $\boldsymbol{C}^{M,N}_{2,2}(u,v)$ that performs two full cycles in both the horizontal and vertical directions, thus creating a diagonally oriented, 2D wave. Of course, the same holds for the corresponding sine functions.

19.1.2 Implementing the 2D DFT

As in the 1D case, we could directly use the definition in Eqn. (19.2) to write a program or procedure that implements the 2D DFT. However, this is not even necessary. A minor rearrangement of Eqn. (19.2) into

$$G(m,n) = \frac{1}{\sqrt{N}} \cdot \sum_{v=0}^{N-1} \underbrace{\left[\frac{1}{\sqrt{M}} \cdot \sum_{u=0}^{M-1} g(u,v) \cdot e^{-\mathrm{i}2\pi\frac{um}{M}}\right]}_{\text{1-dim. DFT of row } g(\cdot,v)} \cdot e^{-\mathrm{i}2\pi\frac{vn}{N}} \tag{19.9}$$

shows that its core contains a *1D* DFT (see Eqn. (18.44)) of the vth row vector $g(\cdot,v)$ that is independent of the "vertical" position v and size N (noting the fact that v and N are placed outside the square brackets in Eqn. (19.9)). If, in a first step, we *replace* each *row* vector $g(\cdot,v)$ of the original image by its 1D Fourier transform,

$$g_{\mathrm{x}}(\cdot,v) \leftarrow \mathsf{DFT}(g(\cdot,v)) \quad \text{for } 0 \leq v < N, \tag{19.10}$$

then we only need to replace each resulting *column* vector by its 1D DFT in a second step:

$$g_{\mathrm{xy}}(u,\cdot) \leftarrow \mathsf{DFT}(g_{\mathrm{x}}(u,\cdot)) \quad \text{for } 0 \leq u < M. \tag{19.11}$$

The resulting function $g''(u,v)$ is precisely the 2D Fourier transform $G(m,n)$. Thus the *2D* DFT can be separated into a sequence of 1D

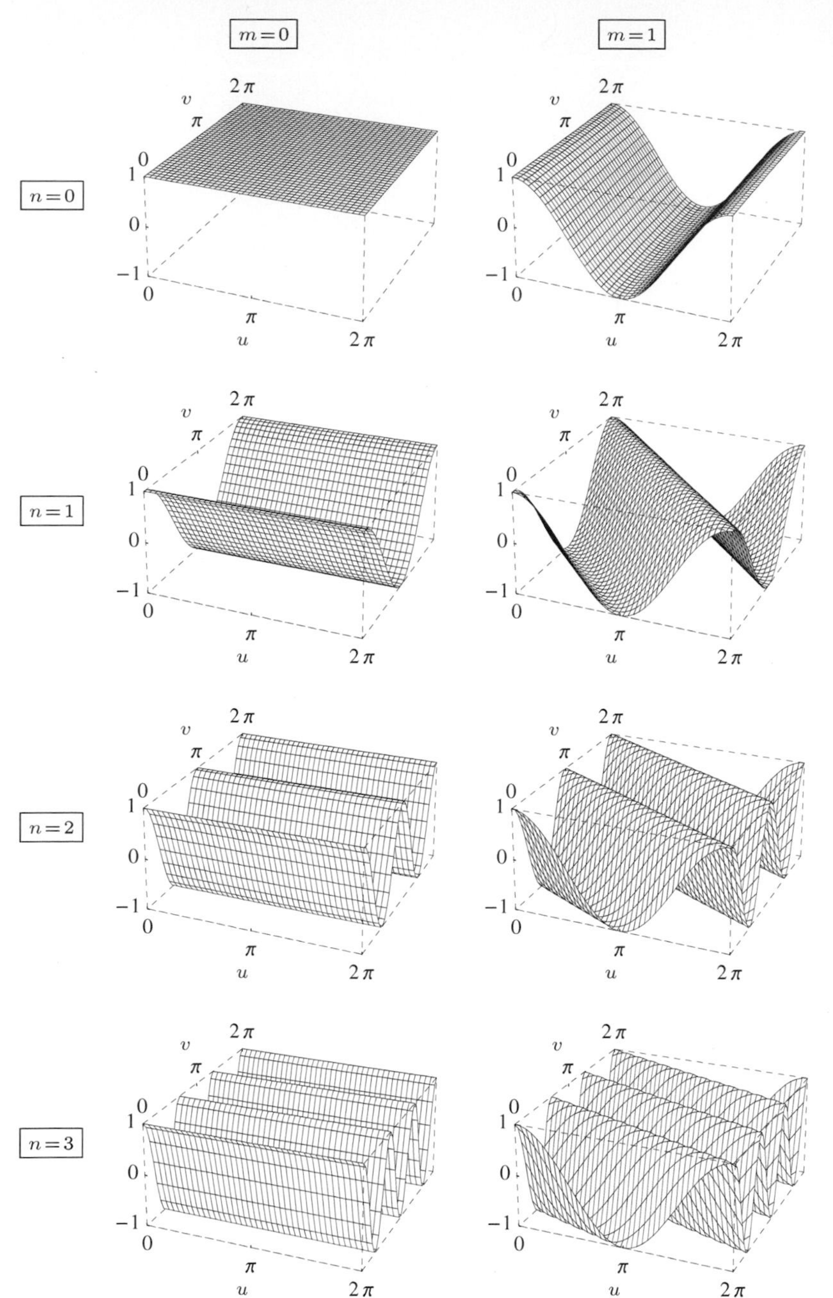

Fig. 19.1
2D cosine functions. $C_{m,n}^{M,N}(u,v) = \cos\left[2\pi\left(\frac{mu}{M}+\frac{nv}{N}\right)\right]$ for $M = N = 16$, $n = 0, \ldots, 3$, $m = 0, 1$.

DFTs over the row and column vectors, respectively, as summarized in Alg. 19.1. The advantage of this is twofold: first, the 2D-DFT can be implemented more efficiently, and second, only a 1D implementation of the DFT (or the 1D FFT, as described in Ch. 18, Sec. 18.4.2) is needed to implement any multidimensional DFT.

As we can see from Eqn. (19.9), the 2D DFT could equally be performed in the opposite way, that is, by first doing a 1D DFT on all *rows* and subsequently on all *columns*. One should also note that all operations in Alg. 19.1 are done "in place", which means that

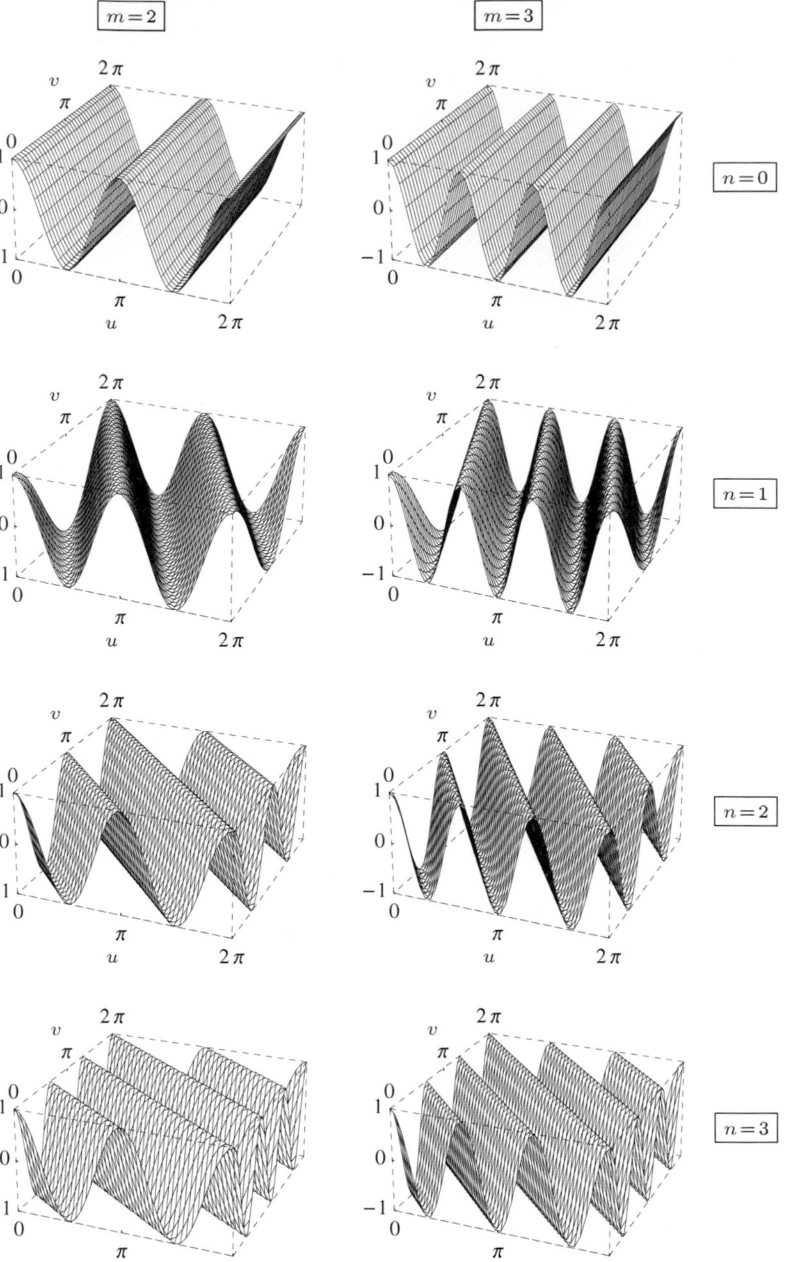

Fig. 19.2 2D cosine functions (*continued*). $C^{M,N}_{m,n}(u, v) = \cos\left[2\pi\left(\frac{um}{M} + \frac{vn}{N}\right)\right]$ for $M = N = 16$, $n = 0, \ldots, 3$, $m = 2, 3$.

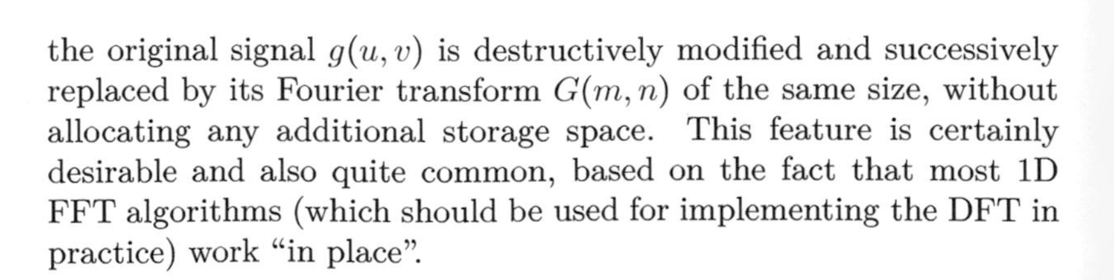

the original signal $g(u, v)$ is destructively modified and successively replaced by its Fourier transform $G(m, n)$ of the same size, without allocating any additional storage space. This feature is certainly desirable and also quite common, based on the fact that most 1D FFT algorithms (which should be used for implementing the DFT in practice) work "in place".

Alg. 19.1
In-place computation of the 2D DFT as a sequence of 1D DFTs on row and column vectors.

```
1: Separable2dDft(g)                                    ▷ g(u,v) ∈ ℂ
       Input: g, a 2D, discrete signal of size M × N, with g(u,v) ∈
       ℂ. Returns the DFT for the 2D function g(u,v). The resulting
       spectrum G(m,n) has the same dimensions as g. The algorithm
       works "in place", i.e., g is modified.
2:     (M,N) ← Size(g)
3:     for v ← 0,...,N−1 do
4:         r ← g(·,v)                     ▷ extract the vth row vector of g
5:         g(·,v) ← DFT(r)                ▷ replace the vth row vector of g
6:     for u ← 0,...,M−1 do
7:         c ← g(u,·)                     ▷ extract the uth column vector of g
8:         g(u,·) ← DFT(c)                ▷ replace the uth column vector of g
       Remark: g(u,v) ≡ G(m,n) now contains the discrete 2D Fourier
       spectrum.
9:     return g
```

19.2 Visualizing the 2D Fourier Transform

Unfortunately, there is no simple method for visualizing 2D complex-valued functions, such as the result of a 2D DFT. One alternative is to display the real and imaginary parts individually as 2D surfaces. Mostly, however, the absolute value of the complex functions is displayed, which in the case of the Fourier transform is $|G(m,n)|$, the *power spectrum* (see Ch. 18, Sec. 18.3.5).

19.2.1 Range of Spectral Values

For most natural images, the "spectral energy" concentrates at the lower frequencies with a clear maximum at wave numbers $(0,0)$; that is, at the co-ordinate center (see also Sec. 19.4). The values of the power spectrum usually cover a wide range, and displaying them linearly often makes the smaller values invisible. To show the full range of spectral values, in particular the smaller values for the high frequencies, it is common to display the square root or the logarithm of the power spectrum, $\sqrt{|G(m,n)|}$ or $\log|G(m,n)|$, respectively.

19.2.2 Centered Representation of the DFT Spectrum

Analogous to the 1D case, the 2D spectrum is a periodic function in both dimensions,

$$G(m,n) = G(m+pM, n+qN), \tag{19.12}$$

for arbitrary $p,q \in \mathbb{Z}$. In the case of a real-valued 2D signal $g(u,v) \in \mathbb{R}$ (see Eqn. (18.57)), the power spectrum is also *symmetric* about the origin, that is,

$$|G(m,n)| = |G(-m,-n)|. \tag{19.13}$$

It is thus common to use a centered representation of the spectrum with coordinates m,n in the ranges

$$-\lfloor\tfrac{M}{2}\rfloor \leq m \leq \lfloor\tfrac{M-1}{2}\rfloor \quad \text{and} \quad -\lfloor\tfrac{N}{2}\rfloor \leq n \leq \lfloor\tfrac{N-1}{2}\rfloor,$$

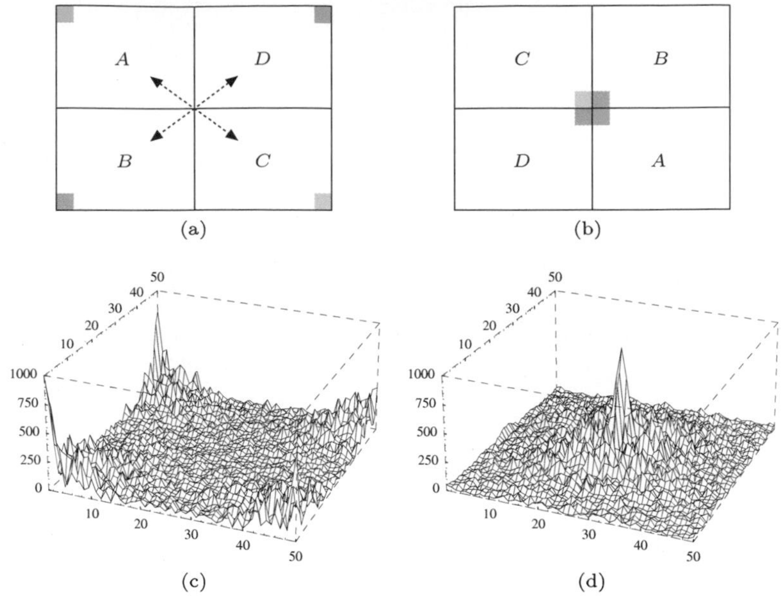

Fig. 19.3
Centering the 2D Fourier spectrum. In the original (noncentered) spectrum, the coordinate center (i.e., the region of low frequencies) is located in the upper left corner and, due to the periodicity of the spectrum, also at all other corners (a). In this case, the coefficients for the highest wave numbers (frequencies) lie at the center. Swapping the quadrants pairwise, as shown in (b), moves all low-frequency coefficients to the center and high frequencies to the periphery. A real 2D power spectrum is shown in its original form in (c) and in centered form in (d).

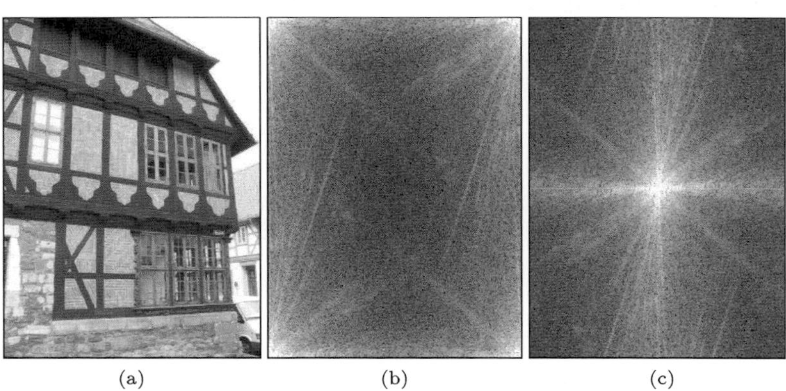

Fig. 19.4
Intensity plot of a 2D power spectrum: original image (a), noncentered spectrum (b), and centered spectrum (c).

respectively. This can be easily accomplished by swapping the four quadrants of the Fourier transform, as illustrated in Fig. 19.3. In the resulting representation, the low-frequency coefficients are found at the center and the high-frequency entries along the outer boundaries. Figure 19.4 shows the plot of a 2D power spectrum as an intensity image in its original and centered form, with the intensity proportional to the logarithm of the spectral values ($\log_{10} |G(m,n)|$).

19.3 Frequencies and Orientation in 2D

19.3.1 Effective Frequency

As we could see in Figs. 19.1 and 19.2, each 2D basis function is an oriented cosine or sine function whose orientation and frequency are determined by its wave numbers m and n for the horizontal and vertical directions, respectively. If we moved along the main direction of such a basis function (i.e., perpendicular to the crest of the waves), we would follow a 1D cosine or sine function of some frequency $\hat{f}$,

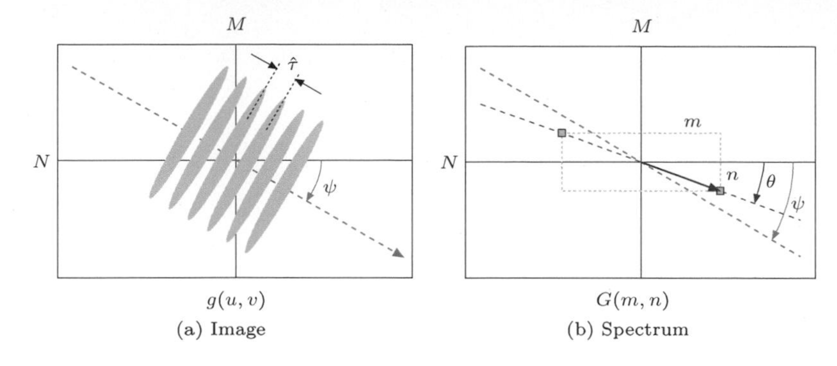

Fig. 19.5
Frequency and orientation in 2D. The image (a) contains a periodic pattern with effective frequency $\hat{f} = 1/\hat{\tau}$ and orientation ψ. The frequency coefficient corresponding to this pattern is found at position $(m, n) = \pm\hat{f} \cdot (M \cos\psi, N \sin\psi)$ (see Eqn. (19.14)) in the 2D Fourier spectrum (b). Thus, if $M \neq N$, the spectral coefficients (m, n) are located at a direction (θ) different to the orientation of the image pattern (ψ).

which we call the *directional* or *effective frequency* of the waveform (see Fig. 19.5).

Recall that the wave numbers m, n specify how many full cycles the associated 2D basis function performs over a distance of M units in the horizontal direction or N units in the vertical direction. Thus, if an image of size $M \times N$ contains a periodic pattern with effective frequency $\hat{f} = 1/\hat{\tau}$ and orientation ψ, the associated frequency coefficients are found at positions

$$\begin{pmatrix} m \\ n \end{pmatrix} = \pm\hat{f} \cdot \begin{pmatrix} M \cdot \cos(\psi) \\ N \cdot \sin(\psi) \end{pmatrix} \tag{19.14}$$

in the corresponding 2D Fourier spectrum (see Fig. 19.5). Given the spectral position (m, n), the effective frequency along the main direction of the wave can be derived (from the 1D case in Eqn. (18.58)) as

$$\hat{f}_{(m,n)} = \tfrac{1}{\tau} \cdot \sqrt{\left(\tfrac{m}{M}\right)^2 + \left(\tfrac{n}{N}\right)^2}, \tag{19.15}$$

where we assume the same spatial sampling interval along the x and y axes (i.e., $\tau = \tau_x = \tau_y$). Thus the *maximum signal frequency* in the directions of the coordinate axes is

$$\hat{f}_{(\pm\frac{M}{2},0)} = \hat{f}_{(0,\pm\frac{N}{2})} = \tfrac{1}{\tau} \cdot \sqrt{\left(\tfrac{1}{2}\right)^2} = \tfrac{1}{2\tau} = \tfrac{1}{2} f_s, \tag{19.16}$$

where $f_s = \frac{1}{\tau}$ denotes the sampling frequency. Notice that the effective signal frequency at the *corners* of the spectrum is

$$\hat{f}_{(\pm\frac{M}{2},\pm\frac{N}{2})} = \tfrac{1}{\tau} \cdot \sqrt{\left(\tfrac{1}{2}\right)^2 + \left(\tfrac{1}{2}\right)^2} = \tfrac{1}{\sqrt{2}\cdot\tau} = \tfrac{1}{\sqrt{2}} f_s, \tag{19.17}$$

which is a factor $\sqrt{2}$ higher than along the coordinate axes (Eqn. (19.16)).

19.3.2 Frequency Limits and Aliasing in 2D

Figure 19.6 illustrates the relationship described in Eqns. (19.16) and (19.17). The highest permissible signal frequencies in any direction lie along the boundary of the centered 2D spectrum of size $M \times N$. Any signal with all frequency components *within* this region complies with the sampling theorem (Nyquist rule) and can thus be reconstructed without aliasing. In contrast, any spectral component

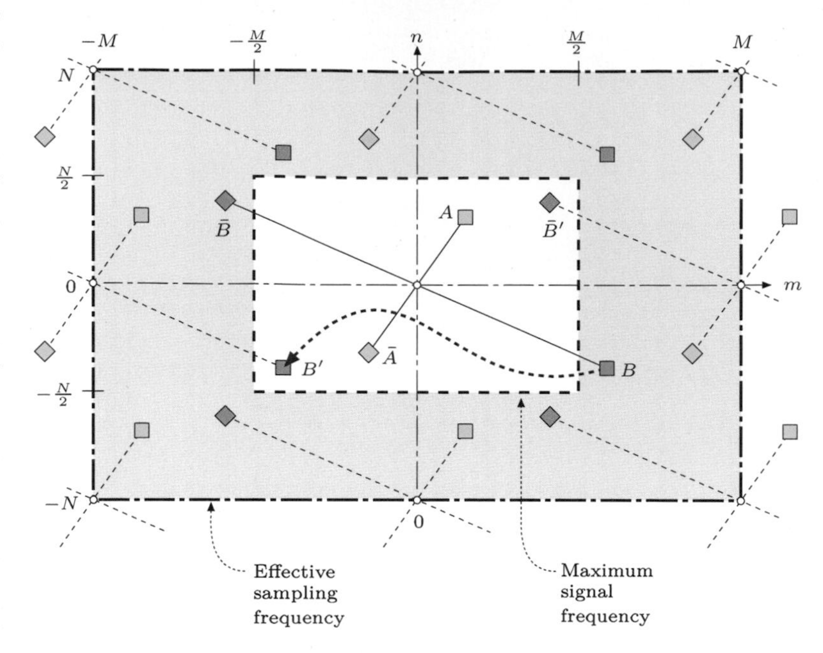

Fig. 19.6 Maximum signal frequencies and aliasing in 2D. The boundary of the $M \times N$ 2D spectrum (inner rectangle) marks the region of permissible signal frequencies along any direction. The outer rectangle corresponds to the effective sampling frequency, which is twice the maximum signal frequency in the same direction. The signal component at spectral position a lies inside the permissible frequency range and thus causes no alasing. In contrast, component b is outside the permissible range. Due to the periodicity of the spectrum, all components repeat (as in the 1D case) at all multiples of the sampling frequency along the m and n axis. This causes the component B to be "aliased" to a lower-frequency position B' (and $\bar{B}$ to $\bar{B}'$) in the visible part of the spectrum. Note that this also changes the *direction* of the corresponding wave in signal space.

outside these limits is reflected across the boundary of this box toward the coordinate center onto lower frequencies, which would appear as visual aliasing in the reconstructed image.

Apparently the lowest effective sampling frequency (Eqn. (19.15)) occurs in the directions of the coordinate axes of the sampling grid. To ensure that a certain image pattern can be reconstructed without aliasing at any orientation, the effective signal frequency $\hat{f}$ of that pattern must be limited to $\frac{f_s}{2} = \frac{1}{2\tau}$ in every direction, again assuming that the sampling interval τ is the same along both coordinate axes.

19.3.3 Orientation

The spatial orientation of a 2D cosine or sine wave with spectral coordinates m, n (wave numbers $0 \leq m < M$, $0 \leq n < N$) is

$$\psi_{(m,n)} = \mathrm{ArcTan}\big(\tfrac{m}{M}, \tfrac{n}{N},\big) = \mathrm{ArcTan}\big(mN, nM\big), \tag{19.18}$$

where $\psi_{(m,n)}$ for $m = n = 0$ is of course undefined.[1] Conversely, a 2D sinusoid with effective frequency $\hat{f}$ and spatial orientation ψ is represented by the spectral coordinates

$$(m, n) = \pm\hat{f} \cdot (M \cos\psi, N \sin\psi)\,, \tag{19.19}$$

as already shown in Fig. 19.5.

19.3.4 Normalizing the Geometry of the 2D Spectrum

From Eqn. (19.19) we can derive that in the special case of a sinusoid with spatial orientation $\psi = 45°$ the corresponding spectral coefficients are found at the frequency coordinates

[1] ArcTan(x, y) in Eqn. (19.18) denotes the inverse tangent function $\tan^{-1}(y/x)$ (also see Sec. F.1.6 in the Appendix).

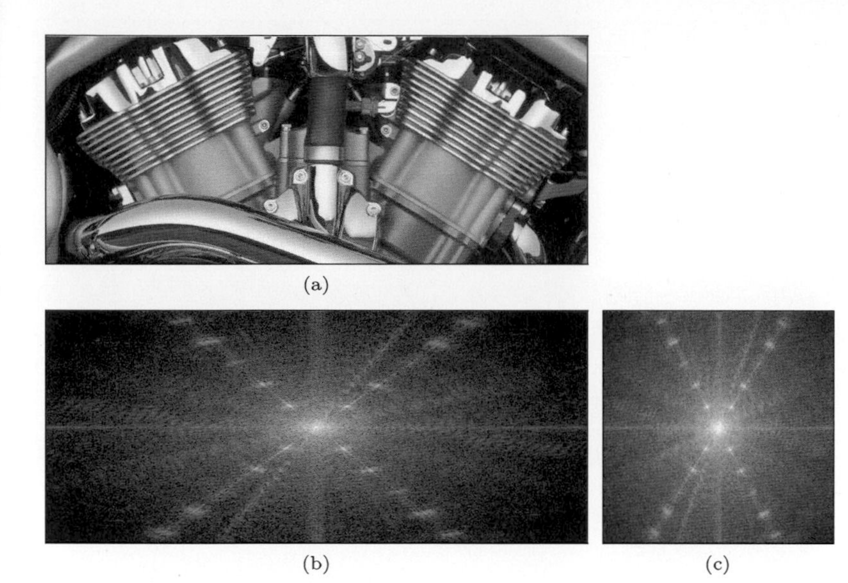

(a)

(b) (c)

Fig. 19.7
Normalizing the 2D spectrum. Original image (a) with dominant oriented patterns that show up as clear peaks in the corresponding spectrum (b). Because the image and the spectrum are not square ($M \neq N$), orientations in the image are not the same as in the actual spectrum (b). After the spectrum is normalized to square proportions (c), we can clearly observe that the cylinders of this (Harley-Davidson *V-Rod*) engine are really arranged at a 60° angle.

$$(m, n) = \pm(\lambda M, \lambda N) \quad \text{for} \; -\tfrac{1}{2} \leq \lambda \leq +\tfrac{1}{2}, \tag{19.20}$$

that is, at the diagonals of the spectrum (see also Eqn. (19.17)). Unless the image (and thus the spectrum) is quadratic ($M = N$), the angle of orientation in the image and in the spectrum are not the same, though they coincide along the directions of the coordinate axes. This means that rotating an image by some angle α does turn the spectrum in the same direction but in general not by the same angle α!

To obtain identical orientations and turning angles in both the image and the spectrum, it is sufficient to scale the spectrum to square size such that the spectral resolution is the same along both frequency axes (as shown in Fig. 19.7).

19.3.5 Effects of Periodicity

When interpreting the 2D DFT of images, one must always be aware of the fact that with any discrete Fourier transform, the signal function is implicitly assumed to be periodic in every dimension. Thus the transitions at the borders between the replicas of the image are also part of the signal, just like the interior of the image itself. If there is a large intensity difference between opposing borders of an image (e.g., between the upper and lower parts of a landscape image), then this causes strong transitions in the resulting periodic signal. Such steep discontinuities are of high bandwidth (i.e., the corresponding signal energy is spread over a wide range along the coordinate axes of the sampling grid; see Fig. 19.8). This broadband energy distribution along the main axes, which is often observed with real images, may lead to a suppression of other relevant signal components in the spectrum.

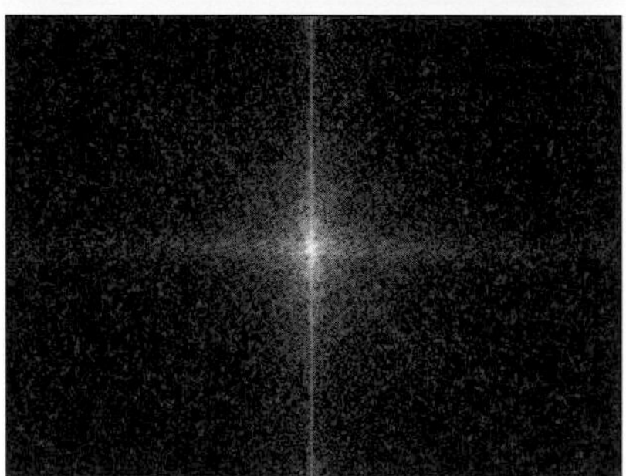

Fig. 19.8
Effects of periodicity in the 2D spectrum. The discrete Fourier transform is computed under the implicit assumption that the image signal is periodic along both dimensions (top). Large differences in intensity at opposite image borders—here most notably in the vertical direction—lead to broad-band signal components that in this case appear as a bright line along the spectrum's vertical axis (bottom).

19.3.6 Windowing

One solution to this problem is to multiply the image function $g(u,v) = I(u,v)$ by a suitable *windowing function* $w(u,v)$, that is,

$$\tilde{g}(u,v) = g(u,v) \cdot w(u,v), \tag{19.21}$$

for $0 \leq u < M$, $0 \leq v < N$, prior to computing the DFT. The windowing function $w(u,v)$ should drop off continuously toward the image borders such that the transitions between image replicas are effectively eliminated. But multiplying the image with $w(u,v)$ has additional effects upon the spectrum. As we already know (from Eqn. (18.26)), a *multiplication* of two functions in signal space corresponds to a *convolution* of the corresponding spectra in frequency space, that is,

$$\tilde{G}(m,n) = G(m,n) * W(m,n). \tag{19.22}$$

To cause the least possible damage to the Fourier transform of the image, the ideal spectrum of $w(u,v)$ would be the impulse function $\delta(m,n)$. Unfortunately, this again corresponds to the constant windowing function $w(u,v) = 1$ with no windowing effect at all. In general, we can say that a broader spectrum of the windowing function

$w(u, v)$ smoothes the resulting spectrum more strongly and individual frequency components are harder to isolate.

Taking a picture is equivalent to cutting out a finite (usually rectangular) region from an infinite image plane, which can be simply modeled as a multiplication with a rectangular pulse function of width M and height N. So, in this case, the spectrum of the original intensity function is convolved with the spectrum of the rectangular pulse (box). The problem is that the spectrum of the rectangular box (see Fig. 19.9(a)) is of extremely high bandwidth and thus far off the ideal narrow impulse function.

These two examples demonstrate a dilemma: windowing functions should for one be as wide as possible to include a maximum part of the original image, and they should also drop off to zero toward the image borders but then again not too steeply to maintain a narrow windowing spectrum.

19.3.7 Common Windowing Functions

Suitable windowing functions should therefore exhibit soft transitions, and many variants have been proposed and analyzed both theoretically and for practical use (see, e.g., [34, Ch. 9, Sec. 9.3], [194, Ch. 10]). Table 19.1 lists the definitions of several popular windowing functions; the corresponding 2D (logarithmic) power spectra are displayed in Figs. 19.9 and 19.10.

The spectrum of the *rectangular pulse* function (Fig. 19.9(a)), which assigns identical weights to all image elements, exhibits a relatively narrow peak at the center, which promises little smoothing in the resulting windowed spectrum. Nevertheless, the spectral energy drops off quite slowly toward the higher frequencies, thus creating a rather wide spectrum. Not surprisingly, the behavior of the *elliptical* windowing function in Fig. 19.9(b) is quite similar. The *Gaussian* window in Fig. 19.9(c) demonstrates how the off-center spectral energy can be significantly suppressed by narrowing the windowing function, however, at the cost of a much wider peak at the center. In fact, none of the functions in Fig. 19.9 makes a good windowing function.

Obviously, the choice of a suitable windowing function is a delicate compromise since even apparently similar functions may exhibit largely different behaviors in the frequency spectrum. For example, good overall results can be obtained with the *Hanning* window (Fig. 19.10(c)) or the *Parzen* window (Fig. 19.10(d)), which are both easy to compute and frequently used in practice.

Figure 19.11 illustrates the effects of selected windowing functions upon the spectrum of an intensity image. As can be seen clearly, narrowing the windowing function leads to a suppression of the artifacts caused by the signal's implicit periodicity. At the same time, however, it also reduces the resolution of the spectrum; the spectrum becomes blurred, and individual peaks are widened.

Table 19.1
2D windowing function definitions. The functions $w(u,v)$ have their maximum values at the image center, $w(M/2,N/2) = 1$. The values r_u, r_v, and $r_{u,v}$ used in the definitions are specified at the top of the table.

Definitions:

$$r_u = \frac{u-M/2}{M/2} = \frac{2u}{M} - 1, \qquad r_v = \frac{v-N/2}{N/2} = \frac{2v}{N} - 1, \qquad r_{u,v} = \sqrt{r_u^2 + r_v^2}$$

Elliptical window:	$w(u,v) = \begin{cases} 1 & \text{for } 0 \leq r_{u,v} \leq 1 \\ 0 & \text{otherwise} \end{cases}$
Gaussian window:	$w(u,v) = e^{\left(\frac{-r_{u,v}^2}{2\sigma^2}\right)}, \quad \sigma = 0.3, \ldots, 0.4$
Super-Gaussian window:	$w(u,v) = e^{\left(\frac{-r_{u,v}^n}{\kappa}\right)}, \quad n = 6,\ \kappa = 0.3, \ldots, 0.4$
Cosine² window:	$w(u,v) = \begin{cases} \cos\left(\frac{\pi}{2} r_u\right) \cdot \cos\left(\frac{\pi}{2} r_v\right) & \text{for } 0 \leq r_u, r_v \leq 1 \\ 0 & \text{otherwise} \end{cases}$
Bartlett window:	$w(u,v) = \begin{cases} 1 - r_{u,v} & \text{for } 0 \leq r_{u,v} \leq 1 \\ 0 & \text{otherwise} \end{cases}$
Hanning window:	$w(u,v) = \begin{cases} 0.5 \cdot \left[\cos(\pi r_{u,v}) + 1\right] & \text{for } 0 \leq r_{u,v} \leq 1 \\ 0 & \text{otherwise} \end{cases}$
Parzen window:	$w(u,v) = \begin{cases} 1 - 6 r_{u,v}^2 + 6 r_{u,v}^3 & \text{for } 0 \leq r_{u,v} < 0.5 \\ 2 \cdot (1 - r_{u,v})^3 & \text{for } 0.5 \leq r_{u,v} < 1 \\ 0 & \text{otherwise} \end{cases}$

19.4 2D Fourier Transform Examples

The following examples demonstrate some basic properties of the 2D DFT on real intensity images. All examples in Figs. 19.12–19.18 show a centered and squared spectrum with logarithmic intensity values (see Sec. 19.2).

Scaling

Figure 19.12 shows that scaling the image in signal space has the opposite effect in frequency space, analogous to the 1D case (see Ch. 18, Fig. 18.4).

Periodic Image Patterns

The images in Fig. 19.13 contain repetitive periodic patterns at various orientations and scales. They appear as distinct peaks at the corresponding positions (see Eqn. (19.19)) in the spectrum.

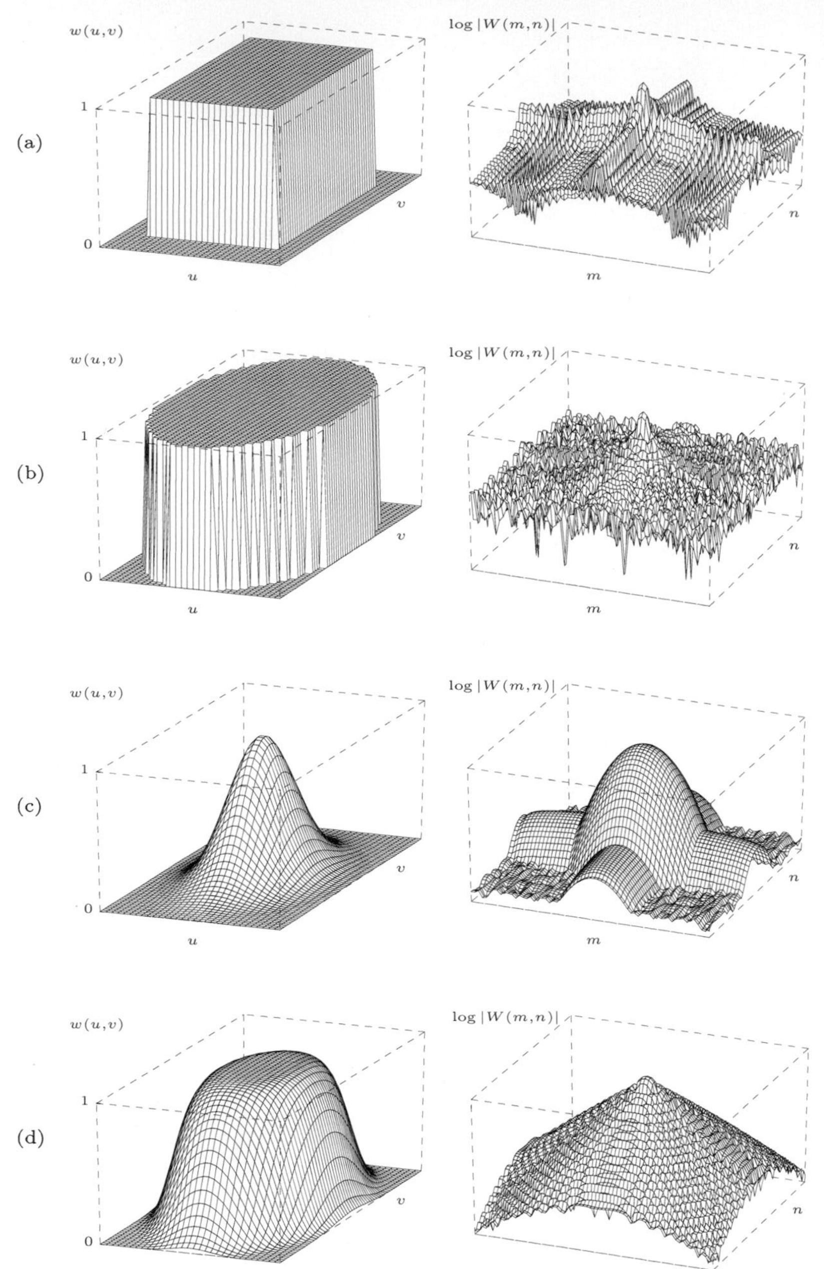

Fig. 19.9
Windowing functions and their logarithmic power spectra. Rectangular pulse (a), elliptical window (b), Gaussian window with $\sigma = 0.3$ (c), and super-Gaussian window of order $n = 6$ and $\kappa = 0.3$ (d). The windowing functions are deliberately of nonsquare size ($M : N = 1 : 2$).

Rotation

Figure 19.14 shows that rotating the image by some angle α rotates the spectrum in the same direction and—if the image is square—by the same angle.

Oriented, elongated structures

Pictures of artificial objects often exhibit regular patterns or elongated structures that appear dominantly in the spectrum. The im-

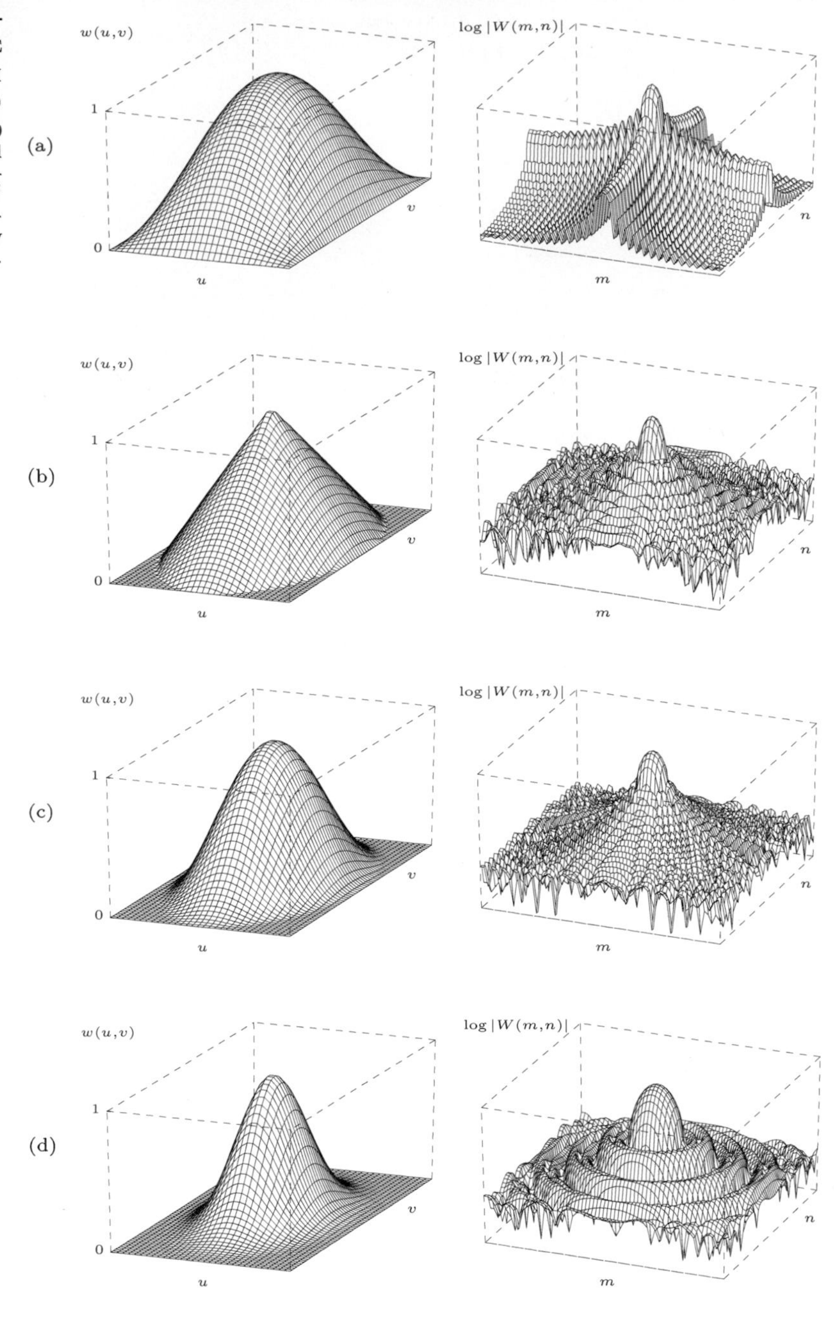

Fig. 19.10
Windowing functions and their logarithmic power spectra (*continued*). Cosine² window (a), Bartlett window (b), Hanning window (c), and Parzen window (d).

ages in Fig. 19.15 show several elongated structures that show up in the spectrum as bright streaks oriented perpendicularly to the main direction of the image patterns.

Natural images

Straight and regular structures are usually less dominant in images of natural objects and scenes, and thus the visual effects in the spectrum are not as obvious as with artificial objects. Some examples of this class of images are shown in Figs. 19.16 and 19.17.

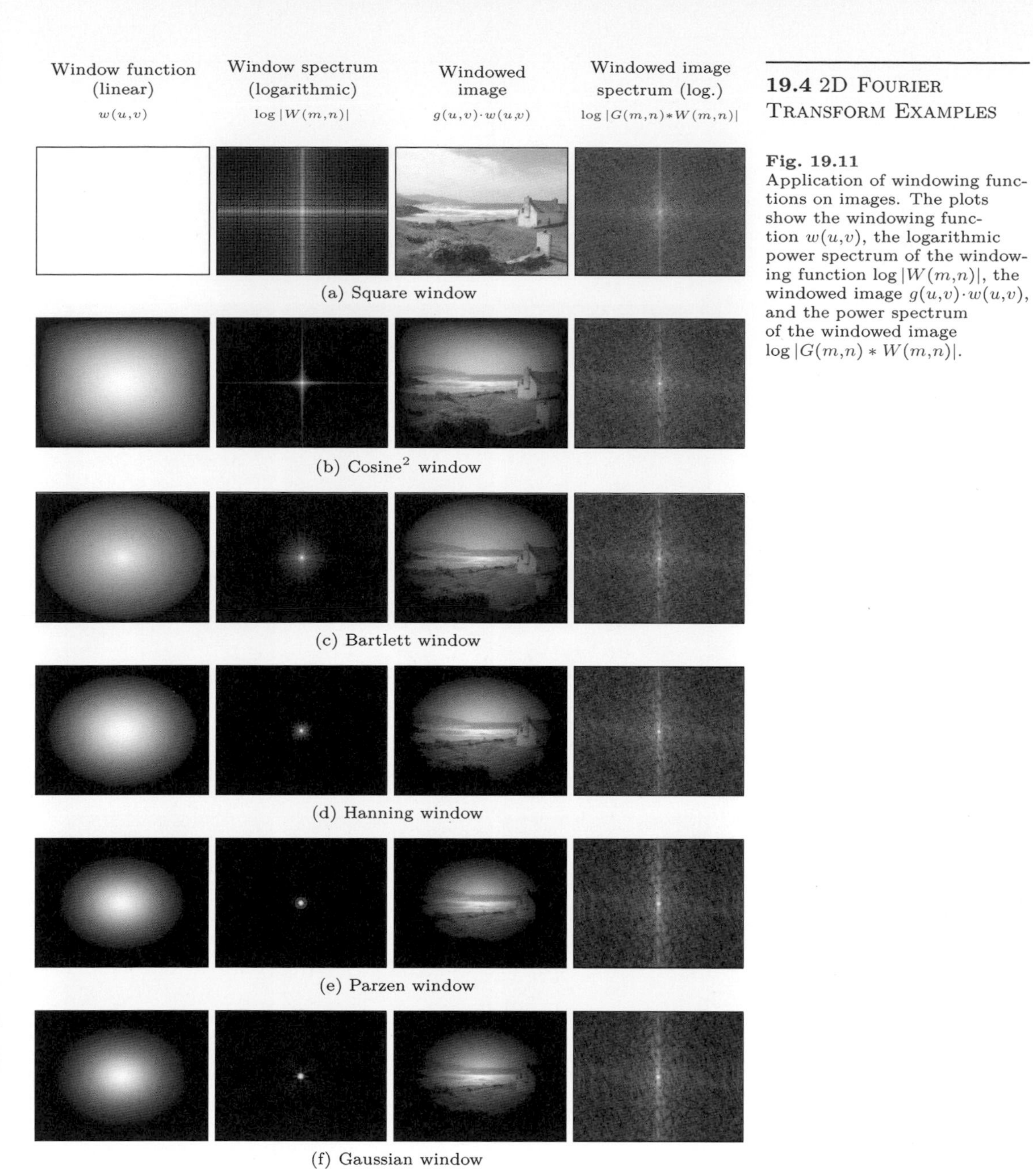

Fig. 19.11
Application of windowing functions on images. The plots show the windowing function $w(u,v)$, the logarithmic power spectrum of the windowing function $\log|W(m,n)|$, the windowed image $g(u,v)\cdot w(u,v)$, and the power spectrum of the windowed image $\log|G(m,n) * W(m,n)|$.

Print patterns

The regular patterns generated by the common raster print techniques (Fig. 19.18) are typical examples for periodic multidirectional structures, which stand out clearly in the corresponding Fourier spectrum.

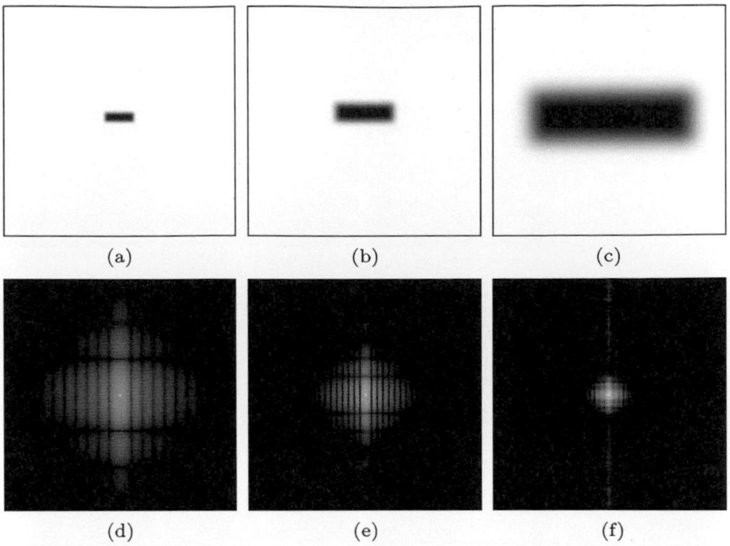

(a) (b) (c)

(d) (e) (f)

Fig. 19.12
DFT under image scaling. The rectangular pulse in the image function (a–c) creates a strongly oscillating power spectrum (d–f), as in the 1D case. Stretching the image causes the spectrum to contract and vice versa.

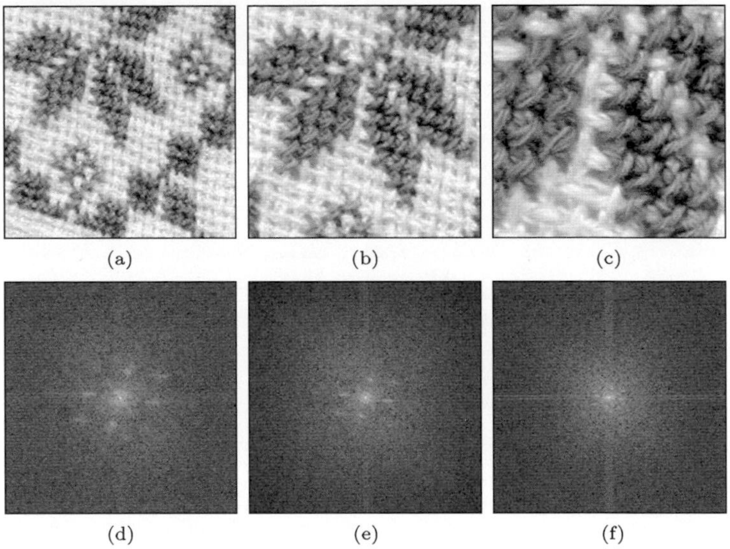

(a) (b) (c)

(d) (e) (f)

Fig. 19.13
DFT of oriented, repetitive patterns. The image function (a–c) contains patterns with three dominant orientations, which appear as pairs of corresponding frequency spots in the spectrum (c–f). Enlarging the image causes the spectrum to contract.

19.5 Applications of the DFT

The Fourier transform and the DFT in particular are important tools in many engineering disciplines. In digital signal and image processing, the DFT (and the FFT) is an indispensable "workhorse" with many applications in image analysis, filtering, and image reconstruction, just to name a few.

19.5.1 Linear Filter Operations in Frequency Space

Performing linear filter operations in frequency space is an interesting option because it provides an efficient way to apply filters of large spatial extent. The approach is based on the *convolution property* of the Fourier transform (see Ch. 18, Sec. 18.1.6), which states that a linear convolution in image space corresponds to a pointwise multiplication in frequency space. Thus the linear convolution $g * h \rightarrow g'$ between

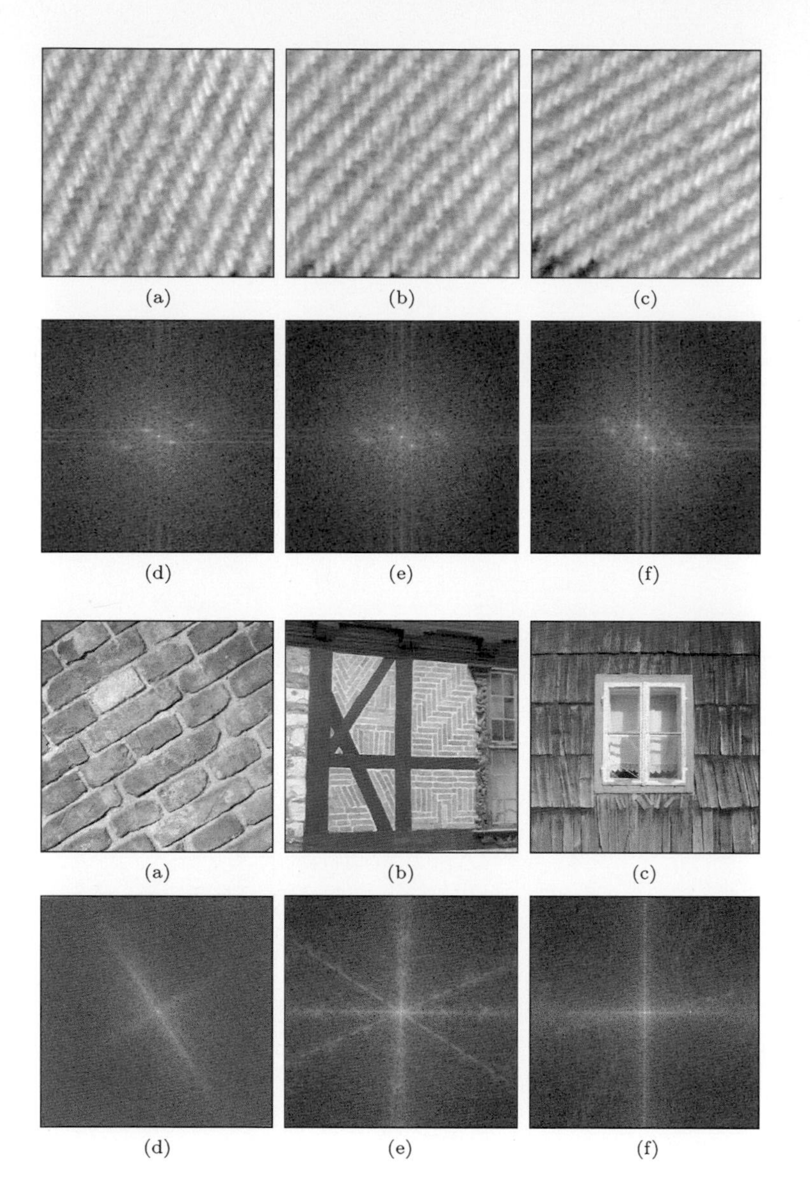

Fig. 19.14
DFT—image rotation. The original image (a) is rotated by 15° (b) and 30° (c). The corresponding (squared) spectrum turns in the same direction and by exactly the same amount (d–f).

Fig. 19.15
DFT—superposition of image patterns. Strong, oriented subpatterns (a–c) are easy to identify in the corresponding spectrum (d–f). Notice the broadband effects caused by straight structures, such as the dark beam on the wall in (b, e).

an image $g(u, v)$ and a filter matrix $h(u, v)$ can be accomplished by the following steps:

$$\begin{array}{lccccc}
\textit{image space:} & g(u,v) & * & h(u,v) & = & g'(u,v) \\
 & \downarrow & & \downarrow & & \uparrow \\
 & \mathsf{DFT} & & \mathsf{DFT} & & \mathsf{DFT}^{-1} \\
 & \downarrow & & \downarrow & & \uparrow \\
\textit{frequency space:} & G(m,n) & \cdot & H(m,n) & \longrightarrow & G'(m,n).
\end{array} \tag{19.23}$$

First, the image g and the filter kernel[2] h are transformed to frequency space using the 2D DFT. The corresponding spectra G and H are then multiplied (pointwise), and the result G' is subsequently

[2] Note that the symbol h is used here for any 1D or 2D filter kernel and H for the corresponding Fourier spectrum. This should not to be confused with the use of h, H for 1D and 2D filter kernels, respectively, in Ch. 5.

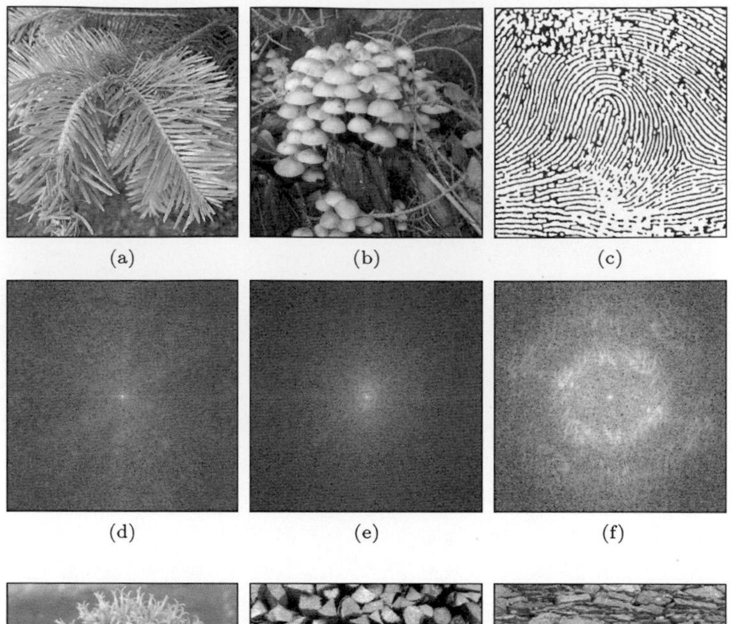

Fig. 19.16 DFT—natural image patterns. Examples of repetitive structures in natural images (a–c) that are also visible in the corresponding spectrum (d–f).

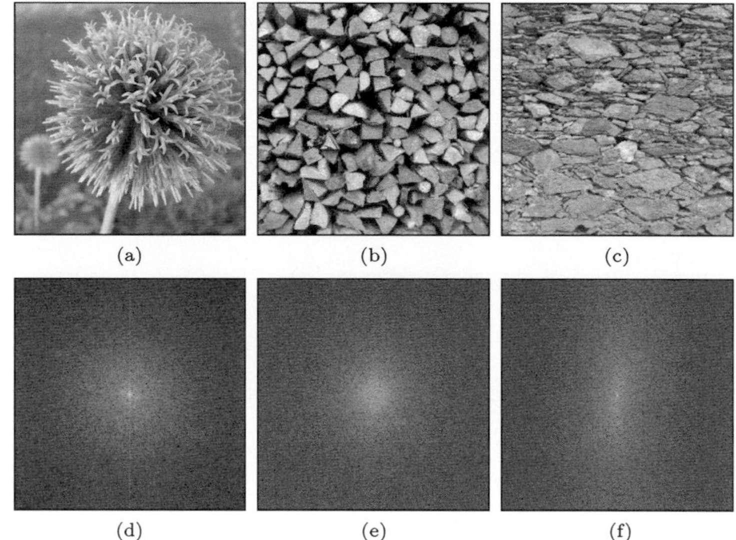

Fig. 19.17 DFT—natural image patterns with no dominant orientation. The repetitive patterns contained in these images (a–c) have no common orientation or sufficiently regular spacing to stand out locally in the corresponding Fourier spectra (d–f).

transformed back to image space using the inverse DFT, thus generating the filtered image g'.

The main advantage of this "detour" lies in its possible efficiency. A direct convolution for an image of size $M \times M$ and a filter matrix of size $N \times N$ requires $\mathcal{O}(M^2 N^2)$ operations. Thus, time complexity increases quadratically with filter size, which is usually no problem for small filters but may render some larger filters too costly to implement. For example, a filter of size 50×50 already requires about 2500 multiplications and additions for every image pixel. In comparison, the transformation from image to frequency space and back can be performed in $\mathcal{O}(M \log_2 M)$ using the FFT, and the pointwise multiplication in frequency space requires M^2 operations, independent of the filter size.

In addition, certain types of filters are easier to specify in frequency space than in image space; for example, an ideal low-pass

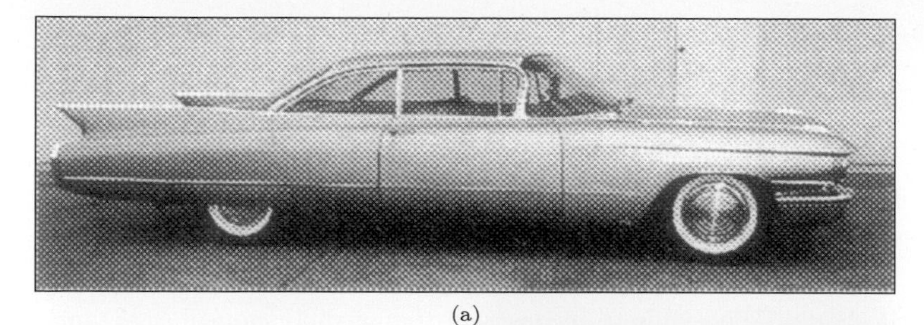

(a)

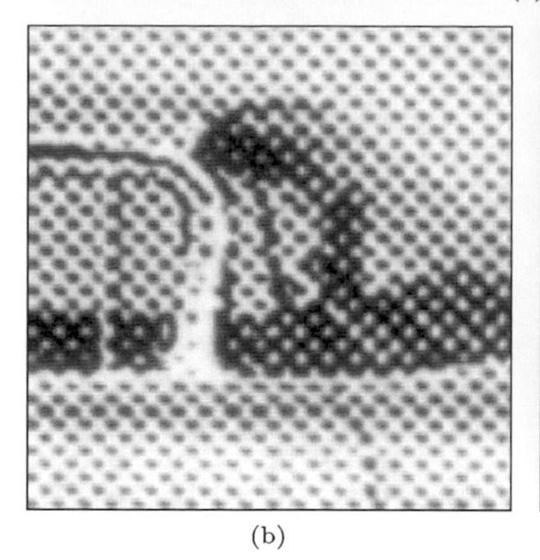

(b)

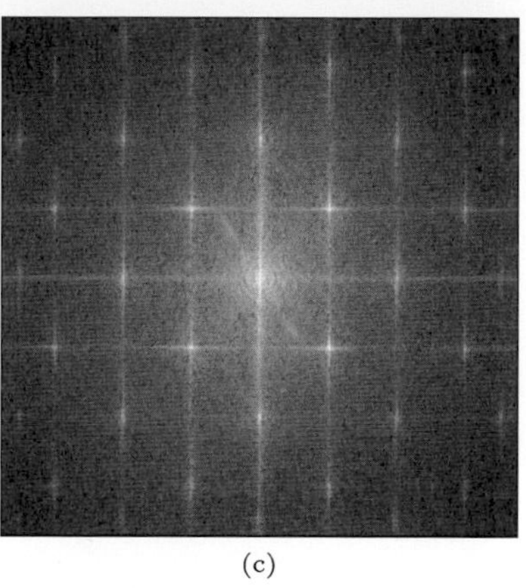

(c)

Fig. 19.18
DFT of a print pattern. The regular diagonally oriented raster pattern (a, b) is clearly visible in the corresponding power spectrum (c). It is possible (at least in principle) to remove such patterns by erasing these peaks in the Fourier spectrum and reconstructing the smoothed image from the modified spectrum using the inverse DFT.

filter, which can be described very compactly in frequency space. Further details on filter operations in frequency space can be found, for example, in [88, Sec. 4.4].

19.5.2 Linear Convolution and Correlation

As discussed in Chapter 5, Sec. 5.3, a linear correlation is the same as a linear convolution with a mirrored filter function. Therefore, the correlation can be computed just like the convolution operation in the frequency domain by following the steps described in Eqn. (19.23). This could be advantageous for comparing images using correlation methods (see Ch. 23, Sec. 23.1.1) because in this case the image and the "filter" matrix (i.e., the second image) are of similar size and thus usually too large to be processed in image space.

Some operations in ImageJ, such as *correlate*, *convolve*, or *deconvolve*, are also implemented in the "Fourier domain" (FD) using the 2D DFT. They can be invoked through the menu Process ▷ FFT ▷ FD Math.

19.5.3 Inverse Filters

Filtering in the frequency domain opens another interesting perspective: reversing the effects of a filter, at least under restricted conditions. In the following, we describe the basic idea only.

Assume we are given an image g_{blur} that has been generated from an original image g_{orig} by some linear filter, for example, motion blur induced by a moving camera. Under the assumption that this image modification can be modeled sufficiently by a linear filter function

h_{blur}, we can state that

$$g_{\text{blur}}(u,v) = (g_{\text{orig}} * h_{\text{blur}})(u,v). \tag{19.24}$$

Recalling that in frequency space this corresponds to a multiplication of the corresponding spectra, that is,

$$G_{\text{blur}}(m,n) = G_{\text{orig}}(m,n) \cdot H_{\text{blur}}(m,n) \tag{19.25}$$

it should be possible to reconstruct the original (non-blurred) image by computing the inverse Fourier transform of the expression

$$G_{\text{orig}}(m,n) = \frac{G_{\text{blur}}(m,n)}{H_{\text{blur}}(m,n)}. \tag{19.26}$$

Unfortunately, this "inverse filter" only works if the spectral coefficients H_{blur} are nonzero, because otherwise the resulting values are infinite. But even small values of H_{blur}, which are typical at high frequencies, lead to large coefficients in the reconstructed spectrum and, as a consequence, large amounts of image noise.

It is also important that the real filter function be accurately approximated, because otherwise the reconstructed image may strongly deviate from the original. The example in Fig. 19.19 shows an image that has been blurred by smooth horizontal motion, whose effect can easily be modeled as a linear convolution. If the filter function that caused the blurring is known exactly, then the reconstruction of the original image can be accomplished without any problems (Fig. 19.19(c)). However, as shown in Fig. 19.19(d), large errors occur if the inverse filter deviates only marginally from the real filter, which quickly renders the method useless.

(a)

(b)

(c)

Fig. 19.19 Removing motion blur by inverse filtering. Original image (a); image blurred by horizontal motion (b); reconstruction using the exact (known) filter function (c); result of the inverse filter when the filter function deviates marginally from the real filter (d).

Beyond this simple idea (which is often referred to as "deconvolution"), better methods for inverse filtering exist, such as the *Wiener filter* and related techniques (see, e.g., [88, Sec. 5.4], [128, Sec. 8.3], [126, Sec. 17.8], [43, Ch. 16]).

19.6 Exercises

Exercise 19.1. Implement the 2D DFT using the 1D DFT, as described in Sec. 19.1.2. Apply the 2D DFT to real intensity images of arbitrary size and display the results (by converting to ImageJ `FloatProcessor` images). Implement the inverse transform and verify that the back-transformed result is identical to the original image.

Exercise 19.2. Assume that the 2D Fourier spectrum of an image with size 640×480 and a spatial resolution of 72 dpi shows a dominant peak at position $\pm(100, 100)$. Determine the orientation and effective frequency (in cycles per cm) of the corresponding image pattern.

Exercise 19.3. An image with size 800×600 contains a wavy intensity pattern with an effective period of 12 pixels, oriented at $30°$. At which frequency coordinates will this pattern manifest itself in the discrete Fourier spectrum?

Exercise 19.4. Generalize Eqn. (19.15) and Eqns. (19.17)–(19.19) for the case where the sampling intervals are *not* identical along the x and y axes (i.e., for $\tau_x \neq \tau_y$).

Exercise 19.5. Implement the *elliptical* and the *super-Gauss* windowing functions (Table 19.1) as ImageJ plugins, and investigate the effects of these windows upon the resulting spectra. Also compare the results to the case where *no* windowing function is used.

20

The Discrete Cosine Transform (DCT)

The Fourier transform and the DFT are designed for processing complex-valued signals, and they always produce a complex-valued spectrum even in the case where the original signal was strictly real-valued. The reason is that neither the real nor the imaginary part of the Fourier spectrum alone is sufficient to represent (i.e., reconstruct) the signal completely. In other words, the corresponding cosine (for the real part) or sine functions (for the imaginary part) alone do not constitute a complete set of basis functions.

On the other hand, we know (see Ch. 18, Eqn. (18.21)) that a real-valued signal has a symmetric Fourier spectrum, so only one half of the spectral coefficients need to be computed without losing any signal information.

There are several spectral transformations that have properties similar to the DFT but do not work with complex function values. The discrete cosine transform (DCT) is a well known example that is particularly interesting in our context because it is frequently used for image and video compression. The DCT uses only cosine functions of various wave numbers as basis functions and operates on real-valued signals and spectral coefficients. Similarly, there is also a discrete sine transform (DST) based on a system of sine functions [128].

20.1 1D DCT

The discrete cosine transform is not, as one may falsely assume, only a "one-half" variant of the discrete Fourier transform. In the 1D case, the *forward* cosine transform for a signal $g(u)$ of length M is defined as

$$G(m) = \sqrt{\tfrac{2}{M}} \cdot \sum_{u=0}^{M-1} g(u) \cdot c_m \cdot \cos\left(\pi \tfrac{m(2u+1)}{2M}\right), \tag{20.1}$$

for $0 \leq m < M$, and the *inverse* transform is

$$g(u) = \sqrt{\tfrac{2}{M}} \cdot \sum_{m=0}^{M-1} G(m) \cdot c_m \cdot \cos\left(\pi \tfrac{m(2u+1)}{2M}\right), \tag{20.2}$$

for $0 \leq u < M$, with

$$c_m = \begin{cases} \frac{1}{\sqrt{2}} & \text{for } m = 0, \\ 1 & \text{otherwise.} \end{cases} \tag{20.3}$$

Note that the index variables (u, m) are used differently in the forward and inverse transforms (Eqns. (20.2)–(20.1)), so the two transforms are—unlike the DFT—*not* symmetric.

20.1.1 DCT Basis Functions

One may ask how it is possible that the DCT can work without any sine functions, while they are essential in the DFT. The trick is to divide all frequencies in half such that they are spaced more densely and thus the frequency resolution in the spectrum is doubled. Comparing the cosine parts of the DFT basis functions (Eqn. (18.49)) and those of the DCT (Eqn. (20.1)),

$$\text{DFT:} \quad \boldsymbol{C}_m^M(u) = \cos\left(2\pi \tfrac{mu}{M}\right), \tag{20.4}$$

$$\text{DCT:} \quad D_m^M(u) = \cos\left(\pi \tfrac{m(2u+1)}{2M}\right) = \cos\left(2\pi \tfrac{m(u+0.5)}{2M}\right), \tag{20.5}$$

one can see that, for a given wave number m, the period $(\tau_m = 2\frac{M}{m})$ of the corresponding DCT basis function is double the period of the DFT basis functions $(\tau_m = \frac{M}{m})$. Notice that the DCT basis functions are also *phase-shifted* by 0.5 units.

Figure 20.1 shows the DCT basis functions $D_m^M(u)$ for the signal length $M = 8$ and wave numbers $m = 0, \ldots, 7$. For example, the basis function $D_7^8(u)$ for wave number $m = 7$ performs seven full cycles over a length of $2M = 16$ units and therefore has the radial frequency $\omega = m/2 = 3.5$.

20.1.2 Implementing the 1D DCT

Since the DCT does not create any complex values and the forward and inverse transforms (Eqns. (20.1) and (20.2)) are almost identical, the whole procedure is fairly easy to implement in Java, as shown in Prog. 20.1. The only notable detail is that the factor c_m in Eqn. (20.1) is independent of the iteration variable u and can thus be calculated outside the inner summation loop (see Prog. 20.1, line 8).

Of course, much more efficient ("fast") DCT algorithms exist. Moreover, the DCT can also be computed in $\mathcal{O}(M \log_2 M)$ time using the FFT [128, p. 152].

20.2 2D DCT

The 2D form of the DCT follows immediately from the the 1D definition (Eqns. (20.1) and (20.2)), resulting in the 2D forward transform

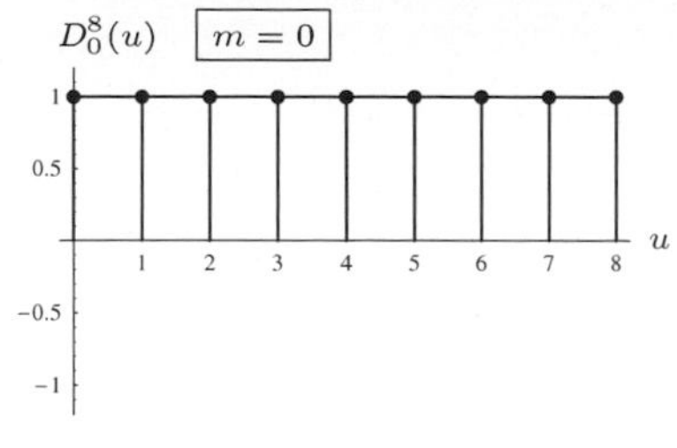

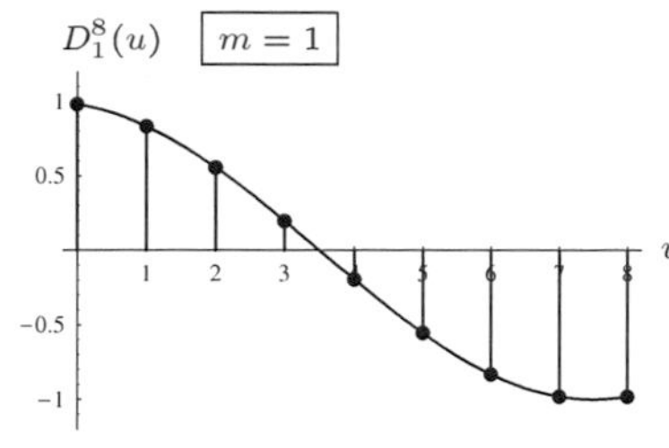

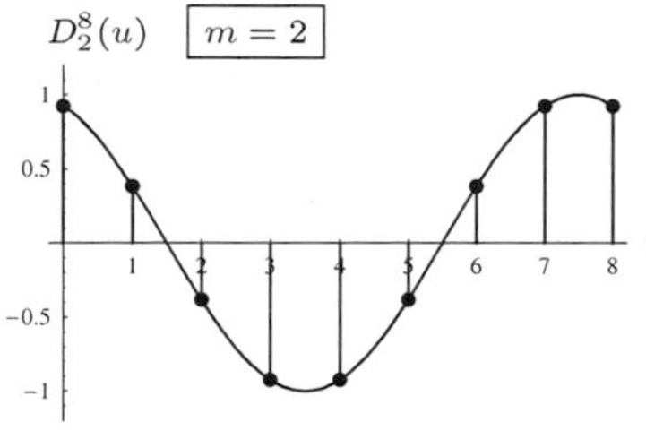

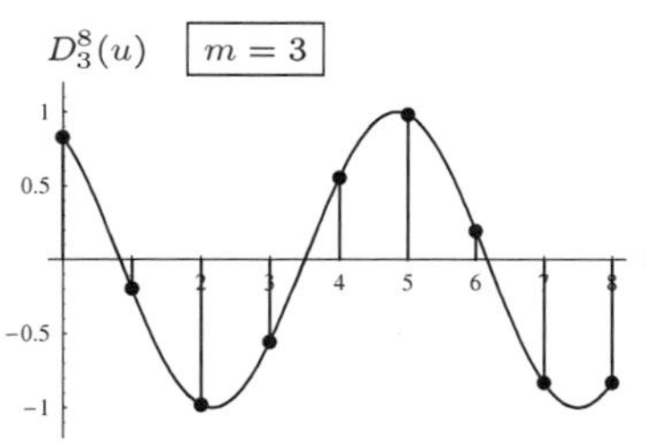

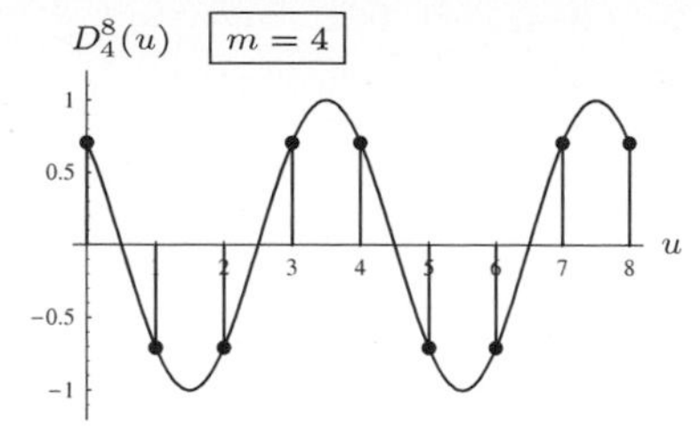

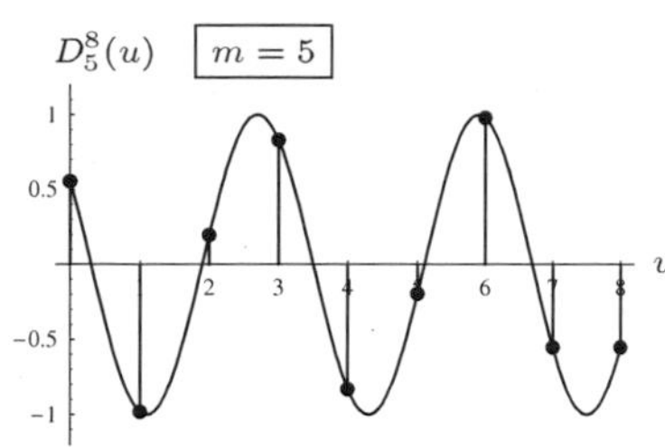

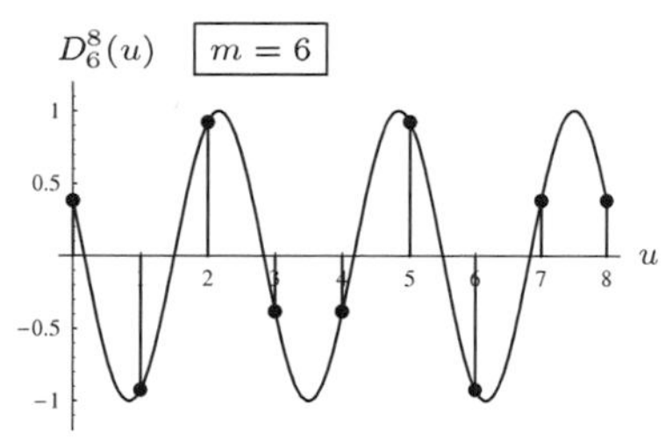

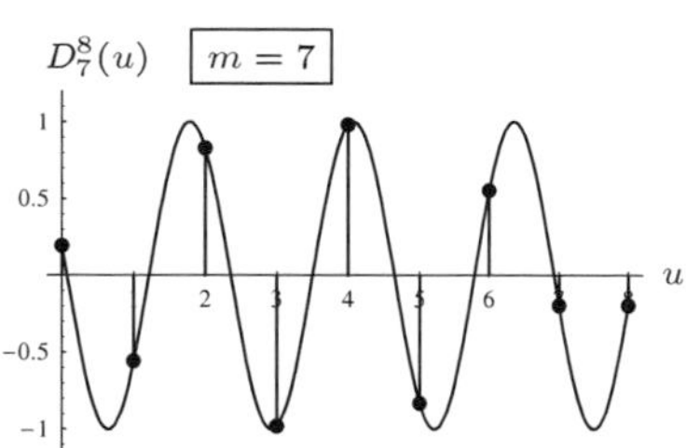

Fig. 20.1
DCT basis functions $D_0^M(u), \ldots, D_7^M(u)$ for $M = 8$. Each plot shows both the discrete function (round dots) and the corresponding continuous function. Compared with the basis functions of the DFT (Figs. 18.11 and 18.12), all frequencies are divided in half and the DCT basis functions are phase-shifted by 0.5 units. All DCT basis functions are thus periodic over the length $2M = 16$ (as compared with M for the DFT).

$$G(m,n) = \frac{2}{\sqrt{MN}} \cdot \sum_{u=0}^{M-1} \sum_{v=0}^{N-1} \Bigl[g(u,v) \cdot c_m \cos\bigl(\tfrac{\pi(2u+1)m}{2M}\bigr) \cdot c_n \cos\bigl(\tfrac{\pi(2v+1)n}{2N}\bigr)\Bigr] \tag{20.6}$$

$$= \frac{2 \cdot c_m \cdot c_n}{\sqrt{MN}} \cdot \sum_{u=0}^{M-1} \sum_{v=0}^{N-1} \cdot \bigl[g(u,v) \cdot D_m^M(u) \cdot D_n^N(v)\bigr], \tag{20.7}$$

for $0 \leq m < M$, $0 \leq n < N$, and the inverse transform

$$g(u,v) = \frac{2}{\sqrt{MN}} \cdot \sum_{m=0}^{M-1} \sum_{n=0}^{N-1} \Bigl[G(m,n) \cdot c_m \cos\bigl(\tfrac{\pi(2u+1)m}{2M}\bigr) \cdot c_n \cos\bigl(\tfrac{\pi(2v+1)n}{2N}\bigr)\Bigr] \tag{20.8}$$

$$= \frac{2}{\sqrt{MN}} \cdot \sum_{m=0}^{M-1} \sum_{n=0}^{N-1} \bigl[G(m,n) \cdot c_m \cdot D_m^M(u) \cdot c_n \cdot D_n^N(v)\bigr], \tag{20.9}$$

for $0 \leq u < M$, $0 \leq v < N$. The coefficients c_m and c_n in Eqns. (20.7) and (20.9) are the same as in the 1D case (Eqn. (20.3)). Notice

Prog. 20.1
1D DCT (Java implementation). The method DCT() computes the forward transform for a real-valued signal vector g of arbitrary length according to the definition in Eqn. (20.1). The method returns the DCT spectrum as a real-valued vector of the same length as the input vector g. The inverse transform iDCT() computes the inverse DCT for the real-valued cosine spectrum G.

```
1  double[] DCT (double[] g) { // forward DCT on signal g
2    int M = g.length;
3    double s = Math.sqrt(2.0 / M); // common scale factor
4    double[] G = new double[M];
5    for (int m = 0; m < M; m++) {
6      double cm = 1.0;
7      if (m == 0)
8        cm = 1.0 / Math.sqrt(2);
9      double sum = 0;
10     for (int u = 0; u < M; u++) {
11       double Phi = Math.PI * m * (2 * u + 1) / (2 * M);
12       sum += g[u] * cm * Math.cos(Phi);
13     }
14     G[m] = s * sum;
15   }
16   return G;
17 }
18
19
20 double[] iDCT (double[] G) { // inverse DCT on spectrum G
21   int M = G.length;
22   double s = Math.sqrt(2.0 / M); //common scale factor
23   double[] g = new double[M];
24   for (int u = 0; u < M; u++) {
25     double sum = 0;
26     for (int m = 0; m < M; m++) {
27       double cm = 1.0;
28       if (m == 0)
29         cm = 1.0 / Math.sqrt(2);
30       double Phi = Math.PI * m * (2 * u + 1) / (2 * M);
31       sum += G[m] * cm * Math.cos(Phi);
32     }
33     g[u] = s * sum;
34   }
35   return g;
36 }
```

that in the forward transform (and only there!) the factors c_m, c_n are independent of the iteration variables u, v and can thus be placed outside the summation (as shown in Eqn. (20.7)).

20.2.1 Examples

Figure 20.2 shows several examples of the DCT in comparison with the results of the DFT. Since the DCT spectrum is (in contrast to the DFT spectrum) not symmetric, it does not get centered but is displayed in its original form with its origin at the upper left corner. The intensity corresponds to the logarithm of the absolute value in the case of the (real-valued) DCT spectrum. Similarly, the usual logarithmic power spectrum is shown for the DFT. Notice that the DCT is not simply a section of the DFT but obviously combines structures from adjacent quadrants of the Fourier spectrum.

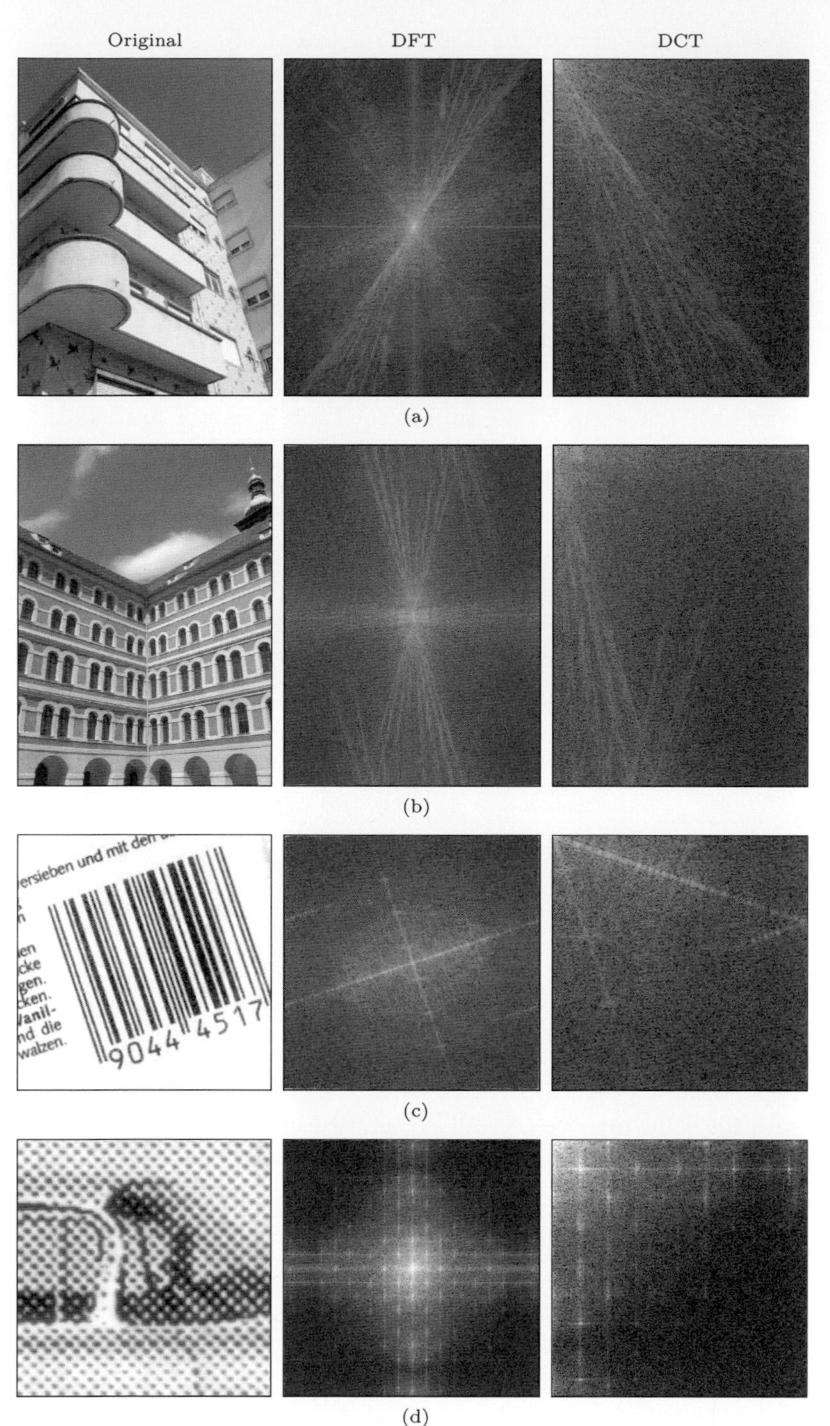

Fig. 20.2
2D DFT versus DCT. Both transforms show the frequency effects of image structures in a similar fashion. In the real-valued DCT spectrum (right), all coefficients are contained in a single quadrant and the frequency resolution is doubled compared with the DFT power spectrum (center). The DFT spectrum is centered as usual, while the origin of the DCT spectrum is located at the upper left corner. Both spectral plots display logarithmic intensity values.

20.2.2 Separability

Similar to the DFT (see Ch. 19, Eqn. (19.9)), the 2D DCT can also be separated into two successive 1D transforms. To make this fact clear, the forward DCT (Eqn. (20.7)) can be expressed in the form

$$G(m,n) = \sqrt{\tfrac{2}{N}} \cdot \sum_{v=0}^{N-1} \Bigl[\underbrace{\sqrt{\tfrac{2}{M}} \cdot \sum_{u=0}^{M-1} g(u,v) \cdot c_m \cdot D_m^M(u)}_{\text{1D DCT of } g(\cdot,v)} \cdot c_n \cdot D_n^N(v) \Bigr]. \tag{20.10}$$

The inner expression in Eqn. (20.10) corresponds to a 1D DCT of the vth line $g(\cdot, v)$ of the 2D signal function. Thus, as with the 2D DFT, one can first apply a 1D DCT to every line of an image and subsequently a DCT to each column. Of course, one could equally follow the reverse order by doing a DCT on the columns first and then on the rows.

The DCT is often used for image compression, in particular for JPEG where the size of the transformed sub-images is fixed at 8×8 and the processing can be highly optimized. Applying the DCT to square images (or sub-images) of size $M \times M$ is indeed an important special case. Here the DCT is commonly expressed in matrix form,

$$\mathbf{G} = \mathbf{A} \cdot \mathbf{g} \cdot \mathbf{A}^{\mathsf{T}}, \tag{20.11}$$

where the matrices $\mathbf{g}$ and $\mathbf{G}$ (both of size $M \times M$) represent the 2D signal and the resulting DCT spectrum, respectively. $\mathbf{A}$ is a quadratic, real-valued transformation matrix with the elements (cf. Eqn. (20.1))

$$A_{i,j} = \sqrt{\tfrac{2}{N}} \cdot c_i \cdot \cos\Bigl(\pi \cdot \frac{i \cdot (2j+1)}{2M}\Bigr), \tag{20.12}$$

with $0 \leq i, j < M$ and c_i as defined in Eqn. (20.3). The x/y separability of the DCT is easy to see in this notation. The matrix $\mathbf{A}$ is real-valued and *orthonormal*, i.e., $\mathbf{A} \cdot \mathbf{A}^{\mathsf{T}} = \mathbf{A}^{\mathsf{T}} \cdot \mathbf{A} = \mathbf{I}$ and so its transpose $\mathbf{A}^{\mathsf{T}}$ is identical to the inverse matrix $\mathbf{A}^{-1}$. Thus the associated inverse transformation from the DCT spectrum $\mathbf{G}$ back to the signal $\mathbf{g}$ can be carried out in the form

$$\mathbf{g} = \mathbf{A}^{\mathsf{T}} \cdot \mathbf{G} \cdot \mathbf{A}, \tag{20.13}$$

with the same matrices $\mathbf{A}$ and $\mathbf{A}^{\mathsf{T}}$ as used in the forward transform. For example, for $M = 4$ the DCT transformation matrix is

$$\mathbf{A} = \begin{pmatrix} A_{0,0} & A_{0,1} & A_{0,2} & A_{0,3} \\ A_{1,0} & A_{1,1} & A_{1,2} & A_{1,3} \\ A_{2,0} & A_{2,1} & A_{2,2} & A_{2,3} \\ A_{3,0} & A_{3,1} & A_{3,2} & A_{3,3} \end{pmatrix} \tag{20.14}$$

$$= \begin{pmatrix} \frac{1}{2}\cos(0) & \frac{1}{2}\cos(0) & \frac{1}{2}\cos(0) & \frac{1}{2}\cos(0) \\ \frac{1}{\sqrt{2}}\cos(\frac{\pi}{8}) & \frac{1}{\sqrt{2}}\cos(\frac{3\pi}{8}) & \frac{1}{\sqrt{2}}\cos(\frac{5\pi}{8}) & \frac{1}{\sqrt{2}}\cos(\frac{7\pi}{8}) \\ \frac{1}{\sqrt{2}}\cos(\frac{2\pi}{8}) & \frac{1}{\sqrt{2}}\cos(\frac{6\pi}{8}) & \frac{1}{\sqrt{2}}\cos(\frac{8\pi}{8}) & \frac{1}{\sqrt{2}}\cos(\frac{10\pi}{8}) \\ \frac{1}{\sqrt{2}}\cos(\frac{3\pi}{8}) & \frac{1}{\sqrt{2}}\cos(\frac{9\pi}{8}) & \frac{1}{\sqrt{2}}\cos(\frac{15\pi}{8}) & \frac{1}{\sqrt{2}}\cos(\frac{21\pi}{8}) \end{pmatrix} \tag{20.15}$$

$$\approx \begin{pmatrix} 0.50000 & 0.50000 & 0.50000 & 0.50000 \\ 0.65328 & 0.27060 & -0.27060 & -0.65328 \\ 0.50000 & -0.50000 & -0.50000 & 0.50000 \\ 0.27060 & -0.65328 & 0.65328 & -0.27060 \end{pmatrix}. \tag{20.16}$$

For the arbitrarily chosen 2D signal (i.e., "image")

$$\mathbf{g} = \begin{pmatrix} 1 & 2 & 3 & 4 \\ 7 & 2 & 0 & 9 \\ 6 & 5 & 2 & 5 \\ 0 & 9 & 8 & 1 \end{pmatrix}, \tag{20.17}$$

for example, the DCT spectrum obtained with Eqn. (20.11) is

$$\mathbf{G} = \mathbf{A}\cdot\mathbf{g}\cdot\mathbf{A}^\mathsf{T} \approx \begin{pmatrix} 16.00000 & -0.95671 & 0.50000 & -2.30970 \\ -2.61313 & -1.81066 & 6.57924 & 0.45711 \\ -2.00000 & -1.65642 & -8.50000 & 1.22731 \\ -1.08239 & 0.95711 & -1.10162 & 0.31066 \end{pmatrix}, \tag{20.18}$$

which is the same as the result from Eqn. (20.7) or, alternatively, Eqn. (20.10).

The matrix notation of the DCT, as shown in Eqn. (20.11) and Eqn. (20.13), is particularly useful for describing the transformation of small, fixed-size sub-images. This is an important component common in most image and video compression methods (including JPEG and MPEG) that calls for efficient implementations.

20.3 Java Implementation

A straightforward Java implementation of the one- and two-dimensional DCT is available online as part of the `imagingbook` library.[1] For efficiency reasons, the following methods generally work "in place", i.e., the supplied data array is destructively modified by the transformation.

`Dct1d` (class)

This class implements the *1D* DCT (see also Prog. 20.1):

`Dct1d (int M)`
: Constructor; `M` denotes the length of the expected signal.

`void DCT (double[] g)`
: Calculates the DCT spectrum of the one-dimensional signal `g`. The array `g` is modified, it's content being replaced by the resulting spectrum.

`void iDCT (double[] G)`
: Reconstructs the original signal from the one-dimensional DCT spectrum `G`. The array `G` is modified, it's content being replaced by the reconstructed signal.

Pre-calculated cosine tables are used in both the forward and inverse transformation for efficient processing.

`Dct2d` (class)

This class implements the *2D* DCT (by using class `Dct1d`):

`Dct2d ()`
: Constructor; in this case no dimension argumens are required.

[1] Package `imagingbook.pub.dct`.

`void DCT (float[][] g)`
Calculates the DCT spectrum of the 2D signal `g`. The array `g` is modified.

`void iDCT (float[][] G)`
Reconstructs the original signal from the two-dimensional DCT spectrum `G`. The array `G` is modified.

`FloatProcessor DCT (FloatProcessor g)`
Calculates the DCT spectrum of the image `g` and returns a new image with the resulting spectrum (`g` is not modified).

`FloatProcessor iDCT (FloatProcessor G)`
Calculates the inverse DCT from the 2D spectrum `G` and returns the reconstructed image (`G` is not modified).

20.4 Other Spectral Transforms

Apparently, the Fourier transform is not the only way to represent a given signal in frequency space; in fact, numerous similar transforms exist. Some of these, such as the discrete cosine transform, also use sinusoidal basis functions, while others, such as the *Hadamard* transform (also known as the *Walsh* transform), build on binary 0/1-functions [43, 126].

All of these transforms are of *global* nature; i.e., the value of any spectral coefficient is equally influenced by all signal values, independent of the spatial position in the signal. Thus a peak in the spectrum could be caused by a high-amplitude event of local extent as well as by a widespread, continuous wave of low amplitude. Global transforms are therefore of limited use for the purpose of detecting or analyzing local events because they are incapable of capturing the spatial position and extent of events in a signal.

A solution to this problem is to use a set of *local*, spatially limited basis functions ("wavelets") in place of the global, spatially fixed basis functions. The corresponding *wavelet transform*, of which several versions have been proposed, allows the simultaneous localization of repetitive signal components in both signal space *and* frequency space [158].

20.5 Exercises

Exercise 20.1. Implement an efficient ("hard-coded") Java method for computing the 1D DCT of length $M = 8$ that operates without iterations (loops) and contains all necessary coefficients as precomputed constants.

Exercise 20.2. Consider how the implementation of the one-dimensional DCT in Prog. 20.1 could be accelerated by using a precalculated table of the cosine values (for a given signal length M). Hint: A table of length $4M$ is required.

Exercise 20.3. Verify by numerical computation that the DCT basis functions $D_m^M(u)$ for $0 \leq m, u < M$ (Eqn. (20.5)) are pairwise orthogonal; i.e., the inner product of the vectors $D_m^M \cdot D_n^M$ is zero for any pair $m \neq n$.

Exercise 20.4. Implement the 2D DCT (Sec. 20.2) as an ImageJ plugin for images of arbitrary size. Make use of the fact that the 2D DCT can be performed as a sequence of 1D transforms (see Eqn. (20.10)).

Exercise 20.5. Verify for the 4×4 DCT example in Eqn. (20.18) that the result of the inverse transformation in Eqn. (20.13) is really identical to the original signal $\mathbf{g}$ in Eqn. (20.17).

Exercise 20.6. Show that the $M \times M$ matrix $\mathbf{A}$ (with elements as defined in Eqn. (20.12)) is really orthonormal, i.e., $\mathbf{A} \cdot \mathbf{A}^\mathsf{T} = \mathbf{I}$.

21

Geometric Operations

Common to all the filters and point operations described so far is the fact that they may change the intensity function of an image but the position of each pixel, and thus the geometry of the image, remains the same. The purpose of geometric operations, which are discussed in this chapter, is to deform an image by altering its geometry. Typical examples are shifting, rotating, or scaling images, as shown in Fig. 21.1. Geometric operations are frequently needed in practical applications, for example, in virtually any modern graphical computer interface. Today we take for granted that windows and images in graphic or video applications can be zoomed continuously to arbitrary size. Geometric image operations are also important in computer graphics where textures, which are usually raster images, are deformed to be mapped onto the corresponding 3D surfaces, possibly in real time. Of course, geometric operations are not as simple as their commonality may suggest. While it is obvious, for example, that an image could be enlarged by some integer factor n simply by replicating each pixel $n \times n$ times, the results would probably not be appealing, and it also gives us no immediate idea how to handle continuous scaling, rotating images, or other image deformations. In general, geometric operations that achieve high-quality results are not trivial to implement and are also computationally demanding, even on today's fast computers.

In principle, a geometric operation transforms a given image I to a new image I' by modifying the *coordinates* of image pixels,

$$I'(x', y') \leftarrow I(x, y), \qquad (21.1)$$

that is, the value of the image function I at the original location (x, y) moves to the new position (x', y') in the transformed image I'. Thus (at least in the continuous case) the *values* of the image elements do not change but only their *positions*.

To model this process, we first need a 2D transformation function or *geometric mapping* T, for example, in the form

$$T : \mathbb{R}^2 \to \mathbb{R}^2, \qquad (21.2)$$

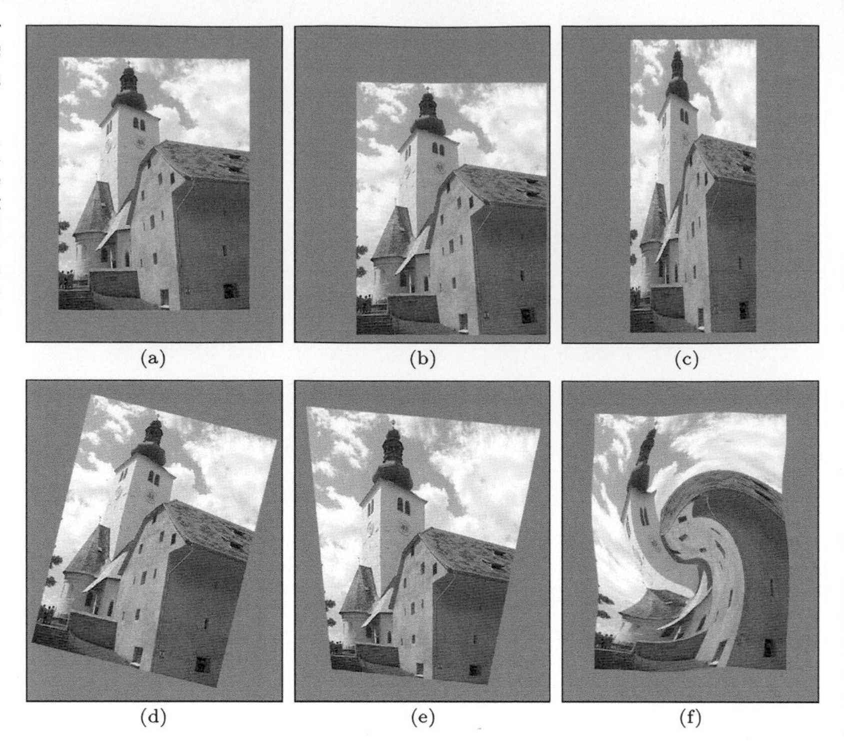

Fig. 21.1
Typical examples for geometric operations: original image (a), translation (b), scaling (contracting or stretching) in x and y directions (c), rotation about the center (d), projective transformation (e), and nonlinear distortion (f).

that specifies for each original 2D coordinate point $\boldsymbol{x} = (x, y)$ the corresponding target point $\boldsymbol{x}' = (x', y')$ in the new image I',

$$(x', y') = T(x, y). \tag{21.3}$$

Notice that the coordinates (x, y) and (x', y') specify points in the *continuous* image plane $\mathbb{R} \times \mathbb{R}$. The main problem in transforming digital images is that the pixels $I(u, v)$ are defined not on a continuous plane but on a *discrete* raster $\mathbb{Z} \times \mathbb{Z}$. Obviously, a transformed coordinate $(u', v') = T(u, v)$ produced by the mapping function $T()$ will, in general, no longer fall onto a discrete raster point. The solution to this problem is to compute intermediate pixel values for the transformed image by a process called *interpolation* (see Ch. 22), which is the second essential element in any geometric operation.

21.1 2D Coordinate Transformations

The mapping function $T()$ in Eqn. (21.3) is an arbitrary continuous function that for reasons of simplicity is often specified as two separate functions,

$$x' = T_{\mathrm{x}}(x, y) \qquad \text{and} \qquad y' = T_{\mathrm{y}}(x, y) \tag{21.4}$$

for the x and y components, respectively.

21.1.1 Simple Geometric Mappings

The simple mapping functions include translation, scaling, shearing, and rotation, defined as follows:

Translation (shift) by a vector (d_x, d_y):

$$\begin{aligned} T_{\mathrm{x}} &: x' = x + d_x \\ T_{\mathrm{y}} &: y' = y + d_y \end{aligned} \quad \text{or} \quad \begin{pmatrix} x' \\ y' \end{pmatrix} = \begin{pmatrix} x \\ y \end{pmatrix} + \begin{pmatrix} d_x \\ d_y \end{pmatrix}. \tag{21.5}$$

Scaling (contracting or stretching) along the x or y axis by the factor s_x or s_y, respectively:

$$\begin{aligned} T_{\mathrm{x}} &: x' = s_x \cdot x \\ T_{\mathrm{y}} &: y' = s_y \cdot y \end{aligned} \quad \text{or} \quad \begin{pmatrix} x' \\ y' \end{pmatrix} = \begin{pmatrix} s_x & 0 \\ 0 & s_y \end{pmatrix} \cdot \begin{pmatrix} x \\ y \end{pmatrix}. \tag{21.6}$$

Shearing along the x and y axis by the factor b_x and b_y, respectively (for shearing in only one direction, the other factor is set to zero):

$$\begin{aligned} T_{\mathrm{x}} &: x' = x + b_x \cdot y \\ T_{\mathrm{y}} &: y' = y + b_y \cdot x \end{aligned} \quad \text{or} \quad \begin{pmatrix} x' \\ y' \end{pmatrix} = \begin{pmatrix} 1 & b_x \\ b_y & 1 \end{pmatrix} \cdot \begin{pmatrix} x \\ y \end{pmatrix}. \tag{21.7}$$

Rotation by an angle α, with the coordinate origin being the center of rotation:

$$\begin{aligned} T_{\mathrm{x}} &: x' = x \cdot \cos\alpha - y \cdot \sin\alpha \\ T_{\mathrm{y}} &: y' = x \cdot \sin\alpha + y \cdot \cos\alpha \end{aligned} \quad \text{or} \tag{21.8}$$

$$\begin{pmatrix} x' \\ y' \end{pmatrix} = \begin{pmatrix} \cos\alpha & -\sin\alpha \\ \sin\alpha & \cos\alpha \end{pmatrix} \cdot \begin{pmatrix} x \\ y \end{pmatrix}. \tag{21.9}$$

Rotating the image by an angle α around an *arbitrary center point* $\boldsymbol{x}_c = (x_c, y_c)$ is accomplished by first translating the image by $(-x_c, -y_c)$, such that $\boldsymbol{x}_c$ coincides with the origin, then rotating the image about the origin (as in Eqn. (21.9)), and finally shifting the image back by (x_c, y_c). The resulting composite transformation is

$$\begin{aligned} T_{\mathrm{x}} &: x' = x_c + (x-x_c) \cdot \cos\alpha - (y-y_c) \cdot \sin\alpha \\ T_{\mathrm{y}} &: y' = y_c + (x-x_c) \cdot \sin\alpha + (y-y_c) \cdot \cos\alpha \end{aligned} \tag{21.10}$$

or

$$\begin{pmatrix} x' \\ y' \end{pmatrix} = \begin{pmatrix} x_c \\ y_c \end{pmatrix} + \begin{pmatrix} \cos\alpha & -\sin\alpha \\ \sin\alpha & \cos\alpha \end{pmatrix} \cdot \begin{pmatrix} x-x_c \\ y-y_c \end{pmatrix}. \tag{21.11}$$

The combination of the operations listed in Eqns. (21.5)–(21.9) constitute the important class of "affine" transformations or *affine mappings* (see also Sec. 21.1.3).

21.1.2 Homogeneous Coordinates

To simplify the concatenation of linear mappings, it is advantageous to specify all operations in the form of vector-matrix multiplications, as in Eqns. (21.6)–(21.9). Note that pure translation Eqn. (21.5), which corresponds to a vector addition, cannot be formulated as a vector-matrix multiplication. Fortunately, this difficulty can be elegantly resolved with so-called *homogeneous coordinates* (see, e.g., [75, p. 204]).[1]

[1] See also Sec. B.5 in the Appendix.

To turn an "ordinary" (i.e., *Cartesian*) coordinate into a homogeneous coordinate, the original vector is simply extended by an additional element with constant value 1. For example, a 2D Cartesian point $\boldsymbol{x} = (x, y)^\mathsf{T}$ converts to a 3D vector,

$$\mathrm{hom}(\boldsymbol{x}) = \mathrm{hom}\begin{pmatrix} x \\ y \end{pmatrix} = \begin{pmatrix} x \\ y \\ 1 \end{pmatrix} = \underline{\boldsymbol{x}}. \tag{21.12}$$

Note that the homogeneous representation is not unique, but any scaled vector $s \cdot \underline{\boldsymbol{x}}$ is an equivalent homogeneous representation of the Cartesian coordinate $\boldsymbol{x}$, that is

$$\boldsymbol{x} = \mathrm{hom}^{-1}(\underline{\boldsymbol{x}}) = \mathrm{hom}^{-1}(s \cdot \underline{\boldsymbol{x}}), \tag{21.13}$$

for any nonzero $s \in \mathbb{R}$. For example, the homogeneous coordinates $\underline{\boldsymbol{x}}_1 = (3, 2, 1)^\mathsf{T}$, $\underline{\boldsymbol{x}}_2 = (-6, -4, -2)^\mathsf{T}$, and $\underline{\boldsymbol{x}}_3 = (30, 20, 10)^\mathsf{T}$ are all equivalent representations of the same Cartesian coordinate $\boldsymbol{x} = (3, 2)^\mathsf{T}$.

The reverse mapping from a 3D homogeneous coordinate $\underline{\boldsymbol{x}} = (\underline{x}, \underline{y}, \underline{z})^\mathsf{T}$ to the corresponding 2D Cartesian coordinate is denoted

$$\mathrm{hom}^{-1}(\underline{\boldsymbol{x}}) = \mathrm{hom}^{-1}\begin{pmatrix} \underline{x} \\ \underline{y} \\ \underline{z} \end{pmatrix} = \frac{1}{\underline{z}} \cdot \begin{pmatrix} \underline{x} \\ \underline{y} \end{pmatrix} = \boldsymbol{x} \tag{21.14}$$

With the help of homogeneous coordinates, we can now define a 2D *translation* (Eqn. (21.5)) as a vector-matrix product in the form

$$\begin{pmatrix} x' \\ y' \end{pmatrix} = \mathrm{hom}^{-1}\Bigl[\begin{pmatrix} 1 & 0 & d_x \\ 0 & 1 & d_y \\ 0 & 0 & 1 \end{pmatrix} \cdot \mathrm{hom}\begin{pmatrix} x \\ y \end{pmatrix}\Bigr] \tag{21.15}$$

$$= \begin{pmatrix} 1 & 0 & d_x \\ 0 & 1 & d_y \end{pmatrix} \cdot \begin{pmatrix} x \\ y \\ 1 \end{pmatrix} = \begin{pmatrix} x + d_x \\ y + d_y \end{pmatrix}, \tag{21.16}$$

which had been our motive for introducing homogeneous coordinates in the first place. As we shall see in the following sections, homogeneous coordinates allow us to write many common 2D coordinate transformations in the form

$$\underline{\boldsymbol{x}}' = \mathbf{A} \cdot \underline{\boldsymbol{x}}\,, \tag{21.17}$$

where $\mathbf{A}$ is a 3×3 matrix. Note that (due to the relation in Eqn. (21.13)) multiplying the matrix $\mathbf{A}$ by some scalar factor s yields the same transformation in terms of Cartesian coordinates, that is,

$$\boldsymbol{x}' = \mathrm{hom}^{-1}[\mathbf{A} \cdot \underline{\boldsymbol{x}}] = \mathrm{hom}^{-1}[s \cdot (\mathbf{A} \cdot \underline{\boldsymbol{x}})] = \mathrm{hom}^{-1}[(s \cdot \mathbf{A}) \cdot \underline{\boldsymbol{x}}], \tag{21.18}$$

for any nonzero $s \in \mathbb{R}$.

21.1.3 Affine (Three-Point) Mapping

In general, and analogous to Eqn. (21.16), we can express any combination of 2D translation, scaling, and rotation as vector-matrix multiplication in homogeneous coordinates in the form

$$\underline{x}' = \mathbf{A}_{\mathsf{affine}} \cdot \underline{x} \tag{21.19}$$

or $\boldsymbol{x}' = \mathrm{hom}^{-1}[\mathbf{A}_{\mathsf{affine}} \cdot \mathrm{hom}(\boldsymbol{x})]$ in Cartesian coordinates, that is,

$$\begin{pmatrix} x' \\ y' \end{pmatrix} = \mathrm{hom}^{-1}\left[\begin{pmatrix} a_{00} & a_{01} & a_{02} \\ a_{10} & a_{11} & a_{12} \\ 0 & 0 & 1 \end{pmatrix} \cdot \begin{pmatrix} x \\ y \\ 1 \end{pmatrix}\right] = \begin{pmatrix} a_{00} & a_{01} & a_{02} \\ a_{10} & a_{11} & a_{12} \end{pmatrix} \cdot \begin{pmatrix} x \\ y \\ 1 \end{pmatrix}. \tag{21.20}$$

This 2D coordinate transformation is called an "affine mapping" with the six parameters $a_{00}, \ldots, a_{12}$, where a_{02}, a_{12} specify the translation (equivalent to d_x, d_y in Eqn. (21.5)) and a_{00}, a_{01}, a_{10}, a_{11} aggregate the scaling, shearing, and rotation coefficients (see Eqns. (21.6)–(21.9)). For example, the affine transformation matrix for a rotation about the origin by an angle α is specified by the matrix

$$\mathbf{A}_{\mathrm{rot}} = \begin{pmatrix} a_{00} & a_{01} & a_{02} \\ a_{10} & a_{11} & a_{12} \\ 0 & 0 & 1 \end{pmatrix} = \begin{pmatrix} \cos\alpha & -\sin\alpha & 0 \\ \sin\alpha & \cos\alpha & 0 \\ 0 & 0 & 1 \end{pmatrix}. \tag{21.21}$$

In this way, compound transformations can be constructed easily by consecutive matrix multiplications (from right to left). For example, the transformation matrix for a rotation by α about a given center point $\boldsymbol{x}_c = (x_c, y_c)^\mathsf{T}$ (see Eqn. (21.11)), composed by a translation to the origin followed by a rotation and another translation, is

$$\mathbf{A} = \underbrace{\begin{pmatrix} 1 & 0 & x_c \\ 0 & 1 & y_c \\ 0 & 0 & 1 \end{pmatrix}}_{\substack{\text{translation by} \\ (x_c, y_c)^\mathsf{T}}} \cdot \underbrace{\begin{pmatrix} \cos\alpha & -\sin\alpha & 0 \\ \sin\alpha & \cos\alpha & 0 \\ 0 & 0 & 1 \end{pmatrix}}_{\substack{\text{rotation by } \alpha \\ \text{(about the origin)}}} \cdot \underbrace{\begin{pmatrix} 1 & 0 & -x_c \\ 0 & 1 & -y_c \\ 0 & 0 & 1 \end{pmatrix}}_{\substack{\text{translation by} \\ (-x_c, -y_c)^\mathsf{T}}} \tag{21.22}$$

$$= \begin{pmatrix} 1 & 0 & x_c \\ 0 & 1 & y_c \\ 0 & 0 & 1 \end{pmatrix} \cdot \begin{pmatrix} \cos\alpha & -\sin\alpha & 0 \\ \sin\alpha & \cos\alpha & 0 \\ 0 & 0 & 1 \end{pmatrix} \cdot \begin{pmatrix} 1 & 0 & x_c \\ 0 & 1 & y_c \\ 0 & 0 & 1 \end{pmatrix}^{-1} \tag{21.23}$$

$$= \begin{pmatrix} \cos\alpha & -\sin\alpha & x_c \cdot (1-\cos\alpha) + y_c \cdot \sin\alpha \\ \sin\alpha & \cos\alpha & y_c \cdot (1-\cos\alpha) - x_c \cdot \sin\alpha \\ 0 & 0 & 1 \end{pmatrix}. \tag{21.24}$$

Of course, the result is the same as in Eqn. (21.10).

Note that multiplying two affine transformation matrices always yields another affine transformation. Also, an affine transformation maps straight lines to straight lines, triangles to triangles, and rectangles to parallelograms, as illustrated in Fig. 21.2. The distance ratio between points on a straight line remains unchanged by this type of mapping function.

Affine transformation parameters from three point pairs

The six parameters of the 2D affine mapping (Eqn. (21.20)) are uniquely determined by three pairs of corresponding points $(\boldsymbol{x}_0, \boldsymbol{x}'_1)$, $(\boldsymbol{x}_1, \boldsymbol{x}'_1)$, $(\boldsymbol{x}_2, \boldsymbol{x}'_2)$, with the first point $\boldsymbol{x}_i = (x_i, y_i)$ of each pair located in the original image and the corresponding point $\boldsymbol{x}'_i = (x'_i, y'_i)$ located in the target image. From these six coordinate values, the

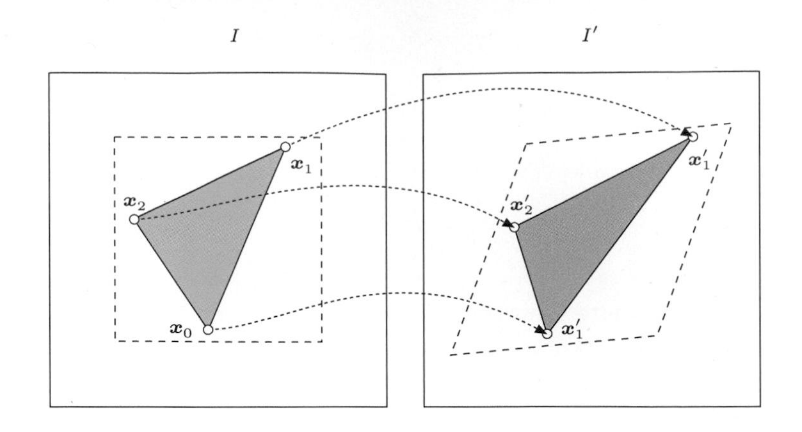

Fig. 21.2
Affine mapping. An affine 2D transformation is uniquely specified by three pairs of corresponding points; for example, $(\boldsymbol{x}_0, \boldsymbol{x}'_1)$, $(\boldsymbol{x}_1, \boldsymbol{x}'_1)$, and $(\boldsymbol{x}_2, \boldsymbol{x}'_2)$.

six transformation parameters $a_{00}, \ldots, a_{12}$ are derived by solving the system of linear equations

$$
\begin{aligned}
x'_0 &= a_{00} \cdot x_0 + a_{01} \cdot y_0 + a_{02}, & y'_0 &= a_{10} \cdot x_0 + a_{11} \cdot y_0 + a_{12}, \\
x'_1 &= a_{00} \cdot x_1 + a_{01} \cdot y_1 + a_{02}, & y'_1 &= a_{10} \cdot x_1 + a_{11} \cdot y_1 + a_{12}, \\
x'_2 &= a_{00} \cdot x_2 + a_{01} \cdot y_2 + a_{02}, & y'_2 &= a_{10} \cdot x_2 + a_{11} \cdot y_2 + a_{12},
\end{aligned}
\tag{21.25}
$$

provided that the points (vectors) $\boldsymbol{x}_0$, $\boldsymbol{x}_1$, $\boldsymbol{x}_2$ are linearly independent (i.e., that they do not lie on a common straight line). Since Eqn. (21.25) consists of two independent sets of linear 3×3 equations for x'_i and y'_i, the solution can be written in closed form as

$$
\begin{aligned}
a_{00} &= \tfrac{1}{d} \cdot [y_0(x'_1 - x'_2) + y_1(x'_2 - x'_0) + y_2(x'_0 - x'_1)], \\
a_{01} &= \tfrac{1}{d} \cdot [x_0(x'_2 - x'_1) + x_1(x'_0 - x'_2) + x_2(x'_1 - x'_0)], \\
a_{10} &= \tfrac{1}{d} \cdot [y_0(y'_1 - y'_2) + y_1(y'_2 - y'_0) + y_2(y'_0 - y'_1)], \\
a_{11} &= \tfrac{1}{d} \cdot [x_0(y'_2 - y'_1) + x_1(y'_0 - y'_2) + x_2(y'_1 - y'_0)], \\
a_{02} &= \tfrac{1}{d} \cdot [x_0(y_2 x'_1 - y_1 x'_2) + x_1(y_0 x'_2 - y_2 x'_0) + x_2(y_1 x'_0 - y_0 x'_1)], \\
a_{12} &= \tfrac{1}{d} \cdot [x_0(y_2 y'_1 - y_1 y'_2) + x_1(y_0 y'_2 - y_2 y'_0) + x_2(y_1 y'_0 - y_0 y'_1)],
\end{aligned}
\tag{21.26}
$$

with $d = x_0(y_2 - y_1) + x_1(y_0 - y_2) + x_2(y_1 - y_0)$.

Inverse affine mapping

The inverse of the affine transformation, which is often required in practice (see Sec. 21.2.2), can be calculated by simply applying the inverse of the transformation matrix $\mathbf{A}_{\text{affine}}$ (Eqn. (21.20)) in homogeneous coordinate space, that is,

$$
\underline{\boldsymbol{x}} = \mathbf{A}_{\text{affine}}^{-1} \cdot \underline{\boldsymbol{x}}' \tag{21.27}
$$

or $\boldsymbol{x} = \text{hom}^{-1}\left[\mathbf{A}_{\text{affine}}^{-1} \cdot \text{hom}(\boldsymbol{x}')\right]$ in Cartesian coordinates, that is,

$$\begin{pmatrix} x \\ y \end{pmatrix} = \mathrm{hom}^{-1}\Bigg[\begin{pmatrix} a_{00} & a_{01} & a_{02} \\ a_{10} & a_{11} & a_{12} \\ 0 & 0 & 1 \end{pmatrix}^{-1} \cdot \begin{pmatrix} x' \\ y' \\ 1 \end{pmatrix}\Bigg] \tag{21.28}$$

$$= \mathrm{hom}^{-1}\Bigg[\underbrace{\frac{1}{a_{00}a_{11}-a_{01}a_{10}} \cdot \begin{pmatrix} a_{11} & -a_{01} & a_{01}a_{12}-a_{02}a_{11} \\ -a_{10} & a_{00} & a_{02}a_{10}-a_{00}a_{12} \\ 0 & 0 & a_{00}a_{11}-a_{01}a_{10} \end{pmatrix}}_{\mathbf{A}_{\mathrm{affine}}^{-1}} \cdot \begin{pmatrix} x' \\ y' \\ 1 \end{pmatrix}\Bigg] \tag{21.29}$$

$$= \frac{1}{a_{00}a_{11}-a_{01}a_{10}} \cdot \begin{pmatrix} a_{11} & -a_{01} & a_{01}a_{12}-a_{02}a_{11} \\ -a_{10} & a_{00} & a_{02}a_{10}-a_{00}a_{12} \end{pmatrix} \cdot \begin{pmatrix} x' \\ y' \end{pmatrix}. \tag{21.30}$$

Since the bottom row of $\mathbf{A}_{\mathrm{affine}}^{-1}$ in Eqn. (21.29) consists of the elements $(0, 0, 1)$, the inverse mapping is again an affine transformation. Of course, the inverse of the affine mapping can also be found directly (i.e., without inverting the transformation matrix) from the given point coordinates $(\boldsymbol{x}_i, \boldsymbol{x}'_i)$ by using Eqns. (21.25) and (21.26) with *interchanged* source and target coordinates.

21.1.4 Projective (Four-Point) Mapping

In contrast to the affine transformation, which provides a mapping between arbitrary triangles, the projective transformation is a linear mapping between arbitrary *quadrilaterals* (Fig. 21.3). This is particularly useful for deforming images controlled by mesh partitioning, as described in Sec. 21.1.7. To map from an arbitrary sequence of four 2D points $(\boldsymbol{x}_0, \boldsymbol{x}_1, \boldsymbol{x}_2, \boldsymbol{x}_3)$ to a set of corresponding points $(\boldsymbol{x}'_0, \boldsymbol{x}'_1, \boldsymbol{x}'_2, \boldsymbol{x}'_3)$, the transformation requires eight degrees of freedom, two more than needed for the affine transformation. Analogous to the affine transformation (Eqn. (21.20)), a projective transformation can be expressed as a linear mapping in homogeneous coordinates,

$$\underline{\boldsymbol{x}}' = \mathbf{A}_{\mathrm{proj}} \cdot \underline{\boldsymbol{x}} \tag{21.31}$$

or $\boldsymbol{x}' = \mathrm{hom}^{-1}[\mathbf{A}_{\mathrm{proj}} \cdot \mathrm{hom}(\boldsymbol{x})]$ in Cartesian coordinates, that is,

$$\begin{pmatrix} x' \\ y' \end{pmatrix} = \mathrm{hom}^{-1}\Bigg[\begin{pmatrix} a_{00} & a_{01} & a_{02} \\ a_{10} & a_{11} & a_{12} \\ a_{20} & a_{21} & 1 \end{pmatrix} \cdot \begin{pmatrix} x \\ y \\ 1 \end{pmatrix}\Bigg] \tag{21.32}$$

$$= \frac{1}{a_{20}\cdot x + a_{21}\cdot y + 1} \cdot \begin{pmatrix} a_{00} & a_{01} & a_{02} \\ a_{10} & a_{11} & a_{12} \end{pmatrix} \cdot \begin{pmatrix} x \\ y \\ 1 \end{pmatrix}, \tag{21.33}$$

with the two additional elements (parameters) a_{20} and a_{21} in the transformation matrix $\mathbf{A}_{\mathrm{proj}}$. Because x, y appear in the denominator of the fraction in Eqn. (21.33), the projective mapping is generally *nonlinear* in Cartesian coordinates. Despite this nonlinearity, straight lines are preserved under this transformation. In fact, this is the most general transformation that maps straight lines to straight lines in 2D, and it actually maps any Nth-order algebraic curve onto another Nth-order algebraic curve. In particular, circles and ellipses always transform into other second-order curves (i.e., conic sections). Unlike the affine transformation, however, parallel lines do not generally map to parallel lines under a projective transformation (cf. Fig.

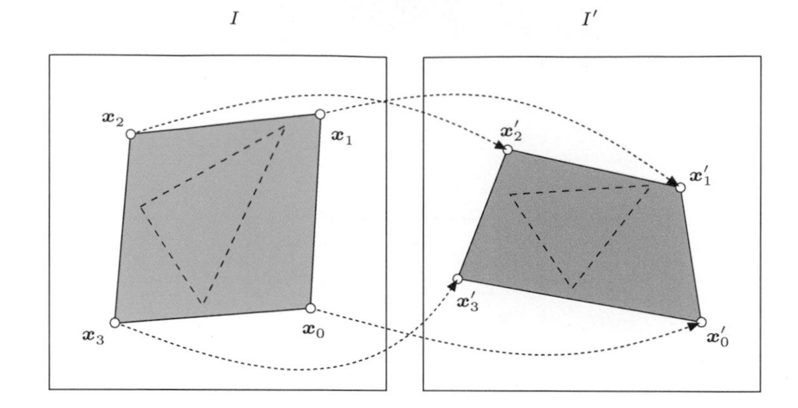

Fig. 21.3
Projective mapping. Four pairs of corresponding 2D points, $(\boldsymbol{x}_0, \boldsymbol{x}'_0)$, $(\boldsymbol{x}_1, \boldsymbol{x}'_1)$, $(\boldsymbol{x}_2, \boldsymbol{x}'_2)$, $(\boldsymbol{x}_3, \boldsymbol{x}'_3)$ uniquely specify a projective transformation. Straight lines are again mapped to straight lines, and a rectangle is mapped to some quadrilateral. In general, neither parallelism between straight lines nor the distance ratio is preserved.

21.3) and the distance ratios between points on a line are not preserved. The projective mapping is therefore sometimes referred to as "perspective" or "pseudo-perspective".

Projective transformation parameters from four point pairs

Given four pairs of corresponding 2D points, $(\boldsymbol{x}_0, \boldsymbol{x}'_0), \ldots, (\boldsymbol{x}_3, \boldsymbol{x}'_3)$, with one point $\boldsymbol{x}_i = (x_i, y_i)^\mathsf{T}$ in the source image and the second point $\boldsymbol{x}'_i = (x'_i, y'_i)^\mathsf{T}$ in the target image, the eight unknown transformation parameters $a_{00}, \ldots, a_{21}$ can be found by solving a system of linear equations. Multiplying Eqn. (21.33) by the common denominator on the right hand side gives

$$
\begin{aligned}
x' \cdot (a_{20} \cdot x + a_{21} \cdot y + 1) &= a_{00} \cdot x + a_{01} \cdot y + a_{02}, \\
y' \cdot (a_{20} \cdot x + a_{21} \cdot y + 1) &= a_{10} \cdot x + a_{11} \cdot y + a_{12},
\end{aligned}
\tag{21.34}
$$

and thus

$$
\begin{aligned}
a_{20} \cdot x \cdot x' + a_{21} \cdot y \cdot x' + x' &= a_{00} \cdot x + a_{01} \cdot y + a_{02}, \\
a_{20} \cdot x \cdot y' + a_{21} \cdot y \cdot y' + y' &= a_{10} \cdot x + a_{11} \cdot y + a_{12},
\end{aligned}
\tag{21.35}
$$

for any pair of corresponding points $\boldsymbol{x} = (x, y)^\mathsf{T}$ and $\boldsymbol{x}' = (x', y')^\mathsf{T}$. By slightly rearranging Eqn. (21.35) and inserting the (known) source and target point coordinates (x_i, y_i) and (x'_i, y'_i), respectively, we obtain one such pair of linear equations

$$
\begin{aligned}
x'_i &= a_{00} \cdot x_i + a_{01} \cdot y_i + a_{02} - a_{20} \cdot x_i \cdot x'_i - a_{21} \cdot y_i \cdot x'_i, \\
y'_i &= a_{10} \cdot x_i + a_{11} \cdot y_i + a_{12} - a_{20} \cdot x_i \cdot y'_i - a_{21} \cdot y_i \cdot y'_i,
\end{aligned}
\tag{21.36}
$$

for each point pair $i = 0, \ldots, 3$ and the eight unknowns $a_{00}, \ldots, a_{21}$. Combining the resulting eight equations in the usual matrix notation yields

$$
\begin{pmatrix} x'_0 \\ y'_0 \\ x'_1 \\ y'_1 \\ x'_2 \\ y'_2 \\ x'_3 \\ y'_3 \end{pmatrix}
=
\begin{pmatrix}
x_0 & y_0 & 1 & 0 & 0 & 0 & -x_0 x'_0 & -y_0 x'_0 \\
0 & 0 & 0 & x_0 & y_0 & 1 & -x_0 y'_0 & -y_0 y'_0 \\
x_1 & y_1 & 1 & 0 & 0 & 0 & -x_1 x'_1 & -y_1 x'_1 \\
0 & 0 & 0 & x_1 & y_1 & 1 & -x_1 y'_1 & -y_1 y'_1 \\
x_2 & y_2 & 1 & 0 & 0 & 0 & -x_2 x'_2 & -y_2 x'_2 \\
0 & 0 & 0 & x_2 & y_2 & 1 & -x_2 y'_2 & -y_2 y'_2 \\
x_3 & y_3 & 1 & 0 & 0 & 0 & -x_3 x'_3 & -y_3 x'_3 \\
0 & 0 & 0 & x_3 & y_3 & 1 & -x_3 y'_3 & -y_3 y'_3
\end{pmatrix}
\cdot
\begin{pmatrix} a_{00} \\ a_{01} \\ a_{02} \\ a_{10} \\ a_{11} \\ a_{12} \\ a_{20} \\ a_{21} \end{pmatrix},
\tag{21.37}
$$

or

$$\boldsymbol{b} = \mathbf{M} \cdot \boldsymbol{a}. \tag{21.38}$$

Note that all elements of the vector $\boldsymbol{b} = (x'_0, \ldots, y'_3)^\mathsf{T}$ and the matrix $\mathbf{M}$ are obtained from the specified point coordinates and are thus constants. The unknown parameters $\boldsymbol{a} = (a_{00}, \ldots, a_{21})^\mathsf{T}$ can be found by solving the system of linear equations in Eqn. (21.38) with standard numerical methods, for example, the Gauss algorithm [35, p. 276]. It is recommended to use proven numerical software for this purpose.[2]

If we want to use *more than four* corresponding point pairs to recover the eight parameters of a projective transformation, the system of linear equations in Eqn. (21.37) becomes overdetermined, that is, the system has more equations than unknowns. In general, n pairs of corresponding points yield a stack of $2n$ equations, so the vector $\boldsymbol{b}$ in Eqn. (21.37) has the length $2n$ and the matrix $\mathbf{M}$ is of size $2n{\times}8$ (vector $\boldsymbol{a}$ remains the same). Overdetermined systems like this can be solved in a least-squares sense (minimizing $\|\mathbf{M} \cdot \boldsymbol{a} - \boldsymbol{b}\|$), for example, using the singular-value (SVD) or QR decomposition of $\mathbf{M}$ [96,145].[3] Other solutions for the multi-point case are discussed later in this section (see p. 524).

Inverse projective mapping

In general, any *linear* transformation of the form $\underline{\boldsymbol{x}}' = \mathbf{A} \cdot \underline{\boldsymbol{x}}$ (in homogeneous coordinates $\underline{\boldsymbol{x}}$, $\underline{\boldsymbol{x}}'$) can be inverted by applying the inverse of the matrix $\mathbf{A}$, that is,

$$\underline{\boldsymbol{x}} = \mathbf{A}^{-1} \cdot \underline{\boldsymbol{x}}' \tag{21.39}$$

provided that $\mathbf{A}$ is nonsingular ($\det(\mathbf{A}) \neq 0$). The inverse of a 3×3 matrix $\mathbf{A}$ is comparatively easy to find in closed form using the relation

$$\mathbf{A}^{-1} = \frac{1}{\det(\mathbf{A})} \cdot \operatorname{adj}(\mathbf{A}), \tag{21.40}$$

with the determinant $\det(\mathbf{A})$ and the *adjugate* matrix $\operatorname{adj}(\mathbf{A})$ (see, e.g., [35, pp. 251, 260], [145, p. 219]). In particular, for a real-valued 3×3 matrix

$$\mathbf{A} = \begin{pmatrix} a_{00} & a_{01} & a_{02} \\ a_{10} & a_{11} & a_{12} \\ a_{20} & a_{21} & a_{22} \end{pmatrix}, \tag{21.41}$$

the determinant can be calculated as

$$\begin{aligned} \det(\mathbf{A}) = \; & a_{00}\, a_{11}\, a_{22} + a_{01}\, a_{12}\, a_{20} + a_{02}\, a_{10}\, a_{21} \\ & - a_{00}\, a_{12}\, a_{21} - a_{01}\, a_{10}\, a_{22} - a_{02}\, a_{11}\, a_{20}, \end{aligned} \tag{21.42}$$

and the 3×3 adjugate matrix is

$$\operatorname{adj}(\mathbf{A}) = \begin{pmatrix} a_{11}a_{22} - a_{12}a_{21} & a_{02}a_{21} - a_{01}a_{22} & a_{01}a_{12} - a_{02}a_{11} \\ a_{12}a_{20} - a_{10}a_{22} & a_{00}a_{22} - a_{02}a_{20} & a_{02}a_{10} - a_{00}a_{12} \\ a_{10}a_{21} - a_{11}a_{20} & a_{01}a_{20} - a_{00}a_{21} & a_{00}a_{11} - a_{01}a_{10} \end{pmatrix}. \tag{21.43}$$

[2] See Sec. B.7.1 in the Appendix.

[3] See Sec. B.7.2 in the Appendix.

In the special case of a projective mapping, the coefficient $a_{22} = 1$ (Eqn. (21.32)), which slightly simplifies the calculation.

Since scalar multiples of homogeneous vectors are all equivalent in Cartesian space (see Eqn. (21.18)), the multiplication by the constant factor $1/\det(\mathbf{A})$ in Eqn. (21.40) can be omitted. Thus, to invert a linear 2D transformation specified by a 3×3 matrix $\mathbf{A}$, we only need to multiply the homogeneous coordinate vector with the adjugate matrix $\operatorname{adj}(\mathbf{A})$, that is,

$$\underline{\boldsymbol{x}} = \mathbf{A}^{-1} \cdot \underline{\boldsymbol{x}}' \equiv \operatorname{adj}(\mathbf{A}) \cdot \underline{\boldsymbol{x}}'. \tag{21.44}$$

Returning to Cartesian coordinates, the inverse transformation can be written as

$$\boldsymbol{x} = \operatorname{hom}^{-1}[\operatorname{adj}(\mathbf{A}) \cdot \operatorname{hom}(\boldsymbol{x}')]. \tag{21.45}$$

This method can be used to invert any linear transformation in 2D, including the affine and projective mapping functions described already. Consequently, the inversion of the *affine* transformation shown earlier (see Eqn. (21.29)) is only a special case of this general method.

Of course, matrix inversion may also be implemented with standard linear algebra software, which is not only less error-prone but also offers better numerical stability (see also Sec. B.1 in the Appendix).

Projective mapping via the unit square

An alternative method for finding the projective mapping parameters for a given set of image points is to use a two-stage mapping through the unit square $\mathcal{S}_1$, which avoids iteratively solving a system of equations [256, p. 55] [105]. The projective mapping, shown in Fig. 21.4, from the four corner points of the unit square $\mathcal{S}_1$ to an arbitrary quadrilateral $\mathcal{Q} = (\boldsymbol{x}'_0, \ldots, \boldsymbol{x}'_3)$ with

$$\begin{aligned} (0,0) &\mapsto \boldsymbol{x}'_0, & (1,1) &\mapsto \boldsymbol{x}'_2, \\ (1,0) &\mapsto \boldsymbol{x}'_1, & (0,1) &\mapsto \boldsymbol{x}'_3, \end{aligned} \tag{21.46}$$

reduces the system of equations in Eqn. (21.37) to

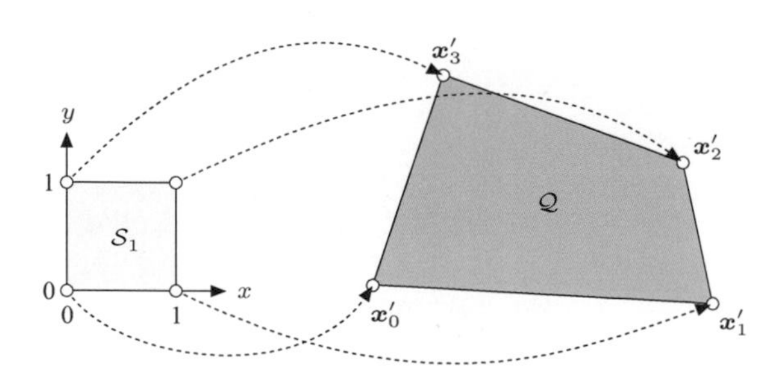

Fig. 21.4 Projective mapping from the unit square $\mathcal{S}_1$ to an arbitrary quadrilateral $\mathcal{Q} = (\boldsymbol{x}'_0, \ldots, \boldsymbol{x}'_3)$.

$$
\begin{aligned}
x'_0 &= a_{02}, \\
y'_0 &= a_{12}, \\
x'_1 &= a_{00} + a_{02} - a_{20} \cdot x'_1, \\
y'_1 &= a_{10} + a_{12} - a_{20} \cdot y'_1, \\
x'_2 &= a_{00} + a_{01} + a_{02} - a_{20} \cdot x'_2 - a_{21} \cdot x'_2, \\
y'_2 &= a_{10} + a_{11} + a_{12} - a_{20} \cdot y'_2 - a_{21} \cdot y'_2, \\
x'_3 &= a_{01} + a_{02} - a_{21} \cdot x'_3, \\
y'_3 &= a_{11} + a_{12} - a_{21} \cdot y'_3.
\end{aligned} \tag{21.47}
$$

This set of equations has the following closed-form solution for the eight unknown transformation parameters $a_{00}, a_{01}, \ldots, a_{21}$:

$$
a_{20} = \frac{(x'_0 - x'_1 + x'_2 - x'_3) \cdot (y'_3 - y'_2) - (y'_0 - y'_1 + y'_2 - y'_3) \cdot (x'_3 - x'_2)}{(x'_1 - x'_2) \cdot (y'_3 - y'_2) - (x'_3 - x'_2) \cdot (y'_1 - y'_2)}, \tag{21.48}
$$

$$
a_{21} = \frac{(y'_0 - y'_1 + y'_2 - y'_3) \cdot (x'_1 - x'_2) - (x'_0 - x'_1 + x'_2 - x'_3) \cdot (y'_1 - y'_2)}{(x'_1 - x'_2) \cdot (y'_3 - y'_2) - (x'_3 - x'_2) \cdot (y'_1 - y'_2)} \tag{21.49}
$$

and

$$
a_{00} = x'_1 - x'_0 + a_{20}\, x'_1, \quad a_{01} = x'_3 - x'_0 + a_{21}\, x'_3, \quad a_{02} = x'_0, \tag{21.50}
$$

$$
a_{10} = y'_1 - y'_0 + a_{20}\, y'_1, \quad a_{11} = y'_3 - y'_0 + a_{21}\, y'_3, \quad a_{12} = y'_0. \tag{21.51}
$$

By calculating the inverse of the corresponding 3×3 transformation matrix (Eqn. (21.40)), the mapping may be *reversed* to transform an arbitrary quadrilateral to the unit square. A mapping T between two arbitrary quadrilaterals,

$$
\mathcal{Q} \xrightarrow{T} \mathcal{Q}',
$$

can thus be implemented by combining a reversed mapping and a forward mapping via the unit square. As illustrated in Fig. 21.5, the transformation of an arbitrary quadrilateral $\mathcal{Q} = (\boldsymbol{x}_0, \boldsymbol{x}_1, \boldsymbol{x}_2, \boldsymbol{x}_3)$ to a second quadrilateral $\mathcal{Q}' = (\boldsymbol{x}'_0, \boldsymbol{x}'_1, \boldsymbol{x}'_2, \boldsymbol{x}'_3)$ is accomplished in two steps involving the linear transformations T_1 and T_2 between the two quadrilaterals and the unit square $\mathcal{S}_1$, that is,

$$
\mathcal{Q} \xleftarrow{T_1} \mathcal{S}_1 \xrightarrow{T_2} \mathcal{Q}'. \tag{21.52}
$$

The parameters for the projective transformations T_1 and T_2 are obtained by inserting the corresponding point coordinates of $\mathcal{Q}$ and $\mathcal{Q}'$ ($\boldsymbol{x}_i$ and $\boldsymbol{x}'_i$, respectively) into Eqns. (21.48)–(21.51). The complete transformation T is then the concatenation of the two transformations T_1^{-1} and T_2, that is,

$$
\boldsymbol{x}' = T(\boldsymbol{x}) = T_2\big(T_1^{-1}(\boldsymbol{x})\big), \tag{21.53}
$$

or, expressed in matrix notation (using homogeneous coordinates),

$$
\underline{\boldsymbol{x}}' = \mathbf{A} \cdot \underline{\boldsymbol{x}} = \mathbf{A}_2 \cdot \mathbf{A}_1^{-1} \cdot \underline{\boldsymbol{x}}\,. \tag{21.54}
$$

Of course, the matrix $\mathbf{A} = \mathbf{A}_2 \cdot \mathbf{A}_1^{-1}$ needs to be calculated only once for a particular transformation and can then be used repeatedly for mapping any other image points $\boldsymbol{x}_i$.

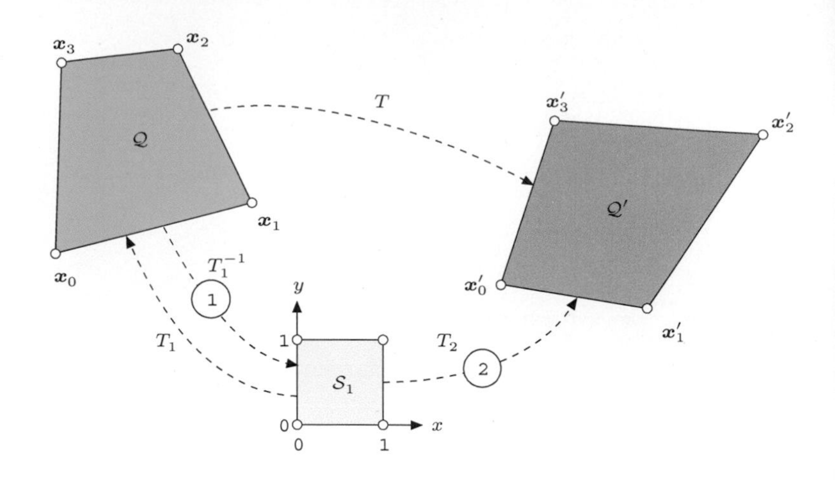

Fig. 21.5 Two-step projective transformation between arbitrary quadrilaterals. In the first step, quadrilateral $\mathcal{Q}$ is transformed to the unit square $\mathcal{S}_1$ by the inverse mapping function T_1^{-1}. In the second step, T_2 transforms the square $\mathcal{S}_1$ to the target quadrilateral $\mathcal{Q}'$. The complete mapping T results from the concatenation of the mappings T_1^{-1} and T_2.

Example

The source and the target quadrilaterals $\mathcal{Q}$ and $\mathcal{Q}'$, respectively, are specified by the following coordinate points:

$$\begin{array}{lllll} \mathcal{Q}: & \boldsymbol{x}_0 = (2,5), & \boldsymbol{x}_1 = (4,6), & \boldsymbol{x}_2 = (7,9), & \boldsymbol{x}_3 = (5,9); \\ \mathcal{Q}': & \boldsymbol{x}'_0 = (4,3), & \boldsymbol{x}'_1 = (5,2), & \boldsymbol{x}'_2 = (9,3), & \boldsymbol{x}'_3 = (7,5). \end{array}$$

Using Eqns. (21.48)–(21.51), the transformation parameters (matrices) for the projective mappings from the unit $\mathcal{S}_1$ square to the quadrilaterals $\mathbf{A}_1 : \mathcal{S}_1 \mapsto \mathcal{Q}$ and $\mathbf{A}_2 : \mathcal{S}_1 \mapsto \mathcal{Q}'$ are obtained as

$$\mathbf{A}_1 = \begin{pmatrix} 3.3\dot{3} & 0.50 & 2.00 \\ 3.00 & -0.50 & 5.00 \\ 0.3\dot{3} & -0.50 & 1.00 \end{pmatrix} \quad \text{and} \quad \mathbf{A}_2 = \begin{pmatrix} 1.00 & -0.50 & 4.00 \\ -1.00 & -0.50 & 3.00 \\ 0.00 & -0.50 & 1.00 \end{pmatrix}.$$

Concatenating the inverse mapping $\mathbf{A}_1^{-1}$ with $\mathbf{A}_2$ (by matrix multiplication), we get the complete mapping $\mathbf{A} = \mathbf{A}_2 \cdot \mathbf{A}_1^{-1}$ with

$$\mathbf{A}_1^{-1} = \begin{pmatrix} 0.60 & -0.45 & 1.05 \\ -0.40 & 0.80 & -3.20 \\ -0.40 & 0.55 & -0.95 \end{pmatrix} \quad \text{and} \quad \mathbf{A} = \begin{pmatrix} -0.80 & 1.35 & -1.15 \\ -1.60 & 1.70 & -2.30 \\ -0.20 & 0.15 & 0.65 \end{pmatrix}.$$

The library method `makeMapping()` in class `ProjectiveMapping` (see Sec. 21.3) is an implementation of this two-step technique.

Projective transformation parameters from *more than four* point pairs

The projective transformation in Eqn. (21.32) describes a mapping between pairs of arbitrary quadrilaterals in the 2D plane. This geometric relation is also known under the terms *projective isomorphism* or *homography*. The concept is frequently encountered in computer vision, because the transformations between two views of a planar 3D point set can be modeled as a homography (with only 8 degrees of freedom) in the 2D image plane, which is important, for example, for camera calibration, and 3D surface reconstruction. In this context, it is often necessary to estimate the homography parameters from a larger set of 2D point matches, for example, from multiple points assumed to be located on a planar 3D surface. This is the same problem as finding the projective mapping between sets of $n > 4$ corresponding point pairs in 2D.

Several approaches to "homography estimation" exist, including linear and (iterative) nonlinear methods. The simplest and most common is the direct linear transform (DLT) method [56, 103], which requires solving a system of $2n$ homogenous linear equations, typically done by singular value decomposition (SVD).

21.1.5 Bilinear Mapping

Similar to the projective transformation (Eqn. (21.32)), the bilinear mapping function

$$\begin{aligned} T_{\mathrm{x}}: \; x' &= a_0 \cdot x + a_1 \cdot y + a_2 \cdot x \cdot y + a_3, \\ T_{\mathrm{y}}: \; y' &= b_0 \cdot x + b_1 \cdot y + b_2 \cdot x \cdot y + b_3, \end{aligned} \tag{21.55}$$

is specified with four pairs of corresponding points and has eight parameters $(a_0, \ldots, a_3, b_0, \ldots, b_3)$. The transformation is nonlinear because of the mixed term $x \cdot y$ and cannot be described by a linear transformation, even with homogeneous coordinates. In contrast to the projective transformation, the straight lines are not preserved in general but map onto quadratic curves. Similarly, circles are not mapped to ellipses by a bilinear transform.

A bilinear mapping is uniquely specified by four corresponding pairs of 2D points $(\boldsymbol{x}_0, \boldsymbol{x}'_0), \ldots, (\boldsymbol{x}_3, \boldsymbol{x}'_3)$. In the general case, for a bilinear mapping between arbitrary quadrilaterals, the coefficients $a_0, \ldots, a_3, b_0, \ldots, b_3$ (Eqn. (21.55)) are found as the solution of two separate systems of equations, each with four unknowns:

$$\begin{pmatrix} x'_0 \\ x'_1 \\ x'_2 \\ x'_3 \end{pmatrix} = \begin{pmatrix} x_0 & y_0 & x_0 \cdot y_0 & 1 \\ x_1 & y_1 & x_1 \cdot y_1 & 1 \\ x_2 & y_2 & x_2 \cdot y_2 & 1 \\ x_3 & y_3 & x_3 \cdot y_3 & 1 \end{pmatrix} \cdot \begin{pmatrix} a_0 \\ a_1 \\ a_2 \\ a_3 \end{pmatrix} \quad \text{or} \quad \boldsymbol{x} = \mathbf{M} \cdot \boldsymbol{a}, \tag{21.56}$$

$$\begin{pmatrix} y'_0 \\ y'_1 \\ y'_2 \\ y'_3 \end{pmatrix} = \begin{pmatrix} x_0 & y_0 & x_0 \cdot y_0 & 1 \\ x_1 & y_1 & x_1 \cdot y_1 & 1 \\ x_2 & y_2 & x_2 \cdot y_2 & 1 \\ x_3 & y_3 & x_3 \cdot y_3 & 1 \end{pmatrix} \cdot \begin{pmatrix} b_0 \\ b_1 \\ b_2 \\ b_3 \end{pmatrix} \quad \text{or} \quad \boldsymbol{y} = \mathbf{M} \cdot \boldsymbol{b}. \tag{21.57}$$

These equations can again be solved using standard numerical methods. In the special case of bilinearly mapping the unit square $\mathcal{S}_1$ to an arbitrary quadrilateral $\mathcal{Q} = (\boldsymbol{x}'_0, \ldots, \boldsymbol{x}'_3)$, the parameters $a_0, \ldots, a_3$ and $b_0, \ldots, b_3$ are found as

$$a_0 = x'_1 - x'_0, \qquad b_0 = y'_1 - y'_0, \tag{21.58}$$

$$a_1 = x'_3 - x'_0, \qquad b_1 = y'_3 - y'_0, \tag{21.59}$$

$$a_2 = x'_0 - x'_1 + x'_2 - x'_3, \qquad b_2 = y'_0 - y'_1 + y'_2 - y'_3, \tag{21.60}$$

$$a_3 = x'_0, \qquad b_3 = y'_0. \tag{21.61}$$

Figure 21.6 shows results of the affine, projective, and bilinear transformations applied to a simple test pattern. The affine transformation (Fig. 21.6(b)) is specified by mapping to the triangle 1-2-3, while the four points of the quadrilateral 1-2-3-4 define the projective and the bilinear transforms (Fig. 21.6(c, d)).

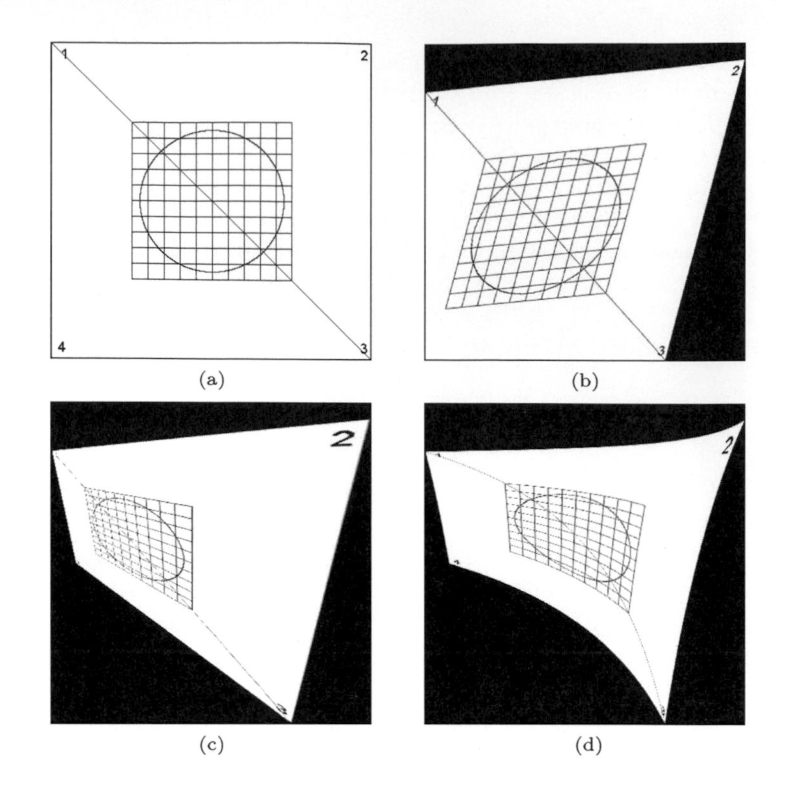

Fig. 21.6
Geometric transformations compared: original image (a), affine transformation with respect to the triangle 1-2-3 (b), projective transformation (c), and bilinear transformation (d).

21.1.6 Other Nonlinear Image Transformations

The bilinear transformation discussed in the previous section is only one example of a nonlinear mapping in 2D that cannot be expressed as a simple matrix-vector multiplication in homogeneous coordinates. Many other types of nonlinear deformations exist; for example, to implement various artistic effects for creative imaging. This type of image deformation is often called "image warping". Depending on the type of transformation used, the derivation of the *inverse* transformation function—which is required for the practical computation of the mapping using *target-to-source mapping* (see Sec. 21.2.2)—is not always easy or may even be impossible. In the following three examples, we therefore look straight at the inverse maps

$$\boldsymbol{x} = T^{-1}(\boldsymbol{x}') \tag{21.62}$$

without really bothering about the corresponding *forward* transformations.

"Twirl" transformation

The twirl mapping causes the image to be rotated around a given anchor point $\boldsymbol{x}_c = (x_c, y_c)$ with a space-variant rotation angle, which has a fixed value α at the center $\boldsymbol{x}_c$ and decreases linearly with the radial distance from the center. The image remains unchanged outside the limiting radius $r_{\max}$. The associated (*inverse*) mapping is defined as

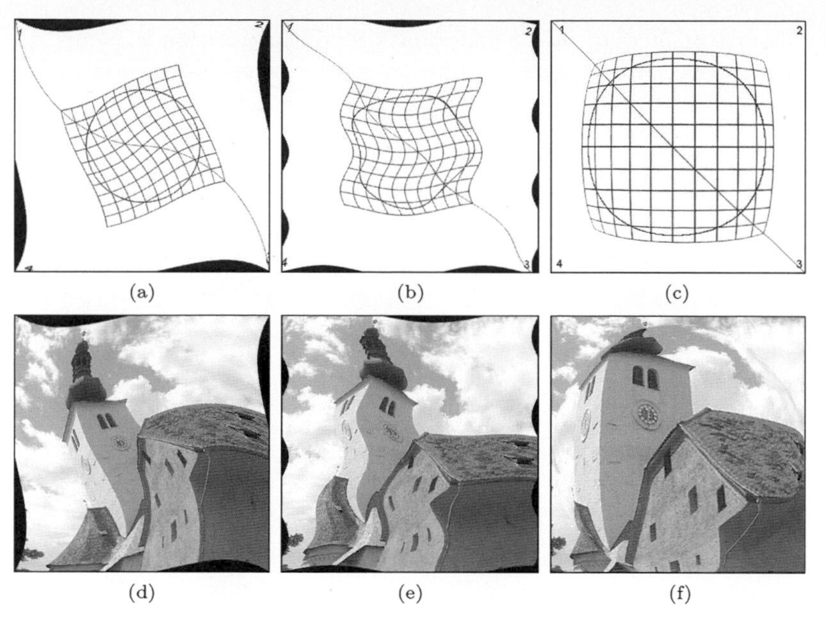

Fig. 21.7
Various nonlinear image deformations: *twirl* (a, d), *ripple* (b, e), and *sphere* (c, f) transformations. The size of the original images is 400 × 400 pixels.

$$T_x^{-1}\colon x = \begin{cases} x_c + r\cdot\cos(\beta) & \text{for } r \le r_{\max},\\ x' & \text{for } r > r_{\max},\end{cases} \tag{21.63}$$

$$T_y^{-1}\colon y = \begin{cases} y_c + r\cdot\sin(\beta) & \text{for } r \le r_{\max},\\ y' & \text{for } r > r_{\max},\end{cases} \tag{21.64}$$

with

$$r = \sqrt{d_x^2 + d_y^2}, \qquad d_x = x' - x_c, \tag{21.65}$$

$$\beta = \operatorname{ArcTan}(d_x, d_y) + \alpha\cdot\left(\tfrac{r_{\max}-r}{r_{\max}}\right), \qquad d_y = y' - y_c. \tag{21.66}$$

Figure 21.7(a, d) shows a twirl mapping with the anchor point $\boldsymbol{x}_c$ placed at the image center. The limiting radius $r_{\max}$ is half the length of the image diagonal, and the rotation angle is $\alpha = 43°$ at the center.

"Ripple" transformation

The ripple transformation causes a local wavelike displacement of the image along both the x and y directions. The parameters of this mapping function are the period lengths $\tau_x, \tau_y \neq 0$ (in pixels) and the corresponding amplitude values a_x, a_y for the displacement in both directions:

$$T_x^{-1}\colon x = x' + a_x\cdot\sin\left(\tfrac{2\pi\cdot y'}{\tau_x}\right), \tag{21.67}$$

$$T_y^{-1}\colon y = y' + a_y\cdot\sin\left(\tfrac{2\pi\cdot x'}{\tau_y}\right). \tag{21.68}$$

An example for the ripple mapping with $\tau_x = 120$, $\tau_y = 250$, $a_x = 10$, and $a_y = 15$ is shown in Fig. 21.7(b, e).

Spherical transformation

The spherical deformation imitates the effect of viewing the image through a transparent hemisphere or lens placed on top of the image.

The parameters of this transformation are the position $\boldsymbol{x}_c = (x_c, y_c)$ of the lens center, the radius of the lens r_{max} and its refraction index ρ. The corresponding mapping functions are defined as

$$T_x^{-1}\colon x = x' - \begin{cases} z \cdot \tan(\beta_x) & \text{for } r \leq r_{\mathrm{max}}, \\ 0 & \text{for } r > r_{\mathrm{max}}, \end{cases} \tag{21.69}$$

$$T_y^{-1}\colon y = y' - \begin{cases} z \cdot \tan(\beta_y) & \text{for } r \leq r_{\mathrm{max}}, \\ 0 & \text{for } r > r_{\mathrm{max}}, \end{cases} \tag{21.70}$$

with

$$\begin{aligned} r &= \sqrt{d_x^2 + d_y^2}\,, & \beta_x &= \big(1 - \tfrac{1}{\rho}\big) \cdot \sin^{-1}\big(\tfrac{d_x}{\sqrt{(d_x^2 + z^2)}}\big), & d_x &= x' - x_c, \\ z &= \sqrt{r_{\mathrm{max}}^2 - r^2}\,, & \beta_y &= \big(1 - \tfrac{1}{\rho}\big) \cdot \sin^{-1}\big(\tfrac{d_y}{\sqrt{(d_y^2 + z^2)}}\big), & d_y &= y' - y_c. \end{aligned} \tag{21.71}$$

Figure 21.7(c, f) shows a spherical transformation with the lens positioned at the image center. The lens radius r_{max} is set to half of the image width, and the refraction index is $\rho = 1.8$.

See Exercise 21.4 for additional examples of nonlinear geometric tarnsformations.

21.1.7 Piecewise Image Transformations

All the geometric transformations discussed so far are *global* (i.e., the same mapping function is applied to all pixels in the given image). It is often necessary to deform an image such that a larger number of n original image points $\boldsymbol{x}_0, \ldots, \boldsymbol{x}_n$ are precisely mapped onto a given set of target points $\boldsymbol{x}'_0, \ldots, \boldsymbol{x}'_n$. For $n = 3$, this problem can be solved with an affine mapping (see Sec. 21.1.3), and for $n = 4$ we could use a projective or bilinear mapping (see Secs. 21.1.4 and 21.1.5). A precise global mapping of $n > 4$ points requires a more complicated function $T(\boldsymbol{x})$ (e.g., a 2D nth-order polynomial or a spline function).

An alternative is to use *local* or *piecewise* transformations, where the image is partitioned into disjoint patches that are transformed separately, applying an individual mapping function to each patch. In practice, it is common to partition the image into a *mesh* of triangles or quadrilaterals, as illustrated in Fig. 21.8.

For a *triangular* mesh partitioning (Fig. 21.8(a)), the transformation between each pair of triangles $\mathcal{D}_i \rightarrow \mathcal{D}'_i$ could be accomplished with an *affine* mapping, whose parameters must be computed individually for every patch. Similarly, the *projective* transformation would be suitable for mapping each patch in a mesh partitioning composed of *quadrilaterals* $\mathcal{Q}_i$ (Fig. 21.8(b)). Since both the affine and the projective transformations preserve the straightness of lines, we can be certain that no holes or overlaps will arise and the deformation will appear continuous between adjacent mesh patches.

Local transformations of this type are frequently used; for example, to register aerial and satellite images or to undistort images for panoramic stitching. In computer graphics, similar techniques are used to map texture images onto polygonal 3D surfaces in the rendered 2D image. Another popular application of this technique is

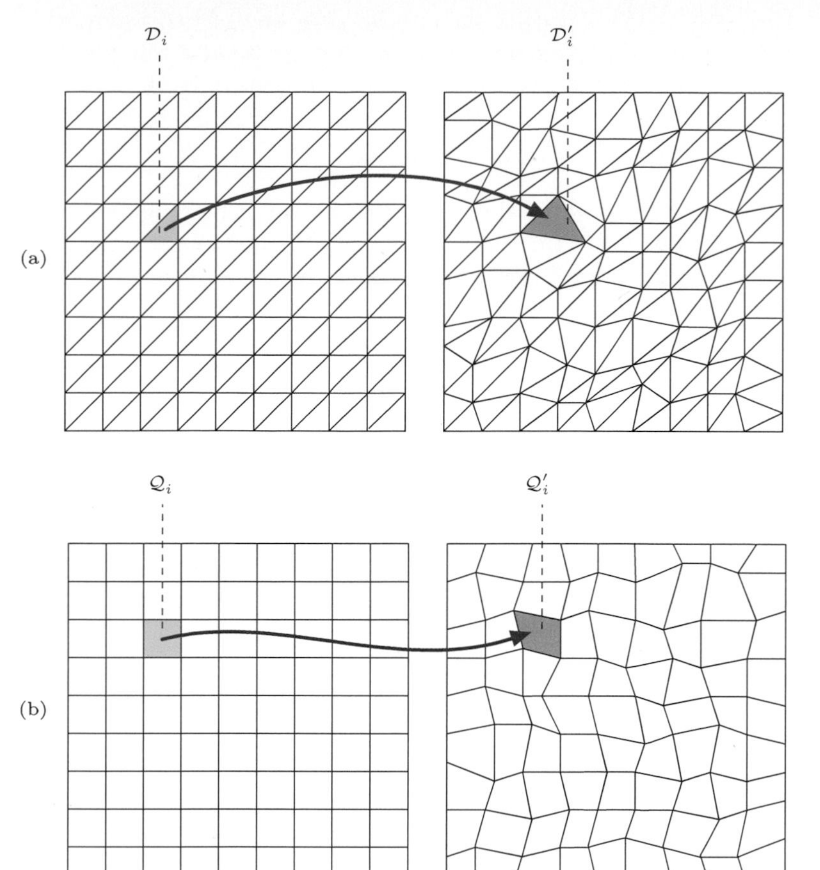

Fig. 21.8
Mesh partitioning examples. Almost arbitrary image deformations can be implemented by partitioning the image plane into nonoverlapping triangles $\mathcal{D}_i, \mathcal{D}'_i$ (a) or quadrilaterals $\mathcal{Q}_i, \mathcal{Q}'_i$ (b) and applying simple local transformations. Every patch in the resulting mesh is transformed separately with the required transformation parameters derived from the corresponding three or four corner points, respectively.

"morphing" [256], which performs a stepwise geometric transformation from one image to another while simultaneously blending their intensity (or color) values.[4]

21.2 Resampling the Image

In the discussion of geometric transformations, we have so far considered the 2D image coordinates as being continuous (i.e., real-valued). In reality, the picture elements in digital images reside at discrete (i.e., integer-valued) coordinates, and thus transferring a discrete image into another discrete image without introducing significant losses in quality is a nontrivial subproblem in the implementation of geometric transformations.

Based on the original image $I(u, v)$ and some (continuous) geometric transformations $T(x, y)$, the aim is to create a transformed image $I'(u', v')$ where all coordinates are discrete (i.e., $u, v \in \mathbb{Z}$ and

[4] Image morphing has also been implemented in ImageJ as a plugin (*iMorph*) by Hajime Hirase (http://rsb.info.nih.gov/ij/plugins/morph.html).

$u', v' \in \mathbb{Z}$).[5] This can be accomplished in one of two ways, which differ by the mapping direction and are commonly referred to as *source-to-target* or *target-to-source* mapping, respectively.

21.2.1 Source-to-Target Mapping

In this approach, which appears quite natural at first sight, we compute for every pixel (u, v) of the original (*source*) image I the corresponding transformed position

$$(x', y') = T(u, v) \tag{21.72}$$

in the target image I'. In general, the result will *not* coincide with any of the raster points, as illustrated in Fig. 21.9. Subsequently, we would have to decide in which pixel in the target image I' the original intensity or color value from $I(u, v)$ should be stored. We could perhaps even think of somehow distributing this value onto all adjacent pixels.

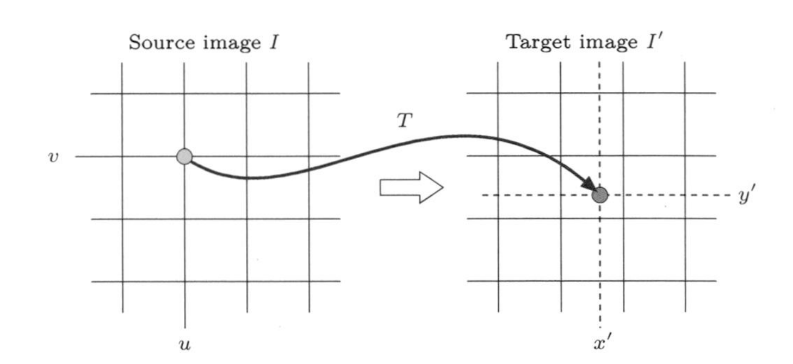

Fig. 21.9 Source-to-target mapping. For each discrete pixel position (u, v) in the source image I, the corresponding (continuous) target position (x', y') is found by applying the geometric transformation $T(u, v)$. In general, the target position (x', y') does not coincide with any discrete raster point. The source pixel value $I(u, v)$ is subsequently transferred to one of the adjacent target pixels.

The problem with the source-to-target method is that, depending on the geometric transformation T, some elements in the target image I' may never be "hit" at all (i.e., never receive a source pixel value)! This happens, for example, when the image is enlarged (even slightly) by the geometric transformation. The resulting holes in the target image would be difficult to close in a subsequent processing step. Conversely, one would have to consider (e.g., when the image is shrunk) that a single element in the target image I' may be hit by multiple source pixels and thus image content may get lost. In the light of all these complications, source-to-target mapping is not really the method of choice.

21.2.2 Target-to-Source Mapping

This method avoids most difficulties encountered in the source-to-target mapping by simply reversing the image generation process. For every discrete pixel position (u', v') in the *target* image, we determine the corresponding (continuous) point

[5] Remark on notation: We mostly use (u, v) or (u', v') to denote *discrete* (integer) coordinates and (x, y) or (x', y') for *continuous* (real-valued) coordinates.

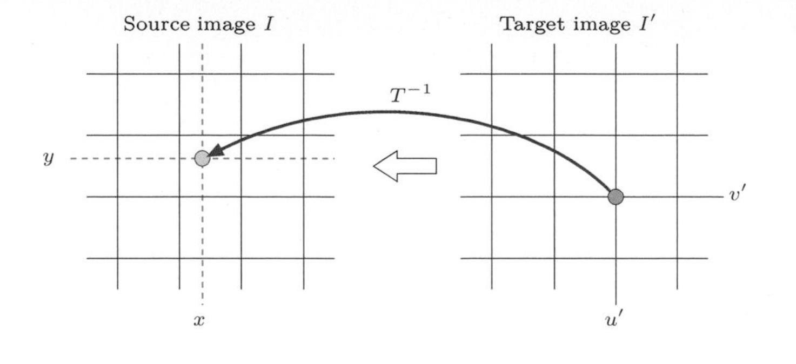

Fig. 21.10
Target-to-source mapping. For each discrete pixel position (u', v') in the target image I', the corresponding continuous source position (x, y) is found by applying the inverse mapping function $T^{-1}(u', v')$. The new pixel value $I'(u', v')$ is determined by interpolating the pixel values in the source image within some neighborhood of (x, y).

$$(x, y) = T^{-1}(u', v') \tag{21.73}$$

in the source image plane using the inverse geometric transformation T^{-1}. Of course, the coordinate (x, y) again does not fall onto a raster point in general (Fig. 21.10) and thus we have to decide from which of the neighboring source pixels to extract the resulting target pixel value. This problem of interpolating among intensity values is discussed in detail in Chapter 22.

The major advantage of the target-to-source method is that all pixels in the target image I' (and only these) are computed and filled exactly once such that no holes or multiple hits can occur. This, however, requires the *inverse* geometric transformation T^{-1} to be available, which is no disadvantage in most cases since the forward transformation T itself is never really needed. Due to its simplicity, which is also demonstrated in Alg. 21.1, *target-to-source* mapping is the common method for geometrically transforming 2D images.

Alg. 21.1
Geometric image transformation using target-to-source mapping. Given are the original (source) image I and the continuous coordinate transformation T. GetInterpolatedValue(I, x, y) returns the interpolated value of the source image I at the continuous position (x, y).

```
1: TransformImage (I, T)
     Input: I, source image; T, continuous mapping ℝ² ↦ ℝ².
     Returns the transformed image.
2:   (M, N) ← Size(I)
3:   I' ← duplicate(I)                       ▷ create the target image
4:   for all (u, v) ∈ M × N do               ▷ loop over all target pixels
5:       (x, y) ← T⁻¹(u, v)
6:       I'(u, v) ← GetInterpolatedValue(I, x, y)
7:   return I'
```

21.3 Java Implementation

In plain ImageJ, only a few simple geometric operations are provided as methods for the `ImageProcessor` class, such as rotation and flipping.[6] This section describes the implementation of the transformations described in this chapter, which is openly available as part of the `imagingbook` library.[7]

[6] Additional operations, including affine transformations, are available as plugin classes as part of the optional `TransformJ` package [162].

[7] Package `imagingbook.pub.geometry.mappings`.

21.3.1 General Mappings (Class Mapping)

The abstract class Mapping is the superclass for all subsequent transformations. All subclasses of Mapping are required to implement the method applyTo(double[] pnt), which applies the associated transformation to a given coordinate point and returns the transformed point. The actual transformations are implemented by its concrete sub-classes. The applyTo() method is defined in multiple versions with different signatures:

double[] applyTo (double[] pnt)
Applies this transformation to the 2D point (of type double[]) and returns the transformed coordinate.

Point2D applyTo (Point2D pnt)
Applies this transformation to the 2D point (of type Point2D) and returns the transformed coordinate.

Point2D[] applyTo (Point2D[] pnts)
Applies this transformation to a sequence of the 2D points (of type Point2D) and returns a sequence of transformed coordinates.

In addition, the Mapping class can also be used to transform entire images:

double[] applyTo (ImageProcessor source, ImageProcessor target, PixelInterpolator.Method im)
Transforms the input image source onto the output image target by target-to-source mapping, using the pixel interpolation method im.

double[] applyTo (ImageProcessor ip, PixelInterpolator.Method im)
Transforms the input image ip destructively, using the pixel interpolation method im.

double[] applyTo (ImageInterpolator source, ImageProcessor target)
Transforms the input image (specified by the interpolator source) onto the output image target by target-to-source mapping.

Other methods defined by class Mapping:

Mapping duplicate ()
Returns a copy of this mapping.

Mapping getInverse ()
Returns the inverse of this mapping if available. Otherwise an UnsupportedOperationException is thrown.

21.3.2 Linear Mappings

Linear transformations are implemented by class LinearMapping,[8] with sub-classes including

AffineMapping, Scaling,
ProjectiveMapping, Shear,
Rotation, Translation.

[8] Package imagingbook.pub.geometry.mappings.linear.

21.3.3 Nonlinear Mappings

Selected nonlinear transformations are implemented by the following subclasses of Mapping:[9]

BilinearMapping, ShereMapping,
RippleMapping, TwirlMapping.

21.3.4 Sample Applications

The following two ImageJ plugins show two simple examples of the use of the classes in Secs. 21.3.2 and 21.3.3 for implementing geometric operations and pixel interpolation (see Ch. 22 for details). Note that these plugins can be applied to any type of image.

Example 1: image rotation

The example in Prog. 21.1 shows a plugin (Transform_Rotate) to rotate an image by 15°. First (in line 16) the geometric mapping object (map) is created as an instance of class Rotation, with the supplied angle being converted from degrees to radians. The actual transformation of the image is performed by invoking the method applyTo() in line 17.

Prog. 21.1
Image rotation example using the Rotation class (ImageJ plugin).

```
1 import ij.ImagePlus;
2 import ij.plugin.filter.PlugInFilter;
3 import ij.process.ImageProcessor;
4 import imagingbook.pub.geometry.interpolators.pixel.
      PixelInterpolator;
5 import imagingbook.pub.geometry.mappings.Mapping;
6 import imagingbook.pub.geometry.mappings.linear.Rotation;
7
8 public class Transform_Rotate implements PlugInFilter {
9   static double angle = 15; // rotation angle (in degrees)
10
11     public int setup(String arg, ImagePlus imp) {
12         return DOES_ALL;
13     }
14
15     public void run(ImageProcessor ip) {
16       Mapping map = new Rotation((2*Math.PI*angle)/360);
17       map.applyTo(ip, PixelInterpolator.Method.Bicubic);
18     }
19 }
```

Example 2: projective transformation

The second example in Prog. 21.2 illustrates the implementation of a projective transformation. The geometric mapping T is defined by two corresponding quadrilaterals P = p0, ..., p3 and Q = q0, ..., q3, respectively. In a real application, these points would probably be specified interactively or given as the result of a mesh partitioning.

[9] Package imagingbook.pub.geometry.mappings.nonlinear.

Prog. 21.2
Projective image transformation example using the `ProjectiveMapping` class (ImageJ plugin).

```
1 import ij.ImagePlus;
2 import ij.plugin.filter.PlugInFilter;
3 import ij.process.ImageProcessor;
4 import imagingbook.pub.geometry.interpolators.pixel.
      PixelInterpolator;
5 import imagingbook.pub.geometry.mappings.Mapping;
6 import imagingbook.pub.geometry.mappings.linear.
      ProjectiveMapping;
7 import java.awt.Point;
8 import java.awt.geom.Point2D;
9
10 public class Transform_Projective implements PlugInFilter {
11
12     public int setup(String arg, ImagePlus imp) {
13         return DOES_ALL;
14     }
15
16     public void run(ImageProcessor ip) {
17       Point2D p0 = new Point(0, 0);
18       Point2D p1 = new Point(400, 0);
19       Point2D p2 = new Point(400, 400);
20       Point2D p3 = new Point(0, 400);
21
22       Point2D q0 = new Point(0, 60);
23       Point2D q1 = new Point(400, 20);
24       Point2D q2 = new Point(300, 400);
25       Point2D q3 = new Point(30, 200);
26
27       Mapping map = new
28         ProjectiveMapping(p0, p1, p2, p3, q0, q1, q2, q3);
29
30       map.applyTo(ip, PixelInterpolator.Method.Bilinear);
31     }
32 }
```

The transformation object `map` (representing the forward transformation T) is created by calling the associated constructor `ProjectiveMapping()` in line 28. The mapping is applied to the input image (line 30), as in the previous example, except for the use of *bilinear* pixel interpolation.

21.4 Exercises

Exercise 21.1. Show that a straight line $y = kx+d$ in 2D is mapped to another straight line under a projective transformation, as defined in Eqn. (21.32).

Exercise 21.2. Show that parallel lines remain parallel under affine transformation (Eqn. (21.20)).

Exercise 21.3. Design a nonlinear geometric transformation similar to the ripple transformation (Eqn. (21.67)) that uses a *sawtooth* function instead of a sinusoid for the distortions in the horizontal

(a) Original image

(b) Radial wave ($a = 10.0$, $\tau = 38$)

(c) Clover ($a = 0.2$, $N = 8$)

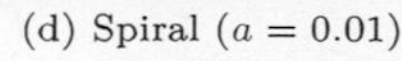

(d) Spiral ($a = 0.01$)

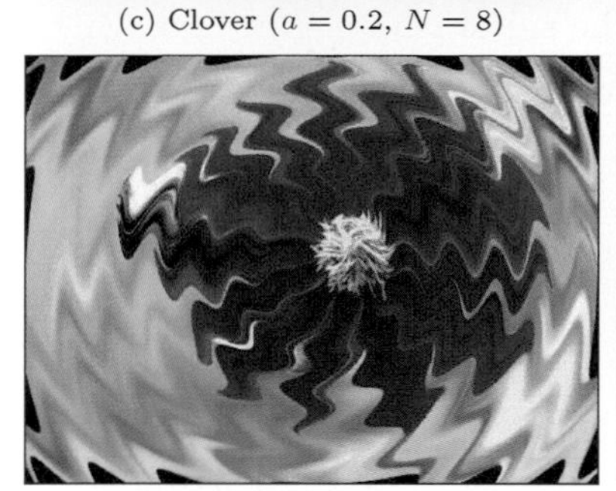

(e) Angular wave ($a = 0.1$, $\tau = 38$)

(f) Tapestry ($a = 5.0$, $\tau_x = \tau_y = 30$)

Fig. 21.11
Examples of the nonlinear geometric transformations defined in Exercise 21.4. The reference point $\boldsymbol{x}_c$ is always taken at the image center.

and vertical directions. Use the class `TwirlMapping` as a template for your implementation.

Exercise 21.4. Implement one or more of the following nonlinear geometric transformations (see Fig. 21.11):

A. **Radial wave** transformation: This transformation simulates an omni-directional wave which originates from a fixed center point $\boldsymbol{x}_c$ (see Fig. 21.11(b)). The inverse transformation (applied to a target image point $\boldsymbol{x}' = (x', y')$) is

$$T^{-1}: \boldsymbol{x} = \begin{cases} \boldsymbol{x}_c & \text{for } r = 0, \\ \boldsymbol{x}_c + \frac{r+\delta}{r} \cdot (\boldsymbol{x}' - \boldsymbol{x}_c) & \text{for } r > 0, \end{cases} \tag{21.74}$$

with $r = \|\boldsymbol{x}' - \boldsymbol{x}_c\|$ and $\delta = a \cdot \sin(2\pi r/\tau)$. Parameter a specifies the *amplitude* (strength) of the distortion and τ is the *period* (width) of the radial wave (in pixel units).

B. **Clover** transformation: This transformation distorts the image in the form of a N-leafed clover shape (see Fig. 21.11(c)). The associated inverse transformation is the same as in Eqn. (21.74) but uses

$$\delta = a \cdot r \cdot \cos(N \cdot \alpha), \quad \text{with } \alpha = \angle(\boldsymbol{x}' - \boldsymbol{x}_{\mathrm{c}}) \tag{21.75}$$

instead. Again $r = \|\boldsymbol{x}' - \boldsymbol{x}_{\mathrm{c}}\|$ is the radius of the target image point $\boldsymbol{x}'$ from the designated center point $\boldsymbol{x}_{\mathrm{c}}$. Parameter a specifies the amplitude of the distortion and N is the number of radial "leaves".

C. **Spiral** transformation: This transformation (see Fig. 21.11(d)) is similar to the *twirl* transformation in Eqns. (21.63)–(21.64), defined by the inverse transformation

$$T^{-1}\colon \boldsymbol{x} = \boldsymbol{x}_{\mathrm{c}} + r \cdot \begin{pmatrix} \cos(\beta) \\ \sin(\beta) \end{pmatrix}, \tag{21.76}$$

with $\beta = \angle(\boldsymbol{x}' - \boldsymbol{x}_{\mathrm{c}}) + a \cdot r$ and $r = \|\boldsymbol{x}' - \boldsymbol{x}_{\mathrm{c}}\|$ denoting the distance from the target point $\boldsymbol{x}'$ and the center point $\boldsymbol{x}_{\mathrm{c}}$. The angle β increases linearly with r; parameter a specifies the "velocity" of the spiral.

D. **Angular wave** transformation: This is another variant of the *twirl* transformation in Eqns. (21.63)–(21.64). Its inverse transformation is the same as for the spiral mapping in Eqn. (21.76), but in this case

$$\beta = \angle(\boldsymbol{x}' - \boldsymbol{x}_{\mathrm{c}}) + a \cdot \sin\left(\tfrac{2\pi r}{\tau}\right). \tag{21.77}$$

Thus the angle β is modified by a sine function with amplitude a (see Fig. 21.11(e)).

E. **Tapestry** transformation: In this case the inverse transformation of a target point $\boldsymbol{x}' = (x', y')$ is

$$T^{-1}\colon \boldsymbol{x} = \boldsymbol{x}' + a \cdot \begin{pmatrix} \sin\left(\frac{2\pi}{\tau_x} \cdot (x' - x_{\mathrm{c}})\right) \\ \sin\left(\frac{2\pi}{\tau_y} \cdot (y' - y_{\mathrm{c}})\right) \end{pmatrix}, \tag{21.78}$$

again with the center point $\boldsymbol{x}_{\mathrm{c}} = (x_{\mathrm{c}}, y_{\mathrm{c}})$. Parameter a specifies the distortion's amplitude and τ_x, τ_y are the wavelengths (measured in pixel units) along the x and y axis, respectively (see Fig. 21.11(f)).

Exercise 21.5. Implement an interactive program (plugin) that performs projective rectification (see Sec. 21.1.4) of a selected quadrilateral, as shown in Fig. 21.12. Make your program perform the following steps:

1. Let the user mark the source quad in the source image I as a polygon-shaped *region of interest* (ROI) with at least four points $\boldsymbol{x}_0, \ldots, \boldsymbol{x}_3$. In ImageJ this is easily done with the built-in polygon selection tool (see Prog. 21.3 for handling ROI points).
2. Create an output image I' of fixed size (i.e., proportional to A4 or Letter paper size).
3. The target rectangle is defined by the four corners $\boldsymbol{x}'_0, \ldots, \boldsymbol{x}'_3$ of the output image. The source and target points are associated 1:1, that is, the four corresponding point pairs are $\langle \boldsymbol{x}_0, \boldsymbol{x}'_0 \rangle, \ldots, \langle \boldsymbol{x}_3, \boldsymbol{x}'_3 \rangle$.

4. From the four point pairs, create an instance of `ProjectiveMapping`, as demonstrated in Prog. 21.2.
5. Test the obtained mapping by applying $\mathbf{A}$ to the specified source points $\boldsymbol{x}_0, \ldots, \boldsymbol{x}_3$. Make sure they project exactly to the specified target points $\boldsymbol{x}'_0, \ldots, \boldsymbol{x}'_3$.
6. Apply the obtained mapping from the source to the target image using the method[10]
```
void applyTo(ImageProcessor source,
    ImageProcessor target, InterpolationMethod im).
```
7. Show the resulting output image.

(a)

(b)

Fig. 21.12
Projective rectification example (see Exercise 21.5). Source image and user-defined selection (a); transformed output image (b).

[10] Defined in class `imagingbook.pub.geometry.mappings.Mapping`.

Prog. 21.3
ImageJ plugin demonstrating the extraction of vertex points from a user-selected polygon-ROI (region of interest). Notice that (in line 21) the region of interest (ROI) is obtained from the associated ImagePlus instance (to which a reference is kept in line 16) and not from the supplied ImageProcessor object. ImageJ's ROI coordinates are integer positions in general.

```
1 import java.awt.Point;
2 import java.awt.Polygon;
3 import java.awt.geom.Point2D;
4
5 import ij.ImagePlus;
6 import ij.gui.PolygonRoi;
7 import ij.gui.Roi;
8 import ij.plugin.filter.PlugInFilter;
9 import ij.process.ImageProcessor;
10
11 public class Get_Roi_Points implements PlugInFilter {
12
13   ImagePlus im = null;
14
15   public int setup(String args, ImagePlus im) {
16     this.im = im;    // keep a reference to im
17     return DOES_ALL + ROI_REQUIRED;
18   }
19
20   public void run(ImageProcessor source) {
21     Roi roi = im.getRoi();
22     if (!(roi instanceof PolygonRoi)) {
23       IJ.error("Polygon selection required!");
24       return;
25     }
26
27     Polygon poly = roi.getPolygon();
28
29     // copy polygon vertices to a point array:
30     Point2D[] pts = new Point2D[poly.npoints];
31     for (int i = 0; i < poly.npoints; i++) {
32       pts[i] = new Point(poly.xpoints[i], poly.ypoints[i]);
33     }
34
35     ... // use the ROI points in pts
36
37   }
38
39 }
```

22

Pixel Interpolation

Interpolation is the process of estimating the intermediate values of a sampled function or signal at continuous positions or the attempt to reconstruct the original continuous function from a set of discrete samples. In the context of geometric operations this task arises from the fact that discrete pixel positions in one image are generally not mapped to discrete raster positions in the other image under some continuous geometric transformation T (or T^{-1}, respectively). The concrete goal is to obtain an optimal estimate for the value of the 2D image function $I(x, y)$ at any continuous position $(x, y) \in \mathbb{R}^2$ to implement the function

$$\mathsf{GetInterpolatedValue}(I, x, y),$$

which we defined in Chapter 21 (see Alg. 21.1). Ideally the interpolated image should preserve as much detail (i.e., sharpness) as possible without causing visible artifacts such as ringing or moiré patterns.

22.1 Simple Interpolation Methods

To illustrate the problem, we first attend to the 1D case (see Fig. 22.1). Several simple, ad-hoc methods exist for interpolating the values of a discrete function $g(u)$, with $u \in \mathbb{Z}$, at arbitrary continuous positions $x \in \mathbb{R}$. The simplest of all interpolation methods is to round the continuous coordinate x to the closest integer u_x and use the associated sample $g(u_x)$ as the interpolated value, that is,

$$\tilde{g}(x) \leftarrow g(u_x), \qquad (22.1)$$

with $u_x = \text{round}(x) = \lfloor x + 0.5 \rfloor$. A typical result of this so-called *nearest-neighbor interpolation* is shown in Fig. 22.2(a).

Another simple method is *linear interpolation*. Here the estimated value is the sum of the two closest samples $g(u_0)$ and $g(u_0 + 1)$, with $u_0 = \lfloor x \rfloor$. The weight of each sample is proportional to its closeness to the continuous position x, that is,

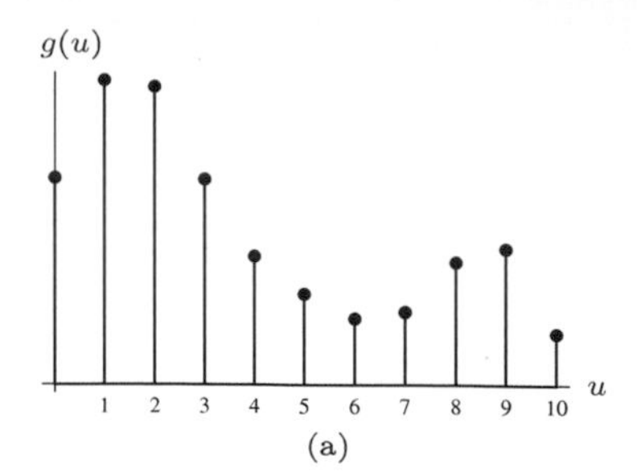

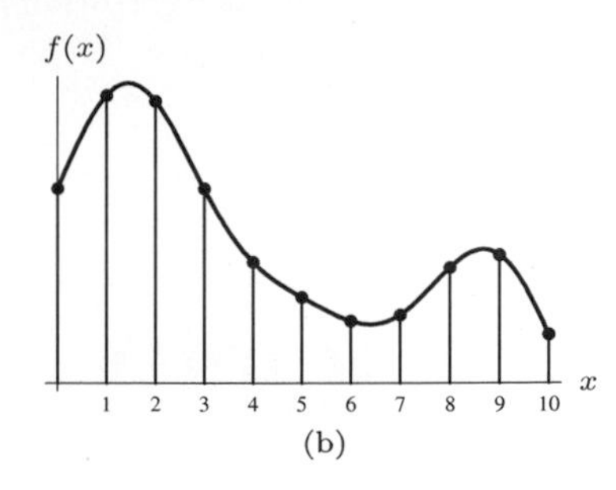

Fig. 22.1
Interpolating a discrete function in 1D. Given the discrete function values $g(u)$ (a), the goal is to estimate the original function $f(x)$ at arbitrary continuous positions $x \in \mathbb{R}$ (b).

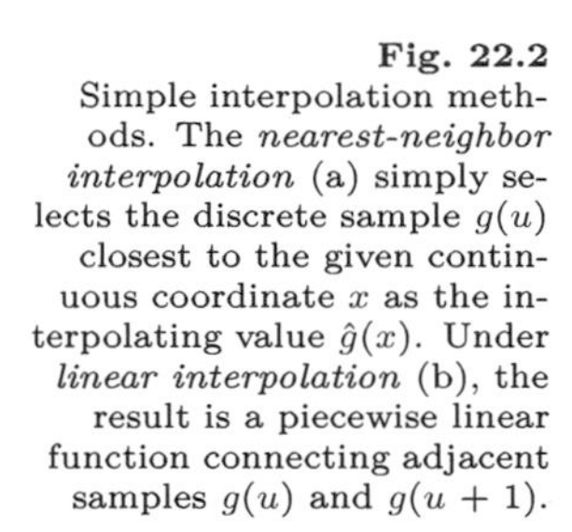

Fig. 22.2
Simple interpolation methods. The *nearest-neighbor interpolation* (a) simply selects the discrete sample $g(u)$ closest to the given continuous coordinate x as the interpolating value $\hat{g}(x)$. Under *linear interpolation* (b), the result is a piecewise linear function connecting adjacent samples $g(u)$ and $g(u+1)$.

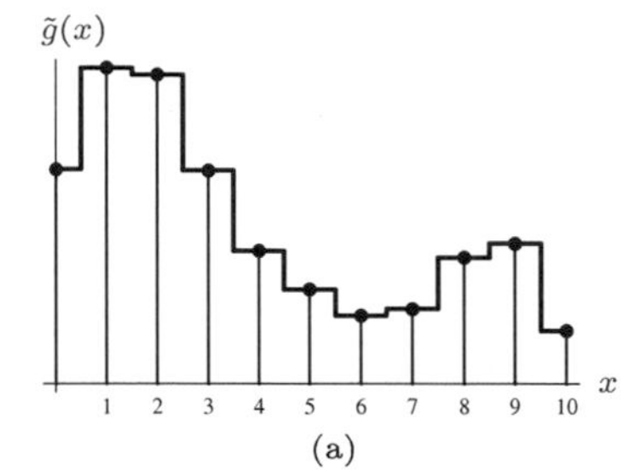

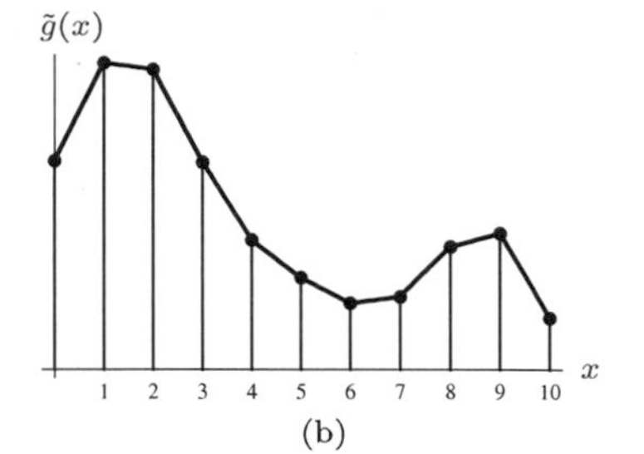

$$
\begin{aligned}
\tilde{g}(x) &= g(u_x) + (x - u_x) \cdot \big(g(u_x + 1) - g(u_x)\big) \\
&= g(u_x) \cdot \big(1 - (x - u_x)\big) + g(u_x + 1) \cdot (x - u_x).
\end{aligned}
\tag{22.2}
$$

As shown in Fig. 22.2(b), the result is a piecewise linear function made up of straight line segments between consecutive sample values.

22.1.1 Ideal Low-Pass Filter

Obviously the results of these simple interpolation methods do not well approximate the original continuous function (Fig. 22.1). But how can we obtain a better approximation from the discrete samples only when the original function is unknown? This may appear hopeless at first, because the discrete samples $g(u)$ could possibly originate from any continuous function $f(x)$ with identical values at the discrete sample positions.

We find an intuitive answer to this question (once again) by looking at the functions in the spectral domain. If the original function $f(x)$ was discretized in accordance with the *sampling theorem* (see Ch. 18, Sec. 18.2.1), then $f(x)$ must have been "band limited"—it could not contain any signal components with frequencies higher than half the sampling frequency ω_s. This means that the reconstructed signal can only contain a limited set of frequencies and thus its trajectory between the discrete sample values is not arbitrary but naturally constrained.

In this context, absolute units of measure are of no concern since in a digital signal all frequencies relate to the sampling frequency. In particular, if we take $\tau_s = 1$ as the (unitless) sampling interval, the resulting sampling frequency is

$$
\omega_s = 2 \cdot \pi \cdot f_s = 2 \cdot \pi \cdot \frac{1}{\tau_s} = 2 \cdot \pi \tag{22.3}
$$

and thus the maximum signal frequency is $\omega_{\max} = \frac{\omega_s}{2} = \pi$. To isolate the frequency range $-\omega_{\max} \ldots \omega_{\max}$ in the corresponding (periodic)

Fourier spectrum, we multiply the spectrum $G(\omega)$ by a square windowing function $\Pi_\pi(\omega)$ of width $\pm\omega_{\max} = \pm\pi$,

$$\tilde{G}(\omega) = G(\omega) \cdot \Pi_\pi(\omega) = G(\omega) \cdot \begin{cases} 1 & \text{for } -\pi \leq \omega \leq \pi, \\ 0 & \text{otherwise.} \end{cases} \tag{22.4}$$

This is called an *ideal low-pass filter*, which cuts off all signal components with frequencies greater than π and keeps all lower-frequency components unchanged. In the signal domain, the operation in Eqn. (22.4) corresponds (see Eqn. (18.27)) to a *linear convolution* with the inverse Fourier transform of the windowing function $\Pi_\pi(\omega)$, which is the *Sinc* function, defined as

$$\mathrm{Sinc}(x) = \frac{\sin(\pi x)}{\pi x}, \tag{22.5}$$

and shown in Fig. 22.3 (see also Ch. 18, Table 18.1). This correspondence, which was already discussed in Chapter 18, Sec. 18.1.6, between convolution in the signal domain and simple multiplication in the frequency domain is summarized in Fig. 22.4.

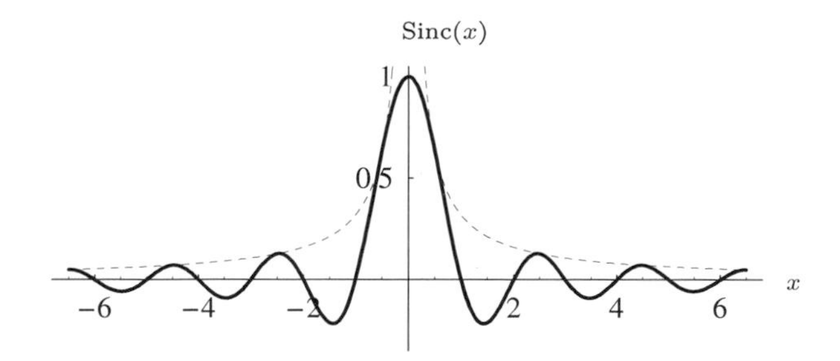

Fig. 22.3
Sinc function in 1D. The function $\mathrm{Sinc}(x)$ has the value 1 at the origin and zero values at all integer positions. The dashed line plots the amplitude $|\frac{1}{\pi x}|$ of the underlying sine function.

So theoretically $\mathrm{Sinc}(x)$ is the ideal interpolation function for reconstructing a frequency-limited continuous signal. To compute the interpolated value for the discrete function $g(u)$ at an arbitrary position x_0, the Sinc function is shifted to x_0 (such that its origin lies at x_0), multiplied with all sample values $g(u)$, with $u \in \mathbb{Z}$, and the results are summed—that is, $g(u)$ and $\mathrm{Sinc}(x)$ are *convolved.* The reconstructed value of the continuous function at position x_0 is thus

$$\tilde{g}(x_0) = [\mathrm{Sinc} * g]\,(x_0) = \sum_{u=-\infty}^{\infty} \mathrm{Sinc}(x_0 - u) \cdot g(u), \tag{22.6}$$

where $*$ is the linear convolution operator (see Ch. 5, Sec. 5.3.1). If the discrete signal $g(u)$ is *finite* with length N (as is usually the case), it is assumed to be *periodic* (i.e., $g(u) = g(u + kN)$ for all $k \in \mathbb{Z}$).[1] In this case, Eqn. (22.6) modifies to

$$\tilde{g}(x_0) = \sum_{u=-\infty}^{\infty} \mathrm{Sinc}(x_0 - u) \cdot g(u \bmod N)\,. \tag{22.7}$$

[1] This assumption is explained by the fact that a discrete Fourier spectrum implicitly corresponds to a periodic signal (also see Ch. 18, Sec. 18.2.2).

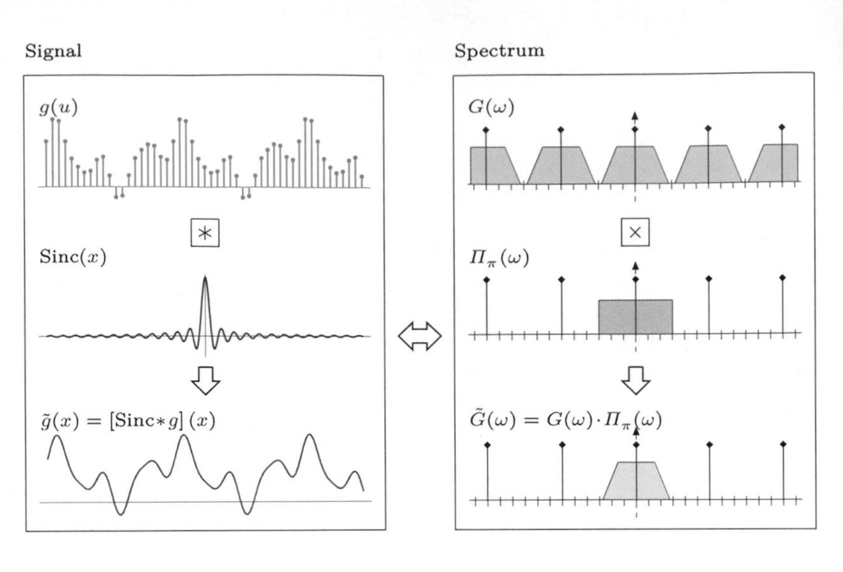

Fig. 22.4 Interpolation of a discrete signal—relation between signal and frequency space. The discrete signal $g(u)$ in signal space (left) corresponds to the periodic Fourier spectrum $G(\omega)$ in frequency space (right). The spectrum $\hat{G}(\omega)$ of the continuous signal is isolated from $G(\omega)$ by point-wise multiplication ($\times$) with the square function $\Pi_\pi(\omega)$, which constitutes an ideal low-pass filter (right). In signal space (left), this operation corresponds to a linear convolution ($*$) with the function Sinc(x).

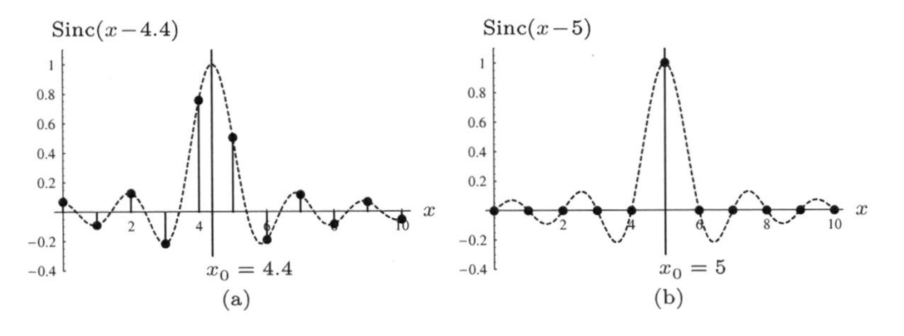

Fig. 22.5 Interpolation by convolving with the Sinc function. The Sinc function is shifted by aligning its origin with the interpolation points $x_0 = 4.4$ (a) and $x_0 = 5$ (b). The values of the shifted Sinc function (dashed curve) at the integral positions are the weights (coefficients) for the corresponding sample values $g(u)$. When the function is interpolated at some *integral* position, such as $x_0 = 5$ (b), only the sample value $g(x_0) = g(5)$ is considered and weighted with 1, while all other samples coincide with the zero positions of the Sinc function and thus do not contribute to the result.

It may be surprising that the ideal interpolation of a discrete function $g(u)$ at a position x_0 apparently involves not only a few neighboring sample points but, in general, *infinitely many* values of $g(u)$ whose weights decrease continuously with their distance from the given interpolation point x_0 (at the rate $|\frac{1}{\pi(x_0-u)}|$). Figure 22.5 shows two examples for interpolating the function $g(u)$ at positions $x_0 = 4.4$ and $x_0 = 5$. If the function is interpolated at some integral position, such as $x_0 = 5$, the sample $g(u)$ at $u = x_0$ receives the weight 1, while all other samples coincide with the zero positions of the Sinc function and are thus ignored. Consequently, the resulting interpolation values are identical to the sample values $g(u)$ at all discrete positions $x = u$.

If a continuous signal is properly frequency limited (by half the sampling frequency $\frac{\omega_s}{2}$), it can be exactly reconstructed from the discrete signal by interpolation with the Sinc function, as Fig. 22.6(a) demonstrates. Problems occur, however, around local high-frequency signal events, such as rapid transitions or pulses, as shown in Fig. 22.6(b, c). In those situations, the Sinc interpolation causes strong overshooting or "ringing" artifacts, which are perceived as visually disturbing. For practical applications, the Sinc function is therefore not suitable as an interpolation kernel—not only because of its infinite extent (and the resulting noncomputability).

A good interpolation function implements a low-pass filter that, on the one hand, introduces minimal blurring by maintaining the

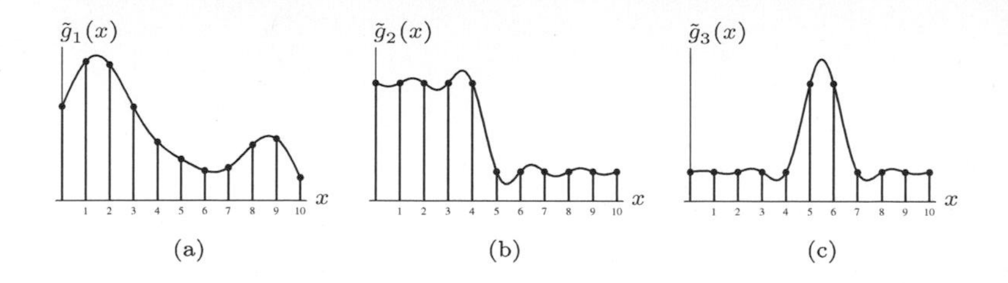

Fig. 22.6
Sinc interpolation applied to various signal types. The reconstructed function in (a) is identical to the continuous, band-limited original. The results for the step function (b) and the pulse function (c) show the strong ringing caused by Sinc (ideal low-pass) interpolation.

maximum signal bandwidth but, on the other hand, also delivers a good reconstruction at rapid signal transitions. In this regard, the Sinc function is an extreme choice—it implements an ideal low-pass filter and thus preserves a maximum bandwidth and signal continuity but gives inferior results at signal transitions. At the opposite extreme, nearest-neighbor interpolation (see Fig. 22.2) can perfectly handle steps and pulses but generally fails to produce a continuous signal reconstruction between sample points. The design of an interpolation function thus always involves a trade-off, and the quality of the results often depends on the particular application and subjective judgment. In the following, we discuss some common interpolation functions that come close to this goal and are therefore frequently used in practice.

22.2 Interpolation by Convolution

As we saw earlier in the context of Sinc interpolation (Eqn. (22.5)), the reconstruction of a continuous signal can be described as a linear convolution operation. In general, we can express interpolation as a convolution of the given discrete function $g(u)$ with some continuous *interpolation kernel* $w(x)$ as

$$\tilde{g}(x_0) = [w * g](x_0) = \sum_{u=-\infty}^{\infty} w(x_0 - u) \cdot g(u). \tag{22.8}$$

The Sinc interpolation in Eqn. (22.6) is obviously only a special case with $w(x) = \mathrm{Sinc}(x)$. Similarly, the 1D *nearest-neighbor interpolation* (Eqn. (22.1), Fig. 22.2(a)) can be expressed as a linear convolution with the kernel

$$w_{\mathrm{nn}}(x) = \begin{cases} 1 & \text{for } -0.5 \leq x < 0.5, \\ 0 & \text{otherwise,} \end{cases} \tag{22.9}$$

and the *linear interpolation* (see Eqn. (22.2), Fig. 22.2(b)) with the kernel

$$w_{\mathrm{lin}}(x) = \begin{cases} 1 - x & \text{for } |x| < 1, \\ 0 & \text{for } |x| \geq 1. \end{cases} \tag{22.10}$$

Both interpolation kernels $w_{\mathrm{nn}}(x)$ and $w_{\mathrm{lin}}(x)$ are shown in Fig. 22.7, and results for various function types are plotted in Fig. 22.8.

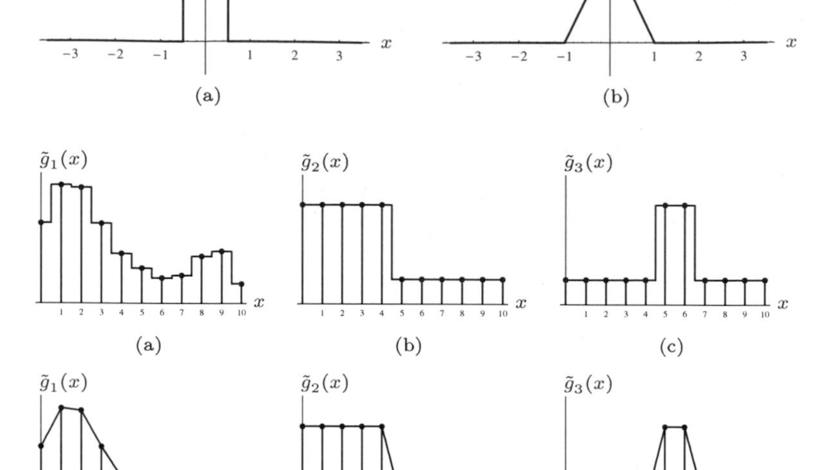

Fig. 22.7
Convolution kernels for the nearest-neighbor interpolation $w_{nn}(x)$ and the linear interpolation $w_{lin}(x)$.

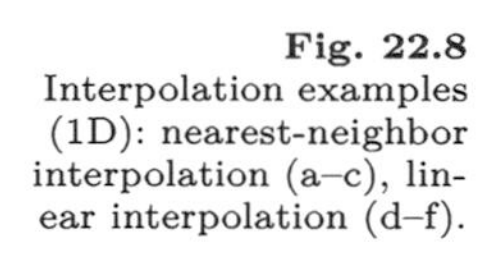

Fig. 22.8
Interpolation examples (1D): nearest-neighbor interpolation (a–c), linear interpolation (d–f).

22.3 Cubic Interpolation

The Sinc function is not a useful interpolation kernel in practice, because of its infinite extent and the ringing artifacts caused by its slowly decaying oscillations. Therefore several interpolation methods employ a truncated version of the Sinc function or an approximation of it, thereby making the convolution kernel more compact and reducing the ringing. A frequently used approximation of a truncated Sinc function is the so-called cubic interpolation, whose convolution kernel is defined as the piecewise cubic polynomial

$$w_{\mathrm{cub}}(x,a) = \begin{cases} (-a+2)\cdot|x|^3 + (a-3)\cdot|x|^2 + 1 & \text{for } 0 \le |x| < 1, \\ -a\cdot|x|^3 + 5a\cdot|x|^2 - 8a\cdot|x| + 4a & \text{for } 1 \le |x| < 2, \\ 0 & \text{for } |x| \ge 2. \end{cases} \tag{22.11}$$

Parameter a can be used to adjust the steepness of the spline function and thus the perceived “sharpness” of the interpolation (see Fig. 22.9(a)). For the standard value $a = 1$, Eqn. (22.11) simplifies to

$$w_{\mathrm{cub}}(x) = \begin{cases} |x|^3 - 2\cdot|x|^2 + 1 & \text{for } 0 \le |x| < 1, \\ -|x|^3 + 5\cdot|x|^2 - 8\cdot|x| + 4 & \text{for } 1 \le |x| < 2, \\ 0 & \text{for } |x| \ge 2. \end{cases} \tag{22.12}$$

The comparison of the Sinc function and the cubic interpolation kernel $w_{\mathrm{cub}}(x) = w_{\mathrm{cub}}(x,-1)$ in Fig. 22.9(b) shows that many high-value coefficients outside $x = \pm 2$ are truncated and thus relatively large errors can be expected. However, because of the compactness of the cubic function, this type of interpolation can be calculated

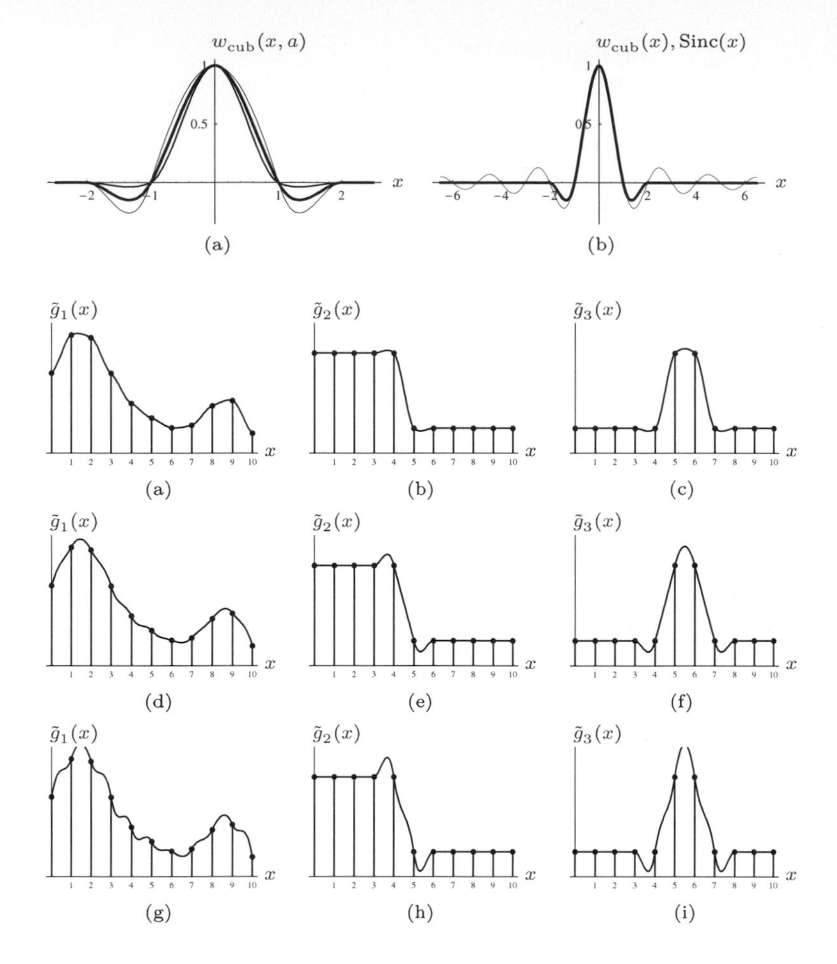

Fig. 22.9
Cubic interpolation kernel. Function $w_{cub}(x, a)$ with control parameter a set to $a = 0.25$ (dashed curve), $a = 1$ (continuous curve), and $a = 1.75$ (dotted curve) (a). Cubic function $w_{cub}(x)$ and Sinc function compared (b).

Fig. 22.10
Cubic interpolation examples. Parameter a in Eqn. (22.11) controls the amount of signal overshoot or perceived sharpness: $a = 0.25$ (a–c), standard setting $a = 1$ (d–f), $a = 1.75$ (g–i). Notice in (d) the ripple effects incurred by interpolating with the standard settings in smooth signal regions.

very efficiently. Since $w_{cub}(x) = 0$ for $|x| \geq 2$, only *four* discrete values $g(u)$ need to be accounted for in the convolution operation (Eqn. (22.8)) at any continuous position $x \in \mathbb{R}$, that is,

$$g(u_0-1),\ g(u_0),\ g(u_0+1),\ g(u_0+2), \quad \text{with } u_0 = \lfloor x_0 \rfloor .$$

This reduces the 1D cubic interpolation to the expression

$$\tilde{g}(x_0) = \sum_{u=\lfloor x_0 \rfloor -1}^{\lfloor x_0 \rfloor +2} w_{cub}(x_0-u) \cdot g(u) . \tag{22.13}$$

Figure 22.10 shows the results of cubic interpolation with different settings of the control parameter a. Notice that the cubic reconstruction obtained with the popular standard setting ($a = 1$) exhibits substantial overshooting at edges as well as strong ripple effects in the continuous parts of the signal (Fig. 22.10(d)). With $a = 0.5$, the expression in Eqn. (22.11) corresponds to a *Catmull-Rom* spline [44] (see also Sec. 22.4), which produces significantly better results than the standard setup (with $a = 1$), particularly in smooth signal regions (see Fig. 22.12(a–c)).

22.4 Spline Interpolation

The cubic interpolation kernel (Eqn. (22.11)) described in the previous section is a piecewise cubic polynomial function, also known as a *cubic spline* in computer graphics. In its general form, this function takes not only one but *two* control parameters (a, b) [164],[2]

$$w_{\mathrm{cs}}(x, a, b) = \tag{22.14}$$
$$\frac{1}{6} \cdot \begin{cases} (-6a - 9b + 12) \cdot |x|^3 \\ \quad + (6a + 12b - 18) \cdot |x|^2 - 2b + 6 & \text{for } 0 \le |x| < 1, \\ (-6a - b) \cdot |x|^3 + (30a + 6b) \cdot |x|^2 \\ \quad + (-48a - 12b) \cdot |x| + 24a + 8b & \text{for } 1 \le |x| < 2, \\ 0 & \text{for } |x| \ge 2. \end{cases}$$

Equation (22.14) describes a family of smooth, C^1-continuous functions (i.e., with continuous first derivatives) with no visible discontinuities or sharp corners. For $b = 0$, the function $w_{\mathrm{cs}}(x, a, b)$ specifies a one-parameter family of so-called *cardinal splines* equivalent to the cubic interpolation function $w_{\mathrm{cub}}(x, a)$ in Eqn. (22.11),

$$w_{\mathrm{cs}}(x, a, 0) = w_{\mathrm{cub}}(x, a), \tag{22.15}$$

and for the standard setting $a = 1$ (Eqn. (22.12)) in particular

$$w_{\mathrm{cs}}(x, 1, 0) = w_{\mathrm{cub}}(x, 1) = w_{\mathrm{cub}}(x). \tag{22.16}$$

Figure 22.11 shows three additional examples of this function type that are important in the context of interpolation: *Catmull-Rom* splines, *cubic B-splines*, and the *Mitchell-Netravali* function. All three functions are briefly described in the following sections. The actual calculation of the interpolated signal follows exactly the same scheme as used for the cubic interpolation described in Eqn. (22.13).

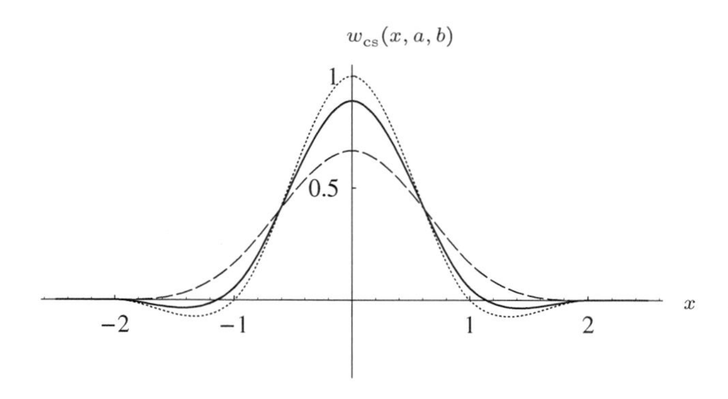

Fig. 22.11 Examples of cubic spline functions as defined in Eqn. (22.14): *Catmull-Rom* spline $w_{\mathrm{cs}}(x, 0.5, 0)$ (dotted line), *cubic B-spline* $w_{\mathrm{cs}}(x, 0, 1)$ (dashed line), and *Mitchell-Netravali* function $w_{\mathrm{cs}}(x, \frac{1}{3}, \frac{1}{3})$ (solid line).

22.4.1 Catmull-Rom Interpolation

With the control parameters set to $a = 0.5$ and $b = 0$, the function in Eqn. (22.14) is a *Catmull-Rom spline* [44], as already mentioned in Sec. 22.3:

[2] In [164], the parameters a and b were originally named C and B, respectively, with $B \equiv b$ and $C \equiv a$.

$$w_{\mathrm{crm}}(x) = w_{\mathrm{cs}}(x, 0.5, 0) \tag{22.17}$$
$$= \frac{1}{2} \cdot \begin{cases} 3 \cdot |x|^3 - 5 \cdot |x|^2 + 2 & \text{for } 0 \le |x| < 1, \\ -|x|^3 + 5 \cdot |x|^2 - 8 \cdot |x| + 4 & \text{for } 1 \le |x| < 2, \\ 0 & \text{for } |x| \ge 2. \end{cases}$$

Examples of signals interpolated with this kernel are shown in Fig. 22.12(a–c). The results are similar to ones produced by cubic interpolation (with $a = 1$, see Fig. 22.10) with regard to sharpness, but the Catmull-Rom reconstruction is clearly superior in smooth signal regions (compare, e.g., Fig. 22.10(d) vs. Fig. 22.12(a)).

22.4.2 Cubic B-spline Approximation

With parameters set to $a = 0$ and $b = 1$, Eqn. (22.14) corresponds to a cubic B-spline function of the form

$$w_{\mathrm{cbs}}(x) = w_{\mathrm{cs}}(x, 0, 1) \tag{22.18}$$
$$= \frac{1}{6} \cdot \begin{cases} 3 \cdot |x|^3 - 6 \cdot |x|^2 + 4 & \text{for } 0 \le |x| < 1, \\ -|x|^3 + 6 \cdot |x|^2 - 12 \cdot |x| + 8 & \text{for } 1 \le |x| < 2, \\ 0 & \text{for } |x| \ge 2. \end{cases}$$

This function is positive everywhere and, when used as an interpolation kernel, causes a pure smoothing effect similar to a Gaussian smoothing filter (see Fig. 22.12(d–f)). The B-spline function in Eqn. (22.18) is C^2-continuous, that is, its first *and* second derivatives are continuous. Notice that—in contrast to all previously described interpolation methods—the reconstructed function does *not* pass through all discrete sample points. Thus, to be precise, the reconstruction with cubic B-splines is not called an *interpolation* but an *approximation* of the signal.

22.4.3 Mitchell-Netravali Approximation

The design of an optimal interpolation kernel is always a trade-off between high bandwidth (sharpness) and good transient response (low ringing). Catmull-Rom interpolation, for example, emphasizes high sharpness, whereas cubic B-spline interpolation blurs but creates no ringing. Based on empirical tests, Mitchell and Netravali [164] proposed a cubic interpolation kernel as described in Eqn. (22.14) with parameter settings $a = \frac{1}{3}$ and $b = \frac{1}{3}$, and the resulting interpolation function

$$w_{\mathrm{mn}}(x) = w_{\mathrm{cs}}\big(x, \tfrac{1}{3}, \tfrac{1}{3}\big) \tag{22.19}$$
$$= \frac{1}{18} \cdot \begin{cases} 21 \cdot |x|^3 - 36 \cdot |x|^2 + 16 & \text{for } 0 \le |x| < 1, \\ -7 \cdot |x|^3 + 36 \cdot |x|^2 - 60 \cdot |x| + 32 & \text{for } 1 \le |x| < 2, \\ 0 & \text{for } |x| \ge 2. \end{cases}$$

This function is the weighted sum of a Catmull-Rom spline in Eqn. (22.17) and a cubic B-spline in Eqn. (22.18).[3] The examples in Fig.

[3] See also Exercise 22.1.

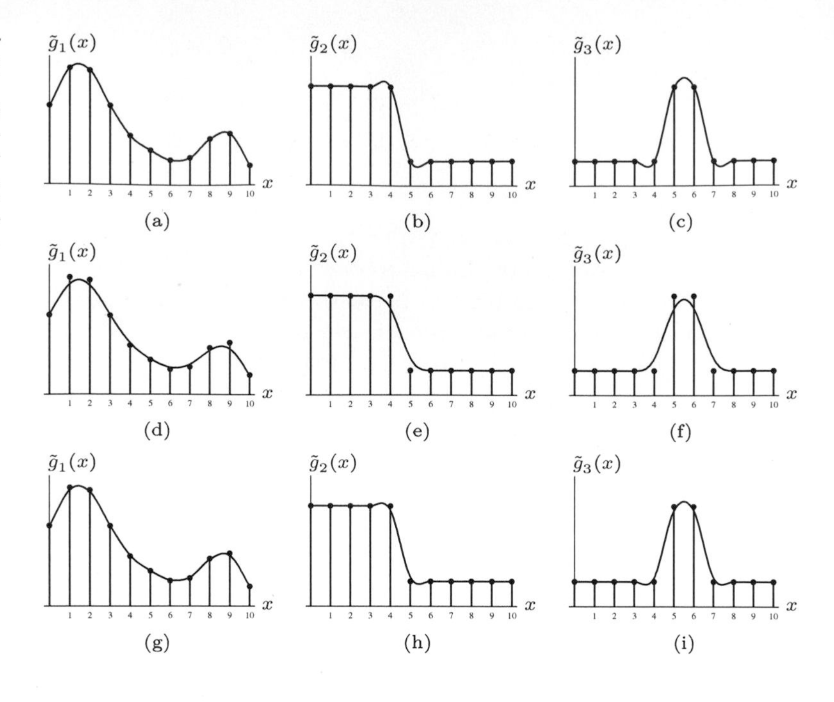

Fig. 22.12 Cardinal spline reconstruction examples: *Catmull-Rom* interpolation (a–c), *cubic B-spline* approximation (d–f), and *Mitchell-Netravali* approximation (g–i).

22.12(g–i) show that this method is a good compromise, creating little overshoot, high edge sharpness, and good signal continuity in smooth regions. Since the resulting function does not pass through the original sample points, the Mitchell-Netravali method is again an *approximation* and not an interpolation.

22.4.4 Lanczos Interpolation

The Lanczos[4] interpolation belongs to the family of "windowed Sinc" methods. In contrast to the methods described in the previous sections, these do *not* use a polynomial (or other) approximation of the Sinc function but the Sinc function *itself* combined with a suitable window function $\psi(x)$; that is, an interpolation kernel of the form

$$w(x) = \psi(x) \cdot \mathrm{Sinc}(x)\,. \tag{22.20}$$

The particular window functions for the Lanczos interpolation are defined as

$$\psi_{\mathrm{L}n}(x) = \begin{cases} 1 & \text{for } |x| = 0, \\ \frac{\sin(\pi x/n)}{\pi x/n} & \text{for } 0 < |x| < n, \\ 0 & \text{for } |x| \geq n, \end{cases} \tag{22.21}$$

where $n \in \mathbb{N}$ denotes the *order* of the filter [176, 237]. Notice that the window function is again a truncated Sinc function! For the Lanczos filters of order $n = 2, 3$, which are the most commonly used in image processing, the corresponding window functions are

[4] Cornelius Lanczos (1893–1974).

$$\psi_{L2}(x) = \begin{cases} 1 & \text{for } |x| = 0, \\ \frac{\sin(\pi x/2)}{\pi x/2} & \text{for } 0 < |x| < 2, \\ 0 & \text{for } |x| \geq 2, \end{cases} \tag{22.22}$$

$$\psi_{L3}(x) = \begin{cases} 1 & \text{for } |x| = 0, \\ \frac{\sin(\pi x/3)}{\pi x/3} & \text{for } 0 < |x| < 3, \\ 0 & \text{for } |x| \geq 3. \end{cases} \tag{22.23}$$

Both window functions are shown in Fig. 22.13(a, b). The 1D interpolation kernels w_{L2} and w_{L3} are obtained as the product of the Sinc function (Eqn. (22.5)) and the associated window function (Eqn. (22.21)), that is,

$$w_{L2}(x) = \begin{cases} 1 & \text{for } |x| = 0, \\ 2 \cdot \frac{\sin(\pi x/2) \cdot \sin(\pi x)}{\pi^2 x^2} & \text{for } 0 < |x| < 2, \\ 0 & \text{for } |x| \geq 2, \end{cases} \tag{22.24}$$

and

$$w_{L3}(x) = \begin{cases} 1 & \text{for } |x| = 0, \\ 3 \cdot \frac{\sin(\pi x/3) \cdot \sin(\pi x)}{\pi^2 x^2} & \text{for } 0 < |x| < 3, \\ 0 & \text{for } |x| \geq 3, \end{cases} \tag{22.25}$$

respectively. In general, for Lanczos interpolation of order n, we get

$$w_{Ln}(x) = \begin{cases} 1 & \text{for } |x| = 0, \\ n \cdot \frac{\sin(\pi x/n) \cdot \sin(\pi x)}{\pi^2 x^2} & \text{for } 0 < |x| < n, \\ 0 & \text{for } |x| \geq n. \end{cases} \tag{22.26}$$

Figure 22.13(c, d) shows the resulting interpolation kernels together with the original Sinc function. The function $w_{L2}(x)$ is quite similar to the Catmull-Rom kernel $w_{crm}(x)$ (Eqn. (22.17), Fig. 22.11), so the results can be expected to be similar as well, as shown in Fig. 22.14(a–c) (cf. Fig. 22.12(a–c)). Notice, however, the relatively poor reconstruction in the smooth signal regions (Fig. 22.14(a)) and the strong ringing introduced in the constant high-amplitude regions (Fig. 22.14(b)). The "3-tap" kernel $w_{L3}(x)$ reduces these artifacts and produces steeper edges, at the cost of increased overshoot (Fig. 22.12(d–f)).

In summary, although Lanczos interpolators have seen revived interest and popularity in recent years, they do not seem to offer much (if any) advantage over other established methods, particularly the cubic, Catmull-Rom, or Mitchell-Netravali interpolations. While these are based on efficiently computable polynomial functions, Lanczos interpolation requires trigonometric functions which are relatively costly to compute, unless some form of tabulation is used.

22.5 Interpolation in 2D

So far we have only looked at interpolating (or reconstructing) *1*D signals from discrete samples. Images are *2*D signals but, as we

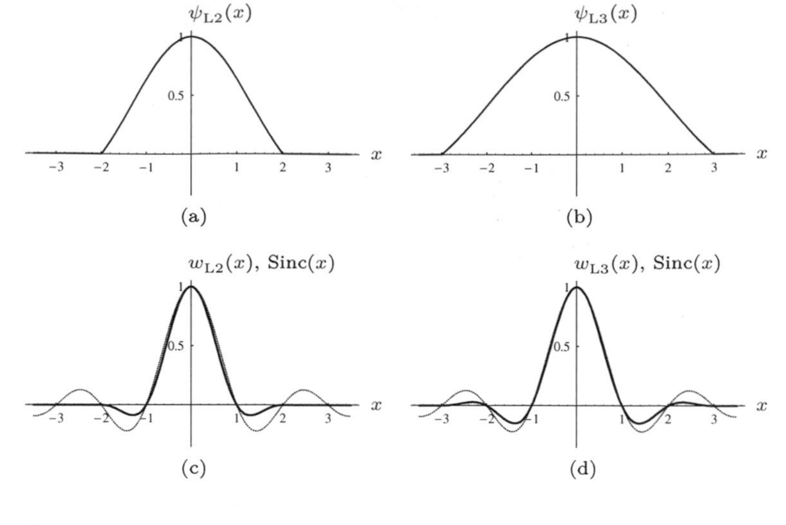

Fig. 22.13 1D Lanczos interpolation kernels. Lanczos window functions ψ_{L2} (a), ψ_{L3} (b), and the corresponding interpolation kernels w_{L2} (c), w_{L3} (d). The original Sinc function (dotted curve) is shown for comparison.

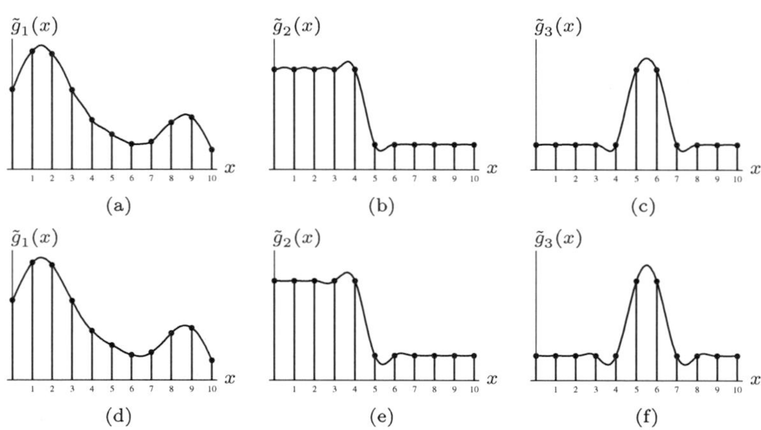

Fig. 22.14 Lanczos interpolation examples: Lanczos-2 (a–c), Lanczos-3 (d–f). Note the ringing in the flat (constant) regions caused by Lanczos-2 interpolation in the left part of (b). The Lanczos-3 interpolator shows less ringing (e) but produces steeper edges at the cost of increased overshoot (e, f).

shall see in this section, the techniques for interpolating images are very similar and can be derived from the 1D approach. In particular, "ideal" (low-pass filter) interpolation requires a 2D Sinc function defined as

$$\text{SINC}(x, y) = \text{Sinc}(x) \cdot \text{Sinc}(y) = \frac{\sin(\pi x)}{\pi x} \cdot \frac{\sin(\pi y)}{\pi y}, \tag{22.27}$$

which is shown in Fig. 22.15(a). Just as in 1D, the 2D Sinc function is not a practical interpolation function for various reasons. In the following, we look at some common interpolation methods for images, particularly the nearest-neighbor, bilinear, bicubic, and Lanczos interpolations, whose 1D versions were described in the previous sections.

22.5.1 Nearest-Neighbor Interpolation in 2D

The position (u_x, v_y) of the pixel closest to a given continuous point (x, y) is found by independently rounding the x and y coordinates to discrete values, that is,

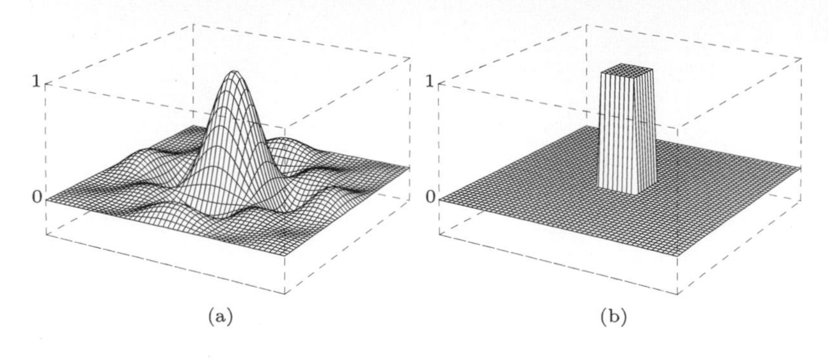

Fig. 22.15 Interpolation kernels in 2D. Sinc kernel $\mathrm{Sinc}(x, y)$ (a) and nearest-neighbor kernel $W_{\mathrm{nn}}(x, y)$ (b) for $-3 \leq x, y \leq 3$.

$$\tilde{I}(x, y) = I(u_x, v_y), \tag{22.28}$$

with $u_x = \mathrm{round}(x) = \lfloor x + 0.5 \rfloor$ und $v_y = \mathrm{round}(y) = \lfloor y + 0.5 \rfloor$.

As in the 1D case, the interpolation in 2D can be described as a linear convolution (linear filter). The 2D kernel for the nearest-neighbor interpolation is, analogous to Eqn. (22.9), defined as

$$W_{\mathrm{nn}}(x, y) = \begin{cases} 1 & \text{for } -0.5 \leq x, y < 0.5, \\ 0 & \text{otherwise.} \end{cases} \tag{22.29}$$

This function is shown in Fig. 22.15(b). Nearest-neighbor interpolation is known for its strong blocking effects (Fig. 22.16(b)) and thus is rarely used for geometric image operations. However, in some situations, this effect may be intended; for example, if an image is to be enlarged by replicating each pixel without any smoothing.

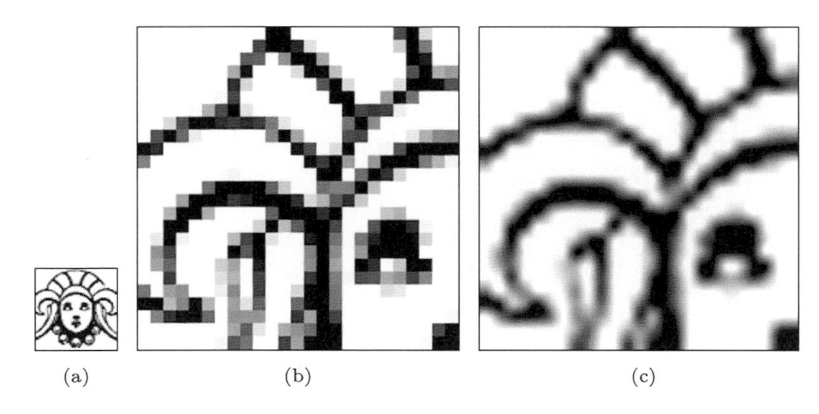

Fig. 22.16 Image enlargement example. Original (a); 8× enlargement using nearest-neighbor interpolation (b) and bicubic interpolation (c).

22.5.2 Bilinear Interpolation

The 2D counterpart to the linear interpolation in 1D (see Sec. 22.1) is the so-called *bilinear* interpolation,[5] whose operation is illustrated in Fig. 22.17. For the given interpolation point (x, y), we first find the four closest (surrounding) pixel vcalues,

$$\begin{aligned} A &= I(u_x, v_y), & B &= I(u_x + 1, v_y), \\ C &= I(u_x, v_y + 1), & D &= I(u_x + 1, v_y + 1), \end{aligned} \tag{22.30}$$

[5] Not to be confused with the bilinear *mapping* (transformation) described in Chapter 21, Sec. 21.1.5.

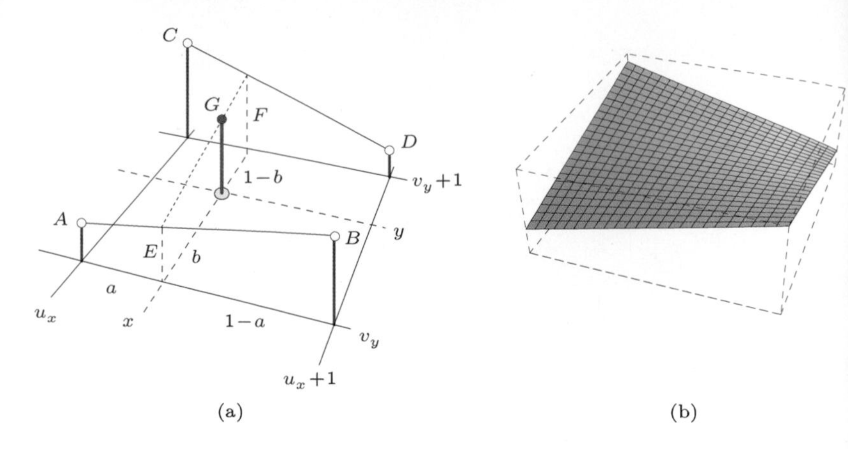

Fig. 22.17
Bilinear interpolation. For a given position (x, y), the interpolated value is computed from the values A, B, C, D of the four closest pixels in two steps (a). First the intermediate values E and F are computed by linear interpolation in the horizontal direction between A, B and C, D, respectively, where $a = x - u_x$ is the distance to the nearest pixel to the left of x. Subsequently, the intermediate values E, F are interpolated in the vertical direction, where $b = y - v_y$ is the distance to the nearest pixel below y. An example for the resulting surface between four adjacent pixels is shown in (b).

where $u_x = \lfloor x \rfloor$ and $v_x = \lfloor y \rfloor$. Then the pixel values A, B, C, D are interpolated in horizontal and subsequently in vertical direction. The intermediate values E, F are calculated from the distance $a = (x - u_x)$ of the specified interpolation position (x, y) from the discrete raster coordinate u_x as

$$E = A + (x - u_x) \cdot (B - A) = A + a \cdot (B - A), \tag{22.31}$$

$$F = C + (x - u_x) \cdot (D - C) = C + a \cdot (D - C), \tag{22.32}$$

and the final interpolation value G is computed from the vertical distance $b = y_0 - v_y$ as

$$\begin{aligned} \tilde{I}(x, y) = G &= E + (y - v_y) \cdot (F - E) = E + b \cdot (F - E) \\ &= (a-1)(b-1)\, A + a(1-b)\, B + (1-a)\, b\, C + a\, b\, D\,. \end{aligned} \tag{22.33}$$

Expressed as a linear convolution filter, the corresponding 2D kernel $W_{\mathrm{bil}}(x, y)$ is the product of the two 1D kernels $w_{\mathrm{lin}}(x)$ and $w_{\mathrm{lin}}(y)$ (Eqn. (22.10)), that is,

$$\begin{aligned} W_{\mathrm{bilin}}(x, y) &= w_{\mathrm{lin}}(x) \cdot w_{\mathrm{lin}}(y) \\ &= \begin{cases} 1 - x - y + x \cdot y & \text{for } 0 \leq |x|, |y| < 1, \\ 0 & \text{otherwise.} \end{cases} \end{aligned} \tag{22.34}$$

In this function (plotted in Fig. 22.18), we can recognize the bilinear term that gives this method its name.

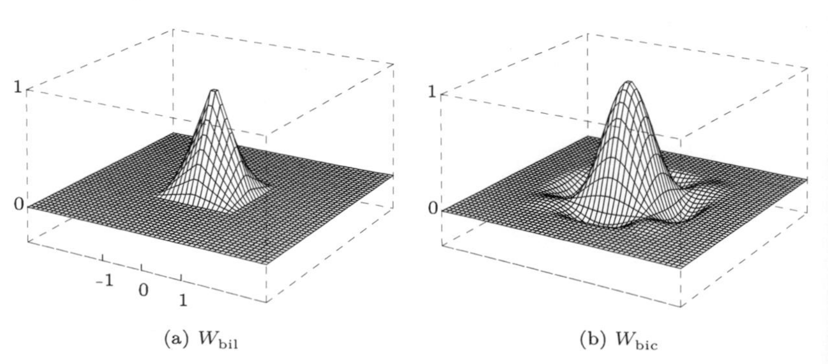

(a) W_{bil} (b) W_{bic}

Fig. 22.18
2D interpolation kernels. bilinear kernel $W_{\mathrm{bil}}(x, y)$ (a) and bicubic kernel $W_{\mathrm{bic}}(x, y)$ (b) for $-3 \leq x, y \leq 3$.

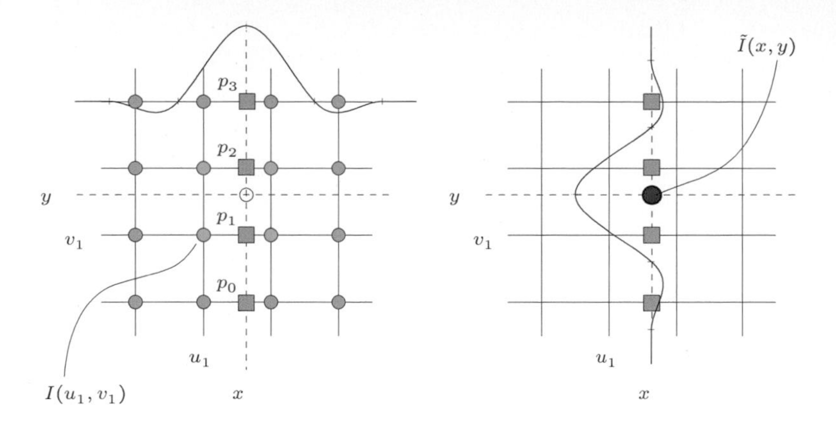

Fig. 22.19
Bicubic interpolation in two steps. The discrete image I (pixel positons correspond to raster lines) is to be interpolated at some continuous position (x, y). In step 1 (left), a 1D interpolation is performed in the horizontal direction with $w_{\text{cub}}(x)$ over four pixels $I(u_i, v_j)$ in four lines. One intermediate result p_j (marked □) is computed for each line j. In step 2 (right), the result $\hat{I}(x_0, y_0)$ is computed by a single cubic interpolation in the vertical direction over the intermediate results $p_0, \ldots, p_3$. In total, $16 + 4 = 20$ interpolation steps are required.

22.5.3 Bicubic and Spline Interpolation in 2D

The convolution kernel for the 2D cubic interpolation is also defined as the product of the corresponding 1D kernels (Eqn. (22.12)),

$$W_{\text{bic}}(x, y) = w_{\text{cub}}(x) \cdot w_{\text{cub}}(y). \tag{22.35}$$

The resulting kernel is plotted in Fig. 22.18(b). Due to the decomposition into 1D kernels (Eqn. (22.13)), the computation of the bicubic interpolation is *separable* in x, y and can thus be expressed as

$$\tilde{I}(x, y) = \sum_{\substack{v= \\ \lfloor y \rfloor - 1}}^{\lfloor y \rfloor + 2} \Biggl[\sum_{\substack{u= \\ \lfloor x \rfloor - 1}}^{\lfloor x \rfloor + 2} I(u, v) \cdot W_{\text{bic}}(x - u, y - v) \Biggr] \tag{22.36}$$

$$= \sum_{j=0}^{3} \Biggl[w_{\text{cub}}(y - v_j) \cdot \underbrace{\sum_{i=0}^{3} I(u_i, v_j) \cdot w_{\text{cub}}(x - u_i)}_{p_j} \Biggr], \tag{22.37}$$

with $u_i = \lfloor x_0 \rfloor - 1 + i$ and $v_j = \lfloor y_0 \rfloor - 1 + j$. The quantity p_j is the intermediate result of the cubic interpolation in the x direction in line j, as illustrated in Fig. 22.19. Equation (22.37) describes a simple and efficient procedure for computing the bicubic interpolation using only a 1D kernel $w_{\text{cub}}(x)$. The interpolation is based on a 4×4 neighborhood of pixels and requires a total of $16 + 4 = 20$ additions and multiplications.

This method, which is summarized in Alg. 22.1, can be used to implement any x/y-separable 2D interpolation kernel of size 4×4, such as the 2D *Catmull-Rom* interpolation (Eqn. (22.17)) with

$$W_{\text{crm}}(x, y) = w_{\text{crm}}(x) \cdot w_{\text{crm}}(y) \tag{22.38}$$

or the *Mitchell-Netravali* interpolation (Eqn. (22.19)) with

$$W_{\text{mn}}(x, y) = w_{\text{mn}}(x) \cdot w_{\text{mn}}(y). \tag{22.39}$$

The corresponding 2D kernels are shown in Fig. 22.20. For interpolation with separable kernels of larger size see the general procedure in Alg. 22.2.

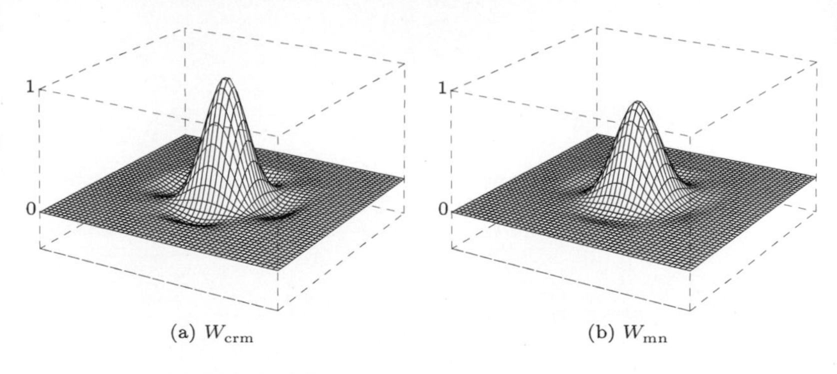

(a) W_{crm} (b) W_{mn}

Fig. 22.20 2D spline interpolation kernels: Catmull-Rom kernel $W_{crm}(x, y)$ (a), Mitchell-Netravali kernel $W_{mn}(x, y)$ (b), for $-3 \leq x, y \leq 3$.

Alg. 22.1 Bicubic interpolation of image I at position (x, y). The 1D cubic function $w_{cub}(\cdot)$ (Eqn. (22.11)) is used for the separate interpolation in the x and y directions based on a neighborhood of 4 × 4 pixels. See Prog. 22.1 for a straightforward implementation in Java.

```
1:  BicubicInterpolation(I, x, y, a)
    Input: I, original image; x, y ∈ ℝ, continuous position; a, con-
    trol parameter. Returns the interpolated image value at position
    (x, y).
2:  q ← 0
3:  for j ← 0, ..., 3 do                        ▷ iterate over 4 lines
4:      v ← ⌊y⌋ − 1 + j
5:      p ← 0
6:      for i ← 0, ..., 3 do                    ▷ iterate over 4 columns
7:          u ← ⌊x⌋ − 1 + i
8:          p ← p + I(u, v) · w_cub(x−u, a)     ▷ see Eq. 22.11
9:      q ← q + p · w_cub(y−v, a)
10: return q
```

22.5.4 Lanczos Interpolation in 2D

The kernels for the 2D Lanczos interpolation are also x/y-separable into 1D kernels (see Eqns. (22.24) and (22.25), respectively), that is,

$$W_{Ln}(x, y) = w_{Ln}(x) \cdot w_{Ln}(y) . \qquad (22.40)$$

The resulting kernels for orders $n = 2$ and $n = 3$ are shown in Fig. 22.21. Because of the separability the 2D Lanczos interpolation can be computed, similar to the bicubic interpolation, separately in the x and y directions. Like the bicubic kernel, the 2-tap Lanczos kernel W_{L2} (Eqn. (22.24)) is *zero* outside the interval $-2 \leq x, y \leq 2$, and thus the procedure described in Eqn. (22.37) and Alg. 22.1 can be used with only a small modification (replace w_{cub} by w_{L2}).

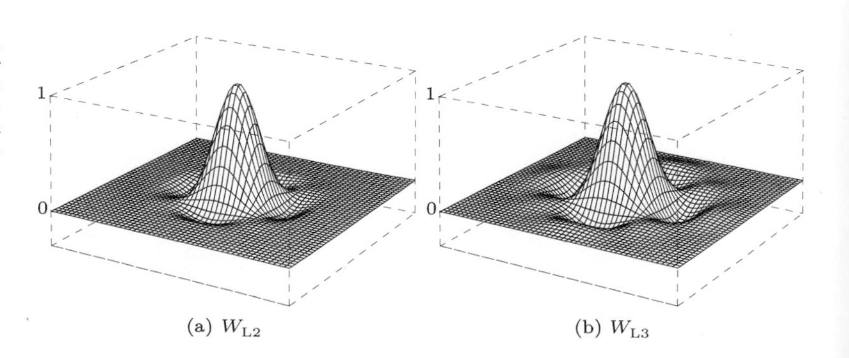

(a) W_{L2} (b) W_{L3}

Fig. 22.21 2D Lanczos kernels for $n = 2$ and $n = 3$: kernels $W_{L2}(x, y)$ (a) and $W_{L3}(x, y)$ (b), with $-3 \leq x, y \leq 3$.

```
1: SeparableInterpolation(I, x, y, w, n)
     Input: I, original image; x, y ∈ ℝ, continuous position; w, a 1D
     interpolation kernel of extent ±n (n ≥ 1).
     Returns the interpolated image value at position (x, y) using the
     composite interpolation kernel W(x, y) = w(x) · w(y).
2:   q ← 0
3:   for j ← 0, ..., 2n−1 do                    ▷ iterate over 2n lines
4:       v ← ⌊y⌋ − n + 1 + j                                    ▷ = v_j
5:       p ← 0
6:       for i ← 0, ..., 2n−1 do              ▷ iterate over 2n columns
7:           u ← ⌊x⌋ − n + 1 + i                                ▷ = u_i
8:           p ← p + I(u, v) · w(x − u)
9:       q ← q + p · w(y − v)
10:  return q
```

Alg. 22.2
General interpolation with a separable interpolation kernel $W(x,y) = w_n(x) \cdot w_n(y)$ of extent $\pm n$ (i.e., the 1D kernel $w_n(x)$ is zero for $x < -n$ and $x > n$, with $n \in \mathbb{N}$). Note that procedure BicubicInterpolation in Alg. 22.1 is a special instance of this algorithm (with $n = 2$).

Compared to Eqn. (22.37), the larger Lanczos kernel W_{L3} (Eqn. (22.25)) requires two additional pixel rows and columns. The calculation of the interpolated pixel value at position (x, y) thus has the form

$$\tilde{I}(x,y) = \sum_{\substack{v=\\ \lfloor y \rfloor - 2}}^{\lfloor y \rfloor + 3} \Biggl[\sum_{\substack{u=\\ \lfloor x \rfloor - 2}}^{\lfloor x \rfloor + 3} I(u,v) \cdot W_{\mathrm{L3}}(x-u, y-v) \Biggr] \tag{22.41}$$

$$= \sum_{j=0}^{5} \Biggl[w_{\mathrm{L3}}(y - v_j) \cdot \sum_{i=0}^{5} I(u_i, v_j) \cdot w_{\mathrm{L3}}(x - u_i) \Biggr], \tag{22.42}$$

with $u_i = \lfloor x \rfloor - 2 + i$ and $v_j = \lfloor y \rfloor - 2 + j$. Thus the L3 Lanczos interpolation in 2D uses a support region of $6 \times 6 = 36$ pixels from the original image, 20 pixels more than the bicubic interpolation.

In general, the expression for a 2D Lanczos interpolator Ln of arbitrary order $n \geq 1$ is

$$\tilde{I}(x,y) = \sum_{\substack{v=\\ \lfloor y \rfloor - n+1}}^{\lfloor y \rfloor + n} \Biggl[\sum_{\substack{u=\\ \lfloor x \rfloor - n+1}}^{\lfloor x \rfloor + n} \bigl[I(u,v) \cdot W_{\mathrm{L}n}(x-u, y-v) \bigr] \Biggr] \tag{22.43}$$

$$= \sum_{j=0}^{2n-1} \Bigl[w_{\mathrm{L}n}(y - v_j) \cdot \sum_{i=0}^{2n-1} \bigl[I(u_i, v_j) \cdot w_{\mathrm{L}n}(x - u_i) \bigr] \Bigr], \tag{22.44}$$

with $u_i = \lfloor x \rfloor - n + 1 + i$ and $v_j = \lfloor y \rfloor - n + 1 + j$. The size of this interpolator's support region is $2n \times 2n$ pixels. How the expression in Eqn. (22.44) could be computed is shown in Alg. 22.2, which actually describes a general interpolation procedure that can be used with any separable interpolation kernel $W(x,y) = w_n(x) \cdot w_n(y)$ of extent $\pm n$.

22.5.5 Examples and Discussion

Figures 22.22 and 22.23 compare the interpolation methods described in this section: nearest-neighbor, bilinear, bicubic Catmull-Rom, cubic B-spline, Mitchell-Netravali, and Lanczos interpolation. In both figures, the original images are rotated counter-clockwise by 15°. A gray background is used to visualize the edge overshoot produced by some of the interpolators.

Nearest-neighbor interpolation (Fig. 22.22(b)) creates no new pixel values but forms, as expected, coarse blocks of pixels with the same intensity.

The effect of the *bilinear* interpolation (Fig. 22.22(c)) is local smoothing over four neighboring pixels. The weights for these four pixels are positive, and thus no result can be smaller than the smallest neighboring pixel value or greater than the greatest neighboring pixel value. In other words, bilinear interpolation cannot create any over- or undershoot at edges.

This is not the case for the *bicubic* interpolation (Fig. 22.22(d)): some of the coefficients in the bicubic interpolation kernel are negative, which makes pixels near edges clearly brighter or darker, respectively, thus increasing the perceived sharpness. In general, bicubic interpolation produces clearly better results than the bilinear method at comparable computing cost, and it is thus widely accepted as the standard technique and used in most image manipulation programs. By adjusting the control parameter a (Eqn. (22.11)), the bicubic kernel can be easily tuned to fit the need of particular applications. For example, the *Catmull-Rom* method (Fig. 22.22(e)) can be implemented with the bicubic interpolation by setting $a = 0.5$ (Eqns. (22.17) and (22.38)).

Results from the 2D *Lanczos* interpolation (Fig. 22.22(h)) using the 2-tap kernel W_{L2} cannot be much better than from the bicubic interpolation, which can be adjusted to give similar results without causing any ringing in flat regions, as seen in Fig. 22.14. The 3-tap Lanczos kernel W_{L3} (Fig. 22.22(i)) on the other hand should produce slightly sharper edges at the cost of increased overshoot (see also Exercise 22.3).

In summary, for high-quality applications one should consider the *Catmull-Rom* (Eqns. (22.17) and (22.38)) or the *Mitchell-Netravali* (Eqns. (22.19) and (22.39)) methods, which offer good reconstruction at the same computational cost as the bicubic interpolation.

22.6 Aliasing

As we described in the main part of this chapter, the usual approach for implementing geometric image transformations can be summarized by the following three steps (Fig. 22.24):

1. Each discrete image point (u', v') of the *target* image is projected by the inverse geometric transformation T^{-1} to the continuous coordinate (x, y) in the source image.
2. The continuous image function $\tilde{I}(x, y)$ is reconstructed from the discrete source image $I(u, v)$ by interpolation (using one of the methods described earlier).
3. The interpolated function is sampled at position (x, y), and the sample value $\tilde{I}(x, y)$ is transferred to the target pixel $I'(u', v')$.

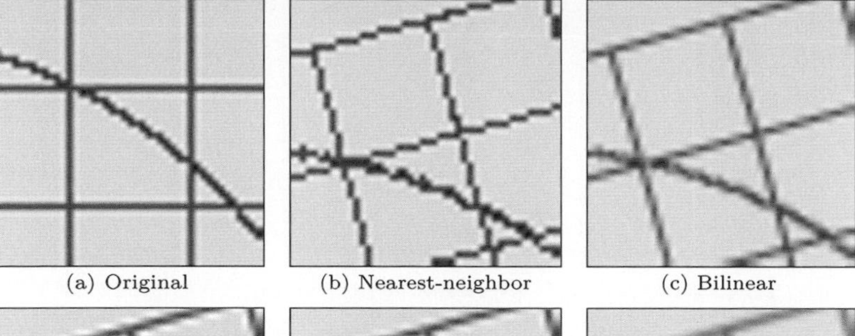

(a) Original (b) Nearest-neighbor (c) Bilinear

(d) Bicubic

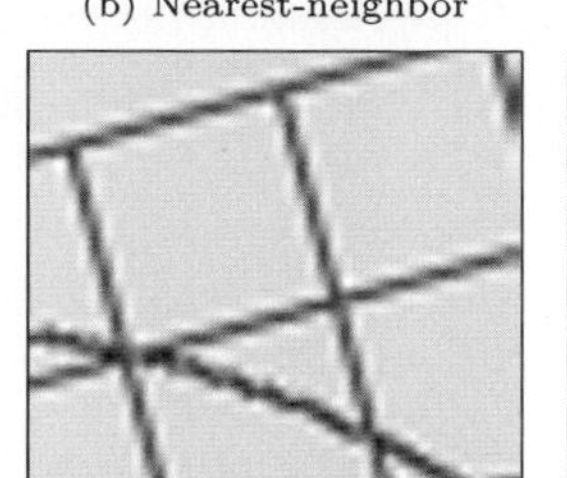

(e) Catmull-Rom

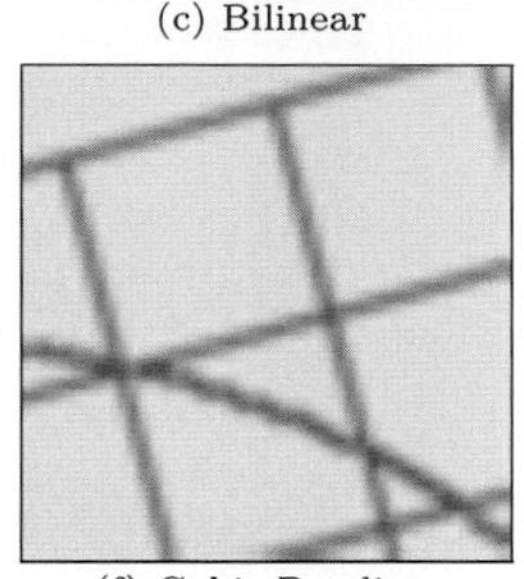

(f) Cubic B-spline

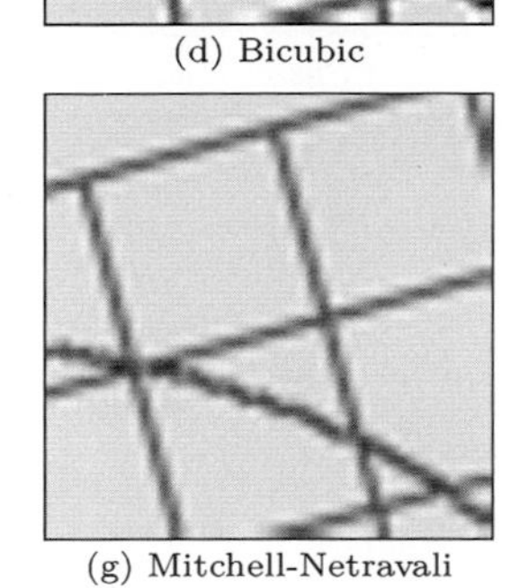

(g) Mitchell-Netravali

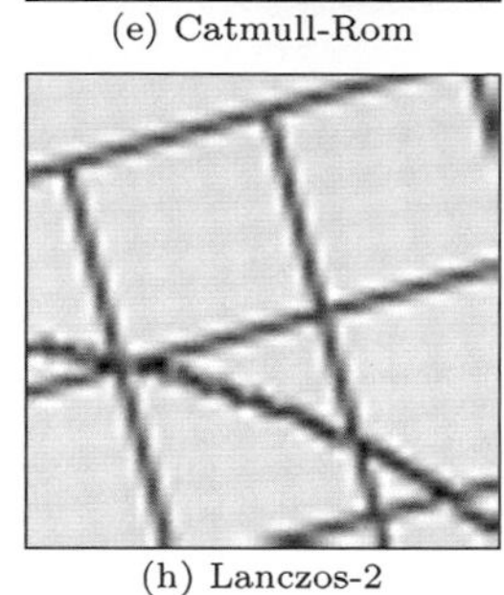

(h) Lanczos-2

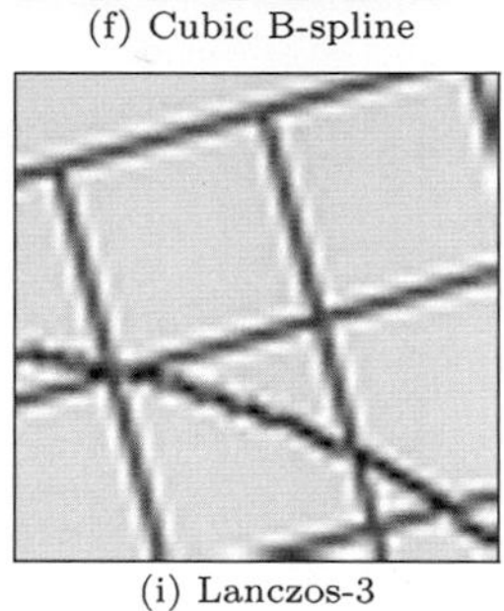

(i) Lanczos-3

Fig. 22.22
Image interpolation methods compared (line art).

22.6.1 Sampling the Interpolated Image

One problem not considered so far concerns the process of sampling the reconstructed, continuous image function in the aforementioned step 3. The problem occurs when the geometric transformation T causes parts of the image to be *contracted.* In this case, the distance between adjacent sample points on the source image is locally *increased* by the corresponding inverse transformation T^{-1}. Now, widening the sampling distance reduces the spatial sampling rate and thus the maximum permissible frequencies in the reconstructed image function $\tilde{I}(x, y)$. Eventually this leads to a violation of the sampling criterion and causes visible aliasing in the transformed image. The problem does not occur when the image is enlarged by the geometric transformation because in this case the sampling interval on the source image is shortened (corresponding to a higher sampling frequency) and no aliasing can occur.

Note that this effect is largely unrelated to the interpolation method, as demonstrated by the examples in Fig. 22.25. The effect is most noticeable under nearest-neighbor interpolation in Fig. 22.25(b), where the thin lines are simply not "hit" by the widened sampling raster and thus disappear in some places. Important image information is thereby lost. The bilinear and bicubic interpolation methods in Fig. 22.25(c, d) have wider interpolation kernels but still

(a) Original (b) Nearest-neighbor (c) Bilinear

(d) Bicubic (e) Catmull-Rom (f) Cubic B-spline

(g) Mitchell-Netravali (h) Lanczos-2 (i) Lanczos-3

Fig. 22.23
Image interpolation methods compared (text image).

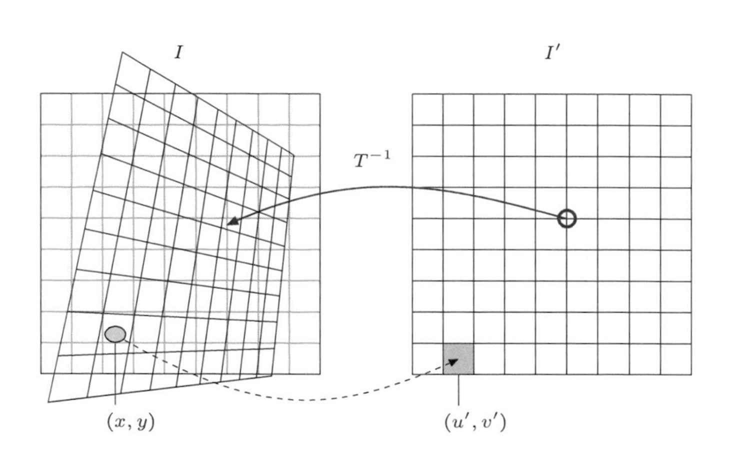

Fig. 22.24
Sampling errors in geometric operations. If the geometric transformation T leads to a local contraction of the image (which corresponds to a local enlargement by T^{-1}), the distance between adjacent sample points in I is increased. This reduces the local sampling frequency and thus the maximum signal frequency allowed in the source image, which eventually leads to aliasing.

cannot avoid the aliasing effect. The problem of course gets worse with increasing reduction factors.

22.6.2 Low-Pass Filtering

One solution to the aliasing problem is to make sure that the interpolated image function is properly frequency-limited before it gets

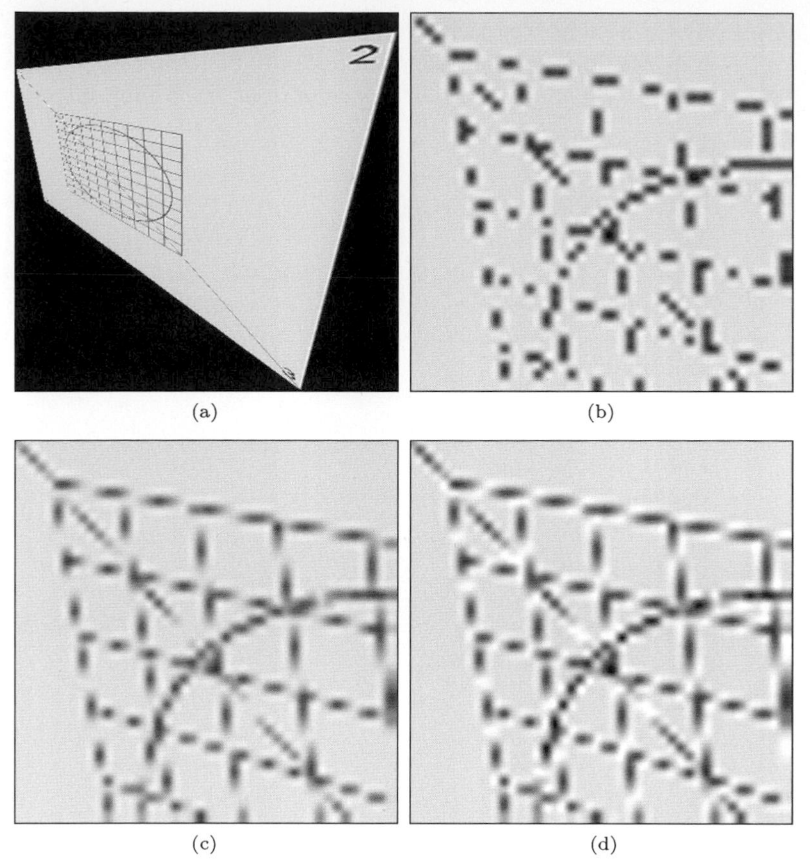

Fig. 22.25
Aliasing caused by local image contraction. Aliasing is caused by a violation of the sampling criterion and is largely unaffected by the interpolation method used: complete transformed image (a), detail using nearest-neighbor interpolation (b), bilinear interpolation (c), and bicubic interpolation (d).

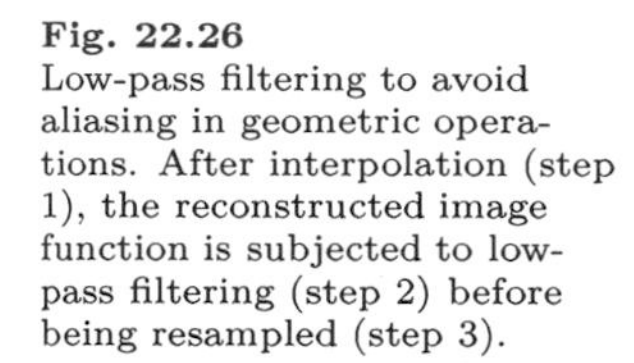

Fig. 22.26
Low-pass filtering to avoid aliasing in geometric operations. After interpolation (step 1), the reconstructed image function is subjected to low-pass filtering (step 2) before being resampled (step 3).

resampled. This can be accomplished with a suitable low-pass filter, as illustrated in Fig. 22.26.

The cutoff frequency of the low-pass filter is determined by the amount of local scale change, which may—depending upon the type of transformation—be different in various parts of the image. In the simplest, case the amount of scale change is the same throughout the image (e.g., under global scaling or affine transformations, where the same filter can be used everywhere in the image). In general, however, the low-pass filter is *space-variant* or *nonhomogeneous*, and the local filter parameters are determined by the transformation T and the current image position. If convolution filters are used for both interpolation and low-pass filtering, they could be combined into a common, space-variant reconstruction filter.

Unfortunately, space-variant filtering is computationally expensive and thus is often avoided, even in professional applications (e.g., Adobe Photoshop). The technique is nevertheless used in certain applications, such as high-quality texture mapping in computer graphics [75, 105, 256]. Integral images, as described in Chapter 3, Sec. 3.8, can be used to implement efficient space-variant smoothing filters.

22.7 Java Implementation

Implementations of most interpolation methods described in this chapter are openly available as part of the **imagingbook** library.[6] The following interpolators are available as subclasses of the abstract class `PixelInterpolator`:

> `BicubicInterpolator`,
> `BilinearInterpolator`,
> `LanczosInterpolator`,
> `NearestNeighborInterpolator`,
> `SplineInterpolator`.

For illustration, the complete implementation of the class `BicubicInterpolator` is shown in Prog. 22.1.

`PixelInterpolator` (class)

This class provides the functionality for interpolating images with scalar pixel values. It defines the following methods:

`static PixelInterpolator create (InterpolationMethod im)`
Factory method which creates and returns a new interpolator. Admissible values for the parameter `im` and associated interpolator types (subclasses of `ScalarInterpolator`) are listed in Table 22.1.

`float getInterpolatedValue (ImageAccessor.Scalar ia, double x, double y)`
Returns the interpolated pixel value at the continuous position `x`, `y` of the scalar-valued image (referenced by the image accessor `ia`).

Table 22.1 Admissible values for `InterpolationMethod` and associated interpolator types returned by the static `create(im)` method of `PixelInterpolator`.

`InterpolationMethod im`	Interpolator Type
`NearestNeighbor`	`NearestNeighborInterpolator()`
`Bilinear`	`BilinearInterpolator()`
`Bicubic`	`BicubicInterpolator(1.00)`
`BicubicSmooth`	`BicubicInterpolator(0.25)`
`BicubicSharp`	`BicubicInterpolator(1.75)`
`CatmullRom`	`SplineInterpolator(0.5, 0.0)`
`CubicBSpline`	`SplineInterpolator(0.0, 1.0)`
`MitchellNetravali`	`SplineInterpolator(1.0/3, 1.0/3)`
`Lanczos2`	`LanczosInterpolator(2)`
`Lanczos3`	`LanczosInterpolator(3)`
`Lanczos4`	`LanczosInterpolator(4)`

[6] Package `imagingbook.lib.interpolation`.

```
 1 package imagingbook.lib.interpolation;
 2
 3 import imagingbook.lib.image.ImageAccessor;
 4 import java.awt.geom.Point2D;
 5
 6 public class BicubicInterpolator
 7             extends PixelInterpolator {
 8
 9   private final double a;     // sharpness value
10
11   public BicubicInterpolator() {
12     this(0.5);
13   }
14   public BicubicInterpolator(double a) {
15     this.a = a;
16   }
17
18   public float getInterpolatedValue(
19           ImageAccessor.Scalar ia, double x, double y) {
20     final int u0 = (int) Math.floor(x);
21     final int v0 = (int) Math.floor(y);
22     double q = 0;
23     for (int j = 0; j <= 3; j++) {
24       int v = v0 - 1 + j;
25       double p = 0;
26       for (int i = 0; i <= 3; i++) {
27         int u = u0 - 1 + i;
28         float pixval = ia.getVal(u, v);
29         p = p + pixval * w_cub(x - u, a);
30       }
31       q = q + p * w_cub(y - v, a);
32     }
33     return (float) q;
34   }
35
36   private final double w_cub(double x, double a) {
37     if (x < 0)
38       x = -x;
39     double z = 0;
40     if (x < 1)
41       z = (-a + 2) * x * x * x + (a - 3) * x * x + 1;
42     else if (x < 2)
43       z = -a * x * x * x + 5 * a * x * x
44           - 8 * a * x + 4 * a;
45     return z;
46   }
47 }
```

Prog. 22.1
Java implementation of bicubic interpolation (class `BicubicInterpolator`), as defined in Alg. 22.1. The class provides two constructors: a default constructor (line 11) with sharpness value $a = 0.5$ and a general constructor for arbitrary a (line 14). The actual pixel interpolation is performed by method `getInterpolatedValue()` in line 18, which implements Alg. 22.1. `w_cub()` in line 36 is the 1D cubic interpolation function (see Eqn. (22.11)).

The class `PixelInterpolator` is primarily used by the methods in class `ImageAccessor`.[7] See Prog. 22.2 for a basic usage example.

[7] The `ImageAccessor` class (in package `imagingbook.lib.image`) provides unified access to all types of images available in ImageJ and also supports pixel interpolation.

Prog. 22.2
Image interpolation example using class ImageAccessor. This ImageJ plugin translates the input image by some (non-integer) distance dx, dy. It uses target-to-source mapping and pixel interpolation of type BicubicSharp (see Table 22.1). The required ImageAccessor (interpolator) object for the source image is created in line 31, another for the target image in line 34. This is followed by an iteration over all pixels of the target image. The source image is interpolated (line 41) at the calculated positions (x, y) and the resulting float[] value is inserted into the target image with setPix() in line 42. Note that this plugin is generic, that is, it works for all image types.

```
1 import ij.ImagePlus;
2 import ij.plugin.filter.PlugInFilter;
3 import ij.process.ImageProcessor;
4 import imagingbook.lib.image.ImageAccessor;
5 import imagingbook.lib.image.OutOfBoundsStrategy;
6 import static imagingbook.lib.image.OutOfBoundsStrategy.*;
7 import imagingbook.lib.interpolation.InterpolationMethod;
8 import static imagingbook.lib.interpolation.
      InterpolationMethod.*;
9
10 public class Interpolator_Demo implements PlugInFilter {
11
12     static double dx =  0.5;  // translation
13     static double dy = -3.5;
14
15     static OutOfBoundsStrategy OBS = NearestBorder;
16     static InterpolationMethod IPM = BicubicSharp;
17
18     public int setup(String arg, ImagePlus imp) {
19         return DOES_ALL + NO_CHANGES;
20     }
21
22     public void run(ImageProcessor source) {
23       final int w = source.getWidth();
24       final int h = source.getHeight();
25
26       // create the target image (same type as source):
27       ImageProcessor target = source.createProcessor(w, h);
28
29       // create an ImageAccessor for the source image:
30       ImageAccessor sA =
31             ImageAccessor.create(source, OBS, IPM);
32
33       // create an ImageAccessor for the target image:
34       ImageAccessor tA = ImageAccessor.create(target);
35
36       // iterate over all pixels of the target image:
37       for (int u = 0; u < w; u++) {
38         for (int v = 0; v < h; v++) {
39           double x = u + dx;   // continuous source position (x,y)
40           double y = v + dy;
41           float[] val = sA.getPix(x, y);
42           tA.setPix(u, v, val);  // update the target pixel
43         }
44       }
45
46       // display the target image:
47       (new ImagePlus("Target", target)).show();
48     }
49 }
```

22.8 Exercises

Exercise 22.1. The 1D interpolation function by Mitchell and Natravali $w_{\mathrm{mn}}(x)$ is defined as a general spline function $w_{\mathrm{cs}}(x, a, b)$ (Eqn. (22.19)). Show that this function can be expressed as the weighted sum of a Catmull-Rom function $w_{\mathrm{crm}}(x)$ (Eqn. (22.17)) and a cubic B-spline $w_{\mathrm{cbs}}(x)$ (Eqn. (22.18)) in the form

$$\begin{aligned} w_{\mathrm{mn}}(x) &= w_{\mathrm{cs}}\big(x, \tfrac{1}{3}, \tfrac{1}{3}\big) \\ &= \tfrac{1}{3} \cdot \big[2 \cdot w_{\mathrm{cs}}(x, 0.5, 0) + w_{\mathrm{cs}}(x, 0, 1)\big] \\ &= \tfrac{1}{3} \cdot \big[2 \cdot w_{\mathrm{crm}}(x) + w_{\mathrm{cbs}}(x)\big]. \end{aligned} \tag{22.45}$$

Exercise 22.2. Implement an "ideal" (low-pass) pixel interpolator based on the Sinc function (see Eqn. (22.5)). Assume that the image function is periodic along both coordinate axes. Determine (by truncating the Sinc function at $\pm N$) the minimum number of samples to include and if the result improves by including additional samples. Use the class `BicubicInterpolator` (Prog. 22.1) as a template for your implementation.

Exercise 22.3. Implement the 2D *Lanczos* interpolation with a W_{L3} kernel, as defined in Eqn. (22.42), as a Java class analogous to the class `BicubicInterpolator` (Prog. 22.1). Compare the results to the bicubic interpolation.

Exercise 22.4. The 1D Lanczos interpolation kernel of order $n = 4$ is (analogous to Eqn. (22.25)) defined as

$$w_{\mathrm{L4}} = \begin{cases} 4 \cdot \frac{\sin(\pi \frac{x}{4}) \cdot \sin(\pi x)}{\pi^2 x^2} & \text{for } 0 \leq |x| < 4, \\ 0 & \text{for } |x| \geq 4. \end{cases} \tag{22.46}$$

Extend the 2D kernel in Eqn. (22.42) to w_{L4} and implement this interpolator as a Java class analogous to `BicubicInterpolator` (Prog. 22.1). How many image pixels does the calculation include at each position? See if there is any noticeable improvement over the bicubic and the Lanczos-3 interpolation (Exercise 22.3).

23

Image Matching and Registration

When we compare two images, we are faced with the following basic question: when are two images the same or similar, and how can this similarity be measured? Of course one could trivially define two images I_1, I_2 as being identical when all pixel values are the same (i.e., the difference $I_1 - I_2$ is zero). Although this kind of definition may be useful in specific applications, such as for detecting changes in successive images under constant lighting and camera conditions, simple pixel differencing is usually too inflexible to be of much practical use. Noise, quantization errors, small changes in lighting, and minute shifts or rotations can all create large numerical pixel differences for pairs of images that would still be perceived as perfectly identical by a human viewer. Obviously, human perception incorporates a much wider concept of similarity and uses cues such as structure and content to recognize similarity between images, even when a direct comparison between individual pixels would not indicate any match. The problem of comparing images at a structural or semantic level is a difficult problem and an interesting research field, for example, in the context of image-based searches on the Internet or database retrieval.

This chapter deals with the much simpler problem of comparing images at the pixel level; in particular, localizing a given subimage—often called a "template"—within some larger image. This task is frequently required, for example, to find matching patches in stereo images, to localize a particular pattern in a scene, or to track a certain pattern through an image sequence. The principal idea behind "template matching" is simple: move the given pattern (template) over the search image, measure the difference against the corresponding subimage at each position, and record those positions where the highest similarity is obtained. But this is not as simple as it may initially sound. After all, what is a suitable distance measure, what total difference is acceptable for a match, and what happens when brightness or contrast changes?

We already touched on this problem of invariance under geometric transformations when we discussed the shape properties of seg-

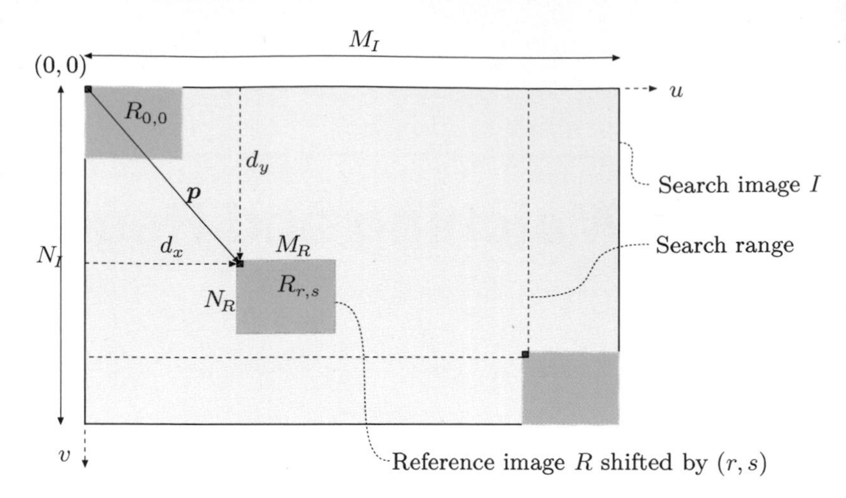

Fig. 23.1
Geometry of template matching. The reference image R is shifted across the search image I by an offset (r, s) using the origins of the two images as the reference points. The dimensions of the search image ($M_I \times N_I$) and the reference image ($M_R \times N_R$) determine the maximal search region for this comparison.

mented regions in Chapter 10, Sec. 10.4.2. However, geometric invariance is not our main concern in the remaining part of this chapter, where we describe only the most basic template-matching techniques: correlation-based methods for intensity images and "chamfer-matching" for binary images.

23.1 Template Matching in Intensity Images

First we look at the problem of localizing a given *reference image* (template) R within a larger intensity (grayscale) image I, which we call the *search image*. The task is to find those positions where the contents of the reference image R and the corresponding subimage of I are either the same or most similar. If we denote by

$$R_{r,s}(u, v) = R(u-r, v-s) \qquad (23.1)$$

the reference image R *shifted* by the distance (r, s) in the horizontal and vertical directions, respectively, then the matching problem (illustrated in Fig. 23.1) can be summarized as follows:

- Given are the search image I and the reference image R. Find the offset $(r, s) \in \mathbb{Z}^2$ such that the similarity between the shifted reference image $R_{r,s}$ and the corresponding subimage of I is a maximum.

To successfully solve this task, several issues need to be addressed, such as determining a minimum similarity value for accepting a match and developing a good search strategy for finding the optimal displacement. First, and most important, a suitable measure of similarity between subimages must be found that is reasonably tolerant against intensity and contrast variations.

23.1.1 Distance between Image Patterns

To quantify the amount of agreement, we compute a "distance" $\mathrm{d}(r, s)$ between the shifted reference image R and the corresponding subimage of I for each offset position (r, s) (Fig. 23.2). Several distance

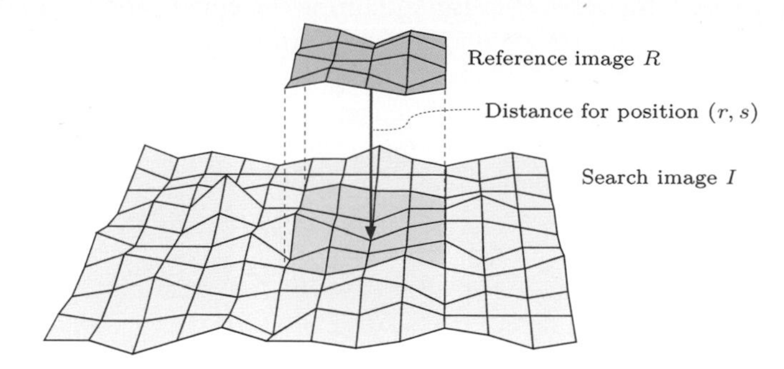

Fig. 23.2
Measuring the distance between 2D image functions. The reference image R is positioned at offset (r, s) on top of the search image I.

measures have been proposed for 2D intensity images, including the following three basic definitions:[1]

Sum of absolute differences:

$$\mathrm{d}_A(r, s) = \sum_{(i,j)\in R} |I(r+i, s+j) - R(i, j)| . \tag{23.2}$$

Maximum difference:

$$\mathrm{d}_M(r, s) = \max_{(i,j)\in R} |I(r+i, s+j) - R(i, j)| . \tag{23.3}$$

Sum of squared differences:

$$\mathrm{d}_E(r, s) = \Big[\sum_{(i,j)\in R} \big(I(r+i, s+j) - R(i, j)\big)^2 \Big]^{1/2} . \tag{23.4}$$

Note that the expression in Eqn. (23.4) is nothing else but the *Euclidean distance* between two N-dimensional vectors of pixels values. Similarly, the sum of differences in Eqn. (23.2) is equivalent to the L_1 distance, and the maximum difference in Eqn. (23.3) equals the L_∞ distance norm.[2]

Distance and correlation

Because of its formal properties, the N-dimensional distance d_E (Eqn. (23.4)) is of special importance and well-known in statistics and optimization. To find the best-matching position between the reference image R and the search image I, it is sufficient to *minimize the square* of d_E (which is always positive), which can be expanded to

$$\mathrm{d}_E^2(r, s) = \sum_{(i,j)\in R} \big(I(r+i, s+j) - R(i, j)\big)^2 \tag{23.5}$$

$$= \underbrace{\sum_{(i,j)\in R} I^2(r+i, s+j)}_{A(r,s)} + \underbrace{\sum_{(i,j)\in R} R^2(i, j)}_{B} - 2 \cdot \underbrace{\sum_{(i,j)\in R} I(r+i, s+j) \cdot R(i, j)}_{C(r,s)} .$$

[1] We use the short notation $(i, j) \in R$ to specify the set of all possible template coordinates, that is, $\{(i, j) \mid 0 \le i < M_R, 0 \le j < N_R\}$.
[2] See also Sec. B.1.2 in the Appendix.

Notice that the term B in Eqn. (23.5) is the sum of the squared pixel values in the reference image R, a constant value (independent of r, s) that can thus be ignored. The term $A(r, s)$ is the sum of the squared values within the subimage of I at the current offset (r, s). $C(r, s)$ is the so-called *linear cross correlation* ($\circledast$) between I and R, which is defined in the general case as

$$(I \circledast R)(r,s) = \sum_{i=-\infty}^{\infty} \sum_{j=-\infty}^{\infty} I(r+i, s+j) \cdot R(i,j), \tag{23.6}$$

which—since R and I are assumed to have zero values outside their boundaries—is, furthermore, equivalent to

$$\sum_{i=0}^{M_R-1} \sum_{j=0}^{N_R-1} I(r+i, s+j) \cdot R(i,j) = \sum_{(i,j)\in R} I(r+i, s+j) \cdot R(i,j) \tag{23.7}$$

and thus the same as $C(r, s)$ in Eqn. (23.5). As we can see in Eqn. (23.6), correlation is in principle the same operation as linear *convolution* (see Ch. 5, Eqn. (5.16)), with the only difference being that the convolution kernel ($R(i, j)$ in this case) is implicitly mirrored.

If we assume for a minute that $A(r, s)$—the "signal energy"— in Eqn. (23.5) is constant throughout the image I, then $A(r, s)$ can also be ignored and the position of maximum cross correlation $C(r, s)$ coincides with the best match between R and I. In this case, the minimum of $d_E^2(r, s)$ (Eqn. (23.5)) can be found by computing the maximum value of the correlation $I \circledast R$ only. This could be interesting for practical reasons if we consider that the linear convolution (and thus the correlation) with large kernels can be computed very efficiently in the frequency domain (see also Ch. 19, Sec. 19.5).

Normalized cross correlation

Unfortunately, the assumption made earlier that $A(r, s)$ is constant does not hold for most images, and thus the result of the cross correlation strongly varies with intensity changes in the image I. The *normalized cross correlation* $C_N(r, s)$ compensates for this dependency by taking into account the energy in the reference image and the current subimage:

$$C_N(r,s) = \frac{C(r,s)}{\sqrt{A(r,s) \cdot B}} = \frac{C(r,s)}{\sqrt{A(r,s)} \cdot \sqrt{B}} \tag{23.8}$$

$$= \frac{\sum_{(i,j)\in R} I(r+i, s+j) \cdot R(i,j)}{\big[\sum_{(i,j)\in R} I^2(r+i, s+j)\big]^{1/2} \cdot \big[\sum_{(i,j)\in R} R^2(i,j)\big]^{1/2}} . \tag{23.9}$$

If the values in the search and reference images are all positive (which is usually the case), then the result of $C_N(r, s)$ is always in the range $[0, 1]$, independent of the remaining contents in I and R. In this case, the result $C_N(r, s) = 1$ indicates a maximum match between R and the current subimage of I at the offset (r, s), while $C_N(r, s) =$

0 signals no agreement. Thus the normalized correlation has the additional advantage of delivering a standardized match value that can be used directly (using a suitable threshold between 0 and 1) to decide about the acceptance or rejection of a match position.

In contrast to the "global" cross correlation in Eqn. (23.6), the expression in Eqn. (23.8) is a "local" distance measure. However, it, too, has the problem of measuring the *absolute* distance between the template and the subimage. If, for example, the overall intensity of the image I is altered, then even the result of the normalized cross correlation $C_N(r,s)$ may also change dramatically.

Correlation coefficient

One solution to this problem is to compare not the original function values but the differences with respect to the average value of R and the average of the current subimage of I. This modification turns Eqn. (23.8) into

$$C_L(r,s) = \frac{\sum\limits_{(i,j)\in R} \bigl(I(r{+}i,s{+}j) - \bar{I}_{r,s}\bigr) \cdot \bigl(R(i,j) - \bar{R}\bigr)}{\Bigl[\sum\limits_{(i,j)\in R} \bigl(I(r{+}i,s{+}j) - \bar{I}_{r,s}\bigr)^2\Bigr]^{1/2} \cdot \underbrace{\Bigl[\sum\limits_{(i,j)\in R} \bigl(R(i,j) - \bar{R}\bigr)^2\Bigr]^{1/2}}_{S_R^2 = K \cdot \sigma_R^2}}, \tag{23.10}$$

with the average values $\bar{I}_{r,s}$ and $\bar{R}$ defined as

$$\bar{I}_{r,s} = \frac{1}{K} \cdot \sum_{(i,j)\in R} I(r{+}i,s{+}j) \quad \text{and} \quad \bar{R} = \frac{1}{K} \cdot \sum_{(i,j)\in R} R(i,j), \tag{23.11}$$

respectively, ($K = |R|$ being the size of the reference image R). In statistics, the expression in Eqn. (23.10) is known as the *correlation coefficient.* However, different from the usual application as a global measure in statistics, $C_L(r,s)$ describes a *local*, piecewise correlation between the template R and the current subimage (at offset r,s) of I. The resulting values of $C_L(r,s)$ are in the range $[-1,1]$ regardless of the contents in R and I. Again a value of 1 indicates maximum agreement between the compared image patterns, while -1 corresponds to a maximum mismatch. The term

$$S_R^2 = K \cdot \sigma_R^2 = \sum_{(i,j)\in R} \bigl(R(i,j) - \bar{R}\bigr)^2 \tag{23.12}$$

in the denominator of Eqn. (23.10) is K times the *variance* (σ_R^2) of the values in the template R, which is constant and thus needs to be computed only once. Due to the fact that $\sigma_R^2 = \frac{1}{K}\sum R^2(i,j) - \bar{R}^2$, the expression in Eqn. (23.12) can be reformulated as

$$S_R^2 = \sum_{(i,j)\in R} R^2(i,j) \;-\; K \cdot \bar{R}^2 \tag{23.13}$$

$$= \sum_{(i,j)\in R} R^2(i,j) \;-\; \frac{1}{K} \cdot \Bigl[\sum_{(i,j)\in R} R(i,j)\Bigr]^2. \tag{23.14}$$

By inserting the results from Eqns. (23.11) and (23.14) we can rewrite Eqn. (23.10) as

$$C_L(r,s) = \frac{\sum\limits_{(i,j)\in R} \big(I(r{+}i,s{+}j)\cdot R(i,j)\big) \;-\; K\cdot \bar{I}_{r,s}\cdot \bar{R}}{\Big[\sum\limits_{(i,j)\in R} I^2(r{+}i,s{+}j) \;-\; K\cdot \bar{I}^2_{r,s}\Big]^{1/2}\cdot S_R}, \tag{23.15}$$

and thereby obtain an efficient way to compute the local correlation coefficient. Since $\bar{R}$ and $S_R = (S_R^2)^{1/2}$ must be calculated only once and the local average of the current subimage $\bar{I}_{r,s}$ is not immediately required for summing up the differences, the whole expression in Eqn. (23.15) can be computed in one common iteration, as shown in Alg. 23.1.

Note that in the calculation of $C_L(r,s)$ in Eqn. (23.15), the denominator becomes zero if any of the two factors is zero. This may happen, for example, if the search image I is locally "flat" and thus has zero variance or if the reference image R is constant. The quantity 1 is added to the denominator in Alg. 23.1 (line 23) to avoid divisions by zero in such cases, which otherwise has no significant effect on the result.

A direct Java implementation of this procedure is shown in Progs. 23.1 and 23.2 in Sec. 23.1.3 (class `CorrCoeffMatcher`).

Examples and discussion

Figure 23.3 compares the performance of the described distance functions in a typical example. The original image (Fig. 23.3(a)) shows a repetitive flower pattern produced under uneven lighting and differences in local brightness. One instance of the repetitive pattern was extracted as the reference image (Fig. 23.3(b)).

- The *sum of absolute differences* (Eqn. (23.2)) in Fig. 23.3(c) shows a distinct peak value at the original template position, as does the *Euclidean distance* (Eqn. (23.4)) in Fig. 23.3(e). Both measures work satisfactorily in this regard but are strongly affected by global intensity changes, as demonstrated in Figs. 23.4 and 23.5.
- The *maximum difference* (Eqn. (23.3)) in Fig. 23.3(d) proves completely useless as a distance measure since it responds more strongly to the lighting changes than to pattern similarity. As expected, the behavior of the *global cross correlation* in Fig. 23.3(f) is also unsatisfactory. Although the result exhibits a *local* maximum at the true template position (hardly visible in the printed image), it is completely dominated by the high-intensity responses in the brighter parts of the image.
- The result from the *normalized cross correlation* in Fig. 23.3(g) appears naturally very similar to the Euclidean distance (Fig. 23.3(e)), because in principle it is the same measure. As expected, the *correlation coefficient* (Eqn. (23.10)) in Fig. 23.3(h) yields the best results. Distinct peaks of similar intensity are produced for all six instances of the template pattern, and the result is unaffected by changing lighting conditions. In this case, the

Alg. 23.1
Calculation of the correlation coefficient. Given is the search image I and the reference image (template) R. In Step 1, the template's average $\bar{R}$ and variance term S_R are computed once. In Step 2, the match function is computed for every template position (r, s) as prescribed by Eqn. (23.15). The result is a map of correlation values $C(r, s) \in [-1, 1]$ that is returned. In line 23 (cf. Eqn. (23.15)) the quantity 1 is added to the denominator to avoid division by zero in the case of zero variance.

```
1:  CorrelationCoefficient (I, R)
        Input: I(u,v), search image; R(i,j), reference image.
        Returns a map C(r,s) containing the values of the correlation
        coefficient between I and R positioned at (r,s).

        Step 1—initialize:
2:      (M_I, N_I) ← Size(I)
3:      (M_R, N_R) ← Size(R)
4:      K ← M_R · N_R
5:      Σ_R ← 0, Σ_R2 ← 0
6:      for i ← 0,...,(M_R−1) do
7:          for j ← 0,...,(N_R−1) do
8:              Σ_R  ← Σ_R  + R(i,j)
9:              Σ_R2 ← Σ_R2 + R²(i,j)
10:     R̄ ← Σ_R/K                                        ▷ Eq. 23.11
11:     S_R ← (Σ_R2 − K·R̄²)^(1/2)                        ▷ Eq. 23.14

        Step 2—compute the correlation map:
12:     Create map C: (M_I−M_R+1) × (N_I−N_R+1) ↦ ℝ
13:     for r ← 0,...,M_I−M_R do                ▷ place R at position (r,s)
14:         for s ← 0,...,N_I−N_R do
                Compute the correlation coefficient for position (r,s):
15:             Σ_I ← 0, Σ_I2 ← 0, Σ_IR ← 0
16:             for i ← 0,...,M_R−1 do
17:                 for j ← 0,...,N_R−1 do
18:                     a_I ← I(r+i, s+j)
19:                     a_R ← R(i,j)
20:                     Σ_I  ← Σ_I  + a_I
21:                     Σ_I2 ← Σ_I2 + a_I²
22:                     Σ_IR ← Σ_IR + a_I · a_R
23:             C(r,s) ← (Σ_IR − Σ_I·R̄) / (1 + sqrt(Σ_I2 − Σ_I²/K) · S_R)
24:     return C                                     ▷ C(r,s) ∈ [−1,1]
```

values range from −1.0 (black) to +1.0 (white), and zero values are shown as gray.

Figure 23.4 compares the results of the *Euclidean distance* against the *correlation coefficient* under globally changing intensity. For this purpose, the intensity of the reference image R is raised by 50 units such that the template is different from any subpattern in the original image. As can be seen clearly, the initially distinct peaks disappear under the Euclidean distance (Fig. 23.4(c)), while the correlation coefficient (Fig. 23.4(d)) naturally remains unaffected by this change.

In summary, the correlation coefficient can be recommended as a reliable measure for template matching in intensity images under realistic lighting conditions. This method proves relatively robust against global changes of brightness or contrast and tolerates small deviations from the reference pattern. Since the resulting values are in the fixed range of $[-1, 1]$, a simple threshold operation can be used to localize the best match points (Fig. 23.6).

Fig. 23.3
Comparison of various distance functions. From the original image I (a), the marked section is used as the reference image R, shown enlarged in (b). In the resulting difference images (c–h), brightness corresponds to the amount of agreement (white equals minimum distance). The position of the true reference point is marked by a red circle.

(a) Original image I

(b) Reference image R

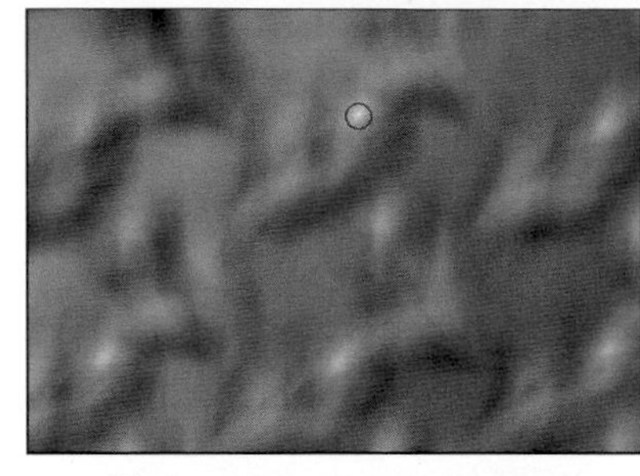

(c) Sum of absolute differences

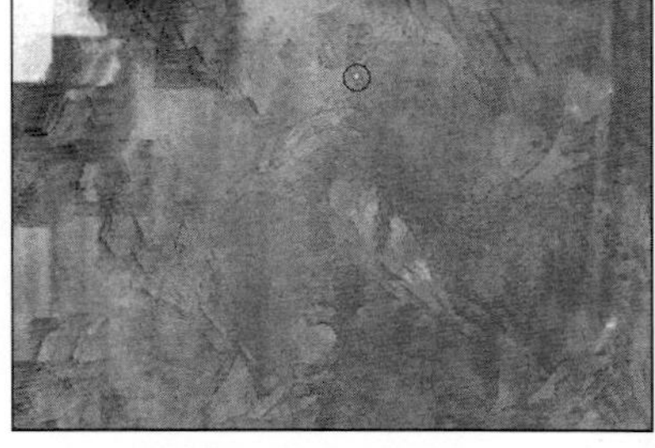

(d) Maximum difference

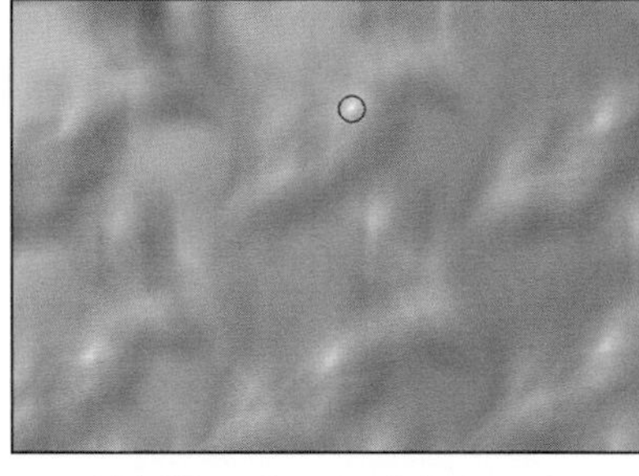

(e) Sum of squared distances

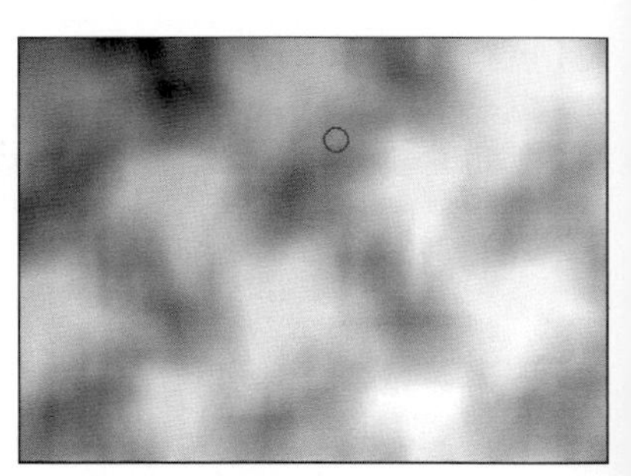

(f) Global cross correlation

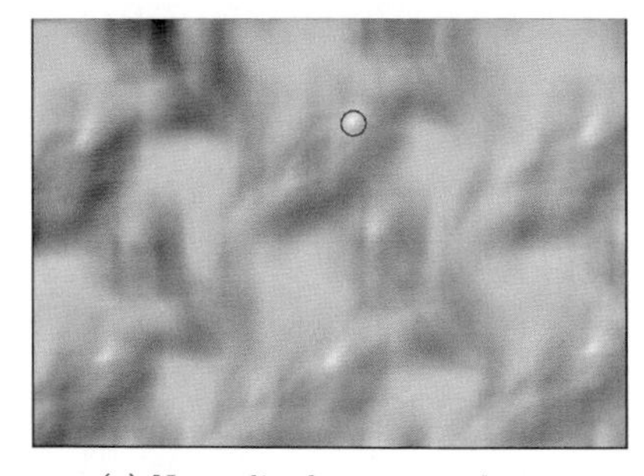

(g) Normalized cross correlation

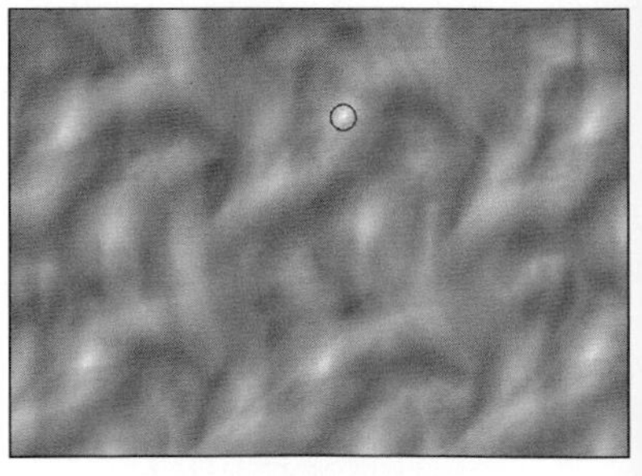

(h) Correlation coefficient

Shape of the template

The shape of the reference image does not need to be rectangular as in the previous examples, although it is convenient for the processing. In some applications, circular, elliptical, or custom-shaped templates may be more applicable than a rectangle. In such a case, the template

Original reference image R

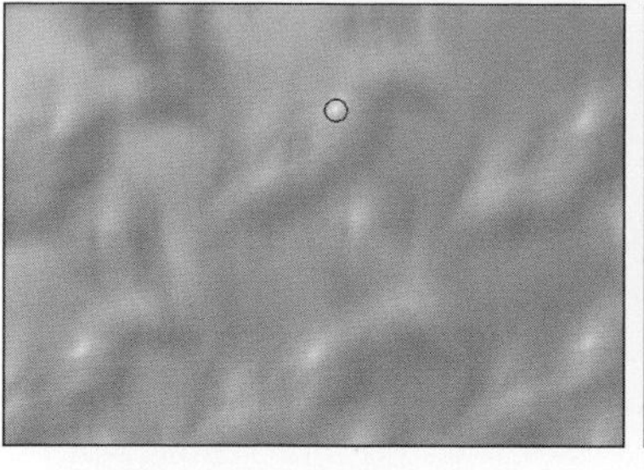

(a) Euclidean distance $d_{\mathrm{E}}(r, s)$ (b) Correlation coefficient $C_L(r, s)$

Modified reference image $R' = R + 50$

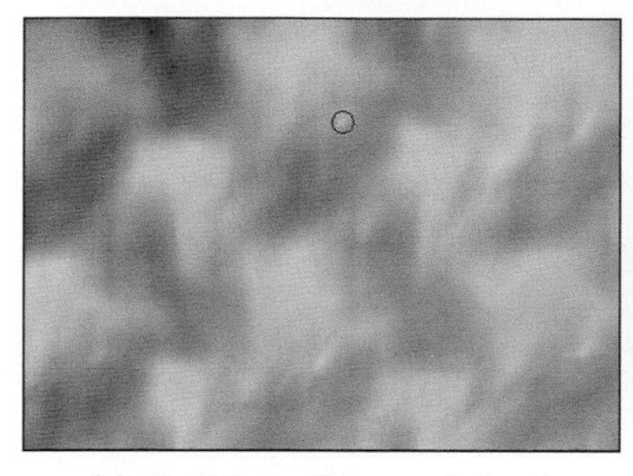

(c) Euclidean distance $d_{\mathrm{E}}(r, s)$

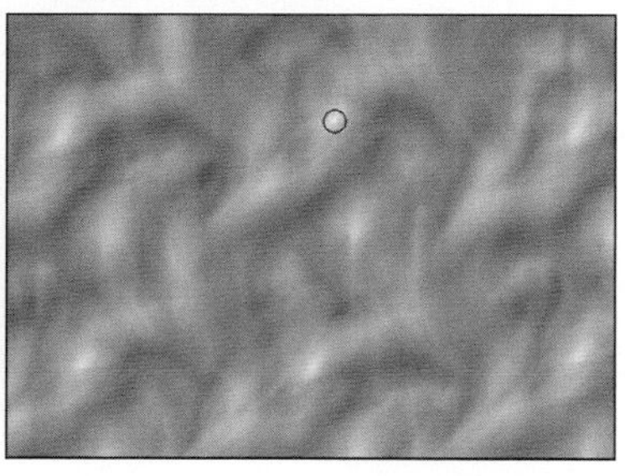

(d) Correlation coefficient $C_L(r, s)$

Fig. 23.4
Effects of changing global brightness. Original reference image R: the results from both the Euclidean distance (a) and the correlation coefficient (b) show distinct peaks at the positions of maximum agreement. Modified reference image $R' = R + 50$: the peak values disappear in the Euclidean distance (c), while the correlation coefficient (d) remains unaffected.

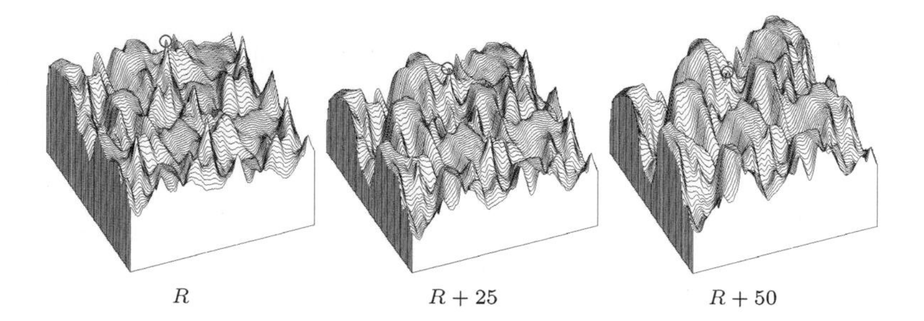

R $R + 25$ $R + 50$

Fig. 23.5
Euclidean distance under global intensity changes. Distance function for the original template R (left), with the template intensity increased by 25 units (center) and 50 units (right). Notice that the local peaks disappear as the template intensity (and thus the total distance between the image and the template) is increased.

(a) (b) (c)

Fig. 23.6
Detection of match points by simple thresholding: correlation coefficient (a), positive values only (b), and values greater than 0.5 (c). The remaining peaks indicate the positions of the six similar (but not identical) tulip patterns in the original image (Fig. 23.3(a)).

may still be stored in a rectangular array, but the relevant pixels must somehow be marked (e.g., using a binary mask).

Even more general is the option to assign individual continuous weights to the template elements such that, for example, the center of a template can be given higher significance in the match than the peripheral regions. Implementing such a "windowed matching" technique should be straightforward and require only minor modifications to the standard approach.

23.1.2 Matching Under Rotation and Scaling

Correlation-based matching methods applied in the way described in this section cannot handle significant rotation or scale differences between the search image and the template. One obvious way to overcome rotation is to match using multiple rotated versions of the template, of course at the price of additional computation time. Similarly, one could try to match using several scaled versions of the template to achieve scale independence to some extent. Although this could be combined by using a set of rotated *and* scaled template patterns, the combinatorially growing number of required matching steps could soon become prohibitive for a practical implementation.

An interesting technique is matching in *logarithmic-polar* space, where rotation and scaling map to translations and can thus be handled with correlation-type methods [267]. However, this requires an initial "anchor point", which again needs to be detected in a rotation and scale invariant way [152, 209, 238]. Another alternative is the popular Lucas-Kanade technique for elastic local matching, which is described at detail in Chapter 24. In principle, given an approximate starting solution, this method cannot only handle rotation and scaling, but arbitrary image transformations or distortions.

23.1.3 Java Implementation

Implementations of most methods described in this chapter are openly available as part of the `imagingbook` library.[3] As an example, the code listed in Progs. 23.1 and 23.2 demonstrates the use of the `CorrCoeffMatcher` class for template matching based on the local correlation coefficient (Eqn. (23.10)). The application assumes that the search image (`I`) and the reference image (`R`) are already available as objects of type `FloatProcessor`. They are used to create a new instance of class `CorrCoeffMatcher`, as shown in the following code segment:

```
FloatProcessor I = ...        // search image
FloatProcessor R = ...        // reference image
CorrCoeffMatcher matcher = new CorrCoeffMatcher(I);
float[][] C = matcher.getMatch(R);
```

The correlation coefficient is computed by the method `getMatch()` and returned as a 2D `float`-array (`C`).

23.2 Matching Binary Images

As became evident in the previous section, the comparison of intensity images based on correlation may not be an optimal solution but is sufficiently reliable and efficient under certain restrictions. If we compare binary images in the same way, by counting the number of identical pixels in the search image and the template, the total difference will only be small when most pixels are in exact agreement.

[3] Package `imagingbook.pub.matching`.

```
1 package imagingbook.pub.matching;
2
3 import ij.process.FloatProcessor;
4
5 class CorrCoeffMatcher {
6
7   private final FloatProcessor I; // search image
8   private final int MI, NI;         // width/height of search image
9
10  private FloatProcessor R;        // reference image
11  private int MR, NR;              // width/height of reference image
12  private int K;
13  private double meanR;            // mean value of reference (R̄)
14  private double varR;             // square root of reference variance
      (σ_R)
15
16  public CorrCoeffMatcher(FloatProcessor I) { // constructor
17    this.I = I;
18    this.MI = this.I.getWidth();
19    this.NI = this.I.getHeight();
20    }
21
22  public float[][] getMatch(FloatProcessor R) {
23    this.R = R;
24    this.MR = R.getWidth();
25    this.NR = R.getHeight();
26    this.K = MR * NR;
27
28    // calculate the mean (R̄) and variance term (S_R) of the template:
29    double sumR = 0;        // Σ_R = ∑ R(i,j)
30    double sumR2 = 0;       // Σ_R2 = ∑ R²(i,j)
31    for (int j = 0; j < NR; j++) {
32      for (int i = 0; i < MR; i++) {
33        float aR = R.getf(i,j);
34        sumR  += aR;
35        sumR2 += aR * aR;
36      }
37    }
38
39    this.meanR = sumR / K;      // R̄ = [∑ R(i,j)]/K
40    this.varR =                 // S_R = [∑ R²(i,j) − K·R̄²]^(1/2)
41        Math.sqrt(sumR2 - K * meanR * meanR);
42
43    float[][] C = new float[MI - MR + 1][NI - NR + 1];
44    for (int r = 0; r <= MI - MR; r++) {
45      for (int s = 0; s <= NI - NR; s++) {
46        float d = (float) getMatchValue(r, s);
47        C[r][s] = d;
48      }
49    }
50    return C;
51  }
52
53  // continued...
```

Prog. 23.1
Implementation of class `CorrCoeffMatcher` (part 1/2). The constructor method (lines 16–20) calculates the mean $\bar{R}$ = `meanR` (Eqn. (23.11)) and the variance S_R = `varR` (Eqn. (23.14)) of the reference image R. The method `getMatch(R)` (lines 22–51) determines the match values between the search image I and the reference image R f for all positions (r, s).

Prog. 23.2 Implementation of class `CorrCoeffMatcher` (part 2/2). The local match value $C(r, s)$ (see Eqn. (23.15)) at the individual position (r, s) is calculated by method `getMatchValue(r,s)` (lines 54–72).

```
54   private double getMatchValue(int r, int s) {
55     double sumI = 0;      // Σ_I = ∑ I(r+i, s+j)
56     double sumI2 = 0;     // Σ_I2 = ∑(I(r+i, s+j))^2
57     double sumIR = 0;     // Σ_IR = ∑ I(r+i, s+j) · R(i, j)
58 
59     for (int j = 0; j < NR; j++) {
60       for (int i = 0; i < MR; i++) {
61         float aI = I.getf(r + i, s + j);
62         float aR = R.getf(i, j);
63         sumI  += aI;
64         sumI2 += aI * aI;
65         sumIR += aI * aR;
66       }
67     }
68 
69     double meanI = sumI / K;  // Ī_r,s = Σ_I/K
70     return (sumIR - K * meanI * meanR) /
71         (1 + Math.sqrt(sumI2 - K * meanI * meanI) * varR);
72   }
73 
74 } // end of class CorrCoeffMatcher
```

Since there is no continuous transition between pixel values, the distribution produced by a simple distance function will generally be ill-behaved (i.e., highly discontinuous with many local extrema; see Fig. 23.7).

23.2.1 Direct Comparison of Binary Images

The problem with directly comparing binary images is that even the smallest deviations between image patterns, such as those caused by a small shift, rotation, or distortion, can create very high distance values. Shifting a thin line drawing by only a single pixel, for example, may be sufficient to switch from full agreement to no agreement at all (i.e., from zero difference to maximum difference). Thus a simple distance function gives no indication how far away and in which direction to search for a better match position.

An interesting question is how matching of binary images can be made more tolerant against small differences of the compared patterns. Thus the goal is not only to detect the single image position, where most foreground pixels in the two images match up, but also (if possible) to obtain a measure indicating how far (in terms of geometry) we are away from this position.

23.2.2 The Distance Transform

A first step in this direction is to record the distance to the closest foreground pixel for every position (u, v) in the search image I. This gives us the minimum distance (though not the direction) for shifting a particular pixel onto a foreground pixel. Starting from a binary image $I(u, v) = I(\boldsymbol{u})$, we denote

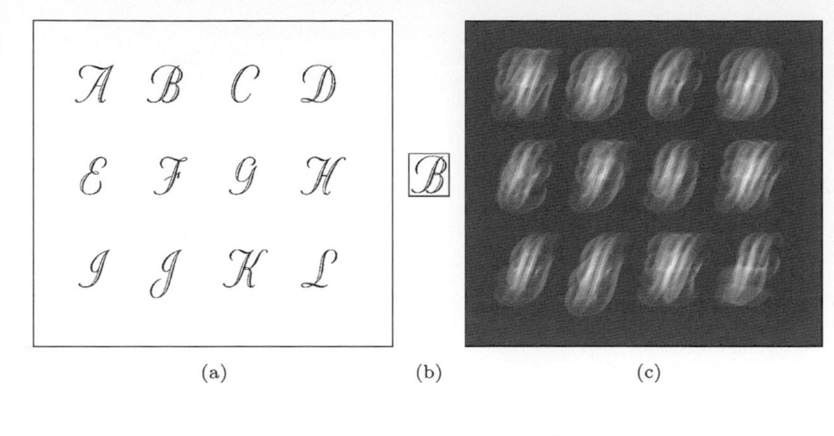

(a) (b) (c)

Fig. 23.7
Direct comparison of binary images. Given are a binary search image (a) and a binary reference image (b). The local similarity value for any template position corresponds to the relative number of matching (black) foreground pixels. High similarity values are shown as bright spots in the result (c). While the maximum similarity is naturally found at the correct position (at the center of the glyph *B*) the match function behaves wildly, with many local maxima.

$$FG(I) = \{\boldsymbol{u} \mid I(\boldsymbol{u}) = 1\}, \tag{23.16}$$
$$BG(I) = \{\boldsymbol{u} \mid I(\boldsymbol{u}) = 0\}, \tag{23.17}$$

as the set of coordinates of the foreground and background pixels, respectively. The so-called distance transform of I, $\mathsf{D}(\boldsymbol{u}) \in \mathbb{R}$, is defined as

$$\mathsf{D}(\boldsymbol{u}) := \min_{\boldsymbol{u}' \in FG(I)} \operatorname{dist}(\boldsymbol{u}, \boldsymbol{u}'), \tag{23.18}$$

for all $\boldsymbol{u} = (u, v)$, where $u = 0, \ldots, M{-}1$, $v = 0, \ldots, N{-}1$ (for image size $M \times N$). The value D at a given position $\boldsymbol{u}$ thus equals the distance between $\boldsymbol{u}$ and the nearest foreground pixel in I. If $I(\boldsymbol{u})$ is a foreground pixel itself (i.e., $x \in FG$), then the distance $D(\boldsymbol{u}) = 0$ since no shift is necessary for moving this pixel onto a foreground pixel.

The function $\operatorname{dist}(\boldsymbol{u}, \boldsymbol{u}')$ in Eqn. (23.18) measures the geometric distance between the two coordinate points $\boldsymbol{u} = (u, v)$ and $\boldsymbol{u}' = (u', v')$. Examples of suitable distance functions are the Euclidean distance (L_2 norm)

$$\mathrm{d}_E(\boldsymbol{u}, \boldsymbol{u}') = \|\boldsymbol{u} - \boldsymbol{u}'\| = \sqrt{(u - u')^2 + (v - v')^2} \;\in \mathbb{R}^+ \tag{23.19}$$

and the *Manhattan distance*[4] (L_1 norm)

$$\mathrm{d}_M(\boldsymbol{u}, \boldsymbol{u}') = |u - u'| + |v - v'| \;\in \mathbb{N}_0. \tag{23.20}$$

Figure 23.8 shows a simple example of a distance transform using the Manhattan distance $\mathrm{d}_M()$.

The direct calculation of the distance transform (following the definition in Eqn. (23.18)) is computationally expensive, because the closest foreground pixel must be found for each pixel position $\boldsymbol{p}$ (unless $I(\boldsymbol{p})$ is a foreground pixel itself).[5]

Chamfer algorithm

The so-called *chamfer* algorithm [30] is an efficient method for computing the distance transform. Similar to the sequential region labeling algorithm (see Ch. 10, Alg. 10.2), the chamfer algorithm traverses

[4] Also called "city block distance".

[5] A simple (brute force) algorithm for the distance transform would perform a full scan over the entire image for each processed pixel, resulting in $\mathcal{O}(N^2 \cdot N^2) = \mathcal{O}(N^4)$ steps for an image of size $N \times N$.

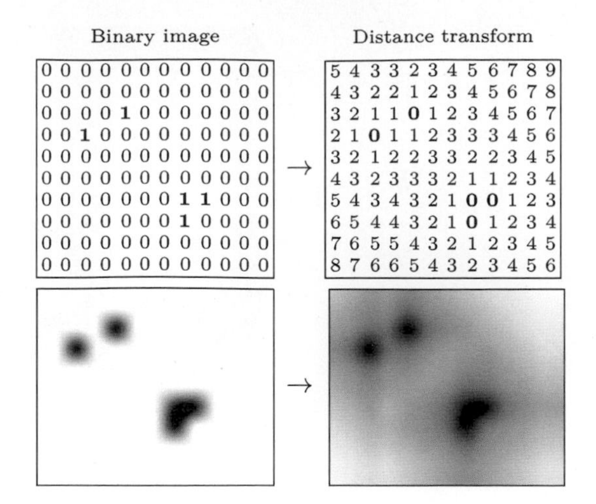

Fig. 23.8
Example of a distance transform of a binary image using the Manhattan distance $\mathrm{d}_M()$. Foreground pixels in the binary image have value 1 (shown inverted).

the image twice by propagating the computed values across the image like a wave. The first traversal starts at the upper left corner of the image and propagates the distance values downward in a diagonal direction. The second traversal proceeds in the opposite direction from the bottom to the top. For each traversal, a "distance mask" is used for the propagation of the distance values; that is,

$$M^L = \begin{bmatrix} m_2 & m_1 & m_2 \\ m_1 & \times & \cdot \\ \cdot & \cdot & \cdot \end{bmatrix} \quad \text{and} \quad M^R = \begin{bmatrix} \cdot & \cdot & \cdot \\ \cdot & \times & m_1 \\ m_2 & m_1 & m_2 \end{bmatrix} \tag{23.21}$$

for the first and second traversals, respectively. The values in M^L and M^R describe the geometric distance between the current pixel (marked $\times$) and the relevant neighboring pixels. They depend upon the distance function $\mathrm{dist}(\boldsymbol{x}, \boldsymbol{x}')$ used. Algorithm 23.2 outlines the chamfer method for computing the distance transform $D(u, v)$ for a binary image $I(u, v)$ using the above distance masks.

For the Manhattan distance, the chamfer algorithm computes the distance transform (Eqn. (23.20)) *exactly* using the masks

$$M^L_M = \begin{bmatrix} 2 & 1 & 2 \\ 1 & \times & \cdot \\ \cdot & \cdot & \cdot \end{bmatrix} \quad \text{and} \quad M^R_M = \begin{bmatrix} \cdot & \cdot & \cdot \\ \cdot & \times & 1 \\ 2 & 1 & 2 \end{bmatrix}. \tag{23.22}$$

Similarly for the Euclidean distance (Eqn. (23.19)) can be calculated with the masks

$$M^L_E = \begin{bmatrix} \sqrt{2} & 1 & \sqrt{2} \\ 1 & \times & \cdot \\ \cdot & \cdot & \cdot \end{bmatrix} \quad \text{and} \quad M^R_E = \begin{bmatrix} \cdot & \cdot & \cdot \\ \cdot & \times & 1 \\ \sqrt{2} & 1 & \sqrt{2} \end{bmatrix}. \tag{23.23}$$

Note that the result obtained with these masks is only an *approximation* of the Euclidean distance to the nearest foreground pixel, which is nevertheless more accurate than the estimate produced by the Manhattan distance. As demonstrated by the examples in Fig. 23.9, the distances obtained with the Euclidean masks are exact along the coordinate axes and the diagonals but are overestimated (i.e., too

Alg. 23.2
Chamfer algorithm for computing the distance transform. From the binary image I, the distance transform D (Eqn. (23.18)) is computed using a pair of distance masks (Eqn. (23.21)) for the first and second passes. Notice that the image borders require special treatment.

```
1:  DistanceTransform(I, norm)
      Input: I, a, binary image; norm ∈ {L1, L2}, distance function.
      Returns the distance transform of I.
    Step 1: initialize
2:    (m1, m2) ← { (1, 2)    for norm = L1
                 { (1, √2)   for norm = L2
3:    (M, N) ← Size(I)
4:    Create map D: M × N ↦ ℝ
5:    for all (u, v) ∈ M × N do
6:        D(u, v) ← { 0   for I(u, v) > 0
                    { ∞   otherwise
    Step 2: L→R pass
7:    for v ← 0, ..., N−1 do                          ▷ top → bottom
8:        for u ← 0, ..., M−1 do                      ▷ left → right
9:            if D(u, v) > 0 then
10:               d1, d2, d3, d4 ← ∞
11:               if u > 0 then
12:                   d1 ← m1 + D(u − 1, v)
13:                   if v > 0 then
14:                       d2 ← m2 + D(u − 1, v − 1)
15:               if v > 0 then
16:                   d3 ← m1 + D(u, v − 1)
17:                   if u < M − 1 then
18:                       d4 ← m2 + D(u + 1, v − 1)
19:               D(u, v) ← min(D(u, v), d1, d2, d3, d4)
    Step 3: R→L pass
20:   for v ← N−1, ..., 0 do                          ▷ bottom → top
21:       for u ← M−1, ..., 0 do                      ▷ right → left
22:           if D(u, v) > 0 then
23:               d1, d2, d3, d4 ← ∞
24:               if u < M−1 then
25:                   d1 ← m1 + D(u + 1, v)
26:                   if v < N−1 then
27:                       d2 ← m2 + D(u + 1, v + 1)
28:               if v < N−1 then
29:                   d3 ← m1 + D(u, v + 1)
30:                   if u > 0 then
31:                       d4 ← m2 + D(u − 1, v + 1)
32:               D(u, v) ← min(D(u, v), d1, d2, d3, d4)
33:   return D
```

high) for all other directions. A more precise approximation can be obtained with distance masks of greater size (e.g., 5×5 pixels; see Exercise 23.3), which include the exact distances to pixels in a larger neighborhood [30]. Furthermore, floating point-operations can be avoided by using distance masks with scaled integer values, such as the masks

$$M^{L}_{E'} = \begin{bmatrix} 4 & 3 & 4 \\ 3 & \times & \cdot \\ \cdot & \cdot & \cdot \end{bmatrix} \quad \text{and} \quad M^{R}_{E'} = \begin{bmatrix} \cdot & \cdot & \cdot \\ \cdot & \times & 3 \\ 4 & 3 & 4 \end{bmatrix} \qquad (23.24)$$

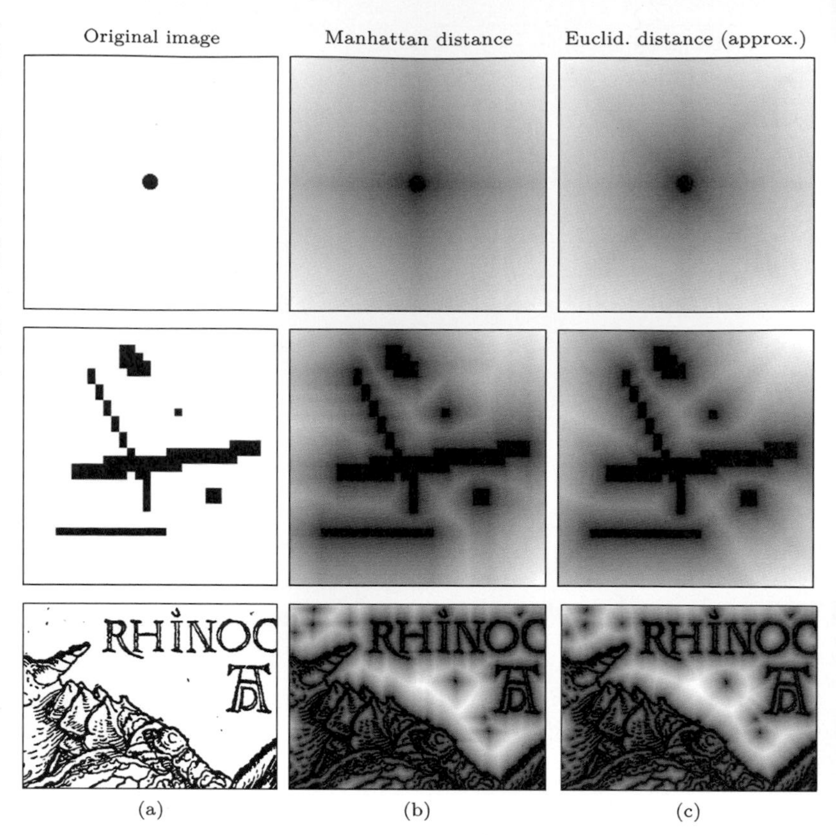

Fig. 23.9
Distance transform with the chamfer algorithm: original image with black foreground pixels (a), and results of distance transforms using the Manhattan distance (b) and the Euclidean distance (c). The brightness (scaled to maximum contrast) corresponds to the estimated distance to the nearest foreground pixel.

for the Euclidean distance. Compared with the original masks (Eqn. (23.23)), the resulting distance values are scaled by about the factor 3.

23.2.3 Chamfer Matching

The chamfer algorithm offers an efficient way to approximate the distance transform for a binary image of arbitrary size. The next step is to use the distance transform for matching binary images. *Chamfer matching* (first described in [19]) uses the distance transform to localize the points of maximum agreement between a binary search image I and a binary reference image (template) R. Instead of counting the overlapping foreground pixels as in the direct approach (see Sec. 23.2.1), chamfer matching uses the accumulated values of the distance transform as the match score Q. At each position (r, s) of the template R, the distance values corresponding to all foreground pixels in R are accumulated, that is,

$$Q(r,s) = \frac{1}{|FG(R)|} \cdot \sum_{\substack{(i,j)\in \\ FG(R)}} D(r+i,\, s+j)\,, \tag{23.25}$$

where $K = |FG(R)|$ denotes the number of foreground pixels in the template R.

The complete procedure for computing the match score Q is summarized in Alg. 23.3. If at some position each foreground pixel in the

```
1:  ChamferMatch (I, R)
        Input: I, binary search image; R, binary reference image.
        Returns a 2D map of match scores.

        Step 1 – initialize:
2:      (M_I, N_I) ← Size(I)
3:      (M_R, N_R) ← Size(R)
4:      D ← DistanceTransform(I)                         ▷ Alg. 23.2
5:      Create map Q: (M_I−M_R+1) × (N_I−N_R+1) ↦ ℝ

        Step 2 – compute match function:
6:      for r ← 0, ..., M_I−M_R do                       ▷ place R at (r, s)
7:          for s ← 0, ..., N_I−N_R do
                Get match score for R placed at (r, s)
8:              q ← 0
9:              n ← 0                      ▷ number of foreground pixels in R
10:             for i ← 0, ..., M_R−1 do
11:                 for j ← 0, ..., N_R−1 do
12:                     if R(i, j) > 0 then          ▷ foreground pixel in R
13:                         q ← q + D(r + i, s + j)
14:                         n ← n + 1
15:             Q(r, s) ← q/n

16:     return Q
```

Alg. 23.3
Chamfer matching (calculation of the match function). Given is a binary search image I and a binary reference image (template) R. In step 1, the distance transform D is computed for the image I using the chamfer algorithm (Alg. 23.2). In step 2, the sum of distance values is accumulated for all foreground pixels in template R for each template position (r, s). The resulting scores are stored in the 2D match map Q, which is returned.

template R coincides with a foreground pixel in the image I, the sum of the distance values is zero, which indicates a perfect match. The more foreground pixels of the template fall onto distance values greater than zero, the larger is the resulting score value Q (sum of distances). The best match is found at the global minimum of Q, that is,

$$\boldsymbol{x}_{\mathrm{opt}} = (r_{\mathrm{opt}}, s_{\mathrm{opt}}) = \underset{(r,s)}{\mathrm{argmin}}(Q(r, s)). \tag{23.26}$$

The example in Fig. 23.10 demonstrates the difference between direct pixel comparison and chamfer matching using the binary image shown in Fig. 23.7. Obviously the match score produced by the chamfer method is considerably smoother and exhibits only a few distinct local maxima. This is of great advantage because it facilitates the detection of optimal match points using simple local search methods. Figure 23.11 shows another example with circles and squares. The circles have different diameters and the medium-sized circle is used as the template. As this example illustrates, chamfer matching is tolerant against small-scale changes between the search image and the template and even in this case yields a smooth score function with distinct peaks.

While chamfer matching is not a "silver bullet", it is efficient and works sufficiently well if the applications and conditions are suitable. It is most suited for matching line or edge images where the percentage of foreground pixels is small, such as for registering aerial images or aligning wide-baseline stereo images. The method tolerates deviations between the image and the template to a small extent but is of course not generally invariant under scaling, rotation, and deformation. The quality of the results deteriorates quickly when images contain random noise ("clutter") or large foreground regions, because

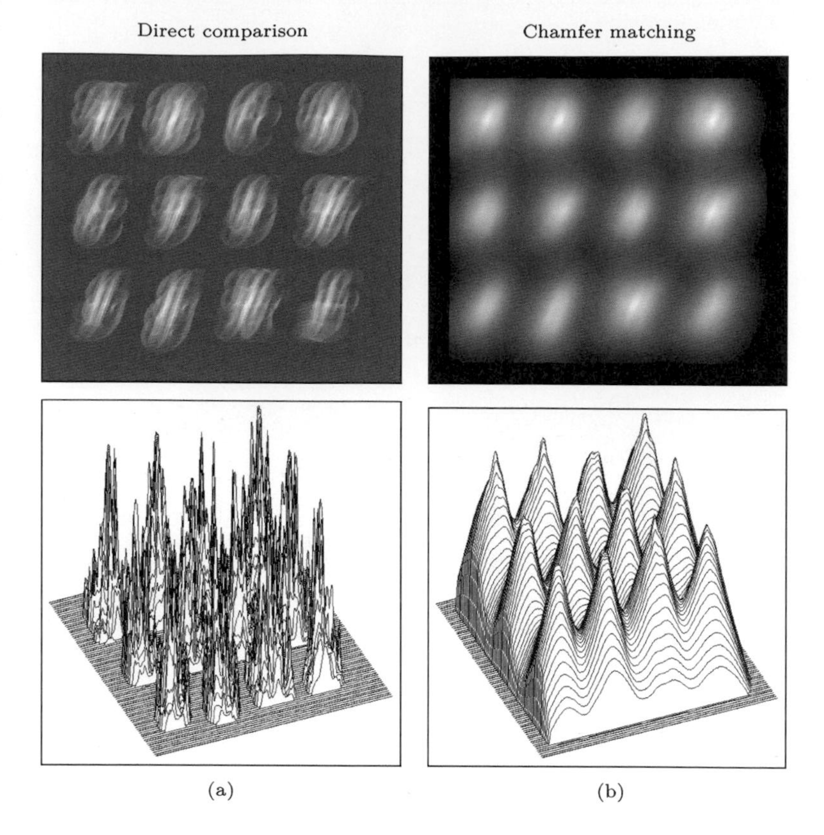

Fig. 23.10
Direct pixel comparison vs. chamfer matching (see original images in Fig. 23.7). Unlike the results of the direct pixel comparison (a), the chamfer match score Q (b) is much smoother. It shows distinct peak values in places of high agreement that are easy to track down with local search methods. The match score Q (Eqn. (23.25)) in (b) is shown inverted for easy comparison.

the method is based on minimizing the distances to foreground pixels. One way to reduce the probability of false matches is not to use a *linear* summation (as in Eqn. (23.25)) but add up the *squared* distances, that is,

$$Q_{rms}(r,s) = \Big[\frac{1}{K} \cdot \sum_{\substack{(i,j)\in\\ FG(R)}} \big(D(r+i,s+i)\big)^2\Big]^{1/2} \tag{23.27}$$

("root mean square" of the distances) as the match score between the template R and the current subimage, as suggested in [30]. Also, hierarchical variants of the chamfer method have been proposed to reduce the search effort as well as to increase robustness [31].

23.2.4 Java Implementation

The calculation of the distance transform, as described in Alg. 23.2, is implemented by the class `DistanceTransform`.[6] Program 23.3 shows the complete code for the class `ChamferMatcher` for comparing binary images with the distance transform, which is a direct implementation of Alg. 23.3. Additional examples (ImageJ plugins) can be found in the on-line code repository.

[6] Package `imagingbook.pub.matching`.

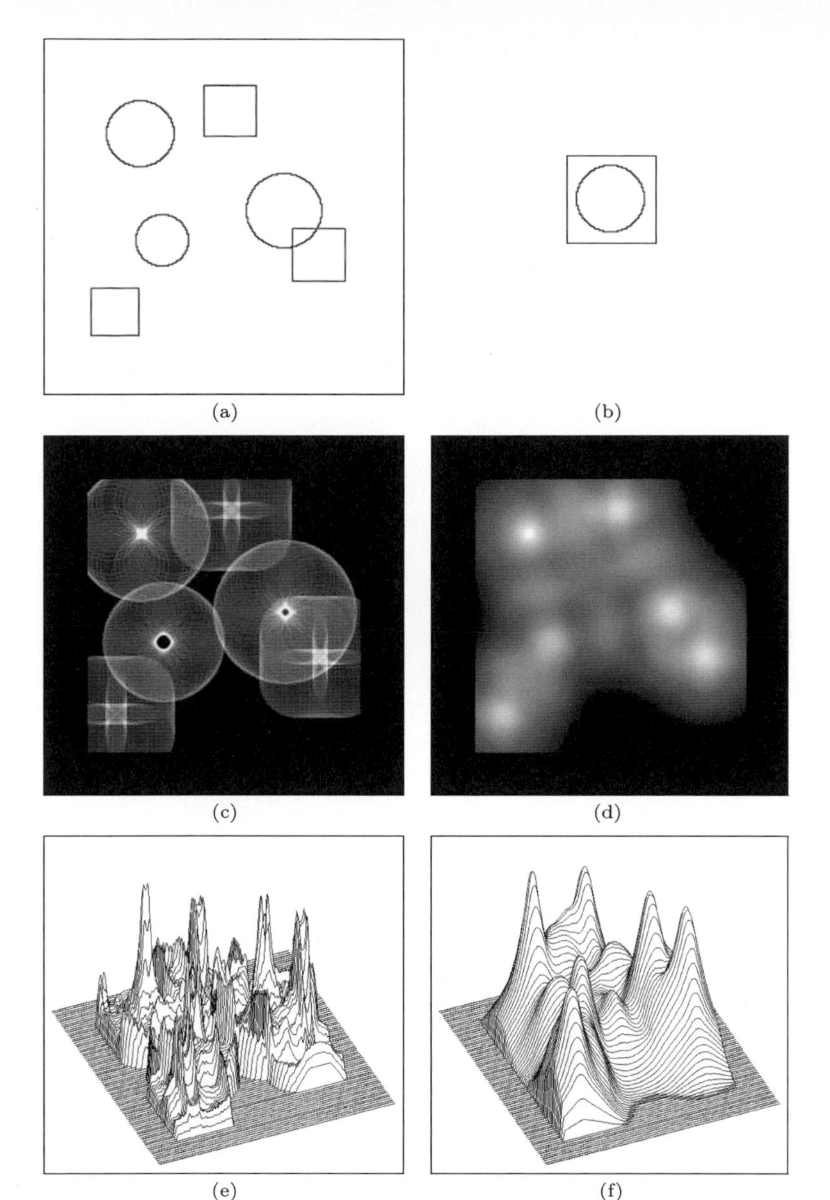

Fig. 23.11
Chamfer matching under varying scales. Binary search image with three circles of different diameters and three identical squares (a). The medium-sized circle at the top is used as the template (b). The result from a direct pixel comparison (c, e) and the result from chamfer matching (d, f). Again the chamfer match produces a much smoother score, which is most notable in the 3D plots shown in the bottom row (e, f). Notice that the three circles and the squares produce high match scores with similar absolute values (f).

23.3 Exercises

Exercise 23.1. Implement the chamfer-matching method (Alg. 23.2) for binary images using the Euclidean distance and the Manhattan distance.

Exercise 23.2. Implement the *exact* Euclidean distance transform using a "brute-force" search for each closest foreground pixel (this may take a while to compute). Compare your results with the approximation obtained with the chamfer method (Alg. 23.2), and compute the maximum deviation (as percentage of the real distance).

Exercise 23.3. Modify the chamfer algorithm for computing the distance transform (Alg. 23.2) by replacing the 3×3 pixel Euclidean distance masks (Eqn. (23.23)) with the following masks of size 5×5:

$$M^L = \begin{bmatrix} \cdot & 2.236 & \cdot & 2.236 & \cdot \\ 2.236 & 1.414 & 1.000 & 1.414 & 2.236 \\ \cdot & 1.000 & \times & \cdot & \cdot \\ \cdot & \cdot & \cdot & \cdot & \cdot \\ \cdot & \cdot & \cdot & \cdot & \cdot \end{bmatrix}, \tag{23.28}$$

$$M^R = \begin{bmatrix} \cdot & \cdot & \cdot & \cdot & \cdot \\ \cdot & \cdot & \cdot & \cdot & \cdot \\ \cdot & \cdot & \times & 1.000 & \cdot \\ 2.236 & 1.414 & 1.000 & 1.414 & 2.236 \\ \cdot & 2.236 & \cdot & 2.236 & \cdot \end{bmatrix}. \tag{23.29}$$

Compare the results with those obtained with the standard masks. Why are no additional mask elements required along the coordinate axes and the diagonals?

Exercise 23.4. Implement the chamfer-matching technique using (a) the linear summation of distances (Eqn. (23.25)) and (b) the summation of squared distances (Eqn. (23.27)) for computing the match score. Select suitable test images to find out if version (b) is really more robust in terms of reducing the number of false matches.

Exercise 23.5. Adapt the template-matching method described in Sec. 23.1 for the comparison of RGB color images.

```
1 package imagingbook.pub.matching;
2 import ij.process.ByteProcessor;
3 import imagingbook.pub.matching.DistanceTransform.Norm;
4
5 public class ChamferMatcher {
6   private final ByteProcessor I;
7   private final int MI, NI;
8   private final float[][] D;          // distance transform of I
9
10  public ChamferMatcher(ByteProcessor I) {
11    this(I, Norm.L2);
12  }
13
14  public ChamferMatcher(ByteProcessor I, Norm norm) {
15    this.I = I;
16    this.MI = this.I.getWidth();
17    this.NI = this.I.getHeight();
18    this.D = (new DistanceTransform(I, norm)).
       getDistanceMap();
19  }
20
21  public float[][] getMatch(ByteProcessor R) {
22    final int MR = R.getWidth();
23    final int NR = R.getHeight();
24    final int[][] Ra = R.getIntArray();
25    float[][] Q = new float[MI - MR + 1][NI - NR + 1];
26    for (int r = 0; r <= MI - MR; r++) {
27      for (int s = 0; s <= NI - NR; s++) {
28        float q = getMatchValue(Ra, r, s);
29        Q[r][s] = q;
30      }
31    }
32    return Q;
33  }
34
35  private float getMatchValue(int[][] R, int r, int s) {
36    float q = 0.0f;
37    for (int i = 0; i < R.length; i++) {
38      for (int j = 0; j < R[i].length; j++) {
39        if (R[i][j] > 0) {  // foreground pixel in reference image
40          q = q + D[r + i][s + j];
41        }
42      }
43    }
44    return q;
45  }
46 }
```

Prog. 23.3
Java implementation of Alg. 23.3 (class `ChamferMatcher`). The distance transform of the binary search image `I` is calculated in the constructor method by an instance of class `DistanceTransform` and stored as a 2D `float` array (line 18). The method `getMatch(R)` in lines 21–45 computes the 2D match function `Q` (again as a `float` array) for the reference image `R`.

24

Non-Rigid Image Matching

The correlation-based registration methods described in Chapter 23 are *rigid* in the sense that they provide for *translation* as the only form of geometric transformation and positioning is limited to whole pixel units. In this chapter we look at methods that are capable of registering a reference image under (almost) arbitrary geometric transformations, such as changes in rotation, scale, and affine distortion, and also to *sub-pixel* accuracy.

At the core of this chapter is a detailed description of the classic Lucas-Kanade algorithm [154] and its efficient implementation. Unlike the methods presented earlier, the algorithms described here typically do not perform a global search over the entire image to find the best match, but start from an initial estimate of the geometric transformation to home in on the optimum position and distortion in an iterative fashion. This is not difficult, for example, in tracking applications, where the approximate location of a particular image patch can be predicted from the observed motion in previous frames. Of course, the global matching methods described in Chapter 23 can be used to find a coarse starting solution.

24.1 The Lucas-Kanade Technique

The basic idea of the Lucas-Kanade technique is best illustrated in the 1D case (see Fig. 24.1(a)).

24.1.1 Registration in 1D

Given two 1D, real-valued functions $f(x)$, $g(x)$, the registration problem is to find the disparity t in the (horizontal) x-direction under the assumption that g is a shifted version of f, that is,

$$g(x) = f(x - t). \tag{24.1}$$

If the function f is linear in a (sufficiently large) neighborhood of some point x with slope $f'(x)$, then

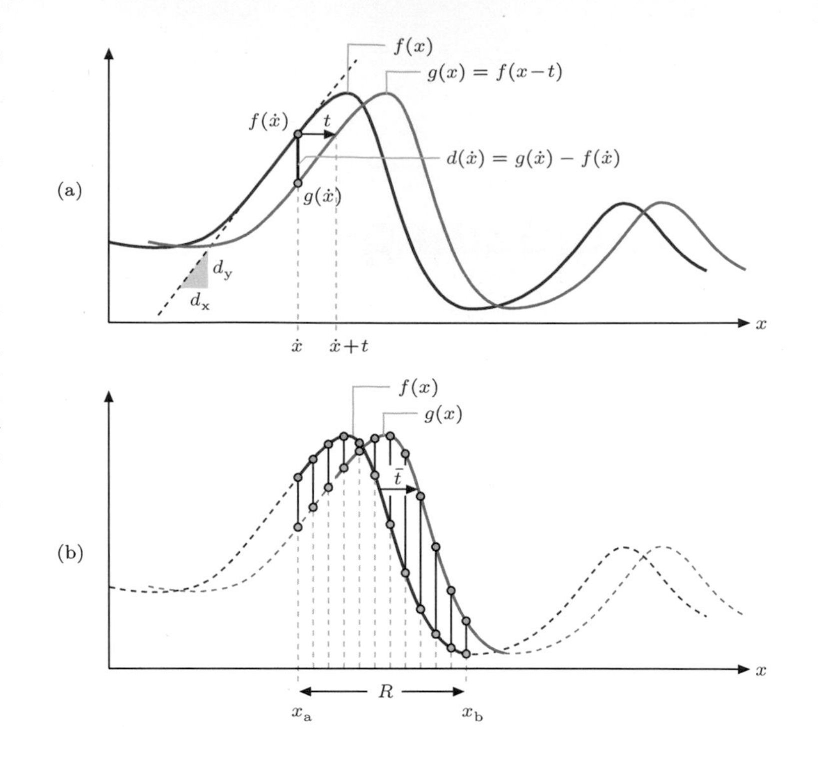

Fig. 24.1 Registering two 1D functions (figure adapted from [154]). The 1D function $g(x)$ is assumed to be a shifted version of $f(x)$. In (a), f is approximately linear at position $\dot{x}$, with slope $f'(\dot{x}) = d_y/d_x$. Under this condition, the horizontal displacement t can be estimated from the difference of the local function values $f(\dot{x})$ and $g(\dot{x})$ as $t \approx (f(\dot{x}) - g(\dot{x}))/f'(\dot{x})$. In (b), the overall displacement $\bar{t}$ is calculated by averaging the individual displacement estimates from multiple samples in the region $R = [x_a, x_b]$.

$$f(x - t) \approx f(x) - t \cdot f'(x) \tag{24.2}$$

and therefore

$$g(x) \approx f(x) - t \cdot f'(x). \tag{24.3}$$

Thus, given the function values $f(x)$, $g(x)$ and the first derivative $f'(x)$ at some point x, the displacement t can be estimated (from Eqn. (24.2)) as

$$t \approx \frac{f(x) - g(x)}{f'(x)}. \tag{24.4}$$

Note that this can be viewed as a first-order Taylor expansion[1] of the function f. Obviously, the estimate of the shift t in Eqn. (24.4) depends only on a single pair of function samples at position x and fails at points where f is either not linear or flat, that is, where the first derivative f' vanishes. To obtain a more robust displacement estimate it appears natural to extend the calculation over a range R of sample values, thereby aligning a complete section of the two functions f and g (see Fig. 24.1(b)). This problem can be formulated as finding the displacement t that minimizes the L_2 distance between the two functions f and g over a range R, that is, finding t such that

$$\mathcal{E}(t) = \sum_{x \in R} [f(x-t) - g(x)]^2 = \sum_{x \in R} [f(x) - t \cdot f'(x) - g(x)]^2 \tag{24.5}$$

[1] See also Sec. C.3.2 in the Appendix.

is a minimum. This can be accomplished by calculating the first derivative of the aforementioned expression (with respect to t) and setting it equal to zero, which gives

$$\frac{\partial \mathcal{E}}{\partial t} = 2 \cdot \sum_{x \in R} f'(x) \cdot \big[f(x) - f'(x) \cdot t - g(x)\big] = 0\,. \qquad (24.6)$$

By solving this equation the optimal shift is found as

$$t_{\text{opt}} = \Big[\sum_{x \in R} [f'(x)]^2\Big]^{-1} \cdot \sum_{x \in R} f'(x) \cdot [f(x) - g(x)]\,. \qquad (24.7)$$

Note that this local estimation works even if the function f is flat at some positions in R, unless $f'(x)$ is zero everywhere R. However, since the estimate is based only on linear (i.e., first-order) prediction, the estimate is generally not accurate. For this purpose, the following iterative optimization scheme is proposed in [154], which is really the basis of the Lucas-Kanade algorithm. With $t^{(0)} = t_{\text{start}}$ as the initial estimate of the displacement (which may be zero), t is successively updated as

$$t^{(k)} = t^{(k-1)} + \Big[\sum_{x \in R} [f'(x)]^2\Big]^{-1} \cdot \sum_{x \in R} f'(x) \cdot [f(x) - g(x)]\,, \qquad (24.8)$$

for $k = 1, 2, \ldots$, until either $t^{(k)}$ converges or a maximum number of steps is reached.

24.1.2 Extension to Multi-Dimensional Functions

As shown in [154], the formulation given in Sec. 24.1.1 can be easily generalized to align multi-dimensional, scalar-valued functions, including 2D images. In general, the involved functions $F(\boldsymbol{x})$ and $G(\boldsymbol{x})$ are now defined over $\mathbb{R}^m$, and thus all coordinates $\boldsymbol{x} = (x_1, \ldots, x_m)$ and spatial shifts $\boldsymbol{t} = (t_1, \ldots, t_m)$ are m-dimensional column vectors. The task is, analogous to Eqn. (24.5), to find the vector $\boldsymbol{t}$ that minimizes the error quantity

$$\mathcal{E}(\boldsymbol{t}) = \sum_{\boldsymbol{x} \in R} [F(\boldsymbol{x} - \boldsymbol{t}) - G(\boldsymbol{x})]^2, \qquad (24.9)$$

where R denotes an m-dimensional region. The linear approximation in Eqn. (24.2) becomes

$$F(\boldsymbol{x} - \boldsymbol{t}) \approx F(\boldsymbol{x}) - \nabla_F(\boldsymbol{x}) \cdot \boldsymbol{t}, \qquad (24.10)$$

where the row vector $\nabla_F(\boldsymbol{x}) = \big(\frac{\partial F}{\partial x_1}(\boldsymbol{x}), \ldots, \frac{\partial F}{\partial x_m}(\boldsymbol{x})\big)$ is the m-dimensional *gradient* of the function F, evaluated at position $\boldsymbol{x}$. Minimizing $\mathcal{E}(\boldsymbol{t})$ over $\boldsymbol{t}$ is again accomplished by solving $\frac{\partial \mathcal{E}}{\partial \boldsymbol{t}} = 0$, that is (analogous to Eqn. (24.6)),

$$2 \cdot \sum_{\boldsymbol{x} \in R} \nabla_F(\boldsymbol{x}) \cdot \big[F(\boldsymbol{x}) - \nabla_F(\boldsymbol{x}) \cdot \boldsymbol{t} - G(\boldsymbol{x})\big] = 0\,. \qquad (24.11)$$

The solution to Eqn. (24.11) is

$$t_{\mathrm{opt}} = \Big[\sum_{\boldsymbol{x}\in R} \nabla_F^{\mathsf{T}}(\boldsymbol{x}) \cdot \nabla_F(\boldsymbol{x})\Big]^{-1} \cdot \Big[\sum_{\boldsymbol{x}\in R} \nabla_F^{\mathsf{T}}(\boldsymbol{x}) \cdot [F(\boldsymbol{x}) - G(\boldsymbol{x})]\Big] \tag{24.12}$$

$$= \mathbf{H}_F^{-1} \cdot \Big[\sum_{\boldsymbol{x}\in R} \nabla_F^{\mathsf{T}}(\boldsymbol{x}) \cdot [F(\boldsymbol{x}) - G(\boldsymbol{x})]\Big], \tag{24.13}$$

where $\mathbf{H}_F$ is an estimate of the $m \times m$ Hessian matrix[2] for the function F over the region R. Note the similarity of Eqn. (24.13) to the 1D version in Eqn. (24.7).

24.2 The Lucas-Kanade Algorithm

Based on the ideas outlined in Sec. 24.1, the Lucas-Kanade algorithm [154] is not only capable of registering 2D images by finding the optimal translation, but works for a range of geometric transformations $T_{\boldsymbol{p}}$ that can be parameterized by a n-dimensional vector $\boldsymbol{p}$. Among others, this includes affine and projective transformations (see Ch. 21) as the most important cases.

The same mathematical notation is used as in Chapter 23, that is, I denotes the *search image* and R is the (typically smaller) *reference image*. The placement and possible distortion of the matching image patch is described by a *geometric transformation* $T_{\boldsymbol{p}}$ (cf. Ch. 21), where $\boldsymbol{p}$ denotes a vector of transformation parameters. The goal of the Lucas-Kanade registration algorithm is to minimize the expression

$$\mathcal{E}(\boldsymbol{p}) = \sum_{\boldsymbol{x}\in R} \big[I(T_{\boldsymbol{p}}(\boldsymbol{x})) - R(\boldsymbol{x})\big]^2 \tag{24.14}$$

with respect to the geometric transformation parameters $\boldsymbol{p}$, where I is the (search) image, R is the reference image (template), and $T_{\boldsymbol{p}}(\boldsymbol{x})$ is a geometric transformation or warp function with parameters $\boldsymbol{p}$. For example, simple 2D translation is described by the transformation

$$T_{\boldsymbol{p}}(\boldsymbol{x}) = \boldsymbol{x} + \boldsymbol{p} = \begin{pmatrix} x + t_{\mathrm{x}} \\ y + t_{\mathrm{y}} \end{pmatrix}, \tag{24.15}$$

where $\boldsymbol{x} = (x, y)^{\mathsf{T}}$ and $\boldsymbol{p} = (t_{\mathrm{x}}, t_{\mathrm{y}})^{\mathsf{T}}$. The task of the alignment process is to find the parameters that describe how to warp the search image I, such that the match between I and R is optimal over the support region R. Figure 24.2 illustrates the corresponding geometry.

In each iteration, the Lucas-Kanade algorithm starts with an estimate of the transformation parameters $\boldsymbol{p}$ and attempts to find the parameter increment $\boldsymbol{q}$ that locally minimizes the expression

$$\mathcal{E}(\boldsymbol{q}) = \sum_{\boldsymbol{x}\in R} \big[I(T_{\boldsymbol{p}+\boldsymbol{q}}(\boldsymbol{x})) - R(\boldsymbol{x})\big]^2. \tag{24.16}$$

After calculating the optimal parameter change $\boldsymbol{q}_{\mathrm{opt}}$, the parameter vector $\boldsymbol{p}$ is updated in the form

$$\boldsymbol{p} \leftarrow \boldsymbol{p} + \boldsymbol{q}_{\mathrm{opt}} \tag{24.17}$$

[2] See Sec. C.2.6 in the Appendix for details.

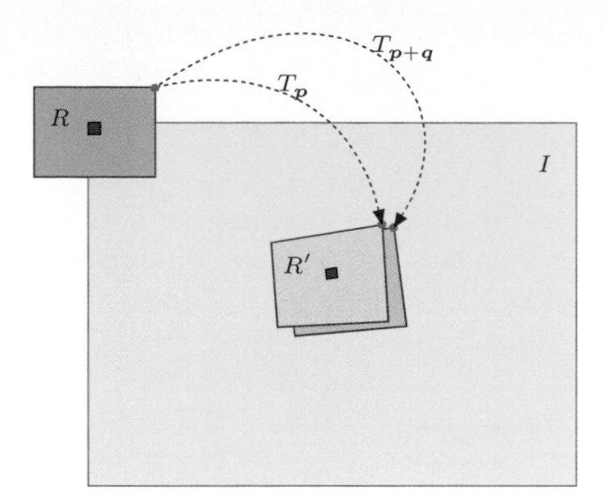

Fig. 24.2
Geometric relations in the (forward) Lucas-Kanade registration algorithm. I denotes the search image and R is the reference image. The mapping $T_{\boldsymbol{p}}$ warps the reference image R from the original position (centered at the origin) to R', with $\boldsymbol{p}$ being the initial parameter estimate. Matching is performed between the search image I and the warped reference image R'. $T_{\boldsymbol{p}+\boldsymbol{q}}$ is the improved warp; the optimal parameter change $\boldsymbol{q}$ is estimated in each iteration.

until the process converges. Typically, the update loop is terminated when the magnitude of the change vector $\boldsymbol{q}_{\mathrm{opt}}$ drops below a predefined threshold.

The expression to be minimized in Eqn. (24.16) depends on the image content and is generally nonlinear with respect to $\boldsymbol{q}$. A locally linear approximation of this function is obtained by the first-order Taylor expansion on I, that is,[3]

$$I(T_{\boldsymbol{p}+\boldsymbol{q}}(\boldsymbol{x})) \approx I(T_{\boldsymbol{p}}(\boldsymbol{x})) + \underbrace{\underbrace{\nabla_I(T_{\boldsymbol{p}}(\boldsymbol{x}))}_{1\times 2} \cdot \underbrace{\mathbf{J}_{T_{\boldsymbol{p}}}(\boldsymbol{x})}_{2\times n} \cdot \underbrace{\boldsymbol{q}}_{n\times 1}}_{\in \mathbb{R}}, \tag{24.18}$$

where the 2D (column) vector

$$\nabla_I(\boldsymbol{x}) = \big(I_{\mathrm{x}}(\boldsymbol{x}), I_{\mathrm{y}}(\boldsymbol{x})\big) \tag{24.19}$$

is the *gradient* of the image I at some position $\boldsymbol{x}$ and $\mathbf{J}_{T_{\boldsymbol{p}}}(\boldsymbol{x})$ denotes the *Jacobian* matrix[4] of the warp function $T_{\boldsymbol{p}}$, also evaluated at position $\boldsymbol{x}$. In general, the Jacobian of a 2D warp function

$$T_{\boldsymbol{p}}(\boldsymbol{x}) = \begin{pmatrix} T_{\mathrm{x},\boldsymbol{p}}(\boldsymbol{x}) \\ T_{\mathrm{y},\boldsymbol{p}}(\boldsymbol{x}) \end{pmatrix} \tag{24.20}$$

with n parameters $\boldsymbol{p} = (p_0, p_1, \ldots, p_{n-1})^{\mathsf{T}}$ is a $2 \times n$ matrix function

$$\mathbf{J}_{T_{\boldsymbol{p}}}(\boldsymbol{x}) = \begin{pmatrix} \frac{\partial T_{\mathrm{x},\boldsymbol{p}}}{\partial p_0}(\boldsymbol{x}) & \frac{\partial T_{\mathrm{x},\boldsymbol{p}}}{\partial p_1}(\boldsymbol{x}) & \cdots & \frac{\partial T_{\mathrm{x},\boldsymbol{p}}}{\partial p_{n-1}}(\boldsymbol{x}) \\ \frac{\partial T_{\mathrm{y},\boldsymbol{p}}}{\partial p_0}(\boldsymbol{x}) & \frac{\partial T_{\mathrm{y},\boldsymbol{p}}}{\partial p_1}(\boldsymbol{x}) & \cdots & \frac{\partial T_{\mathrm{y},\boldsymbol{p}}}{\partial p_{n-1}}(\boldsymbol{x}) \end{pmatrix}. \tag{24.21}$$

With the linear approximation in Eqn. (24.18), the original minimization problem in Eqn. (24.14) can now be written as

[3] In some of the following equations, we distinguish carefully between row and column vectors and the dimensions of vectors and matrices are explicitly displayed (in underbraces) to avoid possible confusion.

[4] The Jacobian $\mathbf{J}$ of a function f is a matrix containing the first partial derivatives of f, that is, it is a matrix of functions (see also Sec. C.2.1 in the Appendix).

$$\mathcal{E}(\boldsymbol{q}) \approx \sum_{\boldsymbol{u}\in R} \bigl[I(T_{\boldsymbol{p}}(\boldsymbol{u})) + \nabla_I(T_{\boldsymbol{p}}(\boldsymbol{u})) \cdot \mathbf{J}_{T_{\boldsymbol{p}}}(\boldsymbol{u}) \cdot \boldsymbol{q} - R(\boldsymbol{u})\bigr]^2 \tag{24.22}$$

$$= \sum_{\boldsymbol{u}\in R} \bigl[I(\acute{\boldsymbol{u}}) + \nabla_I(\acute{\boldsymbol{u}}) \cdot \mathbf{J}_{T_{\boldsymbol{p}}}(\boldsymbol{u}) \cdot \boldsymbol{q} - R(\boldsymbol{u})\bigr]^2, \tag{24.23}$$

with $\acute{\boldsymbol{x}} = T_{\boldsymbol{p}}(\boldsymbol{x})$. Finding the parameters $\boldsymbol{q}$ that give the smallest difference $\mathcal{E}(\boldsymbol{q})$ is a linear least-squares minimization problem, which can be solved by taking the first partial derivative with respect to $\boldsymbol{q}$, that is,

$$\underbrace{\frac{\partial d}{\partial \boldsymbol{q}}}_{n\times 1} \approx \sum_{\boldsymbol{u}\in R} \underbrace{\bigl[\underbrace{\nabla_I(\acute{\boldsymbol{u}})}_{1\times 2} \cdot \underbrace{\mathbf{J}_{T_{\boldsymbol{p}}}(\boldsymbol{u})}_{2\times n}\bigr]^{\mathsf{T}}}_{n\times 1} \cdot \underbrace{\bigl[I(\acute{\boldsymbol{u}}) + \underbrace{\nabla_I(\acute{\boldsymbol{u}})}_{1\times 2} \cdot \underbrace{\mathbf{J}_{T_{\boldsymbol{p}}}(\boldsymbol{u})}_{2\times n} \cdot \underbrace{\boldsymbol{q}}_{n\times 1} - R(\boldsymbol{u})\bigr]^2}_{\in\mathbb{R}}, \tag{24.24}$$

and setting it equal to zero.[5] Solving the resulting equation for the unknown $\boldsymbol{q}$ yields the parameter change minimizing Eqn. (24.24) as

$$\boldsymbol{q}_{\mathrm{opt}} = \bar{\mathbf{H}}^{-1} \cdot \boldsymbol{\delta}_{\boldsymbol{p}}, \tag{24.25}$$

where $\bar{\mathbf{H}}$ is an estimate of the Hessian matrix (see Eqns. (24.29)–(24.30)),

$$\boldsymbol{\delta}_{\boldsymbol{p}} = \sum_{\boldsymbol{u}\in R} \underbrace{\bigl[\nabla_I(\acute{\boldsymbol{u}}) \cdot \mathbf{J}_{T_{\boldsymbol{p}}}(\boldsymbol{u})\bigr]^{\mathsf{T}}}_{\boldsymbol{s}(\boldsymbol{u})\,\in\,\mathbb{R}^n} \cdot \underbrace{\bigl[R(\boldsymbol{u}) - I(\acute{\boldsymbol{u}})\bigr]}_{D(\boldsymbol{u})\,\in\,\mathbb{R}} = \sum_{\boldsymbol{u}\in R} \boldsymbol{s}^{\mathsf{T}}(\boldsymbol{u}) \cdot D(\boldsymbol{u}) \tag{24.26}$$

is a n-dimensional column vector, and

$$D(\boldsymbol{u}) = R(\boldsymbol{u}) - I(\acute{\boldsymbol{u}}) \tag{24.27}$$

is the resulting (scalar-valued) error image. $\boldsymbol{s}(\boldsymbol{u}) = (s_0(\boldsymbol{u}), \ldots, s_{n-1}(\boldsymbol{u}))$ is a n-dimensional row vector, with each element corresponding to one of the parameters in $\boldsymbol{p}$. The 2D *scalar* fields formed by the individual components of the vector field $\boldsymbol{s}(\boldsymbol{u})$,

$$s_0, \ldots, s_{n-1} \colon M_R \times N_R \mapsto \mathbb{R}, \tag{24.28}$$

are called *steepest descent images* for the current transformation parameters $\boldsymbol{p}$.[6] These images are of the same size as the reference image R. Finally, the $n \times n$ matrix

$$\bar{\mathbf{H}} = \sum_{\boldsymbol{u}\in R} \underbrace{\bigl[\underbrace{\nabla_I(\acute{\boldsymbol{u}})}_{1\times 2} \cdot \underbrace{\mathbf{J}_{T_{\boldsymbol{p}}}(\boldsymbol{u})}_{2\times n}\bigr]^{\mathsf{T}}}_{n\times 1} \cdot \underbrace{\bigl[\underbrace{\nabla_I(\acute{\boldsymbol{u}})}_{1\times 2} \cdot \underbrace{\mathbf{J}_{T_{\boldsymbol{p}}}(\boldsymbol{u})}_{2\times n}\bigr]}_{1\times n} \tag{24.29}$$

$$= \sum_{\boldsymbol{u}\in R} \boldsymbol{s}^{\mathsf{T}}(\boldsymbol{u}) \cdot \boldsymbol{s}(\boldsymbol{u}) \approx \begin{pmatrix} \frac{\partial^2 D}{\partial p_0^2}(\boldsymbol{p}) & \cdots & \frac{\partial^2 D}{\partial p_0\, \partial p_{n-1}}(\boldsymbol{p}) \\ \vdots & \ddots & \vdots \\ \frac{\partial^2 D}{\partial p_{n-1}\, \partial p_0}(\boldsymbol{p}) & \cdots & \frac{\partial^2 D}{\partial p_{n-1}^2}(\boldsymbol{p}) \end{pmatrix} \tag{24.30}$$

[5] Note that in Eqn. (24.24) the left factor inside the summation is a n-dimensional column vector, while the right factor is a scalar.

[6] The value $s_k(\boldsymbol{u})$ indicates the optimal change of parameter p_k for the individual pixel position $\boldsymbol{u}$ to achieve a steepest-descent optimization of Eqn. (24.23) (see [13, Sec. 4.3]).

in Eqn. (24.25) is an estimate of the Hessian matrix[7] for the given transformation parameters $\boldsymbol{p}$, calculated over all coordinates $\boldsymbol{x}$ of the reference image R (Eqn. (24.29)).

The inverse of this matrix is used to calculate the optimal parameter change $\boldsymbol{q}_{\mathrm{opt}}$ in Eqn. (24.25). A better alternative to this formulation is to solve

$$\bar{\mathbf{H}} \cdot \boldsymbol{q}_{\mathrm{opt}} = \boldsymbol{\delta}_{\boldsymbol{p}}, \tag{24.31}$$

for $\boldsymbol{q}_{\mathrm{opt}}$ as the unknown, without explicitly calculating $\mathbf{H}_{\boldsymbol{p}}^{-1}$. This is a system of linear equations in the standard form $\boldsymbol{A} \cdot \boldsymbol{x} = \boldsymbol{b}$, which is numerically more stable and efficient to solve than Eqn. (24.25).[8]

24.2.1 Summary of the Algorithm

In order not to get lost after this (quite mathematical) presentation, let us recap the key steps of the Lucas-Kanade method in a more compact form. In summary, given a search image I, a reference image R, a geometric transformation $T_{\boldsymbol{p}}$, an initial parameter estimate $\boldsymbol{p}_{\mathrm{init}}$, and the convergence limit ϵ, the Lucas-Kanade algorithm performs the following steps:

A. **Initialize:**
 1. Calculate the gradient $\nabla_I(\boldsymbol{u})$ of the search image I for all image positions $\boldsymbol{u} \in I$.
 2. Initialize the transformation parameters: $\boldsymbol{p} \leftarrow \boldsymbol{p}_{\mathrm{init}}$.

B. **Repeat:**
 3. Calculate the warped gradient image $\nabla'_I(\boldsymbol{u}) = \nabla_I(T_{\boldsymbol{p}}(\boldsymbol{u}))$, for each position $\boldsymbol{u} \in R$ (by interpolation of ∇_I).
 4. Calculate the $(2 \times n)$ Jacobian matrix $\mathbf{J}_{T_{\boldsymbol{p}}}(\boldsymbol{u}) = \frac{\partial T_{\boldsymbol{p}}}{\partial \boldsymbol{p}}(\boldsymbol{u})$ of the warp function $T_{\boldsymbol{p}}(\boldsymbol{x})$, for each position $\boldsymbol{u} \in R$ and the current parameter vector $\boldsymbol{p}$ (see Eqn. (24.21)).
 5. Compute the n-dim. row vectors $\boldsymbol{s}_{\boldsymbol{u}} = \nabla'_I(\boldsymbol{u}) \cdot \mathbf{J}_{T_{\boldsymbol{p}}}(\boldsymbol{u})$, for each position $\boldsymbol{u} \in R$ (see Eqn. (24.26)).
 6. Compute the cumulative $n \times n$ Hessian matrix as $\bar{\mathbf{H}} = \sum_{\boldsymbol{u} \in R} \boldsymbol{s}_{\boldsymbol{u}}^{\mathsf{T}} \cdot \boldsymbol{s}_{\boldsymbol{u}}$ (see Eqn. (24.29)).
 7. Calculate the error image $D(\boldsymbol{x}) = R(\boldsymbol{u}) - I(T_{\boldsymbol{p}}(\boldsymbol{u}))$, for each position $\boldsymbol{u} \in R$ (by interpolation of I, see Eqn. (24.26)).
 8. Compute the column vector $\boldsymbol{\delta}_{\boldsymbol{p}} = \sum_{\boldsymbol{u} \in R} \boldsymbol{s}_{\boldsymbol{u}}^{\mathsf{T}} \cdot D(\boldsymbol{u})$ (see Eqn. (24.26)).
 9. Calculate the optimal parameter change $\boldsymbol{q}_{\mathrm{opt}} = \bar{\mathbf{H}}^{-1} \cdot \boldsymbol{\delta}_{\boldsymbol{p}}$ (see Eqn. (24.25)).
 10. Update the transformation parameter: $\boldsymbol{p} \leftarrow \boldsymbol{p} + \boldsymbol{q}_{\mathrm{opt}}$ (see Eqn. (24.17)).

 Until $\left\|\boldsymbol{q}_{\mathrm{opt}}\right\| < \epsilon$.

[7] The Hessian matrix of a n-variable, real-valued function f is composed of f's second-order partial derivatives (see also Sec. C.2.6 in the Appendix). The Hessian matrix $\mathbf{H}$ is always symmetric.

[8] Moreover, Eqn. (24.31) may be solvable even if the matrix $\bar{\mathbf{H}}$ is almost singular and thus numerically not invertible [160, p. 164].

Alg. 24.1
Lucas-Kanade ("forward-additive") registration algorithm. The origin of the reference image R is placed at its center. The gradient of the image is calculated only once (line 6), but interpolated in every iteration (line 15). Also, the $n \times n$ Hessian matrix $\bar{\mathbf{H}}$ is calculated and inverted in every iteration. The Jacobian of the warp function T is also evaluated repeatedly (line 16), though this is not an expensive calculation, at least for affine warps (lines 32–33). Procedure $\mathsf{Interpolate}(I, \boldsymbol{x}')$ returns the interpolated value of the image I at the continuous position $\boldsymbol{x}' \in \mathbb{R}^2$ (see Ch. 22 for details and possible implementations).

1: **LucasKanadeForward**$(I, R, T, \boldsymbol{p}_{\text{init}}, \epsilon, i_{\max})$
Input: I, the search image; R, the reference image; T, a 2D warp function that maps any point $\boldsymbol{x} \in \mathbb{R}^2$ to some point $\boldsymbol{x}' = T_{\boldsymbol{p}}(\boldsymbol{x})$, with transformation parameters $\boldsymbol{p} = (p_0, \ldots, p_{n-1})$; $\boldsymbol{p}_{\text{init}}$, initial estimate of the warp parameters; ϵ, the error limit; $i_{\max}$, the maximum number of iterations.
Returns the modified warp parameter vector $\boldsymbol{p}$ for the best fit between I and R, or nil if no match could be found.

2: $(M_R, N_R) \leftarrow \mathsf{Size}(R)$ ▷ size of the reference image R
3: $\boldsymbol{x}_{\text{c}} \leftarrow 0.5 \cdot (M_R-1, N_R-1)$ ▷ center of R
4: $\boldsymbol{p} \leftarrow \boldsymbol{p}_{\text{init}}$ ▷ initial transformation parameters
5: $n \leftarrow \mathsf{Length}(\boldsymbol{p})$ ▷ parameter count
6: $(I_{\text{x}}, I_{\text{y}}) \leftarrow \mathsf{Gradient}(I)$ ▷ calculate the gradient ∇I
7: $i \leftarrow 0$ ▷ iteration counter
8: **do** ▷ main loop
9: $\quad i \leftarrow i+1$
10: $\quad \bar{\mathbf{H}} \leftarrow \mathbf{0}_{n,n}$ ▷ $\bar{\mathbf{H}} \in \mathbb{R}^{n\times n}$, initialized to zero
11: $\quad \boldsymbol{\delta_p} \leftarrow \mathbf{0}_n$ ▷ $s_{\boldsymbol{p}} \in \mathbb{R}^n$, initialized to zero
12: $\quad$ **for** all positions $\boldsymbol{u} \in (M_R \times N_R)$ **do**
13: $\quad\quad \boldsymbol{x} \leftarrow \boldsymbol{u} - \boldsymbol{x}_{\text{c}}$ ▷ position w.r.t. the center of R
14: $\quad\quad \boldsymbol{x}' \leftarrow T_{\boldsymbol{p}}(\boldsymbol{x})$ ▷ warp x to x' by transf. T_p
$\quad\quad$ Estimate the gradient of I at the warped position x':
15: $\quad\quad \nabla \leftarrow \big(\mathsf{Interpolate}(I_{\text{x}}, \boldsymbol{x}'), \mathsf{Interpolate}(I_{\text{y}}, \boldsymbol{x}')\big)$ ▷ 2D row vector
16: $\quad\quad \mathbf{J} \leftarrow \mathsf{Jacobian}(T_{\boldsymbol{p}}, \boldsymbol{x})$ ▷ Jacobian of $T_{\boldsymbol{p}}$ at pos. x
17: $\quad\quad \boldsymbol{s} \leftarrow (\nabla \cdot \mathbf{J})^{\mathsf{T}}$ ▷ $\boldsymbol{s}$ is a column vector of length n
18: $\quad\quad \mathbf{H} \leftarrow \boldsymbol{s} \cdot \boldsymbol{s}^{\mathsf{T}}$ ▷ outer product, $\mathbf{H}$ is of size $n \times n$
19: $\quad\quad \bar{\mathbf{H}} \leftarrow \bar{\mathbf{H}} + \mathbf{H}$ ▷ cumulate the Hessian (Eq. 24.30)
20: $\quad\quad d \leftarrow R(\boldsymbol{u}) - \mathsf{Interpolate}(I, \boldsymbol{x}')$ ▷ pixel difference $d \in \mathbb{R}$
21: $\quad\quad \boldsymbol{\delta_p} \leftarrow \boldsymbol{\delta_p} + \boldsymbol{s} \cdot d$
22: $\quad \boldsymbol{q}_{\text{opt}} \leftarrow \bar{\mathbf{H}}^{-1} \cdot \boldsymbol{\delta_p}$ ▷ Eq. 24.17, or solve $\bar{\mathbf{H}} \cdot \boldsymbol{q}_{\text{opt}} = \boldsymbol{\delta_p}$ (Eq. 24.31)
23: $\quad \boldsymbol{p} \leftarrow \boldsymbol{p} + \boldsymbol{q}_{\text{opt}}$
24: **while** $(\|\boldsymbol{q}_{\text{opt}}\| > \epsilon) \wedge (i < i_{\max})$ ▷ repeat until convergence
25: **if** $i < i_{\max}$ **then**
26: $\quad$ **return** $\boldsymbol{p}$
27: **else**
28: $\quad$ **return** nil

29: **Gradient**(I)
Returns the gradient of I as a pair of maps.
30: $H_{\text{x}} = \frac{1}{8} \cdot \begin{bmatrix} -1 & 0 & 1 \\ -2 & \mathbf{0} & 2 \\ -1 & 0 & 1 \end{bmatrix}, \quad H_{\text{y}} = \frac{1}{8} \cdot \begin{bmatrix} -1 & -2 & -1 \\ 0 & \mathbf{0} & 0 \\ 1 & 2 & 1 \end{bmatrix}$
31: **return** $(I * H_x, I * H_y)$

32: **Jacobian**$(T_{\boldsymbol{p}}, \boldsymbol{x})$
Returns the $2 \times n$ Jacobian matrix of the 2D warp function $T_{\boldsymbol{p}}(\boldsymbol{x}) = (T_{\text{x},\boldsymbol{p}}(\boldsymbol{x}), T_{\text{y},\boldsymbol{p}}(\boldsymbol{x}))$ with parameters $\boldsymbol{p} = (p_0, \ldots, p_{n-1})$ for the spatial position $\boldsymbol{x} \in \mathbb{R}^2$.
33: **return** $\begin{pmatrix} \frac{\partial T_{\text{x},\boldsymbol{p}}}{\partial p_0}(\boldsymbol{x}) & \frac{\partial T_{\text{x},\boldsymbol{p}}}{\partial p_1}(\boldsymbol{x}) & \cdots & \frac{\partial T_{\text{x},\boldsymbol{p}}}{\partial p_{n-1}}(\boldsymbol{x}) \\ \frac{\partial T_{\text{y},\boldsymbol{p}}}{\partial p_0}(\boldsymbol{x}) & \frac{\partial T_{\text{y},\boldsymbol{p}}}{\partial p_1}(\boldsymbol{x}) & \cdots & \frac{\partial T_{\text{y},\boldsymbol{p}}}{\partial p_{n-1}}(\boldsymbol{x}) \end{pmatrix}$ ▷ see Eq. 24.21

The complete specification of the Lucas-Kanade algorithm (referred to as the "forward-additive" algorithm in [13]) is given in Alg. 24.1. In addition to the two images I and R, the procedure requires the assumed type of the geometric transformation T, the estimated initial transformation parameters $\boldsymbol{p}_{\mathrm{init}}$, a convergence limit ϵ and the maximum number of iterations $i_{\max}$. The optimal parameter vector $\boldsymbol{p}$ is returned or nil if the optimization did not converge. For better numerical stability, the origin of the reference image R is placed at its center $\boldsymbol{x}_{\mathrm{c}}$ (see line 3), as is also illustrated in Fig. 24.2. The algorithm shows (unlike the just given summary) that it is sufficient to calculate the Jacobian $\mathbf{J}$ (see line 16) and the Hessian matrix $\bar{\mathbf{H}}$ (see line 18) only for the current position ($\boldsymbol{u}$) in the reference image, which implies relatively modest storage requirements. Additional instructions for calculating the Jacobian and Hessian matrices for specific linear transformations T are described in Sec. 24.4. In the case that $\bar{\mathbf{H}}$ cannot be inverted (because it is singular) in line 22, the algorithm could either stop (and return nil) or continue with a small random perturbation of the transformation parameters $\boldsymbol{p}$.

This so-called forward-additive algorithm performs reliably if the assumed type of geometric transformation is correct and the initial parameter estimate is sufficiently close to the actual parameters. However, it is computationally demanding since it requires repeated warping of the gradient image and the Jacobian $\mathbf{J}_{T_{\boldsymbol{p}}}$ as well as the Hessian matrix $\mathbf{H}$ must be re-calculated in each iteration. Very similar results at greatly improved performance are obtained with the "inverse compositional algorithm" described in Sec. 24.3.

24.3 Inverse Compositional Algorithm

This algorithm, described in [14], exchanges the roles of the search image I and the reference image R. As illustrated in Fig. 24.3, the reference image R remains anchored at the original position, while the geometric transformations are applied to (parts of) the search image I. In particular, the transformation $T_{\boldsymbol{p}}$ now describes the mapping from the warped image I' *back* to the original image I. The advantage of this algorithm is that it avoids re-evaluating the Jacobian and Hessian matrices in every iteration while exhibiting convergence properties similar to the Lucas-Kanade (forward-additive) algorithm described in Sec. 24.2.

In this algorithm, the expression to be minimized in each iteration is (cf. Eqn. (24.16))

$$\mathcal{E}(\boldsymbol{q}) = \sum_{\boldsymbol{u} \in R} \left[R(T_{\boldsymbol{q}}(\boldsymbol{u})) - I(T_{\boldsymbol{p}}(\boldsymbol{u})) \right]^2, \tag{24.32}$$

with respect to the parameter change $\boldsymbol{q}$, producing an optimal change vector $\boldsymbol{q}_{\mathrm{opt}}$. Subsequently, the geometric transformation is updated not by simply *adding* $\boldsymbol{q}_{\mathrm{opt}}$ to the current parameter estimate $\boldsymbol{p}$ (as in Eqn. (24.17)), but by *concatenating* the corresponding warps in the form

$$T_{\boldsymbol{p}'}(\boldsymbol{x}) = (T_{\boldsymbol{q}_{\mathrm{opt}}}^{-1} \circ T_{\boldsymbol{p}})(\boldsymbol{x}) = T_{\boldsymbol{p}}(T_{\boldsymbol{q}_{\mathrm{opt}}}^{-1}(\boldsymbol{x})) \tag{24.33}$$

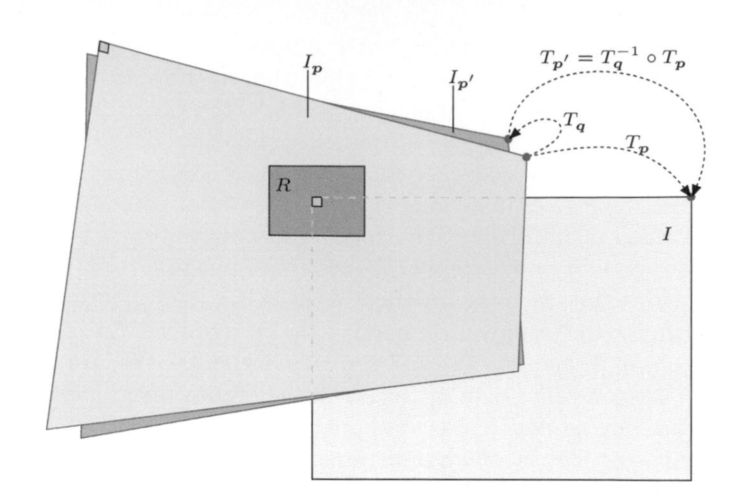

Fig. 24.3
Geometry of the *inverse compositional* registration algorithm. I denotes the search image and R is the reference image. The geometric transformation T_p warps the image I_p *back* to the original search image I, with p being the initial parameter estimate. Matching is performed between the (unwarped) reference image R and the warped search image I_p. Note that the reference image R always remains anchored at the origin. In each iteration, the incremental warp T_q (with parameter vector q) is estimated, mapping the image I_p to image $I_{p'}$. The resulting composite warp $T_{p'}$ (mapping $I_{p'}$ back to I) with parameters p' is obtained by concatenating the transformations T_q^{-1} and T_p.

where $\circ$ denotes the concatenation (successive application) of transformations. In the special (but frequent) case of linear geometric transformations, the concatenation is simply accomplished by multiplying the corresponding transformation matrices $\mathbf{M}_p$, $\mathbf{M}_{q_{\text{opt}}}$, that is,

$$\mathbf{M}_{p'} = \mathbf{M}_p \cdot \mathbf{M}_{q_{\text{opt}}}^{-1} \tag{24.34}$$

(see also Sec. 24.4.4). Also note that the "incremental" transformation $T_{q_{\text{opt}}}$ is *inverted* before it is concatenated with the current warp T_p, to calculate the parameters of the resulting composite warp $T_{p'}$. Thus the geometric transformation T must be invertible, but this is again no problem with linear (affine or projective) warps.

In summary, given a search image I, a reference image R, a geometric transformation T_p, an initial parameter estimate p_{init} and the convergence limit ϵ, the "inverse compositional algorithm" performs the following steps:

A. **Initialize:**
1. Calculate the gradient $\nabla_R(x)$ of the reference image R for all $x \in R$.
2. Calculate the Jacobian $\mathbf{J}(x) = \frac{\partial T_p}{\partial p}(x)$ of the warp function $T_p(x)$ for all $x \in R$, with $p = \mathbf{0}$.
3. Compute $s_x = \nabla_R(x) \cdot \mathbf{J}(x)$ for all $x \in R$.
4. Calculate the Hessian matrix as $\mathbf{H} = \sum_R s_x^\intercal \cdot s_x$ and precalculate its inverse $\mathbf{H}^{-1}$.
5. Initialize the transformation parameters: $p \leftarrow p_{\text{init}}$.

B. **Repeat:**
6. Warp the search image I to I', such that $I'(x) = I(T_p(x))$, for all $x \in R$.
7. Compute the (column) vector $\delta_p = \sum_R s_x \cdot [I'(x) - R(x)]$.
8. Estimate the optimal parameter change $q_{\text{opt}} = \mathbf{H}^{-1} \cdot \delta_p$.
9. Find the warp parameters p', such that $T_{p'} = T_{q_{\text{opt}}}^{-1} \circ T_p$.
10. Update the warp parameter $p \leftarrow p'$.

Until $\|q_{\text{opt}}\| < \epsilon$.

1: **LucasKanadeInverse**($I, R, T, \boldsymbol{p}_{\text{init}}, \epsilon, i_{\max}$)
Input: I, the search image; R, the reference image; T, a 2D warp function that maps any point $\boldsymbol{x} \in \mathbb{R}^2$ to $\boldsymbol{x}' = T_{\boldsymbol{p}}(\boldsymbol{x})$ using parameters $\boldsymbol{p} = (p_0, \ldots, p_{n-1})$; $\boldsymbol{p}_{\text{init}}$, initial estimate of the warp parameters; ϵ, the error limit (typ. $\epsilon = 10^{-3}$); $i_{\max}$, the maximum number of iterations.
Returns the updated warp parameter vector $\boldsymbol{p}$ for the best fit between I and R, or nil if no match could be found.

```
2:   (M_R, N_R) ← Size(R)                     ▷ size of the reference image R
3:   x_c ← 0.5 · (M_R−1, N_R−1)                ▷ center of R
     Initialize:
4:   n ← Length(p)                            ▷ parameter count n
5:   Create map S: (M_R × N_R) ↦ R^n          ▷ n "steepest-descent images"
6:   (R_x, R_y) ← Gradient(R)                 ▷ (R_x(u), R_y(u))^T = ∇_R(u)
7:   H̄ ← 0_{n,n}                              ▷ initialize n × n Hessian matrix to zero
8:   for all positions u ∈ (M_R × N_R) do
9:       x ← u − x_c                          ▷ centered position
10:      ∇_R ← (R_x(u), R_y(u))               ▷ 2-dimensional row vector
11:      J ← Jacobian(T_0(x))                 ▷ Jacob. of T at pos. x with p = 0
12:      s ← (∇_R · J)^T                      ▷ s is a column vector of length n
13:      S(u) ← s                             ▷ keep s for later use
14:      H ← s · s^T                          ▷ outer product, H is of size n×n
15:      H̄ ← H̄ + H                            ▷ cumulate the Hessian (Eq. 24.30)
16:  H̄^{-1} ← Inverse(H̄)
17:  if H̄^{-1} = nil then                     ▷ H̄ could not be inverted
18:      return nil                           ▷ stop
19:  p ← p_init                               ▷ initial parameter estimate
20:  i ← 0                                    ▷ iteration counter
     Main loop:
21:  do
22:      i ← i + 1
23:      δ_p ← 0_n                            ▷ δ_p ∈ R^n, initialized to zero
24:      for all positions u ∈ (M_R × N_R) do
25:          x ← u − x_c                      ▷ centered position
26:          x' ← T_p(x)                      ▷ warp I to I'
27:          d ← Interpolate(I, x') − R(u)    ▷ pixel difference d ∈ R
28:          s ← S(u)                         ▷ get pre-calculated s
29:          δ_p ← δ_p + s · d
30:      q_opt ← H^{-1} · δ_p                 ▷ H^{-1} is pre-calculated in line 16
31:      p' ← determine, such that T_{p'}(x) = T_p(T^{-1}_{q_opt}(x))
32:      p ← p'
33:  while (‖q_opt‖ > ε) ∧ (i < i_max)        ▷ repeat until convergence
34:  return { p    for i < i_max
            { nil  otherwise
```

Alg. 24.2
Inverse compositional registration algorithm. The gradient vectors $\nabla_R(u, v)$ of the reference image R are calculated only once (line 6) using procedure Gradient(), as defined in Alg. 24.1. The Jacobian matrix $\mathbf{J}$ of the warp function $T_{\boldsymbol{p}}$ is also evaluated only once (line 11) for $\boldsymbol{p} = \mathbf{0}$ (i.e., the identity mapping) over all positions of the reference image R. Similarly, the Hessian matrix $\mathbf{H}$ and its inverse $\mathbf{H}^{-1}$ are calculated only once (lines 15, 16). $\mathbf{H}^{-1}$ is used to calculate the optimal parameter change vector $\boldsymbol{q}_{\text{opt}}$ in line 30 of the main loop. Procedure Interpolate() in line 27 is the same as in Alg. 24.1. This algorithm is typically about 5–10 times faster than the original Lucas-Kanade (forward) algorithm (see Alg. 24.1), with similar convergence properties.

One can see clearly that in this variant several steps are performed only once at initialization and do not appear inside the main loop. A detailed and concise listing of the inverse compositional algorithm is given in Alg. 24.2 and concrete setups for various linear transformations are described in Sec. 24.4. Since the Jacobian matrix (for the null parameter vector $\boldsymbol{p} = \mathbf{0}$) and the Hessian matrix are calculated only once during initialization, this algorithm executes significantly faster than the original Lucas-Kanade (forward-additive) algorithm, while offering similar convergence properties.

24.4 Parameter Setups for Various Linear Transformations

The use of linear transformatons for the geometric mapping T is very common. In the following, we describe detailed setups required for the Lucas-Kanade algorithm for various geometric transformations, such as pure translation as well as affine and projective transformations. This should help to reduce the chance of confusion about the content and structure of the involved vectors and matrices. For additional details and concrete implementations of these transformations readers should consult the associated Java source code in the `imagingbook`[9] library.

24.4.1 Pure Translation

In the case of pure 2D translation, we have $n = 2$ parameters t_x, t_y and the geometric transformation is (see Eqn. (24.15))

$$\acute{\boldsymbol{x}} = T_{\boldsymbol{p}}(\boldsymbol{x}) = \boldsymbol{x} + \begin{pmatrix} t_\mathrm{x} \\ t_\mathrm{y} \end{pmatrix}, \tag{24.35}$$

with the parameter vector $\boldsymbol{p} = (p_0, p_1)^\intercal = (t_\mathrm{x}, t_\mathrm{y})^\intercal$ and $\boldsymbol{x} = (x, y)^\intercal$. Thus the two component functions of the transformation (cf. Eqn. (24.18)) are

$$\begin{aligned} T_{\mathrm{x},\boldsymbol{p}}(\boldsymbol{x}) &= x + t_\mathrm{x}, \\ T_{\mathrm{y},\boldsymbol{p}}(\boldsymbol{x}) &= y + t_\mathrm{y}, \end{aligned} \tag{24.36}$$

with the 2×2 Jacobian matrix

$$\mathbf{J}_{T_{\boldsymbol{p}}}(\boldsymbol{x}) = \begin{pmatrix} \frac{\partial T_{\mathrm{x},\boldsymbol{p}}}{\partial t_\mathrm{x}}(\boldsymbol{x}) & \frac{\partial T_{\mathrm{x},\boldsymbol{p}}}{\partial t_\mathrm{y}}(\boldsymbol{x}) \\ \frac{\partial T_{\mathrm{y},\boldsymbol{p}}}{\partial t_\mathrm{x}}(\boldsymbol{x}) & \frac{\partial T_{\mathrm{y},\boldsymbol{p}}}{\partial t_\mathrm{y}}(\boldsymbol{x}) \end{pmatrix} = \begin{pmatrix} 1 & 0 \\ 0 & 1 \end{pmatrix}. \tag{24.37}$$

Note that in this case $\mathbf{J}_{T_{\boldsymbol{p}}}(\boldsymbol{x})$ is constant,[10] that is, independent of the position $\boldsymbol{x}$ and the parameters $\boldsymbol{p}$. The 2D column vector $\boldsymbol{\delta}_{\boldsymbol{p}}$ (Eqn. (24.26)) is calculated as

$$\boldsymbol{\delta}_{\boldsymbol{p}} = \sum_{\boldsymbol{u} \in R} \big[\nabla_I(\underbrace{T_{\boldsymbol{p}}(\boldsymbol{u})}_{\acute{\boldsymbol{u}} \in \mathbb{R}^2}) \cdot \mathbf{J}_{T_{\boldsymbol{p}}}(\boldsymbol{u})\big]^\intercal \cdot \underbrace{\big[R(\boldsymbol{u}) - I(T_{\boldsymbol{p}}(\boldsymbol{u}))\big]}_{D(\boldsymbol{u}) \in \mathbb{R}} \tag{24.38}$$

$$= \sum_{\boldsymbol{u} \in R} \underbrace{\big[(I_\mathrm{x}(\acute{\boldsymbol{u}}), I_\mathrm{y}(\acute{\boldsymbol{u}})) \cdot \left(\begin{smallmatrix} 1 & 0 \\ 0 & 1 \end{smallmatrix}\right)\big]^\intercal}_{\boldsymbol{s}(\boldsymbol{u}) = (s_0(\boldsymbol{u}), s_1(\boldsymbol{u}))} \cdot D(\boldsymbol{u}) = \sum_{\boldsymbol{u} \in R} \begin{pmatrix} I_\mathrm{x}(\acute{\boldsymbol{u}}) \\ I_\mathrm{y}(\acute{\boldsymbol{u}}) \end{pmatrix} \cdot D(\boldsymbol{u}) \tag{24.39}$$

$$= \begin{pmatrix} \sum_{\boldsymbol{u}} I_\mathrm{x}(\acute{\boldsymbol{u}}) \cdot D(\boldsymbol{u}) \\ \sum_{\boldsymbol{u}} I_\mathrm{y}(\acute{\boldsymbol{u}}) \cdot D(\boldsymbol{u}) \end{pmatrix} = \begin{pmatrix} \sum_{\boldsymbol{u}} s_0(\boldsymbol{u}) \cdot D(\boldsymbol{u}) \\ \sum_{\boldsymbol{u}} s_1(\boldsymbol{u}) \cdot D(\boldsymbol{u}) \end{pmatrix} = \begin{pmatrix} \delta_0 \\ \delta_1 \end{pmatrix}, \tag{24.40}$$

[9] Package `imagingbook.pub.geometry.mappings`.

[10] $\mathbf{I}_2$ denotes the 2×2 identity matrix.

where I_x, I_y denote the (estimated) first derivatives of the search image I in x and y-direction, respectively.[11] Thus in this case the *steepest descent images* (Eqn. (24.28)) $s_0(\boldsymbol{x}) = I_x(\acute{\boldsymbol{x}})$ and $s_1(\boldsymbol{x}) = I_y(\acute{\boldsymbol{x}})$ are simply the components of the interpolated gradient of I in the region of the shifted reference image. The associated Hessian matrix (Eqn. (24.29)) is calculated as

$$\bar{\mathbf{H}} = \sum_{\boldsymbol{u}\in R} \big[\nabla_I(T_{\boldsymbol{p}}(\boldsymbol{u})) \cdot \mathbf{J}_{T_{\boldsymbol{p}}}(\boldsymbol{u})\big]^{\mathsf{T}} \cdot \big[\nabla_I(T_{\boldsymbol{p}}(\boldsymbol{u})) \cdot \mathbf{J}_{T_{\boldsymbol{p}}}(\boldsymbol{u})\big] \tag{24.41}$$

$$= \sum_{\boldsymbol{u}\in R} \underbrace{\big[\nabla_I(\acute{\boldsymbol{u}}) \cdot \left(\begin{smallmatrix}1&0\\0&1\end{smallmatrix}\right)\big]^{\mathsf{T}}}_{\boldsymbol{s}(\boldsymbol{u})} \cdot \underbrace{\big[\nabla_I(\acute{\boldsymbol{u}}) \cdot \left(\begin{smallmatrix}1&0\\0&1\end{smallmatrix}\right)\big]}_{\boldsymbol{s}(\boldsymbol{u})} = \sum_{\boldsymbol{u}\in R} \boldsymbol{s}^{\mathsf{T}}(\boldsymbol{u}) \cdot \boldsymbol{s}(\boldsymbol{u}) \tag{24.42}$$

$$= \sum_{\boldsymbol{u}\in R} \nabla_I^{\mathsf{T}}(\acute{\boldsymbol{u}}) \cdot \nabla_I(\acute{\boldsymbol{u}}) = \sum_{\boldsymbol{u}\in R} \begin{pmatrix} I_x(\acute{\boldsymbol{u}}) \\ I_y(\acute{\boldsymbol{u}}) \end{pmatrix} \cdot \big(I_x(\acute{\boldsymbol{u}}),\, I_y(\acute{\boldsymbol{u}})\big) \tag{24.43}$$

$$= \sum_{\boldsymbol{u}\in R} \begin{pmatrix} I_x^2(\acute{\boldsymbol{u}}) & I_x(\acute{\boldsymbol{u}})\cdot I_y(\acute{\boldsymbol{u}}) \\ I_x(\acute{\boldsymbol{u}})\cdot I_y(\acute{\boldsymbol{u}}) & I_y^2(\acute{\boldsymbol{u}}) \end{pmatrix} \tag{24.44}$$

$$= \begin{pmatrix} \Sigma\, I_x^2(\acute{\boldsymbol{u}}) & \Sigma\, I_x(\acute{\boldsymbol{u}})\cdot I_y(\acute{\boldsymbol{u}}) \\ \Sigma\, I_x(\acute{\boldsymbol{u}})\cdot I_y(\acute{\boldsymbol{u}}) & \Sigma\, I_y^2(\acute{\boldsymbol{u}}) \end{pmatrix} = \begin{pmatrix} H_{00} & H_{01} \\ H_{10} & H_{11} \end{pmatrix}, \tag{24.45}$$

again with $\acute{\boldsymbol{u}} = T_{\boldsymbol{p}}(\boldsymbol{u})$. Since $\bar{\mathbf{H}}$ is symmetric ($H_{01} = H_{10}$) and only of size 2×2, its *inverse* can be easily obtained in closed form:

$$\bar{\mathbf{H}}^{-1} = \frac{1}{H_{00}\cdot H_{11} - H_{01}\cdot H_{10}} \cdot \begin{pmatrix} H_{11} & -H_{01} \\ -H_{10} & H_{00} \end{pmatrix} \tag{24.46}$$

$$= \frac{1}{H_{00}\cdot H_{11} - H_{01}^2} \cdot \begin{pmatrix} H_{11} & -H_{01} \\ -H_{01} & H_{00} \end{pmatrix}. \tag{24.47}$$

The resulting optimal parameter increment (see Eqn. (24.25)) is

$$\boldsymbol{q}_{\mathrm{opt}} = \begin{pmatrix} t'_x \\ t'_y \end{pmatrix} = \bar{\mathbf{H}}^{-1} \cdot \boldsymbol{\delta}_{\boldsymbol{p}} = \bar{\mathbf{H}}^{-1} \cdot \begin{pmatrix} \delta_0 \\ \delta_1 \end{pmatrix} \tag{24.48}$$

$$= \frac{1}{H_{11}\cdot H_{22} - H_{12}^2} \cdot \begin{pmatrix} H_{11}\cdot\delta_0 - H_{01}\cdot\delta_1 \\ H_{00}\cdot\delta_1 - H_{01}\cdot\delta_0 \end{pmatrix}, \tag{24.49}$$

with δ_0, δ_1 as defined in Eqn. (24.40). Alternatively the same result could be obtained by solving Eqn. (24.31) for $\boldsymbol{q}_{\mathrm{opt}}$.

24.4.2 Affine Transformation

An affine transformation in 2D can be expressed (for example) with homogeneous coordinates[12] in the form

$$T_{\boldsymbol{p}}(\boldsymbol{x}) = \begin{pmatrix} 1+a & b & t_x \\ c & 1+d & t_y \end{pmatrix} \cdot \begin{pmatrix} x \\ y \\ 1 \end{pmatrix}, \tag{24.50}$$

with $n = 6$ parameters $\boldsymbol{p} = (p_0, \dots, p_5)^{\mathsf{T}} = (a, b, c, d, t_x, t_y)^{\mathsf{T}}$. This parameterization of the affine transformation implies that the *null*

[11] See Sec. C.3.1 in the Appendix for how to estimate gradients of discrete images.

[12] See also Chapter 21, Secs. 21.1.2 and 21.1.3.

parameter vector ($\boldsymbol{p} = \boldsymbol{0}$) corresponds to the *identity* transformation. The component functions of this transformation thus are

$$\begin{aligned} T_{\mathrm{x},\boldsymbol{p}}(\boldsymbol{x}) &= (1+a) \cdot x + b \cdot y + t_{\mathrm{x}}, \\ T_{\mathrm{y},\boldsymbol{p}}(\boldsymbol{x}) &= c \cdot x + (1+d) \cdot y + t_{\mathrm{y}}, \end{aligned} \tag{24.51}$$

and the associated Jacobian matrix at some position $\boldsymbol{x} = (x, y)$ is

$$\mathbf{J}_{T_{\boldsymbol{p}}}(\boldsymbol{x}) = \begin{pmatrix} \frac{\partial T_{\mathrm{x},\boldsymbol{p}}}{\partial a} & \frac{\partial T_{\mathrm{x},\boldsymbol{p}}}{\partial b} & \frac{\partial T_{\mathrm{x},\boldsymbol{p}}}{\partial c} & \frac{\partial T_{\mathrm{x},\boldsymbol{p}}}{\partial d} & \frac{\partial T_{\mathrm{x},\boldsymbol{p}}}{\partial t_{\mathrm{x}}} & \frac{\partial T_{\mathrm{x},\boldsymbol{p}}}{\partial t_{\mathrm{y}}} \\ \frac{\partial T_{\mathrm{y},\boldsymbol{p}}}{\partial a} & \frac{\partial T_{\mathrm{y},\boldsymbol{p}}}{\partial b} & \frac{\partial T_{\mathrm{y},\boldsymbol{p}}}{\partial c} & \frac{\partial T_{\mathrm{y},\boldsymbol{p}}}{\partial d} & \frac{\partial T_{\mathrm{y},\boldsymbol{p}}}{\partial t_{\mathrm{x}}} & \frac{\partial T_{\mathrm{y},\boldsymbol{p}}}{\partial t_{\mathrm{y}}} \end{pmatrix}(\boldsymbol{x}) \tag{24.52}$$

$$= \begin{pmatrix} x & y & 0 & 0 & 1 & 0 \\ 0 & 0 & x & y & 0 & 1 \end{pmatrix}. \tag{24.53}$$

Note that in this case the Jacobian only depends on the position $\boldsymbol{x} = (x, y)$, not on the transformation parameters $\boldsymbol{p}$. It can thus be pre-calculated once for all positions $\boldsymbol{x}$ of the reference image R. The 6-dimensional column vector $\boldsymbol{\delta}_{\boldsymbol{p}}$ (Eqn. (24.26)) is obtained as

$$\boldsymbol{\delta}_{\boldsymbol{p}} = \sum_{\boldsymbol{u} \in R} \underbrace{\big[\nabla_I(T_{\boldsymbol{p}}(\boldsymbol{u})) \cdot \mathbf{J}_{T_{\boldsymbol{p}}}(\boldsymbol{u})\big]^{\mathsf{T}}}_{\boldsymbol{s}(\boldsymbol{u})} \cdot \underbrace{\big[R(\boldsymbol{u}) - I(T_{\boldsymbol{p}}(\boldsymbol{u}))\big]}_{D(\boldsymbol{u})} \tag{24.54}$$

$$= \sum_{\boldsymbol{u} \in R} \left[(I_{\mathrm{x}}(\acute{\boldsymbol{u}}), I_{\mathrm{y}}(\acute{\boldsymbol{u}})) \cdot \begin{pmatrix} x & y & 0 & 0 & 1 & 0 \\ 0 & 0 & x & y & 0 & 1 \end{pmatrix} \right]^{\mathsf{T}} \cdot D(\boldsymbol{u}) \tag{24.55}$$

$$= \sum_{\boldsymbol{u} \in R} \begin{pmatrix} I_{\mathrm{x}}(\acute{\boldsymbol{u}}) \cdot x \\ I_{\mathrm{x}}(\acute{\boldsymbol{u}}) \cdot y \\ I_{\mathrm{y}}(\acute{\boldsymbol{u}}) \cdot x \\ I_{\mathrm{y}}(\acute{\boldsymbol{u}}) \cdot y \\ I_{\mathrm{x}}(\acute{\boldsymbol{u}}) \\ I_{\mathrm{y}}(\acute{\boldsymbol{u}}) \end{pmatrix} \cdot D(\boldsymbol{u}) = \sum_{\boldsymbol{u} \in R} \begin{pmatrix} s_0(\boldsymbol{u}) \\ s_1(\boldsymbol{u}) \\ s_2(\boldsymbol{u}) \\ s_3(\boldsymbol{u}) \\ s_4(\boldsymbol{u}) \\ s_5(\boldsymbol{u}) \end{pmatrix} \cdot D(\boldsymbol{u}) \tag{24.56}$$

$$= \sum_{\boldsymbol{u} \in R} \begin{pmatrix} s_0(\boldsymbol{u}) \cdot D(\boldsymbol{u}) \\ s_1(\boldsymbol{u}) \cdot D(\boldsymbol{u}) \\ s_2(\boldsymbol{u}) \cdot D(\boldsymbol{u}) \\ s_3(\boldsymbol{u}) \cdot D(\boldsymbol{u}) \\ s_4(\boldsymbol{u}) \cdot D(\boldsymbol{u}) \\ s_5(\boldsymbol{u}) \cdot D(\boldsymbol{u}) \end{pmatrix} = \begin{pmatrix} \Sigma\, s_0(\boldsymbol{u}) \cdot D(\boldsymbol{u}) \\ \Sigma\, s_1(\boldsymbol{u}) \cdot D(\boldsymbol{u}) \\ \Sigma\, s_2(\boldsymbol{u}) \cdot D(\boldsymbol{u}) \\ \Sigma\, s_3(\boldsymbol{u}) \cdot D(\boldsymbol{u}) \\ \Sigma\, s_4(\boldsymbol{u}) \cdot D(\boldsymbol{u}) \\ \Sigma\, s_5(\boldsymbol{u}) \cdot D(\boldsymbol{u}) \end{pmatrix}, \tag{24.57}$$

again with $\acute{\boldsymbol{u}} = T_{\boldsymbol{p}}(\boldsymbol{u})$. The corresponding Hessian matrix (of size 6×6) is found as

$$\bar{\mathbf{H}} = \sum_{\boldsymbol{u} \in R} \big[\nabla_I(T_{\boldsymbol{p}}(\boldsymbol{u})) \cdot \mathbf{J}_{T_{\boldsymbol{p}}}(\boldsymbol{u})\big]^{\mathsf{T}} \cdot \big[\nabla_I(T_{\boldsymbol{p}}(\boldsymbol{u})) \cdot \mathbf{J}_{T_{\boldsymbol{p}}}(\boldsymbol{u})\big] \tag{24.58}$$

$$= \sum_{\boldsymbol{x} \in R} \boldsymbol{s}^{\mathsf{T}}(\boldsymbol{u}) \cdot \boldsymbol{s}(\boldsymbol{u}) = \sum_{\boldsymbol{x} \in R} \begin{pmatrix} I_{\mathrm{x}}(\acute{\boldsymbol{u}}) \cdot x \\ I_{\mathrm{x}}(\acute{\boldsymbol{u}}) \cdot y \\ I_{\mathrm{y}}(\acute{\boldsymbol{u}}) \cdot x \\ I_{\mathrm{y}}(\acute{\boldsymbol{u}}) \cdot y \\ I_{\mathrm{x}}(\acute{\boldsymbol{u}}) \\ I_{\mathrm{y}}(\acute{\boldsymbol{u}}) \end{pmatrix}^{\mathsf{T}} \cdot \begin{pmatrix} I_{\mathrm{x}}(\acute{\boldsymbol{u}}) \cdot x \\ I_{\mathrm{x}}(\acute{\boldsymbol{u}}) \cdot y \\ I_{\mathrm{y}}(\acute{\boldsymbol{u}}) \cdot x \\ I_{\mathrm{y}}(\acute{\boldsymbol{u}}) \cdot y \\ I_{\mathrm{x}}(\acute{\boldsymbol{u}}) \\ I_{\mathrm{y}}(\acute{\boldsymbol{u}}) \end{pmatrix} = \tag{24.59}$$

$$\begin{pmatrix}
\Sigma I_{\mathrm{x}}^2(\acute{\boldsymbol{u}})x^2 & \Sigma I_{\mathrm{x}}^2(\acute{\boldsymbol{u}})xy & \Sigma I_{\mathrm{x}}(\acute{\boldsymbol{u}})I_{\mathrm{y}}(\acute{\boldsymbol{u}})x^2 & \Sigma I_{\mathrm{x}}(\acute{\boldsymbol{u}})I_{\mathrm{y}}(\acute{\boldsymbol{u}})xy & \Sigma I_{\mathrm{x}}^2(\acute{\boldsymbol{u}})x & \Sigma I_{\mathrm{x}}(\acute{\boldsymbol{u}})I_{\mathrm{y}}(\acute{\boldsymbol{u}})x \\
\Sigma I_{\mathrm{x}}^2(\acute{\boldsymbol{u}})xy & \Sigma I_{\mathrm{x}}^2(\acute{\boldsymbol{u}})y^2 & \Sigma I_{\mathrm{x}}(\acute{\boldsymbol{u}})I_{\mathrm{y}}(\acute{\boldsymbol{u}})xy & \Sigma I_{\mathrm{x}}(\acute{\boldsymbol{u}})I_{\mathrm{y}}(\acute{\boldsymbol{u}})y^2 & \Sigma I_{\mathrm{x}}^2(\acute{\boldsymbol{u}})y & \Sigma I_{\mathrm{x}}(\acute{\boldsymbol{u}})I_{\mathrm{y}}(\acute{\boldsymbol{u}})y \\
\Sigma I_{\mathrm{x}}(\acute{\boldsymbol{u}})I_{\mathrm{y}}(\acute{\boldsymbol{u}})x^2 & \Sigma I_{\mathrm{x}}(\acute{\boldsymbol{u}})I_{\mathrm{y}}(\acute{\boldsymbol{u}})xy & \Sigma I_{\mathrm{y}}^2(\acute{\boldsymbol{u}})x^2 & \Sigma I_{\mathrm{y}}^2(\acute{\boldsymbol{u}})xy & \Sigma I_{\mathrm{x}}(\acute{\boldsymbol{u}})I_{\mathrm{y}}(\acute{\boldsymbol{u}})x & \Sigma I_{\mathrm{y}}^2(\acute{\boldsymbol{u}})x \\
\Sigma I_{\mathrm{x}}(\acute{\boldsymbol{u}})I_{\mathrm{y}}(\acute{\boldsymbol{u}})xy & \Sigma I_{\mathrm{x}}(\acute{\boldsymbol{u}})I_{\mathrm{y}}(\acute{\boldsymbol{u}})y^2 & \Sigma I_{\mathrm{y}}^2(\acute{\boldsymbol{u}})xy & \Sigma I_{\mathrm{y}}^2(\acute{\boldsymbol{u}})y^2(\acute{\boldsymbol{u}}) & \Sigma I_{\mathrm{x}}(\acute{\boldsymbol{u}})I_{\mathrm{y}}(\acute{\boldsymbol{u}})y & \Sigma I_{\mathrm{y}}^2(\acute{\boldsymbol{u}})y \\
\Sigma I_{\mathrm{x}}^2(\acute{\boldsymbol{u}})x & \Sigma I_{\mathrm{x}}^2(\acute{\boldsymbol{u}})y & \Sigma I_{\mathrm{x}}(\acute{\boldsymbol{u}})I_{\mathrm{y}}(\acute{\boldsymbol{u}})x & \Sigma I_{\mathrm{x}}(\acute{\boldsymbol{u}})I_{\mathrm{y}}(\acute{\boldsymbol{u}})y & \Sigma I_{\mathrm{x}}^2(\acute{\boldsymbol{u}}) & \Sigma I_{\mathrm{x}}(\acute{\boldsymbol{u}})I_{\mathrm{y}}(\acute{\boldsymbol{u}}) \\
\Sigma I_{\mathrm{x}}(\acute{\boldsymbol{u}})I_{\mathrm{y}}(\acute{\boldsymbol{u}})x & \Sigma I_{\mathrm{x}}(\acute{\boldsymbol{u}})I_{\mathrm{y}}(\acute{\boldsymbol{u}})y & \Sigma I_{\mathrm{y}}^2(\acute{\boldsymbol{u}})x & \Sigma I_{\mathrm{y}}^2(\acute{\boldsymbol{u}})y & \Sigma I_{\mathrm{x}}(\acute{\boldsymbol{u}})I_{\mathrm{y}}(\acute{\boldsymbol{u}}) & \Sigma I_{\mathrm{y}}^2(\acute{\boldsymbol{u}})
\end{pmatrix}. \tag{24.60}$$

Finally, the optimal parameter increment (see Eqn. (24.25)) is calculated as

$$\boldsymbol{q}_{\mathrm{opt}} = \left(a', b', c', d', t'_x, t'_y\right)^{\mathsf{T}} = \bar{\mathbf{H}}^{-1} \cdot \boldsymbol{\delta}_{\boldsymbol{p}} \tag{24.61}$$

or, equivalently, by solving $\mathbf{H} \cdot \boldsymbol{q}_{\mathrm{opt}} = \boldsymbol{\delta}_{\boldsymbol{p}}$ (see Eqn. (24.31)). For both approaches, no closed-form solution is possible but numerical methods must be used.

24.4.3 Projective Transformation

A projective transformation[13] can be expressed (for example) with homogeneous coordinates in the form

$$T_{\boldsymbol{p}}(\boldsymbol{x}) = \mathbf{M}_{\boldsymbol{p}} \cdot \boldsymbol{x} = \begin{pmatrix} 1+a & b & t_{\mathrm{x}} \\ c & 1+d & t_{\mathrm{y}} \\ e & f & 1 \end{pmatrix} \cdot \begin{pmatrix} x \\ y \\ 1 \end{pmatrix}, \tag{24.62}$$

with $n = 8$ parameters $\boldsymbol{p} = (p_0, \ldots, p_7) = (a, b, c, d, e, f, t_{\mathrm{x}}, t_{\mathrm{y}})$. Again the null parameter vector corresponds to the identity transformation. In this case, the results need to be converted back to non-homogeneous coordinates (see Ch. 21, Sec. 21.1.2), which yields the transformation's effective (nonlinear) component functions

$$T_{\mathrm{x},\boldsymbol{p}}(\boldsymbol{x}) = \frac{(1+a) \cdot x + b \cdot y + t_{\mathrm{x}}}{e \cdot x + f \cdot y + 1} = \frac{\alpha}{\gamma}, \tag{24.63}$$

$$T_{\mathrm{y},\boldsymbol{p}}(\boldsymbol{x}) = \frac{c \cdot x + (1+d) \cdot y + t_{\mathrm{y}}}{e \cdot x + f \cdot y + 1} = \frac{\beta}{\gamma}, \tag{24.64}$$

with $\boldsymbol{x} = (x, y)$ and

$$\alpha = (1+a) \cdot x + b \cdot y + t_{\mathrm{x}}, \tag{24.65}$$

$$\beta = c \cdot x + (1+d) \cdot y + t_{\mathrm{y}}, \tag{24.66}$$

$$\gamma = e \cdot x + f \cdot y + 1. \tag{24.67}$$

In this case, the associated Jacobian matrix for position $\boldsymbol{x} = (x, y)$,

$$\begin{aligned} \mathbf{J}_{T_{\boldsymbol{p}}}(\boldsymbol{x}) &= \begin{pmatrix} \frac{\partial T_{\mathrm{x},\boldsymbol{p}}}{\partial a} & \frac{\partial T_{\mathrm{x},\boldsymbol{p}}}{\partial b} & \frac{\partial T_{\mathrm{x},\boldsymbol{p}}}{\partial c} & \frac{\partial T_{\mathrm{x},\boldsymbol{p}}}{\partial d} & \frac{\partial T_{\mathrm{x},\boldsymbol{p}}}{\partial e} & \frac{\partial T_{\mathrm{x},\boldsymbol{p}}}{\partial f} & \frac{\partial T_{\mathrm{x},\boldsymbol{p}}}{\partial t_{\mathrm{x}}} & \frac{\partial T_{\mathrm{x},\boldsymbol{p}}}{\partial t_{\mathrm{y}}} \\ \frac{\partial T_{\mathrm{y},\boldsymbol{p}}}{\partial a} & \frac{\partial T_{\mathrm{y},\boldsymbol{p}}}{\partial b} & \frac{\partial T_{\mathrm{y},\boldsymbol{p}}}{\partial c} & \frac{\partial T_{\mathrm{y},\boldsymbol{p}}}{\partial d} & \frac{\partial T_{\mathrm{y},\boldsymbol{p}}}{\partial e} & \frac{\partial T_{\mathrm{y},\boldsymbol{p}}}{\partial f} & \frac{\partial T_{\mathrm{y},\boldsymbol{p}}}{\partial t_{\mathrm{x}}} & \frac{\partial T_{\mathrm{y},\boldsymbol{p}}}{\partial t_{\mathrm{y}}} \end{pmatrix}(\boldsymbol{x}) \\ &= \frac{1}{\gamma} \cdot \begin{pmatrix} x & y & 0 & 0 & -\frac{x \cdot \alpha}{\gamma} & -\frac{y \cdot \alpha}{\gamma} & 1 & 0 \\ 0 & 0 & x & y & -\frac{x \cdot \beta}{\gamma} & -\frac{y \cdot \beta}{\gamma} & 0 & 1 \end{pmatrix}, \end{aligned} \tag{24.68}$$

depends on both the position $\boldsymbol{x}$ as well as the transformation parameters $\boldsymbol{p}$. The setup for the resulting Hessian matrix $\mathbf{H}$ is analogous to Eqns. (24.58)–(24.61).

24.4.4 Concatenating Linear Transformations

The "inverse compositional" algorithm described in Sec. 24.3 requires the concatenation of geometric transformations (see Eqn. (24.33)). In

[13] See also Chapter 21, Sec. 21.1.4.

particular, if $T_{\boldsymbol{p}}, T_{\boldsymbol{q}}$ are *linear* transformations (in homogeneous coordinates, see Eqn. (24.62)), with associated transformation matrices $\mathbf{M}_{\boldsymbol{p}}$ and $\mathbf{M}_{\boldsymbol{q}}$ (such that $T_{\boldsymbol{p}}(\boldsymbol{x}) = \mathbf{M}_{\boldsymbol{p}} \cdot \boldsymbol{x}$ and $T_{\boldsymbol{q}}(\boldsymbol{x}) = \mathbf{M}_{\boldsymbol{q}} \cdot \boldsymbol{x}$, respectively), the matrix for the concatenated transformation,

$$T_{\boldsymbol{p}'}(\boldsymbol{x}) = (T_{\boldsymbol{p}} \circ T_{\boldsymbol{q}})(\boldsymbol{x}) = T_{\boldsymbol{q}}(T_{\boldsymbol{p}}(\boldsymbol{x})) \tag{24.69}$$

is simply the product of the original matrices, that is,

$$\mathbf{M}_{\boldsymbol{p}'} \cdot \boldsymbol{x} = \mathbf{M}_{\boldsymbol{q}} \cdot \mathbf{M}_{\boldsymbol{p}} \cdot \boldsymbol{x}. \tag{24.70}$$

The resulting parameter vector $\boldsymbol{p}'$ for the composite transformation $T_{\boldsymbol{p}'}$ can be simply extracted from the corresponding elements of the matrix $\mathbf{M}_{\boldsymbol{p}'}$ (see Eqn. (24.50) and Eqn. (24.62)), respectively.

24.5 Example

Figure 24.4 shows an example for using the classic Lucas-Kanade (forward-additive) matcher. Initially, a rectangular region Q is selected in the search image I, marked by the green rectangle in Fig. 24.4(a,b), which specifies the approximate position of the reference image. To create the (synthetic) reference image R, all four corners of the rectangle Q were perturbed randomly in x- and y-direction by Gaussian noise (with $\sigma = 2.5$) in x- and y-direction. The resulting quadrilateral Q' (red outline in Fig. 24.4(a,b)) specifies the region in image I where the reference image R was extracted by transformation and interpolation (see Fig. 24.4(d)). The matching process starts from the rectangle Q, which specifies the *initial* warp transformation T_{init}, given by the green rectangle (Q), while the real (but unknown) transformation corresponds to the red quadrilateral (Q'). Each iteration of the matcher updates the warp transformation T. The blue circles in Fig. 24.4(b) mark the corners of the back-projected reference frame under the changing transformation T; the radius of the circles corresponds to the remaining registration error between the reference image R and the current subimage of I.

Figure 24.4(e) shows the steepest-descent images $s_0, \ldots, s_7$ (see Eqn. (24.28)) for the first iteration. Each of these images is of the same size as R and corresponds to one of the 8 parameters $a, b, c, d, e, f, t_{\mathrm{x}}, t_{\mathrm{y}}$ of the projective warp transformation (see Eqn. (24.62)). The value $s_k(u, v)$ in a particular image s_k corresponds to the optimal change of the transformation parameter k with respect to the associated image position (u, v). The actual change of parameter k is calculated by averaging over all positions (u, v) of the reference image R.

The example demonstrates the robustness and fast convergence of the classic Lucas-Kanade matcher, which typically requires only 5–20 iterations. In this case, the matcher performed 7 iterations to converge (with convergence limit $\epsilon = 0.00001$). In comparison, the inverse-compositional matcher typically requires more iterations and is less tolerant to deviations of the initial warp transformation,

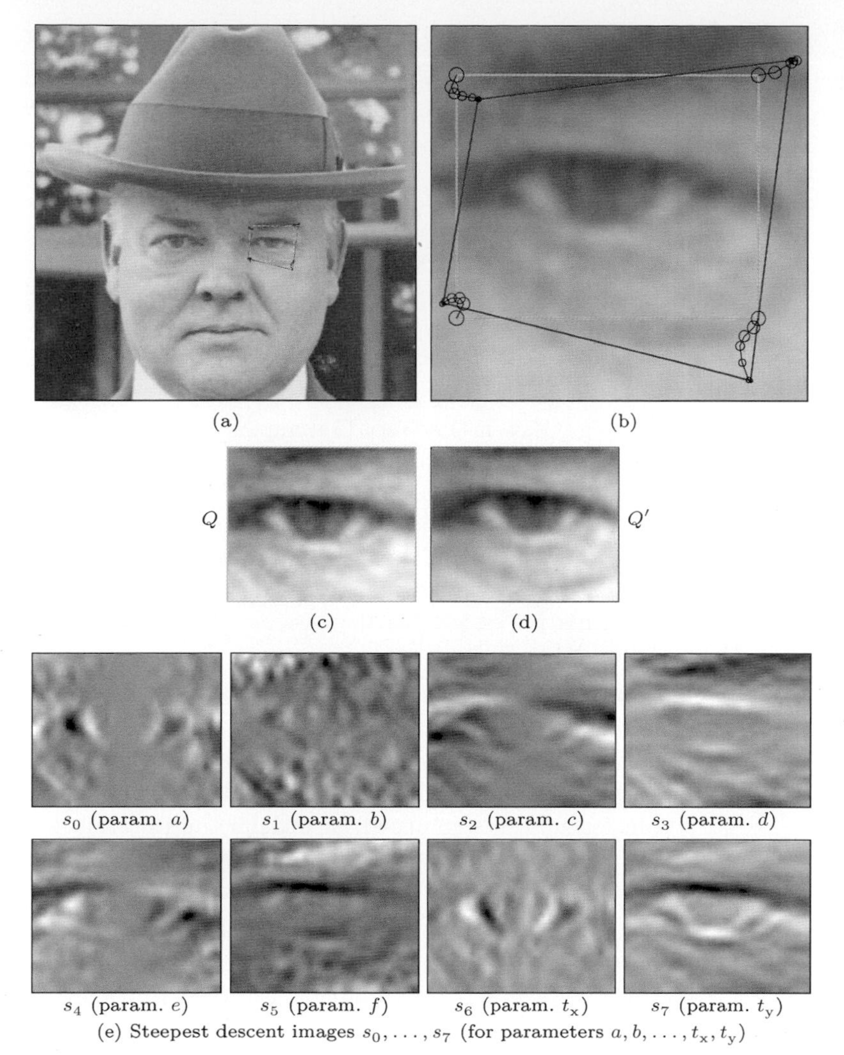

Fig. 24.4
Lucas-Kanade (forward-additive) matcher with projective warp transformation. Original image I (a b); the initial warp transformation T_{init} is visualized by the green rectangle Q, which corresponds to the subimage shown in (c). The actual reference image R (d) has been extracted from the red quadrilateral Q' (by transformation and interpolation). The blue circles mark the corners of the back-projected reference image under the changing transformation $T_{\boldsymbol{p}}$. The radius of each circle corresponds to the registration error between the transformed reference image R and the currently overlapping part of the search image I. The *steepest-descent images* $s_0, \ldots, s_7$ (one for each of the 8 parameters $a, b, c, d, e, f, t_x, t_y$ of the projective transformation) for the first iteration are shown in (e). These images are of the same size as the reference image R.

that is, has a smaller convergence range than the additive-forward algorithm.[14]

24.6 Java Implementation

The algorithms described in this chapter have been implemented in Java, with the source code available as part of the `imagingbook`[15] library on the book's accompanying website. As usual, most Java variables and methods in the online code have been named similarly to the identifiers used in the text for easier understanding.

[14] In fact, the inverse-compositional algorithm does not converge with this particular example.

[15] Package `imagingbook.pub.lucaskanade`.

LucasKanadeMatcher (class)

This is the (abstract) super-class of the concrete matchers (ForwardAdditiveMatcher, InverseCompositionalMatcher) described further. It defines a static inner class Parameters[16] with public parameter fields such as

tolerance ($=\epsilon$, default 0.00001),
maxIterations ($=i_{\text{max}}$, default 100).

In addition, class LucasKanadeMatcher itself provides the following public methods:

LinearMapping getMatch (ProjectiveMapping T)
: Performs a complete match on the given image pair I, R (required by the sub-class constructors), with T used as the initial geometric transformation. The transformation object T may be of any subtype of ProjectiveMapping,[17] including Translation and AffineMapping. The method returns a new transformation object for the optimal match, or null if the matcher did not converge.

ProjectiveMapping iterateOnce (ProjectiveMapping T)
: This method performs a single matching iteration with the current warp transformation T. It is typically invoked repeatedly after an initial call to initializeMatch(). The updated warp transformation is returned, or null if the iteration was unsuccessful (e.g., if the Hessian matrix could not be inverted).

boolean hasConverged ()
: Returns true if (and only if) the minimization criteria (specified by the tolerance parameter) have been reached. This method is typically used to terminate the optimization loop after calling iterateOnce().

Point2D[] getReferencePoints ()
: Returns the four corner points of the bounding rectangle of the reference image R, centered at the origin. All warp transformations (including Tinit and Tp) refer to these coordinates. Note that the returned point coordinates are generally non-integer values; for example, for a reference image size 11×8, the reference corner points are $A = (-5, -3.5)$, $B = (5, -3.5)$, $C = (5, 3.5)$, and $D = (-5, 3.5)$ (see Fig. 24.5).

ProjectiveMapping getReferenceMappingTo (Point2D[] Q)
: Calculates the (linear) geometric transformation between the reference image R (centered at the origin) and the quadrilateral specified by the point sequence Q. The type of the returned mapping depends on the number of points in Q (max. 4).

double getRmsError ()
: Returns the RMS error between images I and R for the most recent iteration (usually called after iterateOnce()).

[16] See the usage example in Prog. 24.1.

[17] Class ProjectiveMapping is described in Chapter 21, Sec. 21.1.4.

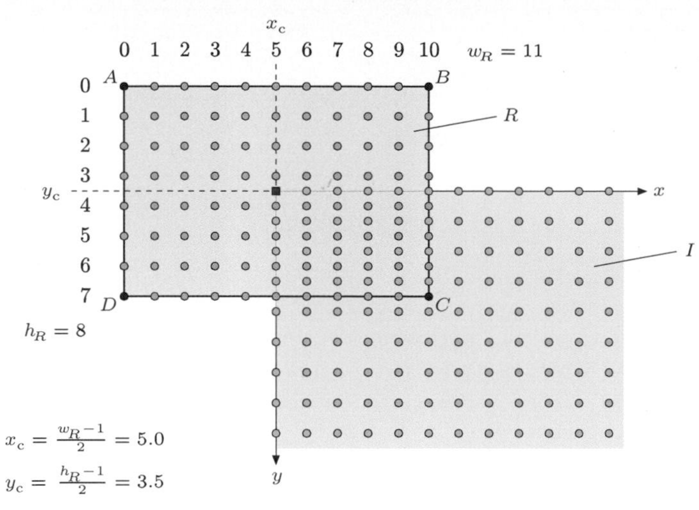

Fig. 24.5
Reference coordinates. The center of the reference image R is aligned with the origin of the search image I (red square), which is taken as the absolute origin. Image samples (indicated by round dots) are assumed to be located at integer positions. In this example, the reference image R is of size $w_R = 11$ and $h_R = 8$, thus the center coordinates are $x_c = 5.0$ and $y_c = 3.5$. In the x/y coordinate frame of I (i.e., absolute coordinates), the four corners of R's bounding rectangle are $A = (-5, -3.5)$, $B = (5, -3.5)$, $C = (5, 3.5)$ and $D = (-5, 3.5)$. All warp transformations refer to these reference points (cf. Figs. 24.2 and 24.3).

`LucasKanadeForwardMatcher` **(class)**

This sub-class of `LucasKanadeMatcher` implements the Lucas-Kanade ("forward-additive") algorithm, as outlined in Alg. 24.1. It provides the aforementioned methods for `LucasKanadeMatcher` and two constructors:

`LucasKanadeForwardMatcher (FloatProcessor I, FloatProcessor R)`
Here `I` is the search image, `R` is the (smaller) reference image. It creates a new instance of `LucasKanadeForwardMatcher` using default parameter values.

`LucasKanadeForwardMatcher (FloatProcessor I, FloatProcessor R, Parameters params)`
Creates a new instance of type `LucasKanadeForwardMatcher` using the specific settings in `params`.

`LucasKanadeInverseMatcher` **(class)**

This sub-class of `LucasKanadeMatcher` implements the "inverse compositional" algorithm, as described in Alg. 24.2. It provides the same methods and constructors as class `LucasKanadeForwardMatcher`:

`LucasKanadeInverseMatcher (FloatProcessor I, FloatProcessor R)`.

`LucasKanadeInverseMatcher (FloatProcessor I, FloatProcessor R, Parameters params)`.

24.6.1 Application Example

The code example in Prog. 24.1 demonstrates the use of the Lucas-Kanade API. The ImageJ plugin is applied to the search image I (the current image) and requires a rectangular ROI to be selected, which is taken as the initial guess for the match region. The reference image is created synthetically by extracting a warped sub-image

Prog. 24.1
Lucas-Kanade code example (ImageJ plugin). This plugin is applied to the search image (I) and assumes that a rectangular ROI is selected whose bounding rectangle and corner points (`Q`) are obtained in lines 22–27. The search image `I` is copied from the current image (as a `FloatProcessor` object) in line 19. The size of the reference image `R` (created in line 24) is defined by the ROI rectangle, whose corner points `Q` also determine the initial parameters of the geometric transformation `Tinit` (line 27 and 37, respectively). The synthetic reference image `R` (with the same size as the ROI) is extracted from the search image by warping from a quadrilateral (`QQ`), which is obtained by randomly perturbing the corner points of the selected ROI (lines 28–29). A new matcher object is created in lines 32–33, in this case of type `LucasKanadeForwardMatcher` (alternatively, `LucasKanadeInverseMatcher` could have been used). The actual match operation is performed in lines 40–44. It consists of a simple `do-while` loop which is terminated if either, the transformation `T` becomes invalid (`null`), the matcher has converged or the maximum number of iterations has been reached. Alternatively, lines 40–44 could have been replaced by the statement `T = matcher.getMatch(Tinit)`. If the matcher has converged, the final transformation `Tp` maps to the best-matching sub-image of `I`.

```
1 import ...
2
3 public class LucasKanade_Demo implements PlugInFilter {
4
5   static int maxIterations = 100;
6
7   public int setup(String args, ImagePlus img) {
8     return DOES_8G + ROI_REQUIRED;
9   }
10
11  public void run(ImageProcessor ip) {
12    Roi roi = img.getRoi();
13    if (roi != null && roi.getType() != Roi.RECTANGLE) {
14      IJ.error("Rectangular selection required!)");
15      return;
16    }
17
18    // Step 1: create the search image I:
19    FloatProcessor I = ip.convertToFloatProcessor();
20
21    // Step 2: create the (empty) reference image R:
22    Rectangle roiR = roi.getBounds();
23    FloatProcessor R =
24        new FloatProcessor(roiR.width, roiR.height);
25
26    // Step 3: perturb the rectangle Q to Q' to extract reference image R:
27    Point2D[] Q  = getCornerPoints(roiR); // = Q
28    Point2D[] QQ = perturbGaussian(Q); // = Q'
29    (new ImageExtractor(I)).extractImage(R, QQ);
30
31    // Step 4: create the Lucas-Kanade matcher (forward or inverse):
32    LucasKanadeMatcher matcher =
33        new LucasKanadeForwardMatcher(I, R);
34
35    // Step 5: calculate the initial mapping T_init:
36    ProjectiveMapping Tinit =
37        matcher.getReferenceMappingTo(Q);
38
39    // Step 6: initialize and run the matching loop:
40    ProjectiveMapping T = Tinit;
41    do {
42      T = matcher.iterateOnce(T);
43    } while (T != null && !matcher.hasConverged() &&
44          matcher.getIteration() < maxIterations);
45
46    // Step 7: evaluate the result:
47    if (T == null || !matcher.hasConverged()) {
48      IJ.log("no match found!");
49      return;
50    }
51    else {
52      ProjectiveMapping Tfinal = T;
53      ...
54  }
55
56 }
```

of I from a random quadrilateral around the selected ROI.[18] The required geometric transformations (such as `ProjectiveMapping`, `AffineMapping`, `Translation` etc.) are described in Chapter 21, Sec. 21.1.

The example demonstrates how the Lucas-Kanade matcher is initialized and called repeatedly inside the optimization loop using a projective transformation. This usage mode is specifically intended for testing purposes, since it allows to retrieve the state of the matcher after every iteration. The same result could be obtained by replacing the whole loop (lines 40–44 in Prog. 24.1) with the single instruction

```
ProjectiveMapping T = matcher.getMatch(Tinit);
```

Moreover, in line 33, the `LucasKanadeForwardMatcher` could be replaced by an instance of `LucasKanadeInverseMatcher` without any additional changes. For further details, see the complete source code on the book's website.

24.7 Exercises

Exercise 24.1. Determine the general structure of the Hessian matrix for the projective transformation (see Sec. 24.4.3), analogous to the affine transformation in Eqns. (24.58)–(24.60).

Exercise 24.2. Create comparative statistics of the convergence properties of the classes `ForwardAdditiveMatcher` and `InverseCompositionalMatcher` by evaluating the number of iterations required including the percentage of failures. Use a test scenario with randomly perturbed reference regions as shown in Prog. 24.1.

Exercise 24.3. It is sometimes suggested to refine the warp transformation step-by-step instead of using the full transformation for the whole matching process. For example, one could first match with a pure translation model, then—starting from the result of the first match—switch to an affine transformation model, and eventually apply a full projective transformation. Explore this idea and find out whether this can yield a more robust matching process.

Exercise 24.4. Adapt the 2D Lucas-Kanade method described in Sec. 24.2 for the registration of discrete *1D* signals under shifting and scaling. Given is a search signal $I(u)$, for $u = 0, \ldots, M_I - 1$, and a reference signal $R(u)$, for $u = 0, \ldots, M_R - 1$. It is assumed that I contains a transformed version of R, which is specified by the mapping $T_{\boldsymbol{p}}(x) = s \cdot x + t$, with the two unknown parameters $\boldsymbol{p} = (s, t)$. A practical application could be the registration of neighboring image lines under perspective distortion.

Exercise 24.5. Use the Lucas-Kanade matcher to design a tracker that follows a given reference patch through a sequence of N images. Hint: In ImageJ, an image sequence (AVI-video or multi-frame TIFF)

[18] The class `ImageExtractor`, used to extract the warped sub-image, is part of the imagingbook library (package `imagingbook.lib.image`).

can be imported as an `ImageStack` and simply processed frame-by-frame. Select the original reference patch in the first frame of the image sequence and use its position to calculate the initial warp transformation to find a match in the second image. Subsequently, take the match obtained in the second image as the initial transformation for the third image, etc. Consider two approaches: (a) use the initial patch as the reference image for *all* frames of the sequence or (b) extract a new reference image for each pair of frames.

25

Scale-Invariant Feature Transform (SIFT)

Many real applications require the localization of reference positions in one or more images, for example, for image alignment, removing distortions, object tracking, 3D reconstruction, etc. We have seen that corner points[1] can be located quite reliably and independent of orientation. However, typical corner detectors only provide the position and strength of each candidate point, they do not provide any information about its characteristic or "identity" that could be used for matching. Another limitation is that most corner detectors only operate at a particular scale or resolution, since they are based on a rigid set of filters.

This chapter describes the *Scale-Invariant Feature Transform* (SIFT) technique for local feature detection, which was originally proposed by D. Lowe [152] and has since become a "workhorse" method in the imaging industry. Its goal is to locate image features that can be identified robustly to facilitate matching in multiple images and image sequences as well as object recognition under different viewing conditions. SIFT employs the concept of "scale space" [151] to capture features at *multiple* scale levels or image resolutions, which not only increases the number of available features but also makes the method highly tolerant to scale changes. This makes it possible, for example, to track features on objects that move towards the camera and thereby change their scale continuously or to stitch together images taken with widely different zoom settings.

Accelerated variants of the SIFT algorithm have been implemented by streamlining the scale space calculation and feature detection or the use of GPU hardware [20, 90, 218].

In principle, SIFT works like a multi-scale corner detector with sub-pixel positioning accuracy and a rotation-invariant feature descriptor attached to each candidate point. This (typically 128-dimensional) feature descriptor summarizes the distribution of the gradient directions in a spatial neighborhood around the corresponding feature point and can thus be used like a "fingerprint". The main steps involved in the calculation of SIFT features are as follows:

[1] See Chapter 7.

1. Extrema detection in a Laplacian-of-Gaussian (LoG) scale space to locate potential interest points.
2. Key point refinement by fitting a continuous model to determine precise location and scale.
3. Orientation assignment by the dominant orientation of the feature point from the directions of the surrounding image gradients.
4. Formation of the feature descriptor by normalizing the local gradient histogram.

These steps are all described in the remaining parts of this chapter. There are several reasons why we explain the SIFT technique here at such great detail. For one, it is by far the most complex algorithm that we have looked at so far, its individual steps are carefully designed and delicately interdependent, with numerous parameters that need to be considered. A good understanding of the inner workings and limitations is thus important for successful use as well as for analyzing problems if the results are not as expected.

25.1 Interest Points at Multiple Scales

The first step in detecting interest points is to find locations with stable features that can be localized under a wide range of viewing conditions and different scales. In the SIFT approach, interest point detection is based on Laplacian-of-Gaussian (LoG) filters, which respond primarily to distinct bright blobs surrounded by darker regions, or vice versa. Unlike the filters used in popular corner detectors,[2] LoG filters are *isotropic*, i.e., insensitive to orientation. To locate interest points over multiple scales, a scale space representation of the input image is constructed by recursively smoothing the image with a sequence of small Gaussian filters. The difference between the images in adjacent scale layers is used to approximate the LoG filter at each scale. Interest points are finally selected by finding the local maxima in the 3D LoG scale space.

25.1.1 The LoG Filter

In this section, we first outline LoG filters and the basic construction of a Gaussian scale space, followed by a detailed description of the actual implementation and the parameters used in the SIFT approach.

The LoG is a so-called *center-surround* operator, which most strongly responds to isolated local intensity peaks, edge, and corner-like image structures. The corresponding filter kernel is based on the second derivative of the Gaussian function, as illustrated in Fig. 25.1 for the 1D case. The 1D Gaussian function of width σ is defined as

$$G_\sigma(x) = \frac{1}{\sqrt{2\pi} \cdot \sigma} \cdot e^{-\frac{x^2}{2\sigma^2}} \qquad (25.1)$$

and its *first* derivative is

[2] See Chapter 7.

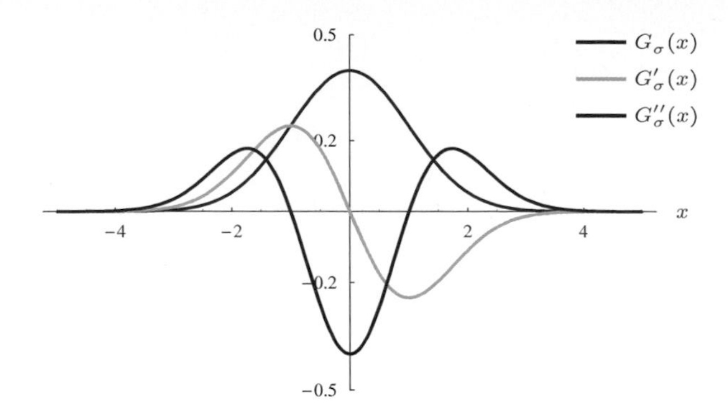

Fig. 25.1
1D Gaussian function $G_\sigma(x)$ with $\sigma = 1$ (black), its first derivative $G'_\sigma(x)$ (green) and second derivative $G''_\sigma(x)$ (blue).

$$G'_\sigma(x) = \frac{\mathrm{d}G_\sigma}{\mathrm{d}\,x}(x) = -\frac{x}{\sqrt{2\pi}\cdot\sigma^3}\cdot e^{-\frac{x^2}{2\sigma^2}}. \tag{25.2}$$

Analogously, the *second* derivative of the 1D Gaussian is

$$G''_\sigma(x) = \frac{\mathrm{d}^2 G_\sigma}{\mathrm{d}\,x^2}(x) = \frac{x^2-\sigma^2}{\sqrt{2\pi}\cdot\sigma^5}\cdot e^{-\frac{x^2}{2\sigma^2}}. \tag{25.3}$$

The *Laplacian* (denoted ∇^2) of a continuous, 2D function $f(x, y)$ is defined as the sum of the second partial derivatives for the x- and y-directions, traditionally written as

$$\left(\nabla^2 f\right)(x,y) = \frac{\partial^2 f}{\partial x^2}(x,y) + \frac{\partial^2 f}{\partial y^2}(x,y). \tag{25.4}$$

Note that, unlike the *gradient*[3] of a 2D function, the result of the Laplacian is not a vector but a *scalar* quantity. Its value is invariant against rotations of the coordinate system, that is, the Laplacian operator has the important property of being *isotropic*.

By applying the *Laplacian* operator to a rotationally symmetric 2D Gaussian,

$$G_\sigma(x,y) = \frac{1}{2\pi\cdot\sigma^2}\cdot e^{-\frac{x^2+y^2}{2\sigma^2}} \tag{25.5}$$

with identical widths $\sigma = \sigma_x = \sigma_y$ in the x/y directions (see Fig. 25.2(a)), we obtain the LoG function

$$\begin{aligned} L_\sigma(x,y) = \left(\nabla^2 G_\sigma\right)(x,y) &= \frac{\partial^2 G_\sigma}{\partial x^2}(x,y) + \frac{\partial^2 G_\sigma}{\partial y^2}(x,y) \\ &= \frac{(x^2-\sigma^2)}{2\pi\cdot\sigma^6}\cdot e^{-\frac{x^2+y^2}{2\cdot\sigma^2}} + \frac{(y^2-\sigma^2)}{2\pi\cdot\sigma^6}\cdot e^{-\frac{x^2+y^2}{2\cdot\sigma^2}} \\ &= \frac{1}{\pi\cdot\sigma^4}\cdot\left(\frac{x^2+y^2-2\sigma^2}{2\cdot\sigma^2}\right)\cdot e^{-\frac{x^2+y^2}{2\cdot\sigma^2}}, \end{aligned} \tag{25.6}$$

as shown in Fig. 25.2(b). The continuous LoG function in Eqn. (25.6) has the absolute value integral

$$\int_{-\infty}^{\infty}\int_{-\infty}^{\infty} |L_\sigma(x,y)|\,\mathrm{d}x\,\mathrm{d}y \;=\; \frac{4}{\sigma^2 e}, \tag{25.7}$$

[3] See Chapter 6, Sec. 6.2.1.

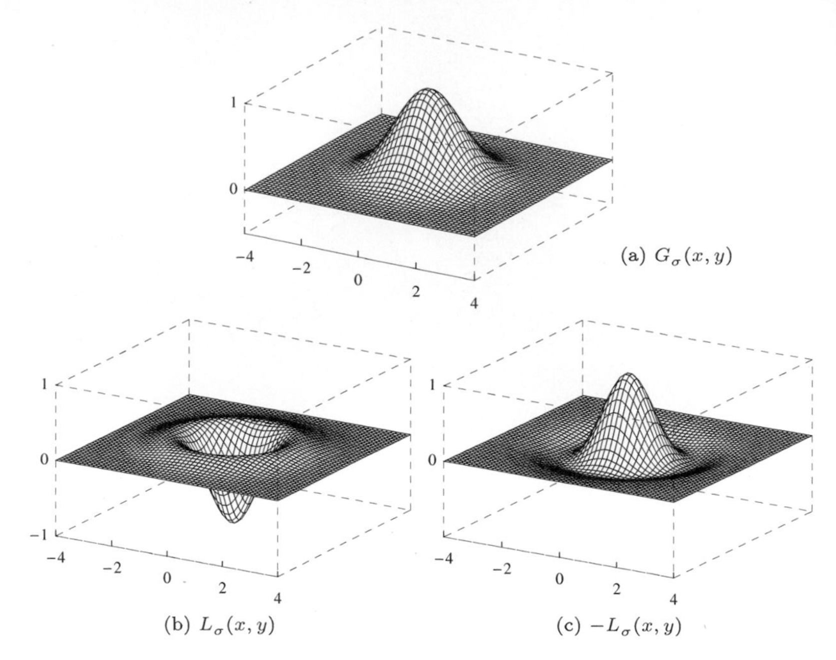

Fig. 25.2
2D Gaussian and LoG. Gaussian function $G_\sigma(x, y)$ with $\sigma = 1$ (a); the corresponding LoG function $L_\sigma(x, y)$ in (b), and the inverted function ("Mexican hat" or "Sombrero" kernel) $-L_\sigma(x, y)$ in (c). For illustration, all three functions are normalized to an absolute value of 1 at the origin.

and zero average, that is,

$$\int_{-\infty}^{\infty}\int_{-\infty}^{\infty} L_\sigma(x, y)\,\mathrm{d}x\,\mathrm{d}y = 0. \tag{25.8}$$

When used as the kernel of a linear filter,[4] the LoG responds maximally to circular spots that are *darker* than the surrounding background and have a radius of approximately σ.[5] Blobs that are *brighter* than the surrounding background are enhanced by filtering with the negative LoG kernel, that is, $-L_\sigma$, which is often referred to as the "Mexican hat" or "Sombrero" filter (see Fig. 25.2). Both types of blobs can be detected simultaneously by simply taking the absolute value of the filter response (see Fig. 25.3).

Since the LoG function is based on derivatives, its magnitude strongly depends on the steepness of the Gaussian slope, which is controlled by σ. To obtain responses of comparable magnitude over multiple scales, a *scale normalized* LoG kernel can be defined in the form [151]

$$\hat{L}_\sigma(x, y) = \sigma^2 \cdot \left(\nabla^2 G_\sigma\right)(x, y) = \sigma^2 \cdot L_\sigma(x, y) \tag{25.9}$$

$$= \frac{1}{\pi\sigma^2} \cdot \left(\frac{x^2 + y^2 - 2\sigma^2}{2\sigma^2}\right) \cdot e^{-\frac{x^2+y^2}{2\sigma^2}}. \tag{25.10}$$

[4] To produce a sufficiently accurate discrete LoG filter kernel, the support radius should be set to at least 4σ (kernel diameter $\geq 8\sigma$).

[5] The LoG is often used as a model for early processes in biological vision systems [161], particularly to describe the center-surround response of receptive fields. In this model, an "on-center" cell is *stimulated* when the center of its receptive field is exposed to light, and is *inhibited* when light falls on its surround. Conversely, an "off-center" cell is stimulated by light falling on its surround. Thus filtering with the original LoG L_σ (Eqn. (25.6)) corresponds to the behavior of off-center cells, while the response to the negative LoG kernel $-L_\sigma$ is that of an on-center cell.

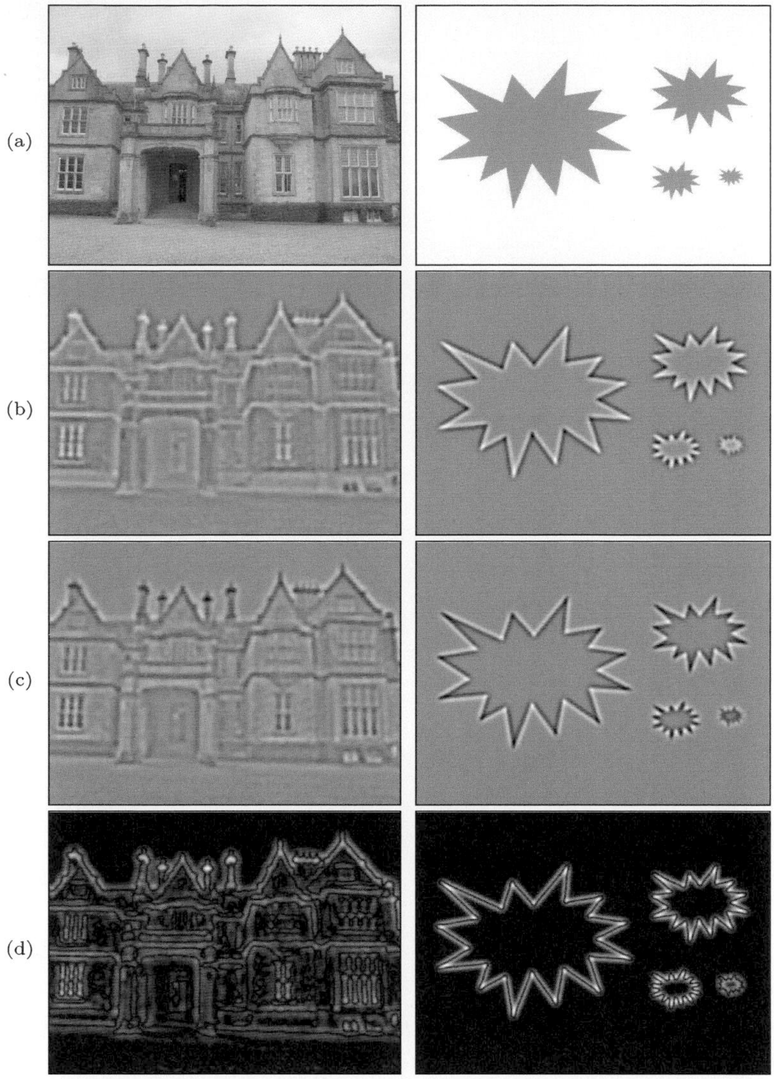

(a)

(b)

(c)

(d)

Fig. 25.3
Filtering with the LoG kernel (with $\sigma = 3$). Original images (a). A linear filter with the LoG kernel $L_\sigma(x, y)$ responds strongest to dark spots in a bright surround (b), while the inverted kernel $-L_\sigma(x, y)$ responds strongest to bright spots in a dark surround (c). In (b, c), zero values are shown as medium gray, negative values are dark, positive values are bright. The absolute value of (b) or (c) combines the responses from both dark and bright spots (d).

Note that the integral of this function,

$$\int_{-\infty}^{\infty}\int_{-\infty}^{\infty} \left|\hat{L}_\sigma(x,y)\right| \mathrm{d}x\, \mathrm{d}y = \frac{4}{e}, \qquad (25.11)$$

is constant and thus (unlike Eqn. (25.7)) independent of the scale parameter σ (see Fig. 25.4).

Approximating the LoG by the difference of two Gaussians (DoG)

Although the LoG is "quasi-separable" [113, 243] and can thus be calculated efficiently, the most common method for implementing the LoG filter is to approximate it by the *difference of two Gaussians* (DoG) of widths σ and $\kappa\sigma$, respectively, that is,

$$L_\sigma(x,y) \approx \lambda \cdot \big[\underbrace{G_{\kappa\sigma}(x,y) - G_\sigma(x,y)}_{= D_{\sigma,\kappa}(x,y)}\big] \qquad (25.12)$$

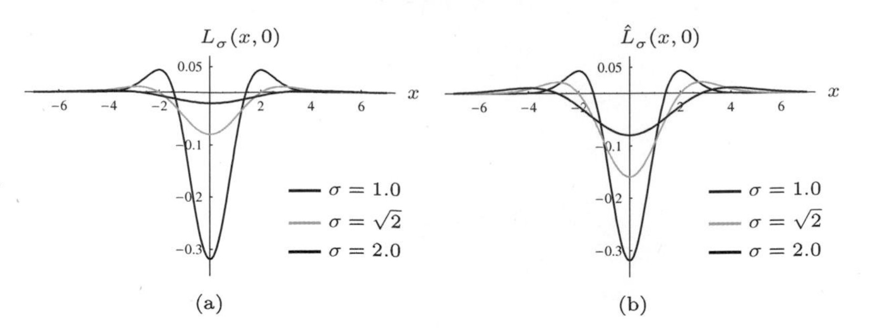

Fig. 25.4 Normalization of the LoG function. Cross section of LoG function $L_\sigma(x, y)$ as defined in Eqn. (25.6) (a); scale-normalized LoG (b) as defined in Eqn. (25.10). $\sigma = 1.0$ (black), $\sigma = \sqrt{2}$ (green), $\sigma = 2.0$ (blue). All three functions in (b) have the same absolute value integral that is independent of σ (see Eqn. (25.11)).

with the parameter $\kappa > 1$ specifying the relative width of the two Gaussians (defined in Eqn. (25.5)). Properly scaled (by some factor λ, see Eqn. (25.13)), the DOG function $D_{\sigma,\kappa}(x, y)$ approximates the LoG function $L_\sigma(x, y)$ in Eqn. (25.6) with arbitrary precision, as κ approaches 1 ($\kappa = 1$ being excluded, of course). In practice, values of κ in the range $1.1, \ldots, 1.3$ yield sufficiently accurate results. As an example, Fig. 25.5 shows the cross-section of the 2D DoG function for $\kappa = 2^{1/3} \approx 1.25992$.[6]

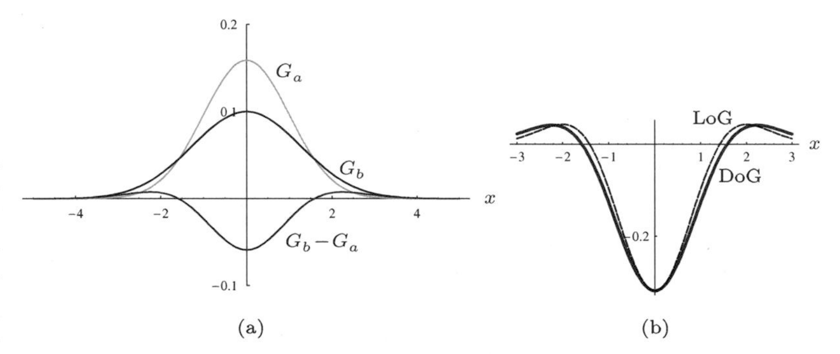

Fig. 25.5 Approximating the LoG by the DoG. The two original Gaussians, $G_a(x)$ with $\sigma_a = 1.0$ and $G_b(x)$ with $\sigma_b = \sigma_a \cdot \kappa = \kappa = 2^{1/3}$, shown by the green and blue curves, respectively (a). The red curve in (a) shows the DoG function $D_{\sigma,\kappa}(x, y) = G_b(x, y) - G_a(x, y)$ for $y = 0$. In (b), the dashed line shows the reference LoG function in comparison to the DoG (red). The DoG is scaled to match the magnitude of the LoG function.

The factor $\lambda \in \mathbb{R}$ in Eqn. (25.12) controls the magnitude of the DoG function; it depends on both the ratio κ and the scale parameter σ. To match the magnitude of the original LoG (Eqn. (25.6)) at the origin, it must be set to

$$\lambda = \frac{2\kappa^2}{\sigma^2 \cdot (\kappa^2 - 1)}. \tag{25.13}$$

Similarly, the *scale-normalized* LoG $\hat{L}_\sigma$ (Eqn. (25.10)) can be approximated by the DoG function $D_{\sigma,\kappa}$ (Eqn. (25.12)) as

$$\begin{aligned} \hat{L}_\sigma(x, y) &= \sigma^2 L_\sigma(x, y) \\ &\approx \underbrace{\sigma^2 \cdot \lambda}_{\hat{\lambda}} \cdot D_{\sigma,\kappa}(x, y) = \frac{2\kappa^2}{\kappa^2 - 1} \cdot D_{\sigma,\kappa}(x, y), \end{aligned} \tag{25.14}$$

[6] The factor $\kappa = 2^{1/3}$ originates from splitting the scale interval 2 (i.e., one scale octave) into 3 equal intervals, as described later on. Another factor mentioned frequently in the literature is 1.6, which, however, does not yield a satisfactory approximation. Possibly that value refers to the ratio of the *variances* σ_2^2/σ_1^2 and not the ratio of the *standard deviations* σ_2/σ_1.

Table 25.1
Scale space-related symbols used in this chapter.

Symbol	Description
$\mathcal{G}(x, y, \sigma)$	continuous Gaussian scale space
$\mathsf{G} = (\mathsf{G}_0, \ldots, \mathsf{G}_{K-1})$	discrete Gaussian scale space with K levels
G_k	single level in a discrete Gaussian scale space
$\mathsf{L} = (\mathsf{L}_0, \ldots, \mathsf{L}_{K-1})$	discrete LoG scale space with K levels
L_k	single level in a LoG scale space
$\mathsf{D} = (\mathsf{D}_0, \ldots, \mathsf{D}_{K-1})$	discrete DoG scale space with K levels
D_k	single level in a DoG scale space
$\mathbf{G} = (\mathbf{G}_0, \ldots, \mathbf{G}_{P-1})$	hierarchical Gaussian scale space with P octaves
$\mathbf{G}_p = (\mathbf{G}_{p,0}, \ldots, \mathbf{G}_{p,Q-1})$	octave in a hier. Gaussian scale space with Q levels
$\mathbf{G}_{p,q}$	single level in a hierarchical Gaussian scale space
$\mathbf{D} = (\mathbf{D}_0, \ldots, \mathbf{D}_{P-1})$	hierarchical DoG scale space with P octaves
$\mathbf{D}_p = (\mathbf{D}_{p,0}, \ldots, \mathbf{D}_{p,Q-1})$	octave in a hierarchical DoG scale space with Q levels
$\mathbf{D}_{p,q}$	single level in a hierarchical DoG scale space
$\mathsf{N}_c(i, j, k)$	$3\times3\times3$ neigborhood in DoG scale space
$\boldsymbol{k} = (p, q, u, v)$	discrete key point position in hierarchical scale space ($p, q, u, v \in \mathbb{Z}$)
$\boldsymbol{k}' = (p, q, x, y)$	continuous (refined) key point position ($x, y \in \mathbb{R}$)

with the factor $\hat{\lambda} = \sigma^2 \cdot \lambda = 2\kappa^2/(\kappa^2 - 1)$ being constant and therefore independent of the scale σ. Thus, as pointed out in [153], with a fixed scale increment κ, the DoG already approximates the scale-normalized LoG up to a constant factor, and thus no additional scaling is required to compare the magnitudes of the DoG responses obtained at different scales.[7]

In the SIFT approach, the DoG is used as an approximation of the (scale-normalized) LoG filter at multiple scales, based on a Gaussian scale space representation of the input image that is described next.[8]

25.1.2 Gaussian Scale Space

The concept of scale space [150] is motivated by the observation that real-world scenes exhibit relevant image features over a large range of sizes and, depending on the particular viewing situation, at various different scales. To relate image structures at different and unknown sizes, it is useful to represent the images simultaneously at different scale levels. The scale space representation of an image adds *scale* as a third coordinate (in addition to the two image coordinates). Thus the scale space is a 3D structure, which can be navigated not only along the x/y positions but also across different scale levels.

Continuous Gaussian scale space

The scale-space representation of an image at a particular scale level is obtained by filtering the image with a kernel that is parameterized to the desired scale. Because of its unique properties [11, 71], the most common type of scale space is based on successive filtering with Gaussian kernels. Conceptually, given a continuous, 2D function $F(x, y)$, its Gaussian scale space representation is a 3D function

[7] See Sec. E.4 in the Appendix for additional details.

[8] See Table 25.1 for a summary of the most important scale space-related symbols used in this chapter.

$$\mathcal{G}(x, y, \sigma) = (F * H^{G,\sigma})(x, y), \tag{25.15}$$

where $H^{G,\sigma} \equiv G_\sigma(x, y)$ is a 2D Gaussian kernel (see Eqn. (25.5)) with unit integral, and $*$ denotes the linear convolution over x, y. Note that $\sigma \geq 0$ serves as both the continuous scale parameter and the width of the corresponding Gaussian filter kernel.

A fully continuous Gaussian scale space $\mathcal{G}(x, y, \sigma)$ covers a 3D volume and represents the original function $F(x, y)$ at varying scales σ. For $\sigma = 0$, the Gaussian kernel $H^{G,0}$ has zero width, which makes it equivalent to an impulse or Dirac function $\delta(x, y)$.[9] This is the neutral element of linear convolution, that is,

$$\mathcal{G}(x, y, 0) = (F * H^{G,0})(x, y) = (F * \delta)(x, y) = F(x, y). \tag{25.16}$$

Thus the base level $\mathcal{G}(x, y, 0)$ of the Gaussian scale space is identical to the input function $F(x, y)$. In general (with $\sigma > 0$), the Gaussian kernel $H^{G,\sigma}$ acts as a low-pass filter with a cutoff frequency proportional to $1/\sigma$ (see Sec. E.3 in the Appendix), the maximum frequency (or bandwidth) of the original "signal" $F(x, y)$ being potentially unlimited.

Discrete Gaussian scale space

This is different for a *discrete* input function $I(u, v)$, whose bandwidth is implicitly limited to half the sampling frequency, as mandated by the sampling theorem to avoid aliasing.[10] Thus, in the discrete case, the lowest level $\mathcal{G}(x, y, 0)$ of the Gaussian scale space is not accessible! To model the implicit bandwidth limitations of the sampling process, the discrete input image $I(u, v)$ is assumed to be pre-filtered (with respect to the underlying continuous signal) with a Gaussian kernel of width $\sigma_s \geq 0.5$ [153], that is,

$$\mathcal{G}(u, v, \sigma_s) \equiv I(u, v). \tag{25.17}$$

Thus the discrete input image $I(u, v)$ is implicitly placed at some initial level σ_s of the Gaussian scale space, and the lower levels with $\sigma < \sigma_s$ are not available.

Any higher level $\sigma_h > \sigma_s$ of the Gaussian scale space can be derived from the original image $I(u, v)$ by filtering with Gaussian kernel $H^{G,\bar{\sigma}}$, that is,

$$\mathcal{G}(u, v, \sigma_h) = (I * H^{G,\bar{\sigma}})(u, v), \quad \text{with } \bar{\sigma} = \sqrt{\sigma_h^2 - \sigma_s^2}\,. \tag{25.18}$$

This is due to the fact that applying two Gaussian filters of widths σ_1 and σ_2, one after the other, is equivalent to a single convolution with a Gaussian kernel of width $\sigma_{1,2}$, that is,[11]

$$\left(I * H^{G,\sigma_1}\right) * H^{G,\sigma_2} \equiv I * H^{G,\sigma_{1,2}}, \tag{25.19}$$

[9] See Chapter 5, Sec. 5.3.4.
[10] See Chapter 18, Sec. 18.2.1.
[11] See Sec. E.1 in the Appendix for additional details on combining Gaussian filters.

with $\sigma_{1,2} = (\sigma_1^2 + \sigma_2^2)^{1/2}$. We define the *discrete Gaussian scale space* representation of an image I as a vector of M images, one for each scale level m:

$$\mathsf{G} = (\mathsf{G}_0, \mathsf{G}_1, \ldots, \mathsf{G}_{M-1}). \tag{25.20}$$

Associated with each level G_m is its absolute scale $\sigma_m > 0$, and each level G_m represents a blurred version of the original image, that is, $\mathsf{G}_m(u,v) \equiv \mathcal{G}(u,v,\sigma_m)$ in the notation introduced in Eqn. (25.15). The scale ratio between adjacent scale levels,

$$\Delta_\sigma = \frac{\sigma_{m+1}}{\sigma_m}, \tag{25.21}$$

is pre-defined and constant. Usually, Δ_σ is specified such that the absolute scale σ_m doubles with a given number of levels Q, called an *octave*. In this case, the resulting scale increment is $\Delta_\sigma = 2^{1/Q}$ with (typically) $Q = 3, \ldots, 6$.

In addition, a *base scale* $\sigma_0 > \sigma_s$ is specified for the initial level G_0, with σ_s denoting the smoothing of the discrete image implied by the sampling process, as discussed already. Based on empirical results, a base scale of $\sigma_0 = 1.6$ is recommended in [153] to achieve reliable interest point detection. Given Q and the base scale σ_0, the absolute scale at an arbitrary scale space level G_m is

$$\sigma_m = \sigma_0 \cdot \Delta_\sigma^m = \sigma_0 \cdot 2^{m/Q}, \tag{25.22}$$

for $m = 0, \ldots, M-1$.

As follows from Eqn. (25.18), each scale level G_m can be obtained directly from the discrete input image I by a filter operation

$$\mathsf{G}_m = I * H^{\mathrm{G},\bar{\sigma}_m}, \tag{25.23}$$

with a Gaussian kernel $H^{\mathrm{G},\bar{\sigma}_m}$ of width

$$\bar{\sigma}_m = \sqrt{\sigma_m^2 - \sigma_s^2} = \sqrt{\sigma_0^2 \cdot 2^{2m/Q} - \sigma_s^2}. \tag{25.24}$$

In particular, the initial scale space level G_0, (with the specified base scale σ_0) is obtained from the discrete input image I by linear filtering using a Gaussian kernel of width

$$\bar{\sigma}_0 = \sqrt{\sigma_0^2 - \sigma_s^2}. \tag{25.25}$$

Alternatively, using the relation $\sigma_m = \sigma_{m-1} \cdot \Delta_\sigma$ (from Eqn. (25.21)), the scale levels $\mathsf{G}_1, \ldots, \mathsf{G}_{M-1}$ could be calculated recursively from the base level G_0 in the form

$$\mathsf{G}_m = \mathsf{G}_{m-1} * H^{\mathrm{G},\sigma'_m}, \tag{25.26}$$

for $m > 0$, with a sequence of Gaussian kernels H^{G,σ'_m} of width

$$\sigma'_m = \sqrt{\sigma_m^2 - \sigma_{m-1}^2} = \sigma_0 \cdot 2^{m/Q} \cdot \sqrt{1 - 1/\Delta_\sigma^2}. \tag{25.27}$$

Table 25.2 lists the resulting kernel widths for $Q = 3$ levels per octave and base scale $\sigma_0 = 1.6$ over a scale range of 6 octaves. The value $\bar{\sigma}_m$ denotes the size of the Gaussian kernel required to compute the image at scale m from the discrete input image I (assumed to be sampled with $\sigma_s = 0.5$). σ'_m is the width of the Gaussian kernel to compute level m recursively from the previous level $m-1$. Apparently (though perhaps unexpectedly), the kernel size required for recursive filtering (σ'_m) grows at the same (exponential) rate as the absolute kernel size $\bar{\sigma}_m$.[12]

Table 25.2
Filter sizes required for calculating Gaussian scale levels G_m for the first 6 octaves. Each octave consists of $Q = 3$ levels, placed at increments of Δ_σ along the scale coordinate. The discrete input image I is assumed to be pre-filtered with σ_s. Column σ_m denotes the absolute scale at level m, starting with the specified base offset scale σ_0. $\bar{\sigma}_m$ is the width of the Gaussian filter required to calculate level G_m directly from the input image I. Values σ'_m are the widths of the Gaussian kernels required to calculate level G_m from the previous level G_{m-1}. Note that the width of the Gaussian kernels needed for recursive filtering (σ'_m) grows at the same exponential rate as the size of the direct filter ($\bar{\sigma}_m$).

m	σ_m	$\bar{\sigma}_m$	σ'_m
18	102.4000	102.3988	62.2908
17	81.2749	81.2734	49.4402
16	64.5080	64.5060	39.2408
15	51.2000	51.1976	31.1454
14	40.6375	40.6344	24.7201
13	32.2540	32.2501	19.6204
12	25.6000	25.5951	15.5727
11	20.3187	20.3126	12.3601
10	16.1270	16.1192	9.8102
9	12.8000	12.7902	7.7864
8	10.1594	10.1471	6.1800
7	8.0635	8.0480	4.9051
6	6.4000	6.3804	3.8932
5	5.0797	5.0550	3.0900
4	4.0317	4.0006	2.4525
3	3.2000	3.1607	1.9466
2	2.5398	2.4901	1.5450
1	2.0159	1.9529	1.2263
0	1.6000	1.5199	—

m ... linear scale index

σ_m ... absolute scale at level m (Eqn. (25.22))

$\bar{\sigma}_m$ ... relative scale at level m w.r.t. the original image (Eqn. (25.24))

σ'_m ... relative scale at level m w.r.t. the previous level $m-1$ (Eqn. (25.27))

$\sigma_s = 0.5$ (sampling scale)

$\sigma_0 = 1.6$ (base scale)

$Q = 3$ (levels per octave)

$\Delta_\sigma = 2^{1/Q} \approx 1.256$

At scale level $m = 16$ and absolute scale $\sigma_{16} = 1.6 \cdot 2^{16/3} \approx 64.5$, for example, the Gaussian filters required to compute G_{16} directly from the input image I has the width $\bar{\sigma}_{16} = (\sigma_{16}^2 - \sigma_s^2)^{1/2} = (64.5080^2 - 0.5^2)^{1/2} \approx 64.5$, while the filter to blur incrementally from the previous scale level has the width $\sigma'_{16} = (\sigma_{16}^2 - \sigma_{15}^2)^{1/2} = (64.5080^2 - 51.1976^2)^{1/2} \approx 39.2$. Since recursive filtering also tends to accrue numerical inaccuracies, this approach does not offer a significant advantage in general. Fortunately, the growth of the Gaussian kernels can be kept small by spatially sub-sampling after each octave, as will be described in Sec. 25.1.4.

The process of constructing a discrete Gaussian scale space using the same parameters as in Table 25.2 is illustrated in Fig. 25.6. Again the input image I is assumed to be pre-filtered at $\sigma_s = 0.5$ due to sampling and the absolute scale of the first level G_0 is set to $\sigma_0 = 1.6$. The scale ratio between successive levels is fixed at $\Delta_\sigma = 2^{1/3} \approx 1.25992$, that is, each octave spans three discrete scale levels. As shown in this figure, each scale level G_m can be calculated either directly from the input image I by filtering with a Gaussian of width $\bar{\sigma}_m$, or recursively from the previous level by filtering with σ'_m.

[12] The ratio of the kernel sizes $\bar{\sigma}_m/\sigma'_m$ converges to $\sqrt{1 - 1/\Delta_\sigma^2}$ (≈ 1.64 for $Q = 3$) and is thus practically constant for larger values of m.

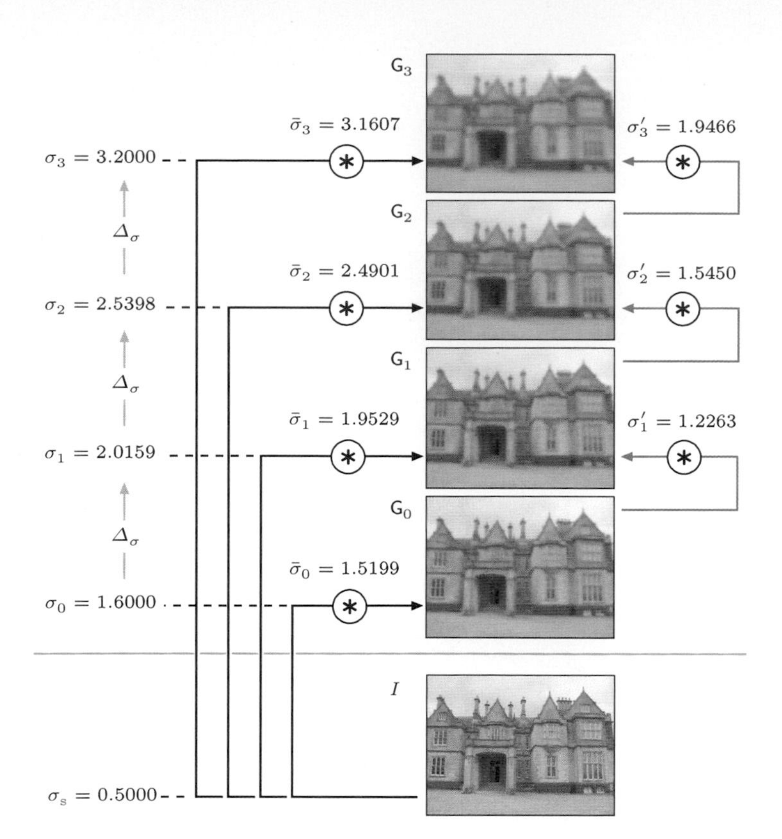

Fig. 25.6
Gaussian scale space construction (first four levels). Parameters are the same as in Table 25.2. The discrete input image I is assumed to be pre-filtered with a Gaussian of width $\sigma_s = 0.5$; the scale of the initial level (base scale offset) is set to $\sigma_0 = 1.6$. The discrete scale space levels $\mathsf{G}_0, \mathsf{G}_1, \ldots$ (at absolute scales $\sigma_0, \sigma_1, \ldots$) are slices through the continuous scale space. Scale levels can either be calculated by filtering directly from the discrete image I with Gaussian kernels of width $\bar{\sigma}_0, \bar{\sigma}_1, \ldots$ (blue arrows) or, alternatively, by recursively filtering with $\sigma'_1, \sigma'_2, \ldots$ (green arrows).

25.1.3 LoG/DoG Scale Space

Interest point detection in the SIFT approach is based on finding local maxima in the output of LoG filters over multiple scales. Analogous to the discrete Gaussian scale space described in Sec. 25.1.2, a LoG scale space representation of an image I can be defined as

$$\mathsf{L} = (\mathsf{L}_0, \mathsf{L}_1, \ldots, \mathsf{L}_{M-1}), \tag{25.28}$$

with levels $\mathsf{L}_m = I * H^{\mathrm{L},\sigma_m}$, where $H^{\mathrm{L},\sigma_m}(x, y) \equiv \hat{L}_{\sigma_m}(x, y)$ is a scale-normalized LoG kernel of width σ_m (see Eqn. (25.10)).

As demonstrated in Eqn. (25.12), the LoG kernel can be approximated by the the difference of two Gaussians whose widths differ by a certain ratio κ. Since pairs of adjacent scale layers in the Gaussian scale space are also separated by a fixed scale ratio, it is straightforward to construct a multi-scale DoG representation,

$$\mathsf{D} = (\mathsf{D}_0, \mathsf{D}_1, \ldots, \mathsf{D}_{M-2}) \tag{25.29}$$

from an existing Gaussian scale space $\mathsf{G} = (\mathsf{G}_0, \mathsf{G}_1, \ldots, \mathsf{G}_{M-1})$. The individual levels in the DoG scale space are defined as

$$\mathsf{D}_m = \hat{\lambda} \cdot (\mathsf{G}_{m+1} - \mathsf{G}_m) \approx \mathsf{L}_m, \tag{25.30}$$

for $m = 0, \ldots, M-2$. The constant factor $\hat{\lambda}$ (defined in Eqn. (25.14)) can be omitted in the aforementioned expression, as the relative width of the involved Gaussians,

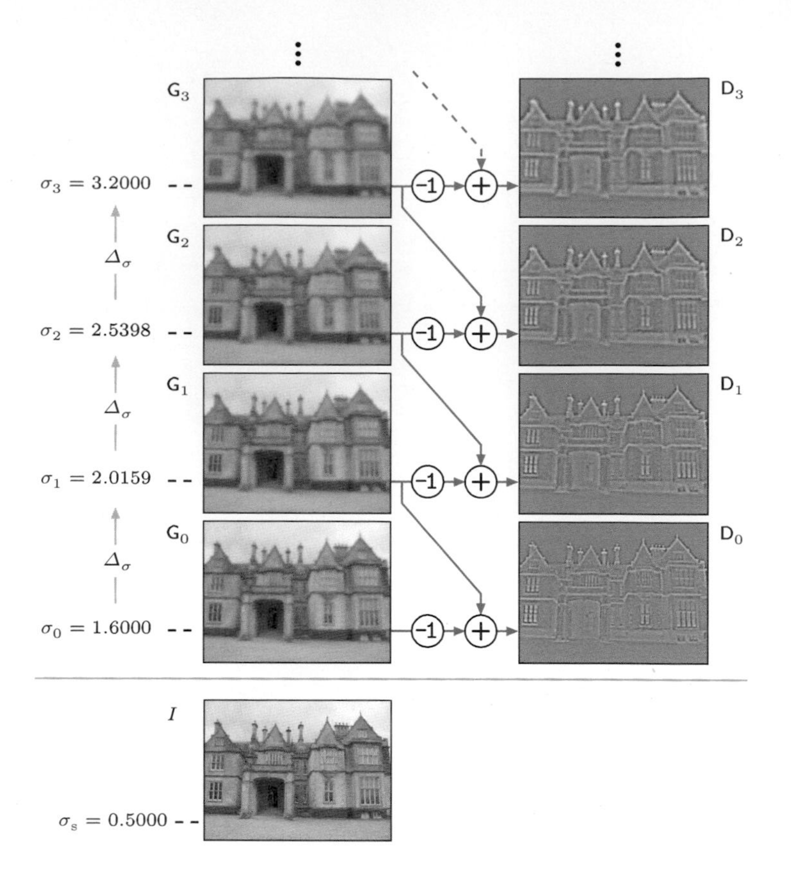

Fig. 25.7
DoG scale-space construction. The differences of successive levels $\mathsf{G}_0, \mathsf{G}_1, \ldots$ of the Gaussian scale space (see Fig. 25.6) are used to approximate a LoG scale space. Each DoG-level D_m is calculated as the point-wise difference $\mathsf{G}_{m+1} - \mathsf{G}_m$ between Gaussian levels G_{m+1} and G_m. The values in $\mathsf{D}_0, \ldots, \mathsf{D}_3$ are scale-normalized (see Eqn. (25.14)) and mapped to a uniform intensity range for viewing.

$$\kappa = \Delta_\sigma = \frac{\sigma_{m+1}}{\sigma_m} = 2^{1/Q}, \tag{25.31}$$

is simply the fixed scale ratio Δ_σ between successive scale space levels. Note that the DoG approximation does not require any additional normalization to approximate a scale-normalized LoG representation (see Eqns. 25.10 and 25.14). The process of calculating a DoG scale space from a discrete Gaussian scale space is illustrated in Fig. 25.7, using the same parameters as in Table 25.2 and Fig. 25.6.

25.1.4 Hierarchical Scale Space

Despite the fact that 2D Gaussian filter kernels are separable into 1D kernels,[13] the size of the required filter grows quickly with increasing scale, regardless if a direct or recursive approach is used (as shown in Table 25.2). However, each Gaussian filter operation reduces the bandwidth of the signal inversely proportional to the width of the kernel (see Sec. E.3 in the Appendix). If the image size is kept constant over all scales, the images become increasingly oversampled at higher scale levels. In other words, the sampling rate in a Gaussian scale space can be reduced with increasing scale without losing relevant signal information.

[13] See also Chapter 5, Sec. 5.3.3.

Octaves and sub-sampling (decimation)

In particular, doubling the scale cuts the bandwidth by half, that is, the signal at scale level 2σ has only half the bandwidth of the signal at level σ. An image signal at scale level 2σ of a Gaussian scale space thus shows only half the bandwidth of the same image at scale level σ. In a Gaussian scale space representation it is thus safe to down-sample the image to half the sample rate after each octave without any loss of information. This suggests a very efficient, "pyramid-like" approach for constructing a DoG scale space, as illustrated in Fig. 25.8.[14]

At the start (bottom) of each octave, the image is down-sampled to half the resolution, that is, each pixel in the new octave covers twice the distance of the pixels in the previous octave in every spatial direction. Within each octave, the same small Gaussian kernels can be used for successive filtering, since their relative widths (with respect to the original sampling lattice) also implicitly double at each octave. To describe these relations formally, we use

$$\mathbf{G} = (\mathbf{G}_0, \mathbf{G}_1, \ldots, \mathbf{G}_{P-1}) \tag{25.32}$$

to denote a *hierarchical Gaussian scale space* consisting of P octaves. Each octave

$$\mathbf{G}_p = \left(\mathbf{G}_{p,0}, \mathbf{G}_{p,1}, \ldots, \mathbf{G}_{p,Q}\right), \tag{25.33}$$

consists of $Q+1$ scale levels $\mathbf{G}_{p,q}$, where $p \in [0, P-1]$ is the octave index and $q \in [0, Q]$ is the level index within the containing octave $\mathbf{G}_p$. With respect to *absolute* scale, a level $\mathbf{G}_{p,q} = \mathbf{G}_p(q)$ in the hierarchical Gaussian scale space corresponds to the level G_m in the non-hierarchical Gaussian scale space (see Eqn. (25.20)) with index

$$m = Q \cdot p + q. \tag{25.34}$$

As follows from Eqn. (25.22), the *absolute scale* at level $\mathbf{G}_{p,q}$ then is

$$\begin{aligned} \sigma_{p,q} &= \sigma_m = \sigma_0 \cdot \Delta_\sigma^m = \sigma_0 \cdot 2^{m/Q} \\ &= \sigma_0 \cdot 2^{(Qp+q)/Q} = \sigma_0 \cdot 2^{p+q/Q}, \end{aligned} \tag{25.35}$$

where $\sigma_0 = \sigma_{0,0}$ denotes the predefined base scale offset (e.g., $\sigma_0 = 1.6$ in Table 25.2). In particular, the absolute scale of the base level $\mathbf{G}_{p,0}$ of *any* octave $\mathbf{G}_p$ is

$$\sigma_{p,0} = \sigma_0 \cdot 2^p. \tag{25.36}$$

The **decimated scale** $\dot{\sigma}_{p,q}$ is the absolute scale $\sigma_{p,q}$ (Eqn. (25.35)) expressed in the coordinate units of octave $\mathbf{G}_p$, that is,

$$\dot{\sigma}_{p,q} = \dot{\sigma}_q = \sigma_{p,q} \cdot 2^{-p} = \sigma_0 \cdot 2^{p+q/Q} \cdot 2^{-p} = \sigma_0 \cdot 2^{q/Q}. \tag{25.37}$$

Note that the decimated scale $\dot{\sigma}_{p,q}$ is independent of the octave index p and therefore $\dot{\sigma}_{p,q} \equiv \dot{\sigma}_q$, for any level index q.

[14] Successive reduction of image resolution by sub-sampling is the core concept of "image pyramid" methods [41].

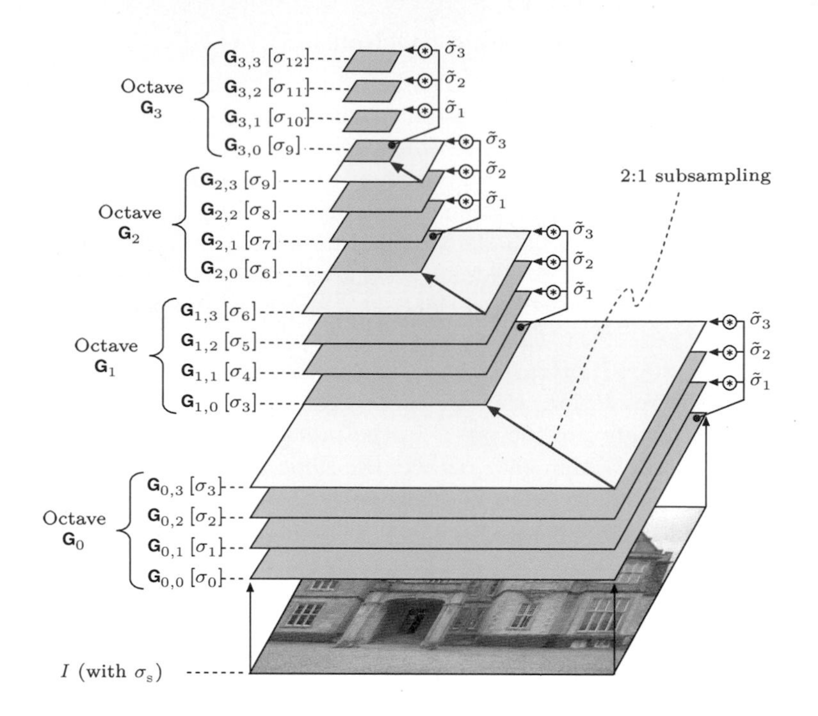

Fig. 25.8
Hierarchical Gaussian scale space. Each octave extends over $Q = 3$ scale steps. The base level $\mathbf{G}_{p,0}$ of each octave $p > 0$ is obtained by 2:1 sub-sampling of the top level $\mathbf{G}_{p-1,3}$ of the next-lower octave. At the transition between octaves, the resolution (image size) is cut in half in the x- and y-direction. The absolute scale at octave level $\mathbf{G}_{p,q}$ is σ_m, with $m = Qp + q$. Within each octave, the same set of Gaussian kernels ($\tilde{\sigma}_1$, $\tilde{\sigma}_2$, $\tilde{\sigma}_3$) is used to calculate the following levels from the octave's base level $\mathbf{G}_{p,0}$.

From the octave's base level $\mathbf{G}_{p,0}$, the subsequent levels in the same octave can be calculated by filtering with relatively small Gaussian kernels. The size of the kernel needed to calculate scale-level $\mathbf{G}_{p,q}$ from the octave's base level $\mathbf{G}_{p,0}$ is obtained from the corresponding decimated scales (Eqn. (25.37)) as

$$\tilde{\sigma}_{p,q} = \sqrt{\dot{\sigma}_{p,q}^2 - \dot{\sigma}_{p,0}^2} = \sqrt{(\sigma_0 \cdot 2^{q/Q})^2 - \sigma_0^2} = \sigma_0 \cdot \sqrt{2^{2q/Q} - 1}\,, \tag{25.38}$$

for $q \geq 0$. Note that $\tilde{\sigma}_q$ is independent of the octave index p and thus the *same* filter kernels can be used at each octave. For example, with $Q = 3$ and $\sigma_0 = 1.6$ (as used in Table 25.2) the resulting kernel widths are

$$\tilde{\sigma}_1 = 1.2263, \qquad \tilde{\sigma}_2 = 1.9725, \qquad \tilde{\sigma}_3 = 2.7713. \tag{25.39}$$

Also note that, instead of filtering all scale levels $\mathbf{G}_{p,q}$ in an octave from the corresponding base level $\mathbf{G}_{p,0}$, we could calculate them recursively from the next-lower level $\mathbf{G}_{p,q-1}$. While this approach requires even smaller Gaussian kernels (and is thus more efficient), recursive filtering tends to accrue numerical inaccuracies. Nevertheless, the method is used frequently in scale-space implementations.

Decimation between successive octaves

With $M \times N$ being the size of the original image I, every sub-sampling step between octaves cuts the size of the image by half, that is,

$$M_{p+1} \times N_{p+1} = \left\lfloor \frac{M_p}{2} \right\rfloor \times \left\lfloor \frac{N_p}{2} \right\rfloor, \tag{25.40}$$

for octaves with index $p \geq 0$. The resulting image size at octave $\mathbf{G}_p$ is thus

$$M_p \times N_p = \left\lfloor \frac{M_0}{2^p} \right\rfloor \times \left\lfloor \frac{N_0}{2^p} \right\rfloor. \tag{25.41}$$

The base level $\mathbf{G}_{p,0}$ of each octave $\mathbf{G}_p$ (with $p > 0$) is obtained by sub-sampling the top level $\mathbf{G}_{p-1,Q}$ of the next-lower octave $\mathbf{G}_{p-1}$ as

$$\mathbf{G}_{p,0} = \mathsf{Decimate}(\mathbf{G}_{p-1,Q}), \tag{25.42}$$

where $\mathsf{Decimate}(G)$ denotes the 2:1 sub-sampling operation, that is,

$$\mathbf{G}_{p,0}(u, v) \leftarrow \mathbf{G}_{p-1,Q}(2u, 2v), \tag{25.43}$$

for each sample position $(u, v) \in [0, M_p - 1] \times [0, N_p - 1]$. Additional low-pass filtering is not required prior to sub-sampling since the Gaussian smoothing performed in each octave also cuts the bandwidth by half.

The main steps involved in constructing a hierarchical Gaussian scale space are summarized in Alg. 25.1. In summary, the input image I is first blurred to scale σ_0 by filtering with a Gaussian kernel of width $\bar{\sigma}_0$. Within each octave $\mathbf{G}_p$, the scale levels $\mathbf{G}_{p,q}$ are calculated from the base level $\mathbf{G}_{p,0}$ by filtering with a set of Gaussian filters of width $\tilde{\sigma}_q$ ($q = 1, \ldots, Q$). Note that the values $\tilde{\sigma}_q$ and the corresponding Gaussian kernels $H^{\mathrm{G},\tilde{\sigma}_q}$ can be pre-calculated once since they are independent of the octave index p (Alg. 25.1, lines 13–14). The base level $\mathbf{G}_{p,0}$ of each higher octave $\mathbf{G}_p$ is obtained by decimating the top level $\mathbf{G}_{p-1,Q}$ of the previous octave $\mathbf{G}_{p-1}$. Typical parameter values are $\sigma_\mathrm{s} = 0.5$, $\sigma_0 = 1.6$, $Q = 3$, $P = 4$.

Spatial positions in the hierarchical scale space

To properly associate the spatial positions of features detected in different octaves of the hierarchical scale space we define the function

$$\boldsymbol{x}_0 \leftarrow \mathsf{AbsPos}(\boldsymbol{x}_p, p),$$

that maps the continuous position $\boldsymbol{x}_p = (x_p, y_p)$ in the local coordinate system of octave p to the corresponding position $\boldsymbol{x} = (x, y)$ in the coordinate system of the original full-resolution image I (octave $p = 0$). The function AbsPos can be defined recursively by relating the positions in successive octaves as

$$\mathsf{AbsPos}(\boldsymbol{x}_p, p) = \begin{cases} \boldsymbol{x}_p & \text{for } p = 0, \\ \mathsf{AbsPos}(2 \cdot \boldsymbol{x}_p, p-1) & \text{for } p > 0, \end{cases} \tag{25.44}$$

which gives $\boldsymbol{x}_0 = \mathsf{AbsPos}(2^p \cdot \boldsymbol{x}_p, 0)$ and thus

$$\mathsf{AbsPos}(\boldsymbol{x}_p, p) = 2^p \cdot \boldsymbol{x}_p. \tag{25.45}$$

Hierarchical LoG/DoG scale space

Analogous to the scheme shown in Fig. 25.7, a *hierarchical* DoG scale space representation is obtained by calculating the difference of adjacent scale levels within each octave of the hierarchical Gaussian scale space, that is,

Alg. 25.1
Building a hierarchical Gaussian scale space. The input image I is first blurred to scale σ_0 by filtering with a Gaussian kernel of width $\bar{\sigma}_0$ (line 3). In each octave $\mathbf{G}_p$, the scale levels $\mathbf{G}_{p,q}$ are calculated from the base level $\mathbf{G}_{p,0}$ by filtering with a set of Gaussian filters of width $\tilde{\sigma}_1, \ldots, \tilde{\sigma}_Q$ (line 13–14). The base level $\mathbf{G}_{p,0}$ of each higher octave is obtained by subsampling the top level $\mathbf{G}_{p-1,Q}$ of the previous octave (line 6).

```
1:  BuildGaussianScaleSpace(I, σ_s, σ_0, P, Q)
      Input: I, source image; σ_s, sampling scale; σ_0, reference scale
      of the first octave; P, number of octaves. Q, number of scale
      steps per octave. Returns a hierarchical Gaussian scale space
      representation G of the image I.
2:    σ̄_0 ← (σ_0² − σ_s²)^{1/2}          ▷ scale to base of 1st octave, Eq. 25.25
3:    G_init ← I ∗ H^{G,σ̄_0}             ▷ apply 2D Gaussian filter of width σ̄_0
4:    G_0 ← MakeGaussianOctave(G_init, 0, Q, σ_0)       ▷ create octave G_0
5:    for p ← 1, ..., P−1 do                           ▷ octave index p
6:        G_next ← Decimate(G_{p−1,Q})      ▷ dec. top level of octave p−1
7:        G_p ← MakeGaussianOctave(G_next, p, Q, σ_0)   ▷ create octave
          G_p
8:    G ← (G_0, ..., G_{P−1})
9:    return G                         ▷ hierarchical Gaussian scale space G

10: MakeGaussianOctave(G_base, p, Q, σ_0)
      Input: G_base, octave base level; p, octave index; Q, number of
      levels per octave; σ_0, reference scale.
11:   G_{p,0} ← G_base
12:   for q ← 1, ..., Q do                              ▷ level index q
13:       σ̃_q ← σ_0 · sqrt(2^{2q/Q} − 1)                 ▷ see Eq. 25.38
14:       G_{p,q} ← G_base ∗ H^{G,σ̃_q} ▷ apply 2D Gaussian filter of width σ̃_q
15:   G_p ← (G_{p,0}, ..., G_{p,Q})
16:   return G_p                               ▷ scale space octave G_p

17: Decimate(G_in)
      Input: G_in, Gaussian scale space level.
18:   (M, N) ← Size(G_in)
19:   M' ← ⌊M/2⌋,   N' ← ⌊N/2⌋                          ▷ decimated size
20:   Create map G_out : M'×N' ↦ ℝ
21:   for all (u, v) ∈ M'×N' do
22:       G_out(u, v) ← G_in(2u, 2v)                    ▷ 2:1 subsampling
23:   return G_out                       ▷ decimated scale level G_out
```

$$\mathbf{D}_{p,q} = \mathbf{G}_{p,q+1} - \mathbf{G}_{p,q} \tag{25.46}$$

for level numbers $q \in [0, Q-1]$. Figure 25.9 shows the corresponding Gaussian and DoG scale levels for the previous example over a range of three octaves. To demonstrate the effects of sub-sampling, the same information is shown in Fig. 25.10 and 25.11, with all level images scaled to the same size. Figure 25.11 also shows the absolute values of the DoG response, which are effectively used for detecting interest points at different scale levels. Note how blob-like features stand out and disappear again as the scale varies from fine to coarse. Analogous results obtained from a different image are shown in Figs. 25.12 and 25.13.

25.1.5 Scale Space Structure in SIFT

In the SIFT approach, the absolute value of the DoG response is used to localize interest points at different scales. For this purpose, local maxima are detected in the 3D space spanned by the spatial x/y-positions and the scale coordinate. To determine local maxima along the scale dimension over a full octave, two additional DoG levels,

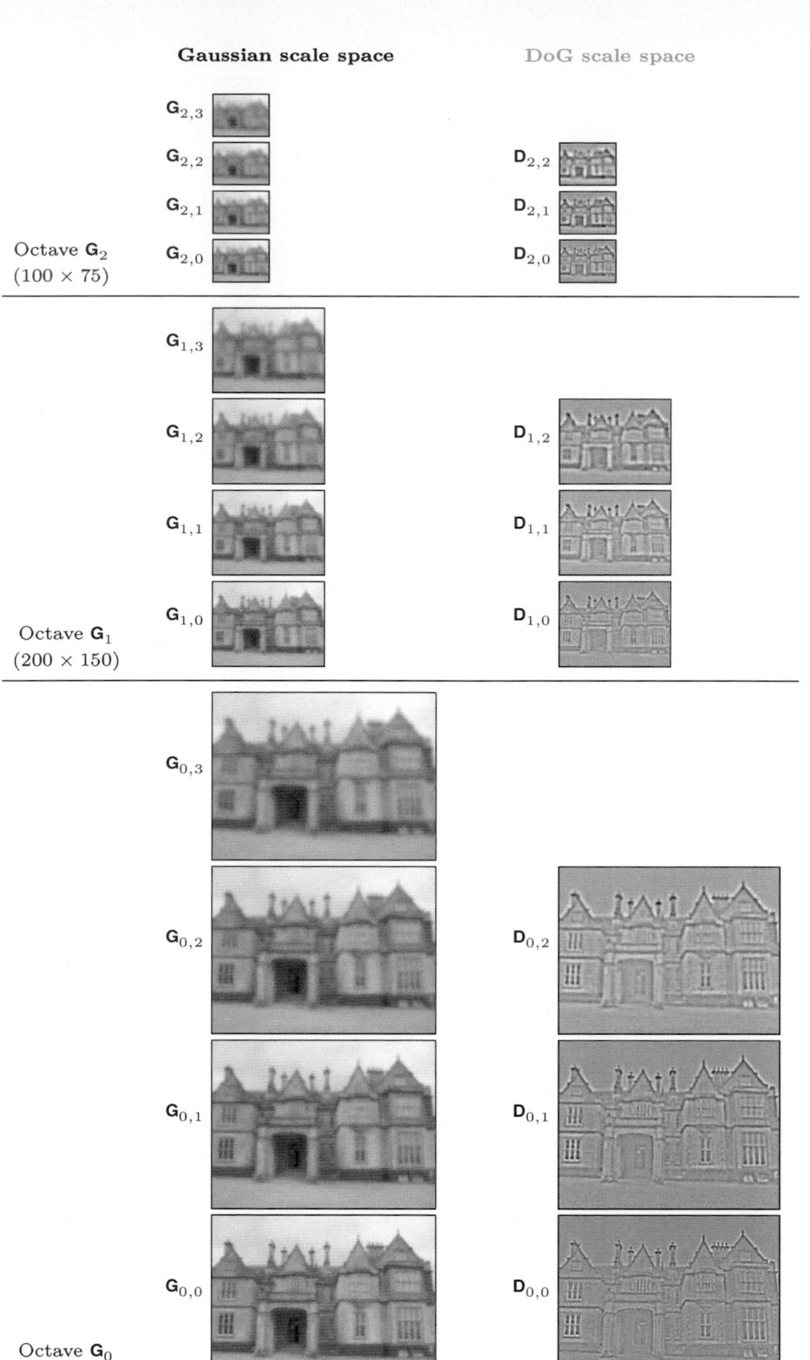

Fig. 25.9
Hierarchical Gaussian and DoG scale space example, with $P = Q = 3$. Gaussian scale space levels $\mathbf{G}_{p,q}$ are shown in the left column, DoG levels $\mathbf{D}_{p,q}$ in the right column. All images are shown at their real scale.

$\mathbf{D}_{p,-1}$ and $\mathbf{D}_{p,Q}$, and two additional Gaussian scale levels, $\mathbf{G}_{p,-1}$ and $\mathbf{G}_{p,Q+1}$, are required in each octave.

In total, each octave $\mathbf{G}_p$ then consists of $Q+3$ Gaussian scale levels $\mathbf{G}_{p,q}$ $(q = -1, \ldots, Q+1)$ and $Q+2$ DoG levels $\mathbf{D}_{p,q}$ $(q = -1, \ldots, Q)$, as shown in Fig. 25.14. For the base level $\mathbf{G}_{0,-1}$, the scale index is $m = -1$ and its absolute scale (see Eqns. (25.22) and (25.35)) is

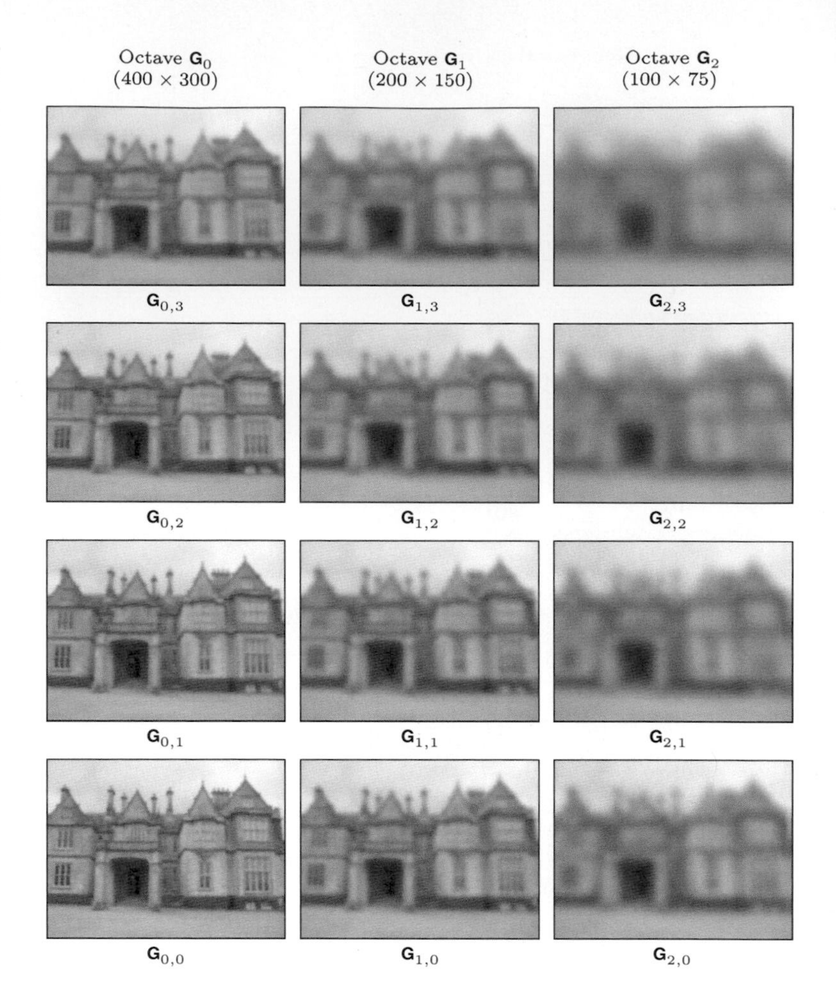

Fig. 25.10 Hierarchical Gaussian scale space example (`castle` image). All images are scaled to the same size. Note that $\mathbf{G}_{1,0}$ is merely a sub-sampled copy of $\mathbf{G}_{0,3}$; analogously, $\mathbf{G}_{2,0}$ is sub-sampled from $\mathbf{G}_{1,3}$.

$$\sigma_{0,-1} = \sigma_0 \cdot 2^{-1/Q} = \sigma_0 \cdot \frac{1}{\Delta_\sigma}. \tag{25.47}$$

Thus, with the usual settings ($\sigma_0 = 1.6$ and $Q = 3$), the *absolute* scale values for the six levels of the first octave are

$$\begin{aligned} \sigma_{0,-1} &= 1.2699, & \sigma_{0,0} &= 1.6000, & \sigma_{0,1} &= 2.0159, \\ \sigma_{0,2} &= 2.5398, & \sigma_{0,3} &= 3.2000, & \sigma_{0,4} &= 4.0317. \end{aligned} \tag{25.48}$$

The complete set of scale values for a SIFT scale space with four octaves ($p = 0, \ldots, 3$) is listed in Table 25.3.

To construct the Gaussian part of the first scale space octave $\mathbf{G}_0$, the initial level $\mathbf{G}_{0,-1}$ is obtained by filtering the input image I with a Gaussian kernel of width

$$\bar{\sigma}_{0,-1} = \sqrt{\sigma_{0,-1}^2 - \sigma_\mathrm{s}^2} = \sqrt{1.2699^2 - 0.5^2} \approx 1.1673 \tag{25.49}$$

For the higher octaves ($p > 0$), the initial level ($q = -1$) is obtained by sub-sampling (decimating) level $Q-1$ of the next-lower octave $\mathbf{G}_{p-1}$, that is,

$$\mathbf{G}_{p,-1} \leftarrow \mathsf{Decimate}(\mathbf{G}_{p-1,Q-1}), \tag{25.50}$$

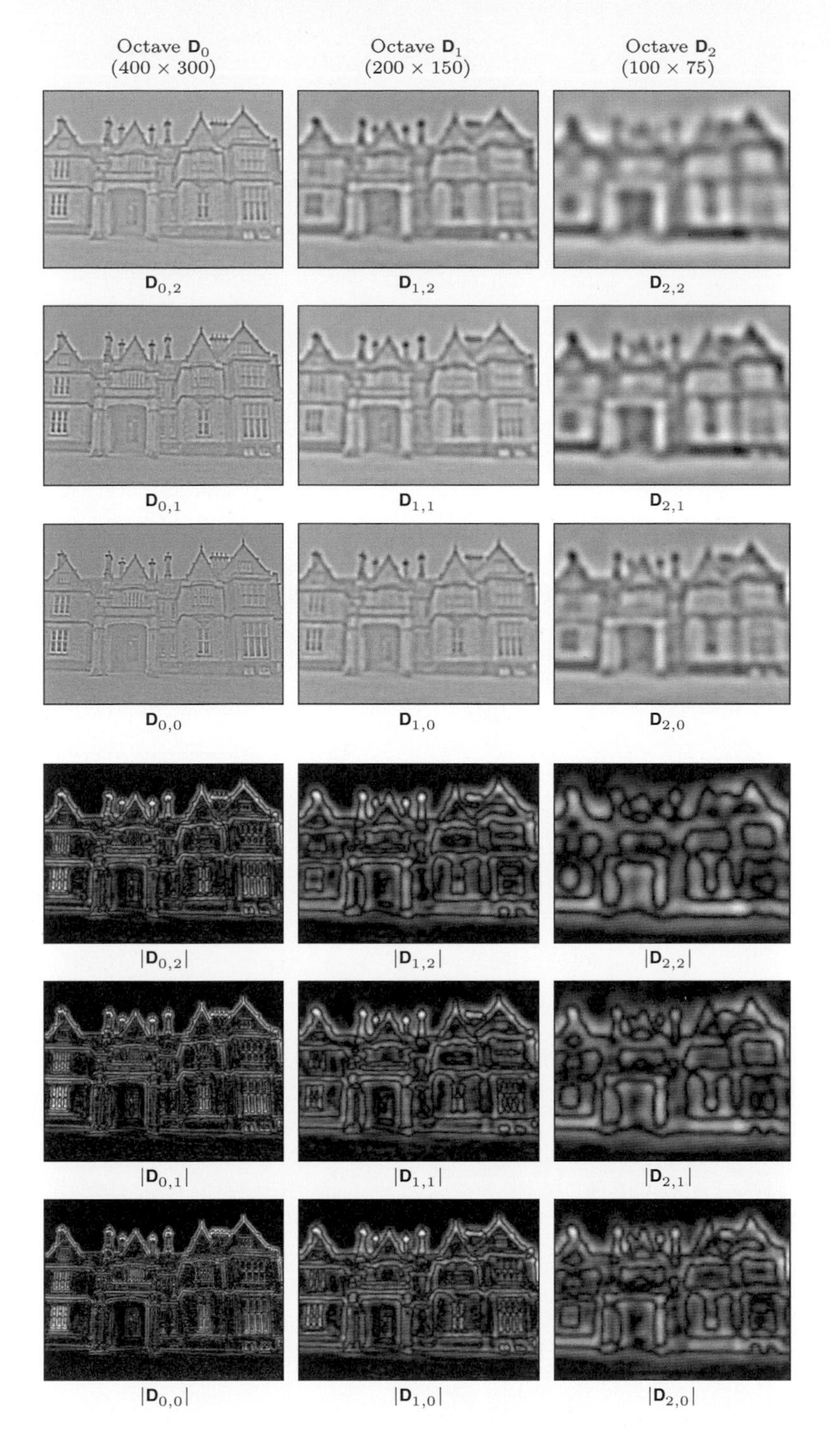

Fig. 25.11
Hierarchical DoG scale space example (`castle` image). The three top rows show the positive and negative DoG values (zero is mapped to intermediate gray). The three bottom rows show the absolute values of the DoG results (zero is mapped to black, maximum values to white). All images are scaled to the size of the original image.

analogous to Eqn. (25.42). The remaining levels $\mathbf{G}_{p,0}, \ldots, \mathbf{G}_{p,Q+1}$ of the octave are either calculated by incremental filtering (as described in Fig. 25.6) or by filtering from the octave's initial level $\mathbf{G}_{p,-1}$ with a Gaussian of width $\tilde{\sigma}_{p,q}$ (see Eqn. (25.38)). The advantage of the direct approach is that numerical errors do not accrue across the scale space; the disadvantage is that the kernels are up to 50 % larger than those needed for the incremental approach ($\tilde{\sigma}_{0,4} = 3.8265$ vs.

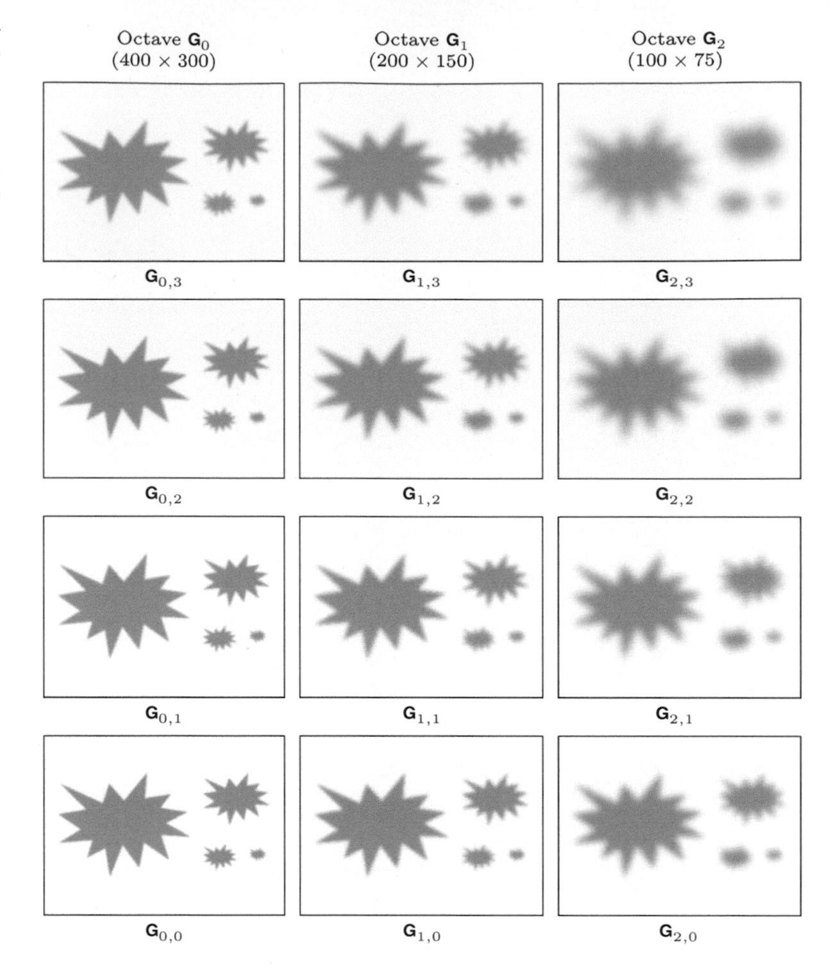

Fig. 25.12
Hierarchical Gaussian scale space example (stars image).

Table 25.3
Absolute and relative scale values for a SIFT scale space with four octaves. Each octave with index $p = 0, \ldots, 3$ consists of 6 Gaussian scale layers $\mathbf{G}_{p,q}$, with $q = -1, \ldots, 4$. For each scale layer, m is the scale index and $\sigma_{p,q}$ is the corresponding *absolute* scale. Within each octave p, $\tilde{\sigma}_{p,q}$ denotes the *relative* scale with respect to the octave's base layer $\mathbf{G}_{p,-1}$. Each base layer $\mathbf{G}_{p,-1}$ is obtained by sub-sampling (decimating) layer $q = Q - 1 = 2$ in the previous octave, i.e., $\mathbf{G}_{p,-1} = \mathsf{Decimate}(\mathbf{G}_{p-1,Q-1})$, for $p > 0$. The base layer $\mathbf{G}_{0,-1}$ in the bottom octave is derived by Gaussian smoothing of the original image. Note that the relative scale values $\tilde{\sigma}_{p,q} = \tilde{\sigma}_q$ are the same inside every octave (independent of p) and thus the same Gaussian filter kernels can be used for calculating all octaves.

p	q	m	d	$\sigma_{p,q}$	$\dot{\sigma}_q$	$\tilde{\sigma}_q$
3	4	13	8	32.2540	4.0317	3.8265
3	3	12	8	25.6000	3.2000	2.9372
3	2	11	8	20.3187	2.5398	2.1996
3	1	10	8	16.1270	2.0159	1.5656
3	0	9	8	12.8000	1.6000	0.9733
3	−1	8	8	10.1594	1.2699	0.0000
2	4	10	4	16.1270	4.0317	3.8265
2	3	9	4	12.8000	3.2000	2.9372
2	2	8	4	10.1594	2.5398	2.1996
2	1	7	4	8.0635	2.0159	1.5656
2	0	6	4	6.4000	1.6000	0.9733
2	−1	5	4	5.0797	1.2699	0.0000
1	4	7	2	8.0635	4.0317	3.8265
1	3	6	2	6.4000	3.2000	2.9372
1	2	5	2	5.0797	2.5398	2.1996
1	1	4	2	4.0317	2.0159	1.5656
1	0	3	2	3.2000	1.6000	0.9733
1	−1	2	2	2.5398	1.2699	0.0000
0	4	4	1	4.0317	4.0317	3.8265
0	3	3	1	3.2000	3.2000	2.9372
0	2	2	1	2.5398	2.5398	2.1996
0	1	1	1	2.0159	2.0159	1.5656
0	0	0	1	1.6000	1.6000	0.9733
0	−1	−1	1	1.2699	1.2699	0.0000

p ... octave index
q ... level index
m ... linear scale index ($m = Qp + q$)
d ... decimation factor ($d = 2^p$)
$\sigma_{p,q}$ ... absolute scale (Eqn. (25.35))
$\dot{\sigma}_q$ ... decimated scale (Eqn. (25.37))
$\tilde{\sigma}_q$ ... relative decimated scale w.r.t. octave's base level $\mathbf{G}_{p,-1}$ (Eqn. (25.38))

$P = 3$ (number of octaves)
$Q = 3$ (levels per octave)
$\sigma_0 = 1.6$ (base scale)

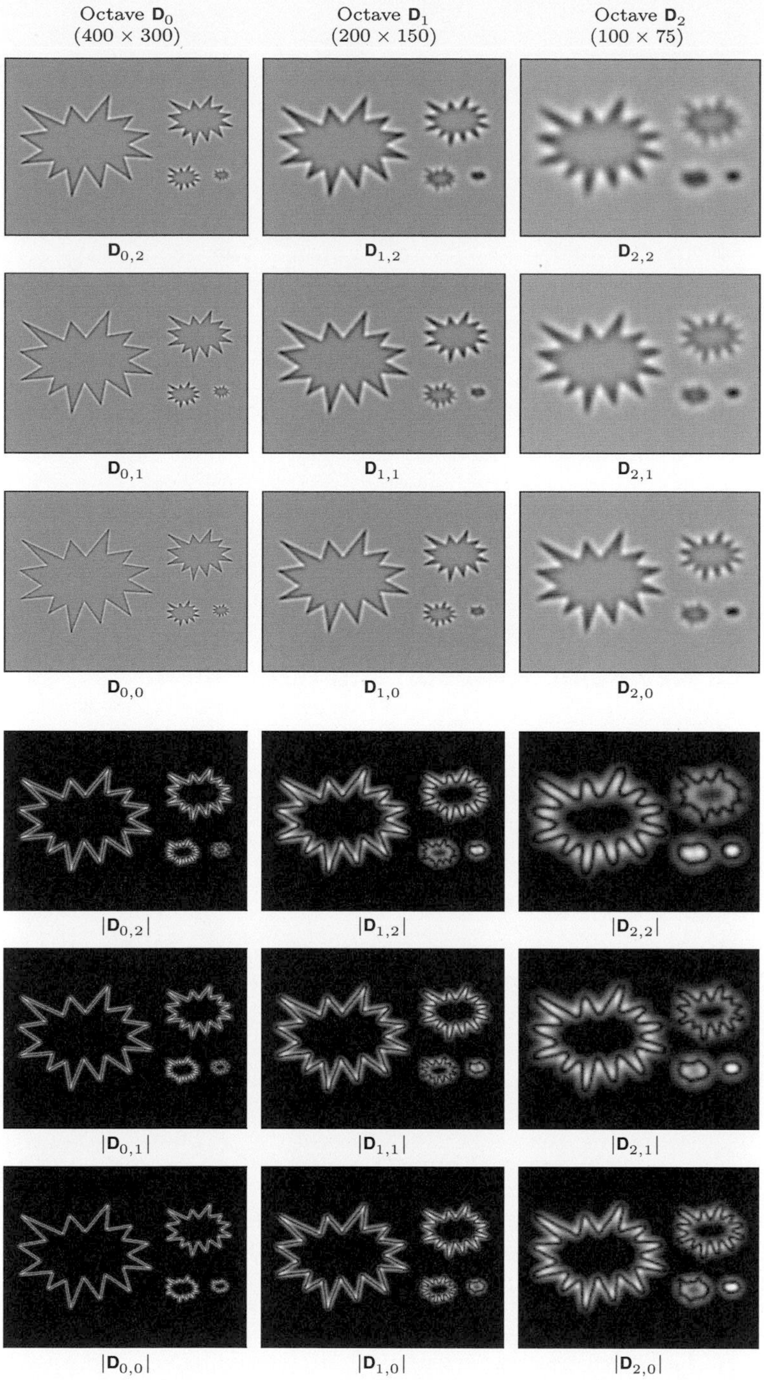

Octave $\mathbf{D}_0$ (400 × 300) | Octave $\mathbf{D}_1$ (200 × 150) | Octave $\mathbf{D}_2$ (100 × 75)

$\mathbf{D}_{0,2}$ $\mathbf{D}_{1,2}$ $\mathbf{D}_{2,2}$

$\mathbf{D}_{0,1}$ $\mathbf{D}_{1,1}$ $\mathbf{D}_{2,1}$

$\mathbf{D}_{0,0}$ $\mathbf{D}_{1,0}$ $\mathbf{D}_{2,0}$

$|\mathbf{D}_{0,2}|$ $|\mathbf{D}_{1,2}|$ $|\mathbf{D}_{2,2}|$

$|\mathbf{D}_{0,1}|$ $|\mathbf{D}_{1,1}|$ $|\mathbf{D}_{2,1}|$

$|\mathbf{D}_{0,0}|$ $|\mathbf{D}_{1,0}|$ $|\mathbf{D}_{2,0}|$

Fig. 25.13
Hierarchical DoG scale space example (stars image). The three top rows show the positive and negative DoG values (zero is mapped to intermediate gray). The three bottom rows show the absolute values of the DoG results (zero is mapped to black, maximum values to white). All images are scaled to the size of the original image.

$\sigma'_{0,4} = 2.4525$). Note that the inner levels $\mathbf{G}_{p,q}$ of all higher octaves (i.e., $p > 0, q \geq 0$) are calculated from the base level $\mathbf{G}_{p,-1}$, using the *same* set of kernels as for the first octave, as listed in Table 25.3. The complete process of building a SIFT scale space is summarized in Alg. 25.2.

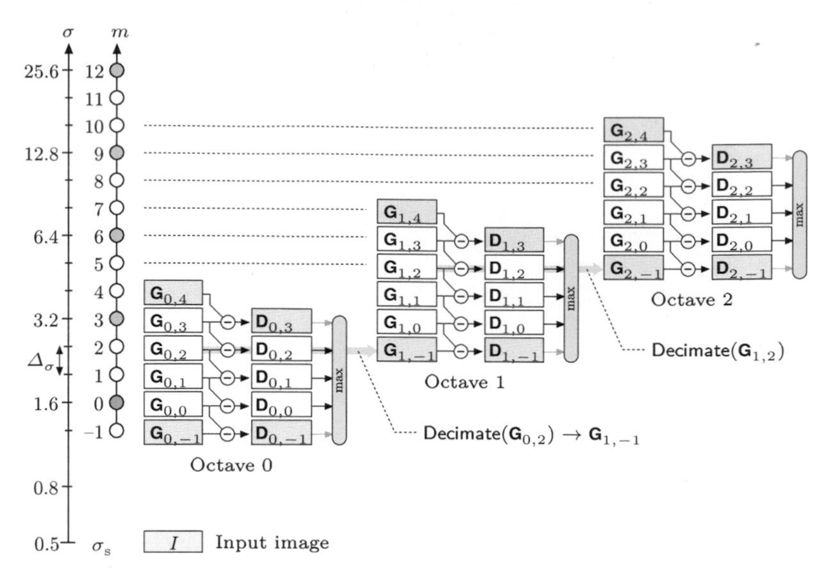

Fig. 25.14 Scale space structure for SIFT with $P = 3$ octaves and $Q = 3$ levels per octave. To perform local maximum detection ("max") over the full octave, $Q+2$ DoG scale space levels $(\mathbf{D}_{p,-1}, \ldots, \mathbf{D}_{p,Q})$ are required. The blue arrows indicate the decimation steps between successive Gaussian octaves. Since the DoG levels are obtained by subtracting pairs of Gaussian scale space levels, $Q+3$ such levels $(\mathbf{G}_{p,-1}, \ldots, \mathbf{G}_{p,Q+1})$ are needed in each octave $\mathbf{G}_p$. The two vertical axes on the left show the absolute scale (σ) and the discrete scale index (m), respectively. Note that the values along the scale axis are logarithmic with constant multiplicative scale increments $\Delta_\sigma = 2^{1/Q}$. The absolute scale of the input image (I) is assumed as $\sigma_s = 0.5$.

25.2 Key Point Selection and Refinement

Key points are identified in three steps: (1) detection of extremal points in the DOG scale space, (2) position refinement by local interpolation, and (3) elimination of edge responses. These steps are detailed in the following and summarized in Algs. 25.3–25.6.

25.2.1 Local Extrema Detection

In the first step, candidate interest points are detected as local extrema in the 3D DoG scale space that we described in the previous section. Extrema detection is performed independently within each octave p. For the sake of convenience we define the 3D scale space coordinate $\boldsymbol{c} = (u, v, q)$, composed of the spatial position (u, v) and the level index q, as well as the function

$$D(\boldsymbol{c}) := \mathbf{D}_{p,q+k}(u, v) \tag{25.51}$$

as a short notation for selecting DoG values from a given octave p. Also, for collecting the DoG values in the 3D neighborhood around a scale space position $\boldsymbol{c}$, we define the map

$$\mathsf{N}_{\boldsymbol{c}}(i, j, k) := D\big(\boldsymbol{c} + i \cdot \mathbf{e}_{\mathrm{i}} + j \cdot \mathbf{e}_{\mathrm{j}} + k \cdot \mathbf{e}_{\mathrm{k}}\big), \tag{25.52}$$

with $i, j, k \in \{-1, 0, 1\}$ and the 3D unit vectors

$$\mathbf{e}_{\mathrm{i}} = (1, 0, 0)^{\mathsf{T}}, \qquad \mathbf{e}_{\mathrm{j}} = (0, 1, 0)^{\mathsf{T}}, \qquad \mathbf{e}_{\mathrm{k}} = (0, 0, 1)^{\mathsf{T}}. \tag{25.53}$$

The neighborhood $\mathsf{N}_{\boldsymbol{c}}$ includes the center value $D(\boldsymbol{c})$ and the 26 values of its immediate neighbors (see Fig. 25.15(a)). These values are used to estimate the 3D gradient vector and the Hessian matrix for the 3D scale space position $\boldsymbol{c}$, as will be described.

A DoG scale space position $\boldsymbol{c}$ is accepted as a local extremum (minimum or maximum) if the associated value $D(\boldsymbol{c}) = \mathsf{N}_{\boldsymbol{c}}(0, 0, 0)$

```
1:  BuildSiftScaleSpace(I, σ_s, σ_0, P, Q)
      Input: I, source image; σ_s, sampling scale; σ_0, reference scale of
      the first octave; P, number of octaves; Q, number of scale steps
      per octave. Returns a SIFT scale space representation ⟨G, D⟩ of
      the image I.
2:    σ_init ← σ_0 · 2^(−1/Q)              ▷ abs. scale at level (0, −1), Eq. 25.47
3:    σ̄_init ← sqrt(σ_init² − σ_s²)        ▷ relative scale w.r.t. σ_s, Eq. 25.49
4:    G_init ← I ∗ H^(G,σ̄_init)            ▷ 2D Gaussian filter with σ̄_init
5:    G_0 ← MakeGaussianOctave(G_init, 0, Q, σ_0)       ▷ Gauss. octave 0
6:    for p ← 1, ..., P−1 do                ▷ for octaves 1, ..., P−1
7:        G_next ← Decimate(G_{p−1,Q−1})     ▷ see Alg. 25.1
8:        G_p ← MakeGaussianOctave(G_next, p, Q, σ_0)   ▷ octave p
9:    G ← (G_0, ..., G_{P−1})               ▷ assemble the Gaussian scale space G
10:   for p ← 0, ..., P−1 do
11:       D_p ← MakeDogOctave(G_p, p, Q)
12:   D ← (D_0, ..., D_{P−1})               ▷ assemble the DoG scale space D
13:   return ⟨G, D⟩

14: MakeGaussianOctave(G_base, p, Q, σ_0)
      Input: G_base, Gaussian base level; p, octave index; Q, scale steps
      per octave, σ_0, reference scale. Returns a new Gaussian octave
      G_p with Q+3 levels levels.
15:   G_{p,−1} ← G_base                     ▷ level q = −1
16:   for q ← 0, ..., Q+1 do                ▷ levels q = −1, ..., Q + 1
17:       σ̃_q ← σ_0 · sqrt(2^(2q/Q) − 2^(−2/Q))   ▷ rel. scale w.r.t base level G_base
18:       G_{p,q} ← G_base ∗ H^(G,σ̃_q)       ▷ 2D Gaussian filter with σ̃_q
19:   G_p ← (G_{p,−1}, ..., G_{p,Q+1})
20:   return G_p

21: MakeDogOctave(G_p, p, Q)
      Input: G_p, Gaussian octave; p, octave index; Q, scale steps per
      octave. Returns a new DoG octave D_p with Q+2 levels.
22:   for q ← −1, ..., Q do
23:       D_{p,q} ← G_{p,q+1} − G_{p,q}      ▷ diff. of Gaussians, Eq. 25.30
24:   D_p ← (D_{p,−1}, D_{p,0}, ..., D_{p,Q})   ▷ levels q = −1, ..., Q
25:   return D_p
```

Alg. 25.2
Building a SIFT scale space. This procedure is an extension of Alg. 25.1 and takes the same parameters. The SIFT scale space (see Fig. 25.14) consists of two components: a hierarchical Gaussian scale space $\mathbf{G} = (\mathbf{G}_0, \ldots, \mathbf{G}_{P-1})$ with P octaves and a (derived) hierarchical DoG scale space $\mathbf{D} = (\mathbf{D}_0, \ldots, \mathbf{D}_{P-1})$. Each Gaussian octave $\mathbf{G}_p$ holds $Q+3$ levels $(\mathbf{G}_{p,-1}, \ldots, \mathbf{G}_{p,Q+1})$. At each Gaussian octave, the lowest level $\mathbf{G}_{p,-1}$ is obtained by decimating level $Q-1$ of the previous octave $\mathbf{G}_{p-1}$ (line 7). Every DoG octave $\mathbf{D}_p$ contains $Q+2$ levels $(\mathbf{D}_{p,-1}, \ldots, \mathbf{D}_{p,Q})$. A DoG level $\mathbf{D}_{p,q}$ is calculated as the pointwise difference of two adjacent Gaussian levels $\mathbf{G}_{p,q+1}$ and $\mathbf{G}_{p,q}$ (line 23). Typical parameter settings are $\sigma_s = 0.5$, $\sigma_0 = 1.6$, $Q = 3$, $P = 4$.

is either *negative* and also *smaller* or *positive* and *greater* than all neighboring values. In addition, a minimum difference $\mathrm{t_{extrm}} \geq 0$ can be specified, indicating how much the center value must at least deviate from the surrounding values. The decision whether a given neighborhood N_c contains a local minimum or maximum can thus be expressed as

$$\mathsf{IsLocalMin}(\mathsf{N}_c) := \mathsf{N}_c(0,0,0) < 0 \;\wedge\; \mathsf{N}_c(0,0,0) + \mathrm{t_{extrm}} < \min_{\substack{(i,j,k)\neq\\(0,0,0)}} \mathsf{N}_c(i,j,k), \tag{25.54}$$

$$\mathsf{IsLocalMax}(\mathsf{N}_c) := \mathsf{N}_c(0,0,0) > 0 \;\wedge\; \mathsf{N}_c(0,0,0) - \mathrm{t_{extrm}} < \max_{\substack{(i,j,k)\neq\\(0,0,0)}} \mathsf{N}_c(i,j,k) \tag{25.55}$$

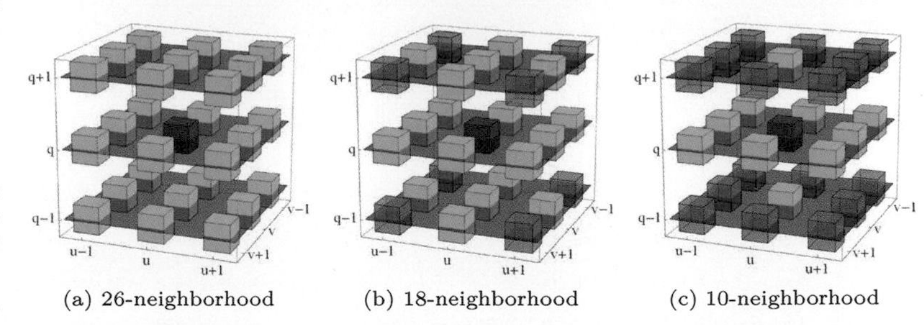

(a) 26-neighborhood (b) 18-neighborhood (c) 10-neighborhood

Fig. 25.15
Different 3D neighborhoods for detecting local extrema in the DoG scale space. The red cube represents the DoG value at the reference coordinate $c = (u, v, q)$ at the spatial position (u, v) at scale level q (within some octave p). Full $3 \times 3 \times 3$ neighborhood with 26 elements (a); other types of neighborhoods with 18 (b) or 10 (c) elements, respectively, are also commonly used. A local maximum/minimum is detected if the DoG value at the center is greater/smaller than all neighboring values (green cubes).

(see procedure $\mathsf{IsExtremum}(\mathsf{N}_c)$ in Alg. 25.5). As illustrated in Fig. 25.15(b–c), alternative 3D neighborhoods with 18 or 10 cells may be specified for extrema detection.

25.2.2 Position Refinement

Once a local extremum is detected in the DoG scale space, only its *discrete* 3D coordinates $\boldsymbol{c} = (u, v, q)$ are known, consisting of the spatial grid position (u, v) and the index (q) of the associated scale level. In the second step, a more accurate, *continuous* position for each candidate key point is estimated by fitting a quadratic function to the local neighborhood, as proposed in [37]. This is particularly important at the higher octaves of the scale space, where the spatial resolution becomes increasingly coarse due to successive decimation. Position refinement is based on a local second-order Taylor expansion of the discrete DoG function, which yields a continuous approximation function whose maximum or minimum can be found analytically. Additional details and illustrative examples are provided in Sec. C.3.2 of the Appendix.

At any extremal position $\boldsymbol{c} = (u, v, q)$ in octave p of the hierarchical DoG scale space $\mathbf{D}$, the corresponding 3×3×3 neighborhood $\mathcal{N}_D(\boldsymbol{c})$ is used to estimate the elements of the continuous 3D gradient, that is,

$$\nabla_D(\boldsymbol{c}) = \begin{pmatrix} d_\mathrm{x} \\ d_\mathrm{y} \\ d_\sigma \end{pmatrix} \approx \frac{1}{2} \cdot \begin{pmatrix} D(\boldsymbol{c}+\mathbf{e}_\mathrm{i}) - D(\boldsymbol{c}-\mathbf{e}_\mathrm{i}) \\ D(\boldsymbol{c}+\mathbf{e}_\mathrm{j}) - D(\boldsymbol{c}-\mathbf{e}_\mathrm{j}) \\ D(\boldsymbol{c}+\mathbf{e}_\mathrm{k}) - D(\boldsymbol{c}-\mathbf{e}_\mathrm{k}) \end{pmatrix}, \tag{25.56}$$

with $D()$ as defined in Eqn. (25.51). Similarly, the 3×3 Hessian matrix for position $\boldsymbol{c}$ is obtained as

$$\mathbf{H}_D(\boldsymbol{c}) = \begin{pmatrix} d_{xx} & d_{xy} & d_{x\sigma} \\ d_{xy} & d_{yy} & d_{y\sigma} \\ d_{x\sigma} & d_{y\sigma} & d_{\sigma\sigma} \end{pmatrix}, \tag{25.57}$$

with the required second order derivatives estimated as

$$\begin{aligned} d_{xx} &= D(\boldsymbol{c}-\mathbf{e}_\mathrm{i}) - 2 \cdot D(\boldsymbol{c}) + D(\boldsymbol{c}+\mathbf{e}_\mathrm{i}), \\ d_{yy} &= D(\boldsymbol{c}-\mathbf{e}_\mathrm{j}) - 2 \cdot D(\boldsymbol{c}) + D(\boldsymbol{c}+\mathbf{e}_\mathrm{j}), \\ d_{\sigma\sigma} &= D(\boldsymbol{c}-\mathbf{e}_\mathrm{k}) - 2 \cdot D(\boldsymbol{c}) + D(\boldsymbol{c}+\mathbf{e}_\mathrm{k}), \\ d_{xy} &= \tfrac{D(\boldsymbol{c}+\mathbf{e}_\mathrm{i}+\mathbf{e}_\mathrm{j}) - D(\boldsymbol{c}-\mathbf{e}_\mathrm{i}+\mathbf{e}_\mathrm{j}) - D(\boldsymbol{c}+\mathbf{e}_\mathrm{i}-\mathbf{e}_\mathrm{j}) + D(\boldsymbol{c}-\mathbf{e}_\mathrm{i}-\mathbf{e}_\mathrm{j})}{4}, \\ d_{x\sigma} &= \tfrac{D(\boldsymbol{c}+\mathbf{e}_\mathrm{i}+\mathbf{e}_\mathrm{k}) - D(\boldsymbol{c}-\mathbf{e}_\mathrm{i}+\mathbf{e}_\mathrm{k}) - D(\boldsymbol{c}+\mathbf{e}_\mathrm{i}-\mathbf{e}_\mathrm{k}) + D(\boldsymbol{c}-\mathbf{e}_\mathrm{i}-\mathbf{e}_\mathrm{k})}{4}, \\ d_{y\sigma} &= \tfrac{D(\boldsymbol{c}+\mathbf{e}_\mathrm{j}+\mathbf{e}_\mathrm{k}) - D(\boldsymbol{c}-\mathbf{e}_\mathrm{j}+\mathbf{e}_\mathrm{k}) - D(\boldsymbol{c}+\mathbf{e}_\mathrm{j}-\mathbf{e}_\mathrm{k}) + D(\boldsymbol{c}-\mathbf{e}_\mathrm{j}-\mathbf{e}_\mathrm{k})}{4}. \end{aligned} \tag{25.58}$$

See the procedures $\mathsf{Gradient}(\mathsf{N}_c)$ and $\mathsf{Hessian}(\mathsf{N}_c)$ in Alg. 25.5 (p. 651) for additional details. From the gradient vector $\nabla_D(\boldsymbol{c})$ and the Hessian matrix $\mathbf{H}_D(\boldsymbol{c})$, the second order Taylor expansion around point $\boldsymbol{c}$ is

$$\tilde{D}_{\boldsymbol{c}}(\boldsymbol{x}) = D(\boldsymbol{c}) + \nabla_D^{\mathsf{T}}(\boldsymbol{c})\cdot(\boldsymbol{x}-\boldsymbol{c}) + \tfrac{1}{2}(\boldsymbol{x}-\boldsymbol{c})^{\mathsf{T}}\cdot\mathbf{H}_D(\boldsymbol{c})\cdot(\boldsymbol{x}-\boldsymbol{c}), \qquad (25.59)$$

for the continuous position $\boldsymbol{x} = (x, y, \sigma)^{\mathsf{T}}$. The scalar-valued function $\tilde{D}_{\boldsymbol{c}}(\boldsymbol{x}) \in \mathbb{R}$, with $\boldsymbol{c} = (u, v, q)^{\mathsf{T}}$ and $\boldsymbol{x} = (x, y, \sigma)^{\mathsf{T}}$, is a local, *continuous* approximation of the discrete DoG function $\mathbf{D}_{p,q}(u, v)$ at octave p, scale level q, and spatial position u, v. This is a quadratic function with an extremum (maximum or minimum) at position

$$\breve{\boldsymbol{x}} = \begin{pmatrix} \breve{x} \\ \breve{y} \\ \breve{\sigma} \end{pmatrix} = \boldsymbol{c} + \boldsymbol{d} = \boldsymbol{c} \underbrace{-\,\mathbf{H}_D^{-1}(\boldsymbol{c}) \cdot \nabla_D(\boldsymbol{c})}_{\boldsymbol{d}=\breve{\boldsymbol{x}}-\boldsymbol{c}} \qquad (25.60)$$

with $\boldsymbol{d} = (x', y', \sigma')^{\mathsf{T}} = \breve{\boldsymbol{x}} - \boldsymbol{c}$, under the assumption that the inverse of the Hessian matrix $\mathbf{H}_D$ exists. By inserting the extremal position $\breve{\boldsymbol{x}}$ into Eqn. (25.59), the peak (minimum or maximum) *value* of the continuous approximation function $\tilde{D}$ is found as[15]

$$\begin{aligned} D_{\mathrm{peak}}(\boldsymbol{c}) = \tilde{D}_{\boldsymbol{c}}(\breve{\boldsymbol{x}}) &= D(\boldsymbol{c}) + \tfrac{1}{2} \cdot \nabla_D^{\mathsf{T}}(\boldsymbol{c}) \cdot (\breve{\boldsymbol{x}} - \boldsymbol{c}) \\ &= D(\boldsymbol{c}) + \tfrac{1}{2} \cdot \nabla_D^{\mathsf{T}}(\boldsymbol{c}) \cdot \boldsymbol{d}, \end{aligned} \qquad (25.61)$$

where $\boldsymbol{d} = \breve{\boldsymbol{x}} - \boldsymbol{c}$ (cf. Eqn. (25.60)) denotes the 3D vector between the neighborhood's discrete center position $\boldsymbol{c}$ and the continuous extremal position $\breve{\boldsymbol{x}}$.

A scale space location $\boldsymbol{c}$ is only retained as a candidate interest point if the estimated magnitude of the DoG exceeds a given threshold $\mathrm{t}_{\mathrm{peak}}$, that is, if

$$|D_{\mathrm{peak}}(\boldsymbol{c})| > \mathrm{t}_{\mathrm{peak}}. \qquad (25.62)$$

If the distance $\boldsymbol{d} = (x', y', \sigma')^{\mathsf{T}}$ from $\boldsymbol{c}$ to the estimated (continuous) peak position $\breve{\boldsymbol{x}}$ in Eqn. (25.60) is greater than a predefined limit (typically 0.5) in any spatial direction, the center point $\boldsymbol{c} = (u, v, q)^{\mathsf{T}}$ is moved to one of the neighboring DoG cells by maximally ± 1 unit steps along the u, v axes, that is,

$$\boldsymbol{c} \leftarrow \boldsymbol{c} + \begin{pmatrix} \min(1, \max(-1, \mathrm{round}(x'))) \\ \min(1, \max(-1, \mathrm{round}(y'))) \\ 0 \end{pmatrix}. \qquad (25.63)$$

The q component of $\boldsymbol{c}$ is not modified in this version, that is, the search continues at the original scale level.[16] Based on the surrounding 3D neighborhood of this new point, a Taylor expansion (Eqn. (25.60)) is again performed to estimate a new peak location. This is repeated until either the peak location is inside the current DoG cell or the allowed number of repositioning steps $\mathrm{n}_{\mathrm{refine}}$ is reached

[15] See Eqn. (C.64) in Sec. C.3.3 in the Appendix for details.

[16] This is handled differently in other SIFT implementations.

(typically $\mathrm{n_{refine}}$ is set to 4 or 5). If successful, the result of this step is a candidate feature point

$$\breve{\boldsymbol{c}} = (\breve{x}, \breve{y}, \breve{q})^{\mathsf{T}} = \boldsymbol{c} + (x', y', 0)^{\mathsf{T}}. \tag{25.64}$$

Notice that (in this implementation) the scale level q remains unchanged even if the 3D Taylor expansion indicates that the estimated peak is located at another scale level. See procedure RefineKeyPosition() in Alg. 25.4 (p. 650) for a concise summary of these steps.

It should be mentioned that the original publication [153] is not particularly explicit about the aforementioned position refinement process and thus slightly different approaches are used in various open-source SIFT implementations. For example, the implementation in *VLFeat*[17] [241] moves to one of the direct neighbors at the same scale level as described earlier, as long as $|x'|$ or $|y'|$ is greater than 0.6. *AutoPano-SIFT*[18] by S. Nowozin calculates the length of the spatial displacement $d = \|(x', y')\|$ and discards the current point if $d > 2$. Otherwise it moves by $\Delta_u = \mathrm{round}(x')$, $\Delta_v = \mathrm{round}(y')$ without limiting the displacement to ± 1. The *Open-Source SIFT Library*[19] [106] used in *OpenCV* also makes full moves in the spatial directions and, in addition, potentially also changes the scale level by $\Delta_q = \mathrm{round}(\sigma')$ in each iteration.

25.2.3 Suppressing Responses to Edge-Like Structures

In the previous step, candidate interest points were selected as those locations in the DoG scale space where the Taylor approximation had a local maximum and the extrapolated DoG value was above a given threshold ($\mathrm{t_{peak}}$). However, the DoG filter also responds strongly to edge-like structures. At such positions, interest points cannot be located with sufficient stability and repeatability. To eliminate the responses near edges, Lowe suggests the use of the principal curvatures of the 2D DoG result along the spatial x, y axes, using the fact that the principal curvatures of a function are proportional to the eigenvalues of the function's Hessian matrix at a given point.

For a particular lattice point $\boldsymbol{c} = (u, v, q)$ in DoG scale space, with neighborhood N_D (see Eqn. (25.52)), the 2×2 Hessian matrix for the spatial coordinates is

$$\mathbf{H}_{xy}(\boldsymbol{c}) = \begin{pmatrix} d_{xx} & d_{xy} \\ d_{xy} & d_{yy} \end{pmatrix}, \tag{25.65}$$

with d_{xx}, d_{xy}, d_{yy} as defined in Eqn. (25.58), that is, these values can be extracted from the corresponding 3×3 Hessian matrix $\mathbf{H}_D(\boldsymbol{c})$ (see Eqn. (25.57)).

The matrix $\mathbf{H}_{xy}(\boldsymbol{c})$ has two eigenvalues λ_1, λ_2, which we define as being ordered, such that λ_1 has the greater magnitude ($|\lambda_1| \geq |\lambda_2|$). If both eigenvalues for a point $\boldsymbol{c}$ are of similar magnitude, the function exhibits a high curvature along two orthogonal directions and in this

[17] http://www.vlfeat.org/overview/sift.html.

[18] http://sourceforge.net/projects/hugin/files/autopano-sift-C/.

[19] http://robwhess.github.io/opensift/.

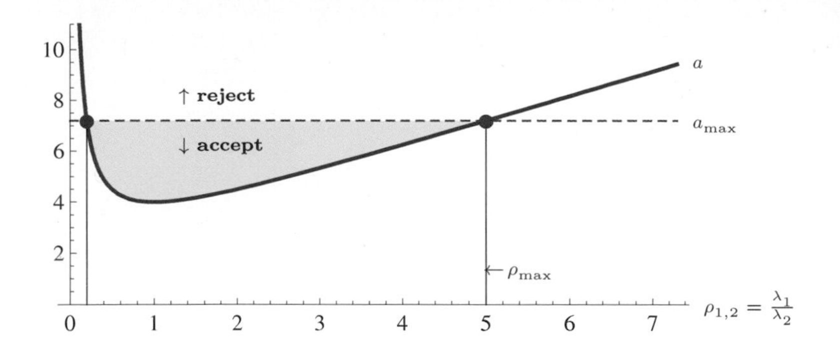

Fig. 25.16
Limiting the ratio of principal curvatures (edge ratio) $\rho_{1,2}$ by specifying a_{max}. The quantity a (blue line) has a minimum when the eigenvalue ratio $\rho_{1,2} = \frac{\lambda_1}{\lambda_2}$ is one, that is, when the two eigenvalues λ_1, λ_2 are equal, indicating a corner-like event. Typically only one of the eigenvalues is dominant in the vicinity of image lines, such that $\rho_{1,2}$ and a values are significantly increased. In this example, the principal curvature ratio $\rho_{1,2}$ is limited to $\rho_{max} = 5.0$ by setting $a_{max} = (5+1)^2/5 = 7.2$ (red line).

case $\boldsymbol{c}$ is likely to be a good reference point that can be located reliably. In the optimal situation (e.g., near a corner), the ratio of the eigenvalues $\rho = \lambda_1/\lambda_2$ is close to 1. Alternatively, if the ratio ρ is high it can be concluded that a single orientation dominates at this position, as is typically the case in the neighborhood of edges.

To estimate the ratio ρ it is not necessary to calculate the eigenvalues themselves. Following the description in [153], the sum and product of the eigenvalues λ_1, λ_2 can be found as

$$\lambda_1 + \lambda_2 = \mathrm{trace}(\mathbf{H}_{xy}(\boldsymbol{c})) = d_{xx} + d_{yy}, \tag{25.66}$$

$$\lambda_1 \cdot \lambda_2 = \det(\mathbf{H}_{xy}(\boldsymbol{c})) = d_{xx} \cdot d_{yy} - d_{xy}^2. \tag{25.67}$$

If the determinant $\det(\mathbf{H}_{xy})$ is *negative*, the principal curvatures of the underlying 2D function have opposite signs and thus point $\boldsymbol{c}$ can be discarded as not being an extremum. Otherwise, if the signs of both eigenvalues λ_1, λ_2 are the *same*, then the ratio

$$\rho_{1,2} = \frac{\lambda_1}{\lambda_2} \tag{25.68}$$

is positive (with $\lambda_1 = \rho_{1,2} \cdot \lambda_2$), and thus the expession

$$a = \frac{[\mathrm{trace}(\mathbf{H}_{xy}(\boldsymbol{c}))]^2}{\det(\mathbf{H}_{xy}(\boldsymbol{c}))} = \frac{(\lambda_1 + \lambda_2)^2}{\lambda_1 \cdot \lambda_2} \tag{25.69}$$

$$= \frac{(\rho_{1,2} \cdot \lambda_2 + \lambda_2)^2}{\rho_{1,2} \cdot \lambda_2^2} = \frac{\lambda_2^2 \cdot (\rho_{1,2} + 1)^2}{\rho_{1,2} \cdot \lambda_2^2} = \frac{(\rho_{1,2} + 1)^2}{\rho_{1,2}} \tag{25.70}$$

depends only on the ratio $\rho_{1,2}$. If the determinant of $\mathbf{H}_{xy}$ is positive, the quantity a has a minimum (4.0) at $\rho_{1,2} = 1$, if the two eigenvalues are equal (see Fig. 25.16). Note that the ratio a is the same for $\rho_{1,2} = \lambda_1/\lambda_2$ or $\rho_{1,2} = \lambda_2/\lambda_1$, since

$$a = \frac{(\rho_{1,2} + 1)^2}{\rho_{1,2}} = \frac{(\frac{1}{\rho_{1,2}} + 1)^2}{\frac{1}{\rho_{1,2}}}. \tag{25.71}$$

To verify that the eigenvalue ratio $\rho_{1,2}$ at a given position $\boldsymbol{c}$ is *below* a specified limit ρ_{max} (making $\boldsymbol{c}$ a good candidate), it is thus sufficient to check the condition

$$a \leq a_{max}, \quad \text{with} \quad a_{max} = \frac{(\rho_{max} + 1)^2}{\rho_{max}}, \tag{25.72}$$

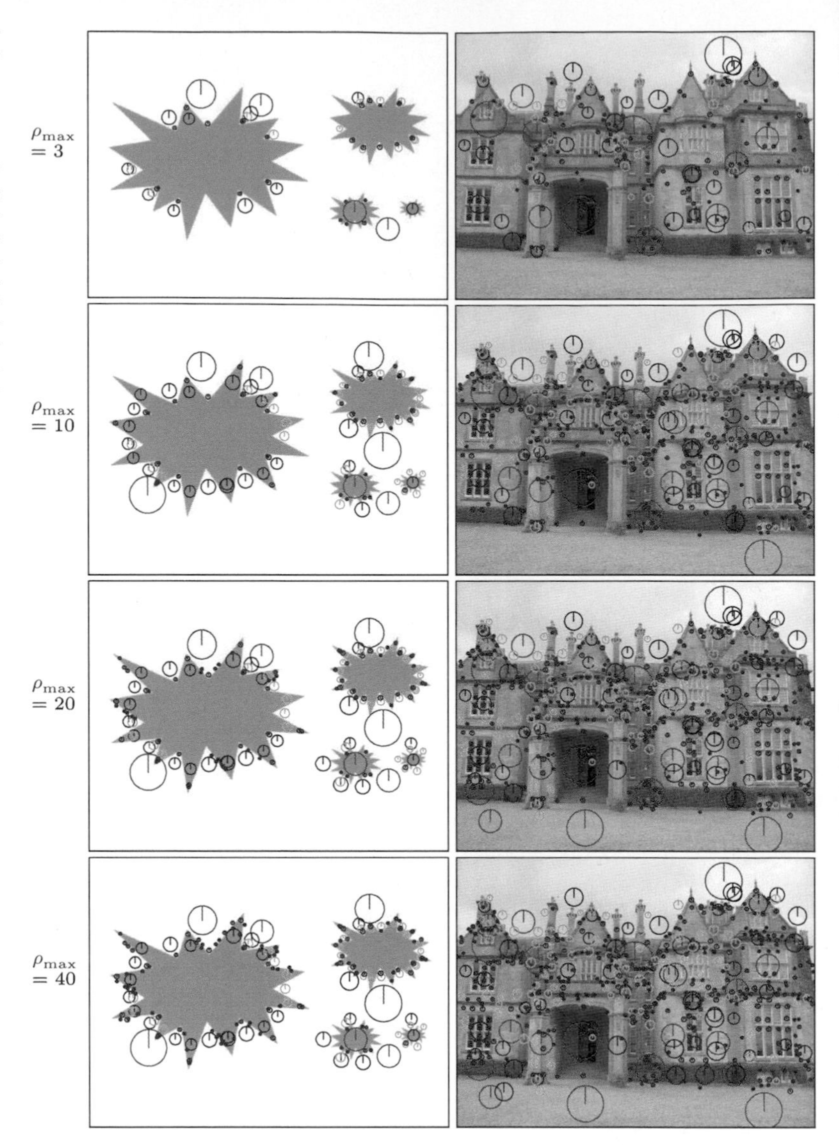

Fig. 25.17
Rejection of edge-like features by controlling the max. curvature ratio ρ_{max}. The size of the circles is proportional to the scale level at which the corresponding key point was detected, the color indicating the containing octave (0 = red, 1 = green, 2 = blue, 3 = magenta).

without the need to actually calculate the individual eigenvalues λ_1 and λ_2.[20] ρ_{max} should be greater than 1 and is typically chosen to be in the range $3, \ldots, 10$ ($\rho_{max} = 10$ is suggested in [153]). The resulting value of a_{max} in Eqn. (25.72) is constant and needs only be calculated once (see Alg. 25.3, line 2). Detection examples for varying values of ρ_{max} are shown in Fig. 25.17. Note that considerably more candidates appear near edges as ρ_{max} is raised from 3 to 40.

25.3 Creating Local Descriptors

For each local maximum detected in the hierarchical DoG scale space, a candidate key point is created, which is subsequently refined to

[20] A similar trick is used in the *Harris* corner detection algorithm (see Chapter 7).

a continuous position following the steps we have just described (see Eqns. (25.56)–(25.64)). Then, for each refined key point $\boldsymbol{k}' = (p, q, x, y)$, one or more (up to four) local descriptors are calculated. Multiple (up to four) descriptors may be created for a position if the local orientation is not unique. This process involves the following steps:

1. Find the *dominant* orientation(s) of the key point $\boldsymbol{k}'$ from the distribution of the gradients at the corresponding Gaussian scale space level.
2. For each dominant orientation, create a separate SIFT *descriptor* at the key point $\boldsymbol{k}'$.

25.3.1 Finding Dominant Orientations

Local orientation from Gaussian scale space

Orientation vectors are obtained by sampling the *gradient* values of the hierarchical Gaussian scale space $\mathbf{G}_{p,q}(u, v)$ (see Eqn. (25.32)). For any lattice position (u, v) at octave p and scale level q, the local gradient is calculated as

$$\nabla_{p,q}(u, v) = \begin{pmatrix} d_\mathrm{x} \\ d_\mathrm{y} \end{pmatrix} = 0.5 \cdot \begin{pmatrix} \mathbf{G}_{p,q}(u{+}1, v) - \mathbf{G}_{p,q}(u{-}1, v) \\ \mathbf{G}_{p,q}(u, v{+}1) - \mathbf{G}_{p,q}(u, v{-}1) \end{pmatrix}. \quad (25.73)$$

From these gradient vectors, the gradient *magnitude* and *orientation* (i.e., polar coordinates) are found as[21]

$$E_{p,q}(u, v) = \left\| \nabla_{p,q}(u, v) \right\| = \sqrt{d_\mathrm{x}^2 + d_\mathrm{y}^2}\,, \quad (25.74)$$

$$\phi_{p,q}(u, v) = \angle \nabla_{p,q}(u, v) = \tan^{-1}(d_\mathrm{y}/d_\mathrm{x}). \quad (25.75)$$

These scalar fields $E_{p,q}$ and $\phi_{p,q}$ are typically pre-calculated for all relevant octaves/levels p, q of the Gaussian scale space $\mathbf{G}$.

Orientation histograms

To find the dominant orientations for a given key point, a histogram h_ϕ of the orientation angles is calculated for the gradient vectors collected from a square window around the key point center. Typically the histogram has $\mathrm{n}_\mathrm{orient} = 36$ bins, that is, the angular resolution is 10°. The orientation histogram is collected from a square region using an isotropic Gaussian weighting function whose width σ_w is proportional to the *decimated scale* $\dot{\sigma}_q$ (see Eqn. (25.37)) of the key point's scale level q. Typically a Gaussian weighting function "with a σ that is 1.5 times that of the scale of the key point" [153] is used, that is,

$$\sigma_\mathrm{w} = 1.5 \cdot \dot{\sigma}_q = 1.5 \cdot \sigma_0 \cdot 2^{q/Q}. \quad (25.76)$$

Note that σ_w is independent of the octave index p and thus the same weighting functions are used in each octave. To calculate the *orientation histogram*, the Gaussian gradients around the given key point are collected from a square region of size $2r_\mathrm{w} \times 2r_\mathrm{w}$, with

[21] See also Chapter 16, Sec. 16.1.

$$r_{\mathrm{w}} = \lceil 2.5 \cdot \sigma_{\mathrm{w}} \rceil \tag{25.77}$$

amply dimensioned to avoid numerical truncation effects. For the parameters listed in Table 25.3 ($\sigma_0 = 1.6$, $Q = 3$), the values for σ_{w} (expressed in the octave's coordinate units) are

q	0	1	2	3
σ_{w}	1.6000	2.0159	2.5398	3.2000
r_{w}	4	5	6	7

(25.78)

In Alg. 25.7, σ_{w} and r_{w} of the Gaussian weighting function are calculated in lines 7 and 8, respectively. At each lattice point (u, v), the gradient vector $\nabla_{p,q}(u, v)$ is calculated in octave p and level q of the Gaussian scale space $\mathbf{G}$ (Alg. 25.7, line 16). From this, the gradient magnitude $E_{p,q}(u, v)$ and orientation $\phi_{p,q}(u, v)$ are obtained (lines 29–30). The corresponding Gaussian weight is calculated (in line 18) from the spatial distance between the grid point (u, v) and the interest point (x, y) as

$$w_{\mathrm{G}}(u, v) = \exp\bigl(-\tfrac{(u-x)^2+(v-y)^2}{2\cdot\sigma_{\mathrm{w}}^2}\bigr). \tag{25.79}$$

For the grid point (u, v), the quantity to be accumulated into the orientation histogram is

$$z = E_{p,q}(u, v) \cdot w_{\mathrm{G}}(u, v), \tag{25.80}$$

that is, the local gradient magnitude weighted by the Gaussian window function (Alg. 25.7, line 19).

The orientation histogram h_ϕ consists of $\mathrm{n_{orient}}$ bins and thus the *continuous* bin number for the angle $\phi(u, v)$ is

$$\kappa_\phi = \frac{\mathrm{n_{orient}}}{2\pi} \cdot \phi(u, v) \tag{25.81}$$

(see Alg. 25.7, line 20). To collect the *continuous* orientations into a histogram with discrete bins, quantization must be performed. The simplest approach is to select the "nearest" bin (by rounding) and to add the associated quantity (denoted z) entirely to the selected bin. Alternatively, to reduce quantization effects, a common technique is to *split* the quantity z onto the two closest bins. Given the continuous bin value κ_ϕ, the indexes of the two closest discrete bins are

$$k_0 = \lfloor \kappa_\phi \rfloor \bmod \mathrm{n_{orient}} \quad \text{and} \quad k_1 = (\lfloor \kappa_\phi \rfloor + 1) \bmod \mathrm{n_{orient}}, \tag{25.82}$$

respectively. The quantity z (Eqn. (25.80)) is then partitioned and accumulated into the neighboring bins k_0, k_1 of the orientation histogram h_ϕ in the form

$$\begin{aligned} \mathsf{h}_\phi(k_0) &\leftarrow \mathsf{h}_\phi(k_0) + (1-\alpha) \cdot z, \\ \mathsf{h}_\phi(k_1) &\leftarrow \mathsf{h}_\phi(k_1) + \alpha \cdot z, \end{aligned} \tag{25.83}$$

with $\alpha = \kappa_\phi - \lfloor \kappa_\phi \rfloor$. This process is illustrated by the example in Fig. 25.18 (see also Alg. 25.7, lines 21–25).

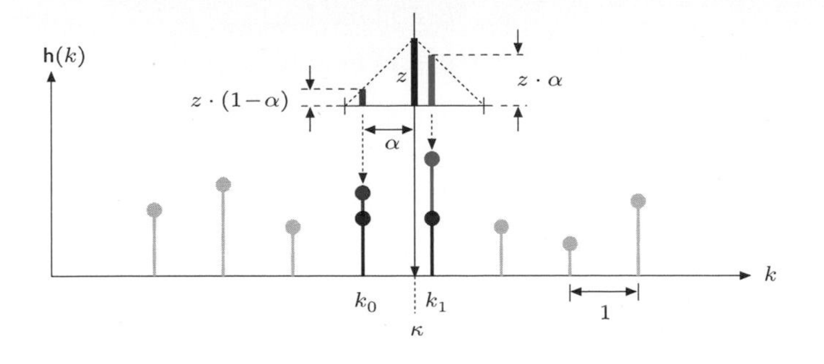

Fig. 25.18
Accumulating into multiple histogram bins by linear interpolation. Assume that some quantity z (blue bar) is to be added to the *discrete* histogram h_ϕ at the *continuous* position κ_ϕ. The histogram bins adjacent to κ_ϕ are $k_0 = \lfloor \kappa_\phi \rfloor$ and $k_1 = \lfloor \kappa_\phi \rfloor + 1$. The fraction of z accumulated into bin k_1 is $z_1 = z \cdot \alpha$ (red bar), with $\alpha = \kappa_\phi - k_0$. Analogously, the quantity added to bin k_0 is $z_0 = z \cdot (1-\alpha)$ (green bar).

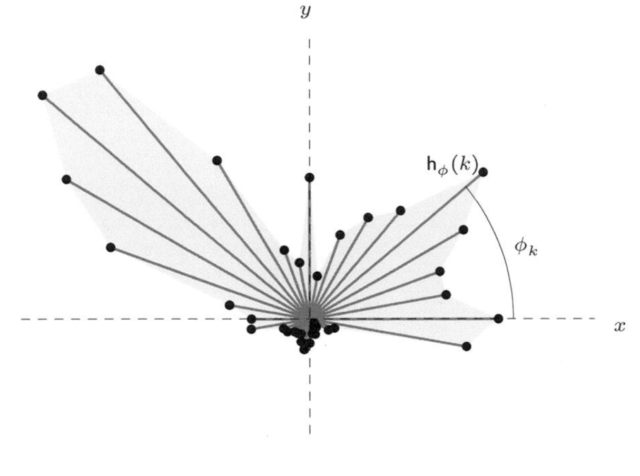

Fig. 25.19
Orientation histogram example. Each of the 36 radial bars corresponds to one entry in the orientation histogram h_ϕ. The *length* (radius) of each radial bar with index k is proportional to the accumulated value in the corresponding bin $\mathsf{h}_\phi(k)$ and its orientation is ϕ_k.

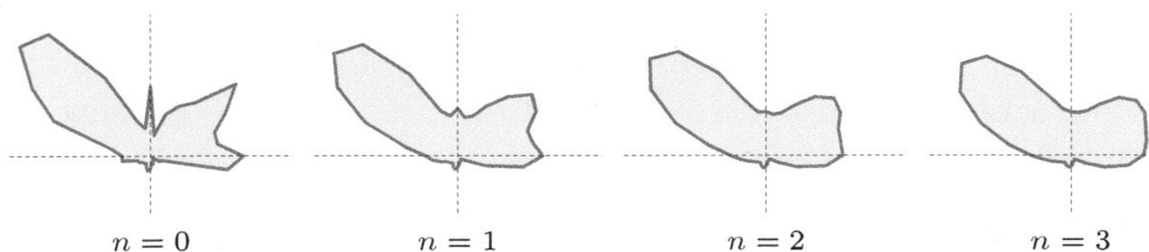

Fig. 25.20
Smoothing the orientation histogram (from Fig. 25.19) by repeatedly applying a circular low-pass filter with the 1D kernel $H = \frac{1}{4} \cdot (1, \mathbf{2}, 1)$.

Orientation histogram smoothing

Figure 25.19 shows a geometric rendering of the orientation histogram that explains the relevance of the cell indexes (discrete angles ϕ_k) and the accumulated quantities (z). Before calculating the dominant orientations, the raw orientation histogram h_ϕ is usually smoothed by applying a (circular) low-pass filter, typically a simple 3-tap Gaussian or box-type filter (see procedure SmoothCircular() in Alg. 25.7, lines 6–16).[22] Stronger smoothing is achieved by applying the filter multiple times, as illustrated in Fig. 25.20. In practice, two to three smoothing iterations appear to be sufficient.

Locating and interpolating orientation peaks

After smoothing the orientation histogram, the next step is to detect the peak entries in h_ϕ. A bin k is considered a significant orientation peak if $\mathsf{h}_\phi(k)$ is a local maximum and its value is not less than a certain fraction of the maximum histogram entry, that is, only if

[22] Histogram smoothing is not mentioned in the original SIFT publication [153] but used in most implementations.

$$\begin{aligned}&\mathsf{h}_\phi(k) > \mathsf{h}_\phi((k-1) \bmod \mathrm{n_{orient}}) \wedge \\ &\mathsf{h}_\phi(k) > \mathsf{h}_\phi((k+1) \bmod \mathrm{n_{orient}}) \wedge \\ &\mathsf{h}_\phi(k) > \mathrm{t_{domor}} \cdot \max_i \mathsf{h}_\phi(i)\,,\end{aligned} \tag{25.84}$$

with $\mathrm{t_{domor}} = 0.8$ as a typical limit.

To achieve a finer angular resolution than provided by the orientation histogram bins (typically spaced at 10° steps) alone, a continuous peak orientation is calculated by quadratic interpolation of the neighboring histogram values. Given a discrete peak index k, the interpolated (continuous) peak position $\breve{k}$ is obtained by fitting a quadratic function to the three successive histogram values $\mathsf{h}_\phi(k-1)$, $\mathsf{h}_\phi(k)$, $\mathsf{h}_\phi(k+1)$ as[23]

$$\breve{k} = k + \frac{\mathsf{h}_\phi(k-1) - \mathsf{h}_\phi(k+1)}{2 \cdot \left[\mathsf{h}_\phi(k-1) - 2\,\mathsf{h}_\phi(k) + \mathsf{h}_\phi(k+1)\right]}\,, \tag{25.85}$$

with all indexes taken modulo $\mathrm{n_{orient}}$. From Eqn. (25.81), the (continuous) dominant orientation angle $\theta \in [0, 2\pi)$ is then obtained as

$$\theta = (\breve{k} \bmod \mathrm{n_{orient}}) \cdot \frac{2\pi}{\mathrm{n_{orient}}}\,, \tag{25.86}$$

mit $\theta \in [0, 2\pi)$. In this way, the dominant orientation can be estimated with accuracy much beyond the coarse resolution of the orientation histogram. Note that, in some cases, multiple histogram peaks are obtained for a given key point (see procedure FindPeakOrientations() in Alg. 25.6, lines 18–31). In this event, individual SIFT descriptors are created for each dominant orientation at the same key point position (see Alg. 25.3, line 8).

Figure 25.21 shows the orientation histograms for a set of detected key points in two different images after applying a varying number of smoothing steps. It also shows the interpolated dominant orientations θ calculated from the orientation histograms (Eqn. (25.86)) by the corresponding vectors.

25.3.2 SIFT Descriptor Construction

For each key point $\boldsymbol{k}' = (p, q, x, y)$ and each dominant orientation θ, a corresponding SIFT descriptor is obtained by sampling the surrounding gradients at octave p and level q of the Gaussian scale space $\mathbf{G}$.

Descriptor geometry

The geometry underlying the calculation of SIFT descriptors is illustrated in Fig. 25.22. The descriptor combines the gradient orientation and magnitude from a square region of size $w_\mathrm{d} \times w_\mathrm{d}$, which is centered at the (continuous) position (x, y) of the associated feature point and aligned with its dominant orientation θ. The side length of the descriptor is set to $w_\mathrm{d} = 10 \cdot \dot{\sigma}_q$, where $\dot{\sigma}_q$ denotes the key point's decimated scale (radius of the inner circle). It depends on the key point's scale level q (see Table 25.4).

[23] See Sec. C.1.2 in the Appendix for details.

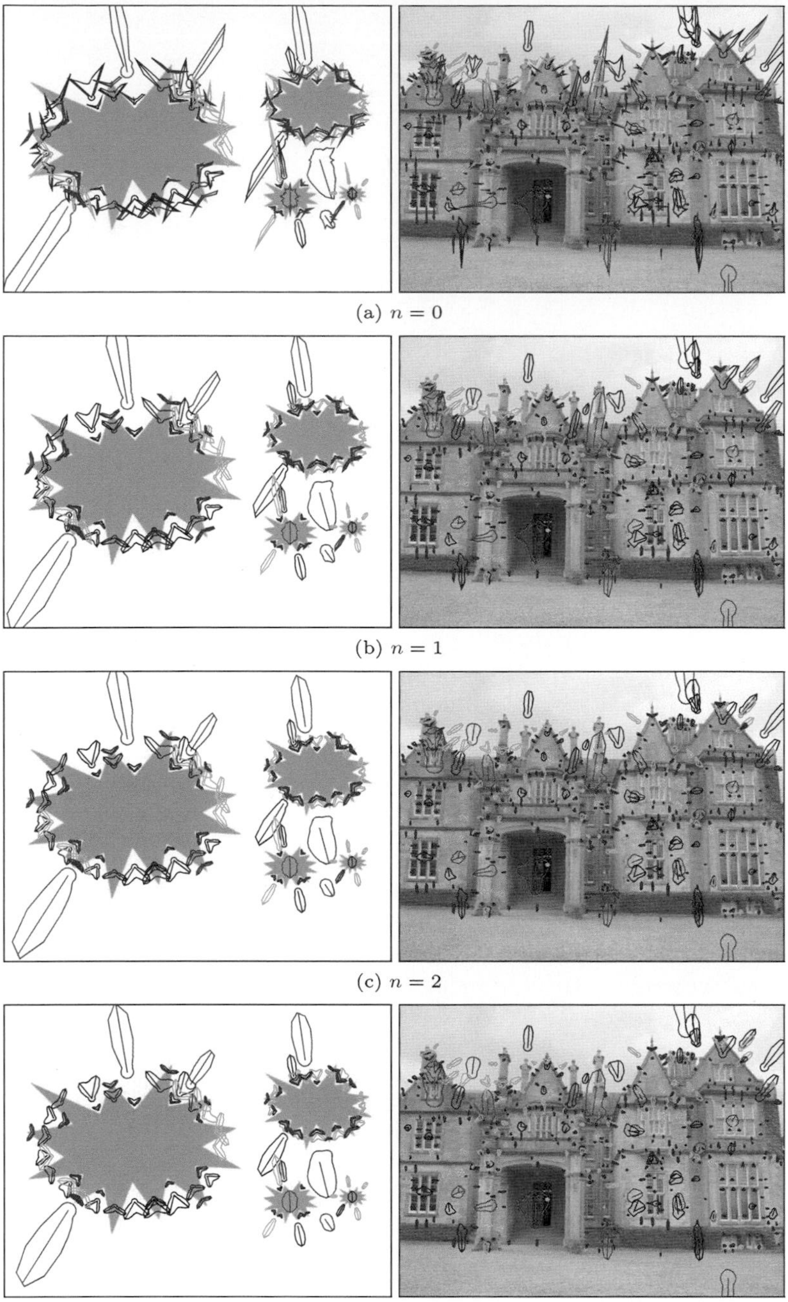

(a) $n = 0$

(b) $n = 1$

(c) $n = 2$

(d) $n = 3$

Fig. 25.21
Orientation histograms and dominant orientations (examples). $n = 0, \ldots, 3$ smoothing iterations were applied to the orientation histograms. The (interpolated) dominant orientations are shown as radial lines that emanate from each feature's center point. The size of the histogram graphs is proportional to the absolute scale ($\sigma_{p,q}$, see Table 25.3) at which the corresponding key point was detected. The colors indicate the index of the containing scale space octave p (red = 0, green = 1, blue = 2, magenta = 3).

The region is partitioned into $\mathrm{n_{spat}} \times \mathrm{n_{spat}}$ sub-squares of identical size; typically $\mathrm{n_{spat}} = 4$ (see Table 25.5). The contribution of each gradient sample is attenuated by a circular Gaussian function of width $\sigma_\mathrm{d} = 0.25 \cdot w_\mathrm{d}$ (blue circle). The weights drop off radially and are practically zero at $r_\mathrm{d} = 2.5 \cdot \sigma_\mathrm{d}$ (green circle in Fig. 25.22). Thus only samples outside this zone need to be included for calculating the descriptor statistics.

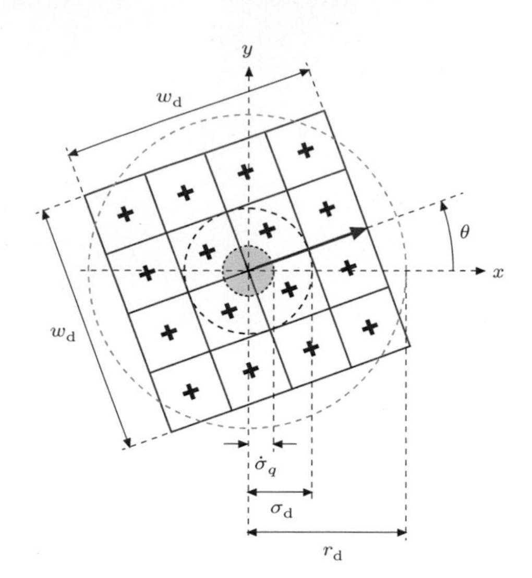

Fig. 25.22
Geometry of a SIFT descriptor. The descriptor is calculated from a square support region that is centered at the key point's position (x, y), aligned to the key point's dominant orientation θ, and partitioned into $n_{spat} \times n_{spat}$ (4×4) sub-squares. The radius of the inner (gray) circle corresponds to the feature point's decimated scale value $(\dot{\sigma}_q)$. The blue circle displays the width (σ_d) of the Gaussian weighting function applied to the gradients; its value is practically zero outside the green circle (r_d).

To achieve rotation invariance, the descriptor region is aligned to the key point's dominant orientation, as determined in the previous steps. To make the descriptor invariant to scale changes, its size w_d (expressed in the grid coordinate units of octave p) is set proportional to the key point's *decimated scale* $\dot{\sigma}_q$ (see Eqn. (25.37)), that is,

$$w_d = s_d \cdot \dot{\sigma}_q = s_d \cdot \sigma_0 \cdot 2^{q/Q}, \tag{25.87}$$

where s_d is a constant size factor. For $s_d = 10$ (see Table 25.5), the descriptor size w_d ranges from 16.0 (at level 0) to 25.4 (at level 2), as listed in Table 25.4. Note that the descriptor size w_d only depends on the scale level index q and is independent of the octave index p. Thus the same descriptor geometry applies to all octaves of the scale space.

Table 25.4
SIFT descriptor dimensions for different scale levels q (for size factor $s_d = 10$ and $Q = 3$ levels per octave). $\dot{\sigma}_q$ is the key point's decimated scale, w_d is the descriptor size, σ_d is the width of the Gaussian weighting function, and r_d is the radius of the descriptor's support region. For $Q = 3$, only scale levels $q = 0, 1, 2$ are relevant. All lengths are expressed in the octave's (i.e., decimated) coordinate units.

q	$\dot{\sigma}_q$	$w_d = s_d \cdot \dot{\sigma}_q$	$\sigma_d = 0.25 \cdot w_d$	$r_d = 2.5 \cdot \sigma_d$
3	3.2000	32.000	8.0000	20.0000
2	2.5398	25.398	6.3495	15.8738
1	2.0159	20.159	5.0398	12.5994
0	1.6000	16.000	4.0000	10.0000
−1	1.2699	12.699	3.1748	7.9369

The descriptor's *spatial resolution* is specified by the parameter n_{spat}. Typically $n_{spat} = 4$ (as shown in Fig. 25.22) and thus the total number of spatial bins is $n_{spat} \times n_{spat} = 16$ (in this case). Each spatial descriptor bin relates to an area of size $(w_d/n_{spat}) \times (w_d/n_{spat})$. For example, at scale level $q = 0$ of any octave, $\dot{\sigma}_0 = 1.6$ and the corresponding descriptor size is $w_d = s_d \cdot \dot{\sigma}_0 = 10 \cdot 1.6 = 16.0$ (see Table 25.4). In this case (illustrated in Fig. 25.23), the descriptor covers 16×16 gradient samples, as suggested in [153]. Figure 25.24 shows an example with M-shaped feature point markers aligned to the dominant orientation and scaled to the descriptor region width w_d of the associated scale level.

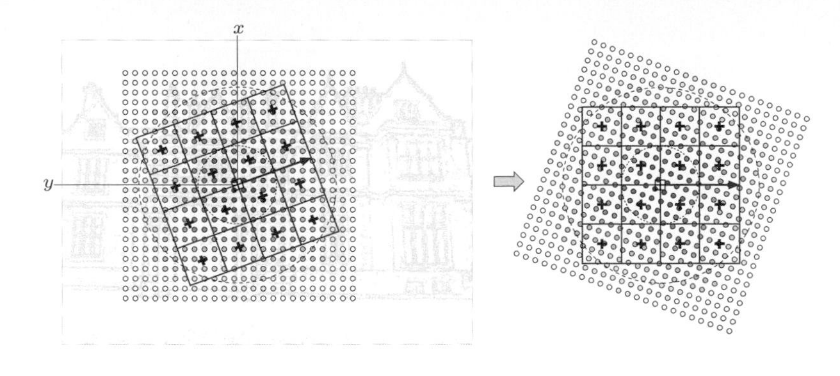

Fig. 25.23
Geometry of the SIFT descriptor in relation to the discrete sample grid of the associated octave (level $q = 0$, parameter $s_d = 10$). In this case, the decimated scale is $\dot{\sigma}_0 = 1.6$ and the width of the descriptor is $w_d = s_d \cdot \dot{\sigma}_0 = 10 \cdot 1.6 = 16.0$.

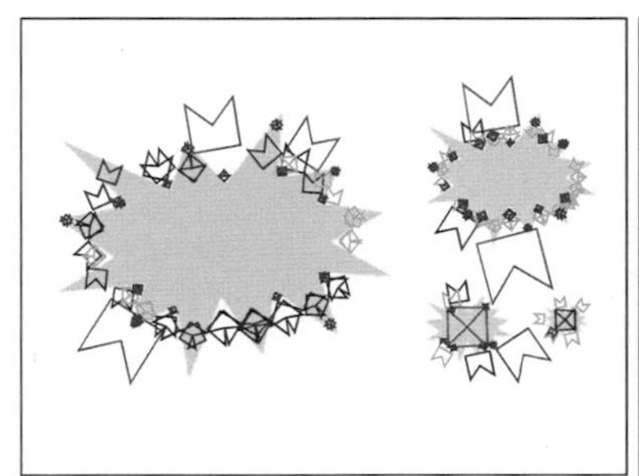

Fig. 25.24
Marked key points aligned to their dominant orientation. Note that multiple feature instances are inserted at key point positions with more than one dominant orientation. The size of the markers is proportional to the absolute scale ($\sigma_{p,q}$, see Table 25.3) at which the corresponding key point was detected. The colors indicate the index of the scale space containing octave p (red = 0, green = 1, blue = 2, magenta = 3).

Gradient features

The actual SIFT descriptor is a feature vector obtained by histogramming the gradient orientations of the Gaussian scale level within the descriptors spatial support region. This requires a 3D histogram $\mathsf{h}_\nabla(i, j, k)$, with two spatial dimensions (i, j) for the $n_{spat} \times n_{spat}$ sub-regions and one additional dimension (k) for n_{angl} gradient orientations. This histogram thus contains $n_{spat} \times n_{spat} \times n_{angl}$ bins.

Figure 25.25 illustrates this structure for the typical setup, with $n_{spat} = 4$ and $n_{angl} = 8$ (see Table 25.5). In this arrangement, eight orientation bins $k = 0, \ldots, 7$ are attached to each of the 16 spatial position bins $(A1, \ldots, D4)$, which makes a total of 128 histogram bins.

For a given key point $\boldsymbol{k}' = (p, q, x, y)$, the histogram h_∇ accumulates the orientations (angles) of the gradients at the Gaussian scale space level $\mathbf{G}_{p,q}$ within the support region around the (continous) center coordinate (x, y). At each grid point (u, v) inside this region, the gradient vector ∇_G is estimated (as described in Eqn. (25.73)), from which the gradient magnitude $E(u, v)$ and orientation $\phi(u, v)$ are calculated (see Eqns. (25.74)–(25.75) and lines 27–31 in Alg. 25.7). For efficiency reasons, $E(u, v)$ and $\phi(u, v)$ are typically pre-calculated for all relevant scale levels.

Each gradient sample contributes to the gradient histogram h_∇ a particular quantity z that depends on the gradient magnitude E and the distance of the sample point (u, v) from the key point's center (x, y). Again a Gaussian weighting function (of width σ_d) is used to attenuate samples with increasing spatial distance; thus the resulting accumulated quantity is

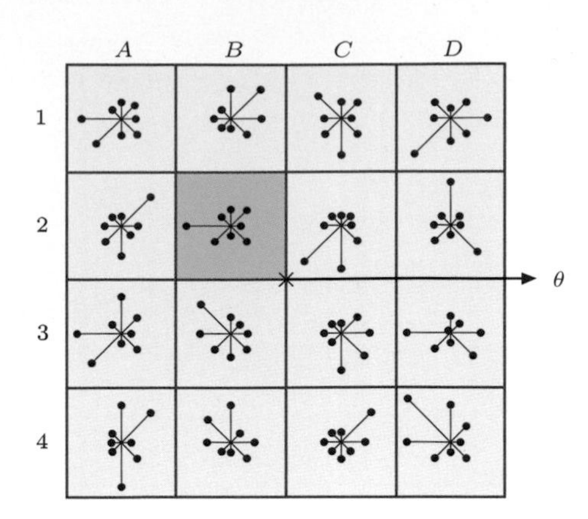

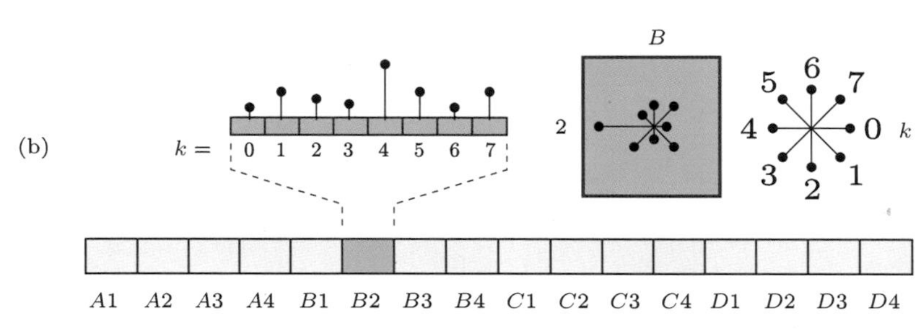

Fig. 25.25 SIFT descriptor structure for $n_{spat} = 4$ and $n_{angl} = 8$. Eight orientation bins $k = 0, \ldots, 7$ are provided for each of the 16 spatial bins $ij = A1, \ldots, D4$. Thus the gradient histogram h_∇ holds 128 cells that are arranged to a 1D feature vector $(A1_0, A1_2 \ldots, D4_6, D4_7)$ as shown in (b).

$$z(u,v) = R(u,v) \cdot w_G = R(u,v) \cdot \exp\left(-\tfrac{(u-x)^2+(v-y)^2}{2\sigma_d^2}\right). \quad (25.88)$$

The width σ_d of the Gaussian function $w_G()$ is proportional to the side length of the descriptor region, with

$$\sigma_d = 0.25 \cdot w_d = 0.25 \cdot \mathsf{s}_d \cdot \dot{\sigma}_q. \quad (25.89)$$

The weighting function drops off radially from the center and is practically zero at distance $r_d = 2.5 \cdot \sigma_d$. Therefore, only gradient samples that are closer to the key point's center than r_d (green circle in Fig. 25.22) need to be considered in the gradient histogram calculation (see Alg. 25.8, lines 7 and 17). For a given key point $\boldsymbol{k}' = (p, q, x, y)$, sampling of the Gaussian gradients can thus be confined to the grid points (u, v) inside the square region bounded by $x \pm r_d$ and $y \pm r_d$ (see Alg. 25.8, lines 8–10 and 15–16). Each sample point (u, v) is then subjected to the affine transformation

$$\begin{pmatrix} u' \\ v' \end{pmatrix} = \frac{1}{w_d} \cdot \begin{pmatrix} \cos(-\theta) & -\sin(-\theta) \\ \sin(-\theta) & \cos(-\theta) \end{pmatrix} \cdot \begin{pmatrix} u-x \\ v-y \end{pmatrix}, \quad (25.90)$$

which performs a rotation by the dominant orientation θ and maps the original (rotated) square of size $w_d \times w_d$ to the unit square with coordinates $u', v' \in [-0.5, +0.5]$ (see Fig. 25.23).

To make feature vectors rotation invariant, the individual gradient orientations $\phi(u, v)$ are rotated by the dominant orientation, that is,

$$\phi'(u,v) = (\phi(u,v) - \theta) \bmod 2\pi, \quad (25.91)$$

with $\phi'(u, v) \in [0, 2\pi)$, such that the relative orientation is preserved.

For each gradient sample, with the continuous coordinates (u', v', ϕ'), the corresponding quantity $z(u, v)$ (Eqn. (25.88)) is accumulated into the 3D gradient histogram h_∇. For a complete description of this step see procedure UpdateGradientHistogram() in Alg. 25.9. It first maps the coordinates (u', v', ϕ') (see Eqn. (25.90)) to the continuous histogram position (i', j', k') by

$$\begin{aligned} i' &= \mathrm{n_{spat}} \cdot u' + 0.5 \cdot (\mathrm{n_{spat}} - 1), \\ j' &= \mathrm{n_{spat}} \cdot v' + 0.5 \cdot (\mathrm{n_{spat}} - 1), \\ k' &= \phi' \cdot \frac{\mathrm{n_{angl}}}{2\pi}, \end{aligned} \tag{25.92}$$

such that $i', j' \in [-0.5, \mathrm{n_{spat}} - 0.5]$ and $k' \in [0, \mathrm{n_{angl}})$.

Analogous to inserting into a continuous position of a 1D histogram by linear interpolation over *two* bins (see Fig. 25.18), the quantity z is distributed over *eight* neighboring histogram bins by *tri-linear* interpolation. The quantiles of z contributing to the individual histogram bins are determined by the distances of the coordinates (i', j', k') from the discrete indexes (i, j, k) of the affected histogram bins. The indexes (i, j, k) are found as the set of possible combinations $\{i_0, i_1\} \times \{j_0, j_1\} \times \{k_0, k_1\}$, with

$$\begin{aligned} i_0 &= \lfloor i' \rfloor, & i_1 &= (i_0 + 1), \\ j_0 &= \lfloor j' \rfloor, & j_1 &= (j_0 + 1), \\ k_0 &= \lfloor k' \rfloor \bmod \mathrm{n_{angl}}, & k_1 &= (k_0 + 1) \bmod \mathrm{n_{angl}}, \end{aligned} \tag{25.93}$$

and the corresponding quantiles (weights) are

$$\begin{aligned} \alpha_0 &= \lfloor i' \rfloor + 1 - i' = i_1 - i', & \alpha_1 &= 1 - \alpha_0, \\ \beta_0 &= \lfloor j' \rfloor + 1 - j' = j_1 - j', & \beta_1 &= 1 - \beta_0, \\ \gamma_0 &= \lfloor k' \rfloor + 1 - k', & \gamma_1 &= 1 - \gamma_0, \end{aligned} \tag{25.94}$$

and the (eight) affected bins of the gradient histogram are finally updated as

$$\begin{aligned} \mathsf{h}_\nabla(i_0, j_0, k_0) &\stackrel{+}{\leftarrow} z \cdot \alpha_0 \cdot \beta_0 \cdot \gamma_0, \\ \mathsf{h}_\nabla(i_1, j_0, k_0) &\stackrel{+}{\leftarrow} z \cdot \alpha_1 \cdot \beta_0 \cdot \gamma_0, \\ \mathsf{h}_\nabla(i_0, j_1, k_0) &\stackrel{+}{\leftarrow} z \cdot \alpha_0 \cdot \beta_1 \cdot \gamma_0, \\ &\;\;\vdots \qquad\qquad \vdots \\ \mathsf{h}_\nabla(i_1, j_1, k_1) &\stackrel{+}{\leftarrow} z \cdot \alpha_1 \cdot \beta_1 \cdot \gamma_1. \end{aligned} \tag{25.95}$$

Attention must be paid to the fact that the coordinate k represents an orientation and must therefore be treated in a *circular* manner, as illustrated in Fig. 25.26 (also see Alg. 25.9, lines 11–12).

For each histogram bin, the range of contributing gradient samples covers half of each neighboring bin, that is, the support regions of neighboring bins overlap, as illustrated in Fig. 25.27.

Normalizing SIFT descriptors

The elements of the gradient histogram h_∇ are the raw material for the SIFT feature vectors $\boldsymbol{f}_{\mathrm{sift}}$. The process of calculating the feature vectors from the gradient histogram is described in Alg. 25.10.

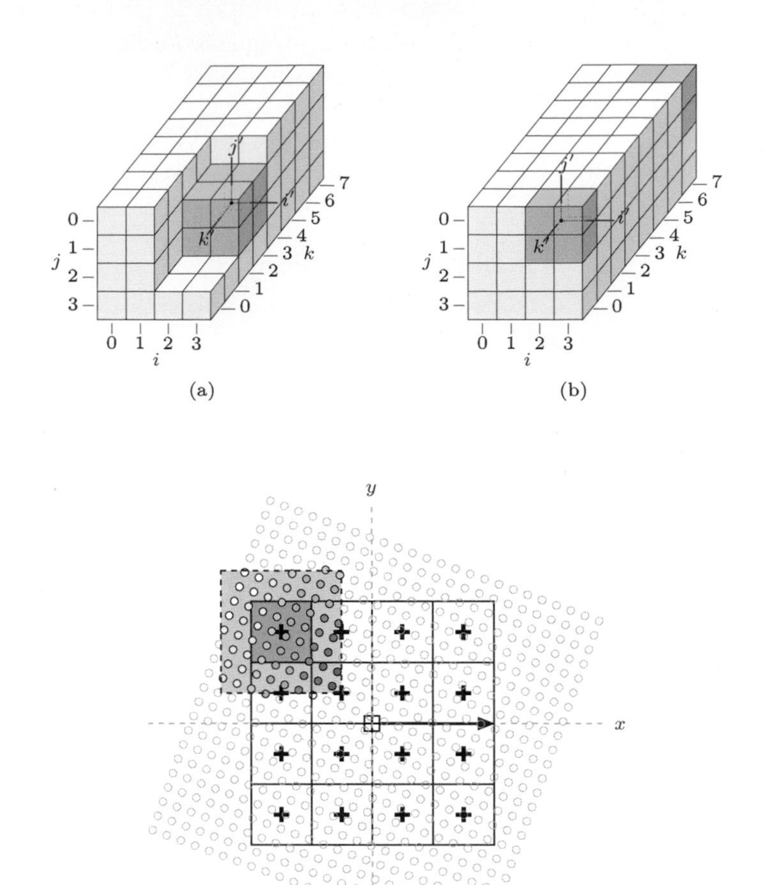

Fig. 25.26
3D structure of the gradient histogram, with $n_{spat} \times n_{spat} = 4 \times 4$ bins for the spatial dimensions (i, j) and $n_{angl} = 8$ bins along the orientation axis (k). For the histogram to accumulate a quantity z into some continuous position (i', j', k'), eight adjacent bins receive different quantiles of z that are determined by tri-linear interpolation (a). Note that the bins along the orientation axis ϕ are treated circularly; for example, bins at $k = 0$ are also considered adjacent to the bins at $k = 7$ (b).

Fig. 25.27
Overlapping support regions in the gradient field. Due to the tri-linear interpolation used in the histogram calculation, the spatial regions associated with the cells of the orientation histogram h_∇ *overlap*. The shading of the circles indicates the weight w_G assigned to each sample by the Gaussian weighting function, whose value depends on the distance of each sample from the key point's center (see Eqn. (25.88)).

Initially, the 3D gradient histogram h_∇ (which contains continuous values) of size $n_{spat} \times n_{spat} \times n_{angl}$ is flattened to a 1D vector $\boldsymbol{f}$ of length $n_{spat}^2 \cdot n_{angl}$ (typ. 128), with

$$\boldsymbol{f}\big((i \cdot n_{spat} + j) \cdot n_{angl} + k\big) \;\leftarrow\; \mathsf{h}_\nabla(i, j, k), \tag{25.96}$$

for $i, j = 0, \ldots, n_{spat}-1$ and $k = 0, \ldots, n_{angl}-1$. The elements in $\boldsymbol{f}$ are thus arranged in the same order as shown in Fig. 25.25, with the orientation index k being the fastest moving and the spatial index i being the slowest (see Alg. 25.10, lines 3–8).[24]

Changes in image contrast have a linear impact upon the gradient magnitude and thus also upon the values of the feature vector $\boldsymbol{f}$. To eliminate these effects, the vector $\boldsymbol{f}$ is subsequently normalized to

$$\boldsymbol{f}(m) \;\leftarrow\; \frac{1}{\|\boldsymbol{f}\|} \cdot \boldsymbol{f}(m), \tag{25.97}$$

for all m, such that $\boldsymbol{f}$ has unit norm (see Alg. 25.10, line 9). Since the gradient is calculated from local pixel differences, changes in absolute

[24] Note that different ordering schemes for arranging the elements of the feature vector are used in various SIFT implementations. For successful matching, the ordering of the elements must be identical, of course.

brightness do not affect the gradient magnitude, unless saturation occurs. Such nonlinear illumination changes tend to produce peak gradient values, which are compensated for by clipping the values of $\boldsymbol{f}$ to a predefined maximum $\mathsf{t}_{\mathrm{fclip}}$, that is,

$$\boldsymbol{f}(m) \leftarrow \min(\boldsymbol{f}(m), \mathsf{t}_{\mathrm{fclip}}), \tag{25.98}$$

with typically $\mathsf{t}_{\mathrm{fclip}} = 0.2$, as suggested in [153] (see Alg. 25.10, line 10). After this step, $\boldsymbol{f}$ is normalized once again, as in Eqn. (25.97). Finally, the real-valued feature vector $\boldsymbol{f}$ is converted to an integer vector by

$$\boldsymbol{f}_{\mathrm{sift}}(m) \leftarrow \min\bigl(\mathrm{round}(\mathsf{s}_{\mathrm{fscale}} \cdot \boldsymbol{f}(m)), 255)\bigr), \tag{25.99}$$

with $\mathsf{s}_{\mathrm{fscale}}$ being a predefined constant (typ. $\mathsf{s}_{\mathrm{fscale}} = 512$). The elements of $\boldsymbol{f}_{\mathrm{sift}}$ are in the range $[0, 255]$ to be conveniently encoded and stored as a byte sequence (see Alg. 25.10, line 12).

The final SIFT descriptor for a given key point $\boldsymbol{k}' = (p, q, x, y)$ is a tuple

$$\boldsymbol{s} = \langle x', y', \sigma, \theta, \boldsymbol{f}_{\mathrm{sift}} \rangle, \tag{25.100}$$

which contains the key point's interpolated position x', y' (in original image coordinates), the absolute scale σ, its dominant orientation θ, and the corresponding integer-valued gradient feature vector $\boldsymbol{f}_{\mathrm{sift}}$ (see Alg. 25.8, line 27). Remember that multiple SIFT descriptors may be produced for different dominant orientations located at the same key point position. These will have the same position and scale values but different θ and $\boldsymbol{f}_{\mathrm{sift}}$ data.

25.4 SIFT Algorithm Summary

This section contains a collection of algorithms that summarizes the SIFT feature extraction process described in the previous sections of this chapter.

Algorithm 25.3 shows the top-level procedure GetSiftFeatures(I), which returns a sequence of SIFT feature descriptors for the given image I. The remaining parts of Alg. 25.3 describe the key point detection as extrema of the DOG scale space. The refinement of key point positions is covered in Alg. 25.4. Algorithm 25.5 contains the procedures used for neighborhood operations, detecting local extrema, and the calculation of the gradient and Hessian matrix in 3D. Algorithm 25.6 covers the operations related to finding the dominant orientations at a given key point location, based on the orientation histogram that is calculated in Alg. 25.7. The final formation of the SIFT descriptors is described in Alg. 25.8, which is based on the procedures defined in Algs. 25.9 and 25.10. The global constants used throughout these algorithms are listed in Table 25.5, together with the corresponding Java identifiers in the associated source code (see Sec. 25.7).

Table 25.5
Predefined constants used in the SIFT algorithms (Algs. 25.3–25.11).

Scale space parameters

Symbol	*Java id.*	*Value*	*Description*
Q	`Q`	3	scale steps (levels) per octave
P	`P`	4	number of scale space octaves
σ_s	`sigma_s`	0.5	sampling scale (nominal smoothing of the input image)
σ_0	`sigma_0`	1.6	base scale of level 0 (base smoothing)

Key-point detection

Symbol	*Java id.*	*Value*	*Description*
n_{orient}	`n_Orient`	36	number of orientation bins (angular resolution) used for calculating the dominant key point orientation
n_{refine}	`n_Refine`	5	max. number of iterations for repositioning a key point
n_{smooth}	`n_Smooth`	2	number of smoothing iterations applied to the orientation histogram
ρ_{max}	`rho_Max`	10.0	max. ratio of principal curvatures $(3, \ldots, 10)$
t_{domor}	`t_DomOr`	0.8	min. value in orientation histogram for selecting dominant orientations (rel. to max. entry)
t_{extrm}	`t_Extrm`	0.0	min. difference w.r.t. any neighbor for extrema detection
t_{mag}	`t_Mag`	0.01	min. DoG magnitude for initial key point candidates
t_{peak}	`t_Peak`	0.01	min. DoG magnitude at interpolated peaks

Feature descriptor

Symbol	*Java id.*	*Value*	*Description*
n_{spat}	`n_Spat`	4	number of spatial descriptor bins along each x/y axis
n_{angl}	`n_Angl`	16	number of angular descriptor bins
s_{d}	`s_Desc`	10.0	spatial size factor of descriptor (relative to feature scale)
s_{fscale}	`s_Fscale`	512.0	scale factor for converting normalized feature values to byte values in $[0, 255]$
t_{fclip}	`t_Fclip`	0.2	max. value for clipping elements of normalized feature vectors

Feature matching

Symbol	*Java id.*	*Value*	*Description*
ρ_{max}	`rho_ax`	0.8	max. ratio of best and second-best matching feature distance

25.5 Matching SIFT Features

Most applications of SIFT features aim at locating corresponding interest points in two or more images of the same scene, for example, for matching stereo pairs, panorama stitching, or feature tracking. Other applications like self-localization or object recognition might use a large database of model descriptors and the task is to match these to the SIFT features detected in a new image or video sequence. All these applications require possibly large numbers of pairs of SIFT features to be compared reliably and efficiently.

25.5.1 Feature Distance and Match Quality

In a typical situation, two sequences of SIFT features $S^{(a)}$ and $S^{(b)}$ are extracted independently from a pair of input images I_a, I_b, that is,

$$S^{(a)} = (\boldsymbol{s}_1^{(a)}, \boldsymbol{s}_2^{(a)}, \ldots, \boldsymbol{s}_{N_a}^{(a)}) \quad \text{and} \quad S^{(b)} = (\boldsymbol{s}_1^{(b)}, \boldsymbol{s}_2^{(b)}, \ldots, \boldsymbol{s}_{N_b}^{(b)}).$$

The goal is to find matching descriptors in the two feature sets. The similarity between a given pair of descriptors, $\boldsymbol{s}_i = \langle x_i, y_i, \sigma_i, \theta_i, \boldsymbol{f}_i \rangle$ and $\boldsymbol{s}_j = \langle x_j, y_j, \sigma_j, \theta_j, \boldsymbol{f}_j \rangle$, is measured by the *distance* between the corresponding feature vectors $\boldsymbol{f}_i, \boldsymbol{f}_j$, that is,

```
1:  GetSiftFeatures(I)
        Input: I, the source image (scalar-valued).
        Returns a sequence of SIFT feature descriptors detected in I.
2:      ⟨G, D⟩ ← BuildSiftScaleSpace(I, σ_s, σ_0, P, Q)         ▷ Alg. 25.2
3:      C ← GetKeyPoints(D)
4:      S ← ()                                ▷ empty list of SIFT descriptors
5:      for all k' ∈ C do                                   ▷ k' = (p, q, x, y)
6:          A ← GetDominantOrientations(G, k')             ▷ Alg. 25.6
7:          for all θ ∈ A do
8:              s ← MakeSiftDescriptor(G, k', θ)           ▷ Alg. 25.8
9:              S ← S ⌣ (s)
10:     return S

11: GetKeypoints(D)
        D: DoG scale space (with P octaves, each containing Q levels).
        Returns a set of key points located in D.
12:     C ← ()                                      ▷ empty list of key points
13:     for p ← 0, ..., P−1 do                            ▷ for all octaves p
14:         for q ← 0, ..., Q−1 do                    ▷ for all scale levels q
15:             E ← FindExtrema(D, p, q)
16:             for all k ∈ E do                           ▷ k = (p, q, u, v)
17:                 k' ← RefineKeyPosition(D, k)           ▷ Alg. 25.4
18:                 if k' ≠ nil then                      ▷ k' = (p, q, x, y)
19:                     C ← C ⌣ (k')              ▷ add refined key point k'
20:     return C

21: FindExtrema(D, p, q)
22:     D_{p,q} ← GetScaleLevel(D, p, q)
23:     (M, N) ← Size(D_{p,q})
24:     E ← ()                                        ▷ empty list of extrema
25:     for u ← 1, ..., M−2 do
26:         for v ← 1, ..., N−2 do
27:             if |D_{p,q}(u, v)| > t_mag then
28:                 k ← (p, q, u, v)
29:                 N_c ← GetNeighborhood(D, k)            ▷ Alg. 25.5
30:                 if IsExtremum(N_c) then                ▷ Alg. 25.5
31:                     E ← E ⌣ (k)                           ▷ add k to E
32:     return E
```

Alg. 25.3
SIFT feature extraction (part 1). Top-level SIFT procedure. Global parameters: σ_s, σ_0, t_{mag}, Q, P (see Table 25.5).

$$\mathrm{dist}(\boldsymbol{s}_i, \boldsymbol{s}_j) := \left\| \boldsymbol{f}_i - \boldsymbol{f}_j \right\|, \tag{25.101}$$

where $\|\cdots\|$ denotes an appropriate norm (typically Euclidean, alternatives will be discussed further).[25]

Note that this distance is measured between individual points distributed in a high-dimensional (typically 128-dimensional) vector space that is only sparsely populated. Since there is *always* a best-matching counterpart for a given descriptor, matches may occur between unrelated features even if the correct feature is not contained in the target set. This is particularly critical if feature matching is used to determine whether two images show any correspondence at all.

Obviously, significant matches should exhibit small feature distances but setting a *fixed limit* on the acceptable feature distance

[25] See also Sec. B.1.2 in the Appendix.

Alg. 25.4
SIFT feature extraction (part 2). Position refinement. Global parameters: $\mathrm{n}_{\mathrm{refine}}$, $\mathrm{t}_{\mathrm{peak}}$, $\rho_{\max}$ (see Table 25.5).

```
1: RefineKeyPosition(D, k)
     Input: D, hierarchical DoG scale space; k = (p, q, u, v), candidate
     (extremal) position.
     Returns a refined key point k' or nil if no proper key point could
     be localized at or near the extremal position k.
2:   a_max ← (ρ_max+1)^2 / ρ_max                       ▷ see Eq. 25.72
3:   k' ← nil                                           ▷ refined key point
4:   done ← false
5:   n ← 1                               ▷ number of repositioning steps
6:   while ¬done ∧ n ≤ n_refine ∧ IsInside(D, k) do
7:     N_c ← GetNeighborhood(D, k)                     ▷ Alg. 25.5
8:     ∇ = (d_x, d_x, d_σ)^T ← Gradient(N_c)           ▷ Alg. 25.5
9:     H_D = ( d_xx d_xy d_xσ
               d_xy d_yy d_yσ
               d_xσ d_yσ d_σσ ) ← Hessian(N_c)         ▷ Alg. 25.5
10:    if det(H_D) = 0 then                 ▷ H_D is not invertible
11:      done ← true                ▷ ignore this point and finish
12:    else
13:      d = (x', y', σ')^T ← −H_D^{-1} · ∇            ▷ Eq. 25.60
14:      if |x'|<0.5 ∧ |y'|<0.5 then ▷ stay in the same DoG cell
15:        done ← true
16:        D_peak ← N_c(0,0,0) + 1/2 · ∇^T·d           ▷ Eq. 25.61
17:        H_xy ← ( d_xx d_xy
                    d_xy d_yy )      ▷ extract 2D Hessian from H_D
18:        if |D_peak| > t_peak ∧ det(H_xy) > 0 then
19:          a ← [trace(H_xy)]^2 / det(H_xy)           ▷ Eq. 25.69
20:          if a < a_max then    ▷ suppress edges, Eq. 25.72
21:            k' ← k + (0, 0, x', y')^T         ▷ refined key point
22:      else
           Move to a neighboring DoG position at same level p, q:
23:        u' ← min(1, max(−1, round(x'))) ▷ move by max. ±1
24:        v' ← min(1, max(−1, round(y'))) ▷ move by max. ±1
25:        k ← k + (0, 0, u', v')^T
26:    n ← n + 1
27:  return k'          ▷ k' is either a refined key point position or nil
```

turns out to be inappropriate in practice, since some descriptors are more discriminative than others. The solution proposed in [153] is to compare the distance obtained for the *best* feature match to that of the *second-best* match. For a given reference descriptor $\boldsymbol{s}_{\mathrm{r}} \in S^{(a)}$, the best match is defined as the descriptor $\boldsymbol{s}_1 \in S^{(b)}$ which has the smallest distance from $\boldsymbol{s}_{\mathrm{r}}$ in the multi-dimensional feature space, that is,

$$\boldsymbol{s}_1 = \operatorname*{argmin}_{\boldsymbol{s}_j \in S^{(b)}} \operatorname{dist}(\boldsymbol{s}_{\mathrm{r}}, \boldsymbol{s}_j), \qquad (25.102)$$

Alg. 25.5
SIFT feature extraction (part 3): Neighborhood operations. Global parameters: Q, $\mathrm{t_{extrm}}$ (see Table 25.5).

1: **IsInside**($\mathbf{D}, \boldsymbol{k}$)
Checks if coordinate $\boldsymbol{k} = (p, q, u, v)$ is inside the DoG scale space $\mathbf{D}$.
2: $(p, q, u, v) \leftarrow \boldsymbol{k}$
3: $(M, N) \leftarrow \mathrm{Size}(\mathrm{GetScaleLevel}(\mathbf{D}, p, q))$
4: **return** $(0 < u < M-1) \wedge (0 < v < N-1) \wedge (0 \leq q < Q)$

5: **GetNeighborhood**($\mathbf{D}, \boldsymbol{k}$) ▷ $\boldsymbol{k} = (p, q, u, v)$
Collects and returns the $3\times 3\times 3$ neighborhood values around position $\boldsymbol{k}$ in the hierarchical DoG scale space $\mathbf{D}$.
6: Create map $\mathsf{N}_c : \{-1, 0, 1\}^3 \mapsto \mathbb{R}$
7: **for all** $(i, j, k) \in \{-1, 0, 1\}^3$ **do** ▷ collect 3×3×3 neighborhood
8: $\quad \mathsf{N}_c(i, j, k) \leftarrow \mathbf{D}_{p,q+k}(u+i, v+j)$
9: **return** N_c

10: **IsExtremum**(N_c) ▷ N_c is a 3×3×3 map
Determines if the center of the 3D neighborhood N_c is either a local minimum or maximum by the threshold $\mathrm{t_{extrm}} \geq 0$. Returns a boolean value (i.e., true or false).
11: $c \leftarrow \mathsf{N}_c(0, 0, 0)$ ▷ center DoG value
12: $\mathit{isMin} \leftarrow c < 0 \wedge (c + \mathrm{t_{extrm}}) < \min\limits_{\substack{(i,j,k)\neq\\(0,0,0)}} \mathsf{N}_c(i, j, k)$ ▷ s. Eq. 25.54
13: $\mathit{isMax} \leftarrow c > 0 \wedge (c - \mathrm{t_{extrm}}) > \max\limits_{\substack{(i,j,k)\neq\\(0,0,0)}} \mathsf{N}_c(i, j, k)$ ▷ s. Eq. 25.55
14: **return** $\mathit{isMin} \vee \mathit{isMax}$

15: **Gradient**(N_c) ▷ N_c is a 3×3×3 map
Returns the estim. gradient vector (∇) for the 3D neighborhood N_c.
16: $d_\mathrm{x} \leftarrow 0.5 \cdot (\mathsf{N}_c(1, 2, 1) - \mathsf{N}_c(1, 0, 1))$
17: $d_\mathrm{y} \leftarrow 0.5 \cdot (\mathsf{N}_c(1, 1, 2) - \mathsf{N}_c(1, 1, 0))$ ▷ see Eq. 25.56
18: $d_\sigma \leftarrow 0.5 \cdot (\mathsf{N}_c(2, 1, 1) - \mathsf{N}_c(0, 1, 1))$
19: $\nabla \leftarrow (d_\mathrm{x}, d_\mathrm{y}, d_\sigma)^\mathsf{T}$
20: **return** ∇

21: **Hessian**(N_c) ▷ N_c is a 3×3×3 map
Returns the estim. Hessian matrix ($\mathbf{H}$) for the neighborhood N_c.
22: $d_{xx} \leftarrow \mathsf{N}_c(-1, 0, 0) - 2{\cdot}\mathsf{N}_c(0, 0, 0) + \mathsf{N}_c(1, 0, 0)$ ▷ see Eq. 25.58
23: $d_{yy} \leftarrow \mathsf{N}_c(0, -1, 0) - 2{\cdot}\mathsf{N}_c(0, 0, 0) + \mathsf{N}_c(0, 1, 0)$
24: $d_{\sigma\sigma} \leftarrow \mathsf{N}_c(0, 0, -1) - 2{\cdot}\mathsf{N}_c(0, 0, 0) + \mathsf{N}_c(0, 0, 1)$
25: $d_{xy} \leftarrow [\,\mathsf{N}_c(1, 1, 0) - \mathsf{N}_c(-1, 1, 0) - \mathsf{N}_c(1, -1, 0) + \mathsf{N}_c(-1, -1, 0)\,]/4$
26: $d_{x\sigma} \leftarrow [\,\mathsf{N}_c(1, 0, 1) - \mathsf{N}_c(-1, 0, 1) - \mathsf{N}_c(1, 0, -1) + \mathsf{N}_c(-1, 0, -1)\,]/4$
27: $d_{y\sigma} \leftarrow [\,\mathsf{N}_c(0, 1, 1) - \mathsf{N}_c(0, -1, 1) - \mathsf{N}_c(0, 1, -1) + \mathsf{N}_c(0, -1, -1)\,]/4$
28: $\mathbf{H} \leftarrow \begin{pmatrix} d_{xx} & d_{xy} & d_{x\sigma} \\ d_{xy} & d_{yy} & d_{y\sigma} \\ d_{x\sigma} & d_{y\sigma} & d_{\sigma\sigma} \end{pmatrix}$
29: **return** $\mathbf{H}$

and the primary distance is $d_{\mathrm{r},1} = \mathrm{dist}(\boldsymbol{s}_\mathrm{r}, \boldsymbol{s}_1)$. Analogously, the second-best matching descriptor is

$$\boldsymbol{s}_2 = \operatorname*{argmin}_{\substack{\boldsymbol{s}_j \in \mathcal{S}^{(b)},\\ \boldsymbol{s}_j \neq \boldsymbol{s}_1}} \mathrm{dist}(\boldsymbol{s}_\mathrm{r}, \boldsymbol{s}_j), \qquad (25.103)$$

and the corresponding distance is $d_{\mathrm{r},2} = \mathrm{dist}(\boldsymbol{s}_\mathrm{r}, \boldsymbol{s}_2)$, with $d_{\mathrm{r},1} \leq d_{\mathrm{r},2}$. Reliable matches are expected to have a distance to the primary

Alg. 25.6
SIFT feature extraction (part 4): Key point orientation assignment. Global parameters: n_{smooth}, t_{domor} (see Table 25.5).

```
1: GetDominantOrientations(G, k')
     Input: G, hierarchical Gaussian scale space; k' = (p, q, x, y), refined key point at octave p, scale level q and spatial position x, y (in octave's coordinates).
     Returns a list of dominant orientations for the key point k'.
2:   h_φ ← GetOrientationHistogram(G, k')                    ▷ Alg. 25.7
3:   SmoothCircular(h_φ, n_smooth)
4:   A ← FindPeakOrientations(h_φ)
5:   return A

6: SmoothCircular(x, n_iter)
     Smooths the real-valued vector x = (x_0, ..., x_{n-1}) circularly using the 3-element kernel H = (h_0, h_1, h_2), with h_1 as the hot-spot. The filter operation is applied n_iter times and "in place", i.e., the vector x is modified.
7:   (h_0, h_1, h_2) ← 1/4 · (1, 2, 1)                       ▷ 1D filter kernel
8:   n ← Size(x)
9:   for i ← 1, ..., n_iter do
10:      s ← x(0)
11:      p ← x(n−1)
12:      for j ← 0, ..., n−2 do
13:          c ← x(j)
14:          x(j) ← h_0·p + h_1·x(j) + h_2·x(j+1)
15:          p ← c
16:      x(n−1) ← h_0·p + h_1·x(n−1) + h_2·s
17:  return

18: FindPeakOrientations(h_φ)
     Returns a (possibly empty) sequence of dominant directions (angles) obtained from the orientation histogram h_φ.
19:  n ← Size(h_φ)
20:  A ← ()
21:  h_max ← max_{0≤i<n} h_φ(i)
22:  for k ← 0, ..., n−1 do
23:      h_c ← h(k)
24:      if h_c > t_domor · h_max then              ▷ only accept dominant peaks
25:          h_p ← h_φ((k−1) mod n)
26:          h_n ← h_φ((k+1) mod n)
27:          if (h_c > h_p) ∧ (h_c > h_n) then      ▷ local max. at index k
28:              k̆ ← k + (h_p − h_n) / (2·(h_p − 2·h_c + h_n))   ▷ quadr. interpol., Eq. 25.85
29:              θ ← (k̆ · 2π/n) mod 2π          ▷ domin. orientation, Eq. 25.86
30:              A ← A ⌣ (θ)
31:  return A
```

feature $\boldsymbol{s}_1$ that is considerably smaller than the distance to any other feature in the target set. In the case of a weak or ambiguous match, on the other hand, it is likely that other matches exist at a distance similar to $d_{r,1}$, including the second-best match $\boldsymbol{s}_2$. Comparing the best and the second-best distances thus provides information about the likelihood of a false match. For this purpose, we define the *feature distance ratio*

$$\rho_{match}(\boldsymbol{s}_r, \boldsymbol{s}_1, \boldsymbol{s}_2) := \frac{d_{r,1}}{d_{r,2}} = \frac{\mathrm{dist}(\boldsymbol{s}_r, \boldsymbol{s}_1)}{\mathrm{dist}(\boldsymbol{s}_r, \boldsymbol{s}_2)}, \tag{25.104}$$

Alg. 25.7
SIFT feature extraction (part 5): Calculation of the orientation histogram and gradients from Gaussian scale levels. Global parameters: n_{orient} (see Table 25.5).

1: **GetOrientationHistogram**$(\mathbf{G}, \boldsymbol{k}')$
Input: $\mathbf{G}$, hierarchical Gaussian scale space; $\boldsymbol{k}' = (p, q, x, y)$, refined key point at octave p, scale level q and relative position x, y.
Returns the gradient orientation histogram for key point $\boldsymbol{k}'$.

2: $\mathbf{G}_{p,q} \leftarrow \text{GetScaleLevel}(\mathbf{G}, p, q)$
3: $(M, N) \leftarrow \text{Size}(\mathbf{G}_{p,q})$
4: Create a new map $\mathsf{h}_\phi : [0, n_{orient}-1] \mapsto \mathbb{R}$. ▷ new histogram h_ϕ
5: **for** $i \leftarrow 0, \ldots, n_{orient}-1$ **do** ▷ initialize h_ϕ to zero
6: $\quad \mathsf{h}_\phi(i) \leftarrow 0$
7: $\sigma_w \leftarrow 1.5 \cdot \sigma_0 \cdot 2^{q/Q}$ ▷ σ of Gaussian weight fun., see Eq. 25.76
8: $r_w \leftarrow \max(1, 2.5 \cdot \sigma_w)$ ▷ rad. of weight fun., see Eq. 25.77
9: $u_{min} \leftarrow \max(\lfloor x - r_w \rfloor, 1)$
10: $u_{max} \leftarrow \min(\lceil x + r_w \rceil, M-2)$
11: $v_{min} \leftarrow \max(\lfloor y - r_w \rfloor, 1)$
12: $v_{max} \leftarrow \min(\lceil y + r_w \rceil, N-2)$
13: **for** $u \leftarrow u_{min}, \ldots, u_{max}$ **do**
14: $\quad$ **for** $v \leftarrow v_{min}, \ldots, v_{max}$ **do**
15: $\quad\quad r^2 \leftarrow (u-x)^2 + (v-y)^2$
16: $\quad\quad$ **if** $r^2 < r_w^2$ **then**
17: $\quad\quad\quad (E, \phi) \leftarrow \text{GetGradientPolar}(\mathbf{G}_{p,q}, u, v)$ ▷ see below
18: $\quad\quad\quad w_G \leftarrow \exp\bigl(-\frac{(u-x)^2+(v-y)^2}{2\sigma_w^2}\bigr)$ ▷ Gaussian weight
19: $\quad\quad\quad z \leftarrow E \cdot w_G$ ▷ quantity to accumulate
20: $\quad\quad\quad \kappa_\phi \leftarrow \frac{n_{orient}}{2\pi} \cdot \phi$ ▷ $\kappa_\phi \in [-\frac{n_{orient}}{2}, +\frac{n_{orient}}{2}]$
21: $\quad\quad\quad \alpha \leftarrow \kappa_\phi - \lfloor \kappa_\phi \rfloor$ ▷ $\alpha \in [0, 1]$
22: $\quad\quad\quad k_0 \leftarrow \lfloor \kappa_\phi \rfloor \bmod n_{orient}$ ▷ lower bin index
23: $\quad\quad\quad k_1 \leftarrow (k_0 + 1) \bmod n_{orient}$ ▷ upper bin index
24: $\quad\quad\quad \mathsf{h}_\phi(k_0) \stackrel{+}{\leftarrow} (1-\alpha) \cdot z$ ▷ update bin k_0
25: $\quad\quad\quad \mathsf{h}_\phi(k_1) \stackrel{+}{\leftarrow} \alpha \cdot z$ ▷ update bin k_1
26: **return** h_ϕ

27: **GetGradientPolar**$(\mathbf{G}_{p,q}, u, v)$
Returns the gradient magnitude (E) and orientation (ϕ) at position (u, v) of the Gaussian scale level $\mathbf{G}_{p,q}$.

28: $\begin{pmatrix} d_x \\ d_y \end{pmatrix} \leftarrow 0.5 \cdot \begin{pmatrix} \mathbf{G}_{p,q}(u+1, v) - \mathbf{G}_{p,q}(u-1, v) \\ \mathbf{G}_{p,q}(u, v+1) - \mathbf{G}_{p,q}(u, v-1) \end{pmatrix}$ ▷ gradient at u, v
29: $E \leftarrow \left(d_x^2 + d_y^2\right)^{1/2}$ ▷ gradient magnitude
30: $\phi \leftarrow \text{ArcTan}(d_x, d_y)$ ▷ gradient orientation $(-\pi \le \phi \le \pi)$
31: **return** (E, ϕ)

such that $\rho_{match} \in [0, 1]$. If the distance $d_{r,1}$ between $\boldsymbol{s}_r$ and the primary feature $\boldsymbol{s}_1$ is small compared to the secondary distance $d_{r,2}$, then the value of ρ_{match} is small as well. Thus, large values of ρ_{match} indicate that the corresponding match (between $\boldsymbol{s}_r$ and $\boldsymbol{s}_1$) is likely to be weak or ambiguous. Matches are only accepted if they are sufficiently distinctive, for example, by enforcing the condition

$$\rho_{match}(\boldsymbol{s}_r, \boldsymbol{s}_1, \boldsymbol{s}_2) \le \rho_{max}\,, \tag{25.105}$$

where $\rho_{max} \in [0, 1]$ is a predefined constant (see Table 25.5). The complete matching process, using the Euclidean distance norm and sequential search, is summarized in Alg. 25.11. Other common options for distance measurement are the L_1 and L_∞ norms.

Alg. 25.8
SIFT feature extraction (part 6): Calculation of SIFT descriptors. Global parameters: Q, σ_0, s_d, n_spat, n_angl (see Table 25.5).

```
1:  MakeSiftDescriptor(G, k', θ)
        Input: G, hierarchical Gaussian scale space; k' = (p, q, x, y), refined key point; θ, dominant orientation.
        Returns a new SIFT descriptor for the key point k'.
2:      G_{p,q} ← GetScaleLevel(G, p, q)
3:      (M, N) ← Size(G_{p,q})
4:      σ̇_q ← σ_0 · 2^{q/Q}                      ▷ decimated scale at level q
5:      w_d ← s_d · σ̇_q            ▷ descriptor size is prop. to key point scale
6:      σ_d ← 0.25 · w_d              ▷ width of Gaussian weighting function
7:      r_d ← 2.5 · σ_d                ▷ cutoff radius of weighting function
8:      u_min ← max(⌊x−r_d⌋, 1)
9:      u_max ← min(⌈x+r_d⌉, M−2)
10:     v_min ← max(⌊y−r_d⌋, 1)
11:     v_max ← min(⌈y+r_d⌉, N−2)
12:     Create map h_∇ : n_spat × n_spat × n_angl ↦ ℝ      ▷ gradient histogram h_∇
13:     for all (i, j, k) ∈ n_spat × n_spat × n_angl do
14:         h_∇(i, j, k) ← 0                         ▷ initialize h_∇ to zero
15:     for u ← u_min, ..., u_max do
16:         for v ← v_min, ..., v_max do
17:             r² ← (u−x)² + (v−y)²
18:             if r² < r_d² then
                    Map to canonical coord. frame, with u', v' ∈ [−1/2, +1/2]:
19:                 (u', v')ᵀ ← (1/w_d) · ( cos(−θ)  −sin(−θ) ; sin(−θ)  cos(−θ) ) · (u−x, v−y)ᵀ
20:                 (E, φ) ← GetGradientPolar(G_{p,q}, u, v)        ▷ Alg. 25.7
21:                 φ' ← (φ − θ) mod 2π            ▷ normalize gradient angle
22:                 w_G ← exp(−r²/(2σ_d²))                  ▷ Gaussian weight
23:                 z ← E · w_G                     ▷ quantity to accumulate
24:                 UpdateGradientHistogram(h_∇, u', v', φ', z)  ▷ Alg. 25.9
25:     f_sift ← MakeFeatureVector(h_∇)                    ▷ see Alg. 25.10
26:     σ ← σ_0 · 2^{p+q/Q}                       ▷ absolute scale, Eq. 25.35
27:     (x', y')ᵀ ← 2^p · (x, y)ᵀ                  ▷ real position, Eq. 25.45
28:     s ← ⟨x', y', σ, θ, f_sift⟩               ▷ create a new SIFT descriptor
29:     return s
```

25.5.2 Examples

The following examples were calculated on pairs of stereographic images taken at the beginning of the 20th century.[26] From each of the two frames of a stereo picture, a sequence of (ca. 1000) SIFT descriptors (marked by blue rectangles) was extracted with identical parameter settings. Matching was done by enumerating all possible descriptor pairs from the left and the right image, calculating their (Euclidean) distance, and showing the 25 closest matches obtained from ca. 1000 detected key points in each frame. Only the

[26] The images used in Figs. 25.28–25.31 are historic stereographs made publicly available by the *Library of Congress* (www.loc.gov).

Alg. 25.9
SIFT feature extraction (part 7): Updating the gradient descriptor histogram. The quantity z pertaining to the continuous position (u', v', ϕ') is to be accumulated into the 3D histogram h_∇ (u', v' are normalized spatial coordinates, ϕ' is the orientation). The quantity z is distributed over up to eight neighboring histogram bins (see Fig. 25.26) by tri-linear interpolation. Note that the orientation coordinate ϕ' receives special treatment because it is circular. Global parameters: $\mathrm{n_{spat}}$, $\mathrm{n_{angl}}$ (see Table 25.5).

```
1: UpdateGradientHistogram(h∇, u', v', φ', z)
     Input: h∇, gradient histogram of size n_spat × n_spat × n_angl, with
     h∇(i,j,k) ∈ ℝ; u', v' ∈ [−0.5, 0.5], normalized spatial position;
     φ' ∈ [0, 2π), normalized gradient orientation; z ∈ ℝ, quantity to
     be accumulated into h∇.
     Returns nothing but modifies the histogram h∇.
2:   i' ← n_spat · u' + 0.5 · (n_spat − 1)              ▷ see Eq. 25.92
3:   j' ← n_spat · v' + 0.5 · (n_spat − 1)      ▷ −0.5 ≤ i', j' ≤ n_spat − 0.5
4:   k' ← n_angl · φ'/(2π)                   ▷ −n_angl/2 ≤ k' ≤ n_angl/2
5:   i0 ← ⌊i'⌋
6:   i1 ← i0 + 1
7:   i ← (i0, i1)                 ▷ see Eq. 25.93; i(0) = i0, i(1) = i1
8:   j0 ← ⌊j'⌋
9:   j1 ← j0 + 1
10:  j ← (j0, j1)                                ▷ j(0) = j0, j(1) = j1
11:  k0 ← ⌊k'⌋ mod n_angl
12:  k1 ← (k0 + 1) mod n_angl
13:  k ← (k0, k1)                                ▷ k(0) = k0, k(1) = k1
14:  α0 ← i1 − i'                                      ▷ see Eq. 25.94
15:  α1 ← 1 − α0
16:  A ← (α0, α1)                                ▷ A(0) = α0, A(1) = α1
17:  β0 ← j1 − j'
18:  β1 ← 1 − β0
19:  B ← (β0, β1)                                ▷ B(0) = β0, B(1) = β1
20:  γ0 ← 1 − (k' − ⌊k'⌋)
21:  γ1 ← 1 − γ0
22:  C ← (γ0, γ1)                                ▷ C(0) = γ0, C(1) = γ1
     Distribute quantity z among (up to) 8 adjacent histogram bins:
23:  for all a ∈ {0, 1} do
24:      i ← i(a)
25:      if (0 ≤ i < n_spat) then
26:          w_a ← A(a)
27:          for all b ∈ {0, 1} do
28:              j ← j(b)
29:              if (0 ≤ j < n_spat) then
30:                  w_b ← B(b)
31:                  for all c ∈ {0, 1} do
32:                      k ← k(c)
33:                      w_c ← C(c)
34:                      h∇(i, j, k) ←+ z·w_a·w_b·w_c       ▷ see Eq. 25.95
35:  return
```

best 25 matches are shown in the examples. Feature matches are numbered according to their goodness, that is, label "1" denotes the best-matching descriptor pair (with the smallest feature distance). Selected details from these results are shown in Fig. 25.29. Unless otherwise noted, all SIFT parameters are set to their default values (see Table 25.5).

Although the use of the Euclidean (L_2) norm for measuring the distances between feature vectors in Eqn. (25.101) is suggested in [153], other norms have been considered [130, 181, 227] to improve

Alg. 25.10
SIFT feature extraction (part 8): Converting the orientation histogram to a SIFT feature vector. Global parameters: n_{spat}, n_{angl}, t_{fclip}, s_{fscale} (see Table 25.5).

```
1:  MakeSiftFeatureVector(h_∇)
        Input: h_∇, gradient histogram of size n_spat × n_spat × n_angl.
        Returns a 1D integer (unsigned byte) vector obtained from h_∇.
2:      Create map f : [0, n_spat^2 · n_angl − 1] ↦ ℝ        ▷ new 1D vector f
3:      m ← 0
4:      for i ← 0, ..., n_spat −1 do                          ▷ flatten h_∇ into f
5:          for j ← 0, ..., n_spat −1 do
6:              for k ← 0, ..., n_angl −1 do
7:                  f(m) ← h_∇(i, j, k)
8:                  m ← m + 1
9:      Normalize(f)
10:     ClipPeaks(f, t_fclip)
11:     Normalize(f)
12:     f_sift ← MapToBytes(f, s_fscale)
13:     return f_sift

14: Normalize(x)
        Scales vector x to unit norm. Returns nothing, but x is modified.
15:     n ← Size(x)
16:     s ← Σ_{i=0}^{n−1} x(i)
17:     for i ← 0, ..., n−1 do
18:         x(i) ← (1/s) · x(i)
19:     return

20: ClipPeaks(x, x_max)
        Limits the elements of x to x_max. Returns nothing, but x is
        modified.
21:     n ← Size(x)
22:     for i ← 0, ..., n−1 do
23:         x(i) ← min(x(i), x_max)
24:     return

25: MapToBytes(x, s)
        Converts the real-valued vector x to an integer (unsigned byte)
        valued vector with elements in [0, 255], using the scale factor
        s > 0.
26:     n ← Size(x)
27:     Create a new map x_int : [0, n−1] ↦ [0, 255]    ▷ new byte vector
28:     for i ← 0, ..., n−1 do
29:         a ← round (s · x(i))                          ▷ a ∈ ℕ_0
30:         x_int(i) ← min(a, 255)                         ▷ x_int(i) ∈ [0, 255]
31:     return x_int
```

the statistical robustness and noise resistance. In Fig. 25.30, matching results are shown using the L_1, L_2, and L_∞ norms, respectively. Note that the resulting sets of top-ranking matches are almost the same with different distance norms, but the ordering of the strongest matches does change.

Figure 25.31 demonstrates the effectiveness of selecting feature matches based on the ratio between the distances to the best and the second-best match (see Eqns. (25.102)–(25.103)). Again the figure shows the 25 top-ranking matches based on the minimum (L_2) feature distance. With the maximum distance ratio ρ_{max} set to 1.0, rejection is practically turned off with the result that several false or ambiguous matches are among the top-ranking feature matches (Fig. 25.31(a)).

Alg. 25.11
SIFT feature matching using Euclidean feature distance and linear search. The returned sequence of SIFT matches is sorted to ascending distance between corresponding feature pairs. Function $\mathsf{Dist}(\boldsymbol{s}_a, \boldsymbol{s}_b)$ demonstrates the calculation of the Euclidean (L_2) feature distance, other options are the L_1 and L_∞ norms.

```
1: MatchDescriptors(S^(a), S^(b), ρ_max)
   Input: S^(a), S^(b), two sets of SIFT descriptors; ρ_max, max. ratio
   of best and second-best matching distance (s. Eq. 25.105).
   Returns a sorted list of matches m_ij = ⟨s_a, s_b, d_ij⟩, with s_a ∈
   S^(a), s_b ∈ S^(b) and d_ij being the distance between s_a, s_b in feature
   space.
2:   M ← ()                                   ▷ empty sequence of matches
3:   for all s_a ∈ S^(a) do
4:       s_1 ← nil,   d_r,1 ← ∞                      ▷ best nearest neighbor
5:       s_2 ← nil,   d_r,2 ← ∞               ▷ second-best nearest neighbor
6:       for all s_b ∈ S^(b) do
7:           d ← Dist(s_a, s_b)
8:           if d < d_r,1 then                   ▷ d is a new 'best' distance
9:               s_2 ← s_1,   d_r,2 ← d_r,1
10:              s_1 ← s_b,   d_r,1 ← d
11:          else
12:              if d < d_r,2 then       ▷ d is a new 'second-best' distance
13:                  s_2 ← s_b,   d_r,2 ← d
14:      if (s_2 ≠ nil) ∧ (d_r,1/d_r,2 ≤ ρ_max) then  ▷ Eqns. (25.104–25.105)
15:          m ← ⟨s_a, s_1, d_r,1⟩                      ▷ add a new match
16:          M ⌣ (m)
17:  Sort(M)                             ▷ sort M to ascending distance d_r,1
18:  return M

19: Dist(s_a, s_b)
   Input: descriptors s_a = ⟨x_a, y_a, σ_a, θ_a, f_a⟩, s_b = ⟨x_b, y_b, σ_b, θ_b,
   f_b⟩. Returns the Euclidean distance between feature vectors f_a
   and f_b.
20:  d ← ‖f_a − f_b‖
21:  return d
```

With $\rho_{\max}$ set to 0.8 and finally 0.5, the number of false matches is effectively reduced (Fig. 25.31(b, c)).[27]

25.6 Efficient Feature Matching

The task of finding the best match based on the minimum distance in feature space is called "nearest-neighbor" search. If performed exhaustively, evaluating all possible matches between two descriptor sets $S^{(a)}$ and $S^{(b)}$ of size N_a and N_b, respectively, requires $N_a \cdot N_b$ feature distance calculations and comparisons. While this may be acceptable for small feature sets (with maybe up to 1000 descriptors each), this linear (brute-force) approach becomes prohibitively expensive for large feature sets with possibly millions of candidates, as required, for example, in the context of image database indexing or robot self-localization. Although efficient methods for exact nearest-neighbor search based on tree structures exist, such as the k-d tree method [80], it has been shown that these methods lose their effectiveness with increasing dimensionality of the search space.

[27] $\rho_{\max} = 0.8$ is recommended in [153].

Fig. 25.28
SIFT feature matching examples on pairs of stereo images. Shown are the 25 best matches obtained with the L_2 feature distance and $\rho_{max} = 0.8$.

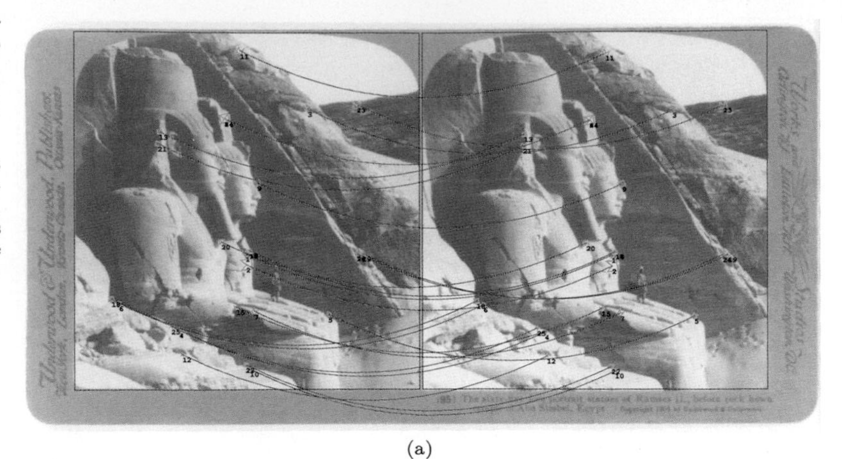

(a)

(b)

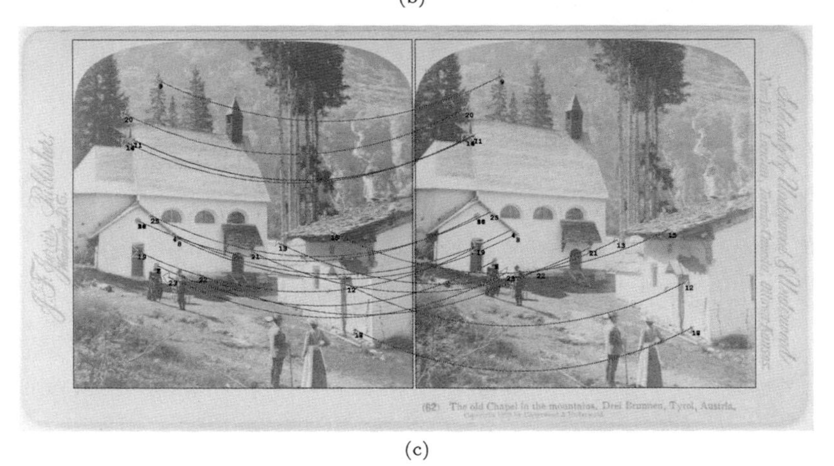

(c)

In fact, no algorithms are known that significantly outperform exhaustive (linear) nearest neighbor search in feature spaces that are more than about 10-dimensional [153]. SIFT feature vectors are 128-dimensional and therefore exact nearest-neighbor search is not a viable option for efficient matching between large descriptor sets.

The approach taken in [21, 153] abandons exact nearest-neighbor search in favor of finding an *approximate* solution with substantially reduced effort, based on ideas described in [9]. This so-called

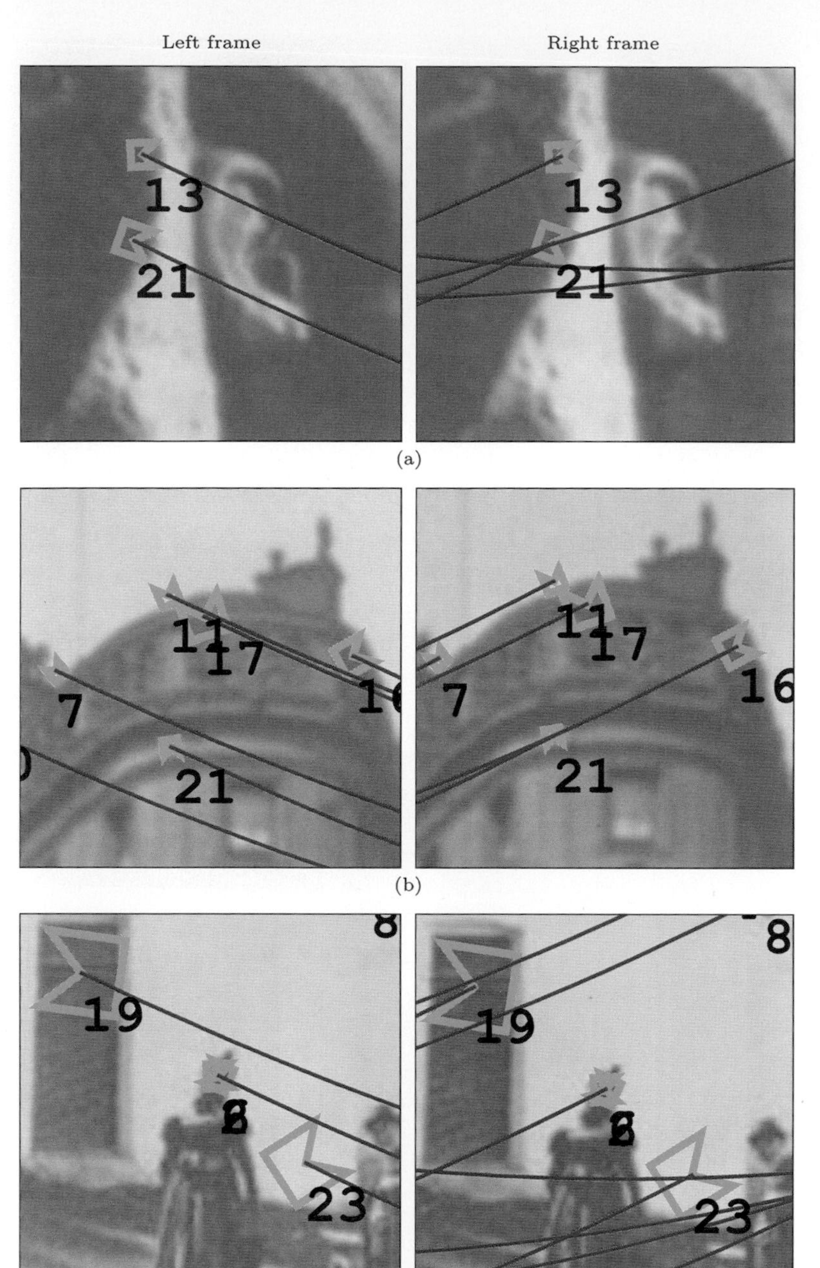

Fig. 25.29
Stereo matching examples (enlarged details from Fig. 25.28).

"best-bin-first" method uses a modified k-d algorithm, which searches neighboring feature space partitions in the order of their closest distance from the given feature vector. To limit the exploration to a small fraction of the feature space, the search is cut off after checking the first 200 candidates, which results in a substantial speedup without compromising the search results, particularly when combined with feature selection based on the ratio of primary and secondary distances (see Eqns. (25.104)–(25.105)). Additional details can be found in [21].

Fig. 25.30
Using different distance norms for feature matching. L_1 (a), L_2 (b), and L_∞ norm (c). All other parameters are set to their default values (see Table 25.5).

(a) L_1-norm

(b) L_2-norm

(c) L_∞-norm

Approximate nearest-neighbor search in high-dimensional spaces is not only essential for practical SIFT matching in real time, but is a general problem with numerous applications in various disciplines and continued research. Open-source implementations of several different methods are available as software libraries.

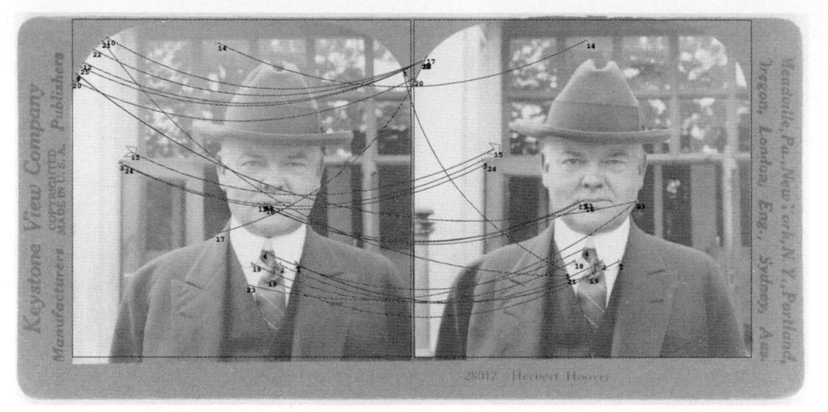

(a) $\rho_{\mathrm{max}} = 1.0$

(b) $\rho_{\mathrm{max}} = 0.8$

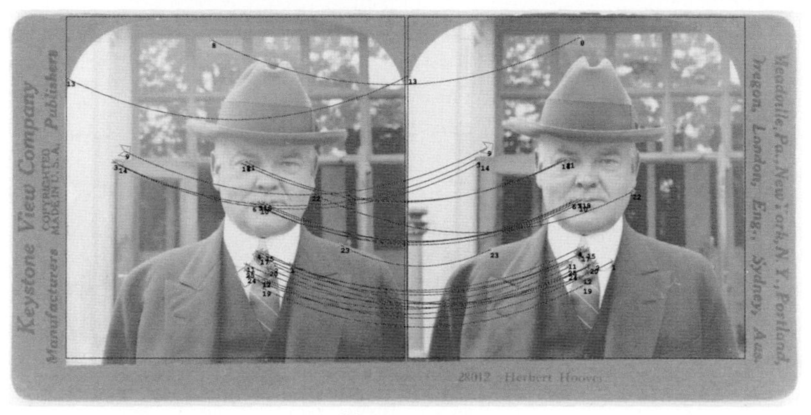

(c) $\rho_{\mathrm{max}} = 0.5$

Fig. 25.31
Rejection of weak or ambiguous matches by limiting the ratio of primary and secondary match distance ρ_{max} (see Eqns. (25.104)–(25.105)).

25.7 Java Implementation

A new and complete Java implementation of the SIFT method has been written from ground up to complement the algorithms described in this chapter. Space limitations do not permit a full listing here, but the entire implementation and additional examples can be found in the source code section of this book's website. Most Java methods are named and structured identically to the procedures listed in the algorithms for easy identification. Note, however, that this implementation is again written for instructional clarity and readability. The code is neither tuned for efficiency nor is it intended to be used in a production environment.

25.7.1 SIFT Feature Extraction

The key class in this Java library is `SiftDetector`, which implements a SIFT detector for a given floating-point image. The following example illustrates its basic use for a given `ImageProcessor` object `ip`:

```
...
FloatProcessor I = ip.convertToFloatProcessor();
SiftDetector sd = new SiftDetector(I);
List<SiftDescriptor> S = sd.getSiftFeatures();
... // process descriptor set S
```

The initial work of setting up the required Gaussian and DoG scale space structures for the given image `I` is accomplished by the constructor in `new SiftDetector(I)`.

The method `getSiftFeatures()` then performs the actual feature detection process and returns a sequence of `SiftDescriptor` objects (`S`) for the image `I`. Each extracted `SiftDescriptor` in `S` holds information about its image position (`x`, `y`), its absolute scale σ (`scale`) and its dominant orientation θ (`orientation`). It also contains an invariant, 128-element, `int`-type feature vector $\boldsymbol{f}_{\mathrm{sift}}$ (see Alg. 25.8).

The SIFT detector uses a large set of parameters that are set to their default values (see Table 25.5) if the simple constructor `new SiftDetector(I)` is used, as in the previous example. All parameters can be adjusted individually by passing a parameter object (of type `SiftDetector.Parameters`) to its constructor, as in the following example, which shows feature extraction from two images `A`, `B` using identical parameters:

```
...
FloatProcessor Ia = A.convertToFloatProcessor();
FloatProcessor Ib = B.convertToFloatProcessor();
...
SiftDetector.Parameters params =
    new SiftDetector.Parameters();
params.sigma_s = 0.5; // modify individual parameters
params.sigma_0 = 1.6;
...
SiftDetector sdA = new SiftDetector(Ia, params);
SiftDetector sdB = new SiftDetector(Ib, params);
List<SiftDescriptor> SA = sda.getSiftFeatures();
List<SiftDescriptor> SB = sdb.getSiftFeatures();
...
// process descriptor sets SA and SB
```

25.7.2 SIFT Feature Matching

Finding matching descriptors from a pair of SIFT descriptor sets `Sa`, `Sb` is accomplished by the class `SiftMatcher`.[28] One descriptor set (`Sa`) is considered the "reference" or "model" set and used to initialize a new `SiftMatcher` object, as shown in the following example. The actual matches are then calculated by invoking the method `matchDescriptors()`, which implements the procedure MatchDescriptors() outlined in Alg. 25.11. It takes the second descriptor set (`Sb`) as the only argument. The following code segment continues from the previous example:

```
...
SiftMatcher.Parameters params =
                new SiftMatcher.Parameters();
// set matcher parameters here (see below)
SiftMatcher matcher = new SiftMatcher(SA, params);
List<SiftMatch> matches = matcher.matchDescriptors(SB);
...
// process matches
```

As noted, certain parameters of class `SiftMatcher` can be set individually, for example,

```
params.norm = FeatureDistanceNorm.L1; // L1, L2, or Linf
params.rmMax = 0.8; // ρ_max, max. ratio of best and second-best match
params.sort = true; // set to true if sorting of matches is desired
```

The method `matchDescriptors()` in this prototypical implementation performs an exhaustive search over all possible descriptor pairs in the two sets `Sa` and `Sb`. To implement efficient approximate nearest-neighbor search (see Sec. 25.6), one would pre-calculate the required search tree structures for the model descriptor set (`Sa`) once inside `SiftMatcher`'s constructor method. The same matcher object could then be reused to match against multiple descriptor sets without the need to recalculate the search tree structure over and over again. This is particularly effective when the given model set is large.

25.8 Exercises

Exercise 25.1. As claimed in Eqn. (25.12), the 2D LoG function $L_\sigma(x,y)$ can be approximated by the DoG in the form $L_\sigma(x,y) \approx \lambda \cdot (G_{\kappa\sigma}(x,y) - G_\sigma(x,y))$. Create a combined plot, similar to the one in Fig. 25.5(b), showing the 1D cross sections of the LoG and DoG functions (with $\sigma = 1.0$ and $y = 0$). Compare both functions by varying the values of $\kappa = 2.00, 1.25, 1.10, 1.05$, and 1.01. How does the approximation change as κ approaches 1, and what happens if κ becomes exactly 1?

Exercise 25.2. Test the performance of the SIFT feature detection and matching on pairs of related images under (a) changes of image brightness and contrast, (b) image rotation, (c) scale changes,

[28] File `imagingbook.sift.SiftMatcher.java`.

(d) adding (synthetic) noise. Choose (or shoot) your own test images, show the results in a suitable way and document the parameters used.

Exercise 25.3. Evaluate the SIFT mechanism for tracking features in video sequences. Search for a suitable video sequence with good features to track and process the images frame-by-frame.[29] Then match the SIFT features detected in pairs of successive frames by connecting the best-matching features, as long as the "match quality" is above a predefined threshold. Visualize the resulting feature trajectories. Could other properties of the SIFT descriptors (such as position, scale, and dominant orientation) be used to improve tracking stability?

[29] In ImageJ, choose an AVI video short enough to fit into main memory and open it as an image stack.

26

Fourier Shape Descriptors

Fourier descriptors are an interesting method for modeling 2D shapes that are described as closed contours. Unlike polylines or splines, which are explicit and local descriptions of the contour, Fourier descriptors are *global* shape representations, that is, each component stands for a particular characteristic of the entire shape. If one component is changed, the whole shape will change. The advantage is that it is possible to capture coarse shape properties with only a few numeric values, and the level of detail can be increased (or decreased) by adding (or removing) descriptor elements. In the following, we describe what is called "cartesian" (or "elliptical") Fourier descriptors, how they can be used to model the shape of closed 2D contours and how they can be adapted to compare shapes in a translation-, scale-, and rotation-invariant fashion.

26.1 Closed Curves in the Complex Plane

Any continuous curve C in the 2D plane can be expressed as a function $f\colon \mathbb{R} \to \mathbb{R}^2$, with

$$f(t) = \begin{pmatrix} x_t \\ y_t \end{pmatrix} = \begin{pmatrix} f_x(t) \\ f_y(t) \end{pmatrix}, \tag{26.1}$$

with the continuous parameter t being varied over the range $[0, t_{\mathrm{max}}]$. If the curve is closed, then $f(0) = f(t_{\mathrm{max}})$ and $f(t) = f(t + t_{\mathrm{max}})$. Note that $f_x(t)$, $f_y(t)$ are independent, real-valued functions, and t is the *path length* along the curve.

26.1.1 Discrete 2D Curves

Sampling a closed curve C at M regularly spaced positions $t_0, t_1, \ldots, t_{M-1}$, with $t_i - t_{i-1} = \Delta_t = \mathrm{Length}(C)/M$, results in a sequence (vector) of discrete 2D coordinates $V = (\boldsymbol{v}_0, \boldsymbol{v}_1, \ldots, \boldsymbol{v}_{M-1})$, with

$$\boldsymbol{v}_k = (x_k, y_k) = f(t_k). \tag{26.2}$$

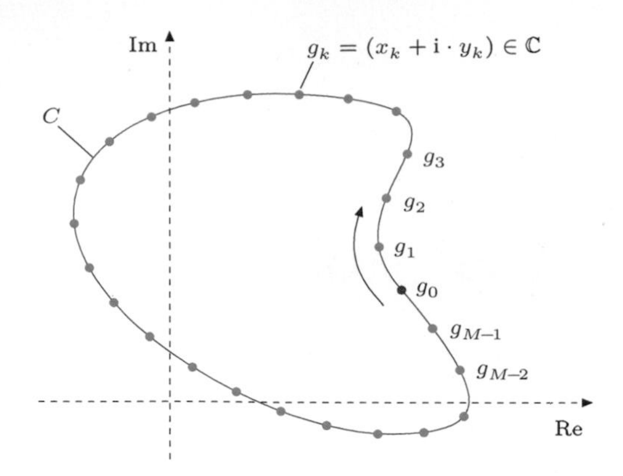

Fig. 26.1
A closed, continuous 2D curve C, represented as a sequence of M uniformly placed samples $\boldsymbol{g} = (g_0, g_1, \ldots, g_{M-1})$ in the complex plane.

Since the curve C is closed, the vector V represents a discrete function that is infinite and periodic, that is,

$$\boldsymbol{v}_k = \boldsymbol{v}_{k+pM}, \tag{26.3}$$

for $0 \leq k < M$ and any $p \in \mathbb{Z}$.

Contour points in the complex plane

Any 2D contour sample $\boldsymbol{v}_k = (x_k, y_k)$ can be interpreted as a point g_k in the complex plane,

$$g_k = x_k + \mathrm{i} \cdot y_k, \tag{26.4}$$

with x_k and y_k taken as the real and imaginary components, respectively.[1] The result is a sequence (vector) of complex values

$$\boldsymbol{g} = (g_0, g_1, \ldots, g_{M-1}), \tag{26.5}$$

representing the discrete 2D contour (see Fig. 26.1).

Regular position sampling

The assumption of input data being obtained by regular sampling is quite fundamental in traditional discrete Fourier analysis. In practice, contours of objects are typically not available as regularly sampled point sequences. For example, if an object has been segmented as a binary region, the coordinates of its boundary pixels could be used as the original contour sequence. However, the number of boundary pixels is usually too large to be used directly and their positions are not strictly uniformly spaced (at least under 8-connectivity). To produce a useful contour sequence from a region boundary, one could choose an arbitrary contour point as the start position $\boldsymbol{x}_0$ and then sample the x/y positions along the contour at regular (equidistant) steps, treating the centers of the boundary pixels as the vertices of a closed polygon. Algorithm 26.1 shows how to calculate a predefined number of contour points on an arbitrary polygon, such that the path

[1] Instead of $g \leftarrow x + \mathrm{i} \cdot y$, we sometimes use the short notation $g \leftarrow (x, y)$ or $g \leftarrow \boldsymbol{v}$ for assigning the components of a 2D vector $\boldsymbol{v} = (x, y) \in \mathbb{R}^2$ to a complex variable $g \in \mathbb{C}$.

```
1:  SamplePolygonUniformly(V, M)
      Input: V = (v_0, ..., v_{N-1}), a sequence of N points representing
      the vertices of a 2D polygon; M, number of desired sample points.
      Returns a sequence g = (g_0, ..., g_{M-1}) of complex values rep-
      resenting points sampled uniformly along the path of the input
      polygon V.
2:    N ← |V|
3:    Δ ← (1/M) · PathLength(V)                 ▷ const. segment length Δ
4:    Create map g: [0, M−1] → ℂ                ▷ complex point sequence g
5:    g(0) ← Complex(V(0))
6:    i ← 0                             ▷ index of polygon segment ⟨v_i, v_{i+1}⟩
7:    k ← 1                             ▷ index of next point to be added to g
8:    α ← 0                             ▷ path position of polygon vertex v_i
9:    β ← Δ                     ▷ path position of next point to be added to g
10:   while (i < N) ∧ (k < M) do
11:       v_A ← V(i)
12:       v_B ← V((i+1) mod N)
13:       δ ← ‖v_B − v_A‖                       ▷ length of segment ⟨v_A, v_B⟩
14:       while (β ≤ α + δ) ∧ (k < M) do
15:           x ← v_A + ((β−α)/δ) · (v_B − v_A)   ▷ linear path interpolation
16:           g(k) ← Complex(x)
17:           k ← k + 1
18:           β ← β + Δ
19:       α ← α + δ
20:       i ← i + 1
21:   return g.

22: PathLength(V)     ▷ returns the path length of the closed polygon V
23:   N ← |V|
24:   L ← 0
25:   for i ← 0, ..., N−1 do
26:       v_A ← V(i)
27:       v_B ← V((i+1) mod N)
28:       L ← L + ‖v_B − v_A‖
29:   return L.
```

Alg. 26.1
Regular sampling of a polygon path. Given a sequence V of 2D points representing the vertices of a closed polygon, SamplePolygonUniformly(V, M) returns a sequence of M complex values $\boldsymbol{g}$ on the polygon V, such that $\boldsymbol{g}(0) \equiv V(0)$ and all remaining points $\boldsymbol{g}(k)$ are uniformly positioned along the polygon path. See Alg. 26.9 for an alternate solution.

length between the sample points is uniform. This algorithm is used in all examples involving contours obtained from binary regions.

Note that if the shape is given as an arbitrary polygon, the corresponding Fourier descriptor can also be calculated directly (and exactly) from the vertices of the polygon, without sub-sampling the polygon contour path at all. This "trigonometric" variant of the Fourier descriptor calculation is described in Sec. 26.3.7.

26.2 Discrete Fourier Transform (DFT)

Fourier descriptors are obtained by applying the 1D Discrete Fourier Transform (DFT)[2] to the complex-valued vector $\boldsymbol{g}$ of 2D contour points (Eqn. (26.5)). The DFT is a transformation of a finite, complex-valued *signal* vector $\boldsymbol{g} = (g_0, g_1, \ldots, g_{M-1})$ to a complex-valued *spec-*

[2] See Chapter 18, Sec. 18.3.

trum $\boldsymbol{G} = (G_0, G_1, \ldots, G_{M-1})$.[3] Both the signal and the spectrum are of the same length (M) and periodic. In the following, we typically use k to denote the index in the time or space domain,[4] and m for a frequency index in the spectral domain.

26.2.1 Forward Fourier Transform

The discrete Fourier spectrum $\boldsymbol{G} = (G_0, G_1, \ldots, G_{M-1})$ is calculated from the discrete, complex-valued signal $\boldsymbol{g} = (g_0, g_1, \ldots, g_{M-1})$ using the forward DFT, defined as[5]

$$G_m = \frac{1}{M} \cdot \sum_{k=0}^{M-1} g_k \cdot e^{-\mathrm{i}\cdot 2\pi m \cdot \frac{k}{M}} = \frac{1}{M} \cdot \sum_{k=0}^{M-1} g_k \cdot e^{-\mathrm{i}\cdot \omega_m \cdot \frac{k}{M}} \tag{26.6}$$

$$= \frac{1}{M} \cdot \sum_{k=0}^{M-1} \underbrace{\left[x_k + \mathrm{i}\cdot y_k\right]}_{g_k} \cdot \left[\cos\bigl(\underbrace{2\pi m}_{\omega_m} \tfrac{k}{M}\bigr) - \mathrm{i}\cdot\sin\bigl(\underbrace{2\pi m}_{\omega_m} \tfrac{k}{M}\bigr)\right] \tag{26.7}$$

$$= \frac{1}{M} \cdot \sum_{k=0}^{M-1} \left[x_k + \mathrm{i}\cdot y_k\right] \cdot \left[\cos\bigl(\omega_m \tfrac{k}{M}\bigr) - \mathrm{i}\cdot\sin\bigl(\omega_m \tfrac{k}{M}\bigr)\right], \tag{26.8}$$

for $0 \leq m < M$.[6] Note that $\omega_m = 2\pi m$ denotes the *angular frequency* for the frequency index m. By applying the usual rules of complex multiplication, we obtain the *real* (Re) and *imaginary* (Im) parts of the spectral coefficients $G_m = (A_m + \mathrm{i} \cdot B_m)$ explicitly as

$$A_m = \mathrm{Re}(G_m) = \frac{1}{M} \sum_{k=0}^{M-1} \left[x_k \cdot \cos\bigl(\omega_m \tfrac{k}{M}\bigr) + y_k \cdot \sin\bigl(\omega_m \tfrac{k}{M}\bigr)\right], \tag{26.9}$$

$$B_m = \mathrm{Im}(G_m) = \frac{1}{M} \sum_{k=0}^{M-1} \left[y_k \cdot \cos\bigl(\omega_m \tfrac{k}{M}\bigr) - x_k \cdot \sin\bigl(\omega_m \tfrac{k}{M}\bigr)\right]. \tag{26.10}$$

The DFT is defined for any signal length $M \geq 1$. If the signal length M is a power of two (that is, $M = 2^n$ for some $n \in \mathbb{N}$), the Fast Fourier Transform (FFT)[7] can be used in place of the DFT for improved performance.

26.2.2 Inverse Fourier Transform (Reconstruction)

The inverse DFT reconstructs the original signal $\boldsymbol{g}$ from a given spectrum $\boldsymbol{G}$. The formulation is almost symmetrical (except for the scale

[3] In most traditional applications of the DFT (e.g. in acoustic processing), the signals are real-valued, that is, the imaginary components of the samples are zero. The Fourier spectrum is generally complex-valued, but it is symmetric for real-valued signals.

[4] We use k instead of the usual i as the running index to avoid confusion with the imaginary constant "i" (despite the deliberate use of different glyphs).

[5] This definition deviates slightly from the one used in Chapter 18, Sec. 18.3 but is otherwise equivalent.

[6] Recall that $z = x + \mathrm{i}y = |z| \cdot (\cos\psi + \mathrm{i} \cdot \sin\psi) = |z| \cdot e^{\mathrm{i}\psi}$, with $\psi = \tan^{-1}(y/x)$.

[7] See Chapter 18, Sec. 18.4.2.

1:	**FourierDescriptorUniform**($\boldsymbol{g}$) Input: $\boldsymbol{g} = (g_0, \ldots, g_{M-1})$, a sequence of M complex values, representing regularly sampled 2D points along a contour path. Returns a Fourier descriptor $\boldsymbol{G}$ of length M.
2:	$M \leftarrow \lvert\boldsymbol{g}\rvert$
3:	Create map $\boldsymbol{G}\colon [0, M-1] \to \mathbb{C}$
4:	**for** $m \leftarrow 0, \ldots, M-1$ **do**
5:	$A \leftarrow 0, \quad B \leftarrow 0$ ▷ real/imag. part of coefficient G_m
6:	**for** $k \leftarrow 0, \ldots, M-1$ **do**
7:	$g \leftarrow \boldsymbol{g}(k)$
8:	$x \leftarrow \mathrm{Re}(g), \quad y \leftarrow \mathrm{Im}(g)$
9:	$\phi \leftarrow 2 \cdot \pi \cdot m \cdot \frac{k}{M}$
10:	$A \leftarrow A + x \cdot \cos(\phi) + y \cdot \sin(\phi)$ ▷ Eq. 26.10
11:	$B \leftarrow B - x \cdot \sin(\phi) + y \cdot \cos(\phi)$
12:	$\boldsymbol{G}(m) \leftarrow \frac{1}{M} \cdot (A + \mathrm{i} \cdot B)$
13:	**return** $\boldsymbol{G}$.

Alg. 26.2
Calculating the Fourier descriptor for a sequence of uniformly sampled contour points. The complex-valued contour points in C represent 2D positions sampled uniformly along the contour path. Applying the DFT to $\boldsymbol{g}$ yields the raw Fourier descriptor $\boldsymbol{G}$.

factor and the different signs in the exponent) to the forward transformation in Eqns. (26.6)–(26.8); its full expansion is

$$g_k = \sum_{m=0}^{M-1} G_m \cdot e^{\mathrm{i} \cdot 2\pi m \cdot \frac{k}{M}} = \sum_{m=0}^{M-1} G_m \cdot e^{\mathrm{i} \cdot \omega_m \cdot \frac{k}{M}} \tag{26.11}$$

$$= \sum_{m=0}^{M-1} \underbrace{\left[\mathrm{Re}(G_m) + \mathrm{i} \cdot \mathrm{Im}(G_m)\right]}_{G_m} \cdot \left[\cos\bigl(\underbrace{2\pi m}_{\omega_m} \tfrac{k}{M}\bigr) + \mathrm{i} \cdot \sin\bigl(\underbrace{2\pi m}_{\omega_m} \tfrac{k}{M}\bigr)\right] \tag{26.12}$$

$$= \sum_{m=0}^{M-1} \left[A_m + \mathrm{i} \cdot B_m\right] \cdot \left[\cos\bigl(\omega_m \tfrac{k}{M}\bigr) + \mathrm{i} \cdot \sin\bigl(\omega_m \tfrac{k}{M}\bigr)\right]. \tag{26.13}$$

Again we can expand Eqn. (26.13) to obtain the real and imaginary parts of the reconstructed signal, that is, the x/y-components of the corresponding curve points $g_k = (x_k, y_k)$ as

$$x_k = \mathrm{Re}(g_k) = \sum_{m=0}^{M-1} \left[\mathrm{Re}(G_m) \cdot \cos\bigl(2\pi m \tfrac{k}{M}\bigr) - \mathrm{Im}(G_m) \cdot \sin\bigl(2\pi m \tfrac{k}{M}\bigr)\right], \tag{26.14}$$

$$y_k = \mathrm{Im}(g_k) = \sum_{m=0}^{M-1} \left[\mathrm{Im}(G_m) \cdot \cos\bigl(2\pi m \tfrac{k}{M}\bigr) + \mathrm{Re}(G_m) \cdot \sin\bigl(2\pi m \tfrac{k}{M}\bigr)\right], \tag{26.15}$$

for $0 \leq k < M$. If *all* coefficients of the spectrum are used, this reconstruction is *exact*, that is, the resulting discrete points g_k are identical to the original contour points.[8]

With the aforementioned formulation we can not only reconstruct the discrete contour points g_k from the DFT spectrum, but also a smooth, interpolating curve as the sum of continuous sine and cosine components. To calculate *arbitrary* points on this curve, we replace the discrete quantity $\frac{k}{M}$ in Eqn. (26.15) by the continuous parameter t in the range $[0, 1)$. We must be careful about the frequencies, though. To achieve the desired *smooth* interpolation, the set of *lowest* possible

[8] Apart from inaccuracies caused by finite floating-point precision.

frequencies ω_m must be used,[9] that is,

$$x(t) = \sum_{m=0}^{M-1} \big[\mathrm{Re}(G_m) \cdot \cos(\omega_m \cdot t) - \mathrm{Im}(G_m) \cdot \sin(\omega_m \cdot t)\big], \quad (26.16)$$

$$y(t) = \sum_{m=0}^{M-1} \big[\mathrm{Im}(G_m) \cdot \cos(\omega_m \cdot t) + \mathrm{Re}(G_m) \cdot \sin(\omega_m \cdot t)\big], \quad (26.17)$$

with $$\omega_m = \begin{cases} 2\pi m & \text{for } m \leq (M \div 2), \\ 2\pi(m-M) & \text{for } m > (M \div 2), \end{cases} \quad (26.18)$$

where $\div$ denotes the quotient (i.e., integer division). Alternatively, we could write Eqn. (26.17) in the form

$$x(t) = \sum_{\substack{m=\\-(M-1)\div 2}}^{M\div 2} [\mathrm{Re}(G_{m \bmod M}) \cdot \cos(2\pi m t) - \mathrm{Im}(G_{m \bmod M}) \cdot \sin(2\pi m t)], \quad (26.19)$$

$$y(t) = \sum_{\substack{m=\\-(M-1)\div 2}}^{M\div 2} [\mathrm{Im}(G_{m \bmod M}) \cdot \cos(2\pi m t) + \mathrm{Re}(G_{m \bmod M}) \cdot \sin(2\pi m t)]. \quad (26.20)$$

This formulation is used for the purpose of shape reconstruction from Fourier descriptors in Alg. 26.4.

Figure (26.2) shows the reconstruction of the discrete contour points as well as the calculation of a continuous outline from the DFT spectrum obtained from a sequence of discrete contour positions. The original sample points were taken at $M = 25$ uniformly spaced positions along the region's contour. The discrete points in Fig. 26.2(b) are exactly reconstructed from the complete DFT spectrum, as specified in Eqn. (26.15). The interpolated (green) outline in Fig. 26.2(c) was calculated with Eqn. (26.15) for continuous positions, based on the frequencies $m = 0, \ldots, M-1$. The oscillations of the resulting curve are explained by the high-frequency components. Note that the curve still passes exactly through each of the original sample points, in fact, these can be perfectly reconstructed from *any* contiguous range of M coefficients and the corresponding harmonic frequencies. The smooth interpolation in Fig. 26.2(d), based on the symmetric low-frequency coefficients $m = -(M-1) \div 2, \ldots, M \div 2$ (see Eqn. (26.20)) shows no such oscillations, since no high-frequency components are included.

26.2.3 Periodicity of the DFT Spectrum

When we apply the DFT, we implicitly assume that both the signal vector $\boldsymbol{g} = (g_0, g_1, \ldots, g_{M-1})$ and the spectral vector $\boldsymbol{G} = (G_0, G_1, \ldots, G_{M-1})$ represent discrete, periodic functions of infinite extent

[9] Due to the periodicity of the discrete spectrum, any summation over M successive frequencies ω_m can be used to reconstruct the original discrete x/y samples. However, a smooth interpolation between the discrete x/y samples can only be obtained from the set of *lowest* frequencies in the range $[-\frac{M}{2}, +\frac{M}{2}]$ centered around the zero frequency, as in Eqns. (26.17) and (26.20).

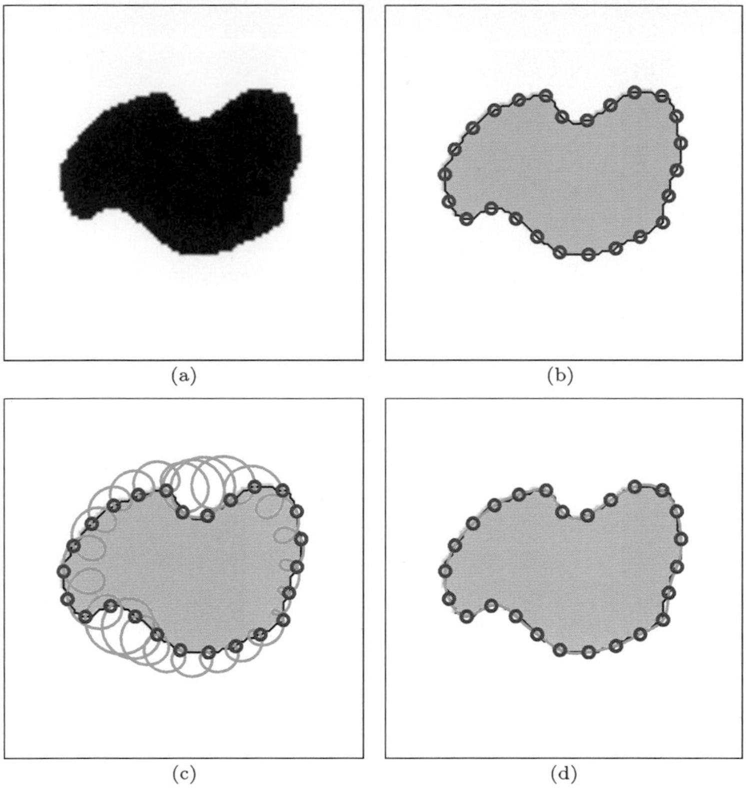

(a) (b)

(c) (d)

Fig. 26.2
Contour reconstruction by inverse DFT. Original image (a), $M = 25$ uniformly spaced sample points on the region's contour (b). Continuous contour (green line) reconstructed by using frequencies ω_m with $m = 0, \ldots, 24$ (c). Note that despite the oscillations introduced by the high frequencies, the continuous contour passes exactly through the original sample points. Smooth interpolation reconstructed with Eqn. (26.17) from the lowest-frequency coefficients in the symmetric range $m = -12, \ldots, +12$ (d).

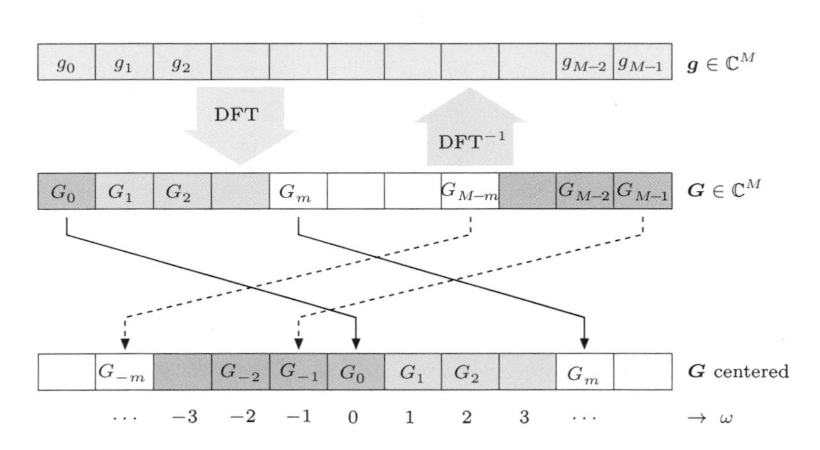

Fig. 26.3
Applying the DFT to a complex-valued vector $\boldsymbol{g}$ of length M yields the complex-valued spectrum $\boldsymbol{G}$ that is also of length M. The DFT spectrum is infinite and periodic with M, thus $G_{-m} = G_{M-m}$, as illustrated by the centered representation of the DFT spectrum (bottom). ω at the bottom denotes the harmonic number (multiple of the fundamental frequency) associated with each coefficient.

(see [39, Ch. 13] for details). Due to this periodicity, $\boldsymbol{G}(0) = \boldsymbol{G}(M)$, $\boldsymbol{G}(1) = \boldsymbol{G}(M+1)$, etc. In general,

$$\boldsymbol{G}(q \cdot M + m) = \boldsymbol{G}(m) \quad \text{and} \quad \boldsymbol{G}(m) = \boldsymbol{G}(m \bmod M), \tag{26.21}$$

for arbitrary integers $q, m \in \mathbb{Z}$. Also, since $(-m \bmod M) = (M-m) \bmod M$, we can state that

$$\boldsymbol{G}(-m) = \boldsymbol{G}(M-m), \tag{26.22}$$

for any $m \in \mathbb{Z}$, such that $\boldsymbol{G}(-1) = \boldsymbol{G}(M-1)$, $\boldsymbol{G}(-2) = \boldsymbol{G}(M-2)$, etc., as illustrated in Fig. 26.3.

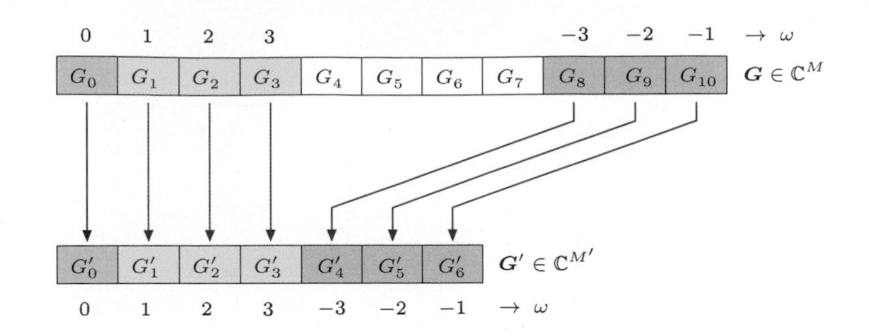

Fig. 26.4
Truncating a DFT spectrum from $M = 11$ to $M' = 7$ coefficients, as specified in Eqns. (26.23) and (26.24). Coefficients $G_4, \ldots, G_7$ are discarded ($M' \div 2 = 3$). Note that the associated harmonic number ω remains the same for each coefficient.

26.2.4 Truncating the DFT Spectrum

In the original formulation in Eqns. (26.6)–(26.8), the DFT is applied to a signal $\boldsymbol{g}$ of length M and yields a discrete Fourier spectrum $\boldsymbol{G}$ with M coefficients. Thus the signal and the spectrum have the same length. For shape representation, it is often useful to work with a truncated spectrum, that is, a reduced number of low-frequency Fourier coefficients.

By truncating a spectrum we mean the removal of coefficients above a certain harmonic number, which are (considering positive and negative frequencies) located around the center of the coefficient vector. Truncating a given spectrum $\boldsymbol{G}$ of length $|\boldsymbol{G}| = M$ to a shorter spectrum $\boldsymbol{G}'$ of length $M' \leq M$ is done as

$$\boldsymbol{G}'(m) \leftarrow \begin{cases} \boldsymbol{G}(m) & \text{for } 0 \leq m \leq M' \div 2, \\ \boldsymbol{G}(M-M'+m) & \text{for } M' \div 2 < m < M', \end{cases} \tag{26.23}$$

or simply

$$\boldsymbol{G}'(m \bmod M') \leftarrow \boldsymbol{G}(m \bmod M), \tag{26.24}$$

for $(M' {\div} 2 - M' {+} 1) \leq m \leq (M' {\div} 2)$. This works for M and M' being even or odd. The example in Fig. 26.4 illustrates how an original DFT spectrum $\boldsymbol{G}$ of length $M = 11$ is truncated to $\boldsymbol{G}'$ with only $M' = 7$ coefficients.

Of course it is also possible to calculate the truncated spectrum directly from the contour samples, without going through the full DFT spectrum. With M being the length of the signal vector $\boldsymbol{g}$ and $M' \leq M$ the desired length of the (truncated) spectrum $\boldsymbol{G}'$, Eqn. (26.6) modifies to

$$\boldsymbol{G}'(m \bmod M') = \frac{1}{M} \cdot \sum_{k=0}^{M-1} g_k \cdot e^{-\mathrm{i} 2\pi m \frac{k}{M}}, \tag{26.25}$$

for m in the same range as in Eqn. (26.24). This approach is more efficient than truncating the complete spectrum, since unneeded coefficients are never calculated. Algorithm 26.3, which is a modified version of Alg. 26.2, summarizes the steps we have described.

Since some of the coefficients are missing, it is not possible to reconstruct the original signal vector $\boldsymbol{g}$ from the truncated DFT spectrum $\boldsymbol{G}'$. However, the calculation of a partial reconstruction is possible, for example, using the formulation in Eqn. (26.20). In this

1:	**FourierDescriptorUniform**$(\boldsymbol{g}, M')$
	Input: $\boldsymbol{g} = (g_0, \ldots, g_{M-1})$, a sequence of M complex values, representing regularly sampled 2D points along a contour path. M', the number of Fourier coefficients ($M' \leq M$). Returns a truncated Fourier descriptor $\boldsymbol{G}$ of length M'.
2:	$M \leftarrow \lvert\boldsymbol{g}\rvert$
3:	Create map $\boldsymbol{G}\colon [0, M'-1] \to \mathbb{C}$
4:	**for** $m \leftarrow (M' \div 2 - M' + 1), \ldots, (M' \div 2)$ **do**
5:	$A \leftarrow 0, \quad B \leftarrow 0$ ▷ real/imag. part of coefficient G_m
6:	**for** $k \leftarrow 0, \ldots, M-1$ **do**
7:	$g \leftarrow \boldsymbol{g}(k)$
8:	$x \leftarrow \mathrm{Re}(g), \quad y \leftarrow \mathrm{Im}(g)$
9:	$\phi \leftarrow 2 \cdot \pi \cdot m \cdot \frac{k}{M}$
10:	$A \leftarrow A + x \cdot \cos(\phi) + y \cdot \sin(\phi)$ ▷ Eq. 26.10
11:	$B \leftarrow B - x \cdot \sin(\phi) + y \cdot \cos(\phi)$
12:	$\boldsymbol{G}(m \bmod M') \leftarrow \frac{1}{M} \cdot (A + \mathrm{i} \cdot B)$
13:	**return** $\boldsymbol{G}$.

Alg. 26.3
Calculating a truncated Fourier descriptor for a sequence of uniformly sampled contour points (adapted from Alg. 26.2). The M complex-valued contour points in $\boldsymbol{g}$ represent 2D positions sampled uniformly along the contour path. The resulting Fourier descriptor $\boldsymbol{G}$ contains only M' coefficients for the M' lowest harmonic frequencies.

case, the discarded (high-frequency) coefficients are simply assumed to have zero values (see Sec. 26.3.6 for more details).

26.3 Geometric Interpretation of Fourier Coefficients

The contour reconstructed by the inverse transformation (Eqn. (26.15)) is the sum of M terms, one for each Fourier coefficient $G_m = (A_m, B_m)$. Each of these M terms represents a particular 2D shape in the spatial domain and the original contour can be obtained by point-wise addition of the individual shapes. So what are the spatial shapes that correspond to the individual Fourier coefficients?

26.3.1 Coefficient G_0 Corresponds to the Contour's Centroid

We first look only at the specific Fourier coefficient G_0 with frequency index $m = 0$. Substituting $m = 0$ and $\omega_0 = 0$ in Eqn. (26.10), we get

$$A_0 = \frac{1}{M} \sum_{k=0}^{M-1} \bigl[x_k \cdot \cos(0) + y_k \cdot \sin(0)\bigr] \tag{26.26}$$

$$= \frac{1}{M} \sum_{k=0}^{M-1} \bigl[x_k \cdot 1 + y_k \cdot 0\bigr] = \frac{1}{M} \sum_{k=0}^{M-1} x_k = \bar{x}, \tag{26.27}$$

$$B_0 = \frac{1}{M} \sum_{k=0}^{M-1} \bigl[y_k \cdot \cos(0) - x_k \cdot \sin(0)\bigr] \tag{26.28}$$

$$= \frac{1}{M} \sum_{k=0}^{M-1} \bigl[y_k \cdot 1 - x_k \cdot 0\bigr] = \frac{1}{M} \sum_{k=0}^{M-1} y_k = \bar{y}. \tag{26.29}$$

Thus $G_0 = (A_0, B_0) = (\bar{x}, \bar{y})$ is simply the average of the x/y-coordinates, that is, the *centroid* of the original contour points g_k (see

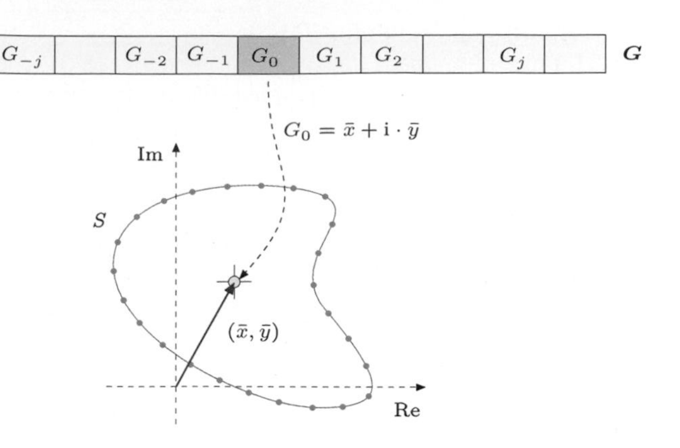

Fig. 26.5
DFT coefficient G_0 corresponds to the centroid of the contour points.

Fig. 26.5).[10] If we apply the *inverse* Fourier transform (Eqn. (26.15)) by ignoring (i.e., zeroing) all coefficients except G_0, we get the *partial reconstruction*[11] of the 2D contour coordinates $g_k^{(0)} = \big(x_k^{(0)}, y_k^{(0)}\big)$ as

$$x_k^{(0)} = \big[A_0 \cdot \cos\big(\omega_0 \tfrac{k}{M}\big) - B_0 \cdot \sin\big(\omega_0 \tfrac{k}{M}\big)\big] \tag{26.30}$$
$$= \bar{x} \cdot \cos(0) - \bar{y} \cdot \sin(0) = \bar{x} \cdot 1 - \bar{y} \cdot 0 = \bar{x}, \tag{26.31}$$
$$y_k^{(0)} = \big[B_0 \cdot \cos\big(\omega_0 \tfrac{k}{M}\big) + A_0 \cdot \sin\big(\omega_0 \tfrac{k}{M}\big)\big] \tag{26.32}$$
$$= \bar{y} \cdot \cos(0) + \bar{x} \cdot \sin(0) = \bar{y} \cdot 1 + \bar{x} \cdot 0 = \bar{y}. \tag{26.33}$$

Thus the contribution of the spectral value G_0 is the *centroid* of the reconstructed shape (see Fig. 26.5). If we perform a partial reconstruction of the contour using only the spectral coefficient G_0, then all contour points

$$g_0^{(0)} = g_1^{(0)} = \ldots = g_k^{(0)} = \ldots = g_{M-1}^{(0)} = (\bar{x}, \bar{y}) \tag{26.34}$$

would have the same (centroid) coordinate. This is because G_0 is the coefficient for the zero frequency and thus the sine and cosine terms in Eqns. (26.27) and (26.29) are constant. Alternatively, if we reconstruct the signal by *omitting* G_0 (i.e., $\boldsymbol{g}^{(1,\ldots,M-1)}$), the resulting contour is identical to the original shape, except that it is centered at the coordinate origin.

26.3.2 Coefficient G_1 Corresponds to a Circle

Next, we look at the geometric interpretation of $G_1 = (A_1, B_1)$, that is, the coefficient with frequency index $m = 1$, which corresponds to the angular frequency $\omega_1 = 2\pi$. Assuming that all coefficients G_m in the DFT spectrum are set to zero, except the single coefficient G_1,

[10] Note that the centroid of a boundary is generally not the same as the centroid of the enclosed region.

[11] We use the notation $\boldsymbol{g}^{(m)} = (g_0^{(m)}, g_1^{(m)}, \ldots, g_{M-1}^{(m)})$ for the *partial reconstruction* of the contour $\boldsymbol{g}$ from only a single Fourier coefficient G_m. For example, $\boldsymbol{g}^{(0)}$ is the reconstruction from the zero-frequency coefficient G_0 only. Analogously, we use $\boldsymbol{g}^{(a,b,c)}$ to denote a partial reconstruction based on selected Fourier coefficients G_a, G_b, G_c.

we get the partially reconstructed contour points $\boldsymbol{g}^{(1)}$ by Eqn. (26.11) as

$$g_k^{(1)} = G_1 \cdot e^{\mathrm{i} \cdot 2\pi \cdot \frac{k}{M}} \tag{26.35}$$
$$= [A_1 + \mathrm{i} \cdot B_1] \cdot [\cos(2\pi \tfrac{k}{M}) + \mathrm{i} \cdot \sin(2\pi \tfrac{k}{M})], \tag{26.36}$$

for $0 \leq k < M$. Remember that the complex values of $e^{\mathrm{i}\varphi}$ describe a *unit circle* in the complex plane that performs one full (counter-clockwise) revolution, as the angle φ runs from $0, \ldots, 2\pi$. Analogously, $e^{\mathrm{i}2\pi t}$ also describes a complete unit circle as t goes from 0 to 1. Since the term $\frac{k}{M}$ (for $0 \leq k < M$) also varies from 0 to 1 in Eqn. (26.36), the M reconstructed contour points are placed on a circle at equal angular steps. Multiplying $e^{\mathrm{i} \cdot 2\pi t}$ by a complex factor z stretches the *radius* of the circle by $|z|$, and also changes the *phase* (starting angle) of the circle by an angle θ, that is,

$$z \cdot e^{\mathrm{i} \cdot \varphi} = |z| \cdot e^{\mathrm{i} \cdot (\varphi + \theta)}, \tag{26.37}$$

with $\theta = \measuredangle z = \arg(z) = \tan^{-1}(\mathrm{Im}(z)/\mathrm{Re}(z))$.

We now see that the points $g_k^{(1)} = G_1 \cdot e^{\mathrm{i}\,2\pi k/M}$, generated by Eqn. (26.36), are positioned uniformly on a circle with radius $r_1 = |G_1|$ and starting angle (phase)

$$\theta_1 = \measuredangle G_1 = \tan^{-1}\Bigl(\frac{\mathrm{Im}(G_1)}{\mathrm{Re}(G_1)}\Bigr) = \tan^{-1}\Bigl(\frac{B_1}{A_1}\Bigr). \tag{26.38}$$

This point sequence is traversed in counter-clockwise direction for $k = 0, \ldots, M-1$ at frequency $m = 1$, that is, the circle performs one full revolution while the contour is traversed once. The circle is centered at the coordinate origin $(0,0)$, its radius is $|G_1|$, and its starting point (Eqn. (26.36) for $k = 0$) is

$$g_0^{(1)} = G_1 \cdot e^{\mathrm{i} \cdot 2\pi m \cdot \frac{k}{M}} = G_1 \cdot e^{\mathrm{i} \cdot 2\pi 1 \cdot \frac{0}{M}} = G_1 \cdot e^0 = G_1, \tag{26.39}$$

as illustrated in Fig. 26.6.

26.3.3 Coefficient G_m Corresponds to a Circle with Frequency m

Based on the aforementioned result for the frequency index $m = 1$, we can easily generalize the geometric interpretation of Fourier coefficients with arbitrary index $m > 0$. Using Eqn. (26.11), the partial reconstruction for the single Fourier coefficient $G_m = (A_m, B_m)$ is the contour $\boldsymbol{g}^{(m)}$, with coordinates

$$g_k^{(m)} = G_m \cdot e^{\mathrm{i} \cdot 2\pi m \cdot \frac{k}{M}} \tag{26.40}$$
$$= [A_m + \mathrm{i} \cdot B_m] \cdot [\cos(2\pi m \tfrac{k}{M}) + \mathrm{i} \cdot \sin(2\pi m \tfrac{k}{M})], \tag{26.41}$$

which again describe a circle with radius $r_m = |G_m|$, phase $\theta_m = \arg(G_m) = \tan^{-1}(B_m/A_m)$, and starting point $g_0^{(m)} = G_m$. In this case, however, the angular velocity is scaled by m, that is, the resulting circle revolves m times faster than the circle for G_1. In other

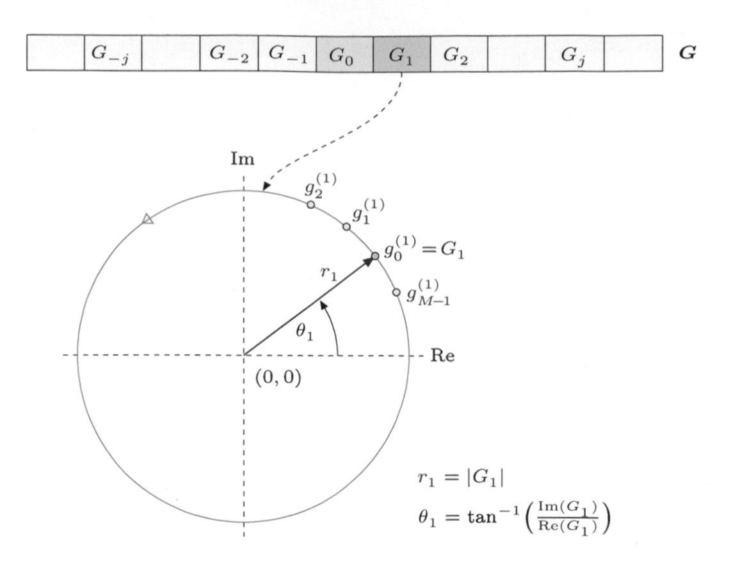

Fig. 26.6
A single DFT coefficient corresponds to a circle. The partial reconstruction from the single DFT coefficient G_m yields a sequence of M points $g_0^{(m)}, \ldots, g_{M-1}^{(m)}$ on a circle centered at the coordinate origin, with radius r_m and starting angle (phase) θ_m.

words, while the contour is traversed once, this circle performs m full revolutions.

Note that G_0 (see Sec. 26.3.1) does not really constitute a special case at all. Formally, it also describes a circle but one that oscillates with zero frequency, that is, all points have the same (constant) position

$$g_k^{(0)} = G_0 \cdot e^{\mathrm{i}\cdot 2\pi m \cdot \frac{k}{M}} = G_0 \cdot e^{\mathrm{i}\cdot 2\pi 0 \cdot \frac{k}{M}} = G_0 \cdot e^0 = G_0, \quad (26.42)$$

for $k = 0, \ldots, M-1$, which is equivalent to the curve's centroid $G_0 = (\bar{x}, \bar{y})$, as shown in Eqns. (26.27)–(26.29). Since the corresponding frequency is zero, the point never moves away from G_0.

26.3.4 Negative Frequencies

The DFT spectrum is periodic and defined for all frequencies $m \in \mathbb{Z}$, including negative frequencies. From Eqn. (26.21) we know that for any DFT coefficient with negative index G_{-m} there is an equivalent coefficient G_n whose index n is in the range $0, \ldots, M-1$. The partial reconstruction of the spectrum with the single coefficient G_{-m} is

$$g_k^{(-m)} = G_{-m} \cdot e^{-\mathrm{i}\cdot 2\pi m \cdot \frac{k}{M}} = G_n \cdot e^{-\mathrm{i}\cdot 2\pi m \cdot \frac{k}{M}}, \quad (26.43)$$

with $n = -m \bmod M$, which is again a sequence of points on the circle with radius $r_{-m} = r_n = |G_n|$ and phase $\theta_{-m} = \theta_n = \arg(G_n)$. The absolute rotation frequency is m, but this circle spins in the opposite, that is, *clockwise* direction, since angles become increasingly negative with growing k.

26.3.5 Fourier Descriptor Pairs Correspond to Ellipses

It follows therefore that the space-domain circles for the Fourier coefficients G_m and G_{-m} rotate with the same absolute frequency m but with different phase angles θ_m, θ_{-m} and in opposite directions. We denote the tuple

$$\text{FP}_m = (G_{-m}, G_{+m})$$

the "Fourier descriptor pair" (or "FD pair") for the frequency index m. If we perform a partial reconstruction from only the two Fourier coefficients G_{-m}, G_{+m} of this FD pair, we obtain the spatial points

$$\begin{aligned} g_k^{(\pm m)} &= g_k^{(-m)} + g_k^{(+m)} \\ &= G_{-m} \cdot e^{-\mathrm{i} \cdot 2\pi m \cdot \frac{k}{M}} + G_m \cdot e^{\mathrm{i} \cdot 2\pi m \cdot \frac{k}{M}} \\ &= G_{-m} \cdot e^{-\mathrm{i} \cdot \omega_m \cdot \frac{k}{M}} + G_m \cdot e^{\mathrm{i} \cdot \omega_m \cdot \frac{k}{M}}. \end{aligned} \tag{26.44}$$

By Eqn. (26.15) we can expand the result from Eqn. (26.44) to cartesian x/y coordinates as[12]

$$\begin{aligned} x_k^{(\pm m)} &= A_{-m} \cdot \cos\bigl(-\omega_m \cdot \tfrac{k}{M}\bigr) - B_{-m} \cdot \sin\bigl(-\omega_m \cdot \tfrac{k}{M}\bigr) + \\ &\quad\; A_m \cdot \cos(\omega_m \cdot \tfrac{k}{M}) - B_m \cdot \sin(\omega_m \cdot \tfrac{k}{M}) \\ &= (A_{-m} + A_m) \cdot \cos(\omega_m \cdot \tfrac{k}{M}) + (B_{-m} - B_m) \cdot \sin(\omega_m \cdot \tfrac{k}{M}), \end{aligned} \tag{26.45}$$

$$\begin{aligned} y_k^{(\pm m)} &= B_{-m} \cdot \cos\bigl(-\omega_m \cdot \tfrac{k}{M}\bigr) + A_{-m} \cdot \sin\bigl(-\omega_m \cdot \tfrac{k}{M}\bigr) + \\ &\quad\; B_m \cdot \cos\bigl(\omega_m \cdot \tfrac{k}{M}\bigr) + A_m \cdot \sin\bigl(\omega_m \cdot \tfrac{k}{M}\bigr) \\ &= (B_{-m} + B_m) \cdot \cos(\omega_m \cdot \tfrac{k}{M}) - (A_{-m} - A_m) \cdot \sin(\omega_m \cdot \tfrac{k}{M}), \end{aligned} \tag{26.46}$$

for $k = 0, \ldots, M-1$. The 2D point sequence $\boldsymbol{g}^{(\pm m)} = (g_0^{(\pm m)}, \ldots, g_{M-1}^{(\pm m)})$, obtained with Eqns. (26.45) and (26.46), describes an oriented *ellipse* that is centered at the origin (see Fig. 26.7). The parametric equation for this ellipse is

$$\begin{aligned} x_t^{(\pm m)} &= (A_{-m} + A_m) \cdot \cos(\omega_m \cdot t) + (B_{-m} - B_m) \cdot \sin(\omega_m \cdot t), \\ &= (A_{-m} + A_m) \cdot \cos(2\pi m t) + (B_{-m} - B_m) \cdot \sin(2\pi m t), \end{aligned} \tag{26.47}$$

$$\begin{aligned} y_t^{(\pm m)} &= (B_{-m} + B_m) \cdot \cos(\omega_m \cdot t) - (A_{-m} - A_m) \cdot \sin(\omega_m \cdot t) \\ &= (B_{-m} + B_m) \cdot \cos(2\pi m t) - (A_{-m} - A_m) \cdot \sin(2\pi m t), \end{aligned} \tag{26.48}$$

for $t = 0, \ldots, 1$.

Ellipse parameters

In general, the parametric equation of an ellipse with radii a, b, centered at (x_c, y_c) and oriented at an angle α is

$$\begin{aligned} x(\psi) &= x_c + a \cdot \cos(\psi) \cdot \cos(\alpha) - b \cdot \sin(\psi) \cdot \sin(\alpha), \\ y(\psi) &= y_c + a \cdot \cos(\psi) \cdot \sin(\alpha) + b \cdot \sin(\psi) \cdot \cos(\alpha), \end{aligned} \tag{26.49}$$

with $\psi = 0, \ldots, 2\pi$. From Eqns. (26.45) and (26.46) we see that the parameters a_m, b_m, α_m of the ellipse for a single Fourier descriptor pair $\text{FP}_m = (G_{-m}, G_{+m})$ are

$$a_m = r_{-m} + r_{+m} = |G_{-m}| + |G_{+m}|, \tag{26.50}$$

$$b_m = |r_{-m} - r_{+m}| = \bigl|\, |G_{-m}| - |G_{+m}| \,\bigr|, \tag{26.51}$$

$$\begin{aligned} \alpha_m &= \frac{1}{2} \cdot \bigl(\underbrace{\sphericalangle G_{-m}}_{\theta_{-m}} + \underbrace{\sphericalangle G_{+m}}_{\theta_{+m}}\bigr) \\ &= \frac{1}{2} \cdot \Bigl[\tan^{-1}\Bigl(\frac{B_{-m}}{A_{-m}}\Bigr) + \tan^{-1}\Bigl(\frac{B_{+m}}{A_{+m}}\Bigr)\Bigr]. \end{aligned} \tag{26.52}$$

[12] Using the relations $\sin(-a) = -\sin(a)$ and $\cos(-a) = \cos(a)$.

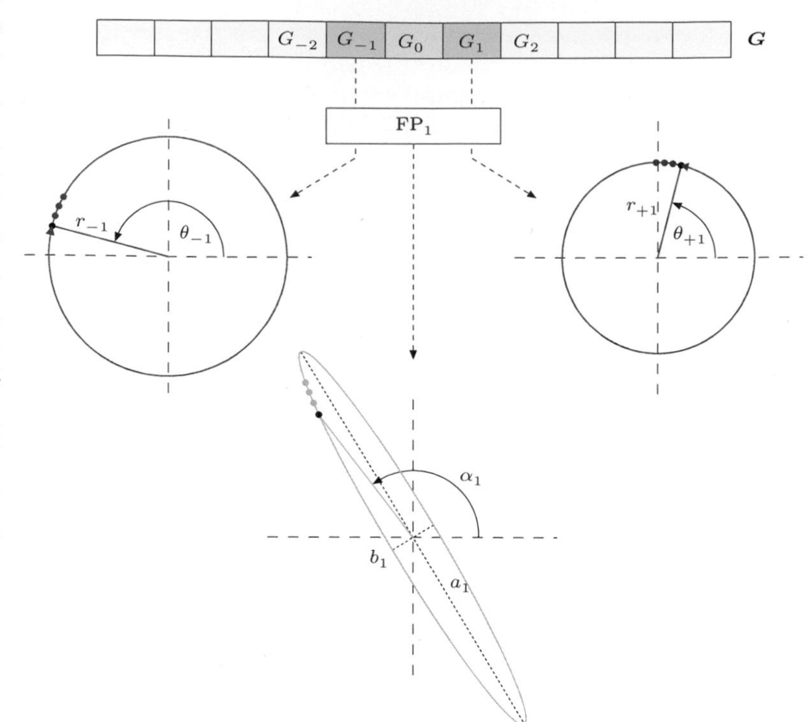

Fig. 26.7
DFT coefficients G_{-m}, G_{+m} form a Fourier descriptor pair FP_m. Each of the two descriptors corresponds to M points on a circle of radius r_{-m}, r_{+m} and phase θ_{-m}, θ_{+m}, respectively, revolving with the same frequency m but in opposite directions. The sum of each point pair is located on an ellipse with radii a_m, b_m and orientation α_m. The orientation α_m of the ellipse's major axis is centered between the starting angles of the circles defined by G_{-m} and G_{+m}; its radii are $a_m = r_{-m} + r_{+m}$ for the major axis and $b_m = |r_{-m} - r_{+m}|$ for the minor axis. The figure shows the situation for $m = 1$.

Like its constituting circles, this ellipse is centered at $(x_c, y_c) = (0, 0)$ and performs m revolutions for one traversal of the contour. G_{-m} specifies the circle

$$z_{-m}(\varphi) = G_{-m} \cdot e^{\mathrm{i}\cdot(-\varphi)} = r_{-m} \cdot e^{\mathrm{i}\cdot(\theta_{-m}-\varphi)}, \tag{26.53}$$

for $\varphi \in [0, 2\pi]$, with starting angle θ_{-m} and radius r_{-m}, rotating in a clockwise direction. Similarly, G_{+m} specifies the circle

$$z_{+m}(\varphi) = G_{+m} \cdot e^{\mathrm{i}\cdot(\varphi)} = r_{+m} \cdot e^{\mathrm{i}\cdot(\theta_{+m}+\varphi)}, \tag{26.54}$$

with starting angle θ_{+m} and radius r_{+m}, rotating in a counter-clockwise direction. Both circles thus rotate at the same angular velocity but in opposite directions, as mentioned before. The corresponding (complex-valued) ellipse points are

$$z_m(\varphi) = z_{-m}(\varphi) + z_{+m}(\varphi). \tag{26.55}$$

The ellipse radius $|z_m(\varphi)|$ is a *maximum* at position $\varphi = \varphi_{\max}$, where the angles on both circles are identical (i.e., the corresponding vectors have the same direction). This occurs when

$$\theta_{-m} - \varphi_{\max} = \theta_{+m} + \varphi_{\max} \qquad \text{or} \qquad \varphi_{\max} = \frac{1}{2} \cdot (\theta_{-m} - \theta_{+m}),$$

that is, at mid-angle between the two starting angles θ_{-m} and θ_{+m}. Therefore, the orientation of the ellipse's major axis is

$$\alpha_m = \theta_{+m} + \frac{\theta_{-m} - \theta_{+m}}{2} = \frac{1}{2} \cdot (\theta_{-m} + \theta_{+m}), \tag{26.56}$$

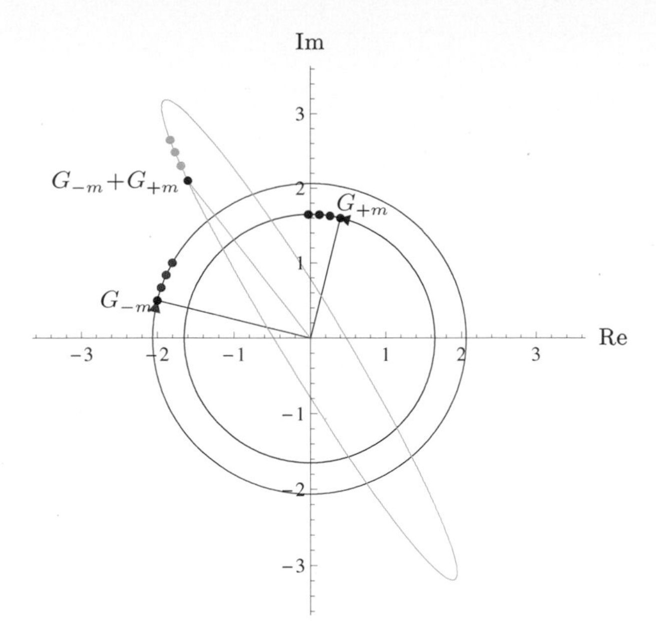

Fig. 26.8 Ellipse created by partial reconstruction from a single Fourier descriptor pair $\mathrm{FP}_m = (G_{-m}, G_{+m})$. The two complex-valued Fourier coefficients $G_{-m} = (-2, 0.5)$ and $G_m = (0.4, 1.6)$ represent circles with starting points G_{-m} and G_{+m}, respectively. The circle for G_{-m} (red) rotates in clockwise direction, the circle for G_{+m} (blue) rotates in counter-clockwise direction. The ellipse (green) is the result of point-wise addition of the two circles, as shown for four successive points, starting with point $G_{-m}+G_{+m}$.

as already stated in Eqn. (26.52). At $\varphi = \varphi_{\max}$ the two radial vectors align, and thus the radius of the ellipse's major axis a_m is the sum of the two circle radii, that is,

$$a_m = r_{-m} + r_{+m} \tag{26.57}$$

(cf. Eqn. (26.50)). Analogously, the ellipse radius is *minimized* at position $\varphi = \varphi_{\min}$, where the $z_{-m}(\varphi_{\min})$ and $z_{+m}(\varphi_{\min})$ lie on opposite sides of the circle. This occurs at angle

$$\varphi_{\min} = \varphi_{\max} + \frac{\pi}{2} = \frac{\pi + \theta_{-m} - \theta_{+m}}{2} \tag{26.58}$$

and the corresponding radius for the ellipse's minor axis is (cf. Eqn. (26.51))

$$b_m = r_{+m} - r_{-m}. \tag{26.59}$$

Figure 26.8 illustrates this situation for a specific Fourier descriptor pair $\mathrm{FP}_m = (G_{-m}, G_{+m}) = (-2 + \mathrm{i}\cdot 0.5, 0.4 + \mathrm{i}\cdot 1.6)$. Note that the ellipse parameters a_m, b_m, α_m (see Eqns. (26.50)–(26.52)) are not explicitly required for reconstructing (drawing) the contour, since the ellipse can also be generated by simply adding the x/y-coordinates of the two counter-revolving circles for the participating Fourier descriptors, as given in Eqn. (26.55). Another example is shown in Fig. 26.9.

26.3.6 Shape Reconstruction from Truncated Fourier Descriptors

Due to the periodicity of the DFT spectrum, the complete reconstruction of the contour points g_k from the Fourier coefficients G_m (see Eqn. (26.11)) could also be written with a different summation range, as long as all spectral coefficients are included, that is,

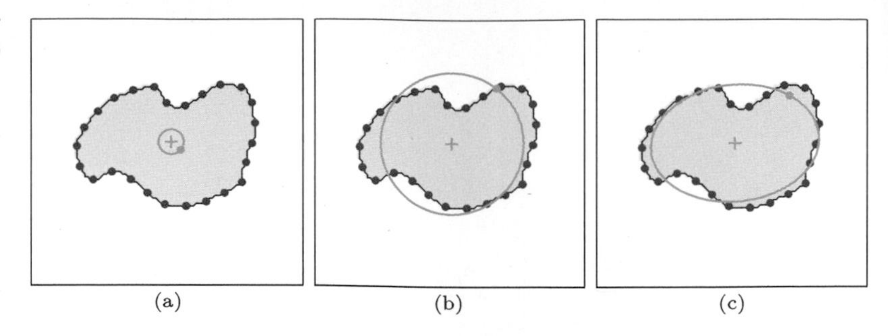

Fig. 26.9 Partial reconstruction from single coefficients and an FD descriptor pair. The two circles reconstructed from DFT coefficient G_{-1} (a) and coefficient G_{+1} (b) are positioned at the centroid of the contour (G_0). The combined reconstruction for $\left(G_{-1}, G_{+1}\right)$ produces the ellipse in (c). The dots on the green curves show the path position for $t = 0$.

$$g_k = \sum_{m=0}^{M-1} G_m \cdot e^{\mathrm{i}\cdot 2\pi m\cdot \frac{k}{M}} = \sum_{m=m_0}^{m_0+M-1} G_m \cdot e^{\mathrm{i}\cdot 2\pi m\cdot \frac{k}{M}}, \tag{26.60}$$

for any start index $m_0 \in \mathbb{Z}$. As a special (though important) case we can perform the summation symmetrically around the zero index and write

$$g_k = \sum_{m=0}^{M-1} G_m \cdot e^{\mathrm{i}\cdot 2\pi m\cdot \frac{k}{M}} = \sum_{\substack{m=\\-(M-1)\div 2}}^{M\div 2} G_m \cdot e^{\mathrm{i}\cdot 2\pi m\cdot \frac{k}{M}}. \tag{26.61}$$

To understand the reconstruction in terms of Fourier descriptor pairs, it is helpful to distinguish if M (the number of contour points and Fourier coefficients) is *even* or *odd*.

Odd number of contour points

If M is *odd*, then the spectrum consists of G_0 (representing the contour's centroid) plus exactly $M \div 2$ Fourier descriptor pairs FP_m, with $m = 1, \ldots, M \div 2$.[13] We can thus rewrite Eqn. (26.60) as

$$\begin{aligned} g_k &= \sum_{m=0}^{M-1} G_m \cdot e^{\mathrm{i}\cdot 2\pi m\cdot \frac{k}{M}} = \underbrace{G_0}_{g_k^{(0)}} + \sum_{m=1}^{M\div 2} \underbrace{\left[G_{-m} \cdot e^{-\mathrm{i}\cdot 2\pi m\cdot \frac{k}{M}} + G_m \cdot e^{\mathrm{i}\cdot 2\pi m\cdot \frac{k}{M}}\right]}_{g_k^{(\pm m)} = g_k^{(-m)} + g_k^{(m)}} \\ &= g_k^{(0)} + \sum_{m=1}^{M\div 2} g_k^{(\pm m)} = g_k^{(0)} + g_k^{(\pm 1)} + g_k^{(\pm 2)} + \ldots + g_k^{(\pm M\div 2)}, \end{aligned} \tag{26.62}$$

where $g_k^{(\pm m)}$ denotes the partial reconstruction from the single Fourier descriptor pair FP_m (see Eqn. (26.44)).

As we already know, the partial reconstruction $g_k^{(\pm m)}$ of an individual Fourier descriptor pair FP_m is a set of points on an ellipse that is centered at the origin $(0, 0)$. The partial reconstruction of the *three* DFT coefficients G_0, G_{-m}, G_{+m} (i.e., FP_m plus the single coefficient G_0) is the point sequence

$$g_k^{(-m,0,m)} = g_k^{(0)} + g_k^{(\pm m)}, \tag{26.63}$$

which is the ellipse for $g_k^{(\pm m)}$ shifted to $g_k^{(0)} = (\bar{x}, \bar{y})$, the centroid of the original contour. For example, the partial reconstruction from the coefficients G_{-1}, G_0, G_{+1},

[13] If M is odd, then $M = 2 \cdot (M \div 2) + 1$.

$$g_k^{(-1,0,1)} = g_k^{(-1,\ldots,1)} = g_k^{(0)} + g_k^{(\pm 1)}, \tag{26.64}$$

yields an ellipse with frequency $m = 1$ that revolves around the (fixed) centroid of the original contour. If we add another Fourier descriptor pair FP_2, the resulting reconstruction is

$$g_k^{(-2,\ldots,2)} = \underbrace{g_k^{(0)} + g_k^{(\pm 1)}}_{\text{ellipse 1}} + \underbrace{g_k^{(\pm 2)}}_{\text{ellipse 2}}. \tag{26.65}$$

The resulting ellipse $g_k^{(\pm 2)}$ has the frequency $m = 2$, but note that it is centered at a moving point on the "slower" ellipse (with frequency $m = 1$), that is, ellipse 2 effectively "rides" on ellipse 1. If we add FP_3, its ellipse is again centered at a point on ellipse 2, and so on. For an illustration, see the examples in Figs. 26.11 and 26.12. In general, the ellipse for descriptor pair FP_j revolves around the (moving) center obtained as the superposition of $j-1$ "slower" ellipses,

$$g_k^{(0)} + \sum_{m=1}^{j-1} g_k^{(\pm m)}. \tag{26.66}$$

Consequently, the curve obtained by the partial reconstruction from descriptor pairs $\mathrm{FP}_1, \ldots, \mathrm{FP}_j$ (for $j \leq M \div 2$) is the point sequence

$$g_k^{(-j,\ldots,j)} = g_k^{(0)} + \sum_{m=1}^{j} g_k^{(\pm m)}, \tag{26.67}$$

for $k = 0, \ldots, M-1$. The fully reconstructed shape is the sum of the centroid (defined by G_0) and $M \div 2$ ellipses, one for each Fourier descriptor pair $\mathrm{FP}_1, \ldots, \mathrm{FP}_{M \div 2}$.

Even number of contour points

If M is *even*,[14] then the reconstructed shape is a superposition of the centroid (defined by G_0), $(M-1) \div 2$ ellipses from the Fourier descriptor pairs $\mathrm{FP}_1, \ldots, \mathrm{FP}_{(M-1)\div 2}$, plus one additional *circle* specified by the single (highest frequency) Fourier coefficient $G_{M \div 2}$. The complete reconstruction from an even-length Fourier descriptor can thus be written as

$$g_k = \sum_{m=0}^{M-1} G_m \cdot e^{\mathrm{i} \cdot 2\pi m \cdot \frac{k}{M}} = \underbrace{g_k^{(0)}}_{\text{center}} + \underbrace{\sum_{m=1}^{(M-1)\div 2} g_k^{(\pm m)}}_{(M-1)\div 2 \text{ ellipses}} + \underbrace{g_k^{(M \div 2)}}_{1 \text{ circle}}. \tag{26.68}$$

The single high-frequency circle associated with $g_k^{(M \div 2)}$ has its (moving) center at the sum of all lower-frequency ellipses that correspond to the Fourier coefficients $G_{-m}, \ldots, G_{+m}$, with $m < (M \div 2)$.

Reconstruction algorithm

Algorithm 26.4 describes the reconstruction of shapes from a Fourier descriptor using only a specified number (M_p) of Fourier descriptor pairs. The number of points on the reconstructed contour (N) can be freely chosen.

[14] In this case, $M = 2 \cdot (M \div 2) = (M-1) \div 2 + 1 + M \div 2$.

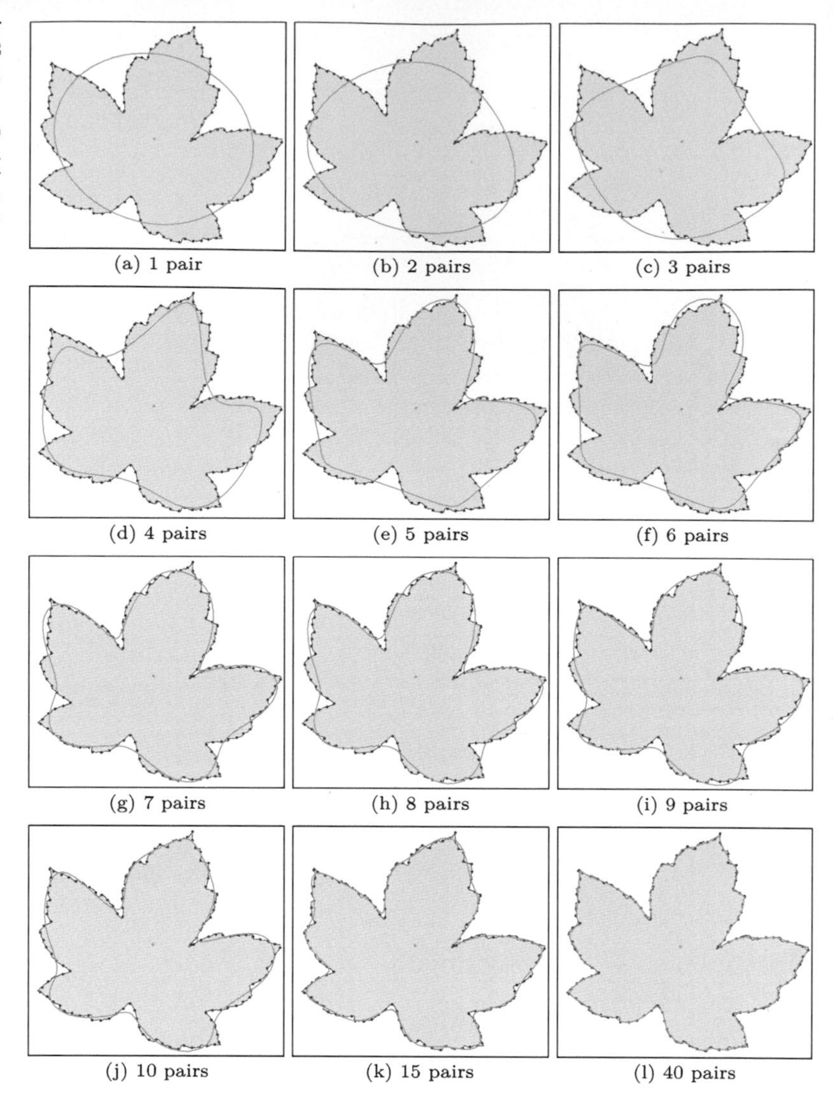

Fig. 26.10
Partial shape reconstruction from a limited set of Fourier descriptor pairs. The full descriptor contains 125 coefficients (G_0 plus 62 FD pairs).

26.3.7 Fourier Descriptors from Unsampled Polygons

The requirement to distribute sample points uniformly along the contour path stems from classical signal processing and Fourier theory, where uniform sampling is a common assumption. However, as shown in [143] (see also [183, 262]), the Fourier descriptors for a polygonal shape can be calculated directly from the original polygon vertices without sub-sampling the contour. This "trigonometric" approach, described in the following, works for arbitrary (convex and non-convex) polygons.

We assume that the shape is specified as a sequence of P points $V = (\boldsymbol{v}_0, \ldots, \boldsymbol{v}_{P-1})$, with $V(i) = \boldsymbol{v}_i = (x_i, y_i)$ representing the 2D vertices of a closed polygon. We define the quantities

$$\boldsymbol{d}(i) = \boldsymbol{v}_{(i+1) \bmod P} - \boldsymbol{v}_i \qquad \text{and} \qquad \lambda(i) = \|\boldsymbol{d}(i)\|, \tag{26.69}$$

for $i = 0, \ldots, P-1$, where $\boldsymbol{d}(i)$ is the vector representing the polygon segment between the vertices $\boldsymbol{v}_i, \boldsymbol{v}_{i+1}$, and $\lambda(i)$ is the length of that

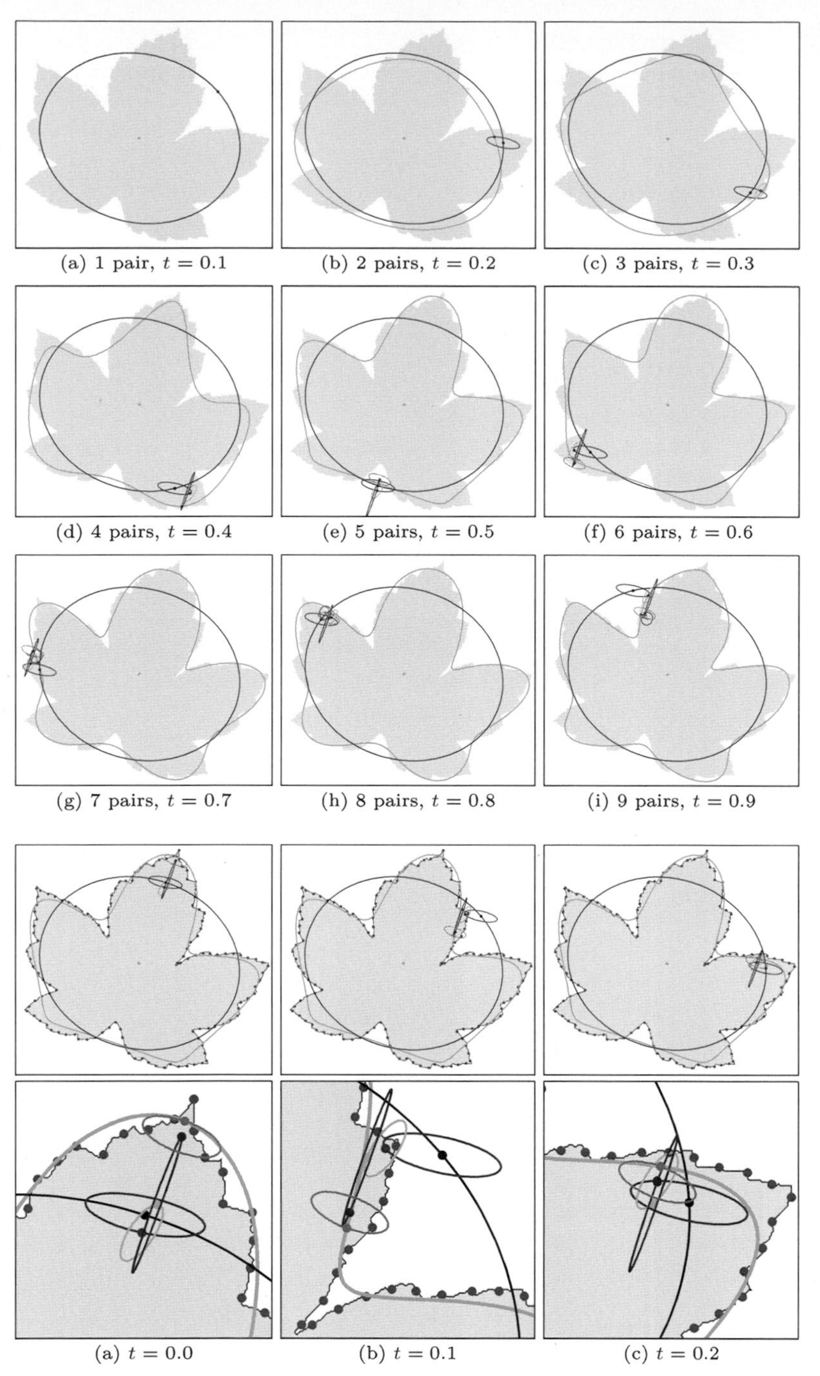

(a) 1 pair, $t = 0.1$ (b) 2 pairs, $t = 0.2$ (c) 3 pairs, $t = 0.3$

(d) 4 pairs, $t = 0.4$ (e) 5 pairs, $t = 0.5$ (f) 6 pairs, $t = 0.6$

(g) 7 pairs, $t = 0.7$ (h) 8 pairs, $t = 0.8$ (i) 9 pairs, $t = 0.9$

Fig. 26.11
Partial reconstruction by ellipse superposition (details). The green curve shows the partial reconstruction from $1, \ldots, 9$ FD pairs. This curve performs one full revolution as the path parameter t runs from 0 to 1. Subfigures (a–i) depict the situation for $1, \ldots, 9$ FD pairs and different path positions $t = 0.1, 0.2, \ldots, 0.9$. Each Fourier descriptor pair corresponds to an ellipse that is centered at the current position t on the previous ellipse. The individual Fourier descriptor pair FP_1 in (a) corresponds to a single ellipse. In (b), the point for $t = 0.2$ on the blue ellipse (for FP_1) is the center of the red ellipse (for FP_2). In (c), the green ellipse (for FP_3) is centered at the point marked on the previous ellipse, and so on. The reconstructed shape is obtained by superposition of all ellipses. See Fig. 26.12 for a detailed view.

(a) $t = 0.0$ (b) $t = 0.1$ (c) $t = 0.2$

Fig. 26.12
Partial reconstruction by ellipse superposition (details). The green curve shows the partial reconstruction from 5 FD pairs $\mathrm{FP}_1, \ldots, \mathrm{FP}_5$. This curve performs one full revolution as the path parameter t runs from 0 to 1. Subfigures (a–c) show the composition of the contour by superposition of the 5 ellipses, each corresponding to one FD pair, at selected positions $t = 0.0, 0.1, 0.2$. The blue ellipse corresponds to FP_1 and revolves once for $t = 0, \ldots, 1$. The blue dot on this ellipse marks the position t, which serves as the center of the next (red) ellipse corresponding to FP_2. This ellipse makes 2 revolutions for $t = 0, \ldots, 1$ and the red dot for position t is again the center of green ellipse (for FP_3), and so on. Position t on the orange ellipse (for FP_1) coincides with the final reconstruction (green curve). The original contour was sampled at 125 equidistant points.

segment. We also define

$$L(i) = \sum_{j=0}^{i-1} \lambda(j), \tag{26.70}$$

for $i = 0, \ldots, P$, which is the cumulative length of the polygon path from the start vertex $\boldsymbol{v}_0$ to vertex $\boldsymbol{v}_i$, such that $L(0)$ is zero and $L(P)$ is the closed path length of the polygon V.

Alg. 26.4
Partial shape reconstruction from a truncated Fourier descriptor $\boldsymbol{G}$. The shape is reconstructed by considering up to M_{p} Fourier descriptor pairs. The resulting sequence of contour points may be of arbitrary length (N). See Figs. 26.10–26.12 for examples.

```
1: GetPartialReconstruction(G, M_p, N)
     Input: G = (G_0, ..., G_{M-1}), Fourier descriptor with M coefficients; M_p, number of Fourier descriptor pairs to consider; N, number of points on the reconstructed shape. Returns the reconstructed contour as a sequence of N complex values.
2:   Create map g: [0, N-1] → ℂ
3:   M ← |G|                                   ▷ total number of Fourier coefficients
4:   M_p ← min(M_p, (M-1) ÷ 2)                 ▷ available Fourier coefficient pairs
5:   for k ← 0, ..., N-1 do
6:       t ← k/N                               ▷ continuous path position t ∈ [0, 1]
7:       g(k) ← GetSinglePoint(G, -M_p, M_p, t)    ▷ see below
8:   return g.

9: GetSinglePoint(G, m_-, m_+, t)
     Returns a single point (as a complex value) on the reconstructed shape for the continuous path position t ∈ [0, 1], based on the Fourier coefficients G(m_-), ..., G(m_+).
10:  M ← |G|
11:  x ← 0,   y ← 0
12:  for m ← m_-, ..., m_+ do
13:      φ ← 2 · π · m · t
14:      G ← G(m mod M)
15:      A ← Re(G),    B ← Im(G)
16:      x ← x + A · cos(φ) − B · sin(φ)
17:      y ← y + A · sin(φ) + B · cos(φ)
18:  return (x + i y).
```

For a (freely chosen) number of Fourier descriptor pairs (M_{p}), the corresponding Fourier descriptor $\boldsymbol{G} = (G_{-M_{\mathrm{p}}}, \ldots, G_0, \ldots, G_{+M_{\mathrm{p}}})$, has $2M_{\mathrm{p}} + 1$ complex-valued coefficients G_m, where

$$G_0 = a_0 + \mathrm{i} \cdot c_0 \tag{26.71}$$

and the remaining coefficients are calculated as

$$G_{+m} = (a_m + d_m) + \mathrm{i} \cdot (c_m - b_m), \tag{26.72}$$
$$G_{-m} = (a_m - d_m) + \mathrm{i} \cdot (c_m + b_m), \tag{26.73}$$

from the "trigonometric coefficients" a_m, b_m, c_m, d_m. As described in [143], these coefficients are obtained directly from the P polygon vertices $\boldsymbol{v}_i$ as

$$\begin{pmatrix} a_0 \\ c_0 \end{pmatrix} = \boldsymbol{v}_0 + \frac{\sum_{i=0}^{P-1} \Bigl[\frac{L^2(i+1) - L^2(i)}{2\,\lambda(i)} \cdot \boldsymbol{d}(i) + \lambda(i) \cdot \sum_{j=0}^{i-1} \boldsymbol{d}(j) - \boldsymbol{d}(i) \cdot \sum_{j=0}^{i-1} \lambda(j) \Bigr]}{L(P)} \tag{26.74}$$

(representing the shape's center), with $\boldsymbol{d}, \lambda, L$ as defined in Eqns. (26.69) and (26.70). This can be simplified to

$$\begin{pmatrix} a_0 \\ c_0 \end{pmatrix} = \boldsymbol{v}_0 + \frac{\sum_{i=0}^{P-1} \Bigl[\Bigl(\frac{L^2(i+1) - L^2(i)}{2\,\lambda(i)} - L(i) \Bigr) \cdot \boldsymbol{d}(i) + \lambda(i) \cdot (\boldsymbol{v}_i - \boldsymbol{v}_0) \Bigr]}{L(P)}. \tag{26.75}$$

```
1:  FourierDescriptorFromPolygon(V, M_p)
    Input: V = (v_0, ..., v_{P-1}), a sequence of P points representing
    the vertices of a closed 2D polygon; M_p, the desired number of
    FD pairs. Returns a new Fourier descriptor of length 2M_p+1.
2:    P ← |V|                                  ▷ number of polygon vertices in V
3:    M ← 2·M_p+1                              ▷ number of Fourier coefficients in G
4:    Create maps d: [0, P−1] → R^2,   λ: [0, P−1] → R,
5:                L: [0, P] → R,       G: [0, M−1] → C
6:    L(0) ← 0
7:    for i ← 0, ..., P−1 do
8:        d(i) ← V((i+1) mod P) − V(i)                         ▷ Eq. 26.69
9:        λ(i) ← ||d(i)||
10:       L(i+1) ← L(i) + λ(i)
11:   (a, c) ← (0, 0)                                          ▷ a = a_0, c = c_0
12:   for i ← 0, ..., P−1 do
13:       s ← (L^2(i+1) − L^2(i)) / (2·λ(i)) − L(i)
14:       (a, c) ← (a, c) + s·d(i) + λ(i)·(V(i) − V(0))        ▷ Eq. 26.75
15:   G(0) ← v_0 + 1/L(P) · (a, c)                             ▷ Eq. 26.71
16:   for m ← 1, ..., M_p do                      ▷ for FD-pairs G_{±1}, ..., G_{±M_p}
17:       (a, c) ← (0, 0),   (b, d) ← (0, 0)                   ▷ a_m, b_m, c_m, d_m
18:       for i ← 0, ..., P−1 do
19:           ω_0 ← 2πm · L(i)/L(P)
20:           ω_1 ← 2πm · L((i+1) mod P)/L(P)
21:           (a, c) ← (a, c) + (cos(ω_1)−cos(ω_0))/λ(i) · d(i)  ▷ Eq. 26.76
22:           (b, d) ← (b, d) + (sin(ω_1)−sin(ω_0))/λ(i) · d(i)  ▷ Eq. 26.77
23:       G(m)         ← L(P)/(2πm)^2 · (a + d, c − b)          ▷ Eq. 26.72
24:       G(−m mod M)  ← L(P)/(2πm)^2 · (a − d, c + b)          ▷ Eq. 26.73
25:   return G.
```

Alg. 26.5 Fourier descriptor from trigonometric data (arbitrary polygons). Parameter M_p specifies the number of Fourier coefficient pairs.

The remaining coefficients a_m, b_m, c_m, d_m ($m = 1, \ldots, M_p$) are calculated as

$$\begin{pmatrix} a_m \\ c_m \end{pmatrix} = \frac{L(P)}{(2\pi m)^2} \cdot \sum_{i=0}^{P-1} \Bigl[\frac{\cos\bigl(2\pi m \frac{L(i+1)}{L(P)}\bigr) - \cos\bigl(2\pi m \frac{L(i)}{L(P)}\bigr)}{\lambda(i)} \cdot \boldsymbol{d}(i) \Bigr], \tag{26.76}$$

$$\begin{pmatrix} b_m \\ d_m \end{pmatrix} = \frac{L(P)}{(2\pi m)^2} \cdot \sum_{i=0}^{P-1} \Bigl[\frac{\sin\bigl(2\pi m \frac{L(i+1)}{L(P)}\bigr) - \sin\bigl(2\pi m \frac{L(i)}{L(P)}\bigr)}{\lambda(i)} \cdot \boldsymbol{d}(i) \Bigr], \tag{26.77}$$

respectively. The complete calculation of a Fourier descriptor from trigonometric coordinates (i.e., from arbitrary polygons) is summarized in Alg. 26.5.

An approximate reconstruction of the original shape can be obtained directly from the trigonometric coefficients a_m, b_m, c_m, d_m de-

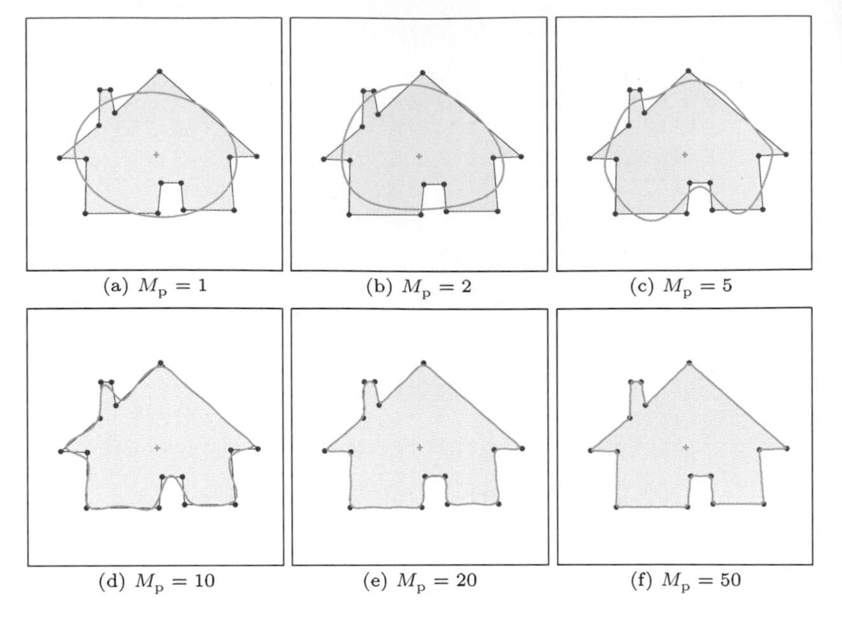

Fig. 26.13 Fourier descriptors calculated from trigonometric data (arbitrary polygons). Shape reconstructions with different numbers of Fourier descriptor pairs (M_p).

fined in Eqns. (26.75) and (26.76) as[15]

$$\boldsymbol{x}(t) = \begin{pmatrix} a_0 \\ c_0 \end{pmatrix} + \sum_{m=1}^{M_p} \Big[\begin{pmatrix} a_m \\ c_m \end{pmatrix} \cdot \cos(2\pi m t) + \begin{pmatrix} b_m \\ d_m \end{pmatrix} \cdot \sin(2\pi m t) \Big], \quad (26.78)$$

for $t = 0, \ldots, 1$. Of course, this reconstruction can also be calculated from the actual DFT coefficients $\boldsymbol{G}$, as described in Eqn. (26.20). Again the reconstruction error is reduced by increasing the number of Fourier descriptor pairs (M_p), as demonstrated in Fig. 26.13.[16] The reconstruction is theoretically perfect as M_p goes to infinity.

Working with the trigonometric technique is an advantage, in particular, if the boundary curvature along the outline varies strongly. For example, the silhouette of a human hand typically exhibits high curvature along the fingertips while other contour sections are almost straight. Capturing the high-curvature parts requires a significantly higher density of samples than in the smooth sections, as illustrated in Fig. 26.14. This figure compares the partial shape reconstructions obtained from Fourier descriptors calculated with uniform and non-uniform contour sampling, using identical numbers of Fourier descriptor pairs (M_p). Note that the coefficients (and thus the reconstructions) are very similar, although considerably fewer samples were used for the trigonometric approach.

[15] Note the analogy to the elliptical reconstruction in Eqns. (26.47) and (26.48).

[16] Most test images used in this chapter were taken from the Kimia dataset [134]. A selected subset of modified images taken from this dataset is available on the book's website.

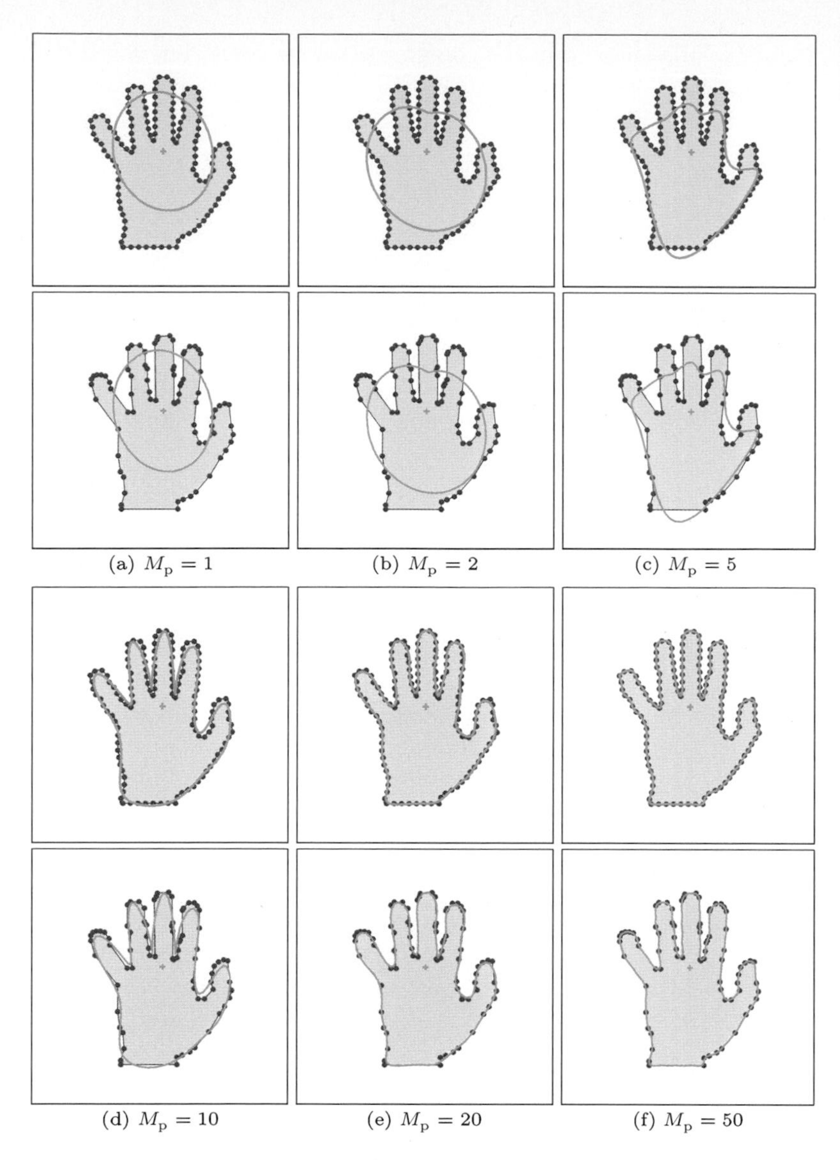

Fig. 26.14
Fourier descriptors from uniformly sampled vs. non-uniformly sampled (trigonometric) contours. Partial constructions from Fourier descriptors obtained from uniformly sampled contours (rows 1, 3) and non-uniformly sampled contours (rows 2, 4), for different numbers of Fourier descriptor pairs (M_p).

26.4 Effects of Geometric Transformations

To be useful for comparing shapes, a representation should be invariant against a certain set of geometric transformations. Typically, a minimal requirement for robust 2D shape matching is invariance to translation, scale changes, and rotation. Fourier shape descriptors in their basic form are *not* invariant under any of these transformations but they can be modified to satisfy these requirements. In this section, we discuss the effects of such transformations upon the corresponding Fourier descriptors. The steps involved for making Fourier descriptors invariant are discussed subsequently in Sec. 26.5.

26.4.1 Translation

As described in Sec. 26.3.1, the coefficient G_0 of a Fourier descriptor $\boldsymbol{G}$ corresponds to the centroid of the encoded contour. Moving the points g_k of a shape $\boldsymbol{g}$ in the complex plane by some constant $z \in \mathbb{C}$,

$$g'_k = g_k + z, \tag{26.79}$$

for $k = 0, \ldots, M-1$, only affects Fourier coefficient G_0, that is,

$$G'_m = \begin{cases} G_m + z & \text{for } m = 0, \\ G_m & \text{for } m \neq 0. \end{cases} \tag{26.80}$$

To make an FD invariant against translation, it is thus sufficient to zero its G_0 coefficient, thereby shifting the shape's center to the origin of the coordinate system. Alternatively, translation invariant matching of Fourier descriptors is achieved by simply ignoring coefficient G_0.

26.4.2 Scale Change

Since the Fourier transform is a linear operation, scaling a 2D shape $\boldsymbol{g}$ uniformly by a real-valued factor s,

$$g'_k = s \cdot g_k, \tag{26.81}$$

also scales the corresponding Fourier spectrum by the same factor, that is,

$$G'_m = s \cdot G_m, \tag{26.82}$$

for $m = 1, \ldots, M-1$. Note that scaling by $s = -1$ (or any other negative factor) corresponds to *reversing* the ordering of the samples along the contour (see also Sec. 26.4.6). Given the fact that the DFT coefficient G_1 represents a circle whose radius $r_1 = |G_1|$ is proportional to the size of the original shape (see Sec. 26.3.2), the Fourier descriptor $\boldsymbol{G}$ could be normalized for scale by setting

$$G^{\mathrm{S}}_m = \frac{1}{|G_1|} \cdot G_m, \tag{26.83}$$

for $m = 1, \ldots, M-1$, such that $\left|G^{\mathrm{S}}_1\right| = 1$. Although it is common to use only G_1 for scale normalization, this coefficient may be relatively small (and thus unreliable) for certain shapes. We therefore prefer to normalize the complete Fourier coefficient vector to achieve scale invariance (see Sec. 26.5.1).

26.4.3 Rotation

If a given shape is rotated about the origin by some angle β, then each contour point $\boldsymbol{v}_k = (x_k, y_k)$ moves to a new position

$$\boldsymbol{v}'_k = \begin{pmatrix} x'_k \\ y'_k \end{pmatrix} = \begin{pmatrix} \cos(\beta) & -\sin(\beta) \\ \sin(\beta) & \cos(\beta) \end{pmatrix} \cdot \begin{pmatrix} x_k \\ y_k \end{pmatrix}. \tag{26.84}$$

If the 2D contour samples are represented as complex values $g_k = x_k + \mathrm{i} \cdot y_k$, this rotation can be expressed as a multiplication

$$g'_k = e^{\mathrm{i}\beta} \cdot g_k, \tag{26.85}$$

with the complex factor $e^{\mathrm{i}\beta} = \cos(\beta) + \mathrm{i} \cdot \sin(\beta)$. As in Eqn. (26.82), we can use the linearity of the DFT to predict the effects of rotating the shape $\boldsymbol{g}$ by angle β as

$$G'_m = e^{\mathrm{i}\beta} \cdot G_m, \tag{26.86}$$

for $m = 0, \ldots, M-1$. Thus, the spatial rotation in Eqn. (26.85) multiplies each DFT coefficient G_m by the *same* complex factor $e^{\mathrm{i}\beta}$, which has unit magnitude. Since

$$e^{\mathrm{i}\beta} \cdot G_m = e^{\mathrm{i}(\theta_m + \beta)} \cdot |G_m|, \tag{26.87}$$

this only rotates the *phase* $\theta_m = \measuredangle G_m$ of each coefficient by the *same* angle β, without changing its *magnitude* $|G_m|$.

26.4.4 Shifting the Sampling Start Position

Despite the implicit periodicity of the boundary sequence and the corresponding DFT spectrum, Fourier descriptors are generally not the same if sampling starts at different positions along the contour. Given a periodic sequence of M discrete contour samples $\boldsymbol{g} = (g_0, g_1, \ldots, g_{M-1})$, we select another sequence $\boldsymbol{g}' = (g'_0, g'_1, \ldots) = (g_{k_s}, g_{k_s+1}, \ldots)$, again of length M, from the same set of samples but starting at point k_s, that is,

$$g'_k = g_{(k+k_s) \bmod M}. \tag{26.88}$$

This is equivalent to *shifting* the original signal $\boldsymbol{g}$ circularly by $-k_s$ positions. The well-known "shift property" of the Fourier transform[17] states that such a change to the "signal" $\boldsymbol{g}$ modifies the corresponding DFT coefficients G_m (for the original contour sequence) to

$$G'_m = e^{\mathrm{i} \cdot m \cdot \frac{2\pi k_s}{M}} \cdot G_m = e^{\mathrm{i} \cdot m \cdot \varphi_s} \cdot G_m, \tag{26.89}$$

where $\varphi_s = \frac{2\pi k_s}{M}$ is a constant phase angle that is obviously proportional to the chosen start position k_s. Note that, in Eqn. (26.89), each DFT coefficient G_m is multiplied by a *different* complex quantity $e^{\mathrm{i} \cdot m \cdot \varphi_s}$, which is of unit magnitude and varies with the frequency index m. In other words, the *magnitude* of any DFT coefficient G_m is again preserved but its *phase* changes individually. The coefficients of any Fourier descriptor pair $\mathrm{FP}_m = (G_{-m}, G_{+m})$ thus become

$$G'_{-m} = e^{-\mathrm{i} \cdot m\varphi_s} \cdot G_{-m} \quad \text{and} \quad G'_{+m} = e^{\mathrm{i} \cdot m\varphi_s} \cdot G_{+m}, \tag{26.90}$$

that is, coefficient G_{-m} is rotated by the angle $-m \cdot \varphi_s$ and G_{+m} is rotated by $m \cdot \varphi_s$. In other words, a circular shift of the signal by $-k_s$ samples rotates the coefficients G_{-m}, G_{+m} by the same angle $m \cdot \varphi_s$ but in *opposite* directions. Therefore, the sum of both angles stays the same, that is,

$$\measuredangle G'_{-m} + \measuredangle G'_{+m} \equiv \measuredangle G_{-m} + \measuredangle G_{+m}. \tag{26.91}$$

[17] See Chapter 18, Sec. 18.1.6.

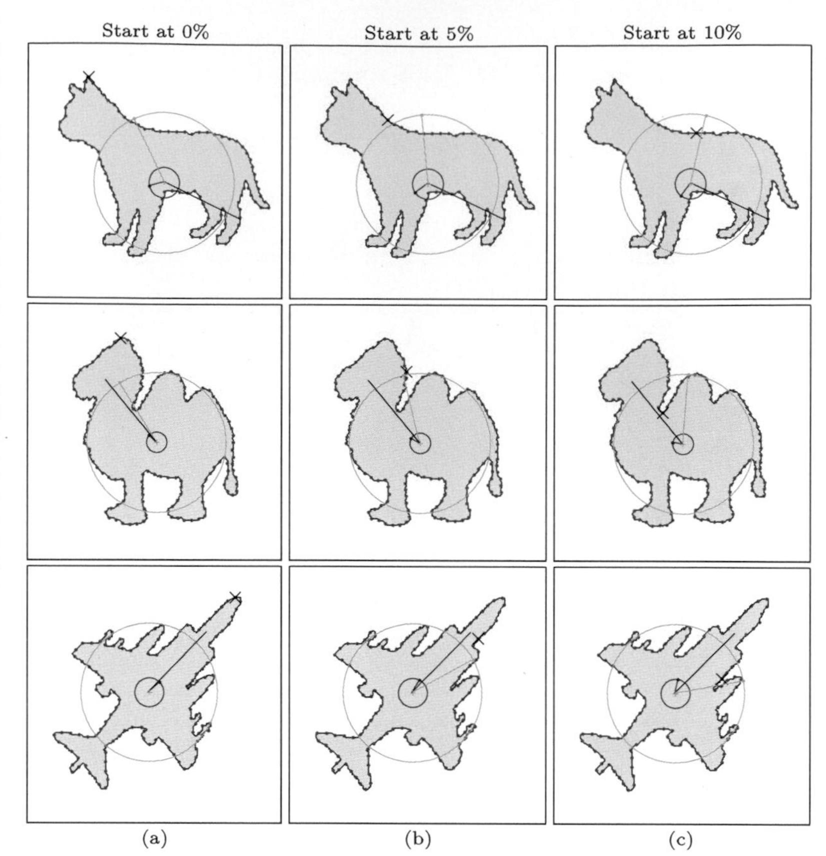

Fig. 26.15
Effects of choosing different start points for contour sampling. The start point (marked $\times$ on the contour) is set to 0%, 5%, 10% of the contour path length. The blue and green circles represent the partial reconstruction from single DFT coefficients G_{-1} and G_{+1}, respectively. The dot on each circle and the associated radial line shows the phase of the corresponding coefficient. The black line indicates the average orientation $(\sphericalangle G_{-1} + \sphericalangle G_{+1})/2$. It can be seen that the phase difference of G_{-1} and G_{+1} is directly related to the start position, but the average *orientation* (black line) remains unchanged.

In particular, we see from Eqn. (26.90) that shifting the start position modifies the coefficients of the *first* descriptor pair $\mathrm{FP}_1 = (G_{-1}, G_{+1})$ to

$$G'_{-1} = e^{-\mathrm{i}\cdot\varphi_{\mathrm{s}}} \cdot G_{-1} \qquad \text{and} \qquad G'_{+1} = e^{\mathrm{i}\cdot\varphi_{\mathrm{s}}} \cdot G_{+1}. \tag{26.92}$$

The resulting *absolute* phase change of the coefficients G_{-1}, G_{+1} is $-\varphi_{\mathrm{s}}, +\varphi_{\mathrm{s}}$, respectively, and thus the change in phase *difference* is $2 \cdot \varphi_{\mathrm{s}}$, that is, the phase difference between the coefficients G_{-1}, G_{+1} is proportional to the chosen start position k_{s} (see Fig. 26.15).

26.4.5 Effects of Phase Removal

As described in the two previous sections, shape rotation (Sec. 26.4.3) and shift of start point (Sec. 26.4.4) both affect the phase of the Fourier coefficients but not their magnitude. The fact that magnitude is preserved suggests a simple solution for rotation invariant shape matching by simply ignoring the phase of the coefficients and comparing only their magnitude (see Sec. 26.6). Although this comes at the price of losing shape descriptiveness, magnitude-only descriptors are often used for shape matching. Clearly, the original shape cannot be reconstructed from a magnitude-only Fourier descriptor, as demonstrated in Fig. 26.16. It shows the reconstruction of shapes from Fourier descriptors with the phase of all coefficients set to zero, except for G_{-1}, G_0 and G_{+1} (to preserve the shape's center and main orientation).

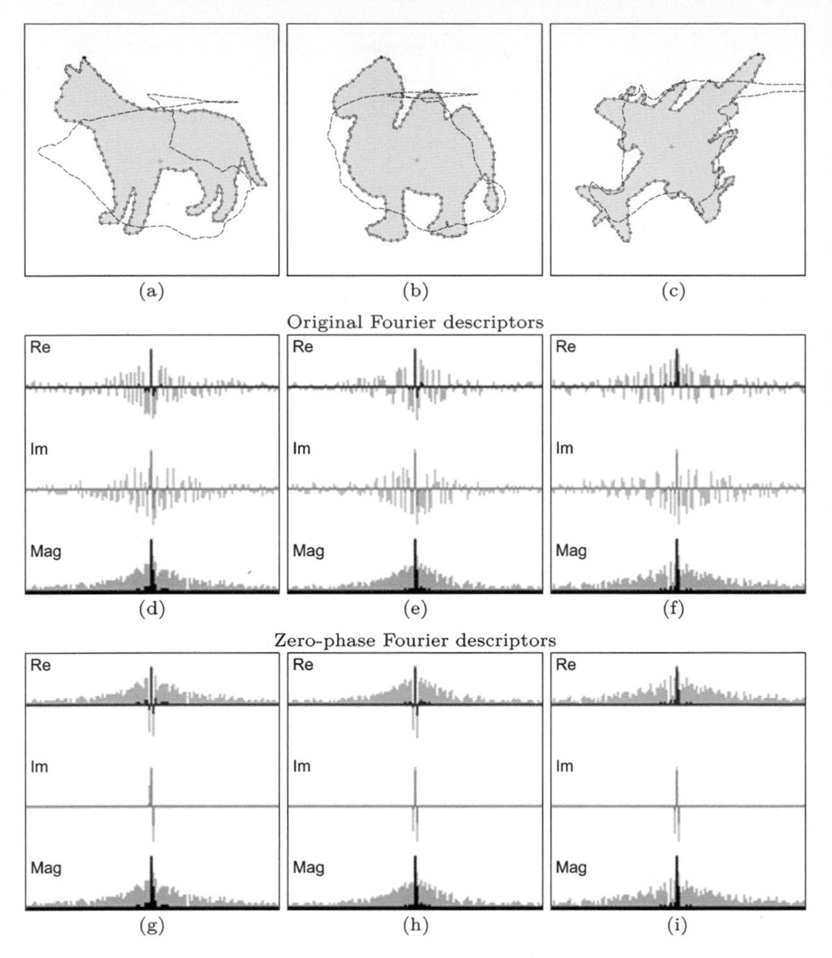

Fig. 26.16
Effects of removing phase information. Original shapes and reconstruction after phase removal (a–c). Original Fourier coefficients (d–f) and zero-phase coefficients (g–i). The red and green plots in (d–i) show the real and imaginary components, respectively; gray plots show the coefficient magnitude. Dark-shaded bars correspond to the actual values, light-shaded bars are logarithmic values. The magnitude of the coefficients in (d–f) is the same as in (g–i).

26.4.6 Direction of Contour Traversal

If the traversal direction of the contour samples is reversed, the coefficients of all Fourier descriptor pairs are exchanged, that is,

$$G'_m = G_{-m \bmod M}. \tag{26.93}$$

This is equivalent to scaling the original shape by $s = -1$, as pointed out in Section 26.4.2. However, this is typically of no relevance in matching, since we can specify all contours to be sampled in either clockwise or counter-clockwise direction.

26.4.7 Reflection (Symmetry)

Mirroring or reflecting a contour about the x-axis is equivalent to replacing each complex-valued point $g_k = x_k + \mathrm{i} \cdot y_k$ by its *complex conjugate* g_k^*, that is,

$$g'_k = g_k^* = x_k - \mathrm{i} \cdot y_k. \tag{26.94}$$

This change to the "signal" results in a modified DFT spectrum with coefficients

$$G'_m = G^*_{-m \bmod M}, \tag{26.95}$$

Table 26.1
Effects of spatial transformations upon the corresponding DFT spectrum. The original contour samples are denoted g_k, the DFT coefficients are G_m.

Operation	Contour samples	DFT coefficients
Forward transformation	g_k, for $k=0,\ldots,M-1$	$G_m = \frac{1}{M} \cdot \sum_{k=0}^{M-1} g_k \cdot e^{-\mathrm{i}2\pi m \frac{k}{M}}$
Inverse transformation	$g_k = \sum_{m=0}^{M-1} G_m \cdot e^{\mathrm{i}2\pi m \frac{k}{M}}$	G_m, for $m=0,\ldots,M-1$
Translation (by $z \in \mathbb{C}$)	$g'_k = g_k + z$	$G'_m = \begin{cases} G_m + z & \text{for } m = 0 \\ G_m & \text{otherwise} \end{cases}$
Uniform scaling (by $s \in \mathbb{R}$)	$g'_k = s \cdot g_k$	$G'_m = s \cdot G_m$
Rotation about the origin (by β)	$g'_k = e^{\mathrm{i} \cdot \beta} \cdot g_k$	$G'_m = e^{\mathrm{i} \cdot \beta} \cdot G_m$
Shift of start position (by k_s)	$g'_k = g_{(k+k_s) \bmod M}$	$G'_m = e^{\mathrm{i} \cdot m \cdot \frac{2\pi k_s}{M}} \cdot G_m$
Direction of contour traversal	$g'_k = g_{-k \bmod M}$	$G'_m = G_{-m \bmod M}$
Reflection about the x-axis	$g'_k = g^*_k$	$G'_m = G^*_{-m \bmod M}$

where G^* denotes the complex conjugate of the original DFT coefficients. Reflections about arbitrary axes can be described in the same way with additional rotations. Fourier descriptors can be made invariant against reflections, such that symmetric contours map to equivalent descriptors [245]. Note, however, that invariance to symmetry is not always desirable, for example, for distinguishing the silhouettes of left and right hands.

The relations between 2D point coordinates and the Fourier spectrum, as well as the effects of the aforementioned geometric shape transformations upon the DFT coefficients are compactly summarized in Table 26.1.

26.5 Transformation-Invariant Fourier Descriptors

As mentioned already, making a Fourier descriptor invariant to *translation* or absolute shape position is easy because the only affected spectral coefficient is G_0. Thus, setting coefficient G_0 to zero implicitly moves the center of the corresponding shape to the coordinate origin and thus creates a descriptor that is invariant to shape translation.

Invariance against a change in *scale* is also a simple issue because it only multiplies the magnitude of all Fourier coefficients by the same real-valued scale factor, which can be easily normalized.

A more challenging task is to make Fourier descriptors invariant against shape *rotation* and shift of the contour *starting point*, because they jointly affect the phase of the Fourier coefficients. If matching is to be based on the complex-valued Fourier descriptors (not on coefficient magnitude only) to achieve better shape discrimination, the phase changes introduced by shape rotation and start point shifts must be eliminated first. However, due to noise and possible ambiguities, this is not a trivial problem (see also [183, 184, 189, 245]).

26.5.1 Scale Invariance

As mentioned in Section 26.4.2, the magnitude G_{+1} is often used as a reference to normalize for scale, since G_{+1} is typically (though not always) the Fourier coefficient with the largest magnitude. Alternatively, one could use the size of the fundamental ellipse, defined by the Fourier descriptor pair FP_1, to measure the overall scale, for example, by normalizing to

$$G_m^{\mathrm{S}} \leftarrow \frac{1}{|G_{-1}| + |G_{+1}|} \cdot G_m, \tag{26.96}$$

which normalizes the *length* of the major axis $a_1 = |G_{-1}| + |G_{+1}|$ (see Eqn. (26.57)) of the fundamental ellipse to unity. Another alternative is

$$G_m^{\mathrm{S}} \leftarrow \frac{1}{(|G_{-1}| \cdot |G_{+1}|)^{1/2}} \cdot G_m, \tag{26.97}$$

which normalizes the *area* of the fundamental ellipse. Since all variants in Eqns. (26.83), (26.96) and (26.97) scale the coefficients G_m by a fixed (real-valued) factor, the shape information contained in the Fourier descriptor remains unchanged.

There are shapes, however, where coefficients G_{+1} and/or G_{-1} are small or almost vanish to zero, such that they are not always a reliable reference for scale. An obvious solution is to include the complete set of Fourier coefficients by standardizing the *norm* of the coefficient vector $\boldsymbol{G}$ to unity in the form

$$G_m^{\mathrm{S}} \leftarrow \frac{1}{\|\boldsymbol{G}\|} \cdot G_m, \tag{26.98}$$

(assuming that $G_0 = 0$). In general, the L_2 norm of a complex-valued vector $Z = (z_0, z_1, \ldots, z_{M-1})$, $z_i \in \mathbb{C}$, is defined as

$$\|Z\| = \Bigl(\sum_{i=1}^{M-1} |z_i|^2\Bigr)^{1/2} = \Bigl(\sum_{i=1}^{M-1} \mathrm{Re}(z_i)^2 + \mathrm{Im}(z_i)^2\Bigr)^{1/2}. \tag{26.99}$$

Scaling the vector Z by the reciprocal of its norm yields a vector with unit norm, that is,

$$\Bigl\| \frac{1}{\|Z\|} \cdot Z \Bigr\| = 1. \tag{26.100}$$

To normalize a given Fourier descriptor $\boldsymbol{G}$, we use all elements except G_0 (which relates to the absolute position of the shape and is not relevant for its shape). The following substitution makes $\boldsymbol{G}$ scale invariant by normalizing the remaining sub-vector $(G_1, G_2, \ldots, G_{M-1})$ to

$$G_m^{\mathrm{S}} \leftarrow \begin{cases} G_m & \text{for } m = 0, \\ \frac{1}{\sqrt{\nu}} \cdot G_m & \text{for } 1 \leq m < M, \end{cases} \quad \text{with } \nu = \sum_{m=1}^{M-1} |G_m|^2. \tag{26.101}$$

See procedure MakeScaleInvariant($\boldsymbol{G}$) in Alg. 26.6 (lines 7–15) for a summary of this step.

26.5.2 Start Point Invariance

As discussed in Sections 26.4.3 and 26.4.4, respectively, shape rotation and shift of start point both affect the phase of the Fourier coefficients in a combined manner, without altering their magnitude. In particular, if the shape is rotated by some angle β (see Eqn. (26.89)) and the start position is shifted by k_s samples (see Eqn. (26.86)), then each Fourier coefficient G_m is modified to

$$G'_m = e^{\mathrm{i}\cdot\beta} \cdot e^{\mathrm{i}\cdot m\cdot\varphi_\mathrm{s}} \cdot G_m = e^{\mathrm{i}\cdot(\beta+m\cdot\varphi_\mathrm{s})} \cdot G_m, \tag{26.102}$$

where $\varphi_\mathrm{s} = 2\pi k_\mathrm{s}/M$ is the corresponding *start point phase*. Thus, the incurred phase shift is not only different for each coefficient but simultaneously depends on the rotation angle β and the start point phase φ_s. Normalization in this case means to remove these phase shifts, which would be straightforward if β and φ_s were known. We derive these two parameters one after the other, starting with the calculation of the start point phase φ_s, which we describe in this section, followed by the estimation of the rotation β, shown subsequently in Section 26.5.3.

To normalize the Fourier descriptor of a particular shape to a "canonical" start point, we need a quantity that can be calculated from the Fourier spectrum and only depends on the start point phase φ_s but is independent of the rotation β. From Eqn. (26.90) and Fig. 26.15 we see that the phase *difference* within any Fourier descriptor pair (G_{-m}, G_{+m}) is proportional to the start point phase φ_s and independent to shape rotation β, since the latter rotates all coefficients by the same angle. Thus, we look for a quantity that depends only on the phase *differences* within Fourier descriptor pairs. This is accomplished, for example, by the function

$$f_\mathrm{p}(\varphi) = \sum_{m=1}^{M_\mathrm{p}} \left[e^{-\mathrm{i}\cdot m\cdot\varphi} \cdot G_{-m}\right] \otimes \left[e^{\mathrm{i}\cdot m\cdot\varphi} \cdot G_m\right], \tag{26.103}$$

where parameter φ is an arbitrary start point phase, M_p is the number of coefficient pairs, and $\otimes$ denotes the "cross product" between two Fourier coefficients.[18] Given a particular start point phase φ, the function in Eqn. (26.103) yields the sum of the cross products of each coefficient pair (G_{-m}, G_m), for $m = 1, \ldots, M_\mathrm{p}$. If each of the complex-valued coefficients is interpreted as a vector in the 2D plane, the magnitude of their cross product is proportional to the *area* of the enclosed parallelogram. The enclosed area is potentially large only if *both* vectors are of significant length, which means that the corresponding ellipse has a distinct eccentricity and orientation. Note that the sign of the cross product may be positive or negative and depends on the relative orientation or "handedness" of the two vectors.

Since the function $f_\mathrm{p}(\varphi)$ is based only on the *relative* orientation (phase) of the involved coefficients, it is invariant to a shape rotation

[18] In analogy to 2D vector notation, we define the "cross product" of two complex quantities $z_1 = (a_1, b_1)$ and $z_2 = (a_2, b_2)$ as $z_1 \otimes z_2 = a_1 \cdot b_2 - b_1 \cdot a_2 = |z_1| \cdot |z_2| \cdot \sin(\theta_2 - \theta_1)$. See also Sec. B.3.3 in the Appendix.

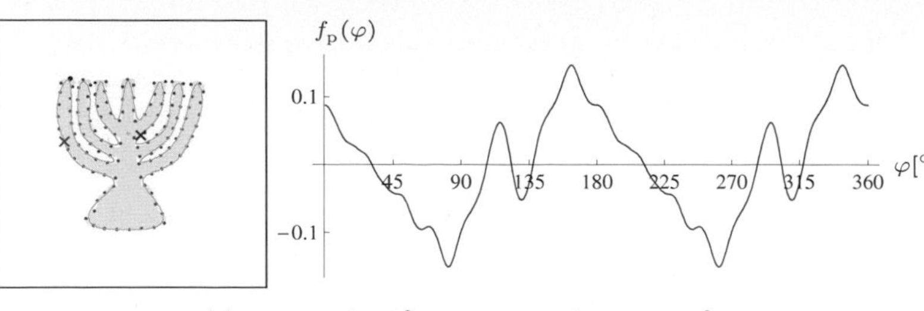

(a) rotation $\theta = 0°$, start point phase $\varphi_s = 0°$

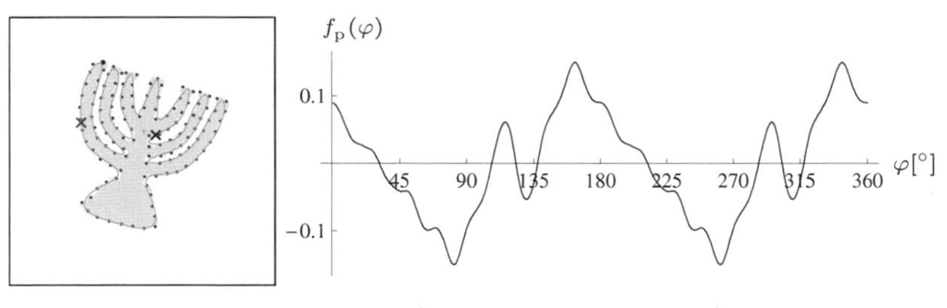

(b) rotation $\theta = 15°$, start point phase $\varphi_s = 0°$

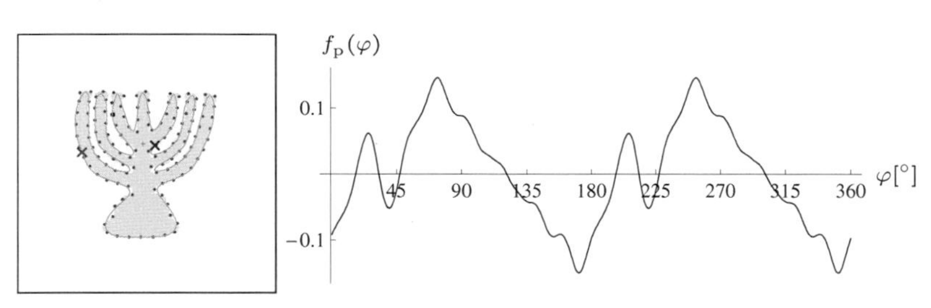

(c) rotation $\theta = 0°$, start point phase $\varphi_s = 90°$

Table 26.2
Plot of the function $f_p(\varphi)$ used for start point normalization. In the figures on the left, the real start point is marked by a black dot. The normalized start points φ_A and $\varphi_B = \varphi_A + \pi$ are marked by a blue and a brown cross, respectively. They correspond to the two peak positions of the function $f_p(\varphi)$, as defined in Eqn. (26.103), separated by a fixed phase shift of $\pi = 180°$ (right). The function is invariant under shape rotation, as demonstrated in (b), where the shape is rotated by 15° but sampled from the same start point as in (a). However, the phase of $f_p(\varphi)$ is proportional to the start point shift, as shown in (c), where the start point is chosen at 25% ($\varphi_s = 90°$) of the boundary path length. The functions were calculated after scale normalization, using $M_p = 25$ Fourier coefficient pairs.

β, which shifts all coefficients by the same angle (see Eqn. (26.86)). As shown in Fig. 26.2, $f_p(\varphi)$ is periodic with π and its phase is proportional to the actual start point shift. We choose the angle φ that *maximizes* $f_p(\varphi)$ as the "canonical" start point phase φ_A, that is,

$$\varphi_A = \underset{0 \le \varphi < \pi}{\operatorname{argmax}} \, f_p(\varphi). \tag{26.104}$$

However, since $f_p(\varphi) = f_p(\varphi + \pi)$, there is also a *second* candidate phase

$$\varphi_B = \varphi_A + \pi, \tag{26.105}$$

displaced by $\pi = 180°$. The two "canonical" start points corresponding to φ_A and φ_B, respectively, are marked on the reconstructed shapes in Fig. 26.2. Although it might seem easy at first to resolve this 180° ambiguity of the start point phase, this turns out to be difficult to achieve in general from the Fourier coefficients alone. Several functions have been proposed for this purpose that work well for certain shapes but fail on others, including the "positive real energy" function suggested in [245]. In particular, any decision based on the magnitude or phase of a *single* coefficient (or a single coefficient pair) must eventually fail, since none of the coefficients is guaranteed to have a significant magnitude. With vanishing coefficient magnitude, phase measurements become unreliable and may be very susceptible to noise.

The complete process of start point normalization is summarized in Alg. 26.7. The start point phase φ_A is found numerically by evaluating the function $f_\mathrm{p}(\varphi)$ at 400 discrete steps for $\varphi = 0, \ldots, \pi$ (lines 6–16). For practical use, this exhaustive method should be substituted by a more efficient and accurate optimization technique (for example, using Brent's method [190, Ch. 10]).[19] Given the estimated start point phase φ_A for the Fourier descriptor $\boldsymbol{G}$, *two* normalized versions $\boldsymbol{G}^\mathrm{A}, \boldsymbol{G}^\mathrm{B}$ are calculated as

$$\begin{aligned} \boldsymbol{G}^\mathrm{A}\colon\; G^\mathrm{A}_m &\leftarrow G_m \cdot e^{\mathrm{i}\cdot m\cdot\varphi_\mathrm{A}}, \\ \boldsymbol{G}^\mathrm{B}\colon\; G^\mathrm{B}_m &\leftarrow G_m \cdot e^{\mathrm{i}\cdot m\cdot(\varphi_\mathrm{A}+\pi)}, \end{aligned} \tag{26.106}$$

for $m = -M_\mathrm{p}, \ldots, M_\mathrm{p}, m \neq 0$. Note that start point normalization does not require the Fourier descriptor $\boldsymbol{G}$ to be normalized for translation and scale (see Sec. 26.5.1).

26.5.3 Rotation Invariance

After normalizing for starting point, the orientation of the fundamental ellipse (formed by the descriptor pair (G_{-1}, G_{+1})) could be assumed to be a reliable reference for global shape rotation. However, for certain shapes (e.g., regular polyhedra with an even number of faces), G_{-1} may vanish. Therefore, we recover the overall shape orientation from the vector obtained as the weighted sum of *all* Fourier coefficients, that is,

$$z = \sum_{m=1}^{M_\mathrm{p}} \frac{1}{m} \cdot (G_{-m} + G_{+m}), \tag{26.107}$$

where the $1/m$ serves as a weighting factor, giving stronger emphasis to the low-frequency coefficients and attenuating the influence of the high-frequency coefficients. The resulting shape orientation estimate is

$$\beta = \measuredangle z = \tan^{-1}\left(\frac{\mathrm{Im}(z)}{\mathrm{Re}(z)}\right). \tag{26.108}$$

To normalize $\boldsymbol{G}^\mathrm{A}, \boldsymbol{G}^\mathrm{B}$ (obtained in Eqn. (26.106)) for shape orientation, we rotate each coefficient (except G_0) by $-\beta$, that is,

$$\begin{aligned} \boldsymbol{G}^\mathrm{A}\colon\; G^\mathrm{A}_m &\leftarrow G^\mathrm{A}_m \cdot e^{-\mathrm{i}\cdot\beta}, \\ \boldsymbol{G}^\mathrm{B}\colon\; G^\mathrm{B}_m &\leftarrow G^\mathrm{B}_m \cdot e^{-\mathrm{i}\cdot\beta}, \end{aligned} \tag{26.109}$$

for $m = -M_\mathrm{p}, \ldots, M_\mathrm{p}, m \neq 0$. For a summary of these steps, see procedure MakeRotationInvariant($\boldsymbol{G}$) in Alg. 26.6 (lines 16–24).

[19] The accompanying Java implementation uses the class `BrentOptimizer` from the *Apache Commons Math library* [4] for this purpose.

```
1:  MakeInvariant(G)
        Input: G, Fourier descriptor with M_p coefficient pairs.
        Returns a pair of normalized Fourier descriptors G^A, G^B, with
        a start point phase offset by 180°.
2:      MakeScaleInvariant(G)                         ▷ see below
3:      (G^A, G^B) ← MakeStartPointInvariant(G)       ▷ see Alg. 26.7
4:      MakeRotationInvariant(G^A)                    ▷ see below
5:      MakeRotationInvariant(G^B)
6:      return (G^A, G^B).

7:  MakeScaleInvariant(G)
        Modifies G by unifying its norm and returns the scale factor ν.
8:      s ← 0                                         ▷ s ∈ ℝ
9:      for m ← 1, ..., M_p do
10:         s ← s + |G(−m)|^2 + |G(m)|^2
11:     ν ← 1/√s
12:     for m ← 1, ..., M_p do
13:         G(−m) ← ν · G(−m)
14:         G(m)  ← ν · G(m)
15:     return ν.

16: MakeRotationInvariant(G)
        Modifies G and returns the estimated rotation angle β.
17:     z ← 0 + i·0                                   ▷ z ∈ ℂ
18:     for m ← 1, ..., M_p do
19:         z ← z + (1/m) · (G(−m) + G(m))            ▷ complex addition!
20:     β ← ∡z
21:     for m ← 1, ..., M_p do                        ▷ rotate all coefficients by −β
22:         G(−m) ← e^(−i·β) · G(−m)
23:         G(m)  ← e^(−i·β) · G(m)
24:     return β.
```

Alg. 26.6 Making Fourier descriptors invariant against scale, shift of start point, and shape rotation. For a given Fourier descriptor G, procedure MakeStartPointInvariant(G) returns a pair of normalized Fourier descriptors (G^A, G^B), one for each normalized start point phase φ_A and $\varphi_B = \varphi_A + \pi$.

26.5.4 Other Approaches

The aforementioned normalization for making Fourier descriptors invariant to geometric transformations deviates from the published "classic" techniques in certain ways, but also adopts some common elements. As representative examples, we briefly discuss two of these techniques (already referenced earlier) in the following.

Persoon and Fu [183,184] proposed (in what they call the "suboptimal" approach) to choose the parameters s (common scale factor), β (shape rotation), and φ_s (start point phase) such that the modified coefficients G'_{-1}, G'_{+1} are both imaginary and $|G_{-1} + G_{+1}| = 1$. As argued in [245], this method leaves a $\pm 180°$ ambiguity for the shape orientation. Also, it requires that both G_{-1}, G_{+1} have significant magnitude, which may not be true for G_{-1} in case of shapes that are circularly symmetric (e.g., equilateral triangles, squares, pentagons etc.).

Wallace and Wintz [245] use $|G_{+1}|$ as the common scale factor, because the coefficient G_{+1} typically has the largest magnitude. The phase of G_{+1}, denoted $\phi_1 = \sphericalangle G_{+1}$, and the phase of another coefficient G_k ($k > 0$) with the second-largest magnitude and phase $\phi_k = \sphericalangle G_k$ are used to compensate for rotation and starting point. Coefficients are phase shifted such that both G'_{+1} and G'_k have zero

Alg. 26.7 Making Fourier descriptors invariant to the shift of start point. Since the result is ambiguous by 180°, two normalized descriptors $(\boldsymbol{G}^{\mathrm{A}}, \boldsymbol{G}^{\mathrm{B}})$ are returned, with the start point phase set to φ_{A} and $\varphi_{\mathrm{A}} + \pi$, respectively.

```
1:  MakeStartPointInvariant(G)
        Input: G, Fourier descriptor with M_p coefficient pairs.
        Returns a pair of new Fourier descriptors G^A, G^B, normalized
        to the start point phase φ_A and φ_A + π, respectively.
2:      φ_A ← GetStartPointPhase(G)                          ▷ see below
3:      G^A ← ShiftStartPointPhase(G, φ_A)                   ▷ see below
4:      G^B ← ShiftStartPointPhase(G, φ_A + π)
5:      return (G^A, G^B).

6:  GetStartPointPhase(G)
        Returns φ maximizing f_p(G, φ), with φ ∈ [0, π). The maximum
        is found by simple brute-force search (for illustration only).
7:      c_max ← −∞
8:      φ_max ← 0
9:      K ← 400                        ▷ do K search steps over 0, ..., π
10:     for k ← 0, ..., K−1 do              ▷ find φ maximizing f_p(G, φ)
11:         φ ← π · k/K
12:         c ← f_p(G, φ)
13:         if c > c_max then
14:             c_max ← c
15:             φ_max ← φ
16:     return φ_max.

17: f_p(G, φ)                                        ▷ see Eq. 26.103
18:     s ← 0
19:     for m ← 1, ..., M_p do
20:         z_1 ← G(−m) · e^(−i·m·φ)
21:         z_2 ← G(m) · e^(i·m·φ)
22:         s ← s + Re(z_1) · Im(z_2) − Im(z_1) · Re(z_2)  ▷ = s + (z_1 ⊗ z_2)
23:     return s.

24: ShiftStartPointPhase(G, φ)              ▷ start-point normalize G by φ
25:     G' ← Duplicate(G)
26:     for m ← 1, ..., M_p do
27:         G'(−m) ← G(−m) · e^(−i·m·φ)
28:         G'(m)  ← G(m) · e^(i·m·φ)
29:     return G'.
```

phase. This is accomplished by multiplying all coefficients in the form

$$G'_m = G_m \cdot e^{\mathrm{i}\cdot[(m-k)\cdot\phi_1 + (1-m)\cdot\phi_k]\cdot(k-1)}, \tag{26.110}$$

for $-\frac{M}{2} + 1 \leq m \leq \frac{M}{2}$ (also used in [189]). Depending on the index k of the second-largest coefficient, there exist $|k-1|$ different orientation/start point combinations to obtain zero-phase in G'_{+1} and G'_k. If $k = 2$, then $|k-1| = 1$, thus the solution is unique and Eqn. (26.110) simplifies to

$$G'_m = G_m \cdot e^{\mathrm{i}\cdot[(m-2)\cdot\phi_1 + (1-m)\cdot\phi_2]}, \tag{26.111}$$

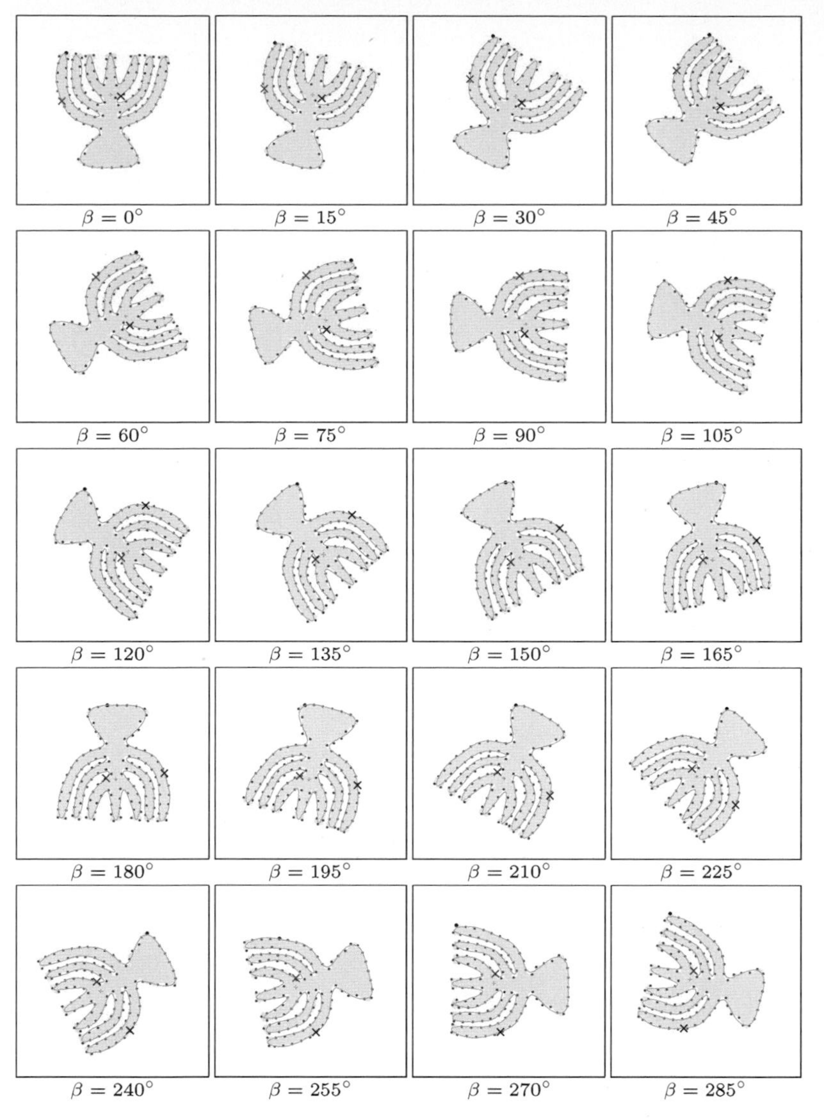

Fig. 26.17
Start point normalization under varying shape rotation (β). The real start point (which varies with shape rotation) is marked by a black dot. The two normalized start points φ_{A} and $\varphi_{\mathrm{B}} = \varphi_{\mathrm{A}} + \pi$ (calculated with the procedure in Alg. 26.7) are marked by a blue and a brown $\times$, respectively. Twenty-five Fourier coefficient pairs are used for the normalization and shape reconstruction. Inaccuracies are due to shape variations caused by the use of nearest-neighbor interpolation for the image rotation.

with $\phi_2 = \sphericalangle G_2$.[20] Otherwise, the ambiguity is resolved by calculating an "ambiguity-resolving" criterion for each of the $|k-1|$ solutions, for example, the amount of "positive real energy",

$$\sum_{m=1}^{N-1} \mathrm{Re}(G'_m) \cdot |\mathrm{Re}(G'_m)| \, ,$$

as defined in [245] (other functions were suggested in [189]). This leaves the problem that, for matching, the normalization of the investigated shape descriptor must be based on the same set of dominant coefficients as the reference descriptor. Alternatively, one could memorize the relevant coefficient indexes for every reference descrip-

[20] Unfortunately, the general use of coefficient G_2 as a phase reference is critical, because the magnitude of G_2 may be small or even zero for certain symmetrical shapes (including all regular polygons with an even number of faces).

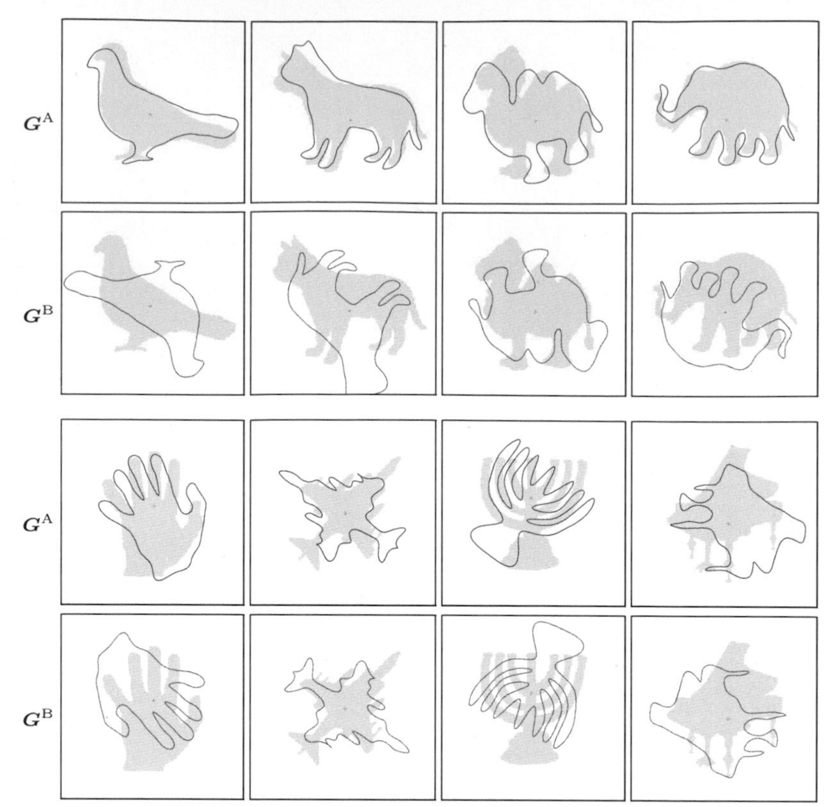

Fig. 26.18 Reconstruction of various shapes from Fourier descriptors normalized for start point shift and shape rotation. The blue shapes (rows 1, 3) correspond to the normalized Fourier descriptors G^A with start point phase φ_A. The brown shapes (rows 2, 4) correspond to the normalized Fourier descriptors G^B with start point phase $\varphi_B = \varphi_A + \pi$. No scale normalization was applied for better visualization.

tor, but then different normalizations must be applied for matching against multiple models in a database.

26.6 Shape Matching with Fourier Descriptors

A typical use of Fourier descriptors is to see if a given shape is identical or similar to an exemplar contained in a database of reference shapes. For this purpose, we need to define a distance measure that quantifies the difference between two Fourier shape descriptors G_1 and G_2. In the following, we assume that the Fourier descriptors G_1, G_2 are at least scale-normalized (as described in Alg. 26.6) and of identical length, each with M_p coefficient pairs.

26.6.1 Magnitude-Only Matching

In the simplest case, we only use the *magnitude* of the Fourier coefficients for comparison and entirely ignore their phase, using the distance function

$$\mathrm{dist}_M(G_1, G_2) = \Bigl[\sum_{\substack{m=-M_p,\\ m\neq 0}}^{M_p} \bigl(|G_1(m)| - |G_2(m)|\bigr)^2\Bigr]^{1/2} \quad (26.112)$$

$$= \Bigl[\sum_{m=1}^{M_p} \bigl(|G_1(-m)| - |G_2(-m)|\bigr)^2 + \bigl(|G_1(m)| - |G_2(m)|\bigr)^2\Bigr]^{1/2},$$

where M_p denotes the number of FD pairs used for matching. Note that Eqn. (26.112) is simply the L_2 norm of the magnitude difference vector, and of course other norms (such as L_1 or L_∞) could be used as well. The advantage of the magnitude-only approach is that no normalization (except for scale) is required. Its drawback is that even highly dissimilar shapes might be mistakenly matched, since the removal of phase naturally eliminates shape information that is possibly essential for discrimination. As demonstrated in Fig. 26.19, a given Fourier magnitude vector may correspond to a great diversity of shapes, and thus the subspace of "equivalent" shapes defined by the magnitude-only distance dist_M is quite large.

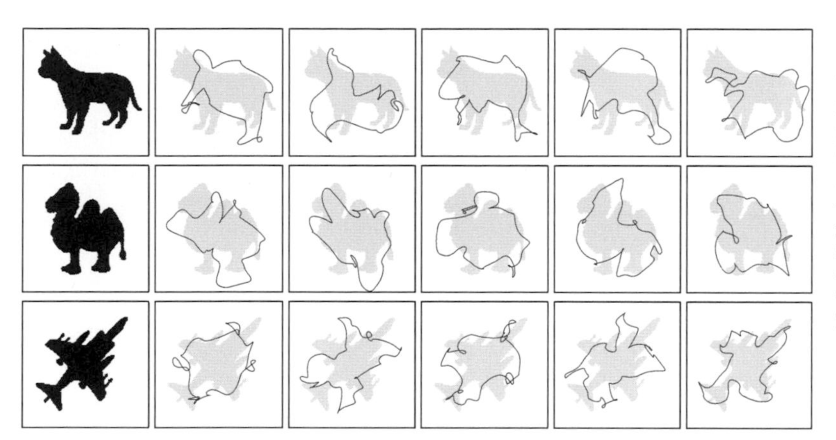

Fig. 26.19
Magnitude-only reconstruction (randomized phase). Reconstruction of shapes from Fourier descriptors with the phase of all coefficients (except G_{-1}, G_0, and G_{+1}) individually randomized. Note that the magnitude of the coefficients is exactly the same for each shape category, so all blue shapes would be considered "equivalent" to the original shape (first column) by a magnitude-only matcher.

Nevertheless, magnitude-only matching may be sufficient in situations where the reference shapes are not too similar. In a sense, the operation of reducing the complex-valued Fourier descriptors to their magnitude vectors can be viewed as a *hash* function. While potentially many different shapes may produce (i.e., "hash to") similar Fourier magnitude vectors, the chance of two real shapes mapping to the same vector (and thus being confused) may be relatively small. Thus, particularly considering its simplicity (only scale-normalization of descriptors is required), magnitude-based matching can be quite effective in practice.

Figure 26.20 shows the pair-wise magnitude-only distances (blue cells, values are $10 \times \mathrm{dist}_M$) between various sample shapes. The corresponding intra-class distances, given in Fig. 26.21, are typically more than one order of magnitude smaller, indicating that shape discrimination based on this measure should be fairly reliable.

26.6.2 Complex (Phase-Preserving) Matching

Assuming that the Fourier descriptors $\boldsymbol{G}_1$ and $\boldsymbol{G}_2$ have been normalized for scale, start point shift, and shape rotation (see Alg. 26.6), we can use the following function to measure their mutual distance:

	bird	cat	camel	elephant	hand	harrier	menora	piano	creature
bird	0.000	4.529	4.482	5.007	5.525	4.314	7.554	5.174	7.076
cat	3.156	0.000	5.788	4.708	5.711	5.701	7.181	5.543	7.677
camel	2.648	3.005	0.000	4.429	5.573	3.726	7.014	4.013	8.480
elephant	3.487	1.933	2.549	0.000	6.100	4.618	5.338	4.369	8.743
hand	4.627	3.146	3.132	2.372	0.000	6.079	8.540	5.580	7.136
harrier	3.712	3.707	2.687	3.553	4.294	0.000	6.818	4.958	8.284
menora	5.835	4.893	4.563	4.162	3.788	5.775	0.000	6.826	11.072
piano	4.037	2.426	2.610	1.876	1.848	3.405	4.315	0.000	7.666
creature	6.030	6.261	5.554	5.492	5.955	5.914	5.190	6.049	0.000

$\mathrm{dist_M}(\boldsymbol{G}_1, \boldsymbol{G}_2)$ $\mathrm{dist_C}(\boldsymbol{G}_1, \boldsymbol{G}_2)$

Fig. 26.20
Inter-class Fourier descriptor distances (magnitude-only and complex-valued). Numbers inside the green fields (lower-left half of the matrix) are the magnitude-only distances $\mathrm{dist_M}$ (see Eqn. (26.112)). Numbers in blue fields (upper-right half of the matrix) are the complex-valued distances $\mathrm{dist_C}$ (see Eqn. (26.114)). Shapes were sampled uniformly at 125 contour positions, with 25 coefficient pairs. Fourier descriptors were normalized for scale, start point and rotation. All distance values are multiplied by 10.

$$\mathrm{dist_C}(\boldsymbol{G}_1, \boldsymbol{G}_2) = \Big(\sum_{\substack{m=-M_\mathrm{p},\\ m\neq 0}}^{M_\mathrm{p}} \big|\boldsymbol{G}_1(m) - \boldsymbol{G}_2(m)\big|^2 \Big)^{1/2} \qquad (26.113)$$

$$= \Big(\sum_{m=1}^{M_\mathrm{p}} \big|\boldsymbol{G}_1(-m) - \boldsymbol{G}_2(-m)\big|^2 + \big|\boldsymbol{G}_1(m) - \boldsymbol{G}_2(m)\big|^2 \Big)^{1/2} \qquad (26.114)$$

$$= \Big(\sum_{\substack{m=-M_\mathrm{p},\\ m\neq 0}}^{M_\mathrm{p}} \big[\mathrm{Re}(\boldsymbol{G}_1(m)) - \mathrm{Re}(\boldsymbol{G}_2(m))\big]^2 + \big[\mathrm{Im}(\boldsymbol{G}_1(m)) - \mathrm{Im}(\boldsymbol{G}_2(m))\big]^2 \Big)^{1/2}. \qquad (26.115)$$

Again, this is simply the L_2 norm of the complex-valued difference vector $\boldsymbol{G}_1 - \boldsymbol{G}_2$ (ignoring the coefficients at $m = 0$), which could be substituted by some other norm. Since the phase of the involved coefficients is fully preserved, a zero distance between two Fourier descriptors means that they represent the very same shape. Thus the set of equivalent shapes defined by the distance function in Eqn. (26.114) is much smaller than the one defined by the magnitude-only distance in Eqn. (26.112). Consequently, the probability of two different shapes being confused for the same is also significantly smaller with this distance measure.

$\alpha =$	0°	17°	34°	51°	68°	85°	102°	119°	136°	153°	170°	187°	204°
dist_M	0.000	0.070	0.126	0.151	0.103	0.058	0.143	0.107	0.195	0.190	0.105	0.078	0.053
dist_C	0.000	0.141	0.222	0.299	0.198	0.111	0.274	0.159	0.313	0.400	0.142	0.162	0.092
dist_M	0.000	0.134	0.144	0.176	0.167	0.055	0.104	0.206	0.227	0.135	0.164	0.083	0.174
dist_C	0.000	0.222	0.214	0.252	0.244	0.081	0.141	0.310	0.339	0.197	0.231	0.157	0.281
dist_M	0.000	0.117	0.346	0.147	0.142	0.141	0.109	0.100	0.125	0.163	0.099	0.147	0.106
dist_C	0.000	0.229	0.728	0.367	0.310	0.386	0.161	0.186	0.202	0.252	0.141	0.191	0.271
dist_M	0.000	0.121	0.195	0.272	0.170	0.057	0.135	0.175	0.216	0.176	0.092	0.112	0.160
dist_C	0.000	0.180	0.317	0.392	0.278	0.080	0.218	0.257	0.307	0.266	0.160	0.198	0.248
dist_M	0.000	0.127	0.138	0.179	0.130	0.048	0.131	0.115	0.329	0.173	0.202	0.109	0.132
dist_C	0.000	0.179	0.186	0.361	0.180	0.085	0.234	0.188	0.496	0.263	0.313	0.182	0.195
dist_M	0.000	0.234	0.171	0.224	0.095	0.090	0.106	0.189	0.228	0.170	0.079	0.121	0.213
dist_C	0.000	0.433	0.290	0.317	0.147	0.129	0.197	0.276	0.344	0.251	0.146	0.197	0.308
dist_M	0.000	0.163	0.148	0.131	0.213	0.116	0.228	0.322	0.334	0.205	0.253	0.108	0.122
dist_C	0.000	0.570	0.330	0.395	0.456	0.169	0.271	0.401	0.465	0.295	0.440	0.149	0.251
dist_M	0.000	0.164	0.186	0.161	0.186	0.101	0.112	0.252	0.159	0.150	0.169	0.104	0.201
dist_C	0.000	0.264	0.362	0.311	0.255	0.175	0.148	0.576	0.230	0.267	0.232	0.142	0.284
dist_M	0.000	0.154	0.190	0.167	0.103	0.084	0.180	0.390	0.210	0.123	0.194	0.084	0.131
dist_C	0.000	0.203	0.260	0.248	0.141	0.108	0.232	0.447	0.308	0.171	0.234	0.120	0.160

Fig. 26.21
Intra-class Fourier descriptor distances (magnitude-only and complex-valued). The reference images (0° column) were rotated by angle α (multiples of 17°), using no (i.e., nearest-neighbor) interpolation. Numbers inside the blue fields are the magnitude-only distances dist_M (see Eqn. (26.112)). Numbers inside the green fields are the complex-valued distances dist_C (see Eqn. (26.114)). Shapes were sampled uniformly at 125 contour positions, with 25 coefficient pairs. Fourier descriptors were normalized for scale, start point shift and shape rotation. All distance values are multiplied by 10. Note that all *intra*-class distances are roughly one order of magnitude smaller than the *inter*-class distances shown in Fig. 26.20.

Complex inter-class and intra-class distance values for the set of sample shapes are listed in Figs. 26.20 and 26.21. Notice that, with the normalization described in Alg. 26.6, the complex intra-class distance values in Fig. 26.21 (which should be as small as possible) are typically about twice as large as the corresponding magnitude-only distance values, but still an order of magnitude smaller than comparable inter-class values in Fig. 26.20, so reliable shape discrimination should be possible.

The price paid for the increased discriminative power is the extra work necessary for normalizing the Fourier descriptors for start point and shape rotation (in addition to scale), as described in Alg. 26.6. Note that this involves the comparison with *two* normalized descriptors to cope with the unresolved 180° ambiguity of the start point normalization (see Eqns. (26.104) and (26.105)). For example, assume we wish to compare two shapes V_1, V_2 with Fourier descriptors $\boldsymbol{G}_1, \boldsymbol{G}_2$, respectively. We first calculate the corresponding invariant descriptors (as described in Alg. 26.6),

$$(G_1^A, G_1^B) \leftarrow \mathsf{MakeInvariant}(G_1)\,,$$
$$(G_2^A, G_2^B) \leftarrow \mathsf{MakeInvariant}(G_2)\,. \tag{26.116}$$

Now we use Eqn. (26.114) to calculate the complex-valued distance as

$$d_{\min} = \min\bigl(\mathrm{dist}_{\mathrm{C}}(G_1^A, G_2^A), \mathrm{dist}_{\mathrm{C}}(G_1^A, G_2^B)\bigr) \tag{26.117}$$

or, alternatively, as

$$d_{\min} = \min\bigl(\mathrm{dist}_{\mathrm{C}}(G_1^A, G_2^A), \mathrm{dist}_{\mathrm{C}}(G_1^B, G_2^A)\bigr). \tag{26.118}$$

Note that, in any case, the resulting distance $d_{\min}$ will be small only if the two shapes V_1, V_2 are really similar. This also means that we only need to store *one* of the two normalized Fourier descriptors—for example, G_{ref}^A—for each reference shape V_{ref} and then (following Eqn. (26.117)) compare it to *both* normalized descriptors G_{new}^A and G_{new}^B of any new shape V_{new}.[21]

To illustrate this idea, Alg. 26.8 shows the construction of a simple Fourier descriptor database from a set of reference shapes and its subsequent use for classifying unknown shapes. First, procedure MakeFdDataBase($\boldsymbol{V}$) returns a map D holding a normalized Fourier descriptor for each of the reference shapes given in $\boldsymbol{V}$. Matching a new shape V_{new} to the entries in the database D is accomplished by procedure FindBestMatch(V_{new}, D, $d_{\max}$), which returns the index of the best-fitting shape in D, or nil if the distance of the closest match exceeds the predefined threshold $d_{\max}$. As common in this situation, we use *squared* distance values (i.e., $\mathrm{dist}_{\mathrm{C}}^2$) for matching in Alg. 26.8 (lines 15–18), thereby avoiding the square root operations in Eqns. (26.112) and (26.114).

26.7 Java Implementation

The algorithms described in this chapter have been implemented as part of the open `imagingbook` library,[22] which is available at the book's accompanying website. As usual, most Java methods are named and structured identically to the procedures defined in the various algorithms for easy identification.

`FourierDescriptor` (class)

This is the main class of this package; it holds all data structures and implements the functionality common to all Fourier descriptors, including methods for shape reconstruction, invariance, and matching, as will be described here.

[21] The justification for keeping only *one* of the two normalized descriptors G_{ref}^A, G_{ref}^B of each reference shape V_{ref} is that if two candidate shapes V_1, V_2 are similar, then the normalization will produce pairs of Fourier descriptors (G_1^A, G_1^B) and (G_2^A, G_2^B) that are also similar but not necessarily in the same order. Therefore G_1^A must only match with *either* G_2^A *or* G_2^B to detect the similarity of V_1 and V_2.

[22] Package `imagingbook.pub.fd`.

Alg. 26.8
Simple shape matching with a database of Fourier descriptors. MakeFdDataBase($\mathsf{V}_{\mathrm{ref}}, M'$) creates and returns a new database (map) R from a sequence of reference shapes $\mathsf{V}_{\mathrm{ref}}$. R can then be passed to FindBestMatch($V_{\mathrm{new}}, M', \mathsf{R}, d_{\max}$) for classifying a new shape $\mathsf{V}_{\mathrm{new}}$, where $d_{\max}$ is a predefined distance threshold.

```
1: MakeFdDataBase(V_ref, M')
     Input: V_ref = (V_0, V_1, ..., V_{N_R}), a sequence of reference shapes;
     M', the number of Fourier coefficients. Returns a sequence of
     model Fourier descriptors for the reference shapes in V_ref.
2:   N_R ← |V_ref|
3:   R ← new map of Fourier descriptors over [0, N_R − 1]
4:   for i ← 0, ..., N_R − 1 do
5:      G ← FourierDescriptorUniform(V_ref(i), M')           ▷ Alg. 26.3
6:      (G^A, G^B) ← MakeInvariant(G)                         ▷ Alg. 26.6
7:      R(i) ← G^A        ▷ store only one normalized descriptor (G^A)
8:   return R.

9: FindBestMatch(V_new, M', R, d_max)
     Input: V_new, a new shape; M', the number of Fourier coefficients;
     R, a sequence of reference Fourier descriptors; d_max, maximum
     squared distance acceptable for a positive match. Returns the
     best-matching shape index i_min or nil if no acceptable match was
     found.
10:  G_new ← FourierDescriptorUniform(V_new, M')               ▷ Alg. 26.3
11:  (G^A_new, G^B_new) ← MakeInvariant(G_new)                  ▷ Alg. 26.6
12:  d_min ← ∞,    i_min ← −1
13:  for i ← 0, ..., |R| − 1 do
14:     G^A_ref ← R(i)
15:     d_2 ← min(D2(G^A_new, G^A_ref), D2(G^B_new, G^A_ref))    ▷ Eq. 26.118
16:     if d_2 < d_min then
17:        d_min ← d_2
18:        i_min ← i
19:  if d_min ≤ d_max then
20:     return i_min                          ▷ best match index is i_min
21:  else
22:     return nil.                     ▷ no matching shape found in R

23: D2(G_1, G_2)
     Returns the squared complex distance dist_C^2(G_1, G_2) between the
     Fourier descriptors G_1, G_2 (see Eq. 26.114).
24:  d ← 0,   M_p ← (min(|G_1|, |G_2|) − 1) ÷ 2
25:  for m ← −M_p, ..., M_p, m ≠ 0 do
26:     d ← d + [Re(G_1(m)) − Re(G_2(m))]^2 +
                [Im(G_1(m)) − Im(G_2(m))]^2
27:  return d.                         ▷ d ≡ (dist_C(G_1, G_2))^2
```

Class `FourierDescriptor` is abstract and thus cannot be instantiated. To create Fourier descriptor objects, one of the concrete subclasses `FourierDescriptorUniform` or `FourierDescriptorFromPolygon` (discussed later in this section) may be used, which provide the appropriate constructors. `FourierDescriptor` provides the following methods for both types of Fourier descriptors.

Access to Fourier coefficients

`Complex[] getCoefficients ()`
Returns the complete vector of complex-valued Fourier coefficients.[23]

[23] The class `Complex` is defined in package `imagingbook.lib.math`.

`Complex getCoefficient (int m)`
Returns the value of the Fourier coefficient $\boldsymbol{G}(\texttt{m} \bmod M)$, with $M = |\boldsymbol{G}|$ as above.

`Complex setCoefficient (int m, Complex z)`
Replaces the Fourier coefficient $\boldsymbol{G}(\texttt{m} \bmod M)$ by the complex value `z`, with $M = |\boldsymbol{G}|$ as above.

`Complex setCoefficient (int m, double a, double b)`
Replaces the Fourier coefficient $\boldsymbol{G}(\texttt{m} \bmod M)$ by the complex value $z = \texttt{a} + \texttt{i} \cdot \texttt{b}$, with $M = |\boldsymbol{G}|$ as above.

`int size ()`
Returns the length (M) of the Fourier descriptor.

`int getMaxNegHarmonic ()`
Returns the max. negative harmonic $m = -(M-1) \div 2$ for this Fourier descriptor (of length M).

`int getMaxPosHarmonic ()`
Returns the max. positive harmonic $m = M \div 2$ for this Fourier descriptor (of length M).

`int getMaxCoefficientPairs ()`
Returns the maximum number of coefficient pairs, $(M-1) \div 2$, for this Fourier descriptor (of length M).

`void truncate (int Mp)`
Truncates this Fourier descriptor to the `Mp` lowest-frequency coefficients (see Eqn. (26.23)).

Comparing Fourier descriptors

`double distanceComplex (FourierDescriptor fd2)`
Returns the complex-valued distance ($\mathrm{dist}_{\mathrm{C}}(\boldsymbol{G}_1, \boldsymbol{G}_2)$, see Eqn. (26.114)) between *this* Fourier descriptor ($\boldsymbol{G}_1$) and another Fourier descriptor `fd2` ($\boldsymbol{G}_2$). The zero-coefficients are ignored.

`double distanceComplex (FourierDescriptor fd2, int Mp)`
As above, but using only `Mp` coefficient pairs (see Eqn. (26.114)).

`double distanceMagnitude (FourierDescriptor fd2)`
Returns the magnitude-only distance ($\mathrm{dist}_{\mathrm{M}}(\boldsymbol{G}_1, \boldsymbol{G}_2)$, see Eqn. (26.112)) between *this* Fourier descriptor ($\boldsymbol{G}_1$) and another Fourier descriptor `fd2` ($\boldsymbol{G}_2$). The zero-coefficients are ignored.

`double distanceMagnitude (FourierDescriptor fd2, int Mp)`
As above, but using only `Mp` coefficient pairs (see Eqn. (26.112)).

Shape reconstruction

`Complex[] getReconstruction (int N)`
Returns the shape reconstructed from the complete Fourier descriptor as a sequence of `N` complex-valued contour points. The contour points are obtained by evaluating `getReconstructionPoint(t)` at uniformly spaced positions $\texttt{t} \in [0, 1)$.

`Complex[] getReconstruction (int N, int Mp)`
Returns a partial shape reconstruction from `Mp` Fourier coefficient pairs as a sequence of `N` complex-valued contour points.

`Complex getReconstructionPoint (double t)`
Returns a single point (as a complex value) on the continuous contour for path parameter $\texttt{t} \in [0,1)$, reconstructed from the complete Fourier descriptor (see Eqn. (26.20)).

`Complex getReconstructionPoint (double t, int Mp)`
Returns a single point (as a complex value) on the continuous contour for path parameter $\texttt{t} \in [0,1)$, reconstructed from `Mp` Fourier coefficient pairs.

Normalization

`FourierDescriptor[] makeInvariant ()`
Returns a pair of Fourier descriptors $(\boldsymbol{G}^{\mathrm{A}}, \boldsymbol{G}^{\mathrm{B}})$ that are normalized for scale, start point shift and shape rotation (see Alg. 26.6).

`double makeRotationInvariant ()`
Normalizes the Fourier descriptor for shape rotation by phase-shifting all coefficients (see Alg. 26.6). Returns the estimated rotation angle β.

`double makeScaleInvariant ()`
Normalizes the Fourier descriptor for scale by multiplying with a common factor, such that the L_2 norm of the resulting vector is 1. Returns the scale factor that was applied for normalization.

`FourierDescriptor[] makeStartPointInvariant ()`
Returns a pair of normalized Fourier descriptors $(\boldsymbol{G}^{\mathrm{A}}, \boldsymbol{G}^{\mathrm{B}})$, one for each start point normalization angles φ_{A} and $\varphi_{\mathrm{B}} = \varphi_{\mathrm{A}} + \pi$, respectively (see Alg. 26.7).

`void makeTranslationInvariant ()`
Modifies this Fourier descriptor by setting the coefficient $\boldsymbol{G}(0)$ to zero. This method is rarely needed because $\boldsymbol{G}(0)$ is ignored for matching.

`FourierDescriptorUniform` (class)

This sub-class of `FourierDescriptor` represents Fourier descriptors obtained from uniformly sampled contours, as described in Alg. 26.2. It provides the constructor methods

`FourierDescriptorUniform (Point2D[] V)`,

`FourierDescriptorUniform (Point2D[] V, int Mp)`,

where `V` is a sequence of M contour points (`Point2D`), assumed to be uniformly sampled. The first constructor creates a full Fourier descriptor with M coefficients (see Alg. 26.2). The second constructor creates a Fourier descriptor with `Mp` coefficient *pairs* (i.e., $2 \cdot \texttt{Mp} + 1$ coefficients), as described in Alg. 26.3

`FourierDescriptorFromPolygon` (class)

This sub-class of `FourierDescriptor` represents Fourier descriptors obtained directly from polygons (without contour sampling, see Alg. 26.5). It provides the single constructor method

`FourierDescriptorFromPolygon (Point2D[] V, int Mp)`,

where V is a sequence of polygon vertices and Mp specifies the number of Fourier coefficient pairs.

`PolygonSampler` (class)

Instances of this utility class can be used to produce uniformly sampled polygons.

`Point2D[] samplePolygonUniformly(Point2D[] V, int M)`
Samples the closed polygon path specified by the vertices in V at M equi-distant positions and returns the resulting point sequence (see Alg. 26.1).

Example

The code example in Prog. 26.1 demonstrates the use of the Fourier descriptor API. It assumes that the binary input image (ip) contains at least one connected foreground region. Region labeling and contour extraction is applied first, using methods provided by the **imagingbook.regions** and **imagingbook.contours** packages.[24] Subsequently, the longest region contour (C) is used to create a Fourier descriptor (fd) with $M_{\mathrm{P}} = 15$ coefficient pairs. A partial reconstruction is calculated from the original Fourier descriptor with 100 sample points along the contour. The last lines show how a pair of invariant descriptors $(\boldsymbol{G}^{\mathrm{A}}, \boldsymbol{G}^{\mathrm{B}})$ is obtained by applying the **makeInvariant()** method. Note that the code fragment in Prog. 26.1 is not complete but would typically be part of the **run()** method in an ImageJ plugin. The full version and additional code examples can be found on the book's website.

26.8 Discussion and Further Reading

The use of Fourier descriptors for shape description and matching dates back to the early 1960's [55, 81], advanced by the work of Zahn and Roskies [262], Granlund [93], Richard and Hemami [196], and Persoon and Fu [183, 184] in the 1970s, particularly in the context of character recognition and aircraft identification. Making Fourier descriptors invariant against various geometric transformations was a key issue from the very beginning, and several relevant contributions were published in the 1980s, including [245], [57] [143], and [189]. Unfortunately, as illustrated in this chapter, to achieve robust invariance and uniqueness of representation in practice is not as easy as sometimes suggested in the literature, despite the simplicity and elegance of the underlying theory. In practice, normalization for descriptor invariance is quite difficult for arbitrary shapes because of possibly vanishing Fourier coefficients and the resulting sensitivity to noise.

Fourier descriptors have nevertheless become popular in a wide range of applications, including geology and, in particular, biological imaging, as documented by the work of Lestrel and others in [146].

[24] See also Chapter 10.

```
1 ...
2 import imagingbook.lib.math.Complex;
3 import imagingbook.pub.fd.*;
4 import imagingbook.pub.regions.*;
5
6 ByteProcessor ip ...;  // assumed to contain a binary image
7
8 // segment ip and select the longest outer region contour:
9 RegionContourLabeling labeling =
10       new RegionContourLabeling(ip);
11 List<Contour> outerContours =
12       labeling.getAllOuterContours(true);
13 Contour contr = outerContours.get(0);  // get the longest contour
14 Point2D[] V = contr.getPointArray();
15
16 // create the Fourier descriptor for V with 15 coefficient pairs:
17 FourierDescriptor fd = new FourierDescriptorUniform(V, 15);
18
19 // reconstruct the corresponding shape with 100 contour points:
20 Complex[] R = fd.getReconstruction(100);
21
22 // create a pair of invariant descriptors (G^A, G^B):
23 FourierDescriptor[] fdAB = fd.makeInvariant();
24 FourierDescriptor fdA = fdAB[0];  // = G^A
25 FourierDescriptor fdB = fdAB[1];  // = G^B
26 ...
```

Prog. 26.1
Fourier descriptor code example. The input image ip is assumed to contain a binary image (line 6). The class RegionContourLabeling is used to find connected regions (line 10). Then the list of outer contours is retrieved (line 12) and the longest contour is assigned to V as an array of type Point2D (lines 13–14). In line 17, the contour V is used to create a Fourier descriptor with 15 coefficient pairs. Alternatively, we could have created a Fourier descriptor of the same length (number of coefficients) as the contour and then truncated it (using the truncate() method) to the specified number of coefficient pairs. A partial reconstruction of the contour (with 100 sample points) is calculated from the Fourier descriptor fd in line 20. Finally, a pair of invariant descriptors (contained in the array fdAB) is calculated in line 23.

Fourier descriptors have been extended to accommodate affine transformations and applied to 3D object identification [5] and stereo matching [257].

Although Fourier descriptors have been investigated to handle open contours and partial shapes [148], they are naturally best suited to dealing with closed contours, as we have described. Of course, this is a limitation if shapes are only partially visible or occluded. The presentation in this chapter was limited to what are frequently called "elliptical" Fourier descriptors [93], since they are most popular and well known. Other types of Fourier descriptors have been proposed, which are not covered here but can be found elsewhere in the literature (see, e.g., [126, p. 534] and [174, Ch. 7]).

26.9 Exercises

Exercise 26.1. Verify that the DFT spectrum is periodic, that is, that $\boldsymbol{G}(-m) = \boldsymbol{G}(M-m)$ holds for arbitrary $m \in \mathbb{Z}$ (as claimed in Eqn. (26.22)).

Exercise 26.2. Algorithm 26.9 shows an alternative solution to uniform polygon sampling. Implement this algorithm and verify that it is equivalent to Alg. 26.1 (implemented as method `samplePolygonUniformly()` in class `PolygonSampler`, see Sec. 26.7).

Exercise 26.3. Assume that the complete outer contour of a binary region is given as a sequence of P boundary pixels with coordinates

Alg. 26.9
Uniform sampling of a polygon path (alternative to Alg. 26.1, proposed by J. Heinzelreiter).

```
1:  SamplePolygonUniformly(V, M)
      Input: V = (v_0, ..., v_{N-1}), a sequence of N points representing
      the vertices of a closed 2D polygon; M, number of desired sample
      points. Returns a new sequence g = (g_0, ..., g_{M-1}) of complex
      values representing sample points sampled uniformly along the
      path of the input polygon V.
2:      N ← |V|
3:      Δ ← (1/M) · PathLength(V)        ▷ segment length Δ, see Alg. 26.1
4:      Create map g: [0, M−1] → ℂ       ▷ complex point sequence g
5:      g(0) ← Complex(V(0))
6:      i ← 0                            ▷ index of path segment ⟨V_i, V_{i+1}⟩
7:      k ← 1                            ▷ index of first unassigned point in g
8:      d_p ← 0                          ▷ path distance between V(i) and V(k−1)
9:      while (i < N) ∧ (k < M) do
10:         v_A ← V(i)
11:         v_B ← V((i+1) mod N)
12:         δ ← ‖v_B − v_A‖                ▷ Euclidean distance
13:         if (Δ − d_p) ≤ δ then
14:             x ← v_A + ((Δ−d_p)/δ) · (v_B − v_A)   ▷ x_k by lin. interpolation
15:             g(k) ← Complex(x)
16:             d_p ← d_p − Δ
17:             k ← k + 1
18:         else
19:             d_p ← d_p + δ
20:             i ← i + 1
21:     return g.
```

$V = (\boldsymbol{p}_0, \ldots, \boldsymbol{p}_{P-1})$. To produce a Fourier descriptor of length $M < P$ there are several options:

1. Sample the original contour V at M uniformly-spaced positions (see Alg. 26.1) and then calculate the Fourier descriptor of length M using Alg. 26.2.
2. Calculate a partial Fourier descriptor of length M' from the original contour V using Alg. 26.3.
3. Calculate the full Fourier descriptor (of length M) from the original contour V (using Alg. 26.2) and subsequently truncate[25] the Fourier descriptor to length M', as described in Eqns. (26.23) and (26.24).
4. Treat the original boundary coordinates V as the vertices of a closed polygon and calculate a Fourier descriptor with $M_{\mathrm{P}} = M \div 2$ coefficient pairs, using the trigonometric method described in Alg. 26.5.

Compare these approaches and discuss their individual merits or disadvantages in terms of efficiency and accuracy.

Exercise 26.4. Test the Fourier descriptor normalization described in Algs. 26.6 and 26.7 (implemented by method `makeInvariant()` in the Java API) for changes in scale, start point shift, and shape rotation on a suitable set of binary shapes (e.g., images from the

[25] See method `truncate(int Mp)` in Sec. 26.7.

KIMIA dataset [134]). See the examples for shape rotation and (implicit) start point shifts in Fig. 26.21. How reliably do the normalized Fourier descriptors of the modified shapes match to their corresponding originals?

Exercise 26.5. Magnitude-only matching (see Sec. 26.6.1) is much simpler than complex-valued matching (see Sec. 26.6.2) of Fourier descriptors, since no normalization for phase (start point shift and shape rotation) is required. However, it can be assumed that different shapes are more likely to be confused if the phase information is ignored. Test this hypothesis on a large number and variety of different shapes. Compare the confusion probability for magnitude-only vs. complex-valued matching.

Appendix A

Mathematical Symbols and Notation

A.1 Symbols

The following symbols are used in the main text primarily with the denotations given here. While some symbols may be used for purposes other than the ones listed, the meaning should always be clear in the particular context.

$(a_0, \ldots, a_{n-1})$ A *vector* or *list*, that is, an ordered sequence of n elements of the same type. Unlike a *set* (see below), a list may contain the same element more than once. If used to denote a *vector*, then $(a_0, \ldots, a_{n-1})$ is usually a *row* vector and $(a_0, \ldots, a_{n-1})^\mathsf{T}$ is the corresponding (transposed) *column* vector.[1] If used to represent a *list*,[2] $()$ represents the *empty* list and (a) is a list with a single element a. $|A|$ is the *length* of the sequence A, that is, the number of contained elements. $A \smile B$ denotes the concatenation of A, B. $A(i)$ or a_i refers to the i-th element of A. $A(i) \leftarrow x$ means that the i-th element of A is set to (i.e., replaced by) the quantity x.

$\{a, b, c, d, \ldots\}$ A *set*, that is, an unordered collection of distinct elements. A particular element x can be contained in a set at most once. $\{\}$ denotes the empty set. $|\mathcal{A}|$ is the size (cardinality) of the set $\mathcal{A}$. $\mathcal{A} \cup \mathcal{B}$ is the union and $\mathcal{A} \cap \mathcal{B}$ is the intersection of two sets $\mathcal{A}, \mathcal{B}$. $x \in \mathcal{A}$ means that the element x is contained in $\mathcal{A}$.

$\langle A, B, C \rangle$ A *tuple*, that is, a fixed-size, ordered sequence of elements, each possibly of a different type.[3]

---

[1] In most programming environments, vectors are implemented as one-dimensional arrays, with elements being referred to by position (index).

[2] Lists are usually implemented with dynamic data structures, such as linked lists. Java's *Collections* framework provides numerous easy-to-use list implementations.

[3] Tuples are typically implemented as *objects* (in Java or C++) or *structures* (in C) with elements being referred to by name.

$[a, b]$ — Numeric interval; $x \in [a, b]$ means $a \leq x \leq b$. Similarly, $x \in [a, b)$ says that $a \leq x < b$.

$|A|$ — Length (number of elements) of a sequence (see above) or size (cardinality) of a set A, that is, $|A| \equiv \text{card}\, A$.

$|\mathbf{A}|$ — Determinant of a matrix $\mathbf{A}$ ($|\mathbf{A}| \equiv \det(\mathbf{A})$).

$|x|$ — Absolute value (magnitude) of a scalar or complex quantity x.

$\|\boldsymbol{x}\|$ — Euclidean (L_2) norm of a vector $\boldsymbol{x}$. $\|\boldsymbol{x}\|_n$ denotes the magnitude of $\boldsymbol{x}$ using a particular norm L_n.

$\lceil x \rceil$ — "Ceil" of x, the smallest integer $z \in \mathbb{Z}$ greater than $x \in \mathbb{R}$. For example, $\lceil 3.141 \rceil = 4$, $\lceil -1.2 \rceil = -1$.

$\lfloor x \rfloor$ — "Floor" of x, the largest integer $z \in \mathbb{Z}$ smaller than $x \in \mathbb{R}$. For example, $\lfloor 3.141 \rfloor = 3$, $\lfloor -1.2 \rfloor = -2$.

$\div$ — Integer division operator: $a \div b$ denotes the quotient of the two integers a, b. For example, $5 \div 3 = 1$ and $-13 \div 4 = -3$ (equivalent to Java's "`/`" operator in the case of integer operands).

$*$ — Linear convolution operator (see Sec. 5.3.1).

$\circledast$ — Linear correlation operator (see Sec. 23.1.1).

$\otimes$ — Outer vector product (see Sec. B.3.2).

$\times$ — Cross product (between vectors or complex quantities (see Sec. B.3.3).

$\oplus$ — Morphological dilation operator (see Sec. 9.2.3).

$\ominus$ — Morphological erosion operator (see Sec. 9.2.4).

$\circ$ — Morphological opening operator (see Sec. 9.3.1).

$\bullet$ — Morphological closing operator (see Sec. 9.3.2).

$\smile$ — Concatenation operator. Given two sequences $A = (a, b, c)$ and $B = (d, e)$, $A \smile B$ denotes the concatenation of A and B, with the result (a, b, c, d, e). Inserting a single element x at the end or front of the list A is written as $A \smile (x)$ or $(x) \smile A$, resulting in (a, b, c, x) or (x, a, b, c), respectively.

$\sim$ — "Similarity" relation used in the context of random variables and statistical distributions.

$\approx$ — "Approximately equal" relation.

$\equiv$ — Equivalence relation.

$\leftarrow$ — Assignment operator: $a \leftarrow \mathit{expr}$ means that expression expr is evaluated and subsequently the result is assigned to the variable a.

$\stackrel{+}{\leftarrow}$ — Incremental assignment operator: $a \stackrel{+}{\leftarrow} b$ is equivalent to $a \leftarrow a + b$.

$:=$ — Function definition operator (used in algorithms). For example, $f(x) := x^2 + 5$ defines a function $f()$ with the bound variable (formal function argument) x.

$.\cdot\cdot$ — "upto" (incrementing) iteration, used in loop constructs like **for** $q \leftarrow 1, .\cdot\cdot, K$ (with $q = 1, 2, \ldots, K-1, K$).

$\cdot\cdot.$ — "downto" (decrementing) iteration, for example, **for** $q \leftarrow K, \cdot\cdot., 1$ (with $q = K, K-1, \ldots, 2, 1$).

$\wedge$	Logical "and" operator.
$\vee$	Logical "or" operator.
∂	Partial derivative operator (see Sec. 6.2.1). For example, $\frac{\partial}{\partial x_i} f$ denotes the *first* derivative of the multi-dimensional function $f(x_1, x_2, \ldots, x_n) : \mathbb{R}^n \to \mathbb{R}$ along variable x_i, $\frac{\partial^2}{\partial x_i^2} f$ is the *second* derivative (i.e., differentiating f twice along variable x_i), etc.
∇	Gradient operator. The gradient of a multi-dimensional function $f(x_1, x_2, \ldots, x_n) : \mathbb{R}^n \to \mathbb{R}$, denoted ∇f (also ∇_f or $\mathrm{grad}\, f$), is the vector of its first partial derivatives (see also Sec. C.2.2).
∇^2	Laplace operator (or *Laplacian*). The Laplacian of a multi-dimensional function $f(x_1, x_2, \ldots, x_n) : \mathbb{R}^n \to \mathbb{R}$, denoted $\nabla^2 f$ (or ∇_f^2), is the sum of its second partial derivatives (see Sec. C.2.5).
$\mathbf{0}$	Zero vector, $\mathbf{0} = (0, \ldots, 0)^\mathsf{T}$.
adj	Adjugate of a square matrix, denoted $\mathrm{adj}(\mathbf{A})$; also called *adjoint* in older texts.
AND	Bitwise "and" operation. Example: $(0011_\mathrm{b}\ \mathrm{AND}\ 1010_\mathrm{b}) = 0010_\mathrm{b}$ (binary) and $(3\ \mathrm{AND}\ 6) = 2$ (decimal).
$\mathrm{ArcTan}(x, y)$	Inverse tangent function. The result of $\mathrm{ArcTan}(x, y)$ is equivalent to $\arctan(\frac{y}{x}) = \tan^{-1}(\frac{y}{x})$ but with two arguments and returning angles in the range $[-\pi, +\pi]$ (i.e., covering all four quadrants). $\mathrm{ArcTan}(x, y)$ is equivalent to the `ArcTan[x,y]` function in *Mathematica* and the `Math.atan2(y,x)` method in Java (but note the reversed arguments!).
$\mathbb{C}$	The set of complex numbers.
card	Size (cardinality) of a set. $\mathrm{card}(\mathcal{A}) = \lvert\mathcal{A}\rvert$ (see also Sec. 3.1).
det	Determinant of a matrix ($\det(\mathbf{A}) = \lvert\mathbf{A}\rvert$).
DFT	Discrete Fourier transform (see Sec. 18.3).
e	Euler's constant.
$\mathbf{e}$	Unit vector. For example, $\mathbf{e}_x = (1, 0)^\mathsf{T}$ denotes the 2D unit vector in x-direction. $\mathbf{e}_\theta = (\cos\theta, \sin\theta)^\mathsf{T}$ is the 2D unit vector oriented at angle θ and $\mathbf{e}_\mathrm{i}, \mathbf{e}_\mathrm{j}, \mathbf{e}_\mathrm{k}$ are the unit vectors along the coordinate axes in 3D.
exp	Exponential function: $\exp(x) = e^x$.
$\mathcal{F}$	Continuous Fourier transform (see Sec. 18.1.4).
false	Boolean constant ($\mathrm{false} = \neg\mathrm{true}$).
grad	Gradient operator (see ∇).
h	Histogram of an image (see Sec. 3.1).
H	Cumulative histogram (see Sec. 3.6).
$\mathbf{H}$	Hessian matrix (see Sec. C.2.6).
hom	Operator for converting Cartesian to homogeneous coordinates. $\mathrm{hom}(\boldsymbol{x}) = \underline{\boldsymbol{x}}$ maps the Cartesian point $\boldsymbol{x}$ to a corresponding homogeneous point $\underline{\boldsymbol{x}}$; the reverse mapping is denoted $\mathrm{hom}^{-1}(\underline{\boldsymbol{x}}) = \boldsymbol{x}$ (see Sec. B.5).
i	Imaginary unit ($\mathrm{i}^2 = -1$), see Sec. A.3.

Appendix A
MATHEMATICAL SYMBOLS AND NOTATION

I Image with scalar pixel values (e.g., an intensity or grayscale image). $I(u, v) \in \mathbb{R}$ is the pixel value at position (u, v)

$\boldsymbol{I}$ Vector-valued image, for example, a RGB color image with 3D color vectors $\boldsymbol{I}(u, v) \in \mathbb{R}^3$ at position (u, v).

$\mathbf{I}_n$ Identity matrix of size $n \times n$. For example, $\mathbf{I}_2 = \left(\begin{smallmatrix} 1 & 0 \\ 0 & 1 \end{smallmatrix}\right)$ is the 2×2 identity matrix.

$\mathbf{J}$ Jacobian matrix (see Sec. C.2.1).

$\mathrm{L}_1, \mathrm{L}_2, \mathrm{L}_\infty$ Common distance measures or *norms* (see Eqns. (15.23)–(15.25)).

$M \times N$ Domain of pixel coordinates (u, v) for an image with M columns (width) and N rows (height); used as a shortcut notation for the set $\{0, \ldots, M-1\} \times \{0, \ldots, N-1\}$.

mod Modulus operator: $(a \bmod b)$ is the remainder of the *integer* division $a \div b$ (see Sec. F.1.2).

μ Arithmetic mean value.

$\mathbb{N}$ The set of natural numbers; $\mathbb{N} = \{1, 2, 3, \ldots\}$, $\mathbb{N}_0 = \{0, 1, 2, \ldots\}$.

nil Null ("nothing") constant, typically used in algorithms to denote an invalid quantity (similar to `null` in Java).

p Discrete probability density function (see Sec. 4.6.1).

P Discrete probability distribution function or cumulative probability density (see Sec. 4.6.1).

$\mathcal{Q}$ Quadrilateral (see Sec. 21.1.4).

$\mathbb{R}$ The set of real numbers.

R, G, B *Red*, *green* and *blue* color components.

rank Rank of a matrix $\mathbf{A}$, denoted by $\operatorname{rank}(\mathbf{A})$.

round Rounding function: returns the integer closest to the scalar $x \in \mathbb{R}$. $\operatorname{round}(x) \equiv \lfloor x + 0.5 \rfloor$.

σ Standard deviation (square root of the *variance* σ^2).

$\mathcal{S}_1$ Unit square (see Sec. 21.1.4).

sgn "Sign" or "signum" function:

$$\operatorname{sgn}(x) = \begin{cases} 1 & \text{for } x > 0 \\ 0 & \text{for } x = 0 \\ -1 & \text{for } x < 0 \end{cases}$$

τ Interval in time or space.

t Continuous time variable.

t Threshold value.

T Transpose of a vector ($\boldsymbol{a}^\mathsf{T}$) or matrix ($\mathbf{A}^\mathsf{T}$).

trace Trace (sum of the diagonal elements) of a matrix, e.g., $\operatorname{trace}(\mathbf{A})$.

true Boolean constant (true $= \neg$false).

$\boldsymbol{u} = (u, v)$ Discrete 2D coordinate variable with $u, v \in \mathbb{Z}$.

$\boldsymbol{x} = (x, y)$ Continuous 2D coordinate variable with $x, y \in \mathbb{R}$.

XOR Bitwise "xor" (exclusive OR) operator. Example: $(0011_\mathrm{b} \text{ XOR } 1010_\mathrm{b}) = 1001_\mathrm{b}$ (binary) and $(3 \text{ XOR } 6) = 5$ (decimal).

$\mathbb{Z}$ The set of integers.

A.2 Set Operators

$\lvert\mathcal{A}\rvert$	The size of the set $\mathcal{A}$ (equal to $\mathrm{card}(\mathcal{A})$).
$\forall_x \ldots$	"All" quantifier (for all x, ...).
$\exists_x \ldots$	"Exists" quantifier (there is some x for which ...).
$\cup$	Set union (e.g., $\mathcal{A} \cup \mathcal{B}$).
$\cap$	Set intersection (e.g., $\mathcal{A} \cap \mathcal{B}$).
$\bigcup_i \mathcal{A}_i$	Union of multiple sets $\mathcal{A}_i$.
$\bigcap_i \mathcal{A}_i$	Intersection over multiple sets $\mathcal{A}_i$.
$\setminus$	Set difference: if $x \in \mathcal{A} \setminus \mathcal{B}$, then $x \in \mathcal{A}$ and $x \notin \mathcal{B}$.

A.3 Complex Numbers

Basic relations:

$$z = a + \mathrm{i} \cdot b \qquad (\text{with } z, \mathrm{i} \in \mathbb{C},\ a, b \in \mathbb{R},\ \mathrm{i}^2 = -1) \tag{A.1}$$

$$s \cdot z = s \cdot a + \mathrm{i} \cdot s \cdot b \qquad (\text{for } s \in \mathbb{R}) \tag{A.2}$$

$$|z| = \sqrt{a^2 + b^2} \tag{A.3}$$

$$|s \cdot z| = s \cdot |z| \tag{A.4}$$

$$z = a + \mathrm{i} \cdot b = |z| \cdot (\cos\psi + \mathrm{i} \cdot \sin\psi) \tag{A.5}$$

$$= |z| \cdot e^{\mathrm{i}\cdot\psi} \qquad (\text{with } \psi = \mathrm{ArcTan}(a, b)) \tag{A.6}$$

$$\mathrm{Re}(a + \mathrm{i} \cdot b) = a \qquad \mathrm{Re}(e^{\mathrm{i}\cdot\varphi}) = \cos\varphi \tag{A.7}$$

$$\mathrm{Im}(a + \mathrm{i} \cdot b) = b \qquad \mathrm{Im}(e^{\mathrm{i}\cdot\varphi}) = \sin\varphi \tag{A.8}$$

$$e^{\mathrm{i}\cdot\varphi} = \cos\varphi + \mathrm{i} \cdot \sin\varphi \tag{A.9}$$

$$e^{-\mathrm{i}\cdot\varphi} = \cos\varphi - \mathrm{i} \cdot \sin\varphi \tag{A.10}$$

$$\cos(\varphi) = \tfrac{1}{2} \cdot \left(e^{\mathrm{i}\cdot\varphi} + e^{-\mathrm{i}\cdot\varphi}\right) \tag{A.11}$$

$$\sin(\varphi) = \tfrac{1}{2i} \cdot \left(e^{\mathrm{i}\cdot\varphi} - e^{-\mathrm{i}\cdot\varphi}\right) \tag{A.12}$$

$$z^* = a - \mathrm{i} \cdot b \qquad (\text{complex conjugate}) \tag{A.13}$$

$$z \cdot z^* = z^* \cdot z = |z|^2 = a^2 + b^2 \tag{A.14}$$

$$z^0 = (a + \mathrm{i} \cdot b)^0 = (1 + \mathrm{i} \cdot 0) = 1 \tag{A.15}$$

Arithmetic operations:

$$z_1 = (a_1 + \mathrm{i} \cdot b_1) = |z_1|\, e^{\mathrm{i}\cdot\varphi_1} \tag{A.16}$$

$$z_2 = (a_2 + \mathrm{i} \cdot b_2) = |z_2|\, e^{\mathrm{i}\cdot\varphi_2} \tag{A.17}$$

$$z_1 + z_2 = (a_1 + a_2) + \mathrm{i} \cdot (b_1 + b_2), \tag{A.18}$$

$$z_1 \cdot z_2 = (a_1 \cdot a_2 - b_1 \cdot b_2) + \mathrm{i} \cdot (a_1 \cdot b_2 + b_1 \cdot a_2) \tag{A.19}$$

$$= |z_1| \cdot |z_2| \cdot e^{\mathrm{i}\cdot(\varphi_1 + \varphi_2)} \tag{A.20}$$

$$\frac{z_1}{z_2} = \frac{a_1 \cdot a_2 + b_1 \cdot b_2}{a_2^2 + b_2^2} + \mathrm{i} \cdot \frac{a_2 \cdot b_1 - a_1 \cdot b_2}{a_2^2 + b_2^2} = \frac{|z_1|}{|z_2|} \cdot e^{\mathrm{i}\cdot(\varphi_1 - \varphi_2)} \tag{A.21}$$

Appendix B

Linear Algebra

This part contains a compact set of elementary tools and concepts from algebra and calculus that are referenced in the main text. Many good textbooks (probably including some of your school books) are available on this subject, for example, [35,36,145,264]. For numerical aspects of linear algebra see [160, 190].

B.1 Vectors and Matrices

Here we describe the basic notation for vectors in two and three dimensions. Let

$$\boldsymbol{a} = \begin{pmatrix} a_0 \\ a_1 \end{pmatrix}, \qquad \boldsymbol{b} = \begin{pmatrix} b_0 \\ b_1 \end{pmatrix} \tag{B.1}$$

denote vectors $\boldsymbol{a}, \boldsymbol{b}$ in 2D, and analogously

$$\boldsymbol{a} = \begin{pmatrix} a_0 \\ a_1 \\ a_2 \end{pmatrix}, \qquad \boldsymbol{b} = \begin{pmatrix} b_0 \\ b_1 \\ b_2 \end{pmatrix} \tag{B.2}$$

vectors in 3D (with $a_i, b_i \in \mathbb{R}$). Vectors are used to describe 2D or 3D points (relative to the origin of the coordinate system) or the displacement between two arbitrary points in the corresponding space.

We commonly use upper-case letters to denote a *matrix*, for example,

$$\mathbf{A} = \begin{pmatrix} A_{0,0} & A_{0,1} \\ A_{1,0} & A_{1,1} \\ A_{2,0} & A_{2,1} \end{pmatrix}. \tag{B.3}$$

This matrix consists of 3 rows and 2 columns; in other words, $\mathbf{A}$ is of size $(3, 2)$. Its individual elements are referenced as $A_{i,j}$, where i is the *row* index (vertical coordinate) and j is the *column* index (horizontal coordinate).[1]

[1] Note that the usual notation for matrix coordinates is (unlike image coordinates) vertical-first!

The *transpose* of $\mathbf{A}$, denoted $\mathbf{A}^\mathsf{T}$, is obtained be exchanging rows and columns, that is,

$$\mathbf{A}^\mathsf{T} = \begin{pmatrix} A_{0,0} & A_{0,1} \\ A_{1,0} & A_{1,1} \\ A_{2,0} & A_{2,1} \end{pmatrix}^\mathsf{T} = \begin{pmatrix} A_{0,0} & A_{1,0} & A_{2,0} \\ A_{0,1} & A_{1,1} & A_{2,1} \end{pmatrix}. \tag{B.4}$$

The *inverse* of a square matrix $\mathbf{A}$ is denoted $\mathbf{A}^{-1}$, such that

$$\mathbf{A} \cdot \mathbf{A}^{-1} = \mathbf{I} \quad \text{and} \quad \mathbf{A}^{-1} \cdot \mathbf{A} = \mathbf{I} \tag{B.5}$$

($\mathbf{I}$ is the identity matrix). Note that not every square matrix has an inverse. Calculation of the inverse can be performed in closed form up to the size $(3, 3)$; for example, see Eqn. (21.29) and Eqn. (24.47). In general, the use of standard numerical methods is recommended (see Sec. B.6).

B.1.1 Column and Row Vectors

For practical purposes, a vector can be considered a special case of a matrix. In particular, a the m-dimensional *column* vector

$$\boldsymbol{a} = \begin{pmatrix} a_0 \\ \vdots \\ a_{m-1} \end{pmatrix} \tag{B.6}$$

corresponds to a matrix of size $(m, 1)$, while its transpose $\boldsymbol{a}^\mathsf{T}$ is a *row* vector and thus like a matrix of size $(1, m)$. By default, and unless otherwise noted, any vector is implicitly assumed to be a *column* vector.

B.1.2 Length (Norm) of a Vector

The *length* or *Euclidean norm* (L_2 norm) of a vector $\boldsymbol{a} = (a_1, \ldots, a_{m-1})^\mathsf{T}$, denoted $\|\boldsymbol{a}\|$, is defined as

$$\|\boldsymbol{a}\| = \Bigl(\sum_{i=0}^{m-1} a_i^2\Bigr)^{1/2}. \tag{B.7}$$

For example, the length of the 3D vector $\boldsymbol{x} = (x, y, z)^\mathsf{T}$ is

$$\|\boldsymbol{x}\| = \sqrt{x^2 + y^2 + z^2}. \tag{B.8}$$

B.2 Matrix Multiplication

B.2.1 Scalar Multiplication

The product of a real-valued matrix and a scalar value $s \in \mathbb{R}$ is defined as

$$s \cdot \mathbf{A} = \mathbf{A} \cdot s = \bigl[s \cdot A_{i,j}\bigr] = \begin{pmatrix} s \cdot A_{0,0} & \cdots & s \cdot A_{0,n-1} \\ \vdots & \ddots & \vdots \\ s \cdot A_{m-1,0} & \cdots & s \cdot A_{m-1,n-1} \end{pmatrix}. \tag{B.9}$$

B.2.2 Product of Two Matrices

We say that a matrix is of size (m, n) if consists of m rows and n columns. Given two matrices $\mathbf{A}$, $\mathbf{B}$ of size (m, n) and (p, q), respectively, the product $\mathbf{A} \cdot \mathbf{B}$ is only defined if $n = p$. Thus the number of columns (n) in $\mathbf{A}$ must always match the number of rows (p) in $\mathbf{B}$. The result is a new matrix $\mathbf{C}$ of size (m, q), that is,

$$\mathbf{C} = \mathbf{A} \cdot \mathbf{B} = \underbrace{\begin{pmatrix} A_{0,0} & \cdots & A_{0,n-1} \\ \vdots & \ddots & \vdots \\ A_{m-1,0} & \cdots & A_{m-1,n-1} \end{pmatrix}}_{(m,n)} \cdot \underbrace{\begin{pmatrix} B_{0,0} & \cdots & B_{0,q-1} \\ \vdots & \ddots & \vdots \\ B_{n-1,0} & \cdots & B_{n-1,q-1} \end{pmatrix}}_{(n,q)}$$
$$= \underbrace{\begin{pmatrix} C_{0,0} & \cdots & C_{0,q-1} \\ \vdots & \ddots & \vdots \\ C_{m-1,0} & \cdots & C_{m-1,q-1} \end{pmatrix}}_{(m,q)}, \tag{B.10}$$

with the elements

$$C_{ij} = \sum_{k=0}^{n-1} A_{i,k} \cdot B_{k,j}, \tag{B.11}$$

for $i = 0, \ldots, m-1$ and $j = 0, \ldots, q-1$. Note that this product is not commutative, that is, $\mathbf{A} \cdot \mathbf{B} \neq \mathbf{B} \cdot \mathbf{A}$ in general.

B.2.3 Matrix-Vector Products

The product $\mathbf{A} \cdot \boldsymbol{x}$ between a matrix $\mathbf{A}$ and a vector $\boldsymbol{x}$ is only a special case of the matrix-matrix multiplication given in Eqn. (B.10). In particular, if $\boldsymbol{x} = (x_0, \ldots, x_{n-1})^\mathsf{T}$ is a n-dimensional *column* vector (i.e., a matrix of size $(n, 1)$), then the multiplication

$$\underbrace{\boldsymbol{y}}_{(m,1)} = \underbrace{\mathbf{A}}_{(m,n)} \cdot \underbrace{\boldsymbol{x}}_{(n,1)} \tag{B.12}$$

is only defined if the matrix $\mathbf{A}$ is of size (m, n), for arbitrary $m \geq 1$. The result $\boldsymbol{y}$ is a *column* vector of length m (equivalent to a matrix of size $(m, 1)$). For example (with $m = 2$, $n = 3$),

$$\mathbf{A} \cdot \boldsymbol{x} = \underbrace{\begin{pmatrix} A & B & C \\ D & E & F \end{pmatrix}}_{(2,3)} \cdot \underbrace{\begin{pmatrix} x \\ y \\ z \end{pmatrix}}_{(3,1)} = \underbrace{\begin{pmatrix} A \cdot x + B \cdot y + C \cdot z \\ D \cdot x + E \cdot y + F \cdot z \end{pmatrix}}_{(2,1)}. \tag{B.13}$$

Here $\mathbf{A}$ operates on the column vector $\boldsymbol{x}$ "from the left", that is, $\mathbf{A} \cdot \boldsymbol{x}$ is the *left-sided* matrix-vector product of $\mathbf{A}$ and $\boldsymbol{x}$.

Similarly, a *right-sided* multiplication of a *row* vector $\boldsymbol{x}^\mathsf{T}$ of length m with a matrix of size (m, n) is performed as

$$\underbrace{\boldsymbol{x}^\mathsf{T}}_{(1,m)} \cdot \underbrace{\mathbf{B}}_{(m,n)} = \underbrace{\boldsymbol{z}}_{(1,n)}, \tag{B.14}$$

where the result $\boldsymbol{z}$ is a n-dimensional *row* vector; for example (again with $m = 2$, $n = 3$),

$$\boldsymbol{x}^\mathsf{T} \cdot \mathbf{B} = \underbrace{(x, y)}_{(1,2)} \cdot \underbrace{\begin{pmatrix} A & B & C \\ D & E & F \end{pmatrix}}_{(2,3)} = \underbrace{(x \cdot A + y \cdot D,\ x \cdot B + y \cdot E,\ x \cdot C + y \cdot F)}_{(1,3)}. \tag{B.15}$$

In general, if $\mathbf{A} \cdot \boldsymbol{x}$ is defined, then

$$\mathbf{A} \cdot \boldsymbol{x} = (\boldsymbol{x}^\mathsf{T} \cdot \mathbf{A}^\mathsf{T})^\mathsf{T} \qquad \text{and} \qquad (\mathbf{A} \cdot \boldsymbol{x})^\mathsf{T} = \boldsymbol{x}^\mathsf{T} \cdot \mathbf{A}^\mathsf{T}. \tag{B.16}$$

Thus, any right-sided matrix-vector product $\mathbf{A} \cdot \boldsymbol{x}$ can also be calculated as a left-sided product $\boldsymbol{x}^\mathsf{T} \cdot \mathbf{A}^\mathsf{T}$ by transposing the corresponding matrix $\mathbf{A}$ and vector $\boldsymbol{x}$.

B.3 Vector Products

Products between vectors are a common cause of confusion, mainly because the same symbol $(\cdot)$ is used to denote widely different operators.

B.3.1 Dot (Scalar) Product

The *dot* product (also called *scalar* or *inner* product) of two vectors $\boldsymbol{a} = (a_0, \ldots, a_{n-1})^\mathsf{T}$, $\boldsymbol{b} = (b_0, \ldots, b_{n-1})^\mathsf{T}$ of the same length n is defined as

$$x = \boldsymbol{a} \cdot \boldsymbol{b} = \sum_{i=0}^{n-1} a_i \cdot b_i. \tag{B.17}$$

Thus the result x is a scalar value (hence the name of this product). If we write this as the product of a row and a column vector, as in Eqn. (B.14),

$$\underbrace{x}_{(1,1)} = \underbrace{\boldsymbol{a}^\mathsf{T}}_{(1,n)} \cdot \underbrace{\boldsymbol{b}}_{(n,1)}, \tag{B.18}$$

we conclude that the result x is a matrix of size $(1, 1)$, that is, a single scalar value. The dot product can be viewed as the *projection* of one vector onto the other, with the relation

$$\boldsymbol{a} \cdot \boldsymbol{b} = \|\boldsymbol{a}\| \cdot \|\boldsymbol{b}\| \cdot \cos(\alpha), \tag{B.19}$$

where α is angle enclosed by the vectors $\boldsymbol{a}$ and $\boldsymbol{b}$. As a consequence, the dot product is *zero* if the two vectors are *orthogonal* to each other.

The dot product of a vector with *itself* gives the square of its length (see Eqn. (B.7)), that is,

$$\boldsymbol{a} \cdot \boldsymbol{a} = \sum_{i=0}^{n-1} a_i^2 = \|\boldsymbol{a}\|^2. \tag{B.20}$$

B.3.2 Outer Product

The outer product of two vectors $\boldsymbol{a} = (a_0, \ldots, a_{m-1})^\mathsf{T}$, $\boldsymbol{b} = (b_0, \ldots, b_{n-1})^\mathsf{T}$ of length m and n, respectively, is defined as

$$\mathbf{M} = \boldsymbol{a} \otimes \boldsymbol{b} = \boldsymbol{a} \cdot \boldsymbol{b}^\mathsf{T} = \begin{pmatrix} a_0 b_0 & a_0 b_1 & \ldots & a_0 b_{n-1} \\ a_1 b_0 & a_1 b_1 & \ldots & a_1 b_{n-1} \\ \vdots & \vdots & \ddots & \vdots \\ a_{m-1} b_0 & a_{m-1} b_1 & \ldots & a_{m-1} b_{n-1} \end{pmatrix}. \qquad (B.21)$$

Thus the result is a *matrix* $\mathbf{M}$ with m rows and n columns and elements $M_{ij} = a_i \cdot b_j$, for $i = 0, \ldots, m-1$ and $j = 1, \ldots, n-1$. Note that $\boldsymbol{a} \cdot \boldsymbol{b}^\mathsf{T}$ in Eqn. (B.21) denotes the ordinary (matrix) product of the column vector $\boldsymbol{a}$ (of size $m \times 1$) and the row vector $\boldsymbol{b}^\mathsf{T}$ (of size $1 \times n$), as defined in Eqn. (B.10). The outer product is a special case of the *Kronecker* product ($\otimes$) which generally operates on pairs of matrices.

B.3.3 Cross Product

Although the cross product ($\times$) is generally defined for n-dimensional vectors, it is almost exclusively used in the 3D case, where the result is geometrically easy to understand. For a pair of 3D vectors, $\boldsymbol{a} = (a_0, a_1, a_2)^\mathsf{T}$ and $\boldsymbol{b} = (b_0, b_1, b_2)^\mathsf{T}$, the *cross product* is defined as

$$\boldsymbol{c} = \boldsymbol{a} \times \boldsymbol{b} = \begin{pmatrix} a_0 \\ a_1 \\ a_2 \end{pmatrix} \times \begin{pmatrix} b_0 \\ b_1 \\ b_2 \end{pmatrix} = \begin{pmatrix} a_1 \cdot b_2 - a_2 \cdot b_1 \\ a_2 \cdot b_0 - a_0 \cdot b_2 \\ a_0 \cdot b_1 - a_1 \cdot b_0 \end{pmatrix}. \qquad (B.22)$$

In the 3D case, the *cross product* is another 3D vector that is perpendicular to both of the original vectors.[2] The magnitude (length) of the vector $\boldsymbol{c}$ relates to the angle θ between $\boldsymbol{a}$ and $\boldsymbol{b}$ as

$$\|\boldsymbol{c}\| = \|\boldsymbol{a} \times \boldsymbol{b}\| = \|\boldsymbol{a}\| \cdot \|\boldsymbol{b}\| \cdot \sin(\theta). \qquad (B.23)$$

The quantity $\|\boldsymbol{a} \times \boldsymbol{b}\|$ corresponds to the area of the parallelogram spanned by the vectors $\boldsymbol{a}$ and $\boldsymbol{b}$.

B.4 Eigenvectors and Eigenvalues

This section gives an elementary introduction to eigenvectors and eigenvalues, which are mentioned at several places in the main text (see also [27, 64]). In general, the eigenvalue problem is to find solutions $\boldsymbol{x} \in \mathbb{R}^n$ and $\lambda \in \mathbb{R}$ for the linear equation

$$\mathbf{A} \cdot \boldsymbol{x} = \lambda \cdot \boldsymbol{x}, \qquad (B.24)$$

with the given square matrix $\mathbf{A}$ of size (n, n). Any non-trivial[3] solution $\boldsymbol{x}$ is an *eigenvector* of $\mathbf{A}$ and the scalar λ (which may be

[2] For dimensions greater than three, the definition (and calculation) of the cross product is considerably more involved.

[3] An obvious but trivial solution is $\boldsymbol{x} = \mathbf{0}$ (where $\mathbf{0}$ denotes the zero-vector).

complex-valued) is the associated *eigenvalue*. Eigenvalue and eigenvectors thus always come in pairs $\langle \lambda_j, \boldsymbol{x}_j \rangle$, usually called *eigenpairs*. Geometrically speaking, applying the matrix $\mathbf{A}$ to an eigenvector only changes the vector's *magnitude* or *length* (by the associated eigenvalue λ), but not its orientation in space. Equation (B.24) can be rewritten as

$$\mathbf{A} \cdot \boldsymbol{x} - \lambda \cdot \boldsymbol{x} = \mathbf{0} \qquad \text{or} \qquad \big(\mathbf{A} - \lambda \cdot \mathbf{I}_n\big) \cdot \boldsymbol{x} = \mathbf{0}\,, \tag{B.25}$$

where $\mathbf{I}_n$ is the (n, n) identity matrix. This homogeneous linear equation has non-trivial solutions only if the matrix $(\mathbf{A} - \lambda \cdot \mathbf{I}_n)$ is *singular*, that is, its rank is *less* than n and thus its determinant det() is zero, that is,

$$\det\big(\mathbf{A} - \lambda \cdot \mathbf{I}_n\big) = 0\,. \tag{B.26}$$

Equation (B.26) is called the "characteristic equation" of the matrix $\mathbf{A}$ and can be expanded to a n-th order polynomial in λ. This polynomial has a maximum of n distinct roots, which are the eigenvalues of $\mathbf{A}$ (that is, solutions to Eqn. (B.26)). A matrix of size (n, n) thus has up to n non-distinct eigenvectors $\boldsymbol{x}_1, \boldsymbol{x}_2, \ldots, \boldsymbol{x}_n$, each with an associated eigenvalue $\lambda_1, \lambda_2, \ldots, \lambda_n$.

If they exist, the eigen*values* of a matrix are *unique*, but the associated eigen*vectors* are not! This results from the fact that, if Eqn. (B.24) is satisfied for a vector $\boldsymbol{x}$ (and the associated eigenvalue λ), it also applies to any *scaled* vector $s\boldsymbol{x}$, that is,

$$\mathbf{A} \cdot s\boldsymbol{x} = \lambda \cdot s\boldsymbol{x}\,, \tag{B.27}$$

for arbitrary $s \in \mathbb{R}$ (and $s \neq 0$). Thus, if $\boldsymbol{x}$ is an eigenvector of $\mathbf{A}$, then $s\boldsymbol{x}$ is also an (equivalent) eigenvector.

Note that the eigenvalues of a real-valued matrix may generally be complex. However, (as an important special case) if the matrix $\mathbf{A}$ is *real* and *symmetric*, *all* its eigenvalues are guaranteed to be *real*.

Example

For the real-valued (non-symmetric) 2×2 matrix

$$\mathbf{A} = \begin{pmatrix} 3 & -2 \\ -4 & 1 \end{pmatrix},$$

the two eigenvalues and their associated eigenvectors are

$$\lambda_1 = 5, \quad \boldsymbol{x}_1 = s \cdot \begin{pmatrix} 4 \\ -4 \end{pmatrix}, \qquad \text{and} \qquad \lambda_2 = -1, \quad \boldsymbol{x}_2 = s \cdot \begin{pmatrix} -2 \\ -4 \end{pmatrix},$$

for any nonzero $s \in \mathbb{R}$. The result can be easily verified by inserting pairs $\langle \lambda_1, \boldsymbol{x}_1 \rangle$ and $\langle \lambda_2, \boldsymbol{x}_2 \rangle$, respectively, into Eqn. (B.24).

B.4.1 Calculation of Eigenvalues

Special case: 2×2 matrix

For the special (but frequent) case of $n = 2$, the solution can be found in closed form (and without any software libraries). In this case, the characteristic equation (Eqn. (B.26)) reduces to

Alg. B.1
Calculating the real eigenvalues and eigenvectors for a 2×2 real-valued matrix $\mathbf{A}$. If the matrix has real eigenvalues, an ordered sequence of two "eigenpairs" $\langle \lambda_i, \boldsymbol{x}_i \rangle$, each containing the eigenvalue λ_i and the associated eigenvector $\boldsymbol{x}_i$, is returned ($i = 1, 2$). The resulting sequence is ordered by decreasing eigenvalues. nil is returned if $\mathbf{A}$ has no real eigenvalues.

```
1: RealEigenValues2x2 (A, B, C, D)
   Input: A, B, C, D ∈ ℝ, the elements of a real-valued 2×2 matrix
   A = (A B; C D). Returns an ordered sequence of real-valued
   eigenpairs ⟨λ_i, x_i⟩ for A, or nil if the matrix has no real-valued
   eigenvalues.
2:   R ← (A+D)/2
3:   S ← (A−D)/2
4:   if (S^2 + B·C) < 0 then
5:       return nil                  ▷ A has no real-valued eigenvalues
6:   else
7:       T ← sqrt(S^2 + B·C)
8:       λ_1 ← R + T                              ▷ eigenvalue λ_1
9:       λ_2 ← R − T                              ▷ eigenvalue λ_2
10:      if (A − D) ≥ 0 then
11:          x_1 ← (S + T, C)^T                   ▷ eigenvector x_1
12:          x_2 ← (B, −S − T)^T                  ▷ eigenvector x_2
13:      else
14:          x_1 ← (B, −S + T)^T                  ▷ eigenvector x_1
15:          x_2 ← (S − T, C)^T                   ▷ eigenvector x_2
16:      return (⟨λ_1, x_1⟩, ⟨λ_2, x_2⟩)             ▷ λ_1 ≥ λ_2
```

$$\det(\mathbf{A} - \lambda \cdot \mathbf{I}_2) = \left| \begin{pmatrix} A & B \\ C & D \end{pmatrix} - \lambda \begin{pmatrix} 1 & 0 \\ 0 & 1 \end{pmatrix} \right| = \begin{vmatrix} A-\lambda & B \\ C & D-\lambda \end{vmatrix} \tag{B.28}$$

$$= \lambda^2 - (A+D) \cdot \lambda + (AD - BC) = 0 \,. \tag{B.29}$$

The two possible solutions to this quadratic equation,

$$\begin{aligned} \lambda_{1,2} &= \frac{A+D}{2} \pm \left[\left(\frac{A+D}{2} \right)^2 - (AD - BC) \right]^{1/2} \\ &= \frac{A+D}{2} \pm \left[\left(\frac{A-D}{2} \right)^2 + BC \right]^{1/2} \\ &= R \pm \sqrt{S^2 + BC}, \end{aligned} \tag{B.30}$$

are the eigenvalues of the matrix $\mathbf{A}$, with

$$\begin{aligned} \lambda_1 &= R + \sqrt{S^2 + B \cdot C}, \\ \lambda_2 &= R - \sqrt{S^2 + B \cdot C}. \end{aligned} \tag{B.31}$$

Both λ_1, λ_2 are real-valued if the term under the square root is positive, that is, if

$$S^2 + B \cdot C = \left(\frac{A-D}{2} \right)^2 + B \cdot C \geq 0 \,. \tag{B.32}$$

In particular, if the matrix is *symmetric* (i.e., $B = C$), this condition is guaranteed (because $B \cdot C \geq 0$). In this case, $\lambda_1 \geq \lambda_2$. Algorithm B.1[4] summarizes the closed-form computation of the eigenvalues and eigenvectors of a 2×2 matrix.

[4] See [27] and its reprint in [28, Ch. 5].

General case: $n \times n$

In general, proven numerical software should be used for eigenvalue calculations. See the example using the Apache Commons Math library in Sec. B.6.5.

B.5 Homogeneous Coordinates

Homogeneous coordinates are an alternative representation of points in multi-dimensional space. They are commonly used in 2D and 3D geometry because they can greatly simplify the description of certain transformations. For example, affine and projective transformations become matrices with homogeneous coordinates and the composition of transformations can be performed by simple matrix multiplication.[5]

To convert a given n-dimensional *Cartesian* point $\boldsymbol{x} = (x_0, \ldots, x_{n-1})^\mathsf{T}$ to *homogeneous* coordinates $\underline{\boldsymbol{x}}$, we use the notation[6]

$$\mathrm{hom}(\boldsymbol{x}) = \underline{\boldsymbol{x}}. \tag{B.33}$$

This operation increases the dimensionality of the original vector by one by inserting the additional element 1, that is,

$$\mathrm{hom}\begin{pmatrix} x_0 \\ \vdots \\ x_{n-1} \end{pmatrix} = \begin{pmatrix} x_0 \\ \vdots \\ x_{n-1} \\ 1 \end{pmatrix} = \begin{pmatrix} \underline{x}_0 \\ \vdots \\ \underline{x}_{n-1} \\ \underline{x}_n \end{pmatrix}. \tag{B.34}$$

Note that the homogeneous representation of a Cartesian vector is not unique, but every multiple of the homogeneous vector is an equivalent representation of $\boldsymbol{x}$. Thus any scaled homogeneous vector $\underline{\boldsymbol{x}}' = s \cdot \underline{\boldsymbol{x}}$ (with $s \in \mathbb{R}$, $s \neq 0$) corresponds to the *same* Cartesian vector (see also Eqn. (B.39)).

To convert a given homogeneous point $\underline{\boldsymbol{x}} = (\underline{x}_0, \ldots, \underline{x}_n)^\mathsf{T}$ back to Cartesian coordinates $\boldsymbol{x}$ we simply write

$$\mathrm{hom}^{-1}(\underline{\boldsymbol{x}}) = \boldsymbol{x}. \tag{B.35}$$

This operation can be easily derived as

$$\mathrm{hom}^{-1}\begin{pmatrix} \underline{x}_0 \\ \vdots \\ \underline{x}_{n-1} \\ \underline{x}_n \end{pmatrix} = \frac{1}{\underline{x}_n} \cdot \begin{pmatrix} \underline{x}_0 \\ \vdots \\ \underline{x}_{n-1} \end{pmatrix} = \begin{pmatrix} x_0 \\ \vdots \\ x_{n-1} \end{pmatrix}, \tag{B.36}$$

provided that $\underline{x}_n \neq 0$. Two homogeneous points $\underline{\boldsymbol{x}}_1, \underline{\boldsymbol{x}}_2$ are considered *equivalent* ($\equiv$), if they represent the same Cartesian point, that is,

$$\underline{\boldsymbol{x}}_1 \equiv \underline{\boldsymbol{x}}_2 \;\Leftrightarrow\; \mathrm{hom}^{-1}(\underline{\boldsymbol{x}}_1) = \mathrm{hom}^{-1}(\underline{\boldsymbol{x}}_2). \tag{B.37}$$

It follows from Eqn. (B.36) that

[5] See Chapter 21, Sec. 21.1.2.

[6] The operator hom( ) is introduced here for convenience and clarity.

$$\text{hom}^{-1}(\underline{\boldsymbol{x}}) = \text{hom}^{-1}(s \cdot \underline{\boldsymbol{x}}) \tag{B.38}$$

for any nonzero factor $s \in \mathbb{R}$. Thus, as mentioned earlier, any scaled homogeneous point corresponds to the same Cartesian point, that is,

$$\underline{\boldsymbol{x}} \equiv s \cdot \underline{\boldsymbol{x}}. \tag{B.39}$$

For example, for the Cartesian point $\boldsymbol{x} = (3, 7, 2)^\mathsf{T}$, the homogeneous coordinates

$$\text{hom}(\boldsymbol{x}) = \begin{pmatrix} 3 \\ 7 \\ 2 \\ 1 \end{pmatrix} \equiv \begin{pmatrix} -3 \\ -7 \\ -2 \\ -1 \end{pmatrix} \equiv \begin{pmatrix} 9 \\ 31 \\ 6 \\ 3 \end{pmatrix} \equiv \begin{pmatrix} 30 \\ 70 \\ 20 \\ 10 \end{pmatrix} \cdots \tag{B.40}$$

are all equivalent. Homogeneous coordinates can be used for vector spaces of arbitrary dimension, including 2D coordinates.

B.6 Basic Matrix-Vector Operations with the *Apache Commons Math* Library

It is recommended to use proven standard software, such as the *Apache Commons Math*[7] (ACM) library, for any non-trivial linear algebra calculation.

B.6.1 Vectors and Matrices

The basic data structures for representing vectors and matrices are `RealVector` and `RealMatrix`, respectively. The following ACM examples show the conversion from and to simple Java arrays of element-type `double`:

```
import org.apache.commons.math3.linear.MatrixUtils;
import org.apache.commons.math3.linear.RealMatrix;
import org.apache.commons.math3.linear.RealVector;

// Data given as simple arrays:
double[]   xa = {1, 2, 3};
double[][] Aa = {{2, 0, 1}, {0, 2, 0}, {1, 0, 2}};

// Conversion to vectors and matrices:
RealVector x = MatrixUtils.createRealVector(xa);
RealMatrix A = MatrixUtils.createRealMatrix(Aa);

// Get a single matrix element A_{i,j}:
int i, j; // specify row (i) and column (j)
double aij = A.getEntry(i, j);

// Set a single matrix element to a new value:
double value;
A.setEntry(i, j, value);

// Extract data to arrays again:
double[]   xb = x.toArray();
double[][] Ab = A.getData();
```

[7] http://commons.apache.org/math/.

```
// Transpose the matrix A:
RealMatrix At = A.transpose();
```

B.6.2 Matrix-Vector Multiplication

The following examples show how to implement the various matrix-vector products described in Sec. B.2.3.

```
RealMatrix A = ...; // matrix A of size (m,n)
RealMatrix B = ...; // matrix B of size (p,q), with p = n
RealVector x = ...; // vector x of length n

// Scalar multiplication C ← s · A:
double s = ...;
RealMatrix C = A.scalarMultiply(s);

// Product of two matrices: C ← A · B:
RealMatrix C = A.multiply(B); // C is of size (m,q)

// Left-sided matrix-vector product: y ← A · x:
RealVector y = A.operate(x);

// Right-sided matrix-vector product: y ← x^T · A:
RealVector y = A.preMultiply(x);
```

B.6.3 Vector Products

The following code segments show the use of the ACM library for calculating various vector products described in Sec. B.3.

```
RealVector a, b;  // vectors a, b (both of length n)

// Multiplication by a scalar c ← s · a:
double s;
RealVector c = a.mapMultiply(s);

// Dot (scalar) product x ← a · b:
double x = a.dotProduct(b);

// Outer product M ← a ⊗ b:
RealMatrix M = a.outerProduct(b);
```

B.6.4 Inverse of a Square Matrix

The following example shows the inversion of a square matrix:

```
RealMatrix A  = ... ;    // a square matrix
RealMatrix Ai = MatrixUtils.inverse(A);
```

B.6.5 Eigenvalues and Eigenvectors

The following code segment illustrates the calculation of eigenvalues and eigenvalues of a square matrix `A` using the class `EigenDecomposition` of the Apache Commons Math API. Note that the eigenvalues returned by getRealEigenvalues() are sorted in non-increasing order. The same ordering applies to the associated eigenvectors.


```
import org.apache.commons.math3.linear.EigenDecomposition;
...

RealMatrix A = MatrixUtils.createRealMatrix(new double[][]
    {{2, 0, 1},
     {0, 2, 0},
     {1, 0, 2}});

EigenDecomposition ed = new EigenDecomposition(A);

if (ed.hasComplexEigenvalues()) {
  System.out.println("A has complex Eigenvalues!");
}
else {
  // get all real eigenvalues:
  double[] lambda = ed.getRealEigenvalues();  // = (3, 2, 1)
  // get the associated eigenvectors:
  for (int i = 0; i < lambda.length; i++) {
    RealVector x = ed.getEigenvector(i);
    ...
  }
}
```

B.7 Solving Systems of Linear Equations

This section describes standard methods for solving systems of linear equations. Such systems appear widely and frequently in all sorts of engineering problems. Identifying them and knowing about standard solution methods is thus quite important and may save much time in any development process. In addition, the solution techniques presented here are very mature and numerically stable. Note that this section is supposed to give only a brief summary of the topic and practical implementations using the Apache Commons Math library. Further details and the underlying theory can be found in most linear algebra textbooks (e.g., [145, 190]).

Systems of linear equations generally come in the form

$$\begin{pmatrix} A_{0,0} & A_{0,1} & \cdots & A_{0,n-1} \\ A_{1,0} & A_{1,1} & \cdots & A_{1,n-1} \\ A_{2,0} & A_{2,1} & \cdots & A_{2,n-1} \\ \vdots & \vdots & \ddots & \vdots \\ A_{m-1,0} & A_{m-1,1} & \cdots & A_{m-1,n-1} \end{pmatrix} \cdot \begin{pmatrix} x_0 \\ x_1 \\ \vdots \\ x_{n-1} \end{pmatrix} = \begin{pmatrix} b_0 \\ b_1 \\ b_2 \\ \vdots \\ b_{m-1} \end{pmatrix}, \tag{B.41}$$

or, in the standard notation,

$$\mathbf{A} \cdot \boldsymbol{x} = \boldsymbol{b}, \tag{B.42}$$

where the (known) matrix $\mathbf{A}$ is of size (m, n), the *unknown* vector $\boldsymbol{x}$ is n-dimensional, and the (known) vector $\boldsymbol{b}$ is m-dimensional. Thus n corresponds to the number of unknowns and m to the number of equations. Each row i of the matrix $\mathbf{A}$ thus represents a single equation

$$A_{i,0} \cdot x_0 + A_{i,1} \cdot x_1 + \ldots + A_{i,n-1} \cdot x_{n-1} = b_i \tag{B.43}$$

$$\text{or} \quad \sum_{j=0}^{n-1} A_{i,j} \cdot x_j = b_i \,, \tag{B.44}$$

for $i = 0, \ldots, m-1$. Depending on m and n, the following situations may occur:

- If $m = n$ (i.e., $\mathbf{A}$ is square) the number of unknowns matches the number of equations and the system typically (but not always, of course) has a unique solution (see Sec. B.7.1 below).
- If $m < n$, we have more unknowns than equations. In this case no unique solution exists (but possibly infinitely many).
- With $m > n$ the system is said to be *over-determined* and thus not solvable in general. Nevertheless, this is a frequent case that is typically handled by calculating a minimum least squares solution (see Sec. B.7.2).

B.7.1 Exact Solutions

If the number of equations (m) is equal to the number of unknowns (n) and the resulting (square) matrix $\mathbf{A}$ is non-singular and of full rank $m = n$, the system $\mathbf{A} \cdot \boldsymbol{x} = \boldsymbol{b}$ can be expected to have a unique solution for $\boldsymbol{x}$. For example, the system[8]

$$\begin{aligned} 2 \cdot x_0 + 3 \cdot x_1 - 2 \cdot x_2 &= 1, \\ -x_0 + 7 \cdot x_1 + 6 \cdot x_2 &= -2, \\ 4 \cdot x_0 - 3 \cdot x_1 - 5 \cdot x_2 &= 1, \end{aligned} \tag{B.45}$$

with

$$\mathbf{A} = \begin{pmatrix} 2 & 3 & -2 \\ -1 & 7 & 6 \\ 4 & -3 & -5 \end{pmatrix}, \quad \boldsymbol{x} = \begin{pmatrix} x_0 \\ x_1 \\ x_2 \end{pmatrix}, \quad \boldsymbol{b} = \begin{pmatrix} 1 \\ -2 \\ 1 \end{pmatrix}, \tag{B.46}$$

has the unique solution $\boldsymbol{x} = (-0.3698, 0.1780, -0.6027)^{\mathsf{T}}$. The following code segment shows how the previous example is solved using class `LUDecomposition` of the ACM library:

```
import org.apache...linear.DecompositionSolver;
import org.apache...linear.LUDecomposition;

RealMatrix A = MatrixUtils.createRealMatrix(new double[][]
        {{ 2,  3, -2},
         {-1,  7,  6},
         { 4, -3, -5}});
RealVector b = MatrixUtils.createRealVector(new double[]
         {1, -2, 1});
DecompositionSolver solver =
              new LUDecomposition(A).getSolver();
RealVector x = solver.solve(b);
```

An exception is thrown if the matrix `A` is non-square or singular.

[8] Example taken from the *Apache Commons Math User Guide* [4].

B.7.2 Over-Determined System (Least-Squares Solutions)

If a system of linear equations has more equations than unknowns (i.e., $m > n$) it is over-determined and thus has no exact solution. In other words, there is no vector $\boldsymbol{x}$ that satisfies $\mathbf{A} \cdot \boldsymbol{x} = \boldsymbol{b}$ or

$$\mathbf{A} \cdot \boldsymbol{x} - \boldsymbol{b} = \mathbf{0}. \tag{B.47}$$

Instead, *any* $\boldsymbol{x}$ plugged into Eqn. (B.47) yields some non-zero "residual" vector $\boldsymbol{\epsilon}$, such that

$$\mathbf{A} \cdot \boldsymbol{x} - \boldsymbol{b} = \boldsymbol{\epsilon}. \tag{B.48}$$

A "best" solution is commonly found by minimizing the squared norm of this residual, that is, by searching for $\boldsymbol{x}$ such that

$$\|\mathbf{A} \cdot \boldsymbol{x} - \boldsymbol{b}\|^2 = \|\boldsymbol{\epsilon}\|^2 \rightarrow \min. \tag{B.49}$$

Several matrix decompositions can be used for calculating the "least-squares solution" of an over-determined system of linear equations. As a simple example, we add a fourth line ($m = 4$) to the system in Eqns. (B.45) and (B.46) to

$$\mathbf{A} = \begin{pmatrix} 2 & 3 & -2 \\ -1 & 7 & 6 \\ 4 & -3 & -5 \\ 2 & -2 & -1 \end{pmatrix}, \quad \boldsymbol{x} = \begin{pmatrix} x_0 \\ x_1 \\ x_2 \end{pmatrix}, \quad \boldsymbol{b} = \begin{pmatrix} 1 \\ -2 \\ 1 \\ 0 \end{pmatrix}, \tag{B.50}$$

without changing the number of unknowns ($n = 3$). The least-squares solution to this over-determined system is (approx.) $\boldsymbol{x} = (-0.2339, 0.1157, -0.4942)^\mathsf{T}$. The following code segment shows the calculation using the `SingularValueDecomposition` class of the ACM library:

```
import org.apache...linear.DecompositionSolver;
import org.apache...linear.SingularValueDecomposition;

RealMatrix A = MatrixUtils.createRealMatrix(new double[][]
        {{ 2,  3, -2},
         {-1,  7,  6},
         { 4, -3, -5},
         { 2, -2, -1}});
RealVector b = MatrixUtils.createRealVector(new double[]
         {1, -2, 1, 0});
DecompositionSolver solver =
          new SingularValueDecomposition(A).getSolver();
RealVector x = solver.solve(b);
```

Alternatively, an instance of `QRDecomposition` could be used for calculating the least-squares solution. If an *exact* solution exists (see Sec. B.7.1), it is the same as the least-squares solution (with zero residual $\boldsymbol{\epsilon} = \mathbf{0}$).

Appendix C

Calculus

This part outlines selected topics from calculus that may serve as a useful supplement to Chapters 6, 16, 17, 24, and 25, in particular.

C.1 Parabolic Fitting

Given a single-variable (1D), discrete function $g\colon \mathbb{Z} \mapsto \mathbb{R}$, it is sometimes useful to locally fit a quadratic (parabolic) function, for example, for precisely locating a maximum or minimum position.

C.1.1 Fitting a Parabolic Function to Three Sample Points

For a quadratic function (second-order polynomial)

$$y = f(x) = a \cdot x^2 + b \cdot x + c \tag{C.1}$$

with parameters a, b, c to pass through a given set of three sample points $\boldsymbol{p}_i = (x_i, y_i)$, $i = 1, 2, 3$, means that the following three equations must be satisfied:

$$\begin{aligned} y_1 &= a \cdot x_1^2 + b \cdot x_1 + c, \\ y_2 &= a \cdot x_2^2 + b \cdot x_2 + c, \\ y_3 &= a \cdot x_3^2 + b \cdot x_3 + c. \end{aligned} \tag{C.2}$$

Written in the standard matrix form $\boldsymbol{A} \cdot \boldsymbol{x} = \boldsymbol{b}$, or

$$\begin{pmatrix} x_1^2 & x_1 & 1 \\ x_2^2 & x_2 & 1 \\ x_3^2 & x_3 & 1 \end{pmatrix} \cdot \begin{pmatrix} a \\ b \\ c \end{pmatrix} = \begin{pmatrix} y_1 \\ y_2 \\ y_3 \end{pmatrix}, \tag{C.3}$$

the unknown coefficient vector $\boldsymbol{x} = (a, b, c)^\mathsf{T}$ is directly found as

$$\boldsymbol{x} = \boldsymbol{A}^{-1} \cdot \boldsymbol{b} = \begin{pmatrix} x_1^2 & x_1 & 1 \\ x_2^2 & x_2 & 1 \\ x_3^2 & x_3 & 1 \end{pmatrix}^{-1} \cdot \begin{pmatrix} y_1 \\ y_2 \\ y_3 \end{pmatrix}, \tag{C.4}$$

assuming that the matrix $\boldsymbol{A}$ has a non-zero determinant. Geometrically this means that the points $\boldsymbol{p}_i$ must not be *collinear*.

Example:

Fitting the sample points $\boldsymbol{p}_1 = (-2, 5)^\mathsf{T}$, $\boldsymbol{p}_2 = (-1, 6)^\mathsf{T}$, $\boldsymbol{p}_3 = (3, -10)^\mathsf{T}$ to a quadratic function, the equation to solve is (analogous to Eqn. (C.3))

$$\begin{pmatrix} 4 & -2 & 1 \\ 1 & -1 & 1 \\ 9 & 3 & 1 \end{pmatrix} \cdot \begin{pmatrix} a \\ b \\ c \end{pmatrix} = \begin{pmatrix} 5 \\ 6 \\ -10 \end{pmatrix}, \tag{C.5}$$

with the solution

$$\begin{pmatrix} a \\ b \\ c \end{pmatrix} = \begin{pmatrix} 4 & -2 & 1 \\ 1 & -1 & 1 \\ 9 & 3 & 1 \end{pmatrix}^{-1} \cdot \begin{pmatrix} 5 \\ 6 \\ -10 \end{pmatrix} = \frac{1}{20} \cdot \begin{pmatrix} 4 & -5 & 1 \\ -8 & 5 & 3 \\ -12 & 30 & 2 \end{pmatrix} \cdot \begin{pmatrix} 5 \\ 6 \\ -10 \end{pmatrix} = \begin{pmatrix} -1 \\ -2 \\ 5 \end{pmatrix}.$$

Thus $a = -1$, $b = -2$, $c = 5$, and the equation of the quadratic fitting function is $y = -x^2 - 2x + 5$. The result for this example is shown graphically in Fig. C.1.

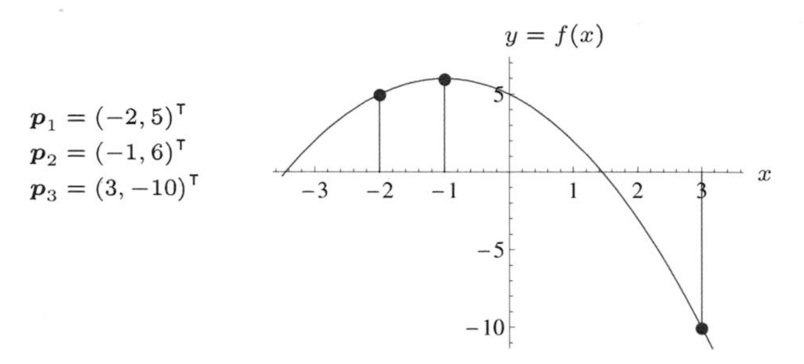

Fig. C.1
Fitting a quadratic function to three arbitrary sample points.

C.1.2 Locating Extrema by Quadratic Interpolation

A special situation is when the given points are positioned at $x_1 = -1$, $x_2 = 0$, and $x_3 = +1$. This is useful, for example, to estimate a continuous extremum position from successive discrete function values defined on a regular lattice. Again the objective is to fit a quadratic function (as in Eqn. (C.1)) to pass through the points $\boldsymbol{p}_1 = (-1, y_1)^\mathsf{T}$, $\boldsymbol{p}_2 = (0, y_2)^\mathsf{T}$, and $\boldsymbol{p}_3 = (1, y_3)^\mathsf{T}$. In this case, the simultaneous equations in Eqn. (C.2) simplify to

$$\begin{aligned} y_1 &= a - b + c, \\ y_2 &= \phantom{a - b +{}} c, \\ y_3 &= a + b + c, \end{aligned} \tag{C.6}$$

with the solution

$$a = \frac{y_1 - 2 \cdot y_2 + y_3}{2}, \qquad b = \frac{y_3 - y_1}{2}, \qquad c = y_2. \tag{C.7}$$

To estimate a local extremum position, we take the first derivative of the quadratic fitting function (Eqn. (C.1)), which is the linear function $f'(x) = 2a \cdot x + b$, and find the position $\breve{x}$ of its (single) root by solving

$$2a \cdot x + b = 0. \tag{C.8}$$

With a, b taken from Eqn. (C.7), the extremal position is thus found as

$$\breve{x} = \frac{-b}{2a} = \frac{y_1 - y_3}{2 \cdot (y_1 - 2y_2 + y_3)} \,. \tag{C.9}$$

The corresponding extremal *value* can then be found by evaluating the quadratic function $f()$ at position $\breve{x}$, that is,

$$\breve{y} = f(\breve{x}) = a \cdot \breve{x}^2 + b \cdot \breve{x} + c, \tag{C.10}$$

with a, b, c as defined in Eqn. (C.7). Figure C.2 shows an example with sample points $\boldsymbol{p}_1 = (-1, -2)^\mathsf{T}$, $\boldsymbol{p}_2 = (0, 7)^\mathsf{T}$, $\boldsymbol{p}_3 = (1, 6)^\mathsf{T}$. In this case, the interpolated maximum position is at $\breve{x} = 0.4$ and the corresponding maximum value is $f(\breve{x}) = 7.8$.

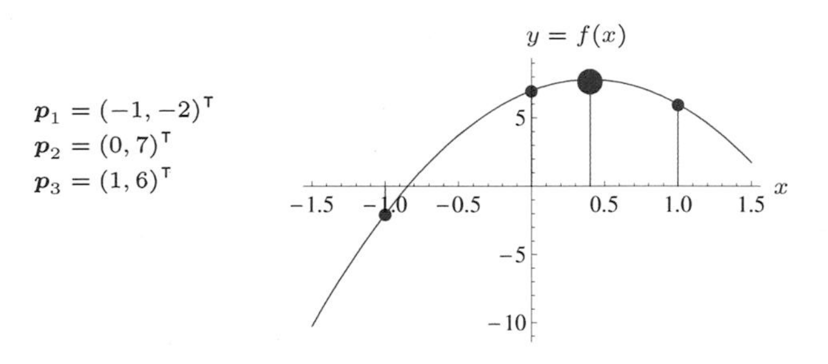

Fig. C.2
Fitting a quadratic function to three reference points at positions $x_1 = -1, x_2 = 0, x_3 = +1$. The interpolated, continuous curve has a maximum at the continuous position $\breve{x} = 0.4$ (large circle).

Using the above scheme, we can interpolate any triplet of successive sample values centered around some position $u \in \mathbb{Z}$, that is, $\boldsymbol{p}_1 = (u-1, y_1)^\mathsf{T}$, $\boldsymbol{p}_2 = (u, y_2)^\mathsf{T}$, $\boldsymbol{p}_3 = (u+1, y_3)^\mathsf{T}$, with arbitrary values y_1, y_2, y_3. In this case the estimated position of the extremum is simply (from Eqn. (C.9))

$$\breve{x} = u + \frac{y_1 - y_3}{2 \cdot (y_1 - 2 \cdot y_2 + y_3)} \,. \tag{C.11}$$

The application of quadratic interpolation to multi-variable functions is described in Sec. C.3.3.

C.2 Scalar and Vector Fields

An RGB color image $\boldsymbol{I}(u, v) = (I_R(u, v), I_G(u, v), I_B(u, v))$ can be considered a 2D function whose values are 3D vectors. Mathematically, this is a special case of a vector-valued function $\boldsymbol{f}\colon \mathbb{R}^n \mapsto \mathbb{R}^m$,

$$\boldsymbol{f}(\boldsymbol{x}) = \boldsymbol{f}(x_0, \ldots, x_{n-1}) = \begin{pmatrix} f_0(\boldsymbol{x}) \\ \vdots \\ f_{m-1}(\boldsymbol{x}) \end{pmatrix}, \tag{C.12}$$

which is composed of m scalar-valued functions $f_i\colon \mathbb{R}^n \mapsto \mathbb{R}$, each being defined on the domain of n-dimensional vectors.

A multi-variable, scalar-valued function $f\colon \mathbb{R}^n \mapsto \mathbb{R}$ is called a *scalar field*, while a vector-valued function $f\colon \mathbb{R}^n \mapsto \mathbb{R}^m$ is referred to as a *vector field*.

C.2.1 The Jacobian Matrix

Assuming that the function $\boldsymbol{f}(\boldsymbol{x}) = (f_0(\boldsymbol{x}), \ldots, f_{m-1}(\boldsymbol{x}))^\mathsf{T}$ is differentiable, the so-called *functional* or *Jacobian* matrix at a specific point $\dot{\boldsymbol{x}} = (\dot{x}_0, \ldots, \dot{x}_{n-1})$ is defined as

$$\mathbf{J}_{\boldsymbol{f}}(\dot{\boldsymbol{x}}) = \begin{pmatrix} \frac{\partial}{\partial x_0} f_0(\dot{\boldsymbol{x}}) & \cdots & \frac{\partial}{\partial x_{n-1}} f_0(\dot{\boldsymbol{x}}) \\ \vdots & \ddots & \vdots \\ \frac{\partial}{\partial x_0} f_{m-1}(\dot{\boldsymbol{x}}) & \cdots & \frac{\partial}{\partial x_{n-1}} f_{m-1}(\dot{\boldsymbol{x}}) \end{pmatrix}. \tag{C.13}$$

The Jacobian matrix is of size $m \times n$ and composed of the first derivatives of the m component functions $f_0, \ldots, f_{m-1}$ with respect to each of the n independent variables $x_0, \ldots, x_{n-1}$. Thus each of its elements $\frac{\partial}{\partial x_j} f_i(\dot{\boldsymbol{x}})$ quantifies how much the value of the scalar-valued component function $f_i(\boldsymbol{x}) = f_i(x_0, \ldots, x_{n-1})$ changes when only variable x_j is varied and all other variables remain fixed. Note that the matrix $\mathbf{J}_{\boldsymbol{f}}(\boldsymbol{x})$ is not constant for a given function $\boldsymbol{f}$ but is different at each position $\dot{\boldsymbol{x}}$. In general, the Jacobian matrix is neither square (unless $m = n$) nor symmetric.

C.2.2 Gradients

Gradient of a scalar field

The gradient of a *scalar field* $f \colon \mathbb{R}^n \mapsto \mathbb{R}$, with $f(\boldsymbol{x}) = f(x_0, \ldots, x_{n-1})$, at a given position $\dot{\boldsymbol{x}} \in \mathbb{R}^n$ is defined as

$$(\nabla f)(\dot{\boldsymbol{x}}) = (\operatorname{grad} f)(\dot{\boldsymbol{x}}) = \begin{pmatrix} \frac{\partial}{\partial x_0} f(\dot{\boldsymbol{x}}) \\ \vdots \\ \frac{\partial}{\partial x_{n-1}} f(\dot{\boldsymbol{x}}) \end{pmatrix}. \tag{C.14}$$

The resulting vector-valued function quantifies the amount of output change with respect to changing any of the input variables $x_0, \ldots, x_{n-1}$ at position $\dot{\boldsymbol{x}}$. Thus the gradient of a scalar field is a vector field.

The *directional* gradient of a scalar field describes how the (scalar) function value changes when the coordinates are modified along a particular direction, specified by the unit vector $\mathbf{e}$. We denote the directional gradient as $\nabla_{\mathbf{e}} f$ and define

$$(\nabla_{\mathbf{e}} f)(\dot{\boldsymbol{x}}) = (\nabla f)(\dot{\boldsymbol{x}}) \cdot \mathbf{e}\,, \tag{C.15}$$

where $\cdot$ is the scalar product (see Sec. B.3.1). The result is a scalar value that can be interpreted as the slope of the tangent on the n-dimensional surface of the scalar field at position $\dot{\boldsymbol{x}}$ along the direction specified by the n-dimensional unit vector $\mathbf{e} = (e_0, \ldots, e_{n-1})^\mathsf{T}$.

Gradient of a vector field

To calculate the gradient of a *vector field* $\boldsymbol{f} \colon \mathbb{R}^n \mapsto \mathbb{R}^m$, we note that each row i in the $m \times n$ Jacobian matrix $\mathbf{J}_{\boldsymbol{f}}$ (Eqn. (C.13)) is the transposed gradient vector of the corresponding component function f_i, that is,

$$\mathbf{J}_{\boldsymbol{f}}(\dot{\boldsymbol{x}}) = \begin{pmatrix} (\nabla f_0)(\dot{\boldsymbol{x}})^\mathsf{T} \\ \vdots \\ (\nabla f_{m-1})(\dot{\boldsymbol{x}})^\mathsf{T} \end{pmatrix}, \tag{C.16}$$

and thus the Jacobian matrix is equivalent to the gradient of the vector field $\boldsymbol{f}$,

$$(\mathrm{grad}\,\boldsymbol{f})(\dot{\boldsymbol{x}}) \equiv \mathbf{J}_{\boldsymbol{f}}(\dot{\boldsymbol{x}}). \tag{C.17}$$

Analogous to Eqn. (C.15), the *directional* gradient of the vector field is then defined as

$$(\mathrm{grad}_{\mathbf{e}}\,\boldsymbol{f})(\dot{\boldsymbol{x}}) \equiv \mathbf{J}_{\boldsymbol{f}}(\dot{\boldsymbol{x}}) \cdot \mathbf{e}, \tag{C.18}$$

where $\mathbf{e}$ is again a unit vector specifying the gradient direction and $\cdot$ is the ordinary matrix-vector product. In this case the resulting gradient is a m-dimensional vector with one element for each component function in $\boldsymbol{f}$.

C.2.3 Maximum Gradient Direction

In case of a scalar field $f(\boldsymbol{x})$, a resulting non-zero gradient vector $(\nabla f)(\dot{\boldsymbol{x}})$ (Eqn. (C.14)) is also the direction of the steepest ascent of $f(\boldsymbol{x})$ at position $\dot{\boldsymbol{x}}$.[1] In this case, the L_2 norm (see Sec. B.1.2) of the gradient vector, that is, $\|(\nabla f)(\dot{\boldsymbol{x}})\|$, corresponds to the maximum slope of f at point $\dot{\boldsymbol{x}}$.

In case of a vector field $\boldsymbol{f}(\boldsymbol{x})$, the direction of maximum slope cannot be obtained directly, since the gradient is not a n-dimensional vector but its $m \times n$ Jacobian matrix. In this case, the direction of maximum change in the function $\boldsymbol{f}$ is found as the eigenvector $\boldsymbol{x}_k$ of the square $(n \times n)$ matrix

$$\mathbf{M} = \mathbf{J}_{\boldsymbol{f}}^\mathsf{T}(\dot{\boldsymbol{x}}) \cdot \mathbf{J}_{\boldsymbol{f}}(\dot{\boldsymbol{x}}) \tag{C.19}$$

that corresponds to its largest eigenvalue λ_k (see also Sec. B.4).

C.2.4 Divergence of a Vector Field

If the vector field maps to the same vector space (i.e., $\boldsymbol{f}\colon \mathbb{R}^n \mapsto \mathbb{R}^n$), its *divergence* (div) is defined as

$$(\mathrm{div}\,\boldsymbol{f})(\dot{\boldsymbol{x}}) = \tfrac{\partial}{\partial x_0} f_0(\dot{\boldsymbol{x}}) + \cdots + \tfrac{\partial}{\partial x_{n-1}} f_{n-1}(\dot{\boldsymbol{x}}) \tag{C.20}$$

$$= \sum_{i=0}^{n-1} \tfrac{\partial}{\partial x_i} f_i(\dot{\boldsymbol{x}}) \in \mathbb{R}, \tag{C.21}$$

for a given point $\dot{\boldsymbol{x}}$. The result is a scalar value and thus $(\mathrm{div}\,\boldsymbol{f})(\dot{\boldsymbol{x}})$ yields a scalar field $\mathbb{R}^n \mapsto \mathbb{R}$. Note that, in this case, the Jacobian matrix $\mathbf{J}_{\boldsymbol{f}}$ in Eqn. (C.13) is square (of size $n \times n$) and $\mathrm{div}\,\boldsymbol{f}$ is equivalent to the trace of $\mathbf{J}_{\boldsymbol{f}}$, that is,

$$(\mathrm{div}\,\boldsymbol{f})(\dot{\boldsymbol{x}}) \equiv \mathrm{trace}(\mathbf{J}_{\boldsymbol{f}}(\dot{\boldsymbol{x}})). \tag{C.22}$$

[1] If the gradient vector is *zero*, that is, if $(\nabla f)(\dot{\boldsymbol{x}}) = \mathbf{0}$, the direction of the gradient is undefined at position $\dot{\boldsymbol{x}}$.

C.2.5 Laplacian Operator

The *Laplacian* (or Laplace operator) of a scalar field $f\colon \mathbb{R}^n \mapsto \mathbb{R}$ is a linear differential operator, commonly denoted Δ or ∇^2. The result of applying ∇^2 to the scalar field $f\colon \mathbb{R}^n \mapsto \mathbb{R}$ generates another scalar field that consists of the sum of all unmixed second-order partial derivatives of f (if existent), that is,

$$(\nabla^2 f)(\dot{\boldsymbol{x}}) = \tfrac{\partial^2}{\partial x_0^2} f(\dot{\boldsymbol{x}}) + \cdots + \tfrac{\partial^2}{\partial x_{n-1}^2} f(\dot{\boldsymbol{x}}) = \sum_{i=0}^{n-1} \tfrac{\partial^2}{\partial x_i^2} f(\dot{\boldsymbol{x}}). \tag{C.23}$$

The result is a scalar value that is equivalent to the *divergence* (see Eqn. (C.21)) of the *gradient* (see Eqn. (C.14)) of the scalar field f, that is,

$$(\nabla^2 f)(\dot{\boldsymbol{x}}) = (\mathrm{div}\nabla f)s(\dot{\boldsymbol{x}}). \tag{C.24}$$

The *Laplacian* is also found as the *trace* of the function's Hessian matrix $\mathbf{H}_f$ (see Sec. C.2.6).

For a *vector*-valued function $\boldsymbol{f}\colon \mathbb{R}^n \mapsto \mathbb{R}^m$, the Laplacian at point $\dot{\boldsymbol{x}}$ is again a vector field $\mathbb{R}^n \mapsto \mathbb{R}^m$,

$$(\nabla^2 \boldsymbol{f})(\dot{\boldsymbol{x}}) = \begin{pmatrix} (\nabla^2 f_0)(\dot{\boldsymbol{x}}) \\ (\nabla^2 f_2)(\dot{\boldsymbol{x}}) \\ \vdots \\ (\nabla^2 f_{m-1})(\dot{\boldsymbol{x}}) \end{pmatrix} \in \mathbb{R}^m, \tag{C.25}$$

that is obtained by applying the Laplacian to the individual (scalar-valued) component functions.

C.2.6 The Hessian Matrix

The Hessian matrix of a n-variable, real-valued function $f\colon \mathbb{R}^n \mapsto \mathbb{R}$ is the $n \times n$ square matrix composed of its second-order partial derivatives (assuming they all exist), that is,

$$\mathbf{H}_f = \begin{pmatrix} H_{0,0} & H_{0,1} & \cdots & H_{0,n-1} \\ H_{1,0} & H_{1,1} & \cdots & H_{1,n-1} \\ \vdots & \vdots & \ddots & \vdots \\ H_{n-1,0} & H_{n-1,1} & \cdots & H_{n-1,n-1} \end{pmatrix} \tag{C.26}$$

$$= \begin{pmatrix} \frac{\partial^2}{\partial x_0^2} f & \frac{\partial^2}{\partial x_0\, \partial x_1} f & \cdots & \frac{\partial^2}{\partial x_0\, \partial x_{n-1}} f \\ \frac{\partial^2}{\partial x_1\, \partial x_0} f & \frac{\partial^2}{\partial x_1^2} f & \cdots & \frac{\partial^2}{\partial x_1\, \partial x_{n-1}} f \\ \vdots & \vdots & \ddots & \vdots \\ \frac{\partial^2}{\partial x_{n-1}\, \partial x_0} f & \frac{\partial^2}{\partial x_{n-1}\, \partial x_1} f & \cdots & \frac{\partial^2}{\partial x_{n-1}^2} f \end{pmatrix}. \tag{C.27}$$

Since the order of differentiation does not matter (i.e., $H_{i,j} = H_{j,i}$), $\mathbf{H}_f$ is symmetric. Note that the Hessian is a matrix of *functions*. To evaluate the Hessian at a particular point $\dot{\boldsymbol{x}} \in \mathbb{R}^n$, we write

$$\mathbf{H}_f(\dot{\boldsymbol{x}}) = \begin{pmatrix} \frac{\partial^2}{\partial x_0^2} f(\dot{\boldsymbol{x}}) & \cdots & \frac{\partial^2}{\partial x_0\, \partial x_{n-1}} f(\dot{\boldsymbol{x}}) \\ \vdots & \ddots & \vdots \\ \frac{\partial^2}{\partial x_{n-1}\, \partial x_0} f(\dot{\boldsymbol{x}}) & \cdots & \frac{\partial^2}{\partial x_{n-1}^2} f(\dot{\boldsymbol{x}}) \end{pmatrix}, \qquad \text{(C.28)}$$

which is a scalar-valued matrix of size $n \times n$. As mentioned already, the *trace* of the Hessian matrix is the *Laplacian* ∇^2 of the function f, that is,

$$\nabla^2 f = \operatorname{trace}(\mathbf{H}_f) = \sum_{i=0}^{n-1} \frac{\partial^2}{\partial x_i^2} f\,. \qquad \text{(C.29)}$$

Example

Given a 2D, continuous, grayscale image or scalar-valued intensity function $I(x, y)$, the corresponding Hessian matrix (of size 2×2) contains all second derivatives along the coordinates x, y, that is,

$$\mathbf{H}_I = \begin{pmatrix} \frac{\partial^2}{\partial x^2} I & \frac{\partial^2}{\partial x \partial y} I \\ \frac{\partial^2}{\partial y \partial x} I & \frac{\partial^2}{\partial y^2} I \end{pmatrix} = \begin{pmatrix} I_{xx} & I_{xy} \\ I_{yx} & I_{yy} \end{pmatrix}, \qquad \text{(C.30)}$$

The elements of $\mathbf{H}_I$ are 2D, scalar-valued functions over x, y and thus scalar fields again. Evaluating the Hessian matrix at a particular point $\dot{\boldsymbol{x}}$ yields the values of the second partial derivatives of I at this position,

$$\mathbf{H}_I(\dot{\boldsymbol{x}}) = \begin{pmatrix} \frac{\partial^2}{\partial x^2} I(\dot{\boldsymbol{x}}) & \frac{\partial^2}{\partial x \partial y} I(\dot{\boldsymbol{x}}) \\ \frac{\partial^2}{\partial y \partial x} I(\dot{\boldsymbol{x}}) & \frac{\partial^2}{\partial y^2} I(\dot{\boldsymbol{x}}) \end{pmatrix} = \begin{pmatrix} I_{xx}(\dot{\boldsymbol{x}}) & I_{xy}(\dot{\boldsymbol{x}}) \\ I_{yx}(\dot{\boldsymbol{x}}) & I_{yy}(\dot{\boldsymbol{x}}) \end{pmatrix}, \qquad \text{(C.31)}$$

that is, a matrix with scalar-valued elements.

C.3 Operations on Multi-Variable, Scalar Functions (Scalar Fields)

C.3.1 Estimating the Derivatives of a Discrete Function

Images are typically discrete functions (i.e., $I\colon \mathbb{N}^2 \mapsto \mathbb{R}$) and thus not differentiable. The derivatives can nevertheless be estimated by calculating finite differences from the pixel values in a 3×3 neighborhood, which can be expressed as a linear filter or convolution operation ($*$). In particular, the *first*-order derivatives $I_x = \partial I/\partial x$ and $I_y = \partial I/\partial y$ are usually estimated in the form

$$I_x \approx I * \begin{bmatrix} -0.5 & \mathbf{0} & 0.5 \end{bmatrix}, \qquad I_y \approx I * \begin{bmatrix} -0.5 \\ \mathbf{0} \\ 0.5 \end{bmatrix}, \qquad \text{(C.32)}$$

the second-order derivatives $I_{xx} = \partial^2 I/\partial x^2$ and $I_{yy} = \partial^2 I/\partial y^2$ as

$$I_{xx} \approx I * \begin{bmatrix} 1 & \mathbf{-2} & 1 \end{bmatrix}, \qquad I_{yy} \approx I * \begin{bmatrix} 1 \\ \mathbf{-2} \\ 1 \end{bmatrix}, \qquad \text{(C.33)}$$

and the mixed derivative

$$\frac{\partial^2 I}{\partial x \partial y} = I_{xy} = I_{yx}$$
$$\approx I * \begin{bmatrix} -0.5 & \mathbf{0} & 0.5 \end{bmatrix} * \begin{bmatrix} -0.5 \\ \mathbf{0} \\ 0.5 \end{bmatrix} = I * \begin{bmatrix} 0.25 & 0 & -0.25 \\ 0 & \mathbf{0} & 0 \\ -0.25 & 0 & 0.25 \end{bmatrix}. \quad \text{(C.34)}$$

C.3.2 Taylor Series Expansion of Functions

Single-variable functions

The Taylor series expansion (of degree d) of a single-variable function $f \colon \mathbb{R} \mapsto \mathbb{R}$ about a reference point a is

$$f(x) = f(a) + f'(a) \cdot (x-a) + f''(a) \cdot \frac{(x-a)^2}{2} + \cdots$$
$$\cdots + f^{(d)}(a) \cdot \frac{(x-a)^d}{d!} + R_d \quad \text{(C.35)}$$
$$= f(a) + \sum_{i=1}^{d} f^{(i)}(a) \cdot \frac{(x-a)^i}{i!} + R_d \quad \text{(C.36)}$$
$$= \sum_{i=0}^{d} f^{(i)}(a) \cdot \frac{(x-a)^i}{i!} + R_d, \quad \text{(C.37)}$$

where R_d is the residual term.[2] This means that if the value $f(a)$ and the first d derivatives $f'(a), f''(a), \ldots, f^{(d)}(a)$ exist and are known at some position a, the value of f at *another* point $\dot{x}$ can be estimated (up to the residual R_d) only from the values at point a, without actually evaluating $f(x)$. Omitting the remainder R_d, the result is an *approximation* for $f(\dot{x})$, that is,

$$f(x) \approx \sum_{i=0}^{d} f^{(i)}(a) \cdot \frac{(x-a)^i}{i!}, \quad \text{(C.38)}$$

whose accuracy depends upon d and the distance $x - a$.

Multi-variable functions

In general, for a real-valued function of n variables,

$$f(\boldsymbol{x}) = f(x_0, x_2, \ldots, x_{n-1}) \in \mathbb{R},$$

the full Taylor series expansion about a reference point $\boldsymbol{a} = (a_0, \ldots, a_{n-1})^\mathsf{T}$ is

$$f(x_0, \ldots, x_{n-1}) = f(\boldsymbol{a}) + \quad \text{(C.39)}$$
$$\sum_{i_0=1}^{\infty} \cdots \sum_{i_{n-1}=1}^{\infty} \Big[\frac{\partial^{i_0}}{\partial x_0^{i_0}} \cdots \frac{\partial^{i_{n-1}}}{\partial x_{n-1}^{i_{n-1}}} \Big] f(\boldsymbol{a}) \cdot \frac{(x_0 - a_0)^{i_0} \cdots (x_{n-1} - a_{n-1})^{i_{n-1}}}{i_1! \cdots i_n!}$$
$$= \sum_{i_1=0}^{\infty} \cdots \sum_{i_n=0}^{\infty} \Big[\frac{\partial^{i_0}}{\partial x_0^{i_0}} \cdots \frac{\partial^{i_{n-1}}}{\partial x_{n-1}^{i_{n-1}}} \Big] f(\boldsymbol{a}) \cdot \frac{(x_0 - a_0)^{i_0} \cdots (x_{n-1} - a_{n-1})^{i_{n-1}}}{i_0! \cdots i_{n-1}!}.$$

[2] Note that $f^{(0)} = f$, $f^{(1)} = f'$, $f^{(2)} = f''$ etc., and $1! = 1$.

In Eqn. (C.39),[3] the term

$$\left[\frac{\partial^{i_0}}{\partial x_0^{i_0}} \cdots \frac{\partial^{i_{n-1}}}{\partial x_{n-1}^{i_{n-1}}}\right] f(\boldsymbol{a}) \tag{C.40}$$

is the value of the function f, after applying a sequence of n partial derivatives, at the n-dimensional position $\boldsymbol{a}$. The operator $\frac{\partial^i}{\partial x_k^i}$ denotes the i-th partial derivative on the variable x_k.

To formulate Eqn. (C.39) in a more compact fashion, we define the index vector

$$\boldsymbol{i} = (i_0, i_1, \ldots, i_{n-1}), \tag{C.41}$$

(with $i_k \in \mathbb{N}_0$ and thus $\boldsymbol{i} \in \mathbb{N}_0^n$), and the associated operations

$$\begin{aligned} \boldsymbol{i}! &= i_0! \cdot i_1! \cdot \ldots \cdot i_{n-1}!, \\ \boldsymbol{x}^{\boldsymbol{i}} &= x_1^{i_0} \cdot x_2^{i_1} \cdot \ldots \cdot x_{n-1}^{i_{n-1}}, \\ \Sigma\boldsymbol{i} &= i_0 + i_1 + \ldots + i_{n-1}. \end{aligned} \tag{C.42}$$

As a shorthand notation for the combined partial derivative operator in Eqn. (C.40) we define

$$\mathrm{D}^{\boldsymbol{i}} := \frac{\partial^{i_0}}{\partial x_0^{i_0}} \frac{\partial^{i_1}}{\partial x_1^{i_1}} \cdots \frac{\partial^{i_{n-1}}}{\partial x_{n-1}^{i_{n-1}}} = \frac{\partial^{i_0+i_1+\ldots+i_{n-1}}}{\partial x_0^{i_0} \partial x_1^{i_1} \cdots \partial x_{n-1}^{i_{n-1}}}. \tag{C.43}$$

With these definitions, the full Taylor expansion of a multi-variable function about a point $\boldsymbol{a}$, as given in Eqn. (C.39), can be elegantly written in the form

$$f(\boldsymbol{x}) = \sum_{\boldsymbol{i} \in \mathbb{N}_0^n} \mathrm{D}^{\boldsymbol{i}} f(\boldsymbol{a}) \cdot \frac{(\boldsymbol{x}-\boldsymbol{a})^{\boldsymbol{i}}}{\boldsymbol{i}!}. \tag{C.44}$$

Note that $\mathrm{D}^{\boldsymbol{i}} f$ is again a n-dimensional function $\mathbb{R}^n \mapsto \mathbb{R}$, and thus $[\mathrm{D}^{\boldsymbol{i}} f](\boldsymbol{a})$ in Eqn. (C.44) is the scalar quantity obtained by evaluating the function $[\mathrm{D}^{\boldsymbol{i}} f]$ at the n-dimensional point $\boldsymbol{a}$.

To obtain a Taylor *approximation* of order d, the sum of the indices $i_1, \ldots, i_n$ is limited to d, that is, the summation is constrained to index vectors $\boldsymbol{i}$, with $\Sigma\boldsymbol{i} \leq d$. The resulting formulation,

$$f(\boldsymbol{x}) \approx \sum_{\substack{\boldsymbol{i} \in \mathbb{N}_0^n \\ \Sigma\boldsymbol{i} \leq d}} \mathrm{D}^{\boldsymbol{i}} f(\boldsymbol{a}) \cdot \frac{(\boldsymbol{x}-\boldsymbol{a})^{\boldsymbol{i}}}{\boldsymbol{i}!}, \tag{C.45}$$

is obviously analogous to the 1D case in Eqn. (C.38).

Example: two-variable (2D) function

This example demonstrates the second-order ($d = 2$) Taylor expansion of a 2D ($n = 2$) function $f\colon \mathbb{R}^2 \mapsto \mathbb{R}$ around a point $\boldsymbol{a} = (x_a, y_a)$. By inserting into Eqn. (C.44), we get

[3] Note that symbols $x_0, \ldots, x_{n-1}$ denote the individual variables, while $\dot{x}_0, \ldots, \dot{x}_{n-1}$ are the coordinates of a specific point in n-dimensional space.

$$f(x,y) \approx \sum_{\substack{\boldsymbol{i} \in \mathbb{N}_0^2 \\ \Sigma \boldsymbol{i} \leq 2}} \mathrm{D}^{\boldsymbol{i}} f(x_a, y_a) \cdot \frac{1}{\boldsymbol{i}!} \cdot \begin{pmatrix} x - x_a \\ y - y_a \end{pmatrix}^{\boldsymbol{i}} \tag{C.46}$$

$$= \sum_{\substack{0 \leq i,j \leq 2 \\ (i+j) \leq 2}} \frac{\partial^{i+j}}{\partial x^i \, \partial y^j} f(x_a, y_a) \cdot \frac{(x - x_a)^i \cdot (y - y_a)^j}{i! \cdot j!}. \tag{C.47}$$

Since $d = 2$, the six permissible index vectors $\boldsymbol{i} = (i, j)$, with $\Sigma \boldsymbol{i} \leq 2$, are $(0,0)$, $(1,0)$, $(0,1)$, $(1,1)$, $(2,0)$, and $(0,2)$. Inserting into Eqn. (C.47), we obtain the corresponding Taylor approximationat position $(\dot{x}, \dot{y})$ as

$$\begin{aligned} f(x,y) \approx\; & \tfrac{\partial^0}{\partial x^0 \, \partial y^0} f(x_a, y_a) \cdot \frac{(x - x_a)^0 \cdot (y - y_a)^0}{1 \cdot 1} \\ & + \tfrac{\partial^1}{\partial x^1 \, \partial y^0} f(x_a, y_a) \cdot \frac{(x - x_a)^1 \cdot (y - y_a)^0}{1 \cdot 1} \\ & + \tfrac{\partial^1}{\partial x^0 \, \partial y^1} f(x_a, y_a) \cdot \frac{(x - x_a)^0 \cdot (y - y_a)^1}{1 \cdot 1} \\ & + \tfrac{\partial^2}{\partial x^1 \, \partial y^1} f(x_a, y_a) \cdot \frac{(x - x_a)^1 \cdot (y - y_a)^1}{1 \cdot 1} \\ & + \tfrac{\partial^2}{\partial x^2 \, \partial y^0} f(x_a, y_a) \cdot \frac{(x - x_a)^2 \cdot (y - y_a)^0}{2 \cdot 1} \\ & + \tfrac{\partial^2}{\partial x^0 \, \partial y^2} f(x_a, y_a) \cdot \frac{(x - x_a)^0 \cdot (y - y_a)^2}{1 \cdot 2} \end{aligned} \tag{C.48}$$

$$\begin{aligned} =\; & f(x_a, y_a) \\ & + \tfrac{\partial}{\partial x} f(x_a, y_a) \cdot (x - x_a) + \tfrac{\partial}{\partial y} f(x_a, y_a) \cdot (y - y_a) \\ & + \tfrac{\partial^2}{\partial x \, \partial y} f(x_a, y_a) \cdot (x - x_a) \cdot (y - y_a) \\ & + \frac{1}{2} \cdot \tfrac{\partial^2}{\partial x^2} f(x_a, y_a) \cdot (x - x_a)^2 + \frac{1}{2} \cdot \tfrac{\partial^2}{\partial y^2} f(x_a, y_a) \cdot (y - y_a)^2. \end{aligned} \tag{C.49}$$

It is assumed that the required derivatives of f exist, that is, f is differentiable at point (x_a, y_a) with respect to x and y up to the second order. By slightly rearranging Eqn. (C.49) to

$$\begin{aligned} f(x,y) \approx\; & f(x_a, y_a) + \tfrac{\partial}{\partial x} f(x_a, y_a) \cdot (x - x_a) + \tfrac{\partial}{\partial y} f(x_a, y_a) \cdot (y - y_a) \\ & + \frac{1}{2} \cdot \Big[\tfrac{\partial^2}{\partial x^2} f(x_a, y_a) \cdot (x - x_a)^2 + 2 \cdot \tfrac{\partial^2}{\partial x \, \partial y} f(x_a, y_a) \cdot (x - x_a) \cdot (y - y_a) \\ & \quad + \tfrac{\partial^2}{\partial y^2} f(x_a, y_a) \cdot (y - y_a)^2 \Big] \end{aligned} \tag{C.50}$$

we can now write the Taylor expansion in matrix-vector notation as

$$\begin{aligned} f(x,y) \approx \tilde{f}(x,y) =\; & f(x_a, y_a) + \Big(\tfrac{\partial}{\partial x} f(x_a, y_a), \tfrac{\partial}{\partial y} f(x_a, y_a) \Big) \cdot \begin{pmatrix} x - x_a \\ y - y_a \end{pmatrix} \\ & + \frac{1}{2} \cdot \left[(x - x_a, y - y_a) \cdot \begin{pmatrix} \tfrac{\partial^2}{\partial x^2} f(x_a, y_a) & \tfrac{\partial^2}{\partial x \, \partial y} f(x_a, y_a) \\ \tfrac{\partial^2}{\partial x \, \partial y} f(x_a, y_a) & \tfrac{\partial^2}{\partial y^2} f(x_a, y_a) \end{pmatrix} \cdot \begin{pmatrix} x - x_a \\ y - y_a \end{pmatrix} \right] \end{aligned} \tag{C.51}$$

or, even more compactly, in the form

$$\tilde{f}(\boldsymbol{x}) = f(\boldsymbol{a}) + \nabla_f^{\mathsf{T}}(\boldsymbol{a}) \cdot (\boldsymbol{x} - \boldsymbol{a}) + \tfrac{1}{2} \cdot (\boldsymbol{x} - \boldsymbol{a})^{\mathsf{T}} \cdot \mathbf{H}_f(\boldsymbol{a}) \cdot (\boldsymbol{x} - \boldsymbol{a}). \tag{C.52}$$

Here $\nabla_f^{\mathsf{T}}(\boldsymbol{a})$ denotes the (transposed) *gradient* vector of the function f at point $\boldsymbol{a}$ (see Sec. C.2.2), and $\mathbf{H}_f$ is the 2×2 *Hessian* matrix of f (see Sec. C.2.6),

$$\mathbf{H}_f(\boldsymbol{a}) = \begin{pmatrix} H_{00} & H_{01} \\ H_{10} & H_{11} \end{pmatrix} = \begin{pmatrix} \frac{\partial^2}{\partial x^2} f(\boldsymbol{a}) & \frac{\partial^2}{\partial x\, \partial y} f(\boldsymbol{a}) \\ \frac{\partial^2}{\partial x\, \partial y} f(\boldsymbol{a}) & \frac{\partial^2}{\partial y^2} f(\boldsymbol{a}) \end{pmatrix}. \tag{C.53}$$

If the function f is *discrete*, for example, a scalar-valued image I, the required partial derivatives at some lattice point $\boldsymbol{a} = (u_a, v_a)^{\mathsf{T}}$ can be estimated from its 3×3 neighborhood, as described in Sec. C.3.1.

Example: three-variable (3D) function

For a 3D function $f\colon \mathbb{R}^3 \mapsto \mathbb{R}$, the second-order Taylor expansion $(d = 2)$ is analogous to Eqns. (C.51–C.52) for the 2D case, except that now the positions $\boldsymbol{x} = (x, y, z)^{\mathsf{T}}$ and $\boldsymbol{a} = (x_a, y_a, z_a)^{\mathsf{T}}$ are 3D vectors. The associated (transposed) gradient vector is

$$\nabla_f^{\mathsf{T}}(\boldsymbol{a}) = \left(\tfrac{\partial}{\partial x} f(\boldsymbol{a}), \tfrac{\partial}{\partial y} f(\boldsymbol{a}), \tfrac{\partial}{\partial z} f(\boldsymbol{a})\right), \tag{C.54}$$

and the Hessian, composed of all second-order partial derivatives, is the 3×3 matrix

$$\mathbf{H}_f(\boldsymbol{a}) = \begin{pmatrix} \frac{\partial^2}{\partial x^2} f(\boldsymbol{a}) & \frac{\partial^2}{\partial x \partial y} f(\boldsymbol{a}) & \frac{\partial^2}{\partial x \partial z} f(\boldsymbol{a}) \\ \frac{\partial^2}{\partial y \partial x} f(\boldsymbol{a}) & \frac{\partial^2}{\partial y^2} f(\boldsymbol{a}) & \frac{\partial^2}{\partial y \partial z} f(\boldsymbol{a}) \\ \frac{\partial^2}{\partial z \partial x} f(\boldsymbol{a}) & \frac{\partial^2}{\partial z \partial y} f(\boldsymbol{a}) & \frac{\partial^2}{\partial z^2} f(\boldsymbol{a}) \end{pmatrix}. \tag{C.55}$$

Note that the order of differentiation is not relevant since, for example, $\frac{\partial^2}{\partial x\, \partial y} = \frac{\partial^2}{\partial y\, \partial x}$, and therefore $\mathbf{H}_f$ is always symmetric.

This can be easily generalized to the n-dimensional case, though things become considerably more involved for Taylor expansions of higher orders $(d > 2)$.

C.3.3 Finding the Continuous Extremum of a Multi-Variable Discrete Function

In Sec. C.1.2 we described how the position of a local extremum can be determined by fitting a quadratic function to the neighboring samples of a *1D* function. This section shows how this technique can be extended to n-dimensional, scalar-valued functions $f : \mathbb{R}^n \mapsto \mathbb{R}$.

Without loss of generality we can assume that the Taylor expansion of the function $f(\boldsymbol{x})$ is carried out around the point $\boldsymbol{a} = \mathbf{0} = (0, \ldots, 0)$, which clearly simplifies the remaining formulation. The Taylor approximation function (see Eqn. (C.52)) for this point can be written as

$$\tilde{f}(\boldsymbol{x}) = f(\mathbf{0}) + \nabla_f^{\mathsf{T}}(\mathbf{0}) \cdot \boldsymbol{x} + \tfrac{1}{2} \cdot \boldsymbol{x}^{\mathsf{T}} \cdot \mathbf{H}_f(\mathbf{0}) \cdot \boldsymbol{x}, \tag{C.56}$$

with the gradient ∇_f and the Hessian matrix $\mathbf{H}_f$ evaluated at position $\mathbf{0}$. The vector of the first derivative of this function is

$$\tilde{f}'(\boldsymbol{x}) = \nabla_f(\mathbf{0}) + \tfrac{1}{2} \cdot \left[(\boldsymbol{x}^{\mathsf{T}} \cdot \mathbf{H}_f(\mathbf{0}))^{\mathsf{T}} + \mathbf{H}_f(\mathbf{0}) \cdot \boldsymbol{x}\right]. \tag{C.57}$$

Since $(\boldsymbol{x}^\mathsf{T} \cdot \mathbf{H}_f)^\mathsf{T} = (\mathbf{H}_f^\mathsf{T} \cdot \boldsymbol{x})$ and because the Hessian matrix $\mathbf{H}_f$ is symmetric (i.e., $\mathbf{H}_f = \mathbf{H}_f^\mathsf{T}$), this simplifies to

$$\tilde{f}'(\boldsymbol{x}) = \nabla_f(\mathbf{0}) + \tfrac{1}{2}\cdot(\mathbf{H}_f(\mathbf{0}) \cdot \boldsymbol{x} + \mathbf{H}_f(\mathbf{0}) \cdot \boldsymbol{x}) \tag{C.58}$$

$$= \nabla_f(\mathbf{0}) + \mathbf{H}_f(\mathbf{0}) \cdot \boldsymbol{x}. \tag{C.59}$$

A local maximum or minimum is found where all first derivatives $\tilde{f}'$ are zero, so we need to solve

$$\nabla_f(\mathbf{0}) + \mathbf{H}_f(\mathbf{0})\cdot\breve{\boldsymbol{x}} = \mathbf{0}, \tag{C.60}$$

for the unknown position $\breve{\boldsymbol{x}}$. By multiplying both sides with $\mathbf{H}_f^{-1}$ (assuming that the inverse of $\mathbf{H}_f(\mathbf{0})$ exists), the solution is

$$\breve{\boldsymbol{x}} = -\mathbf{H}_f^{-1}(\mathbf{0}) \cdot \nabla_f(\mathbf{0}), \tag{C.61}$$

for the specific expansion point $\boldsymbol{a} = \mathbf{0}$ (Eqn. (C.63)). Analogously, for an arbitrary expansion point $\boldsymbol{a}$, the extremum position is

$$\breve{\boldsymbol{x}} = \boldsymbol{a} - \mathbf{H}_f^{-1}(\boldsymbol{a}) \cdot \nabla_f(\boldsymbol{a}). \tag{C.62}$$

Note that the inverse Hessian matrix $\mathbf{H}_f^{-1}$ is again symmetric.

The estimated extremal *value* of the approximation function $\tilde{f}$ is found by replacing $\boldsymbol{x}$ in Eqn. (C.56) with the extremal position $\breve{\boldsymbol{x}}$ (calculated in Eqn. (C.61)) as

$$\begin{aligned} \tilde{f}_{\text{extrm}} &= \tilde{f}(\breve{\boldsymbol{x}}) = f(\mathbf{0}) + \nabla_f^\mathsf{T}(\mathbf{0}) \cdot \breve{\boldsymbol{x}} + \tfrac{1}{2} \cdot \breve{\boldsymbol{x}}^\mathsf{T} \cdot \mathbf{H}_f(\mathbf{0}) \cdot \breve{\boldsymbol{x}} \\ &= f(\mathbf{0}) + \nabla_f^\mathsf{T}(\mathbf{0}) \cdot \breve{\boldsymbol{x}} + \tfrac{1}{2} \cdot \breve{\boldsymbol{x}}^\mathsf{T} \cdot \mathbf{H}_f(\mathbf{0}) \cdot (-\mathbf{H}_f^{-1}(\mathbf{0})) \cdot \nabla_f(\mathbf{0}) \\ &= f(\mathbf{0}) + \nabla_f^\mathsf{T}(\mathbf{0}) \cdot \breve{\boldsymbol{x}} - \tfrac{1}{2} \cdot \breve{\boldsymbol{x}}^\mathsf{T} \cdot \mathbf{I} \cdot \nabla_f(\mathbf{0}) \\ &= f(\mathbf{0}) + \nabla_f^\mathsf{T}(\mathbf{0}) \cdot \breve{\boldsymbol{x}} - \tfrac{1}{2} \cdot \nabla_f^\mathsf{T}(\mathbf{0}) \cdot \breve{\boldsymbol{x}} \\ &= f(\mathbf{0}) + \tfrac{1}{2} \cdot \nabla_f^\mathsf{T}(\mathbf{0}) \cdot \breve{\boldsymbol{x}}, \end{aligned} \tag{C.63}$$

again for the expansion point $\boldsymbol{a} = \mathbf{0}$.

$$\tilde{f}_{\text{extrm}} = \tilde{f}(\breve{\boldsymbol{x}}) = f(\boldsymbol{a}) + \tfrac{1}{2} \cdot \nabla_f^\mathsf{T}(\boldsymbol{a}) \cdot (\breve{\boldsymbol{x}} - \boldsymbol{a})\,. \tag{C.64}$$

Note that $\tilde{f}_{\text{extrm}}$ may be a local minimum or maximum, but could also be a *saddle point* where the first derivatives of the function are zero as well.

Local extrema in 2D

The aforementioned scheme can be applied to n-dimensional functions. In the special case of a 2D function $f\colon \mathbb{R}^2 \mapsto \mathbb{R}$ (e.g., a 2D image), the gradient vector and the Hessian matrix for the given expansion point $\boldsymbol{a} = (x_a, y_a)^\mathsf{T}$ can be noted as

$$\nabla_f(\boldsymbol{a}) = \begin{pmatrix} d_x \\ d_y \end{pmatrix} \quad \text{and} \quad \mathbf{H}_f(\boldsymbol{a}) = \begin{pmatrix} H_{00} & H_{01} \\ H_{01} & H_{11} \end{pmatrix}, \tag{C.65}$$

for a given expansion point $\boldsymbol{a} = (x_a, y_a)^\mathsf{T}$. In this case, the inverse of the Hessian matrix is

$$\mathbf{H}_f^{-1} = \frac{1}{H_{01}^2 - H_{00} \cdot H_{11}} \cdot \begin{pmatrix} -H_{11} & H_{01} \\ H_{01} & -H_{00} \end{pmatrix} \tag{C.66}$$

and the resulting *position* of the extremal point is (see Eqn. (C.62))

$$\breve{\boldsymbol{x}} = \begin{pmatrix} x_a \\ y_a \end{pmatrix} - \frac{1}{H_{01}^2 - H_{00} \cdot H_{11}} \cdot \begin{pmatrix} -H_{11} & H_{01} \\ H_{01} & -H_{00} \end{pmatrix} \cdot \begin{pmatrix} d_x \\ d_y \end{pmatrix} \tag{C.67}$$

$$= \begin{pmatrix} x_a \\ y_a \end{pmatrix} - \frac{1}{H_{01}^2 - H_{00} \cdot H_{11}} \cdot \begin{pmatrix} H_{01} \cdot d_y - H_{11} \cdot d_x \\ H_{01} \cdot d_x - H_{00} \cdot d_y \end{pmatrix}. \tag{C.68}$$

The extremal position is only defined if the denominator in Eqn. (C.68), $H_{01}^2 - H_{00} \cdot H_{11}$ (equivalent to the determinant of $\mathbf{H}_f$), is non-zero, indicating that the Hessian matrix $\mathbf{H}_f$ is non-singular and thus has an inverse. The associated *value* of $\tilde{f}$ at the estimated extremal position $\breve{\boldsymbol{x}} = (\breve{x}, \breve{y})^\mathsf{T}$ can be now calculated using Eqn. (C.64) as

$$\tilde{f}(\breve{x}, \breve{y}) = f(x_a, y_a) + \tfrac{1}{2} \cdot (d_x, d_y) \cdot \begin{pmatrix} \breve{x} - x_a \\ \breve{y} - y_a \end{pmatrix}$$

$$= f(x_a, y_a) + \frac{d_x \cdot (\breve{x} - x_a) + d_y \cdot (\breve{y} - y_a)}{2}. \tag{C.69}$$

Numeric 2D example

The following example shows how a local extremum can be found in a discrete 2D image with sub-pixel accuracy using a second-order Taylor approximation. Assume we are given a grayscale image $I\colon \mathbb{Z} \times \mathbb{Z} \mapsto \mathbb{R}$ with the sample values

	u_a-1	u_a	u_a+1
v_a-1	8	11	7
v_a	15	16	9
v_a+1	14	12	10

(C.70)

in the 3×3 neighborhood of position $\boldsymbol{a} = (u_a, v_a)^\mathsf{T}$. Obviously, the discrete center value $f(\boldsymbol{a}) = 16$ is a local maximum but (as we shall see) the maximum of the *continuous* approximation function is *not* at the center. The gradient vector ∇_I and the Hessian Matrix $\mathbf{H}_I$ at the expansion point $\boldsymbol{a}$ are calculated from local finite differences (see Sec. C.3.1) as

$$\nabla_I(\boldsymbol{a}) = \begin{pmatrix} d_x \\ d_y \end{pmatrix} = 0.5 \cdot \begin{pmatrix} 9-15 \\ 12-11 \end{pmatrix} = \begin{pmatrix} -3 \\ 0.5 \end{pmatrix} \quad \text{and} \tag{C.71}$$

$$\mathbf{H}_I(\boldsymbol{a}) = \begin{pmatrix} H_{11} & H_{12} \\ H_{12} & H_{22} \end{pmatrix} = \begin{pmatrix} 9-2\cdot16+15 & 0.25\cdot(8-14-7+10) \\ 0.25\cdot(8-14-7+10) & 11-2\cdot16+12 \end{pmatrix}$$

$$= \begin{pmatrix} -8.00 & -0.75 \\ -0.75 & -9.00 \end{pmatrix}, \tag{C.72}$$

respectively. The resulting second-order Taylor expansion about the point $\boldsymbol{a}$ is the continuous function (see Eqn. (C.52))

$$\tilde{f}(\boldsymbol{x}) = f(\boldsymbol{a}) + \nabla_I^\mathsf{T}(\boldsymbol{a}) \cdot (\boldsymbol{x}-\boldsymbol{a}) + \tfrac{1}{2} \cdot (\boldsymbol{x}-\boldsymbol{a})^\mathsf{T} \cdot \mathbf{H}_I(\boldsymbol{a}) \cdot (\boldsymbol{x}-\boldsymbol{a})$$

$$= 16 + (-3, 0.5) \cdot \begin{pmatrix} x-u_a \\ y-v_a \end{pmatrix} \tag{C.73}$$

$$+ \tfrac{1}{2} \cdot (x-u_a, y-v_a) \cdot \begin{pmatrix} -8.00 & -0.75 \\ -0.75 & -9.00 \end{pmatrix} \cdot \begin{pmatrix} x-u_a \\ y-v_a \end{pmatrix}.$$

We use the inverse of the 2×2 Hessian matrix at position $\boldsymbol{a}$ (see Eqn. (C.66)),

$$\mathbf{H}_I^{-1}(\boldsymbol{a}) = \begin{pmatrix} -8.00 & -0.75 \\ -0.75 & -9.00 \end{pmatrix}^{-1} = \begin{pmatrix} -0.125984 & 0.010499 \\ 0.010499 & -0.111986 \end{pmatrix}, \tag{C.74}$$

to calculate the *position* of the local extremum $\breve{\boldsymbol{x}}$ (see Eqn. (C.68)) as

$$\begin{aligned} \breve{\boldsymbol{x}} &= \boldsymbol{a} - \mathbf{H}_I^{-1}(\boldsymbol{a}) \cdot \nabla_I(\boldsymbol{a}) \\ &= \begin{pmatrix} u_a \\ v_a \end{pmatrix} - \begin{pmatrix} -0.125984 & 0.010499 \\ 0.010499 & -0.111986 \end{pmatrix} \cdot \begin{pmatrix} -3 \\ 0.5 \end{pmatrix} = \begin{pmatrix} u_a - 0.3832 \\ v_a + 0.0875 \end{pmatrix}. \end{aligned} \tag{C.75}$$

Finally, the extremal *value* (see Eqn. (C.64)) is found as

$$\begin{aligned} \tilde{f}(\breve{\boldsymbol{x}}) &= f(\boldsymbol{a}) + \tfrac{1}{2} \cdot \nabla_f^{\mathsf{T}}(\boldsymbol{a}) \cdot (\breve{\boldsymbol{x}} - \boldsymbol{a}) \\ &= 16 + \tfrac{1}{2} \cdot (-3, 0.5) \cdot \begin{pmatrix} u_a - 0.3832 - u_a \\ v_a + 0.0875 - v_a \end{pmatrix} \\ &= 16 + \tfrac{1}{2} \cdot (3 \cdot 0.3832 + 0.5 \cdot 0.0875) = 16.5967\,. \end{aligned} \tag{C.76}$$

Figure (C.3) illustrates the aforementioned example, with the expansion point set to $\boldsymbol{a} = (u_a, v_a)^{\mathsf{T}} = (0, 0)^{\mathsf{T}}$.

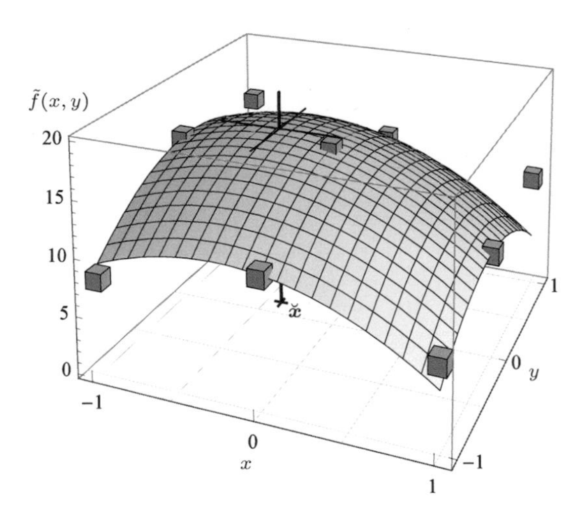

Fig. C.3
Continuous Taylor approximation of a discrete 2D image function for determining the local extremum position with sub-pixel accuracy. The cubes represent the discrete image samples in a 3 × 3 neighborhood around the reference coordinate (0, 0), which is a local maximum of the discrete image function (see Eqn. (C.70) for the concrete values). The parabolic surface shows the continuous approximation $\tilde{f}(x, y)$ obtained by second-order Taylor expansion about the center position $\boldsymbol{a} = (0, 0)$. The vertical line marks the position of the local maximum $\tilde{f}(\breve{\boldsymbol{x}}) = 16.5967$ at $\breve{\boldsymbol{x}} = (-0.3832, 0.0875)$.

Local extrema in 3D

In the case of a three-variable, scalar function $f : \mathbb{R}^3 \mapsto \mathbb{R}$, with a given expansion point $\boldsymbol{a} = (x_a, y_a, z_a)^{\mathsf{T}}$ and

$$\nabla_f(\boldsymbol{a}) = \begin{pmatrix} d_x \\ d_y \\ d_z \end{pmatrix} \quad \text{and} \quad \mathbf{H}_f(\boldsymbol{a}) = \begin{pmatrix} H_{00} & H_{01} & H_{02} \\ H_{01} & H_{11} & H_{12} \\ H_{02} & H_{12} & H_{22} \end{pmatrix} \tag{C.77}$$

being the gradient vector and the Hessian matrix of f at point $\boldsymbol{a}$, respectively, the estimated extremal *position* is

$$\breve{\boldsymbol{x}} = (\breve{x}, \breve{y}, \breve{z})^{\mathsf{T}} = \boldsymbol{a} - \mathbf{H}_f^{-1}(\boldsymbol{a}) \cdot \nabla_f(\boldsymbol{a}) \tag{C.78}$$

$$= \begin{pmatrix} x_a \\ y_a \\ z_a \end{pmatrix} - \frac{1}{H_{02}^2 \cdot H_{11} + H_{01}^2 \cdot H_{22} + H_{00} \cdot H_{12}^2 - H_{00} \cdot H_{11} \cdot H_{22} - 2 \cdot H_{01} \cdot H_{02} \cdot H_{12}}$$

$$\cdot \begin{pmatrix} H_{12}^2 - H_{11} \cdot H_{22} & H_{01} \cdot H_{22} - H_{02} H_{12} & H_{02} \cdot H_{11} - H_{01} \cdot H_{12} \\ H_{01} \cdot H_{22} - H_{02} \cdot H_{12} & H_{02}^2 - H_{00} \cdot H_{22} & H_{00} \cdot H_{12} - H_{01} \cdot H_{02} \\ H_{02} \cdot H_{11} - H_{01} \cdot H_{12} & H_{00} \cdot H_{12} - H_{01} \cdot H_{02} & H_{01}^2 - H_{00} \cdot H_{11} \end{pmatrix} \cdot \begin{pmatrix} d_x \\ d_y \\ d_z \end{pmatrix}.$$

Note that the inverse of the 3×3 Hessian matrix $\mathbf{H}_f^{-1}$ is again symmetric and can be calculated in closed form (as shown in Eqn. (C.78)).[4]

Again using Eqn. (C.64), the estimated extremal *value* at position $\breve{\boldsymbol{x}} = (\breve{x}, \breve{y}, \breve{z})^{\mathsf{T}}$ is found as

$$\tilde{f}(\breve{\boldsymbol{x}}) = f(\boldsymbol{a}) + \tfrac{1}{2} \cdot \nabla_f^{\mathsf{T}}(\boldsymbol{a}) \cdot (\breve{\boldsymbol{x}} - \boldsymbol{a}) \tag{C.79}$$

$$= f(\boldsymbol{a}) + \frac{d_x \cdot (\breve{x} - x_a) + d_y \cdot (\breve{y} - y_a) + d_z \cdot (\breve{z} - z_a)}{2}. \tag{C.80}$$

[4] Nevertheless, the use of standard numerical methods is recommended.

Appendix D

Statistical Prerequisites

This part summarizes some essential statistical concepts for vector-valued data, intended as a supplement particularly to Chapters 11 and 17.

D.1 Mean, Variance, and Covariance

For the following definitions we assume a sequence $X = (\boldsymbol{x}_0, \boldsymbol{x}_1, \ldots, \boldsymbol{x}_{n-1})$ of n vector-valued, m-dimensional measurements, with "samples"

$$\boldsymbol{x}_i = (x_{i,0}, x_{i,1}, \ldots, x_{i,m-1})^{\mathsf{T}} \in \mathbb{R}^m. \tag{D.1}$$

D.1.1 Mean

The n-dimensional *sample mean vector* is defined as

$$\boldsymbol{\mu}(X) = (\mu_0, \mu_1, \ldots, \mu_{m-1})^{\mathsf{T}} \tag{D.2}$$

$$= \frac{1}{n} \cdot (\boldsymbol{x}_0 + \boldsymbol{x}_1 + \ldots + \boldsymbol{x}_{n-1}) = \frac{1}{n} \cdot \sum_{i=0}^{n-1} \boldsymbol{x}_i. \tag{D.3}$$

Geometrically speaking, the vector $\boldsymbol{\mu}(X)$ corresponds to the *centroid* of the sample vectors $\boldsymbol{x}_i$ in m-dimensional space. Each scalar element μ_p is the mean of the associated component (also called *variate* or *dimension*) p over all n samples, that is

$$\mu_p = \frac{1}{n} \cdot \sum_{i=0}^{n-1} x_{i,p}\,, \tag{D.4}$$

for $p = 0, \ldots, m-1$.

D.1.2 Variance and Covariance

The *covariance* quantifies the strength of interaction between a pair of components p, q in the sample X, defined as

$$\sigma_{p,q}(X) = \frac{1}{n} \cdot \sum_{i=0}^{n-1} (x_{i,p} - \mu_p) \cdot (x_{i,q} - \mu_q). \tag{D.5}$$

For efficient calculation, this expression can be rewritten in the form

$$\sigma_{p,q}(X) = \frac{1}{n} \cdot \Bigl[\underbrace{\sum_{i=0}^{n-1} (x_{i,p} \cdot x_{i,q})}_{S_{p,q}(X)} - \frac{1}{n} \cdot \bigl(\underbrace{\sum_{i=0}^{n-1} x_{i,p}}_{S_p(X)}\bigr) \cdot \bigl(\underbrace{\sum_{i=0}^{n-1} x_{i,q}}_{S_q(X)}\bigr)\Bigr], \tag{D.6}$$

which does not require the explicit calculation of μ_p and μ_q. In the special case of $p = q$, we get

$$\sigma_{p,p}(X) = \sigma_p^2(X) = \frac{1}{n} \cdot \sum_{i=0}^{n-1} (x_{i,p} - \mu_p)^2 \tag{D.7}$$

$$= \frac{1}{n} \cdot \Bigl[\sum_{i=0}^{n-1} x_{i,p}^2 - \frac{1}{n} \cdot \bigl(\sum_{i=0}^{n-1} x_{i,p}\bigr)^2\Bigr], \tag{D.8}$$

which is the *variance within* the component p. This corresponds to the ordinary (one-dimensional) variance $\sigma_p^2(X)$ of the n scalar sample values $x_{0,p}, x_{1,p}, \ldots, x_{n-1,p}$ (see also Sec. 3.7.1).

D.1.3 Biased vs. Unbiased Variance

If the variance (or covariance) of some population is estimated from a small set of random samples, the results obtained by the formulation given in the previous section are known to be statistically *biased.*[1] The most common form of correcting for this bias is to use the factor $1/(n-1)$ instead of $1/n$ in the variance calculations. For example, Eqn. (D.5) would change to

$$\breve{\sigma}_{p,q}(X) = \frac{1}{n-1} \cdot \sum_{i=0}^{n-1} (x_{i,p} - \mu_p) \cdot (x_{i,q} - \mu_q) \tag{D.9}$$

to yield an *unbiased* sample variance. In the following (and throughout the text), we ignore the bias issue and consistently use the factor $1/n$ for all variance calculations. Note, however, that many software packages[2] use the bias-corrected factor $1/(n-1)$ by default and thus may return different results (which can be easily scaled for comparison).

D.2 The Covariance Matrix

The *covariance matrix* Σ for the m-dimensional sample X is a square matrix of size $m \times m$ that is composed of the covariance values $\sigma_{p,q}$ for all pairs (p, q) of components, that is,

[1] Note that the estimation of the mean by the *sample mean* (Eqn. (D.3)) is not affected by this bias problem.

[2] For example, *Apache Commons Math*, *Matlab*, *Mathematica*.

$$\Sigma(X) = \begin{pmatrix} \sigma_{0,0} & \sigma_{0,1} & \cdots & \sigma_{0,m-1} \\ \sigma_{1,0} & \sigma_{1,1} & \cdots & \sigma_{1,m-1} \\ \vdots & \vdots & \ddots & \vdots \\ \sigma_{m-1,0} & \sigma_{m-1,1} & \cdots & \sigma_{m-1,m-1} \end{pmatrix} \tag{D.10}$$

$$= \begin{pmatrix} \sigma_0^2 & \sigma_{0,1} & \cdots & \sigma_{0,m-1} \\ \sigma_{1,0} & \sigma_1^2 & \cdots & \sigma_{1,m-1} \\ \vdots & \vdots & \ddots & \vdots \\ \sigma_{m-1,0} & \sigma_{m-1,1} & \cdots & \sigma_{m-1}^2 \end{pmatrix}. \tag{D.11}$$

Note that any diagonal element of $\Sigma(X)$ is the ordinary (scalar) variance $\sigma_p^2(X)$ (see Eqn. (D.7)), for $p = 0, \ldots, m-1$, which can never be negative. All other entries of a covariance matrix may be positive or negative in general. Since $\sigma_{p,q} = \sigma_{q,p}$, a covariance matrix is always symmetric, with up to $(m^2 + m)/2$ unique elements. Thus, any covariance matrix has the important property of being *positive semidefinite*, which implies that all its eigenvalues (see Sec. B.4) are positive (i.e., non-negative). The covariance matrix can also be written in the form

$$\Sigma(X) = \frac{1}{n} \cdot \sum_{i=0}^{n-1} \underbrace{[\boldsymbol{x}_i - \boldsymbol{\mu}(X)] \cdot [\boldsymbol{x}_i - \boldsymbol{\mu}(X)]^\intercal}_{= [\boldsymbol{x}_i - \boldsymbol{\mu}(X)] \otimes [\boldsymbol{x}_i - \boldsymbol{\mu}(X)]}, \tag{D.12}$$

where $\otimes$ denotes the outer (vector) product.

The *trace* (sum of the diagonal elements) of the covariance matrix,

$$\sigma_{\text{total}}(X) = \operatorname{trace}(\Sigma(X)), \tag{D.13}$$

is called the *total variance* of the multivariate sample. Alternatively, the (Frobenius) *norm* of the covariance matrix $\Sigma(X)$, defined as

$$\|\Sigma(X)\|_2 = \Bigl(\sum_{i=0}^{m-1} \sum_{j=0}^{m-1} \sigma_{i,j}^2\Bigr)^{1/2}, \tag{D.14}$$

can be used to quantify the overall variance in the sample data.

D.2.1 Example

Assume that the sample X consists of the following set of four 3D vectors (i.e., $m = 3$ and $n = 4$)

$$\boldsymbol{x}_0 = \begin{pmatrix} 75 \\ 37 \\ 12 \end{pmatrix}, \quad \boldsymbol{x}_1 = \begin{pmatrix} 41 \\ 27 \\ 20 \end{pmatrix}, \quad \boldsymbol{x}_2 = \begin{pmatrix} 93 \\ 81 \\ 11 \end{pmatrix}, \quad \boldsymbol{x}_3 = \begin{pmatrix} 12 \\ 48 \\ 52 \end{pmatrix},$$

with each $\boldsymbol{x}_i = (x_{i,\mathrm{R}}, x_{i,\mathrm{G}}, x_{i,\mathrm{B}})^\intercal$ representing a particular RGB color. The resulting *sample mean vector* (see Eqn. (D.3)) is

$$\boldsymbol{\mu}(X) = \begin{pmatrix} \mu_R \\ \mu_G \\ \mu_B \end{pmatrix} = \frac{1}{4} \cdot \begin{pmatrix} 75 + 41 + 93 + 12 \\ 37 + 27 + 81 + 48 \\ 12 + 20 + 11 + 52 \end{pmatrix} = \frac{1}{4} \cdot \begin{pmatrix} 221 \\ 193 \\ 95 \end{pmatrix} = \begin{pmatrix} 55.25 \\ 48.25 \\ 23.75 \end{pmatrix},$$

and the associated *covariance matrix* (Eqn. (D.11)) is

$$\Sigma(X) = \begin{pmatrix} 972.188 & 331.938 & -470.438 \\ 331.938 & 412.688 & -53.188 \\ -470.438 & -53.188 & 278.188 \end{pmatrix}.$$

As predicted, this matrix is symmetric and all diagonal elements are non-negative. Note that *no* sample bias-correction (see Sec. D.1.3) was used in this example. The *total variance* (Eqn. (D.13)) of the sample set is

$$\sigma_{\mathrm{total}}(X) = \mathrm{trace}\,(\Sigma(X)) = 972.188 + 412.688 + 278.188 \approx 1663.06,$$

and the *Froebenius norm* of the covariance matrix (see Eqn. (D.14)) is $\|\Sigma(X)\|_2 \approx 1364.36$.

D.2.2 Practical Calculation

The calculation of covariance matrices is implemented in almost any software package for statistical analysis or linear algebra. For example, with the *Apache Commons Math* library this could be accomplished as follows:

```
import org.apache.commons.math3.stat.correlation.Covariance;
...
double[][] X;          // X[i] is the i-th sample vector
Covariance cov = new Covariance(X, false);  // no bias correction
RealMatrix S = cov.getCovarianceMatrix();
...
```

D.3 Mahalanobis Distance

The Mahalanobis distance[3] [157] is used to measure distances in multi-dimensional distributions. Unlike the Euclidean distance it takes into account the amount of scatter in the distribution and the correlation between features. In particular, the Mahalanobis distance can be used to measure distances in distributions, where the individual components substantially differ in scale. Depending on their scale, a few components (or even a single component) may dominate the ordinary (Euclidean) distance outcome and the "smaller" components have no influence whatsoever.

D.3.1 Definition

Given a distribution of m-dimensional samples $X = (\boldsymbol{x}_0, \ldots, \boldsymbol{x}_{n-1})$, with $\boldsymbol{x}_k \in \mathbb{R}^m$, the Mahalanobis distance between two samples $\boldsymbol{x}_a$, $\boldsymbol{x}_b$ is defined as

$$\mathrm{d_M}(\boldsymbol{x}_a, \boldsymbol{x}_b) = \|\boldsymbol{x}_a - \boldsymbol{x}_b\|_{\mathrm{M}} = \sqrt{(\boldsymbol{x}_a - \boldsymbol{x}_b)^{\mathsf{T}} \cdot \boldsymbol{\Sigma}^{-1} \cdot (\boldsymbol{x}_a - \boldsymbol{x}_b)}, \quad \text{(D.15)}$$

where $\boldsymbol{\Sigma}$ is the $m \times m$ *covariance matrix* of the distribution X, as described in Sec. D.2.[4]

[3] http://en.wikipedia.org/wiki/Mahalanobis_distance.

[4] Note that the expression under the root in Eqn. (D.15) is the (dot) product of a row vector and a column vector, that is, the result is a non-negative scalar value.

The Mahalanobis distance normalizes each feature component to *zero mean* and *unit variance*. This makes the distance calculation independent of the scale of the individual components, that is, all components are "treated fairly" even if their range is many orders of magnitude different. In other words, no component can dominate the others even if its magnitude is disproportionally large.

D.3.2 Relation to the Euclidean Distance

Recall that the Euclidean distance between two points $\boldsymbol{x}_a, \boldsymbol{x}_b$ in $\mathbb{R}^m$ is equivalent to the (L2) norm of the difference vector $\boldsymbol{x}_a - \boldsymbol{x}_b$, which can be written in the form

$$\mathrm{d}_{\mathrm{E}}(\boldsymbol{x}_a, \boldsymbol{x}_b) = \|\boldsymbol{x}_a - \boldsymbol{x}_b\|_2 = \sqrt{(\boldsymbol{x}_a - \boldsymbol{x}_b)^{\mathsf{T}} \cdot (\boldsymbol{x}_a - \boldsymbol{x}_b)}. \qquad \text{(D.16)}$$

Note the structural similarity with the definition of the Mahalanobis distance in Eqn. (D.15), the only difference being the missing matrix $\boldsymbol{\Sigma}^{-1}$. This becomes even clearer if we analogously insert the identity matrix $\mathbf{I}$ into Eqn. (D.16), that is,

$$\mathrm{d}_{\mathrm{E}}(\boldsymbol{x}_a, \boldsymbol{x}_b) = \|\boldsymbol{x}_a - \boldsymbol{x}_b\|_2 = \sqrt{(\boldsymbol{x}_a - \boldsymbol{x}_b)^{\mathsf{T}} \cdot \mathbf{I} \cdot (\boldsymbol{x}_a - \boldsymbol{x}_b)}\,, \qquad \text{(D.17)}$$

which obviously does not change the outcome. The purpose of $\boldsymbol{\Sigma}^{-1}$ in Eqn. (D.15) is to map the difference vectors (and thus the involved vectors $\boldsymbol{x}_a, \boldsymbol{x}_b$) into a transformed (scaled and rotated) space, where the actual distance measurement is performed. In contrast, with the Euclidean distance, all components contribute equally to the distance measure, without any scaling or other transformation.

D.3.3 Numerical Aspects

For calculating the Mahalobis distance (Eqn. (D.15)) the *inverse* of the covariance matrix (Sec. D.2) is needed. By definition, a covariance matrix $\boldsymbol{\Sigma}$ is symmetric and its diagonal values are non-negative. Similarly (at least in theory), its inverse $\boldsymbol{\Sigma}^{-1}$ should also be symmetric with non-negative diagonal values. This is necessary to ensure that the quantities under the square root in Eqn. (D.15) are always positive.

Unfortunately, $\boldsymbol{\Sigma}$ is often ill-conditioned because of diagonal values that are very small or even zero. In this case, $\boldsymbol{\Sigma}$ is not positive-definite (as it should be), that is, one or more of its eigenvalues are negative, the inversion becomes numerically unstable and the resulting $\boldsymbol{\Sigma}^{-1}$ is non-symmetric. A simple remedy to this problem is to add a small quantity to the diagonal of the original covariance matrix $\boldsymbol{\Sigma}$, that is,

$$\tilde{\boldsymbol{\Sigma}} = \boldsymbol{\Sigma} + \epsilon \cdot \mathbf{I}, \qquad \text{(D.18)}$$

to enforce positive definiteness, and to use $\tilde{\boldsymbol{\Sigma}}^{-1}$ in Eqn. (D.15).

A possible alternative is to calculate the *Eigen decomposition*[5] of $\boldsymbol{\Sigma}$ in the form

[5] See http://mathworld.wolfram.com/EigenDecomposition.html and the class `EigenDecomposition` in the *Apache Commons Math* library.

$$\boldsymbol{\Sigma} = \mathbf{V} \cdot \boldsymbol{\Lambda} \cdot \mathbf{V}^{\mathsf{T}} \tag{D.19}$$

where $\boldsymbol{\Lambda}$ is a diagonal matrix containing the eigenvalues of $\boldsymbol{\Sigma}$ (which may be zero or negative). From this we create a modified diagonal matrix $\tilde{\boldsymbol{\Lambda}}$ by substituting all non-positive eigenvalues with a small positive quantity ϵ, that is,

$$\tilde{\boldsymbol{\Lambda}}_{i,i} = \min(\boldsymbol{\Lambda}_{i,i}, \epsilon). \tag{D.20}$$

(typically $\epsilon \approx 10^{-6}$) and finally calculate the modified covariance matrix as

$$\tilde{\boldsymbol{\Sigma}} = \mathbf{V} \cdot \tilde{\boldsymbol{\Lambda}} \cdot \mathbf{V}^{\mathsf{T}}, \tag{D.21}$$

which should be positive definite. The (symmetric) inverse $\tilde{\boldsymbol{\Sigma}}^{-1}$ is then used in Eqn. (D.15).

D.3.4 Pre-Mapping Data for Efficient Mahalanobis Matching

Assume that we have a large set of sample vectors ("data base") $X = (\boldsymbol{x}_0, \ldots, \boldsymbol{x}_{n-1})$ which shall be frequently queried for the instance most similar (i.e., closest) to a given search sample $\boldsymbol{x}_{\mathrm{s}}$. Assuming that the search through X is performed linearly, we would need to calculate $d_{\mathrm{M}}(\boldsymbol{x}_{\mathrm{s}}, \boldsymbol{x}_i)$—using Eqn. (D.15)—for all elements of $\boldsymbol{x}_i$ in X.

One way to accelerate the matching is to perform the transformation defined by $\boldsymbol{\Sigma}^{-1}$ to the entire data set only once, such that the Euclidean norm alone can be used for the distance calculation. For the sake of simplicity we write

$$\mathrm{d}^2_{\mathrm{M}}(\boldsymbol{x}_a, \boldsymbol{x}_b) = \|\boldsymbol{x}_a - \boldsymbol{x}_b\|^2_{\mathrm{M}} = \|\boldsymbol{y}\|^2_{\mathrm{M}} \tag{D.22}$$

with the difference vector $\boldsymbol{y} = \boldsymbol{x}_a - \boldsymbol{x}_b$, such that Eqn. (D.15) becomes

$$\|\boldsymbol{y}\|^2_{\mathrm{M}} = \boldsymbol{y}^{\mathsf{T}} \cdot \boldsymbol{\Sigma}^{-1} \cdot \boldsymbol{y}\,. \tag{D.23}$$

The goal is to find a transformation $\mathbf{U}$ such that we can calculate the Mahalanobis distance from the transformed vectors directly as

$$\hat{\boldsymbol{y}} = \mathbf{U} \cdot \boldsymbol{y}, \tag{D.24}$$

by using the ordinary *Euclidean* norm $\|\cdot\|_2$ instead, that is, in the form

$$\begin{aligned} \|\boldsymbol{y}\|^2_{\mathrm{M}} &= \|\hat{\boldsymbol{y}}\|^2_2 = \hat{\boldsymbol{y}}^{\mathsf{T}} \cdot \hat{\boldsymbol{y}} && \text{(D.25)} \\ &= (\mathbf{U} \cdot \boldsymbol{y})^{\mathsf{T}} \cdot (\mathbf{U} \cdot \boldsymbol{y}) = (\boldsymbol{y}^{\mathsf{T}} \cdot \mathbf{U}^{\mathsf{T}}) \cdot (\mathbf{U} \cdot \boldsymbol{y}) && \text{(D.26)} \\ &= \boldsymbol{y}^{\mathsf{T}} \cdot \mathbf{U}^{\mathsf{T}} \cdot \mathbf{U} \cdot \boldsymbol{y} = \boldsymbol{y}^{\mathsf{T}} \cdot \boldsymbol{\Sigma}^{-1} \cdot \boldsymbol{y}\,. && \text{(D.27)} \end{aligned}$$

While we do not know the matrix $\mathbf{U}$ yet, we see from Eqn. (D.27) that it must satisfy

$$\mathbf{U}^{\mathsf{T}} \cdot \mathbf{U} = \boldsymbol{\Sigma}^{-1}. \tag{D.28}$$

Fortunately, since $\boldsymbol{\Sigma}^{-1}$ is symmetric and positive definite, such a decomposition of $\boldsymbol{\Sigma}^{-1}$ always exists.

The standard method for calculating $\mathbf{U}$ in Eqn. (D.28) is by the Cholesky decomposition,[6] which can factorize any symmetric, positive definite matrix $\mathbf{A}$ in the form

$$\mathbf{A} = \mathbf{L} \cdot \mathbf{L}^{\mathsf{T}} \qquad \text{or} \qquad \mathbf{A} = \mathbf{U}^{\mathsf{T}} \cdot \mathbf{U}, \tag{D.29}$$

where $\mathbf{L}$ is a *lower-triangular* matrix or, alternatively, $\mathbf{U}$ is an *upper-triangular* matrix (the second variant is the one we need).[7] Since the transformation of the difference vectors $\boldsymbol{y} \rightarrow \mathbf{U} \cdot \boldsymbol{y}$ is a linear operation, the result is the same if we apply the transformation individually to the original vectors, that is,

$$\hat{\boldsymbol{y}} = \mathbf{U} \cdot \boldsymbol{y} = \mathbf{U} \cdot (\boldsymbol{x}_a - \boldsymbol{x}_b) = \mathbf{U} \cdot \boldsymbol{x}_a - \mathbf{U} \cdot \boldsymbol{x}_b \,. \tag{D.30}$$

This means that, given the transformation $\mathbf{U}$, we can obtain the Mahalanobis distance between two points $\boldsymbol{x}_a, \boldsymbol{x}_b$ (as defined in Eqn. (D.15)) by simply calculating the Euclidean distance in the form

$$\mathrm{d}_{\mathrm{M}}(\boldsymbol{x}_a, \boldsymbol{x}_b) = \|\mathbf{U} \cdot (\boldsymbol{x}_a - \boldsymbol{x}_b)\|_2 = \|\mathbf{U} \cdot \boldsymbol{x}_a - \mathbf{U} \cdot \boldsymbol{x}_b\|_2 \,. \tag{D.31}$$

In summary, this suggests the following solution to a large-database Mahalanobis matching problem:

1. Calculate the covariance matrix $\boldsymbol{\Sigma}$ for the original dataset $X = (\boldsymbol{x}_0, \ldots, \boldsymbol{x}_{n-1})$.
2. Condition $\boldsymbol{\Sigma}$, such that it is positive definite (see Sec. D.3.3).
3. Find the matrix $\mathbf{U}$, such that $\mathbf{U}^{\mathsf{T}} \cdot \mathbf{U} = \boldsymbol{\Sigma}^{-1}$ (by Cholesky decomposition of $\boldsymbol{\Sigma}^{-1}$).
4. Transform all samples of the original data set $X = (\boldsymbol{x}_0, \ldots, \boldsymbol{x}_{n-1})$ to $\hat{X} = (\hat{\boldsymbol{x}}_0, \ldots, \hat{\boldsymbol{x}}_{n-1})$, with $\hat{\boldsymbol{x}}_k = \mathbf{U} \cdot \boldsymbol{x}_k$. This now becomes the actual "database".
5. Apply the same transformation to the search sample $\boldsymbol{x}_{\mathrm{s}}$, that is, calculate $\hat{\boldsymbol{x}}_s = \mathbf{U} \cdot \boldsymbol{x}_{\mathrm{s}}$.
6. Find the index l of the best-matching element in X (in terms of the Mahalanobis distance) by calculating the *Euclidean* (!) distance between the transformed vectors, that is

$$l = \operatorname*{argmin}_{0 \leq k < n} \|\hat{\boldsymbol{x}}_s - \hat{\boldsymbol{x}}_k\|^2 \,. \tag{D.32}$$

Since the matching is now performed with the ordinary Euclidean distance and the Mahalanobis calculation is not required during the search, the savings should be substantial. Also, this opens an easy path to the use of advanced, tree-based matching techniques, such as the common k-nearest neighbor methods.

[6] See http://mathworld.wolfram.com/CholeskyDecomposition.html.

[7] The Cholesky decomposition (CD) requires that the supplied matrix $\mathbf{A}$ is symmetric and positive definite, otherwise the decomposition will fail. In fact, the CD itself is commonly used to test if a given matrix is positive definite. It is implemented by class `CholeskyDecomposition` of the *Apache Commons Math* library.

D.4 The Gaussian Distribution

The Gaussian distribution plays a major role in decision theory, pattern recognition, and statistics in general, because of its convenient analytical properties. A continuous, scalar quantity X is said to be subject to a Gaussian distribution, if the probability of observing a particular value x is

$$p(X=x) = p(x) = \frac{1}{\sqrt{2\pi\sigma^2}} \cdot e^{-\frac{(x-\mu)^2}{2\cdot\sigma^2}}. \qquad \text{(D.33)}$$

The Gaussian distribution is completely defined by its mean μ and variance σ^2. The Gaussian distribution, also called a "normal" distribution, is commonly denoted in the form

$$p(x) \sim \mathcal{N}(X \,|\, \mu, \sigma^2) \qquad \text{or} \qquad X \sim \mathcal{N}(\mu, \sigma^2), \qquad \text{(D.34)}$$

saying that "X is normally distributed with parameters μ and σ^2." As required for any valid probability distribution,

$$\mathcal{N}(X \,|\, \mu, \sigma^2) > 0 \qquad \text{and} \qquad \int_{-\infty}^{\infty} \mathcal{N}(X \,|\, \mu, \sigma^2)\, \mathrm{d}x = 1. \qquad \text{(D.35)}$$

Thus the area under the probability distribution curve is always one, that is, $\mathcal{N}()$ is normalized. The Gaussian function in Eqn. (D.33) has its maximum height (called "mode") at position $x = \mu$, where its value is

$$p(x=\mu) = \frac{1}{\sqrt{2\pi\sigma^2}}. \qquad \text{(D.36)}$$

If a random variable X is normally distributed with mean μ and variance σ^2, then the result of a linear mapping of the kind $X' = aX + b$ is again a random variable that is normally distributed, with parameters $\bar{\mu} = a\cdot\mu + b$ and $\bar{\sigma}^2 = a^2\cdot\sigma^2$:

$$X \sim \mathcal{N}(\mu, \sigma^2) \;\Rightarrow\; a\cdot X + b \sim \mathcal{N}(a\cdot\mu + b, a^2\cdot\sigma^2), \qquad \text{(D.37)}$$

for $a, b \in \mathbb{R}$.

Moreover, if X_1, X_2 are statistically *independent*, normally distributed random variables with means μ_1, μ_2 and variances σ_1^2, σ_2^2, respectively, then a linear combination of the form $a_1 X_1 + a_2 X_2$ is again normally distributed with $\mu_{12} = a_1\cdot\mu_1 + a_2\cdot\mu_2$ and $\sigma_{12} = a_1^2\cdot\sigma_1^2 + a_2^2\cdot\sigma_2^2$, that is,

$$(a_1 X_1 + a_2 X_2) \sim \mathcal{N}(a_1\cdot\mu_1 + a_2\cdot\mu_2, a_1^2\cdot\sigma_1^2 + a_2^2\cdot\sigma_2^2). \qquad \text{(D.38)}$$

D.4.1 Maximum Likelihood Estimation

The probability density function $p(x)$ of a statistical distribution tells us how probable it is to observe the result x for some fixed distribution parameters, such as μ and σ, in case of a normal distribution. If these parameters are *unknown* and need to be estimated,[8] it is interesting to ask the reverse question:

[8] As required, for example, for "minimum error thresholding" in Chapter 11, Sec. 11.1.6.

> How likely are particular parameter values for a given set of empirical observations (assuming a certain type of distribution)?

This is (in a casual sense) what the term "likelihood" stands for. In particular, a distribution's *likelihood function* quantifies the probability that a given (fixed) set of observations was generated by some varying distribution parameters.

Note that the probability of observing the outcome x from the normal distribution,

$$p(x) = p(x \,|\, \mu, \sigma^2), \tag{D.39}$$

is really a *conditional* probability, stating how probable it is to observe the value x from a given normal distribution with known parameters μ and σ^2. Conversely, a likelihood function for the normal distribution could be viewed as a conditional function

$$L(\mu, \sigma^2 \,|\, x), \tag{D.40}$$

which quantifies the likelihood of (μ, σ^2) being the correct distribution parameters for a given observation x. The maximum likelihood method tries to find optimal parameters by *maximizing* the value of a distribution's likelihood function L.

If we draw two independent[9] samples x_a, x_b that are subjected to the same distribution, their *joint probability* (i.e., the probability of x_a *and* x_b occurring together in the sample) is the product of their individual probabilities, that is,

$$p(x_a \wedge x_b) = p(x_a) \cdot p(x_b)\,. \tag{D.41}$$

In general, if we are given a vector of m independent observations $X = (x_1, x_2, \ldots, x_m)$ from the same distribution, the probability of observing exactly this set of values is

$$\begin{aligned} p(X) &= p(x_0 \wedge x_1 \wedge \ldots \wedge x_{m-1}) \\ &= p(x_0) \cdot p(x_1) \cdot \ldots \cdot p(x_{m-1}) = \prod_{i=0}^{m-1} p(x_i)\,. \end{aligned} \tag{D.42}$$

Thus, if the sample X originates from a normal distribution $\mathcal{N}$, a suitable likelihood function is

$$L(\mu, \sigma^2 \,|\, X) = p(X \,|\, \mu, \sigma^2) \tag{D.43}$$

$$= \prod_{i=0}^{m-1} \mathcal{N}(x_i \,|\, \mu, \sigma^2) = \prod_{i=0}^{m-1} \frac{1}{\sqrt{2\pi\sigma^2}} \cdot e^{-\frac{(x_i-\mu)^2}{2\cdot\sigma^2}}. \tag{D.44}$$

The parameters $(\hat{\mu}, \hat{\sigma}^2)$, for which $L(\mu, \sigma^2 \,|\, X)$ is a maximum, are called the maximum-likelihood estimate for X.

Note that it is not necessary for a likelihood function to be a proper (i.e., normalized) probability distribution, since it is only necessary to calculate whether a particular set of distribution parameters

[9] Although this assumption is often violated, independence is important to keep statistical problems simple and tractable. In particular, the values of adjacent image pixels are usually not independent.

is more probable than another. Thus the likelihood function L may be any monotonic function of the corresponding probability p in Eqn. (D.43), in particular its *logarithm*, which is commonly used to avoid multiplying small values.

D.4.2 Gaussian Mixtures

In practice, probabilistic models are often too complex to be described by a single Gaussian (or other standard) distribution. Without losing the mathematical convenience of Gaussian models, highly complex distributions can be modeled as combinations of multiple Gaussian distributions with different parameters. Such a Gaussian *mixture model* is a linear superposition of K Gaussian distributions of the form

$$p(x) = \sum_{j=0}^{K-1} \pi_j \cdot \mathcal{N}(x \,|\, \mu_j, \sigma_j^2), \tag{D.45}$$

where the weights ("mixing coefficients") π_j express the probability that an event x was generated by the jth component (with $\sum_{j=0}^{K-1} \pi_j = 1$).[10] The interpretation of this mixture model is, that there are K independent Gaussian "components" (each with its parameters μ_j, σ_j) that contribute to a common stream of events x_i. If a particular value x is observed, it is assumed to be the result of exactly *one* of the K components, but the identity of that component is unknown.

Assume, as a special case, that a probability distribution $p(x)$ is the superposition (mixture) of *two* Gaussian distributions, that is,

$$p(x) = \pi_a \cdot \mathcal{N}(x \,|\, \mu_a, \sigma_a^2) + \pi_b \cdot \mathcal{N}(x \,|\, \mu_b, \sigma_b^2). \tag{D.46}$$

Any observed value x is assumed to be generated by either the first component (with μ_a, σ_a^2 and prior probability π_a) or the second component (with μ_b, σ_b^2 and prior probability π_b). These parameters as well as the prior probabilities are unknown but can be estimated by maximimizing the likelihood function L. Note that, in general, the unknown parameters cannot be calculated in closed form but only with numerical methods. For further details and solution techniques see [24, 64, 228], for example.

D.4.3 Creating Gaussian Noise

Synthetic Gaussian noise is often used for testing in image processing, particularly for assessing the quality of smoothing filters. While the generation of pseudo-random values that follow a Gaussian distribution is not a trivial task in general,[11] it is readily implemented in Java by the standard class `Random`. For example, the Java method `addGaussianNoise()` in Prog. D.1 adds Gaussian noise with zero mean ($\mu = 0$) and standard deviation `sigma` (σ) to a grayscale image `I` of type `FloatProcessor` (ImageJ). The random values produced

[10] The weight π_j is also called the *prior* probability of the component j.

[11] Typically the so-called *polar method* is used for generating Gaussian random values [138, Sec. 3.4.1].

by successive calls to the method `nextGaussian()` in line 10 follow a Gaussian distribution $\mathcal{N}(0,1)$, with mean $\mu = 0$ and variance $\sigma^2 = 1$. As implied by Eqn. (D.37),

$$X \sim \mathcal{N}(0,1) \;\Rightarrow\; a + s \cdot X \sim \mathcal{N}(a, s^2), \tag{D.47}$$

and thus scaling the results from `nextGaussian()` by `s` and additive shifting by a makes the resulting random variable `noise` normally distributed with $\mathcal{N}(a, s^2)$.

```
1 import java.util.Random;
2
3 void addGaussianNoise (FloatProcessor I, double sigma) {
4   int w = I.getWidth();
5   int h = I.getHeight();
6   Random rnd = new Random();
7   for (int v = 0; v < h; v++) {
8     for (int u = 0; u < w; u++) {
9       float val = I.getf(u, v);
10      float noise = (float) (rnd.nextGaussian() * sigma);
11      I.setf(u, v, val + noise);
12    }
13  }
14 }
```

Prog. D.1
Java method for adding Gaussian noise to an image of type `FloatProcessor`.

Appendix E

Gaussian Filters

This part supplements the material presented in Ch. 25 (SIFT).

E.1 Cascading Gaussian Filters

To compute a Gaussian scale space efficiently (as used in the SIFT method, for example), the scale layers are usually not obtained directly from the input image by smoothing with Gaussians of increasing size. Instead, each layer can be calculated recursively from the previous layer by filtering with relatively small Gaussians. Thus, the entire scale space is implemented as a concatenation or "cascade" of smaller Gaussian filters.[1]

If Gaussian filters of sizes σ_1, σ_2 are applied successively to the same image, the resulting smoothing effect is identical to using a single larger Gaussian filter H^G_σ, that is,

$$\left(I * H^{\mathrm{G}}_{\sigma_1}\right) * H^{G}_{\sigma_2} = I * \left(H^{\mathrm{G}}_{\sigma_1} * H^{G}_{\sigma_2}\right) = I * H^{\mathrm{G}}_{\sigma}, \qquad \text{(E.1)}$$

with $\sigma = \sqrt{\sigma_1^2 + \sigma_2^2}$ being the size of the resulting combined Gaussian filter H^{G}_σ [129, Sec. 4.5.4]. Put in other words, the *variances* (squares of the σ values) of successive Gaussian filters add up, that is,

$$\sigma^2 = \sigma_1^2 + \sigma_2^2. \qquad \text{(E.2)}$$

In the special case of the *same* Gaussian filter being applied twice ($\sigma_1 = \sigma_2$), the effective width of the combined filter is $\sigma = \sqrt{2} \cdot \sigma_1$.

E.2 Gaussian Filters and Scale Space

In a Gaussian scale space, the scale corresponding to each level is proportional to the width (σ) of the Gaussian filter required to derive this level from the original (completely unsmoothed) image. Given an image that is already pre-smoothed by a Gaussian filter of width

[1] See Chapter 25, Sec. 25.1.1 for details.

σ_1 and should be smoothed to some target scale $\sigma_2 > \sigma_1$, the required width of the additional Gaussian filter is

$$\sigma_d = \sqrt{\sigma_2^2 - \sigma_1^2}. \tag{E.3}$$

Usually the neighboring layers of the scale space differ by a constant scale factor (κ) and the transformation from one scale level to another can be accomplished by successively applying Gaussian filters. Despite the constant scale factor, however, the width of the required filters is *not* constant but depends on the image's initial scale. In particular, if we want to transform an image with scale σ_0 by a factor κ to a new scale $\kappa \cdot \sigma_0$, then (from Eqn. (E.2)) for σ_d the relation

$$(\kappa \cdot \sigma_0)^2 = \sigma_0^2 + \sigma_d^2 \tag{E.4}$$

must hold. Thus, the width σ_d of the required Gaussian smoothing filter is

$$\sigma_d = \sigma_0 \cdot \sqrt{\kappa^2 - 1}. \tag{E.5}$$

For example, doubling the scale ($\kappa = 2$) of an image that is pre-smoothed with σ_0 requires a Gaussian filter of width $\sigma_d = \sigma_0 \cdot (2^2 - 1)^{1/2} = \sigma_0 \cdot \sqrt{3} \approx \sigma_0 \cdot 1.732$.

E.3 Effects of Gaussian Filtering in the Frequency Domain

For the 1D Gaussian function

$$g_\sigma(x) = \frac{1}{\sigma\sqrt{2\pi}} \cdot e^{-\frac{x^2}{2\sigma^2}} \tag{E.6}$$

the continuous Fourier transform[2] $\mathcal{F}(g_\sigma)$ is

$$G_\sigma(\omega) = \frac{1}{\sqrt{2\pi}} \cdot e^{-\frac{\omega^2\sigma^2}{2}}. \tag{E.7}$$

Doubling the width (σ) of a Gaussian filter corresponds to cutting the bandwidth by half. If σ is doubled, the Fourier transform becomes

$$G_{2\sigma}(\omega) = \frac{1}{\sqrt{2\pi}} \cdot e^{-\frac{\omega^2(2\sigma)^2}{2}} = \frac{1}{\sqrt{2\pi}} \cdot e^{-\frac{4\omega^2\sigma^2}{2}} \tag{E.8}$$

$$= \frac{1}{\sqrt{2\pi}} \cdot e^{-\frac{(2\omega)^2\sigma^2}{2}} = G_\sigma(2\omega) \tag{E.9}$$

and, in general, when scaling the filter by a factor k,

$$G_{k\sigma}(\omega) = G_\sigma(k\omega). \tag{E.10}$$

That is, if σ is *increased* (or the kernel widened) by a factor k, the corresponding Fourier transform gets *contracted* by the same factor. In terms of linear filtering this means that widening the kernel by some factor k decimates the resulting signal bandwidth by $\frac{1}{k}$.

[2] See also Chapter 18, Sec. 18.1.

E.4 LoG-Approximation by the DoG

The 2D LoG kernel (see Ch. 25, Sec. 25.1.1),

$$L_\sigma(x,y) = \left(\nabla^2 g_\sigma\right)(x,y) = \frac{1}{\pi\sigma^4}\left(\frac{x^2+y^2-2\sigma^2}{2\sigma^2}\right)\cdot e^{-\frac{x^2+y^2}{2\sigma^2}}, \tag{E.11}$$

has a (negative) peak at the origin with the associated function value

$$L_\sigma(0,0) = -\frac{1}{\pi\sigma^4}. \tag{E.12}$$

Thus, the *scale normalized* LoG kernel, defined in Eqn. (25.10) as

$$\hat{L}_\sigma(x,y) = \sigma^2 \cdot L_\sigma(x,y), \tag{E.13}$$

has the peak value

$$\hat{L}_\sigma(0,0) = -\frac{1}{\pi\sigma^2} \tag{E.14}$$

at the origin. In comparison, for a given scale factor κ, the unscaled DoG function

$$\begin{aligned}\mathrm{DoG}_{\sigma,\kappa}(x,y) &= G_{\kappa\sigma}(x,y) - G_\sigma(x,y)\\ &= \frac{1}{2\pi\kappa^2\sigma^2}\cdot e^{-\frac{x^2+y^2}{2\kappa^2\sigma^2}} - \frac{1}{2\pi\sigma^2}\cdot e^{-\frac{x^2+y^2}{2\sigma^2}},\end{aligned} \tag{E.15}$$

has a peak value

$$\mathrm{DoG}_{\sigma,\kappa}(0,0) = -\frac{\kappa^2-1}{2\pi\kappa^2\sigma^2}. \tag{E.16}$$

By scaling the DoG function by some factor λ to match the LoG's center peak value, such that $L_\sigma(0,0) = \lambda \cdot \mathrm{DoG}_{\sigma,\kappa}(0,0)$, the original LoG (Eqn. (E.11)) is approximated by the DoG in the form

$$L_\sigma(x,y) \approx \frac{2\kappa^2}{\sigma^2(\kappa^2-1)} \cdot \mathrm{DoG}_{\sigma,\kappa}(x,y). \tag{E.17}$$

Similarly, the scale-normalized LoG (Eqn. (E.13)) is approximated by the DoG as[3]

$$\hat{L}_\sigma(x,y) \approx \frac{2\kappa^2}{\kappa^2-1} \cdot \mathrm{DoG}_{\sigma,\kappa}(x,y). \tag{E.18}$$

Since the factor in Eqn. (E.18) depends on κ only, the DoG approximation is (for a constant size ratio κ) implicitly proportional to the scale normalized LoG for any scale σ.

[3] A different formulation, $\hat{L}_\sigma(x,y) \approx \frac{1}{\kappa-1} \cdot \mathrm{DoG}_{\sigma,\kappa}(x,y)$, is given in [153], which is the same as Eqn. (E.18) for $\kappa \to 1$, but not for $\kappa > 1$. The essence is that the leading factor is constant and independent of σ, and can thus be ignored when comparing the magnitude of the filter responses at varying scales.

Appendix F

Java Notes

As a text for undergraduate engineering curricula, this book assumes basic programming skills in a procedural language, such as Java, C#, or C. The examples in the main text should be easy to understand with the help of an introductory book on Java or one of the many online tutorials. Experience shows, however, that difficulties with some basic Java concepts pertain and often cause complications, even at higher levels. The following sections address some of these typical problem spots.

F.1 Arithmetic

Java is a "strongly typed" programming language, which means in particular that any variable has a fixed type that cannot be altered dynamically. Also, the result of an expression is determined by the types of the involved operands and *not* (in the case of an assignment) by the type of the "receiving" variable.

F.1.1 Integer Division

Division involving integer operands is a frequent cause of errors. If the variables `a` and `b` are both of type `int`, then the expression `a / b` is evaluated according to the rules of integer division. The result—the number of times `b` is contained in `a`—is again of type `int`. For example, after the Java statements

```
int a = 2;
int b = 5;
double c = a / b;     // resulting value of c is zero!
```

the value of `c` is *not* 0.4 but 0.0 because the expression `a / b` on the right yields the `int`-value 0, which is then automatically converted to the `double` value 0.0.

If we wanted to evaluate `a / b` as a *floating-point* operation (as most pocket calculators do), at least one of the involved operands must be converted to a floating-point value, such as by an explicit type cast, for example,

```
double c = (double) a / b;      // value of c is 0.4
```

or alternatively

```
double c = a / (double) b;      // value of c is 0.4
```

Example

Assume, for example, that we want to scale any pixel value a of an image such that the maximum pixel value a_{max} is mapped to 255 (see Ch. 4). In mathematical notation, the scaling of the pixel values is simply expressed as

$$c \leftarrow \frac{a_i}{a_{\mathrm{max}}} \cdot 255$$

and it may be tempting to convert this 1:1 into Java code, such as

```
int a_max = ip.getMaxValue();
for ... {
  int a = ip.getPixel(u,v);
  int c = (a / a_max) * 255;   // ← problem!
  ip.putPixel(u, v, c);
}
...
```

As we can easily predict, the resulting image will be all black (zero values), except those pixels whose value was `a_max` originally (they are set to 255). The reason is again that the division `a / a_max` has two operands of type `int`, and the result is thus zero whenever the denumerator (`a_max`) is greater than the numerator (`a`).

Of course, the entire operation could be performed in the floating-point domain by converting one of the operands (as we have shown), but this is not even necessary in this case. Instead, we may simply swap the order of operations and start with the multiplication:

```
int c = a * 255 / a_max;
```

Why does this work now? The subexpression `a * 255` is evaluated first,[1] generating large intermediate values that pose no problem for the subsequent (integer) division. Nevertheless, *rounding* should always be considered to obtain more accurate results when computing fractions of integers (see Sec. F.1.5).

F.1.2 Modulus Operator

The result of the modulus operator $a \bmod b$ (used in several places in the main text) is defined [92, p. 82] as the remainder of the "floored" division a/b,

$$a \bmod b \;\equiv\; \begin{cases} a & \text{for } b = 0, \\ a - b \cdot \lfloor a/b \rfloor & \text{otherwise,} \end{cases} \tag{F.1}$$

[1] In Java, expressions at the same level are always evaluated in left-to-right order, and therefore no parentheses are required in this example (though they would do no harm either).

for $a, b \in \mathbb{R}$. This type of operator or library method was not available in the standard Java API until recently.[2] The following Java method implements the mod operation according to the definition in Eqn. (F.1):[3]

```
int Mod(int a, int b) {
  if (b == 0)
    return a;
  if (a * b >= 0)
    return a - b * (a / b);
  else
    return a - b * (a / b - 1);
}
```

Note that the *remainder* operator %, defined as

$$a \,\%\, b \equiv a - b \cdot \mathrm{truncate}(a/b), \qquad \text{for } b \neq 0, \tag{F.2}$$

is often used in this context, but yields the same results only for *positive* operands $a \geq 0$ and $b > 0$. For example,

$$\begin{array}{rlrcr} 13 & \mathrm{mod} & 4 & = & 1 \\ 13 & \mathrm{mod} & -4 & = & -3 \\ -13 & \mathrm{mod} & 4 & = & 3 \\ -13 & \mathrm{mod} & -4 & = & -1 \end{array} \quad \text{vs.} \quad \begin{array}{rlrcr} 13 & \% & 4 & = & 1 \\ 13 & \% & -4 & = & 1 \\ -13 & \% & 4 & = & -1 \\ -13 & \% & -4 & = & -1 \end{array}$$

F.1.3 Unsigned Byte Data

Most grayscale and indexed images in Java and ImageJ are composed of pixels of type `byte`, and the same holds for the individual components of most color images. A single byte consists of eight bits and can thus represent $2^8 = 256$ different bit patterns or values, usually mapped to the numeric range $0, \ldots, 255$. Unfortunately, Java (unlike C and C++) does *not* provide a suitable "unsigned" 8-bit data type. The primitive Java type `byte` is "signed", using one of its eight bits for the $\pm$ sign, and is intended to hold values in the range $-128, \ldots, +127$.

Java's `byte` data can still be used to represent the values 0 to 255, but conversions must take place to perform proper arithmetic computations. For example, after execution of the statements

```
int  a = 200;
byte b = (byte) p;
```

the variables `a` (32-bit `int`) and `b` (8-bit `byte`) contain the binary patterns

```
a = 00000000000000000000000011001000
b =                         11001000
```

Interpreted as a (signed) `byte` value, with the leftmost bit[4] as the sign bit, the variable `b` has the decimal value -56. Thus after the statement

[2] Starting with Java version 1.8 the mod operation (as defined in Eqn. (F.1)) is implemented by the standard method `Math.floorMod(a, b)`.

[3] The definition in Eqn. (F.1) is not restricted to integer operands.

[4] Java uses the standard "2s-complement" representation, where a sign bit = 1 stands for a negative value.

```
int  a1 = b;                  // a1 == -56
```

the value of the new `int` variable `a1` is -56! To (ab-)use signed `byte` data as *unsigned* data, we can circumvent Java's standard conversion mechanism by disguising the content of `b` as a logic (i.e., nonarithmetic) *bit pattern*; for example, by

```
int  a2 = (0xff & b);         // a2 == 200
```

where `0xff` (in hexadecimal notation) is an `int` value with the binary bit pattern `00000000000000000000000011111111` and `&` is the bitwise AND operator. Now the variable `a2` contains the right integer value (200) and we thus have a way to use Java's (signed) `byte` data type for storing *unsigned* values. Within ImageJ, access to pixel data is routinely implemented in this way, which is considerably faster than using the convenience methods `getPixel()` and `putPixel()`.

F.1.4 Mathematical Functions in Class Math

Java provides most standard mathematical functions as static methods in class `Math`, as listed in Table F.1. The `Math` class is part of the `java.lang` package and thus requires no explicit import to be used. Most `Math` methods accept arguments of type `double` and also return values of type `double`. As a simple example, a typical use of the cosine function $y = \cos(x)$ is

```
double x;
double y = Math.cos(x);
```

Similarly, the `Math` class defines some common numerical constants as static variables; for example, the value of π could be obtained by

```
double pi = Math.PI;
```

Table F.1
Mathematical methods and constants defined by Java's Math class.

double	abs(double a)	double	max(double a, double b)
int	abs(int a)	float	max(float a, float b)
float	abs(float a)	int	max(int a, int b)
long	abs(long a)	long	max(long a, long b)
double	ceil(double a)	double	min(double a, double b)
double	floor(double a)	float	min(float a, float b)
int	floorMod(int a, int b)	int	min(int a, int b)
long	floorMod(long a, long b)	long	min(long a, long b)
double	rint(double a)		
long	round(double a)	double	random()
int	round(float a)		
double	toDegrees(double rad)	double	toRadians(double deg)
double	sin(double a)	double	asin(double a)
double	cos(double a)	double	acos(double a)
double	tan(double a)	double	atan(double a)
double	atan2(double y, double x)		
double	log(double a)	double	exp(double a)
double	sqrt(double a)	double	pow(double a, double b)
double	E	double	PI

F.1.5 Numerical Rounding

Java's `Math` class (confusingly) offers three different methods for rounding floating-point values:

```
double rint(double x)
long   round(double x)
int    round(float x)
```

For example, a `double` value `x` can be rounded to `int` in any of the following ways:

```
double x; int k;
k = (int) Math.rint(x);
k = (int) Math.round(x);
k = Math.round((float) x);
```

If the operand `x` is known to be positive (as is typically the case with pixel values) rounding can be accomplished without using any method calls by

```
k = (int) (x + 0.5);     // only if x >= 0
```

In this case, the expression `(x + 0.5)` is first computed as a floating-point (`double`) value, which is then truncated (toward zero) by the explicit `(int)` typecast.

F.1.6 Inverse Tangent Function

The inverse tangent function $\varphi = \tan^{-1}(a)$ or $\varphi = \arctan(a)$ is used in several places in the main text. This function is implemented by the method `atan(double a)` in Java's `Math` class (Table F.1). The return value of `atan()` is in the range $[-\frac{\pi}{2}, \ldots, \frac{\pi}{2}]$ and thus restricted to only two of the four quadrants. Without any additional constraints, the resulting angle is ambiguous. In many practical situations, however, a is given as the ratio of two catheti $(\Delta x, \Delta y)$ of a right-angled triangle in the form

$$\varphi = \arctan\left(\tfrac{y}{x}\right), \tag{F.3}$$

for which we introduced the two-parameter function

$$\varphi = \mathrm{ArcTan}(x, y) \tag{F.4}$$

in the main text. The function $\mathrm{ArcTan}(x, y)$ is implemented by the standard method `atan2(dy,dx)` in Java's `Math` class (note the reversed parameters though) and returns an unambiguous angle φ in the range $[-\pi, \ldots, \pi]$; that is, in any of the four quadrants of the unit circle.[5] Also, the `atan2()` method returns a useful value even if both arguments are zero.

[5] The function `atan2(dy,dx)` is available in most current programming languages, including Java, C, and C++.

F.1.7 Classes Float and Double

The representation of floating-point numbers in Java follows the IEEE standard, and thus the types float and double include the values

```
Float.MIN_VALUE,              Double.MIN_VALUE,
Float.MAX_VALUE,              Double.MAX_VALUE,
Float.POSITIVE_INFINITY,      Double.POSITIVE_INFINITY,
Float.NEGATIVE_INFINITY,      Double.NEGATIVE_INFINITY,
Float.NaN,                    Double.NaN.
```

These values are defined as constants in the corresponding wrapper classes Float and Double, respectively. If any INFINITY or NaN[6] value occurs in the course of a computation (e.g., as the result of dividing by zero),[7] Java continues without raising an error, so incorrect values may ripple through a whole chain of calculations, making the actual bugs difficult to locate.

F.1.8 Testing Floating-Point Values Against Zero

Comparing floating-point values or testing them for zero is a nontrivial issue and a frequent cause of errors. In particular, one should *never* write

```
if (x == 0.0) {...}    ← problem!
```

if x is a floating-point variable. This is often needed, for example, to make sure that it is safe to divide another quantity by x. The aforementioned test, however, is not sufficient since x may be nonzero but still too small as a divisor.

A much better alternative is to test if x is "close" to zero, that is, within some small positive/negative (*epsilon*) interval. While the proper choice of this interval depends on the specific situation, the following settings are usually sufficient for safe operation:[8]

```
static final float  EPSILON_FLOAT  = 1e-7f;
static final double EPSILON_DOUBLE = 2e-16;

float x;
double y;

if (Math.abs(x) < EPSILON_FLOAT) {
  ... // x is practically zero
}

if (Math.abs(y) < EPSILON_DOUBLE) {
  ... // y is practically zero
}
```

[6] NaN stands for "not a number".

[7] In Java, this only holds for floating-point operations, whereas integer division by zero always causes an *exception*.

[8] These settings account for the limited *machine accuracy* (ϵ_{m}) of the IEEE 754 standard types float ($\epsilon_{\mathrm{m}} \approx 1.19 \cdot 10^{-7}$) and double ($\epsilon_{\mathrm{m}} \approx 2.22 \cdot 10^{-16}$) [190, Ch. 1, Sec. 1.1.2].

F.2 Arrays in Java

F.2.1 Creating Arrays

Unlike in most traditional programming languages (such as FORTRAN or C), arrays in Java can be created *dynamically*, meaning that the size of an array can be specified at runtime using the value of some variable or arithmetic expression. For example:

```
int N = 20;
int[] A = new int[N];
int[] B = new int[N * N];
```

Once allocated, however, the size of any Java array is fixed and cannot be subsequently altered.[9] Note that Java arrays may be of length *zero*!

After its definition, an array variable can be assigned any other compatible array or the constant value `null`, for example, [10]

```
A = B;      // A now references the data in B
B = null;
```

With the assignment `A = B`, the array initially referenced by `A` becomes unaccessible and thus turns into *garbage*. In contrast to C and C++, where unnecessary storage needs to be *de*allocated explicitly, this is taken care of in Java by its built-in "garbage collector". It is also convenient that newly created arrays of numerical element types (`int`, `float`, `double`, etc.) are automatically initialized to zero.

F.2.2 Array Size

Since an array may be created dynamically, it is important that its actual size can be determined at runtime. This is done by accessing the `length` attribute[11]

```
int k = A.length;  // number of elements in A
```

The size is a property of the array itself and can therefore be obtained inside any method from array arguments passed to it. Thus (unlike in C, for example) it is not necessary to pass the size of an array as a separate function argument.

If an array has more than one dimension, the size (`length`) along every dimension must be queried separately (see Sec. F.2.4). Also arrays are not necessarily rectangular; for example, the rows of a 2D array may have different lengths (including zero).

F.2.3 Accessing Array Elements

In Java, the index of the first array element is always 0 and the index of the last element is $N-1$ for an array with a total of N elements. To iterate through a 1D array `A` of arbitrary size, one would typically use a construct like

[9] For additional flexibility, Java provides a number of universal container classes (e.g., the classes `Set` and `List`) for a wide range of applications.

[10] This is not possible if the array variable was defined with the `final` attribute.

[11] Notice that the `length` attribute of an array is not a method!

```
for (int i = 0; i < A.length; i++) {
    // do something with A[i]
}
```

Alternatively, if only the array *values* are relevant and the array *index* (i) is not needed, one could use to following (even simpler) loop construct:

```
for (int a : A) {
    // do something with array values a
}
```

In both cases, the Java compiler can generate very efficient runtime code, since the source code makes obvious that the `for` loop does not access any elements outside the array limits and thus no explicit boundary checking is needed at execution time. This fact is very important for implementing efficient image processing programs in Java.

Images in Java and ImageJ are usually stored as 1D arrays (accessible through the `ImageProcessor` method `getPixels()` in ImageJ), with pixels arranged in row-first order.[12] Statistical calculations and most point operations can thus be efficiently implemented by directly accessing the underlying 1D array. For example, the `run` method of the contrast enhancement plugin in Prog. 4.1 (see Chapter 4, p. 58) could also be implemented in the following manner:

```
public void run(ImageProcessor ip) {
  // ip is assumed to be of type ByteProcessor
  byte[] pixels = (byte[]) ip.getPixels();
  for (int i = 0; i < pixels.length; i++) {
    int a = 0xFF & pixels[i];              // direct read operation
    int b = (int) (a * 1.5 + 0.5);
    if (b > 255)
      b = 255;
    pixels[i] = (byte) (0xFF & b);         // direct write operation
  }
}
```

F.2.4 2D Arrays

Multidimensional arrays are a frequent source of confusion. In Java, all arrays are 1D in principle, and multi-dimensional arrays are implemented as 1D arrays of arrays etc. (see Fig. F.1). If, for example, the 3×3 matrix

$$A = \begin{bmatrix} a_{0,0} & a_{0,1} & a_{0,2} \\ a_{1,0} & a_{1,1} & a_{1,2} \\ a_{2,0} & a_{2,1} & a_{2,2} \end{bmatrix} = \begin{bmatrix} 1 & 2 & 3 \\ 4 & 5 & 6 \\ 7 & 8 & 9 \end{bmatrix} \quad \text{(F.5)}$$

is defined as a 2D `int` array,

```
int[][] A = {{1,2,3},
             {4,5,6},
             {7,8,9}};
```

[12] This means that horizontally adjacent image pixels are stored next to each other in computer memory.

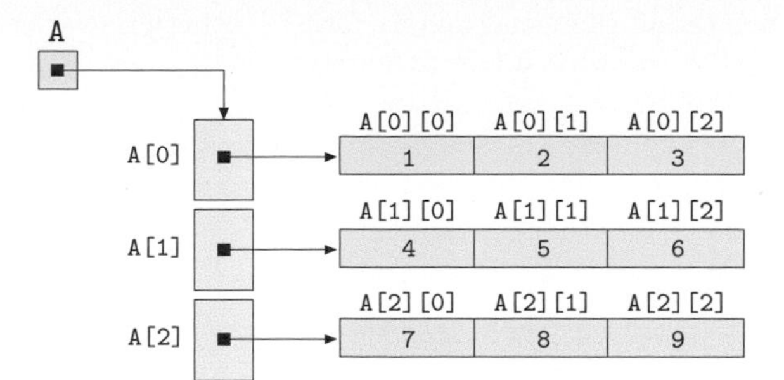

Fig. F.1
Layout of elements of a 2D Java array (corresponding to Eqn. (F.5)). In Java, multidimensional arrays are generally implemented as *1D* arrays whose elements are again 1D arrays.

then `A` is actually a *1D* array with three elements, each of which is again a 1D array. The elements `A[0]`, `A[1]` and `A[2]` are of type `int[]` and correspond to the three rows of the matrix A (see Fig. F.1).

The usual assumption is that the array elements are arranged in *row-first* order, as illustrated in Fig. F.1. The first index thus corresponds to the *row* number r and the second index corresponds to the *column* number c, that is,

$$a_{r,c} \equiv \texttt{A}[r]\,[c]\,. \tag{F.6}$$

This conforms to the mathematical convention and makes the array definition in the code segment above look exactly the same as the original matrix in Eqn. (F.5). Note that in this scheme the first array index corresponds to the *vertical* coordinate and the second index to the *horizontal* coordinate.

However, if an array is used to specify the contents of an *image* $I(u, v)$ or a *filter kernel* $H(i, j)$, we usually assume that the first index (u or i, respectively) is associated with the horizontal x-coordinate and the second index (v bzw. j) with the vertical y-coordinate. For example, if we represent the filter kernel

$$H = \begin{bmatrix} h_{0,0} & h_{1,0} & h_{2,0} \\ h_{0,1} & h_{1,1} & h_{2,1} \\ h_{0,2} & h_{1,2} & h_{2,2} \end{bmatrix} = \begin{bmatrix} -1 & -2 & 0 \\ -2 & 0 & 2 \\ 0 & 2 & 1 \end{bmatrix}$$

as a 2D Java array,

```
double[][] H = {{-1,-2, 0},
                {-2, 0, 2},
                { 0, 2, 1}};
```

then the row and column indexes must be *reversed* in order to access the correct elements. In this case we have the relation

$$h_{i,j} \equiv \texttt{H}[j]\,[i], \tag{F.7}$$

that is, the ordering of the indexes for array `H` is not the same as for the i/j coordinates of the filter kernel. In this case the *first* array index (j) corresponds to the *vertical* coordinate and the *second* index (i) to the *horizontal* coordinate. The advantage is that (as shown in the aforementioned code segment) the definition of the filter kernel can be written in the usual matrix form[13] (otherwise we would have to specify the transposed kernel matrix).

If a 2D array is merely used as an image container (whose contents are never defined in matrix form) any convention can be used for the ordering of the indexes. For example, the ImageJ method `getFloatArray()` of class `ImageProcessor`, when called in the form

```
float[][] I = ip.getFloatArray();
```

returns the image as a 2D array (`I`), whose indexes are arranged in the usual x/y order, that is,

$$I(x,y) \equiv \mathtt{I}[x][y]. \tag{F.8}$$

In this case, the image pixels are arranged in column-order, that is, *vertically* adjacent elements are stored next to each other in memory.

Size of multi-dimensional arrays

The size of a multi-dimensional array can be obtained by querying the size of its sub-arrays. For example, given the following 3D array with dimensions $P \times Q \times R$,

```
int A[][][] = new int[P][Q][R];
```

the size of `A` along its three dimensions is obtained by the statements

```
int p = A.length;           // = P
int q = A[0].length;        // = Q
int r = A[0][0].length;     // = R
```

This at least works for "rectangular" Java arrays, that is, multi-dimensional arrays with all sub-arrays at the same level having *identical* lengths, which is warranted by the array initialization in the aforementioned case. However, every 1D sub-array of `A` may be replaced by a suitable 1D array of *different* length,[14] for example, by the statement

```
A[0][0] = new int[0];
```

To avoid "index-out-of-bounds" errors, the length of each sub-array should be determined dynamically. The following example shows a "bullet-proof" iteration over all elements of a 3D array `A` whose sub-arrays may have different lengths or may even be empty:

```
int A[][][];
...
for (int i = 0; i < A.length; i++) {
  for (int j = 0; j < A[i].length; j++) {
    for (int k = 0; k < A[i][j].length; k++) {
      // safely access A[i][j][k]
    }
  }
}
```

[13] This scheme is used, for example, in the implementation of the 3×3 filter plugin in Prog. 5.2 (Chapter 5, p. 95).

[14] Even if the array `A` was originally declared `final`, the structure and contents of its sub-arrays may be modified any time.

F.2.5 Arrays of Objects

In Java, as mentioned earlier, we can create arrays dynamically; that is, the size of an array can be specified at runtime. This is convenient because we can adapt the size of the arrays to the given problem. For example, we could write

```
Corner[] corners = new Corner[n];
```

to create an array that can hold n objects of type Corner (as defined in Chapter 7, Sec. 7.3). Note that the new array corners is not filled with corners yet but initialized with null references, so the newly created array holds no objects at all. We can insert a Corner object into its first (or any other) cell, for example, by

```
corners[0] = new Corner(10, 20, 6789.0f);
```

F.2.6 Searching for Minimum and Maximum Values

Unfortunately, the standard Java API does not provide methods for retrieving the minimum and maximum values of a numeric array. Although these values are easily found by iterating over all elements of the sequence, care must be taken regarding the initialization.

For example, finding the extreme values of a sequence of int-values could be accomplished as follows:[15]

```
int[] A = ...
int minval = Integer.MAX_VALUE;
int maxval = Integer.MIN_VALUE;
for (int val : A) {
  minval = Math.min(minval, val);
  maxval = Math.max(maxval, val);
}
```

Note the use of the constants MIN_VALUE and MAX_VALUE, which are defined for any numeric Java type.

However, in the case of *floating-point* values, these are not the proper values for initialization.[16] Instead, POSITIVE_INFINITY and NEGATIVE_INFINITY should be used, as shown in the following code segment:

```
double[] B = ...
double minval = Double.POSITIVE_INFINITY;
double maxval = Double.NEGATIVE_INFINITY;
for (double val : B) {
  minval = Math.min(minval, val);
  maxval = Math.max(maxval, val);
}
```

[15] Alternatively, one could initialize minval and maxval with the first array element A[0].

[16] Because Double.MIN_VALUE and Float.MIN_VALUE specify to the smallest *positive* values.

F.2.7 Sorting Arrays

Arrays can be sorted efficiently with the standard method

```
Arrays.sort(type[] arr)
```

in class `java.util.Arrays`, where `arr` can be any array of primitive *`type`* (`int`, `float`, etc.) or an array of objects. In the latter case, the array may not have `null` entries. Also, the class of every contained object must implement the `Comparable` interface, that is, provide a public method `compareTo()` that returns an `int` value of -1, 0, or 1, depending upon the intended ordering relation. For example, the class `Corner` defines the `compareTo()` method as follows:

```
public class Corner implements Comparable<Corner> {
  float x, y, q;
  ...
  public int compareTo(Corner other) {
    if (this.q > other.q) return -1;
    else if (this.q < other.q) return 1;
    else return 0;
  }
}
```

References

1. Adobe Systems. "Adobe RGB (1998) Color Space Specification" (2005). http://www.adobe.com/digitalimag/pdfs/AdobeRGB1998.pdf.
2. M. Ahmed and R. Ward. A rotation invariant rule-based thinning algorithm for character recognition. *IEEE Transactions on Pattern Analysis and Machine Intelligence* **24**(12), 1672–1678 (2002).
3. L. Alvarez, P.-L. Lions, and J.-M. Morel. Image selective smoothing and edge detection by nonlinear diffusion (II). *SIAM Journal on Numerical Analysis* **29**(3), 845–866 (1992).
4. Apache Software Foundation. "Commons Math: The Apache Commons Mathematics Library". http://commons.apache.org/math/index.html.
5. K. Arbter, W. E. Snyder, H. Burkhardt, and G. Hirzinger. Application of affine-invariant Fourier descriptors to recognition of 3-D objects. *IEEE Transactions on Pattern Analysis and Machine Intelligence* **12**(7), 640–647 (1990).
6. G. R. Arce, J. Bacca, and J. L. Paredes. Nonlinear filtering for image analysis and enhancement. In A. Bovik, editor, "Handbook of Image and Video Processing", pp. 109–133. Academic Press, New York, second ed. (2005).
7. C. Arcelli and G. Sanniti di Baja. A one-pass two-operation process to detect the skeletal pixels on the 4-distance transform. *IEEE Transactions on Pattern Analysis and Machine Intelligence* **11**(4), 411–414 (1989).
8. K. Arnold, J. Gosling, and D. Holmes. "The Java Programming Language". Prentice Hall, fifth ed. (2012).
9. S. Arya, D. M. Mount, N. S. Netanyahu, R. Silverman, and A. Y. Wu. An optimal algorithm for approximate nearest neighbor searching in fixed dimensions. *Journal of the ACM* **45**(6), 891–923 (1998).
10. J. Astola, P. Haavisto, and Y. Neuvo. Vector median filters. *Proceedings of the IEEE* **78**(4), 678–689 (1990).
11. J. Babaud, A. P. Witkin, M. Baudin, and R. O. Duda. Uniqueness of the Gaussian kernel for scale-space filtering. *IEEE Transactions on Pattern Analysis and Machine Intelligence* **8**(1), 26–33 (1986).
12. W. Bailer. "Writing ImageJ Plugins—A Tutorial" (2003). http://www.imagingbook.com.
13. S. Baker and I. Matthews. Lucas-Kanade 20 years on: A unifying framework: Part 1. Technical Report CMU-RI-TR-02-16, Robotics Institute, Carnegie Mellon University (2003).
14. S. Baker and I. Matthews. Lucas-Kanade 20 years on: A unifying framework. *International Journal of Computer Vision* **56**(3), 221–255 (2004).
15. D. H. Ballard and C. M. Brown. "Computer Vision". Prentice Hall, Englewood Cliffs, NJ (1982).
16. D. Barash. Fundamental relationship between bilateral filtering, adaptive smoothing, and the nonlinear diffusion equation. *IEEE*

Transactions on Pattern Analysis and Machine Intelligence **24**(6), 844–847 (2002).

17. C. B. Barber, D. P. Dobkin, and H. Huhdanpaa. The quickhull algorithm for convex hulls. *ACM Transactions on Mathematical Software* **22**(4), 469–483 (1996).
18. M. Barni. A fast algorithm for 1-norm vector median filtering. *IEEE Transactions on Image Processing* **6**(10), 1452–1455 (1997).
19. H. G. Barrow, J. M. Tenenbaum, R. C. Bolles, and H. C. Wolf. Parametric correspondence and chamfer matching: two new techniques for image matching. In R. Reddy, editor, "Proceedings of the 5th International Joint Conference on Artificial Intelligence", pp. 659–663, Cambridge, MA (1977). William Kaufmann, Los Altos, CA.
20. H. Bay, A. Ess, T. Tuytelaars, and L. Van Gool. SURF: Speeded up robust features. *Computer Vision, Graphics, and Image Processing: Image Understanding* **110**(3), 346–359 (2008).
21. J. S. Beis and D. G. Lowe. Shape indexing using approximate nearest-neighbour search in high-dimensional spaces. In "Proceedings of the 1997 Conference on Computer Vision and Pattern Recognition (CVPR'97)", pp. 1000–1006, Puerto Rico (June 1997).
22. R. Bencina and M. Kaltenbrunner. The design and evolution of fiducials for the reacTIVision system. In "Proceedings of the 3rd International Conference on Generative Systems in the Electronic Arts", Melbourne (2005).
23. J. Bernsen. Dynamic thresholding of grey-level images. In "Proceedings of the International Conference on Pattern Recognition (ICPR)", pp. 1251–1255, Paris (October 1986). IEEE Computer Society.
24. C. M. Bishop. "Pattern Recognition and Machine Learning". Springer, New York (2006).
25. R. E. Blahut. "Fast Algorithms for Digital Signal Processing". Addison-Wesley, Reading, MA (1985).
26. I. Blayvas, A. Bruckstein, and R. Kimmel. Efficient computation of adaptive threshold surfaces for image binarization. *Pattern Recognition* **39**(1), 89–101 (2006).
27. J. Blinn. Consider the lowly 2×2 matrix. *IEEE Computer Graphics and Applications* **16**(2), 82–88 (1996).
28. J. Blinn. "Jim Blinn's Corner: Notation, Notation, Notation". Morgan Kaufmann (2002).
29. J. Bloch. "Effective Java". Addison-Wesley, second ed. (2008).
30. G. Borgefors. Distance transformations in digital images. *Computer Vision, Graphics and Image Processing* **34**, 344–371 (1986).
31. G. Borgefors. Hierarchical chamfer matching: a parametric edge matching algorithm. *IEEE Transactions on Pattern Analysis and Machine Intelligence* **10**(6), 849–865 (1988).
32. A. I. Borisenko and I. E. Tarapov. "Vector and Tensor Analysis with Applications". Dover Publications, New York (1979).
33. J. E. Bresenham. A linear algorithm for incremental digital display of circular arcs. *Communications of the ACM* **20**(2), 100–106 (1977).
34. E. O. Brigham. "The Fast Fourier Transform and Its Applications". Prentice Hall, Englewood Cliffs, NJ (1988).
35. I. N. Bronstein and K. A. Semendjajew. "Handbook of Mathematics". Springer-Verlag, Berlin, third ed. (2007).
36. I. N. Bronstein, K. A. Semendjajew, G. Musiol, and H. Mühlig. "Taschenbuch der Mathematik". Verlag Harri Deutsch, fifth ed. (2000).
37. M. Brown and D. Lowe. Invariant features from interest point groups. In "Proceedings of the British Machine Vision Conference", pp. 656–665 (2002).

38. H. BUNKE AND P. S.-P. WANG, editors. "Handbook of Character Recognition and Document Image Analysis". World Scientific, Singapore (2000).
39. W. BURGER AND M. J. BURGE. "Digital Image Processing—An Algorithmic Introduction using Java". Texts in Computer Science. Springer, New York (2008).
40. W. BURGER AND M. J. BURGE. "ImageJ Short Reference for Java Developers" (2008). http://www.imagingbook.com.
41. P. J. BURT AND E. H. ADELSON. The Laplacian pyramid as a compact image code. *IEEE Transactions on Communications* **31**(4), 532–540 (1983).
42. J. F. CANNY. A computational approach to edge detection. *IEEE Transactions on Pattern Analysis and Machine Intelligence* **8**(6), 679–698 (1986).
43. K. R. CASTLEMAN. "Digital Image Processing". Prentice Hall, Upper Saddle River, NJ (1995).
44. E. E. CATMULL AND R. ROM. A class of local interpolating splines. In R. E. BARNHILL AND R. F. RIESENFELD, editors, "Computer Aided Geometric Design", pp. 317–326. Academic Press, New York (1974).
45. F. CATTÉ, P.-L. LIONS, J.-M. MOREL, AND T. COLL. Image selective smoothing and edge detection by nonlinear diffusion. *SIAM Journal on Numerical Analysis* **29**(1), 182–193 (1992).
46. C. I. CHANG, Y. DU, J. WANG, S. M. GUO, AND P. D. THOUIN. Survey and comparative analysis of entropy and relative entropy thresholding techniques. *IEE Proceedings—Vision, Image and Signal Processing* **153**(6), 837–850 (2006).
47. F. CHANG, C. J. CHEN, AND C. J. LU. A linear-time component-labeling algorithm using contour tracing technique. *Computer Vision, Graphics, and Image Processing: Image Understanding* **93**(2), 206–220 (2004).
48. P. CHARBONNIER, L. BLANC-FERAUD, G. AUBERT, AND M. BARLAUD. Two deterministic half-quadratic regularization algorithms for computed imaging. In "Proceedings IEEE International Conference on Image Processing (ICIP-94)", vol. 2, pp. 168–172, Austin (November 1994).
49. Y. CHEN AND G. LEEDHAM. Decompose algorithm for thresholding degraded historical document images. *IEE Proceedings—Vision, Image and Signal Processing* **152**(6), 702–714 (2005).
50. H. D. CHENG, X. H. JIANG, Y. SUN, AND J. WANG. Color image segmentation: advances and prospects. *Pattern Recognition* **34**(12), 2259–2281 (2001).
51. P. R. COHEN AND E. A. FEIGENBAUM. "The Handbook of Artificial Intelligence". William Kaufmann, Los Altos, CA (1982).
52. B. COLL, J. L. LISANI, AND C. SBERT. Color images filtering by anisotropic diffusion. In "Proceedings of the IEEE International Conference on Systems, Signals, and Image Processing (IWSSIP)", pp. 305–308, Chalkida, Greece (2005).
53. D. COMANICIU AND P. MEER. Mean shift: A robust approach toward feature space analysis. *IEEE Transactions on Pattern Analysis and Machine Intelligence* **24**(5), 603–619 (2002).
54. T. H. CORMEN, C. E. LEISERSON, R. L. RIVEST, AND C. STEIN. "Introduction to Algorithms". MIT Press, Cambridge, MA, second ed. (2001).
55. R. L. COSGRIFF. Identification of shape. Technical Report 820-11, Antenna Laboratory, Ohio State University, Department of Electrical Engineering, Columbus, Ohio (December 1960).

56. A. CRIMINISI, I. D. REID, AND A. ZISSERMAN. A plane measuring device. *Image and Vision Computing* **17**(8), 625–634 (1999).
57. T. R. CRIMMINS. A complete set of Fourier descriptors for two-dimensional shapes. *IEEE Transactions on Systems, Man, and Cybernetics* **12**(6), 848–855 (1982).
58. F. C. CROW. Summed-area tables for texture mapping. *SIGGRAPH Computer Graphics* **18**(3), 207–212 (1984).
59. A. CUMANI. Edge detection in multispectral images. *Computer Vision, Graphics and Image Processing* **53**(1), 40–51 (1991).
60. A. CUMANI. Efficient contour extraction in color images. In "Proceedings of the Third Asian Conference on Computer Vision", ACCV, pp. 582–589, Hong Kong (January 1998). Springer.
61. L. S. DAVIS. A survey of edge detection techniques. *Computer Graphics and Image Processing* **4**, 248–270 (1975).
62. R. DERICHE. Using Canny's criteria to derive a recursively implemented optimal edge detector. *International Journal of Computer Vision* **1**(2), 167–187 (1987).
63. S. DI ZENZO. A note on the gradient of a multi-image. *Computer Vision, Graphics and Image Processing* **33**(1), 116–125 (1986).
64. R. O. DUDA, P. E. HART, AND D. G. STORK. "Pattern Classification". Wiley, New York (2001).
65. F. DURAND AND J. DORSEY. Fast bilateral filtering for the display of high-dynamic-range images. In "Proceedings of the 29th annual conference on Computer graphics and interactive techniques (SIGGRAPH'02)", pp. 257–266, San Antonio, Texas (July 2002).
66. B. ECKEL. "Thinking in Java". Prentice Hall, Englewood Cliffs, NJ, fourth ed. (2006). Earlier versions available online.
67. M. ELAD. On the origin of the bilateral filter and ways to improve it. *IEEE Transactions on Image Processing* **11**(10), 1141–1151 (2002).
68. A. FERREIRA AND S. UBEDA. Computing the medial axis transform in parallel with eight scan operations. *IEEE Transactions on Pattern Analysis and Machine Intelligence* **21**(3), 277–282 (1999).
69. N. I. FISHER. "Statistical Analysis of Circular Data". Cambridge University Press (1995).
70. D. FLANAGAN. "Java in a Nutshell". O'Reilly, Sebastopol, CA, fifth ed. (2005).
71. L. M. J. FLORACK, B. M. TER HAAR ROMENY, J. J. KOENDERINK, AND M. A. VIERGEVER. Scale and the differential structure of images. *Image and Vision Computing* **10**(6), 376–388 (1992).
72. J. FLUSSER. On the independence of rotation moment invariants. *Pattern Recognition* **33**(9), 1405–1410 (2000).
73. J. FLUSSER. Moment forms invariant to rotation and blur in arbitrary number of dimensions. *IEEE Transactions on Pattern Analysis and Machine Intelligence* **25**(2), 234–246 (2003).
74. J. FLUSSER, B. ZITOVA, AND T. SUK. "Moments and Moment Invariants in Pattern Recognition". John Wiley & Sons (2009).
75. J. D. FOLEY, A. VAN DAM, S. K. FEINER, AND J. F. HUGHES. "Computer Graphics: Principles and Practice". Addison-Wesley, Reading, MA, second ed. (1996).
76. A. FORD AND A. ROBERTS. "Colour Space Conversions" (1998). http://www.poynton.com/PDFs/coloureq.pdf.
77. W. FÖRSTNER AND E. GÜLCH. A fast operator for detection and precise location of distinct points, corners and centres of circular features. In A. GRÜN AND H. BEYER, editors, "Proceedings, International Society for Photogrammetry and Remote Sensing Intercommission Conference on the Fast Processing of Photogrammetric Data", pp. 281–305, Interlaken (June 1987).

78. D. A. Forsyth and J. Ponce. "Computer Vision—A Modern Approach". Prentice Hall, Englewood Cliffs, NJ (2003).
79. H. Freeman. Computer processing of line drawing images. *ACM Computing Surveys* **6**(1), 57–97 (1974).
80. J. H. Friedman, J. L. Bentley, and R. A. Finkel. An algorithm for finding best matches in logarithmic expected time. *ACM Transactions on Mathematical Software* **3**(3), 209–226 (1977).
81. D. L. Fritzsche. A systematic method for character recognition. Technical Report 1222-4, Antenna Laboratory, Ohio State University, Department of Electrical Engineering, Columbus, Ohio (November 1961).
82. M. Gervautz and W. Purgathofer. A simple method for color quantization: octree quantization. In A. Glassner, editor, "Graphics Gems I", pp. 287–293. Academic Press, New York (1990).
83. T. Gevers, A. Gijsenij, J. van de Weijer, and J.-M. Geusebroek. "Color in Computer Vision". Wiley (2012).
84. T. Gevers and H. Stokman. Classifying color edges in video into shadow-geometry, highlight, or material transitions. *IEEE Transactions on Multimedia* **5**(2), 237–243 (2003).
85. T. Gevers, J. Van De Weijer, and H. Stokman. Color feature detection. In R. Lukac and K. N. Plataniotis, editors, "Color Image Processing: Methods and Applications", pp. 203–226. CRC Press (2006).
86. C. A. Glasbey. An analysis of histogram-based thresholding algorithms. *Computer Vision, Graphics, and Image Processing: Graphical Models and Image Processing* **55**(6), 532–537 (1993).
87. A. S. Glassner. "Principles of Digital Image Synthesis". Morgan Kaufmann Publishers, San Francisco (1995).
88. R. C. Gonzalez and R. E. Woods. "Digital Image Processing". Addison-Wesley, Reading, MA (1992).
89. R. C. Gonzalez and R. E. Woods. "Digital Image Processing". Pearson Prentice Hall, Upper Saddle River, NJ, third ed. (2008).
90. M. Grabner, H. Grabner, and H. Bischof. Fast approximated SIFT. In "Proceedings of the 7th Asian Conference of Computer Vision", pp. 918–927 (2006).
91. R. L. Graham. An efficient algorithm for determining the convex hull of a finite planar set. *Information Processing Letters* **1**, 132–133 (1972).
92. R. L. Graham, D. E. Knuth, and O. Patashnik. "Concrete Mathematics: A Foundation for Computer Science". Addison-Wesley, Reading, MA, second ed. (1994).
93. G. H. Granlund. Fourier preprocessing for hand print character recognition. *IEEE Transactions on Computers* **21**(2), 195–201 (1972).
94. P. Green. Colorimetry and colour differences. In P. Green and L. MacDonald, editors, "Colour Engineering", ch. 3, pp. 40–77. Wiley, New York (2002).
95. F. Guichard, L. Moisan, and J.-M. Morel. A review of P.D.E. models in image processing and image analysis. *J. Phys. IV France* **12**(1), 137–154 (2002).
96. W. W. Hager. "Applied Numerical Linear Algebra". Prentice Hall (1988).
97. E. L. Hall. "Computer Image Processing and Recognition". Academic Press, New York (1979).
98. A. Hanbury. Circular statistics applied to colour images. In "Proceedings of the 8th Computer Vision Winter Workshop", pp. 55–60, Valtice, Czech Republic (February 2003).

99. J. C. Hancock. "An Introduction to the Principles of Communication Theory". McGraw-Hill (1961).
100. I. Hannah, D. Patel, and R. Davies. The use of variance and entropic thresholding methods for image segmentation. *Pattern Recognition* **28**(4), 1135–1143 (1995).
101. W. W. Harman. "Principles of the Statistical Theory of Communication". McGraw-Hill (1963).
102. C. G. Harris and M. Stephens. A combined corner and edge detector. In C. J. Taylor, editor, "4th Alvey Vision Conference", pp. 147–151, Manchester (1988).
103. R. Hartley and A. Zisserman. "Multiple View Geometry in Computer Vision". Cambridge University Press, 2 ed. (2013).
104. P. S. Heckbert. Color image quantization for frame buffer display. *Computer Graphics* **16**(3), 297–307 (1982).
105. P. S. Heckbert. Fundamentals of texture mapping and image warping. Master's thesis, University of California, Berkeley, Dept. of Electrical Engineering and Computer Science (1989).
106. R. Hess. An open-source SIFT library. In "Proceedings of the International Conference on Multimedia, MM'10", pp. 1493–1496, Firenze, Italy (October 2010).
107. J. Holm, I. Tastl, L. Hanlon, and P. Hubel. Color processing for digital photography. In P. Green and L. MacDonald, editors, "Colour Engineering", ch. 9, pp. 179–220. Wiley, New York (2002).
108. C. M. Holt, A. Stewart, M. Clint, and R. H. Perrott. An improved parallel thinning algorithm. *Communications of the ACM* **30**(2), 156–160 (1987).
109. V. Hong, H. Palus, and D. Paulus. Edge preserving filters on color images. In "Proceedings Int'l Conf. on Computational Science, ICCS", pp. 34–40, Kraków, Poland (2004).
110. B. K. P. Horn. "Robot Vision". MIT-Press, Cambridge, MA (1982).
111. P. V. C. Hough. Method and means for recognizing complex patterns. US Patent 3,069,654 (1962).
112. M. K. Hu. Visual pattern recognition by moment invariants. *IEEE Transactions on Information Theory* **8**, 179–187 (1962).
113. A. Huertas and G. Medioni. Detection of intensity changes with subpixel accuracy using Laplacian-Gaussian masks. *IEEE Transactions on Pattern Analysis and Machine Intelligence* **8**(5), 651–664 (1986).
114. R. W. G. Hunt. "The Reproduction of Colour". Wiley, New York, sixth ed. (2004).
115. J. Hutchinson. Culture, communication, and an information age madonna. *IEEE Professional Communications Society Newsletter* **45**(3), 1, 5–7 (2001).
116. J. Illingworth and J. Kittler. Minimum error thresholding. *Pattern Recognition* **19**(1), 41–47 (1986).
117. J. Illingworth and J. Kittler. A survey of the Hough transform. *Computer Vision, Graphics and Image Processing* **44**, 87–116 (1988).
118. International Color Consortium. "Specification ICC.1:2010-12 (Profile Version 4.3.0.0): Image Technology Colour Management—Architecture, Profile Format, and Data Structure" (2010). http://www.color.org.
119. International Electrotechnical Commission, IEC, Geneva. "IEC 61966-2-1: Multimedia Systems and Equipment—Colour Measurement and Management, Part 2-1: Colour Management—Default RGB Colour Space—sRGB" (1999). http://www.iec.ch.
120. International Organization for Standardization, ISO, Geneva. "ISO 13655:1996, Graphic Technology—Spectral Measurement and Colorimetric Computation for Graphic Arts Images" (1996).

121. International Organization for Standardization, ISO, Geneva. "ISO 15076-1:2005, Image Technology Colour Management—Architecture, Profile Format, and Data Structure: Part 1" (2005). Based on ICC.1:2004-10.
122. International Telecommunications Union, ITU, Geneva. "ITU-R Recommendation BT.709-3: Basic Parameter Values for the HDTV Standard for the Studio and for International Programme Exchange" (1998).
123. International Telecommunications Union, ITU, Geneva. "ITU-R Recommendation BT.601-5: Studio Encoding Parameters of Digital Television for Standard 4:3 and Wide-Screen 16:9 Aspect Ratios" (1999).
124. K. Jack. "Video Demystified—A Handbook for the Digital Engineer". LLH Publishing, Eagle Rock, VA, third ed. (2001).
125. B. Jähne. "Practical Handbook on Image Processing for Scientific Applications". CRC Press, Boca Raton, FL (1997).
126. B. Jähne. "Digitale Bildverarbeitung". Springer-Verlag, Berlin, fifth ed. (2002).
127. B. Jähne. "Digital Image Processing". Springer-Verlag, Berlin, sixth ed. (2005).
128. A. K. Jain. "Fundamentals of Digital Image Processing". Prentice Hall, Englewood Cliffs, NJ (1989).
129. R. Jain, R. Kasturi, and B. G. Schunck. "Machine Vision". McGraw-Hill, Boston (1995).
130. Y. Jia and T. Darrell. Heavy-tailed distances for gradient based image descriptors. In "Proceedings of the Twenty-Fifth Annual Conference on Neural Information Processing Systems (NIPS)", Grenada, Spain (December 2011).
131. X. Y. Jiang and H. Bunke. Simple and fast computation of moments. *Pattern Recognition* **24**(8), 801–806 (1991).
132. L. Jin and D. Li. A switching vector median filter based on the CIELAB color space for color image restoration. *Signal Processing* **87**(6), 1345–1354 (2007).
133. J. N. Kapur, P. K. Sahoo, and A. K. C. Wong. A new method for gray-level picture thresholding using the entropy of the histogram. *Computer Vision, Graphics, and Image Processing* **29**, 273–285 (1985).
134. B. Kimia. A large binary image database. Technical Report, LEMS Vision Group, Brown University (2002).
135. J. King. Engineering color at Adobe. In P. Green and L. MacDonald, editors, "Colour Engineering", ch. 15, pp. 341–369. Wiley, New York (2002).
136. R. A. Kirsch. Computer determination of the constituent structure of biological images. *Computers in Biomedical Research* **4**, 315–328 (1971).
137. L. Kitchen and A. Rosenfeld. Gray-level corner detection. *Pattern Recognition Letters* **1**, 95–102 (1982).
138. D. E. Knuth. "The Art of Computer Programming, Volume 2: Seminumerical Algorithms". Addison-Wesley, third ed. (1997).
139. J. J. Koenderink. The structure of images. *Biological Cybernetics* **50**(5), 363–370 (1984).
140. A. Koschan and M. A. Abidi. Detection and classification of edges in color images. *IEEE Signal Processing Magazine* **22**(1), 64–73 (2005).
141. A. Koschan and M. A. Abidi. "Digital Color Image Processing". Wiley (2008).
142. P. Kovesi. Arbitrary Gaussian filtering with 25 additions and 5 multiplications per pixel. Technical Report UWA-CSSE-09-002, The

University of Western Australia, School of Computer Science and Software Engineering (2009).

143. F. P. Kuhl and C. R. Giardina. Elliptic Fourier features of a closed contour. *Computer Graphics and Image Processing* **18**(3), 236–258 (1982).
144. M. Kuwahara, K. Hachimura, S. Eiho, and M. Kinoshita. Processing of RI-angiocardiographic image. In K. Preston and M. Onoe, editors, “Digital Processing of Biomedical Images”, pp. 187–202. Plenum, New York (1976).
145. D. C. Lay. “Linear Algebra and Its Applications”. Pearson, Boston, third ed. (2006).
146. P. E. Lestrel, editor. “Fourier Descriptors and Their Applications in Biology”. Cambridge University Press, New York (1997).
147. P.-S. Liao, T.-S. Chen, and P.-C. Chung. A fast algorithm for multilevel thresholding. *Journal of Information Science and Engineering* **17**, 713–727 (2001).
148. C. C. Lin and R. Chellappa. Classification of partial 2-D shapes using Fourier descriptors. *IEEE Transactions on Pattern Analysis and Machine Intelligence* **9**(5), 686–690 (1987).
149. B. J. Lindbloom. Accurate color reproduction for computer graphics applications. *SIGGRAPH Computer Graphics* **23**(3), 117–126 (1989).
150. T. Lindeberg. “Scale-Space Theory in Computer Vision”. Kluwer Academic Publishers (1994).
151. T. Lindeberg. Feature detection with automatic scale selection. *International Journal of Computer Vision* **30**(2), 77–116 (1998).
152. D. G. Lowe. Object recognition from local scale-invariant features. In “Proceedings of the 7th IEEE International Conference on Computer Vision”, vol. 2 of “ICCV’99”, pp. 1150–1157, Kerkyra, Corfu, Greece (1999).
153. D. G. Lowe. Distinctive image features from scale-invariant keypoints. *International Journal of Computer Vision* **60**, 91–110 (2004).
154. B. D. Lucas and T. Kanade. An iterative image registration technique with an application to stereo vision. In P. J. Hayes, editor, “Proceedings of the 7th International Joint Conference on Artificial Intelligence IJCAI’81”, pp. 674–679, Vancouver, BC (1981). William Kaufmann, Los Altos, CA.
155. R. Lukac, B. Smolka, and K. N. Plataniotis. Sharpening vector median filters. *Signal Processing* **87**(9), 2085–2099 (2007).
156. R. Lukac, B. Smolka, K. N. Plataniotis, and A. N. Venetsanopoulos. Vector sigma filters for noise detection and removal in color images. *Journal of Visual Communication and Image Representation* **17**(1), 1–26 (2006).
157. P. C. Mahalanobis. On the generalised distance in statistics. *Proceedings of the National Institute of Sciences of India* **2**(1), 49–55 (1936).
158. S. Mallat. “A Wavelet Tour of Signal Processing”. Academic Press, New York (1999).
159. C. Mancas-Thillou and B. Gosselin. Color text extraction with selective metric-based clustering. *Computer Vision, Graphics, and Image Processing: Image Understanding* **107**(1-2), 97–107 (2007).
160. M. J. Maron and R. J. Lopez. “Numerical Analysis”. Wadsworth Publishing, third ed. (1990).
161. D. Marr and E. Hildreth. Theory of edge detection. *Proceedings of the Royal Society of London, Series B* **207**, 187–217 (1980).
162. E. H. W. Meijering, W. J. Niessen, and M. A. Viergever. Quantitative evaluation of convolution-based methods for medical image interpolation. *Medical Image Analysis* **5**(2), 111–126 (2001).

163. J. MIANO. "Compressed Image File Formats". ACM Press, Addison-Wesley, Reading, MA (1999).
164. D. P. MITCHELL AND A. N. NETRAVALI. Reconstruction filters in computer-graphics. In R. J. BEACH, editor, "Proceedings of the 15th Annual Conference on Computer Graphics and Interactive Techniques, SIGGRAPH'88", pp. 221–228, Atlanta, GA (1988). ACM Press, New York.
165. P. A. MLSNA AND J. J. RODRIGUEZ. Gradient and Laplacian-type edge detection. In A. BOVIK, editor, "Handbook of Image and Video Processing", pp. 415–431. Academic Press, New York (2000).
166. P. A. MLSNA AND J. J. RODRIGUEZ. Gradient and Laplacian-type edge detection. In A. BOVIK, editor, "Handbook of Image and Video Processing", pp. 415–431. Academic Press, New York, second ed. (2005).
167. J. MOROVIC. "Color Gamut Mapping". Wiley (2008).
168. J. D. MURRAY AND W. VANRYPER. "Encyclopedia of Graphics File Formats". O'Reilly, Sebastopol, CA, second ed. (1996).
169. M. NADLER AND E. P. SMITH. "Pattern Recognition Engineering". Wiley, New York (1993).
170. M. NAGAO AND T. MATSUYAMA. Edge preserving smoothing. *Computer Graphics and Image Processing* **9**(4), 394–407 (1979).
171. S. K. NAIK AND C. A. MURTHY. Standardization of edge magnitude in color images. *IEEE Transactions on Image Processing* **15**(9), 2588–2595 (2006).
172. W. NIBLACK. "An Introduction to Digital Image Processing". Prentice-Hall (1986).
173. M. NITZBERG AND T. SHIOTA. Nonlinear image filtering with edge and corner enhancement. *IEEE Transactions on Pattern Analysis and Machine Intelligence* **14**(8), 826–833 (1992).
174. M. NIXON AND A. AGUADO. "Feature Extraction and Image Processing". Academic Press, second ed. (2008).
175. W. OH AND W. B. LINDQUIST. Image thresholding by indicator kriging. *IEEE Transactions on Pattern Analysis and Machine Intelligence* **21**(7), 590–602 (1999).
176. A. V. OPPENHEIM, R. W. SHAFER, AND J. R. BUCK. "Discrete-Time Signal Processing". Prentice Hall, Englewood Cliffs, NJ, second ed. (1999).
177. N. OTSU. A threshold selection method from gray-level histograms. *IEEE Transactions on Systems, Man, and Cybernetics* **9**(1), 62–66 (1979).
178. N. R. PAL AND S. K. PAL. A review on image segmentation techniques. *Pattern Recognition* **26**(9), 1277–1294 (1993).
179. S. PARIS AND F. DURAND. A fast approximation of the bilateral filter using a signal processing approach. *International Journal of Computer Vision* **81**(1), 24–52 (2007).
180. T. PAVLIDIS. "Algorithms for Graphics and Image Processing". Computer Science Press / Springer-Verlag, New York (1982).
181. O. PELE AND M. WERMAN. A linear time histogram metric for improved SIFT matching. In "Proceedings of the 10th European Conference on Computer Vision (ECCV'08)", pp. 495–508, Marseille, France (October 2008).
182. P. PERONA AND J. MALIK. Scale-space and edge detection using anisotropic diffusion. *IEEE Transactions on Pattern Analysis and Machine Intelligence* **12**(4), 629–639 (1990).
183. E. PERSOON AND K.-S. FU. Shape discrimination using Fourier descriptors. *IEEE Transactions on Systems, Man and Cybernetics* **7**(3), 170–179 (1977).

184. E. Persoon and K.-S. Fu. Shape discrimination using Fourier descriptors. *IEEE Transactions on Pattern Analysis and Machine Intelligence* **8**(3), 388–397 (1986).
185. T. Q. Pham and L. J. van Vliet. Separable bilateral filtering for fast video preprocessing. In "Proceedings IEEE International Conference on Multimedia and Expo", pp. CD1–4, Los Alamitos, USA (July 2005). IEEE Computer Society.
186. K. N. Plataniotis and A. N. Venetsanopoulos. "Color Image Processing and Applications". Springer (2000).
187. F. Porikli. Constant time O(1) bilateral filtering. In "Proceedings IEEE Conf. on Computer Vision and Pattern Recognition (CVPR)", pp. 1–8, Anchorage (June 2008).
188. C. A. Poynton. "Digital Video and HDTV Algorithms and Interfaces". Morgan Kaufmann Publishers, San Francisco (2003).
189. S. Prakash and F. V. D. Heyden. Normalisation of Fourier descriptors of planar shapes. *Electronics Letters* **19**(20), 828–830 (1983).
190. W. H. Press, S. A. Teukolsky, W. T. Vetterling, and B. P. Flannery. "Numerical Recipes". Cambridge University Press, third ed. (2007).
191. J. Prewitt. Object enhancement and extraction. In B. Lipkin and A. Rosenfeld, editors, "Picture Processing and Psychopictorics", pp. 415–431. Academic Press (1970).
192. R. R. Rakesh, P. Chaudhuri, and C. A. Murthy. Thresholding in edge detection: a statistical approach. *IEEE Transactions on Image Processing* **13**(7), 927–936 (2004).
193. W. S. Rasband. "ImageJ". U.S. National Institutes of Health, MD (1997–2007). http://rsb.info.nih.gov/ij/.
194. C. E. Reid and T. B. Passin. "Signal Processing in C". Wiley, New York (1992).
195. D. Rich. Instruments and methods for colour measurement. In P. Green and L. MacDonald, editors, "Colour Engineering", ch. 2, pp. 19–48. Wiley, New York (2002).
196. C. W. Richard and H. Hemami. Identification of three-dimensional objects using Fourier descriptors of the boundary curve. *IEEE Transactions on Systems, Man, and Cybernetics* **4**(4), 371–378 (1974).
197. I. E. G. Richardson. "H.264 and MPEG-4 Video Compression". Wiley, New York (2003).
198. T. W. Ridler and S. Calvard. Picture thresholding using an iterative selection method. *IEEE Transactions on Systems, Man, and Cybernetics* **8**(8), 630–632 (1978).
199. L. G. Roberts. Machine perception of three-dimensional solids. In J. T. Tippet, editor, "Optical and Electro-Optical Information Processing", pp. 159–197. MIT Press, Cambridge, MA (1965).
200. G. Robinson. Edge detection by compass gradient masks. *Computer Graphics and Image Processing* **6**(5), 492–501 (1977).
201. P. I. Rockett. An improved rotation-invariant thinning algorithm. *IEEE Transactions on Pattern Analysis and Machine Intelligence* **27**(10), 1671–1674 (2005).
202. A. Rosenfeld and J. L. Pfaltz. Sequential operations in digital picture processing. *Journal of the ACM* **12**, 471–494 (1966).
203. J. C. Russ. "The Image Processing Handbook". CRC Press, Boca Raton, FL, third ed. (1998).
204. P. K. Sahoo, S. Soltani, A. K. C. Wong, and Y. C. Chen. A survey of thresholding techniques. *Computer Vision, Graphics and Image Processing* **41**(2), 233–260 (1988).
205. G. Sapiro. "Geometric Partial Differential Equations and Image Analysis". Cambridge University Press (2001).

206. G. SAPIRO AND D. L. RINGACH. Anisotropic diffusion of multivalued images with applications to color filtering. *IEEE Transactions on Image Processing* **5**(11), 1582–1586 (1996).
207. J. SAUVOLA AND M. PIETIKÄINEN. Adaptive document image binarization. *Pattern Recognition* **33**(2), 1135–1143 (2000).
208. H. SCHILDT. "Java: A Beginner's Guide". Mcgraw-Hill Osborne Media (2014).
209. C. SCHMID AND R. MOHR. Local grayvalue invariants for image retrieval. *IEEE Transactions on Pattern Analysis and Machine Intelligence* **19**(5), 530–535 (1997).
210. C. SCHMID, R. MOHR, AND C. BAUCKHAGE. Evaluation of interest point detectors. *International Journal of Computer Vision* **37**(2), 151–172 (2000).
211. Y. SCHWARZER, editor. "Die Farbenlehre Goethes". Westerweide Verlag, Witten (2004).
212. M. SEUL, L. O'GORMAN, AND M. J. SAMMON. "Practical Algorithms for Image Analysis". Cambridge University Press, Cambridge (2000).
213. M. SEZGIN AND B. SANKUR. Survey over image thresholding techniques and quantitative performance evaluation. *Journal of Electronic Imaging* **13**(1), 146–165 (2004).
214. L. G. SHAPIRO AND G. C. STOCKMAN. "Computer Vision". Prentice Hall, Englewood Cliffs, NJ (2001).
215. G. SHARMA AND H. J. TRUSSELL. Digital color imaging. *IEEE Transactions on Image Processing* **6**(7), 901–932 (1997).
216. F. Y. SHIH AND S. CHENG. Automatic seeded region growing for color image segmentation. *Image and Vision Computing* **23**(10), 877–886 (2005).
217. N. SILVESTRINI AND E. P. FISCHER. "Farbsysteme in Kunst und Wissenschaft". DuMont, Cologne (1998).
218. S. N. SINHA, J.-M. FRAHM, M. POLLEFEYS, AND Y. GENC. Feature tracking and matching in video using programmable graphics hardware. *Machine Vision and Applications* **22**(1), 207–217 (2011).
219. Y. SIRISATHITKUL, S. AUWATANAMONGKOL, AND B. UYYANONVARA. Color image quantization using distances between adjacent colors along the color axis with highest color variance. *Pattern Recognition Letters* **25**, 1025–1043 (2004).
220. S. M. SMITH AND J. M. BRADY. SUSAN—a new approach to low level image processing. *International Journal of Computer Vision* **23**(1), 45–78 (1997).
221. B. SMOLKA, M. SZCZEPANSKI, K. N. PLATANIOTIS, AND A. N. VENETSANOPOULOS. Fast modified vector median filter. In "Proceedings of the 9th International Conference on Computer Analysis of Images and Patterns", CAIP'01, pp. 570–580, London, UK (2001). Springer-Verlag.
222. M. SONKA, V. HLAVAC, AND R. BOYLE. "Image Processing, Analysis and Machine Vision". PWS Publishing, Pacific Grove, CA, second ed. (1999).
223. M. SPIEGEL AND S. LIPSCHUTZ. "Schaum's Outline of Vector Analysis". McGraw-Hill, New York, second ed. (2009).
224. M. STOKES AND M. ANDERSON. "A Standard Default Color Space for the Internet—sRGB". Hewlett-Packard, Microsoft, www.w3.org/Graphics/Color/sRGB.html (1996).
225. S. SÜSSTRUNK. Managing color in digital image libraries. In P. GREEN AND L. MACDONALD, editors, "Colour Engineering", ch. 17, pp. 385–419. Wiley, New York (2002).
226. B. TANG, G. SAPIRO, AND V. CASELLES. Color image enhancement via chromaticity diffusion. *IEEE Transactions on Image Processing* **10**(5), 701–707 (2001).

227. C.-Y. Tang, Y.-L. Wu, M.-K. Hor, and W.-H. Wang. Modified SIFT descriptor for image matching under interference. In "Proceedings of the International Conference on Machine Learning and Cybernetics (ICMLC)", pp. 3294–3300, Kunming, China (July 2008).
228. S. Theodoridis and K. Koutroumbas. "Pattern Recognition". Academic Press, New York (1999).
229. C. Tomasi and R. Manduchi. Bilateral filtering for gray and color images. In "Proceedings Int'l Conf. on Computer Vision", ICCV'98, pp. 839–846, Bombay (1998).
230. F. Tomita and S. Tsuji. Extraction of multiple regions by smoothing in selected neighborhoods. *IEEE Transactions on Systems, Man, and Cybernetics* **7**, 394–407 (1977).
231. Ø. D. Trier and T. Taxt. Evaluation of binarization methods for document images. *IEEE Transactions on Pattern Analysis and Machine Intelligence* **17**(3), 312–315 (1995).
232. E. Trucco and A. Verri. "Introductory Techniques for 3-D Computer Vision". Prentice Hall, Englewood Cliffs, NJ (1998).
233. D. Tschumperlé. "PDEs Based Regularization of Multivalued Images and Applications". PhD thesis, Université de Nice, Sophia Antipolis, France (2005).
234. D. Tschumperlé. Fast anisotropic smoothing of multi-valued images using curvature-preserving PDEs. *International Journal of Computer Vision* **68**(1), 65–82 (2006).
235. D. Tschumperlé and R. Deriche. Diffusion PDEs on vector-valued images: local approach and geometric viewpoint. *IEEE Signal Processing Magazine* **19**(5), 16–25 (2002).
236. D. Tschumperlé and R. Deriche. Vector-valued image regularization with PDEs: A common framework for different applications. *IEEE Transactions on Pattern Analysis and Machine Intelligence* **27**(4), 506–517 (2005).
237. K. Turkowski. Filters for common resampling tasks. In A. Glassner, editor, "Graphics Gems I", pp. 147–165. Academic Press, New York (1990).
238. T. Tuytelaars and L. J. Van Gool. Matching widely separated views based on affine invariant regions. *International Journal of Computer Vision* **59**(1), 61–85 (2004).
239. J. van de Weijer. "Color Features and Local Structure in Images". PhD thesis, University of Amsterdam (2005).
240. M. I. Vardavoulia, I. Andreadis, and P. Tsalides. A new vector median filter for colour image processing. *Pattern Recognition Letters* **22**(6-7), 675–689 (2001).
241. A. Vedaldi and B. Fulkerson. VLFeat: An open and portable library of computer vision algorithms. http://www.vlfeat.org/ (2008).
242. F. R. D. Velasco. Thresholding using the ISODATA clustering algorithm. *IEEE Transactions on Systems, Man, and Cybernetics* **10**(11), 771–774 (1980).
243. D. Vernon. "Machine Vision". Prentice Hall (1999).
244. P. Viola and M. Jones. Robust real-time face detection. *International Journal of Computer Vision* **57**(2), 137–154 (2004).
245. T. P. Wallace and P. A. Wintz. An efficient three-dimensional aircraft recognition algorithm using normalized Fourier descriptors. *Computer Vision, Graphics and Image Processing* **13**(2), 99–126 (1980).
246. D. Wallner. Color management and transformation through ICC profiles. In P. Green and L. MacDonald, editors, "Colour Engineering", ch. 11, pp. 247–261. Wiley, New York (2002).

247. A. WATT. "3D Computer Graphics". Addison-Wesley, Reading, MA, third ed. (1999).
248. A. WATT AND F. POLICARPO. "The Computer Image". Addison-Wesley, Reading, MA (1999).
249. J. WEICKERT. "Anisotropic Diffusion in Image Processing". PhD thesis, Universität Kaiserslautern, Fachbereich Mathematik (1996).
250. J. WEICKERT. A review of nonlinear diffusion filtering. In B. M. TER HAAR ROMENY, L. FLORACK, J. J. KOENDERINK, AND M. A. VIERGEVER, editors, "Proceedings First International Conference on Scale-Space Theory in Computer Vision, Scale-Space'97", Lecture Notes in Computer Science, pp. 3–28, Utrecht (July 1997). Springer.
251. J. WEICKERT. Coherence-enhancing diffusion filtering. *International Journal of Computer Vision* **31**(2/3), 111–127 (1999).
252. J. WEICKERT. Coherence-enhancing diffusion of colour images. *Image and Vision Computing* **17**(3/4), 201–212 (1999).
253. B. WEISS. Fast median and bilateral filtering. *ACM Transactions on Graphics* **25**(3), 519–526 (2006).
254. M. WELK, J. WEICKERT, F. BECKER, C. SCHNÖRR, C. FEDDERN, AND B. BURGETH. Median and related local filters for tensor-valued images. *Signal Processing* **87**(2), 291–308 (2007).
255. P. WENDYKIER. "High Performance Java Software for Image Processing". PhD thesis, Emory University (2009).
256. G. WOLBERG. "Digital Image Warping". IEEE Computer Society Press, Los Alamitos, CA (1990).
257. M.-F. WU AND H.-T. SHEU. Contour-based correspondence using Fourier descriptors. *IEE Proceedings—Vision, Image and Signal Processing* **144**(3), 150–160 (1997).
258. G. WYSZECKI AND W. S. STILES. "Color Science: Concepts and Methods, Quantitative Data and Formulae". Wiley–Interscience, New York, second ed. (2000).
259. Q. YANG, K.-H. TAN, AND N. AHUJA. Real-time O(1) bilateral filtering. In "Proceedings IEEE Conf. on Computer Vision and Pattern Recognition (CVPR)", pp. 557–564, Miami (2009).
260. S. D. YANOWITZ AND A. M. BRUCKSTEIN. A new method for image segmentation. *Computer Vision, Graphics, and Image Processing* **46**(1), 82–95 (1989).
261. G. W. ZACK, W. E. ROGERS, AND S. A. LATT. Automatic measurement of sister chromatid exchange frequency. *Journal of Histochemistry and Cytochemistry* **25**(7), 741–753 (1977).
262. C. T. ZAHN AND R. Z. ROSKIES. Fourier descriptors for plane closed curves. *IEEE Transactions on Computers* **21**(3), 269–281 (1972).
263. P. ZAMPERONI. A note on the computation of the enclosed area for contour-coded binary objects. *Signal Processing* **3**(3), 267–271 (1981).
264. E. ZEIDLER, editor. "Teubner-Taschenbuch der Mathematik". B. G. Teubner Verlag, Leipzig, second ed. (2002).
265. T. Y. ZHANG AND C. Y. SUEN. A fast parallel algorithm for thinning digital patterns. *Communications of the ACM* **27**(3), 236–239 (1984).
266. S.-Y. ZHU, K. N. PLATANIOTIS, AND A. N. VENETSANOPOULOS. Comprehensive analysis of edge detection in color image processing. *Optical Engineering* **38**(4), 612–625 (1999).
267. S. ZOKAI AND G. WOLBERG. Image registration using log-polar mappings for recovery of large-scale similarity and projective transformations. *IEEE Transactions on Image Processing* **14**(10), 1422–1434 (2005).

Index

Symbols

A

G

H

M

N

Q

R

S

T

U

V

W

X

Y

Z